Fifth Edition

Introductory and Intermediate Algebra

Margaret L. Lial
American River College

John Hornsby
University of New Orleans

Terry McGinnis

PEARSON

Boston   Columbus   Indianapolis   New York   San Francisco   Upper Saddle River
Amsterdam   Cape Town   Dubai   London   Madrid   Milan   Munich   Paris   Montréal   Toronto
Delhi   Mexico City   São Paulo   Sydney   Hong Kong   Seoul   Singapore   Taipei   Tokyo

Editorial Director	Christine Hoag
Editor in Chief	Maureen O'Connor
Executive Content Editor	Kari Heen
Content Editor	Christine Whitlock
Senior Content Editor	Lauren Morse
Assistant Editor	Rachel Haskell
Senior Managing Editor	Karen Wernholm
Senior Production Project Manager	Kathleen A. Manley
Digital Assets Manager	Marianne Groth
Supplements Production Coordinator	Kerri Consalvo
Media Producer	Stephanie Green
Software Development	Eric Gregg, MathXL;
	Mary Durnwald, TestGen
Marketing Manager	Rachel Ross
Senior Author Support/Technology Specialist	Joe Vetere
Rights and Permissions Advisor	Cheryl Besenjak
Image Manager	Rachel Youdelman
Procurement Manager	Evelyn Beaton
Procurement Specialist	Debbie Rossi
Media Procurement Specialist	Ginny Michaud
Associate Director of Design	Andrea Nix
Senior Designer	Barbara Atkinson
Text Design, Production Coordination,	Cenveo Publisher Services/
Composition, and Illustrations	Nesbitt Graphics, Inc.
Cover Image	*Meadow Stream* © Lorraine Cota Manley

For permission to use copyrighted material, grateful acknowledgment is made to the copyright holders on page I-8, which is hereby made part of this copyright page.

Many of the designations used by manufacturers and sellers to distinguish their products are claimed as trademarks. Where those designations appear in this book, and Pearson Education was aware of a trademark claim, the designations have been printed in initial caps or all caps.

Library of Congress Cataloging-in-Publication Data
Lial, Margaret L.
  Introductory and intermediate algebra/Margaret L. Lial, John Hornsby, Terry McGinnis.—5th ed.
    p. cm.
  ISBN 978-0-321-86553-3
  1. Algebra—Textbooks. I. Hornsby, E. John. II. McGinnis, Terry. III. Title.
  QA152.3.L56 2013
  512.9—dc23   2012013815

2 3 4 5 6 7 8 9 10—CRK—16 15 14

www.pearsonhighered.com

ISBN 13: 978-0-321-86553-3
ISBN 10: 0-321-86553-7

To Margaret L. Lial

On March 16, 2012, the mathematics education community lost one of its most influential members with the passing of our beloved mentor, colleague, and friend Marge Lial. On that day, Marge lost her long battle with ALS. Throughout her illness, Marge showed the remarkable strength and courage that characterized her entire life.

We would like to share a few comments from among the many messages we received from friends, colleagues, and others whose lives were touched by our beloved Marge:

"What a lady"   "A great friend"
"A remarkable person"   "Sorely missed but so fondly remembered"
"Gracious to everyone"   "Even though our crossed path was narrow, she made an impact
"One of a kind"   and I will never forget her."
"Truly someone special"   "There is talent and there is Greatness. Marge was truly Great."
"A loss in the mathematical world"   "Her true impact is almost more than we can imagine."

In the world of college mathematics publishing, Marge Lial was a rock star. People flocked to her, and she had a way of making everyone feel like they truly mattered. And to Marge, they did. She and Chuck Miller began writing for Scott Foresman in 1970. Just three years ago she told us that she could no longer continue because "just getting from point A to point B" had become too challenging. That's our Marge—she even gave a geometric interpretation to her illness.

It has truly been an honor and a privilege to work with Marge Lial these past twenty years. While we no longer have her wit, charm, and loving presence to guide us, so much of who we are as mathematics educators has been shaped by her influence. We will continue doing our part to make sure that the Lial name represents excellence in mathematics education. And we remember daily so many of the little ways she impacted us, including her special expressions, "Margisms" as we like to call them. She often ended emails with one of them—the single word "Onward."

We conclude with a poem penned by another of Marge's coauthors, Callie Daniels.

Your courage inspires me
Your strength...impressive
Your wit humors me
Your vision...progressive

Your determination motivates me
Your accomplishments pave my way
Your vision sketches images for me
Your influence will forever stay.

Thank you, dearest Marge.
Knowing you and working with you has been a divine gift.

Onward.

John Hornsby
Terry McGinnis

Contents

Preface   vii

CHAPTER R Prealgebra Review   1

R.1 Fractions   1
R.2 Decimals and Percents   15
Study Skills Using Your Math Textbook   25
Study Skills Reading Your Math Textbook   26

CHAPTER 1 The Real Number System   27

1.1 Exponents, Order of Operations, and
     Inequality   28
Study Skills Taking Lecture Notes   36
1.2 Variables, Expressions, and Equations   37
Study Skills Tackling Your Homework   45
1.3 Real Numbers and the Number Line   46
Study Skills Using Study Cards   55
1.4 Adding Real Numbers   56
1.5 Subtracting Real Numbers   64
1.6 Multiplying and Dividing Real Numbers   73
Summary Exercises *Performing Operations with*
Real Numbers   85
1.7 Properties of Real Numbers   87
Study Skills Reviewing a Chapter   97
1.8 Simplifying Expressions   98
Study Skills Taking Math Tests   104

Summary 105 ▾ Review Exercises 109 ▾ Test 114

Math in the Media The Magic Number in Sports   116

CHAPTER 2 Equations, Inequalities,
               and Applications   117

2.1 The Addition Property of Equality   118
Study Skills Managing Your Time   125
2.2 The Multiplication Property of Equality   126
Study Skills Analyzing Your Test Results   132
2.3 More on Solving Linear Equations   133
Study Skills Using Study Cards Revisited   145
Summary Exercises *Applying Methods for*
Solving Linear Equations   146

2.4 An Introduction to Applications
     of Linear Equations   148
2.5 Formulas and Additional Applications
     from Geometry   162
2.6 Ratio, Proportion, and Percent   174
Summary Exercises *Applying Problem-Solving*
Techniques   186
2.7 Solving Linear Inequalities   188

Summary 203 ▾ Review Exercises 207 ▾ Test 211
Cumulative Review Exercises Chapters R–2   213

CHAPTER 3 Graphs of Linear Equations and
               Inequalities in Two Variables;
               Functions   215

3.1 Linear Equations in Two Variables;
     The Rectangular Coordinate System   216
Study Skills Preparing for Your Math Final Exam   230
3.2 Graphing Linear Equations in Two Variables   231
3.3 The Slope of a Line   244
3.4 Writing and Graphing Equations of Lines   260
Summary Exercises *Applying Graphing and*
Equation-Writing Techniques for Lines   276
3.5 Graphing Linear Inequalities in Two Variables   278
3.6 Introduction to Relations and Functions   284
3.7 Function Notation and Linear Functions   293

Summary 301 ▾ Review Exercises 305 ▾ Test 309
Cumulative Review Exercises Chapters R–3   311

CHAPTER 4 Systems of Linear Equations
               and Inequalities   313

4.1 Solving Systems of Linear Equations
     by Graphing   314
4.2 Solving Systems of Linear Equations
     by Substitution   323
4.3 Solving Systems of Linear Equations
     by Elimination   331
Summary Exercises *Applying Techniques for*
Solving Systems of Linear Equations   338
4.4 Applications of Linear Systems   340
4.5 Solving Systems of Linear Inequalities   352

Summary 357 ▾ Review Exercises 360 ▾ Test 363
Cumulative Review Exercises Chapters R–4   365

CHAPTER 5	Exponents and Polynomials	367
5.1	Adding and Subtracting Polynomials	368
5.2	The Product Rule and Power Rules for Exponents	377
5.3	Multiplying Polynomials	385
5.4	Special Products	392
5.5	Integer Exponents and the Quotient Rule	399

Summary Exercises **Applying the Rules for Exponents** 408

5.6	Dividing a Polynomial by a Monomial	410
5.7	Dividing a Polynomial by a Polynomial	414
5.8	An Application of Exponents: Scientific Notation	420

Summary 427 ▼ Review Exercises 430 ▼ Test 434
Cumulative Review Exercises Chapters R–5 436

Math in the Media Floods, Hurricanes, and Earthquakes, Oh My! 438

CHAPTER 6	Factoring and Applications	439
6.1	Factors; The Greatest Common Factor	440
6.2	Factoring Trinomials	450
6.3	Factoring Trinomials by Grouping	457
6.4	Factoring Trinomials by Using the FOIL Method	461
6.5	Special Factoring Techniques	467
6.6	A General Approach to Factoring	477
6.7	Solving Quadratic Equations by Factoring	481
6.8	Applications of Quadratic Equations	490

Summary 501 ▼ Review Exercises 504 ▼ Test 509
Cumulative Review Exercises Chapters R–6 511

CHAPTER 7	Rational Expressions and Functions	513
7.1	Rational Expressions and Functions; Multiplying and Dividing	514
7.2	Adding and Subtracting Rational Expressions	525
7.3	Complex Fractions	534
7.4	Equations with Rational Expressions and Graphs	540

Summary Exercises **Simplifying Rational Expressions vs. Solving Rational Equations** 548

| 7.5 | Applications of Rational Expressions | 550 |
| 7.6 | Variation | 565 |

Summary 575 ▼ Review Exercises 579 ▼ Test 583
Cumulative Review Exercises Chapters R–7 585

CHAPTER 8	Equations, Inequalities, and Systems Revisited	587
8.1	Review of Solving Linear Equations and Inequalities	588
8.2	Set Operations and Compound Inequalities	602
8.3	Absolute Value Equations and Inequalities	611

Summary Exercises **Solving Linear and Absolute Value Equations and Inequalities** 624

| 8.4 | Review of Systems of Linear Equations in Two Variables | 626 |
| 8.5 | Systems of Linear Equations in Three Variables; Applications | 635 |

Summary 647 ▼ Review Exercises 652 ▼ Test 656
Cumulative Review Exercises Chapters R–8 658

Math in the Media To Pass or to Fail? That Is the Question 660

CHAPTER 9	Roots, Radicals, and Root Functions	661
9.1	Radical Expressions and Graphs	662
9.2	Rational Exponents	675
9.3	Simplifying Radical Expressions	685
9.4	Adding and Subtracting Radical Expressions	698
9.5	Multiplying and Dividing Radical Expressions	702

Summary Exercises **Performing Operations with Radicals and Rational Exponents** 711

| 9.6 | Solving Equations with Radicals | 713 |
| 9.7 | Complex Numbers | 722 |

Summary 731 ▼ Review Exercises 735 ▼ Test 739
Cumulative Review Exercises Chapters R–9 741

CHAPTER 10	Quadratic Equations, Inequalities, and Functions	743
10.1	Solving Quadratic Equations by the Square Root Property	744
10.2	Solving Quadratic Equations by Completing the Square	751
10.3	Solving Quadratic Equations by the Quadratic Formula	759
10.4	Equations Quadratic in Form	767

Summary Exercises **Applying Methods for Solving Quadratic Equations** 779

10.5	Formulas and Applications	781
10.6	Graphs of Quadratic Functions	789
10.7	More about Parabolas and Their Applications	799
10.8	Polynomial and Rational Inequalities	811

Summary 821 ▼ Review Exercises 825 ▼ Test 829
Cumulative Review Exercises Chapters R–10   831

CHAPTER 11 **Inverse, Exponential, and Logarithmic Functions**   833

11.1 Operations on Functions and Composition   834
11.2 Inverse Functions   843
11.3 Exponential Functions   851
11.4 Logarithmic Functions   859
11.5 Properties of Logarithms   867
11.6 Common and Natural Logarithms   875
11.7 Exponential and Logarithmic Equations and Their Applications   881

Summary 890 ▼ Review Exercises 894 ▼ Test 898
Cumulative Review Exercises Chapters R–11   900

Math in the Media So, Did the Scarecrow Really Get a Brain?   902

CHAPTER 12 **Nonlinear Functions, Conic Sections, and Nonlinear Systems**   903

12.1 Additional Graphs of Functions   904
12.2 The Circle and the Ellipse   911

12.3 The Hyperbola and Other Functions Defined by Radicals   921
12.4 Nonlinear Systems of Equations   929
12.5 Second-Degree Inequalities and Systems of Inequalities   936

Summary 942 ▼ Review Exercises 946 ▼ Test 949
Cumulative Review Exercises Chapters R–12   951

Appendix A Review of Exponents, Polynomials, and Factoring (Transition from Introductory to Intermediate Algebra)   953

Appendix B Solving Systems of Linear Equations by Matrix Methods   960

Appendix C Synthetic Division   967

Answers to Selected Exercises   A-1

Solutions to Selected Exercises   S-1

Index   I-1

Credits   I-8

Preface

In the fifth edition of *Introductory and Intermediate Algebra,* we have addressed the diverse needs of today's students by creating a tightly coordinated text and technology package that includes integrated activities to help students improve their study skills, an attractive design, updated applications and graphs, helpful features, and careful explanations of concepts. We have also expanded the supplements and study aids. We have revamped the video series into a complete Lial Video Library with expanded video coverage and new, easier navigation. And we have added the new Lial MyWorkBook. We have also responded to the suggestions of users and reviewers and have added many new examples and exercises based on their feedback.

Students who have never studied algebra—as well as those who require further review of basic algebraic concepts before taking additional courses in mathematics, business, science, nursing, or other fields—will benefit from the text's student-oriented approach. Of particular interest to students and instructors will be the new guided solutions in margin problems and exercises, the new Concept Check exercises, and the enhanced Study Skills activities.

This text is part of a series that also includes the following books:

- *Basic College Mathematics,* Ninth Edition, by Lial, Salzman, and Hestwood
- *Essential Mathematics,* Fourth Edition, by Lial and Salzman
- *Prealgebra,* Fifth Edition, by Lial and Hestwood
- *Introductory Algebra,* Tenth Edition, by Lial, Hornsby, and McGinnis
- *Intermediate Algebra,* Tenth Edition, by Lial, Hornsby, and McGinnis
- *Prealgebra and Introductory Algebra,* Fourth Edition, by Lial, Hestwood, Hornsby, and McGinnis
- *Developmental Mathematics: Basic Mathematics and Algebra,* Third Edition, by Lial, Hornsby, McGinnis, Salzman, and Hestwood

WHAT'S NEW IN THIS EDITION

We are pleased to offer the following new textbook features and supplements.

▶ *Engaging Chapter Openers* The new Chapter Openers portray real life situations that reflect the mathematical content and are relevant to students. Each opener also includes an expanded outline of the chapter contents. (See pp. 27, 117, and 215—Chapters 1, 2, and 3.)

▶ *Revised Exposition* As each section of the text was revised, we paid special attention to the exposition, which has been tightened and polished. (See Section 1.3, Real Numbers and the Number Line, for example.) We believe this has improved discussions and presentations of topics.

▶ *Guided Solutions* Selected exercises in the margins and in the exercise sets, marked with a GS icon, step students through solutions as they learn new concepts or procedures. (See p. 119, margin problem, and p. 157, Exercises 23 and 24.)

▶ *Concept Check Exercises* Each section exercise set now begins with a group of these exercises, designated CONCEPT CHECK. They are designed to facilitate students' mathematical thinking and conceptual understanding. Many of these emphasize vocabulary. (See pp. 238 and 290.) Additional Concept Check problems are sprinkled throughout the exercise sets and ask students to apply mathematical processes and concepts or to identify What Went Wrong? in incorrect solutions. (See pp. 103 and 719.)

▶ *Essential Study Skills* Poor study skills are a major reason why students do not succeed in mathematics. Each one page long, these eleven enhanced activities provide helpful information, tips, and strategies on a variety of essential study skills, including *Reading Your Math Textbook, Tackling Your Homework, Taking Math Tests,* and *Managing Your Time.* Each has been designed independently to allow flexible use with individuals or small groups of students, or as a source of material for in-class discussions. (See pp. 55 and 104.)

▶ *Helpful Teaching Tips* All new *Teaching Tips*, located in the margins of the *Annotated Instructor's Edition,* provide helpful suggestions, emphasize common student trouble spots, and offer other pertinent information that instructors, especially those new to teaching this course, may find helpful. (See pp. 100 and 677.)

▶ *Lial Video Library* The Lial Video Library, available in MyMathLab and on the Video Resources DVD, provides students with a wealth of video resources to help them navigate the road to success. All video resources in the library include optional captions in English and Spanish. The Lial Video Library includes Section Lecture Videos, Solutions Clips, Quick Review Lectures, and Chapter Test Prep Videos. The Chapter Test Prep Videos are also available on YouTube (searchable using author name and book title), or by scanning the QR Code® on the inside back cover for easy access.

▶ *MyWorkBook* This new workbook provides Guided Examples and corresponding Now Try Exercises for each text objective. The extra practice exercises for every section of the text, with ample space for students to show their work, are correlated to Examples, Lecture Videos, and Exercise Solution Clips, to give students the help they need to successfully complete problems. Additionally, MyWorkBook lists the learning objectives and key vocabulary terms for every text section, along with vocabulary practice problems.

CONTENT CHANGES

The scope and sequence of topics in *Introductory and Intermediate Algebra* has stood the test of time and rates highly with our reviewers. Therefore, the table of contents remains intact, making the transition to the new edition easier. Specific content changes include the following:

▶ We gave the exercise sets special attention. There are approximately 585 new and updated exercises, including problems that check conceptual understanding, focus on skill development, and provide review. We also worked to improve the even-odd pairing of exercises.

▶ Real-world data in over 200 applications in the examples and exercises has been updated.

▶ We increased the emphasis on checking solutions and answers, as indicated by a new **CHECK** tag and ✓ in the exposition and examples.

▶ In preparation for Chapter 3 and the material on forms of linear equations, Section 2.5 provides a new example and corresponding exercises on solving a linear equation in two variables x and y for y.

▶ Section 2.6 includes entirely new discussion and examples on percent, percent equations, and percent applications, plus corresponding exercises.

▶ Section 3.4 on writing and graphing equations of lines provides increased development and coverage of the slope-intercept form, including two new examples. There is an enhanced objective on writing equations of horizontal and vertical lines.

▶ The former section Introduction to Functions has been divided into two sections for more flexible coverage. Section 3.6 has increased discussion of determining domain, and Section 3.7 includes more on finding function values from a graph. Both exercise sets have been expanded.

► Sections 6.2–6.4 include three new examples on factoring trinomials.

► The discussion of recognizing the graph of a rational function in Section 7.4 has been expanded.

► Section 9.6 includes a new objective on using the power rule to solve a formula for a specified variable.

► The discussion of operations with and composition of functions has been moved to precede that of inverse functions. These topics now appear in new Section 11.1.

► The following topics are among those that have been enhanced and/or expanded:

Using and completing tables when solving applications (throughout)

Dividing real numbers involving zero (Section 1.6)

Applying the properties of real numbers (Section 1.7)

Solving applications involving consecutive integers and angle measures (Section 2.4)

Using interval notation and solving linear inequalities (Section 2.7)

Graphing linear equations in two variables using intercepts (Section 3.2)

Solving systems of equations with decimal coefficients (Section 4.2)

Finding numerical values of rational expressions and values for which rational expressions are undefined (Section 7.1)

Solving quadratic equations by using the zero-factor property (Section 10.1)

Emphasizing identification of the vertex, axis of symmetry, domain, and range of graphs of quadratic functions (Section 10.6)

Graphing ellipses and hyperbolas whose centers are translated away from the origin (Sections 12.2 and 12.3)

HALLMARK FEATURES

We have enhanced the following popular features, each of which is designed to increase ease-of-use by students and/or instructors.

► *Emphasis on Problem-Solving*   We introduce our six-step problem-solving method in Chapter 2 and integrate it throughout the text. The six steps, *Read, Assign a Variable, Write an Equation, Solve, State the Answer,* and *Check,* are emphasized in boldface type and repeated in examples and exercises to reinforce the problem-solving process for students. (See pp. 148 and 340.) We also provide students with   Problem-Solving Hint   boxes that feature helpful problem-solving tips and strategies. (See pp. 150 and 153.)

► *Helpful Learning Objectives*   We begin each section with clearly stated, numbered objectives, and the included material is directly keyed to these objectives so that students and instructors know exactly what is covered in each section. (See pp. 28 and 148.)

► *Popular Cautions and Notes*   One of the most popular features of previous editions, we include information marked   CAUTION   and   Note   to warn students about common errors and emphasize important ideas throughout the exposition. The updated text design makes them easy to spot. (See pp. 28 and 47.)

► *Comprehensive Examples*   The new edition features a multitude of step-by-step, worked-out examples that include pedagogical color, helpful side comments, and special pointers. We give increased attention to checking example solutions—more checks, designated using a special **CHECK** tag and ✓, are included than in past editions. (See pp. 120 and 541.)

▶ *More Pointers*   Well received by both students and instructors in the previous edition, we incorporate more pointers in examples and discussions throughout this edition of the text. They provide students with important on-the-spot reminders and warnings about common pitfalls. (See pp. 127 and 535.)

▶ *Ample Margin Problems*   Margin problems, with answers immediately available at the bottom of the page, are found in every section of the text. (See pp. 30 and 537.) This expanded key feature allows students to immediately practice the material covered in the examples in preparation for the exercise sets. Many include new guided solutions.

▶ *Updated Figures, Photos, and Hand-Drawn Graphs*   Today's students are more visually oriented than ever. As a result, we have made a concerted effort to include appealing mathematical figures, diagrams, tables, and graphs, including a "hand-drawn" style of graphs, whenever possible. (See pp. 216 and 544.) We have incorporated new depictions of well-known mathematicians as well as new photos to accompany applications in examples and exercises. (See pp. 221 and 562.)

▶ *Optional Calculator Tips*   These tips, marked ▦, offer helpful information and instruction for students using calculators in the course. (See pp. 31 and 663.)

▶ *Relevant Real-Life Applications*   We include many new or updated applications from fields such as business, pop culture, sports, technology, and the health sciences that show the relevance of algebra to daily life. (See pp. 183 and 562.)

▶ *Extensive and Varied Exercise Sets*   The text contains a wealth of exercises to provide students with opportunities to practice, apply, connect, review, and extend the skills they are learning. Numerous illustrations, tables, graphs, and photos help students visualize the problems they are solving. Problem types include skill building, writing, and calculator exercises, as well as applications, matching, true/false, multiple-choice, and fill-in-the-blank problems. (See pp. 290–292 and 670–674.)

In the Annotated Instructor's Edition of the text, the writing exercises are marked with an icon ▱ so that instructors may assign these problems at their discretion. Exercises suitable for calculator work are marked in both the student and instructor editions with a calculator icon ▦. Students can watch an instructor work through the complete solution for all exercises marked with a Play Button icon ▶ on the Videos on DVD or in MyMathLab.

▶ *Flexible Relating Concepts Exercises*   These help students tie concepts together and develop higher level problem-solving skills as they compare and contrast ideas, identify and describe patterns, and extend concepts to new situations. (See pp. 259 and 337.) These exercises, now located at the end of selected exercise sets, make great collaborative activities for pairs or small groups of students.

▶ *Special Summary Exercises*   We include a set of these popular in-chapter exercises in many of the chapters, beginning in Chapter 1. They provide students with the all-important *mixed review problems* they need to master topics and often include summaries of solution methods and/or additional examples. (See pp. 146 and 276.)

▶ *Math in the Media*   Each of these one-page activities presents a relevant look at how mathematics is used in the media. Designed to help instructors answer the often-asked question, "When will I ever use this stuff?," these activities ask students to read and interpret data from newspaper articles, the Internet, and other familiar, real-world sources. (See pp. 116 and 438.) The activities are well-suited to collaborative work or they can be completed by individuals or used for open-ended class discussions.

▶ *Step-by-Step Solutions to Selected Exercises*   Exercise numbers enclosed in a blue square, such as **11.**, indicate that a worked-out solution for the problem is included at the back of the text. These solutions are given for selected exercises that most commonly cause students difficulty. (See pp. S-1 through S-22.)

▶ *Extensive Review Opportunities*   We conclude each chapter with the following important review components:

A **Chapter Summary** that features a helpful list of **Key Terms,** organized by section, **New Symbols, Test Your Word Power** vocabulary quiz (with answers immediately following), and a **Quick Review** of each section's contents, complete with additional examples (See pp. 357–359.)

A comprehensive set of **Chapter Review Exercises,** keyed to individual sections for easy student reference, as well as a set of **Mixed Review Exercises** that helps students further synthesize concepts (See pp. 360–362.)

A **Chapter Test** that students can take under test conditions to see how well they have mastered the chapter material (See pp. 363–364.)

A set of **Cumulative Review Exercises** (beginning in Chapter 2) that covers material going back to Chapter R (See pp. 365–366.)

STUDENT SUPPLEMENTS

Student's Solutions Manual
- By Jeffery A. Cole, Anoka-Ramsey Community College
- Provides detailed solutions to the odd-numbered section-level exercises and to all margin, Relating Concepts, Summary, Chapter Review, Chapter Test, and Cumulative Review Exercises
  ISBNs: 0-321-86558-8, 978-0-321-86558-8

NEW MyWorkBook
- Provides Guided Examples and corresponding Now Try Exercises for each text objective
- Refers students to correlated Examples, Lecture Videos, and Exercise Solution Clips
- Includes extra practice exercises for every section of the text with ample space for students to show their work
- Lists learning objectives and key vocabulary terms for every text section, along with vocabulary practice problems
  ISBNs: 0-321-87262-2, 978-0-321-87262-3

NEW Lial Video Library
The Lial Video Library, available in MyMathLab and on the Video Resources DVD, provides students with a wealth of video resources to help them navigate the road to success. All video resources in the library include optional captions in English and Spanish. The Lial Video Library includes the following resources:
- **Section Lecture Videos** offer a new navigation menu that allows students to easily focus on the key examples and exercises that they need to review in each section.
- **Solutions Clips** show an instructor working through the complete solutions to selected exercises from the text. Exercises with a solution clip are marked in the text and e-book with a Play Button icon ▶.
- **Quick Review Lectures** provide a short summary lecture of each key concept from the Quick Reviews at the end of every chapter in the text.
- **Chapter Test Prep Videos** let students watch instructors work through step-by-step solutions to all the Chapter Test exercises from the textbook. Chapter Test Prep videos are also available on YouTube™ (search using author name and book title) and in MyMathLab, or by scanning the QR Code® on the inside back cover for easy access.

INSTRUCTOR SUPPLEMENTS

Annotated Instructor's Edition
- Provides answers to all text exercises in color next to the corresponding problems
- Includes all NEW Teaching Tips located in the margins
- Identifies writing 🖊 and calculator ▦ exercises
  ISBNs: 0-321-86556-1, 978-0-321-86556-4

Instructor's Solutions Manual (Download only)
- By Jeffery A. Cole, Anoka-Ramsey Community College
- Provides complete solutions to all exercises in the text
- Available for download at www.pearsonhighered.com
  ISBNs: 0-321-86560-X, 978-0-321-86560-1

Instructor's Resource Manual with Tests and Mini-Lectures (Download only)
- Contains a test bank with two diagnostic pretests, six free-response and two multiple-choice test forms per chapter, and two final exams
- Contains a mini-lecture for each section of the text with objectives, key examples, and teaching tips
- Includes a correlation guide from the fourth to the fifth edition and phonetic spellings for all key terms in the text
- Includes resources to help both new and adjunct faculty with course preparation and classroom management, by offering helpful teaching tips correlated to the sections of the text
- Available for download at www.pearsonhighered.com
  ISBNs: 0-321-86561-8, 978-0-321-86561-8

ADDITIONAL MEDIA SUPPLEMENTS

MyMathLab® **MyMathLab® Online Course (access code required)**
MyMathLab from Pearson is the world's leading online resource in mathematics, integrating interactive homework, assessment, and media in a flexible, easy-to-use format. MyMathLab delivers **proven results** in helping individual students succeed. It provides **engaging experiences** that personalize, stimulate, and measure learning for each student. And, it comes from an **experienced partner** with educational expertise and an eye on the future.

To learn more about how MyMathLab combines proven learning applications with powerful assessment, visit **www.mymathlab.com** or contact your Pearson representative.

MyMathLab® Ready to Go Course (access code required)

These new Ready to Go courses provide students with all the same great MyMathLab features, but make it easier for instructors to get started. Each course includes pre-assigned homework and quizzes to make creating a course even simpler. Ask your Pearson representative about the details for this particular course or to see a copy of this course.

MyMathLab® Plus/MyStatLab™ Plus

MyLabsPlus combines proven results and engaging experiences from MyMathLab® and MyStatLab™ with convenient management tools and a dedicated services team. Designed to support growing math and statistics programs, it includes additional features such as:

- **Batch Enrollment:** Your school can create the login name and password for every student and instructor, so everyone can be ready to start class on the first day. Automation of this process is also possible through integration with your school's Student Information System.
- **Login from your campus portal:** Students and instructors can link directly from their campus portal into MyLabsPlus courses. A Pearson service team works with each institution to create a single sign-on experience for instructors and students.
- **Advanced Reporting:** MyLabsPlus's advanced reporting allows instructors to review and analyze students' strengths and weaknesses by tracking their performance on tests, assignments, and tutorials. Administrators can review grades and assignments across all courses on your MyLabsPlus campus for a broad overview of program performance.
- **24/7 Support:** Students and instructors receive 24/7 support, 365 days a year, by email or online chat.

MyLabsPlus is available to qualified adopters. For more information, visit our website at *www.mylabsplus.com* or contact your Pearson representative.

MathXL® Online Course (access code required)

MathXL® is the homework and assessment engine that runs MyMathLab. (MyMathLab is MathXL plus a learning management system.)

With MathXL, instructors can:
- Create, edit, and assign online homework and tests using algorithmically generated exercises correlated at the objective level to the textbook.
- Create and assign their own online exercises and import TestGen tests for added flexibility.
- Maintain records of all student work tracked in MathXL's online gradebook.

With MathXL, students can:
- Take chapter tests in MathXL and receive personalized study plans and/or personalized homework assignments based on their test results.
- Use the study plan and/or the homework to link directly to tutorial exercises for the objectives they need to study.
- Access supplemental animations and video clips directly from selected exercises.

MathXL is available to qualified adopters. For more information, visit our website at *www.mathxl.com,* or contact your Pearson representative.

TestGen®

TestGen® (*www.pearsoned.com/testgen*) enables instructors to build, edit, print, and administer tests using a computerized bank of questions developed to cover all the objectives of the text. TestGen is algorithmically based, allowing instructors to create multiple but equivalent versions of the same question or test with the click of a button. Instructors can also modify test bank questions or add new questions. The software and testbank are available for download from Pearson Education's online catalog.

PowerPoint® Lecture Slides

- Present key concepts and definitions from the text
- Available for download at *www.pearsonhighered.com* or in MyMathLab

ACKNOWLEDGMENTS

The comments, criticisms, and suggestions of users, nonusers, instructors, and students have positively shaped this textbook over the years, and we are most grateful for the many responses we have received. The feedback gathered for this revision of the text was particularly helpful, and we especially wish to thank the following individuals who provided invaluable suggestions for this and the previous edition:

Mary Kay Abbey, *Montgomery College*
Randall Allbritton, *Daytona State College*
Theresa Allen, *University of Idaho*
Sonya Armstrong, *West Virginia State College*
Jannette Avery, *Monroe Community College*
Linda Beattie, *Western New Mexico University*
Linda Beller, *Brevard Community College*
Carla J. Bissell, *University of Nebraska at Omaha*
Jean Bolyard, *Fairmont State University*
Vernon Bridges, *Durham Technical Community College*
Tim C. Caldwell, *Meridian Community College*
Russell Campbell, *Fairmont State University*
Dawn Cox, *Cochise College*
Julie Dewan, *Mohawk Valley Community College*
Bill Dunn, *Las Positas College*
Lucy Edwards, *Las Positas College*
Rob Farinelli, *Community College of Allegheny–Boyce Campus*
Scott Fallstrom, *Shoreline Community College*
J. Lloyd Harris, *Gulf Coast Community College*
Terry Haynes, *Eastern Oklahoma State College*
Edith Hays, *Texas Woman's University*
Anthony Hearn, *Community College of Philadelphia*
Karen Heavin, *Morehead State University*
Elizabeth Heston, *Monroe Community College*
Sharon Jackson, *Brookhaven College*
Susitha Karunaratne, *Purdue University—North Central*
Harriet Kiser, *Floyd College*
Jeffrey Kroll, *Brazosport College*
Barbara Krueger, *Cochise College*

Valerie Lazzara, *Palm Beach State College*
Christine Heinecke Lehmann, *Purdue University—North Central*
Sandy Lofstock, *California Lutheran University*
Valerie H. Maley, *Cape Fear Community College*
Susan McClory, *San Jose State University*
Pam Miller, *Phoenix College*
Jeffrey Mills, *Ohio State University*
Linda J. Murphy, *Northern Essex Community College*
Celia Nippert, *Western Oklahoma State College*
Elizabeth Olgilvie, *Horry-Georgetown Technical College*
Ted Panitz, *Cape Cod Community College*
Claire Peacock, *Chattanooga State Technical Community College*
Faith Peters, *Miami Dade College*
Larry Pontaski, *Pueblo Community College*
Serban Raianu, *California State University—Dominguez Hills*
Janice Rech, *University of Nebraska at Omaha*
Diann Robinson, *Ivy Tech State College—Lafayette*
Rachael Schettenhelm, *Southern Connecticut State University*
Dwight Smith, *Big Sandy Community and Technical College*
Lee Ann Spahr, *Durham Technical Community College*
Theresa Stalder, *University of Illinois–Chicago*
Carol Stewart, *Fairmont State University*
Mark Tom, *College of the Sequoias*
Cora S. West, *Florida State College at Jacksonville*
Leigh Ann Wheeler, *Greenville Technical College*
Johanna Windmueller, *Seminole State College*
Gabriel Yimesghen, *Community College of Philadelphia*

Our sincere thanks go to the dedicated individuals at Pearson who have worked hard to make this revision a success: Maureen O'Connor, Kathy Manley, Barbara Atkinson, Michelle Renda, Rachel Ross, Kari Heen, Christine Whitlock, Lauren Morse, Stephanie Green, and Rachel Haskell.

Abby Tanenbaum did an excellent job updating the real-data applications, as well as helping us with manuscript preparation. We are also grateful to Marilyn Dwyer and Kathy Diamond of Cenveo/Nesbitt Graphics, for their excellent production work; David Abel, for supplying his copyediting expertise; Beth Anderson, for her fine photo research; Lucie Haskins, for producing a useful index; Lisa Collette, for checking the index; Jeff Cole, for writing the solutions manuals; and Janis Cimperman, Chris Heeren, Paul Lorczak, and Sarah Sponholz for timely accuracy checking of the manuscript and page proofs.

As an author team, we are committed to providing the best possible text and supplements package to help students succeed and instructors teach. As we continue to work toward this goal, we would welcome any comments or suggestions you might have via e-mail to *math@pearson.com*.

John Hornsby
Terry McGinnis

Prealgebra Review

R.1 Fractions
R.2 Decimals and Percents
Study Skills *Using Your Math Textbook*
Study Skills *Reading Your Math Textbook*

R.1 Fractions

The numbers used most often in everyday life are the **whole numbers,**

$$0, 1, 2, 3, 4, 5, \ldots$$

and **fractions,** such as

$$\frac{1}{3}, \quad \frac{5}{4}, \quad \text{and} \quad \frac{11}{12}.$$

The parts of a fraction are named as follows.

$$\text{Fraction bar} \rightarrow \frac{4}{7} \begin{array}{l} \leftarrow \text{Numerator} \\ \leftarrow \text{Denominator} \end{array}$$

The fraction bar represents division $\left(\frac{a}{b} = a \div b\right)$.

A fraction is classified as being either a **proper fraction** or an **improper fraction.**

Proper fractions	$\dfrac{1}{5}, \dfrac{2}{7}, \dfrac{9}{10}, \dfrac{23}{25}$	Numerator is **less than** denominator. Value is less than 1.
Improper fractions	$\dfrac{3}{2}, \dfrac{5}{5}, \dfrac{11}{7}, \dfrac{28}{4}$	Numerator is **greater than or equal to** denominator. Value is greater than or equal to 1.

OBJECTIVE 1 Identify prime numbers. In work with fractions, we will need to write the numerators and denominators as products. A **product** is the answer to a multiplication problem. When 12 is written as the product 2×6, for example, 2 and 6 are **factors** of 12. Other factors of 12 are 1, 3, 4, and 12. A whole number is **prime** if it has exactly two different factors (itself and 1).

2, 3, 5, 7, 11, 13, 17, 19, 23, 29, 31, 37   First dozen prime numbers

A whole number greater than 1 that is not prime is a **composite number.**

4, 6, 8, 9, 10, 12, 14, 15, 16, 18, 20, 21   First dozen composite numbers

By agreement, the number 1 is neither prime nor composite.

OBJECTIVES

1. Identify prime numbers.
2. Write numbers in prime factored form.
3. Write fractions in lowest terms.
4. Convert between improper fractions and mixed numbers.
5. Multiply and divide fractions.
6. Add and subtract fractions.
7. Solve applied problems that involve fractions.
8. Interpret data in a circle graph.

❶ Tell whether each number is *prime* or *composite*.

(a) 12

(b) 13

(c) 27

(d) 59

(e) 1806

❷ Write each number in prime factored form.

(a) 70

(b) 72

(c) 693

(d) 97

EXAMPLE 1 **Distinguishing between Prime and Composite Numbers**

Decide whether each number is *prime* or *composite*.

(a) 33   Since 33 has factors of 3 and 11, as well as 1 and 33, it is composite.

(b) 43   Since there are no numbers other than 1 and 43 itself that divide *evenly* into 43, the number 43 is prime.

(c) 9832   Since 9832 can be divided by 2, giving 2×4916, it is composite.

◀ **Work Problem** ❶ **at the Side.**

OBJECTIVE ❷ **Write numbers in prime factored form.** We factor a number by writing it as the product of two or more numbers.

Multiplication	Factoring	
$6 \cdot 3 = 18$	$18 = 6 \cdot 3$	Factoring is the reverse of multiplying two numbers to get the product.

Factors Product   Product Factors

In algebra, a dot $\cdot$ is used instead of the $\times$ symbol to indicate multiplication because $\times$ may be confused with the letter x. A composite number written using factors that are all prime numbers is in **prime factored form.**

EXAMPLE 2 **Writing Numbers in Prime Factored Form**

Write each number in prime factored form.

(a) 35   We factor 35 using the prime factors 5 and 7 as $35 = 5 \cdot 7$.

(b) 24   We use a factor tree, as shown below. The prime factors are circled.

Divide by the least prime factor of 24, which is 2.	$24 = 2 \cdot 12$	② · 12
Divide 12 by 2 to find two factors of 12.	$24 = 2 \cdot 2 \cdot 6$	② · 6
Now factor 6 as $2 \cdot 3$.	$24 = \underline{2 \cdot 2 \cdot 2 \cdot 3}$	② · ③

All factors are prime.

◀ **Work Problem** ❷ **at the Side.**

Note

When factoring, we need not start with the least prime factor. No matter which prime factor we start with, we will *always* obtain the same prime factorization. Verify this in **Example 2(b)** by starting with 3 instead of 2.

OBJECTIVE ❸ **Write fractions in lowest terms.** A fraction is in **lowest terms** when the numerator and denominator have no factors in common (other than 1). The following properties are useful.

Properties of 1

Any nonzero number divided by itself is equal to 1. For example, $\frac{3}{3} = 1$.

Any number multiplied by 1 remains the same. For example, $7 \cdot 1 = 7$.

Writing a Fraction in Lowest Terms

Step 1 Write the numerator and denominator in factored form.

Step 2 Replace each pair of factors common to the numerator and denominator with 1.

Step 3 Multiply the remaining factors in the numerator and in the denominator.

(This procedure is sometimes called **"simplifying the fraction."**)

EXAMPLE 3   **Writing Fractions in Lowest Terms**

Write each fraction in lowest terms.

(a) $\dfrac{10}{15} = \dfrac{2 \cdot 5}{3 \cdot 5} = \dfrac{2}{3} \cdot \dfrac{5}{5} = \dfrac{2}{3} \cdot 1 = \dfrac{2}{3}$ Use the first property of 1 to replace $\frac{5}{5}$ with 1.

(b) $\dfrac{15}{45}$

By inspection, the greatest common factor of 15 and 45 is 15.

$$\dfrac{15}{45} = \dfrac{15}{3 \cdot 15} = \dfrac{1}{3 \cdot 1} = \dfrac{1}{3}$$ Remember to write 1 in the numerator.

If the greatest common factor is not obvious, factor the numerator and denominator into prime factors.

$$\dfrac{15}{45} = \dfrac{3 \cdot 5}{3 \cdot 3 \cdot 5} = \dfrac{1 \cdot 1}{3 \cdot 1 \cdot 1} = \dfrac{1}{3}$$ The same answer results.

(c) $\dfrac{150}{200} = \dfrac{3 \cdot 50}{4 \cdot 50} = \dfrac{3}{4} \cdot 1 = \dfrac{3}{4}$ $\frac{50}{50} = 1$

Note

When writing a fraction in lowest terms, look for the greatest common factor in the numerator and the denominator. If none is obvious, factor the numerator and the denominator into prime factors. *Any* common factor can be used and the fraction can be simplified in stages.

$$\dfrac{150}{200} = \dfrac{15 \cdot 10}{20 \cdot 10} = \dfrac{3 \cdot 5 \cdot 10}{4 \cdot 5 \cdot 10} = \dfrac{3}{4}$$ Example 3(c)

Work Problem ❸ at the Side. ▶

OBJECTIVE ▶ ④ Convert between improper fractions and mixed numbers.
A **mixed number** is a single number that represents the sum of a natural number and a proper fraction.

$$\text{Mixed number} \rightarrow 5\dfrac{3}{4} = 5 + \dfrac{3}{4}$$

Any improper fraction whose value is not a whole number can be rewritten as a mixed number, and any mixed number can be rewritten as an improper fraction.

---

❸ Write each fraction in lowest terms.

(a) $\dfrac{8}{14}$

(b) $\dfrac{35}{42}$

(c) $\dfrac{72}{120}$

Answers

3. (a) $\dfrac{4}{7}$ (b) $\dfrac{5}{6}$ (c) $\dfrac{3}{5}$

4 Write $\frac{92}{5}$ as a mixed number.

5 Write $11\frac{2}{3}$ as an improper fraction.

6 Find each product, and write it in lowest terms.

(a) $\frac{5}{8} \cdot \frac{2}{10}$

(b) $\frac{1}{10} \cdot \frac{12}{5}$

(c) $\frac{7}{9} \cdot \frac{12}{14}$

(d) $3\frac{1}{3} \cdot 1\frac{3}{4}$

Answers

4. $18\frac{2}{5}$

5. $\frac{35}{3}$

6. (a) $\frac{1}{8}$ **(b)** $\frac{6}{25}$ **(c)** $\frac{2}{3}$ **(d)** $\frac{35}{6}$, or $5\frac{5}{6}$

| EXAMPLE 4 | Converting an Improper Fraction to a Mixed Number |

Write $\frac{59}{8}$ as a mixed number.

We divide the numerator of the improper fraction by the denominator.

Denominator of fraction (divisor) → $8\overline{)59}$ ← Numerator of fraction (dividend)

7 ← Quotient

$\underline{56}$

3 ← Remainder

$$\frac{59}{8} = 7\frac{3}{8}$$

◀ **Work Problem 4 at the Side.**

| EXAMPLE 5 | Converting a Mixed Number to an Improper Fraction |

Write $6\frac{4}{7}$ as an improper fraction.

We multiply the denominator of the fraction by the whole number and add the numerator to get the numerator of the improper fraction.

$$7 \cdot 6 + 4 = 42 + 4 = 46$$

The denominator of the improper fraction is the same as the denominator in the mixed number, which is **7** here. Thus, $6\frac{4}{7} = \frac{46}{7}$.

◀ **Work Problem 5 at the Side.**

OBJECTIVE **5** **Multiply and divide fractions.**

Multiplying Fractions

To multiply two fractions, multiply the numerators to get the numerator of the product, and multiply the denominators to get the denominator of the product. *The product should be written in lowest terms.*

| EXAMPLE 6 | Multiplying Fractions |

Find each product, and write it in lowest terms.

(a) $\frac{3}{8} \cdot \frac{4}{9}$

$= \frac{3 \cdot 4}{8 \cdot 9}$ Multiply numerators. Multiply denominators.

$= \frac{3 \cdot 4}{2 \cdot 4 \cdot 3 \cdot 3}$ Factor the denominator.

$= \frac{1}{2 \cdot 3}$ $\frac{3}{3} = 1$ and $\frac{4}{4} = 1$; Remember to write 1 in the numerator.

$= \frac{1}{6}$ Write in lowest terms.

(b) $2\frac{1}{3} \cdot 5\frac{1}{2}$

$= \frac{7}{3} \cdot \frac{11}{2}$ Write each mixed number as an improper fraction.

$= \frac{77}{6}$, or $12\frac{5}{6}$ Multiply numerators and denominators. Write as a mixed number.

◀ **Work Problem 6 at the Side.**

Two fractions are **reciprocals** of each other if their product is 1. See the table in the margin. Because division is the opposite (or inverse) of multiplication, we use reciprocals to divide fractions.

Number	Reciprocal
$\frac{3}{4}$	$\frac{4}{3}$
$\frac{11}{7}$	$\frac{7}{11}$
$\frac{1}{5}$	5, or $\frac{5}{1}$
9, or $\frac{9}{1}$	$\frac{1}{9}$

A number and its reciprocal have a product of 1. For example,

$$\frac{3}{4} \cdot \frac{4}{3} = \frac{12}{12} = 1.$$

> **Dividing Fractions**
>
> To divide two fractions, multiply the first fraction by the reciprocal of the second. The result or **quotient** should be written in lowest terms.

As an example of why this works, we know that

$$20 \div 10 = 2 \quad \text{and also that} \quad 20 \cdot \frac{1}{10} = 2.$$

EXAMPLE 7 Dividing Fractions

Find each quotient, and write it in lowest terms.

(a) $\dfrac{3}{4} \div \dfrac{8}{5}$

$= \dfrac{3}{4} \cdot \dfrac{5}{8}$ Multiply by the reciprocal of the second fraction.

$= \dfrac{3 \cdot 5}{4 \cdot 8}$ Multiply numerators.
            Multiply denominators.

$= \dfrac{15}{32}$

(b) $\dfrac{3}{4} \div \dfrac{5}{8}$

$= \dfrac{3}{4} \cdot \dfrac{8}{5}$ Multiply by the reciprocal.

$= \dfrac{3 \cdot 4 \cdot 2}{4 \cdot 5}$ Multiply and factor.

$= \dfrac{6}{5}, \quad \text{or} \quad 1\dfrac{1}{5}$

(c) $\dfrac{5}{8} \div 10$

$= \dfrac{5}{8} \cdot \dfrac{1}{10}$ Multiply by the reciprocal.

$= \dfrac{5 \cdot 1}{8 \cdot 2 \cdot 5}$ Multiply and factor.

$= \dfrac{1}{16}$ Remember to write 1 in the numerator.

(d) $1\dfrac{2}{3} \div 4\dfrac{1}{2}$

$= \dfrac{5}{3} \div \dfrac{9}{2}$ Write as improper fractions.

$= \dfrac{5}{3} \cdot \dfrac{2}{9}$ Multiply by the reciprocal of the second fraction.

$= \dfrac{10}{27}$ Multiply numerators and denominators.

· **Work Problem ❼ at the Side.** ▶

❼ Find each quotient, and write it in lowest terms.

(a) $\dfrac{3}{10} \div \dfrac{2}{7}$

(b) $\dfrac{3}{4} \div \dfrac{7}{16}$

(c) $\dfrac{4}{3} \div 6$

(d) $3\dfrac{1}{4} \div 1\dfrac{2}{5}$

Answers

7. **(a)** $\dfrac{21}{20}$, or $1\dfrac{1}{20}$ **(b)** $\dfrac{12}{7}$, or $1\dfrac{5}{7}$ **(c)** $\dfrac{2}{9}$

    (d) $\dfrac{65}{28}$, or $2\dfrac{9}{28}$

8 Add. Write sums in lowest terms.

(a) $\dfrac{3}{5} + \dfrac{4}{5}$

Add and subtract fractions. The result of adding two numbers is the **sum** of the numbers. For example, since $2 + 3 = 5$, the sum of 2 and 3 is 5.

Adding Fractions

To find the sum of two fractions with the *same* denominator, add their numerators and *keep the same denominator.*

EXAMPLE 8 Adding Fractions with the Same Denominator

Add. Write sums in lowest terms.

(a) $\dfrac{3}{7} + \dfrac{2}{7}$

$= \dfrac{3 + 2}{7}$, or $\dfrac{5}{7}$ Add numerators.
Keep the same denominator.

(b) $\dfrac{2}{10} + \dfrac{3}{10}$

$= \dfrac{2 + 3}{10}$ Add numerators.
Keep the same denominator.

$= \dfrac{5}{10}$, or $\dfrac{1}{2}$ Write in lowest terms.

◀ Work Problem **8** at the Side.

(b) $\dfrac{5}{14} + \dfrac{3}{14}$

If the fractions to be added do not have the same denominator, we must first rewrite them with a common denominator. For example, to rewrite $\frac{3}{4}$ as a fraction with a denominator of 32, think as follows.

$$\dfrac{3}{4} = \dfrac{?}{32}$$

We must find the number that can be multiplied by 4 to give 32. Since $4 \cdot 8 = 32$, by the second property of 1, we multiply the numerator and the denominator by 8.

$$\dfrac{3}{4} = \dfrac{3}{4} \cdot 1 = \dfrac{3}{4} \cdot \dfrac{8}{8} = \dfrac{3 \cdot 8}{4 \cdot 8} = \dfrac{24}{32}$$

$\frac{3}{4}$ and $\frac{24}{32}$ are equivalent fractions.

This process is the reverse of writing a fraction in lowest terms.

Finding the Least Common Denominator (LCD)

Step 1   Factor all denominators to prime factored form.

Step 2   The LCD is the product of every (different) factor that appears in any of the factored denominators. If a factor is repeated, use the greatest number of repeats as factors of the LCD.

Step 3   Write each fraction with the LCD as the denominator, using the second property of 1.

Answers

8. (a) $\dfrac{7}{5}$, or $1\dfrac{2}{5}$ (b) $\dfrac{4}{7}$

EXAMPLE 9	Adding Fractions with Different Denominators

Add. Write sums in lowest terms.

(a) $\dfrac{4}{15} + \dfrac{5}{9}$

Step 1 To find the LCD, factor the denominators to prime factored form.

$$15 = 5 \cdot 3 \quad \text{and} \quad 9 = 3 \cdot 3$$

3 is a factor of both denominators.

Step 2 $\qquad$ LCD $= 5 \cdot 3 \cdot 3 = 45$

In this example, the LCD needs one factor of 5 and two factors of 3 because the second denominator has two factors of 3.

Step 3 Now we can use the second property of 1 to write each fraction with 45 as the denominator.

$$\frac{4}{15} = \frac{4}{15} \cdot \frac{3}{3} = \frac{12}{45} \quad \text{and} \quad \frac{5}{9} = \frac{5}{9} \cdot \frac{5}{5} = \frac{25}{45}$$

At this stage, the fractions are *not* in lowest terms.

Now add the two equivalent fractions to get the sum.

$$\frac{4}{15} + \frac{5}{9}$$

$$= \frac{12}{45} + \frac{25}{45} \qquad \text{Use a common denominator.}$$

$$= \frac{37}{45} \qquad \text{The sum is in lowest terms.}$$

(b) $3\dfrac{1}{2} + 2\dfrac{3}{4}$

Method 1 $\qquad 3\dfrac{1}{2} + 2\dfrac{3}{4}$

$$= \frac{7}{2} + \frac{11}{4} \qquad \text{Write each mixed number as an improper fraction.}$$

Think: $\frac{7 \cdot 2}{2 \cdot 2} = \frac{14}{4}$

$$= \frac{14}{4} + \frac{11}{4} \qquad \text{Find a common denominator. The LCD is 4.}$$

$$= \frac{25}{4}, \quad \text{or} \quad 6\frac{1}{4} \qquad \text{Add. Write as a mixed number.}$$

Method 2 $\qquad \begin{aligned} 3\frac{1}{2} &= 3\frac{2}{4} \\ + \ 2\frac{3}{4} &= 2\frac{3}{4} \end{aligned}$ Write $3\frac{1}{2}$ as $3\frac{2}{4}$. Then add vertically. Add the whole numbers and the fractions separately.

$$5\frac{5}{4} = 5 + 1\frac{1}{4} = 6\frac{1}{4}, \quad \text{or} \quad \frac{25}{4} \qquad \begin{array}{l}\text{The same}\\ \text{answer results.}\end{array}$$

Work Problem 9 at the Side. ▶

9 Add. Write sums in lowest terms.

(a) $\dfrac{7}{30} + \dfrac{2}{45}$

(b) $\dfrac{17}{10} + \dfrac{8}{27}$

(c) $2\dfrac{1}{8} + 1\dfrac{2}{3}$

(d) $132\dfrac{4}{5} + 28\dfrac{3}{4}$

Answers

9. **(a)** $\dfrac{5}{18}$ **(b)** $\dfrac{539}{270}$, or $1\dfrac{269}{270}$

(c) $\dfrac{91}{24}$, or $3\dfrac{19}{24}$ **(d)** $161\dfrac{11}{20}$

⑩ Subtract.

(a) $\dfrac{9}{11} - \dfrac{3}{11}$

(b) $\dfrac{13}{15} - \dfrac{5}{6}$

(c) $2\dfrac{3}{8} - 1\dfrac{1}{2}$

(d) $50\dfrac{1}{4} - 32\dfrac{2}{3}$

The **difference** between two numbers is found by subtracting the numbers.

Subtracting Fractions

To find the difference between two fractions with the *same* denominator, subtract their numerators and *keep the same denominator.*

If the fractions have *different* denominators, write them with a common denominator first.

EXAMPLE 10 **Subtracting Fractions**

Subtract. Write differences in lowest terms.

(a) $\dfrac{15}{8} - \dfrac{3}{8}$

$= \dfrac{15 - 3}{8}$ Subtract numerators.
Keep the same denominator.

$= \dfrac{12}{8}$

$= \dfrac{3}{2}$,   or $1\dfrac{1}{2}$ Write in lowest terms or as a mixed number.

(b) $\dfrac{15}{16} - \dfrac{4}{9}$

$= \dfrac{15}{16} \cdot \dfrac{9}{9} - \dfrac{4}{9} \cdot \dfrac{16}{16}$ Since 16 and 9 have no common factors greater than 1, the LCD is $16 \cdot 9 = 144$.

$= \dfrac{135}{144} - \dfrac{64}{144}$ Write equivalent fractions.

$= \dfrac{71}{144}$ Subtract numerators.
Keep the common denominator.

(c) $4\dfrac{1}{2} - 1\dfrac{3}{4}$

Method 1 $4\dfrac{1}{2} - 1\dfrac{3}{4}$

$= \dfrac{9}{2} - \dfrac{7}{4}$ Write each mixed number as an improper fraction.

Think: $\frac{9 \cdot 2}{2 \cdot 2} = \frac{18}{4}$ ➤ $= \dfrac{18}{4} - \dfrac{7}{4}$ Find a common denominator. The LCD is 4.

$= \dfrac{11}{4}$,   or $2\dfrac{3}{4}$ Subtract. Write as a mixed number.

Method 2 $4\dfrac{1}{2} = 4\dfrac{2}{4} = 3\dfrac{6}{4}$ The LCD is 4.
$4\dfrac{2}{4} = 3 + 1 + \dfrac{2}{4} = 3 + \dfrac{4}{4} + \dfrac{2}{4} = 3\dfrac{6}{4}$

$ - 1\dfrac{3}{4} = 1\dfrac{3}{4} = 1\dfrac{3}{4}$

$\rule{3cm}{0.4pt}$

$ 2\dfrac{3}{4}$,   or $\dfrac{11}{4}$ The same answer results.

◀ **Work Problem ⑩ at the Side.**

Answers

10. (a) $\dfrac{6}{11}$ **(b)** $\dfrac{1}{30}$ **(c)** $\dfrac{7}{8}$ **(d)** $17\dfrac{7}{12}$

OBJECTIVE ▶ **7** Solve applied problems that involve fractions.

EXAMPLE 11 Solving an Applied Problem with Fractions

The diagram in **Figure 1** appears in the book *Woodworker's 39 Sure-Fire Projects*. Find the height of the desk to the top of the writing surface.

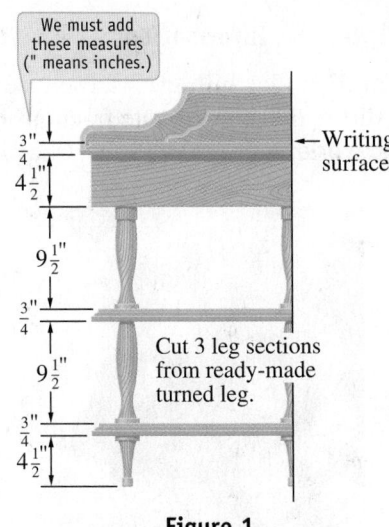

Figure 1

$$\frac{3}{4} \rightarrow \frac{3}{4}$$

$$4\frac{1}{2} = 4\frac{2}{4}$$

$$9\frac{1}{2} = 9\frac{2}{4}$$

$$\frac{3}{4} \rightarrow \frac{3}{4}$$

$$9\frac{1}{2} = 9\frac{2}{4}$$

$$\frac{3}{4} \rightarrow \frac{3}{4}$$

$$+ \; 4\frac{1}{2} = 4\frac{2}{4}$$

$$\overline{\qquad 26\frac{17}{4}}$$

Use Method 2 from **Example 9(b).** The common denominator is 4.

Because $\frac{17}{4}$ is an improper fraction, this is not the final answer.

Since $\frac{17}{4} = 4\frac{1}{4}$, $26\frac{17}{4} = 26 + 4\frac{1}{4} = 30\frac{1}{4}$. The height is $30\frac{1}{4}$ in.

Work Problem 11 at the Side. ▶

EXAMPLE 12 Solving an Applied Problem with Fractions

An upholsterer needs $2\frac{1}{4}$ yd from a bolt of fabric to cover a chair. How many chairs can be covered with $23\frac{2}{3}$ yd of fabric?

To better understand the problem, we replace the fractions with whole numbers. Suppose each chair requires 2 yd, and we have 24 yd of fabric. Dividing 24 by 2 gives 12, the number of chairs that can be covered. To solve the original problem, we must divide $23\frac{2}{3}$ by $2\frac{1}{4}$.

$$23\frac{2}{3} \div 2\frac{1}{4}$$

$$= \frac{71}{3} \div \frac{9}{4} \qquad \text{Convert the mixed numbers to improper fractions.}$$

$$= \frac{71}{3} \cdot \frac{4}{9} \qquad \text{Multiply by the reciprocal.}$$

$$= \frac{284}{27}, \quad \text{or} \quad 10\frac{14}{27} \qquad \text{Multiply numerators and multiply denominators. Write as a mixed number.}$$

Thus, 10 chairs can be covered, with some fabric left over.

Work Problem 12 at the Side. ▶

11 Solve the problem.
To make a three-piece outfit from the same fabric, Wei Jen needs $1\frac{1}{4}$ yd for the blouse, $1\frac{2}{3}$ yd for the skirt, and $2\frac{1}{2}$ yd for the jacket. How much fabric does she need?

12 Solve the problem.
A gallon of paint covers 500 ft². (ft² means square feet.) To paint his house, Tram needs enough paint to cover 4200 ft². How many gallons of paint should he buy?

Answers

11. $5\frac{5}{12}$ yd

12. $8\frac{2}{5}$ gal are needed, so he should buy 9 gal.

13 Refer to the circle graph in **Figure 2.**

(a) Which region had the second-largest number of Internet users in March 2011?

(b) Estimate the number of Internet users in Europe.

(c) How many actual Internet users were there in Europe?

OBJECTIVE **8** **Interpret data in a circle graph.** In a **circle graph,** or **pie chart,** a circle is used to indicate the total of all the data categories represented. The circle is divided into *sectors,* or wedges, whose sizes show the relative magnitudes of the categories. The sum of all the fractional parts must be 1 (for 1 whole circle).

EXAMPLE 13 Using a Circle Graph to Interpret Information

In March 2011, there were about 2100 million (2.1 billion) Internet users worldwide. The circle graph in **Figure 2** shows the approximate fractions of these users living in various regions of the world.

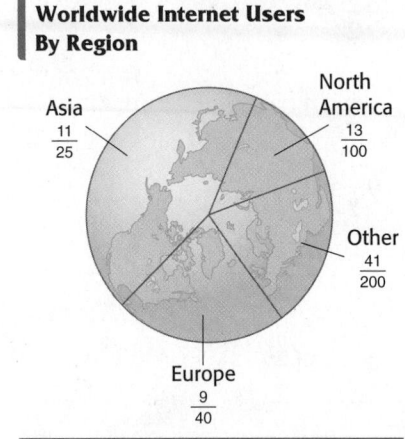

Worldwide Internet Users By Region

Asia $\frac{11}{25}$

North America $\frac{13}{100}$

Other $\frac{41}{200}$

Europe $\frac{9}{40}$

Source: www.internetworldstats.com

Figure 2

(a) Which region had the largest share of Internet users in March 2011? What was that share?

In the circle graph, the sector for Asia is the largest, so Asia had the largest share of Internet users, $\frac{11}{25}$.

(b) Estimate the number of Internet users in North America.

A share of $\frac{13}{100}$ can be rounded to $\frac{10}{100}$, or $\frac{1}{10}$, and the total number of Internet users, 2100 million, can be rounded to 2000 million (2 billion).

$$\frac{1}{10} \cdot 2000 = 200 \text{ million}$$
Multiply to estimate the number of Internet users in North America.

(c) How many actual Internet users were there in North America?

$$\frac{13}{100} \cdot 2100$$
Multiply the actual fraction from the graph for North America by the number of users.

$$= \frac{13}{100} \cdot \frac{2100}{1}$$
$a = \frac{a}{1}$ for all a.

$$= \frac{27,300}{100}$$
Multiply numerators and multiply denominators.

$$= 273$$
Divide.

Thus, 273 million, or 273,000,000 people in North America used the Internet.

◀ **Work Problem 13** at the Side.

Answers

13. (a) Europe (b) 500 million
   (c) 472.5 million

R.1 Exercises

   Download the MyDashBoard App MyMathLab®

CONCEPT CHECK *Decide whether each statement is* true *or* false. *If it is* false, *say why.*

1. In the fraction $\frac{3}{7}$, 3 is the numerator and 7 is the denominator.

2. The mixed number equivalent of $\frac{41}{5}$ is $8\frac{1}{5}$.

3. The fraction $\frac{7}{7}$ is proper.

4. The number 1 is prime.

5. The fraction $\frac{17}{51}$ is in lowest terms.

6. The reciprocal of $\frac{8}{2}$ is $\frac{4}{1}$.

7. The product of 8 and 2 is 10.

8. The difference between 12 and 2 is 6.

Identify each number as prime *or* composite. *See Example 1.*

9. 19

10. 99

11. 52

12. 61

13. 2468

14. 3125

15. 97

16. 83

Write each number in prime factored form. See Example 2.

17. 30

18. 40

19. 252

20. 168

21. 124

22. 165

23. 29

24. 31

Write each fraction in lowest terms. See Example 3.

25. $\frac{8}{16}$

26. $\frac{4}{12}$

27. $\frac{15}{18}$

28. $\frac{16}{20}$

29. $\frac{15}{75}$

30. $\frac{24}{64}$

31. $\frac{144}{120}$

32. $\frac{132}{77}$

Write each improper fraction as a mixed number. ***See Example 4.***

33. $\dfrac{12}{7}$ **34.** $\dfrac{28}{5}$ **35.** $\dfrac{77}{12}$ **36.** $\dfrac{101}{15}$ **37.** $\dfrac{83}{11}$ **38.** $\dfrac{67}{13}$

Write each mixed number as an improper fraction. ***See Example 5.***

39. $2\dfrac{3}{5}$ **40.** $5\dfrac{6}{7}$ **41.** $10\dfrac{3}{8}$ **42.** $12\dfrac{2}{3}$ **43.** $10\dfrac{1}{5}$ **44.** $18\dfrac{1}{6}$

45. CONCEPT CHECK For the fractions $\dfrac{p}{q}$ and $\dfrac{r}{s}$, which can serve as a common denominator?

A. $q \cdot s$ **B.** $q + s$ **C.** $p \cdot r$ **D.** $p + r$

46. CONCEPT CHECK Which fraction is *not* equal to $\dfrac{5}{9}$?

A. $\dfrac{15}{27}$ **B.** $\dfrac{30}{54}$ **C.** $\dfrac{40}{74}$ **D.** $\dfrac{55}{99}$

Find each product or quotient, and write it in lowest terms. ***See Examples 6 and 7.***

47. $\dfrac{4}{5} \cdot \dfrac{6}{7}$ **48.** $\dfrac{5}{9} \cdot \dfrac{10}{7}$ **49.** $\dfrac{1}{10} \cdot \dfrac{12}{5}$ **50.** $\dfrac{6}{11} \cdot \dfrac{2}{3}$

51. $\dfrac{15}{4} \cdot \dfrac{8}{25}$ **52.** $\dfrac{4}{7} \cdot \dfrac{21}{8}$ **53.** $2\dfrac{2}{3} \cdot 5\dfrac{4}{5}$ **54.** $3\dfrac{3}{5} \cdot 7\dfrac{1}{6}$

55. $\dfrac{5}{4} \div \dfrac{3}{8}$ **56.** $\dfrac{7}{6} \div \dfrac{9}{10}$ **57.** $\dfrac{32}{5} \div \dfrac{8}{15}$ **58.** $\dfrac{24}{7} \div \dfrac{6}{21}$

59. $\dfrac{3}{4} \div 12$ **60.** $\dfrac{2}{5} \div 30$ **61.** $2\dfrac{5}{8} \div 1\dfrac{15}{32}$ **62.** $2\dfrac{3}{10} \div 7\dfrac{4}{5}$

Find each sum or difference, and write it in lowest terms. ***See Examples 8–10.***

63. $\dfrac{7}{12} + \dfrac{1}{12}$

64. $\dfrac{3}{16} + \dfrac{5}{16}$

65. $\dfrac{5}{9} + \dfrac{1}{3}$

66. $\dfrac{4}{15} + \dfrac{1}{5}$

67. $3\dfrac{1}{8} + \dfrac{1}{4}$

68. $5\dfrac{3}{4} + \dfrac{2}{3}$

69. $\dfrac{7}{12} - \dfrac{1}{9}$

70. $\dfrac{11}{16} - \dfrac{1}{12}$

71. $6\dfrac{1}{4} - 5\dfrac{1}{3}$

72. $8\dfrac{4}{5} - 7\dfrac{4}{9}$

73. $\dfrac{5}{3} + \dfrac{1}{6} - \dfrac{1}{2}$

74. $\dfrac{7}{15} + \dfrac{1}{6} - \dfrac{1}{10}$

Solve each applied problem. ***See Examples 11 and 12.***

Use the chart to answer the questions in Exercises 75 and 76.

75. How many cups of water would be needed for eight microwave servings of Quaker Quick Grits?

76. How many teaspoons of salt would be needed for five stove top servings? (*Hint:* 5 is halfway between 4 and 6.)

	Microwave		Stove Top	
Servings	**1**	**1**	**4**	**6**
Water	$\frac{3}{4}$ cup	1 cup	3 cups	4 cups
Grits	3 Tbsp	3 Tbsp	$\frac{3}{4}$ cup	1 cup
Salt (optional)	Dash	Dash	$\frac{1}{4}$ tsp	$\frac{1}{2}$ tsp

Source: Package of Quaker Quick Grits.

77. A piece of property has an irregular shape, with five sides as shown in the figure. Find the total distance around the piece of property. This is the **perimeter** of the figure.

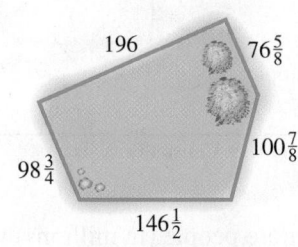

Measurements in feet

78. A triangle has sides of lengths $5\frac{1}{4}$ ft, $7\frac{1}{2}$ ft, and $10\frac{1}{8}$ ft. Find the perimeter of the triangle. **See Exercise 77.**

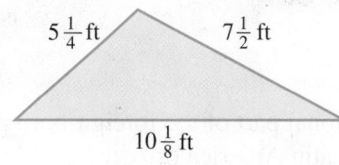

79. A hardware store sells a 40-piece socket wrench set. The measure of the largest socket is $\frac{3}{4}$ in., while the measure of the smallest socket is $\frac{3}{16}$ in. What is the difference between these measures?

80. Two sockets in a socket wrench set have measures of $\frac{9}{16}$ in. and $\frac{3}{8}$ in. What is the difference between these two measures?

81. Under current standards, adopted in 2001, most of the holes (called "eyes") in Grade A Swiss cheese must have a diameter between $\frac{3}{8}$ in. and $\frac{13}{16}$ in. How much less is $\frac{3}{8}$ in. than $\frac{13}{16}$ in.? (*Source*: U.S. Department of Agriculture.)

82. The Pride Golf Tee Company, the only U.S. manufacturer of wooden golf tees, has created the Professional Tee System. Two lengths of tees are the ProLength Max and the Shortee, as shown in the figure. How much longer is the ProLength Max than the Shortee? (*Source: The Gazette.*)

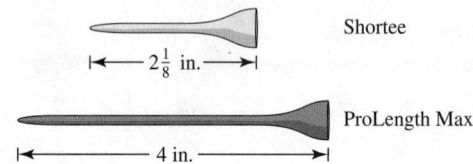

83. It takes $2\frac{3}{8}$ yd from a bolt of fabric to make a costume for a school play. How much fabric would be needed for seven costumes?

84. A cookie recipe calls for $2\frac{2}{3}$ cups of sugar. How much sugar would be needed to make four batches of cookies?

85. A cake recipe calls for $1\frac{3}{4}$ cups of sugar. A caterer has $15\frac{1}{2}$ cups of sugar on hand. How many cakes can he make?

86. Carlotta Valdez needs $2\frac{1}{4}$ yd of fabric to cover a chair. How many chairs can she cover with $23\frac{2}{3}$ yd of fabric?

Approximately 37.5 million people living in the United States in 2010 were born in other countries. The circle graph gives the fractional number from each region of birth for these people. Use the graph to answer each question. See Example 13.

87. What fractional part of the foreign-born population was from other regions?

88. What fractional part of the foreign-born population was from Latin America or Asia?

U.S. Foreign-Born Population By Region of Birth

Other

Latin America $\frac{41}{75}$

Asia $\frac{4}{15}$

Europe $\frac{3}{25}$

Source: U.S. Census Bureau.

89. How many people (in millions) were born in Europe?

90. How many more people (in millions) were born in Latin America than in Asia?

R.2 Decimals and Percents

Fractions are one way to represent parts of a whole. Another way is with a **decimal fraction** or **decimal,** a number written with a decimal point, such as 9.4. Each digit in a decimal number has a place value, as shown below.

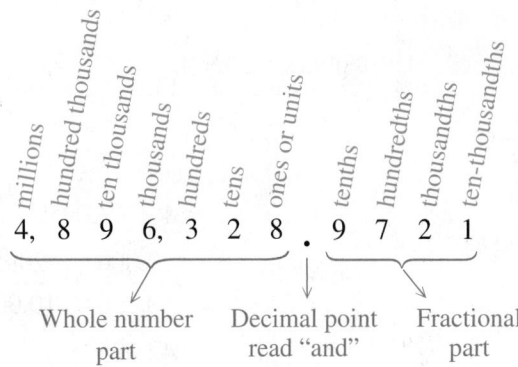

Each successive place value is ten times greater than the place value to its right and is one-tenth as great as the place value to its left.

Prices are often written as decimals. The price $14.75 means 14 dollars and 75 cents, or 14 dollars and $\frac{75}{100}$ of a dollar.

OBJECTIVE 1 Write decimals as fractions. Place value is used to write a decimal number as a fraction. For example, since the last digit (that is, the digit farthest to the right) of 0.67 is in the *hundredths* place,

$$0.67 = \frac{67}{100}.$$

Similarly, $0.9 = \frac{9}{10}$ and $0.25 = \frac{25}{100}$. Digits to the left of the decimal point indicate whole numbers, so 12.342 is the sum of 12 and 0.342.

$$12.342$$

$$= 12 + 0.342 \qquad \text{Write as the sum of a whole number and a decimal.}$$

$$= 12 + \frac{342}{1000} \qquad \text{Write the decimal as a fraction.}$$

$$= \frac{12{,}000}{1000} + \frac{342}{1000} \qquad \text{Write with the common denominator 1000.}$$

$$= \frac{12{,}342}{1000} \qquad \text{Add the numerators and keep the same denominator.}$$

These examples suggest the following rule.

Converting a Decimal to a Fraction

Read the name of the decimal using the correct place value. Write it in fraction form just as we read it. The denominator will be a **power of 10,** a number like 10, 100, 1000, and so on.

The same thing is accomplished by counting the number of digits to the right of the decimal point, and then writing the given number without a decimal point over a denominator of 1 followed by that number of zeros.

❶ Write each decimal as a fraction. Do not write in lowest terms.

(a) 0.8

(b) 0.431

(c) 20.58

❷ Add or subtract as indicated.

(a)
$$\begin{array}{r} 68.9 \\ 42.72 \\ +\ 8.973 \end{array}$$

(b)
$$\begin{array}{r} 32.5 \\ -\ 21.72 \end{array}$$

(c) 42.83 + 71 + 3.074

(d) 351.8 − 2.706

Answers

1. (a) $\frac{8}{10}$ (b) $\frac{431}{1000}$ (c) $\frac{2058}{100}$

2. (a) 120.593 (b) 10.78
   (c) 116.904 (d) 349.094

EXAMPLE 1   Writing Decimals as Fractions

Write each decimal as a fraction. Do not write in lowest terms.

(a) 0.95
We read 0.95 as "ninety-five hundredths," so the fraction form is $\frac{95}{100}$. Using the shortcut method, since there are two places to the right of the decimal point, there will be two zeros in the denominator.

$$0.95 = \frac{95}{100}$$

2 places   2 zeros

(b) $0.056 = \frac{56}{1000}$

3 places   3 zeros

We read 0.056 as "fifty-six thousandths."

(c) 4.2095

4 places

$$= 4 + 0.2095$$

$$= \frac{40{,}000}{10{,}000} + \frac{2095}{10{,}000} \quad \text{The LCD is 10,000.}$$

$$= \frac{42{,}095}{10{,}000}$$

4 zeros

◀ Work Problem ❶ at the Side.

OBJECTIVE ❷ Add and subtract decimals.

EXAMPLE 2   Adding and Subtracting Decimals

Add or subtract as indicated.

(a) 6.92 + 14.8 + 3.217
Place the digits of the numbers in columns, so that tenths are in one column, hundredths in another column, and so on.

Be sure to line up decimal points.

$$\begin{array}{r} 6.92 \\ 14.8 \\ +\ 3.217 \\ \hline 24.937 \end{array}$$ Decimal points are aligned.

To avoid errors, attach zeros to make all the numbers the same length.

$$\begin{array}{r} 6.92 \\ 14.8 \\ +\ 3.217 \end{array} \quad \text{becomes} \quad \begin{array}{r} 6.920 \\ 14.800 \\ +\ 3.217 \\ \hline 24.937 \end{array} \quad \text{Attach zeros.}$$

(b) 47.6 − 32.509

$$\begin{array}{r} 47.6 \\ -\ 32.509 \end{array} \quad \text{becomes} \quad \begin{array}{r} 47.600 \\ -\ 32.509 \\ \hline 15.091 \end{array} \quad \text{Write the numbers in columns, attaching zeros to 47.6.}$$

(c) 3 − 0.253

$$\begin{array}{r} 3.000 \\ -\ 0.253 \\ \hline 2.747 \end{array} \quad \text{A whole number is assumed to have the decimal point at the right of the number. Write 3 as 3.000.}$$

◀ Work Problem ❷ at the Side.

OBJECTIVE ▸ 3 **Multiply and divide decimals.**

Multiplying Decimals

Step 1 Ignore the decimal points and multiply as if the numbers were whole numbers.

Step 2 Add the number of **decimal places** (digits to the *right* of the decimal point) in each number being multiplied. Place the decimal point in the answer that many digits from the right.

EXAMPLE 3 **Multiplying Decimals**

Multiply.

(a) 29.3×4.52

$$
\begin{array}{r}
29.3 \\
\times\ \ 4.52 \\
\hline
586 \\
1465 \\
1172 \\
\hline
132.436
\end{array}
$$

1 decimal place in first number
2 decimal places in second number
$1 + 2 = 3$

3 decimal places in answer

(b) 7.003×55.8

$$
\begin{array}{r}
7.003 \\
\times\ \ \ 55.8 \\
\hline
56024 \\
35015 \\
35015 \\
\hline
390.7674
\end{array}
$$

3 decimal places
1 decimal place
$3 + 1 = 4$

4 decimal places

(c) 31.42×65

$$
\begin{array}{r}
31.42 \\
\times\ \ \ \ 65 \\
\hline
15710 \\
18852 \\
\hline
2042.30
\end{array}
$$

2 decimal places
0 decimal places
$2 + 0 = 2$

2 decimal places

The final 0 can be dropped and the result can be expressed as 2042.3.

Work Problem 3 at the Side. ▶

To divide decimals, follow these steps.

Dividing Decimals

Step 1 Change the **divisor** (the number we are dividing *by*) into a whole number by moving the decimal point as many places as necessary to the right.

Step 2 Move the decimal point in the **dividend** (the number we are dividing *into*) to the right by the same number of places.

Step 3 Move the decimal point straight up and then divide as with whole numbers.

$$
\text{Divisor} \longrightarrow 25\overline{)125} \xleftarrow{} \text{Quotient}
$$

↑
Dividend

Remember this terminology for the parts of a division problem.

---

3 Multiply.

(a) 2.13×0.05

(b) 9.32×1.4

(c) 300.2×0.052

(d) $42,001 \times 0.012$

Answers

3. **(a)** 0.1065 **(b)** 13.048 **(c)** 15.6104
   (d) 504.012

4 Divide.

(a) $14.9\overline{)451.47}$

(b) $0.37\overline{)5.476}$

(c) $375.1 \div 3.001$
(Round the answer to two decimal places.)

Answers

4. (a) 30.3   (b) 14.8
(c) 124.99 (rounded)

EXAMPLE 4   **Dividing Decimals**

Divide.

(a) $233.45 \div 11.5$
Write the problem as follows.

$$11.5\overline{)233.45}$$

To change 11.5 into a whole number, move the decimal point one place to the right. Move the decimal point in 233.45 the *same* number of places to the right, to get 2334.5.

$$11.5\overline{)233.4\,5} \qquad \text{Move each decimal point one place to the right.}$$

To see why this works, write the division in fraction form and multiply by $\frac{10}{10}$, or 1.

$$\frac{233.45}{11.5} \cdot \frac{10}{10} = \frac{2334.5}{115}$$

The result is the same as when we moved the decimal point one place to the right in the divisor and the dividend.
Move the decimal point straight up and divide as with whole numbers.

$$
\begin{array}{r}
20.3 \\
115\overline{)2334.5} \\
230 \\
\hline
345 \\
345 \\
\hline
0
\end{array}
\qquad \text{Move the decimal point straight up.}
$$

In the second step of the division, 115 does not divide into 34, so we used zero as a placeholder in the quotient.

(b) $73.85\overline{)1852.882}$ (Round the answer to two decimal places.)
Move the decimal point two places to the right in 73.85, to get 7385. Do the same thing with 1852.882, to get 185288.2.

$$73.85\,\overline{)1852.88\,2}$$

Move the decimal point straight up and divide as with whole numbers.

$$
\begin{array}{r}
25.089 \\
7385\overline{)185288.200} \\
14770 \\
\hline
37588 \\
36925 \\
\hline
66320 \\
59080 \\
\hline
72400 \\
66465 \\
\hline
5935 \\
\end{array}
$$

We carried out the division to three decimal places so that we could round to two decimal places, obtaining the quotient 25.09.

◀ **Work Problem** **4** **at the Side.**

A shortcut can be used when multiplying or dividing by powers of 10.

Multiplying or Dividing by Powers of 10

To *multiply* by a power of 10, *move the decimal point to the right* as many places as the number of zeros.

To *divide* by a power of 10, *move the decimal point to the left* as many places as the number of zeros.

In both cases, insert 0s as placeholders if necessary.

EXAMPLE 5 **Multiplying and Dividing by Powers of 10**

Multiply or divide as indicated.

(a) 48.731×100
   $= 48.73 \ 1$ or 4873.1 Move the decimal point two places to the right because 100 has two zeros.

(b) $48.7 \div 1000$
   $= 048.7$ or 0.0487 Move the decimal point three places to the left because 1000 has three zeros. Insert a zero in front of the 4 to do this.

························ Work Problem **5** at the Side. ▶

To avoid misplacing the decimal point, check your work by estimating the answer. *To get a quick estimate, round numbers so that only the first digit is not zero,* using the rule for rounding. For more accurate estimates, numbers could be rounded to the first two or three nonzero digits.

Rule for Rounding

If the digit to become 0 or be dropped is 5 or more, round up by adding 1 to the final digit to be kept.

If the digit to become 0 or be dropped is 4 or less, do not round up.

For example, to estimate the answer to **Example 2(a)** from earlier in this section $(6.92 + 14.8 + 3.217)$, round as follows.

6.92 to **7**,   14.8 to **10**,   and   3.217 to **3**

↑   ↑   ↑

5 or more   4 or less   4 or less

Since $7 + 10 + 3 = 20$, our earlier answer of 24.937 is reasonable. In **Example 4(a)**, round 233.45 to 200 and 11.5 to 10. Since $200 \div 10 = 20$, the answer of 20.3 is reasonable.

OBJECTIVE ▶ **4** **Write fractions as decimals.**

Writing a Fraction as a Decimal

Because a fraction bar indicates division, write a fraction as a decimal by dividing the numerator by the denominator.

---

5 Multiply or divide as indicated.

(a) 294.72×10

(b) 19.5×1000

(c) $4.793 \div 100$

(d) $960.1 \div 10$

Answers

5. (a) 2947.2 **(b)** 19,500 **(c)** 0.04793
  (d) 96.01

6 Convert to decimals. For repeating decimals, write the answer two ways: using the bar notation and rounding to the nearest thousandth.

(a) $\dfrac{2}{9}$

(b) $\dfrac{17}{20}$

(c) $\dfrac{1}{11}$

EXAMPLE 6 Writing Fractions as Decimals

Write each fraction as a decimal.

(a) $\dfrac{19}{8}$

$$
\begin{array}{r}
2.375 \\
8\overline{)19.000} \\
16 \\
\hline
30 \\
24 \\
\hline
60 \\
56 \\
\hline
40 \\
40 \\
\hline
0
\end{array}
$$

Divide 19 by 8. Add a decimal point and as many 0s as necessary.

$$\dfrac{19}{8} = 2.375$$

(b) $\dfrac{2}{3}$

$$
\begin{array}{r}
0.6666\ldots \\
3\overline{)2.0000\ldots} \\
18 \\
\hline
20 \\
18 \\
\hline
20 \\
18 \\
\hline
20 \\
18 \\
\hline
20
\end{array}
$$

$$\dfrac{2}{3} = 0.6666\ldots$$

The remainder in the division in part (b) is never 0. Because 2 is always left after the subtraction, this quotient is a **repeating decimal.** A convenient notation for a repeating decimal is a bar over the digit (or digits) that repeats.

$$\dfrac{2}{3} = 0.6666\ldots, \quad \text{or} \quad 0.\overline{6}$$

We often round repeating decimals to as many places as needed.

$$\dfrac{2}{3} \approx 0.667 \qquad \text{An approximation to the nearest thousandth}$$
$$(\approx \text{ means "approximately equal to")}$$

> **CAUTION**
> When rounding, be careful to distinguish between *thousandths* and *thousands* or between *hundredths* and *hundreds*, and so on.

◀ **Work Problem 6 at the Side.**

OBJECTIVE 5 **Convert percents to decimals and decimals to percents.** An important application of decimals is in work with percents. The word **percent** means "per one hundred." Percent is written with the symbol %. *One percent means "one per one hundred," or "one one-hundredth."*

$$1\% = 0.01, \quad \text{or} \quad 1\% = \dfrac{1}{100}$$

EXAMPLE 7 Converting Percents and Decimals

(a) Write 73% as a decimal.
  Since 1% = 0.01, we convert as follows.

$$73\% = 73 \cdot 1\% = 73 \times 0.01 = 0.73$$

Also, 73% can be written as a decimal using the fraction form $1\% = \frac{1}{100}$.

$$73\% = 73 \cdot 1\% = 73 \cdot \dfrac{1}{100} = \dfrac{73}{100} = 0.73$$

Continued on Next Page

Answers

6. **(a)** $0.\overline{2}, 0.222$ **(b)** 0.85 **(c)** $0.\overline{09}, 0.091$

(b) Write 125% as a decimal.

$$125\% = 125 \cdot 1\% = 125 \times 0.01 = 1.25$$

> A percent greater than 100 represents a number greater than 1.

(c) Write $3\frac{1}{2}\%$ as a decimal.

First write the fractional part as a decimal.

$$3\frac{1}{2}\% = (3 + 0.5)\% = 3.5\%$$

Now change the percent to decimal form.

$$3.5\% = 3.5 \times 0.01 = 0.035 \qquad 1\% = 0.01$$

(d) Write 0.32 as a percent.

Since 0.32 means 32 hundredths, write 0.32 as 32×0.01. Finally, replace 0.01 with 1%.

$$0.32 = 32 \times 0.01 = 32 \times 1\% = 32\%$$

(e) Write 2.63 as a percent.

$$2.63 = 263 \times 0.01 = 263 \times 1\% = 263\%$$

> A number greater than 1 is more than 100%.

❼ Convert as indicated.

(a) 23% to a decimal

(b) 310% to a decimal

(c) 0.71 to a percent

(d) 1.32 to a percent

(e) 0.685 to a percent

(f) 0.006 to a percent

Note

A quick way to change from a percent to a decimal is to move the decimal point two places to the left and drop the % symbol. To change from a decimal to a percent, move the decimal point two places to the right and attach the percent symbol.

Divide by 100.
← Move 2 places left.

Decimal   **Percent**

Multiply by 100.
Move 2 places right. →

EXAMPLE 8 Converting Percents and Decimals by Moving the Decimal Point

Convert each percent to a decimal and each decimal to a percent.

(a) $45\% = 0.45$

(b) $250\% = 2.50$

(c) $0.57 = 57\%$

(d) $1.5 = 1.50 = 150\%$

(e) $0.327 = 32.7\%$

(f) $0.007 = 0.007 = 0.7\%$

·· **Work Problem ❼ at the Side.** ▶

In this book, we use 0 in the ones place for decimals between 0 and 1, such as 0.5 and 0.72. Some calculators, including graphing calculators, do *not* show 0 in the ones place. Either way is correct.

Answers

7. **(a)** 0.23 **(b)** 3.10 **(c)** 71%
(d) 132% **(e)** 68.5% **(f)** 0.6%

8 Solve each problem.

(a) Chris found a pair of jeans, regularly priced at $50, on sale at 20% off. If he buys the jeans on sale, how much will he save? What is the sale price?

(b) Miguel was earning $7.50 per hour on his job and received a $33\frac{1}{3}\%$ raise. How much was his hourly raise? What was his new hourly pay?

OBJECTIVE ▶ **6** **Use fraction, decimal, and percent equivalents.** The fraction $\frac{1}{2}$ can be written as the decimal 0.5, or as the percent 50%. *These are three different ways of writing the same number.*

FRACTION, DECIMAL, AND PERCENT EQUIVALENTS

Fraction (or whole number)	Decimal	Percent
$\frac{1}{100}$	0.01	1%
$\frac{1}{50}$	0.02	2%
$\frac{1}{20}$	0.05	5%
$\frac{1}{10}$	0.1	10%
$\frac{1}{8}$	0.125	$12\frac{1}{2}\%$, or 12.5%
$\frac{1}{6}$	$0.1\overline{6}$	$16\frac{2}{3}\%$, or $16.\overline{6}\%$
$\frac{1}{5}$	0.2	20%
$\frac{1}{4}$	0.25	25%
$\frac{1}{3}$	$0.\overline{3}$	$33\frac{1}{3}\%$, or $33.\overline{3}\%$
$\frac{1}{2}$	0.5	50%
$\frac{2}{3}$	$0.\overline{6}$	$66\frac{2}{3}\%$, or $66.\overline{6}\%$
$\frac{3}{4}$	0.75	75%
1	1.0	100%

EXAMPLE 9 **Using Percent and Fraction Equivalents**

Solve each problem.

(a) Marissa bought a jacket with a regular price of $80, on sale at 25% off. How much money will she save? What is the sale price?

From the table, $25\% = \frac{1}{4}$, and $\frac{1}{4}$ of $80 is **$20**, so she will save $20.

$$\$80 - \$20 = \$60 \quad \text{Original price } - \text{ discount } = \text{ sale price}$$

(b) Caleb pays $600 per month in rent. If his landlord raises the rent by 15%, how much more will he pay each month? What will his new rent be?

$$15\% = 10\% + 5\%$$

We know that $10\% = \frac{1}{10}$. Thus, 10%, or $\frac{1}{10}$, of $600 is **$60**. Also, 5% is half of 10%, so since 10% of $600 is $60, it follows that 5% of $600 is half of $60, or **$30**. Caleb's new monthly rent will be $600 plus these increases.

$$\$600 + \$60 + \$30 = \$690 \quad \text{New monthly rent}$$
$$\underbrace{10\% + 5\%}_{15\%}$$

He will pay $60 + $30 = $90 more each month.

◀ **Work Problem 8** at the Side.

R.2 Exercises

FOR EXTRA HELP

Download the MyDashBoard App

MyMathLab®

CONCEPT CHECK *In Exercises 1–8, provide the correct response.*

1. In the decimal 367.9412, name the digit that has each place value.

   (a) tens   (b) tenths   (c) thousandths

   (d) ones or units   (e) hundredths

2. Write a number that has 5 in the thousands place, 0 in the tenths place, and 4 in the ten-thousandths place.

3. For the decimal number 46.249, round to the place value indicated.

   (a) hundredths   (b) tenths

   (c) ones or units   (d) tens

4. Round each decimal to the nearest thousandth.

   (a) $0.\overline{8}$ (b) $0.\overline{5}$

   (c) 0.9762   (d) 0.8642

5. For the sum $35.89 + 24.1$, which is the best estimate?

   A. 40   **B.** 50   **C.** 60   **D.** 70

6. For the difference $109.83 - 32.4$, which is the best estimate?

   A. 40   **B.** 50   **C.** 60   **D.** 70

7. For the product 84.9×98.3, which is the best estimate?

   A. 7000   **B.** 8000   **C.** 80,000   **D.** 70,000

8. For the quotient $9845.3 \div 97.2$, which is the best estimate?

   A. 10   **B.** 1000   **C.** 100   **D.** 10,000

Write each decimal as a fraction. Do not write in lowest terms. **See Example 1.**

9. 0.4   10. 0.6   11. 0.64   12. 0.82

13. 0.138   14. 0.104   15. 3.805   16. 5.166

Add or subtract as indicated. Make sure that your answer is reasonable by estimating first. Give your estimate, and then give the exact answer. **See Example 2** *and the rules for rounding in this section.*

17. $25.32 + 109.2 + 8.574$ 18. $90.527 + 32.43 + 589.83 + 399.327$

19. $28.73 - 3.12$ 20. $46.88 - 13.45$ 21. $43.5 - 28.17$ 22. $345.1 - 56.31$

23. $32.56 + 47.356 + 1.8$ 24. $75.2 + 123.96 + 3.897$ 25. $18 - 2.789$ 26. $29 - 8.582$

Multiply or divide as indicated. Make sure that your answer is reasonable by estimating first. Give your estimate, and then give the exact answer. See **Examples 3–5** *and the rules for rounding in this section.*

27. 0.2×0.03 28. 0.07×0.004 29. 12.8×9.1 30. 34.04×0.56

31. $78.65 \div 11$ 32. $73.36 \div 14$ 33. $19.967 \div 9.74$ 34. $44.4788 \div 5.27$

35. 57.116×100 **36.** 0.094×1000 **37.** $1.62 \div 10$ **38.** $24.03 \div 100$

Write each fraction as a decimal. For repeating decimals, write the answer two ways: using bar notation and rounding to the nearest thousandth. See Example 6.

39. $\dfrac{1}{8}$ **40.** $\dfrac{7}{8}$ **41.** $\dfrac{5}{4}$ **42.** $\dfrac{9}{5}$

43. $\dfrac{5}{9}$ **44.** $\dfrac{8}{9}$ **45.** $\dfrac{1}{6}$ **46.** $\dfrac{5}{6}$

47. In your own words, explain how to convert a decimal to a percent.

48. In your own words, explain how to convert a percent to a decimal.

Convert each percent to a decimal. See Examples 7(a)–(c), 8(a), and 8(b).

49. 54% **50.** 39% **51.** 117% **52.** 189% **53.** 2.4%

54. 3.1% **55.** $6\dfrac{1}{4}\%$ **56.** $5\dfrac{1}{2}\%$ **57.** 0.8% **58.** 0.9%

Convert each decimal to a percent. See Examples 7(d), 7(e), and 8(c)–(f).

59. 0.73 **60.** 0.83 **61.** 0.004 **62.** 0.005 **63.** 1.28

64. 2.35 **65.** 0.3 **66.** 0.6 **67.** 6 **68.** 10

*One method of converting a fraction to a percent is to first convert the fraction to a decimal, as shown in **Example 6,** and then convert the decimal to a percent, as shown in **Examples 7 and 8.** Convert each fraction to a percent in this way.*

69. $\dfrac{4}{5}$ **70.** $\dfrac{3}{25}$ **71.** $\dfrac{2}{11}$ **72.** $\dfrac{4}{9}$

73. $\dfrac{7}{4}$ **74.** $\dfrac{11}{8}$ **75.** $\dfrac{13}{6}$ **76.** $\dfrac{31}{9}$

Solve each problem mentally. See Example 9.

77. Liam, Shannon, Chris, and Kayla went out for dinner together. The total cost of their meals was $75. If they decide to leave a 20% tip, how much should the tip be?

78. Where Lan lives, a sales tax of 5% is added to all purchases. If he buys a Blu-ray player that costs $80, how much will he pay for the Blu-ray player, including the tax?

79. Nicole took an algebra test that was worth 50 points. If she earned 35 points on the test, what percent did she score?

80. In a survey of 2400 eligible voters, 400 said that they planned to vote in an upcoming election. What percent of those surveyed planned to vote?

Study Skills
USING YOUR MATH TEXTBOOK

Your textbook is a valuable resource. You will learn more if you fully make use of the features it offers.

General Features

Locate each feature, and complete the blanks as indicated.

▶ **Table of Contents**   This is located at the front of the text. Find it and mark the chapters and sections you will cover, as noted on your course syllabus.

▶ **Answer Section**   At the back of the book, answers to odd-numbered section exercises are provided. Answers to ALL Concept Check, writing, Relating Concepts, summary, chapter review, test, and cumulative review exercises are given. Tab this section so you can refer to it when doing homework.

▶ **Solution Section**   At the back of the book, step-by-step solutions are provided to selected exercises that have a blue screen over the exercise number, such as **31**. In Section 1.1, Exercise _____ indicates a solution in the Solution Section. This solution is located on page _____.

▶ **List of Formulas**   Inside the back cover of the text is a helpful list of geometric formulas, along with review information on triangles and angles. Use these for reference throughout the course. The formula for the volume of a cube is _____.

Specific Features

▶ **Objectives**   The objectives are listed at the beginning of each section and again within the section as the corresponding material is presented. Once you finish a section, ask yourself if you have accomplished them.

▶ **Margin Problems**   These exercises allow you to immediately practice the material covered in the examples and prepare you for the exercise set. Check your results using the answers at the bottom of the page.

▶ **Pointers**   These small shaded balloons provide on-the-spot warnings and reminders, point out key steps, and give other helpful tips.

▶ **Cautions**   These provide warnings about common errors that students often make or trouble spots to avoid.

▶ **Notes**   These provide additional explanations or emphasize important ideas.

▶ **Problem-Solving Hints**   These give helpful tips or strategies to use when you work applications.

▶ **Calculator Tips**   Marked with 🖩, these tips provide helpful information about using a calculator. A 🖩 beside an exercise is a recommendation to solve the exercise using a calculator.

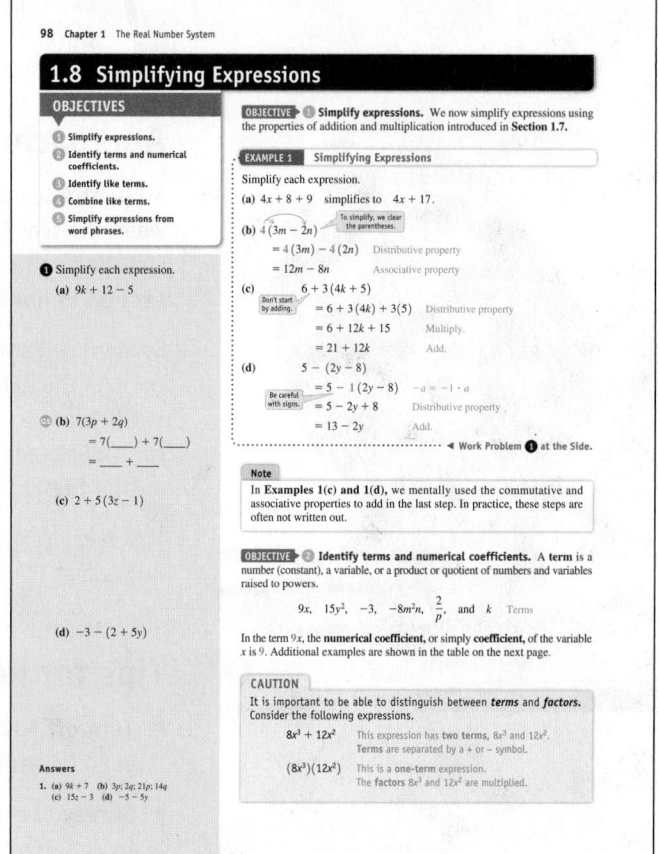

Complete each of the following.

Write Objective 5 from Section 3.1. _____

Write the answer to Section 1.7, Margin Problem 1(a). _____

Write the information given by a pointer on page 30. _____

List a page number for each of the following.

A Caution in Section 2.3 _____

A Note in Section 3.3 _____

A Problem-Solving Hint in Section 2.4 _____

A Calculator Tip in Section 1.1 _____

Study Skills
READING YOUR MATH TEXTBOOK

*T*ake time to read each section and its examples before doing your homework. You will learn more and be better prepared to work the exercises your instructor assigns.

Approaches to Reading Your Math Textbook

Student A learns best by listening to her teacher explain things. She "gets it" when she sees the instructor work problems. She previews the section before the lecture, so she knows generally what to expect. **Student A carefully reads the section in her text *AFTER* she hears the classroom lecture on the topic.**

Student B learns best by reading on his own. He reads the section and works through the examples before coming to class. That way, he knows what the teacher is going to talk about and what questions he wants to ask. **Student B carefully reads the section in his text *BEFORE* he hears the classroom lecture on the topic.**

Which reading approach works best for you—that of Student A or Student B? _____

Tips for Reading Your Math Textbook

▶ **Turn off your cell phone.** You will be able to concentrate more fully on what you are reading.

▶ **Survey the material.** Take a few minutes to glance over the assigned material to get an idea of the "big picture." Look at the list of objectives to see what you will be learning.

▶ **Read slowly.** Read only one section—or even part of a section—at a sitting, with paper and pencil in hand.

▶ **Pay special attention to important information given in colored boxes or set in boldface type.** Highlight any additional information you find especially helpful.

▶ **Study the examples carefully.** Pay particular attention to the blue side comments and any pointers.

▶ **Do the margin problems in the workspace provided or on separate paper as you go.** These mirror the examples and prepare you for the exercise set. Check your answers with those given at the bottom of the page.

▶ **Make study cards as you read.** Make cards for new vocabulary, rules, procedures, formulas, and sample problems.

▶ **Mark anything you don't understand. *ASK QUESTIONS* in class—everyone will benefit. Follow up with your instructor, as needed.**

Mark two or three reading tips to try this week.

1

The Real Number System

Positive and negative numbers, shown here to indicate gains and losses, are examples of real numbers, the subject of this chapter.

1.1 Exponents, Order of Operations, and Inequality

Study Skills Taking Lecture Notes

1.2 Variables, Expressions, and Equations

Study Skills Tackling Your Homework

1.3 Real Numbers and the Number Line

Study Skills Using Study Cards

1.4 Adding Real Numbers

1.5 Subtracting Real Numbers

1.6 Multiplying and Dividing Real Numbers

Summary Exercises Performing Operations with Real Numbers

1.7 Properties of Real Numbers

Study Skills Reviewing a Chapter

1.8 Simplifying Expressions

Study Skills Taking Math Tests

Math in the Media The Magic Number in Sports

1.1 Exponents, Order of Operations, and Inequality

OBJECTIVES

1. Use exponents.
2. Use the rules for order of operations.
3. Use more than one grouping symbol.
4. Know the meanings of $\neq$, $<$, $>$, $\leq$, and $\geq$.
5. Translate word statements to symbols.
6. Write statements that change the direction of inequality symbols.

OBJECTIVE ▶ 1 Use exponents. In **Chapter R,** we factored a number as the product of its prime factors. For example,

81   can be written as $3 \cdot 3 \cdot 3 \cdot 3$, $\cdot$ indicates multiplication.

where the factor 3 appears four times. Repeated factors are written in an abbreviated form by using an *exponent.*

$$\underbrace{3 \cdot 3 \cdot 3 \cdot 3}_{\text{4 factors of 3}} = 3^{\overset{\displaystyle\longleftarrow \text{Exponent}}{4}}_{\longrightarrow \text{Base}}$$

The number 4 is the **exponent,** or **power,** and 3 is the **base** in the **exponential expression** 3^4. The exponent tells how many times the base is used as a factor. We read 3^4 as **"3 to the fourth power,"** or simply **"3 to the fourth."** *A number raised to the first power is simply that number.* For example,

$$6^1 = 6 \quad \text{and} \quad (2.5)^1 = 2.5. \quad \text{In general, } a^1 = a.$$

EXAMPLE 1   Evaluating Exponential Expressions

Find the value of each exponential expression.

(a) 5^2 means $\underbrace{5 \cdot 5,}$ which equals   25.

$\qquad\qquad$ 5 is used as a factor 2 times.

Read 5^2 as "5 to the second power" or, more commonly, "5 squared."

(b) 6^3 means $\underbrace{6 \cdot 6 \cdot 6,}$ which equals   216.

$\qquad\qquad$ 6 is used as a factor 3 times.

Read 6^3 as "6 to the third power" or, more commonly, "6 cubed."

(c) 2^5 means $2 \cdot 2 \cdot 2 \cdot 2 \cdot 2$,   which equals   32.   2 is used as a factor 5 times.
Read 2^5 as "2 to the fifth power."

(d) $\left(\dfrac{2}{3}\right)^3$ means $\dfrac{2}{3} \cdot \dfrac{2}{3} \cdot \dfrac{2}{3}$,   which equals $\dfrac{8}{27}$. $\frac{2}{3}$ is used as a factor 3 times.

(e) $(0.3)^2$ means $0.3\,(0.3)$,   which equals   0.09.   0.3 is used as a factor 2 times.

◀ **Work Problem** ❶ **at the Side.**

CAUTION

Squaring, or raising a number to the second power, is not the same as doubling the number. **For example,**

$$3^2 \quad \textbf{means} \quad 3 \cdot 3, \quad \textit{not} \quad 2 \cdot 3.$$

Thus $3^2 = 9$, not 6. Similarly, cubing, or raising a number to the third power, does *not* mean tripling the number.

---

❶ Find the value of each exponential expression.

(a) 6^2 means ____ $\cdot$ ____, which equals ____.

(b) 3^5

(c) $\left(\dfrac{3}{4}\right)^2$ means ____ $\cdot$ ____, which equals ____.

(d) $\left(\dfrac{1}{2}\right)^4$

(e) $(0.4)^3$ means _____, which equals ____.

Answers

1. **(a)** 6; 6; 36   **(b)** 243   **(c)** $\dfrac{3}{4}; \dfrac{3}{4}; \dfrac{9}{16}$

$\quad$ **(d)** $\dfrac{1}{16}$ **(e)** $0.4\,(0.4)\,(0.4)$; 0.064

OBJECTIVE **2** **Use the rules for order of operations.** When a problem involves more than one operation, we often use **grouping symbols.** If no grouping symbols are used, we apply the rules for order of operations.

Consider the expression $5 + 2 \cdot 3$. To show that the multiplication should be performed before the addition, we use parentheses to group $2 \cdot 3$.

$$5 + (2 \cdot 3) \quad \text{equals} \quad 5 + 6, \quad \text{or} \quad 11.$$

If addition is to be performed first, the parentheses should group $5 + 2$.

$$(5 + 2) \cdot 3 \quad \text{equals} \quad 7 \cdot 3, \quad \text{or} \quad 21.$$

Other grouping symbols are brackets [ ], braces { }, and fraction bars. (For example, in $\frac{8-2}{3}$, the expression $8 - 2$ is considered to be grouped in the numerator.)

To work problems with more than one operation, use the following rules for **order of operations.** This order is used by most calculators and computers.

Order of Operations

If grouping symbols are present, simplify within them, innermost first (and above and below fraction bars separately), in the following order.

Step 1 Apply all **exponents.**

Step 2 Do any **multiplications** or **divisions** in the order in which they occur, working from left to right.

Step 3 Do any **additions** or **subtractions** in the order in which they occur, working from left to right.

If no grouping symbols are present, start with Step 1.

Another way to show multiplication besides with a raised dot is with parentheses.

$$3(7) \quad \text{means} \quad 3 \cdot 7, \quad \text{or} \quad 21.$$

$$3(4 + 5) \quad \text{means} \quad \text{"3 times the sum of 4 and 5."}$$

When simplifying $3(4 + 5)$, the sum in parentheses must be found first, then the product.

EXAMPLE 2 **Using the Rules for Order of Operations**

Find the value of each expression.

(a) $24 - 12 \div 3$ — Be careful. Divide first.

$= 24 - 4$ Divide.

$= 20$ Subtract.

(b) $9(6 + 11)$

$= 9(17)$ Work inside parentheses.

$= 153$ Multiply.

(c) $6 \cdot 8 + 5 \cdot 2$

$= 48 + 10$ Multiply, working from left to right.

$= 58$ Add.

Continued on Next Page

2 Label the order in which each
GS expression should be evaluated.
Then find the value of each
expression.

(a) $7 + 3 \cdot 8$

② ①

$= 7 + \underline{\qquad}$

$= \underline{\qquad}$

(b) $2 \cdot 9 + 7 \cdot 3$

○ ○ ○

$= \underline{\qquad} + \underline{\qquad}$

$= \underline{\qquad}$

(c) $7 \cdot 6 - 3(8 + 1)$

○ ○○ ○

(d) $2 + 3^2 - 5$

○○○

3 Find the value of each
expression.

(a) $9[(4 + 8) - 3]$

(b) $\dfrac{2(7 + 8) + 2}{3 \cdot 5 + 1}$

(d)

$$2(5 + 6) + 7 \cdot 3$$

Start here.

$= 2(11) + 7 \cdot 3$ Work inside parentheses.

$= 22 + 21$ Multiply.

$= 43$ Add.

$2^3 = 2 \cdot 2 \cdot 2,$ not $2 \cdot 3$

(e) $\qquad 9 + 2^3 - 5$

$= 9 + 8 - 5$ Apply the exponent.

$= 12$ Add, and then subtract.

◀ **Work Problem ❷ at the Side.**

OBJECTIVE ▶ 3 Use more than one grouping symbol. In an expression
such as $2(8 + 3(6 + 5))$, we often use brackets, [ ], in place of the outer
pair of parentheses.

EXAMPLE 3   **Using Brackets and Fraction Bars as Grouping Symbols**

Find the value of each expression.

Start here.

(a) $2[8 + 3(6 + 5)]$

$= 2[8 + 3(11)]$ Add inside parentheses.

$= 2[8 + 33]$ Multiply inside brackets.

$= 2[41]$ Add inside brackets.

$= 82$ Multiply.

(b) $\dfrac{4(5 + 3) + 3}{2(3) - 1}$ Simplify the numerator and denominator separately.

$= \dfrac{4(8) + 3}{2(3) - 1}$ Work inside parentheses.

$= \dfrac{32 + 3}{6 - 1}$ Multiply.

$= \dfrac{35}{5}$ Add and subtract.

$= 7$ Divide.

◀ **Work Problem ❸ at the Side.**

Note

The expression $\dfrac{4(5 + 3) + 3}{2(3) - 1}$ in **Example 3(b)** can be written as a quotient.

$$[4(5 + 3) + 3] \div [2(3) - 1]$$

The fraction bar "groups" the numerator and denominator separately.

Answers

2. **(a)** 24; 31   **(b)** ①, ③, ②; 18; 21; 39

   (c) ②, ④, ③, ①; 15

   (d) ②, ①, ③; 6

3. **(a)** 81   **(b)** 2

▦ **Calculator Tip**

Calculators follow the order of operations. Try some of the examples to see that your calculator gives the same answers. Use the parentheses keys to insert parentheses where they are needed, such as around the numerator and the denominator in **Example 3(b).**

OBJECTIVE ▶ ④ **Know the meanings of ≠, <, >, ≤, and ≥.** So far, we have used only the symbols of arithmetic, such as $+, -, \cdot,$ and $\div$ and the equality symbol $=$. The equality symbol with a slash through it means "is *not* equal to."

$$7 \neq 8 \quad \text{7 is not equal to 8.}$$

If two numbers are not equal, then one of the numbers must be less than the other. The symbol $<$ represents "is less than."

$$7 < 8 \quad \text{7 is less than 8.}$$

The symbol $>$ means "is greater than."

$$8 > 2 \quad \text{8 is greater than 2.}$$

To keep the meanings of the symbols $<$ and $>$ clear, remember that the symbol always points to the lesser number.

$$\text{Lesser number} \rightarrow 8 < 15$$

$$15 > 8 \leftarrow \text{Lesser number}$$

Work Problem ④ at the Side. ▶

The symbol $\leq$ means "is less than or equal to."

$$5 \leq 9 \quad \text{5 is less than or equal to 9.}$$

If either the $<$ part or the $=$ part is true, then the inequality $\leq$ is true. The statement $5 \leq 9$ is true. Also, $8 \leq 8$ is true because $8 = 8$ is true.

The symbol $\geq$ means "is greater than or equal to."

$$9 \geq 5 \quad \text{9 is greater than or equal to 5.}$$

EXAMPLE 4 Using the Symbols $\leq$ and $\geq$

Determine whether each statement is *true* or *false*.

(a) $15 \leq 20$ The statement $15 \leq 20$ is true because $15 < 20$.

(b) $12 \geq 12$ Since $12 = 12$, this statement is true.

(c) $15 \leq 20 \cdot 2$ The statement $15 \leq 20 \cdot 2$ is true, because $15 < 40$.

(d) $\dfrac{6}{15} \geq \dfrac{2}{3}$

$\dfrac{6}{15} \geq \dfrac{10}{15}$ Find a common denominator.

The statements $\frac{6}{15} > \frac{10}{15}$ and $\frac{6}{15} = \frac{10}{15}$ are false. Because at least one of them is false, $\frac{6}{15} \geq \frac{2}{3}$ is also false.

Work Problem ⑤ at the Side. ▶

④ Write each statement in words. Then decide whether it is *true* or *false*.

(a) $7 < 5$

(b) $12 > 6$

(c) $4 \neq 10$

⑤ Determine whether each statement is *true* or *false*.

(a) $30 \leq 40$

(b) $25 \geq 10$

(c) $40 \leq 10$

(d) $21 \leq 21$

(e) $9 \cdot 3 \geq 28$

(f) $\dfrac{4}{7} \leq \dfrac{5}{8}$

Answers

4. **(a)** Seven is less than five. False
   (b) Twelve is greater than six. True
   (c) Four is not equal to ten. True
5. **(a)** true **(b)** true **(c)** false
   (d) true **(e)** false **(f)** true

6 Write each word statement in symbols.

(a) Nine is equal to eleven minus two.

(b) Seventeen is less than thirty.

(c) Eight is not equal to ten.

(d) Fourteen is greater than twelve.

(e) Thirty is less than or equal to fifty.

(f) Two is greater than or equal to two.

7 Write each statement as another true statement with the inequality symbol reversed.

(a) $8 < 10$

(b) $3 > 1$

(c) $9 \leq 15$

(d) $6 \geq 2$

OBJECTIVE **5** **Translate word statements to symbols.**

EXAMPLE 5 Translating from Words to Symbols

Write each word statement in symbols.

(a) Twelve **is equal to** ten **plus** two. $12 = 10 + 2$

(b) Nine **is less than** ten. $9 < 10$
Compare this with "9 less than 10," which is written $10 - 9$.

(c) Fifteen **is not equal to** eighteen. $15 \neq 18$

(d) Seven **is greater than** four. $7 > 4$

(e) Thirteen **is less than or equal to** forty. $13 \leq 40$

(f) Six **is greater than or equal to** six. $6 \geq 6$

◀ **Work Problem** **6** **at the Side.**

OBJECTIVE **6** **Write statements that change the direction of inequality symbols.** Any statement with $<$ can be converted to one with $>$, and any statement with $>$ can be converted to one with $<$. *We do this by reversing both the order of the numbers and the direction of the symbol.*

$$6 < 10 \quad \text{becomes} \quad 10 > 6.$$

Interchange numbers.

Reverse symbol.

EXAMPLE 6 Converting between Inequality Symbols

Parts (a)–(c) show the same statements written in two equally correct ways. In each inequality, the symbol points toward the lesser number.

(a) $5 > 2$, $2 < 5$ (b) $3 \leq 8$, $8 \geq 3$ (c) $12 \geq 5$, $5 \leq 12$

◀ **Work Problem** **7** **at the Side.**

Symbol	Meaning	Example
$=$	Is equal to	$0.5 = \frac{1}{2}$ means 0.5 is equal to $\frac{1}{2}$.
$\neq$	Is not equal to	$3 \neq 7$ means 3 is not equal to 7.
$<$	Is less than	$6 < 10$ means 6 is less than 10.
$>$	Is greater than	$15 > 14$ means 15 is greater than 14.
$\leq$	Is less than or equal to	$4 \leq 8$ means 4 is less than or equal to 8.
$\geq$	Is greater than or equal to	$1 \geq 0$ means 1 is greater than or equal to 0.

CAUTION

Equality and inequality symbols are used to write mathematical *sentences,* while operation symbols ($+$, $-$, $\cdot$, and $\div$) are used to write mathematical *expressions* that represent a number. Compare the following.

Sentence: $4 < 10$ ← Gives the relationship between 4 and 10

Expression: $4 + 10$ ← Tells how to operate on 4 and 10 to get 14

1.1 Exercises

FOR EXTRA HELP

  MyMathLab®

CONCEPT CHECK *Decide whether each statement is* true *or* false. *If it is* false, *explain why.*

1. An exponent tells how many times its base is used as a factor.

2. Some grouping symbols are $+$, $-$, $\cdot$, and $\div$.

3. When evaluated, $4 + 3(8 - 2)$ is equal to 42.

4. $3^3 = 9$

5. The statement "4 is 12 less than 16" is interpreted $4 = 12 - 16$.

6. The statement "6 is 4 less than 10" is interpreted $6 < 10 - 4$.

7. **CONCEPT CHECK** When evaluating $(4^2 + 3^3)^4$, what is the *last* exponent that would be applied? Why?

8. **CONCEPT CHECK** Which are not grouping symbols—parentheses, brackets, fraction bars, exponents?

Find the value of each exponential expression. **See Example 1.**

9. 7^2

10. 4^2

11. 12^2

12. 14^2

13. 4^3

14. 5^3

15. 10^3

16. 11^3

17. 3^4

18. 6^4

19. 4^5

20. 3^5

21. $\left(\dfrac{2}{3}\right)^4$

22. $\left(\dfrac{3}{4}\right)^3$

23. $(0.04)^3$

24. $(0.05)^4$

Find the value of each expression. **See Examples 2 and 3.**

25. $13 + 9 \cdot 5$
   $\bigcirc \bigcirc$
   $= 13 + \underline{\quad}$
   $= \underline{\quad}$

26. $11 + 7 \cdot 6$
   $\bigcirc \bigcirc$
   $= 11 + \underline{\quad}$
   $= \underline{\quad}$

27. $20 - 4 \cdot 3 + 5$
   $\bigcirc \bigcirc \bigcirc$
   $= 20 - \underline{\quad} + 5$
   $= \underline{\quad} + 5$
   $= \underline{\quad}$

28. $18 - 7 \cdot 2 + 6$
   $\bigcirc \bigcirc \bigcirc$
   $= 18 - \underline{\quad} + 6$
   $= \underline{\quad} + 6$
   $= \underline{\quad}$

29. $9 \cdot 5 - 13$
   $\bigcirc \bigcirc$

30. $7 \cdot 6 - 11$
   $\bigcirc \bigcirc$

31. $18 - 2 + 3$
   $\bigcirc \bigcirc$

32. $22 - 8 + 9$
   $\bigcirc \bigcirc$

33. $\dfrac{1}{4} \cdot \dfrac{2}{3} + \dfrac{2}{5} \cdot \dfrac{11}{3}$

34. $\dfrac{9}{4} \cdot \dfrac{2}{3} + \dfrac{4}{5} \cdot \dfrac{5}{3}$

35. $9 \cdot 4 - 8 \cdot 3$

36. $11 \cdot 4 + 10 \cdot 3$

37. $2.5(1.9) + 4.3(7.3)$

38. $4.3(1.2) + 2.1(8.5)$

39. $10 + 40 \div 5 \cdot 2$

40. $12 + 35 \div 7 \cdot 3$

41. $18 - 2(3 + 4)$

42. $30 - 3(4 + 2)$

43. $5[3 + 4(2^2)]$

44. $4^2[(13 + 4) - 8]$

45. $\left(\frac{3}{2}\right)^2\left[\left(11 + \frac{1}{3}\right) - 6\right]$

46. $6\left[\frac{3}{4} + 8\left(\frac{1}{2}\right)^3\right]$

47. $\dfrac{8 + 6(3^2 - 1)}{3 \cdot 2 - 2}$

48. $\dfrac{8 + 2(8^2 - 4)}{4 \cdot 3 - 10}$

49. $\dfrac{4(7 + 2) + 8(8 - 3)}{6(4 - 2) - 2^2}$

50. $\dfrac{6(5 + 1) - 9(1 + 1)}{5(8 - 4) - 2^3}$

CONCEPT CHECK *Insert one pair of parentheses so that the left side of each equation is equal to the right side.*

51. $3 \cdot 6 + 4 \cdot 2 = 60$

52. $2 \cdot 8 - 1 \cdot 3 = 42$

53. $10 - 7 - 3 = 6$

54. $15 - 10 - 2 = 7$

55. $8 + 2^2 = 100$

56. $4 + 2^2 = 36$

Tell whether each statement is true *or* false. *In Exercises 59–68, first simplify each expression involving an operation.* **See Example 4.**

57. $8 \geq 17$

58. $10 \geq 41$

59. $17 \leq 18 - 1$

60. $12 \geq 10 + 2$

61. $6 \cdot 8 + 6 \cdot 6 \geq 0$

62. $4 \cdot 20 - 16 \cdot 5 \geq 0$

63. $6[5 + 3(4 + 2)] \leq 70$

64. $6[2 + 3(2 + 5)] \leq 135$

65. $\dfrac{9(7 - 1) - 8 \cdot 2}{4(6 - 1)} > 3$

66. $\dfrac{2(5 + 3) + 2 \cdot 2}{2(4 - 1)} > 1$

67. $8 \leq 4^2 - 2^2$

68. $10^2 - 8^2 > 6^2$

Write each word statement in symbols. **See Example 5.**

69. Fifteen is equal to five plus ten.

70. Twelve is equal to twenty minus eight.

71. Nine is greater than five minus four.

72. Ten is greater than six plus one.

73. Sixteen is not equal to nineteen.

74. Three is not equal to four.

75. Two is less than or equal to three.

76. Five is less than or equal to nine.

Write each statement in words and decide whether it is true *or* false. *(Hint: To compare fractions, write them with the same denominator.)*

77. $7 < 19$

78. $9 < 10$

79. $\dfrac{1}{3} \neq \dfrac{3}{10}$

80. $\dfrac{10}{7} \neq \dfrac{3}{2}$

81. $8 \geq 11$

82. $4 \leq 2$

Write each statement as another true statement with the inequality symbol reversed.
See Example 6.

83. $5 < 30$

84. $8 > 4$

85. $12 \geq 3$

86. $25 \leq 41$

87. $2.5 \geq 1.3$

88. $4.1 \leq 5.3$

89. $\dfrac{4}{5} > \dfrac{3}{4}$

90. $\dfrac{8}{3} < \dfrac{11}{4}$

One way to measure a person's cardio fitness is to calculate how many METs, or metabolic units, he or she can reach at peak exertion. One MET is the amount of energy used when sitting quietly. To calculate ideal METs, we can use the following expressions.

$$14.7 - \text{age} \cdot 0.13 \quad \text{For women}$$

$$14.7 - \text{age} \cdot 0.11 \quad \text{For men}$$

(*Source: New England Journal of Medicine.*)

91. A 40-yr-old woman wishes to calculate her ideal MET.

   (a) Write the expression, using her age.

   (b) Calculate her ideal MET. (*Hint:* Use the rules for order of operations.)

   (c) Researchers recommend that a person reach approximately 85% of his or her MET when exercising. Calculate 85% of the ideal MET from part (b). Then refer to the following table. What activity can the woman do that is approximately this value?

Activity	METs	Activity	METs
Golf (with cart)	2.5	Skiing (water or downhill)	6.8
Walking (3 mph)	3.3	Swimming	7.0
Mowing lawn (power)	4.5	Walking (5 mph)	8.0
Ballroom or square dancing	5.5	Jogging	10.2
Cycling	5.7	Skipping rope	12.0

Source: Harvard School of Public Health.

   (d) Repeat parts (a)–(c) for a 55-yr-old man.

92. Repeat parts (a)–(c) of **Exercise 91** for your age and gender. For yourself 5 yr from now.

The table shows the number of pupils per teacher in U.S. public schools in selected states during the 2009–2010 school year. Use this table to answer the questions in Exercises 93–96.

State	Pupils per Teacher
Alaska	15.3
Texas	14.7
California	22.2
Wyoming	12.7
Maine	11.8
Idaho	18.4
Missouri	14.1

93. Which states had a number greater than 14.1?

94. Which states had a number that was at most 14.7?

95. Which states had a number not less than 14.1?

96. Which states had a number greater than 22.2?

Source: National Center for Education Statistics.

Study Skills
TAKING LECTURE NOTES

Study the set of sample math notes given here.

▶ **Include the date and title** of the day's lecture topic.

▶ **Include definitions,** written here in parentheses—don't trust your memory.

▶ **Skip lines and write neatly** to make reading easier.

▶ **Emphasize direction words** (like *simplify*) with their explanations.

▶ **Mark important concepts with stars, underlining, etc.**

▶ **Use two columns,** which allows an example and its explanation to be close together.

▶ **Use brackets and arrows** to clearly show steps, related material, etc.

January 12   *Exponents*

Exponents used to show repeated multiplication.

$3 \cdot 3 \cdot 3 \cdot 3$ can be written 3^4 *exponent* (how many times it's multiplied)

base (the number being multiplied)

Read 3^2 as 3 to the 2nd power or 3 squared

3^3 as 3 to the 3rd power or 3 cubed

3^4 as 3 to the 4th power

etc.

Simplifying an expression with exponents

→ actually do the repeated multiplication

2^3 means $2 \cdot 2 \cdot 2$ and $2 \cdot 2 \cdot 2 = 8$

★ Careful! [5^2 means $5 \cdot 5$ NOT $5 \cdot 2$

so $5^2 = 5 \cdot 5 = 25$ BUT $5^2 \neq 10$

Example	*Explanation*
simplify $(2^4) \cdot (3^2)$	Exponents mean multiplication.
$2 \cdot 2 \cdot 2 \cdot 2 \cdot 3 \cdot 3$	Use 2 as a factor 4 times. Use 3 as a factor 2 times.
16 · 9	$2 \cdot 2 \cdot 2 \cdot 2$ is 16 $\rangle$ 16 · 9 is 144 $3 \cdot 3$ is 9
144	Simplified result is 144 (no exponents left)

Now Try This

With a partner or in a small group, compare lecture notes.

1 What are you doing to show main points in your notes (such as boxing, using stars, etc.)? _____

2 In what ways do you set off explanations from worked problems and subpoints (such as indenting, using arrows, circling, etc.)? _____

3 What new ideas did you learn by examining your classmates' notes? _____

4 What new techniques will you try in your notes? _____

1.2 Variables, Expressions, and Equations

A **variable** is a symbol, usually a letter, used to represent an unknown number. A **constant** is a fixed, unchanging number.

x, y, or z Variables | 5, $\frac{3}{4}$, $8\frac{1}{2}$, 10.8 Constants

An **algebraic expression** is a collection of constants, variables, operation symbols, and grouping symbols, such as parentheses, square brackets, or fraction bars.

$x + 5$, $2m - 9$, $8p^2 + 6(p - 2)$ Algebraic expressions

$2m$ means $2 \cdot m$, the product of 2 and m.

$6(p - 2)$ means the product of 6 and $p - 2$.

OBJECTIVE ❶ **Evaluate algebraic expressions, given values for the variables.** An algebraic expression can have different numerical values for different values of the variables.

EXAMPLE 1   Evaluating Expressions

Find the value of each algebraic expression for $m = 5$ and then for $m = 9$.

(a) $8m$

$= 8 \cdot 5$ Let $m = 5$.
$= 40$ Multiply.

$8m$

$= 8 \cdot 9$ Let $m = 9$.
$= 72$ Multiply.

(b) $3m^2$

$5^2 = 5 \cdot 5$

$= 3 \cdot 5^2$ Let $m = 5$.
$= 3 \cdot 25$ Square 5.
$= 75$ Multiply.

$3m^2$

$9^2 = 9 \cdot 9$

$= 3 \cdot 9^2$ Let $m = 9$.
$= 3 \cdot 81$ Square 9.
$= 243$ Multiply.

Work Problem ❶ **at the Side.** ▶

CAUTION

$3m^2$ means $3 \cdot m^2$, **not** $3m \cdot 3m$.   See **Example 1(b)**.

Unless parentheses are used, the exponent refers only to the variable or number just before it. We use parentheses to write $3m \cdot 3m$ with exponents as $(3m)^2$.

EXAMPLE 2   Evaluating Expressions

Find the value of each expression for $x = 5$ and $y = 3$.

(a) $2x + 5y$

We could write this with parentheses as $2(5) + 5(3)$.

Follow the rules for order of operations.

$= 2 \cdot 5 + 5 \cdot 3$ Replace x with 5 and y with 3.
$= 10 + 15$ Multiply.
$= 25$ Add.

Continued on Next Page

OBJECTIVES

❶ Evaluate algebraic expressions, given values for the variables.

❷ Translate word phrases to algebraic expressions.

❸ Identify solutions of equations.

❹ Translate sentences to equations.

❺ Distinguish between *equations* and *expressions*.

❶ Find the value of each expression for $p = 3$.

(a) $6p$

$= 6 \cdot \underline{\quad}$
$= \underline{\quad}$

(b) $p + 12$

$= \underline{\quad} + 12$
$= \underline{\quad}$

(c) $5p^2$

$= 5 \cdot \underline{\quad}^2$
$= 5 \cdot \underline{\quad}$
$= \underline{\quad}$

(d) $16p$

(e) $2p^3$

Answers

1. **(a)** 3; 18   **(b)** 3; 15
   (c) 3; 9; 45   **(d)** 48   **(e)** 54

❷ Find the value of each expression for $x = 6$ and $y = 9$.

(GS) **(a)** $4x + 7y$

$$= 4 \cdot \underline{} + 7 \cdot \underline{}$$

$$= \underline{} + \underline{}$$

$$= \underline{}$$

(b) $\dfrac{4x - 2y}{x + 1}$

(c) $2x^2 + y^2$

(b) $\dfrac{9x - 8y}{2x - y}$

$$= \frac{9 \cdot 5 - 8 \cdot 3}{2 \cdot 5 - 3} \qquad \text{Replace } x \text{ with 5 and } y \text{ with 3.}$$

$$= \frac{45 - 24}{10 - 3} \qquad \text{Multiply.}$$

$$= \frac{21}{7} \qquad \text{Subtract.}$$

$$= 3 \qquad \text{Divide.}$$

(c) $x^2 - 2y^2$

$\boxed{3^2 = 3 \cdot 3}$

$$= 5^2 - 2 \cdot 3^2 \qquad \text{Replace } x \text{ with 5 and } y \text{ with 3.}$$

$\boxed{5^2 = 5 \cdot 5}$

$$= 25 - 2 \cdot 9 \qquad \text{Apply the exponents.}$$

$$= 25 - 18 \qquad \text{Multiply.}$$

$$= 7 \qquad \text{Subtract.}$$

◀ **Work Problem ❷ at the Side.**

OBJECTIVE **❷** **Translate word phrases to algebraic expressions.**

Problem-Solving Hint

Sometimes variables must be used to change word phrases into algebraic expressions. This process will be important in later chapters when we solve applied problems.

EXAMPLE 3 **Using Variables to Write Word Phrases as Algebraic Expressions**

Write each word phrase as an algebraic expression, using x as the variable.

(a) The **sum** of a number and 9

$\qquad x + 9, \quad \text{or} \quad 9 + x \qquad$ "Sum" is the answer to an addition problem.

(b) 7 **minus** a number

$\qquad\qquad 7 - x \qquad$ "Minus" indicates subtraction.

$x - 7$ is incorrect. We cannot subtract in either order and get the same result.

(c) A number **subtracted from 12**

$$12 - x \; \longleftarrow \boxed{\text{Be careful with order.}}$$

Compare this result with "12 subtracted from a number," which is $x - 12$.

(d) The **product** of 11 and a number

$\qquad\qquad 11 \cdot x, \quad \text{or} \quad 11x$

Continued on Next Page

Answers

2. **(a)** $6; 9; 24; 63; 87$ **(b)** $\dfrac{6}{7}$ **(c)** 153

(e) 5 **divided by** a number

$$5 \div x, \quad \text{or} \quad \frac{5}{x} \quad \boxed{\frac{x}{5} \text{ is } not \text{ correct here.}}$$

(f) The **product of** 2 and the **difference** between a number and 8

We are multiplying 2 times "something." This "something" is the difference between a number and 8, written $x - 8$. We use parentheses around this difference.

$$2 \cdot (x - 8), \quad \text{or} \quad 2(x - 8)$$

$\boxed{8 - x, \text{ which means the difference between 8 and a number, is not correct.}}$

.... **Work Problem 3 at the Side.** ▶

OBJECTIVE ▶ 3 Identify solutions of equations. An **equation** is a statement that two expressions are equal. *An equation always includes the equality symbol, =.*

$$x + 4 = 11, \qquad 2y = 16, \qquad 4p + 1 = 25 - p,$$

$$\frac{3}{4}x + \frac{1}{2} = 0, \qquad z^2 = 4, \qquad 4(m - 0.5) = 2m \Bigg\} \text{Equations}$$

To **solve** an equation, we must find all values of the variable that make the equation true. Such values of the variable are the **solutions** of the equation.

EXAMPLE 4   Deciding Whether a Number Is a Solution of an Equation

Decide whether the given number is a solution of the equation.

(a) $5p + 1 = 36; \quad 7$

$$5p + 1 = 36$$
$$5 \cdot 7 + 1 \overset{?}{=} 36 \qquad \text{Let } p = 7.$$
$$35 + 1 \overset{?}{=} 36 \qquad \text{Multiply.}$$

$\boxed{\text{Be careful. Multiply first.}}$ $\qquad 36 = 36 \checkmark \qquad$ True—the left side of the equation equals the right side.

The number 7 is a solution of the equation.

(b) $9m - 6 = 32; \quad \frac{14}{3}$

$$9m - 6 = 32$$
$$9 \cdot \frac{14}{3} - 6 \overset{?}{=} 32 \qquad \text{Let } m = \frac{14}{3}.$$
$$42 - 6 \overset{?}{=} 32 \qquad \text{Multiply.}$$
$$36 = 32 \qquad \text{False—the left side does } not \text{ equal the right side.}$$

The number $\frac{14}{3}$ is not a solution of the equation.

.... **Work Problem 4 at the Side.** ▶

3 Write each word phrase as an algebraic expression. Use x as the variable.

(a) The sum of 5 and a number

(b) A number minus 4

(c) A number subtracted from 48

(d) The product of 6 and a number

(e) 9 multiplied by the sum of a number and 5

4 Decide whether the given number is a solution of the equation.

(a) $p - 1 = 3; \quad 2$

(b) $2k + 3 = 15; \quad 7$

(c) $7p - 11 = 5; \quad \frac{16}{7}$

Answers

3. **(a)** $5 + x$ **(b)** $x - 4$ **(c)** $48 - x$
(d) $6x$ **(e)** $9(x + 5)$
4. **(a)** no **(b)** no **(c)** yes

❺ Write each word sentence as an equation. Use x as the variable.

ᴳˢ (a) Three times the sum of a number and 13 is 19.

_____ (_____ + _____) = _____

(b) Five times a number subtracted from 21 is 15.

(c) Five less than six times a number is equal to nineteen.

❻ Decide whether each is an *equation* or an *expression*.

(a) $2x + 5y - 7$

(b) $\dfrac{3x - 1}{5}$

(c) $2x + 5 = 7$

(d) $\dfrac{x}{x - 3} = 4x$

OBJECTIVE ▶ 4 Translate sentences to equations. Sentences given in words are translated as equations.

EXAMPLE 5 **Translating Sentences to Equations**

Write each word sentence as an equation. Use x as the variable.

(a) Twice the sum of a number and four is six.

"Twice" means two times. The word *is* suggests equals. With x representing the number, translate as follows.

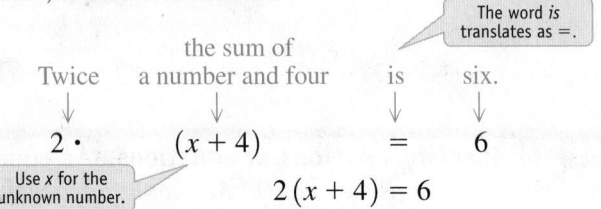

$$2(x + 4) = 6$$

(b) Nine more than five times a number is 49.

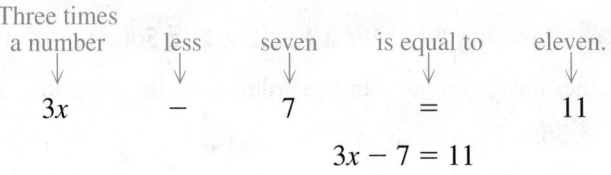

$$5x + 9 = 49 \qquad 5 \cdot x = 5x$$

(c) Seven less than three times a number is equal to eleven.

Three times a number	less	seven	is equal to	eleven.
↓	↓	↓	↓	↓
$3x$	$-$	7	$=$	11

$$3x - 7 = 11$$

◀ **Work Problem ❺ at the Side.**

OBJECTIVE ▶ 5 Distinguish between equations and expressions. Students often have trouble distinguishing between equations and expressions. *An equation is a sentence—it has something on the left side, an = symbol, and something on the right side. An expression is a phrase that represents a number.*

$$\underbrace{4x + 5}_{\text{Left side}} \underset{\text{Right side}}{=} 9 \qquad \underset{\text{Expression}}{4x + 5}$$

Equation

EXAMPLE 6 **Distinguishing between Equations and Expressions**

Decide whether each is an *equation* or an *expression*.

(a) $2x - 3$ — Ask, "Is there an equality symbol?" The answer is no, so this is an expression.

(b) $2x - 3 = 8$ — Because there is an equality symbol with something on either side of it, this is an equation.

(c) $5x^2 + 2y^2$ — There is no equality symbol. This is an expression.

◀ **Work Problem ❻ at the Side.**

1.2 Exercises

  MyMathLab®

CONCEPT CHECK *Choose the letter(s) of the correct response.*

1. The expression $8x^2$ means _____.

   A. $8 \cdot x \cdot 2$ **B.** $8 \cdot x \cdot x$ **C.** $8 + x^2$ **D.** $8x^2 \cdot 8x^2$

2. If $x = 2$ and $y = 1$, then the value of xy is _____.

   A. $\dfrac{1}{2}$ **B.** 1 **C.** 2 **D.** 3

3. The sum of 15 and a number x is represented by _____.

   A. $15 + x$ **B.** $15 - x$ **C.** $x - 15$ **D.** $15x$

4. Which of the following are expressions?

   A. $6x = 7$ **B.** $6x + 7$ **C.** $6x - 7$ **D.** $6x - 7 = 0$

CONCEPT CHECK *Complete each statement.*

5. For $x = 3$, the value of $x + 8$ is _____.

6. For $x = 1$ and $y = 2$, the value of $5xy$ is _____.

7. "The sum of 13 and x" is represented by the expression _____. For $x = 3$, the value of this expression is _____.

8. $2x + 6$ is an (*equation / expression*), while $2x + 6 = 8$ is an (*equation / expression*).

In Exercises 9–12, give a short explanation.

9. If the words *more than* in **Example 5(b)** were changed to *less than*, how would the equation be changed?

10. Why is $2x^3$ not the same as $2x \cdot 2x \cdot 2x$? Explain, using an exponent to write $2x \cdot 2x \cdot 2x$.

11. There are many pairs of values of x and y for which $2x + y$ can be evaluated. Name two such pairs and then evaluate the expression.

12. When evaluating the expression $4x^2$ for $x = 3$, explain why 3 must be squared *before* multiplying by 4.

Find the value of each expression for (a) $x = 4$ and (b) $x = 6$. See Example 1.

13. $4x^2$

(a)   (b)

14. $5x^2$

(a)   (b)

15. $\dfrac{3x - 5}{2x}$

(a)   (b)

16. $\dfrac{4x - 1}{3x}$

(a)   (b)

17. $\dfrac{6.459x}{2.7}$ (to the nearest thousandth)

(a)   (b)

18. $\dfrac{0.74x^2}{0.85}$ (to the nearest thousandth)

(a)   (b)

19. $3x^2 + x$

(a)   (b)

20. $2x + x^2$

(a)   (b)

Find the value of each expression for (a) $x = 2$ and $y = 1$ and (b) $x = 1$ and $y = 5$.
See Example 2.

21. $3(x + 2y)$

(a)   (b)

22. $2(2x + y)$

(a)   (b)

23. $x + \dfrac{4}{y}$

(a)   (b)

24. $y + \dfrac{8}{x}$

(a)   (b)

25. $\dfrac{x}{2} + \dfrac{y}{3}$

(a)   (b)

26. $\dfrac{x}{5} + \dfrac{y}{4}$

(a)   (b)

27. $\dfrac{2x + 4y - 6}{5y + 2}$

(a)   (b)

28. $\dfrac{4x + 3y - 1}{2x + y}$

(a)   (b)

29. $2y^2 + 5x$

(a)   (b)

30. $6x^2 + 4y$

(a)   (b)

31. $\dfrac{3x + y^2}{2x + 3y}$

(a)   (b)

32. $\dfrac{x^2 + 1}{4x + 5y}$

(a)   (b)

33. $0.841x^2 + 0.32y^2$

(a)   (b)

34. $0.941x^2 + 0.2y^2$

(a)   (b)

Write each word phrase as an algebraic expression, using x as the variable.
See Example 3.

35. Twelve times a number

36. Thirteen added to a number

37. Two subtracted from a number

38. Eight subtracted from a number

39. One-third of a number, subtracted from seven

40. One-fifth of a number, subtracted from fourteen

41. The difference between twice a number and 6

42. The difference between 6 and half a number

43. 12 divided by the sum of a number and 3

44. The difference between a number and 5, divided by 12

45. The product of 6 and four less than a number

46. The product of 9 and five more than a number

47. Suppose that the directions on a quiz read, "Evaluate each equation for $x = 3$." How can you justify politely correcting the person who wrote these directions?

48. Suppose that the directions on a test read, "Solve the following expressions." How can you justify politely correcting the person who wrote these directions?

Decide whether the given number is a solution of the equation. **See Example 4.**

49. Is 7 a solution of $x - 5 = 12$?

50. Is 10 a solution of $x + 6 = 15$?

51. Is 1 a solution of $5x + 2 = 7$?

52. Is 1 a solution of $3x + 5 = 8$?

53. Is $\frac{1}{5}$ a solution of $6x + 4x + 9 = 11$?

54. Is $\frac{12}{5}$ a solution of $2x + 3x + 8 = 20$?

55. Is 3 a solution of $2y + 3(y - 2) = 14$?

56. Is 2 a solution of $6x + 2(x + 3) = 14$?

57. Is $\frac{1}{3}$ a solution of $\frac{z+4}{2-z} = \frac{13}{5}$?

58. Is $\frac{13}{4}$ a solution of $\frac{x+6}{x-2} = \frac{37}{5}$?

59. Is 4.3 a solution of $3r^2 - 2 = 53.47$?

60. Is 3.7 a solution of $2x^2 + 1 = 28.38$?

Write each word sentence as an equation. Use x as the variable. ***See Example 5.***

61. The sum of a number and 8 is 18.

62. A number minus three equals 1.

63. Five more than twice a number is 5.

64. The product of 2 and the sum of a number and 5 is 14.

65. Sixteen minus three-fourths of a number is 13.

66. The sum of six-fifths of a number and 2 is 14.

67. Three times a number is equal to 8 more than twice the number.

68. Twelve divided by a number equals $\frac{1}{3}$ times that number.

Identify each as an expression *or an* equation. ***See Example 6.***

69. $3x + 2(x - 4)$

70. $5y - (3y + 6)$

71. $7t + 2(t + 1) = 4$

72. $9r + 3(r - 4) = 2$

73. $x + y = 9$

74. $x + y - 9$

75. $\frac{3x - 8}{2}$

76. $\frac{4x + 3}{2} = 11$

Relating Concepts (Exercises 77–80) For Individual or Group Work

A **mathematical model** is an equation that describes the relationship between two quantities. For example, the life expectancy of Americans at birth can be approximated by the equation

$$y = 0.174x - 270,$$

where x is a year between 1950 and 2008 and y is age in years. *(Source: Centers for Disease Control and Prevention.)*

Use this model to approximate life expectancy *(to the nearest year)* in each of the following years.

77. 1950

78. 1975

79. 1995

80. 2008

Study Skills

TACKLING YOUR HOMEWORK

You are ready to do your homework **AFTER** you have read the corresponding textbook section and worked through the examples and margin problems.

Homework Tips

▶ **Survey the exercise set.** Take a few minutes to glance over the problems that your instructor has assigned to get a general idea of the types of exercises you will be working. Skim directions, and note any references to section examples.

▶ **Work problems neatly.** Use pencil and write legibly, so others can read your work. Skip lines between steps. Clearly separate problems from each other.

▶ **Show all your work.** It is tempting to take shortcuts. Include ALL steps.

▶ **Check your work frequently to make sure you are on the right track.** It is hard to unlearn a mistake. For all odd-numbered problems and other selected exercises, answers are given in the back of the book.

▶ **If you have trouble with a problem, refer to the corresponding worked example in the section.** The exercise directions will often reference specific examples to review. Pay attention to every line of the worked example to see how to get from step to step.

▶ **If you are having trouble with an even-numbered problem, work the corresponding odd-numbered problem.** Check your answer in the back of the book, and apply the same steps to work the even-numbered problem.

▶ **Does the problem or a similar problem have a blue screen around the problem number, such as 11.?** If it does, refer to the worked-out solution in the selected solutions section at the back of the book. Study this solution.

▶ **Do some homework problems every day.** This is a good habit, even if your math class does not meet each day.

▶ **Mark any problems you don't understand.** Ask your instructor about them.

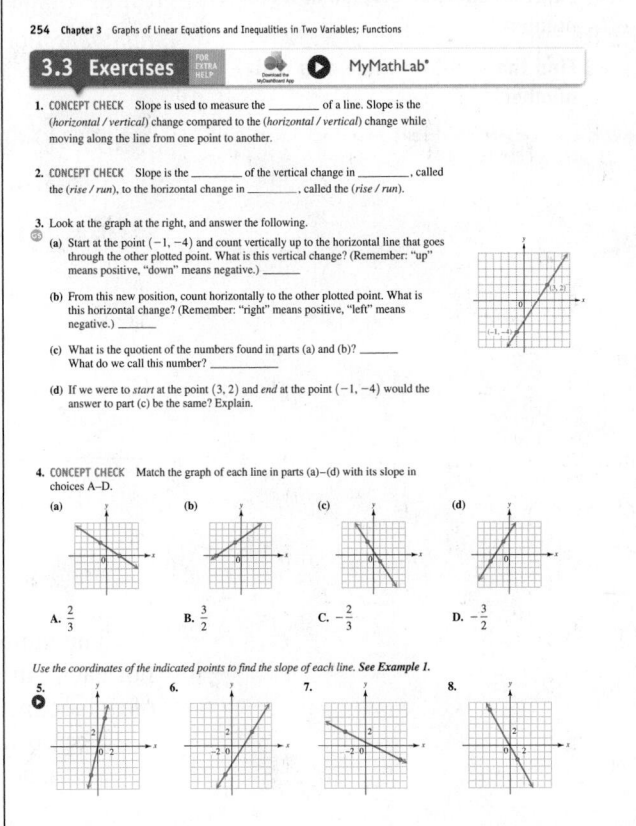

Now Try This

Think through and answer each of the following.

1 What are your biggest homework concerns?

_____

_____

2 List one or more of the homework tips to try.

_____

_____

1.3 Real Numbers and the Number Line

OBJECTIVES

1 Classify numbers and graph them on number lines.

2 Tell which of two real numbers is less than the other.

3 Find the additive inverse of a real number.

4 Find the absolute value of a real number.

A **set** is a collection of objects. In mathematics, these objects are usually numbers. The objects that belong to the set are its **elements.** They are written between braces.

$$\{1, 2, 3, 4, 5\}$$ The set of numbers 1, 2, 3, 4, and 5

OBJECTIVE ▶ 1 **Classify numbers and graph them on number lines.** The set of numbers used for counting is the *natural numbers*. The set of *whole numbers* includes 0 with the natural numbers.

Natural Numbers and Whole Numbers

$\{1, 2, 3, 4, 5, \dots\}$ is the set of **natural numbers** (or **counting numbers**).

$\{0, 1, 2, 3, 4, 5, \dots\}$ is the set of **whole numbers.**

We can represent numbers on a **number line** like the one in **Figure 1.**

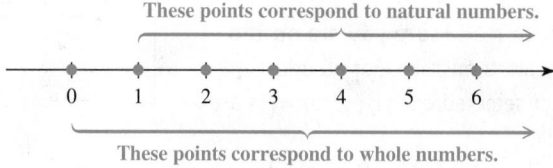

To draw a number line, choose any point on the line and label it 0. Then choose any point to the right of 0 and label it 1. Use the distance between 0 and 1 as the scale to locate, and then label, other points.

Figure 1

The natural numbers are located to the right of 0 on the number line. For each natural number, we can place a corresponding number to the left of 0, labeling the points -1, -2, -3, and so on, as shown in **Figure 2.** Each is the **opposite,** or **negative,** of a natural number. The natural numbers, their opposites, and 0 form the set of *integers*.

Integers

$\{\dots, -3, -2, -1, 0, 1, 2, 3, \dots\}$ is the set of **integers.**

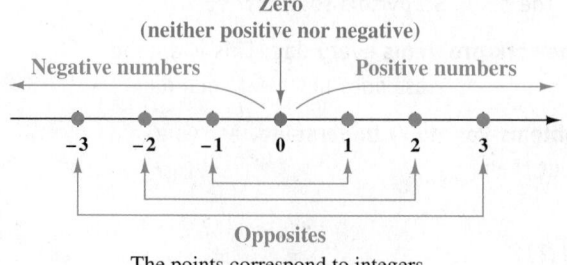

Figure 2

Positive numbers and *negative numbers* are **signed numbers.**

> **EXAMPLE 1**   Using Negative Numbers in Applications
>
> Use an integer to express the boldface italic number in each application.
>
> **(a)** The lowest Fahrenheit temperature ever recorded in meteorological records was *129°* below zero at Vostok, Antarctica, on July 21, 1983. (*Source: World Almanac and Book of Facts.*)
> Use −129 because "below zero" indicates a negative number.
>
> **(b)** The shore surrounding the Dead Sea is *1348* ft below sea level. (*Source: World Almanac and Book of Facts.*)
> Again, "below sea level" indicates a negative number, −1348.
>
> .... **Work Problem ❶ at the Side.** ▶

❶ Use an integer to express the boldface italic number(s) in each application.

(a) Erin discovers that she has spent *$53* more than she has in her checking account.

Fractions, reviewed in **Chapter R,** are *rational numbers.*

Rational Numbers

$\{x \mid x$ is a quotient of two integers, with denominator not $0\}$ is the set of **rational numbers.**
 (Read the part in the braces as "the set of all numbers x such that x is a quotient of two integers, with denominator not 0.")

(b) The record high Fahrenheit temperature in the United States was *134°* in Death Valley, California, on July 10, 1913. (*Source: World Almanac and Book of Facts.*)

Note

The set symbolism used in the definition of rational numbers,

$$\{x \mid x \text{ has a certain property}\},$$

is called **set-builder notation.** This notation is convenient to use when it is not possible to list all the elements of a set.

Since any number that can be written as the quotient of two integers (that is, as a fraction) is a rational number, *all integers, mixed numbers, terminating (or ending) decimals, and repeating decimals are rational.* The table gives examples.

(c) A football team gained *5* yd, then lost *10* yd on the next play.

Rational Number	Equivalent Quotient of Two Integers
−5	$\frac{-5}{1}$ (means $-5 \div 1$)
$1\frac{3}{4}$	$\frac{7}{4}$ (means $7 \div 4$)
0.23 (terminating decimal)	$\frac{23}{100}$ (means $23 \div 100$)
0.3333…, or $0.\overline{3}$ (repeating decimal)	$\frac{1}{3}$ (means $1 \div 3$)
4.7	$\frac{47}{10}$ (means $47 \div 10$)

To **graph** a number, we place a dot on the number line at the point that corresponds to the number. The number is the **coordinate** of the point. Think of the graph of a set of numbers as a picture of the set.

Answers

1. **(a)** −53 **(b)** 134 **(c)** 5, −10

2 Graph each number on the number line.

$$-3, \quad \frac{17}{8}, \quad -2.75, \quad 1\frac{1}{2}, \quad -\frac{3}{4}$$

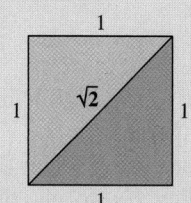

EXAMPLE 2   **Graphing Numbers**

Graph each number on a number line.

$$-\frac{3}{2}, \quad -\frac{2}{3}, \quad \frac{1}{2}, \quad 1\frac{1}{3}, \quad \frac{23}{8}, \quad 3\frac{1}{4}$$

To locate the improper fractions on the number line, write them as mixed numbers or decimals. The graph is shown in **Figure 3.**

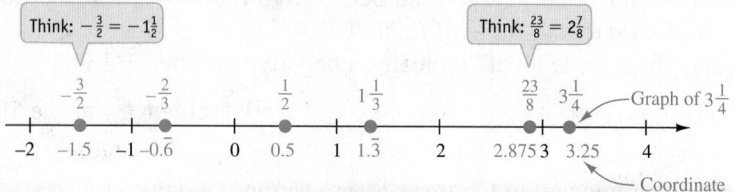

Figure 3

◄ **Work Problem 2 at the Side.**

The square root of 2, written $\sqrt{2}$, cannot be written as a quotient of two integers. Because of this, $\sqrt{2}$ is an *irrational number.* (See **Figure 4.**)

Irrational Numbers

$\{x \mid x$ is a nonrational number represented by a point on the number line$\}$ is the set of **irrational numbers.**

The decimal form of an irrational number neither terminates nor repeats.

Both rational and irrational numbers can be represented by points on the number line. Together, they form the **real numbers.** See **Figure 5.**

Real Numbers

$\{x \mid x$ is a rational or an irrational number$\}$ is the set of **real numbers.**

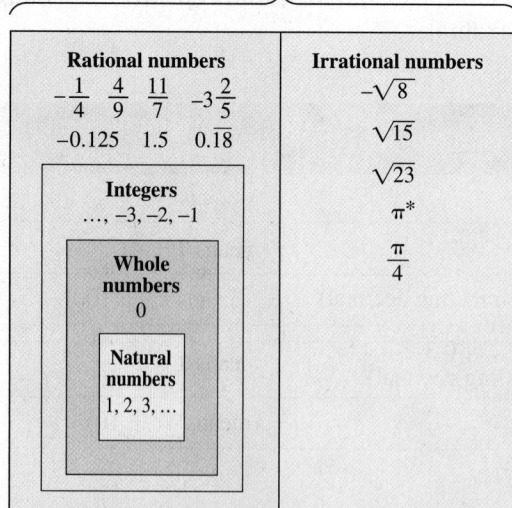

Figure 5

*The value of π (pi) is approximately 3.141592654. The decimal digits continue forever with no repeated pattern.

This square has diagonal of length $\sqrt{2}$. The number $\sqrt{2}$ is an irrational number.

Figure 4

Answer

2.
$$\begin{array}{c} -2.75 \quad -\frac{3}{4} \qquad 1\frac{1}{2} \ \ \frac{17}{8} \\ \end{array}$$

EXAMPLE 3   Determining Whether a Number Belongs to a Set

List the numbers in the following set that belong to each set of numbers.

$$\left\{ -5, \ -\frac{2}{3}, \ 0, \ 0.\overline{6}, \ \sqrt{2}, \ 3\frac{1}{4}, \ 5, \ 5.8 \right\}$$

(a) Natural numbers: 5

(b) Whole numbers: 0 and 5
The whole numbers consist of the natural (counting) numbers and 0.

(c) Integers: $-5, 0,$ and 5

(d) Rational numbers: $-5, -\frac{2}{3}, 0, 0.\overline{6}\left(\text{or } \frac{2}{3}\right), 3\frac{1}{4}\left(\text{or } \frac{13}{4}\right), 5,$ and $5.8\left(\text{or } \frac{58}{10}\right)$
Each of these numbers can be written as the quotient of two integers.

(e) Irrational numbers: $\sqrt{2}$

(f) Real numbers: All the numbers in the set are real numbers.

·································· **Work Problem ❸ at the Side.** ▶

OBJECTIVE ❷ Tell which of two real numbers is less than the other.
Given any two positive integers, you probably can tell which number is less than the other. Positive numbers decrease as the corresponding points on the number line go to the left. For example, $8 < 12$ because 8 is to the left of 12 on the number line. This ordering is extended to all real numbers by definition.

> **Ordering of Real Numbers**
>
> For any two real numbers a and b, ***a* is less than *b*** if a is to the left of b on a number line.
>
>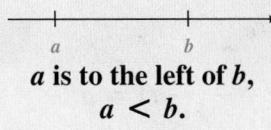
>
> ***a* is to the left of *b*,**
> $a < b.$

We can also say that, for any real numbers a and b, ***a* is greater than *b*** if a is to the right of b on the number line. Any negative number is less than 0, and any negative number is less than any positive number. Also, 0 is less than any positive number.

EXAMPLE 4   Determining the Order of Real Numbers

Is the statement $-3 < -1$ *true* or *false*?
Locate -3 and -1 on a number line, as shown in **Figure 6.** Because -3 is to the left of -1 on the number line, -3 is less than -1. The statement $-3 < -1$ is true.

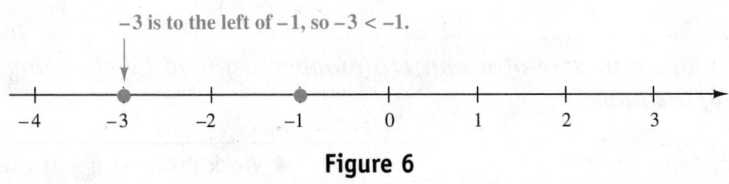

-3 is to the left of -1, so $-3 < -1$.

Figure 6

·································· **Work Problem ❹ at the Side.** ▶

---

❸ List the numbers in the following set that belong to each set of numbers.

$$\left\{ -7, \ -\frac{4}{5}, \ 0, \ \sqrt{3}, \ 2.7, \ 13 \right\}$$

(a) Whole numbers

(b) Integers

(c) Rational numbers

(d) Irrational numbers

❹ Tell whether each statement is *true* or *false*.

GS **(a)** $-2 < 4$
-2 is to the (*left / right*) of 4 on a number line, so the statement is (*true / false*).

GS **(b)** $6 > -3$
6 is to the (*left / right*) of -3 on a number line, so the statement is (*true / false*).

(c) $-9 < -12$

(d) $-4 \geq -1$

(e) $-6 \leq 0$

Answers

3. **(a)** $0, 13$ **(b)** $-7, 0, 13$
   (c) $-7, -\frac{4}{5}, 0, 2.7, 13$ **(d)** $\sqrt{3}$

4. **(a)** left; true   **(b)** right; true   **(c)** false
   (d) false   **(e)** true

5 Find the additive inverse of each number.

(a) 15

(b) −9

(c) −12

(d) $\dfrac{1}{2}$

(e) 0

OBJECTIVE ▶ **3** **Find the additive inverse of a real number.** By a property of the real numbers, for any real number x (except 0), there is exactly one number on a number line the same distance from 0 as x, but on the *opposite* side of 0. See **Figure 7.** Such pairs of numbers are *additive inverses,* or *opposites,* of each other.

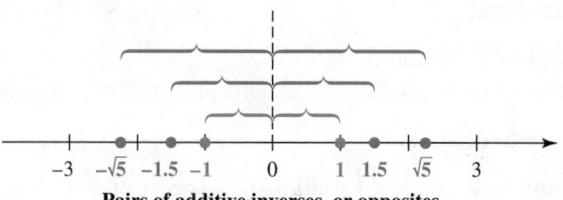

Pairs of additive inverses, or opposites

Figure 7

> **Additive Inverse**
>
> The **additive inverse** of a number x is the number that is the same distance from 0 on a number line as x, but on the *opposite* side of 0. (The number 0 is its own additive inverse.)

We indicate the additive inverse of a number by writing the symbol − in front of the number. For example, the additive inverse of 7 is −7 (read "negative 7"). We could write the additive inverse of −4 as −(−4), but we know that 4 is the opposite of −4. Since a number can have only one additive inverse, −(−4) and 4 must represent the same number, so

$$-(-4) = 4.$$

This idea can be generalized.

> **Double Negative Rule**
>
> For any real number a, $-(-a) = a.$

The following table shows several numbers and their additive inverses.

Number	Additive Inverse
−4	−(−4), or 4
0	0
5	−5
$-\dfrac{2}{3}$	$\dfrac{2}{3}$
0.52	−0.52

The table suggests the following rule.

> *The additive inverse of a nonzero number is found by changing the sign of the number.*

◀ **Work Problem** **5** **at the Side.**

OBJECTIVE ④ **Find the absolute value of a real number.** Because additive inverses are the same distance from 0 on a number line, a number and its additive inverse have the same *absolute value*. The **absolute value** of a real number x, written $|x|$ and read **"the absolute value of x,"** can be defined as the distance between 0 and the number on a number line.

$$|2| = 2 \quad \text{The distance between 2 and 0 on a number line is 2 units.}$$

$$|-2| = 2 \quad \text{The distance between } -2 \text{ and 0 on a number line is also 2 units.}$$

Distance is a physical measurement, which is never negative. *Therefore, the absolute value of a number is never negative.*

Absolute Value

For any real number x,

$$|x| = \begin{cases} x & \text{if } x \geq 0 \\ -x & \text{if } x < 0. \end{cases}$$

By this definition, if x is a positive number or 0, then its absolute value is x itself. For example, since 8 is a positive number,

$$|8| = 8 \quad \text{and} \quad |0| = 0.$$

If x is a negative number, then its absolute value is the additive inverse of x.

$$|-8| = -(-8) = 8 \quad \text{The additive inverse of } -8 \text{ is } 8.$$

EXAMPLE 5   **Finding the Absolute Value**

Simplify by finding the absolute value.

(a) $|0| = 0$ **(b)** $|5| = 5$ **(c)** $|-5| = -(-5) = 5$

(d) $-|5| = -(5) = -5$ **(e)** $-|-5| = -(5) = -5$

(f) $|8-2| = |6| = 6$ **(g)** $-|8-2| = -|6| = -6$

Parts (f) and (g) show that absolute value bars are grouping symbols. We perform any operations inside absolute value symbols *before* finding the absolute value.

·························· **Work Problem ⑥ at the Side.** ▶

---

⑥ Simplify by finding the absolute value.

(a) $|-6|$

(b) $|9|$

(c) $-|15|$

GS **(d)** $-|-9|$

$$= -\underline{\quad}$$

$$= \underline{\quad}$$

(e) $|9-4|$

GS **(f)** $-|32-2|$

$$= -|\underline{\quad}|$$

$$= \underline{\quad}$$

Answers

6. (a) 6 **(b)** 9 **(c)** -15
(d) 9; -9 **(e)** 5 **(f)** 30; -30

1.3 Exercises

FOR EXTRA HELP

 Download the MyDashBoard App

 MyMathLab®

CONCEPT CHECK   *Complete each statement.*

1. The number _____ is a whole number, but not a natural number.

2. The natural numbers, their additive inverses, and 0 form the set of _____.

3. The additive inverse of every negative number is a (*negative / positive*) number.

4. If *x* and *y* are real numbers with $x > y$, then *x* lies to the (*left / right*) of *y* on a number line.

5. A rational number is the _____ of two integers, with the _____ not equal to 0.

6. Decimals that neither terminate nor repeat are _____ numbers.

Use an integer to express each boldface italic number representing a change in the following applications. **See Example 1.**

7. Between July 1, 2008, and July 1, 2009, the population of the United States increased by approximately **2,632,000.** (*Source:* U.S. Census Bureau.)

8. From 2010 to 2011, the mean SAT critical reading score for Illinois students increased by **14,** while the mathematics score increased by **17.** (*Source:* The College Board.)

9. From 2005 to 2009, the paid circulation of daily newspapers in the United States went from 53,345 thousand to 46,278 thousand, representing a decrease of **7067** thousand. (*Source: Editor and Publisher International Yearbook,* 2010.)

10. In 1935, there were 15,295 banks in the United States. By 2010, the number was 7821, representing a decrease of **7474** banks. (*Source:* Federal Deposit Insurance Corporation.)

Graph each group of numbers on a number line. **See Example 2.**

11. $0, 3, -5, -6$

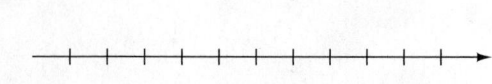

12. $2, 6, -2, -1$

13. $-2, -6, -4, 3, 4$

14. $-5, -3, -2, 0, 4$

15. $\frac{1}{4}, 2\frac{1}{2}, -3\frac{4}{5}, -4, -\frac{13}{8} \doteq 1\frac{5}{8}$

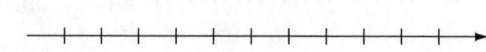

16. $5\frac{1}{4}, \frac{41}{9}, -2\frac{1}{3}, 0, -3\frac{2}{5}$

List all numbers from each set that are **(a)** *natural numbers,* **(b)** *whole numbers,* **(c)** *integers,* **(d)** *rational numbers,* **(e)** *irrational numbers,* **(f)** *real numbers.* **See Example 3.**

17. $\left\{ -9, -\sqrt{7}, -1\frac{1}{4}, -\frac{3}{5}, 0, \sqrt{5}, 3, 5.9, 7 \right\}$

18. $\left\{ -5.3, -5, -\sqrt{3}, -1, -\frac{1}{9}, 0, 1.2, 4, \sqrt{12} \right\}$

19. $\left\{ \dfrac{7}{9}, -2.\overline{3}, \sqrt{3}, 0, -8\dfrac{3}{4}, 11, -6, \pi \right\}$ **20.** $\left\{ 1\dfrac{5}{8}, -0.\overline{4}, \sqrt{6}, 9, -12, 0, \sqrt{10}, 0.026 \right\}$

CONCEPT CHECK *In Exercises 21–26, give a number that satisfies the given condition.*

21. An integer between 3.6 and 4.6

22. A rational number between 2.8 and 2.9

23. A whole number that is not positive and is less than 1

24. A whole number greater than 3.5

25. An irrational number that is between $\sqrt{12}$ and $\sqrt{14}$

26. A real number that is neither negative nor positive

CONCEPT CHECK *In Exercises 27–32, decide whether each statement is* true *or* false.

27. Every natural number is positive.

28. Every whole number is positive.

29. Every integer is a rational number.

30. Every rational number is a real number.

31. Some numbers are both rational and irrational.

32. Every terminating decimal is a rational number.

CONCEPT CHECK *Give three numbers between* −6 *and* 6 *that satisfy each given condition.*

33. Positive real numbers but not integers

34. Real numbers but not positive numbers

35. Real numbers but not whole numbers

36. Rational numbers but not integers

37. Real numbers but not rational numbers

38. Rational numbers but not negative numbers

Select the lesser number in each pair. **See Example 4.**

39. $-11, -4$ **40.** $-9, -16$ **41.** $-21, 1$ **42.** $-57, 3$

43. $0, -100$ **44.** $-215, 0$ **45.** $-\dfrac{2}{3}, -\dfrac{1}{4}$ **46.** $-\dfrac{3}{8}, -\dfrac{9}{16}$

Decide whether each statement is true *or* false. **See Example 4.**

47. $8 < -16$ **48.** $12 < -24$ **49.** $-3 < -2$ **50.** $-10 < -9$

For each number, (a) find its additive inverse and (b) find its absolute value.

51. −2
▶ (a)   (b)

52. −8
(a)   (b)

53. 6
(a)   (b)

54. 11
(a)   (b)

55. $-\dfrac{3}{4}$
▶
(a)   (b)

56. $-\dfrac{1}{3}$
(a)   (b)

57. 4.95
(a)   (b)

58. 0.007
(a)   (b)

59. **CONCEPT CHECK**   Match each expression in Column I with its value in Column II. Choices in Column II may be used once, more than once, or not at all.

I	II
(a) $\lvert -9 \rvert$	A. 9
(b) $-(-9)$	B. −9
(c) $-\lvert -9 \rvert$	C. Neither A nor B
(d) $-\lvert -(-9) \rvert$	D. Both A and B

60. **CONCEPT CHECK**   Fill in the blanks with the correct values: The opposite of −5 is ____, while the absolute value of −5 is ____. The additive inverse of −5 is ____, while the additive inverse of the absolute value of −5 is ____.

Simplify by finding the absolute value. ***See Example 5.***

61. $\lvert -7 \rvert$

62. $\lvert -3 \rvert$

63. $-\lvert 12 \rvert$

64. $-\lvert 23 \rvert$

65. $-\left\lvert -\dfrac{2}{3} \right\rvert$

66. $-\left\lvert -\dfrac{4}{5} \right\rvert$

67. $\lvert 13 - 4 \rvert$

68. $\lvert 8 - 7 \rvert$

Decide whether each statement is true *or* false. ***See Examples 4 and 5.***

69. $\lvert -8 \rvert < 7$

70. $\lvert -6 \rvert \geq -\lvert 6 \rvert$

71. $4 \leq \lvert 4 \rvert$

72. $-\lvert -3 \rvert > 2$

The table shows the change in the Consumer Price Index (CPI) for selected categories of goods and services from 2008 to 2009 and from 2009 to 2010. Use the table to answer Exercises 73–76.

73. Which category for which period represents the greatest decrease?

Category	Change from 2008 to 2009	Change from 2009 to 2010
Apparel	1.2	−0.6
Food	3.9	1.6
Energy	−43.6	18.3
Medical care	11.5	12.8
Transportation	−16.2	14.1

Source: U.S. Bureau of Labor Statistics.

74. Which category for which period represents the least change?

75. Which has lesser absolute value, the change for Transportation from 2008 to 2009 or from 2009 to 2010?

76. Which has greater absolute value, the change for Energy from 2008 to 2009 or from 2009 to 2010?

Study Skills
USING STUDY CARDS

You may have used "flash cards" in other classes. In math, "study cards" can help you remember terms and definitions, procedures, and concepts. Use study cards to

▶ Help you understand and learn the material;

▶ Quickly review when you have a few minutes;

▶ Review before a quiz or test.

One of the advantages of study cards is that you learn while you are making them.

Vocabulary Cards

Put the word and a page reference on the front of the card. On the back, write the definition, an example, any related words, and a sample problem (if appropriate).

Front of Card

Integers   *p. 46*

Back of Card

Def: The natural numbers   {1, 2, 3, 4, ...}
their opposites   {-1, -2, -3, -4, ...}
and 0.   {0}

Integers   { ... , -3, -2, -1, 0, 1, 2, 3, ...}

→ No fractions, decimals, roots
→ Related word: rational numbers

Procedure ("Steps") Cards

Write the name of the procedure on the front of the card. Then write each step in words. On the back of the card, put an example showing each step.

Front of Card

Evaluating Absolute Value (Simplifying)   *p. 51*

1. Work inside absolute value bars first (like working inside parentheses).
2. Find the absolute value (*never* negative).
3. A negative sign *in front of* the absolute value bar is NOT affected, so keep it!

Back of Card

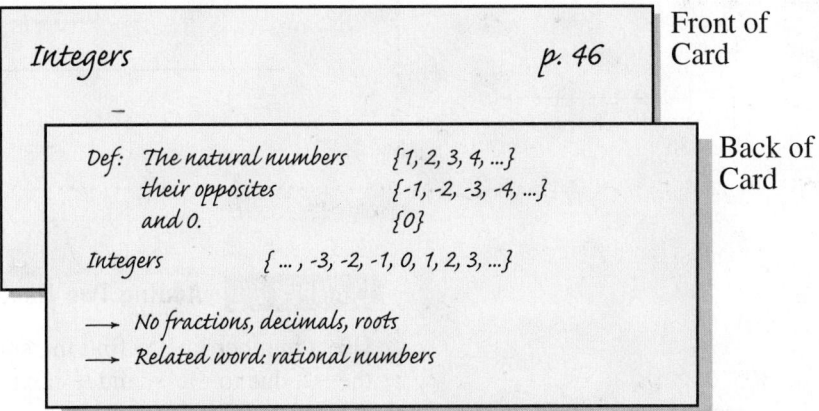

Examples:

Simplify |10 − 6|
|4| = 4   Work inside: 10 − 6 = 4
          Absolute value of 4 is 4.

Simplify   −|−12|
     −12   Absolute value of −12 is 12.
          Keep negative sign that was in front.

Make a vocabulary card and a procedure card for material you are learning now.

1.4 Adding Real Numbers

OBJECTIVES

1. Add two numbers with the same sign.
2. Add numbers with different signs.
3. Add mentally.
4. Use the rules for order of operations with real numbers.
5. Translate words and phrases that indicate addition.

OBJECTIVE ▶ **1** **Add two numbers with the same sign.** Recall that the answer to an addition problem is the **sum.** A number line can be used to add real numbers.

EXAMPLE 1 **Adding Two Positive Numbers on a Number Line**

Use a number line to find the sum $2 + 3$.

Step 1 Start at 0 and draw an arrow 2 units to the *right*. See **Figure 8**.

Step 2 From the right end of that arrow, draw another arrow 3 units to the right.

The number below the end of this second arrow is 5, so $2 + 3 = 5$.

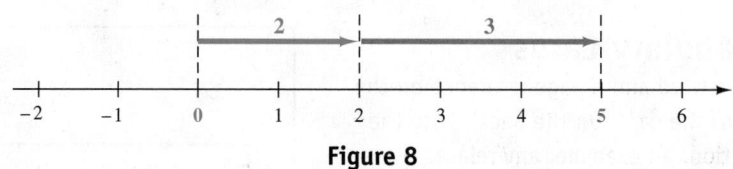

Figure 8

◀ Work Problem **1** at the Side.

EXAMPLE 2 **Adding Two Negative Numbers on a Number Line**

Use a number line to find the sum $-2 + (-4)$. (We put parentheses around the -4 due to the $+$ and $-$ next to each other.)

Step 1 Start at 0 and draw an arrow 2 units to the *left*. See **Figure 9**.

Step 2 From the left end of the first arrow, draw a second arrow 4 units to the *left* to represent the addition of a *negative* number.

The number below the end of this second arrow is -6, so $-2 + (-4) = -6$.

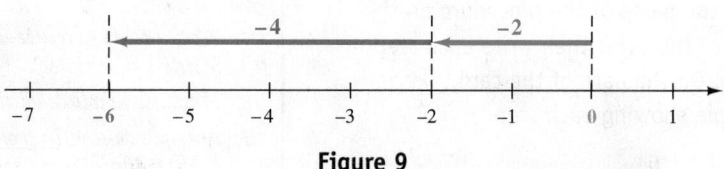

Figure 9

◀ Work Problem **2** at the Side.

In **Example 2,** we found that the sum of the two negative numbers -2 and -4 is a negative number whose distance from 0 is the sum of the distance of -2 from 0 and the distance of -4 from 0. *That is, **the sum of two negative numbers is the opposite of the sum of their absolute values.***

$$-2 + (-4)$$
$$= -(|-2| + |-4|)$$
$$= -(2 + 4)$$
$$= -6$$

---

1 Use a number line to find the sum $1 + 4$.

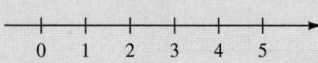

2 Use a number line to find the sum $-2 + (-5)$.

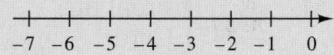

Answers

1. $1 + 4 = 5$

2. $-2 + (-5) = -7$

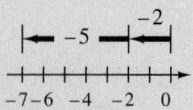

Adding Real Numbers with the Same Sign

To add two numbers with the same sign, add the absolute values of the numbers. Give the result the same sign as the numbers being added.

Example: $-4 + (-3) = -7$

EXAMPLE 3 Adding Two Negative Numbers

Find each sum.

(a) $-2 + (-9)$	(b) $-8 + (-12)$	(c) $-15 + (-3)$
$= -11$	$= -20$	$= -18$

The sum of two negative numbers is negative.

· Work Problem **3** at the Side. ▶

OBJECTIVE **2** **Add numbers with different signs.**

EXAMPLE 4 Adding Numbers with Different Signs

Use the number line to find the sum $-2 + 5$.

Step 1 Start at 0 and draw an arrow 2 units to the left. See **Figure 10**.

Step 2 From the left end of this arrow, draw a second arrow 5 units to the right.

The number below the end of the second arrow is 3, so $-2 + 5 = 3$.

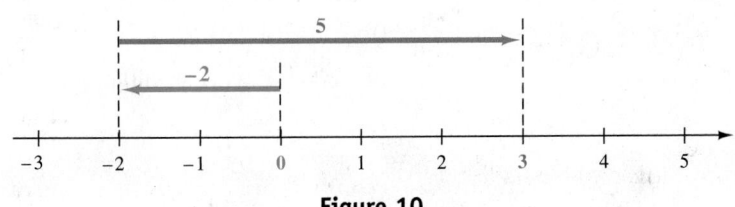

Figure 10

· Work Problem **4** at the Side. ▶

Adding Real Numbers with Different Signs

To add two numbers with different signs, find the absolute values of the numbers. Subtract the lesser absolute value from the greater. Give the answer the same sign as the number with the greater absolute value.

Example: $-12 + 6 = -6$

EXAMPLE 5 Adding Numbers with Different Signs

Find the sum $-12 + 5$.

Find the absolute value of each number.

$$|-12| = 12 \quad \text{and} \quad |5| = 5$$

Then find the difference between these absolute values: $12 - 5 = 7$. The sum will be negative, since $|-12| > |5|$.

$$-12 + 5 = -7$$

· Work Problem **5** at the Side. ▶

---

3 Find each sum.

(a) $-7 + (-3)$

(b) $-12 + (-18)$

(c) $-15 + (-4)$

4 Use a number line to find each sum.

(a) $6 + (-3)$

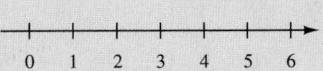

(b) $-5 + 1$

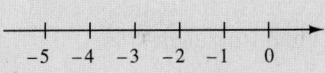

5 Find each sum.

(a) $-24 + 11$

(b) $30 + (-8)$

Answers

3. (a) -10 (b) -30 (c) -19
4. (a) $6 + (-3) = 3$

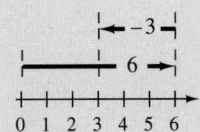

(b) $-5 + 1 = -4$

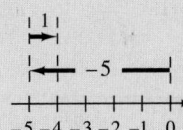

5. (a) -13 (b) 22

6 Check each answer by adding mentally. If necessary, use a number line.

(a) $-8 + 2 = -6$

(b) $-15 + 4 = -11$

(c) $17 + (-10) = 7$

(d) $\frac{3}{4} + \left(-1\frac{3}{8}\right) = -\frac{5}{8}$

(e) $-9.5 + 3.8 = -5.7$

(f) $37 + (-37) = 0$

7 Label the order in which each

GS expression should be evaluated. Then find each sum.

(a) $2 + [7 + (-3)]$
   ② ①
   $= 2 + \underline{\quad}$
   $= \underline{\quad}$

(b) $6 + [(-2 + 5) + 7]$
   ○   ○ ○
   $= 6 + [\underline{\quad} + 7]$
   $= 6 + \underline{\quad}$
   $= \underline{\quad}$

(c) $-9 + [-4 + (-8 + 6)]$
   ○   ○   ○

Answers

6. All are correct.

7. (a) 4; 6   (b) ③,①,②; 3; 10; 16
   (c) ③,②,①; −15

☐ **Calculator Tip**

The ⊝ or ⊕⁄⊝ key is used to input a negative number in some scientific calculators.

OBJECTIVE 3 Add mentally.

EXAMPLE 6   Adding a Positive Number and a Negative Number

Check each answer by adding mentally. If necessary, use a number line.

(a) $7 + (-4)$
   $= 3$

(b) $-8 + 12$
   $= 4$

(c) $-\frac{1}{2} + \frac{1}{8}$
   $= -\frac{4}{8} + \frac{1}{8}$
   $= -\frac{3}{8}$

(d) $\frac{5}{6} + \left(-1\frac{1}{3}\right)$
   $= \frac{5}{6} + \left(-\frac{4}{3}\right)$
   $= \frac{5}{6} + \left(-\frac{8}{6}\right)$
   $= -\frac{3}{6}$, or $-\frac{1}{2}$

(e) $-4.6 + 8.1$
   $= 3.5$

(f) $-16 + 16$
   $= 0$

(g) $42 + (-42)$
   $= 0$

◀ **Work Problem 6 at the Side.**

Adding Signed Numbers

Same sign   Add the absolute values of the numbers. Give the sum the same sign as the numbers being added.

Different signs   Find the absolute values of the numbers, and subtract the lesser absolute value from the greater. Give the answer the sign of the number having the greater absolute value.

OBJECTIVE 4 Use the rules for order of operations with real numbers. When a problem involves square brackets, [ ], we do the calculations inside the brackets until a single number is obtained.

EXAMPLE 7   Adding with Brackets

Find each sum.   Start here.

(a) $-3 + [4 + (-8)]$
   $= -3 + (-4)$
   $= -7$

(b) $8 + [(-2 + 6) + (-3)]$
   $= 8 + [4 + (-3)]$
   $= 8 + 1$
   $= 9$

Follow the order of operations.

◀ **Work Problem 7 at the Side.**

OBJECTIVE 5 **Translate words and phrases that indicate addition.** Problem solving requires translating words and phrases into symbols. We began this process in **Section 1.1.** The table lists key words and phrases that indicate addition.

Word or Phrase Indicating Addition	Example	Numerical Expression and Simplification
Sum of	The *sum of* −3 and 4	−3 + 4, which equals 1
Added to	5 *added to* −8	−8 + 5, which equals −3
More than	12 *more than* −5	(−5) + 12, which equals 7
Increased by	−6 *increased by* 13	−6 + 13, which equals 7
Plus	3 *plus* 14	3 + 14, which equals 17

EXAMPLE 8   **Translating Words and Phrases (Addition)**

Write a numerical expression for each phrase, and simplify the expression.

(a) The *sum of* −8 and 4 and 6

$-8 + 4 + 6$ simplifies to $-4 + 6$,   which equals   2.

> Add in order from left to right.

(b) 3 *more than* −5, **increased by** 12

$(-5 + 3) + 12$ simplifies to $-2 + 12$,   which equals   10.

Here we *simplified* each expression by performing the operations.

··· **Work Problem 8 at the Side.** ▶

In applications, gains may be interpreted as positive numbers and losses as negative numbers.

EXAMPLE 9   **Interpreting Gains and Losses**

A football team gained 3 yd on first down, lost 12 yd on second down, and then gained 13 yd on third down. How many yards did the team gain or lose altogether on these plays?

$3 + (-12) + 13$ Gains plus losses

$= \left[3 + (-12)\right] + 13$ Add from left to right.

$= (-9) + 13$

$= 4$ Add.

The team gained 4 yd altogether on these plays.

·· **Work Problem 9 at the Side.** ▶

8 Write a numerical expression for each phrase, and simplify the expression.

(a) 4 more than −12

(b) The sum of 6 and −7

(c) −12 added to −31

(d) 7 increased by the sum of 8 and −3

9 Solve the problem.
   A football team lost 8 yd on first down, lost 5 yd on second down, and then gained 7 yd on third down. How many yards did the team gain or lose altogether on these plays?

Answers

8. **(a)** $-12 + 4$; -8 **(b)** $6 + (-7)$; -1
   (c) $-31 + (-12)$; -43
   (d) $7 + \left[8 + (-3)\right]$; 12
9. The team lost 6 yd.

1.4 Exercises

 MyMathLab®

CONCEPT CHECK *Complete each of the following.*

1. The sum of two negative numbers will always be a (*positive / negative*) number.
   ▶ Give a number-line illustration using the sum $-2 + (-3) = $ ____.

2. The sum of a number and its opposite will always be _____.

3. When adding a positive number and a negative number, where the negative
   ▶ number has the greater absolute value, the sum will be a (*positive / negative*)
   number. Give a number-line illustration using the sum $-4 + 2 = $ ____.

4. To simplify the expression $8 + [-2 + (-3 + 5)]$, one should begin by adding
   ____ and ____, according to the rules for order of operations.

CONCEPT CHECK *By the rules for order of operations, what is the first step you would use to simplify each expression?*

5. $4[3(-2 + 5) - 1]$

6. $[-4 + 7(-6 + 2)]$

7. $9 + ([-1 + (-3)] + 5)$

8. $[(-8 + 4) + (-6)] + 5$

Find each sum. **See Examples 1–7.**

9. $6 + (-4)$

10. $8 + (-5)$

11. $12 + (-15)$

12. $4 + (-8)$

13. $-7 + (-3)$

14. $-11 + (-4)$

15. $-10 + (-3)$

16. $-16 + (-7)$

17. $-12.4 + (-3.5)$

18. $-21.3 + (-2.5)$

19. $10 + [-3 + (-2)]$
    = $10 + ($____$)$
    = ____

20. $13 + [-4 + (-5)]$
    = $13 + ($____$)$
    = ____

21. $5 + [14 + (-6)]$

22. $7 + [3 + (-14)]$

23. $-3 + [5 + (-2)]$

24. $-7 + [10 + (-3)]$

25. $-8 + [3 + (-1) + (-2)]$

26. $-7 + [5 + (-8) + 3]$

27. $\dfrac{9}{10} + \left(-\dfrac{3}{5}\right)$

28. $\dfrac{5}{8} + \left(-\dfrac{17}{12}\right)$

29. $-\dfrac{1}{6} + \dfrac{2}{3}$

30. $-\dfrac{6}{25} + \dfrac{19}{20}$

31. $2\dfrac{1}{2} + \left(-3\dfrac{1}{4}\right)$

32. $6\dfrac{1}{2} + \left(-4\dfrac{3}{8}\right)$

33. $7.8 + (-9.4)$

34. $14.7 + (-10.1)$

35. $-7.1 + \left[3.3 + (-4.9)\right]$

36. $-9.5 + \left[-6.8 + (-1.3)\right]$

37. $\left[-8 + (-3)\right] + \left[-7 + (-7)\right]$
$= \underline{} + (\underline{})$
$= \underline{}$

38. $\left[-5 + (-4)\right] + \left[9 + (-2)\right]$
$= \underline{} + \underline{}$
$= \underline{}$

39. $\left(-\dfrac{1}{2} + 0.25\right) + \left(-\dfrac{3}{4} + 0.75\right)$

40. $\left(-\dfrac{3}{2} + 0.75\right) + \left(-\dfrac{1}{2} + 2.25\right)$

Perform each operation, and then determine whether the statement is true *or* false.
Try to do all work mentally. **See Examples 6 and 7.**

41. $-11 + 13 = 13 + (-11)$

42. $16 + (-9) = -9 + 16$

43. $-10 + 6 + 7 = -3$

44. $-12 + 8 + 5 = -1$

45. $\dfrac{7}{3} + \left(-\dfrac{1}{3}\right) + \left(-\dfrac{6}{3}\right) = 0$

46. $-\dfrac{3}{2} + 1 + \dfrac{1}{2} = 0$

47. $\left|-8 + 10\right| = -8 + (-10)$

48. $\left|-4 + 6\right| = -4 + (-6)$

49. $2\dfrac{1}{5} + \left(-\dfrac{6}{11}\right) = -\dfrac{6}{11} + 2\dfrac{1}{5}$

50. $-1\dfrac{1}{2} + \dfrac{5}{8} = \dfrac{5}{8} + \left(-1\dfrac{1}{2}\right)$

51. $-7 + \left[-5 + (-3)\right] = \left[(-7) + (-5)\right] + 3$

52. $6 + \left[-2 + (-5)\right] = \left[(-4) + (-2)\right] + 5$

Write a numerical expression for each phrase, and simplify the expression.
See Example 8.

53. The sum of -5 and 12 and 6

54. The sum of -3 and 5 and -12

55. 14 added to the sum of -19 and -4

56. -2 added to the sum of -18 and 11

57. The sum of -4 and -10, increased by 12

58. The sum of -7 and -13, increased by 14

59. $\frac{2}{7}$ more than the sum of $\frac{5}{7}$ and $-\frac{9}{7}$

60. 0.85 more than the sum of -1.25 and -4.75

Solve each problem. ***See Example 9.***

61. Nathaniel owed his older sister Jenna $24 for his share of the bill when they took their mother out to dinner for her birthday. He later borrowed $38 from his younger sister Ilana to buy two DVDs. What positive or negative number represents Nathaniel's financial situation with his siblings?

62. Bonika's checking account balance is $54.00. She then takes a gamble by writing a check for $89.00. What is her new balance? (Write the balance as a signed number.)

63. The surface, or rim, of a canyon is at altitude 0. On a hike down into the canyon, a party of hikers stops for a rest at 130 m below the surface. They then descend another 54 m. What is their new altitude? (Write the altitude as a signed number.)

64. A pilot announces to the passengers that the current altitude of their plane is 34,000 ft. Because of some unexpected turbulence, the pilot is forced to descend 2100 ft. What is the new altitude of the plane? (Write the altitude as a signed number.)

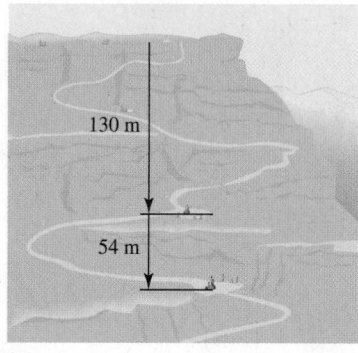

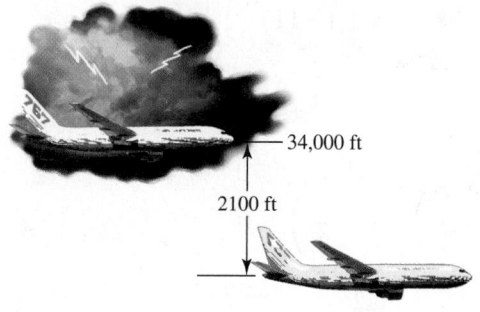

65. J. D. Patin enjoys playing Triominoes every Wednesday night. Last Wednesday, on four successive turns, his scores were -19, 28, -5, and 13. What was his final score for the four turns?

66. Gay Aguillard also enjoys playing Triominoes. On five successive turns, her scores were -13, 15, -12, 24, and 14. What was her total score for the five turns?

67. On three consecutive passes, a quarterback passed for a gain of 6 yd, was sacked for a loss of 12 yd, and passed for a gain of 43 yd. What positive or negative number represents the total net yardage for the plays?

68. On a series of three consecutive running plays, a running back gained 4 yd, lost 3 yd, and lost 2 yd. What positive or negative number represents his total net yardage for the series of plays?

69. The lowest temperature ever recorded in Arkansas was −29°F. The highest temperature ever recorded there was 149°F more than the lowest. What was this highest temperature? (*Source: National Climatic Data Center.*)

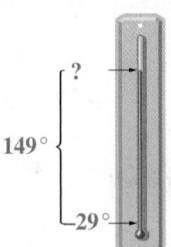

70. On January 23, 1943, the temperature rose 49°F in two minutes in Spearfish, South Dakota. If the starting temperature was −4°F, what was the temperature two minutes later?

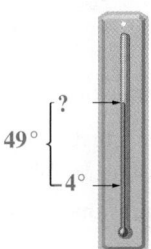

71. Dana Weightman owes $153 to a credit card company. She makes a $14 purchase with the card, and then pays $60 on the account. What is her current balance as a signed number?

72. A female polar bear weighed 660 lb when she entered her winter den. She lost 45 lb during each of the first two months of hibernation, and another 205 lb before leaving the den with her two cubs in March. How much did she weigh when she left the den?

73. Based on the 2010 Census, New York lost 2 seats in the U.S. House of Representatives. Illinois, Iowa, Missouri, and Massachusetts were among the states that each lost 1 seat. Write a signed number that represents the total change in number of seats for these five states. (*Source: U.S. Census Bureau.*)

74. Based on the 2010 Census, Ohio and New York lost the most seats in the U.S. House of Representatives, each losing 2 seats. The states gaining the most seats were Texas with 4 and Florida with 2. Write a signed number that represents the algebraic sum of these changes. (*Source: U.S. Census Bureau.*)

1.5 Subtracting Real Numbers

OBJECTIVE **1** **Subtract two numbers on a number line.** Recall that the answer to a subtraction problem is a **difference.** In the subtraction $x - y$, x is the **minuend** and y is the **subtrahend.**

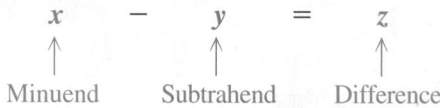

EXAMPLE 1 **Subtracting Numbers on a Number Line**

Use a number line to find the difference $7 - 4$.

Step 1 Start at 0 and draw an arrow 7 units to the *right*. See **Figure 11.**

Step 2 From the right end of the first arrow, draw a second arrow 4 units to the *left* to represent the subtraction of a positive number.

The number below the end of this second arrow is 3, so $7 - 4 = 3$.

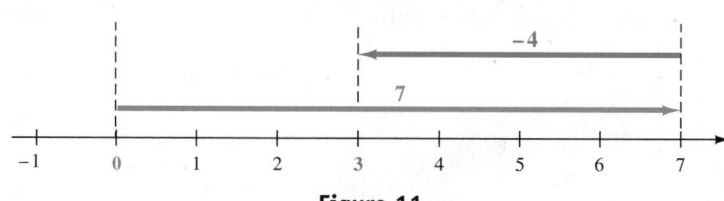

Figure 11

◀ Work Problem **1** at the Side.

OBJECTIVE **2** **Use the definition of subtraction.** The procedure used in **Example 1** to find $7 - 4$ is the same procedure for finding $7 + (-4)$.

$$7 - 4 \quad \text{is equal to} \quad 7 + (-4).$$

This shows that *subtracting* a positive number from a larger positive number is the same as *adding* the opposite of the smaller number to the larger.

Definition of Subtraction

For any real numbers a and b,

$$a - b \quad \text{is defined as} \quad a + (-b).$$

To subtract b from a, add the additive inverse (or opposite) of b to a. In words, change the subtrahend to its opposite and add.

Example: $4 - 9$
$$= 4 + (-9)$$
$$= -5$$

Subtracting Signed Numbers

Step 1 Change the subtraction symbol to an addition symbol, and change the sign of the subtrahend.

Step 2 Add, as in **Section 1.4.**

---

1 Use a number line to find each difference.

(a) $5 - 1$

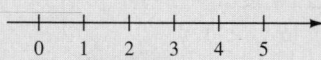

(b) $6 - 2$

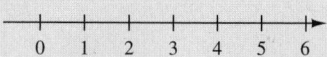

Answers

1. **(a)** $5 - 1 = 4$

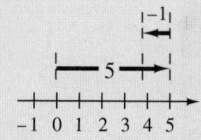

(b) $6 - 2 = 4$

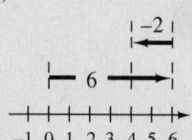

EXAMPLE 2 Using the Definition of Subtraction

Subtract.

No change ⟶ | Change − to +. | Opposite of 3

> −7 has the greater absolute value, so the sum is negative.

(a) $12 - 3 = 12 + (-3) = 9$ **(b)** $5 - 7 = 5 + (-7) = -2$

(c) $-8 - 15 = -8 + (-15) = -23$

No change ⟶ | Change − to +. | Opposite of −5

(d) $-3 - (-5) = -3 + (5) = 2$

(e) $\dfrac{3}{8} - \left(-\dfrac{4}{5}\right)$

$= \dfrac{15}{40} - \left(-\dfrac{32}{40}\right)$ Find a common denominator.

$= \dfrac{15}{40} + \dfrac{32}{40}$ Use the definition of subtraction.

$= \dfrac{47}{40}$ Add the fractions.

···· **Work Problem ② at the Side.** ▶

Uses of the Symbol −

We use the symbol − for three purposes.

1. It can represent **subtraction,** as in $9 - 5 = 4$.

2. It can represent **negative numbers,** such as -10, -2, and -3.

3. It can represent **the opposite (or additive inverse) of a number,** as in "the opposite (or additive inverse) of 8 is -8."

We may see more than one use in the same problem, such as $-6 - (-9)$, where -9 is subtracted from -6. The meaning of the symbol depends on its position in the algebraic expression.

OBJECTIVE ▶ **③ Work subtraction problems that involve brackets.** As before, first perform any operations inside the parentheses and brackets.

EXAMPLE 3 Subtracting with Grouping Symbols

Perform each operation.

(a) $-6 - [2 - (8 + 3)]$ ← Work from the inside out.

$= -6 - [2 - 11]$ Add inside the parentheses.

$= -6 - [2 + (-11)]$ Definition of subtraction

$= -6 - (-9)$ Add.

$= -6 + 9$ Definition of subtraction

$= 3$ Add.

···· **Continued on Next Page**

---

② Subtract.

GS (a) $6 - 10$

$= 6 + (\underline{})$

$= \underline{}$

(b) $-2 - 4$

GS (c) $3 - (-5)$

$= 3 + \underline{}$

$= \underline{}$

(d) $-8 - (-12)$

(e) $\dfrac{5}{4} - \left(-\dfrac{3}{7}\right)$

Answers

2. (a) -10; -4 **(b)** -6 **(c)** 5; 8

(d) 4 **(e)** $\dfrac{47}{28}$

❸ Label the order in which each
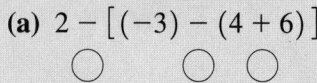 expression should be evaluated.
Then perform each operation.

(a) $2 - [(-3) - (4 + 6)]$

○ ○ ○

(b) $[(5 - 7) + 3] - 8$

○ ○ ○

(c) $6 - [(-1 - 4) - 2]$

○ ○ ○

(b) $5 - \left[\left(-\dfrac{1}{3} - \dfrac{1}{2} \right) - (4 - 1) \right]$ ◁ Work from the inside out.

$= 5 - \left[\left(-\dfrac{1}{3} + \left(-\dfrac{1}{2} \right) \right) - 3 \right]$ | Start within each set of parentheses inside the brackets.

$= 5 - \left[\left(-\dfrac{5}{6} \right) - 3 \right]$ | Use 6 as the common denominator; $-\frac{1}{3} + \left(-\frac{1}{2} \right) = -\frac{2}{6} + \left(-\frac{3}{6} \right) = -\frac{5}{6}$

$= 5 - \left[\left(-\dfrac{5}{6} \right) + (-3) \right]$ | Definition of subtraction

$= 5 - \left[\left(-\dfrac{5}{6} \right) + \left(-\dfrac{18}{6} \right) \right]$ | Use 6 as the common denominator; $-\frac{3}{1} \cdot \frac{6}{6} = -\frac{18}{6}$

$= 5 - \left(-\dfrac{23}{6} \right)$ | Add inside the brackets.

$= 5 + \dfrac{23}{6}$ | Definition of subtraction

$= \dfrac{30}{6} + \dfrac{23}{6}$ | Write 5 as an improper fraction.

$= \dfrac{53}{6}$ | Add fractions.

◁ **Work Problem ❸ at the Side.**

OBJECTIVE ❹ Translate words and phrases that indicate subtraction.
The table lists words and phrases that indicate subtraction in problem solving.

Word, Phrase, or Sentence Indicating Subtraction	Example	Numerical Expression and Simplification
Difference between	The *difference between* −3 and −8	$-3 - (-8)$ simplifies to $-3 + 8$, which equals 5
Subtracted from*	12 *subtracted from* 18	$18 - 12$, which equals 6
From ..., subtract	*From* 12, *subtract* 8.	$12 - 8$ simplifies to $12 + (-8)$, which equals 4
Less	6 *less* 5	$6 - 5$, which equals 1
Less than*	6 *less than* 5	$5 - 6$ simplifies to $5 + (-6)$, which equals −1
Decreased by	9 *decreased by* −4	$9 - (-4)$ simplifies to $9 + 4$, which equals 13
Minus	8 *minus* 5	$8 - 5$, which equals 3

*Be careful with order when translating.

CAUTION

When subtracting two numbers, be careful to write them in the correct order, because, in general,

$$x - y \neq y - x.$$ For example, $5 - 3 \neq 3 - 5$.

Think carefully before interpreting an expression involving subtraction.

Answers

3. **(a)** ③, ②, ①; 15

(b) ①, ②, ③; −7

(c) ③, ①, ②; 13

EXAMPLE 4 Translating Words and Phrases (Subtraction)

Write a numerical expression for each phrase, and simplify the expression.

(a) The **difference between** −8 and 5

When "difference between" is used, write the numbers in the order they are given.

$$-8 - 5 \quad \text{simplifies to} \quad -8 + (-5), \quad \text{which equals} \quad -13.$$

(b) 4 **subtracted from** the sum of 8 and −3

Here the operation of addition is also used, as indicated by the word *sum*. First, add 8 and −3. Next, subtract 4 from this sum. The expression is

$$[8 + (-3)] - 4 \quad \text{simplifies to} \quad 5 - 4, \quad \text{which equals} \quad 1.$$

(c) 4 **less than** −6

Here 4 must be taken *from* −6, so write −6 first.

> Be careful with order.

$$-6 - 4 \quad \text{simplifies to} \quad -6 + (-4), \quad \text{which equals} \quad -10.$$

Notice that "4 less than −6" differs from "4 *is less than* −6." The statement "4 is less than −6" is symbolized as $4 < -6$ (which is a false statement).

(d) 8, **decreased by** 5 **less than** 12

First, write "5 less than 12" as $12 - 5$. Next, subtract $12 - 5$ from 8.

$$8 - (12 - 5) \quad \text{simplifies to} \quad 8 - 7, \quad \text{which equals} \quad 1.$$

·············· **Work Problem ④ at the Side.** ▶

EXAMPLE 5 Solving an Applied Problem Involving Subtraction

The record high temperature of 134°F in the United States was recorded at Death Valley, California, in 1913. The record low was −80°F, at Prospect Creek, Alaska, in 1971. See **Figure 12.** What is the difference between these highest and lowest temperatures? (*Source:* National Climatic Data Center.)

We must subtract the lowest temperature from the highest temperature.

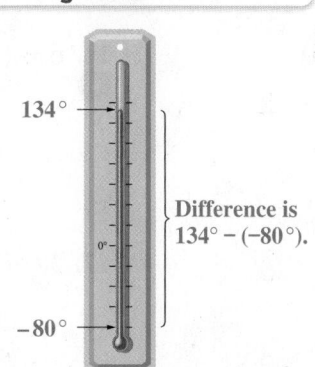

> Order of numbers matters in subtraction.

$$134 - (-80)$$
$$= 134 + 80 \qquad \text{Use the definition of subtraction.}$$
$$= 214 \qquad \text{Add.}$$

The difference between the two temperatures is 214°F.

Figure 12

·············· **Work Problem ⑤ at the Side.** ▶

---

④ Write a numerical expression for each phrase, and simplify the expression.

(a) The difference between −5 and −12

(b) −2 subtracted from the sum of 4 and −4

(c) 7 less than −2

(d) 9, decreased by 10 less than 7

⑤ Solve the problem.

The highest elevation in Argentina is Mt. Aconcagua, which is 6960 m above sea level. The lowest point in Argentina is the Valdes Peninsula, 40 m below sea level. Find the difference between the highest and lowest elevations.

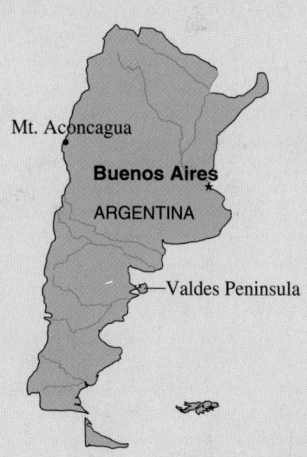

Answers

4. **(a)** $-5 - (-12)$; 7
(b) $[4 + (-4)] - (-2)$; 2
(c) $-2 - 7$; −9
(d) $9 - (7 - 10)$; 12
5. 7000 m

1.5 Exercises  MyMathLab®

CONCEPT CHECK *Fill in each blank with the correct response.*

1. By the definition of subtraction, in order to perform the subtraction $-6 - (-8)$, we must add the opposite of _____ to _____ to get _____.

2. By the rules for order of operations, to simplify $8 - [3 - (-4 - 5)]$, the first step is to subtract _____ from _____.

3. "The difference between 7 and 12" translates as _____, while "the difference between 12 and 7" translates as _____.

4. To subtract b from a, add the _____ _____, or _____, of b to a.

5. $-8 - 4$ can be rewritten as $-8 + (\_\_\_)$.

6. $-19 - 22$ can be rewritten as $-19 + (\_\_\_)$.

Perform the indicated operations. **See Examples 1–3.**

7. $-7 - 3$
$= -7 + (\_\_\_)$
$= \_\_\_$

8. $-12 - 5$
$= -12 + (\_\_\_)$
$= \_\_\_$

9. $-10 - 6$

10. $-13 - 16$

11. $7 - (-4)$
$= 7 + \_\_\_$
$= \_\_\_$

12. $9 - (-6)$
$= 9 + \_\_\_$
$= \_\_\_$

13. $6 - (-13)$

14. $13 - (-3)$

15. $-7 - (-3)$

16. $-8 - (-6)$

17. $3 - (4 - 6)$

18. $6 - (7 - 14)$

19. $-3 - (6 - 9)$

20. $-4 - (5 - 12)$

21. $\dfrac{1}{2} - \left(-\dfrac{1}{4}\right)$

22. $\dfrac{1}{3} - \left(-\dfrac{4}{3}\right)$

23. $-\dfrac{3}{4} - \dfrac{5}{8}$

24. $-\dfrac{5}{6} - \dfrac{1}{2}$

25. $\dfrac{5}{8} - \left(-\dfrac{1}{2} - \dfrac{3}{4}\right)$

26. $\dfrac{9}{10} - \left(-\dfrac{1}{8} - \dfrac{3}{10}\right)$

27. $4.4 - (-9.2)$

28. $6.7 - (-12.6)$

29. $-7.4 - 4.5$

30. $-5.4 - 9.6$

31. $8 - (-3) - 9 + 6$

32. $12 - (-8) - 25 + 9$

33. $-5.2 - (8.4 - 10.8)$

34. $-9.6 - (3.5 - 12.6)$

35. $[(-3.1) - 4.5] - (0.8 - 2.1)$

36. $[(-7.8) - 9.3] - (0.6 - 3.5)$

37. $-12 - [(9 - 2) - (-6 - 3)]$

38. $-4 - [(6 - 9) - (-7 - 4)]$

39. $\left(-\dfrac{3}{8} - \dfrac{2}{3}\right) - \left(-\dfrac{9}{8} - 3\right)$

40. $\left(-\dfrac{3}{4} - \dfrac{5}{2}\right) - \left(-\dfrac{1}{8} - 1\right)$

41. $[-12.25 - (8.34 + 3.57)] - 17.88$

42. $[-34.99 + (6.59 - 12.25)] - 8.33$

Write a numerical expression for each phrase and simplify. ***See Example 4.***

43. The difference between 4 and -8

44. The difference between 7 and -14

45. 8 less than -2

46. 9 less than -13

47. The sum of 9 and -4, decreased by 7

48. The sum of 12 and -7, decreased by 14

49. 12 less than the difference between 8 and -5

50. 19 less than the difference between 9 and -2

Solve each problem. ***See Example 5.***

51. The lowest temperature ever recorded in Illinois was $-36°$F on January 5, 1999. The lowest temperature ever recorded in Utah was on February 1, 1985, and was $33°$F lower than Illinois's record low. What is the record low temperature for Utah? (*Source:* National Climatic Data Center.)

52. The lowest temperature ever recorded in South Carolina was $-19°$F on January 21, 1985. The lowest temperature ever recorded in Wisconsin was $36°$ lower than South Carolina's record low. What is the record low temperature in Wisconsin? (*Source:* National Climatic Data Center.)

53. The top of Mount Whitney, visible from Death Valley, has an altitude of 14,494 ft above sea level. The bottom of Death Valley is 282 ft below sea level. Using 0 as sea level, find the difference between these two elevations. (*Source: World Almanac and Book of Facts.*)

54. The height of Mount Pumasillo in Peru is 20,492 ft. The depth of the Peru-Chile Trench in the Pacific Ocean is 26,457 ft below sea level. Find the difference between these two elevations. (*Source: World Almanac and Book of Facts.*)

55. Samir owed his brother $10. He later borrowed $70. What positive or negative number represents his present financial status?

56. Francesca has $15 in her purse, and Emilio has a debt of $12. Find the difference between these amounts.

57. A chemist is running an experiment under precise conditions. At first, she runs it at $-174.6°$F. She then lowers the temperature by $2.3°$F. What is the new temperature for the experiment?

58. At 2:00 A.M., a plant worker found that a dial reading was 7.904. At 3:00 A.M., she found the reading to be -3.291. Find the difference between these two readings.

59. In August, Kari Heen began with a checking account balance of $904.89. Her checks and deposits for August are given below.

Checks	Deposits
$35.84	$85.00
$26.14	$120.76
$3.12	

Assuming no other transactions, what was her account balance at the end of August?

60. In September, Derek Bowen began with a checking account balance of $904.89. His checks and deposits for September are given below.

Checks	Deposits
$41.29	$80.59
$13.66	$276.13
$84.40	

Assuming no other transactions, what was his account balance at the end of September?

61. A certain Greek mathematician was born in 426 B.C. His father was born 43 years earlier. In what year was his father born?

62. A certain Roman philosopher was born in 325 B.C. Her mother was born 35 years earlier. In what year was her mother born?

63. Kim Falgout owes $870.00 on her MasterCard account. She returns two items costing $35.90 and $150.00 and receives credits for these on the account. Next, she makes a purchase of $82.50, and then two more purchases of $10.00 each. She makes a payment of $500.00. She then incurs a finance charge of $37.23. How much does she still owe?

64. Charles Vosburg owes $679.00 on his Visa account. He returns three items costing $36.89, $29.40, and $113.55 and receives credits for these on the account. Next, he makes purchases of $135.78 and $412.88, and two purchases of $20.00 each. He makes a payment of $400. He then incurs a finance charge of $24.57. How much does he still owe?

65. José Martinez enjoys diving in Lake Okoboji. He dives to 34 ft below the surface of the lake. His partner, Sean O'Malley, dives to 40 ft below the surface, but then ascends 20 ft. What is the vertical distance between José and Sean?

66. Rhonda Alessi also enjoys diving. She dives to 12 ft below the surface of False River. Her sister, Sandy, dives to 20 ft below the surface, but then ascends 10 ft. What is the vertical distance between Rhonda and Sandy?

67. The height of Mt. Foraker is 17,400 ft, while the depth of the Java Trench is 23,376 ft. What is the vertical distance between the top of Mt. Foraker and the bottom of the Java Trench? (*Source: World Almanac and Book of Facts.*)

68. The height of Mt. Wilson in Colorado is 14,246 ft, while the depth of the Cayman Trench is 24,721 ft. What is the vertical distance between the top of Mt. Wilson and the bottom of the Cayman Trench? (*Source: World Almanac and Book of Facts.*)

69. In 2005, Americans saved −0.5% of their after-tax incomes, the first negative personal savings rate since 1933. In the fourth quarter of 2011, they saved 3.9%. Find the difference between the two amounts for 2011 and 2005.

70. Refer to **Exercise 69.** How is it possible that Americans had a negative personal savings rate in 2005?

71. In 2000, the federal budget had a surplus of $236 billion. In 2010, the federal budget had a deficit of $1294 billion. Find the difference between the amounts for 2000 and 2010. (*Source:* U.S. Office of Management and Budget.)

72. In 1998, undergraduate college students had an average (mean) credit card balance of $1879. The average balance increased $869 by 2000, then dropped $579 by 2004, and then increased $1004 by 2008. What was the average credit card balance of undergraduate college students in 2008? (*Source:* Sallie Mae.)

Median sales prices for existing single-family homes in the United States for the years 2005 through 2009 are shown in the table. Complete the table, determining the change from one year to the next by subtraction.

	Year	Median Sales Price	Change from Previous Year
	2005	$219,900	——
73.	2006	$221,900	
74.	2007	$219,000	
75.	2008	$198,100	
76.	2009	$175,500	

Source: National Association of Realtors.

The table shows the heights of some selected mountains. Use the information given to answer Exercises 77 and 78.

77. How much higher is Mt. Wilson than Pikes Peak?

78. If Mt. Wilson and Pikes Peak were stacked one on top of the other, how much higher would they be than Mt. Foraker?

Mountain	Height (in feet)
Foraker	17,400
Wilson	14,246
Pikes Peak	14,110

Source: World Almanac and Book of Facts.

CONCEPT CHECK *In Exercises 79–84, suppose that x represents a positive number and y represents a negative number. Determine whether the given expression must represent a positive number or a negative number.*

79. $y - x$

80. $x - y$

81. $x + |y|$

82. $y - |x|$

83. $|x| + |y|$

84. $-(|x| + |y|)$

1.6 Multiplying and Dividing Real Numbers

The result of multiplication is the **product.** We know that the product of two positive numbers is positive. We also know that the product of 0 and any positive number is 0, so we extend that property to all real numbers.

Multiplication Property of 0

For any real number a, the following hold.
$$a \cdot 0 = 0 \quad \text{and} \quad 0 \cdot a = 0$$

OBJECTIVE ▶ **1** **Find the product of a positive number and a negative number.** Observe the following pattern.

$$3 \cdot 5 = 15$$
$$3 \cdot 4 = 12$$
$$3 \cdot 3 = 9$$
$$3 \cdot 2 = 6$$
$$3 \cdot 1 = 3$$
$$3 \cdot 0 = 0$$
$$3 \cdot (-1) = ?$$

The products decrease by 3.

What should $3(-1)$ equal? Since multiplication can also be considered repeated addition, the product $3(-1)$ represents the sum

$$-1 + (-1) + (-1), \quad \text{which equals} \quad -3,$$

so the product should be -3, which fits the pattern. Also, $3(-2)$ represents the sum

$$-2 + (-2) + (-2), \quad \text{which equals} \quad -6.$$

Work Problem ❶ at the Side. ▶

These results suggest the following rule.

Multiplying Numbers with Different Signs

The product of a positive number and a negative number is negative.

Examples: $6(-3) = -18$ and $-3(6) = -18$

EXAMPLE 1 **Multiplying a Positive Number and a Negative Number**

Find each product using the multiplication rule given in the box.

(a) $8(-5)$
$= -(8 \cdot 5)$
$= -40$

(b) $-9\left(\dfrac{1}{3}\right)$
$= -3$

(c) $-6.2(4.1)$
$= -25.42$

Work Problem ❷ at the Side. ▶

OBJECTIVES

1. Find the product of a positive number and a negative number.

2. Find the product of two negative numbers.

3. Use the reciprocal of a number to apply the definition of division.

4. Use the rules for order of operations when multiplying and dividing signed numbers.

5. Evaluate expressions involving variables.

6. Translate words and phrases involving multiplication and division.

7. Translate simple sentences into equations.

---

❶ Find each product by finding the sum of three numbers.

(a) $3(-3)$

(b) $3(-4)$

(c) $3(-5)$

❷ Find each product.

(a) $2(-6)$

(b) $7(-8)$

(c) $-9(2)$

(d) $-16\left(\dfrac{5}{32}\right)$

(e) $4.56(-10)$

Answers
1. (a) -9 (b) -12 (c) -15
2. (a) -12 (b) -56 (c) -18
   (d) $-\dfrac{5}{2}$ (e) -45.6

③ Find each product.

(a) $-5(-6)$

(b) $-7(-3)$

(c) $-8(-5)$

(d) $-11(-2)$

(e) $-17(3)(-7)$

(f) $-41(2)(-13)$

OBJECTIVE ❷ **Find the product of two negative numbers.** The product of two positive numbers is positive, and the product of a positive number and a negative number is negative. What about the product of two negative numbers? Look at another pattern.

$$-5(4) = -20$$
$$-5(3) = -15$$
$$-5(2) = -10$$
$$-5(1) = -5$$
$$-5(0) = 0$$
$$-5(-1) = ?$$

The products increase by 5.

The numbers in color on the left side of the equality symbols decrease by 1 for each step down the list. The products on the right increase by 5 for each step down the list. To maintain this pattern, $-5(-1)$ should be 5 more than $-5(0)$, or 5 more than 0, so

$$-5(-1) = 5.$$

The pattern continues with

$$-5(-2) = 10$$
$$-5(-3) = 15$$
$$-5(-4) = 20$$
$$-5(-5) = 25, \quad \text{and so on.}$$

This pattern suggests the next rule.

Multiplying Two Negative Numbers

The product of two negative numbers is positive.

Example: $-5(-4) = 20$

EXAMPLE 2 **Multiplying Two Negative Numbers**

Find each product using the multiplication rule given in the box.

(a) $-9(-2)$
$\quad = 18$

(b) $-6(-12)$
$\quad = 72$

(c) $-2(4)(-1)$
$\quad = -8(-1)$
$\quad = 8$

(d) $3(-5)(-2)$
$\quad = -15(-2)$
$\quad = 30$

◀ Work Problem **③** at the Side.

Multiplying Signed Numbers

The product of two numbers having the *same* sign is *positive.*

The product of two numbers having *different* signs is *negative.*

Answers

3. **(a)** 30 **(b)** 21 **(c)** 40 **(d)** 22
  (e) 357 **(f)** 1066

OBJECTIVE ▶ ③ **Use the reciprocal of a number to apply the definition of division.** Recall that the result of division is the **quotient.** The *quotient* of two numbers involves multiplying by the *reciprocal* of the second number, which is the *divisor.*

Reciprocals

Pairs of numbers whose product is 1 are **reciprocals, or multiplicative inverses,** of each other.

The following table shows several numbers and their reciprocals.

Number	Reciprocal
4	$\frac{1}{4}$
-5	$\frac{1}{-5}$, or $-\frac{1}{5}$
0.3, or $\frac{3}{10}$	$\frac{10}{3}$
$-\frac{5}{8}$	$-\frac{8}{5}$

A number and its reciprocal have a product of 1. For example,

$$4 \cdot \tfrac{1}{4} = \tfrac{4}{4}, \text{ or } 1.$$

0 has no reciprocal because the product of 0 and any number is 0, and cannot be 1.

Work Problem ④ at the Side. ▶

The quotient of *a* and *b* is the product of *a* and the reciprocal of *b.*

Definition of Division

The quotient $\frac{a}{b}$ of real numbers *a* and *b*, with $b \neq 0$, is

$$\frac{a}{b} = a \cdot \frac{1}{b}.$$

Example: $\dfrac{8}{-4}$ means $8\left(-\dfrac{1}{4}\right)$, which equals -2.

Since division is defined in terms of multiplication, the rules for multiplying signed numbers also apply to dividing them.

EXAMPLE 3 Using the Definition of Division

Find each quotient using the definition of division.

(a) $\dfrac{12}{3}$

$= 12 \cdot \dfrac{1}{3} \qquad \dfrac{a}{b} = a \cdot \dfrac{1}{b}$

$= 4$

(b) $\dfrac{5(-2)}{2}$

$= -10 \cdot \dfrac{1}{2}$

$= -5$

(c) $\dfrac{-1.47}{-7}$

$= -1.47 \cdot \left(-\dfrac{1}{7}\right)$

$= 0.21$

(d) $-\dfrac{2}{3} \div \left(-\dfrac{5}{4}\right)$

$= -\dfrac{2}{3} \cdot \left(-\dfrac{4}{5}\right)$

$= \dfrac{8}{15}$

Work Problem ⑤ at the Side. ▶

④ Complete the table.

	Number	Reciprocal
(a)	6	_____
(b)	-2	_____
(c)	$\frac{2}{3}$	_____
(d)	$-\frac{1}{4}$	_____
(e)	0.75	_____
(f)	0	_____

⑤ Find each quotient.

(a) $\dfrac{42}{7}$

(b) $\dfrac{-36}{(-2)(-3)}$

(c) $\dfrac{-12.56}{-0.4}$

(d) $\dfrac{10}{7} \div \left(-\dfrac{24}{5}\right)$

Answers

4. (a) $\dfrac{1}{6}$ (b) $\dfrac{1}{-2}$, or $-\dfrac{1}{2}$ (c) $\dfrac{3}{2}$

(d) -4 (e) $\dfrac{4}{3}$ (f) none

5. (a) 6 (b) -6 (c) 31.4

(d) $-\dfrac{25}{84}$

6 Decide whether each quotient is undefined or equal to 0.

(a) $\dfrac{-3}{0}$

(b) $\dfrac{0}{-53}$

7 Find each quotient.

(a) $\dfrac{-8}{-2}$

(b) $\dfrac{-16.4}{2.05}$

(c) $\dfrac{1}{4} \div \left(-\dfrac{2}{3}\right)$

(d) $\dfrac{12}{-4}$

In **Example 3(a)**, $\dfrac{12}{3} = 4$, since $4 \cdot 3 = 12$. ◁ Multiply to check a division problem.

This relationship between multiplication and division allows us to investigate division involving 0. Consider the quotient $\frac{0}{3}$.

$$\frac{0}{3} = 0, \quad \text{since} \quad 0 \cdot 3 = 0.$$

Now consider $\frac{3}{0}$.

$$\frac{3}{0} = ?$$

We need to find a number that when multiplied by 0 will equal 3, that is, $? \cdot 0 = 3$. *No* real number satisfies this equation, since the product of any real number and 0 must be 0. ***Thus, division by 0 is undefined.***

Division Involving 0

For any real number x, with $x \neq 0$,

$$\frac{0}{x} = 0 \quad \text{and} \quad \frac{x}{0} \text{ is undefined.}$$

Examples: $\dfrac{0}{-10} = 0$ and $\dfrac{-10}{0}$ is undefined.

◁ **Work Problem 6** at the Side.

When dividing fractions, multiplying by the reciprocal of the divisor works well. However it is often easier to divide in the usual way, and then determine the sign of the answer.

Dividing Signed Numbers

The quotient of two numbers having the *same* sign is *positive*.
The quotient of two numbers having *different* signs is *negative*.

Examples: $\dfrac{-15}{-5} = 3$, $\dfrac{15}{-5} = -3$, and $\dfrac{-15}{5} = -3$

EXAMPLE 4 **Dividing Signed Numbers**

Find each quotient.

(a) $\dfrac{8}{-2} = -4$ **(b)** $\dfrac{-10}{2} = -5$ **(c)** $\dfrac{-4.5}{-0.09} = 50$

(d) $-\dfrac{1}{8} \div \left(-\dfrac{3}{4}\right)$

$\quad = -\dfrac{1}{8} \cdot \left(-\dfrac{4}{3}\right)$ Multiply by the reciprocal of the divisor.

$\quad = \dfrac{1}{6}$ Write in lowest terms.

◁ **Work Problem 7** at the Side.

Answers

6. (a) undefined **(b)** 0

7. (a) 4 **(b)** -8 **(c)** $-\dfrac{3}{8}$ **(d)** -3

From the definitions of multiplication and division of real numbers,

$$\frac{-40}{8} = -5 \quad \text{and} \quad \frac{40}{-8} = -5, \quad \text{so} \quad \frac{-40}{8} = \frac{40}{-8}.$$

Based on this example, the quotient of a positive number and a negative number can be written in any of the following three forms.

> **Equivalent Forms**
>
> For any positive real numbers a and b, the following equivalences hold.
>
> $$\frac{-a}{b} = \frac{a}{-b} = -\frac{a}{b}$$

Similarly, the quotient of two negative numbers can be expressed as the quotient of two positive numbers.

> **Equivalent Forms**
>
> For any positive real numbers a and b, the following equivalence holds.
>
> $$\frac{-a}{-b} = \frac{a}{b}$$

OBJECTIVE ▶ **4** **Use the rules for order of operations when multiplying and dividing signed numbers.**

> **EXAMPLE 5** Using the Rules for Order of Operations

Simplify.

(a) $-9(2) - (-3)(2)$

$\quad = -18 - (-6)$ Multiply.

$\quad = -18 + 6$ Definition of subtraction

$\quad = -12$ Add.

(b) $-6(-2) - 3(-4)$

$\quad = 12 - (-12)$ Multiply.

$\quad = 12 + 12$ Definition of subtraction

$\quad = 24$ Add.

(c) $\dfrac{5(-2) - 3(4)}{2(1 - 6)}$

$\quad = \dfrac{-10 - 12}{2(-5)}$ Simplify the numerator and denominator separately.

$\quad = \dfrac{-22}{-10}$ Subtract in the numerator.
Multiply in the denominator.

$\quad = \dfrac{11}{5}$ Write in lowest terms.

·········· **Work Problem** **8** **at the Side.** ▶

8 Label the order in which each expression should be evaluated. Then perform the indicated operations.

GS **(a)** $-3(4) - 2(6)$

    ◯   ◯◯

$= \underline{\quad} - \underline{\quad}$

$= \underline{\quad}$

GS **(b)** $-8[-1 - (-4)(-5)]$

    ◯   ◯   ◯

$= -8[-1 - \underline{\quad}]$

$= -8[\underline{\quad}]$

$= \underline{\quad}$

(c) $\dfrac{6(-4) - 2(5)}{3(2 - 7)}$

(d) $\dfrac{-6(-8) + 3(9)}{-2[4 - (-3)]}$

Answers

8. **(a)** ①, ③, ②; -12; 12; -24

    (b) ③, ②, ①; 20; -21; 168

    (c) $\dfrac{34}{15}$ **(d)** $-\dfrac{75}{14}$

9 Evaluate each expression.

(GS) **(a)** $2x - 7(y + 1)$,
for $x = -4$ and $y = 3$

$2x - 7(y + 1)$

$= 2(\underline{\quad}) - 7(\underline{\quad} + 1)$

$= \underline{\quad} - 7(\underline{\quad})$

$= \underline{\quad} - \underline{\quad}$

$= \underline{\quad}$

(b) $2x^2 - 4y^2$,
for $x = -2$ and $y = -3$

(c) $\dfrac{4x - 2y}{-3x}$,
for $x = 2$ and $y = -1$

(d) $\left(\dfrac{2}{5}x - \dfrac{5}{6}y\right)\left(-\dfrac{1}{3}z\right)$,
for $x = 10$, $y = 6$, and
$z = -9$

Answers

9. **(a)** -4; 3; -8; 4; -8; 28; -36

   (b) -28 **(c)** $-\dfrac{5}{3}$ **(d)** -3

OBJECTIVE **5** **Evaluate expressions involving variables.** To *evaluate* an expression means to find its *value*.

EXAMPLE 6   **Evaluating Expressions for Numerical Values**

Evaluate each expression for $x = -1$, $y = -2$, and $m = -3$.

(a) $(3x + 4y)(-2m)$ 〔Use parentheses around substituted negative values to avoid errors.〕

$= [3(-1) + 4(-2)][-2(-3)]$ Substitute the given values for the variables.

$= [-3 + (-8)][6]$ Multiply.

$= [-11]6$ Add inside the brackets.

$= -66$ Multiply.

(b) $2x^2 - 3y^2$ 〔Think: $(-2)^2 = -2(-2)$〕

$= 2(-1)^2 - 3(-2)^2$ Substitute.   〔Think: $(-1)^2 = -1(-1)$〕

$= 2(1) - 3(4)$ Apply the exponents.

$= 2 - 12$ Multiply.

$= -10$ Subtract.

(c) $\dfrac{4y^2 + x}{m}$

$= \dfrac{4(-2)^2 + (-1)}{-3}$ Substitute.

$= \dfrac{4(4) + (-1)}{-3}$ Apply the exponent.

$= \dfrac{16 + (-1)}{-3}$ Multiply.

$= \dfrac{15}{-3}$,   or -5 Add, and then divide.

(d) $\left(\dfrac{3}{4}x + \dfrac{5}{8}y\right)\left(-\dfrac{1}{2}m\right)$

$= \left[\dfrac{3}{4}(-1) + \dfrac{5}{8}(-2)\right] \cdot \left[-\dfrac{1}{2}(-3)\right]$ Substitute.

$= \left[-\dfrac{3}{4} + \left(-\dfrac{5}{4}\right)\right] \cdot \left[\dfrac{3}{2}\right]$ Multiply inside the brackets.

$= \left[-\dfrac{8}{4}\right] \cdot \left[\dfrac{3}{2}\right]$ Add inside the brackets.

$= (-2)\left(\dfrac{3}{2}\right)$ Divide.

$= -3$ Multiply.

◀ **Work Problem** **9** at the Side.

OBJECTIVE ▶ 6 Translate words and phrases involving multiplication and division. The table gives words and phrases that indicate multiplication.

Word or Phrase Indicating Multiplication	Example	Numerical Expression and Simplification
Product of	The *product of* −5 and −2	−5(−2), which equals 10
Times	13 *times* −4	13(−4), which equals −52
Twice (meaning "2 times")	*Twice* 6	2(6), which equals 12
Of (used with fractions)	$\frac{1}{2}$ *of* 10	$\frac{1}{2}$(10), which equals 5
Percent of	12% *of* −16	0.12(−16), which equals −1.92
As much as	$\frac{2}{3}$ *as much as* 30	$\frac{2}{3}$(30), which equals 20

EXAMPLE 7 Translating Words and Phrases (Multiplication)

Write a numerical expression for each phrase, and simplify the expression.

(a) The **product of** 12 and the sum of 3 and −6

$12[3 + (−6)]$ simplifies to $12[−3]$, which equals −36.

(b) **Twice** the difference between 8 and −4

$2[8 − (−4)]$ simplifies to $2[12]$, which equals 24.

(c) Two-thirds **of** the sum of −5 and −3

$\frac{2}{3}[−5 + (−3)]$ simplifies to $\frac{2}{3}[−8]$, which equals $−\frac{16}{3}$.

(d) 15% **of** the difference between 14 and −2

$0.15[14 − (−2)]$ simplifies to $0.15[16]$, which equals 2.4.

Remember that 15% = 0.15.

(e) **Double** the product of 3 and 4

$2 \cdot (3 \cdot 4)$ simplifies to $2(12)$, which equals 24.

Work Problem ⑩ at the Side. ▶

The word *quotient* refers to the answer in a division problem. In algebra, a quotient is usually represented with a fraction bar. The table gives some phrases associated with division.

Phrase Indicating Division	Example	Numerical Expression and Simplification
Quotient of	The *quotient of* −24 and 3	$\frac{−24}{3}$, which equals −8
Divided by	−16 *divided by* −4	$\frac{−16}{−4}$, which equals 4
Ratio of	The *ratio of* 2 to 3	$\frac{2}{3}$

When translating a phrase involving division, we write the first number named as the numerator and the second as the denominator.

⑩ Write a numerical expression for each phrase, and simplify the expression.

(a) The product of 6 and the sum of −5 and −4

(b) Three times the difference between 4 and −6

(c) Three-fifths of the sum of 2 and −7

(d) 20% of the sum of 9 and −4

(e) Triple the product of 5 and 6

Answers

10. (a) $6[(−5) + (−4)]$; −54
(b) $3[4 − (−6)]$; 30
(c) $\frac{3}{5}[2 + (−7)]$; −3
(d) $0.20[9 + (−4)]$; 1
(e) $3(5 \cdot 6)$; 90

11 Write a numerical expression for each phrase, and simplify the expression.

(a) The quotient of 20 and the sum of 8 and −3

(b) The product of −9 and 2, divided by the difference between 5 and −1

EXAMPLE 8 Translating Words and Phrases (Division)

Write a numerical expression for each phrase, and simplify the expression.

(a) The **quotient of** 14 and the sum of −9 and 2

$$\frac{14}{-9+2} \quad \text{simplifies to} \quad \frac{14}{-7}, \quad \text{which equals} \quad -2.$$

"Quotient" indicates division.

(b) The product of 5 and −6, **divided by** the difference between −7 and 8

$$\frac{5(-6)}{-7-8} \quad \text{simplifies to} \quad \frac{-30}{-15}, \quad \text{which equals} \quad 2.$$

◀ Work Problem **11** at the Side.

OBJECTIVE ▶ 7 Translate simple sentences into equations. We can use words and phrases to translate sentences into equations.

---

12 Write each sentence in symbols, using x to represent the number.

(a) Twice a number is −6.

(b) The difference between −8 and a number is −11.

(c) The sum of 5 and a number is 8.

(d) The quotient of a number and −2 is 6.

EXAMPLE 9 Translating Sentences into Equations

Write each sentence in symbols, using x to represent the number.

(a) Three **times** a number **is** −18.

The word *times* indicates multiplication. The word *is* translates as =.

$$3 \cdot x = -18, \quad \text{or} \quad 3x = -18 \qquad 3 \cdot x = 3x$$

(b) The **sum** of a number and 9 **is** 12.

$$x + 9 = 12$$

(c) The **difference between** a number and 5 **is** 0.

$$x - 5 = 0$$

(d) The **quotient of** 24 and a number **is** −2.

$$\frac{24}{x} = -2$$

◀ Work Problem **12** at the Side.

CAUTION

In **Examples 7 and 8,** the *phrases* translate as *expressions,* while in **Example 9,** the *sentences* translate as *equations.* **An expression is a phrase. An equation is a sentence with something on the left side, an = symbol, and something on the right side.**

$$\underbrace{\frac{5(-6)}{-7-8}}_{\text{Expression}} \qquad \underbrace{3x = -18}_{\text{Equation}}$$

Answers

11. (a) $\dfrac{20}{8+(-3)}$; 4 (b) $\dfrac{-9(2)}{5-(-1)}$; −3

12. (a) $2x = -6$ (b) $-8 - x = -11$

  (c) $5 + x = 8$ (d) $\dfrac{x}{-2} = 6$

1.6 Exercises

 Download the MyDashBoard App   ▶ MyMathLab®

CONCEPT CHECK   *Fill in each blank with one of the following:*

greater than 0,   less than 0,   *or*   equal to 0.

1. The product or the quotient of two numbers with the same sign is _____.

2. The product or the quotient of two numbers with different signs is _____.

3. If three negative numbers are multiplied together, the product is _____.

4. If two negative numbers are multiplied and then their product is divided by a negative number, the result is _____.

5. If a negative number is squared and the result is added to a positive number, the final answer is _____.

6. The reciprocal of a negative number is _____.

Find each product. See Examples 1 and 2.

7. $-7(4)$

8. $-8(5)$

9. $-5(-6)$

10. $-4(-20)$

11. $-8(0)$

12. $0(-12)$

13. $-\dfrac{3}{8}\left(-\dfrac{20}{9}\right)$

14. $-\dfrac{5}{4}\left(-\dfrac{6}{25}\right)$

15. $-6.8(0.35)$

16. $-4.6(0.24)$

17. $-6\left(-\dfrac{1}{4}\right)$

18. $-8\left(-\dfrac{1}{2}\right)$

Find each quotient. See Examples 3 and 4 and the discussion of division involving 0.

19. $\dfrac{-15}{5}$

20. $\dfrac{-18}{6}$

21. $\dfrac{20}{-10}$

22. $\dfrac{28}{-4}$

23. $\dfrac{-160}{-10}$

24. $\dfrac{-260}{-20}$

25. $\dfrac{0}{-3}$

26. $\dfrac{0}{-5}$

27. $\dfrac{-10.252}{0}$

28. $\dfrac{-29.584}{0}$

29. $\left(-\dfrac{3}{4}\right) \div \left(-\dfrac{1}{2}\right)$

30. $\left(-\dfrac{3}{16}\right) \div \left(-\dfrac{5}{8}\right)$

31. CONCEPT CHECK Which expression is undefined?

A. $\dfrac{5-5}{5+5}$ B. $\dfrac{5+5}{5+5}$ C. $\dfrac{5-5}{5-5}$ D. $\dfrac{5-5}{5}$

32. CONCEPT CHECK Which expression is undefined?

A. $13 \div 0$ B. $13 \div 13$ C. $0 \div 13$ D. $13 \cdot 0$

Perform each indicated operation. **See Example 5.**

33. $\dfrac{-5(-6)}{9-(-1)}$

$= \dfrac{}{}$

$= \underline{}$

34. $\dfrac{-12(-5)}{7-(-5)}$

$= \dfrac{}{}$

$= \underline{}$

35. $\dfrac{-21(3)}{-3-6}$

36. $\dfrac{-40(3)}{-2-3}$

37. $\dfrac{-10(2)+6(2)}{-3-(-1)}$

38. $\dfrac{8(-1)+6(-2)}{-6-(-1)}$

39. $\dfrac{-27(-2)-(-12)(-2)}{-2(3)-2(2)}$

40. $\dfrac{-13(-4)-(-8)(-2)}{(-10)(2)-4(-2)}$

41. $\dfrac{3^2-4^2}{7(-8+9)}$

42. $\dfrac{5^2-7^2}{2(3+3)}$

43. $\dfrac{4(2^3-5)-5(-3^3+21)}{3[6-(-2)]}$

44. $\dfrac{-3(-2^4+10)+4(2^5-12)}{-2[8-(-7)]}$

Evaluate each expression for $x = 6$, $y = -4$, and $a = 3$. **See Example 6.**

45. $6x - 5y + 4a$

46. $5x - 2y + 3a$

47. $(5x - 2y)(-2a)$

48. $(2x + y)(3a)$

49. $\left(\dfrac{5}{6}x + \dfrac{3}{2}y\right)\left(-\dfrac{1}{3}a\right)$

50. $\left(\dfrac{1}{3}x - \dfrac{4}{5}y\right)\left(-\dfrac{1}{5}a\right)$

51. $(6 - x)(5 + y)(3 + a)$

52. $(-5 + x)(-3 + y)(3 - a)$

53. $5x - 4a^2$

54. $-2y^2 + 3a$

55. $\dfrac{xy + 9a}{x + y - 2}$

56. $\dfrac{2y^2 - x}{a - 3}$

*Write a numerical expression for each phrase and simplify. **See Examples 7 and 8.***

57. The product of 4 and -7, added to -12

58. The product of -9 and 2, added to 9

59. Twice the product of -8 and 2, subtracted from -1

60. Twice the product of -1 and 6, subtracted from -4

61. The product of -3 and the difference between 3 and -7

62. The product of 12 and the difference between 9 and -8

63. Three-tenths of the sum of -2 and -28

64. Four-fifths of the sum of -8 and -2

65. The quotient of -20 and the sum of -8 and -2

66. The quotient of -12 and the sum of -5 and -1

67. The sum of -18 and -6, divided by the product of 2 and -4

68. The sum of 15 and -3, divided by the product of 4 and -3

69. The product of $-\frac{2}{3}$ and $-\frac{1}{5}$, divided by $\frac{1}{7}$

70. The product of $-\frac{1}{2}$ and $\frac{3}{4}$, divided by $-\frac{2}{3}$

Write each sentence in symbols, using x to represent the number. **See Example 9.**

71. Nine times a number is −36.

72. Seven times a number is −42.

73. The quotient of a number and 4 is −1.

74. The quotient of a number and 3 is −3.

75. $\frac{9}{11}$ less than a number is 5.

76. $\frac{1}{2}$ less than a number is 2.

77. When 6 is divided by a number, the result is −3.

78. When 15 is divided by a number, the result is −5.

In 2011, the following question and expression appeared on boxes of Swiss Miss Chocolate: On average, how many mini-marshmallows are in one serving?

$$3 + 2 \times 4 \div 2 - 3 \times 7 - 4 + 47$$

79. The box gave 92 as the answer. What is the *correct* answer? (*Source:* Swiss Miss Chocolate box.)

80. Explain the error that somebody at the company made in calculating the answer.

Relating Concepts (Exercises 81–86) For Individual or Group Work

*To find the **average** of a group of numbers, we add the numbers and then divide this sum by the number of terms added. **Work Exercises 81–84 in order,** to find the average of 23, 18, 13, −4, and −8. Then find the averages in **Exercises 85 and 86.***

81. Find the sum of the given group of numbers.

82. How many numbers are in the group?

83. Divide your answer for **Exercise 81** by your answer for **Exercise 82.** Give the quotient as a mixed number.

84. What is the average of the given group of numbers?

85. What is the average of all integers between −10 and 14, including both −10 and 14?

86. What is the average of all integers between −15 and −10, including both −15 and −10?

Summary Exercises *Performing Operations with Real Numbers*

Operations with Signed Numbers

Addition

Same sign   Add the absolute values of the numbers. The sum has the same sign as the numbers.

Different signs   Find the absolute values of the numbers, and subtract the lesser absolute value from the greater. Give the sum the sign of the number having the greater absolute value.

Subtraction

Add the opposite of the subtrahend to the minuend.

Multiplication and Division

Same sign   The product or quotient of two numbers with the same sign is positive.

Different signs   The product or quotient of two numbers with different signs is negative.

Division by 0 is undefined.

Perform each indicated operation.

1. $14 - 3 \cdot 10$

2. $-3(8) - 4(-7)$

3. $(3 - 8)(-2) - 10$

4. $-6(7 - 3)$

5. $7 - (-3)(2 - 10)$

6. $-4[(-2)(6) - 7]$

7. $(-4)(7) - (-5)(2)$

8. $-5[-4 - (-2)(-7)]$

9. $40 - (-2)[8 - 9]$

10. $\dfrac{5(-4)}{-7 - (-2)}$

11. $\dfrac{-3 - (-9 + 1)}{-7 - (-6)}$

12. $\dfrac{5(-8 + 3)}{13(-2) + (-7)(-3)}$

13. $\dfrac{6^2 - 8}{-2(2) + 4(-1)}$

14. $\dfrac{16(-8 + 5)}{15(-3) + (-7 - 4)(-3)}$

15. $\dfrac{9(-6) - 3(8)}{4(-7) + (-2)(-11)}$

16. $\dfrac{2^2 + 4^2}{5^2 - 3^2}$

17. $\dfrac{(2 + 4)^2}{(5 - 3)^2}$

18. $\dfrac{4^3 - 3^3}{-5(-4 + 2)}$

19. $\dfrac{-9(-6) + (-2)(27)}{3(8 - 9)}$

20. $|-4(9)| - |-11|$

21. $\dfrac{6(-10 + 3)}{15(-2) - 3(-9)}$

22. $\dfrac{(-9)^2 - 9^2}{3^2 - 5^2}$

23. $\dfrac{(-10)^2 + 10^2}{-10(5)}$

24. $-\dfrac{3}{4} \div \left(-\dfrac{5}{8}\right)$

25. $\dfrac{1}{2} \div \left(-\dfrac{1}{2}\right)$

26. $\dfrac{8^2 - 12}{(-5)^2 + 2(6)}$

27. $\left[\dfrac{5}{8} - \left(-\dfrac{1}{16}\right)\right] + \dfrac{3}{8}$

28. $\left(\dfrac{1}{2} - \dfrac{1}{3}\right) - \dfrac{5}{6}$

29. $-0.9(-3.7)$

30. $-5.1(-0.2)$

31. $-3^2 - 2^2$

32. $|-2(3) + 4| - |-2|$

33. $40 - (-2)[-5 - 3]$

Evaluate each expression for $x = -2$, $y = 3$, and $a = 4$.

34. $-x + y - 3a$

35. $(x + 6)^3 - y^3$

36. $(x - y) - (a - 2y)$

37. $\left(\dfrac{1}{2}x + \dfrac{2}{3}y\right)\left(-\dfrac{1}{4}a\right)$

38. $\dfrac{2x + 3y}{a - xy}$

39. $\dfrac{x^2 - y^2}{x^2 + y^2}$

40. $-x^2 + 3y$

41. $\dfrac{-x + 2y}{2x + a}$

42. $\dfrac{2x + a}{-x + 2y}$

1.7 Properties of Real Numbers

In the basic properties covered in this section, a, b, and c represent real numbers.

OBJECTIVE ▶ **1** **Use the commutative properties.** The word *commute* means to go back and forth. Many people commute to work or to school. If you travel from home to work and follow the same route from work to home, you travel the same distance each time.

The **commutative properties** say that if two numbers are added or multiplied in either order, the result is the same.

OBJECTIVES

1 Use the commutative properties.
2 Use the associative properties.
3 Use the identity properties.
4 Use the inverse properties.
5 Use the distributive property.

Commutative Properties	
$a + b = b + a$	Addition
$ab = ba$	Multiplication

EXAMPLE 1 **Using the Commutative Properties**

Use a commutative property to complete each statement.

(a) $-8 + 5 = 5 +$ _?_ | Notice that the "order" changed. |

$-8 + 5 = 5 + (-8)$ Commutative property of addition

(b) $(-2)7 =$ _?_ (-2)

$-2(7) = 7(-2)$ Commutative property of multiplication

························· **Work Problem 1 at the Side.** ▶

OBJECTIVE ▶ **2** **Use the associative properties.** When we *associate* one object with another, we think of those objects as being grouped together.

The **associative properties** say that when we add or multiply three numbers, we can group the first two together or the last two together and get the same answer.

Associative Properties	
$(a + b) + c = a + (b + c)$	Addition
$(ab)c = a(bc)$	Multiplication

EXAMPLE 2 **Using the Associative Properties**

Use an associative property to complete each statement.

(a) $-8 + (1 + 4) = (-8 +$ _?_ $) + 4$ | The "order" is the same. The "grouping" changed. |

$-8 + (1 + 4) = (-8 + 1) + 4$ Associative property of addition

(b) $[2 \cdot (-7)] \cdot 6 = 2 \cdot$ _?_

$[2 \cdot (-7)] \cdot 6 = 2 \cdot [(-7) \cdot 6]$ Associate property of multiplication

························· **Work Problem 2 at the Side.** ▶

1 Complete each statement. Use a commutative property.

(a) $x + 9 = 9 +$ ____

(b) $-12(4) =$ ____ (-12)

(c) $5x = x \cdot$ ____

2 Complete each statement. Use an associative property.

(a) $(9 + 10) + (-3)$
$= 9 + [$ ____ $+ (-3)]$

(b) $-5 + (2 + 8)$
$= ($ _____ $) + 8$

(c) $10 \cdot [-8 \cdot (-3)]$
$=$ _____

Answers

1. **(a)** x **(b)** 4 **(c)** 5
2. **(a)** 10 **(b)** $-5 + 2$
   (c) $[10 \cdot (-8)] \cdot (-3)$

3 Decide whether each statement is an example of a commutative property, an associative property, or both.

(a) $2 \cdot (4 \cdot 6) = (2 \cdot 4) \cdot 6$

(b) $(2 \cdot 4) \cdot 6 = (4 \cdot 2) \cdot 6$

(c) $(2 + 4) + 6 = 4 + (2 + 6)$

4 Find each sum or product.

(a) $5 + 18 + 29 + 31 + 12$

(b) $5(37)(20)$

By the associative property, the sum (or product) of three numbers will be the same no matter how the numbers are "associated" in groups. Parentheses can be left out if a problem contains only addition (or multiplication). For example,

$$(-1 + 2) + 3 \quad \text{and} \quad -1 + (2 + 3) \quad \text{can be written as} \quad -1 + 2 + 3.$$

EXAMPLE 3 **Distinguishing between Properties**

Decide whether each statement is an example of a commutative property, an associative property, or both.

(a) $(2 + 4) + 5 = 2 + (4 + 5)$

The order of the three numbers is the same on both sides of the equality symbol. The only change is in the *grouping*, or association, of the numbers. This is an example of the associative property.

(b) $6 \cdot (3 \cdot 10) = 6 \cdot (10 \cdot 3)$

The same numbers, 3 and 10, are grouped on each side. On the left, the 3 appears first, but on the right, the 10 appears first. Since the only change involves the *order* of the numbers, this is an example of the commutative property.

(c) $(8 + 1) + 7 = 8 + (7 + 1)$

Both the order and the grouping are changed. On the left, the order of the three numbers is 8, 1, and 7. On the right, it is 8, 7, and 1. On the left, the 8 and 1 are grouped. On the right, the 7 and 1 are grouped. Therefore, *both* properties are used.

◀ **Work Problem 3** at the Side.

Example 4 illustrates that we can sometimes use the commutative and associative properties to rearrange and regroup numbers to simplify calculations.

EXAMPLE 4 **Using the Commutative and Associative Properties**

Find each sum or product.

(a) $23 + 41 + 2 + 9 + 25$
$= (41 + 9) + (23 + 2) + 25$
$= 50 + 25 + 25$ Use the commutative and associative properties.
$= 100$

(b) $25(69)(4)$
$= 25(4)(69)$
$= 100(69)$
$= 6900$

◀ **Work Problem 4** at the Side.

OBJECTIVE 3 **Use the identity properties.** If a child wears a costume on Halloween, the child's appearance is changed, but his or her *identity* is unchanged. The identity of a real number is left unchanged when identity properties are applied.

The **identity properties** say that the sum of 0 and any number equals that number, and the product of 1 and any number equals that number.

Identity Properties

$$a + 0 = a \quad \text{and} \quad 0 + a = a \quad \text{Addition}$$

$$a \cdot 1 = a \quad \text{and} \quad 1 \cdot a = a \quad \text{Multiplication}$$

The number 0 leaves the identity, or value, of any real number unchanged by addition, so 0 is the **identity element for addition**, or the **additive identity.** Since multiplication by 1 leaves any real number unchanged, 1 is the **identity element for multiplication**, or the **multiplicative identity.**

> **EXAMPLE 5** Using the Identity Properties
>
> Use an identity property to complete each statement.
>
> **(a)** $-3 + \underline{\quad?\quad} = -3$
>
> $-3 + \mathbf{0} = -3$
>
> Identity property for addition
>
> **(b)** $\dfrac{?}{} \cdot \dfrac{1}{2} = \dfrac{1}{2}$
>
> $\mathbf{1} \cdot \dfrac{1}{2} = \dfrac{1}{2}$
>
> Identity property for multiplication

Work Problem ⑤ at the Side. ▶

We use the identity property for multiplication to write fractions in lowest terms and to find common denominators.

> **EXAMPLE 6** Using the Identity Property for Multiplication
>
> In part (a), write in lowest terms. In part (b), perform the operation.
>
> **(a)** $\dfrac{49}{35}$
>
> $= \dfrac{7 \cdot 7}{5 \cdot 7}$ Factor.
>
> $= \dfrac{7}{5} \cdot \dfrac{7}{7}$ Write as a product.
>
> $= \dfrac{7}{5} \cdot 1$ Property of 1
>
> $= \dfrac{7}{5}$ Identity property
>
> **(b)** $\dfrac{3}{4} + \dfrac{5}{24}$
>
> $= \dfrac{3}{4} \cdot 1 + \dfrac{5}{24}$ Identity property
>
> $= \dfrac{3}{4} \cdot \dfrac{6}{6} + \dfrac{5}{24}$ Use $1 = \dfrac{6}{6}$ to get a common denominator.
>
> $= \dfrac{18}{24} + \dfrac{5}{24}$ Multiply.
>
> $= \dfrac{23}{24}$ Add.

Work Problem ⑥ at the Side. ▶

OBJECTIVE ▶ ④ Use the inverse properties. Each day before you go to work or school, you probably put on your shoes. Before you go to sleep at night, you probably take them off. These operations from everyday life are examples of *inverse* operations.

The **inverse properties** of addition and multiplication lead to the additive and multiplicative identities, respectively. Recall that $-a$ is the **additive inverse**, or **opposite**, of a and $\frac{1}{a}$ is the **multiplicative inverse**, or **reciprocal**, of the nonzero number a. The sum of the numbers a and $-a$ is 0, and the product of the nonzero numbers a and $\frac{1}{a}$ is 1.

> **Inverse Properties**
>
> $a + (-a) = 0$ and $-a + a = 0$ Addition
>
> $a \cdot \dfrac{1}{a} = 1$ and $\dfrac{1}{a} \cdot a = 1$ $(a \neq 0)$ Multiplication

⑤ Use an identity property to complete each statement.

(a) $9 + 0 = $ _____

(b) _____ $+ (-7) = -7$

(c) _____ $\cdot 1 = 5$

⑥ In part (a), write in lowest terms. In part (b), perform the operation.

(a) $\dfrac{85}{105}$

(b) $\dfrac{9}{10} - \dfrac{53}{50}$

Answers

5. (a) 9   (b) 0   (c) 5

6. (a) $\dfrac{17}{21}$ (b) $-\dfrac{4}{25}$

7 Complete each statement so that it is an example of either an identity property or an inverse property. Tell which property is used.

(a) $-6 +$ _____ $= 0$

(b) $\dfrac{4}{3} \cdot$ _____ $= 1$

(c) $-\dfrac{1}{9} \cdot ($ _____ $) = 1$

(d) $275 +$ _____ $= 275$

(e) $-0.75 + \dfrac{3}{4} =$ _____

(f) $0.2(5) =$ _____

EXAMPLE 7 **Using the Inverse Properties**

Use an inverse property to complete each statement.

(a) _?_ $+ \dfrac{1}{2} = 0$

$-\dfrac{1}{2} + \dfrac{1}{2} = 0$

(b) $4 +$ _?_ $= 0$

$4 + (-4) = 0$

(c) $-0.75 + \dfrac{3}{4} =$ _?_

$-0.75 + \dfrac{3}{4} = 0$

The inverse property for addition is used in parts (a)–(c).

(d) _?_ $\cdot \dfrac{5}{2} = 1$

$\dfrac{2}{5} \cdot \dfrac{5}{2} = 1$

(e) $-5($ _?_ $) = 1$

$-5\left(-\dfrac{1}{5}\right) = 1$

(f) $4(0.25) =$ _?_

$4(0.25) = 1$

The inverse property for multiplication is used in parts (d)–(f).

◀ **Work Problem ⑦ at the Side.**

OBJECTIVE ▶ ⑤ Use the distributive property. The word *distribute* means "to give out from one to several." Consider the following expressions.

$$2(5+8) \quad \text{equals} \quad 2(13), \quad \text{or} \quad 26.$$
$$2(5) + 2(8) \quad \text{equals} \quad 10 + 16, \quad \text{or} \quad 26.$$

Since both expressions equal 26,

$$2(5+8) = 2(5) + 2(8).$$

This result is an example of the *distributive property of multiplication with respect to addition,* the only property involving *both* addition and multiplication. With this property, a product can be changed to a sum. This idea is illustrated by the divided rectangle in **Figure 13.**

The area of the left part is $2(5) = 10$.
The area of the right part is $2(8) = 16$.
The total area is $2(5 + 8) = 26$ or the total area is
$2(5) + 2(8) = 10 + 16 = 26$.
Thus, $2(5 + 8) = 2(5) + 2(8)$.

Figure 13

The **distributive property** says that multiplying a number a by a sum of numbers $b + c$ gives the same result as multiplying a by b and a by c and then adding the two products.

Distributive Property

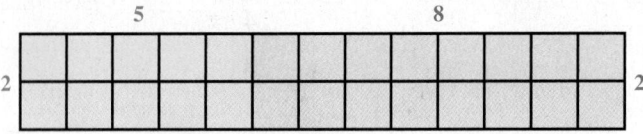

$$a(b + c) = ab + ac \quad \text{and} \quad (b + c)a = ba + ca$$

As the arrows show, the a outside the parentheses is "distributed" over the b and c inside.

The distributive property is also valid for subtraction.

$$a(b - c) = ab - ac \quad \text{and} \quad (b - c)a = ba - ca$$

The distributive property also can be extended to more than two numbers.

$$a(b + c + d) = ab + ac + ad$$

The distributive property can also be written "in reverse."

$$ab + ac = a(b + c)$$

We will use this form in the next section to combine like terms.

EXAMPLE 8 **Using the Distributive Property**

Use the distributive property to rewrite each expression.

(a) $5(9 + 6)$

$$= 5 \cdot 9 + 5 \cdot 6 \quad \text{Distributive property}$$

$$= 45 + 30 \quad \text{Multiply.}$$

$$= 75 \quad \boxed{\text{Multiply first.}} \quad \text{Add.}$$

(b) $4(x + 5 + y)$

$$= 4x + 4 \cdot 5 + 4y \quad \text{Distributive property}$$

$$= 4x + 20 + 4y \quad \text{Multiply.}$$

(c) $-2(x + 3)$

$$= -2x + (-2)(3) \quad \text{Distributive property}$$

$$= -2x + (-6) \quad \text{Multiply.}$$

$$= -2x - 6 \quad \text{Definition of subtraction}$$

(d) $3(k - 9)$ $\boxed{\text{Be careful here.}}$

$$= 3[k + (-9)] \quad \text{Definition of subtraction}$$

$$= 3k + 3(-9) \quad \text{Distributive property}$$

$$= 3k - 27 \quad \text{Multiply.}$$

(e) $8(3r + 11t + 5z)$

$$= 8(3r) + 8(11t) + 8(5z) \quad \text{Distributive property}$$

$$= (8 \cdot 3)r + (8 \cdot 11)t + (8 \cdot 5)z \quad \text{Associative property}$$

$$= 24r + 88t + 40z \quad \text{Multiply.}$$

(f) $6 \cdot 8 + 6 \cdot 2$

$$= 6(8 + 2) \quad \text{Distributive property in reverse}$$

$$= 6(10) \quad \text{Add.}$$

$$= 60 \quad \text{Multiply.}$$

(g) $4x - 4m$

$$= 4(x - m) \quad \text{Distributive property in reverse}$$

Work Problem 8 at the Side. ▶

---

8 Use the distributive property to rewrite each expression.

(a) $2(p + 5)$

$$= 2 \cdot \underline{\quad} + 2 \cdot \underline{\quad}$$

$$= \underline{\quad} + \underline{\quad}$$

(b) $-4(y + 7)$

(c) $5(m - 4)$

(d) $7(2y + 7k - 9m)$

(e) $9 \cdot k + 9 \cdot 5$

$$= \underline{\quad} (\underline{\quad} + \underline{\quad})$$

(f) $3a - 3b$

Answers

8. **(a)** p; 5; $2p$; 10 **(b)** $-4y - 28$
(c) $5m - 20$ **(d)** $14y + 49k - 63m$
(e) 9; k; 5 **(f)** $3(a - b)$

9 Write each expression without parentheses.

(GS) (a) $-(3k-5)$

$$= \underline{\quad} \cdot (3k-5)$$

$$= \underline{\quad} \cdot 3k + (-1)(\underline{\quad})$$

$$= \underline{\quad} + \underline{\quad}$$

(b) $-(2-r)$

(c) $-(-5y+8)$

(d) $-(-z+4)$

(e) $-(-t-4u+5v)$

The symbol $-a$ may be interpreted as $-1 \cdot a$. Using this result and the distributive property, we can remove (or clear) parentheses from some expressions.

EXAMPLE 9 Using the Distributive Property to Remove (Clear) Parentheses

Write each expression without parentheses.

(a) $-(2y+3)$

> The $-$ symbol indicates a factor of -1.

$$= -1 \cdot (2y+3) \qquad -a = -1 \cdot a$$

$$= -1 \cdot 2y + (-1) \cdot 3 \qquad \text{Distributive property}$$

$$= -2y - 3 \qquad \text{Multiply.}$$

(b) $-(-9w-2)$

$$= -1(-9w-2)$$

$$= -1(-9w) - 1(-2)$$

$$= 9w + 2$$

> We can also interpret the negative sign in front of the parentheses to mean the *opposite* of each of the terms within the parentheses.

(c) $-(-x-3y+6z)$

$$= -1(-1x-3y+6z) \qquad \boxed{\text{Be careful with signs.}}$$

$$= -1(-1x) - 1(-3y) - 1(6z) \qquad \text{Distributive property}$$

$$= x + 3y - 6z \qquad -1(-1x) = 1x = x$$

◀ **Work Problem 9** at the Side.

Here is a summary of the basic properties of real numbers.

Properties of Addition and Multiplication

For any real numbers a, b, and c, the following properties hold.

Commutative properties $\qquad a + b = b + a \qquad ab = ba$

Associative properties $\qquad (a + b) + c = a + (b + c)$

$$(ab)c = a(bc)$$

Identity properties $\qquad$ There is a real number 0 such that

$$a + 0 = a \quad \text{and} \quad 0 + a = a.$$

There is a real number 1 such that

$$a \cdot 1 = a \quad \text{and} \quad 1 \cdot a = a.$$

Inverse properties $\qquad$ For each real number a, there is a single real number $-a$ such that

$$a + (-a) = 0 \quad \text{and} \quad (-a) + a = 0.$$

For each nonzero real number a, there is a single real number $\frac{1}{a}$ such that

$$a \cdot \frac{1}{a} = 1 \quad \text{and} \quad \frac{1}{a} \cdot a = 1.$$

Distributive property $\qquad a(b + c) = ab + ac$

$$(b + c)a = ba + ca$$

Answers

9. (a) $-1; -1; -5; -3k; 5$ **(b)** $-2 + r$
(c) $5y - 8$ **(d)** $z - 4$
(e) $t + 4u - 5v$

1.7 Exercises

 Download the MyDashBoard App   MyMathLab®

1. CONCEPT CHECK Match each item in Column I with the correct choice(s) from Column II. Choices may be used once, more than once, or not at all.

I

(a) Identity element for addition

(b) Identity element for multiplication

(c) Additive inverse of a

(d) Multiplicative inverse, or reciprocal, of the nonzero number a

(e) The number that is its own additive inverse

(f) The two numbers that are their own multiplicative inverses

(g) The only number that has no multiplicative inverse

(h) An example of the associative property

(i) An example of the commutative property

(j) An example of the distributive property

II

A. $(5 \cdot 4) \cdot 3 = 5 \cdot (4 \cdot 3)$

B. 0

C. $-a$

D. -1

E. $5 \cdot 4 \cdot 3 = 60$

F. 1

G. $(5 \cdot 4) \cdot 3 = 3 \cdot (5 \cdot 4)$

H. $5(4 + 3) = 5 \cdot 4 + 5 \cdot 3$

I. $\dfrac{1}{a}$

2. CONCEPT CHECK Fill in the blanks: The commutative property allows us to change the _____ of the terms in a sum or the factors in a product. The associative property allows us to change the _____ of the terms in a sum or the factors in a product.

CONCEPT CHECK *Tell whether or not the following everyday activities are commutative.*

3. Washing your face and brushing your teeth

4. Putting on your left sock and putting on your right sock

5. Preparing a meal and eating a meal

6. Starting a car and driving away in a car

7. Putting on your socks and putting on your shoes

8. Getting undressed and taking a shower

9. CONCEPT CHECK Use parentheses to show how the associative property can be used to give two different meanings to the phrase "foreign sales clerk."

10. CONCEPT CHECK Use parentheses to show how the associative property can be used to give two different meanings to the phrase "defective merchandise counter."

Use the commutative or the associative property to complete each statement. State which property is used. **See Examples 1 and 2.**

11. $-15 + 9 = 9 + $ _____

12. $6 + (-2) = -2 + $ _____

13. $-8 \cdot 3 = $ _____ $\cdot (-8)$

14. $-12 \cdot 4 = 4 \cdot $ _____

15. $(3 + 6) + 7 = 3 + ($ _____ $+ 7)$

16. $(-2 + 3) + 6 = -2 + ($ _____ $+ 6)$

17. $7 \cdot (2 \cdot 5) = ($ _____ $\cdot 2) \cdot 5$

18. $8 \cdot (6 \cdot 4) = (8 \cdot $ _____ $) \cdot 4$

19. CONCEPT CHECK Evaluate $25 - (6 - 2)$ and evaluate $(25 - 6) - 2$. Do you think subtraction is associative?

20. CONCEPT CHECK Evaluate $180 \div (15 \div 3)$ and evaluate $(180 \div 15) \div 3$. Do you think division is associative?

21. CONCEPT CHECK Complete the table and the statements beside it.

Number	Additive Inverse	Multiplicative Inverse
5		
-10		
$-\frac{1}{2}$		
$\frac{3}{8}$		
x		
$-y$		

In general, a number and its additive inverse have (*the same / opposite*) signs.

A number and its multiplicative inverse have (*the same / opposite*) signs.

22. CONCEPT CHECK The following conversation actually took place between one of the authors of this book and his son, Jack, when Jack was 4 years old.

DADDY: "Jack, what is $3 + 0$?"
JACK: "3."
DADDY: "Jack, what is $4 + 0$?"
JACK: "4. And Daddy, *string* plus zero equals *string*!"

What property of addition did Jack recognize?

Decide whether each statement is an example of a commutative, associative, identity, or inverse property, or of the distributive property. **See Examples 1, 2, 3, and 5–8.**

23. $\frac{2}{3}(-4) = -4\left(\frac{2}{3}\right)$

24. $6\left(-\frac{5}{6}\right) = \left(-\frac{5}{6}\right)6$

25. $-6 + (12 + 7) = (-6 + 12) + 7$

26. $(-8 + 13) + 2 = -8 + (13 + 2)$

27. $-6 + 6 = 0$

28. $12 + (-12) = 0$

29. $\left(\frac{2}{3}\right)\left(\frac{3}{2}\right) = 1$

30. $\left(\frac{5}{8}\right)\left(\frac{8}{5}\right) = 1$

31. $2.34 \cdot 1 = 2.34$

32. $-8.456 \cdot 1 = -8.456$

33. $(4 + 17) + 3 = 3 + (4 + 17)$

34. $(-8 + 4) + (-12) = -12 + (-8 + 4)$

35. $6(x + y) = 6x + 6y$

36. $14(t + s) = 14t + 14s$

37. $-\dfrac{5}{9} = -\dfrac{5}{9} \cdot \dfrac{3}{3} = -\dfrac{15}{27}$

38. $\dfrac{13}{12} = \dfrac{13}{12} \cdot \dfrac{7}{7} = \dfrac{91}{84}$

39. $5(2x) + 5(3y) = 5(2x + 3y)$

40. $3(5t) - 3(7r) = 3(5t - 7r)$

41. CONCEPT CHECK Suppose that a student simplifies the expression $-3(4 - 6)$ as shown.

$$-3(4 - 6)$$
$$= -3(4) - 3(6)$$
$$= -12 - 18$$
$$= -30$$

What Went Wrong? Work the problem correctly.

42. Explain how the procedure of changing $\frac{3}{4}$ to $\frac{9}{12}$ requires the use of the multiplicative identity element, 1.

*Write a new expression that is equal to the given expression, using the given property. Then simplify the new expression if possible. **See Examples 1, 2, 5, 7, and 8.***

43. $r + 7$;   commutative

44. $t + 9$;   commutative

45. $s + 0$;   identity

46. $w + 0$;   identity

47. $-6(x + 7)$;   distributive

48. $-5(y + 2)$;   distributive

49. $(w + 5) + (-3)$;   associative

50. $(b + 8) + (-10)$;   associative

*Use the properties of this section to perform the operations. **See Example 4.***

51. $50(67)2$

52. $26 + 8 - 26 + 12$

53. $-\dfrac{3}{8} + \dfrac{2}{5} + \dfrac{8}{5} + \dfrac{3}{8}$

54. $-\dfrac{5}{12} - \dfrac{3}{7} + \dfrac{5}{12} + \dfrac{10}{7}$

55. $\left(-\dfrac{8}{5}\right)(0.77)\left(-\dfrac{5}{8}\right)$

56. $\dfrac{9}{7}(-0.38)\left(\dfrac{7}{9}\right)$

Use the distributive property to rewrite each expression. Simplify if possible.
See Example 8.

57. $4(t + 3)$

58. $5(w + 4)$

59. $-8(r + 3)$

60. $-11(x + 4)$

61. $-5(y - 4)$
$$= -5(\underline{\quad}) - 5(\underline{\quad})$$
$$= \underline{\qquad}$$

62. $-9(g - 4)$
$$= -9(\underline{\quad}) - 9(\underline{\quad})$$
$$= \underline{\qquad}$$

63. $-\dfrac{4}{3}(12y + 15z)$

64. $-\dfrac{2}{5}(10b + 20a)$

65. $8 \cdot z + 8 \cdot w$

66. $4 \cdot s + 4 \cdot r$

67. $5 \cdot 3 + 5 \cdot 17$

68. $15 \cdot 6 + 5 \cdot 6$

69. $7(2v) + 7(5r)$

70. $13(5w) + 13(4p)$

71. $8(3r + 4s - 5y)$

72. $2(5u - 3v + 7w)$

73. $-3(8x + 3y + 4z)$

74. $-5(2x - 5y + 6z)$

Use the distributive property to write each expression without parentheses.
See Example 9.

75. $-(4t + 5m)$

76. $-(9x + 12y)$

77. $-(-5c - 4d)$

78. $-(-13x - 15y)$

79. $-(-3q + 5r - 8s)$

80. $-(-4z + 5w - 9y)$

Study Skills

Your textbook provides material to help you prepare for quizzes or tests in this course. Refer to the **Chapter Summary** as you read through the following techniques.

Chapter Reviewing Techniques

▶ **Review the Key Terms.** Make a study card for each. Include the given definition, an example, a sketch (if appropriate), and a section or page reference.

▶ **Take the Test Your Word Power quiz** to check your understanding of new vocabulary. The answers immediately follow.

▶ **Read the Quick Review.** Pay special attention to the headings. Study the explanations and examples given for each concept. Try to think about the whole chapter.

▶ **Reread your lecture notes.** Focus on what your instructor has emphasized in class, and review that material in your text.

▶ **Work the Review Exercises.** They are grouped by section.

  ✓ Pay attention to direction words, such as *simplify*, *solve*, and *estimate*.

  ✓ After you've done each section of exercises, check your answers in the answer section.

  ✓ Are your answers exact and complete? Did you include the correct labels, such as $, cm², ft, etc.?

  ✓ Make study cards for difficult problems.

▶ **Work the Mixed Review Exercises.** They are in mixed-up order. Check your answers in the answer section.

▶ **Take the Chapter Test under test conditions.**

  ✓ Time yourself.

  ✓ Use a calculator or notes (if your instructor permits them on tests).

  ✓ Take the test in one sitting.

  ✓ Show all your work.

  ✓ Check your answers in the back of the book.

Reviewing a chapter will take some time. Avoid rushing through your review in one night. Use the suggestions over a few days or evenings to better understand the material and remember it longer.

Follow these reviewing techniques for your next test. After the test, evaluate how they worked for you. What will you do differently when reviewing for your next test?

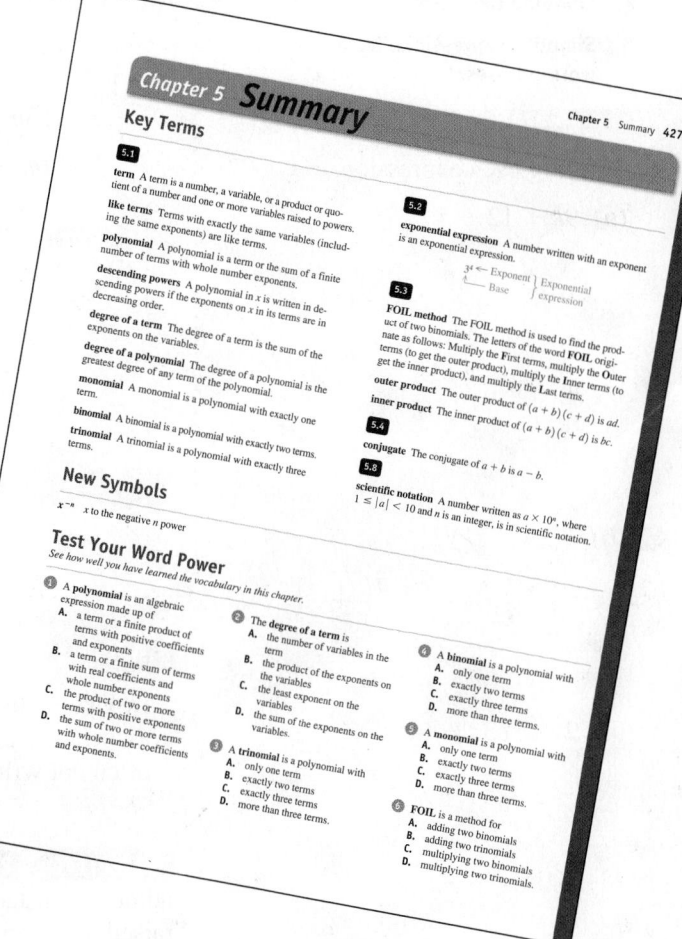

1.8 Simplifying Expressions

OBJECTIVES

1. Simplify expressions.
2. Identify terms and numerical coefficients.
3. Identify like terms.
4. Combine like terms.
5. Simplify expressions from word phrases.

OBJECTIVE ① **Simplify expressions.** We now simplify expressions using the properties of addition and multiplication introduced in **Section 1.7.**

EXAMPLE 1 Simplifying Expressions

Simplify each expression.

(a) $4x + 8 + 9$ simplifies to $4x + 17$.

(b) $4(3m - 2n)$ | To simplify, we clear the parentheses.

$= 4(3m) - 4(2n)$ Distributive property

$= 12m - 8n$ Associative property

(c) $6 + 3(4k + 5)$

Don't start by adding.

$= 6 + 3(4k) + 3(5)$ Distributive property

$= 6 + 12k + 15$ Multiply.

$= 21 + 12k$ Add.

(d) $5 - (2y - 8)$

Be careful with signs.

$= 5 - 1(2y - 8)$ $-a = -1 \cdot a$

$= 5 - 2y + 8$ Distributive property

$= 13 - 2y$ Add.

◄ **Work Problem** ① **at the Side.**

---

① Simplify each expression.

(a) $9k + 12 - 5$

GS (b) $7(3p + 2q)$

$= 7(\underline{\hspace{0.5cm}}) + 7(\underline{\hspace{0.5cm}})$

$= \underline{\hspace{0.5cm}} + \underline{\hspace{0.5cm}}$

(c) $2 + 5(3z - 1)$

(d) $-3 - (2 + 5y)$

Note

In **Examples 1(c) and 1(d),** we mentally used the commutative and associative properties to add in the last step. In practice, these steps are often not written out.

OBJECTIVE ② **Identify terms and numerical coefficients.** A **term** is a number (constant), a variable, or a product or quotient of numbers and variables raised to powers.

$$9x, \quad 15y^2, \quad -3, \quad -8m^2n, \quad \frac{2}{p}, \quad \text{and} \quad k \quad \text{Terms}$$

In the term $9x$, the **numerical coefficient,** or simply **coefficient,** of the variable x is 9. Additional examples are shown in the table on the next page.

CAUTION

It is important to be able to distinguish between **_terms_** and **_factors_**. Consider the following expressions.

$8x^3 + 12x^2$ This expression has **two terms,** $8x^3$ and $12x^2$. **Terms** are separated by a + or − symbol.

$(8x^3)(12x^2)$ This is a **one-term** expression. The **factors** $8x^3$ and $12x^2$ are multiplied.

Answers

1. **(a)** $9k + 7$ **(b)** $3p; 2q; 21p; 14q$
(c) $15z - 3$ **(d)** $-5 - 5y$

Term	Numerical Coefficient
$-7y$	-7
$34r^3$	34
$-26x^5yz^4$	-26
$-k$, or $-1 \cdot k$	-1
r, or $1r$	1
$\frac{3x}{8} = \frac{3}{8}x$	$\frac{3}{8}$

Work Problem ② at the Side. ▶

OBJECTIVE ▶ ③ Identify like terms. Terms with exactly the same variables that have the same exponents are **like terms**. Here are some examples.

Like Terms	**Unlike Terms**	
$9t$ and $4t$	$4y$ and $7t$	Different variables
$6x^2$ and $-5x^2$	$17x$ and $-8x^2$	Different exponents
$-2pq$ and $11pq$	$4xy^2$ and $4xy$	Different exponents
$3x^2y$ and $5x^2y$	$-7wz^3$ and $2xz^3$	Different variables

Work Problem ③ at the Side. ▶

OBJECTIVE ▶ ④ Combine like terms. Recall the distributive property.

$x(y + z) = xy + xz$ can also be written "in reverse" as $xy + xz = x(y + z)$.

This last form, which may be used to find the sum or difference of like terms, provides justification for **combining like terms.**

EXAMPLE 2 Combining Like Terms

Combine like terms in each expression.

(a) $-9m + 5m$

$= (-9 + 5)m$

$= -4m$

(b) $6r + 3r + 2r$

$= (6 + 3 + 2)r$

$= 11r$

(c) $4x + x$

$= 4x + 1x$ $x = 1x$

$= (4 + 1)x$

$= 5x$

(d) $16y^2 - 9y^2$

$= (16 - 9)y^2$

$= 7y^2$

(e) $32y + 10y^2$ These unlike terms cannot be combined.

············· **Work Problem ④ at the Side.** ▶

Simplifying an Expression

An expression has been simplified when the following conditions have been met.

- All grouping symbols have been removed.

- All like terms have been combined.

- Operations have been performed, when possible.

---

② Give the numerical coefficient of each term.

(a) $15q$

(b) $-2m^3$

(c) $-18m^7q^4$

(d) $-r$

(e) $\dfrac{5x}{4}$

③ Identify each pair of terms as *like* or *unlike*.

(a) $9x, 4x$

(b) $-8y^3, 12y^2$

(c) $5x^2y^4, 5x^4y^2$

(d) $7x^2y^4, -7x^2y^4$

(e) $13kt, 4tk$

④ Combine like terms.

(a) $4k + 7k$

(b) $4r - r$

(c) $5z + 9z - 4z$

(d) $8p + 8p^2$

Answers

2. (a) 15 **(b)** -2 **(c)** -18
(d) -1 **(e)** $\dfrac{5}{4}$

3. (a) like **(b)** unlike **(c)** unlike
(d) like **(e)** like

4. (a) $11k$ **(b)** $3r$ **(c)** $10z$
(d) cannot be combined

5 Simplify.

(GS) (a) $10p + 3(5 + 2p)$

$= 10p + \_\_\_(5) + 3(\_\_\_)$

$= 10p + \_\_\_ + \_\_\_$

$= _____$

(b) $7z - 2 - (1 + z)$

(c) $-(3k^2 + 5k) + 7(k^2 - 4k)$

(d) $-\dfrac{3}{4}(x - 8) - \dfrac{1}{3}x$

6 Translate each phrase into a mathematical expression and simplify.

(a) Three times a number, subtracted from the sum of the number and 8

(b) Twice a number added to the sum of 6 and the number

Answers

5. (a) 3; $2p$; 15; $6p$; $16p + 15$ **(b)** $6z - 3$

(c) $4k^2 - 33k$ **(d)** $-\dfrac{13}{12}x + 6$

6. (a) $(x + 8) - 3x$; $-2x + 8$

(b) $(6 + x) + 2x$; $3x + 6$

> **CAUTION**
>
> *Remember that only like terms may be combined.*

EXAMPLE 3 Simplifying Expressions Involving Like Terms

Simplify each expression.

(a) $14y + 2(6 + 3y)$

$= 14y + 2(6) + 2(3y)$ Distributive property

$= 14y + 12 + 6y$ Multiply.

$= 20y + 12$ Combine like terms.

(b) $9k - 6 - 3(2 - 5k)$ [Be careful with signs.]

$= 9k - 6 - 3(2) - 3(-5k)$ Distributive property

$= 9k - 6 - 6 + 15k$ Multiply.

$= 24k - 12$ Combine like terms.

(c) $\quad -(2 - r) + 10r$

$= -1(2 - r) + 10r$ $-(2 - r) = -1(2 - r)$

$= -1(2) - 1(-r) + 10r$ Distributive property [Be careful with signs.]

$= -2 + r + 10r$ Multiply.

$= -2 + 11r$ Combine like terms.

(d) $-\dfrac{2}{3}(x - 6) - \dfrac{1}{6}x$

$= -\dfrac{2}{3}x - \dfrac{2}{3}(-6) - \dfrac{1}{6}x$ Distributive property

$= -\dfrac{2}{3}x + 4 - \dfrac{1}{6}x$ Multiply.

$= -\dfrac{4}{6}x + 4 - \dfrac{1}{6}x$ Get a common denominator.

$= -\dfrac{5}{6}x + 4$ Combine like terms.

◀ **Work Problem 5 at the Side.**

OBJECTIVE ▶ 5 Simplify expressions from word phrases.

EXAMPLE 4 Translating Words into a Mathematical Expression

Translate the phrase into a mathematical expression and simplify.

> The sum of 9, five times a number, four times the number, and six times the number

The word "sum" indicates that the terms should be added. Use x for the number.

$9 + 5x + 4x + 6x$ simplifies to $9 + 15x$. Combine like terms.

[This is an expression to be simplified, *not* an equation to be solved.]

◀ **Work Problem 6 at the Side.**

1.8 Exercises

FOR EXTRA HELP

 MyMathLab®

CONCEPT CHECK *In Exercises 1–4, choose the letter of the correct response.*

1. Which expression is a simplified form of $-(6x - 3)$?
   A. $-6x - 3$ **B.** $-6x + 3$
   C. $6x - 3$ **D.** $6x + 3$

2. Which is an example of a pair of like terms?
   A. $6t, 6w$ **B.** $-8x^2 y, 9xy^2$
   C. $5ry, 6yr$ **D.** $-5x^2, 2x^3$

3. Which is an example of a term with numerical coefficient 5?
   A. $5x^3 y^7$ **B.** x^5
   C. $\dfrac{x}{5}$ **D.** $5^2 xy^3$

4. Which is a correct translation for "six times a number, subtracted from the product of eleven and the number" (if x represents the number)?
   A. $6x - 11x$ **B.** $11x - 6x$
   C. $(11 + x) - 6x$ **D.** $6x - (11 + x)$

Simplify each expression. **See Examples 1 and 2.**

5. $3x + 12x$

6. $4y + 9y$

7. $12b + b$

8. $19x + x$

9. $4r + 19 - 8$

10. $7t + 18 - 4$

11. $5 + 2(x - 3y)$

12. $8 + 3(s - 6t)$

13. $-2 - (5 - 3p)$

14. $-10 - (7 - 14r)$

15. $-\dfrac{5}{6}(y + 12) - \dfrac{1}{2}y$

16. $-\dfrac{2}{3}(w + 15) - \dfrac{1}{4}w$

Give the numerical coefficient. **See Objective 2.**

17. $-12k$

18. $-23y$

19. $5m^2$

20. $-3n^6$

21. xw

22. pq

23. $-x$

24. $-t$

25. 74

26. 98

27. $-\dfrac{3}{8}x$

28. $-\dfrac{5}{4}z$

29. $\dfrac{x}{2}$

30. $\dfrac{x}{6}$

31. $\dfrac{2x}{5}$

32. $\dfrac{8x}{9}$

33. $-1.28r^2$

34. $-2.985t^3$

Identify each group of terms as like *or* unlike. **See Objective 3.**

35. $8r, -13r$

36. $-7a, 12a$

37. $5z^4, 9z^3$

38. $8x^5, -10x^3$

39. $4, 9, -24$

40. $7, 17, -83$

41. x, y

42. t, s

*Simplify each expression. **See Examples 2 and 3.***

43. $-5 - 2(x - 3)$

$= -5 - 2(\underline{}) - 2(\underline{})$

$= -5 - \underline{} + \underline{}$

$= \underline{}$

44. $-8 - 3(2x + 4)$

$= -8 - 3(\underline{}) - 3(\underline{})$

$= -8 - \underline{} - \underline{}$

$= \underline{}$

45. $-\dfrac{4}{3} + 2t + \dfrac{1}{3}t - 8 - \dfrac{8}{3}t$

46. $-\dfrac{5}{6} + 8x + \dfrac{1}{6}x - 7 - \dfrac{7}{6}$

47. $-5.3r + 4.9 - (2r + 0.7) + 3.2r$

48. $2.7b + 5.8 - (3b + 0.5) - 4.4b$

49. $2y^2 - 7y^3 - 4y^2 + 10y^3$

50. $9x^4 - 7x^6 + 12x^4 + 14x^6$

51. $13p + 4(4 - 8p)$

52. $5x + 3(7 - 2x)$

53. $\dfrac{1}{2}(2x + 4) - \dfrac{1}{3}(9x - 6)$

54. $\dfrac{1}{4}(8x + 16) - \dfrac{1}{5}(20x - 15)$

55. $-\dfrac{2}{3}(5x + 7) - \dfrac{1}{3}(4x + 8)$

56. $-\dfrac{3}{4}(7x + 9) - \dfrac{1}{4}(5x + 7)$

57. $-\dfrac{4}{3}(y - 12) - \dfrac{1}{6}y$

58. $-\dfrac{7}{5}(t - 15) - \dfrac{1}{2}t$

59. $-5(5y - 9) + 3(3y + 6)$

60. $-3(2t + 4) + 8(2t - 4)$

61. $-3(2r - 3) + 2(5r + 3)$

62. $-4(5y - 7) + 3(2y - 5)$

63. $8(2k - 1) - (4k - 3)$

64. $6(3p - 2) - (5p + 1)$

65. $-2(-3k + 2) - (5k - 6) - 3k - 5$

66. $-2(3r - 4) - (6 - r) + 2r - 5$

67. $-4(-3x + 3) - (6x - 4) - 2x + 1$

68. $-5(8x + 2) - (5x - 3) - 3x + 17$

69. $-7.5(2y + 4) - 2.9(3y - 6)$

70. $8.4(6t - 6) + 2.4(9 - 3t)$

Write each phrase as a mathematical expression. Use x to represent the number.
Combine like terms when possible. ***See Example 4.***

71. Five times a number, added to the sum of the number and three

72. Six times a number, added to the sum of the number and six

73. A number multiplied by −7, subtracted from the sum of 13 and six times the number

74. A number multiplied by 5, subtracted from the sum of 14 and eight times the number

75. Six times a number added to −4, subtracted from twice the sum of three times the number and 4

76. Nine times a number added to 6, subtracted from triple the sum of 12 and 8 times the number

77. Write the expression $9x - (x + 2)$ using words, as in **Exercises 71–76.**

78. Write the expression $2(3x + 5) - 2(x + 4)$ using words, as in **Exercises 71–76.**

79. CONCEPT CHECK A student simplified the expression $7x - 2(3 - 2x)$ as shown.

$$7x - 2(3 - 2x)$$
$$= 7x - 2(3) - 2(2x)$$
$$= 7x - 6 - 4x$$
$$= 3x - 6$$

What Went Wrong? Find the correct simplified answer.

80. CONCEPT CHECK A student simplified the expression $3 + 2(4x - 5)$ as shown.

$$3 + 2(4x - 5)$$
$$= 5(4x - 5)$$
$$= 5(4x) + 5(-5)$$
$$= 20x - 25$$

What Went Wrong? Find the correct simplified answer.

Relating Concepts (Exercises 81–84) For Individual or Group Work

A manufacturer has fixed costs of $1000 *to produce widgets. Each widget costs* $5 *to make. The fixed cost to produce gadgets is* $750, *and each gadget costs* $3 *to make.*
Work Exercises 81–84 in order.

81. Write an expression for the cost to make x widgets. (*Hint:* The cost will be the sum of the fixed cost and the cost per item times the number of items.)

82. Write an expression for the cost to make y gadgets.

83. Write an expression for the total cost to make x widgets and y gadgets.

84. Simplify the expression you wrote in **Exercise 83.**

Study Skills
TAKING MATH TESTS

Techniques To Improve Your Test Score	Comments
Come prepared with a pencil, eraser, paper, and calculator, if allowed.	Working in pencil lets you erase, keeping your work neat and readable.
Scan the entire test, note the point values of different problems, and plan your time accordingly.	To do 20 problems in 50 minutes, allow $50 \div 20 = 2.5$ minutes per problem. Spend less time on the easier problems.
Do a "knowledge dump" when you get the test. Write important notes, such as formulas, to yourself in a corner of the test.	Writing down tips and things that you've memorized at the beginning allows you to relax later.
Read directions carefully, and circle any significant words. When you finish a problem, read the directions again to make sure you did what was asked.	Pay attention to announcements written on the board or made by your instructor. Ask if you don't understand.
Show all your work. Many teachers give partial credit if some steps are correct, even if the final answer is wrong. **Write neatly.**	If your teacher can't read your writing, you won't get credit for it. If you need more space to work, ask to use extra paper.
Write down anything that might help solve a problem: a formula, a diagram, etc. If you can't get it, circle the problem and come back to it later. Do *not* erase anything you wrote down.	If you know even a little bit about the problem, write it down. The answer may come to you as you work on it, or you may get partial credit. Don't spend too long on any one problem.
If you can't solve a problem, make a guess. Do not change it unless you find an obvious mistake.	Have a good reason for changing an answer. Your first guess is often your best bet.
Check that the answer to an application problem is reasonable and makes sense. Read the problem again to make sure you've answered the question.	Use common sense. Can the father really be seven years old? Would a month's rent be $32,140? Label your answer: $, years, inches, etc.
Check for careless errors. Rework the problem without looking at your previous work. Compare the two answers.	Reworking the problem from the beginning forces you to rethink it. If possible, use a different method to solve the problem.

Mark several tips to try when you take your next math test. After the test, evaluate how they worked for you. What will you do differently when taking your next test?

Chapter 1 *Summary*

Key Terms

1.1

exponent An exponent, or **power,** is a number that indicates how many times a factor is repeated.

$$3^4 \xleftarrow{\text{Exponent}} \left.\vphantom{\begin{array}{c}a\\b\end{array}}\right\} \begin{array}{l}\text{Exponential}\\\text{expression}\end{array}$$

Base

base The base is the number that is a repeated factor when written with an exponent.

exponential expression A number written with an exponent is an exponential expression.

1.2

variable A variable is a symbol, usually a letter, used to represent an unknown number.

constant A constant is a fixed, unchanging number.

algebraic expression An algebraic expression is a collection of numbers (constants), variables, operation symbols, and grouping symbols.

equation An equation is a statement that says two expressions are equal.

solution A solution of an equation is any value of the variable that makes the equation true.

1.3

natural numbers The set of natural numbers is $\{1, 2, 3, 4, \ldots\}$.

whole numbers The set of whole numbers is $\{0, 1, 2, 3, 4, \ldots\}$.

number line A number line shows the ordering of real numbers on a line.

additive inverse The additive inverse of a number a is the number that is the same distance from 0 on the number line as a, but on the opposite side of 0. This number is also called the opposite of a.

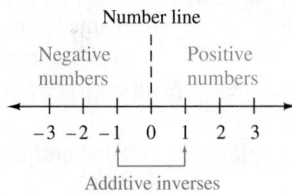

Number line

integers The set of integers is $\{\ldots, -3, -2, -1, 0, 1, 2, 3, \ldots\}$.

negative number A negative number is located to the *left* of 0 on the number line.

positive number A positive number is located to the *right* of 0 on the number line.

signed numbers Positive numbers and negative numbers are signed numbers.

rational numbers A rational number is a number that can be written as the quotient of two integers, with denominator not 0.

set-builder notation Set-builder notation uses a variable and a description to describe a set. It is often used to describe sets whose elements cannot easily be listed.

coordinate The number that corresponds to a point on the number line is the coordinate of that point.

irrational numbers An irrational number is a real number that is not a rational number.

real numbers Real numbers are numbers that can be represented by points on the number line (that is, all rational and irrational numbers).

absolute value The absolute value of a number is the distance between 0 and the number on a number line.

1.4

sum The answer to an addition problem is the sum.

1.5

minuend In the operation $a - b$, a is the minuend.

subtrahend In the operation $a - b$, b is the subtrahend.

difference The answer to a subtraction problem is the difference.

$$15 \quad - \quad 7 = 8 \xleftarrow{} \text{Difference}$$

Minuend   Subtrahend

1.6

product The answer to a multiplication problem is the product.

quotient The answer to a division problem is the quotient.

reciprocal Pairs of numbers whose product is 1 are reciprocals, or **multiplicative inverses,** of each other.

(continued)

1.7

identity element for addition When the identity element for addition, which is 0, is added to a number, the number is unchanged.

identity element for multiplication When a number is multiplied by the identity element for multiplication, which is 1, the number is unchanged.

1.8

term A term is a number (constant), a variable, or a product or quotient of a number and one or more variables raised to powers.

$$\underset{\text{Term}}{\underbrace{\overset{\overset{\text{Numerical}}{\overset{\text{coefficient}}{\downarrow}}}{-7x^2}}}$$

numerical coefficient The numerical factor of a term is its numerical coefficient, or **coefficient.**

like terms Terms with exactly the same variables (including the same exponents) are like terms.

New Symbols

a^n	n factors of a	$a(b)$, $(a)b$, $(a)(b)$, $a \cdot b$, or ab	a times b
$=$	is equal to	$\dfrac{a}{b}$, a/b, or $a \div b$	a divided by b
$\neq$	is not equal to	$\{\ \}$	set braces
		$\{x \mid x \text{ has a certain property}\}$	set-builder notation
$<$	is less than	$[\ ]$	square brackets
$\leq$	is less than or equal to	$\lvert x \rvert$	absolute value of x
$>$	is greater than	$-x$	additive inverse, or opposite, of x
$\geq$	is greater than or equal to	$\dfrac{1}{x}$	multiplicative inverse, or reciprocal, of x (where $x \neq 0$)

Test Your Word Power

See how well you have learned the vocabulary in this chapter.

1 A **product** is
A. the answer in an addition problem
B. the answer in a multiplication problem
C. one of two or more numbers that are added to get another number
D. one of two or more numbers that are multiplied to get another number.

2 An **exponent** is
A. a symbol that tells how many numbers are being multiplied
B. a number raised to a power
C. a number that tells how many times a factor is repeated
D. one of two or more numbers that are multiplied.

3 A **variable** is
A. a symbol used to represent an unknown number
B. a value that makes an equation true
C. a solution of an equation
D. the answer in a division problem.

4 An **integer** is
A. a positive or negative number
B. a natural number, its opposite, or zero
C. any number that can be graphed on a number line
D. the quotient of two numbers.

5 A **coordinate** is
A. the number that corresponds to a point on a number line
B. the graph of a number
C. any point on a number line
D. the distance from 0 on a number line.

6 The **absolute value** of a number is
A. the graph of the number
B. the reciprocal of the number
C. the opposite of the number
D. the distance between 0 and the number on a number line.

7 A **term** is
A. a numerical factor
B. a number, a variable, or a product or quotient of numbers and variables raised to powers
C. one of several variables with the same exponents
D. a sum of numbers and variables raised to powers.

8 The **subtrahend** in $a - b = c$ is
A. a
B. b
C. c
D. $a - b$.

Answers to Test Your Word Power

1. B; *Example:* The product of 2 and 5, or 2 times 5, is 10.

2. C; *Example:* In 2^3, the number 3 is the exponent (or power), so 2 is a factor three times; $2^3 = 2 \cdot 2 \cdot 2 = 8$.

3. A; *Examples: x, y, z*

4. B; *Examples:* $-9, 0, 6$

5. A; *Example:* The point graphed three units to the right of 0 on a number line has coordinate 3.

6. D; *Examples:* $|2| = 2$ and $|-2| = 2$

7. B; *Examples:* $6, \frac{x}{2}, -4ab^2$

8. B; *Example:* In $5 - 3 = 2$, 5 is the minuend, 3 is the subtrahend, and 2 is the difference.

Quick Review

Concepts	Examples
1.1 Exponents, Order of Operations, and Inequality	$$\frac{9(2+6)}{2} - 2(2^3 + 3)$$
Order of Operations	
Simplify within any parentheses or brackets and above and below fraction bars, using the following steps.	$= \dfrac{9(8)}{2} - 2(8+3)$ Add inside parentheses. Apply the exponent.
Step 1 Apply all exponents.	$= 36 - 2(8+3)$ Multiply and divide.
Step 2 Multiply or divide from left to right.	$= 36 - 2(11)$ Add inside parentheses.
Step 3 Add or subtract from left to right.	$= 36 - 22$ Multiply.
	$= 14$ Subtract.
1.2 Variables, Expressions, and Equations	Evaluate $2x + y^2$ for $x = 3$ and $y = -4$.
To evaluate an expression means to find its value. Evaluate an expression with a variable by substituting a given number for the variable.	$2x + y^2$
	$= 2(3) + (-4)^2$ Substitute.
	$= 6 + 16$ Multiply and apply the exponent.
	$= 22$ Add.
Values of a variable that make an equation true are solutions of the equation.	Is 2 a solution of $5x + 3 = 18$?
	$5(2) + 3 \stackrel{?}{=} 18$ Let $x = 2$.
	$13 = 18$ False
	2 is not a solution.
1.3 Real Numbers and the Number Line	
Ordering Real Numbers	
a is less than b if a is to the left of b on a number line.	$-2 < 3 \qquad 3 > 0 \qquad 1 < 3$
The additive inverse, or opposite, of a is $-a$.	$-(5) = -5 \quad -(-7) = 7 \qquad -0 = 0$
The absolute value of a, written $|a|$, is the distance between a and 0 on a number line.	$|13| = 13 \qquad |0| = 0 \qquad |-5| = 5$
1.4 Adding Real Numbers	
To add two numbers with the *same sign*, add their absolute values. The sum has that same sign.	$9 + 4 = 13$
	$-8 + (-5) = -13$
To add two numbers with *different signs*, subtract the lesser absolute value from the greater. The sum has the sign of the number with greater absolute value.	$7 + (-12) = -5$
	$-5 + 13 = 8$

Concepts	Examples

1.5 Subtracting Real Numbers

For any real numbers a and b,

$$a - b \quad \text{is defined as} \quad a + (-b).$$

$$
\begin{array}{lll}
5 - (-2) & -3 - 4 & -2 - (-6) \\
= 5 + 2 & = -3 + (-4) & = -2 + 6 \\
= 7 & = -7 & = 4
\end{array}
$$

1.6 Multiplying and Dividing Real Numbers

The product (or quotient) of two numbers having the *same* sign is *positive*.

$$
6 \cdot 5 = 30 \qquad -7(-8) = 56 \qquad \frac{-24}{-6} = 4
$$

The product (or quotient) of two numbers having *different* signs is *negative*.

$$
-6(5) = -30 \qquad 6(-5) = -30
$$

To divide a by b, multiply a by the reciprocal of b.

$$
\begin{array}{lll}
-18 \div 9 & 49 \div (-7) & 10 \div \dfrac{2}{3} \\[2mm]
= \dfrac{-18}{9} & = \dfrac{49}{-7} & = 10 \cdot \dfrac{3}{2} \\[2mm]
= -2 & = -7 & = 15
\end{array}
$$

0 divided by a nonzero number is 0.
Division by 0 is undefined.

$$
\frac{0}{5} = 0 \qquad \frac{5}{0} \text{ is undefined.}
$$

1.7 Properties of Real Numbers

Commutative Properties

$$
\begin{aligned}
a + b &= b + a \\
ab &= ba
\end{aligned}
$$

$$
\begin{aligned}
7 + (-1) &= -1 + 7 \\
5(-3) &= (-3)5
\end{aligned}
$$

Associative Properties

$$
\begin{aligned}
(a + b) + c &= a + (b + c) \\
(ab)c &= a(bc)
\end{aligned}
$$

$$
\begin{aligned}
(3 + 4) + 8 &= 3 + (4 + 8) \\
[-2(6)]4 &= -2[6(4)]
\end{aligned}
$$

Identity Properties

$$
\begin{aligned}
a + 0 &= a & 0 + a &= a \\
a \cdot 1 &= a & 1 \cdot a &= a
\end{aligned}
$$

$$
\begin{aligned}
-7 + 0 &= -7 & 0 + (-7) &= -7 \\
9 \cdot 1 &= 9 & 1 \cdot 9 &= 9
\end{aligned}
$$

Inverse Properties

$$
\begin{aligned}
a + (-a) &= 0 & -a + a &= 0 \\
a \cdot \frac{1}{a} &= 1 & \frac{1}{a} \cdot a &= 1 \quad (a \neq 0)
\end{aligned}
$$

$$
\begin{aligned}
7 + (-7) &= 0 & -7 + 7 &= 0 \\
-2\left(-\frac{1}{2}\right) &= 1 & -\frac{1}{2}(-2) &= 1
\end{aligned}
$$

Distributive Properties

$$
\begin{aligned}
a(b + c) &= ab + ac \\
(b + c)a &= ba + ca \\
a(b - c) &= ab - ac
\end{aligned}
$$

$$
\begin{aligned}
5(4 + 2) &= 5(4) + 5(2) \\
(4 + 2)5 &= 4(5) + 2(5) \\
9(5 - 4) &= 9(5) - 9(4)
\end{aligned}
$$

1.8 Simplifying Expressions

Only like terms may be combined. We use the distributive property.

$$
\begin{aligned}
& 4(3 + 2x) - 6(5 - x) \\
& = 12 + 8x - 30 + 6x \\
& = 14x - 18
\end{aligned}
$$

Chapter 1 *Review Exercises*

If you need help with any of these Review Exercises, look in the section indicated.

1.1 *Find the value of each exponential expression.*

1. 5^4 **2.** $(0.03)^4$ **3.** 0.21^3 **4.** $\left(\dfrac{5}{2}\right)^3$

Find the value of each expression.

5. $8 \cdot 5 - 13$ **6.** $5[4^2 + 3(2^3)]$ **7.** $\dfrac{7(3^2 - 5)}{16 - 2 \cdot 6}$ **8.** $\dfrac{5(6^2 - 2^4)}{3 \cdot 5 + 10}$

Write each word sentence in symbols.

9. Thirteen is less than seventeen.

10. Five plus two is not equal to ten.

11. Write $6 < 15$ in words.

12. Construct a false statement that involves addition on the left side, the symbol $\geq$, and division on the right side.

1.2 *Evaluate each expression for $x = 6$ and $y = 3$.*

13. $2x + 6y$ **14.** $4(3x - y)$ **15.** $\dfrac{x}{3} + 4y$ **16.** $\dfrac{x^2 + 3}{3y - x}$

Change each word phrase to an algebraic expression. Use x to represent the number.

17. Six added to a number

18. A number subtracted from eight

19. Nine subtracted from six times a number

20. Three-fifths of a number added to 12

Decide whether the given number is a solution of the equation.

21. $5x + 3(x + 2) = 22;$ 2

22. $\dfrac{x + 5}{3x} = 1;$ 6

Change each word sentence to an equation. Use x to represent the number.

23. Six less than twice a number is 10.

24. The product of a number and 4 is 8.

Identify each of the following as either an equation *or an* expression.

25. $5r - 8(r + 7) = 2$

26. $2y + (5y - 9) + 2$

1.3 *Graph each group of numbers on a number line.*

27. $-4, -\dfrac{1}{2}, 0, 2.5, 5$

28. $-3\dfrac{1}{4}, \dfrac{14}{5}, -1\dfrac{1}{8}, \dfrac{5}{6}$

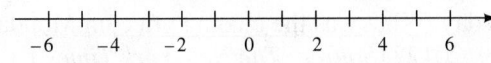

Select the lesser number in each pair.

29. $-10, 5$
30. $-8, -9$
31. $-\dfrac{2}{3}, -\dfrac{3}{4}$
32. $0, -|23|$

Decide whether each statement is true *or* false.

33. $12 > -13$
34. $0 > -5$
35. $-9 < -7$
36. $-13 > -13$

Simplify by finding the absolute value.

37. $-|3|$
38. $-|-19|$
39. $-|9 - 2|$
40. $|15 - 6|$

1.4 *Find each sum.*

41. $-10 + 4$
42. $14 + (-18)$
43. $-8 + (-9)$
44. $\dfrac{4}{9} + \left(-\dfrac{5}{4}\right)$

45. $[-6 + (-8) + 8] + [9 + (-13)]$
46. $(-4 + 7) + (-11 + 3) + (-15 + 1)$

Write a numerical expression for each phrase, and simplify the expression.

47. 19 added to the sum of -31 and 12

48. 13 more than the sum of -4 and -8

Solve each problem.

49. Like many people, Otis Taylor neglects to keep up his checkbook balance. When he finally balanced his account, he found that the balance was $-\$23.75$, so he deposited $\$50.00$. What is his new balance?

50. The low temperature in Yellowknife, in the Canadian Northwest Territories, one January day was $-26°$F. It rose $16°$ to its high that day. What was the high temperature?

1.5 *Find each difference.*

51. $-7 - 4$
52. $-12 - (-11)$
53. $5 - (-2)$
54. $-\dfrac{3}{7} - \dfrac{4}{5}$

55. $2.56 - (-7.75)$
56. $(-10 - 4) - (-2)$
57. $(-3 + 4) - (-1)$
58. $|5 - 9| - |-3 + 6|$

Write a numerical expression for each phrase, and simplify the expression.

59. The difference between -4 and -6

60. Five less than the sum of 4 and -8

61. The difference between 18 and -23, decreased by 15

62. Nineteen, decreased by 12 less than -7

Solve each problem.

63. The quaterback for the Chicago Bears passed for a gain of 8 yd, was sacked for a loss of 12 yd, and then threw a 42 yd touchdown pass. What positive or negative number represents the total net yardage for the plays?

64. On Tuesday, January 10, 2012, the Dow Jones Industrial Average closed at 12,462.47, up 69.78 from the previous day. What was the closing price on Monday, January 9, 2012? (*Source: The New York Times.*)

The table shows the number of people naturalized in the United States (that is, made citizens of the United States) for the years 2003 through 2010. In Exercises 65–68, use a signed number to represent the change in the number of people naturalized for each time period.

Year	Number of People (in thousands)
2003	463
2004	337
2005	604
2006	702
2007	1383
2008	526
2009	570
2010	710

Source: U.S. Department of Homeland Security.

65. 2003–2004

66. 2004–2005

67. 2006–2007

68. 2007–2008

1.6 *Perform the indicated operations.*

69. $(-12)(-3)$

70. $15(-7)$

71. $-\dfrac{4}{3}\left(-\dfrac{3}{8}\right)$

72. $-4.8(-2.1)$

73. $5(8-12)$

74. $(5-7)(8-3)$

75. $2(-6)-(-4)(-3)$

76. $3(-10)-5$

77. $\dfrac{-36}{-9}$

78. $\dfrac{220}{-11}$

79. $-\dfrac{1}{2}\div\dfrac{2}{3}$

80. $-33.9\div(-3)$

81. $\dfrac{-5(3)-1}{8-4(-2)}$

82. $\dfrac{5(-2)-3(4)}{-2[3-(-2)]+10}$

83. $\dfrac{10^2-5^2}{8^2+3^2-(-2)}$

84. $\dfrac{4^2-8\cdot2}{(-1.2)^2-(-0.56)}$

Evaluate each expression for $x=-5$, $y=4$, and $z=-3$.

85. $6x-4z$

86. $5x+y-z$

87. $5x^2$

88. $z^2(3x-8y)$

Write a numerical expression for each phrase, and simplify the expression.

89. Nine less than the product of -4 and 5

90. Five-sixths of the sum of 12 and -6

91. The quotient of 12 and the sum of 8 and -4

92. The product of -20 and 12, divided by the difference between 15 and -15

Translate each sentence to an equation, using x to represent the number.

93. The quotient of a number and the sum of the number and 5 is -2.

94. 3 less than 8 times a number is -7.

1.7 *Decide whether each statement is an example of a commutative, associative, identity, or inverse property, or of the distributive property.*

95. $6 + 0 = 6$

96. $5 \cdot 1 = 5$

97. $-\dfrac{2}{3}\left(-\dfrac{3}{2}\right) = 1$

98. $17 + (-17) = 0$

99. $3x + 3y = 3(x + y)$

100. $w(xy) = (wx)y$

101. $5 + (-9 + 2) = \left[5 + (-9)\right] + 2$

102. $(1 + 2) + 3 = 3 + (1 + 2)$

Use the distributive property to rewrite each expression. Simplify if possible.

103. $7y + y$

104. $-12(4 - t)$

105. $3(2s) + 3(4y)$

106. $-(-4r + 5s)$

1.8 *Use the distributive property as necessary and combine like terms.*

107. $16p^2 - 8p^2 + 9p^2$

108. $4r^2 - 3r + 10r + 12r^2$

109. $-8(5k - 6) + 3(7k + 2)$

110. $2s - (-3s + 6)$

111. $-7(2t - 4) - 4(3t + 8) - 19(t + 1)$

112. $3.6t^2 + 9t - 8.1(6t^2 + 4t)$

Mixed Review Exercises*

Perform the indicated operations.

113. $\dfrac{15}{2} \cdot \left(-\dfrac{4}{5}\right)$

114. $\left(-\dfrac{5}{6}\right)^2$

115. $-\left|(-7)(-4)\right| - (-2)$

116. $\dfrac{6(-4) + 2(-12)}{5(-3) + (-3)}$

117. $\dfrac{3}{8} - \dfrac{5}{12}$

118. $\dfrac{12^2 + 2^2 - 8}{10^2 - (-4)(-15)}$

119. $\dfrac{8^2 + 6^2}{7^2 + 1^2}$

120. $-16(-3.5) - 7.2(-3)$

121. $2\dfrac{5}{6} - 4\dfrac{1}{3}$

122. $-8 + \left[(-4 + 17) - (-3 - 3)\right]$

123. $-\dfrac{12}{5} \div \dfrac{9}{7}$

124. $(-8 - 3) - 5(2 - 9)$

125. $\left[-7 + (-2) - (-3)\right] + \left[8 + (-13)\right]$

126. $\left[(-2) + 7 - (-5)\right] + \left[-4 - (-10)\right]$

Solve each problem.

127. The highest temperature ever recorded in Iowa was 118°F at Keokuk on July 20, 1934. The lowest temperature ever recorded in the state was at Elkader on February 3, 1996, and was 165° lower than the highest temperature. What is the record low temperature for Iowa? (*Source:* National Climatic Data Center.)

128. Humpback whales love to heave their 45-ton bodies out of the water. This is called *breaching*. Chantelle, a researcher based on the island of Maui, noticed that one of her favorite whales, "Pineapple," breached 15 ft above the surface of the ocean while her mate cruised 12 ft below the surface. What is the difference between these two heights?

15 ft

12 ft

---

*The order of exercises in this final group does not correspond to the order in which topics occur in the chapter. This random ordering should help you prepare for the chapter test in yet another way.

Chapter 1 *Test*

The Chapter Test Prep Videos with test solutions are available on DVD, in MyMathLab, and on YouTube—search "LialCombinedAlg" and click on "Channels."

Decide whether each statement is true *or* false.

1. $4[-20 + 7(-2)] \leq -135$

2. $\left(\dfrac{1}{2}\right)^2 + \left(\dfrac{2}{3}\right)^2 = \left(\dfrac{1}{2} + \dfrac{2}{3}\right)^2$

3. Graph the numbers $-1, -3, |-4|,$ and $|-1|$ on the number line.

$\xleftarrow{}\underset{\substack{-3 \;-2 \;-1 \;\; 0 \;\;\; 1 \;\;\; 2 \;\;\; 3 \;\;\; 4}}{+\!+\!+\!+\!+\!+\!+\!+\!}\xrightarrow{}$

Select the lesser number from each pair.

4. $6, -|-8|$

5. $-0.742, -1.277$

6. Write in symbols: The quotient of -6 and the sum of 2 and -8. Simplify the expression.

7. If a and b are both negative, is $\dfrac{a+b}{a \cdot b}$ positive or negative?

Perform the indicated operations whenever possible.

8. $-2 - (5 - 17) + (-6)$

9. $-5\dfrac{1}{2} + 2\dfrac{2}{3}$

10. $4^2 + (-8) - (2^3 - 6)$

11. $-6.2 - [-7.1 + (2.0 - 3.1)]$

12. $(-5)(-12) + 4(-4) + (-8)^2$

13. $\dfrac{-7 - |-6 + 2|}{-5 - (-4)}$

14. $\dfrac{30(-1-2)}{-9[3 - (-2)] - 12(-2)}$

In Exercises 15 and 16, evaluate each expression for $x = -2$ and $y = 4$.

15. $3x - 4y^2$

16. $\dfrac{5x + 7y}{3(x+y)}$

Solve each problem.

17. The highest Fahrenheit temperature ever recorded in Idaho was 118°F, while the lowest was −60°F. What is the difference between these highest and lowest temperatures? (*Source: World Almanac and Book of Facts.*)

18. For a certain system of rating relief pitchers, 3 points are awarded for a save, 3 points are awarded for a win, 2 points are subtracted for a loss, and 2 points are subtracted for a blown save. If a pitcher has 4 saves, 3 wins, 2 losses, and 1 blown save, how many points does he have?

19. For 2009, the U.S. federal government collected $2.10 trillion in revenues, but spent $3.52 trillion. Write the federal budget deficit as a signed number. (*Source: The Gazette.*)

Match each statement in Column I with the property it illustrates in Column II.

I	**II**
20. $3x + 0 = 3x$	**A.** Commutative property
21. $(5 + 2) + 8 = 8 + (5 + 2)$	**B.** Associative property
22. $-3(x + y) = -3x + (-3y)$	**C.** Inverse property
23. $-5 + (3 + 2) = (-5 + 3) + 2$	**D.** Identity property
24. $-\dfrac{5}{3}\left(-\dfrac{3}{5}\right) = 1$	**E.** Distributive property

Simplify.

25. $8x + 4x - 6x + x + 14x$ **26.** $-2(3x^2 + 4) - 3(x^2 + 2x)$

27. Which properties are used to show that $-(3x + 1) = -3x - 1$?

28. Consider the expression $-6\left[5 + (-2)\right]$.

   (a) Evaluate it by first working within the brackets.

   (b) Evaluate it by using the distributive property.

   (c) Why must the answers in parts (a) and (b) be the same?

Math in the Media

THE MAGIC NUMBER IN SPORTS

The climax of any sports season is the playoffs. Baseball fans eagerly debate predictions of which team will win the pennant for their division. The *magic number* for each first-place team is often reported in media outlets. The **magic number** (sometimes called the **elimination number**) is the combined number of wins by the first-place team and losses by the second-place team that would clinch the title for the first-place team.

American League

East	W	L	PCT	GB
New York	87	57	.604	—
Boston	85	60	.586	2.5
Tampa Bay	80	64	.556	7.0
Toronto	73	73	.500	15.0
Baltimore	58	86	.403	29.0

Central	W	L	PCT	GB
Detroit	83	62	.572	—
Chicago	73	71	.507	9.5
Cleveland	71	72	.497	11.0
Kansas City	61	86	.415	23.0
Minnesota	59	86	.407	24.0

West	W	L	PCT	GB
Texas	82	64	.562	—
Los Angeles	80	65	.552	1.5
Oakland	66	79	.455	15.5
Seattle	61	84	.421	20.5

Source: mlb.com

To calculate the magic number, consider the following conditions.

The number of wins for the first-place team (W_1) plus the magic number (M) is one more than the sum of the number of wins to date (W_2) and the number of games remaining in the season (N_2) for the second-place team.

1. First, use the variable definitions to write an equation involving the magic number. Second, solve the equation for the magic number and write a formula for it.

2. The American League standings on September 10, 2011, are shown above. There were 162 regulation games in the 2011 season. Find the magic number for each team. The number of games remaining in the season for the second-place team is calculated as

$$N_2 = 162 - (W_2 + L_2),$$

where L_2 represents the number of losses for the second-place team.

   (a) AL East: New York vs Boston

   Magic Number _____

   (b) AL Central: Detroit vs Chicago

   Magic Number _____

   (c) AL West: Texas vs Los Angeles

   Magic Number _____

3. Try to calculate the magic number for Seattle vs Texas. (Treat Seattle as if it were the second-place team.) How can we interpret the result?

2

Equations, Inequalities, and Applications

Solving *linear equations*, the subject of this chapter, can be thought of in terms of the concept of balance.

2.1 The Addition Property of Equality

Study Skills *Managing Your Time*

2.2 The Multiplication Property of Equality

Study Skills *Analyzing Your Test Results*

2.3 More on Solving Linear Equations

Study Skills *Using Study Cards Revisited*

Summary Exercises *Applying Methods for Solving Linear Equations*

2.4 An Introduction to Applications of Linear Equations

2.5 Formulas and Additional Applications from Geometry

2.6 Ratio, Proportion, and Percent

Summary Exercises *Applying Problem-Solving Techniques*

2.7 Solving Linear Inequalities

117

2.1 The Addition Property of Equality

OBJECTIVES

1. **Identify linear equations.**
2. **Use the addition property of equality.**
3. **Simplify, and then use the addition property of equality.**

An *equation* is a statement asserting that two algebraic expressions are equal. In this chapter, we *solve* equations.

CAUTION

Remember that an equation always includes an equality symbol.

Equation	Expression
$\downarrow$	$\downarrow$
Left side → $x - 5 = 2$ ← Right side	$x - 5$
An equation can be solved.	An expression **cannot** be solved. (It can often be *evaluated* or *simplified*.)

OBJECTIVE 1 Identify linear equations.

Linear Equation in One Variable

A **linear equation in one variable** can be written in the form

$$Ax + B = C,$$

where A, B, and C are real numbers, with $A \neq 0$.

$$4x + 9 = 0, \quad 2x - 3 = 5, \quad \text{and} \quad x = 7 \qquad \text{Linear equations}$$

$$x^2 + 2x = 5, \quad x^3 - x = 6, \quad \text{and} \quad |2x + 6| = 0 \qquad \textit{Non}\text{linear equations}$$

Recall that a **solution** of an equation is a number that makes the equation true when it replaces the variable. An equation is solved by finding its **solution set,** the set of all solutions. Equations that have exactly the same solution sets are **equivalent equations.** A linear equation in x is solved by using a series of steps to produce a simpler equivalent equation of the form

$$x = \textbf{a number} \quad \text{or} \quad \textbf{a number} = x.$$

OBJECTIVE 2 Use the addition property of equality. In the equation $x - 5 = 2$, both $x - 5$ and 2 represent the same number because that is the meaning of the equality symbol. To solve the equation, we change the left side from $x - 5$ to just x, as follows.

$x - 5 = 2$	Given equation
$x - 5 + 5 = 2 + 5$	Add 5 to *each* side to keep them equal.
$x + 0 = 7$	Additive inverse property
$x = 7$	Additive identity property

Add 5. It is the opposite (additive inverse) of -5, and $-5 + 5 = 0$.

The solution is 7. We check by replacing x with 7 in the original equation.

CHECK	$x - 5 = 2$	Original equation
	$7 - 5 \overset{?}{=} 2$	Let $x = 7$.
	$2 = 2 \checkmark$	True

The left side equals the right side.

Since the final equation is true, 7 checks as the solution and $\{7\}$ is the solution set. We write a solution set using set braces $\{ \ \}$.

To solve the equation $x - 5 = 2$, we added the same number, 5, to each side. The **addition property of equality** justifies this step.

Addition Property of Equality

If A, B, and C represent real numbers, then the equations

$$A = B \quad \text{and} \quad A + C = B + C$$

are equivalent equations.

In words, we can add the same number to each side of an equation without changing the solution set.

In this property, any quantity that represents a real number C can be added to each side of an equation to obtain an equivalent equation.

Note

Equations can be thought of in terms of a balance. Thus, adding the *same* quantity to each side does not affect the balance. See **Figure 1**.

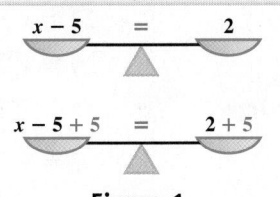

Figure 1

EXAMPLE 1   Using the Addition Property of Equality

Solve $x - 16 = 7$.

Our goal is to get an equivalent equation of the form $x = $ **a number**.

$$x - 16 = 7$$

$$x - 16 + 16 = 7 + 16 \qquad \text{Add 16 to each side.}$$

$$x = 23 \qquad \text{Combine like terms.}$$

CHECK

$$x - 16 = 7 \qquad \text{Original equation}$$

$$23 - 16 \overset{?}{=} 7 \qquad \text{Substitute 23 for } x.$$

> 7 is *not* the solution.

$$7 = 7 \checkmark \qquad \text{True}$$

Since a true statement results, **23** is the solution and $\{23\}$ is the solution set.

············· **Work Problem ❶ at the Side.** ▶

EXAMPLE 2   Using the Addition Property of Equality

Solve $x - 2.9 = -6.4$.

> Our goal is to isolate x.

$$x - 2.9 = -6.4$$

$$x - 2.9 + 2.9 = -6.4 + 2.9 \qquad \text{Add 2.9 to each side.}$$

$$x = -3.5$$

CHECK

$$x - 2.9 = -6.4 \qquad \text{Original equation}$$

$$-3.5 - 2.9 \overset{?}{=} -6.4 \qquad \text{Let } x = -3.5.$$

$$-6.4 = -6.4 \checkmark \qquad \text{True}$$

Since a true statement results, the solution set is $\{-3.5\}$.

············· **Work Problem ❷ at the Side.** ▶

---

❶ Solve.

(GS) **(a)** Fill in the blanks.

$$x - 12 = 9$$

$$x - 12 + \underline{\quad} = 9 + \underline{\quad}$$

$$x = \underline{\quad}$$

CHECK $x - 12 = 9$

$$\underline{\quad} - 12 \overset{?}{=} 9$$

$$\underline{\quad} = 9 \quad (\textit{True / False})$$

The solution set is $\underline{\quad}$.

(b) $x - 25 = -18$

(c) $x - 8 = -13$

❷ Solve.

(a) $x - 3.7 = -8.1$

(b) $a - 4.1 = 6.3$

Answers

1. (a) 12; 12; 21; 21; 9; True; $\{21\}$
  (b) $\{7\}$ **(c)** $\{-5\}$
2. (a) $\{-4.4\}$ **(b)** $\{10.4\}$

3 Solve.

(a) $-3 = a + 2$

Our goal is to isolate ____.
On each side of the equation,
(*add / subtract*) 2.
Now solve.

In **Section 1.5,** subtraction was defined as addition of the opposite. Thus, we can also use the following rule when solving an equation.

> **The same number may be *subtracted* from each side of an equation without changing the solution set.**

For example, to solve the equation $x + 4 = 10$, we *subtract* 4 from each side, which is the same as adding -4. The result is $x = 6$.

EXAMPLE 3 **Using the Addition Property of Equality**

Solve $-7 = x + 22$.

Here the variable x is on the right side of the equation.

$$-7 = x + 22$$ ◁ The variable can be isolated on *either* side.

$$-7 - 22 = x + 22 - 22$$ Subtract 22 from each side.

$$-29 = x, \quad \text{or} \quad x = -29$$ Rewrite; a number $= x$, or $x =$ a number.

CHECK $$-7 = x + 22$$ Original equation

$$-7 \stackrel{?}{=} -29 + 22$$ Let $x = -29$.

$$-7 = -7 \checkmark$$ True

The check confirms that the solution set is $\{-29\}$.

◀ **Work Problem 3 at the Side.**

(b) $22 = -16 + r$

Note

In **Example 3,** what happens if we subtract $-7 - 22$ incorrectly, obtaining $x = -15$ (instead of $x = -29$) as the last line of the solution? A check should indicate an error.

CHECK $$-7 = x + 22$$ Original equation from **Example 3**

$$-7 \stackrel{?}{=} -15 + 22$$ Let $x = -15$.

The left side does *not* equal the right side. ▷ $$-7 = 7$$ False

The false statement indicates that -15 is *not* a solution of the equation. If this happens, rework the problem.

EXAMPLE 4 **Subtracting a Variable Term**

Solve $\frac{3}{5}k + 15 = \frac{8}{5}k$.

$$\frac{3}{5}k + 15 = \frac{8}{5}k$$ ◁ We must get the terms with k on the same side of the $=$ symbol.

$$\frac{3}{5}k + 15 - \frac{3}{5}k = \frac{8}{5}k - \frac{3}{5}k$$ Subtract $\frac{3}{5}k$ from each side.

From now on we will skip this step. ▷ $$15 = 1k$$ $\frac{3}{5}k - \frac{3}{5}k = 0; \frac{8}{5}k - \frac{3}{5}k = \frac{5}{5}k = 1k$

$$15 = k$$ Multiplicative identity property

Check by substituting 15 in the original equation. The solution set is $\{15\}$.

Answers

3. **(a)** a; subtract; $\{-5\}$ **(b)** $\{38\}$

What happens if we solve the equation in **Example 4** by first subtracting $\frac{8}{5}k$ from each side?

$$\frac{3}{5}k + 15 = \frac{8}{5}k \qquad \text{Original equation from Example 4}$$

$$\frac{3}{5}k + 15 - \frac{8}{5}k = \frac{8}{5}k - \frac{8}{5}k \qquad \text{Subtract } \tfrac{8}{5}k \text{ from each side.}$$

$$15 - k = 0 \qquad \tfrac{3}{5}k - \tfrac{8}{5}k = -\tfrac{5}{5}k = -1k = -k; \ \tfrac{8}{5}k - \tfrac{8}{5}k = 0$$

$$15 - k - 15 = 0 - 15 \qquad \text{Subtract 15 from each side.}$$

$$-k = -15 \qquad \text{Combine like terms; additive inverse}$$

This result gives the value of $-k$, but not of k itself. However, it does say that the additive inverse of k is -15, which means that k must be 15.

$$k = 15 \qquad \text{Same result as in Example 4}$$

(This result can also be justified using the multiplication property of equality, covered in **Section 2.2.**) We can make the following generalization.

If a is a number and $-x = a$, then $x = -a$.

Work Problem 4 at the Side. ▶

> **EXAMPLE 5**   Using the Addition Property of Equality Twice

Solve $8 - 6p = -7p + 5$.

We must get all terms with variables on the same side of the equation and all terms without variables on the other side of the equation.

$$8 - 6p = -7p + 5$$

$$8 - 6p + 7p = -7p + 5 + 7p \qquad \text{Add } 7p \text{ to each side.}$$

$$8 + p = 5 \qquad \text{Combine like terms.}$$

$$8 + p - 8 = 5 - 8 \qquad \text{Subtract 8 from each side.}$$

$$p = -3 \qquad \text{Combine like terms.}$$

CHECK   Substitute -3 for p in the original equation.

$$8 - 6p = -7p + 5 \qquad \text{Original equation}$$

$$8 - 6(-3) \stackrel{?}{=} -7(-3) + 5 \qquad \text{Let } p = -3.$$

$$8 + 18 \stackrel{?}{=} 21 + 5 \qquad \text{Multiply.}$$

Use parentheses when substituting to avoid errors.

$$26 = 26 \ \checkmark \qquad \text{True}$$

The check results in a true statement, so the solution set is $\{-3\}$.

····· Work Problem 5 at the Side. ▶

> **Note**
>
> ***There are often several equally correct ways to solve an equation.*** In **Example 5,** we could begin by adding $6p$, instead of $7p$, to each side. Combining like terms and subtracting 5 from each side gives $3 = -p$. (Try this.) If $3 = -p$, then $-3 = p$, and the variable has been isolated on the right side of equation. The same solution results.

---

4 (a) Solve $5m + 4 = 6m$.

(b) Solve $\dfrac{7}{2}m + 1 = \dfrac{9}{2}m$.

(c) What is the solution set of $-x = 6$?

(d) What is the solution set of $-x = -12$?

5 Solve.

GS (a) $\qquad 10 - a = -2a + 9$

$$10 - a + \underline{\quad} = -2a + 9 + 2a$$

$$10 + a = 9$$

$$10 + a - \underline{\quad} = 9 - \underline{\quad}$$

$$a = \underline{\quad}$$

The solution set is $\underline{\quad}$.

(b) $6x - 8 = 12 + 5x$

(c) $8t + 6 = 6 + 7t$

Answers

4. (a) $\{4\}$ (b) $\{1\}$ (c) $\{-6\}$ (d) $\{12\}$
5. (a) $2a$; 10; 10; -1; $\{-1\}$
     (b) $\{20\}$ (c) $\{0\}$

6 Solve.

(a) $4x + 6 + 2x - 3$
$= 9 + 5x - 4$

(b) $9r + 4r + 6 - 2$
$= 9r + 4 + 3r$

7 Solve.

(a) $4(r + 1) - (3r + 5) = 1$

GS (b) $2(5 + 2m) - 3(m - 4) = 29$
$2(5) + 2(2m) - \underline{\quad} m$
$- \underline{\quad}(\underline{\quad}) = 29$
$10 + 4m - \underline{\quad} + \underline{\quad} = 29$
Now complete the solution.

OBJECTIVE **3** Simplify, and then use the addition property of equality.

EXAMPLE 6 Combining Like Terms When Solving

Solve $3t - 12 + t + 2 = 5 + 3t + 2$.

$3t - 12 + t + 2 = 5 + 3t + 2$	Combine like terms on each side.
$4t - 10 = 7 + 3t$	
$4t - 10 - 3t = 7 + 3t - 3t$	Subtract $3t$ from each side.
$t - 10 = 7$	Combine like terms.
$t - 10 + 10 = 7 + 10$	Add 10 to each side.
$t = 17$	Combine like terms.

CHECK

$3t - 12 + t + 2 = 5 + 3t + 2$	Original equation
$3(17) - 12 + 17 + 2 \stackrel{?}{=} 5 + 3(17) + 2$	Let $t = 17$.
$51 - 12 + 17 + 2 \stackrel{?}{=} 5 + 51 + 2$	Multiply.
$58 = 58$ ✓	True

The check results in a true statement, so the solution set is $\{17\}$.

◀ Work Problem **6** at the Side.

CAUTION

The final line of the check does not give the solution to the equation. It gives a confirmation that the solution found is correct.

EXAMPLE 7 Using the Distributive Property When Solving

Solve $3(2 + 5x) - (1 + 14x) = 6$.

$3(2 + 5x) - (1 + 14x) = 6$ — Be sure to distribute to *all* terms within the parentheses.

$3(2 + 5x) - 1(1 + 14x) = 6$	$-(1 + 14x) = -1(1 + 14x)$
$3(2) + 3(5x) - 1(1) - 1(14x) = 6$	Distributive property
$6 + 15x - 1 - 14x = 6$	Multiply.
$x + 5 = 6$	Combine like terms.
$x + 5 - 5 = 6 - 5$	Subtract 5 from each side.
$x = 1$	Combine like terms.

Check by substituting 1 for x in the original equation. The solution set is $\{1\}$.

◀ Work Problem **7** at the Side.

CAUTION

Be careful to apply the distributive property correctly in a problem like that in **Example 7**, or a sign error may result.

2.1 Exercises

FOR EXTRA HELP  Download the MyDashBoard App   ▶   **MyMathLab®**

CONCEPT CHECK *In Exercises 1–4, fill in each blank with the correct response.*

1. An equation includes a(n) _____ symbol, while a(n) _____ does not.

2. A(n) _____ equation in one variable can be written in the form $Ax + B$ ____ C.

3. Equations that have exactly the same solution set are _____ equations.

4. The addition property of equality states that the same expression may be _____ or _____ each side of an equation without changing the solution set.

5. CONCEPT CHECK Substitute to determine which number is a solution of the equation $2x - 5 = 3x$.

   A. 0   **B.** 5   **C.** −5   **D.** −1

6. CONCEPT CHECK Which of the following are *not* linear equations in one variable?

   A. $x^2 - 5x + 6 = 0$ **B.** $x^3 = x$

   C. $3x - 4 = 0$ **D.** $7x - 6x = 3 + 9x$

7. CONCEPT CHECK Decide whether each is an *expression* or an *equation*. If it is an expression, simplify it. If it is an equation, solve it.

   (a) $5x + 8 - 4x + 7$ **(b)** $-6m + 12 + 7m - 5$

   (c) $5x + 8 - 4x = 7$ **(d)** $-6m + 12 + 7m = -5$

8. Explain how to check a solution of an equation.

Solve each equation, and check your solution. **See Examples 1–5.**

9. $x - 4 = 8$ **10.** $x - 8 = 9$ **11.** $x - 5 = -8$ **12.** $x - 7 = -9$

13. $r + 9 = 13$ **14.** $t + 6 = 10$ **15.** $x + 26 = 17$ **16.** $x + 45 = 24$

17. $x + \dfrac{1}{4} = -\dfrac{1}{2}$ **18.** $x + \dfrac{2}{3} = -\dfrac{1}{6}$ **19.** $x - 8.4 = -2.1$ **20.** $z - 15.5 = -5.1$

21. $t + 12.3 = -4.6$ **22.** $x + 21.5 = -13.4$ **23.** $7 + r = -3$ **24.** $8 + k = -4$

25. $2 = p + 15$ **26.** $3 = z + 17$ **27.** $-\dfrac{1}{3} = x - \dfrac{3}{5}$ **28.** $-\dfrac{1}{4} = x - \dfrac{2}{3}$

29. $3x = 2x + 7$ **30.** $5x = 4x + 9$ **31.** $8x - 3 = 9x$ **32.** $4x - 1 = 5x$

33. $10x + 4 = 9x$ **34.** $8t + 5 = 7t$ **35.** $\dfrac{9}{7}r - 3 = \dfrac{2}{7}r$ **36.** $\dfrac{8}{5}w - 6 = \dfrac{3}{5}w$

37. $5.6x + 2 = 4.6x$ **38.** $9.1x + 5 = 8.1x$ **39.** $3p + 6 = 10 + 2p$ **40.** $8x + 4 = -6 + 7x$

41. $5 - x = -2x - 11$ **42.** $3 - 8x = -9x - 1$ **43.** $-4z + 7 = -5z + 9$ **44.** $-6q + 3 = -7q + 10$

Solve each equation, and check your solution. **See Examples 6 and 7.**

45. $3x + 6 - 10 = 2x - 2$ **46.** $8k - 4 + 6 = 7k + 1$

47. $6x + 5 + 7x + 3 = 12x + 4$ **48.** $4x - 3 - 8x + 1 = -5x + 9$

49. $10x + 5x + 7 - 8 = 12x + 3 + 2x$ **50.** $7p + 4p + 13 - 7 = 7p + 9 + 3p$

51. $5.2q - 4.6 - 7.1q = -0.9q - 4.6$ **52.** $-4.0x + 2.7 - 1.6x = -4.6x + 2.7$

53. $\dfrac{5}{7}x + \dfrac{1}{3} = \dfrac{2}{5} - \dfrac{2}{7}x + \dfrac{2}{5}$ **54.** $\dfrac{6}{7}s - \dfrac{3}{4} = \dfrac{4}{5} - \dfrac{1}{7}s + \dfrac{1}{6}$

55. $(5x + 6) - (3 + 4x) = 10$ **56.** $(8r - 3) - (7r + 1) = -6$

57. $2(p + 5) - (9 + p) = -3$ **58.** $4(k - 6) - (3k + 2) = -5$

59. $-6(2x + 1) + (13x - 7) = 0$ **60.** $-5(3w - 3) + (16w + 1) = 0$

61. $10(-2x + 1) = -19(x + 1)$ **62.** $2(-3r + 2) = -5(r - 3)$

63. $-2(8p + 2) - 3(2 - 7p) = 2(4 + 2p)$ **64.** $4(3 - z) - 5(1 - 2z) = 7(3 + z)$

65. **CONCEPT CHECK** Write an equation that requires the use of the addition property of equality, in which 6 must be added to each side to solve the equation and the solution is a negative number.

66. **CONCEPT CHECK** Write an equation that requires the use of the addition property of equality, in which $\frac{1}{2}$ must be subtracted from each side and the solution is a positive number.

Study Skills
MANAGING YOUR TIME

Many college students juggle a difficult schedule and multiple responsibilities, including school, work, and family demands.

Time Management Tips

▶ **Read the syllabus for each class.** Understand class policies, such as attendance, late homework, and make-up tests. Find out how you are graded.

▶ **Make a semester or quarter calendar.** Put test dates and major due dates for *all* your classes on the *same* calendar. Try using a different color pen for each class.

▶ **Make a weekly schedule.** After you fill in your classes and other regular responsibilities, block off some study periods. Aim for 2 hours of study for each 1 hour in class.

▶ **Make "to-do" lists.** Number tasks in order of importance. Cross off tasks as you complete them.

▶ **Break big assignments into smaller chunks.** Make deadlines for each smaller chunk so that you stay on schedule.

▶ **Choose a regular study time and place** (such as the campus library). Routine helps.

▶ **Keep distractions to a minimum.** Get the most out of the time you have set aside for studying by limiting interruptions. Turn off your cell phone. Take a break from social media. Avoid studying in front of the TV.

▶ **Take breaks when studying.** Do not try to study for hours at a time. Take a 10-minute break each hour or so.

▶ **Ask for help when you need it.** Talk with your instructor during office hours. Make use of the learning center, tutoring center, counseling office, or other resources available at your school.

Now Try This

Think through and answer each question.

1 How many hours do you have available for studying this week?

_____

2 Which two or three of the above suggestions will you try this week to improve your time management?

_____

3 Once the week is over, evaluate how these suggestions worked. What will you do differently next week?

_____

2.2 The Multiplication Property of Equality

OBJECTIVES

1 Use the multiplication property of equality.

2 Simplify, and then use the multiplication property of equality.

1 Check that 5 is the solution
GS of $3x = 15$.

$$3x = 15$$
$$3(\_\_\_) \stackrel{?}{=} 15$$
$$\_\_\_ = 15 \quad (\textit{True / False})$$

The solution set is ___.

OBJECTIVE **1** **Use the multiplication property of equality.** The addition property of equality is not enough to solve some equations.

$$3x + 2 = 17$$
$$3x + 2 - 2 = 17 - 2 \quad \text{Subtract 2 from each side.}$$
$$3x = 15 \quad \text{Combine like terms.}$$

The coefficient of x here is **3**, not 1 as desired. The **multiplication property of equality** is needed to change $3x = 15$ to an equation of the form

$$x = \text{a number.}$$

Since $3x = 15$, both $3x$ and 15 must represent the same number. Multiplying both $3x$ and 15 by the same number will result in an equivalent equation.

> **Multiplication Property of Equality**
>
> If A, B, and C (where $C \neq 0$) represent real numbers, then the equations
>
> $$A = B \quad \text{and} \quad AC = BC$$
>
> are equivalent equations.
>   In words, we can multiply each side of an equation by the same nonzero number without changing the solution set.

In $3x = 15$, we must change $3x$ to $1x$, or x. To do this, we multiply each side of the equation by $\frac{1}{3}$, the reciprocal of 3, because $\frac{1}{3} \cdot 3 = \frac{3}{3} = 1$.

$$3x = 15$$

$$\frac{1}{3}(3x) = \frac{1}{3}(15) \quad \text{Multiply each side by } \frac{1}{3}.$$

$$\left(\frac{1}{3} \cdot 3\right)x = \frac{1}{3}(15) \quad \text{Associative property}$$

The product of a number and its reciprocal is 1.

$$1x = 5 \quad \text{Multiplicative inverse property}$$

$$x = 5 \quad \text{Multiplicative identity property}$$

The solution is 5. We can check this result in the original equation.

◀ **Work Problem** **1** **at the Side.**

Just as the addition property of equality permits *subtracting* the same number from each side of an equation, the multiplication property of equality permits *dividing* each side of an equation by the same nonzero number.

$$3x = 15$$

$$\frac{3x}{3} = \frac{15}{3} \quad \text{Divide each side by 3.}$$

$$x = 5 \quad \text{Same result as above}$$

> We can divide each side of an equation by the same nonzero number without changing the solution. ***Do not, however, divide each side by a variable, since the variable might be equal to 0.***

Answer

1. 5; 15; True; {5}

Note

In practice, it is usually easier to multiply on each side if the coefficient of the variable is a fraction, and divide on each side if the coefficient is an integer or a decimal. For example, to solve

$$\frac{3}{4}x = 12, \quad \text{it is easier to multiply by } \frac{4}{3} \text{ than to divide by } \frac{3}{4}.$$

On the other hand, to solve

$$5x = 20, \quad \text{it is easier to divide by 5 than to multiply by } \frac{1}{5}.$$

EXAMPLE 1 Dividing Each Side of an Equation by a Nonzero Number

Solve $5x = 60$.

$$5x = 60 \quad \boxed{\text{Our goal is to isolate } x.}$$

$$\frac{5x}{5} = \frac{60}{5} \qquad \text{Divide each side by 5, the coefficient of } x.$$

$\boxed{\text{Dividing by 5 is the same as multiplying by } \frac{1}{5}.}$

$$x = 12 \qquad \frac{5x}{5} = \frac{5}{5}x = 1x = x$$

CHECK Substitute 12 for x in the original equation.

$$5x = 60 \qquad \text{Original equation}$$

$$5(12) \overset{?}{=} 60 \qquad \text{Let } x = 12.$$

$\boxed{\text{60 is } not \text{ the solution.}}$

$$60 = 60 \checkmark \qquad \text{True}$$

Since a true statement results, the solution set is $\{12\}$.

·········· **Work Problem ② at the Side.** ▶

EXAMPLE 2 Using the Multiplication Property of Equality

Solve $-25p = 30$.

$$-25p = 30$$

$$\frac{-25p}{-25} = \frac{30}{-25} \qquad \text{Divide each side by } -25, \text{ the coefficient of } p.$$

$$p = -\frac{30}{25} \qquad \frac{a}{-b} = -\frac{a}{b}$$

$$p = -\frac{6}{5} \qquad \text{Write in lowest terms.}$$

CHECK $\qquad -25p = 30 \qquad \text{Original equation}$

$$\frac{-25}{1}\left(-\frac{6}{5}\right) \overset{?}{=} 30 \qquad \text{Let } p = -\frac{6}{5}.$$

$$30 = 30 \checkmark \qquad \text{True}$$

The check confirms that the solution set is $\left\{-\frac{6}{5}\right\}$.

·········· **Work Problem ③ at the Side.** ▶

---

② Solve.

(a) Fill in the blanks.

$$7x = 91$$

$$\frac{7x}{\underline{\quad}} = \frac{91}{\underline{\quad}}$$

$$x = \underline{\quad}$$

CHECK $\qquad 7x = 91$

$$7(\underline{\quad}) \overset{?}{=} 91$$

$$\underline{\quad} = 91 \quad (True / False)$$

The solution set is $\underline{\quad}$.

(b) $3r = -12$

(c) $15x = 75$

---

③ Solve.

(a) $2m = 15$

(b) $-6x = 14$

(c) $10z = -45$

Answers

2. **(a)** 7; 7; 13; 13; 91; True; $\{13\}$
  (b) $\{-4\}$ **(c)** $\{5\}$

3. **(a)** $\left\{\frac{15}{2}\right\}$ **(b)** $\left\{-\frac{7}{3}\right\}$ **(c)** $\left\{-\frac{9}{2}\right\}$

4 Solve.

(a) $-0.7m = -5.04$

(b) $-63.75 = 12.5k$

EXAMPLE 3   Using the Multiplication Property of Equality

Solve $6.09 = 2.1x$.

$$6.09 = 2.1x \quad \boxed{\text{Isolate } x \text{ on the right.}}$$

$$\frac{6.09}{2.1} = \frac{2.1x}{2.1} \qquad \text{Divide each side by 2.1.}$$

$$2.9 = x, \quad \text{or} \quad x = 2.9 \qquad \text{Use a calculator if necessary.}$$

Check to confirm that the solution set is $\{2.9\}$.

◀ **Work Problem 4 at the Side.**

---

5 Solve.

(a) $\dfrac{x}{5} = 5$

What is the coefficient of x here? ____ By what number should we multiply each side? ____ Now solve.

(b) $\dfrac{p}{4} = -6$

EXAMPLE 4   Using the Multiplication Property of Equality

Solve $\frac{x}{4} = 3$.

$$\frac{x}{4} = 3$$

$$\frac{1}{4}x = 3 \qquad \frac{x}{4} = \frac{1x}{4} = \frac{1}{4}x$$

$$4 \cdot \frac{1}{4}x = 4 \cdot 3 \qquad \begin{array}{l}\text{Multiply each side by 4, the}\\ \text{reciprocal of } \frac{1}{4}.\end{array}$$

$$\boxed{4 \cdot \frac{1}{4}x = 1x = x} \; x = 12 \qquad \begin{array}{l}\text{Multiplicative inverse property;}\\ \text{multiplicative identity property}\end{array}$$

CHECK

$$\frac{x}{4} = 3 \qquad \text{Original equation}$$

$$\frac{12}{4} \stackrel{?}{=} 3 \qquad \text{Let } x = 12.$$

$$3 = 3 \; \checkmark \qquad \text{True}$$

Since a true statement results, the solution set is $\{12\}$.

◀ **Work Problem 5 at the Side.**

---

6 Solve.

(a) $-\dfrac{5}{6}t = -15$

By what number should we multiply each side? ____ Now solve.

(b) $\dfrac{3}{5}k = -21$

EXAMPLE 5   Using the Multiplication Property of Equality

Solve $\frac{3}{4}h = 6$.

$$\frac{3}{4}h = 6$$

$$\frac{4}{3} \cdot \frac{3}{4}h = \frac{4}{3} \cdot 6 \qquad \begin{array}{l}\text{Multiply each side by } \frac{4}{3},\\ \text{the reciprocal of } \frac{3}{4}.\end{array}$$

$$1 \cdot h = \frac{4}{3} \cdot \frac{6}{1} \qquad \text{Multiplicative inverse property}$$

$$h = 8 \qquad \begin{array}{l}\text{Multiplicative identity property;}\\ \text{multiply fractions.}\end{array}$$

Check to confirm that the solution set is $\{8\}$.

◀ **Work Problem 6 at the Side.**

Answers

4. (a) $\{7.2\}$ (b) $\{-5.1\}$

5. (a) $\frac{1}{5}$; 5; $\{25\}$ (b) $\{-24\}$

6. (a) $-\frac{6}{5}$; $\{18\}$ (b) $\{-35\}$

In **Section 2.1,** we obtained the equation $-k = -15$ in our alternative solution to **Example 4.** We reasoned that since this equation says that the additive inverse (or opposite) of k is -15, then k must equal 15. We can also use the multiplication property of equality to obtain the same result.

EXAMPLE 6 Using the Multiplication Property of Equality When the Coefficient of the Variable Is -1

Solve $-k = -15$.

$$-k = -15$$

$$-1 \cdot k = -15 \qquad -k = -1 \cdot k$$

$$-1(-1 \cdot k) = -1(-15) \qquad \text{Multiply each side by } -1.$$

$$[-1(-1)] \cdot k = 15 \qquad \text{Associative property; multiply.}$$

$$1 \cdot k = 15 \qquad \text{Multiplicative inverse property}$$

$$k = 15 \qquad \text{Multiplicative identity property}$$

CHECK $\qquad -k = -15 \qquad$ Original equation

$$-(15) \overset{?}{=} -15 \qquad \text{Let } k = 15.$$

$$-15 = -15 \checkmark \qquad \text{True}$$

The solution, 15, checks, so the solution set is $\{15\}$.

·········· **Work Problem 7 at the Side.** ▶

OBJECTIVE ▶ **2** **Simplify, and then use the multiplication property of equality.**

EXAMPLE 7 Combining Like Terms When Solving

Solve $5m + 6m = 33$.

$$5m + 6m = 33$$

$$11m = 33 \qquad \text{Combine like terms.}$$

$$\frac{11m}{11} = \frac{33}{11} \qquad \text{Divide each side by 11.}$$

$$m = 3 \qquad \text{Multiplicative identity property}$$

CHECK $\qquad 5m + 6m = 33 \qquad$ Original equation

$$5(3) + 6(3) \overset{?}{=} 33 \qquad \text{Let } m = 3.$$

$$15 + 18 \overset{?}{=} 33 \qquad \text{Multiply.}$$

$$33 = 33 \checkmark \qquad \text{True}$$

Since a true statement results, the solution set is $\{3\}$.

·········· **Work Problem 8 at the Side.** ▶

---

7 Solve.

(a) $-m = 2$

(b) $-p = -7$

8 Solve.

(a) $7m - 5m = -12$

(b) $4r - 9r = 20$

Answers

7. (a) $\{-2\}$ (b) $\{7\}$
8. (a) $\{-6\}$ (b) $\{-4\}$

2.2 Exercises

 FOR EXTRA HELP   Download the MyDashBoard App   ▶ MyMathLab®

1. CONCEPT CHECK Tell whether you would use the addition or multiplication property of equality to solve each equation. *Do not actually solve.*

(a) $3x = 12$ (b) $3 + x = 12$

(c) $-x = 4$ (d) $-12 = 6 + x$

2. CONCEPT CHECK Which equation does *not* require the use of the multiplication property of equality?

A. $3x - 5x = 6$ **B.** $-\dfrac{1}{4}x = 12$

C. $5x - 4x = 7$ **D.** $\dfrac{x}{3} = -2$

CONCEPT CHECK *By what number is it necessary to multiply each side of each equation in order to isolate x on the left side? Do not actually solve.*

3. $\dfrac{2}{3}x = 8$

4. $\dfrac{4}{5}x = 6$

5. $\dfrac{x}{10} = 3$

6. $\dfrac{x}{100} = 8$

7. $-\dfrac{9}{2}x = -4$

8. $-\dfrac{8}{3}x = -11$

9. $-x = 0.36$

10. $-x = 0.29$

CONCEPT CHECK *By what number is it necessary to divide each side of each equation in order to isolate x on the left side? Do not actually solve.*

11. $6x = 5$

12. $7x = 10$

13. $-4x = 13$

14. $-13x = 6$

15. $0.12x = 48$

16. $0.21x = 63$

17. $-x = 23$

18. $-x = 49$

19. CONCEPT CHECK In the solution of a linear equation, the next-to-the-last step reads "$-x = -\dfrac{3}{4}$." Which of the following would be the solution of this equation?

A. $-\dfrac{3}{4}$ **B.** $\dfrac{3}{4}$ **C.** -1 **D.** $\dfrac{4}{3}$

20. CONCEPT CHECK Which of the following is the solution of the equation

$$-x = -24?$$

A. 24   **B.** -24 **C.** 1   **D.** -1

Solve each equation, and check your solution. **See Examples 1–7.**

21. $5x = 30$

22. $7x = 56$

23. $2m = 15$

24. $3m = 10$

25. $3a = -15$

26. $5k = -70$

27. $10t = -36$

28. $4s = -34$

29. $-6x = -72$

30. $-8x = -64$

31. $2r = 0$

$$\dfrac{2r}{\rule{1cm}{0.4pt}} = \dfrac{0}{\rule{1cm}{0.4pt}}$$

$r = \rule{1cm}{0.4pt}$

Solution set: $\rule{1cm}{0.4pt}$

32. $5x = 0$

$$\dfrac{5x}{\rule{1cm}{0.4pt}} = \dfrac{0}{\rule{1cm}{0.4pt}}$$

$x = \rule{1cm}{0.4pt}$

Solution set: $\rule{1cm}{0.4pt}$

33. $-x = 12$

34. $-t = 14$

35. $0.2t = 8$

36. $0.9x = 18$

37. $-2.1m = 25.62$

38. $-3.9a = 31.2$

39. $\frac{1}{4}x = -12$

40. $\frac{1}{5}p = -3$

41. $\frac{z}{6} = 12$

42. $\frac{x}{5} = 15$

43. $\frac{x}{7} = -5$

44. $\frac{k}{8} = -3$

45. $\frac{2}{7}p = 4$

46. $\frac{3}{8}x = 9$

47. $-\frac{7}{9}c = \frac{3}{5}$

48. $-\frac{5}{6}d = \frac{4}{9}$

49. $4x + 3x = 21$

50. $9x + 2x = 121$

51. $3r - 5r = 10$

52. $9p - 13p = 24$

53. $\frac{2}{5}x - \frac{3}{10}x = 2$

54. $\frac{2}{3}x - \frac{5}{9}x = 4$

55. $5m + 6m - 2m = 63$

56. $11r - 5r + 6r = 168$

57. $x + x - 3x = 12$

(*Hint:* What is the coefficient of each x? ____)

58. $z - 3z + z = -16$

(*Hint:* What is the coefficient of each z? ____)

59. $-6x + 4x - 7x = 0$

60. $-5x + 4x - 8x = 0$

61. $0.9w - 0.5w + 0.1w = -3$

62. $0.5x - 0.6x + 0.3x = -1$

63. CONCEPT CHECK Write an equation that requires the use of the multiplication property of equality, where each side must be multiplied by $\frac{2}{3}$ and the solution is a negative number.

64. CONCEPT CHECK Write an equation that requires the use of the multiplication property of equality, where each side must be divided by 100 and the solution is not an integer.

Write an equation using the information given in the problem. Use x as the variable. Then solve the equation.

65. When a number is multiplied by -4, the result is 10. Find the number.

66. When a number is multiplied by 4, the result is 6. Find the number.

67. When a number is divided by -5, the result is 2. Find the number.

68. If twice a number is divided by 5, the result is 4. Find the number.

Study Skills
ANALYZING YOUR TEST RESULTS

An exam is a learning opportunity—learn from your mistakes. After a test is returned, do the following:

▶ **Note what you got wrong and why you had points deducted.**

▶ **Figure out how to solve the problems you missed.** Check your textbook or notes, or ask your instructor. Rework the problems correctly.

▶ **Keep all quizzes and tests that are returned to you.** Use them to study for future tests and the final exam.

Typical Reasons for Errors on Math Tests

These are test taking errors. They are easy to correct if you read carefully, show all your work, proofread, and double-check units and labels.

1. You read the directions wrong.
2. You read the question wrong or skipped over something.
3. You made a computation error.
4. You made a careless error. (For example, you incorrectly copied a correct answer onto a separate answer sheet.)
5. Your answer is not complete.
6. You labeled your answer wrong. (For example, you labeled an answer "ft" instead of "ft^2".)
7. You didn't show your work.

These are test preparation errors. You must practice the kinds of problems that you will see on tests.

8. You didn't understand a concept.
9. You were unable to set up the problem (in an application).
10. You were unable to apply a procedure.

Below are sample charts for tracking your test taking progress. Use them to find out if you tend to make certain kinds of errors on tests. Check the appropriate box when you've made an error in a particular category.

Test Taking Errors

Test	Read directions wrong	Read question wrong	Computation error	Not exact or accurate	Not complete	Labeled wrong	Didn't show work
1							
2							
3							

Test Preparation Errors

Test	Didn't understand concept	Didn't set up problem correctly	Couldn't apply concept to new situation
1			
2			
3			

What will you do to avoid these kinds of errors on your next test?

2.3 More on Solving Linear Equations

In this section, we solve linear equations using *both* properties of equality introduced in **Sections 2.1 and 2.2.**

Work Problem **1** at the Side. ▶

OBJECTIVE ▶ **1** **Learn and use the four steps for solving a linear equation.** We use the following method.

Solving a Linear Equation

Step 1 **Simplify each side separately.** Clear (eliminate) parentheses, fractions, and decimals, using the distributive property as needed, and combine like terms.

Step 2 **Isolate the variable term on one side.** Use the addition property so that the variable term is on one side of the equation and a number is on the other.

Step 3 **Isolate the variable.** Use the multiplication property to get the equation in the form $x = $ a number, or a number $= x$. (Other letters may be used for the variable.)

Step 4 **Check.** Substitute the proposed solution into the original equation to see if a true statement results. If not, rework the problem.

Remember that when we solve an equation, our primary goal is to isolate the variable on one side of the equation.

EXAMPLE 1 **Using Both Properties of Equality to Solve a Linear Equation**

Solve $-6x + 5 = 17$.

Step 1 There are no parentheses, fractions, or decimals in this equation, so this step is not necessary.

> Our goal is to isolate x. $\quad -6x + 5 = 17$

Step 2 $\quad -6x + 5 - 5 = 17 - 5 \quad$ Subtract 5 from each side.

$\quad\quad\quad -6x = 12 \quad$ Combine like terms.

Step 3 $\quad\quad \dfrac{-6x}{-6} = \dfrac{12}{-6} \quad$ Divide each side by -6.

$\quad\quad\quad x = -2$

Step 4 Check by substituting -2 for x in the original equation.

CHECK $\quad\quad -6x + 5 = 17 \quad$ Original equation

$\quad\quad -6(-2) + 5 \overset{?}{=} 17 \quad$ Let $x = -2$.

$\quad\quad 12 + 5 \overset{?}{=} 17 \quad$ Multiply.

> 17 is *not* the solution. $\quad 17 = 17 ✓ \quad$ True

The solution, -2, checks, so the solution set is $\{-2\}$.

Work Problem **2** at the Side. ▶

OBJECTIVES

1 Learn and use the four steps for solving a linear equation.

2 Solve equations that have no solution or infinitely many solutions.

3 Solve equations with fractions or decimals as coefficients.

4 Write expressions for two related unknown quantities.

1 As a review, tell whether you would use the addition or multiplication property of equality to solve each equation. *Do not actually solve.*

(a) $7 + x = -9$

(b) $-13x = 26$

(c) $-x = \dfrac{3}{4}$

(d) $-12 = x - 4$

2 Solve.

(a) $-5p + 4 = 19$

(b) $7 + 2m = -3$

Answers

1. (a) and **(d):** addition property of equality
(b) and **(c):** multiplication property of equality

2. (a) $\{-3\}$ **(b)** $\{-5\}$

3 Solve.

(a) $5 - 8k = 2k - 5$

Begin with Step 2.

$5 - 8k - 2k = 2k - 5 - \underline{\quad}$

$5 - \underline{\quad} = -5$

$5 - 10k - \underline{\quad} = -5 - \underline{\quad}$

$-10k = -10$

$\dfrac{-10k}{\underline{\quad}} = \dfrac{-10}{\underline{\quad}}$

$k = \underline{\quad}$

Check by substituting $\underline{\quad}$ for k in the original equation.

The solution set is $\underline{\quad}$.

(b) $2q + 3 = 4q - 9$

(c) $6x + 7 = -8 + 3x$

EXAMPLE 2 **Using Both Properties of Equality to Solve a Linear Equation**

Solve $3x + 2 = 5x - 8$.

Step 1 There are no parentheses, fractions, or decimals in the equation, so we begin with Step 2.

$3x + 2 = 5x - 8$ — Our goal is to isolate x.

Step 2 $3x + 2 - 5x = 5x - 8 - 5x$ Subtract $5x$ from each side.

$-2x + 2 = -8$ Combine like terms.

$-2x + 2 - 2 = -8 - 2$ Subtract 2 from each side.

$-2x = -10$ Combine like terms.

Step 3 $\dfrac{-2x}{-2} = \dfrac{-10}{-2}$ Divide each side by -2.

$x = 5$

Step 4 Check by substituting 5 for x in the original equation.

CHECK $3x + 2 = 5x - 8$ Original equation

$3(5) + 2 \overset{?}{=} 5(5) - 8$ Let $x = 5$.

$15 + 2 \overset{?}{=} 25 - 8$ Multiply.

$17 = 17 ✓$ True

The solution, 5, checks, so the solution set is $\{5\}$.

Note

Remember that a variable can be isolated on either side of an equation. In **Example 2**, x will be isolated on the right if we begin by subtracting $3x$, instead of $5x$, from each side of the equation.

$3x + 2 = 5x - 8$ Equation from **Example 2**

$3x + 2 - 3x = 5x - 8 - 3x$ Subtract $3x$ from each side.

$2 = 2x - 8$ Combine like terms.

$2 + 8 = 2x - 8 + 8$ Add 8 to each side.

$10 = 2x$ Combine like terms.

$\dfrac{10}{2} = \dfrac{2x}{2}$ Divide each side by 2.

$5 = x$ The same solution results.

There are often several equally correct ways to solve an equation.

◀ **Work Problem** **3** at the Side.

Answers

3. **(a)** $2k$; $10k$; 5; 5; -10; -10; 1; 1; $\{1\}$
    (b) $\{6\}$ **(c)** $\{-5\}$

EXAMPLE 3 **Solving a Linear Equation**

Solve $4(k - 3) - k = k - 6$.

Step 1 Clear the parentheses using the distributive property.

$$4(k - 3) - k = k - 6$$

$4(k) + 4(-3) - k = k - 6$	Distributive property
$4k - 12 - k = k - 6$	Multiply.
$3k - 12 = k - 6$	Combine like terms.

Step 2

$3k - 12 - k = k - 6 - k$	Subtract k.
$2k - 12 = -6$	Combine like terms.
$2k - 12 + 12 = -6 + 12$	Add 12.
$2k = 6$	Combine like terms.

Step 3

$\dfrac{2k}{2} = \dfrac{6}{2}$	Divide by 2.
$k = 3$	

Step 4 Check by substituting 3 for k in the original equation.

CHECK

$4(k - 3) - k = k - 6$	Original equation
$4(3 - 3) - 3 \overset{?}{=} 3 - 6$	Let $k = 3$.
$4(0) - 3 \overset{?}{=} 3 - 6$	Work inside the parentheses.
$-3 = -3 \checkmark$	True

The solution, 3, checks, so the solution set is $\{3\}$.

· **Work Problem ④ at the Side. ▶**

EXAMPLE 4 **Solving a Linear Equation**

Solve $8a - (3 + 2a) = 3a + 1$.

Step 1 $8a - (3 + 2a) = 3a + 1$

$8a - 1(3 + 2a) = 3a + 1$	Multiplicative identity property
$8a - 3 - 2a = 3a + 1$	Distributive property
$6a - 3 = 3a + 1$	Combine like terms.

> Be careful with signs.

Step 2

$6a - 3 - 3a = 3a + 1 - 3a$	Subtract $3a$.
$3a - 3 = 1$	Combine like terms.
$3a - 3 + 3 = 1 + 3$	Add 3.
$3a = 4$	Combine like terms.

Step 3

$\dfrac{3a}{3} = \dfrac{4}{3}$	Divide by 3.
$a = \dfrac{4}{3}$	

Step 4 Check that the solution set is $\left\{\dfrac{4}{3}\right\}$.

④ Solve.

(GS) (a) $7(p - 2) + p = 2p + 4$

Step 1 ____ the parentheses.

$$\_\_\_(p) + 7(\_\_\_) + p = 2p + 4$$

$$\_\_\_ - \_\_\_ + p = 2p + 4$$

$$\_\_\_ - 14 = 2p + 4$$

Now complete the solution.
Give the solution set.

(b) $11 + 3(x + 1) = 5x + 16$

Answers

4. (a) Clear; 7; -2; $7p$; 14; $8p$;
The solution set is $\{3\}$.
(b) $\{-1\}$

5 Solve.

(a) $7m - (2m - 9) = 39$

(b) $5x - (x + 9) = x - 4$

6 Solve.

(a) $2(4 + 3r) = 3(r + 1) + 11$

(b) $2 - 3(2 + 6z)$
  $= 4(z + 1) - 8$

> **CAUTION**
>
> In an expression such as $8a - (3 + 2a)$ in **Example 4,** the $-$ sign acts like a factor of -1 and affects the sign of *every* term within the parentheses.
>
> $$8a - (3 + 2a) \longleftarrow \text{Left side of the equation in \textbf{Example 4}}$$
> $$= 8a - \mathbf{1}(3 + 2a)$$
> $$= 8a + (-\mathbf{1})(3 + 2a)$$
> $$= 8a - 3 - 2a$$
>
> Change to $-$ in *both* terms.

◀ **Work Problem 5 at the Side.**

EXAMPLE 5   **Solving a Linear Equation**

Solve $4(4 - 3x) = 32 - 8(x + 2)$. [Do *not* subtract 8 from 32 here.]

Step 1 $\quad 4(4 - 3x) = 32 - 8(x + 2)$ [Be careful with signs.]

$16 - 12x = 32 - 8x - 16$ Distributive property

$16 - 12x = 16 - 8x$ Combine like terms.

Step 2 $\quad 16 - 12x + \mathbf{8x} = 16 - 8x + \mathbf{8x}$ Add $8x$.

$16 - 4x = 16$ Combine like terms.

$16 - 4x - \mathbf{16} = 16 - \mathbf{16}$ Subtract 16.

$-4x = 0$ Combine like terms.

Step 3 $\quad \dfrac{-4x}{-4} = \dfrac{0}{-4}$ Divide by -4.

$x = 0$

Step 4 CHECK $\quad 4(4 - 3x) = 32 - 8(x + 2)$ Original equation

$4[4 - 3(0)] \stackrel{?}{=} 32 - 8(0 + 2)$ Let $x = 0$.

$4(4 - 0) \stackrel{?}{=} 32 - 8(2)$ Work inside the brackets and parentheses.

$4(4) \stackrel{?}{=} 32 - 16$ Subtract and multiply.

$16 = 16$ ✓   True

Since the solution 0 checks, the solution set is $\{0\}$.

◀ **Work Problem 6 at the Side.**

OBJECTIVE 2 Solve equations that have no solution or infinitely many solutions. Each equation so far has had exactly one solution. An equation with exactly one solution is a **conditional equation** because it is only true under certain conditions. Some equations have no solution or infinitely many solutions.

EXAMPLE 6 Solving an Equation That Has Infinitely Many Solutions

Solve $5x - 15 = 5(x - 3)$.

$$5x - 15 = 5(x - 3)$$

$$5x - 15 = 5x - 15 \qquad \text{Distributive property}$$

$$5x - 15 - 5x = 5x - 15 - 5x \qquad \text{Subtract } 5x.$$

> Notice that the variable "disappeared."

$$-15 = -15 \qquad \text{Combine like terms.}$$

$$-15 + 15 = -15 + 15 \qquad \text{Add 15.}$$

$$0 = 0 \qquad \text{True}$$

Solution set: **{all real numbers}**

Since the last statement $(0 = 0)$ is true, *any* real number is a solution. We could have predicted this from the second line in the solution,

$$5x - 15 = 5x - 15 \; \longleftarrow \; \text{This is true for } any \text{ value of } x.$$

Try several values for x in the original equation to see that they all satisfy it.

An equation with both sides exactly the same, like $0 = 0$, is an **identity.** An identity is true for all replacements of the variables. As shown above, we write the solution set as **{all real numbers}**.

CAUTION

In **Example 6,** do not write $\{0\}$ as the solution set of the equation. While 0 is a solution, there are infinitely many other solutions. *For* $\{0\}$ *to be the solution set, the last line must include a variable, such as x, and read x = 0 (as in Example 5), not 0 = 0.*

EXAMPLE 7 Solving an Equation That Has No Solution

Solve $2x + 3(x + 1) = 5x + 4$.

$$2x + 3(x + 1) = 5x + 4$$

$$2x + 3x + 3 = 5x + 4 \qquad \text{Distributive property}$$

$$5x + 3 = 5x + 4 \qquad \text{Combine like terms.}$$

$$5x + 3 - 5x = 5x + 4 - 5x \qquad \text{Subtract } 5x.$$

> Again, the variable "disappeared."

$$3 = 4 \qquad \text{False}$$

There is no solution. **Solution set:** $\varnothing$

A false statement $(3 = 4)$ results. A **contradiction** is an equation that has no solution. Its solution set is the **empty set,** or **null set,** symbolized $\varnothing$.

··· Work Problem **7** at the Side. ▶

CAUTION

Do not write $\{\varnothing\}$ to represent the empty set.

OBJECTIVE ▶ **3** Solve equations with fractions or decimals as coefficients. To avoid messy computations, we clear an equation of fractions by multiplying each side by the least common denominator (LCD) of all the fractions in the equation. Doing this will give an equation with only *integer* coefficients.

7 Solve.

(a) $2(x - 6) = 2x - 12$

(b) $3x + 6(x + 1) = 9x - 4$

(c) $-4x + 12 = 3 - 4(x - 3)$

Answers

7. (a) {all real numbers}   **(b)** $\varnothing$ **(c)** $\varnothing$

8 Solve.

(a) $\dfrac{1}{4}x - 4 = \dfrac{3}{2}x + \dfrac{3}{4}x$

(b) $\dfrac{1}{2}x + \dfrac{5}{8}x = \dfrac{3}{4}x - 6$

EXAMPLE 8 Solving an Equation with Fractions as Coefficients

Solve $\dfrac{2}{3}x - \dfrac{1}{2}x = -\dfrac{1}{6}x - 2$.

Step 1

$$\dfrac{2}{3}x - \dfrac{1}{2}x = -\dfrac{1}{6}x - 2$$

> Pay particular attention here.

$$6\left(\dfrac{2}{3}x - \dfrac{1}{2}x\right) = 6\left(-\dfrac{1}{6}x - 2\right)$$ Multiply each side by 6, the LCD.

$$6\left(\dfrac{2}{3}x\right) + 6\left(-\dfrac{1}{2}x\right) = 6\left(-\dfrac{1}{6}x\right) + 6(-2)$$ Distributive property; multiply *each* term inside the parentheses by 6.

> The fractions have been cleared.

$$4x - 3x = -x - 12$$ Multiply.

$$x = -x - 12$$ Combine like terms.

Step 2 $$x + x = -x - 12 + x$$ Add x.

$$2x = -12$$ Combine like terms.

Step 3 $$\dfrac{2x}{2} = \dfrac{-12}{2}$$ Divide by 2.

$$x = -6$$

Step 4   **CHECK** $$\dfrac{2}{3}x - \dfrac{1}{2}x = -\dfrac{1}{6}x - 2$$ Original equation

$$\dfrac{2}{3}(-6) - \dfrac{1}{2}(-6) \overset{?}{=} -\dfrac{1}{6}(-6) - 2$$ Let $x = -6$.

$$-4 + 3 \overset{?}{=} 1 - 2$$ Multiply.

$$-1 = -1 \checkmark$$ True

The solution, -6, checks. The solution set is $\{-6\}$.

◀ **Work Problem 8 at the Side.**

> **CAUTION**
>
> *When clearing an equation of fractions, be sure to multiply every term on each side of the equation by the LCD.*

EXAMPLE 9 Solving an Equation with Fractions as Coefficients

Solve $\dfrac{1}{3}(x + 5) - \dfrac{3}{5}(x + 2) = 1$.

Step 1

$$\dfrac{1}{3}(x + 5) - \dfrac{3}{5}(x + 2) = 1$$

$$15\left[\dfrac{1}{3}(x + 5) - \dfrac{3}{5}(x + 2)\right] = 15(1)$$ To clear the fractions, multiply by 15, the LCD.

$$15\left[\dfrac{1}{3}(x + 5)\right] + 15\left[-\dfrac{3}{5}(x + 2)\right] = 15(1)$$ Distributive property

$$5(x + 5) - 9(x + 2) = 15$$ Multiply.

> $15\left[\frac{1}{3}(x+5)\right]$
> $= 15 \cdot \frac{1}{3} \cdot (x+5)$
> $= 5(x+5)$

$$5x + 25 - 9x - 18 = 15$$ Distributive property

$$-4x + 7 = 15$$ Combine like terms.

Answers

8. (a) $\{-2\}$ (b) $\{-16\}$

········ **Continued on Next Page**

Step 2 $-4x + 7 - 7 = 15 - 7$ Subtract 7.

                           $-4x = 8$ Combine like terms.

Step 3 $\dfrac{-4x}{-4} = \dfrac{8}{-4}$ Divide by -4.

                            $x = -2$

Step 4 Check to confirm that $\{-2\}$ is the solution set.

·· **Work Problem** ❾ **at the Side.** ▶

⑨ Solve.

$$\frac{1}{4}(x + 3) - \frac{2}{3}(x + 1) = -2$$

> **CAUTION**
>
> Be sure you understand how to multiply by the LCD to clear an equation of fractions. *Study Step 1 in Examples 8 and 9 carefully.*

EXAMPLE 10   **Solving an Equation with Decimals as Coefficients**

Solve $0.1t + 0.05(20 - t) = 0.09(20)$.

Step 1   The decimals here are expressed as tenths (0.1) and hundredths $(0.05$ and $0.09)$. We choose the least exponent on 10 needed to eliminate the decimal points. In this equation, we use $10^2 = 100$.

$$\mathbf{0.1}t + 0.05(20 - t) = 0.09(20)$$

$$\mathbf{0.10}t + 0.05(20 - t) = 0.09(20) \qquad 0.1 = 0.10$$

$$\mathbf{100}[0.10t + 0.05(20 - t)] = \mathbf{100}[0.09(20)] \qquad \text{Multiply by 100.}$$

$$\mathbf{100}(0.10t) + \mathbf{100}[0.05(20 - t)] = \mathbf{100}[0.09(20)] \qquad \text{Distributive property}$$

$$10t + 5(20 - t) = 9(20) \qquad \text{Multiply.}$$

$$10t + 5(20) + 5(-t) = 180 \qquad \text{Distributive property}$$

$$10t + 100 - 5t = 180 \qquad \text{Multiply.}$$

$$5t + 100 = 180 \qquad \text{Combine like terms.}$$

Step 2 $5t + 100 - \mathbf{100} = 180 - \mathbf{100}$ Subtract 100.

                     $5t = 80$ Combine like terms.

Step 3 $\dfrac{5t}{5} = \dfrac{80}{5}$ Divide by 5.

                      $t = 16$

Step 4 Check to confirm that $\{16\}$ is the solution set.

··· **Work Problem** ❿ **at the Side.** ▶

⑩ Solve.

$$0.06(100 - x) + 0.04x$$
$$= 0.05(92)$$

> **Note**
>
> In **Example 10,** multiplying by 100 is accomplished by moving the decimal point two places to the right.
>
> $$0.\underset{\llcorner\uparrow}{1}0t + 0.\underset{\llcorner\uparrow}{0}5(20 - t) = 0.\underset{\llcorner\uparrow}{0}9(20)$$
>
> $$10t + 5(20 - t) = 9(20) \qquad \text{Multiply by 100.}$$

11 Perform each translation.

(a) Two numbers have a sum of 36. One of the numbers is represented by r. Write an expression for the other number.

(b) Two numbers have a product of 18. One of the numbers is represented by x. Write an expression for the other number.

The table summarizes the solution sets of the equations in this section.

Type of Equation	Final Equation in Solution	Number of Solutions	Solution Set
Conditional (See Examples 1–5, 8–10.)	$x =$ a number	One	$\{$a number$\}$
Identity (See Example 6.)	A true statement with no variable, such as $0 = 0$	Infinite	$\{$all real numbers$\}$
Contradiction (See Example 7.)	A false statement with no variable, such as $3 = 4$	None	$\varnothing$

OBJECTIVE ▶ **4** **Write expressions for two related unknown quantities.**

Problem-Solving Hint

When we solve applied problems, we must often write *expressions* to relate unknown quantities. We then use these expressions to write the *equation* needed to solve the application. The next example provides preparation for doing this in **Section 2.4.**

EXAMPLE 11   Translating Phrases into Algebraic Expressions

Perform each translation.

(a) Two numbers have a sum of 23. If one of the numbers is represented by x, find an expression for the other number.

First, suppose that the sum of two numbers is 23, and one of the numbers is 10. To find the other number, we would subtract 10 from 23.

$$23 - 10 \leftarrow \text{This gives 13 as the other number.}$$

Instead of using 10 as one of the numbers, we use x. The other number would be obtained in the same way—by subtracting x from 23.

$$23 - x \quad \boxed{x - 23 \text{ is } not \text{ correct. Subtraction is } not \text{ commutative.}}$$

To check, we find the sum of the two numbers.

$$x + (23 - x) = 23, \quad \text{as required.}$$

(b) Two numbers have a product of 24. If one of the numbers is represented by x, find an expression for the other number.

Suppose that one of the numbers is 4. To find the other number, we would divide 24 by 4.

$$\frac{24}{4} \leftarrow \text{This gives 6 as the other number. The product } 6 \cdot 4 \text{ is 24.}$$

In the same way, if x is one of the numbers, then we divide 24 by x to find the other number.

$$\frac{24}{x} \leftarrow \text{The other number}$$

◀ **Work Problem 11 at the Side.**

Answers

11. (a) $36 - r$ **(b)** $\dfrac{18}{x}$

2.3 Exercises

MyMathLab®

CONCEPT CHECK *Based on the methods of this section, fill in each blank to indicate what we should do first to solve the given equation. Do not actually solve.*

1. $7x + 8 = 1$

Use the _____ property of equality to _____ 8 from each side.

2. $7x - 5x + 15 = 8 + x$

On the _____ side, combine _____ terms.

3. $3(2t - 4) = 20 - 2t$

Use the _____ property to clear _____ on the left side of the equation.

4. $\frac{3}{4}z = -15$

Use the _____ property of equality to multiply each side by _____ to obtain z on the left.

5. $\frac{2}{3}x - \frac{1}{6} = \frac{3}{2}x + 1$

Clear _____ by multiplying by _____, the LCD.

6. $0.9x + 0.3(x + 12) = 6$

Clear _____ by multiplying each side by _____ to obtain $9x$ as the first term on the left.

7. CONCEPT CHECK Suppose that when solving three linear equations, we obtain the final results shown in parts (a)–(c). Fill in the blanks in parts (a)–(c), and then match each result with the solution set in choices A–C for the *original* equation.

(a) $6 = 6$ (The original equation is a(n) _____.)

(b) $x = 0$ (The original equation is a(n) _____ equation.)

(c) $-5 = 0$ (The original equation is a(n) _____.)

A. $\{0\}$

B. $\{\text{all real numbers}\}$

C. $\emptyset$

CONCEPT CHECK *In Exercises 8–10, give the letter of the correct choice.*

8. Which linear equation does *not* have all real numbers as solutions?

A. $5x = 4x + x$ **B.** $2(x + 6) = 2x + 12$ **C.** $\frac{1}{2}x = 0.5x$ **D.** $3x = 2x$

9. The expression $12\left[\frac{1}{6}(x + 2) - \frac{2}{3}(x + 1)\right]$ is equivalent to which of the following?

A. $-6x - 4$ **B.** $2x - 8$ **C.** $-6x + 1$ **D.** $-6x + 4$

10. The expression $100\left[0.03(x - 10)\right]$ is equivalent to which of the following?

A. $0.03x - 0.3$ **B.** $3x - 3$ **C.** $3x - 10$ **D.** $3x - 30$

Solve each equation, and check your solution. **See Examples 1–7.**

11. $3x + 2 = 14$ **12.** $4x + 3 = 27$ **13.** $-5z - 4 = 21$ **14.** $-7w - 4 = 10$

15. $4p - 5 = 2p$ **16.** $6q - 2 = 3q$ **17.** $5m + 8 = 7 + 3m$ **18.** $4r + 2 = r - 6$

19. $10p + 6 = 12p - 4$ **20.** $-5x + 8 = -3x + 10$ **21.** $7r - 5r + 2 = 5r - r$ **22.** $9p - 4p + 6 = 7p - p$

23. $x + 3 = -(2x + 2)$ **24.** $2x + 1 = -(x + 3)$ **25.** $3(4x - 2) + 5x = 30 - x$

26. $5(2m + 3) - 4m = 8m + 27$ **27.** $-2p + 7 = 3 - (5p + 1)$ **28.** $4x + 9 = 3 - (x - 2)$

29. $11x - 5(x + 2) = 6x + 5$ **30.** $6x - 4(x + 1) = 2x + 4$ **31.** $-(8x - 2) + 5x - 6 = -4$

32. $-(7x - 5) + 4x - 7 = -2$ **33.** $4(2x - 1) = -6(x + 3)$ **34.** $6(3w + 5) = 2(10w + 10)$

35. $6(4x - 1) = 12(2x + 3)$ **36.** $6(2x + 8) = 4(3x - 6)$ **37.** $3(2x - 4) = 6(x - 2)$

38. $3(6 - 4x) = 2(-6x + 9)$ **39.** $24 - 4(7 - 2t) = 4(t - 1)$ **40.** $8 - 2(2 - x) = 4(x + 1)$

Solve each equation, and check your solution. **See Examples 8–10.**

41. $-\dfrac{2}{7}r + 2r = \dfrac{1}{2}r + \dfrac{17}{2}$

(*Hint:* Clear the fractions by multiplying each side of the equation by the LCD, ____.)

42. $\dfrac{3}{5}t - \dfrac{1}{10}t = t - \dfrac{5}{2}$

(*Hint:* Clear the fractions by multiplying each side of the equation by the LCD, ____.)

43. $\dfrac{3}{4}x - \dfrac{1}{3}x + 5 = \dfrac{5}{6}x$

44. $\dfrac{1}{5}x - \dfrac{2}{3}x - 2 = -\dfrac{2}{5}x$

45. $\dfrac{1}{7}(3x + 2) - \dfrac{1}{5}(x + 4) = 2$

46. $\dfrac{1}{4}(3x - 1) + \dfrac{1}{6}(x + 3) = 3$

47. $\frac{1}{9}(x + 18) + \frac{1}{3}(2x + 3) = x + 3$

48. $-\frac{1}{4}(x - 12) + \frac{1}{2}(x + 2) = x + 4$

49. $-\frac{5}{6}q - \left(q - \frac{1}{2}\right) = \frac{1}{4}(q + 1)$

50. $\frac{2}{3}k - \left(k + \frac{1}{4}\right) = \frac{1}{12}(k + 4)$

51. $0.3(30) + 0.15x = 0.2(30 + x)$

(*Hint:* Clear the decimals by multiplying each side of the equation by ____.)

52. $0.2(60) + 0.05x = 0.1(60 + x)$

(*Hint:* Clear the decimals by multiplying each side of the equation by ____.)

53. $0.92x + 0.98(12 - x) = 0.96(12)$

54. $1.00x + 0.05(12 - x) = 0.10(63)$

55. $0.02(5000) + 0.03x = 0.025(5000 + x)$

56. $0.06(10,000) + 0.08x = 0.072(10,000 + x)$

Solve each equation, and check your solution. ***See Examples 1–10.***

57. $-3(5z + 24) + 2 = 2(3 - 2z) - 4$

58. $-2(2s - 4) - 8 = -3(4s + 4) - 1$

59. $-(6k - 5) - (-5k + 8) = -3$

60. $-(4x + 2) - (-3x - 5) = 3$

61. $8(t - 3) + 4t = 6(2t + 1) - 10$

62. $9(v + 1) - 3v = 2(3v + 1) - 8$

63. $4(x + 3) = 2(2x + 8) - 4$

64. $4(x + 8) = 2(2x + 6) + 20$

65. $\frac{1}{3}(x + 3) + \frac{1}{6}(x - 6) = x + 3$

66. $\frac{1}{2}(x + 2) + \frac{3}{4}(x + 4) = x + 5$

67. $0.3(x + 15) + 0.4(x + 25) = 25$

68. $0.1(x + 80) + 0.2x = 14$

Write the answer to each problem in terms of the variable. ***See Example 11.***

69. Two numbers have a sum of 12. One of the numbers is q. What expression represents the other number?

70. Two numbers have a sum of 26. One of the numbers is r. What expression represents the other number?

71. The product of two numbers is 9. One of the numbers is z. What expression represents the other number?

72. The product of two numbers is 13. One of the numbers is k. What expression represents the other number?

73. A football player gained x yards rushing. On the next down, he gained 29 yd. What expression represents the number of yards he gained altogether?

74. A football player gained y yards on a punt return. On the next return, he gained 25 yd. What expression represents the number of yards he gained altogether?

75. Monica is a years old. What expression represents her age 12 yr from now?   2 yr ago?

76. Chandler is b years old. What expression represents his age 3 yr ago?   5 yr from now?

77. Tom has r quarters. Express the value of the quarters in cents.

78. Jean has y dimes. Express the value of the dimes in cents.

79. A bank teller has t dollars, all in $5 bills. What expression represents the number of $5 bills the teller has?

80. A clerk has v dollars, all in $10 bills. What expression represents the number of $10 bills the clerk has?

81. A plane ticket costs x dollars for an adult and y dollars for a child. Find an expression that represents the total cost for 3 adults and 2 children.

82. A concert ticket costs p dollars for an adult and q dollars for a child. Find an expression that represents the total cost for 4 adults and 6 children.

Study Skills
USING STUDY CARDS REVISITED

Two additional types of study cards follow. As with vocabulary and procedure cards introduced earlier, use tough problem and practice quiz cards to do the following.

▶ To help you understand and learn the material

▶ To quickly review when you have a few minutes

▶ To review before a quiz or test

Tough Problem Cards

When you are doing your homework and encounter a "difficult" problem, write the procedure to work the problem on the front of a card in words. Include special notes or tips (like what *not* to do). On the back of the card, work an example. Show all steps, and label what you are doing.

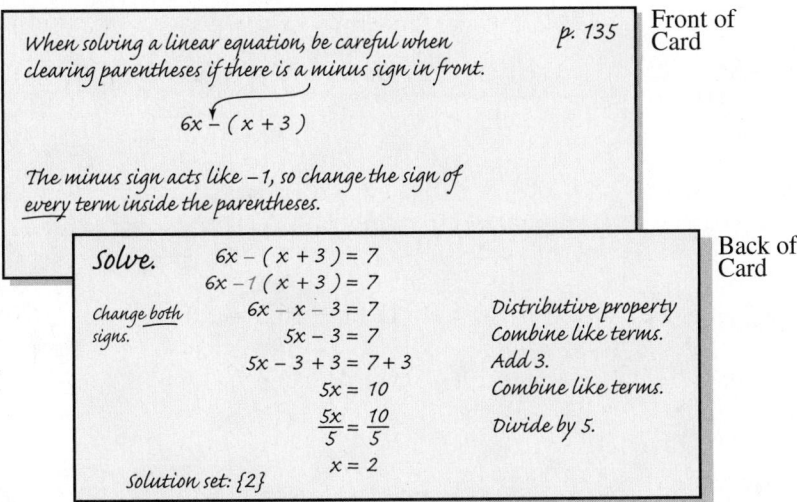

Front of Card

Back of Card

Practice Quiz Cards

Write a problem with direction words (like *solve, simplify*) on the front of a card, and work the problem on the back. Make one for each type of problem you learn.

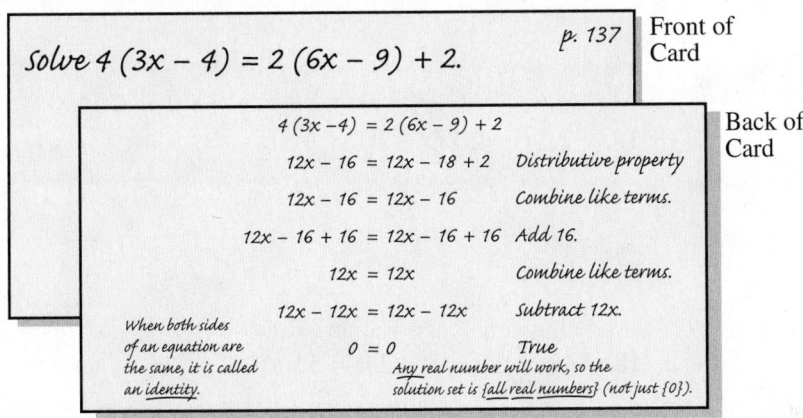

Front of Card

Back of Card

Make a tough problem card and a practice quiz card for material you are learning now.

Summary Exercises   *Applying Methods for Solving Linear Equations*

These exercises provide mixed practice in solving all the types of linear equations introduced in **Sections 2.1–2.3.** Refer back to the examples in these sections to review the various solution methods.

Solve each equation, and check your solution.

1. $a + 2 = -3$

2. $2m + 8 = 16$

3. $16.5k = -84.15$

4. $-x = -25$

5. $\dfrac{4}{5}x = -20$

6. $9x - 7x = -12$

7. $5x - 9 = 4(x - 3)$

8. $\dfrac{a}{-2} = 8$

9. $5x - 9 = 3(x - 3)$

10. $\dfrac{2}{3}x + 8 = \dfrac{1}{4}x$

11. $4x + 2(3 - 2x) = 6$

12. $x - 16.2 = 7.5$

13. $-3(t - 5) + 2(7 + 2t) = 36$

14. $2 - (m + 4) = 3m - 2$

15. $0.08x + 0.06(x + 9) = 1.24$

16. $-0.3x + 2.1(x - 4) = -6.6$

17. $-x = 16$

18. $3(m + 5) - 1 + 2m = 5(m + 2)$

19. $10m - (5m - 9) = 39$

20. $7(p - 2) + p = 2(p + 2)$

21. $-2t + 5t - 9 = 3(t - 4) - 5$

22. $-9z = -21$

23. $0.02(50) + 0.08r = 0.04(50 + r)$

24. $2.3x + 13.7 = 1.3x + 2.9$

25. $2(3 + 7x) - (1 + 15x) = 2$

26. $6q - 9 = 12 + 3q$

27. $2(5 + 3x) = 3(x + 1) + 13$

28. $r + 9 + 7r = 4(3 + 2r) - 3$

29. $\dfrac{5}{6}x + \dfrac{1}{3} = 2x + \dfrac{2}{3}$

30. $0.06x + 0.09(15 - x) = 0.07(15)$

31. $\dfrac{3}{4}(a - 2) - \dfrac{1}{3}(5 - 2a) = -2$

32. $2 - (m + 4) = 3m + 8$

33. $5.2x - 4.6 - 7.1x = -2.1 - 1.9x - 2.5$

34. $9(2m - 3) - 4(5 + 3m) = 5(4 + m) - 3$

2.4 An Introduction to Applications of Linear Equations

OBJECTIVES

1. Learn the six steps for solving applied problems.
2. Solve problems involving unknown numbers.
3. Solve problems involving sums of quantities.
4. Solve problems involving consecutive integers.
5. Solve problems involving complementary and supplementary angles.

OBJECTIVE ① Learn the six steps for solving applied problems. While there is not one specific method, we suggest the following.

Solving an Applied Problem

Step 1   **Read** the problem, several times if necessary. What information is given? What is to be found?

Step 2   **Assign a variable** to represent the unknown value. Use a sketch, diagram, or table, as needed. Express any other unknown values in terms of the variable.

Step 3   **Write an equation** using the variable expression(s).

Step 4   **Solve** the equation.

Step 5   **State the answer.** Label it appropriately. Does the answer seem reasonable?

Step 6   **Check** the answer in the words of the *original* problem.

OBJECTIVE ② Solve problems involving unknown numbers.

EXAMPLE 1   Finding the Value of an Unknown Number

The product of 4, and a number decreased by 7, is 100. What is the number?

Step 1   **Read** the problem carefully. We are asked to find a number.

Step 2   **Assign a variable** to represent the unknown quantity.

$$\text{Let } x = \text{the number}.$$

> Writing a "word equation" is often helpful.

Step 3   **Write an equation.**

The product of 4,	and	a number	decreased by	7,	is	100.
$4 \cdot$		$(x$	$-$	$7)$	$=$	100

> Note the careful use of parentheses.

> *Is, are, was,* and *were* translate as =.

Because of the commas in the given problem, writing the equation as $4x - 7 = 100$ is *incorrect*. The equation $4x - 7 = 100$ corresponds to the statement "The product of 4 and a number, decreased by 7, is 100."

Step 4   **Solve** the equation.

$$4(x - 7) = 100 \qquad \text{Equation from Step 3}$$

$$4x - 28 = 100 \qquad \text{Distributive property}$$

$$4x - 28 + 28 = 100 + 28 \qquad \text{Add 28.}$$

$$4x = 128 \qquad \text{Combine like terms.}$$

$$\frac{4x}{4} = \frac{128}{4} \qquad \text{Divide by 4.}$$

$$x = 32$$

Continued on Next Page

Step 5 **State the answer.** The number is 32.

Step 6 **Check.** The number 32 decreased by 7 is 25. The product of 4 and 25 is 100, as required. The answer, 32, is correct.

·· **Work Problem ❶ at the Side.** ▶

OBJECTIVE ▶ ❸ **Solve problems involving sums of quantities.**

> **Problem-Solving Hint**
>
> In general, to solve problems involving sums of quantities, choose a variable to represent one of the unknowns. **Then represent the other quantity in terms of the same variable.** (See **Example 11** in **Section 2.3.**)

EXAMPLE 2 **Finding Numbers of Olympic Medals**

At the 2010 Winter Olympics in Vancouver, Canada, the United States won 14 more medals than Norway. The two countries won a total of 60 medals. How many medals did each country win? (*Source: World Almanac and Book of Facts.*)

Step 1 **Read** the problem. We are given the total number of medals and asked to find the number each country won.

Step 2 **Assign a variable.**

Let x = the number of medals Norway won.

Then $x + 14$ = the number of medals the United States won.

Step 3 **Write an equation.**

The total	is	the number of medals Norway won	plus	the number of medals the U.S. won.
↓	↓	↓	↓	↓
60	=	x	+	$(x + 14)$

Step 4 **Solve** the equation.

$$60 = 2x + 14 \qquad \text{Combine like terms.}$$
$$60 - 14 = 2x + 14 - 14 \qquad \text{Subtract 14.}$$
$$46 = 2x \qquad \text{Combine like terms.}$$
$$\frac{46}{2} = \frac{2x}{2} \qquad \text{Divide by 2.}$$
$$23 = x, \quad \text{or} \quad x = 23$$

Step 5 **State the answer.** The variable x represents the number of medals Norway won, so Norway won 23 medals. Then the number of medals the United States won is

$$x + 14 = 23 + 14, \quad \text{or} \quad 37.$$

Step 6 **Check.** Since the United States won 37 medals and Norway won 23, the total number of medals was $37 + 23 = 60$. Because $37 - 23 = 14$, the United States won 14 more medals than Norway. This information agrees with what is given in the problem, so the answer checks.

❶ Solve each problem.

(a) If 5 is added to a number, the result is 7 less than 3 times the number. Find the number.

Step 1
We must find a _____.

Step 2
Let x = the _____.

Step 3

If 5 is added to a number,	the result is	7 less than 3 times the number.
↓	↓	↓
_____	=	_____

Complete Steps 4–6 to solve the problem. Give the answer.

(b) If 5 is added to the product of 9 and a number, the result is 19 less than the number. Find the number.

Answers

1. **(a)** number; number; $x + 5$ (or $5 + x$);
   $3x - 7$; The number is 6.
   (b) -3

❷ Solve each problem.

GS **(a)** The 150-member Iowa legislature includes 18 fewer Democrats than Republicans. (No other parties are represented.) How many Democrats and Republicans are there in the legislature? (*Source:* www.legis.iowa.gov)

Step 1

We must find the number of Democrats and _____.

Step 2

Let x = the number of Republicans.

Then _____ = the number of _____.

Complete Steps 3–6 to solve the problem. Give the equation and the answer.

(b) At the 2010 Winter Olympics, Germany won 19 more medals than China. The two countries won a total of 41 medals. How many medals did each country win? (*Source: World Almanac and Book of Facts.*)

❸ Solve the problem.

At a club meeting, each member brought two nonmembers. If a total of 27 people attended, how many were members and how many were nonmembers?

Meeting

| Members x | Nonmembers $2x$ | = 27 |

Answers

2. **(a)** Republicans; $x - 18$; Democrats; $x + (x - 18) = 150$; Republicans: 84; Democrats: 66
   (b) Germany: 30; China: 11
3. members: 9; nonmembers: 18

Problem-Solving Hint

The problem in **Example 2** could also be solved by letting x represent the number of medals the United States won. Then $x - 14$ would represent the number of medals Norway won. The equation would be different.

$$60 = x + (x - 14) \quad \text{Alternative equation for Example 2}$$

The solution of this equation is 37, which is the number of U.S. medals. The number of Norwegian medals would be $37 - 14 = 23$. **The answers are the same,** whichever approach is used, even though the equation and its solution are different.

◀ **Work Problem ❷ at the Side.**

EXAMPLE 3 **Analyzing a Gasoline/Oil Mixture**

A lawn trimmer uses a mixture of gasoline and oil. The mixture contains 16 oz of gasoline for each 1 ounce of oil. If the tank holds 68 oz of the mixture, how many ounces of oil and how many ounces of gasoline does it require when it is full?

Step 1 **Read** the problem. We must find how many ounces of oil and gasoline are needed to fill the tank.

Step 2 **Assign a variable.**

Let x = the number of ounces of oil required.

Then $16x$ = the number of ounces of gasoline required.

A diagram like the following is sometimes helpful.

Tank

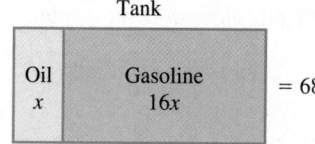

Step 3 **Write an equation.**

| Amount of gasoline | plus | amount of oil | is | total amount in tank. |
| $16x$ | $+$ | x | $=$ | 68 |

Step 4 **Solve.**

$$17x = 68 \quad \text{Combine like terms.}$$

$$\frac{17x}{17} = \frac{68}{17} \quad \text{Divide by 17.}$$

$$x = 4$$

Step 5 **State the answer.** When full, the lawn trimmer requires 4 oz of oil and $16x = 16(4)$ oz, or 64 oz of gasoline.

Step 6 **Check.** Since $4 + 64 = 68$, and $64 = 16(4)$, the answer checks.

◀ **Work Problem ❸ at the Side.**

Problem-Solving Hint

Sometimes we must find three unknown quantities. *When the three unknowns are compared in pairs, let the variable represent the unknown found in both pairs.*

EXAMPLE 4   Dividing a Board into Pieces

A project calls for three pieces of wood. The longest piece must be twice the length of the middle-sized piece. The shortest piece must be 10 in. shorter than the middle-sized piece. If a board 70 in. long is to be used, how long must each piece be?

Step 1   Read the problem. Three lengths must be found.

Step 2   Assign a variable. Since the middle-sized piece appears in both pairs of comparisons, let x represent the length, in inches, of the middle-sized piece.

Let x = the length of the middle-sized piece.

Then $2x$ = the length of the longest piece,

and $x - 10$ = the length of the shortest piece.

A sketch is helpful here. See **Figure 2.**

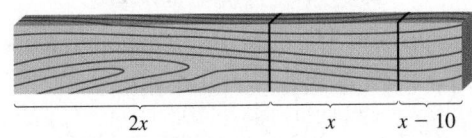

$$2x \qquad x \quad x - 10$$

Figure 2

Step 3   Write an equation.

Longest	plus	middle-sized	plus	shortest	is	total length.
↓	↓	↓	↓	↓	↓	↓
$2x$	$+$	x	$+$	$(x - 10)$	$=$	70

Step 4   Solve.

$$4x - 10 = 70 \qquad \text{Combine like terms.}$$
$$4x - 10 + 10 = 70 + 10 \qquad \text{Add 10.}$$
$$4x = 80 \qquad \text{Combine like terms.}$$
$$\frac{4x}{4} = \frac{80}{4} \qquad \text{Divide by 4.}$$
$$x = 20$$

Step 5   State the answer. The middle-sized piece is 20 in. long, the longest piece is $2(20) = 40$ in. long, and the shortest piece is $20 - 10 = 10$ in. long.

Step 6   Check. The sum of the lengths is 70 in. All conditions of the problem are satisfied.

·· **Work Problem 4 at the Side.** ▶

---

4 Solve each problem.

(a) Over a 6-hr period, a basketball player spent twice as much time lifting weights as practicing free throws and 2 hr longer watching game films than practicing free throws. How many hours did he spend on each task?

Steps 1 and 2
The unknown found in both pairs of comparisons is the time spent _____.

Let x = the time spent practicing free throws.

Then _____ = the time spent lifting weights, and

_____ = the time spent watching game films.

Step 3
Write an equation.

_____ $= 6$

Complete Steps 4–6 to solve the problem. Give the answer.

(b) A piece of pipe is 50 in. long. It is cut into three pieces. The longest piece is 10 in. longer than the middle-sized piece, and the shortest piece measures 5 in. less than the middle-sized piece. Find the lengths of the three pieces.

Answers

4. (a) practicing free throws;
    $2x$; $x + 2$; $x + 2x + (x + 2)$;
    practicing free throws: 1 hr;
    lifting weights: 2 hr;
    watching game films: 3 hr
(b) longest: 25 in.; middle: 15 in.;
    shortest: 10 in.

5 Solve the problem.

Two pages that face each other in this book have a sum of 569. What are the page numbers?

integers that differ by 1 are **consecutive integers.** For example, 3 and 4, 6 and 7, and −2 and −1 are pairs of consecutive integers. See **Figure 3.**

Consecutive integers

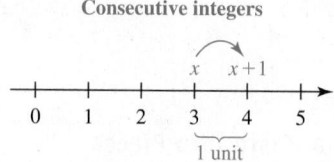

Figure 3

In general, if x represents an integer, then x + 1 represents the next greater consecutive integer.

EXAMPLE 5 **Finding Consecutive Integers**

Two pages that face each other in this book have 305 as the sum of their page numbers. What are the page numbers?

Step 1 **Read** the problem. Because the two pages face each other, they must have page numbers that are consecutive integers.

Step 2 **Assign a variable.**

Let x = the lesser page number.

Then $x + 1$ = the greater page number.

Step 3 **Write an equation.** The sum of the page numbers is 305.

$$x + (x + 1) = 305$$

Step 4 **Solve.** 　$2x + 1 = 305$ 　　Combine like terms.

$$2x = 304$$ 　　Subtract 1.

$$x = 152$$ 　　Divide by 2.

Step 5 **State the answer.** The lesser page number is 152, and the greater is $152 + 1 = 153$. (Your book is opened to these two pages.)

Step 6 **Check.** The sum of 152 and 153 is 305. The answer is correct.

◀ Work Problem **5** at the Side.

Consecutive *even* integers, such as 2 and 4, and 8 and 10, differ by 2. Similarly, **consecutive *odd* integers,** such as 1 and 3, and 9 and 11, also differ by 2. See **Figure 4.**

Consecutive even integers

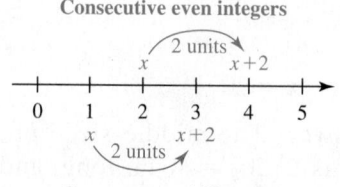

Consecutive odd integers

Figure 4

In general, if x represents an even (or odd) integer, then x + 2 represents the next greater consecutive even (or odd) integer, respectively.

Answer

5. 284, 285

In this book, we list consecutive integers in increasing order.

> **Problem-Solving Hint**
>
> When solving consecutive integer problems, if x = the lesser integer, then the following apply.
>
> For two consecutive integers, use $\qquad$ **x, $x + 1$.**
>
> For two consecutive *even* integers, use $\qquad$ **x, $x + 2$.**
>
> For two consecutive *odd* integers, use $\qquad$ **x, $x + 2$.**

EXAMPLE 6   Finding Consecutive Odd Integers

If the lesser of two consecutive odd integers is doubled, the result is 7 more than the greater of the two integers. Find the two integers.

Step 1   **Read** the problem. We must find two consecutive odd integers.

Step 2   **Assign a variable.**

Let $\qquad x$ = the lesser consecutive odd integer.

Then $\quad x + 2$ = the greater consecutive odd integer.

Step 3   **Write an equation.**

If the lesser is doubled, the result is 7 more than the greater.

$$2x = 7 + x + 2$$

Step 4   **Solve.** $2x = 9 + x$ Combine like terms.

$\qquad\qquad\qquad x = 9$ Subtract x.

Step 5   **State the answer.** The lesser integer is 9. The greater is $9 + 2 = 11$.

Step 6   **Check.** When 9 is doubled, we get 18, which is 7 more than the greater odd integer, 11. The answers are correct.

·· **Work Problem ❻ at the Side.** ▶

OBJECTIVE ▶ ⑤ Solve problems involving complementary and supplementary angles. An angle can be measured by a unit called the degree (°), which is $\frac{1}{360}$ of a complete rotation. Two angles whose sum is 90° are *complements* of each other, or **complementary.** An angle that measures 90° is a **right angle.** Two angles whose sum is 180° are *supplements* of each other, or **supplementary.** One angle *supplements* the other to form a **straight angle** of 180°. See **Figure 5.**

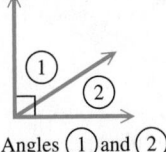

Angles ① and ② are complementary. They form a right angle, indicated by ⌐.

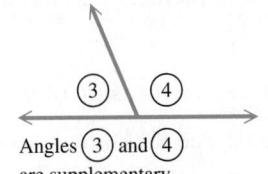

Angles ③ and ④ are supplementary. They form a straight angle.

180°

Straight angle

Figure 5

❻ Solve each problem.

GS (a) The sum of two consecutive even integers is 254. Find the integers.

Step 1
We must find two consecutive _____ .

Step 2
Let x = the lesser of the two _____ even integers.

Then _____ = the greater of the two consecutive even integers.

Step 3

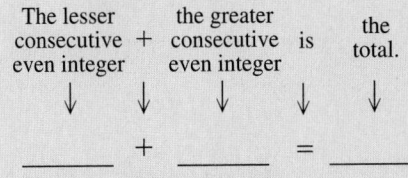

The lesser consecutive even integer + the greater consecutive even integer is the total.

_____ + _____ = _____

Complete Steps 4–6 to solve the problem. Give the answer.

(b) Find two consecutive odd integers such that the sum of twice the lesser and three times the greater is 191.

7 Fill in the blank below each figure. Then solve.

(a) Find the complement of an angle that measures 26°.

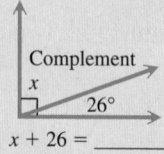

Complement
x
26°

$x + 26 =$ _____

(b) Find the supplement of an angle that measures 92°.

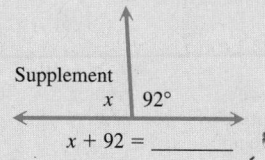

Supplement
x 92°

$x + 92 =$ _____

8 Solve each problem. Give the equation and the answer. (Let $x =$ the degree measure of the angle.)

(a) Find the measure of an angle whose complement is eight times its measure.

(b) Find the measure of an angle whose supplement is twice its measure.

Problem-Solving Hint

If x represents the degree measure of an angle, then

$90 - x$ represents the degree measure of its complement,

and $180 - x$ represents the degree measure of its supplement.

◀ **Work Problem 7 at the Side.**

EXAMPLE 7 Finding the Measure of an Angle

Find the measure of an angle whose complement is five times its measure.

Step 1 **Read** the problem. We must find the measure of an angle, given information about the measure of its complement.

Step 2 **Assign a variable.**

Let $x =$ the degree measure of the angle.

Then $90 - x =$ the degree measure of its complement.

Step 3 **Write an equation.**

Measure of the complement ... is ... 5 times the measure of the angle.

$$90 - x = 5x$$

Step 4 **Solve.** $90 - x + x = 5x + x$ Add x.

$$90 = 6x$$ Combine like terms.

$$\frac{90}{6} = \frac{6x}{6}$$ Divide by 6.

$$15 = x, \quad \text{or} \quad x = 15$$

Step 5 **State the answer.** The measure of the angle is 15°.

Step 6 **Check.** If the angle measures 15°, then its complement measures $90° - 15° = 75°$, which is equal to five times 15°, as required.

◀ **Work Problem 8 at the Side.**

EXAMPLE 8 Finding the Measure of an Angle

Find the measure of an angle whose supplement is 10° more than twice its complement.

Step 1 **Read** the problem. We must find the measure of an angle, given information about its complement and its supplement.

Step 2 **Assign a variable.**

Let $x =$ the degree measure of the angle.

Then $90 - x =$ the degree measure of its complement,

and $180 - x =$ the degree measure of its supplement.

················· **Continued on Next Page**

We can visualize this information using a sketch. See **Figure 6.**

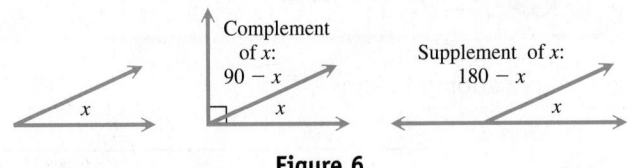

Figure 6

Step 3 **Write an equation.**

Supplement is 10 more than twice its complement.

$$180 - x = 10 + 2 \cdot (90 - x)$$

Step 4 **Solve.**

$$180 - x = 10 + 2(90 - x)$$

> Be sure to use parentheses here.

$$180 - x = 10 + 180 - 2x \quad \text{Distributive property}$$

$$180 - x = 190 - 2x \quad \text{Combine like terms.}$$

$$180 - x + 2x = 190 - 2x + 2x \quad \text{Add } 2x.$$

$$180 + x = 190 \quad \text{Combine like terms.}$$

$$180 + x - 180 = 190 - 180 \quad \text{Subtract } 180.$$

$$x = 10$$

Step 5 **State the answer.** The measure of the angle is 10°.

Step 6 **Check.** The complement of 10° is 80° and the supplement of 10° is 170°. Also, 170° is equal to 10° more than twice 80° (that is, 170 = 10 + 2(80) is true). Therefore, the answer is correct.

· **Work Problem ❾ at the Side.** ▶

---

❾ Solve each problem.

🅖🅢 **(a)** Find the measure of an angle such that twice its complement is 30° less than its supplement.

Let x = the degree measure of the angle.

Then _____ = the degree measure of its complement,

and _____ = the degree measure of its supplement.

Complete the solution. Give the equation and the answer.

(b) Find the measure of an angle whose supplement is 46° less than three times its complement.

2.4 Exercises

FOR EXTRA HELP

Download the MyDashBoard App

MyMathLab®

1. Give the six steps introduced in this section for solving application problems.

   Step 1: _____   *Step 4:* _____

   Step 2: _____   *Step 5:* _____

   Step 3: _____   *Step 6:* _____

2. **CONCEPT CHECK**   List some of the words that translate as "=" when writing an equation to solve an applied problem.

CONCEPT CHECK   *In Exercises 3–6, which choice would **not** be a reasonable answer? Justify your response.*

3. A problem requires finding the number of cars on a dealer's lot.

   A. 0   **B.** 45   **C.** 1   **D.** $6\frac{1}{2}$

4. A problem requires finding the number of hours a light bulb is on during a day.

   A. 0   **B.** 4.5   **C.** 13   **D.** 25

5. A problem requires finding the distance traveled in miles.

   A. −10   **B.** 1.8   **C.** $10\frac{1}{2}$ **D.** 50

6. A problem requires finding the time in minutes.

   A. 0   **B.** 10.5   **C.** −5   **D.** 90

CONCEPT CHECK   *Fill in each blank with the correct response.*

7. Consecutive integers differ by _____, such as 15 and _____, and −8 and _____.

8. Consecutive odd integers are _____ integers that differ by _____, such as _____ and 13. Consecutive even integers are _____ integers that differ by _____, such as 12 and _____.

9. Two angles whose measures sum to 90° are _____ angles. Two angles whose measures sum to 180° are _____ angles.

10. A right angle has measure _____. A straight angle has measure _____.

CONCEPT CHECK   *Answer each question.*

11. Is there an angle that is equal to its supplement? Is there an angle that is equal to its complement? If the answer is yes to either question, give the measure of the angle.

12. If *x* represents an integer, how can you express the next *smaller* consecutive integer in terms of *x*? The next *smaller* even integer?

Solve each problem. In each case, give the equation using x as the variable, and give the answer. See Example 1.

13. The product of 8, and a number increased by 6, is 104. What is the number?

14. The product of 5, and 3 more than twice a number, is 85. What is the number?

15. If 2 is added to five times a number, the result is equal ⏵ to 5 more than four times the number. Find the number.

16. If four times a number is added to 8, the result is three times the number, added to 5. Find the number.

17. Two less than three times a number is equal to 14 ⏵ more than five times the number. What is the number?

$$3x \quad = 5x + 14$$

18. Nine more than five times a number is equal to 3 less than seven times the number. What is the number?

19. If 2 is subtracted from a number and this difference is tripled, the result is 6 more than the number. Find the number.

20. If 3 is added to a number and this sum is doubled, the result is 2 more than the number. Find the number.

21. The sum of three times a number and 7 more than the number is the same as the difference between -11 and twice the number. What is the number?

$$3x +$$

22. If 4 is added to twice a number and this sum is multiplied by 2, the result is the same as if the number is multiplied by 3 and 4 is added to the product. What is the number?

Ⓖ *In Exercises 23 and 24, complete the six problem-solving steps to solve each problem. See Example 2.*

23. Pennsylvania and New York were among the states with the most remaining drive-in movie screens in 2012. Pennsylvania had 3 more screens than New York, and there were 59 screens total in the two states. How many drive-in movie screens remained in each state? (*Source:* www.Drive-Ins.com)

Step 1 **Read.** What are we asked to find?

_____

_____

Step 2 **Assign a variable.** Let x = the number of screens in New York.

Then $x + 3$ = _____

_____

Step 3 **Write an equation.**

____ + ____ = 59

Step 4 **Solve** the equation.

x = ____

Step 5 **State the answer.** New York had ____ screens. Pennsylvania had ____ + 3 = ____ screens.

Step 6 **Check.** The number of screens in Pennsylvania was ____ more than the number of _____. The total number of screens was 28 + ____ = ____.

24. Two of the most watched episodes in television were the final episodes of *M*A*S*H* and *Cheers*. The total number of viewers for these two episodes was about 92 million, with 8 million more people watching the *M*A*S*H* episode than the *Cheers* episode. How many people watched each episode? (*Source:* Nielsen Media Research.)

Step 1 **Read.** What are we asked to find?

_____

_____

Step 2 **Assign a variable.** Let x = the number of people who watched the *Cheers* episode.

Then $x + 8$ = _____

_____

Step 3 **Write an equation.**

____ + ____ = ____

Step 4 **Solve** the equation.

x = ____

Step 5 **State the answer.** ____ million people watched the *Cheers* episode. ____ + 8 = ____ million watched the *M*A*S*H** episode.

Step 6 **Check.** The number of people who watched the *M*A*S*H** episode was ____ million more than the number who watched the *Cheers* episode. The total number of viewers in millions was 42 + ____ = ____.

Solve each problem. **See Example 2.**

25. The total number of Democrats and Republicans in the U.S. House of Representatives during the 112th session (2011–2012) was 435. There were 49 more Republicans than Democrats. How many members of each party were there? (*Source: World Almanac and Book of Facts.*)

26. During the 112th session, the U.S. Senate had a total of 98 Democrats and Republicans. There were 4 more Democrats than Republicans. How many Democrats and Republicans were there in the Senate? (*Source: World Almanac and Book of Facts.*)

27. Bon Jovi and Roger Waters had the two top-grossing North American concert tours in 2010, together generating $197.7 million in ticket sales. If Roger Waters took in $18.7 million less than Bon Jovi, how much did each tour generate? (*Source:* Pollstar.)

28. In the United States in 2010, Honda Accord sales were 30 thousand less than Toyota Camry sales, and 596 thousand of these two cars were sold. How many of each make of car were sold? (*Source:* www.wikipedia.org)

29. In the 2010–2011 NBA regular season, the Los Angeles Lakers won 7 more than twice as many games as they lost. The Lakers played 82 games. How many wins and losses did the team have? (*Source:* www.nba.com)

30. In the 2011 regular MLB season, the Detroit Tigers won 39 fewer than twice as many games as they lost. The Tigers played 162 regular season games. How many wins and losses did the team have? (*Source:* www.mlb.com)

31. A one-cup serving of orange juice contains 3 mg less than four times the amount of vitamin C as a one-cup serving of pineapple juice. Servings of the two juices contain a total of 122 mg of vitamin C. How many milligrams of vitamin C are in a serving of each type of juice? (*Source:* U.S. Agriculture Department.)

32. A one-cup serving of pineapple juice has 9 more than three times as many calories as a one-cup serving of tomato juice. Servings of the two juices contain a total of 173 calories. How many calories are in a serving of each type of juice? (*Source:* U.S. Agriculture Department.)

Solve each problem. **See Example 3.**

33. U.S. five-cent coins are made from a combination of nickel and copper. For every 1 lb of nickel, 3 lb of copper are used. How many pounds of copper would be needed to make 560 lb of five-cent coins? (*Source:* The United States Mint.)

34. The recipe for a special whole-grain bread calls for 1 oz of rye flour for every 4 oz of whole-wheat flour. How many ounces of each kind of flour should be used to make a loaf of bread weighing 32 oz?

35. A medication contains 9 mg of active ingredients for every 1 mg of inert ingredients. How much of each kind of ingredient would be contained in a single 250-mg caplet?

36. A recipe for salad dressing uses 2 oz of olive oil for each 1 oz of red wine vinegar. If 42 oz of salad dressing are needed, how many ounces of each ingredient should be used?

37. The value of a "Mint State-63" (uncirculated) 1950 Jefferson nickel minted at Denver is $\frac{4}{3}$ the value of a similar condition 1944 nickel minted at Philadelphia. Together, the value of the two coins is $28.00. What is the value of each coin? (*Source:* Yeoman, R., *A Guide Book of United States Coins.*)

38. In one day, a store sold $\frac{8}{5}$ as many DVDs as Blu-ray discs. The total number of DVDs and Blu-ray discs sold that day was 273. How many DVDs were sold?

39. The world's largest taco contained approximately 1 kg of onion for every 6.6 kg of grilled steak. The total weight of these two ingredients was 617.6 kg. To the nearest tenth of a kilogram, how many kilograms of each ingredient were used? (*Source: Guinness World Records.*)

40. As of mid-2011, the combined population of China and India was estimated at 2.5 billion. If there were about 88% as many people living in India as China, what was the population of each country, to the nearest tenth of a billion? (*Source: U.S. Census Bureau.*)

*Solve each problem. **See Example 4.***

41. An office manager booked 55 airline tickets. He booked 7 more tickets on American Airlines than United Airlines. On Southwest Airlines, he booked 4 more than twice as many tickets as on United. How many tickets did he book on each airline?

42. A mathematics textbook editor spent 7.5 hr making telephone calls, writing e-mails, and attending meetings. She spent twice as much time attending meetings as making telephone calls and 0.5 hr longer writing e-mails than making telephone calls. How many hours did she spend on each task?

middle = x +5   long = x   short = x+9

43. Nagaraj Nanjappa has a party-length submarine sandwich 59 in. long. He wants to cut it into three pieces so that the middle piece is 5 in. longer than the shortest piece and the shortest piece is 9 in. shorter than the longest piece. How long should the three pieces be?

44. A three-foot-long deli sandwich must be split into three pieces so that the middle piece is twice as long as the shortest piece and the shortest piece is 8 in. shorter than the longest piece. How long should the three pieces be? (*Hint:* How many inches are in 3 ft?)

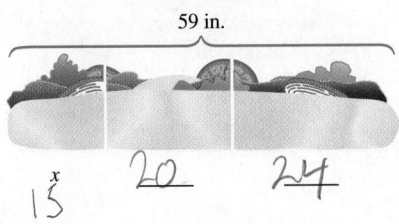

59 in.

x
15 *20 24*

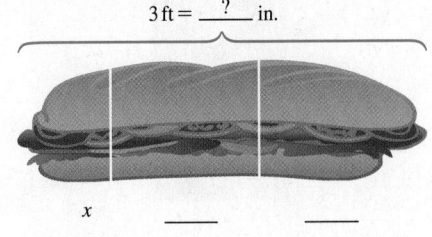

3 ft = ___?___ in.

x   ___   ___

45. The United States earned 104 medals at the 2012 Summer Olympics. The number of silver medals earned was the same as the number of bronze medals. The number of gold medals was 17 more than the number of silver medals. How many of each kind of medal did the United States earn? (*Source:* www.london2012.com)

46. China earned 88 medals at the 2012 Summer Olympics. The number of gold medals earned was 15 more than the number of bronze medals. The number of silver medals earned was 4 more than the number of bronze medals. How many of each kind of medal did China earn? (*Source:* www.london2012.com)

47. Venus is 31.2 million mi farther from the sun than Mercury, while Earth is 57 million mi farther from the sun than Mercury. If the total of the distances from these three planets to the sun is 196.2 million mi, how far away from the sun is Mercury? (All distances given here are mean (*average*) distances.) (*Source: The New York Times Almanac.*)

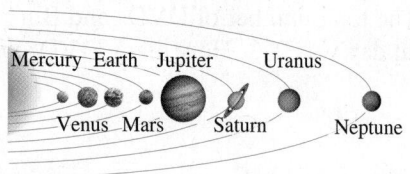

48. Saturn, Jupiter, and Uranus have a total of 153 known satellites (moons). Jupiter has 2 more satellites than Saturn, and Uranus has 35 fewer satellites than Saturn. How many known satellites does Uranus have? (*Source:* http://solarsystem.nasa.gov)

49. The sum of the measures of the angles of any triangle is 180°. In triangle *ABC*, angles *A* and *B* have the same measure, while the measure of angle *C* is 60° greater than each of *A* and *B*. What are the measures of the three angles?

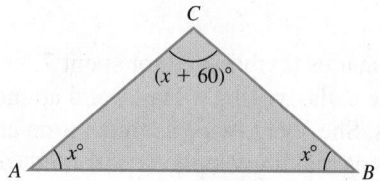

50. In triangle *ABC*, the measure of angle *A* is 141° more than the measure of angle *B*. The measure of angle *B* is the same as the measure of angle *C*. Find the measure of each angle. (*Hint:* See **Exercise 49.**)

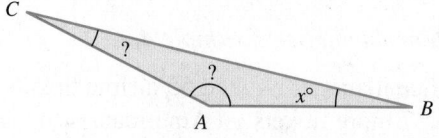

Solve each problem. **See Examples 5 and 6.**

51. The numbers on two consecutively numbered gym lockers have a sum of 137. What are the locker numbers?

52. The sum of two consecutive check numbers is 357. Find the numbers.

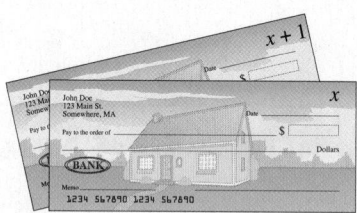

53. Two pages that are back-to-back in this book have 293 as the sum of their page numbers. What are the page numbers?

54. Two hotel rooms have room numbers that are consecutive integers. The sum of the numbers is 515. What are the two room numbers?

55. Find two consecutive even integers such that the lesser added to three times the greater gives a sum of 46.

56. Find two consecutive even integers such that six times the lesser added to the greater gives a sum of 86.

57. Find two consecutive odd integers such that 59 more than the lesser is four times the greater.

58. Find two consecutive odd integers such that twice the greater is 17 more than the lesser.

59. When the lesser of two consecutive integers is added to three times the greater, the result is 43. Find the integers.

60. If five times the lesser of two consecutive integers is added to three times the greater, the result is 59. Find the integers.

61. If the sum of three consecutive even integers is 60, what is the first of the three even integers? (*Hint:* If x and $x + 2$ represent the first two consecutive even integers, how would we represent the third consecutive even integer?)

62. If the sum of three consecutive odd integers is 69, what is the third of the three odd integers? (*Hint:* If x and $x + 2$ represent the first two consecutive odd integers, how would we represent the third consecutive odd integer?)

*Solve each problem. **See Examples 7 and 8.***

63. Find the measure of an angle whose complement is four times its measure.

64. Find the measure of an angle whose complement is five times its measure.

65. Find the measure of an angle whose supplement is eight times its measure.

66. Find the measure of an angle whose supplement is three times its measure.

67. Find the measure of an angle whose supplement measures 39° more than twice its complement.

68. Find the measure of an angle whose supplement measures 38° less than three times its complement.

69. Find the measure of an angle such that the difference between the measures of its supplement and three times its complement is 10°.

70. Find the measure of an angle such that the sum of the measures of its complement and its supplement is 160°.

2.5 Formulas and Additional Applications from Geometry

OBJECTIVES

1 Solve a formula for one variable, given the values of the other variables.

2 Use a formula to solve an applied problem.

3 Solve problems involving vertical angles and straight angles.

4 Solve a formula for a specified variable.

A **formula** is an equation in which variables are used to describe a relationship. For example, formulas exist for finding perimeters and areas of geometric figures, for calculating money earned on bank savings, and for converting among measurements.

$$P = 4s, \quad A = \pi r^2, \quad I = prt, \quad F = \frac{9}{5}C + 32 \quad \text{Formulas}$$

Many of the formulas used in this book are given inside the back cover.

OBJECTIVE ▶ 1 **Solve a formula for one variable, given the values of the other variables.** In **Example 1,** we use the idea of *area.* The **area** of a plane (two-dimensional) geometric figure is a measure of the surface covered by the figure. Area is measured in square units.

EXAMPLE 1 Using Formulas to Evaluate Variables

Find the value of the remaining variable in each formula.

(a) $A = LW$; $A = 64, L = 10$

This formula gives the area of a rectangle. See **Figure 7.** Substitute the given values for A and L into the formula.

$$A = LW \quad \boxed{\text{Solve for } W.}$$

$$64 = 10W \quad \text{Let } A = 64 \text{ and } L = 10.$$

$$\frac{64}{10} = \frac{10W}{10} \quad \text{Divide by 10.}$$

$$6.4 = W$$

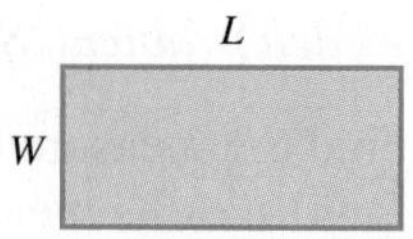

L

W

Rectangle
$A = LW$

Figure 7

The width is **6.4.** Since $10(6.4) = 64$, the given area, the answer checks.

(b) $A = \frac{1}{2}h(b + B)$; $A = 210, B = 27, h = 10$

This formula gives the area of a trapezoid. See **Figure 8.**

$$A = \frac{1}{2}h(b + B)$$

$$210 = \frac{1}{2}(10)(b + 27) \quad \boxed{\text{Solve for } b.} \quad \text{Let } A = 210, h = 10, B = 27.$$

$$210 = 5(b + 27) \quad \text{Multiply } \tfrac{1}{2}(10).$$

$$210 = 5b + 135 \quad \text{Distributive property}$$

$$210 - 135 = 5b + 135 - 135 \quad \text{Subtract 135.}$$

$$75 = 5b \quad \text{Combine like terms.}$$

$$\frac{75}{5} = \frac{5b}{5} \quad \text{Divide by 5.}$$

$$15 = b$$

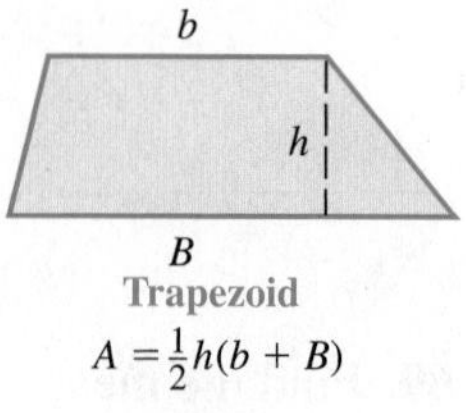

b

h

B
Trapezoid
$A = \frac{1}{2}h(b + B)$

Figure 8

The length of the shorter parallel side, b, is **15.** Since $\frac{1}{2}(10)(15 + 27) = 210$, the given area, the answer checks.

---

1 Find the value of the remaining variable in each formula.

(a) $A = bh$
(area of a parallelogram);
$A = 96, h = 8$

Given values for A and h, solve for the value of ____.

GS (b) $I = prt$ (simple interest);
$I = 246, r = 0.06$ (that is, 6%), $t = 2$

$$I = prt$$

$$\_\_\_\_ = p(\_\_\_)(\_\_\_)$$

$$246 = \_\_\_ p$$

$$\frac{246}{\_\_\_} = \frac{0.12p}{0.12}$$

$$\_\_\_ = p$$

(c) $P = 2L + 2W$
(perimeter of a rectangle);
$P = 126, W = 25$

Answers

1. **(a)** $b; b = 12$
   (b) 246; 0.06; 2; 0.12; 0.12; 2050
   (c) $L = 38$

◀ **Work Problem 1** at the Side.

OBJECTIVE ▶ 2 Use a formula to solve an applied problem. **Examples 2 and 3** use the idea of *perimeter*. The **perimeter** of a plane (two-dimensional) geometric figure is the distance around the figure. For a polygon (e.g., a rectangle, square, or triangle), it is the sum of the lengths of its sides.

EXAMPLE 2 Finding the Dimensions of a Rectangular Yard

Kari Heen's backyard is in the shape of a rectangle. The length is 5 m less than twice the width, and the perimeter is 80 m. Find the dimensions of the yard.

Step 1 **Read** the problem. We must find the dimensions of the yard.

Step 2 **Assign a variable.** Let W = the width of the lot, in meters. Since the length is 5 m less than twice the width, the length is given by $L = 2W - 5$. See **Figure 9.**

Step 3 **Write an equation.** Use the formula for the perimeter of a rectangle.

$$P = 2L + 2W$$

Perimeter $= 2 \cdot$ Length $+ 2 \cdot$ Width

$$80 = 2(2W - 5) + 2W \qquad \text{Substitute } 2W - 5 \text{ for length } L.$$

Step 4 **Solve.**

$80 = 4W - 10 + 2W$	Distributive property
$80 = 6W - 10$	Combine like terms.
$80 + 10 = 6W - 10 + 10$	Add 10.
$90 = 6W$	Combine like terms.
$\dfrac{90}{6} = \dfrac{6W}{6}$	Divide by 6.
$15 = W$	

Step 5 **State the answer.** The width is 15 m. The length is $2(15) - 5 = 25$ m.

Step 6 **Check.** If the width of the yard is 15 m and the length is 25 m, the perimeter is $2(25) + 2(15) = 50 + 30 = 80$ m, as required.

························ **Work Problem ❷ at the Side.** ▶

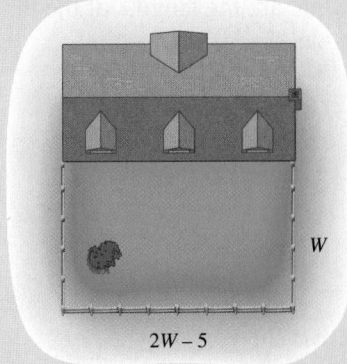

Figure 9

❷ Solve the problem.
A farmer has 800 m of fencing material to enclose a rectangular field. The width of the field is 175 m. Find the length of the field.

EXAMPLE 3 Finding the Dimensions of a Triangle

The longest side of a triangle is 3 ft longer than the shortest side. The medium side is 1 ft longer than the shortest side. If the perimeter of the triangle is 16 ft, what are the lengths of the three sides?

Step 1 **Read** the problem. We are given the perimeter of a triangle and must find the lengths of the three sides.

Step 2 **Assign a variable.** The shortest side is mentioned in each pair of comparisons.

Let s = the length of the shortest side, in feet.

Then $s + 1$ = the length of the medium side, in feet,

and $s + 3$ = the length of the longest side, in feet.

It is a good idea to draw a sketch. See **Figure 10.**

················· **Continued on Next Page**

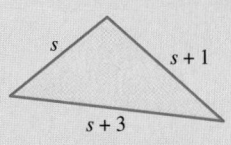

Figure 10

Answer

2. 225 m

❸ Solve the problem.

The longest side of a triangle is 1 in. longer than the medium side. The medium side is 5 in. longer than the shortest side. If the perimeter is 32 in., what are the lengths of the three sides?

Step 3 **Write an equation.** Use the formula for the perimeter of a triangle.

$$P = a + b + c \qquad \text{Perimeter of a triangle}$$

$$16 = s + (s + 1) + (s + 3) \qquad \text{Substitute.}$$

Step 4 **Solve.** $\quad 16 = 3s + 4 \qquad$ Combine like terms.

$$12 = 3s \qquad \text{Subtract 4.}$$

$$4 = s \qquad \text{Divide by 3.}$$

Step 5 **State the answer.** Since s represents the length of the shortest side, its measure is **4** ft.

$$s + 1 = 4 + 1 = 5 \text{ ft} \qquad \text{Length of the medium side}$$

$$s + 3 = 4 + 3 = 7 \text{ ft} \qquad \text{Length of the longest side}$$

Step 6 **Check.** The medium side, 5 ft, is 1 ft longer than the shortest side, and the longest side, 7 ft, is 3 ft longer than the shortest side. Furthermore, the perimeter is

$$4 + 5 + 7 = 16 \text{ ft}, \quad \text{as required.}$$

◀ **Work Problem ❸ at the Side.**

EXAMPLE 4 Finding the Height of a Triangular Sail

The area of a triangular sail of a sailboat is 126 ft². (Recall that ft² means "square feet.") The base of the sail is 12 ft. Find the height of the sail.

Step 1 **Read** the problem. We must find the height of the triangular sail.

Step 2 **Assign a variable.** Let $h =$ the height of the sail, in feet. See **Figure 11.**

❹ Solve the problem.

The area of a triangle is 120 m². The height is 24 m. Find the length of the base of the triangle.

Figure 11

Step 3 **Write an equation.** Use the formula for the area of a triangle.

$$A = \frac{1}{2}bh \qquad \begin{array}{l} A \text{ is the area, } b \text{ is the base,} \\ \text{and } h \text{ is the height.} \end{array}$$

$$126 = \frac{1}{2}(12)h \qquad \text{Substitute } A = 126, b = 12.$$

Step 4 **Solve.** $\quad 126 = 6h \qquad$ Multiply.

$$21 = h \qquad \text{Divide by 6.}$$

Step 5 **State the answer.** The height of the sail is 21 ft.

Step 6 **Check** to see that the values $A = 126$, $b = 12$, and $h = 21$ satisfy the formula for the area of a triangle.

Answers

3. 7 in.; 12 in.; 13 in.

4. 10 m

◀ **Work Problem ❹ at the Side.**

OBJECTIVE ▸ **3** **Solve problems involving vertical angles and straight angles.** **Figure 12** shows two intersecting lines forming angles that are numbered ①, ②, ③, and ④. Angles ① and ③ lie "opposite" each other. They are **vertical angles.** Another pair of vertical angles is ② and ④. ***Vertical angles have equal measures.***

Now look at angles ① and ②. When their measures are added, we get 180°, the measure of a **straight angle.** There are three other angle pairs that form straight angles: ② and ③, ③ and ④, and ① and ④.

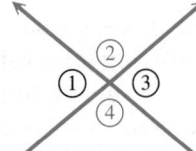

Figure 12

EXAMPLE 5 **Finding Angle Measures**

Refer to the appropriate figure in each part.

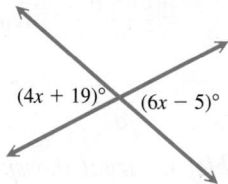

Figure 13

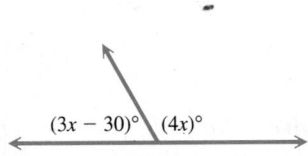

Figure 14

(a) Find the measure of each marked angle in **Figure 13.**
Since the marked angles are vertical angles, they have equal measures.

$$4x + 19 = 6x - 5 \quad \text{Set } 4x + 19 \text{ equal to } 6x - 5.$$
$$19 = 2x - 5 \quad \text{Subtract } 4x.$$
$$24 = 2x \quad \text{Add 5.}$$
$$12 = x \quad \text{Divide by 2.}$$

> This is *not* the answer.

Since $x = 12$, one angle has measure $4(12) + 19 = \mathbf{67}$ degrees. The other has the same measure, since $6(12) - 5 = \mathbf{67}$ as well. Each angle measures 67°.

(b) Find the measure of each marked angle in **Figure 14.**
The measures of the marked angles must add to 180° because together they form a straight angle. (They are also *supplements* of each other.)

$$(3x - 30) + 4x = 180 \quad \text{Supplementary angles sum to } 180°.$$
$$7x - 30 = 180 \quad \text{Combine like terms.}$$
$$7x = 210 \quad \text{Add 30.}$$
$$x = 30 \quad \text{Divide by 7.}$$

> Don't stop here!

Replace x with 30 in the measure of each marked angle.

$$3x - 30 = 3(30) - 30 = 90 - 30 = \mathbf{60}$$
$$4x = 4(30) = \mathbf{120}$$

The measures of the angles add to 180°, as required.

The two angle measures are **60°** and **120°**.

CAUTION

In **Example 5,** the answer is *not* the value of *x.* ***Remember to substitute the value of the variable into the expression given for each angle.***

Work Problem 5 at the Side. ▸

---

5 Find the measure of each marked angle.

(a)

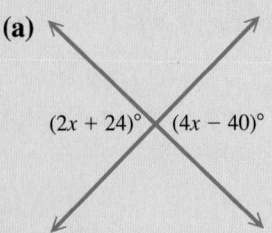

(b)

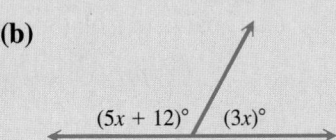

GS **(c)**

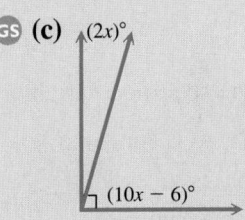

Because of the ⌐ symbol, the marked angles are _____ angles and have a sum of ____. The equation to use is

_____ = 90.

Solve this equation to find that $x =$ ____. Then substitute to find the measure of each angle.

$$2x = 2(\_\_\_)$$
$$= \_\_\_$$
$$10x - 6 = 10(\_\_\_) - 6$$
$$= \_\_\_$$

The two angles measure
_____.

Answers

5. **(a)** Both measure 88°. **(b)** 117° and 63°
   (c) complementary; 90°; $2x + (10x - 6)$;
   8; 8; 16; 8; 74; 16° and 74°

❻ Solve each formula for the specified variable.

(a) $W = Fd$ for F

ⓖⓢ (b) $I = prt$ for t
Our goal is to isolate ____.

$$I = prt$$

$$\frac{I}{\rule{1.2cm}{0.4pt}} = \frac{prt}{\rule{1.2cm}{0.4pt}}$$

$$\frac{\rule{1.2cm}{0.4pt}}{} = t$$

❼ Solve for the specified variable.

ⓖⓢ (a) $Ax + By = C$ for A
Our goal is to isolate ____.
To do this, subtract ____
from each side. Then ____
each side by ____.

Show these steps and write
the formula solved for A.

(b) $Ax + By = C$ for B

Answers

6. **(a)** $F = \dfrac{W}{d}$ **(b)** $t; pr; pr; \dfrac{I}{pr}$

7. **(a)** $A; By;$ divide; $x; A = \dfrac{C - By}{x}$

   (b) $B = \dfrac{C - Ax}{y}$

OBJECTIVE ▶ ④ Solve a formula for a specified variable. Sometimes we want to rewrite a formula in terms of a *different* variable in the formula. For example, consider $A = LW$, the formula for the area of a rectangle.

How can we rewrite $A = LW$ in terms of W?

The process whereby we do this is called **solving for a specified variable,** or **solving a literal equation.**

To solve a formula for a specified variable, we use the *same* steps that we used to solve an equation with just one variable. Consider the parallel reasoning to solve each of the following for x.

$3x + 4 = 13$		$ax + b = c$	
$3x + 4 - 4 = 13 - 4$	Subtract 4.	$ax + b - b = c - b$	Subtract b.
$3x = 9$		$ax = c - b$	
$\dfrac{3x}{3} = \dfrac{9}{3}$	Divide by 3.	$\dfrac{ax}{a} = \dfrac{c - b}{a}$	Divide by a.
$x = 3$		$x = \dfrac{c - b}{a}$	

When we solve a formula for a specified variable, we treat the specified variable as if it were the ONLY variable in the equation, and treat the other variables as if they were numbers.

EXAMPLE 6 **Solving for a Specified Variable**

Solve $A = LW$ for W.

Think of undoing what has been done to W. Since W is multiplied by L, undo the multiplication by dividing each side of $A = LW$ by L.

$$A = LW \quad \text{← Our goal is to isolate } W.$$

$$\frac{A}{L} = \frac{LW}{L} \qquad \text{Divide by } L.$$

$$\frac{A}{L} = W, \quad \text{or} \quad W = \frac{A}{L} \qquad \frac{LW}{L} = \frac{L}{L} \cdot W = 1 \cdot W = W$$

◀ **Work Problem ❻ at the Side.**

EXAMPLE 7 **Solving for a Specified Variable**

Solve $P = 2L + 2W$ for L.

$$P = 2L + 2W \quad \text{← Our goal is to isolate } L.$$

$$P - 2W = 2L + 2W - 2W \qquad \text{Subtract } 2W.$$

$$P - 2W = 2L \qquad \text{Combine like terms.}$$

$$\frac{P - 2W}{2} = \frac{2L}{2} \qquad \text{Divide by 2.}$$

$$\frac{P - 2W}{2} = L, \quad \text{or} \quad L = \frac{P - 2W}{2} \qquad \frac{2L}{2} = \frac{2}{2} \cdot L = 1 \cdot L = L$$

◀ **Work Problem ❼ at the Side.**

EXAMPLE 8 Solving for a Specified Variable

Solve $F = \frac{9}{5}C + 32$ for C.

> Our goal is to isolate C.

$$F = \frac{9}{5}C + 32 \qquad \text{This is the formula for converting temperatures from Celsius to Fahrenheit.}$$

$$F - 32 = \frac{9}{5}C + 32 - 32 \qquad \text{Subtract 32.}$$

> Be sure to use parentheses.

$$F - 32 = \frac{9}{5}C$$

$$\frac{5}{9}(F - 32) = \frac{5}{9} \cdot \frac{9}{5}C \qquad \text{Multiply by } \frac{5}{9}.$$

$$\frac{5}{9}(F - 32) = C \qquad \frac{5}{9} \cdot \frac{9}{5}C = 1C = C$$

$$C = \frac{5}{9}(F - 32) \qquad \text{This is the formula for converting temperatures from Fahrenheit to Celsius.}$$

Work Problem ❽ at the Side. ▶

❽ Solve each formula for the specified variable.

(a) $A = p + prt$ for t

(b) $x = u + zs$ for z

EXAMPLE 9 Solving for a Specified Variable

Solve each equation for y.

(a) $\quad 2x - y = 7 \quad$ ◁ Our goal is to isolate y.

$$2x - y - 2x = 7 - 2x \qquad \text{Subtract } 2x.$$

$$-y = 7 - 2x \qquad \text{Combine like terms.}$$

$$-1(-y) = -1(7 - 2x) \qquad \text{Multiply by } -1.$$

$$y = -7 + 2x \qquad \text{Multiply; distributive property}$$

$$y = 2x - 7 \qquad -a + b = b - a$$

We could have added y and subtracted 7 from each side of the equation to isolate y on the right, giving $2x - 7 = y$, a different form of the same result. There is often more than one way to solve for a specified variable.

(b) $\qquad -3x + 2y = 6$

$$-3x + 2y + 3x = 6 + 3x \qquad \text{Add } 3x.$$

$$2y = 3x + 6 \qquad \text{Combine like terms; commutative property}$$

$$\frac{2y}{2} = \frac{3x + 6}{2} \qquad \text{Divide by 2.}$$

> Be careful here.

$$y = \frac{3x}{2} + \frac{6}{2} \qquad \frac{a+b}{c} = \frac{a}{c} + \frac{b}{c}$$

> $\frac{3x}{2} = \frac{3}{2} \cdot \frac{x}{1} = \frac{3}{2}x$

$$y = \frac{3}{2}x + 3 \qquad \text{Simplify.}$$

Although we could have given our answer as $y = \frac{3x+6}{2}$, in preparation for our work in **Chapter 3,** we simplified further as shown in the last two lines of the solution.

Work Problem ❾ at the Side. ▶

❾ Solve each equation for y.

(a) $5x + y = 3$

(b) $x - 2y = 8$

Answers

8. **(a)** $t = \dfrac{A - p}{pr}$ **(b)** $z = \dfrac{x - u}{s}$

9. **(a)** $y = -5x + 3$ **(b)** $y = \dfrac{1}{2}x - 4$

2.5 Exercises

  MyMathLab®

CONCEPT CHECK *Give a one-sentence definition of each term.*

1. Perimeter of a plane geometric figure

2. Area of a plane geometric figure

CONCEPT CHECK *Decide whether perimeter or area would be used to solve a problem concerning the measure of the quantity.*

3. Sod for a lawn

4. Carpeting for a bedroom

5. Baseboards for a living room

6. Fencing for a yard

7. Fertilizer for a garden

8. Tile for a bathroom

9. Determining the cost of planting rye grass in a lawn for the winter

10. Determining the cost of replacing a linoleum floor with a wood floor

In the following exercises a formula is given, along with the values of all but one of the variables in the formula. Find the value of the variable that is not given. In Exercises 21–24, use 3.14 as an approximation for π (pi). **See Example 1.**

11. $P = 2L + 2W$ (perimeter of a rectangle); $L = 8$, $W = 5$

12. $P = 2L + 2W$; $L = 6$, $W = 4$

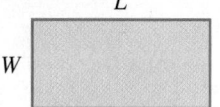

13. $A = \dfrac{1}{2}bh$ (area of a triangle); $b = 8$, $h = 16$

14. $A = \dfrac{1}{2}bh$; $b = 10$, $h = 14$

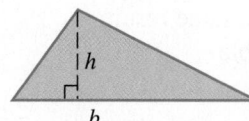

15. $P = a + b + c$ (perimeter of a triangle); $P = 12$, $a = 3$, $c = 5$

16. $P = a + b + c$; $P = 15$, $a = 3$, $b = 7$

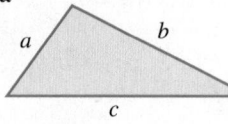

17. $d = rt$ (distance formula); $d = 252$, $r = 45$

18. $d = rt$; $d = 100$, $t = 2.5$

19. $A = \dfrac{1}{2}h(b + B)$ (area of a trapezoid); $A = 91$, $b = 12$, $B = 14$

20. $A = \dfrac{1}{2}h(b + B)$; $A = 75$, $b = 19$, $B = 31$

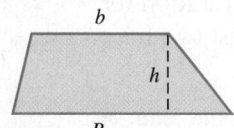

21. $C = 2\pi r$ (circumference of a circle);
▦ $C = 16.328$

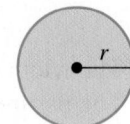

22. $C = 2\pi r$; $C = 8.164$
▦

23. $A = \pi r^2$ (area of a circle); $r = 4$
▦

24. $A = \pi r^2$; $r = 12$
▦

*The **volume** of a three-dimensional object is a measure of the space occupied by the object. For example, we would need to know the volume of a gasoline tank in order to know how many gallons of gasoline it would take to completely fill the tank. Volume is measured in cubic units.*

In Exercises 25–30, a formula for the volume (V) of a three-dimensional object is given, along with values for the other variables. Evaluate V. (Use 3.14 as an approximation for π.) **See Example 1.**

25. $V = LWH$ (volume of a rectangular box);
$L = 10$, $W = 5$, $H = 3$

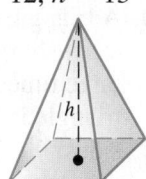

26. $V = LWH$; $L = 12$, $W = 8$, $H = 4$

27. $V = \dfrac{1}{3}Bh$ (volume of a pyramid); $B = 12$, $h = 13$

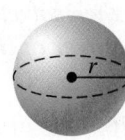

28. $V = \dfrac{1}{3}Bh$; $B = 36$, $h = 4$

29. $V = \dfrac{4}{3}\pi r^3$ (volume of a sphere); $r = 12$
▦

30. $V = \dfrac{4}{3}\pi r^3$; $r = 6$
▦

▦ *Simple interest I in dollars is calculated using the following formula.*

$$I = prt \quad \text{Simple interest formula}$$

Here, p represents the principal, or amount, in dollars that is invested or borrowed, r represents the annual interest rate, expressed as a percent, and t represents time, in years.

In Exercises 31–36, find the value of the remaining variable in the simple interest formula. **See Example 1.** *(Hint: Write percents as decimals. See **Section R.2** to review how to do this if necessary.)*

31. $p = \$7500$, $r = 4\%$, $t = 2$ yr

32. $p = \$3600$, $r = 3\%$, $t = 4$ yr

33. $I = \$33$, $r = 2\%$, $t = 3$ yr

34. $I = \$270$, $r = 5\%$, $t = 6$ yr

35. $I = \$180$, $p = \$4800$, $r = 2.5\%$
(*Hint:* 2.5% written as a decimal is _____.)

36. $I = \$162$, $p = \$2400$, $r = 1.5\%$
(*Hint:* 1.5% written as a decimal is _____.)

*Solve each perimeter problem. **See Examples 2 and 3.***

37. The length of a rectangle is 9 in. more than the width. The perimeter is 54 in. Find the length and the width of the rectangle.

38. The width of a rectangle is 3 ft less than the length. The perimeter is 62 ft. Find the length and the width of the rectangle.

39. The perimeter of a rectangle is 36 m. The length is 2 m more than three times the width. Find the length and the width of the rectangle.

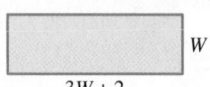

$3W + 2$ · W

40. The perimeter of a rectangle is 36 yd. The width is 18 yd less than twice the length. Find the length and the width of the rectangle.

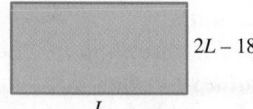

$2L - 18$ · L

41. The longest side of a triangle is 3 in. longer than the shortest side. The medium side is 2 in. longer than the shortest side. If the perimeter of the triangle is 20 in., what are the lengths of the three sides?

42. The perimeter of a triangle is 28 ft. The medium side is 4 ft longer than the shortest side, while the longest side is twice as long as the shortest side. What are the lengths of the three sides?

43. Two sides of a triangle have the same length. The third side measures 4 m less than twice that length. The perimeter of the triangle is 24 m. Find the lengths of the three sides.

44. A triangle is such that its medium side is twice as long as its shortest side and its longest side is 7 yd less than three times its shortest side. The perimeter of the triangle is 47 yd. What are the lengths of the three sides?

🔢 *Use a formula to write an equation for each application, and then solve. (Use 3.14 as an approximation for π.)* **Formulas are found inside the back cover of this book. See Examples 2–4.**

45. One of the largest fashion catalogues in the world was published in Hamburg, Germany. Each of the 212 pages in the catalogue measured 1.2 m by 1.5 m. What was the perimeter of a page? What was the area? (*Source: Guinness World Records.*)

46. One of the world's largest mandalas (sand paintings) measures 12.24 m by 12.24 m. What is the perimeter of the sand painting? To the nearest hundredth of a square meter, what is the area? (*Source: Guinness World Records.*)

47. The area of a triangular road sign is 70 ft². If the base of the sign measures 14 ft, what is the height of the sign?

48. The area of a triangular advertising banner is 96 ft². If the height of the banner measures 12 ft, find the measure of the base.

49. A prehistoric ceremonial site dating to about 3000 B.C. was discovered at Stanton Drew in southwestern England. The site, which is larger than Stonehenge, is a nearly perfect circle, consisting of nine concentric rings that probably held upright wooden posts. Around this timber temple is a wide, encircling ditch enclosing an area with a diameter of 443 ft. Find this enclosed area to the nearest thousand square feet. (*Source: Archaeology.*)

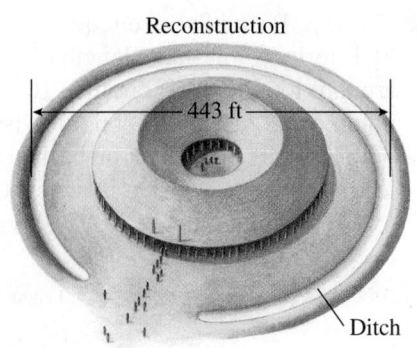

50. The Rogers Centre in Toronto, Canada, is the first stadium with a hard-shell, retractable roof. The steel dome is 630 ft in diameter. To the nearest foot, what is the circumference of this dome? (*Source:* www.ballparks.com)

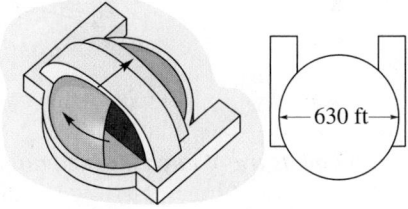

51. One of the largest drums ever constructed was made from Japanese cedar and cowhide, with radius 7.87 ft. What was the area of the circular face of the drum? What was the circumference of the drum? Round your answers to the nearest hundredth. (*Source: Guinness World Records.*)

52. A drum played at the Royal Festival Hall in London had radius 6.5 ft. What was the area of the circular face of the drum? What was the circumference of the drum? (*Source: Guinness World Records.*)

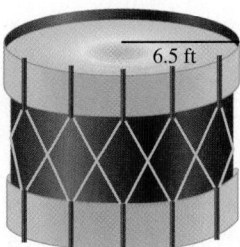

53. The survey plat depicted here shows two lots that form a trapezoid. The measures of the parallel sides are 115.80 ft and 171.00 ft. The height of the trapezoid is 165.97 ft. Find the combined area of the two lots. Round your answer to the nearest hundredth of a square foot.

54. Lot A in the figure is in the shape of a trapezoid. The parallel sides measure 26.84 ft and 82.05 ft. The height of the trapezoid is 165.97 ft. Find the area of Lot A. Round your answer to the nearest hundredth of a square foot.

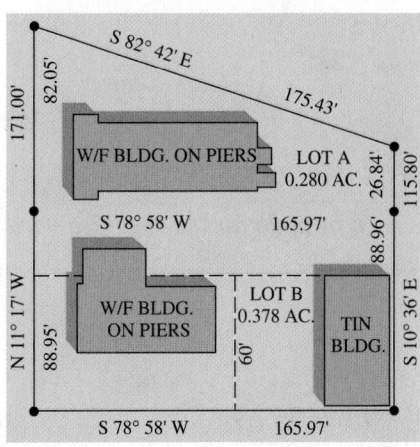

Source: Property survey in New Roads, Louisiana.

55. The U.S. Postal Service requires that any box sent by Priority Mail® have length plus girth (distance around) totaling no more than 108 in. The maximum volume that meets this condition is contained by a box with a square end 18 in. on each side. What is the length of the box? What is the maximum volume? (*Source:* United States Postal Service.)

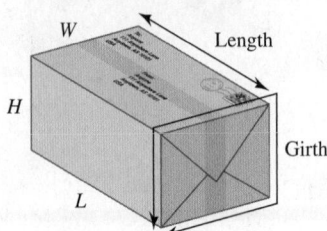

56. One of the world's largest sandwiches, made by Wild Woody's Chill and Grill in Roseville, Michigan, was 12 ft long, 12 ft wide, and $17\frac{1}{2}$ in. $\left(1\frac{11}{24}\text{ ft}\right)$ thick. What was the volume of the sandwich? (*Source: Guinness World Records.*)

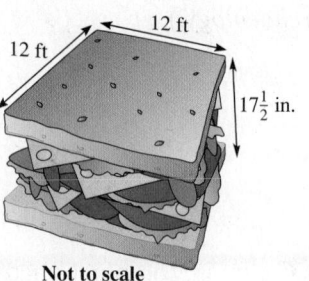

Find the measure of each marked angle. See Example 5.

57.

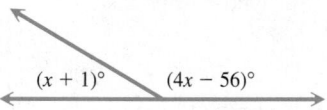

$(x + 1)°$ $(4x − 56)°$

58.

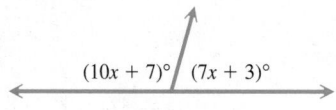

$(10x + 7)°$ $(7x + 3)°$

59.

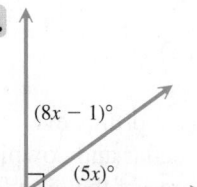

$(8x − 1)°$
$(5x)°$

60.

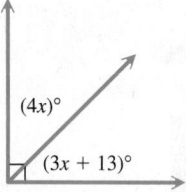

$(4x)°$
$(3x + 13)°$

61.

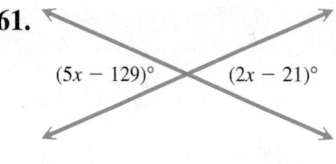

$(5x − 129)°$ $(2x − 21)°$

62.

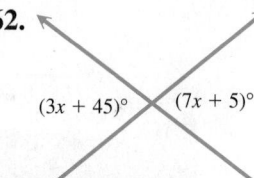

$(3x + 45)°$ $(7x + 5)°$

63.

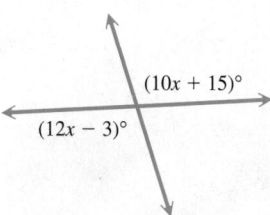

$(10x + 15)°$
$(12x − 3)°$

64.

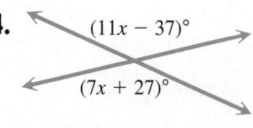

$(11x − 37)°$
$(7x + 27)°$

Solve each formula for the specified variable. See Examples 6–8.

65. $d = rt$ for t

66. $d = rt$ for r

67. $V = LWH$ for H

68. $V = LWH$ for L

69. $P = a + b + c$ for b

70. $P = a + b + c$ for c

71. $C = 2\pi r$ for r

72. $C = \pi d$ for d

73. $I = prt$ for r

74. $I = prt$ for p

75. $A = \dfrac{1}{2}bh$ for h

76. $A = \dfrac{1}{2}bh$ for b

77. $V = \dfrac{1}{3}\pi r^2 h$ for h

78. $V = \pi r^2 h$ for h

79. $P = 2L + 2W$ for W

80. $A = p + prt$ for r

81. $y = mx + b$ for m

82. $y = mx + b$ for x

83. $Ax + By = C$ for y

84. $Ax + By = C$ for x

85. $M = C(1 + r)$ for r

86. $C = \dfrac{5}{9}(F - 32)$ for F

87. $P = 2(a + b)$ for a

88. $P = 2(a + b)$ for b

Solve each equation for y. **See Example 9.**

89. $6x + y = 4$

90. $3x + y = 6$

91. $5x - y = 2$

92. $4x - y = 1$

93. $-3x + 5y = -15$

94. $-2x + 3y = -9$

95. $x - 3y = 12$

96. $x - 5y = 10$

2.6 Ratio, Proportion, and Percent

OBJECTIVES

1. Write ratios.
2. Solve proportions.
3. Solve applied problems using proportions.
4. Find percents and percentages.

OBJECTIVE ▶ 1 **Write ratios.** A **ratio** is a comparison of two quantities using a quotient.

> **Ratio**
>
> The ratio of the number a to the number b (where $b \neq 0$) is written as follows.
>
> $$a \text{ to } b, \quad a:b, \quad \text{or} \quad \frac{a}{b}$$

Writing a ratio as a quotient $\frac{a}{b}$ is most common in algebra.

EXAMPLE 1   Writing Word Phrases as Ratios

Write a ratio for each word phrase.

(a) 5 hr to 3 hr $\dfrac{5 \text{ hr}}{3 \text{ hr}} = \dfrac{5}{3}$

(b) 6 hr to 3 days

First convert 3 days to hours.

$$3 \text{ days} = 3 \cdot 24 = 72 \text{ hr} \quad 1 \text{ day} = 24 \text{ hr}$$

Now write the ratio using the common unit of measure, hours.

$$\frac{6 \text{ hr}}{3 \text{ days}} = \frac{6 \text{ hr}}{72 \text{ hr}} = \frac{6}{72}, \quad \text{or} \quad \frac{1}{12} \quad \text{Write in lowest terms.}$$

◀ **Work Problem 1 at the Side.**

An application of ratios is in *unit pricing,* to see which size of an item offered in different sizes produces the best price per unit.

EXAMPLE 2   Finding Price per Unit

A Jewel-Osco supermarket charges the following prices for a jar of extra crunchy peanut butter.

 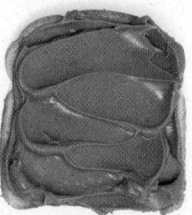

PEANUT BUTTER

Size	Price
18 oz	$3.49
28 oz	$4.99
40 oz	$6.79

Which size is the best buy? That is, which size has the lowest unit price?

Continued on Next Page

---

1 Write each ratio.

(a) 9 women to 5 women

(GS) (b) 4 in. to 1 ft

First convert 1 ft to _____ in.

Then write a ratio.

(c) 8 months to 2 yr

Answers

1. **(a)** $\dfrac{9}{5}$ **(b)** 12; $\dfrac{4}{12}$, or $\dfrac{1}{3}$

(c) $\dfrac{8}{24}$, or $\dfrac{1}{3}$

To find the best buy, write ratios comparing the price for each size jar to the number of units (ounces) per jar.

Size	Unit Price (dollars per ounce)
18 oz	$\frac{\$3.49}{18} = \$0.194$
28 oz	$\frac{\$4.99}{28} = \$0.178$
40 oz	$\frac{\$6.79}{40} = \$0.170$ ← The best buy

(Results are rounded to the nearest thousandth.)

Because the 40-oz size produces the lowest unit price, it is the best buy. Buying the largest size does not always provide the best buy, although it often does, as in this case.

·········· **Work Problem 2 at the Side.** ▶

OBJECTIVE ▶ 2 Solve proportions. A ratio is used to compare two numbers or amounts. A **proportion** says that two ratios are equal. For example, the proportion

$$\frac{3}{4} = \frac{15}{20}$$

A proportion is a special type of equation.

says that the ratios $\frac{3}{4}$ and $\frac{15}{20}$ are equal. In the proportion

$$\frac{a}{b} = \frac{c}{d} \quad (\text{where } b, d \neq 0),$$

a, b, c, and d are the **terms** of the proportion. The a and d terms are called the **extremes,** and the b and c terms are called the **means.** We read the proportion $\frac{a}{b} = \frac{c}{d}$ as "*a is to b as c is to d.*" Multiplying each side of this proportion by the common denominator, bd, gives the following.

$$bd \cdot \frac{a}{b} = bd \cdot \frac{c}{d} \quad \text{Multiply each side by } bd.$$

$$\frac{b}{b}(d \cdot a) = \frac{d}{d}(b \cdot c) \quad \text{Associative and commutative properties}$$

$$ad = bc \quad \text{Commutative and identity properties}$$

We can also find the products ad and bc by multiplying diagonally.

$$ad = bc$$

$$\frac{a}{b} = \frac{c}{d}$$

For this reason, ad and bc are called **cross products of the proportion.**

Cross Products of a Proportion

If $\frac{a}{b} = \frac{c}{d}$, then the cross products ad and bc are equal—*the product of the extremes equals the product of the means.*

Also, if $ad = bc$, then $\frac{a}{b} = \frac{c}{d}$ (where $b, d \neq 0$).

2 Solve each problem.

(a) A supermarket charges the following prices for a popular brand of pork and beans.

PORK AND BEANS

Size	Price
8 oz	$0.76
28 oz	$1.00
53 oz	$1.99

Calculate the unit price to the nearest thousandth of an ounce for each size.

8 oz: $\dfrac{\$0.76}{\underline{\quad}} = \underline{\quad}$

28 oz: $\dfrac{\underline{\quad}}{28} = \underline{\quad}$

53 oz: $\dfrac{\underline{\quad}}{\underline{\quad}} = \underline{\quad}$

Which size is the best buy? What is the unit price for that size?

(b) A supermarket charges the following prices for a certain brand of liquid detergent.

LAUNDRY DETERGENT

Size	Price
75 oz	$8.94
100 oz	$13.97
150 oz	$19.97

Which size is the best buy? What is the unit price to the nearest thousandth for that size?

Answers

2. (a) 8; $0.095; $1.00; $0.036; $1.99; 53; $0.038; 28 oz; $0.036 per oz
   (b) 75 oz; $0.119 per oz

❸ Solve each proportion.

(a) $\dfrac{y}{6} = \dfrac{35}{42}$

$y \cdot \underline{\hspace{1cm}} = \underline{\hspace{1cm}} \cdot 35$

$\underline{\hspace{1cm}} y = \underline{\hspace{1cm}}$

$y = \underline{\hspace{1cm}}$

The solution set is $\underline{\hspace{1cm}}$.

(b) $\dfrac{a}{24} = \dfrac{15}{16}$

❹ Solve each equation.

(a) $\dfrac{z}{2} = \dfrac{z+1}{3}$

(b) $\dfrac{p+3}{3} = \dfrac{p-5}{4}$

Note

If $\frac{a}{c} = \frac{b}{d}$, then $ad = cb$, or $ad = bc$. This means that the two proportions are equivalent, and the proportion

$$\frac{a}{b} = \frac{c}{d} \quad \text{can also be written as} \quad \frac{a}{c} = \frac{b}{d}.$$

Sometimes one form is more convenient to work with than the other.

Four numbers are used in a proportion. If any three of these numbers are known, the fourth can be found.

EXAMPLE 3 Finding an Unknown in a Proportion

Solve the proportion $\frac{5}{9} = \frac{x}{63}$.

$$\frac{5}{9} = \frac{x}{63} \quad \boxed{\text{Solve for } x.}$$

$$5 \cdot 63 = 9 \cdot x \quad \text{Cross products must be equal.}$$

$$315 = 9x \quad \text{Multiply.}$$

$$35 = x \quad \text{Divide by 9.}$$

Check by substituting 35 for x in the proportion. The solution set is $\{35\}$.

◀ **Work Problem ❸ at the Side.**

CAUTION

The cross product method cannot be used directly if there is more than one term on either side of the equality symbol.

EXAMPLE 4 Solving an Equation by Using Cross Products

Solve the equation $\frac{m-2}{5} = \frac{m+1}{3}$.

$$\frac{m-2}{5} = \frac{m+1}{3} \quad \boxed{\text{Be sure to use parentheses.}}$$

$$3(m-2) = 5(m+1) \quad \text{Find the cross products.}$$

$$3m - 6 = 5m + 5 \quad \text{Distributive property}$$

$$3m = 5m + 11 \quad \text{Add 6.}$$

$$-2m = 11 \quad \text{Subtract } 5m.$$

$$m = -\frac{11}{2} \quad \text{Divide by } -2.$$

A check confirms that the solution is $-\frac{11}{2}$, so the solution set is $\left\{-\frac{11}{2}\right\}$.

◀ **Work Problem ❹ at the Side.**

Note

When we set cross products equal to each other, we are really multiplying each ratio in the proportion by a common denominator.

Answers

3. **(a)** 42; 6; 42; 210; 5; $\{5\}$ **(b)** $\left\{\dfrac{45}{2}\right\}$

4. **(a)** $\{2\}$ **(b)** $\{-27\}$

OBJECTIVE ▶ ③ **Solve applied problems using proportions.**

EXAMPLE 5 Applying Proportions

After Lee Ann Spahr pumped 5.0 gal of gasoline, the display showing the price read $23.20. When she finished pumping the gasoline, the price display read $67.28. How many gallons did she pump?

To solve this problem, set up a proportion, with prices in the numerators and gallons in the denominators. Let x = the number of gallons pumped.

$$\text{Price} \rightarrow \frac{\$23.20}{5.0} = \frac{\$67.28}{x} \begin{array}{l}\leftarrow \text{Price} \\ \leftarrow \text{Gallons}\end{array}$$

> Be sure that numerators represent the *same* quantities and denominators represent the *same* quantities.

$$23.20x = 5.0\,(67.28) \quad \text{Cross products}$$
$$23.20x = 336.40 \quad \text{Multiply.}$$
$$x = 14.5 \quad \text{Divide by 23.20.}$$

She pumped a total of 14.5 gal. Check this answer. Notice that the way the proportion was set up uses the fact that the unit price is the same, no matter how many gallons are purchased.

🖩 **Calculator Tip**

Using a calculator to perform the arithmetic in **Example 5** reduces the possibility of errors.

Work Problem ⑤ at the Side. ▶

OBJECTIVE ▶ ④ **Find percents and percentages.** *A percent is a ratio where the second number is always 100.*

50% represents the ratio of 50 to 100, that is, $\frac{50}{100}$, or, 0.50.

27% represents the ratio of 27 to 100, that is, $\frac{27}{100}$, or, 0.27.

125% represents the ratio of 125 to 100, that is, $\frac{125}{100}$, or, 1.25.

5% represents the ratio of 5 to 100, that is, $\frac{5}{100}$, or, 0.05.

Recall from **Section R.2** that *the decimal point can be moved two places to the left and the percent symbol dropped to change a percent to a decimal.*

🖩 **Calculator Tip**

Many calculators have a percent key that does this automatically.

We can solve a percent problem involving $x\%$ by writing it as a proportion.

$$\frac{\textit{amount}}{\textit{base}} = \frac{x}{100}$$

The amount, or **percentage,** is compared to the **base** (the whole amount). We can also write this proportion as follows.

$$\frac{\text{amount}}{\text{base}} = \text{percent (as a decimal)} \qquad \frac{x}{100} \text{ is equivalent to percent as a decimal.}$$

or **amount = percent (as a decimal) · base** Basic percent equation

---

⑤ Solve each problem.

(a) Twelve gal of gasoline costs $57.48. To the nearest cent, how much would 16.5 gal of the same fuel cost?

(b) Eight quarts of oil cost $14.00. How much do 5 qt of oil cost?

6 Solve each problem.

(a) What is 20% of 70?

(b) 40% of what number is 130?

(c) 121 is what percent of 484?

| EXAMPLE 6 | Solving Percent Equations |

Solve each problem.

(a) What is 15% of 600?

Let n = the number. The word *of* indicates multiplication.

What   is   15%   of   600? ◄ Translate each word or phrase to write the equation.

$$n = 0.15 \cdot 600 \quad \text{Write the percent equation.}$$

$$n = 90 \quad \text{(Write 15% as a decimal.)} \quad \text{Multiply.}$$

Thus, **90** is 15% of 600.

(b) 32% of what number is 64?

32%   of   what number   is   64?

$$0.32 \cdot n = 64 \quad \text{Write the percent equation.}$$

(Write 32% as a decimal.)

$$n = \frac{64}{0.32} \quad \text{Divide by 0.32.}$$

$$n = 200 \quad \text{Simplify. Use a calculator.}$$

32% of **200** is 64.

(c) 90 is what percent of 360?

90   is   what percent   of   360?

$$90 = p \cdot 360 \quad \text{Write the percent equation.}$$

$$\frac{90}{360} = p \quad \text{Divide by 360.}$$

$$0.25 = p, \quad \text{or} \quad 25\% = p \quad \text{Simplify. Write 0.25 as a percent.}$$

Thus, 90 is **25%** of 360.

◄ **Work Problem 6 at the Side.**

| EXAMPLE 7 | Solving Applied Percent Problems |

Solve each problem.

(a) A DVD with a regular price of $18 is on sale this week at 22% off. Find the amount of the discount and the sale price of the disc.

The discount is 22% of 18, so we must find the number that is 22% of 18.

What number   is   22%   of   18?

$$n = 0.22 \cdot 18 \quad \text{Write the percent equation.}$$

$$n = 3.96 \quad \text{Multiply.}$$

The discount is $3.96, so the sale price is found by subtracting.

$$\$18.00 - \$3.96 = \$14.04 \quad \text{Original price} - \text{discount} = \text{sale price}$$

Continued on Next Page

Answers

6. (a) 14 (b) 325 (c) 25%

(b) A newspaper ad offered a set of tires at a sale price of $258. The regular price was $300. What percent of the regular price was the savings?

The savings amounted to $300 − $258 = $42. We can now restate the problem: What percent of 300 is 42?

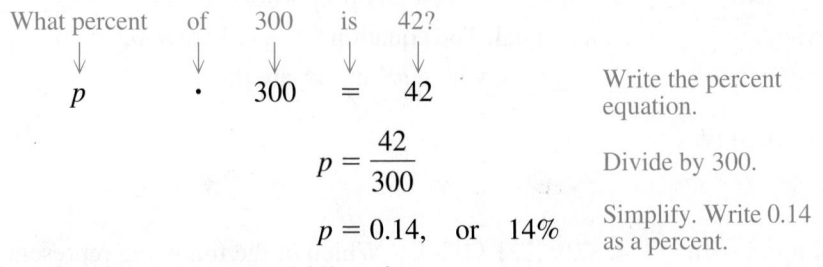

$$p \cdot 300 = 42 \qquad \text{Write the percent equation.}$$

$$p = \frac{42}{300} \qquad \text{Divide by 300.}$$

$$p = 0.14, \quad \text{or} \quad 14\% \qquad \text{Simplify. Write 0.14 as a percent.}$$

The sale price represents a 14% savings.

··· **Work Problem ➐ at the Side.** ▶

➐ Solve each problem.

(a) A television set with a regular price of $950 is on sale at 25% off. Find the amount of the discount and the sale price of the set.

(b) A winter coat is on a clearance sale for $48. The regular price is $120. What percent of the regular price is the savings?

GS (c) Mark scored 34 points on a test, which was 85% of the possible points. How many possible points were on the test?

Write the percent equation. Give the percent as a decimal.

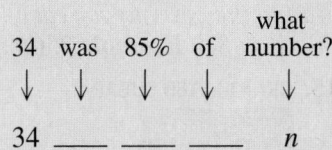

$$34 \ \underline{\quad} \ \underline{\quad} \ \underline{\quad} \ n$$

Complete the solution, and give the answer to the problem.

2.6 Exercises

FOR EXTRA HELP

 Download the MyDashBoard App

 MyMathLab®

1. CONCEPT CHECK Ratios are used to _____ two numbers or quantities. Which of the following indicate the ratio of *a* to *b*?

A. $\dfrac{a}{b}$ **B.** $\dfrac{b}{a}$ **C.** $a \cdot b$ **D.** $a : b$

2. CONCEPT CHECK A proportion says that two _____ are equal. The equation $\frac{a}{b} = \frac{c}{d}$ (where $b, d, \neq 0$) is a _____ , where *ad* and *bc* are the _____ .

3. CONCEPT CHECK Match each ratio in Column I with the ratio equivalent to it in Column II.

I	II
(a) 75 to 100	**A.** 80 to 100
(b) 5 to 4	**B.** 50 to 100
(c) $\dfrac{1}{2}$	**C.** 3 to 4
(d) 4 to 5	**D.** 15 to 12

4. CONCEPT CHECK Which of the following represent a ratio of 4 days to 2 weeks?

A. $\dfrac{4}{2}$ **B.** $\dfrac{4}{7}$ **C.** $\dfrac{4}{14}$

D. $\dfrac{2}{1}$ **E.** $\dfrac{2}{7}$ **F.** $\dfrac{1}{2}$

Write a ratio for each word phrase. Write fractions in lowest terms. **See Example 1.**

5. 60 ft to 70 ft

6. 30 mi to 40 mi

7. 72 dollars to 220 dollars

8. 80 people to 120 people

9. 30 in. to 8 ft

10. 8 ft to 20 yd

11. 16 min to 1 hr

12. 24 min to 2 hr

13. 2 yd to 60 in.

14. 3 days to 40 hr

72

Find the best buy for each item. Give the unit price to the nearest thousandth for that size. **See Example 2.** (*Source:* Jewel-Osco and HyVee.)

15. GRANULATED SUGAR

Size	Price
4 lb	$3.29
10 lb	$7.49

16. APPLESAUCE

Size	Price
23 oz	$1.99
48 oz	$3.49

17. ORANGE JUICE

Size	Price
64 oz	$2.99
89 oz	$4.79
128 oz	$6.49

18. SALAD DRESSING

Size	Price
8 oz	$1.69
16 oz	$1.97
36 oz	$5.99

19. MAPLE SYRUP

Size	Price
8.5 oz	$5.79
12.5 oz	$7.99
32 oz	$16.99

20. MOUTHWASH

Size	Price
16.9 oz	$3.39
33.8 oz	$3.49
50.7 oz	$5.29

21. TOMATO KETCHUP

Size	Price
32 oz	$1.79
36 oz	$2.69
40 oz	$2.49
64 oz	$4.38

22. GRAPE JELLY

Size	Price
1 lb	$0.79
2 lb	$1.49
5 lb	$3.59
20 lb	$12.99

Solve each equation. ***See Examples 3 and 4.***

23. $\dfrac{k}{4} = \dfrac{175}{20}$ ⏵

24. $\dfrac{x}{6} = \dfrac{18}{4}$

25. $\dfrac{49}{56} = \dfrac{z}{8}$

26. $\dfrac{20}{100} = \dfrac{z}{80}$

27. $\dfrac{x}{4} = \dfrac{12}{30}$

28. $\dfrac{x}{6} = \dfrac{5}{21}$

29. $\dfrac{z}{4} = \dfrac{z+1}{6}$

30. $\dfrac{m}{5} = \dfrac{m-2}{2}$

31. $\dfrac{3y-2}{5} = \dfrac{6y-5}{11}$ ⏵

32. $\dfrac{2r+8}{4} = \dfrac{3r-9}{3}$

33. $\dfrac{5k+1}{6} = \dfrac{3k-2}{3}$ ⏵

34. $\dfrac{x+4}{6} = \dfrac{x+10}{8}$

35. $\dfrac{2p+7}{3} = \dfrac{p-1}{4}$

36. $\dfrac{3m-2}{5} = \dfrac{4-m}{3}$

🖩 *Solve each problem.* ***See Example 5.***

37. If 16 candy bars cost $20.00, how much do 24 candy bars cost?

38. If 12 ring tones cost $30.00, how much do 8 ring tones cost?

39. If 6 gal of premium gasoline cost $22.74, how much would it cost to completely fill a 15-gal tank?

40. If sales tax on a $16.00 DVD is $1.32, how much would the sales tax be on a $120.00 DVD player?

41. Biologists tagged 500 fish in Grand Bay. At a later date, they found 7 tagged fish in a sample of 700. Estimate the total number of fish in Grand Bay to the nearest hundred.

42. Researchers at West Okoboji Lake tagged 840 fish. A later sample of 1000 fish contained 18 that were tagged. Approximate the fish population in West Okoboji Lake to the nearest hundred.

43. The distance between Kansas City, Missouri, and Denver is 600 mi. On a certain wall map, this is represented by a length of 2.4 ft. On the map, how many feet would there be between Memphis and Philadelphia, two cities that are actually 1000 mi apart?

44. The distance between Singapore and Tokyo is 3300 mi. On a certain wall map, this distance is represented by a length of 11 in. The actual distance between Mexico City and Cairo is 7700 mi. How far apart are they on the same map?

45. In 2009, there were 87 licensed drivers for every
ⓖⓢ 100 U.S. residents of driving age (16 and over). How many of the 241 million residents of driving age were licensed drivers? (*Source:* Bureau of Transportation Statistics.)

    Complete the proportion, solve it, and give the answer.

$$\dfrac{\underline{\quad}}{\underline{\quad}} = \dfrac{x}{241} \begin{array}{l} \leftarrow \text{Drivers (in millions)} \\ \leftarrow \text{Residents (in millions)} \end{array}$$

46. In 2009, there were 7 licensed drivers for every
ⓖⓢ 10 registered vehicles in California. There were 33.8 million registered vehicles. How many licensed drivers were there? (*Source:* Bureau of Transportation Statistics.)

    Complete the proportion (using x as the variable), solve it, and give the answer.

$$\begin{array}{l} \text{Drivers} \rightarrow \\ \text{Vehicles} \rightarrow \end{array} \dfrac{7}{10} = \dfrac{\underline{\quad}}{\underline{\quad}}$$

47. According to the directions on the label of a bottle of Armstrong® Concentrated Floor Cleaner, for routine cleaning, $\frac{1}{4}$ cup of cleaner should be mixed with 1 gal of warm water. How much cleaner should be mixed with $10\frac{1}{2}$ gal of water?

48. The directions on the bottle mentioned in **Exercise 47** also specify that, for extra-strength cleaning, $\frac{1}{2}$ cup of cleaner should be used for each 1 gal of water. For extra-strength cleaning, how much cleaner should be mixed with $15\frac{1}{2}$ gal of water?

49. On January 3, 2012, the exchange rate between euros and U.S. dollars was 1 euro to $1.3051. Ashley went to Rome and exchanged her U.S. currency for euros, receiving 300 euros. How much in U.S. dollars did she exchange? (*Source:* www.xe.com/ucc)

50. On the same date as in **Exercise 49**, 12 U.S. dollars could be exchanged for 163.8 Mexican pesos. At that exchange rate, how many pesos could Carlos obtain for $100? (*Source:* www.xe.com/ucc)

*Two triangles are **similar** if they have the same shape (but not necessarily the same size). Similar triangles have sides that are proportional. The figure shows two similar triangles. Notice that the ratios of the corresponding sides all equal $\frac{3}{2}$.*

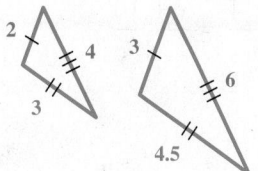

$$\frac{3}{2} = \frac{3}{2} \qquad \frac{4.5}{3} = \frac{3}{2} \qquad \frac{6}{4} = \frac{3}{2}$$

If we know that two triangles are similar, we can set up a proportion to solve for the length of an unknown side.

   In Exercises 51–56, find the lengths x and y as needed in each pair of similar triangles.

51.
GS

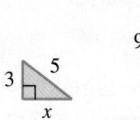 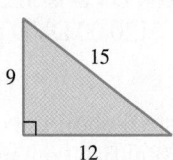

Complete the proportion and then solve for *x*.

$$\frac{3}{\underline{\quad}} = \frac{x}{12}$$

52.
GS

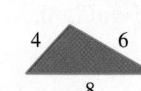 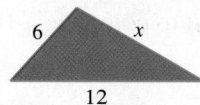

Complete the proportion and then solve for *x*.

$$\frac{x}{6} = \frac{12}{\underline{\quad}}$$

53.

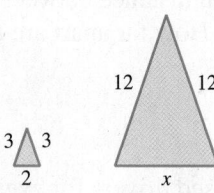

54.

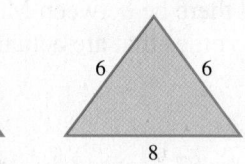

55.

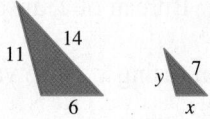

56.

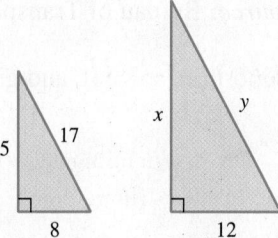

GS Use the information in each problem to complete the diagram of similar triangles. Then write a proportion and solve the problem.

57. An enlarged version of the chair used by George Washington at the Constitutional Convention casts a shadow 18 ft long at the same time a vertical pole 12 ft high casts a shadow 4 ft long. How tall is the chair?

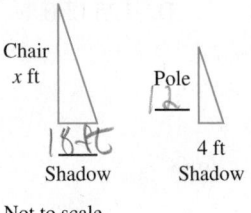

Chair
x ft

Pole

18 ft

4 ft

Shadow   Shadow

Not to scale

58. One of the tallest candles ever constructed was exhibited at the 1897 Stockholm Exhibition. If it cast a shadow 5 ft long at the same time a vertical pole 32 ft high cast a shadow 2 ft long, how tall was the candle?

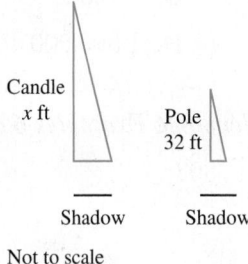

Candle
x ft

Pole
32 ft

Shadow   Shadow

Not to scale

GS Children are often given antibiotics in liquid form, called an oral suspension. Pharmacists make up these suspensions by mixing the medication in powder form with water. They use proportions to calculate the volume of the suspension for the amount of medication that has been prescribed. In Exercises 59 and 60, do each of the following.

(a) Find the total amount of medication in milligrams to be given over the full course of treatment.

(b) Write a proportion that can be solved to find the total volume of the liquid suspension that the pharmacist will prepare. Use x as the variable.

(c) Solve the proportion to determine the total volume of the oral suspension.

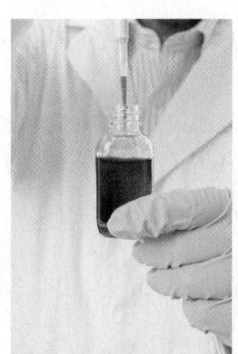

59. Amoxil is a common antibiotic used to treat a variety of infections. Logan's pediatric nurse practitioner has prescribed 375 mg of amoxil a day for 7 days to treat his ear infection. The pharmacist uses 125 mg of amoxil in each 5 mL of the suspension. (*Source:* www.drugs.com)

60. An amoxil oral suspension can also be made by using 250 mg for each 5 mL of suspension. Ava's pediatrician prescribed 900 mg a day for 10 days to treat her bronchitis. (*Source:* www.drugs.com)

The Consumer Price Index (CPI) provides a means of determining the purchasing power of the U.S. dollar from one year to the next. Using the period from 1982 to 1984 as a measure of 100.0, the CPI for selected years from 1998 to 2010 is shown in the table. To use the CPI to predict a price in a particular year, we set up a proportion and compare it with a known price in another year.

$$\frac{\text{price in year } A}{\text{index in year } A} = \frac{\text{price in year } B}{\text{index in year } B}$$

Use the CPI figures in the table to find the amount that would be charged for using the same amount of electricity that cost $225 in 1998. Give answers to the nearest dollar.

Year	Consumer Price Index
1998	163.0
2000	172.2
2002	179.9
2004	188.9
2006	201.6
2008	215.3
2010	218.1

Source: U.S. Bureau of Labor Statistics.

61. in 2000   **62.** in 2002   **63.** in 2006   **64.** in 2010

CONCEPT CHECK   *Use mental techniques to answer each question.*

65. The 2010 U.S. Census gave the population of Alabama as 4,780,000, where 26.0% were African-Americans. What is the best estimate of the African-American population in Alabama? (*Source:* U.S. Census Bureau.)

    A. 500,000   **B.** 750,000

    C. 1,250,000   **D.** 1,500,000

66. The same census gave the population of New Mexico as 2,059,000, with 46.3% being Hispanic. What is the best estimate of the Hispanic population in New Mexico? (*Source:* U.S. Census Bureau.)

    A. 950,000   **B.** 95,000

    C. 1,250,000   **D.** 125,000

*Solve each problem. **See Examples 6 and 7.***

67. What is 14% of 780?

68. What is 26% of 480?

69. 42% of what number is 294?

70. 18% of what number is 108?

71. 120% of what number is 510?

72. 140% of what number is 315?

73. 4 is what percent of 50?

74. 8 is what percent of 64?

75. What percent of 30 is 36?

76. What percent of 48 is 96?

77. Find the discount on a leather recliner with a regular price of $795 if the recliner is 15% off. What is the sale price of the recliner?

78. A laptop computer with a regular price of $597 is on sale at 20% off. Find the amount of the discount and the sale price of the computer.

79. Clayton earned 48 points on a 60-point geometry project. What percent of the total points did he earn?

80. On a 75-point algebra test, Grady scored 63 points. What percent of the total points did he score?

81. Anna saved $1950, which was 65% of the amount she needed for a used car. What was the total amount she needed for the car?

82. Bryn had $525, which was 70% of the total amount she needed for a deposit on an apartment. What was the total deposit she needed?

*Work each percent problem. Round all money amounts to the nearest dollar and percents to the nearest tenth, as needed. **See Examples 5–7.***

83. A family of four with a monthly income of $3800 plans to spend 8% of this amount on entertainment. How much will be spent on entertainment?

84. Quinhon Dac Ho earns $3200 per month. He saves 12% of this amount. How much does he save?

85. In 2010, the U.S. civilian labor force consisted of 153,889,000 persons. Of this total, 14,825,000 were unemployed. What was the percent of unemployment? (*Source:* U.S. Bureau of Labor Statistics.)

The number unemployed	was	what percent	of	the total civilian labor force?
↓	↓	↓	↓	↓
_____	=	p	____	_____

86. In 2010, the U.S. civilian labor force consisted of 124,073,000 persons. Of this total, 14,715,000 were union members. What percent were union members? (*Source:* U.S. Bureau of Labor Statistics.)

What percent	of	the total U.S. labor force	was	the number of union members?
↓	↓	↓	↓	↓
p	____	_____	=	_____

87. As of February 2011, U.S. households owned 377,410,000 pets. Of these, 78,200,000 were dogs. What percent of the pets were *not* dogs? (*Source:* American Pet Product Manufacturers Association.)

88. In the fall of 2010, total public and private school enrollment in the United States was 62.4 million. Of this total, 10.9% of the students were enrolled in private schools. How many students were enrolled in *public* schools? (*Source:* National Center for Education Statistics.)

89. The 1916 dime minted in Denver is quite rare. The 1979 edition of *A Guide Book of United States Coins* listed its value in Extremely Fine condition as $625. The 2012 value had increased to $6200. What was the percent increase in the value of this coin? (*Hint:* First subtract to find the increase in value. Then write a percent equation that uses this increase.)

90. Here is a common business problem: If the sales tax rate is 6.5% and we have collected $3400 in sales tax, how much were sales? (*Hint:* To solve using a percent equation, ask "6.5% of what number is $3400?" Write 6.5% as a decimal.)

Relating Concepts (Exercises 91–94) For Individual or Group Work

Work Exercises 91–94 in order. The steps justify the method of solving a proportion by cross products.

91. What is the LCD of the fractions in the following equation?

$$\frac{x}{6} = \frac{2}{5}$$

92. Solve the equation in **Exercise 91** as follows.

(a) Multiply each side by the LCD. What equation results?

(b) Solve the equation from part (a) by dividing each side by the coefficient of *x*.

93. Solve the equation in **Exercise 91** using cross products.

94. Compare your answers from **Exercises 92(b) and 93.** What do you notice?

Summary Exercises *Applying Problem-Solving Techniques*

The following problems are of the various types discussed in this chapter. Solve each problem.

1. On an algebra test, the highest grade was 42 points more than the lowest grade. The sum of the two grades was 138. Find the lowest grade.

2. Find the measure of an angle whose measure is 70° more than its complement.

3. The perimeter of a certain square is seven times the length of a side, decreased by 12. Find the length of a side.

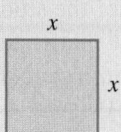

4. Find the measures of the marked angles.

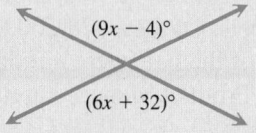

5. If 2 is added to five times a number, the result is equal to 5 more than four times the number. Find the number.

6. Find two consecutive even integers such that four times the greater added to the lesser is 98.

7. Find the measures of the marked angles.

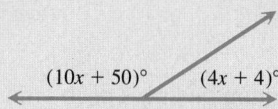

8. A store has 39 qt of milk, some in pint cartons and some in quart cartons. There are six times as many quart cartons as pint cartons. How many quart cartons are there? (*Hint:* 1 qt = 2 pt.)

9. Grant Wood painted his most famous work, *American Gothic*, in 1930 on composition board with perimeter 108.44 in. If the width of the painting is 29.88 in., find the length. Then find the area. Round your answers to the nearest hundredth. (*Source: The Gazette.*)

10. The interest in 1 yr on a deposit of $11,000 was $682. What percent interest was paid?

11. Soccer teams Barcelona and Real Madrid along with baseball's New York Yankees were the best-paid global sports teams during the 2009–2010 season. Barcelona's average pay was $0.5 million more than Real Madrid's. The Yankees average pay was $0.6 million less than Real Madrid's. Average pay for the three teams totaled $22.1 million. What was the average pay for each team? (*Source:* www.sportingintelligence.com)

12. A fully inflated professional basketball has a circumference of 78 cm. What is the radius of a circular cross section through the center of the ball? (Use 3.14 as the approximation for π.) Round your answer to the nearest hundredth.

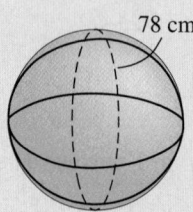

13. A music player that normally sells for $90 is on sale for $75. What is the percent discount on the player?

14. Two slices of bacon contain 85 calories. How many calories are there in twelve slices of bacon?

15. Athletes in vigorous training programs can eat 50 calories per day for every 2.2 lb of body weight. To the nearest hundred, how many calories can a 175 lb athlete consume per day? (*Source: The Gazette.*)

16. In the 2011 Masters Golf Tournament in Augusta, Georgia, Adam Scott finished with a score of 2 more than the tournament winner, Charl Schwartzel. The sum of their scores was 550. Find their scores. (*Source:* www.golf.about.com)

17. Find the best buy (based on price per unit). Give the unit price to the nearest thousandth for that size. (*Source:* HyVee.)

SPAGHETTI SAUCE

Size	Price
14 oz	$1.79
24 oz	$1.77
48 oz	$3.65

18. In the 2010 Winter Olympics in Vancouver, Canada, athletes from the host country earned 26 medals. Seven of every 13 medals were gold. How many gold medals did the Canadian team earn? (*Source: World Almanac and Book of Facts.*)

19. The average cost of a traditional Thanksgiving dinner for 10 people, featuring turkey, stuffing, cranberries, pumpkin pie, and trimmings, was $43.47 in 2010. This price increased 13.2% in 2011. What was the price of this traditional Thanksgiving dinner in 2011? Round your answer to the nearest cent. (*Source:* American Farm Bureau.)

20. Refer to **Exercise 19.** The cost of a traditional Thanksgiving dinner in 1986, the year this information was first collected, was $28.74. What percent increase is the 2011 cost over the 1986 cost? Round your answer to the nearest tenth of a percent. (*Source:* American Farm Bureau.)

2.7 Solving Linear Inequalities

OBJECTIVES

1. Graph intervals on a number line.
2. Use the addition property of inequality.
3. Use the multiplication property of inequality.
4. Solve linear inequalities by using both properties of inequality.
5. Solve linear inequalities with three parts.
6. Use inequalities to solve applied problems.

An **inequality** relates algebraic expressions using the symbols

$<$ "is less than," $\leq$ "is less than or equal to,"

$>$ "is greater than," $\geq$ "is greater than or equal to."

In each case, the interpretation is based on reading the symbol from left to right.

Linear Inequality in One Variable

A **linear inequality in one variable** can be written in the form

$$Ax + B < C, \quad Ax + B \leq C, \quad Ax + B > C, \quad \text{or} \quad Ax + B \geq C,$$

where A, B, and C represent real numbers, and $A \neq 0$.

Some examples of linear inequalities in one variable follow.

$$x + 5 < 2, \quad z - \frac{3}{4} \geq 5, \quad \text{and} \quad 2k + 5 \leq 10 \qquad \text{Linear inequalities}$$

We solve a linear inequality by finding all of its real number solutions. For example, the set

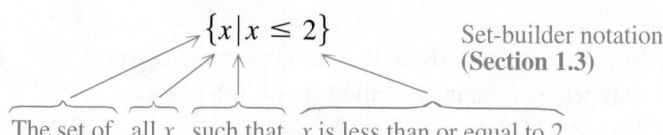

Set-builder notation **(Section 1.3)**

The set of   all x such that x is less than or equal to 2

includes *all real numbers* that are less than or equal to 2, not just the *integers* less than or equal to 2.

OBJECTIVE **1** **Graph intervals on a number line.** Graphing is a good way to show the solution set of an inequality. To graph all real numbers belonging to the set $\{x \mid x \leq 2\}$, we place a square bracket at 2 on a number line and draw an arrow extending from the bracket to the left (since all numbers *less than* 2 are also part of the graph). See **Figure 15.**

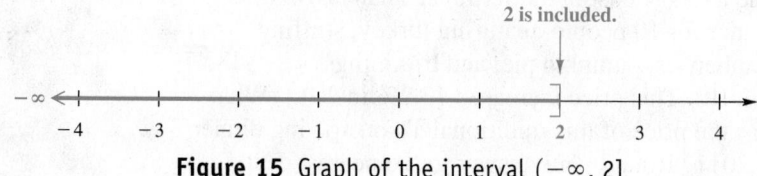

Figure 15 Graph of the interval $(-\infty, 2]$

The set of numbers less than or equal to 2 is an example of an **interval** on the number line. We can write this interval using **interval notation,** which uses the **infinity symbols ∞ or $-\infty$.**

$$(-\infty, 2] \qquad \text{Interval notation}$$

The negative infinity symbol $-\infty$ here does not indicate a number, but shows that the interval includes *all* real numbers less than 2. Again, the square bracket indicates that 2 is part of the solution.

EXAMPLE 1 Graphing an Interval on a Number Line

Graph $x > -5$.

The statement $x > -5$ says that x can represent any value greater than -5 but cannot equal -5, written $(-5, \infty)$. We graph this interval by placing a parenthesis at -5 and drawing an arrow to the right, as in **Figure 16**. The parenthesis at -5 indicates that -5 is *not* part of the graph.

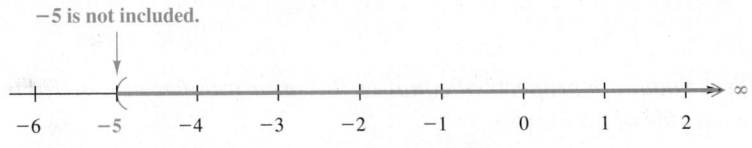

Figure 16 Graph of the interval $(-5, \infty)$

.... **Work Problem ① at the Side.** ▶

Keep the following important concepts regarding interval notation in mind.

1. A parenthesis indicates that an endpoint is *not included* in a solution set.
2. A bracket indicates that an endpoint is *included* in a solution set.
3. A parenthesis is *always* used next to an infinity symbol, $-\infty$ or ∞.
4. The set of all real numbers is written in interval notation as $(-\infty, \infty)$.

EXAMPLE 2 Graphing an Interval on a Number Line

Graph $3 > x$.

The statement $3 > x$ means the same as $x < 3$. *The inequality symbol continues to point to the lesser value.* The graph of $x < 3$, written in interval notation as $(-\infty, 3)$, is shown in **Figure 17.**

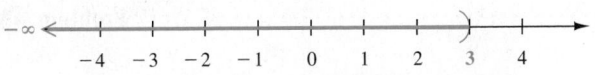

Figure 17 Graph of the interval $(-\infty, 3)$

.... **Work Problem ② at the Side.** ▶

The table summarizes solution sets of linear inequalities.

Set-Builder Notation	Interval Notation	Graph
$\{x \mid x < a\}$	$(-\infty, a)$	
$\{x \mid x \le a\}$	$(-\infty, a]$	
$\{x \mid x > a\}$	(a, ∞)	
$\{x \mid x \ge a\}$	$[a, \infty)$	
$\{x \mid x \text{ is a real number}\}$	$(-\infty, \infty)$	

OBJECTIVE ② Use the addition property of inequality. Consider the true inequality $2 < 5$. If 4 is added to each side, the result is

$$2 + 4 < 5 + 4 \quad \text{Add 4.}$$

$$6 < 9, \quad \text{True}$$

also a true sentence. This suggests the **addition property of inequality.**

① Write each inequality in interval notation, and graph the interval.

(a) $x \le 3$

(b) $x > -4$

(c) $x \le -\dfrac{3}{4}$

② Write each inequality in interval notation and graph the interval.

(a) $-4 \ge x$

(b) $0 < x$

Answers

1. (a) $(-\infty, 3]$

(b) $(-4, \infty)$

(c) $\left(-\infty, -\dfrac{3}{4}\right]$

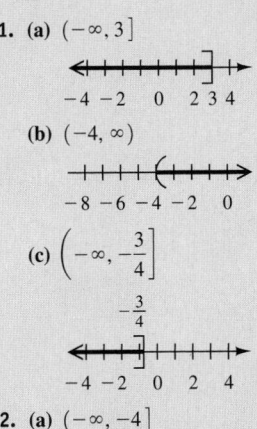

2. (a) $(-\infty, -4]$

(b) $(0, \infty)$

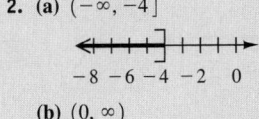
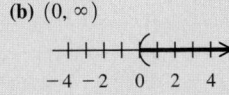

❸ Solve each inequality, and graph the solution set.

(a) $-1 + 8r < 7r + 2$

(b) $5 + 5x \geq 4x + 3$

Answers

3. (a) $(-\infty, 3)$

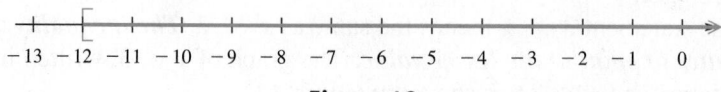

$-4 \;\; -2 \quad 0 \quad 2 \; 3 \; 4$

(b) $[-2, \infty)$

$-4 \; -2 \quad 0 \quad 2 \quad 4$

Addition Property of Inequality

For any real numbers A, B, and C, the inequalities

$$A < B \quad \text{and} \quad A + C < B + C$$

have exactly the same solutions.

In words, the same number may be added to each side of an inequality without changing the solutions.

As with the addition property of equality, the same number may be subtracted from each side of an inequality.

EXAMPLE 3 **Using the Addition Property of Inequality**

Solve $7 + 3k \geq 2k - 5$, and graph the solution set.

$$7 + 3k \geq 2k - 5$$

$7 + 3k - 2k \geq 2k - 5 - 2k$	Subtract $2k$.
$7 + k \geq -5$	Combine like terms.
$7 + k - 7 \geq -5 - 7$	Subtract 7.
$k \geq -12$	Combine like terms.

The solution set is $[-12, \infty)$. Its graph is shown in **Figure 18.**

$-13 \;\; -12 \;\; -11 \;\; -10 \;\; -9 \;\; -8 \;\; -7 \;\; -6 \;\; -5 \;\; -4 \;\; -3 \;\; -2 \;\; -1 \quad 0$

Figure 18

◀ **Work Problem ❸ at the Side.**

Note

Because an inequality has many solutions, we cannot check all of them by substitution as we did with the single solution of an equation. To check the solutions in the **Example 3** solution set $[-12, \infty)$, we use a multistep process. First, we substitute -12 for k in the related *equation.*

CHECK	$7 + 3k = 2k - 5$	Related equation
	$7 + 3(-12) \stackrel{?}{=} 2(-12) - 5$	Let $k = -12$.
	$7 - 36 \stackrel{?}{=} -24 - 5$	Multiply.
	$-29 = -29 \;\checkmark$	True

A true statement results, so -12 is indeed the "boundary" point. Now we test a number other than -12 from the interval $[-12, \infty)$. We choose 0 since it is easy to substitute.

CHECK	$7 + 3k \geq 2k - 5$	Original inequality
	$7 + 3(0) \stackrel{?}{\geq} 2(0) - 5$	Let $k = 0$.
0 is easy to substitute.	$7 \geq -5 \;\checkmark$	True

Again, a true statement results, so the checks confirm that solutions to the inequality are in the interval $[-12, \infty)$. Any number "outside" the interval $[-12, \infty)$, that is, any number in $(-\infty, -12)$, will give a false statement when tested. (Try this.)

OBJECTIVE 3 Use the multiplication property of inequality. Consider the true inequality $3 < 7$. Multiply each side by the positive number 2.

$$3 < 7$$
$$2(3) < 2(7) \qquad \text{Multiply by 2.}$$
$$6 < 14 \qquad \text{True}$$

Now multiply each side of $3 < 7$ by the negative number -5.

$$3 < 7$$
$$-5(3) < -5(7) \qquad \text{Multiply by } -5.$$
$$-15 < -35 \qquad \text{False}$$

To get a true statement when multiplying each side by -5, *we must reverse the direction of the inequality symbol.*

$$3 < 7$$
$$-5(3) > -5(7) \qquad \text{Multiply by } -5. \text{ Reverse}$$
$$\qquad \qquad \qquad \text{the direction of the symbol.}$$
$$-15 > -35 \qquad \text{True}$$

Work Problem 4 at the Side. ▶

These examples suggest the **multiplication property of inequality.**

Multiplication Property of Inequality

Let A, B, and C (where $C \neq 0$) be real numbers.

1. If C is *positive,* then the inequalities

$$A < B \quad \text{and} \quad AC < BC$$

have exactly the same solutions.

2. If C is *negative,* then the inequalities

$$A < B \quad \text{and} \quad AC > BC$$

have exactly the same solutions.

In words, each side of an inequality may be multiplied by the same positive number without changing the solutions. *If the multiplier is negative, we must reverse the direction of the inequality symbol.*

As with the multiplication property of equality, the same nonzero number may be divided into each side.

Note the following differences for positive and negative numbers.

1. When each side of an inequality is multiplied or divided by a *positive number,* the direction of the inequality symbol *does not change.*

2. When each side of an inequality is multiplied or divided by a *negative number,* reverse the direction of the inequality symbol.

4 Work each of the following.

(a) Multiply each side of

$$-3 < 7$$

by 2 and then by -5. Reverse the direction of the inequality symbol if necessary to make a true statement.

(b) Multiply each side of

$$3 > -7$$

by 2 and then by -5. Reverse the direction of the inequality symbol if necessary to make a true statement.

(c) Multiply each side of

$$-7 < -3$$

by 2 and then by -5. Reverse the direction of the inequality symbol if necessary to make a true statement.

Answers

4. (a) $-6 < 14$; $15 > -35$
(b) $6 > -14$; $-15 < 35$
(c) $-14 < -6$; $35 > 15$

⑤ Solve each inequality. Graph the solution set.

GS **(a)** $9x < -18$

$$\frac{9x}{9} \, (</>) \, \frac{-18}{\_\_}$$

$$x \, (</>) \, \_\_$$

The solution set is ___.

———————————————→

GS **(b)** $-2r > -12$

$$\frac{-2r}{\_\_} \, (</>) \, \frac{-12}{\_\_}$$

$$r \, (</>) \, \_\_$$

The solution set is ___.

———————————————→

(c) $-5p \le 0$

———————————————→

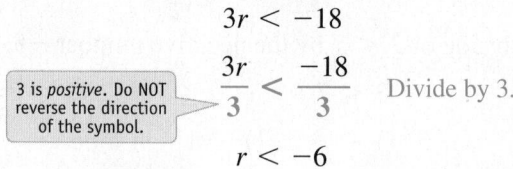

EXAMPLE 4 **Using the Multiplication Property of Inequality**

Solve $3r < -18$, and graph the solution set.

We divide each side by 3, a positive number, so the direction of the inequality symbol *does not* change. *(It does not matter that the number on the right side of the inequality is negative.)*

$$3r < -18$$

3 is positive. Do NOT reverse the direction of the symbol. $\dfrac{3r}{3} < \dfrac{-18}{3}$ Divide by 3.

$$r < -6$$

The graph of the solution set, $(-\infty, -6)$, is shown in **Figure 19**.

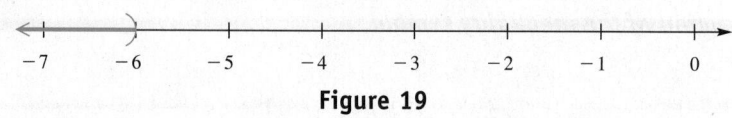

Figure 19

EXAMPLE 5 **Using the Multiplication Property of Inequality**

Solve $-4t \ge 8$, and graph the solution set.

Here each side of the inequality must be divided by -4, a negative number, which *does* require changing the direction of the inequality symbol.

$$-4t \ge 8$$

−4 is negative. Change ≥ to ≤. $\dfrac{-4t}{-4} \le \dfrac{8}{-4}$ Divide by -4.
Reverse the symbol.

$$t \le -2$$

The solution set, $(-\infty, -2]$, is graphed in **Figure 20**.

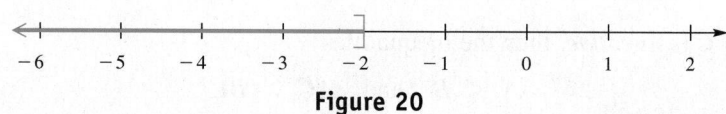

Figure 20

· ◀ **Work Problem ⑤ at the Side.**

OBJECTIVE ▶ **④** **Solve linear inequalities by using both properties of inequality.** The steps to solve a linear inequality are summarized below.

Solving a Linear Inequality

Step 1 **Simplify each side separately.** Use the distributive property to clear parentheses and combine like terms on each side as needed.

Step 2 **Isolate the variable term on one side.** Use the addition property of inequality to get all terms with variables on one side of the inequality and all numbers on the other side.

Step 3 **Isolate the variable.** Use the multiplication property of inequality to write the inequality as $x <$ a number or $x >$ a number.

Remember: Reverse the direction of the inequality symbol only when multiplying or dividing each side of an inequality by a negative number.

EXAMPLE 6	Solving a Linear Inequality

Solve $3x + 2 - 5 > -x + 7 + 2x$. Graph the solution set.

Step 1 Combine like terms and simplify.

$$3x + 2 - 5 > -x + 7 + 2x$$

$$3x - 3 > x + 7$$

Step 2 Use the addition property of inequality.

$$3x - 3 - x > x + 7 - x \qquad \text{Subtract } x.$$

$$2x - 3 > 7 \qquad \text{Combine like terms.}$$

$$2x - 3 + 3 > 7 + 3 \qquad \text{Add 3.}$$

$$2x > 10 \qquad \text{Combine like terms.}$$

Step 3 Use the multiplication property of inequality.

Because 2 is positive, keep the symbol $>$.

$$\frac{2x}{2} > \frac{10}{2} \qquad \text{Divide by 2.}$$

$$x > 5$$

The solution set is $(5, \infty)$. Its graph is shown in **Figure 21.**

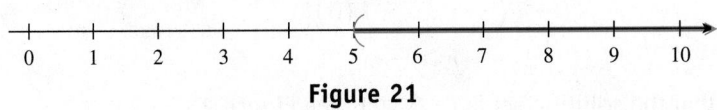

Figure 21

·········· Work Problem **6** at the Side. ▶

EXAMPLE 7	Solving a Linear Inequality

Solve $5(k - 3) - 7k \geq 4(k - 3) + 9$. Graph the solution set.

Step 1 $5(k - 3) - 7k \geq 4(k - 3) + 9$ ◀ Start by clearing parentheses.

$$5k - 15 - 7k \geq 4k - 12 + 9 \qquad \text{Distributive property}$$

$$-2k - 15 \geq 4k - 3 \qquad \text{Combine like terms.}$$

Step 2 $-2k - 15 - 4k \geq 4k - 3 - 4k$ Subtract $4k$.

$$-6k - 15 \geq -3 \qquad \text{Combine like terms.}$$

$$-6k - 15 + 15 \geq -3 + 15 \qquad \text{Add 15.}$$

$$-6k \geq 12 \qquad \text{Combine like terms.}$$

Step 3 $\dfrac{-6k}{-6} \leq \dfrac{12}{-6}$ Divide by -6. Reverse the symbol.

Because -6 is negative, change $\geq$ to $\leq$.

$$k \leq -2$$

The solution set is $(-\infty, -2]$. Its graph is shown in **Figure 22.**

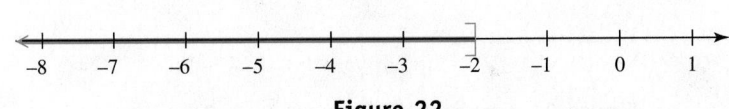

Figure 22

·········· Work Problem **7** at the Side. ▶

6 Solve.

$$7x - 6 + 1 \geq 5x - x + 2$$

Graph the solution set.

⟶

7 Solve.

$$-15 - (2x + 1) \geq 4(x - 1) - 3x$$

Graph the solution set.

⟶

Answers

6. $\left[\dfrac{7}{3}, \infty\right)$

$$\frac{7}{3}$$

‹+++++[+++›
  -2   0   2 3 4

7. $(-\infty, -4]$

‹++]+++++›
 -6 -4 -2   0

8 Solve, check, and graph the solution set of each inequality.

(a) $\frac{3}{4}(x-2) + \frac{1}{2} > \frac{1}{5}(x-8)$

(b) $\frac{1}{4}(m+3) + 2 \leq \frac{3}{4}(m+8)$

9 Write each inequality in interval notation and graph the interval.

(a) $-7 < x < -2$

(b) $-6 < x \leq -4$

Answers

8. (a) $\left(-\frac{12}{11}, \infty\right)$

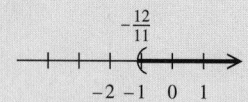

(b) $\left[-\frac{13}{2}, \infty\right)$

9. (a) $(-7, -2)$

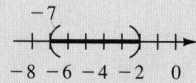

(b) $(-6, -4]$

EXAMPLE 8 **Solving a Linear Inequality with Fractions**

Solve $-\frac{2}{3}(r-3) - \frac{1}{2} < \frac{1}{2}(5-r)$, and graph the solution set.

$$-\frac{2}{3}(r-3) - \frac{1}{2} < \frac{1}{2}(5-r)$$

Step 1 $\quad 6\left[-\frac{2}{3}(r-3) - \frac{1}{2}\right] < 6\left[\frac{1}{2}(5-r)\right]$ Multiply by 6, the LCD.

$$6\left[-\frac{2}{3}(r-3)\right] - 6\left(\frac{1}{2}\right) < 6\left[\frac{1}{2}(5-r)\right]$$ Distributive property

Be careful here.

$$-4(r-3) - 3 < 3(5-r)$$ Multiply.

$$-4r + 12 - 3 < 15 - 3r$$ Distributive property

$$-4r + 9 < 15 - 3r$$ Combine like terms.

Step 2 $\quad -4r + 9 + 3r < 15 - 3r + 3r$ Add 3r.

$$-r + 9 < 15$$ Combine like terms.

$$-r + 9 - 9 < 15 - 9$$ Subtract 9.

$$-r < 6$$ Combine like terms.

Step 3 $\quad -1(-r) > -1(6)$ Multiply by -1. Change $<$ to $>$.

$$r > -6$$

Check that the solution set is $(-6, \infty)$. See **Figure 23.**

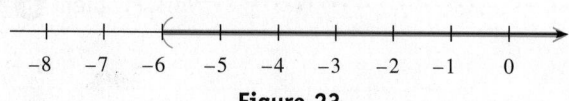

Figure 23

◄ **Work Problem 8 at the Side.**

OBJECTIVE ▶ 5 **Solve linear inequalities with three parts.** An inequality that says that one number is *between* two other numbers is a **three-part inequality.** For example,

$$-3 < 5 < 7 \quad \text{says that} \quad 5 \quad \text{is between} \quad -3 \text{ and } 7.$$

EXAMPLE 9 **Graphing a Three-Part Inequality**

Write the inequality $-3 \leq x < 2$ in interval notation, and graph the interval.

The statement is read "-3 is less than or equal to *x and x* is less than 2." We want the set of numbers that are *between* -3 and 2, with -3 included and 2 excluded. In interval notation, we write $[-3, 2)$, using a square bracket at -3 because -3 is part of the graph and a parenthesis at 2 because 2 is not part of the graph. See **Figure 24.**

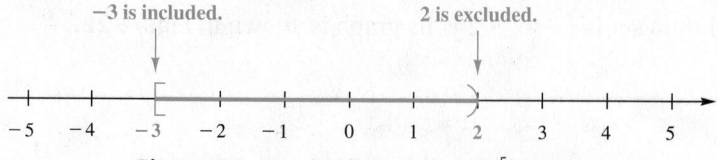

Figure 24 Graph of the interval $[-3, 2)$

◄ **Work Problem 9 at the Side.**

The three-part inequality $3 < x + 2 < 8$ says that $x + 2$ is between 3 and 8. We solve this inequality as follows.

$$3 < \quad x + 2 \quad < 8$$
$$3 - 2 < x + 2 - 2 < 8 - 2 \qquad \text{Subtract 2 from } each \text{ part.}$$
$$1 < \quad x \quad < 6$$

The idea is to get the inequality in the form

a number $< x <$ **another number.**

> ### CAUTION
> *Three-part inequalities are written so that the symbols point in the same direction and both point toward the lesser number.* It would be *wrong* to write an inequality as $8 < x + 2 < 3$, since this would imply that $8 < 3$, a false statement.

The table summarizes solution sets of three-part inequalities.

Set-Builder Notation	Interval Notation	Graph
$\{x \mid a < x < b\}$	(a, b)	
$\{x \mid a < x \le b\}$	$(a, b]$	
$\{x \mid a \le x < b\}$	$[a, b)$	
$\{x \mid a \le x \le b\}$	$[a, b]$	

EXAMPLE 10 Solving a Three-Part Inequality

Solve each inequality, and graph the solution set.

(a)
$$4 \le \quad 3x - 5 \quad < 10$$
$$4 + 5 \le 3x - 5 + 5 < 10 + 5 \qquad \text{Add 5 to each part.}$$
$$9 \le \quad 3x \quad < 15$$

Remember to divide all *three* parts by 3.

$$\frac{9}{3} \le \quad \frac{3x}{3} \quad < \frac{15}{3} \qquad \text{Divide each part by 3.}$$
$$3 \le \quad x \quad < 5$$

The solution set is $[3, 5)$. Its graph is shown in **Figure 25.**

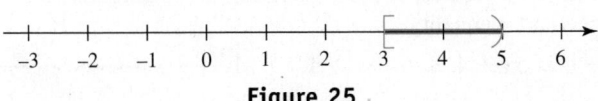

Figure 25

Continued on Next Page

10 Solve each inequality, and graph the solution set.

(a) $2 \le 3x - 1 \le 8$

_____→

(b) $-4 \le \dfrac{3}{2}x - 1 \le 0$

_____→

(b) $-4 \le \dfrac{2}{3}m - 1 < 8$ ◁ Work with all three parts at the same time.

$$3(-4) \le 3\left(\dfrac{2}{3}m - 1\right) < 3(8)$$ Multiply each part by 3 to clear the fraction.

$$-12 \le 2m - 3 < 24$$ Multiply. Use the distributive property.

$$-12 + 3 \le 2m - 3 + 3 < 24 + 3$$ Add 3 to each part.

$$-9 \le 2m < 27$$

$$\dfrac{-9}{2} \le \dfrac{2m}{2} < \dfrac{27}{2}$$ Divide each part by 2.

$$-\dfrac{9}{2} \le m < \dfrac{27}{2}$$

The solution set is $\left[-\dfrac{9}{2}, \dfrac{27}{2}\right)$. Its graph is shown in **Figure 26.**

Think: $-\dfrac{9}{2} = -4\dfrac{1}{2}$ Think: $\dfrac{27}{2} = 13\dfrac{1}{2}$

$-\dfrac{9}{2}$ $\dfrac{27}{2}$

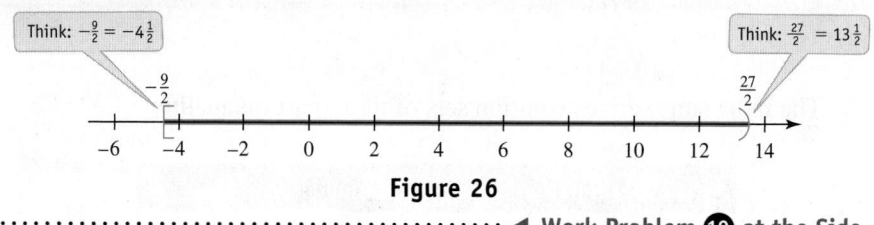

Figure 26

◀ **Work Problem 10 at the Side.**

Note

The inequality in **Example 10(b)**, $-4 \le \dfrac{2}{3}m - 1 < 8$, can also be solved by first adding 1 to each part and then multiplying each part by $\dfrac{3}{2}$. Try this.

Types of solution sets for linear equations and linear inequalities are summarized here.

SOLUTIONS OF LINEAR EQUATIONS AND INEQUALITIES

Equation or Inequality	Typical Solution Set	Graph of Solution Set
Linear equation $5x + 4 = 14$	$\{2\}$	• 2
Linear inequality $5x + 4 < 14$	$(-\infty, 2)$	←———) 2
$5x + 4 > 14$	$(2, \infty)$	(———→ 2
Three-part inequality $-1 \le 5x + 4 \le 14$	$[-1, 2]$	[———] −1 2

Answers

10. (a) $[1, 3]$

0 1   3

(b) $\left[-2, \dfrac{2}{3}\right]$

$\dfrac{2}{3}$

−3 −2 −1 0 1 2 3

OBJECTIVE ▶ **6** **Use inequalities to solve applied problems.** The table gives some common phrases that suggest inequality.

Phrase/Word	Example	Inequality
Is greater than	A number *is greater than* 4	$x > 4$
Is less than	A number *is less than* -12	$x < -12$
Exceeds	A number *exceeds* 3.5	$x > 3.5$
Is at least	A number *is at least* 6	$x \geq 6$
Is at most	A number *is at most* 8	$x \leq 8$

Work Problem 11 at the Side. ▶

The next example uses the idea of finding the average of a number of scores. ***In general, to find the average of n numbers, add the numbers and divide by n.*** We use the six problem-solving steps from **Section 2.4,** changing Step 3 to "Write an inequality."

EXAMPLE 11 **Finding an Average Test Score**

Brent has scores of 86, 88, and 78 (out of a possible 100) on each of his first three tests in geometry. If he wants an average of at least 80 after his fourth test, what are the possible scores he can make on that test?

Step 1 **Read** the problem again.

Step 2 **Assign a variable.**

Let x = Brent's score on his fourth test.

Step 3 **Write an inequality.**

$$\underset{\text{Average}}{\underbrace{\frac{86 + 88 + 78 + x}{4}}} \underset{\substack{\text{is at}\\ \text{least }80.}}{\geq \quad 80}$$

To find his average after four tests, add the test scores and divide by 4.

Step 4 **Solve.**

$$\frac{252 + x}{4} \geq 80 \qquad \text{Add the known scores.}$$

$$4\left(\frac{252 + x}{4}\right) \geq 4\,(80) \qquad \text{Multiply by 4.}$$

$$252 + x \geq 320$$

$$252 + x - 252 \geq 320 - 252 \qquad \text{Subtract 252.}$$

$$x \geq 68 \qquad \text{Combine like terms.}$$

Step 5 **State the answer.** He must score 68 or more on the fourth test to have an average of *at least* 80.

Step 6 **Check.** $\dfrac{86 + 88 + 78 + 68}{4} = \dfrac{320}{4} = 80$

To complete the check, also show that any number greater than 68 (but less than or equal to 100) makes the average greater than 80.

·············· **Work Problem 12 at the Side.** ▶

11 Translate each statement into an inequality, using x as the variable.

(a) The total cost is less than $10.

(b) Chicago received at most 5 in. of snow.

(c) The car's speed exceeded 60 mph.

(d) You must be at least 18 yr old to vote.

12 Solve each problem.

(a) Kristine has grades of 98 and 85 on her first two tests in algebra. If she wants an average of at least 92 after her third test, what score must she make on that test?

(b) Maggie has scores of 98, 86, and 88 on her first three tests in algebra. If she wants an average of at least 90 after her fourth test, what score must she make on her fourth test?

Answers

11. **(a)** $x < 10$ **(b)** $x \leq 5$
(c) $x > 60$ **(d)** $x \geq 18$
12. **(a)** 93 or more **(b)** 88 or more

13 Solve each problem.

(a) A rental company charges $10 to rent a leaf blower, plus $7.50 per hr. Marge Ruhberg can spend no more than $40 to blow leaves from her driveway and pool deck. What is the maximum amount of time she can use the rented leaf blower?

Step 1
Find the (*minimum / maximum*) amount of time.

Step 2
Let h = the number of _____ she can use the leaf blower.

Step 3
Write an inequality.

_____

Solve the inequality.

$h\,(\leq / \geq)\,$_____

The maximum amount of time she can use the blower is _____ .

(b) A local health club charges a $40 one-time enrollment fee, plus $35 per month for a membership. Sara can spend no more than $355 on this exercise expense. What is the maximum number of months that Sara can belong to this health club?

| EXAMPLE 12 | Using a Linear Inequality to Solve a Rental Problem |

A rental company charges $20 to rent a chain saw, plus $9 per hr. Tom Ruhberg can spend no more than $65 to clear some logs from his yard. What is the maximum amount of time he can use the rented saw?

Step 1 **Read** the problem again.

Step 2 **Assign a variable.** Let h = the number of hours he can rent the saw.

Step 3 **Write an inequality.** He must pay $20, plus $9h$, to rent the saw for h hours, and this amount must be *no more than* $65.

Cost of renting	is no more than	65 dollars.
$20 + 9h$	$\leq$	65

Step 4 **Solve.** $9h \leq 45$ Subtract 20.

$h \leq 5$ Divide by 9.

Step 5 **State the answer.** He can use the saw for a maximum of 5 hr. (He may use it for less time, as indicated by the inequality $h \leq 5$.)

Step 6 **Check.** If Tom uses the saw for **5** hr, he will spend

$$20 + 9\,(5) = 65 \text{ dollars, the maximum amount.}$$

◀ **Work Problem 13** at the Side.

Answers

13. (a) maximum; hours; $10 + 7.50h \leq 40$; $\leq$; 4; 4 hr
   (b) 9 months

2.7 Exercises

Download the MyDashBoard App

MyMathLab®

CONCEPT CHECK *Work each problem.*

1. When graphing an inequality, use a parenthesis if the inequality symbol is _____ or _____. Use a square bracket if the inequality symbol is _____ or _____.

2. *True* or *false*? In interval notation, a square bracket is sometimes used next to an infinity symbol.

3. In interval notation, the set $\{x \mid x > 0\}$ is written _____.

4. In interval notation, the set of all real numbers is written _____.

CONCEPT CHECK *Write an inequality using the variable x that corresponds to each graph of solutions on a number line.*

5.
$$-4\ -3\ -2\ -1\ \ 0\ \ 1\ \ 2\ \ 3$$

6.
$$-4\ -3\ -2\ -1\ \ 0\ \ 1\ \ 2\ \ 3$$

7.
$$-2\ -1\ \ 0\ \ 1\ \ 2\ \ 3\ \ 4\ \ 5$$

8.
$$-2\ -1\ \ 0\ \ 1\ \ 2\ \ 3\ \ 4\ \ 5$$

9.
$$-1\ \ \ \ 0\ \ \ \ 1\ \ \ \ 2$$

10.
$$-1\ \ \ \ 0\ \ \ \ 1\ \ \ \ 2$$

Write each inequality in interval notation, and graph the interval. **See Examples 1, 2, and 9.**

11. $k \leq 4$

12. $r \leq -10$

13. $x > -3$

14. $x > 3$

15. $8 \leq x \leq 10$

16. $3 \leq x \leq 5$

17. $0 < x \leq 10$

18. $-3 \leq x < 5$

Solve each inequality. Write the solution set in interval notation, and graph it. **See Example 3.**

19. $z - 8 \geq -7$

20. $p - 3 \geq -11$

21. $2k + 3 \geq k + 8$

22. $3x + 7 \geq 2x + 11$

23. $3n + 5 < 2n - 1$

24. $5x - 2 < 4x - 5$

25. Under what conditions must the inequality symbol be reversed when using the multiplication property of inequality?

26. Explain the steps you would use to solve the inequality $-5x > 20$.

Solve each inequality. Write the solution set in interval notation, and graph it.
See Examples 4 and 5.

27. $3x < 18$

28. $5x < 35$

29. $2x \geq -20$

30. $6m \geq -24$

31. $-8t > 24$

32. $-7x > 49$

33. $-x \geq 0$

34. $-k < 0$

35. $-\dfrac{3}{4}r < -15$

36. $-\dfrac{7}{8}t < -14$

37. $-0.02x \leq 0.06$

38. $-0.03v \geq -0.12$

Solve each inequality. Write the inequality in interval notation, and graph it.
See Examples 3–8.

39. $8x + 9 \leq -15$

40. $6x + 7 \leq -17$

41. $-4x - 3 < 1$

42. $-5x - 4 < 6$

43. $5r + 1 \geq 3r - 9$

44. $6t + 3 < 3t + 12$

45. $6x + 3 + x < 2 + 4x + 4$

46. $-4w + 12 + 9w \geq w + 9 + w$

47. $-x + 4 + 7x \leq -2 + 3x + 6$

48. $14x - 6 + 7x > 4 + 10x - 10$

49. $5(t - 1) > 3(t - 2)$

50. $7(m - 2) < 4(m - 4)$

51. $5(x + 3) - 6x \leq 3(2x + 1) - 4x$

52. $2(x - 5) + 3x < 4(x - 6) + 1$

53. $\dfrac{2}{3}(p + 3) > \dfrac{5}{6}(p - 4)$

54. $\dfrac{7}{9}(n - 4) \leq \dfrac{4}{3}(n + 5)$

55. $\frac{2}{3}(3k - 1) \geq \frac{3}{2}(2k - 3)$

56. $\frac{7}{5}(10m - 1) < \frac{2}{3}(6m + 5)$

57. $-\frac{1}{4}(p + 6) + \frac{3}{2}(2p - 5) < 10$

58. $\frac{3}{5}(k - 2) - \frac{1}{4}(2k - 7) \leq 3$

Solve each inequality. Write the solution set in interval notation, and graph it.
See Example 10.

59. $-5 \leq 2x - 3 \leq 9$

60. $-7 \leq 3x - 4 \leq 8$

61. $10 < 7p + 3 < 24$

62. $-8 \leq 3r - 1 \leq -1$

63. $-12 < -1 + 6m \leq -5$

64. $-14 \leq 1 + 5q < 3$

65. $-12 \leq \frac{1}{2}z + 1 \leq 4$

66. $-6 \leq 3 + \frac{1}{3}x \leq 5$

67. $1 \leq \frac{2}{3}p + 3 \leq 7$

68. $2 < \frac{3}{4}x + 6 < 12$

CONCEPT CHECK *Translate each statement into an inequality. Use x as the variable.*

69. You must be at least 16 yr old to drive.

70. Less than 1 in. of rain fell.

71. Denver received more than 8 in. of snow.

72. A full-time student must take at least 12 credits.

73. Tracy could spend at most $20 on a gift.

74. The wind speed exceeded 40 mph.

Solve each problem. See Examples 11 and 12.

75. John Douglas has grades of 84 and 98 on his first two history tests. What must he score on his third test so that his average is at least 90?

76. Elizabeth Gainey has scores of 74 and 82 on her first two algebra tests. What must she score on her third test so that her average is at least 80?

77. A student has scores of 87, 84, 95, and 79 on four quizzes. What must she score on the fifth quiz to have an average of at least 85?

78. Another student has scores of 82, 93, 94, and 86 on four quizzes. What must he score on the fifth quiz to have an average of at least 90?

79. When 2 is added to the difference between six times a number and 5, the result is greater than 13 added to 5 times the number. Find all such numbers.

80. When 8 is subtracted from the sum of three times a number and 6, the result is less than 4 more than the number. Find all such numbers.

81. The formula for converting Celsius temperature to Fahrenheit is $F = \frac{9}{5}C + 32$. The Fahrenheit temperature of Providence, Rhode Island, has never exceeded 104°. How would you describe this using Celsius temperature?

82. The formula for converting Fahrenheit temperature to Celsius is $C = \frac{5}{9}(F - 32)$. If the Celsius temperature on a certain day in San Diego, California, is never more than 25°, how would you describe the corresponding Fahrenheit temperature?

83. For what values of x would the rectangle have perimeter of at least 400?

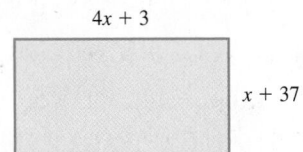

84. For what values of x would the triangle have perimeter of at least 72?

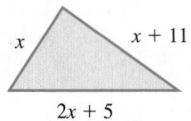

85. An international phone call costs $2.00, plus $0.30 per minute or fractional part of a minute. If x represents the number of minutes of the length of the call, then $2 + 0.30x$ represents the cost of the call. If Jorge has $5.60 to spend on a call, what is the maximum total time he can use the phone?

86. At the Speedy Gas 'n Go, a car wash costs $4.50, and gasoline is selling for $3.40 per gal. Terri Hoelker has $43.60 to spend, and her car is so dirty that she must have it washed. What is the maximum number of gallons of gasoline that she can purchase?

Relating Concepts (Exercises 87–90) For Individual or Group Work

Work Exercises 87–90 in order, *to see the connection between the solution of an equation and the solutions of the corresponding inequalities.*
In Exercises 87 and 88, solve and graph the solutions.

87. $3x + 2 = 14$

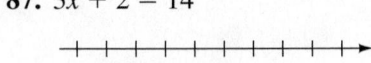

88. (a) $3x + 2 < 14$

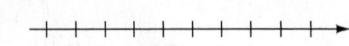

(b) $3x + 2 > 14$

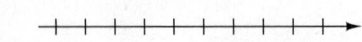

89. Now graph all the solutions together on the following number line.

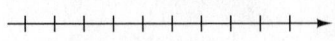

Describe the graph.

90. Based on the results from **Exercises 87–89,** if we were to graph the solutions of

$$-4x + 3 = -1, \quad -4x + 3 > -1,$$
$$\text{and} \quad -4x + 3 < -1$$

on the same number line, describe the graph. Give this solution set using interval notation.

Chapter 2 *Summary*

Key Terms

2.1

linear equation A linear equation in one variable is an equation that can be written in the form $Ax + B = C$, where A, B, and C are real numbers, with $A \neq 0$.

solution set The set of all solutions of an equation is its solution set.

equivalent equations Equations that have exactly the same solution sets are equivalent equations.

2.3

conditional equation A conditional equation is an equation that is true for some values of the variable and false for others.

identity An identity is an equation that is true for all values of the variable.

contradiction A contradiction is an equation with no solution.

2.4

consecutive integers Two integers that differ by 1 are consecutive integers.

consecutive even (or odd) integers Two even (or odd) integers that differ by 2 are consecutive even (or odd) integers.

complementary angles Two angles whose measures have a sum of 90° are complementary angles.

right angle A right angle measures 90°.

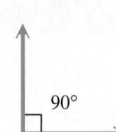

supplementary angles Two angles whose measures have a sum of 180° are supplementary angles.

straight angle A straight angle measures 180°.

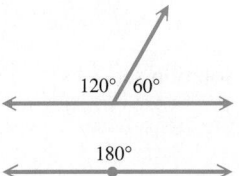

2.5

formula A formula is an equation in which variables are used to describe a relationship.

area The area of a plane geometric figure is a measure of the surface covered by the figure.

perimeter The perimeter of a plane geometric figure is the distance around the figure.

vertical angles Vertical angles are angles formed by intersecting lines. They have the same measure. (In the figure, ① and ③ are vertical angles, as are ② and ④.)

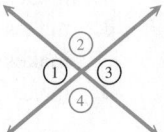

2.6

ratio A ratio is a comparison of two quantities using a quotient.

proportion A proportion is a statement that two ratios are equal.

cross products of a proportion The method of cross products provides a way of determining whether a proportion is true.

$\dfrac{a}{b} = \dfrac{c}{d}$ *ad* and *bc* are cross products.

terms In the proportion $\frac{a}{b} = \frac{c}{d}$, a, b, c, and d are the terms. The a and d terms are the **extremes,** and the b and c terms are the **means**.

2.7

inequality Inequalities are statements relating algebraic expressions using $<$, $\leq$, $>$, or $\geq$.

linear inequality A linear inequality in one variable can be written in the form $Ax + B < C$, $Ax + B \leq C$, $Ax + B > C$, or $Ax + B \geq C$, where A, B, and C are real numbers, with $A \neq 0$.

interval An interval is a portion of a number line.

interval notation Interval notation is a special notation that uses parentheses ( ) and/or brackets [ ] to describe an interval on a number line.

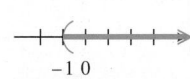

The interval $(-1, \infty)$

three-part inequality An inequality that says that one number is between two other numbers is a three-part inequality.

New Symbols

$\emptyset$ empty (null) set	a to b, $a:b$, or $\frac{a}{b}$	the ratio of a to b	∞ infinity
$1°$ one degree	(a, b)	interval notation for $a < x < b$	$-\infty$ negative infinity
$\urcorner$ right angle	$[a, b]$	interval notation for $a \leq x \leq b$	$(-\infty, \infty)$ set of all real numbers

Test Your Word Power

See how well you have learned the vocabulary in this chapter.

1 A **solution** of an equation is a number that
- **A.** makes an expression undefined
- **B.** makes the equation false
- **C.** makes the equation true
- **D.** makes an expression equal to 0.

2 **Complementary angles** are angles
- **A.** formed by two parallel lines
- **B.** whose sum is 90°
- **C.** whose sum is 180°
- **D.** formed by perpendicular lines.

3 **Supplementary angles** are angles
- **A.** formed by two parallel lines
- **B.** whose sum is 90°
- **C.** whose sum is 180°
- **D.** formed by perpendicular lines.

4 A **ratio**
- **A.** compares two quantities using a quotient
- **B.** says that two quotients are equal
- **C.** is a product of two quantities
- **D.** is a difference between two quantities.

5 A **proportion**
- **A.** compares two quantities using a quotient
- **B.** says that two ratios are equal
- **C.** is a product of two quantities
- **D.** is a difference between two quantities.

Answers to Test Your Word Power

1. C; *Example:* 8 is the solution of $2x + 5 = 21$.

2. B; *Example:* Angles with measures 35° and 55° are complementary angles.

3. C; *Example:* Angles with measures 112° and 68° are supplementary angles.

4. A; *Example:* $\frac{7 \text{ in.}}{12 \text{ in.}} = \frac{7}{12}$

5. B; *Example:* $\frac{2}{3} = \frac{8}{12}$

Quick Review

Concepts	*Examples*
2.1 The Addition Property of Equality The same number may be added to (or subtracted from) each side of an equation without changing the solution set.	Solve. $\qquad x - 6 = 12$ $x - 6 + 6 = 12 + 6 \qquad$ Add 6. $\qquad\qquad x = 18 \qquad$ Combine like terms. Solution set: $\{18\}$
2.2 The Multiplication Property of Equality Each side of an equation may be multiplied (or divided) by the same nonzero number without changing the solution set.	Solve. $\qquad \dfrac{3}{4}x = -9$ $\dfrac{4}{3} \cdot \dfrac{3}{4}x = \dfrac{4}{3}(-9) \qquad$ Multiply by $\frac{4}{3}$. $\qquad\qquad x = -12$ Solution set: $\{-12\}$

Concepts	Examples

2.3 More on Solving Linear Equations

Step 1 Simplify each side separately.

Step 2 Isolate the variable term on one side.

Step 3 Isolate the variable.

Step 4 Check.

Solve.
$$2x + 2(x + 1) = 14 + x$$
$$2x + 2x + 2 = 14 + x \quad \text{Distributive property}$$
$$4x + 2 = 14 + x \quad \text{Combine like terms.}$$
$$4x + 2 - x - 2 = 14 + x - x - 2$$
$$\text{Subtract } x. \text{ Subtract 2.}$$
$$3x = 12 \quad \text{Combine like terms.}$$
$$\frac{3x}{3} = \frac{12}{3} \quad \text{Divide by 3.}$$
$$x = 4$$

CHECK $2(4) + 2(4 + 1) \stackrel{?}{=} 14 + 4$ Let $x = 4$.
$$18 = 18 \checkmark \quad \text{True}$$
Solution set: $\{4\}$

2.4 An Introduction to Applications of Linear Equations

Step 1 Read.

Step 2 Assign a variable.

Step 3 Write an equation.
Step 4 Solve the equation.

Step 5 State the answer.
Step 6 Check.

One number is 5 more than another. Their sum is 21. What are the numbers?

We are looking for two numbers.

Let $\quad x = $ the lesser number.
Then $\quad x + 5 = $ the greater number.
$$x + (x + 5) = 21$$
$$2x + 5 = 21 \quad \text{Combine like terms.}$$
$$2x = 16 \quad \text{Subtract 5.}$$
$$x = 8 \quad \text{Divide by 2.}$$
The numbers are 8 and 13.

13 is 5 more than 8, and $8 + 13 = 21$. The answer checks.

2.5 Formulas and Additional Applications from Geometry

To find the value of one of the variables in a formula, given values for the others, substitute the known values into the formula.

To solve a formula for one of the variables, isolate that variable by treating the other variables as numbers and using the steps for solving equations.

Find L if $A = LW$, given that $A = 24$ and $W = 3$.
$$24 = L \cdot 3 \quad A = 24, W = 3$$
$$\frac{24}{3} = \frac{L \cdot 3}{3} \quad \text{Divide by 3.}$$
$$8 = L$$

Solve $P = 2a + 2b$ for b.
$$P - 2a = 2a + 2b - 2a \quad \text{Subtract } 2a.$$
$$P - 2a = 2b \quad \text{Combine like terms.}$$
$$\frac{P - 2a}{2} = \frac{2b}{2} \quad \text{Divide by 2.}$$
$$\frac{P - 2a}{2} = b, \quad \text{or} \quad b = \frac{P - 2a}{2}$$

Concepts	Examples

2.6 Ratio, Proportion, and Percent

To write a ratio, express quantities in the same units.

4 ft to 8 in. can be written **48 in.** to 8 in., which is the ratio

$$\frac{48}{8}, \quad \text{or} \quad \frac{6}{1}.$$

To solve a proportion, use the method of cross products.

Solve. $\dfrac{x}{12} = \dfrac{35}{60}$

$60x = 12 \cdot 35$ Cross products

$60x = 420$ Multiply.

$x = 7$ Divide by 60.

Solution set: $\{7\}$

To solve a percent problem, use the percent equation.

amount = percent (as a decimal) · base

65   is   what percent   of   325?

$\downarrow$ $\downarrow$ $\downarrow$ $\downarrow$ $\downarrow$

65   = p ·   325

$$\frac{65}{325} = p$$

$$0.2 = p, \quad \text{or} \quad 20\% = p$$

65 is **20%** of 325.

2.7 Solving Linear Inequalities

Step 1 Simplify each side separately.

Solve and graph the solution set.

$3(1 - x) + 5 - 2x > 9 - 6$

$3 - 3x + 5 - 2x > 9 - 6$ Distributive property

$8 - 5x > 3$ Combine like terms.

Step 2 Isolate the variable term on one side.

$8 - 5x - 8 > 3 - 8$ Subtract 8.

$-5x > -5$ Combine like terms.

Step 3 Isolate the variable.

$\dfrac{-5x}{-5} < \dfrac{-5}{-5}$ Divide by -5.

               Change $>$ to $<$.

Be sure to reverse the direction of the inequality symbol when multiplying or dividing by a negative number.

$x < 1$

Solution set: $(-\infty, 1)$

To solve a three-part inequality such as

$$4 < 2x + 6 < 8,$$

work with all three parts at the same time.

Solve. $4 < 2x + 6 < 8$

$4 - 6 < 2x + 6 - 6 < 8 - 6$ Subtract 6.

$-2 < 2x < 2$

$\dfrac{-2}{2} < \dfrac{2x}{2} < \dfrac{2}{2}$ Divide by 2.

$-1 < x < 1$

Solution set: $(-1, 1)$

Chapter 2 Review Exercises

2.1–2.3 *Solve each equation. Check the solution.*

1. $x - 7 = 2$

2. $4r - 6 = 10$

3. $5x + 8 = 4x + 2$

4. $8t = 7t + \dfrac{3}{2}$

5. $(4r - 8) - (3r + 12) = 0$

6. $7(2x + 1) = 6(2x - 9)$

7. $-\dfrac{6}{5}y = -18$

8. $\dfrac{1}{2}r - \dfrac{1}{6}r + 3 = 2 + \dfrac{1}{6}r + 1$

9. $3x - (-2x + 6) = 4(x - 4) + x$

10. $0.10(x + 80) + 0.20x = 8 + 0.30x$

2.4 *Solve each problem.*

11. If 7 is added to five times a number, the result is equal to three times the number. Find the number.

12. If 4 is subtracted from twice a number, the result is 36. Find the number.

13. The land area of Hawaii is 5213 mi² greater than that of Rhode Island. Together, the areas total 7637 mi². What is the area of each state?

14. The height of Seven Falls in Colorado is $\frac{5}{2}$ the height (in feet) of Twin Falls in Idaho. The sum of the heights is 420 ft. Find the height of each.

15. The supplement of an angle measures 10 times the measure of its complement. What is the measure of the angle (in degrees)?

16. Find two consecutive odd integers such that when the lesser is added to twice the greater, the result is 24 more than the greater integer.

2.5 *A formula is given in each exercise, along with the values for all but one of the variables. Find the value of the variable that is not given. (For Exercises 19 and 20, use 3.14 as an approximation for π.)*

17. $A = \dfrac{1}{2}bh; A = 44, b = 8$

18. $A = \dfrac{1}{2}h(b + B); b = 3, B = 4, h = 8$

19. $C = 2\pi r; C = 29.83$

20. $V = \dfrac{4}{3}\pi r^3; r = 9$

Solve each formula for the specified variable.

21. $A = bh$ for h

22. $A = \dfrac{1}{2}h(b + B)$ for h

23. Solve each equation for y.

 (a) $x + y = 11$ **(b)** $3x - 2y = 12$

Find the measure of each marked angle.

24.

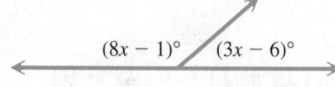

25.

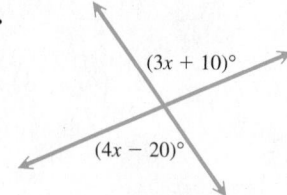

▦ *Solve each application of geometry.*

26. A cinema screen in Indonesia has length 92.75 ft and width 70.5 ft. What is the perimeter? What is the area? (*Source: Guinness World Records.*)

27. A Montezuma cypress in Mexico is 137 ft tall and has a circumference of about 146.9 ft. What is the diameter of the tree? What is its radius? Use 3.14 as an approximation for π. Round answers to the nearest hundredth. (*Source: Guinness World Records.*)

2.6 *Write a ratio for each word phrase, giving fractions in lowest terms.*

28. 60 cm to 40 cm

29. 5 days to 2 weeks

30. 90 in. to 10 ft

Solve each equation.

31. $\dfrac{p}{21} = \dfrac{5}{30}$

32. $\dfrac{5 + x}{3} = \dfrac{2 - x}{6}$

33. $\dfrac{y}{5} = \dfrac{6y - 5}{11}$

34. Explain how 40% can be expressed as a ratio of two whole numbers.

Solve each problem.

35. If 2 lb of fertilizer will cover 150 ft² of lawn, how many pounds would be needed to cover 500 ft²?

36. If 8 oz of medicine must be mixed with 20 oz of water, how many ounces of medicine must be mixed with 90 oz of water?

37. The distance between two cities on a road map is 32 cm. The two cities are actually 150 km apart. The distance on the map between two other cities is 80 cm. How far apart are these cities?

38. Find the best buy. Give the unit price to the nearest thousandth for that size. (*Source:* Jewel-Osco.)

CEREAL	
Size	Price
9 oz	$3.49
14 oz	$3.99
18 oz	$4.49

39. What is 8% of 75?

40. What percent of 12 is 21?

41. 6 is what percent of 18?

42. 36% of what number is 900?

43. Nicholas paid $28,790, including sales tax, for his 2012 Prius v in the base model. The sales tax rate where he lives is 6%. What was the actual price of the car to the nearest dollar? (*Source:* www.toyota.com)

44. Maureen, from the mathematics editorial division of Pearson Education, took the mathematics faculty from a community college out to dinner. The bill was $304.75. Maureen added a 15% tip, and paid for the meal with her corporate credit card. What was the total price she paid to the nearest cent?

2.7 *Write each inequality in interval notation, and graph it.*

45. $p \geq -4$

46. $x < 7$

47. $-5 \leq k < 6$

48. $r \geq \dfrac{1}{2}$

Solve each inequality. Write the solution set in interval notation, and graph it.

49. $x + 6 \geq 3$

50. $5t < 4t + 2$

51. $-6x \leq -18$

52. $8(k - 5) - (2 + 7k) \geq 4$

53. $4x - 3x > 10 - 4x + 7x$

54. $3(2w + 5) + 4(8 + 3w) < 5(3w + 2) + 2w$

55. $-3 \leq 2x + 1 < 4$

56. $8 < 3x + 5 \leq 20$

57. Justin Sudak has grades of 94 and 88 on his first two calculus tests. What possible scores on a third test will give him an average of at least 90?

58. If nine times a number is added to 6, the result is at most 3. Find all such numbers.

Mixed Review Exercises

Solve.

59. $\dfrac{x}{7} = \dfrac{x - 5}{2}$

60. $d = 2r$ for r

61. $-2x > -4$

62. $2k - 5 = 4k + 13$

63. $0.05x + 0.02x = 4.9$

64. $2 - 3(t - 5) = 4 + t$

65. $9x - (7x + 2) = 3x + (2 - x)$

66. $\dfrac{1}{3}s + \dfrac{1}{2}s + 7 = \dfrac{5}{6}s + 5 + 2$

67. In 2010, the top-selling frozen pizza brands, DiGiorno and Red Baron, together had sales of $937.5 million. Red Baron's sales were $399.9 million less than DiGiorno's. What were sales in millions for each brand? (*Source:* www.aibonline.org)

68. On a world globe, the distance between Capetown and Bangkok, two cities that are actually 10,080 km apart, is 12.4 in. The actual distance between Moscow and Berlin is 1610 km. How far apart are Moscow and Berlin on this globe, to the nearest inch?

69. Of the 26 medals earned by Canada during the 2010 Winter Olympic games, there were two times as many gold as silver medals and 9 more gold than bronze medals. How many of each medal did Canada earn? (*Source: World Almanac and Book of Facts.*)

70. In triangle *DEF*, the measure of angle *E* is twice the measure of angle *D*. Angle *F* has measure 18° less than six times the measure of angle *D*. Find the measure of each angle.

71. The perimeter of a triangle is 96 m. One side is twice as long as another, and the third side is 30 m long. What is the length of the longest side?

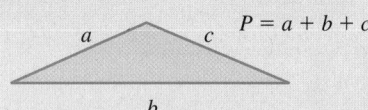

72. The perimeter of a rectangle is 288 ft. The length is 4 ft longer than the width. Find the width.

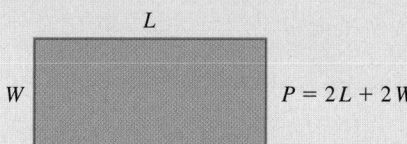

73. Find the best buy. Give the unit price to the nearest thousandth for that size. (*Source:* Jewel-Osco.)

LAUNDRY DETERGENT

Size	Price
50 oz	$3.99
100 oz	$7.29
160 oz	$9.99

74. Find the measure of each marked angle.

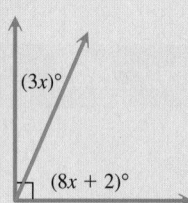

75. Latarsha has grades of 82 and 96 on her first two English tests. What must she make on her third test so that her average will be at least 90?

76. If nine pairs of jeans cost $355.50, find the cost of five pairs. (Assume all are equally priced.)

Solve each equation, and check your solution.

1. $3x - 7 = 11$

2. $5x + 9 = 7x + 21$

3. $2 - 3(x - 5) = 3 + (x + 1)$

4. $2.3x + 13.7 = 1.3x + 2.9$

5. $-\dfrac{4}{7}x = -12$

6. $-8(2x + 4) = -4(4x + 8)$

7. $0.06(x + 20) + 0.08(x - 10) = 4.6$

8. $7 - (m - 4) = -3m + 2(m + 1)$

Solve each problem.

9. The Chicago Bulls finished with the best record for the 2010–2011 NBA regular season. They won 2 more than three times as many games as they lost. They played 82 games. How many games did they win and lose? (*Source:* www.nba.com)

10. Three islands in the Hawaiian island chain are Hawaii (the Big Island), Maui, and Kauai. Together, their areas total 5300 mi^2. The island of Hawaii is 3293 mi^2 larger than the island of Maui, and Maui is 177 mi^2 larger than Kauai. What is the area of each island?

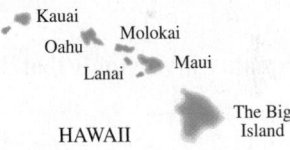

11. If the lesser of two consecutive even integers is tripled, the result is 20 more than twice the greater integer. Find the two integers.

12. Find the measure of an angle if its supplement measures 10° more than three times its complement.

13. The formula for the perimeter of a rectangle is $P = 2L + 2W$.

   (a) Solve for W.

   (b) If $P = 116$ and $L = 40$, find the value of W.

14. Solve the following equation for y.
$$5x - 4y = 8$$

Find the measure of each marked angle.

15.

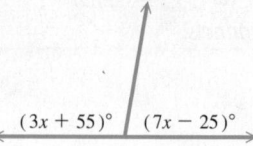

$(3x + 55)°$ | $(7x - 25)°$

16.

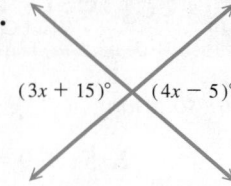

$(3x + 15)°$ $(4x - 5)°$

Solve each equation.

17. $\dfrac{z}{8} = \dfrac{12}{16}$

18. $\dfrac{x + 5}{3} = \dfrac{x - 3}{4}$

Solve each problem.

19. Find the best buy. Give the unit price to the nearest thousandth for that size. (*Source:* Jewel-Osco.)

PROCESSED CHEESE SLICES

Size	Price
8 oz	$2.99
16 oz	$3.99
48 oz	$14.69

20. The distance between Milwaukee and Boston is 1050 mi. On a certain map, this distance is represented by 42 in. On the same map, Seattle and Cincinnati are 92 in. apart. What is the actual distance between Seattle and Cincinnati?

21. An eReader regularly priced at $149 is on sale at 18% off. Find the amount of the discount and the sale price of the eReader.

22. What percent of 65 is 26?

23. Write an inequality involving x that describes the numbers graphed.

(a)

$-2 \;-1 \quad 0 \quad 1 \quad 2 \quad 3$

(b)

$-2 \;-1 \quad 0 \quad 1 \quad 2 \quad 3$

Solve each inequality. Write the solution set in interval notation, and graph it.

24. $-3x > -33$

25. $-0.04x \leq 0.12$

26. $-4x + 2(x - 3) \geq 4x - (3 + 5x) - 7$

27. $-10 < 3x - 4 \leq 14$

28. Shania Johnson has scores of 76 and 81 on her first two algebra tests. If she wants an average of at least 80 after her third test, what score must she make on her third test?

Chapters R–2 Cumulative Review Exercises

Write each fraction in lowest terms.

1. $\dfrac{15}{40}$

2. $\dfrac{108}{144}$

Perform the indicated operations.

3. $\dfrac{5}{6} + \dfrac{1}{4} + \dfrac{7}{15}$

4. $16\dfrac{7}{8} - 3\dfrac{1}{10}$

5. $\dfrac{9}{8} \cdot \dfrac{16}{3}$

6. $\dfrac{3}{4} \div \dfrac{5}{8}$

7. $4.8 + 12.5 + 16.73$

8. $56.3 - 28.99$

9. $67.8\,(0.45)$

10. $236.46 \div 4.2$

11. In making dresses, Earth Works uses $\frac{5}{8}$ yd of trim per dress. How many yards of trim would be used to make 56 dresses?

12. A cook wants to increase a recipe for Quaker Quick Grits that serves 4 to make enough for 10 people. The recipe calls for 3 cups of water. How much water will be needed to serve 10?

13. First published in 1953, the digest-sized *TV Guide* changed to a full-sized magazine in 2005. See the figure. The new magazine is 3 in. wider than the old guide. What is the difference in their heights? (*Source: TV Guide.*)

14. A small business owner bought 3 business laptop computers for $529.99, $599.99, and $629.99 and 3 ergonomic office chairs for $279.99 each. What was the final bill (without tax)? (*Source:* www.staples.com)

Tell whether each inequality is true *or* false.

15. $\dfrac{8\,(7) - 5\,(6 + 2)}{3 \cdot 5 + 1} \geq 1$

16. $\dfrac{4\,(9 + 3) - 8\,(4)}{2 + 3 - 3} \geq 2$

Perform the indicated operations.

17. $-11 + 20 + (-2)$

18. $13 + (-19) + 7$

19. $9 - (-4)$

20. $-2(-5)(-4)$

21. $\dfrac{4 \cdot 9}{-3}$

22. $\dfrac{8}{7 - 7}$

23. $(-5 + 8) + (-2 - 7)$

24. $(-7 - 1)(-4) + (-4)$

25. $\dfrac{-3 - (-5)}{1 - (-1)}$

26. $\dfrac{6\,(-4) - (-2)(12)}{3^2 + 7^2}$

27. $\dfrac{(-3)^2 - (-4)(2^4)}{5 \cdot 2 - (-2)^3}$

28. $\dfrac{-2\,(5^3) - 6}{4^2 + 2\,(-5) + (-2)}$

Find the value of each expression for x = −2, y = −4, and z = 3.

29. $xz^3 - 5y^2$

30. $\dfrac{xz - y^3}{-4z}$

Name the property illustrated by each equation.

31. $7(k + m) = 7k + 7m$

32. $3 + (5 + 2) = 3 + (2 + 5)$

33. $7 + (-7) = 0$

34. $3.5(1) = 3.5$

Simplify each expression.

35. $4p - 6 + 3p - 8$

36. $-4(k + 2) + 3(2k - 1)$

Solve each equation, and check the solution.

37. $2r - 6 = 8$

38. $2(p - 1) = 3p + 2$

39. $4 - 5(a + 2) = 3(a + 1) - 1$

40. $2 - 6(z + 1) = 4(z - 2) + 10$

41. $-(m - 1) = 3 - 2m$

42. $\dfrac{x - 2}{3} = \dfrac{2x + 1}{5}$

43. $\dfrac{2x + 3}{5} = \dfrac{x - 4}{2}$

44. $\dfrac{2}{3}x + \dfrac{3}{4}x = -17$

Solve each formula for the indicated variable.

45. $P = a + b + c + B$ for c

46. $P = 4s$ for s

Solve each inequality. Write the solution set in interval notation, and graph it.

47. $-5z \geq 4z - 18$

48. $6(r - 1) + 2(3r - 5) < -4$

Solve each problem.

49. Abby Tanenbaum bought textbooks at the college bookstore for $244.33, including 6% sales tax. What did the books cost before tax?

50. A used car has a price of $9500. For trading in her old car, Shannon d'Hemecourt will get 25% off. Find the price of the car with the trade-in.

51. The perimeter of a rectangle is 98 cm. The width is 19 cm. Find the length.

?
19 cm

52. The area of a triangle is 104 in.². The base is 13 in. Find the height.

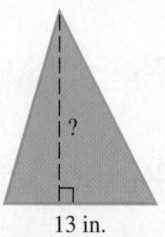

?

13 in.

3 Graphs of Linear Equations and Inequalities in Two Variables; Functions

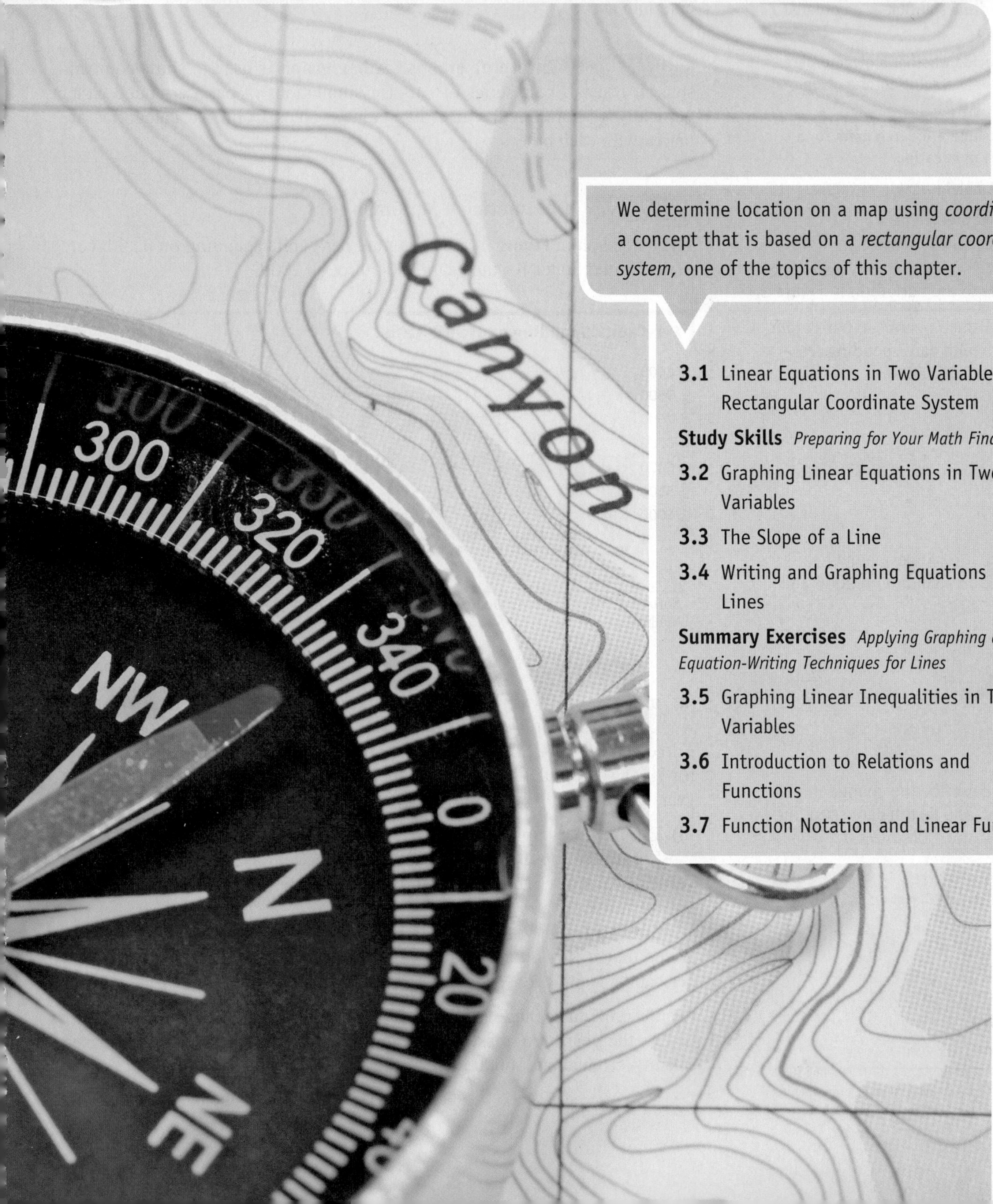

We determine location on a map using *coordinates*, a concept that is based on a *rectangular coordinate system*, one of the topics of this chapter.

3.1 Linear Equations in Two Variables; The Rectangular Coordinate System

Study Skills *Preparing for Your Math Final Exam*

3.2 Graphing Linear Equations in Two Variables

3.3 The Slope of a Line

3.4 Writing and Graphing Equations of Lines

Summary Exercises *Applying Graphing and Equation-Writing Techniques for Lines*

3.5 Graphing Linear Inequalities in Two Variables

3.6 Introduction to Relations and Functions

3.7 Function Notation and Linear Functions

3.1 Linear Equations in Two Variables; The Rectangular Coordinate System

OBJECTIVES

1. Interpret graphs.

2. Write a solution as an ordered pair.

3. Decide whether a given ordered pair is a solution of a given equation.

4. Complete ordered pairs for a given equation.

5. Complete a table of values.

6. Plot ordered pairs.

---

1 Refer to the bar graph in **Figure 1.**

(a) Which years had per-capita health care spending less than $7000?

(b) Estimate per-capita health care spending in 2006 and 2007.

(c) Describe the change in per-capita health care spending from 2006 to 2007.

Answers

1. (a) 2004, 2005, 2006
  (b) 2006: about $6750; 2007: about $7000
  (c) Per-capita health care spending increased by about $250 from 2006 to 2007.

As we saw in **Section R.1,** circle graphs (pie charts) provide a convenient way to organize and communicate information. Along with *bar graphs* and *line graphs*, they can be used to analyze data, make predictions, or simply to entertain us.

OBJECTIVE ▶ 1 Interpret graphs. A **bar graph** is used to show comparisons. It consists of a series of bars (or simulations of bars) arranged either vertically or horizontally. In a bar graph, values from two categories are paired with each other.

EXAMPLE 1 Interpreting a Bar Graph

The bar graph in **Figure 1** shows annual per-capita spending on health care in the United States for the years 2004 through 2009.

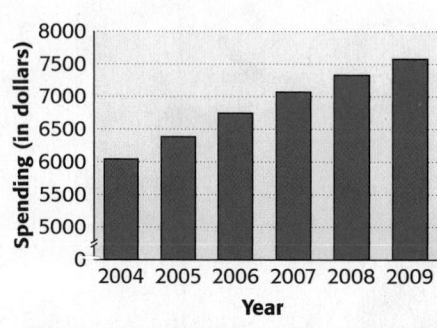

Per-Capita Spending on Health Care

Source: U.S. Centers for Medicare and Medicaid Services.

Figure 1

(a) In what years was per-capita health care spending greater than $7000?
  Locate 7000 on the vertical axis and follow the line across to the right. Three years—2007, 2008, and 2009—have bars that extend above the line for 7000, so per-capita health care spending was greater than $7000 in those years.

(b) Estimate per-capita health care spending in 2004 and 2008.
  Locate the top of the bar for 2004 and move horizontally across to the vertical scale to see that it is about 6000. Per-capita health care spending for 2004 was about $6000.
  Similarly, follow the top of the bar for 2008 across to the vertical scale to see that it lies a little more than halfway between 7000 and 7500, so per-capita health care spending in 2008 was about $7300.

(c) Describe the change in per-capita spending as the years progressed.
  As the years progressed, per-capita spending on health care increased steadily, from about $6000 in 2004 to about $7500 in 2009.

◀ **Work Problem 1** at the Side.

A **line graph** is used to show changes or trends in data over time. To form a line graph, we connect a series of points representing data with line segments.

EXAMPLE 2 Interpreting a Line Graph

The line graph in **Figure 2** shows average prices of a gallon of regular unleaded gasoline in the United States for the years 2004 through 2011.

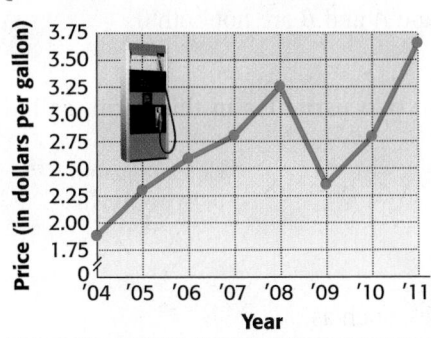

Average U.S. Gasoline Prices

Source: U.S. Department of Energy.

Figure 2

(a) Between which years did the average price of a gallon of gasoline decrease?

The line between 2008 and 2009 falls, so the average price of a gallon of gasoline decreased from 2008 to 2009.

(b) What was the general trend in the average price of a gallon of gasoline from 2004 through 2008?

The line graph rises from 2004 to 2008, so the average price of a gallon of gasoline increased over those years.

(c) Estimate the average price of a gallon of gasoline in 2004 and 2008. About how much did the price increase between 2004 and 2008?

Move up from 2004 on the horizontal scale to the point plotted for 2004. This point is about halfway between the lines on the vertical scale for $1.75 and $2.00. Halfway between $1.75 and $2.00 would be about $1.88. So, it cost about $1.88 for a gallon of gasoline in 2004.

Similarly, locate the point plotted for 2008. Moving across to the vertical scale, the graph indicates that the price for a gallon of gasoline in 2008 was about $3.25.

Between 2004 and 2008, the average price of a gallon of gasoline increased by about

$$\$3.25 - \$1.88 = \$1.37.$$

···················· **Work Problem** ② **at the Side.** ▶

The line graph in **Figure 2** relates years to average prices for a gallon of gasoline. We can also represent these two related quantities using a table of data, as shown in the margin. In table form, we can see more precise data rather than estimating it. Trends in the data are easier to see from the graph, which gives a "picture" of the data.

---

② Refer to the line graph in **Figure 2.**

(a) Estimate the average price of a gallon of regular unleaded gasoline in 2005.

(b) About how much did the average price of a gallon of gasoline increase from 2005 to 2008?

(c) About how much did the average price of a gallon of gasoline decrease from 2008 to 2009?

Year	Average Price (in dollars per gallon)
2004	1.88
2005	2.30
2006	2.59
2007	2.80
2008	3.25
2009	2.35
2010	2.79
2011	3.65

Source: U.S. Department of Energy.

Answers

2. (a) about $2.30   **(b)** about $0.95
   (c) about $0.90

❸ Write each solution as an ordered pair.

GS (a) $x = 5$ and $y = 7$

$$\underset{\downarrow}{\text{x-value}} \quad \underset{\downarrow}{\text{y-value}}$$

$$(\underline{\quad}, \underline{\quad})$$

GS (b) $y = 6$ and $x = -1$

$$\underset{\downarrow}{\text{x-value}} \quad \underset{\downarrow}{\text{y-value}}$$

$$(\underline{\quad}, \underline{\quad})$$

(c) $y = 4$ and $x = -3$

(d) $x = \dfrac{2}{3}$ and $y = -12$

(e) $y = 1.5$ and $x = -2.4$

(f) $x = 0$ and $y = 0$

We can extend these ideas to the subject of this chapter, *linear equations in two variables*. A linear equation in two variables, one for each of the quantities being related, can be used to represent the data in a table or graph. ***The graph of a linear equation in two variables is a line.***

Linear Equation in Two Variables

A **linear equation in two variables** is an equation that can be written in the form

$$Ax + By = C,$$

where A, B, and C are real numbers and A and B are not both 0.

Some examples of linear equations in two variables in this form, called *standard form*, follow.

$$3x + 4y = 9, \quad x - y = 0, \quad \text{and} \quad x + 2y = -8 \qquad \substack{\text{Linear equations} \\ \text{in two variables}}$$

Note

Other linear equations in two variables, such as

$$y = 4x + 5 \quad \text{and} \quad 3x = 7 - 2y,$$

are not written in standard form, but could be algebraically rewritten in this form. We discuss the forms of linear equations in **Section 3.4.**

OBJECTIVE ❷ Write a solution as an ordered pair. Recall from **Section 1.2** that a *solution* of an equation is a number that makes the equation true when it replaces the variable. For example, the linear equation in *one* variable $x - 2 = 5$ has solution 7, since replacing x with 7 gives a true statement.

A solution of a linear equation in two variables requires two numbers, one for each variable. For example, a true statement results when we replace x with 2 and y with 13 in the equation $y = 4x + 5$ since

$$13 = 4(2) + 5. \qquad \text{Let } x = 2 \text{ and } y = 13.$$

The pair of numbers $x = 2$ and $y = 13$ gives one solution of the equation $y = 4x + 5$. The phrase "$x = 2$ and $y = 13$" is abbreviated

$$\underset{\text{x-value}}{\underbrace{(2,}} \underset{\text{y-value}}{\underbrace{13)}}$$

Ordered pair

with the x-value, 2, and the y-value, 13, given as a pair of numbers written inside parentheses. ***The x-value is always given first.*** A pair of numbers such as $(2, 13)$ is called an **ordered pair.**

CAUTION

The ordered pairs $(2, 13)$ and $(13, 2)$ are *not* the same. In the first pair, $x = 2$ and $y = 13$. In the second pair, $x = 13$ and $y = 2$. ***The order in which the numbers are written in an ordered pair is important.*** For two ordered pairs to be equal, their x-values must be equal and their y-values must be equal.

Answers

3. **(a)** $5; 7$ **(b)** $-1; 6$ **(c)** $(-3, 4)$

   (d) $\left(\dfrac{2}{3}, -12\right)$ **(e)** $(-2.4, 1.5)$

   (f) $(0, 0)$

◀ Work Problem **❸** at the Side.

OBJECTIVE ③ **Decide whether a given ordered pair is a solution of a given equation.** We substitute the x- and y-values of an ordered pair into a linear equation in two variables to see whether the ordered pair is a solution.

EXAMPLE 3 Deciding Whether Ordered Pairs Are Solutions

Decide whether each ordered pair is a solution of the equation $2x + 3y = 12$.

(a) $(3, 2)$

Substitute 3 for x and 2 for y in the given equation $2x + 3y = 12$.

$$2x + 3y = 12$$
$$2(3) + 3(2) \stackrel{?}{=} 12 \qquad \text{Let } x = 3 \text{ and } y = 2.$$
$$6 + 6 \stackrel{?}{=} 12 \qquad \text{Multiply.}$$
$$12 = 12 \checkmark \qquad \text{True}$$

This result is true, so $(3, 2)$ is a solution of $2x + 3y = 12$.

(b) $(-2, -7)$

$$2x + 3y = 12$$
$$2(-2) + 3(-7) \stackrel{?}{=} 12 \qquad \text{Let } x = -2 \text{ and } y = -7.$$
$$-4 + (-21) \stackrel{?}{=} 12 \qquad \text{Multiply.}$$
$$-25 = 12 \qquad \text{False}$$

> Use parentheses to avoid errors.

This result is false, so $(-2, -7)$ is *not* a solution of $2x + 3y = 12$.

· ▶ **Work Problem ④ at the Side.** ▶

OBJECTIVE ④ **Complete ordered pairs for a given equation.** We substitute a number for one variable to find the value of the other variable.

EXAMPLE 4 Completing Ordered Pairs

Complete each ordered pair for the equation $y = 4x + 5$.

(a) $(7, \underline{\quad})$

> The x-value always comes first.

In this ordered pair, $x = 7$. Replace x with 7 in the given equation.

$$y = 4x + 5$$

> Solve for the value of y.

$$y = 4(7) + 5 \qquad \text{Let } x = 7.$$
$$y = 28 + 5 \qquad \text{Multiply.}$$
$$y = 33 \qquad \text{Add.}$$

The ordered pair is $(7, 33)$.

(b) $(\underline{\quad}, -3)$

In this ordered pair, $y = -3$. Replace y with -3 in the given equation.

$$y = 4x + 5$$

> Solve for the value of x.

$$-3 = 4x + 5 \qquad \text{Let } y = -3.$$
$$-8 = 4x \qquad \text{Subtract 5 from each side.}$$
$$-2 = x \qquad \text{Divide each side by 4.}$$

The ordered pair is $(-2, -3)$.

· ▶ **Work Problem ⑤ at the Side.** ▶

---

④ Decide whether each ordered pair is a solution of the equation $5x + 2y = 20$.

GS **(a)** $(0, 10)$

In this ordered pair,

$x = \underline{\quad}$ and $y = \underline{\quad}$.

$$5x + 2y = 20$$
$$5(\underline{\quad}) + 2(\underline{\quad}) \stackrel{?}{=} 20$$
$$\underline{\quad} + 20 \stackrel{?}{=} 20$$
$$\underline{\quad} = 20$$

Is $(0, 10)$ a solution?
(yes / no)

(b) $(2, -5)$

(c) $(3, 2)$

(d) $(-4, 20)$

⑤ Complete each ordered pair for the equation $y = 2x - 9$.

GS **(a)** $(5, \underline{\quad})$

In this ordered pair,

$x = \underline{\quad}$.

We must find the corresponding value of $\underline{\quad}$.

$$y = 2x - 9$$
$$y = 2(\underline{\quad}) - 9$$
$$y = \underline{\quad} - 9$$
$$y = \underline{\quad}$$

The ordered pair is $\underline{\quad}$.

(b) $(2, \underline{\quad})$

(c) $(\underline{\quad}, 7)$

(d) $(\underline{\quad}, -13)$

Answers

4. (a) 0; 10; 0; 10; 0; 20; yes
   (b) no **(c)** no **(d)** yes
5. (a) 5; y; 5; 10; 1; $(5, 1)$
   (b) $(2, -5)$ **(c)** $(8, 7)$ **(d)** $(-2, -13)$

6 Complete the table of values for each equation. Write the results as ordered pairs.

GS **(a)** $2x - 3y = 12$

x	y
0	
	0
3	
	-3

From the first row of the table, let $x =$ ____.

$$2x - 3y = 12$$
$$2(\underline{\quad}) - 3y = 12$$
$$\underline{\quad} - 3y = 12$$
$$-3y = 12$$
$$\frac{-3y}{\underline{\quad}} = \frac{12}{\underline{\quad}}$$
$$y = \underline{\quad}$$

Write ____ for y in the first row of the table. Repeat this process to complete the rest of the table.

(b) $x = -1$ **(c)** $y = 4$

x	y
	-4
	0
-1	2

x	y
-3	4
2	
5	

Answers

6. **(a)** 0; 0; 0; -3; -3; -4; -4

x	y
0	-4
6	0
3	-2
$\frac{3}{2}$	-3

$(0, -4), (6, 0), (3, -2), \left(\frac{3}{2}, -3\right)$

(b)
x	y
-1	-4
-1	0
-1	2

(c)
x	y
-3	4
2	4
5	4

$(-1, -4), (-1, 0),$ $(-3, 4), (2, 4),$
$(-1, 2)$ $(5, 4)$

OBJECTIVE **5** **Complete a table of values.** Ordered pairs are often displayed in a **table of values.** Although we usually write tables of values vertically, they may be written horizontally.

EXAMPLE 5 **Completing Tables of Values**

Complete the table of values for each equation. Then write the results as ordered pairs.

(a) $x - 2y = 8$

x	y
2	
10	
	0
	-2

Ordered Pairs
$(2, \underline{\quad})$
$(10, \underline{\quad})$
$(\underline{\quad}, 0)$
$(\underline{\quad}, -2)$

From the first row of the table, let $x = 2$ in the equation. From the second row of the table, let $x = 10$.

If	$x = 2,$	If	$x = 10,$
then	$x - 2y = 8$	then	$x - 2y = 8$
becomes	$2 - 2y = 8$	becomes	$10 - 2y = 8$
	$-2y = 6$		$-2y = -2$
	$y = -3.$		$y = 1.$

The first two ordered pairs are $(2, -3)$ and $(10, 1)$. From the third and fourth rows of the table, let $y = 0$ and $y = -2$, respectively.

If	$y = 0,$	If	$y = -2,$
then	$x - 2y = 8$	then	$x - 2y = 8$
becomes	$x - 2(0) = 8$	becomes	$x - 2(-2) = 8$
	$x - 0 = 8$		$x + 4 = 8$
	$x = 8.$		$x = 4.$

The last two ordered pairs are $(8, 0)$ and $(4, -2)$. The completed table of values and corresponding ordered pairs follow.

> Write y-values in the second column.

x	y
2	-3
10	1
8	0
4	-2

Ordered Pairs
$(2, -3)$
$(10, 1)$
$(8, 0)$
$(4, -2)$

> Write x-values in the first column.

Each ordered pair is a solution of the given equation $x - 2y = 8$.

(b) $x = 5$

x	y
	-2
	6
	3

The given equation is $x = 5$. No matter which value of y is chosen, the value of x is *always* 5.

x	y
5	-2
5	6
5	3

Ordered Pairs
$(5, -2)$
$(5, 6)$
$(5, 3)$

◀ **Work Problem** **6** **at the Side.**

Note

We can think of $x = 5$ in **Example 5(b)** as an equation in two variables by rewriting $x = 5$ as

$$x + 0y = 5.$$

This form shows that, for any value of y, the value of x is 5. Similarly, $y = 4$ in **Margin Problem 6(c)** on the preceding page is the same as

$$0x + y = 4.$$

OBJECTIVE ▸ **6** **Plot ordered pairs.** In **Section 2.3,** we saw that linear equations in *one* variable have either one, zero, or an infinite number of real number solutions. These solutions can be graphed on *one* number line. For example, the linear equation in one variable $x - 2 = 5$ has solution 7, which is graphed on the number line in **Figure 3.**

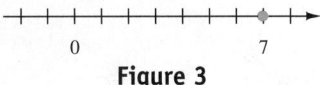

Figure 3

Every linear equation in *two* variables has an infinite number of ordered pairs as solutions. To graph these solutions, represented as the ordered pairs (x, y), we need *two* number lines, one for each variable. These two number lines are drawn at right angles as in **Figure 4.** The horizontal number line is the **x-axis**, and the vertical line is the **y-axis**. Together, the x-axis and y-axis form a **rectangular coordinate system.** The point at which the x-axis and y-axis intersect is the **origin.**

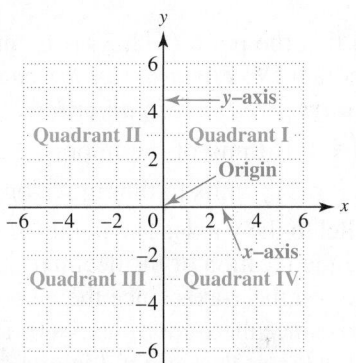

Rectangular coordinate system

Figure 4

The rectangular coordinate system is divided into four regions, or **quadrants.** These quadrants are numbered counterclockwise. See **Figure 4.**

Work Problem **7** at the Side. ▸

The x-axis and y-axis determine a **plane**—a flat surface (as illustrated by a sheet of paper). By referring to the two axes, every point in the plane can be associated with an ordered pair. The numbers in the ordered pair are the **coordinates** of the point.

Note

In a plane, *both* numbers in the ordered pair are needed to locate a point. The ordered pair is a name for the point.

7 Name the quadrant in which each point in the figure is located.

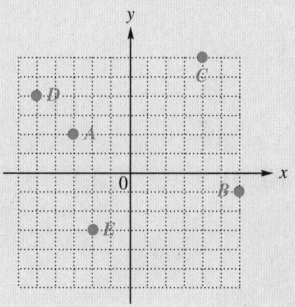

René Descartes (1596–1650)

The rectangular coordinate system is also called the **Cartesian coordinate system,** in honor of René Descartes, the French mathematician credited with its invention.

Answer

7. A: II; B: IV; C: I; D: II; E: III

8 Plot the given points in a coordinate system.

(a) $(3, 5)$ (b) $(-2, 6)$

(c) $(-4.5, 0)$ (d) $(-5, -2)$

(e) $(6, -2)$ (f) $(0, -6)$

(g) $(0, 0)$ (h) $\left(-3, \dfrac{5}{2}\right)$

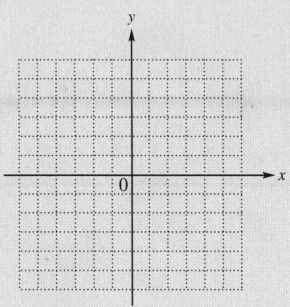

Indicate the points that lie in each quadrant.

quadrant I: _____

quadrant II: _____

quadrant III: _____

quadrant IV: _____

no quadrant: _____

Answers

8.

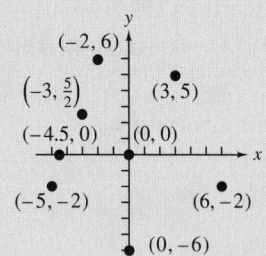

quadrant I: $(3, 5)$

quadrant II: $(-2, 6), \left(-3, \dfrac{5}{2}\right)$

quadrant III: $(-5, -2)$

quadrant IV: $(6, -2)$

no quadrant: $(-4.5, 0), (0, 0), (0, -6)$

EXAMPLE 6 Plotting Ordered Pairs

Plot the given points in a coordinate system.

(a) $(2, 3)$ (b) $(-1, -4)$ (c) $(-2, 3)$ (d) $(3, -2)$ (e) $\left(\dfrac{3}{2}, 2\right)$

(f) $(4, -3.75)$ (g) $(5, 0)$ (h) $(0, -3)$ (i) $(0, 0)$

The point $(2, 3)$ from part (a) is **plotted** (graphed) in **Figure 5.** The other points are plotted in **Figure 6.** In each case, we begin at the origin.

Step 1 Move right or left the number of units that corresponds to the x-coordinate in the ordered pair—*right if the x-coordinate is positive or left if it is negative.*

Step 2 Then turn and move up or down the number of units that corresponds to the y-coordinate—*up if the y-coordinate is positive or down if it is negative.*

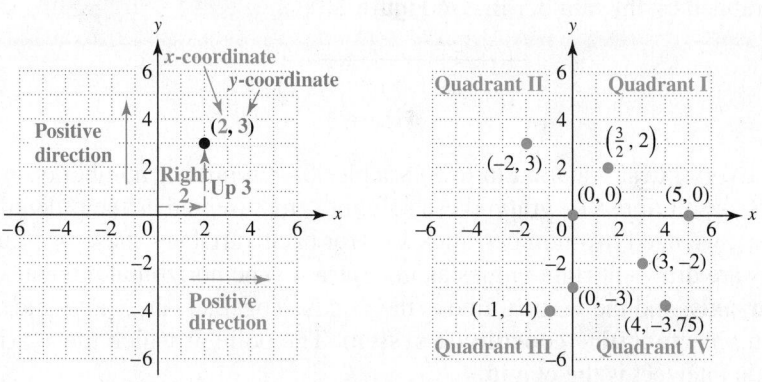

Figure 5   **Figure 6**

Notice in **Figure 6** that the point $(-2, 3)$ is in quadrant II, whereas the point $(3, -2)$ is in quadrant IV. *The order of the coordinates is important. The x-coordinate is always given first in an ordered pair.*

To plot the point $\left(\dfrac{3}{2}, 2\right)$, think of the improper fraction $\dfrac{3}{2}$ as the mixed number $1\dfrac{1}{2}$ and move $\dfrac{3}{2}$ $\left(\text{or } 1\dfrac{1}{2}\right)$ units to the right along the x-axis. Then turn and go 2 units up, parallel to the y-axis. The point $(4, -3.75)$ is plotted similarly, by approximating the location of the decimal y-coordinate.

The point $(5, 0)$ lies on the x-axis since the y-coordinate is 0. The point $(0, -3)$ lies on the y-axis since the x-coordinate is 0. The point $(0, 0)$ is at the origin. *Points on the axes themselves are not in any quadrant.*

◀ Work Problem **8** at the Side.

We can use a linear equation in two variables to mathematically describe, or *model*, certain real-life situations, as shown in the next example.

EXAMPLE 7 Using a Linear Model

The annual number of twin births in the United States from 2001 through 2007 can be approximated by the linear equation

Number of twin births ——┐   ┌—— Year

$$y = 2.929x - 5738,$$

which relates x, the year, and y, the number of twin births in thousands. (*Source:* National Center for Health Statistics.)

········· **Continued on Next Page**

(a) Complete the table of values for the given linear equation.

x (Year)	y (Number of Twin Births, in thousands)
2001	
2004	
2007	

To find y for $x = 2001$, substitute into the equation.

$$y = 2.929x - 5738$$

$$y = 2.929(\mathbf{2001}) - 5738 \qquad \text{Let } x = 2001.$$

> ≈ means "is approximately equal to."

$$y \approx 123 \qquad\qquad \text{Use a calculator.}$$

In 2001, there were about 123 thousand (or 123,000) twin births.

Work Problem ❾ at the Side. ▶

Including the results from **Margin Problem 9** gives the completed table that follows.

x (Year)	y (Number of Twin Births, in thousands)	Ordered Pairs (x, y)	Here each year x is paired with a number of twin births y (in thousands).
2001	123	⟶ (2001, 123)	
2004	132	⟶ (2004, 132)	
2007	141	⟶ (2007, 141)	

(b) Graph the ordered pairs found in part (a).
See **Figure 7.** A graph of ordered pairs of data is a **scatter diagram.**

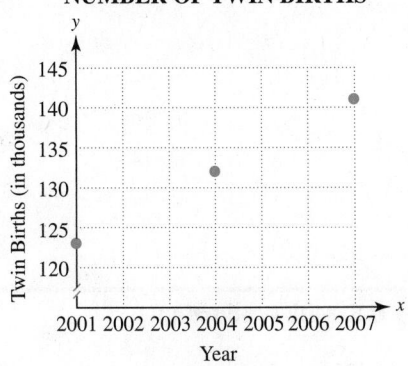

NUMBER OF TWIN BIRTHS

Notice the axis labels and scales. Each square represents 1 unit in the horizontal direction and 5 units in the vertical direction. We show a break in the y-axis, to indicate the jump from 0 to 120.

Figure 7

A scatter diagram can indicate whether two quantities are related. In **Figure 7,** the plotted points could be connected to approximate a straight *line,* so the variables x (year) and y (number of twin births) have a *line*ar relationship. The increase in the number of twin births is also reflected.

---

❾ Refer to the linear equation in **Example 7.** Round to the nearest whole number.

(a) Find the y-value for $x = 2004$.

$$y = 2.929x - 5738$$

$$y = 2.929(\underline{\quad}) - \underline{\quad}$$

$$y \approx \underline{\quad}$$

(b) Find the y-value for $x = 2007$. Interpret your result.

Answers

9. **(a)** 2004; 5738; 132
   (b) 141; In 2007, there were about 141 thousand (or 141,000) twin births in the United States.

3.1 Exercises

CONCEPT CHECK *Complete each statement.*

1. The symbol (x, y) (*does / does not*) represent an ordered pair, while the symbols $[x, y]$ and $\{x, y\}$ (*do / do not*) represent ordered pairs.

2. The point whose graph has coordinates $(-4, 2)$ is in quadrant _____.

3. The point whose graph has coordinates $(0, 5)$ lies on the _____-axis.

4. The ordered pair $(4, \_\_\_\_)$ is a solution of the equation $y = 3$.

5. The ordered pair $(\_\_\_\_, -2)$ is a solution of the equation $x = 6$.

6. The ordered pair $(3, 2)$ is a solution of the equation $2x - 5y = \_\_\_\_$.

7. Do $(4, -1)$ and $(-1, 4)$ represent the same ordered pair? Explain.

8. Explain why it would be easier to find the corresponding y-value for $x = \frac{1}{3}$ in the equation $y = 6x + 2$ than it would be for $x = \frac{1}{7}$.

CONCEPT CHECK *Fill in each blank with the word* positive *or the word* negative.

The point with coordinates (x, y) is in

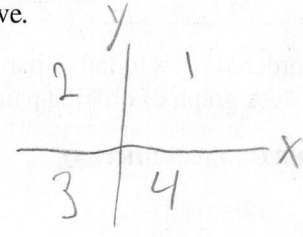

9. quadrant III if x is _____ and y is _____.

10. quadrant II if x is _____ and y is _____.

11. quadrant IV if x is _____ and y is _____.

12. quadrant I if x is _____ and y is _____.

The bar graph shows total U.S. milk production in billions of pounds for the years 2004 through 2010. Use the bar graph to work Exercises 13–16. **See Example 1.**

13. In what years was U.S. milk production greater than 185 billion pounds?

14. In what years was U.S. milk production about the same?

15. Estimate U.S. milk production in 2004 and 2010.

16. Describe the change in U.S. milk production from 2004 to 2010.

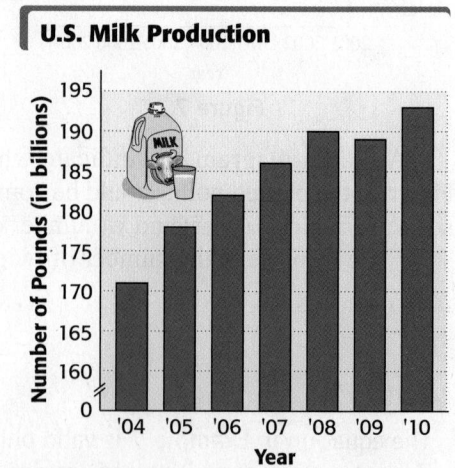

U.S. Milk Production

Source: U.S. Department of Agriculture.

*The line graph shows the number of new and used passenger cars (in millions) imported into the United States over the years 2005 through 2010. Use the line graph to work Exercises 17–20. **See Example 2.***

17. Over which two consecutive years did the number of imported cars increase the most? About how much was this increase?

18. Estimate the number of cars imported during 2007, 2008, and 2009.

19. Describe the trend in car imports from 2006 to 2009.

20. During which year(s) were fewer than 6 millions cars imported into the United States?

Passenger Cars Imported into the U.S.

Source: U.S. Census Bureau.

*Decide whether each ordered pair is a solution of the given equation. **See Example 3.***

21. $x + y = 9$; $(0, 9)$

22. $x + y = 8$; $(0, 8)$

23. $2x - y = 6$; $(4, 2)$

24. $2x + y = 5$; $(3, -1)$

25. $4x - 3y = 6$; $(2, 1)$

26. $5x - 3y = 15$; $(5, 2)$

27. $y = \dfrac{2}{3}x$; $(-6, -4)$

28. $y = -\dfrac{1}{4}x$; $(-8, 2)$

29. $x = -6$; $(5, -6)$

30. $x = -9$; $(8, -9)$

31. $y = 2$; $(2, 4)$

32. $y = 7$; $(7, -2)$

*Complete each ordered pair for the equation $y = 2x + 7$. **See Example 4.***

33. $(2, \underline{\quad})$

34. $(0, \underline{\quad})$

35. $(\underline{\quad}, 0)$

36. $(\underline{\quad}, -3)$

*Complete each ordered pair for the equation $y = -4x - 4$. **See Example 4.***

37. $(0, \underline{\quad})$

38. $(\underline{\quad}, 0)$

39. $(\underline{\quad}, 16)$

40. $(\underline{\quad}, 24)$

*Complete each table of values. Write the results as ordered pairs. **See Example 5.***

41. $2x + 3y = 12$

x	y
0	
	0
	8

42. $4x + 3y = 24$

x	y
0	
	0
	4

43. $3x - 5y = -15$

x	y
0	
	0
	-6

44. $4x - 9y = -36$

x	y
	0
0	
	8

45. $x = -9$

x	y
	6
	2
	-3

46. $x = 12$

x	y
	3
	8
	0

47. $y = -6$

x	y
8	
4	
-2	

48. $y = -10$

x	y
4	
0	
-4	

49. $x - 8 = 0$

x	y
	8
	3
	0

50. $x + 4 = 0$

x	y
	4
	0
	-4

51. $y + 2 = 0$

x	y
9	
2	
0	

52. $y + 1 = 0$

x	y
1	
0	
-1	

Give the ordered pairs for the points labeled A–F in the figure. (All coordinates are integers.) Tell the quadrant in which each point is located. ***See Example 6.***

53. A

54. B

55. C

56. D

57. E

58. F

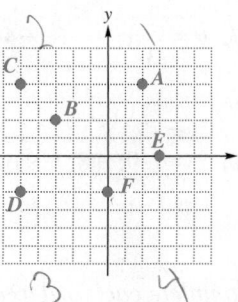

59. A point (x, y) has the property that $xy < 0$. In which quadrant(s) must the point lie? Explain.

60. A point (x, y) has the property that $xy > 0$. In which quadrant(s) must the point lie? Explain.

Plot each ordered pair on the rectangular coordinate system provided. ***See Example 6.***

61. $(6, 2)$

62. $(5, 3)$

63. $(-4, 2)$

64. $(-3, 5)$

65. $\left(-\dfrac{4}{5}, -1\right)$

66. $\left(-\dfrac{3}{2}, -4\right)$

67. $(3, -1.75)$

68. $(5, -4.25)$

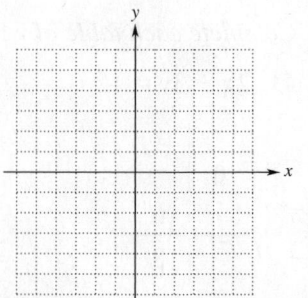

69. $(0, 4)$

70. $(0, -3)$

71. $(4, 0)$

72. $(-3, 0)$

*Complete each table of values, and then plot the ordered pairs. **See Examples 5 and 6.***

73. $x - 2y = 6$

x	y
0	
	0
2	
	−1

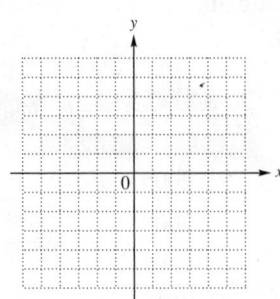

74. $2x - y = 4$

x	y
0	
	0
1	
	−6

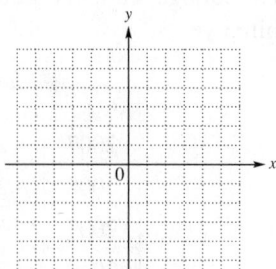

75. $3x - 4y = 12$

x	y
0	
	0
−4	
	−4

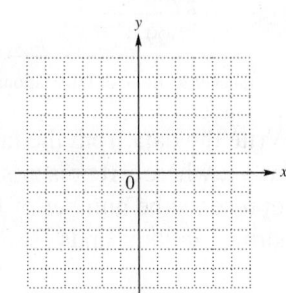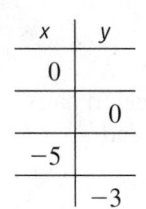

76. $2x - 5y = 10$

x	y
0	
	0
−5	
	−3

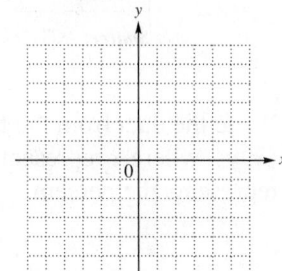

77. $y + 4 = 0$

x	y
0	
5	
−2	
−3	

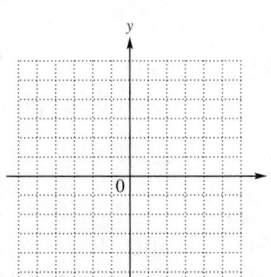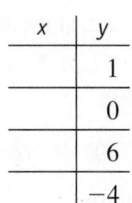

78. $x - 5 = 0$

x	y
	1
	0
	6
	−4

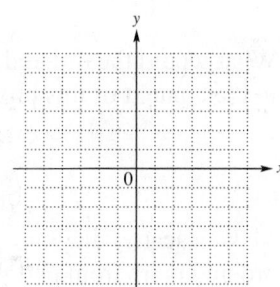

79. Describe the pattern indicated by the plotted points in **Exercises 73–78.**

80. Answer each question.

  (a) A line through the plotted points in **Exercise 77** would be horizontal. What do you notice about the *y*-coordinates of the ordered pairs?

  (b) A line through the plotted points in **Exercise 78** would be vertical. What do you notice about the *x*-coordinates of the ordered pairs?

*Work each problem. **See Example 7.***

81. Suppose that it costs a flat fee of $20 plus $5 per day to rent a pressure washer. Therefore, the cost *y* in dollars to rent the pressure washer for *x* days is given by

$$y = 5x + 20.$$

Express each of the following as an ordered pair.

  (a) When the washer is rented for 5 days, the cost is $45. (*Hint:* What does *x* represent? What does *y* represent?)

  (b) We paid $50 when we returned the washer, so we must have rented it for 6 days.

82. Suppose that it costs $5000 to start up a business selling snow cones. Furthermore, it costs $0.50 per cone in labor, ice, syrup, and overhead. Then the cost *y* in dollars to make *x* snow cones is given by

$$y = 0.50x + 5000.$$

Express each of the following as an ordered pair.

  (a) When 100 snow cones are made, the cost is $5050. (*Hint:* What does *x* represent? What does *y* represent?)

  (b) When the cost is $6000, the number of snow cones made is 2000.

83. The table shows the rate (in percent) at which 2-year college students (public) complete a degree within 3 years.

Year	Percent
2007	27.1
2008	29.3
2009	28.3
2010	28.0
2011	26.9

Source: ACT.

(a) Write the data from the table as ordered pairs (x, y), where x represents the year and y represents the percent.

(b) What would the ordered pair (2000, 32.4) mean in the context of this problem?

(c) Make a scatter diagram of the data using the ordered pairs from part (a).

2-YEAR COLLEGE STUDENTS COMPLETING A DEGREE WITHIN 3 YEARS

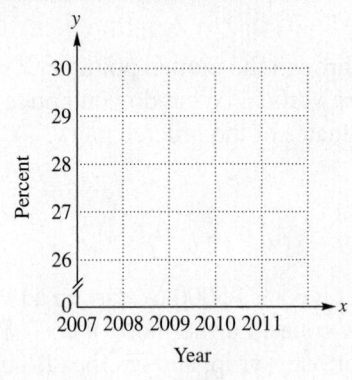

(d) Describe the pattern indicated by the points on the scatter diagram. What is happening to the rates at which 2-year college students complete a degree within 3 years?

84. The table shows the number of U.S. students studying abroad (in thousands) for recent academic years.

Academic Year	Number of Students (in thousands)
2005	224
2006	242
2007	262
2008	260
2009	271

Source: Institute of International Education.

(a) Write the data from the table as ordered pairs (x, y), where x represents the year and y represents the number of U.S. students studying abroad, in thousands.

(b) What does the ordered pair (2000, 154) mean in the context of this problem?

(c) Make a scatter diagram of the data using the ordered pairs from part (a).

U.S. STUDENTS STUDYING ABROAD

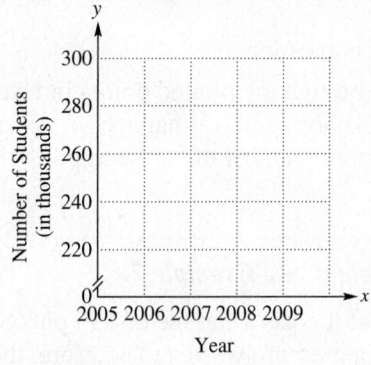

(d) Describe the pattern indicated by the points on the scatter diagram. What is the trend in the number of U.S. students studying abroad?

85. The maximum benefit for the heart from exercising occurs if the heart rate is in the target heart rate zone. The lower limit of this target zone can be approximated by the linear equation

$$y = -0.5x + 108,$$

where x represents age and y represents heartbeats per minute. (*Source:* www.fitresource.com)

(a) Complete the table of values for this linear equation.

Age	Heartbeats (per minute)
20	
40	
60	
80	

(b) Write the data from the table of values as ordered pairs.

(c) Make a scatter diagram of the data. Do the points lie in a linear pattern?

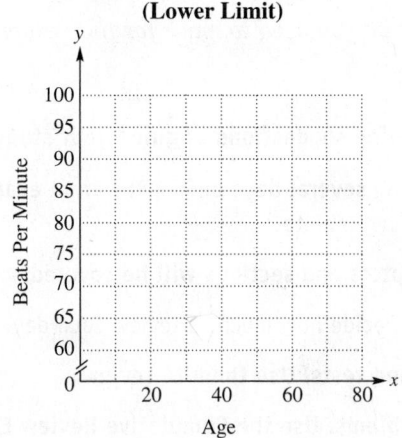

TARGET HEART RATE ZONE
(Lower Limit)

86. (See **Exercise 85.**) The upper limit of the target heart rate zone can be approximated by the linear equation

$$y = -0.8x + 173,$$

where x represents age and y represents heartbeats per minute. (*Source:* www.fitresource.com)

(a) Complete the table of values for this linear equation.

Age	Heartbeats (per minute)
20	
40	
60	
80	

(b) Write the data from the table of values as ordered pairs.

(c) Make a scatter diagram of the data. Describe the pattern indicated by the data.

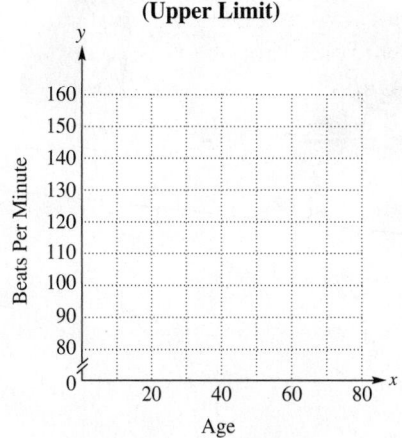

TARGET HEART RATE ZONE
(Upper Limit)

87. See **Exercises 85 and 86.** What is the target heart rate zone for age 20? Age 40?

88. See **Exercises 85 and 86.** What is the target heart rate zone for age 60? Age 80?

Study Skills
PREPARING FOR YOUR MATH FINAL EXAM

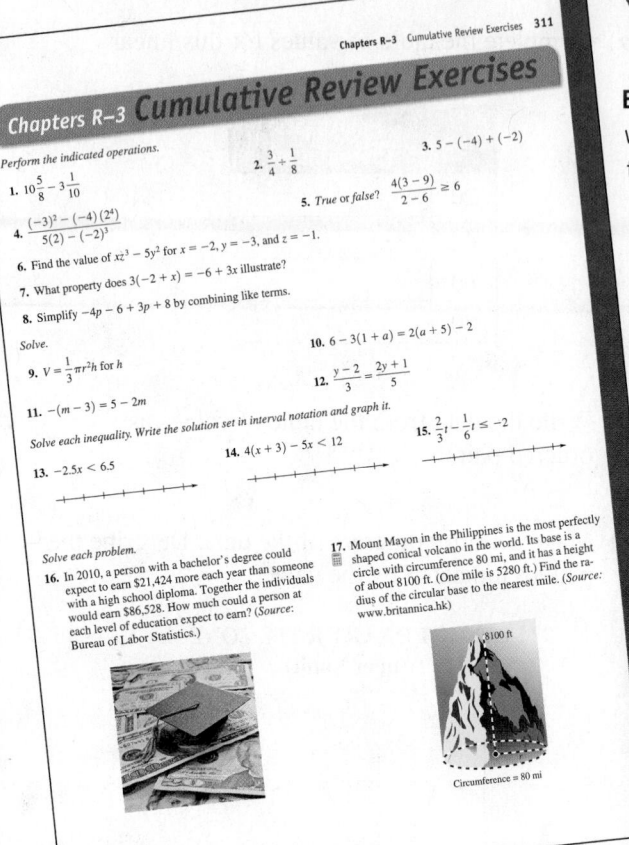

Chapters R–3 Cumulative Review Exercises

Perform the indicated operations.

1. $10\frac{5}{8} - 3\frac{1}{10}$

2. $\frac{3}{4} + \frac{1}{8}$

3. $5 - (-4) + (-2)$

4. $\frac{(-3)^2 - (-4)(2^4)}{5(2) - (-2)^3}$

5. *True or false?* $\frac{4(3-9)}{2-6} \ge 6$

6. Find the value of $xz^3 - 5y^2$ for $x = -2$, $y = -3$, and $z = -1$.

7. What property does $3(-2 + x) = -6 + 3x$ illustrate?

8. Simplify $-4p - 6 + 3p + 8$ by combining like terms.

Solve.

9. $V = \frac{1}{3}\pi r^2 h$ for h

10. $6 - 3(1 + a) = 2(a + 5) - 2$

11. $-(m - 3) = 5 - 2m$

12. $\frac{y-2}{3} = \frac{2y+1}{5}$

Solve each inequality. Write the solution set in interval notation and graph it.

13. $-2.5x < 6.5$

14. $4(x + 3) - 5x < 12$

15. $\frac{2}{3}t - \frac{1}{6}t \le -2$

Solve each problem.

16. In 2010, a person with a bachelor's degree could expect to earn \$21,424 more each year than someone with a high school diploma. Together the individuals would earn \$86,528. How much could a person at each level of education expect to earn? (*Source:* Bureau of Labor Statistics.)

17. Mount Mayon in the Philippines is the most perfectly shaped conical volcano in the world. Its base is a circle with circumference 80 mi, and it has a height of about 8100 ft. (One mile is 5280 ft.) Find the radius of the circular base to the nearest mile. (*Source:* www.britannica.hk)

8100 ft

Circumference = 80 mi

Your math final exam is likely to be a comprehensive exam, which means it will cover material from the entire term. **One way to prepare for it now is by working a set of Cumulative Review Exercises** each time your class finishes a chapter. This continual review will help you remember concepts and procedures as you progress through the course.

Final Exam Preparation Suggestions

1. **Figure out the grade you need to earn on the final exam to get the course grade you want.** Check your course syllabus for grading policies, or ask your instructor if you are not sure.

   How many points do you need to earn on your math final exam to get the grade you want?

2. **Create a final exam week plan.** Set priorities that allow you to spend extra time studying. This may mean making adjustments, in advance, in your work schedule or enlisting extra help with family responsibilities.

   What adjustments do you need to make for final exam week? List two or three here.

3. **Use the following suggestions to guide your studying.**

   ▶ **Begin reviewing several days before the final exam.** DON'T wait until the last minute.

   ▶ **Know exactly which chapters and sections will be covered.**

   ▶ **Divide up the chapters.** Decide how much to review each day.

   ▶ **Keep returned quizzes and tests. Use them to review.**

   ▶ **Practice all types of problems. Use the Cumulative Review Exercises** at the end of each chapter in your textbook beginning in Chapter 2. All answers are given in the answer section.

   ▶ **Review or rewrite your notes** to create summaries of key information.

   ▶ **Make study cards for all types of problems.** Carry the cards with you, and review them whenever you have a few minutes.

   ▶ **Take plenty of short breaks as you study to reduce physical and mental stress.** Exercising, listening to music, and enjoying a favorite activity are effective stress busters.

Finally, *DON'T* stay up all night the night before an exam—*get a good night's sleep.*

Which of these suggestions will you use as you study for your math final exam?

3.2 Graphing Linear Equations in Two Variables

OBJECTIVES

OBJECTIVE ▶ 1 Graph linear equations by plotting ordered pairs. There are infinitely many ordered pairs that satisfy an equation in two variables. We find these ordered-pair solutions by choosing as many values of x (or y) as we wish and then completing each ordered pair.

For example, consider the equation $x + 2y = 7$. If we choose $x = 1$, then we can substitute to find the corresponding value of y.

$$x + 2y = 7 \quad \text{Given equation}$$
$$1 + 2y = 7 \quad \text{Let } x = 1.$$
$$2y = 6 \quad \text{Subtract 1.}$$
$$y = 3 \quad \text{Divide by 2.}$$

If $x = 1$, then $y = 3$, and the ordered pair $(1, 3)$ is a solution of the equation.

$$1 + 2(3) = 7 \; ✓ \quad (1, 3) \text{ is a solution.}$$

Work Problem ❶ at the Side. ▶

Figure 8 shows a graph of all the ordered-pair solutions found above and in **Margin Problem 1** for $x + 2y = 7$.

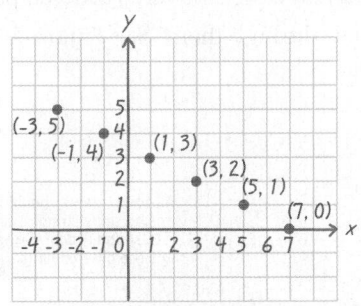

Figure 8

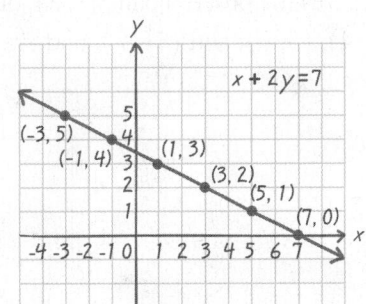

Figure 9

Notice that the points plotted in **Figure 8** all appear to lie on a straight line, as shown in **Figure 9.** In fact, the following is true.

Every point on the line represents a solution of the equation $x + 2y = 7$, and every solution of the equation corresponds to a point on the line.

This line gives a "picture" of all the solutions of the equation $x + 2y = 7$. The line extends indefinitely in both directions, as suggested by the arrowhead on each end, and is the **graph** of the equation $x + 2y = 7$. The process of plotting the ordered pairs and drawing the line through the corresponding points is called **graphing.**

The preceding discussion can be generalized.

Graph of a Linear Equation
The graph of any linear equation in two variables is a straight line.

Notice that the word ***line*** appears in the term "*line*ar equation."

OBJECTIVES

1. Graph linear equations by plotting ordered pairs.
2. Find intercepts.
3. Graph linear equations of the form $Ax + By = 0$.
4. Graph linear equations of the form $y = b$ or $x = a$.
5. Use a linear equation to model data.

❶ Complete each ordered pair to find additional solutions of the equation $x + 2y = 7$.

GS (a) $(-3, \underline{\quad})$

Substitute ____ for x in the equation and solve for ____.

(b) $(-1, \underline{\quad})$

(c) $(3, \underline{\quad})$

(d) $(5, \underline{\quad})$

(e) $(7, \underline{\quad})$

Answers
1. **(a)** $-3; y; (-3, 5)$ **(b)** $(-1, 4)$
   (c) $(3, 2)$ **(d)** $(5, 1)$ **(e)** $(7, 0)$

② Graph the linear equation.

GS $x + y = 6$

x	y
0	
	0
2	

From the first row of the table,
let $x = $ ____ .

$$x + y = 6$$
$$\underline{\quad} + y = 6$$
$$y = \underline{\quad}$$

Write ____ for the y-value in the first row of the table.

The first ordered pair is ____ .

Repeat this process to complete the table. Then plot the corresponding points, and draw the ____ through them.

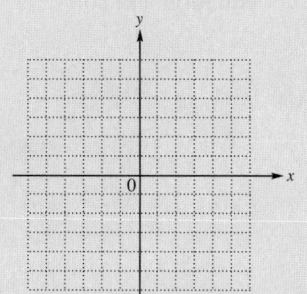

EXAMPLE 1 **Graphing a Linear Equation**

Graph $4x - 5y = 20$.

At least two different points are needed to draw the graph. First let $x = 0$ and then let $y = 0$ to determine two ordered pairs.

$4x - 5y = 20$		$4x - 5y = 20$	
$4(0) - 5y = 20$	0 is easy to substitute.	$4x - 5(0) = 20$	0 is easy to substitute.
$0 - 5y = 20$	Multiply.	$4x - 0 = 20$	Multiply.
$-5y = 20$	Subtract.	$4x = 20$	Subtract.
$y = -4$	Divide by -5.	$x = 5$	Divide by 4.

The ordered pairs are $(0, -4)$ and $(5, 0)$. Find a third ordered pair (as a check) by choosing a number *other than 0* for x or y. We choose $y = 2$.

$$4x - 5y = 20$$
$$4x - 5(2) = 20 \qquad \text{We arbitrarily let } y = 2. \text{ Other numbers could be used for } y, \text{ or for } x, \text{ instead.}$$
$$4x - 10 = 20 \qquad \text{Multiply.}$$
$$4x = 30 \qquad \text{Add 10.}$$
$$x = \frac{30}{4}, \quad \text{or} \quad \frac{15}{2} \qquad \text{Divide by 4. Write in lowest terms.}$$

This gives the ordered pair $\left(\frac{15}{2}, 2\right)$, or $\left(7\frac{1}{2}, 2\right)$. We plot the three ordered pairs $(0, -4)$, $(5, 0)$, and $\left(7\frac{1}{2}, 2\right)$, and draw a line through them. See **Figure 10.**

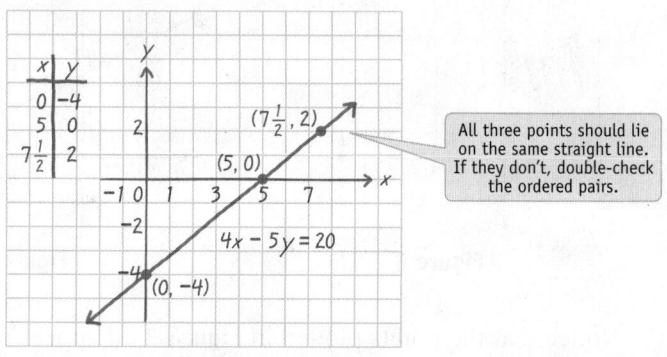

All three points should lie on the same straight line. If they don't, double-check the ordered pairs.

Figure 10

⊲ **Work Problem ②** at the Side.

OBJECTIVE **②** **Find intercepts.** In **Figure 10,** the graph crosses, or intersects, the y-axis at $(0, -4)$ and the x-axis at $(5, 0)$. For this reason, $(0, -4)$ is called the **y-intercept,** and $(5, 0)$ is called the **x-intercept** of the graph. The intercepts are particularly useful for graphing linear equations.

Answer

2. 0; 0; 6; 6; $(0, 6)$; line

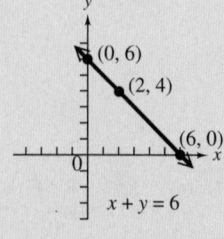

Finding Intercepts

To find the x-intercept, let $y = 0$ in the given equation and solve for x. Then $(x, 0)$ is the x-intercept.

To find the y-intercept, let $x = 0$ in the given equation and solve for y. Then $(0, y)$ is the y-intercept.

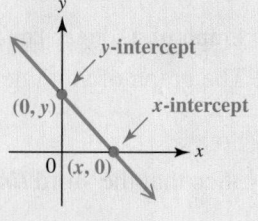

EXAMPLE 2 Graphing a Linear Equation Using Intercepts

Graph $2x + y = 4$ using intercepts.

To find the y-intercept, let $x = 0$.	To find the x-intercept, let $y = 0$.
$2x + y = 4$	$2x + y = 4$
$2(0) + y = 4$ Let $x = 0$.	$2x + 0 = 4$ Let $y = 0$.
$0 + y = 4$ Multiply.	$2x = 4$ Add.
$y = 4$ y-intercept is $(0, 4)$.	$x = 2$ x-intercept is $(2, 0)$.

The intercepts are $(0, 4)$ and $(2, 0)$. To find a third point (as a check), we let $x = 1$.

$$2x + y = 4$$
$$2(1) + y = 4 \quad \text{Let } x = 1.$$
$$2 + y = 4 \quad \text{Multiply.}$$
$$y = 2 \quad \text{Subtract 2.}$$

This gives the ordered pair $(1, 2)$. The graph, with the two intercepts in red, is shown in **Figure 11.**

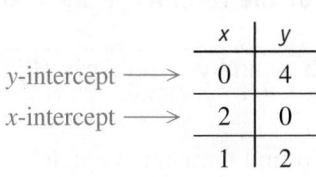

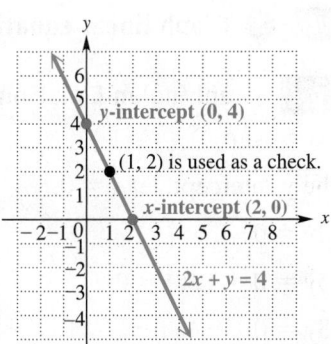

Figure 11

······ **Work Problem ③ at the Side.** ▶

EXAMPLE 3 Graphing a Linear Equation Using Intercepts

Graph $y = -\frac{3}{2}x + 3$.

    Although this linear equation is not in standard form $Ax + By = C$, it *could* be written in that form. The intercepts can be found by first letting $x = 0$ and then letting $y = 0$.

$y = -\dfrac{3}{2}x + 3$	$y = -\dfrac{3}{2}x + 3$
$y = -\dfrac{3}{2}(0) + 3$ Let $x = 0$.	$0 = -\dfrac{3}{2}x + 3$ Let $y = 0$.
$y = 0 + 3$ Multiply.	$\dfrac{3}{2}x = 3$ Add $\frac{3}{2}x$.
$y = 3$ Add.	$x = 2$ Multiply by $\frac{2}{3}$.

This gives the intercepts $(0, 3)$ and $(2, 0)$.

·········· **Continued on Next Page**

③ Graph the linear equation using
Ⓖⓢ intercepts. (Be sure to get a third point as a check.)

$$5x + 2y = 10$$

Find the y-intercept by letting $x = \underline{\quad}$ in the equation.

The y-intercept is $\underline{\quad}$. Find the x-intercept by letting $y = \underline{\quad}$ in the equation.

The x-intercept is $\underline{\quad}$.

Find a third point with $x = 4$ as a check, and graph the equation.

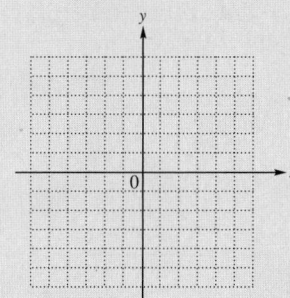

Answer

3. 0; $(0, 5)$; 0; $(2, 0)$

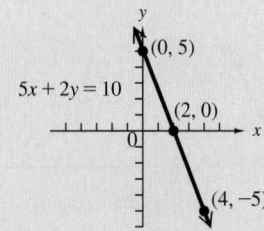

❹ Graph the linear equation using intercepts.

$$y = \frac{2}{3}x - 2$$

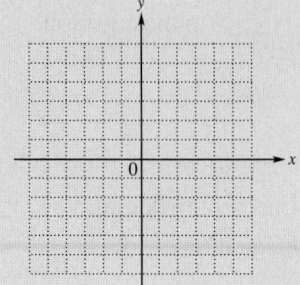

❺ Graph $2x - y = 0$.

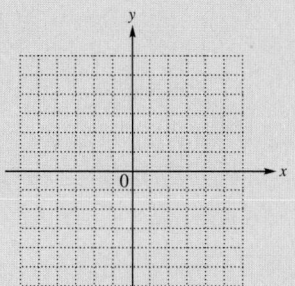

To find a third point, we let $x = -2$.

$$y = -\frac{3}{2}x + 3$$

> Choosing a multiple of 2 makes multiplying by $-\frac{3}{2}$ easier.

$$y = -\frac{3}{2}(-2) + 3 \quad \text{Let } x = -2.$$

$$y = 3 + 3 \qquad\qquad \text{Multiply.}$$

$$y = 6 \qquad\qquad \text{Add.}$$

This gives the ordered pair $(-2, 6)$. We plot this ordered pair and both intercepts and draw a line through them, as shown in **Figure 12**.

x	y
0	3
2	0
-2	6

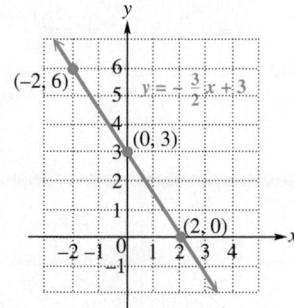

Figure 12

◀ **Work Problem ❹ at the Side.**

OBJECTIVE ▸ ③ Graph linear equations of the form $Ax + By = 0$.

EXAMPLE 4 Graphing an Equation with x- and y-Intercepts $(0, 0)$

Graph $x - 3y = 0$.

To find the y-intercept, let $x = 0$.

$$x - 3y = 0$$
$$0 - 3y = 0 \quad \text{Let } x = 0.$$
$$-3y = 0 \quad \text{Subtract.}$$
$$y = 0 \quad y\text{-intercept is } (0, 0).$$

To find the x-intercept, let $y = 0$.

$$x - 3y = 0$$
$$x - 3(0) = 0 \quad \text{Let } y = 0.$$
$$x - 0 = 0 \quad \text{Multiply.}$$
$$x = 0 \quad x\text{-intercept is } (0, 0).$$

The x- and y-intercepts are the *same* point, $(0, 0)$. We must select *two other values* for x or y to find two other points. We choose $x = 6$ and $x = -6$.

$$x - 3y = 0$$
$$6 - 3y = 0 \quad \text{Let } x = 6.$$
$$-3y = -6 \quad \text{Subtract 6.}$$
$$y = 2 \quad \text{Gives } (6, 2)$$

$$x - 3y = 0$$
$$-6 - 3y = 0 \quad \text{Let } x = -6.$$
$$-3y = 6 \quad \text{Add 6.}$$
$$y = -2 \quad \text{Gives } (-6, -2)$$

We use $(-6, -2)$, $(0, 0)$, and $(6, 2)$ to draw the graph in **Figure 13**.

x	y
0	0
6	2
-6	-2

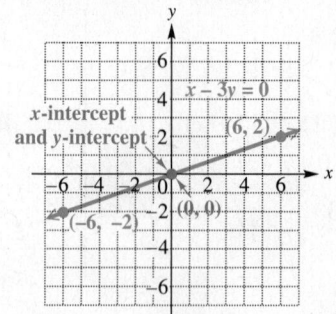

Figure 13

◀ **Work Problem ❺ at the Side.**

Answers

4.

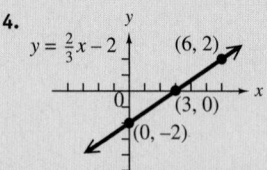

5.

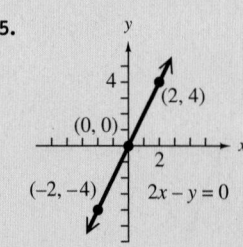

Line through the Origin

The graph of a linear equation of the form

$$Ax + By = 0,$$

where A and B are nonzero real numbers, passes through the origin $(0, 0)$.

OBJECTIVE **4** **Graph linear equations of the form $y = b$ or $x = a$.**
Consider the following linear equations.

| $y = -4$ | can be written as | $0x + y = -4.$ |
| $x = 3$ | can be written as | $x + 0y = 3.$ |

When the coefficient of x or y is 0, the graph is a horizontal or vertical line.

EXAMPLE 5 **Graphing a Horizontal Line ($y = b$)**

Graph $y = -4$.

For any value of x, y is always equal to -4. Three ordered pairs that satisfy the equation are shown in the table of values. Drawing a line through these points gives the **horizontal line** shown in **Figure 14.** The y-intercept is $(0, -4)$. There is no x-intercept.

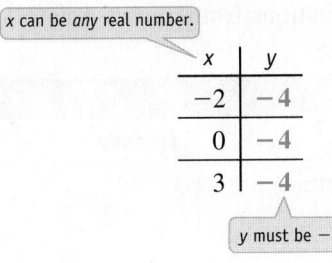

x can be *any* real number.

x	y
-2	-4
0	-4
3	-4

y must be -4.

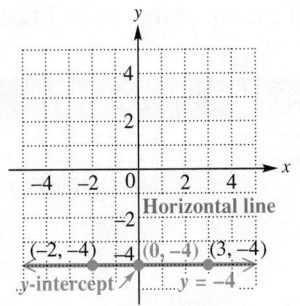

Figure 14

Work Problem **6** at the Side. ▶

EXAMPLE 6 **Graphing a Vertical Line ($x = a$)**

Graph $x - 3 = 0$.

First we add 3 to each side of the equation $x - 3 = 0$ to get the equivalent equation $x = 3$. All ordered-pair solutions of this equation have x-coordinate 3. Any number can be used for y. We show three ordered pairs that satisfy the equation in the table of values. The graph is the **vertical line** shown in **Figure 15.** The x-intercept is $(3, 0)$. There is no y-intercept.

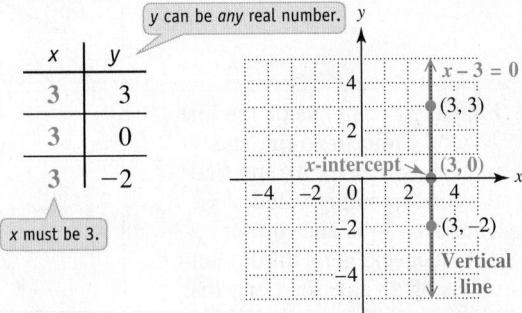

y can be *any* real number.

x	y
3	3
3	0
3	-2

x must be 3.

Figure 15

Work Problem **7** at the Side. ▶

6 Graph $y = -5$.

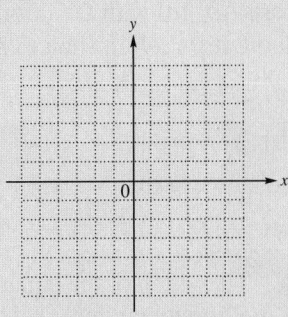

This is the graph of a _____ line. There is no x-intercept.

7 Graph $x + 4 = 6$.

First subtract ____ from each side to obtain the equivalent equation $x = $ ____. Now graph this equation.

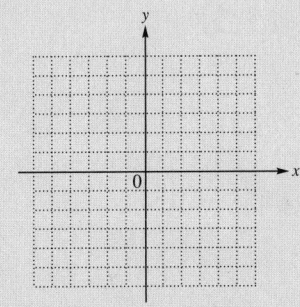

This is the graph of a _____ line. There is no y-intercept.

Answers

6.

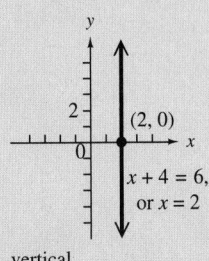

$(0, -5)$ $y = -5$

horizontal

7. 4; 2

$(2, 0)$

$x + 4 = 6,$ or $x = 2$

vertical

8 Match each linear equation in parts (a)–(d) with the information about its graph in choices A–D.

(a) $x = 5$

(b) $2x - 5y = 8$

(c) $y - 2 = 3$

(d) $x + 4y = 0$

A. The graph of the equation is a horizontal line.

B. The graph of the equation passes through the origin.

C. The graph of the equation is a vertical line.

D. The graph of the equation passes through $(9, 2)$.

From **Examples 5 and 6,** we make the following observations.

Horizontal and Vertical Lines

The graph of the linear equation $y = b$, where b is a real number, is the **horizontal line** with y-intercept $(0, b)$. There is no x-intercept (unless the horizontal line is the x-axis itself).

The graph of the linear equation $x = a$, where a is a real number, is the **vertical line** with x-intercept $(a, 0)$. There is no y-intercept (unless the vertical line is the y-axis itself).

The x-axis is the horizontal line given by the equation $y = 0$, and the y-axis is the vertical line given by the equation $x = 0$.

CAUTION

The equations of horizontal and vertical lines are often confused with each other. The graph of $y = b$ is parallel to the x-axis and the graph of $x = a$ is parallel to the y-axis (for $a \neq 0$ and $b \neq 0$).

A summary of the forms of linear equations from this section follows.

Graphing a Linear Equation

Equation	To Graph	Example
$y = b$	Draw a horizontal line, through $(0, b)$.	$y = -2$ $(0, -2)$
$x = a$	Draw a vertical line, through $(a, 0)$.	$(4, 0)$ $x = 4$
$Ax + By = 0$	The graph passes through $(0, 0)$. To find additional points that lie on the graph, choose any value for x or y, except 0.	$x = 2y$ $(0, 0)$ $(4, 2)$
$Ax + By = C$ (**where** A, B, $C \neq 0$)	Find any two points on the line. A good choice is to find the intercepts. Let $x = 0$, and find the corresponding value of y. Then let $y = 0$, and find x. As a check, get a third point by choosing a value for x or y that has not yet been used.	$(4, 3)$ $(2, 0)$ $3x - 2y = 6$ $(0, -3)$

Answers

8. **(a)** C **(b)** D **(c)** A **(d)** B

◀ **Work Problem 8 at the Side.**

OBJECTIVE ▶ ⑤ **Use a linear equation to model data.**

EXAMPLE 7 | **Using a Linear Equation to Model Credit Card Debt**

The amount of credit card debt y in billions of dollars in the United States from 2000 through 2008 can be modeled by the linear equation

$$y = 32.0x + 684,$$

where $x = 0$ represents the year 2000, $x = 1$ represents 2001, and so on. (*Source:* The Nilson Report.)

(a) Use the equation to approximate credit card debt in the years 2000, 2004, and 2008.

Substitute the appropriate value for each year x to find credit card debt in that year.

For 2000: $y = 32.0\,(0) + 684$ Replace x with 0.
                  $y = 684$ billion dollars   Multiply, and then add.

For 2004: $y = 32.0\,(4) + 684$ $2004 - 2000 = 4$
                  $y = 812$ billion dollars   Replace x with 4.

For 2008: $y = 32.0\,(8) + 684$ $2008 - 2000 = 8$
                  $y = 940$ billion dollars   Replace x with 8.

(b) Write the information from part (a) as three ordered pairs, and use them to graph the given linear equation.

Since x represents the year and y represents the debt, the ordered pairs are $(0, 684)$, $(4, 812)$, and $(8, 940)$. See **Figure 16.** (Arrowheads are not included with the graphed line, since the data are for the years 2000 to 2008 only—that is, from $x = 0$ to $x = 8$.)

U.S. CREDIT CARD DEBT

Debt (in billions of dollars)

$y = 32.0x + 684$

Year

Figure 16

(c) Use the graph and the equation to approximate credit card debt in 2002.

For 2002, $x = 2$. On the graph, find 2 on the horizontal axis, move up to the graphed line and then across to the vertical axis. It appears that credit card debt in 2002 was about 750 billion dollars.

To use the equation, substitute 2 for x.

     $y = 32.0x + 684$ Given linear equation

     $y = 32.0\,(2) + 684$ Let $x = 2$.

     $y = 748$ billion dollars   Multiply, and then add.

This result for 2002 is close to our estimate of 750 billion dollars from the graph.

······· **Work Problem** ⑨ **at the Side.** ▶

⑨ Use **(a)** the graph and **(b)** the equation in **Example 7** to approximate credit card debt in 2006.

Answers

9. (a) about 875 billion dollars
    (b) 876 billion dollars

3.2 Exercises

CONCEPT CHECK *Fill in each blank with the correct response.*

1. A linear equation in two variables can be written in the form $Ax +$ _____ $=$ _____, where A, B, and C are real numbers and A and B are not both _____.

2. The graph of any linear equation in two variables is a straight _____. Every point on the line represents a _____ of the equation.

3. **CONCEPT CHECK** Match the information about each graph in Column I with the correct linear equation in Column II.

	I		II
(a)	The graph of the equation has y-intercept $(0, -4)$.	**A.**	$3x + y = -4$
(b)	The graph of the equation has $(0, 0)$ as x-intercept and y-intercept.	**B.**	$x - 4 = 0$
(c)	The graph of the equation does not have an x-intercept.	**C.**	$y = 4x$
(d)	The graph of the equation has x-intercept $(4, 0)$.	**D.**	$y = 4$

4. **CONCEPT CHECK** Which of these equations have a graph with only one intercept?

   A. $x + 8 = 0$ **B.** $x - y = 3$ **C.** $x + y = 0$ **D.** $y = 4$

CONCEPT CHECK *Identify the intercepts of each graph. (All coordinates are integers.)*

5.
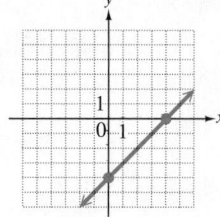

6.
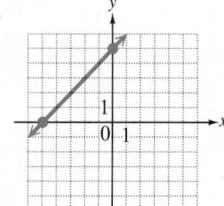

7.
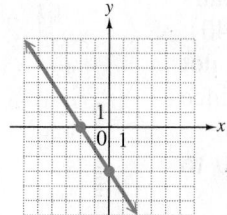

8.
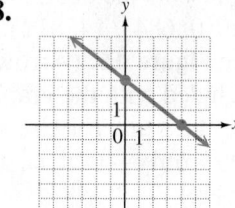

Complete the given ordered pairs for each equation. Then graph each equation by plotting the points and drawing the line through them. **See Examples 1 and 2.**

9. $x + y = 5$

   $(0, \underline{\quad}), (\underline{\quad}, 0), (2, \underline{\quad})$

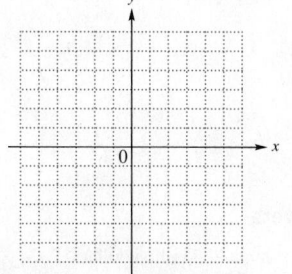

10. $x - y = 2$

   $(0, \underline{\quad}), (\underline{\quad}, 0), (5, \underline{\quad})$

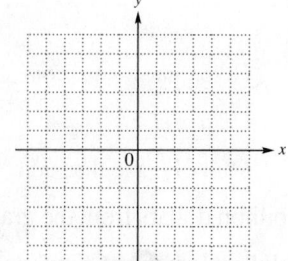

11. $y = \dfrac{2}{3}x + 1$

   $(0, \underline{\quad}), (3, \underline{\quad}), (-3, \underline{\quad})$

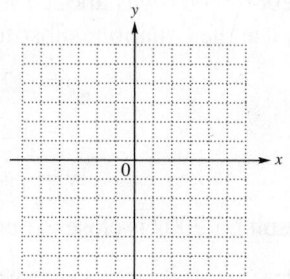

12. $y = -\dfrac{3}{4}x + 2$

$(0, \underline{\quad}), (4, \underline{\quad}), (-4, \underline{\quad})$

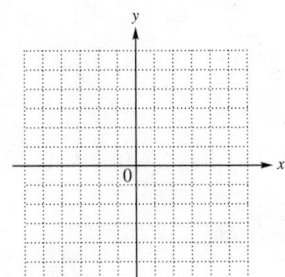

13. $3x = -y - 6$

$(0, \underline{\quad}), (\underline{\quad}, 0), \left(-\dfrac{1}{3}, \underline{\quad}\right)$

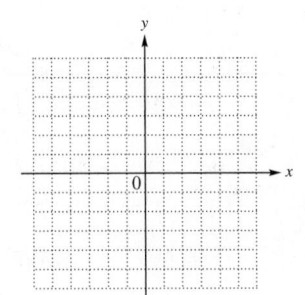

14. $x = 2y + 3$

$(\underline{\quad}, 0), (0, \underline{\quad}), \left(\underline{\quad}, \dfrac{1}{2}\right)$

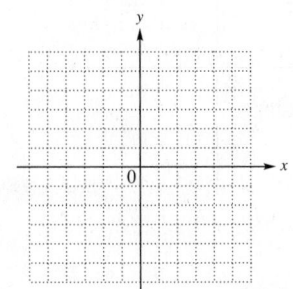

Find the intercepts for the graph of each equation. ***See Example 2.***

15. $x - y = 8$

To find the y-intercept,
let $x = \underline{\qquad}$.
The y-intercept is $\underline{\qquad}$.
To find the x-intercept,
let $y = \underline{\qquad}$.
The x-intercept is $\underline{\qquad}$.

16. $x - y = 10$

To find the y-intercept,
let $x = \underline{\qquad}$.
The y-intercept is $\underline{\qquad}$.
To find the x-intercept,
let $y = \underline{\qquad}$.
The x-intercept is $\underline{\qquad}$.

17. $2x - 3y = 24$

y-intercept: $\underline{\qquad}$
x-intercept: $\underline{\qquad}$

18. $-3x + 8y = 48$

y-intercept: $\underline{\qquad}$
x-intercept: $\underline{\qquad}$

19. $x + 6y = 0$

y-intercept: $\underline{\qquad}$
x-intercept: $\underline{\qquad}$

20. $3x - y = 0$

y-intercept: $\underline{\qquad}$
x-intercept: $\underline{\qquad}$

Graph each linear equation using intercepts. ***See Examples 1–6.***

21. $y = x - 2$

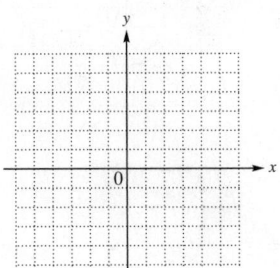

22. $y = -x + 6$

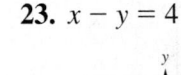
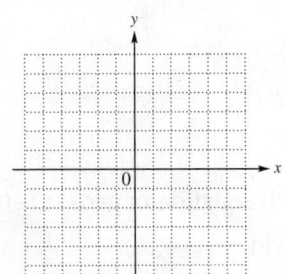

23. $x - y = 4$

24. $x - y = 5$

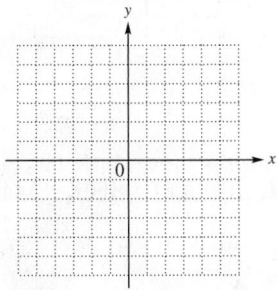

25. $2x + y = 6$

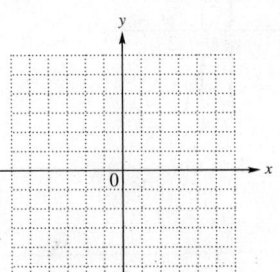

26. $-3x + y = -6$

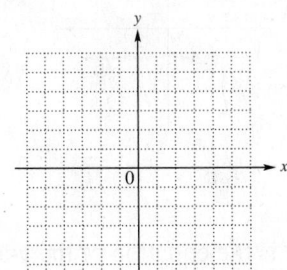

27. $y = 2x - 5$

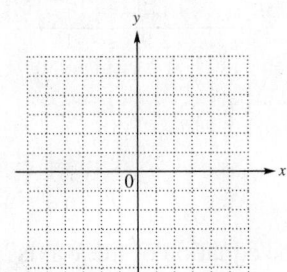

28. $y = 4x + 3$

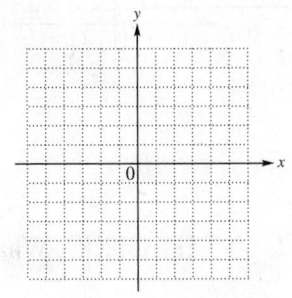

29. $3x + 7y = 14$

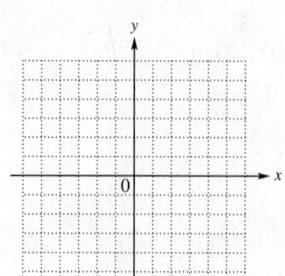

30. $6x - 5y = 18$

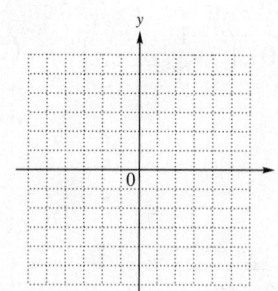

31. $y - 2x = 0$

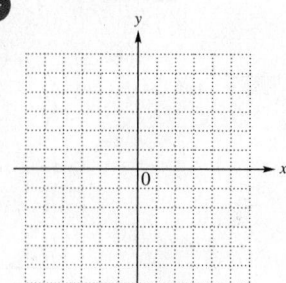

32. $y + 3x = 0$

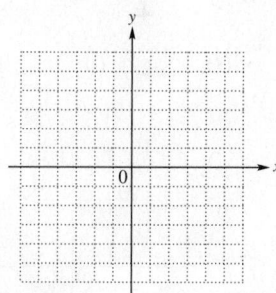

33. $y = -6x$

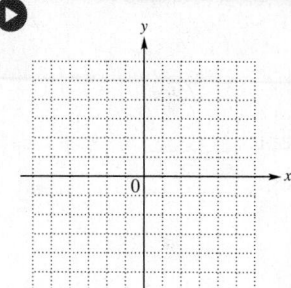

34. $y = 4x$

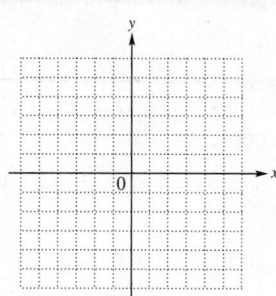

35. $x = -2$

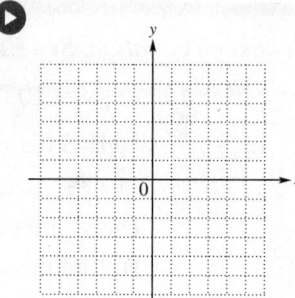

36. $x = 4$

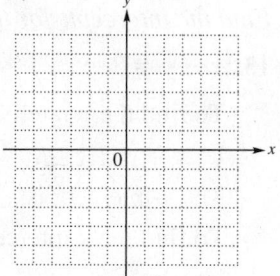

37. $y - 3 = 0$

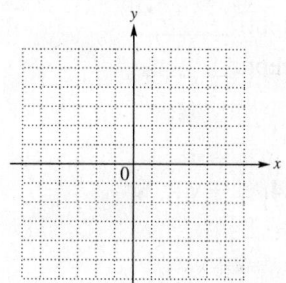

38. $y + 1 = 0$

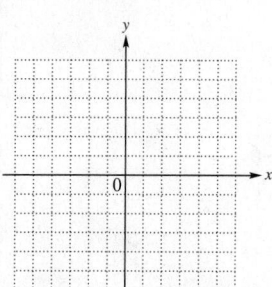

39. $-3y = 15$

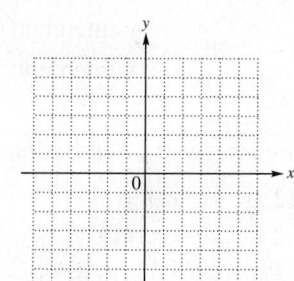

40. $-2y = 12$

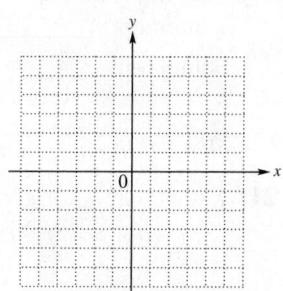

41. CONCEPT CHECK Match each equation in parts (a)–(d) with its graph in choices A–D.

(a) $x = -2$ **(b)** $y = -2$ **(c)** $x = 2$ **(d)** $y = 2$

A.

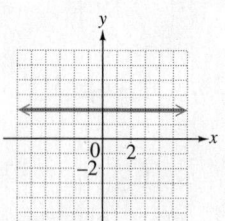

B.

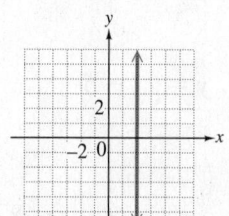

C.

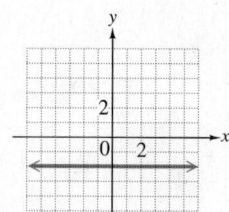

D.

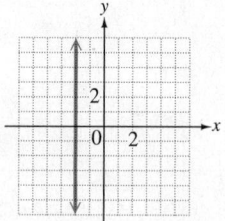

42. CONCEPT CHECK What is the equation of the x-axis? What is the equation of the y-axis?

In Exercises 43–46, describe what the graph of each linear equation will look like on the coordinate plane. (Hint: Rewrite the equation if necessary so that it is in a more recognizable form.)

43. $3x = y - 9$

44. $x - 10 = 1$

45. $3y = -6$

46. $2x = 4y$

Solve each problem. See Example 7.

47. The height y of a woman is related to the length of her radius bone x and is approximated by the linear equation

$$y = 3.9x + 73.5,$$

where both x and y are in centimeters.

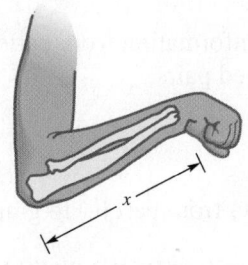

(a) Use the equation to find the approximate heights of women with radius bones of lengths 20 cm, 22 cm, and 26 cm.

(b) Write the information from part (a) as three ordered pairs.

(c) Graph the equation using the data from part (b).

HEIGHTS OF WOMEN

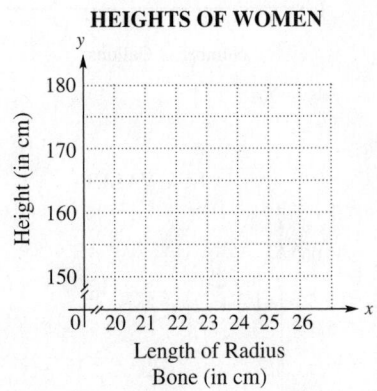

(d) Use the graph to estimate the length of the radius bone in a woman who is 167 cm tall. Then use the equation to find the length of this radius bone to the nearest centimeter. (*Hint:* Substitute for y in the equation.)

48. The weight y (in pounds) of a man taller than 60 in. can be roughly approximated by the linear equation

$$y = 5.5x - 220,$$

where x is the height of the man in inches.

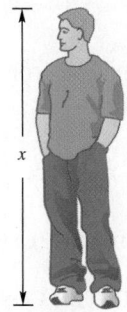

(a) Use the equation to approximate the weights of men whose heights are 62 in., 66 in., and 72 in.

(b) Write the information from part (a) as three ordered pairs.

(c) Graph the equation using the data from part (b).

WEIGHTS OF MEN

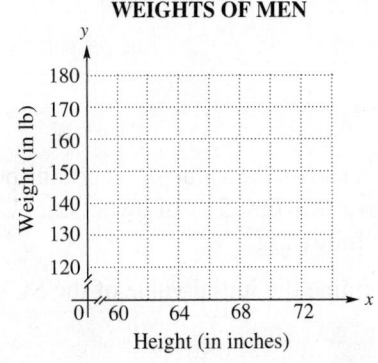

(d) Use the graph to estimate the height of a man who weighs 155 lb. Then use the equation to find the height of this man to the nearest inch. (*Hint:* Substitute for y in the equation.)

49. As a fundraiser, a school club is selling posters. The printer charges a $25 set-up fee, plus $0.75 for each poster. The cost y in dollars to print x posters is given by the linear equation

$$y = 0.75x + 25.$$

(a) What is the cost y in dollars to print 50 posters? To print 100 posters?

(b) Find the number of posters x if the printer billed the club for costs of $175.

(c) Write the information from parts (a) and (b) as three ordered pairs.

(d) Use the data from part (c) to graph the equation.

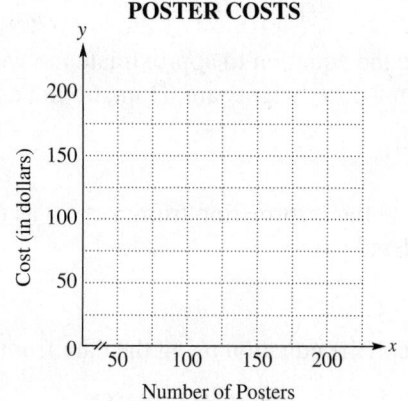

POSTER COSTS

50. A gas station is selling gasoline for $4.50 per gallon and charges $7 for a car wash. The cost y in dollars for x gallons of gasoline and a car wash is given by the linear equation

$$y = 4.50x + 7.$$

(a) What is the cost y in dollars for 9 gallons of gasoline and a car wash? For 4 gallons of gasoline and a car wash?

(b) Find the number of gallons of gasoline x if the cost for the gasoline and a car wash is $43.

(c) Write the information from parts (a) and (b) as three ordered pairs.

(d) Use the data from part (c) to graph the equation.

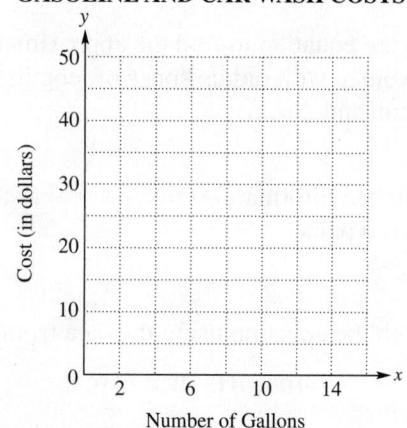

GASOLINE AND CAR WASH COSTS

51. The graph shows the value of a certain sport-utility vehicle over the first 5 yr of ownership. Use the graph to do the following.

(a) Determine the initial value of the SUV.

(b) Find the **depreciation** (loss in value) from the original value after the first 3 yr.

(c) What is the annual or yearly depreciation in each of the first 5 yr?

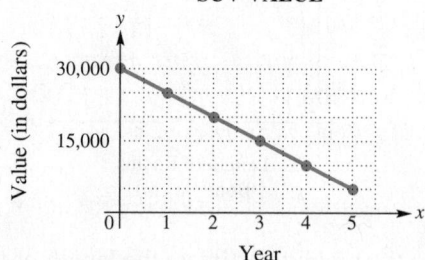

SUV VALUE

(d) What does the ordered pair (5, 5000) mean in the context of this problem?

52. Demand for an item is often closely related to its price. As price increases, demand decreases, and as price decreases, demand increases. Suppose demand for a video game is 2000 units when the price is $40, and demand is 2500 units when the price is $30.

(a) Let x be the price and y be the demand for the game. Graph the two given pairs of prices and demands.

VIDEO GAME PRICE/DEMAND

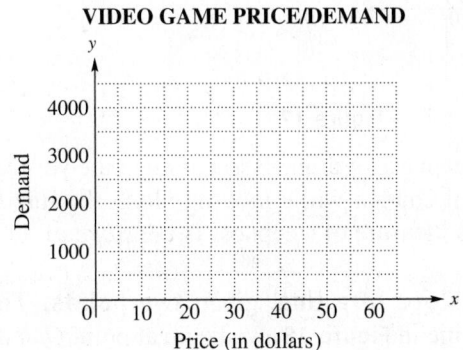

(b) Assume the relationship is linear. Draw a line through the two points from part (a). From your graph, estimate the demand if the price drops to $20.

(c) Use the graph to estimate the price if the demand is 3500 units.

(d) Write the prices and demands from parts (b) and (c) as ordered pairs.

(e) What does the ordered pair $(50, 1500)$ mean in the context of this problem?

53. U.S. per capita consumption of cheese increased for the years 1980 through 2009 as shown in the graph. If $x = 0$ represents 1980, $x = 5$ represents 1985, and so on, per capita consumption y in pounds can be modeled by the linear equation

$$y = 0.5249x + 18.54.$$

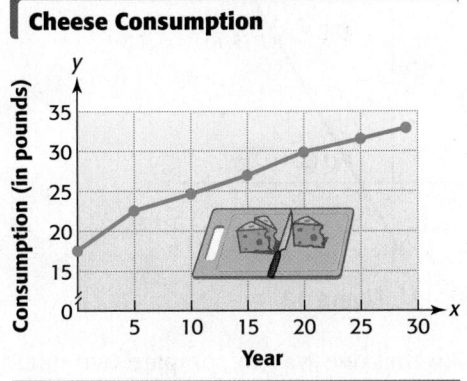

Cheese Consumption

Source: U.S. Department of Agriculture.

(a) Use the equation to approximate cheese consumption in 1990, 2000, and 2009 to the nearest tenth.

(b) Use the graph to estimate cheese consumption for the same years.

(c) How do the approximations using the equation compare to the estimates from the graph?

54. The number of U.S. marathon finishers y (in thousands) from 1990 through 2010 are shown in the graph and modeled by the linear equation

$$y = 13.36x + 220.8,$$

where $x = 0$ corresponds to 1990, $x = 5$ corresponds to 1995, and so on.

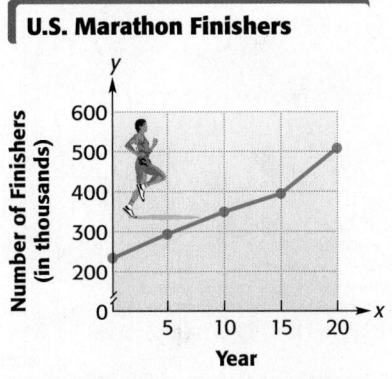

U.S. Marathon Finishers

Source: Running U.S.A.

(a) Use the equation to approximate the number of U.S. marathon finishers in 1990, 2000, and 2010 to the nearest thousand.

(b) Use the graph to estimate the number of U.S. marathon finishers for the same years.

(c) How do the approximations using the equation compare to the estimates from the graph?

3.3 The Slope of a Line

OBJECTIVES

1. Find the slope of a line, given two points.

2. Find the slope from the equation of a line.

3. Use slope to determine whether two lines are parallel, perpendicular, or neither.

4. Solve problems involving average rate of change.

An important characteristic of the lines we graphed in the previous section is their slant or "steepness," as viewed from *left to right*. See **Figure 17.**

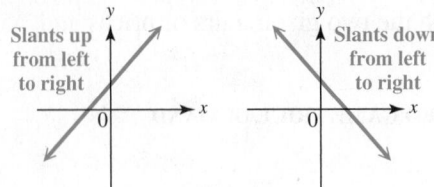

Figure 17

One way to measure the steepness of a line is to compare the vertical change in the line to the horizontal change while moving along the line from one fixed point to another. This measure of steepness is the *slope* of the line.

OBJECTIVE ➊ **Find the slope of a line, given two points.** To find the steepness, or slope, of the line in **Figure 18**, we begin at point Q and move to point P. The vertical change, or **rise,** is the change in the y-values, which is the difference

$$6 - 1 = 5 \text{ units.}$$

The horizontal change, or **run,** from Q to P is the change in the x-values, which is the difference

$$5 - 2 = 3 \text{ units.}$$

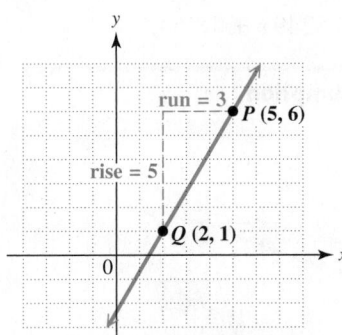

Figure 18

Remember from **Section 2.6** that one way to compare two numbers is by using a ratio. **Slope** is the ratio of the vertical change in y to the horizontal change in x. The line in **Figure 18** has

$$\text{slope} = \frac{\text{vertical change in } y \text{ (rise)}}{\text{horizontal change in } x \text{ (run)}} = \frac{5}{3}.$$

To confirm this ratio, we can count grid squares. We start at point Q in **Figure 18** and count *up* 5 grid squares to find the vertical change (rise). To find the horizontal change (run) and arrive at point P, we count to the *right* 3 grid squares. The slope is $\frac{5}{3}$, as found above.

> *Slope is a single number that allows us to determine the direction in which a line is slanting from left to right, as well as how much slant there is to the line.*

EXAMPLE 1 Finding the Slope of a Line

Find the slope of the line in **Figure 19.**

We use the two points shown on the line. The vertical change is the difference in the y-values, or $-1 - 3 = -4$, and the horizontal change is the difference in the x-values, or $6 - 2 = 4$. Thus, the line has

$$\text{slope} = \frac{\text{change in } y \text{ (rise)}}{\text{change in } x \text{ (run)}} = \frac{-4}{4}, \text{ or } -1.$$

Counting grid squares, we begin at point P and count *down* 4 grid squares. Because we counted down, we write the vertical change as a negative number, -4 here. Then we count to the *right* 4 grid squares to reach point Q. The slope is $\frac{-4}{4}$, or -1.

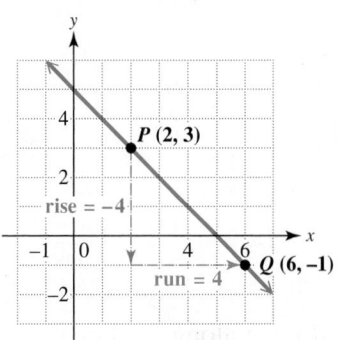

Figure 19

············· Work Problem **1** at the Side. ▶

Note

The slope of a line is the same for any two points on the line. To see this, refer to **Figure 19.** Find two points, say $(3, 2)$ and $(5, 0)$, which also lie on the line. If we start at $(3, 2)$ and count *down* 2 units and then to the *right* 2 units, we arrive at $(5, 0)$. The slope is $\frac{-2}{2}$, or -1, the same slope we found in **Example 1.**

The concept of slope is used in many everyday situations. See **Figure 20.** A highway with a 10%, or $\frac{1}{10}$, grade (or slope) rises 1 m for every 10 m horizontally. Architects specify the pitch of a roof by using slope. A $\frac{5}{12}$ roof means that the roof rises 5 ft vertically for every 12 ft that it runs horizontally. The slope of a stairwell indicates the ratio of the vertical rise to the horizontal run. The slope of the stairwell is $\frac{8}{12}$, or $\frac{2}{3}$.

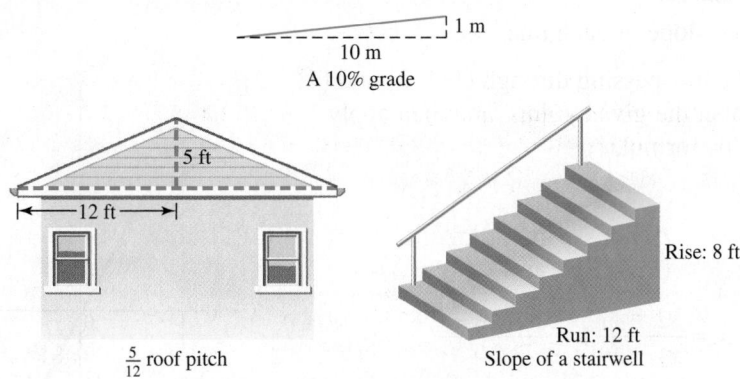

Figure 20

We can generalize the preceding discussion and find the slope of a line through two nonspecific points (x_1, y_1) and (x_2, y_2). This notation is called **subscript notation.** Read x_1 as "**x-sub-one**" and x_2 as "**x-sub-two.**" See **Figure 21** on the next page.

1 Find the slope of each line.

GS **(a)**

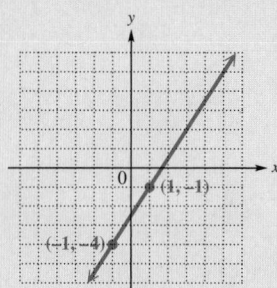

Begin at $(-1, -4)$.

Count (*up* / *down*) ____ units.

Then count to the (*right* / *left*) ____ units.

$$\text{slope} = \frac{\text{vertical change in } y}{\text{horizontal change in } x} = \frac{}{}$$

The slope is ____ .

(b)

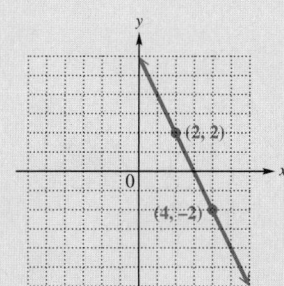

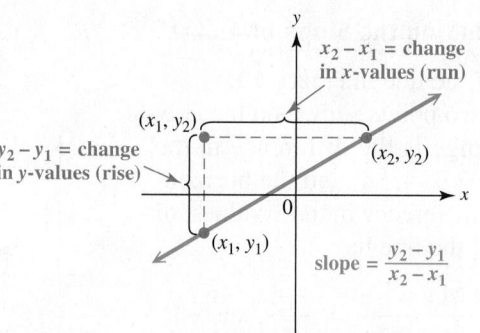

Figure 21

Moving along the line from the point (x_1, y_1) to the point (x_2, y_2), we see that y changes by $y_2 - y_1$ units. This is the vertical change (rise). Similarly, x changes by $x_2 - x_1$ units, which is the horizontal change (run). The slope of the line is the ratio of $y_2 - y_1$ to $x_2 - x_1$.

> **Note**
>
> Subscript notation is used to identify a point. It does *not* indicate any operation. Note the difference between x_2, a nonspecific value, and x^2, which means $x \cdot x$. Read x_2 as "x-sub-two," *not* "x squared."

> **Slope Formula**
>
> The **slope m** of the line passing through the points (x_1, y_1) and (x_2, y_2) is defined as follows. (Traditionally, the letter m represents slope.)
>
> $$m = \frac{\textbf{change in } y}{\textbf{change in } x} = \frac{y_2 - y_1}{x_2 - x_1} \qquad (\text{where } x_1 \neq x_2)$$

The slope gives the change in y for each unit of change in x.

EXAMPLE 2 **Finding Slopes of Lines Given Two Points**

Find the slope of each line.

(a) The line passing through $(-4, 7)$ and $(1, -2)$

Label the given points, and then apply the slope formula.

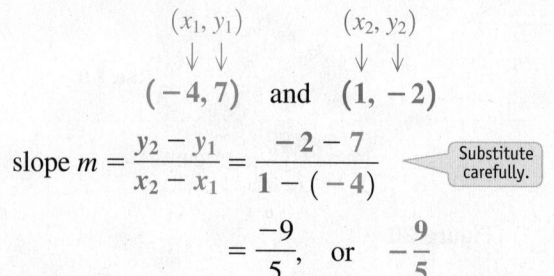

$$\text{slope } m = \frac{y_2 - y_1}{x_2 - x_1} = \frac{-2 - 7}{1 - (-4)} \quad \boxed{\text{Substitute carefully.}}$$

$$= \frac{-9}{5}, \quad \text{or} \quad -\frac{9}{5}$$

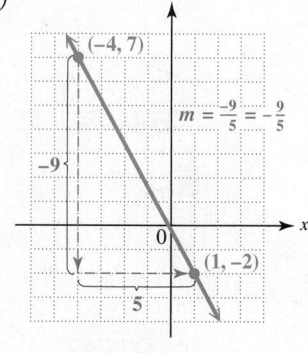

Figure 22

Count grid squares in **Figure 22** to confirm that the slope is $-\frac{9}{5}$.

Continued on Next Page

(b) The line passing through $(-9, -2)$ and $(12, 5)$

Label the points, and then apply the slope formula.

$$
\underset{\downarrow\ \downarrow}{(x_1, y_1)} \qquad \underset{\downarrow\ \downarrow}{(x_2, y_2)}
$$

$$
(-9, -2) \quad \text{and} \quad (12, 5)
$$

$$
\text{slope } m = \frac{y_2 - y_1}{x_2 - x_1} = \frac{5 - (-2)}{12 - (-9)}
$$

$$
= \frac{7}{21}, \quad \text{or} \quad \frac{1}{3} \quad \begin{array}{l}\text{Write in}\\ \text{lowest terms.}\end{array}
$$

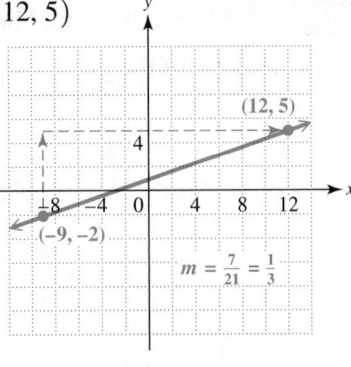

$$m = \frac{7}{21} = \frac{1}{3}$$

Figure 23

Confirm this calculation using **Figure 23.**

The same slope is obtained if we label the points in reverse order. *It makes no difference which point is identified as (x_1, y_1) or (x_2, y_2).*

$$
\underset{\downarrow\ \downarrow}{(x_2, y_2)} \qquad \underset{\downarrow\ \downarrow}{(x_1, y_1)}
$$

$$
(-9, -2) \quad \text{and} \quad (12, 5)
$$

$$
\text{slope } m = \frac{y_2 - y_1}{x_2 - x_1} = \frac{-2 - 5}{-9 - 12} \qquad \text{Substitute.}
$$

$$
= \frac{-7}{-21}, \quad \text{or} \quad \frac{1}{3} \qquad \text{The same slope results.}
$$

················· **Work Problem ❷ at the Side.** ▶

The lines in **Figures 22 and 23** suggest the following generalization.

Orientation of Lines with Positive and Negative Slopes

A line with positive slope rises (slants up) from left to right.

A line with negative slope falls (slants down) from left to right.

EXAMPLE 3 Finding the Slope of a Horizontal Line

Find the slope of the line passing through $(-5, 4)$ and $(2, 4)$.

$$
\underset{\downarrow\ \downarrow}{(x_1, y_1)} \qquad \underset{\downarrow\ \downarrow}{(x_2, y_2)}
$$

$$
(-5, 4) \quad \text{and} \quad (2, 4) \qquad \text{Label the points.}
$$

$$
m = \frac{y_2 - y_1}{x_2 - x_1} = \frac{4 - 4}{2 - (-5)} \qquad \text{Substitute in the slope formula.}
$$

$$
= \frac{0}{7}, \quad \text{or} \quad 0 \quad \text{Slope 0}
$$

As shown in **Figure 24,** the line through these two points is horizontal, with equation $y = 4$. *All horizontal lines have slope 0,* since the difference in their y-values is always 0.

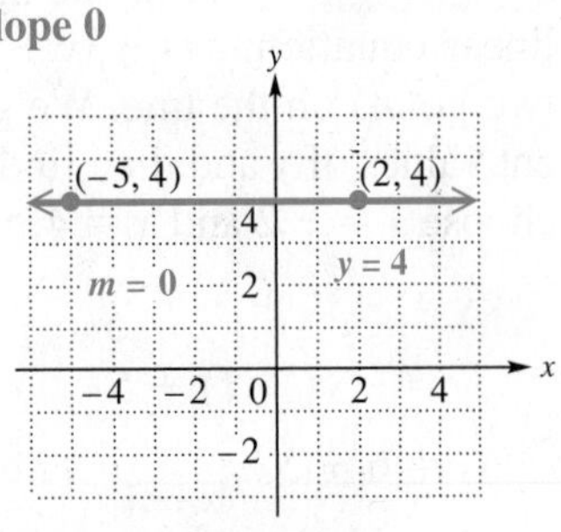

Figure 24

---

❷ Find the slope of each line.

(a) The line passing through $(6, -2)$ and $(5, 4)$

Label the points.

$$
\underset{\downarrow\ \downarrow}{(x_1, y_1)} \qquad \underset{\downarrow\ \downarrow}{(\_\_, \_\_)}
$$

$$
(6, -2) \quad \text{and} \quad (5, \ \ 4)
$$

$$
\text{slope } m = \frac{y_2 - y_1}{x_2 - x_1}
$$

$$
= \frac{\_\_ - (\_\_)}{\_\_ - 6}
$$

$$
= \frac{6}{\_\_}, \quad \text{or} \quad \_\_
$$

(b) The line passing through $(-3, 5)$ and $(-4, -7)$

(c) The line passing through $(6, -8)$ and $(-2, 4)$

(Find this slope in two different ways as in **Example 2(b)**.)

Answers

2. (a) $x_2; y_2; 4; -2; 5; -1; -6$

(b) 12 **(c)** $-\frac{3}{2}; -\frac{3}{2}$

3 Find the slope of each line.

   (a) The line passing through $(2, 5)$ and $(-1, 5)$

   (b) The line passing through $(3, 1)$ and $(3, -4)$

EXAMPLE 4   **Finding the Slope of a Vertical Line**

Find the slope of the line passing through $(6, 2)$ and $(6, -4)$.

$$\begin{array}{cc} (x_1, y_1) & (x_2, y_2) \\ \downarrow\ \downarrow & \downarrow\ \downarrow \\ (6, 2) \quad \text{and} & (6, -4) \end{array} \quad \text{Label the points.}$$

$$m = \frac{y_2 - y_1}{x_2 - x_1} = \frac{-4 - 2}{6 - 6} \quad \text{Substitute in the slope formula.}$$

$$= \frac{-6}{0} \quad \text{Undefined slope}$$

Because division by 0 is undefined, this line has undefined slope. (This is why the slope formula at the beginning of this section had the restriction $x_1 \neq x_2$.) The graph in **Figure 25** shows that this line is vertical, with equation $x = 6$. All points on a vertical line have the same x-value, so ***the slope of any vertical line is undefined.***

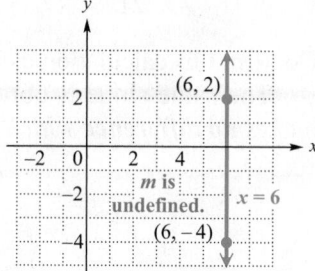

Figure 25

$\blacktriangleleft$ **Work Problem 3 at the Side.**

---

4 Find the slope of each line.

GS

   (a) The line with equation $y = -1$

     This is the equation of a (*horizontal* / *vertical*) line. It has (0 / *undefined*) slope.

   (b) The line with equation $x - 4 = 0$

     Rewrite this equation in an equivalent form.

$$x - 4 = 0$$

$$x = \underline{\quad\quad}$$

     This is the equation of a (*horizontal* / *vertical*) line. It has (0 / *undefined*) slope.

> **Slopes of Horizontal and Vertical Lines**
>
> **Horizontal lines,** which have equations of the form $y = k$ (where k is a constant (number)), have **slope 0.**
>
> **Vertical lines,** which have equations of the form $x = k$ (where k is a constant (number)), have **undefined slope.**

$\blacktriangleleft$ **Work Problem 4 at the Side.**

Figure 26 summarizes the four cases for slopes of lines.

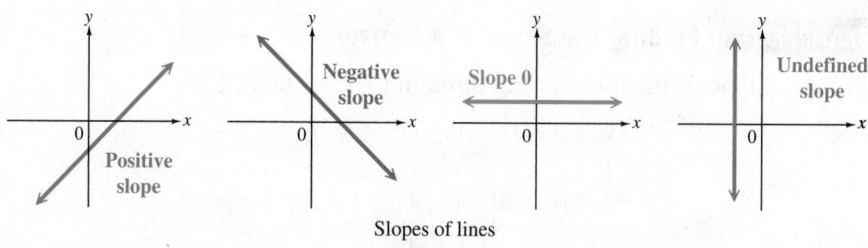

Slopes of lines

Figure 26

OBJECTIVE **2** **Find the slope from the equation of a line.** Consider the linear equation $y = -3x + 5$. We can find the slope of the line using any two points on the line. We get these two points by first choosing two different values of x and then finding the corresponding values of y. We arbitrarily choose $x = -2$ and $x = 4$.

$y = -3x + 5$	$y = -3x + 5$
$y = -3(-2) + 5$ Let $x = -2$.	$y = -3(4) + 5$ Let $x = 4$.
$y = 6 + 5$ Multiply.	$y = -12 + 5$ Multiply.
$y = 11$ Add.	$y = -7$ Add.

The ordered pairs are $(-2, 11)$ and $(4, -7)$.

Answers

3. **(a)** 0   **(b)** undefined
4. **(a)** horizontal; 0
   (b) 4; vertical; undefined

Now we apply the slope formula using the two points $(-2, 11)$ and $(4, -7)$.

$$m = \frac{-7 - 11}{4 - (-2)} = \frac{-18}{6} = -3$$

The slope, $m = -3$, is the same number as the coefficient of x in the equation $y = -3x + 5$. It can be shown that this always happens, *as long as the equation is solved for y.* This fact is used to find the slope of a line from its equation.

Finding the Slope of a Line from Its Equation

Step 1 Solve the equation for y. (See **Section 2.5.**)

Step 2 The slope is given by the coefficient of x.

EXAMPLE 5 Finding Slopes from Equations

Find the slope of each line.

(a) $2x - 5y = 4$

Step 1 Solve the equation for y.

$$2x - 5y = 4 \quad \boxed{\text{Isolate } y \text{ on one side.}}$$

$$-5y = 4 - 2x \qquad \text{Subtract } 2x.$$

$$-5y = -2x + 4 \qquad \text{Commutative property}$$

$$\boxed{\tfrac{-2x}{-5} = \tfrac{-2}{-5}x = \tfrac{2}{5}x}$$

$$y = \frac{2}{5}x - \frac{4}{5} \qquad \text{Divide } each \text{ term by } -5.$$

$$\uparrow$$
Slope

Step 2 The slope is given by the coefficient of x, so the slope is $\frac{2}{5}$.

(b)
$$8x + 4y = 1$$

$$\boxed{\text{Solve for } y.} \quad 4y = 1 - 8x \qquad \text{Subtract } 8x.$$

$$4y = -8x + 1 \qquad \text{Commutative property}$$

$$y = -2x + \frac{1}{4} \qquad \text{Divide } each \text{ term by } 4.$$

The slope of this line is given by the coefficient of x, which is -2.

(c)
$$3y + x = -3 \quad \boxed{\text{We leave out the step showing the commutative property.}}$$

$$3y = -x - 3 \qquad \text{Subtract } x.$$

$$y = \frac{-x}{3} - 1 \qquad \text{Divide } each \text{ term by } 3.$$

$$\boxed{\text{The slope is } -\tfrac{1}{3}, \textit{not } \tfrac{-x}{3} \text{ or } -\tfrac{x}{3}.}$$

$$y = -\frac{1}{3}x - 1 \qquad \frac{-x}{3} = \frac{-1x}{3} = -\frac{1}{3}x$$

The coefficient of x is $-\frac{1}{3}$, so the slope of this line is $-\frac{1}{3}$.

⸻⸻ **Work Problem ❺ at the Side.** ▶

❺ Find the slope of each line.

GS (a) $y = -\dfrac{7}{2}x + 1$

Since this equation is already solved for y, Step 1 is not needed. The slope, which is given by the _____ of ____, can be read directly from the equation. The slope is ____ .

GS (b) $4y = 4x - 3$

Solve for y here by dividing each term by ____ to get the equation

$$y = \text{_____} .$$

The coefficient of x is ____, so the slope is ____ .

(c) $3x + 2y = 9$

(d) $5y - x = 10$

Answers

5. (a) coefficient; x; $-\dfrac{7}{2}$ **(b)** 4; $x - \dfrac{3}{4}$; 1; 1

(c) $-\dfrac{3}{2}$ **(d)** $\dfrac{1}{5}$

OBJECTIVE **3** **Use slope to determine whether two lines are parallel, perpendicular, or neither.** Two lines in a plane that never intersect are **parallel.** We use slopes to tell whether two lines are parallel.

Figure 27 shows the graphs of $x + 2y = 4$ and $x + 2y = -6$. These lines appear to be parallel. We solve each equation for y to find the slope.

$x + 2y = 4$

$2y = -x + 4$ Subtract x.

$y = \dfrac{-x}{2} + 2$ Divide by 2.

$y = -\dfrac{1}{2}x + 2$

$\dfrac{-x}{2} = \dfrac{-1x}{2} = -\dfrac{1}{2}x$

↑ Slope

$x + 2y = -6$

$2y = -x - 6$ Subtract x.

$y = \dfrac{-x}{2} - 3$ Divide by 2.

$y = -\dfrac{1}{2}x - 3$

The slope is $-\dfrac{1}{2}$, not $-\dfrac{x}{2}$.

↑ Slope

Each line has slope $-\dfrac{1}{2}$. *Nonvertical parallel lines always have equal slopes.*

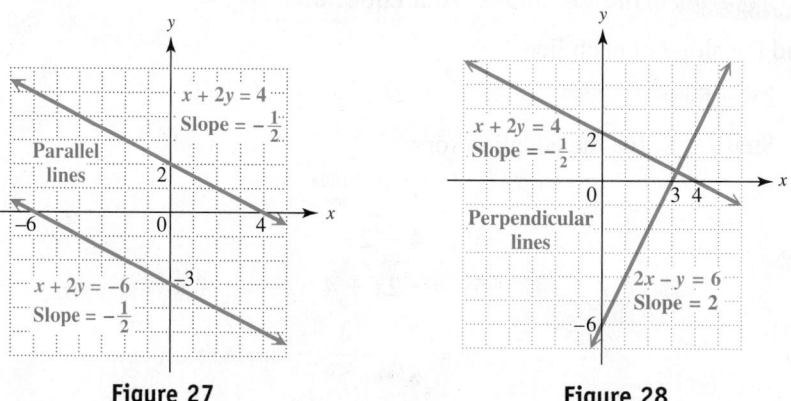

Figure 27   **Figure 28**

Figure 28 shows the graphs of $x + 2y = 4$ and $2x - y = 6$. These lines appear to be **perpendicular** (that is, they intersect at a 90° angle). As shown on the left above, solving $x + 2y = 4$ for y gives $y = -\dfrac{1}{2}x + 2$, with slope $-\dfrac{1}{2}$. We must solve $2x - y = 6$ for y.

$2x - y = 6$

$-y = -2x + 6$ Subtract $2x$.

$y = 2x - 6$ Multiply by -1.

↑ Slope

The product of the two slopes $-\dfrac{1}{2}$ and 2 is

$$-\frac{1}{2}(2) = -1.$$

This condition is true in general. *The product of the slopes of two perpendicular lines, neither of which is vertical, is always* **−1.** This means that the slopes of perpendicular lines are negative (or opposite) reciprocals—if one slope is the nonzero number a, then the other is $-\dfrac{1}{a}$. The table in the margin shows several examples.

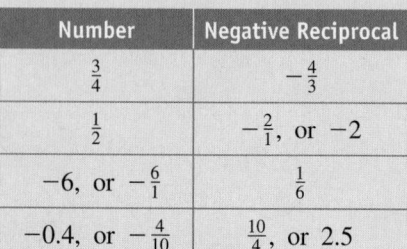

Number	Negative Reciprocal
$\frac{3}{4}$	$-\frac{4}{3}$
$\frac{1}{2}$	$-\frac{2}{1}$, or -2
-6, or $-\frac{6}{1}$	$\frac{1}{6}$
-0.4, or $-\frac{4}{10}$	$\frac{10}{4}$, or 2.5

The product of each number and its negative reciprocal is **−1.**

Slopes of Parallel and Perpendicular Lines

Two lines with the same slope are parallel.

Two lines whose slopes have a product of -1 are perpendicular.

EXAMPLE 6 **Deciding Whether Two Lines Are Parallel or Perpendicular**

Decide whether each pair of lines is *parallel*, *perpendicular*, or *neither*.

(a) $x + 2y = 7$ and $-2x + y = 3$

Find the slope of each line by first solving each equation for y.

$x + 2y = 7$ $-2x + y = 3$

$2y = -x + 7$ Subtract x. $y = 2x + 3$ Add $2x$.

$y = -\dfrac{1}{2}x + \dfrac{7}{2}$ Divide by 2.

The slope is $-\frac{1}{2}$.   The slope is 2.

Because the slopes are not equal, the lines are not parallel. Check the product of the slopes: $-\frac{1}{2}(2) = -1$. The two lines are perpendicular because the product of their slopes is -1. See **Figure 29.**

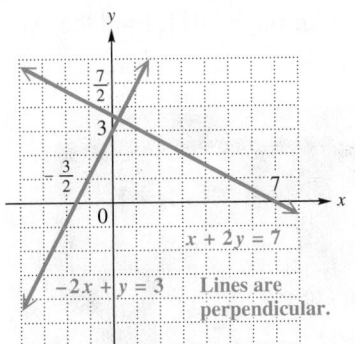

Figure 29

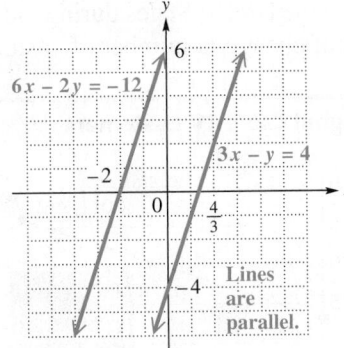

Figure 30

(b) $3x - y = 4$ Solve each $y = 3x - 4$
$6x - 2y = -12$ equation for y. $y = 3x + 6$

Both lines have slope 3, so the lines are parallel. See **Figure 30.**

(c) $4x + 3y = 6$ Solve each $y = -\dfrac{4}{3}x + 2$

$2x - y = 5$ equation for y. $y = 2x - 5$

The slopes are $-\frac{4}{3}$ and 2. Because $-\frac{4}{3} \neq 2$ and $-\frac{4}{3}(2) \neq -1$, these lines are neither parallel nor perpendicular.

(d) $5x - y = 1$ Solve each $y = 5x - 1$

$x - 5y = -10$ equation for y. $y = \dfrac{1}{5}x + 2$ $5\left(\frac{1}{5}\right) = 1$, *not* -1.

Here the slopes are 5 and $\frac{1}{5}$. The lines are not parallel, nor are they perpendicular.

⋯⋯⋯⋯⋯⋯⋯⋯⋯⋯ **Work Problem** ❻ **at the Side.** ▶

❻ Decide whether each pair of lines is *parallel, perpendicular,* or *neither.*

(a) $x + y = 6$
  $x + y = 1$

(b) $3x - y = 4$
  $x + 3y = 9$

(c) $2x - y = 5$
  $2x + y = 3$

(d) $3x - 7y = 35$
  $7x - 3y = -6$

Answers

6. **(a)** parallel **(b)** perpendicular
  (c) neither **(d)** neither

7 There were approximately 40 million digital cable TV customers in the United States in 2008. (*Source:* SNL Kagan.)

(a) Using this number for 2008 and the data for 2011 from the graph in **Figure 32**, find the average rate of change in number of customers per year from 2008 to 2011.

OBJECTIVE ▶ **4** **Solve problems involving average rate of change.** The slope formula applied to any two points on a line gives the **average rate of change** in *y* per unit change in *x*, where the value of *y* depends on the value of *x*.

For example, suppose the height of a boy increased from 60 to 68 in. between the ages of 12 and 16, as shown in **Figure 31**.

Change in height $y \rightarrow \dfrac{68 - 60}{16 - 12} = \dfrac{8}{4} = 2$ **in. per yr**
Change in age $x \longrightarrow$

The boy may have grown more than 2 in. during some years and less than 2 in. during others. If we plotted ordered pairs (age, height) for those years and drew a line connecting any two of the points, the average rate of change would likely be slightly different than that found above. However using the data for ages 12 and 16, the boy's *average* change in height was 2 in. per year over these years.

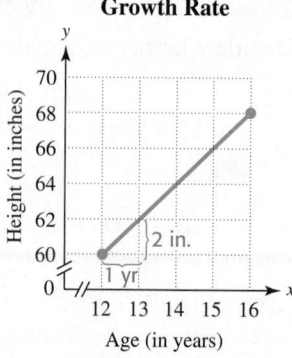

Growth Rate

Figure 31

EXAMPLE 7   **Interpreting Slope as Average Rate of Change**

The graph in **Figure 32** approximates the number of digital cable TV customers in the United States during the years 2006 through 2011. Find the average rate of change in number of customers per year.

(b) How does the average rate of change from part (a) compare to the average rate of change from 2006 to 2011 found in **Example 7?**

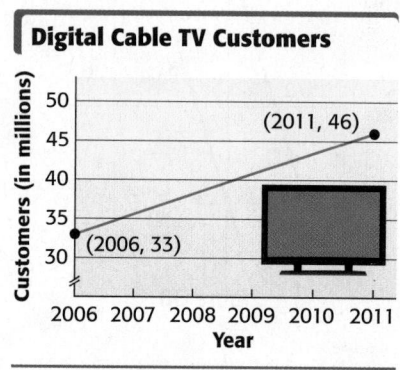

Digital Cable TV Customers

(2011, 46)

(2006, 33)

Source: SNL Kagan.

Figure 32

To find the average rate of change, we need two pairs of data. From the graph, we have the ordered pairs (2006, 33) and (2011, 46). We use the slope formula.

average rate of change $= \dfrac{46 - 33}{2011 - 2006} = \dfrac{13}{5} = 2.6$ ◀ A positive slope indicates an increase.

This means that the number of digital cable TV customers *increased* by an average of 2.6 million customers per year from 2006 to 2011.

◀ **Work Problem 7** at the Side.

Answers

7. (a) 2 million customers per yr
  (b) It is less than the average rate of change for 2006–2011.

EXAMPLE 8 Interpreting Slope as Average Rate of Change

In 2006, there were 65 million basic cable TV customers in the United States. There were 58 million such customers in 2011. Find the average rate of change in the number of customers per year. (*Source:* SNL Kagan.)

To use the slope formula, we let one ordered pair be $(2006, 65)$ and the other be $(2011, 58)$.

$$\text{average rate of change} = \frac{58 - 65}{2011 - 2006} = \frac{-7}{5} = -1.4$$

A negative slope indicates a decrease.

The graph in **Figure 33** confirms that the line through the ordered pairs falls from left to right and therefore has negative slope. Thus, the number of basic cable TV customers *decreased* by an average of 1.4 million customers per year from 2006 to 2011.

The negative sign in -1.4 denotes the *decrease*. (We say "The number of customers decreased by 1.4 million per year." It is *incorrect* to say "The number of customers decreased by -1.4 million per year.")

Basic Cable TV Customers

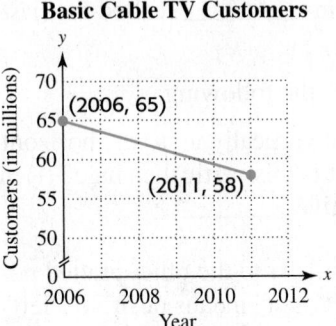

Figure 33

······· Work Problem **8** at the Side. ▶

8 In 2000, 943 million compact discs were sold in the United States. In 2010, 226 million CDs were sold. Find the average rate of change in CDs sold per year. (*Source:* Recording Industry Association of America.)

3.3 Exercises

 MyMathLab®

Download the MyDashBoard App

1. **CONCEPT CHECK** Slope is used to measure the _____ of a line. Slope is the (*horizontal / vertical*) change compared to the (*horizontal / vertical*) change while moving along the line from one point to another.

2. **CONCEPT CHECK** Slope is the _____ of the vertical change in _____, called the (*rise / run*), to the horizontal change in _____, called the (*rise / run*).

3. Look at the graph at the right, and answer the following.

   (a) Start at the point $(-1, -4)$ and count vertically up to the horizontal line that goes through the other plotted point. What is this vertical change? (Remember: "up" means positive, "down" means negative.) _____

   (b) From this new position, count horizontally to the other plotted point. What is this horizontal change? (Remember: "right" means positive, "left" means negative.) _____

   (c) What is the quotient of the numbers found in parts (a) and (b)? _____ What do we call this number? _____

   (d) If we were to *start* at the point $(3, 2)$ and *end* at the point $(-1, -4)$ would the answer to part (c) be the same? Explain.

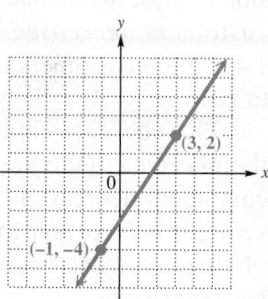

4. **CONCEPT CHECK** Match the graph of each line in parts (a)–(d) with its slope in choices A–D.

   (a)   **(b)**   **(c)**   **(d)**

 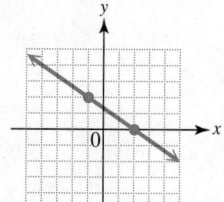 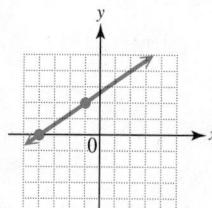 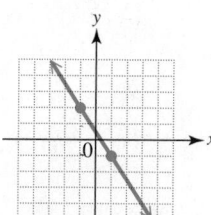 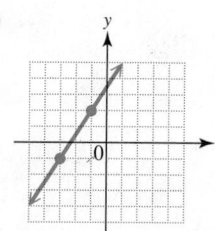

   A. $\dfrac{2}{3}$ **B.** $\dfrac{3}{2}$ **C.** $-\dfrac{2}{3}$ **D.** $-\dfrac{3}{2}$

*Use the coordinates of the indicated points to find the slope of each line. **See Example 1.***

5.   **6.**   **7.**   **8.**

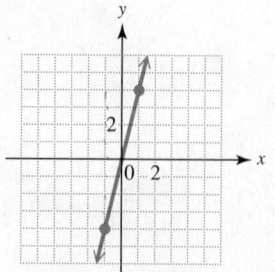 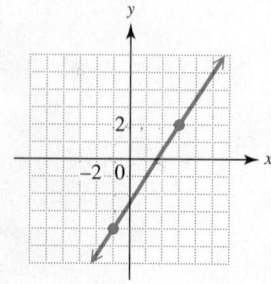 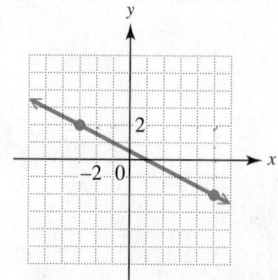 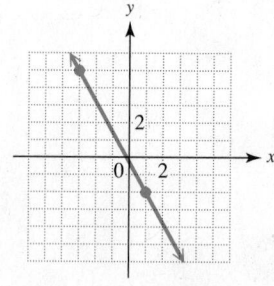

9. CONCEPT CHECK On the given axes, sketch the graph of a straight line with the indicated slope.

(a) Negative   (b) Positive   (c) Undefined   (d) Zero

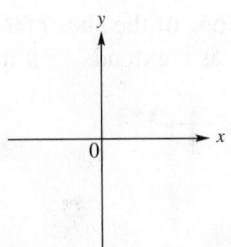

10. CONCEPT CHECK Decide whether the line with the given slope *m rises from left to right, falls from left to right*, is *horizontal*, or is *vertical*.

(a) $m = -4$ (b) $m = 0$ (c) *m* is undefined.   (d) $m = \dfrac{3}{7}$

CONCEPT CHECK *The figure at the right shows a line that has a positive slope (because it rises from left to right) and a positive y-value for the y-intercept (because it intersects the y-axis above the origin).*

For each figure in Exercises 11–16, decide whether (a) the slope is positive, negative, *or* 0 *and whether (b) the y-value of the y-intercept is* positive, negative, *or* 0.

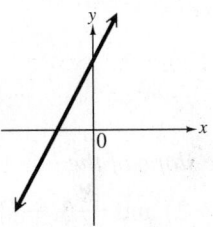

11. (a) _____

     (b) _____

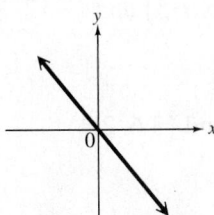

12. (a) _____

     (b) _____

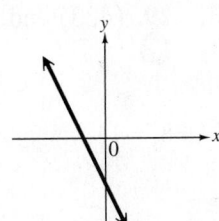

13. (a) _____

     (b) _____

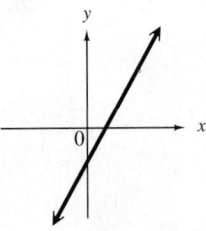

14. (a) _____

     (b) _____

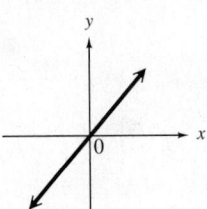

15. (a) _____

     (b) _____

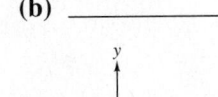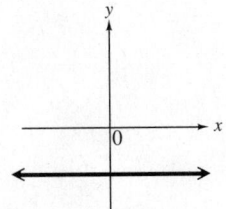

16. (a) _____

     (b) _____

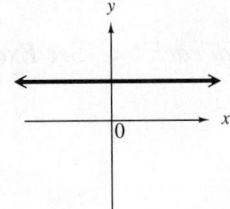

CONCEPT CHECK *Answer each question.*

17. What is the slope (or grade) of this hill?

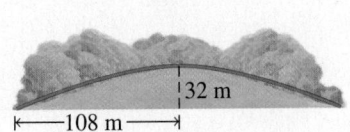

18. What is the slope (or pitch) of this roof?

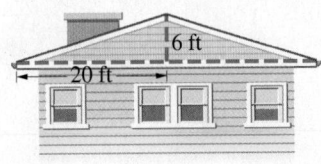

19. What is the slope of the slide? (*Hint:* The slide *drops* 8 ft vertically as it extends 12 ft horizontally.)

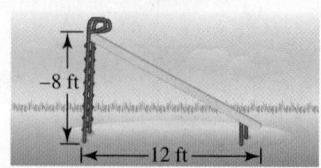

20. What is the slope (or grade) of this ski slope? (*Hint:* The ski slope *drops* 25 ft vertically as it extends 100 ft horizontally.)

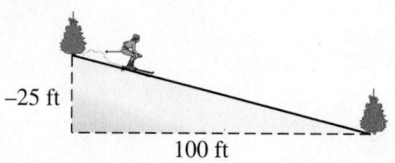

21. CONCEPT CHECK A student found the slope of the line through the points $(2, 5)$ and $(-1, 3)$ as follows.

$$\frac{3-5}{2-(-1)} = \frac{-2}{3}, \quad \text{or} \quad -\frac{2}{3}$$

What Went Wrong? Give the correct slope.

22. CONCEPT CHECK A student found the slope of the line through the points $(-2, 4)$ and $(6, -1)$ as follows.

$$\frac{6-(-2)}{-1-4} = \frac{8}{-5}, \quad \text{or} \quad -\frac{8}{5}$$

What Went Wrong? Give the correct slope.

Find the slope of the line passing through each pair of points. ***See Examples 2–4.***

23. $(1, -2)$ and $(-3, -7)$
24. $(4, -1)$ and $(-2, -8)$
25. $(0, 3)$ and $(-2, 0)$
26. $(-8, 0)$ and $(0, -5)$

27. $(-2, 4)$ and $(-3, 7)$
28. $(-4, -5)$ and $(-5, -8)$
29. $(4, 3)$ and $(-6, 3)$
30. $(6, -5)$ and $(-12, -5)$

31. $(-12, 3)$ and $(-12, -7)$
32. $(-8, 6)$ and $(-8, -1)$
33. $(4.8, 2.5)$ and $(3.6, 2.2)$

34. $(3.1, 2.6)$ and $(1.6, 2.1)$
35. $\left(-\frac{7}{5}, \frac{3}{10}\right)$ and $\left(\frac{1}{5}, -\frac{1}{2}\right)$
36. $\left(-\frac{4}{3}, \frac{1}{2}\right)$ and $\left(\frac{1}{3}, -\frac{5}{6}\right)$

Find the slope of each line. ***See Example 5.***

37. $y = 5x + 12$
38. $y = 2x - 3$
39. $4y = x + 1$

40. $2y = -x + 4$
41. $3x - 2y = 3$
42. $-6x + 4y = 4$

43. $-2y - 3x = 5$
44. $-4y - 3x = 2$
45. $y = 6$
46. $y = 4$

47. $x = -2$
48. $x = 5$
49. $x - y = 0$
50. $x + y = 0$

For each pair of equations, give the slope of each line, and then determine whether the two lines are parallel, perpendicular, *or* neither. *See Example 6.*

51. $-4x + 3y = 4$
$-8x + 6y = 0$

52. $2x + 5y = 4$
$4x + 10y = 1$

53. $5x - 3y = -2$
$3x - 5y = -8$

54. $8x - 9y = 6$
$8x + 6y = -5$

55. $3x - 5y = -1$
$5x + 3y = 2$

56. $3x - 2y = 6$
$2x + 3y = 3$

Figure A *gives the percent of first-time full-time freshmen at 4-year colleges and universities who planned to major in the Biological Sciences.* **Figure B** *shows the percent of the same group of students who planned to major in Business.*

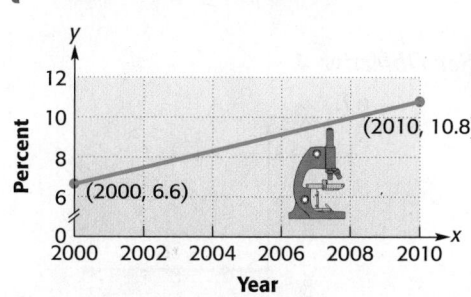

Source: Higher Education Research Institute, UCLA.

Figure A

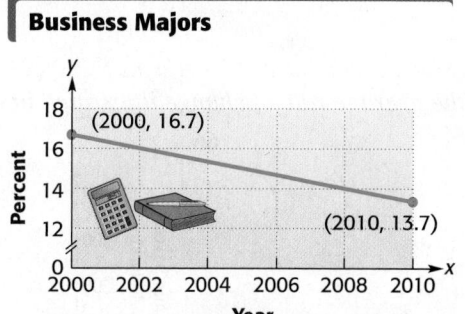

Source: Higher Education Research Institute, UCLA.

Figure B

Use **Figures A and B** *to work Exercises 57–62 in order.*

57. Use the given ordered pairs to find the slope of the line in **Figure A.**

58. The slope of the line in **Figure A** is (*positive / negative*). This means that during the period represented, the percent of freshmen planning to major in the Biological Sciences (*increased / decreased*).

59. The slope of a line represents the *rate of change.* Based on **Figure A,** what was the increase in the percent of freshmen *per year* who planned to major in the Biological Sciences during the period shown?

60. Use the given ordered pairs to find the slope of the line in **Figure B.**

61. The slope of the line in **Figure B** is (*positive / negative*). This means that during the period represented, the percent of freshmen planning to major in Business (*increased / decreased*).

62. Based on **Figure B,** what was the decrease in the percent of freshmen *per year* who planned to major in Business?

Solve each problem.

63. The upper deck at U.S. Cellular Field in Chicago has produced, among other complaints, displeasure with its steepness. It is 160 ft from home plate to the front of the upper deck and 250 ft from home plate to the back. The top of the upper deck is 63 ft above the bottom. What is its slope? (Consider the slope as a positive number.)

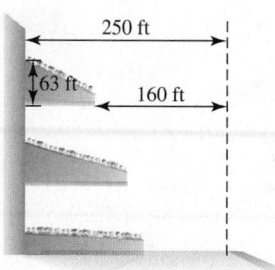

64. When designing the TD Bank North Garden arena in Boston, architects designed the ramps leading up to the entrances so that circus elephants would be able to march up the ramps. The maximum grade (or slope) that an elephant will walk on is 13%. Suppose that such a ramp were constructed with a horizontal run of 150 ft. What would be the maximum vertical rise the architects could use?

Find and interpret the average rate of change illustrated in each graph. **See Objective 4.**

65.

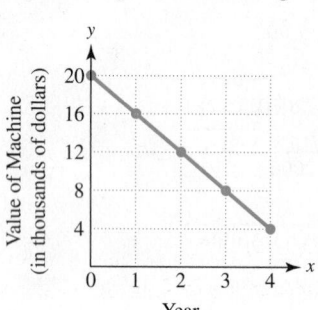

66.

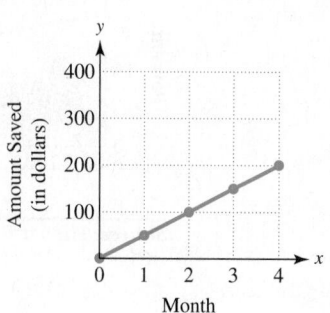

67.

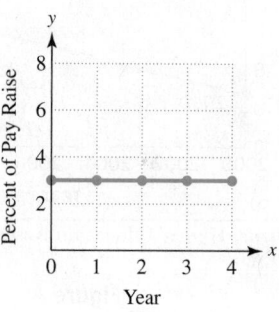

68. CONCEPT CHECK If the graph of a linear equation rises from left to right, then the average rate of change is (*positive/negative*). If the graph of a linear equation falls from left to right, then the average rate of change is (*positive/negative*).

Solve each problem. **See Examples 7 and 8.**

69. The graph provides a good approximation of the number of drive-in theaters in the United States from 2005 through 2011.

(a) Use the given ordered pairs to find the average rate of change in the number of drive-in theaters per year during this period.

(b) Explain how a negative slope is interpreted in this situation.

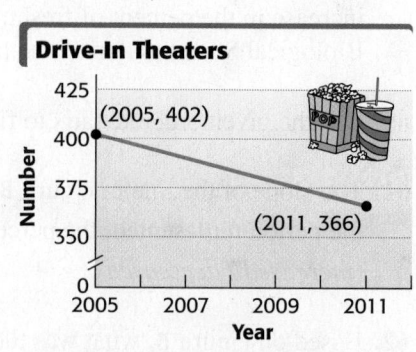

Source: United Drive-In Theatre Owners Association.

70. The graph provides a good approximation of the number of mobile homes (in thousands) placed in use in the United States during 2005–2011.

    (a) Use the given ordered pairs to find the average rate of change in the number of mobile homes per year during this period.

    (b) Interpret what a negative slope means in this situation.

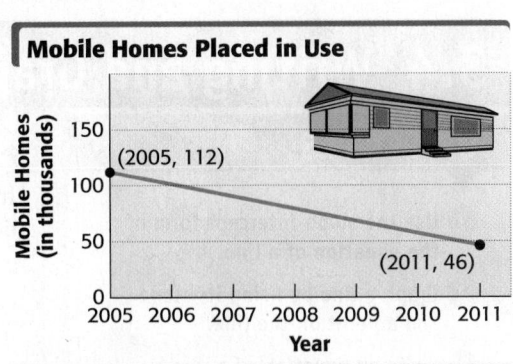

Mobile Homes Placed in Use

(2005, 112)

(2011, 46)

Source: U.S. Census Bureau.

71. The graph shows the number of cellular phone subscribers (in millions) in the United States from 2006 to 2010.

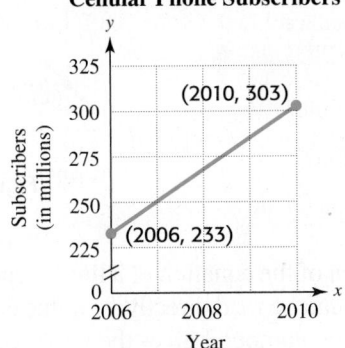

Cellular Phone Subscribers

(2010, 303)

(2006, 233)

Year

Source: CTIA-The Wireless Association.

    (a) Use the given ordered pairs to find the slope of the line.

    (b) Interpret the slope in the context of this problem.

72. The graph shows spending on personal care products (in billions of dollars) in the United States from 2005 to 2008.

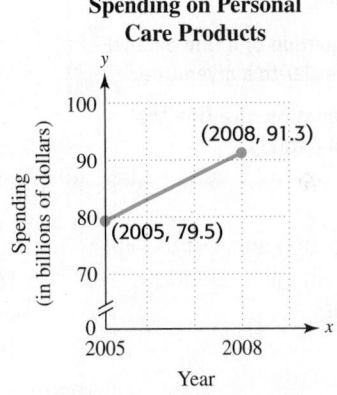

Spending on Personal Care Products

(2008, 91.3)

(2005, 79.5)

Year

Source: U.S. Department of Commerce.

    (a) Use the given ordered pairs to find the slope of the line to the nearest tenth.

    (b) Interpret the slope in the context of this problem.

Relating Concepts (Exercises 73–78) For Individual or Group Work

*In these exercises we investigate a method of determining whether three points lie on the same straight line. Such points are said to be **collinear**. The points we consider are* $A(3, 1)$, $B(6, 2)$, *and* $C(9, 3)$. ***Work Exercises 73–78 in order.***

73. Find the slope of segment AB.

74. Find the slope of segment BC.

75. Find the slope of segment AC.

76. If slope of AB = slope of BC = slope of AC, then A, B, and C are collinear. Use the results of **Exercises 73–75** to show that this statement is satisfied.

77. Use the slope formula to determine whether the points $(1, -2)$, $(3, -1)$, and $(5, 0)$ are collinear.

78. Repeat **Exercise 77** for the points $(0, 6)$, $(4, -5)$, and $(-2, 12)$.

3.4 Writing and Graphing Equations of Lines

OBJECTIVES

1. Use the slope-intercept form of the equation of a line.

2. Graph a line by using its slope and a point on the line.

3. Write an equation of a line by using its slope and any point on the line.

4. Write an equation of a line by using two points on the line.

5. Write equations of horizontal and vertical lines.

6. Write an equation of a line parallel or perpendicular to a given line.

7. Write an equation of a line that models real data.

1 Identify the slope and y-intercept of the line with each equation.

(a) $y = 2x - 6$

(b) $y = -\frac{3}{5}x - 9$

(c) $y = -\frac{x}{3} + \frac{7}{3}$

(d) $y = -x$

OBJECTIVE **1** **Use the slope-intercept form of the equation of a line.**
In **Section 3.3,** we found the slope (steepness) of a line by solving the equation of the line for y. In that form, the slope is the coefficient of x. For example, the graph of the equation

$$y = 2x + 3 \quad \text{has slope} \quad 2.$$

What does the number **3** represent? To find out, suppose a line has slope m and y-intercept $(0, b)$. We can find an equation of this line by choosing another point (x, y) on the line, as shown in **Figure 34.** Then we use the slope formula.

$$m = \frac{y - b}{x - 0} \quad \leftarrow \text{Change in } y\text{-values}$$
$$\phantom{m = \frac{y - b}{x - 0}} \leftarrow \text{Change in } x\text{-values}$$

$$m = \frac{y - b}{x} \quad \begin{array}{l}\text{Subtract in the}\\\text{denominator.}\end{array}$$

$$mx = y - b \quad \text{Multiply by } x.$$

$$mx + b = y \quad \text{Add } b.$$

$$y = mx + b \quad \text{Interchange sides.}$$

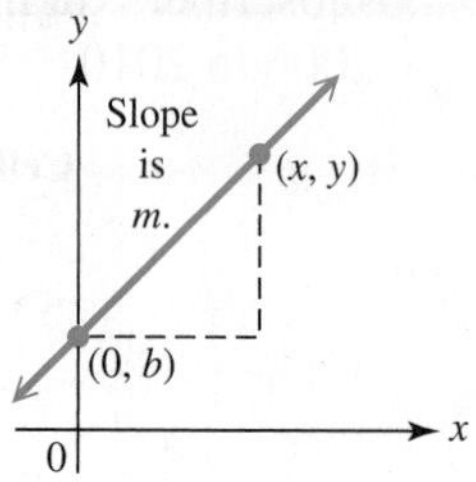

Figure 34

This result is the *slope-intercept form* of the equation of a line, because both the slope and the y-intercept of the line can be read directly from the equation. For the line with equation $y = 2x + 3$, the number 3 gives the y-intercept $(0, \mathbf{3})$.

Slope-Intercept Form

The **slope-intercept form** of the equation of a line with slope m and y-intercept $(0, b)$ is given as follows.

$$y = mx + b$$

Slope $\nearrow$ $\qquad$ $\nwarrow$ $(0, b)$ is the y-intercept.

The intercept given is the y-intercept.

EXAMPLE 1 **Identifying Slopes and y-Intercepts**

Identify the slope and y-intercept of the line with each equation.

(a) $y = -4x + 1$
Slope $\nearrow$ $\qquad$ $\nwarrow$ y-intercept $(0, 1)$

(b) $y = x - 8$ can be written as $y = 1x + (-8)$.
$\qquad\qquad\qquad\qquad\qquad$ Slope $\nearrow$ $\qquad$ $\nwarrow$ y-intercept $(0, -8)$

(c) $y = 6x$ can be written as $y = 6x + 0$.
$\qquad\qquad\qquad\quad$ Slope $\nearrow$ $\qquad$ $\nwarrow$ y-intercept $(0, 0)$

(d) $y = \frac{x}{4} - \frac{3}{4}$ can be written as $y = \frac{1}{4}x + \left(-\frac{3}{4}\right)$.
$\qquad\qquad\qquad\qquad\quad$ Slope $\nearrow$ $\qquad$ $\nwarrow$ y-intercept $\left(0, -\frac{3}{4}\right)$

$\cdots\cdots\cdots\cdots\cdots\cdots\cdots\cdots$ ◀ **Work Problem** **1** at the Side.

Answers

1. (a) slope: 2; y-intercept: $(0, -6)$
(b) slope: $-\frac{3}{5}$; y-intercept: $(0, -9)$
(c) slope: $-\frac{1}{3}$; y-intercept: $\left(0, \frac{7}{3}\right)$
(d) slope: -1; y-intercept: $(0, 0)$

EXAMPLE 2 Writing an Equation of a Line

Write an equation of the line with slope $\frac{2}{3}$ and y-intercept $(0, -1)$.

Here, $m = \frac{2}{3}$ and $b = -1$, so we can write the following equation.

Slope $\longrightarrow$ $\quad$ $\longleftarrow$ y-intercept $(0, b)$

$$y = mx + b \qquad \text{Slope-intercept form}$$

$$y = \frac{2}{3}x + (-1), \quad \text{or} \quad y = \frac{2}{3}x - 1 \quad \text{Substitute for } m \text{ and } b.$$

$\cdots\cdots\cdots\cdots\cdots\cdots\cdots\cdots\cdots\cdots\cdots\cdots\cdots$ Work Problem ❷ at the Side. ▶

OBJECTIVE ❷ **Graph a line by using its slope and a point on the line.**
We can use the slope and y-intercept to graph a line.

Graphing a Line by Using Its Slope and y-Intercept

Step 1 Write the equation in slope-intercept form $y = mx + b$, if necessary, by solving for y.

Step 2 Identify the y-intercept. Graph the point $(0, b)$.

Step 3 Identify slope m of the line. Use the geometric interpretation of slope ("rise over run") to find another point on the graph by counting from the y-intercept.

Step 4 Join the two points with a line to obtain the graph.

EXAMPLE 3 Graphing Lines by Using Slopes and y-Intercepts

Graph each equation by using the slope and y-intercept.

(a) $y = \frac{2}{3}x - 1$

Step 1 The equation is in slope-intercept form.

$$y = \frac{2}{3}x - 1$$

$\qquad\qquad$ ↑ $\quad$ ↑
$\qquad$ Slope $\quad$ Value of b in y-intercept $(0, b)$

Step 2 The y-intercept is $(0, -1)$. Graph this point. See **Figure 35**.

Step 3 The slope is $\frac{2}{3}$. By definition,

$$\text{slope } m = \frac{\text{change in } y \text{ (rise)}}{\text{change in } x \text{ (run)}} = \frac{2}{3}.$$

From the y-intercept, count up 2 units and to the right 3 units to obtain the point $(3, 1)$.

Step 4 Draw the line through the points $(0, -1)$ and $(3, 1)$ to obtain the graph in **Figure 35**.

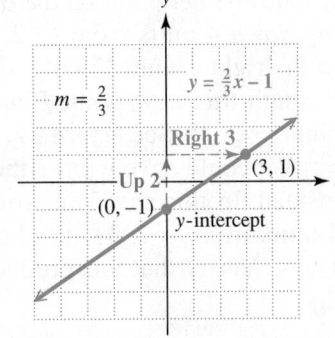

Figure 35

$\cdots\cdots\cdots\cdots\cdots\cdots$ **Continued on Next Page**

---

❷ Write an equation of the line with the given slope and y-intercept.

(a) slope $\frac{1}{2}$; y-intercept $(0, -4)$

(b) slope -1; y-intercept $(0, 8)$

(c) slope 3; y-intercept $(0, 0)$

(d) slope 0; y-intercept $(0, 2)$

(e) slope 1; y-intercept $(0, 0.75)$

Answers

2. **(a)** $y = \frac{1}{2}x - 4$ **(b)** $y = -x + 8$
(c) $y = 3x$ **(d)** $y = 2$ **(e)** $y = x + 0.75$

❸ Graph $3x - 4y = 8$ by using the slope and y-intercept.

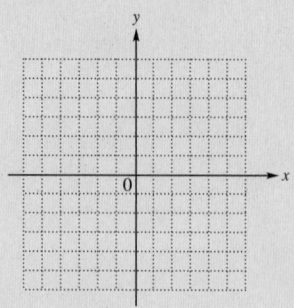

(b) $3x + 4y = 8$

Step 1   Solve for y to write the equation in slope-intercept form.

$$3x + 4y = 8$$

Isolate y on one side. ⟶ $4y = -3x + 8$ Subtract $3x$.

Slope-intercept form ⟶ $y = -\dfrac{3}{4}x + 2$ Divide *each* term by 4.

Step 2   The y-intercept is $(0, 2)$. Graph this point. See **Figure 36.**

Step 3   The slope is $-\frac{3}{4}$, which can be written as either $\frac{-3}{4}$ or $\frac{3}{-4}$. We use $\frac{-3}{4}$ here.

$$m = \frac{\text{change in } y \text{ (rise)}}{\text{change in } x \text{ (run)}} = \frac{-3}{4}$$

From the y-intercept, count *down* 3 units (because of the negative sign) and to the right 4 units, to obtain the point $(4, -1)$.

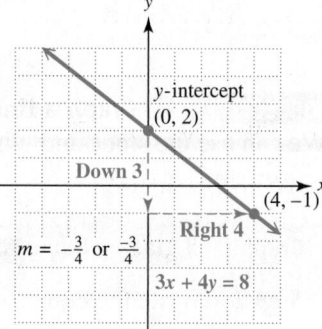

Figure 36

Step 4   Draw the line through the two points $(0, 2)$ and $(4, -1)$ to obtain the graph in **Figure 36.**

◄ **Work Problem ❸ at the Side.**

❹ Graph the line passing through the point $(2, -3)$, with slope $-\frac{1}{3}$.

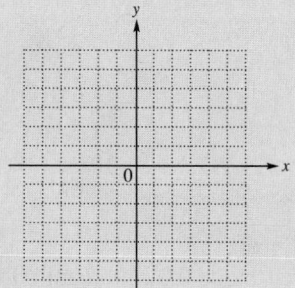

Note

In Step 3 of **Example 3(b)**, we could use $\frac{3}{-4}$ for the slope. From the y-intercept, count up 3 units and to the *left* 4 units (because of the negative sign) to obtain the point $(-4, 5)$. Verify that this produces the same line.

EXAMPLE 4   Graphing a Line by Using Its Slope and a Point

Graph the line passing through the point $(-2, 3)$, with slope -4.

First, locate the point $(-2, 3)$. See **Figure 37.** Then write the slope -4 as

$$\text{slope } m = \frac{\text{change in } y}{\text{change in } x} = -4 = \frac{-4}{1}.$$

Locate another point on the line by counting *down* 4 units from $(-2, 3)$ and then to the right 1 unit. Finally, draw the line through this new point P and the given point $(-2, 3)$. See **Figure 37.**

We could have written the slope as $\frac{4}{-1}$ instead. In this case, we would move up 4 units from $(-2, 3)$ and then to the *left* 1 unit. Verify that this produces the same line.

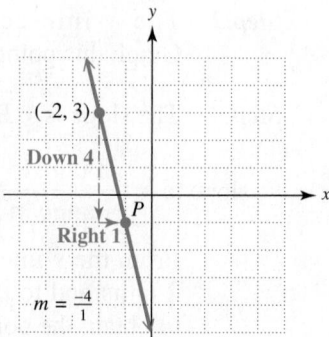

Figure 37

◄ **Work Problem ❹ at the Side.**

Answers

3.

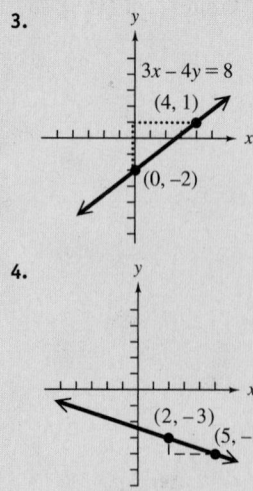

4.

OBJECTIVE ▶ 3 Write an equation of a line by using its slope and any point on the line. We can use the slope-intercept form to do this.

EXAMPLE 5 Using the Slope-Intercept Form to Write an Equation of a Line

Write an equation, in slope-intercept form, of the line having slope 4 passing through the point $(2, 5)$.

Since the line passes through the point $(2, 5)$, we can substitute $x = 2$, $y = 5$, and the given slope $m = 4$ into $y = mx + b$ and solve for b.

$$y = mx + b \qquad \text{Slope-intercept form}$$

$$5 = 4(2) + b \qquad \text{Let } x = 2, y = 5, \text{ and } m = 4.$$

$$5 = 8 + b \qquad \text{Multiply.}$$

> $(0, b)$ is the y-intercept. Don't stop here.

$$-3 = b \qquad \text{Subtract 8.}$$

Now substitute the values of m and b into slope-intercept form.

$$y = mx + b \qquad \text{Slope-intercept form}$$

$$y = 4x - 3 \qquad \text{Let } m = 4 \text{ and } b = -3.$$

············ **Work Problem ⑤ at the Side.** ▶

Note

Slope-intercept form is an especially useful form for a linear equation because of the information we can determine from it. It is the form used by graphing calculators and the one that describes *a linear function.*

There is another form that can be used to write the equation of a line. To develop this form, let m represent the slope of a line and let (x_1, y_1) represent a given point on the line. Let (x, y) represent any other point on the line. See **Figure 38.**

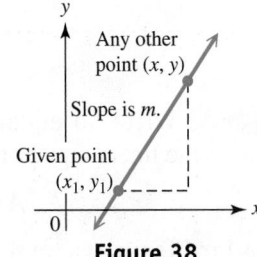

Figure 38

$$m = \frac{y - y_1}{x - x_1} \qquad \text{Definition of slope}$$

$$m(x - x_1) = y - y_1 \qquad \text{Multiply each side by } x - x_1.$$

$$y - y_1 = m(x - x_1) \qquad \text{Interchange sides.}$$

This result is the **point-slope form** of the equation of a line.

Point-Slope Form

The point-slope form of the equation of a line with slope m passing through the point (x_1, y_1) is given as follows.

$$\overset{\text{Slope}}{y - y_1 = m(x - x_1)}$$
$$\underset{\text{Given point}}{}$$

---

⑤ Write an equation, in slope-intercept form, of each line.

GS (a) The line having slope -2 passing through the point $(-1, 4)$

Substitute values and solve for b.

$$y = mx + b$$

$$\underline{\quad} = \underline{\quad} (\underline{\quad}) + b$$

$$4 = \underline{\quad} + b$$

$$\underline{\quad} = b$$

An equation of the line is $y = \underline{\quad\quad}$.

(b) The line having slope 3 passing through the point $(-2, 1)$

6 Write an equation of each line. Give the final answer in slope-intercept form.

GS **(a)** The line passing through $(-1, 3)$, with slope -2

$$y - y_1 = m(x - x_1)$$

$$y - \underline{\hspace{0.5cm}} = \underline{\hspace{0.5cm}} [x - (\underline{\hspace{0.5cm}})]$$

$$y - 3 = -2(x + \underline{\hspace{0.5cm}})$$

$$y - 3 = -2x - \underline{\hspace{0.5cm}}$$

$$y = \underline{\hspace{1.5cm}}$$

(b) The line passing through $(5, 2)$, with slope $-\frac{1}{3}$

7 Which equation is written in standard form as specified in the discussion?

A. $y = -4x - 7$

B. $-3x + 4y = 12$

C. $x - 6y = 3$

D. $\frac{1}{2}x + y = 0$

E. $6x - 2y = 10$

Write the equations not written in standard form in that form.

EXAMPLE 6   Using Point-Slope Form to Write Equations

Write an equation of each line. Give the final answer in slope-intercept form.

(a) The line passing through $(-2, 4)$, with slope -3

The given point is $(-2, 4)$ so $x_1 = -2$ and $y_1 = 4$. Also, $m = -3$. Substitute these values into the point-slope form.

Only y_1, m, and x_1 are replaced with numbers.

$$y - y_1 = m(x - x_1) \qquad \text{Point-slope form}$$

$$y - 4 = -3[x - (-2)] \qquad \text{Let } x_1 = -2,\ y_1 = 4,\ m = -3.$$

$$y - 4 = -3(x + 2) \qquad \text{Definition of subtraction}$$

$$y - 4 = -3x - 6 \qquad \text{Distributive property}$$

$$y = -3x - 2 \qquad \text{Add 4.}$$

(b) The line passing through $(4, 2)$, with slope $\frac{3}{5}$

$$y - y_1 = m(x - x_1) \qquad \text{Point-slope form}$$

$$y - 2 = \frac{3}{5}(x - 4) \qquad \text{Let } x_1 = 4,\ y_1 = 2,\ m = \frac{3}{5}.$$

$$y - 2 = \frac{3}{5}x - \frac{12}{5} \qquad \text{Distributive property}$$

$$y = \frac{3}{5}x - \frac{12}{5} + \frac{10}{5} \qquad \text{Add } 2 = \frac{10}{5}.$$

$$y = \frac{3}{5}x - \frac{2}{5} \qquad \text{Combine like terms.}$$

We did not clear fractions after the substitution step because we want the equation in slope-intercept form—that is, solved for y.

◀ **Work Problem 6 at the Side.**

OBJECTIVE 4 Write an equation of a line by using two points on the line. Many of the linear equations in **Sections 3.1–3.3** were given in the form

$$Ax + By = C,$$

called **standard form**, where A, B, and C are real numbers and A and B are not both 0. In most cases, A, B, and C are rational numbers. For consistency in this book, we give answers so that A, B, and C are integers with greatest common factor 1 and $A \geq 0$. (If $A = 0$, then we give $B > 0$.)

Note

The definition of standard form is not the same in all texts. A linear equation can be written in many different, yet equally correct, ways. For example,

$$3x + 4y = 12, \quad 6x + 8y = 24, \quad \text{and} \quad -9x - 12y = -36$$

all represent the same set of ordered pairs. When giving answers, $3x + 4y = 12$ is preferable to the other forms because the greatest common factor of 3, 4, and 12 is 1 and $A \geq 0$.

◀ **Work Problem 7 at the Side.**

Answers

6. (a) 3; -2; -1; 1; 2; $-2x + 1$

 (b) $y = -\frac{1}{3}x + \frac{11}{3}$

7. C is in standard form;

 A: $4x + y = -7$; B: $3x - 4y = -12$;

 D: $x + 2y = 0$; E: $3x - y = 5$

EXAMPLE 7 **Writing an Equation of a Line by Using Two Points**

Write an equation of the line passing through the points $(3, 4)$ and $(-2, 5)$. Give the final answer in slope-intercept form and then in standard form.

First, find the slope of the line.

$$\overset{(x_1, y_1)}{(3, 4)} \quad \text{and} \quad \overset{(x_2, y_2)}{(-2, 5)} \qquad \text{Label the points.}$$

$$\text{slope } m = \frac{y_2 - y_1}{x_2 - x_1} = \frac{5 - 4}{-2 - 3} \qquad \text{Apply the slope formula.}$$

$$= \frac{1}{-5}, \quad \text{or} \quad -\frac{1}{5} \qquad \text{Simplify the fraction.}$$

Now use (x_1, y_1), here $(3, 4)$, and either slope-intercept or point-slope form.

$$y - y_1 = m(x - x_1) \qquad \text{We choose point-slope form.}$$

$$y - 4 = -\frac{1}{5}(x - 3) \qquad \text{Let } x_1 = 3, y_1 = 4, m = -\frac{1}{5}.$$

$$y - 4 = -\frac{1}{5}x + \frac{3}{5} \qquad \text{Distributive property}$$

$$y = -\frac{1}{5}x + \frac{3}{5} + \frac{20}{5} \qquad \text{Add } 4 = \frac{20}{5} \text{ to each side.}$$

Slope-intercept form $\longrightarrow y = -\frac{1}{5}x + \frac{23}{5} \qquad$ Combine like terms.

$$5y = -x + 23 \qquad \text{Multiply by 5 to clear fractions.}$$

Standard form $\longrightarrow x + 5y = 23 \qquad$ Add x.

The same result would be found using $(-2, 5)$ for (x_1, y_1).

· Work Problem **8** at the Side. ▶

---

8 Write an equation of the line passing through each pair of points. Give the final answer in slope-intercept form and then in standard form.

(a) $(2, 5)$ and $(-1, 6)$

(b) $(-3, 1)$ and $(2, 4)$

OBJECTIVE ▶ **5** **Write equations of horizontal and vertical lines.** A horizontal line has slope 0. Using point-slope form, we can find the equation of a horizontal line through the point (a, b).

$$y - y_1 = m(x - x_1) \qquad \text{Point-slope form}$$

$$y - b = 0(x - a) \qquad \text{Let } y_1 = b, m = 0, x_1 = a.$$

$$y - b = 0 \qquad \text{Multiplication property of 0}$$

Horizontal line $\rightarrow y = b \qquad$ Add b.

The point-slope form does not apply to a vertical line, since the slope of a vertical line is undefined. A vertical line through the point (a, b) has equation $x = a$.

In summary, horizontal and vertical lines have the following equations.

Equations of Horizontal and Vertical Lines

The horizontal line through the point (a, b) has equation $y = b$.

The vertical line through the point (a, b) has equation $x = a$.

Answers

8. (a) $y = -\frac{1}{3}x + \frac{17}{3}; x + 3y = 17$

(b) $y = \frac{3}{5}x + \frac{14}{5}; 3x - 5y = -14$

9 Write an equation of the line that satisfies the given conditions.

(a) Through $(4, -4)$; undefined slope

(b) Through $(4, -4)$; slope 0

(c) Through $(8, -2)$; $m = 0$

(d) The vertical line through $(3, 5)$

Answers

9. **(a)** $x = 4$ **(b)** $y = -4$
   (c) $y = -2$ **(d)** $x = 3$

EXAMPLE 8   Writing Equations of Horizontal and Vertical Lines

Write an equation of the line passing through the point $(-3, 3)$ that satisfies the given condition.

(a) The line has slope 0.

Since the slope is 0, this is a horizontal line. A horizontal line through the point (a, b) has equation $y = b$. In $(-3, \mathbf{3})$, the y-coordinate is $\mathbf{3}$, so the equation is $y = 3$.

(b) The line has undefined slope.

This is a vertical line, since the slope is undefined. A vertical line through the point (a, b) has equation $x = a$. In $(\mathbf{-3}, 3)$, the x-coordinate is -3, so the equation is $x = -3$.

Both lines are graphed in **Figure 39**.

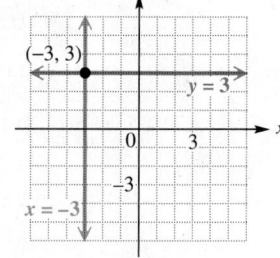

Figure 39

◄ Work Problem **9** at the Side.

OBJECTIVE ▶ 6 **Write an equation of a line parallel or perpendicular to a given line.** Recall that parallel lines have the same slope and perpendicular lines have slopes that are negative reciprocals.

EXAMPLE 9   Writing Equations of Parallel or Perpendicular Lines

Write an equation in slope-intercept form of the line passing through the point $(-3, 6)$ that satisfies the given condition.

(a) The line is parallel to the line $2x + 3y = 6$.

We must find the slope of the given line by solving for y.

$$2x + 3y = 6$$
$$3y = -2x + 6 \qquad \text{Subtract } 2x.$$
$$y = -\frac{2}{3}x + 2 \qquad \text{Divide by 3.}$$
$$\underset{\text{Slope}}{\uparrow}$$

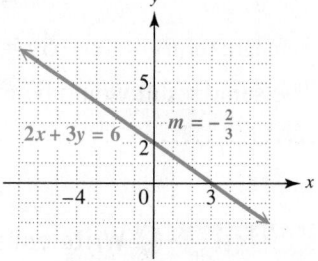

Figure 40

The slope is given by the coefficient of x, so $m = -\frac{2}{3}$. See **Figure 40**.

The required equation of the line through $(-3, 6)$ and parallel to $2x + 3y = 6$ must also have slope $-\frac{2}{3}$. To find this equation, we use the point-slope form, with $(x_1, y_1) = (-3, 6)$ and $m = -\frac{2}{3}$.

$$y - 6 = -\frac{2}{3}[x - (-3)] \qquad \begin{array}{l} y_1 = 6,\, m = \frac{2}{3}, \\ x_1 = -3 \end{array}$$
$$y - 6 = -\frac{2}{3}(x + 3) \qquad \begin{array}{l} \text{Definition of} \\ \text{subtraction} \end{array}$$
$$y - 6 = -\frac{2}{3}x - 2 \qquad \begin{array}{l} \text{Distributive} \\ \text{property} \end{array}$$
$$y = -\frac{2}{3}x + 4 \qquad \text{Add 6.}$$

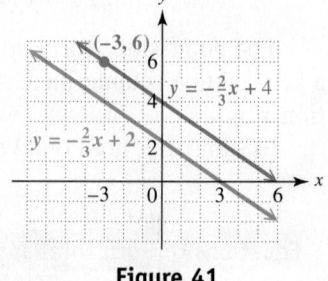

Figure 41

We did not clear the fraction here because we want the equation in slope-intercept form—that is, solved for y. Both lines are shown in **Figure 41**.

·········· **Continued on Next Page**

(b) The line is perpendicular to the line $2x + 3y = 6$.

In part (a), we wrote the equation of the given line $2x + 3y = 6$ in slope-intercept form.

$$y = -\frac{2}{3}x + 2$$

$\uparrow$ Slope

To be perpendicular to the line $2x + 3y = 6$, a line must have slope $\frac{3}{2}$, the negative reciprocal of $-\frac{2}{3}$.

We use $(-3, 6)$ and slope $\frac{3}{2}$ in the point-slope form to find the equation of the perpendicular line shown in **Figure 42.**

$y - 6 = \frac{3}{2}\left[x - (-3)\right]$ $\quad y_1 = 6, m = \frac{3}{2},$ $x_1 = -3$

$y - 6 = \frac{3}{2}(x + 3)$ $\quad$ Definition of subtraction

$y - 6 = \frac{3}{2}x + \frac{9}{2}$ $\quad$ Distributive property

$y = \frac{3}{2}x + \frac{21}{2}$ $\quad$ Add $6 = \frac{12}{2}$.

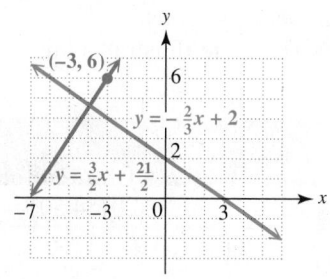

Figure 42

· ▸ **Work Problem ⑩ at the Side.** ▸

Summary of Forms of Linear Equations		
Equation	*Description*	*When to Use*
$y = mx + b$	**Slope-Intercept Form** Slope is m. y-intercept is $(0, b)$.	The slope and y-intercept can be easily identified and used to quickly graph the equation.
$y - y_1 = m(x - x_1)$	**Point-Slope Form** Slope is m. Line passes through the point (x_1, y_1).	This form is ideal for finding the equation of a line if the slope and a point on the line or two points on the line are known.
$Ax + By = C$	**Standard Form** (A, B, and C integers, with $A \geq 0$) Slope is $-\frac{A}{B}$ ($B \neq 0$). x-intercept is $\left(\frac{C}{A}, 0\right)$ ($A \neq 0$). y-intercept is $\left(0, \frac{C}{B}\right)$ ($B \neq 0$).	The x- and y-intercepts can be found quickly and used to graph the equation. The slope must be calculated.
$y = b$	**Horizontal Line** Slope is 0. y-intercept is $(0, b)$.	If the graph intersects only the y-axis, then y is the only variable in the equation.
$x = a$	**Vertical Line** Slope is undefined. x-intercept is $(a, 0)$.	If the graph intersects only the x-axis, then x is the only variable in the equation.

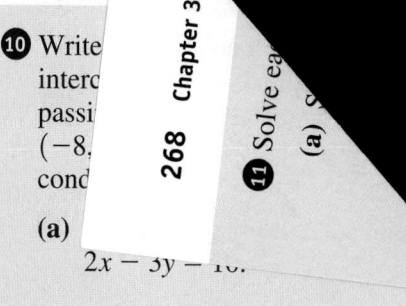

⑩ Write interc passi $(-8,$ conc

(a)

$2x - 3y$ ̶ ̶1̶0̶.

⑪ Solve e

(a) S

(b) The line is perpendicular to the line $2x - 3y = 10$.

Answers

10. (a) $y = \frac{2}{3}x + \frac{25}{3}$ **(b)** $y = -\frac{3}{2}x - 9$

ch problem.

Suppose it costs $0.10 per minute to make a long-distance call. Write an equation to describe the cost y to make an x-minute call.

(b) Suppose there is a flat rate of $0.20 plus a charge of $0.10 per minute to make a call. Write an equation that gives the cost y for a call of x minutes.

(c) Interpret the ordered pair $(15, 1.7)$ in relation to the equation from part (b).

OBJECTIVE ⑦ **Write an equation of a line that models real data.** If a given set of data changes at a fairly constant rate, the data may fit a linear pattern, where the rate of change is the slope of the line.

EXAMPLE 10 Writing a Linear Equation to Describe Data

A local gasoline station is selling 89-octane gas for $4.50 per gal.

(a) Write an equation that describes the cost y to buy x gallons of gas.

The total cost is determined by the number of gallons we buy multiplied by the price per gallon (in this case, $4.50). As the gas is pumped, two sets of numbers spin by: the number of gallons pumped and the cost of that number of gallons.

The table illustrates this situation.

Number of Gallons Pumped	Cost of This Number of Gallons
0	0 ($4.50) = $ 0.00
1	1 ($4.50) = $ 4.50
2	2 ($4.50) = $ 9.00
3	3 ($4.50) = $13.50
4	4 ($4.50) = $18.00

If we let x denote the number of gallons pumped, then the total cost y in dollars can be found by using the following linear equation.

Total price ⟶   ⟵ Number of gallons

$$y = 4.50x$$

Theoretically, there are infinitely many ordered pairs (x, y) that satisfy this equation, but here we are limited to nonnegative values for x, since we cannot have a negative number of gallons. There is also a practical maximum value for x in this situation, which varies from one car to another. What determines this maximum value?

(b) A car wash at this gas station costs an additional $3.00. Write an equation that defines the cost of gas and a car wash.

The cost will be $4.50x + 3.00$ dollars for x gallons of gas and a car wash.

$$y = 4.5x + 3 \quad \text{Final 0's need not be included.}$$

(c) Interpret the ordered pairs $(5, 25.5)$ and $(10, 48)$ in relation to the equation from part (b).

The ordered pair $(5, 25.5)$ indicates that 5 gal of gas and a car wash cost $25.50. Similarly, the ordered pair $(10, 48)$ indicates that 10 gal of gas and a car wash cost $48.00.

◀ **Work Problem** ⑪ **at the Side.**

Note

In **Example 10(a),** the ordered pair $(0, 0)$ satisfied the equation, so the linear equation has the form $y = mx$, where $b = 0$. If a situation involves an initial charge b plus a charge per unit m as in **Example 10(b),** the equation has the form $y = mx + b$, where $b \neq 0$.

Answers

11. **(a)** $y = 0.1x$ (Note: $0.10x = 0.1x$)
   (b) $y = 0.1x + 0.2$
   (c) The ordered pair $(15, 1.7)$ indicates that the price of a 15-minute call is $1.70.

> ### EXAMPLE 11   Writing an Equation of a Line That Models Data

Average annual tuition and fees for in-state students at public four-year colleges are shown in the table for selected years and graphed as ordered pairs of points in the **scatter diagram** in **Figure 43**, where $x = 0$ represents 2005, $x = 1$ represents 2006, and so on, and y represents the cost in dollars.

Year	Cost (in dollars)
2005	6350
2006	6443
2007	6715
2008	6770
2009	7396
2010	7889
2011	8244

Source: The College Board.

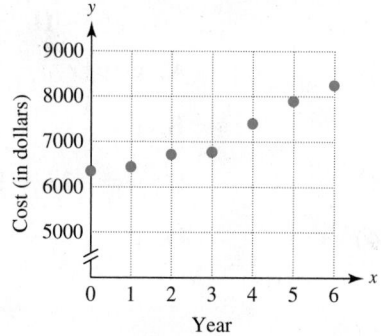

Figure 43

(a) Write an equation that models the data.

Since the points in **Figure 43** lie approximately on a straight line, we can write a linear equation that models the relationship between year x and cost y. We choose two data points, $(0, 6350)$ and $(5, 7889)$, to find the slope of the line.

Start with the x- and y-values of the same point.

$$m = \frac{7889 - 6350}{5 - 0} = \frac{1539}{5} = 307.8$$

The slope 307.8 indicates that the cost of tuition and fees increased by about $307.80 per year from 2005 to 2011. We use this slope, the y-intercept $(0, \mathbf{6350})$, and slope-intercept form to write an equation of the line.

$$y = \mathbf{307.8}x + \mathbf{6350}$$

(b) Use the equation from part (a) to predict the approximate cost of tuition and fees at public four-year colleges in 2013.

The value $x = 8$ corresponds to the year 2013.

$$y = 307.8x + 6350 \qquad \text{Equation from part (a)}$$
$$y = 307.8\,(\mathbf{8}) + 6350 \qquad \text{Substitute 8 for } x.$$
$$y \approx 8812 \qquad \text{Multiply, and then add.}$$

According to the model, average tuition and fees for in-state students at public four-year colleges in 2013 will be about $8812.

··· **Work Problem 12 at the Side.** ▶

> ### Note
>
> Choosing different data points in **Example 11** would result in a slightly different line (particularly in regard to its slope) and, hence, a slightly different equation. However, all such equations should yield similar results.

12 The percentage of the U.S. population aged 25 yr and older with at least a high school diploma is shown in the table for selected years.

Year	Percent
1950	34.3
1960	41.1
1970	52.3
1980	66.5
1990	77.6
2000	84.1
2010	87.1

Source: U.S. Census Bureau.

(a) Let $x = 0$ represent 1950, $x = 10$ represent 1960, and so on. Use the data for 1950 and 2000 to write an equation that models the data.

(b) Use the equation from part (a) to approximate the percentage, to the nearest tenth, of the U.S. population aged 25 yr and older who were at least high school graduates in 1995.

3.4 Exercises

FOR EXTRA HELP Download the MyDashBoard App   ▶   MyMathLab®

CONCEPT CHECK *Work each problem.*

1. Match each description of a line in Column I with the correct equation in Column II.

          I   **II**

  (a) Slope -2, passes through the point $(4, 1)$ **A.** $y = 4x$

  (b) Slope -2, y-intercept $(0, 1)$ **B.** $y = \dfrac{1}{4}x$

  (c) Passes through the points $(0, 0)$ and $(4, 1)$ **C.** $y = -2x + 1$

  (d) Passes through the points $(0, 0)$ and $(1, 4)$ **D.** $y - 1 = -2(x - 4)$

2. Which equations are equivalent to $2x - 3y = 6$?

  A. $y = \dfrac{2}{3}x - 2$ **B.** $-2x + 3y = -6$ **C.** $y = -\dfrac{3}{2}x + 3$ **D.** $y - 2 = \dfrac{2}{3}(x - 6)$

3. Match each equation with the graph that would most closely resemble its graph.

  (a) $y = x + 3$ **(b)** $y = -x + 3$ **(c)** $y = x - 3$ **(d)** $y = -x - 3$

  A. 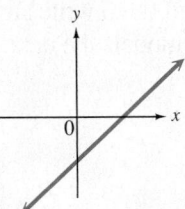   **B.** 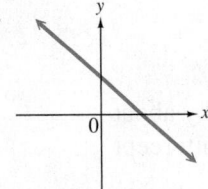   **C.** 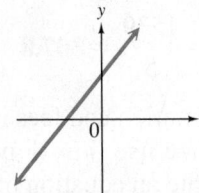   **D.**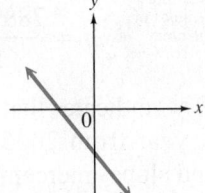

4. **(a)** What is the common name given to the vertical line whose x-intercept is the origin?

  (b) What is the common name given to the line with slope 0 whose y-intercept is the origin?

Identify the slope and y-intercept of the line with each equation. **See Example 1.**

5. $y = \dfrac{5}{2}x - 4$ 6. $y = \dfrac{7}{3}x - 6$ 7. $y = -x + 9$

8. $y = x + 1$ 9. $y = \dfrac{x}{5} - \dfrac{3}{10}$ 10. $y = \dfrac{x}{7} - \dfrac{5}{14}$

Write the equation of the line with the given slope and y-intercept. **See Example 2.**

11. slope 4, ▶ y-intercept $(0, -3)$ 12. slope -5, y-intercept $(0, 6)$ 13. $m = -1$, y-intercept $(0, -7)$ 14. $m = 1$, y-intercept $(0, -9)$

15. slope 0,
    y-intercept (0, 3)

16. slope 0,
    y-intercept (0, −4)

17. undefined slope,
    y-intercept (0, −2)

18. undefined slope,
    y-intercept (0, 5)

CONCEPT CHECK *Use the geometric interpretation of slope (rise divided by run, from* **Section 3.3***) to find the slope of each line. Then, by identifying the y-intercept from the graph, write the slope-intercept form of the equation of the line.*

19.

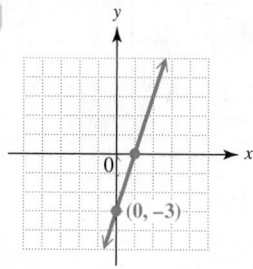

20.

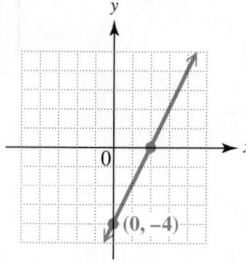

21.

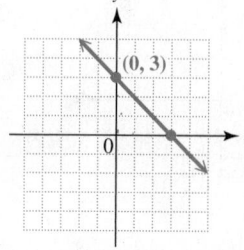

22.

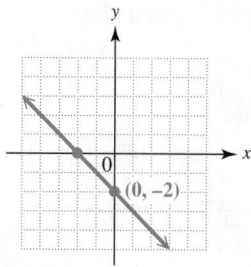

23.

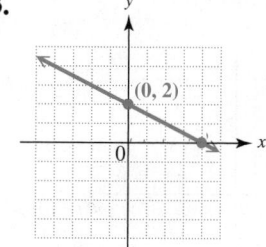

24.

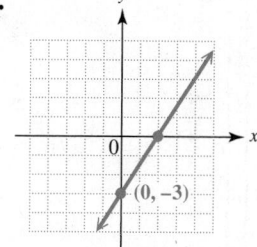

Graph each equation by identifying the slope and y-intercept, and using their definitions to find two points on the line. **See Example 3.**

25. $y = 3x + 2$

26. $y = 4x - 4$

27. $y = \dfrac{3}{4}x - 1$

28. $y = \dfrac{3}{2}x + 2$

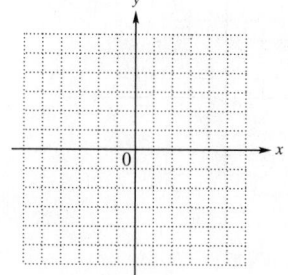

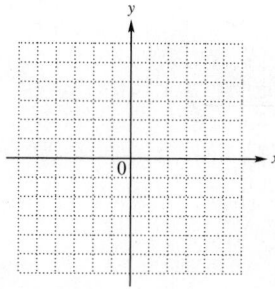

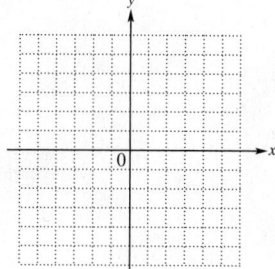

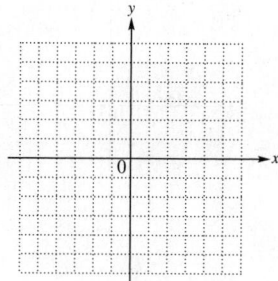

29. $2x + y = -5$

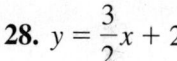

30. $3x + y = -2$

31. $x + 2y = 4$

32. $x + 3y = 12$

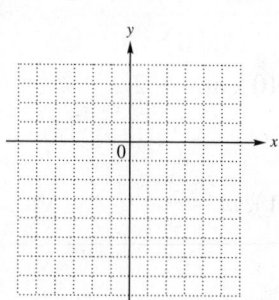

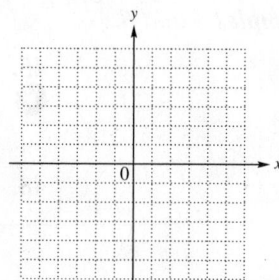

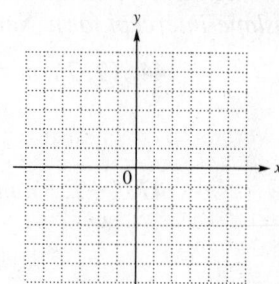

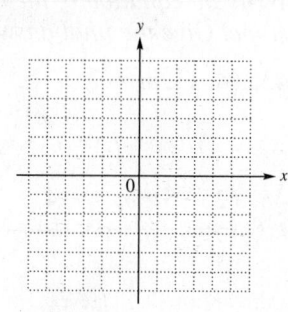

Graph each line passing through the given point and having the given slope. (In Exercises 37–40, recall the types of lines having slope 0 and undefined slope.) Give the slope-intercept form of the equation of the line if possible. **See Examples 4 and 5.**

33. $(-2, 3)$, $m = \dfrac{1}{2}$

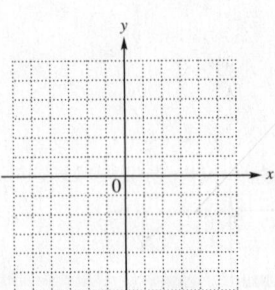

34. $(-4, -1)$, $m = \dfrac{3}{4}$

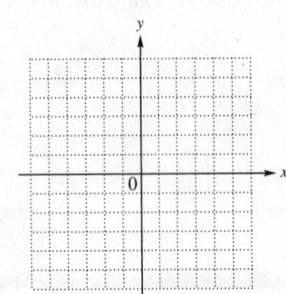

35. $(1, -5)$, $m = -\dfrac{2}{5}$

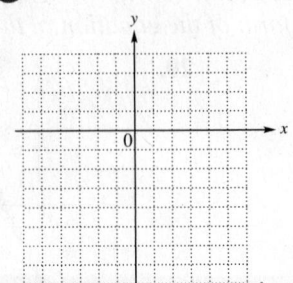

36. $(2, -1)$, $m = -\dfrac{1}{3}$

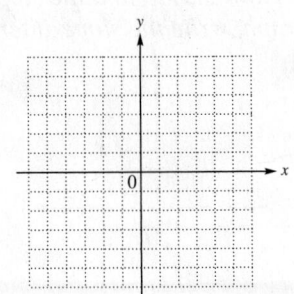

37. $(3, 2)$, $m = 0$

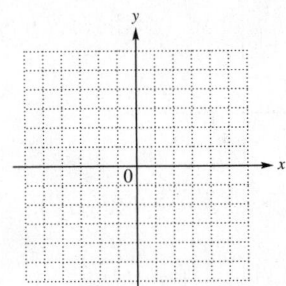

38. $(-2, 3)$, $m = 0$

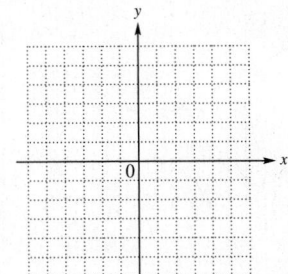

39. $(3, -2)$, undefined slope

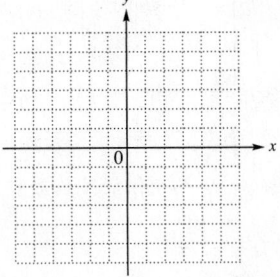

40. $(2, 4)$, undefined slope

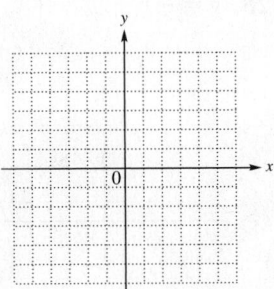

41. $(0, 0)$, $m = \dfrac{2}{3}$

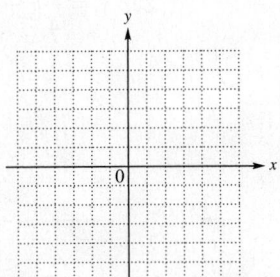

42. $(0, 0)$, $m = \dfrac{5}{2}$

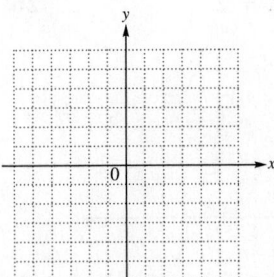

Write an equation of the line passing through the given point and having the given slope. Give the final answer in slope-intercept form. **See Examples 5 and 6.**

43. $(4, 1)$, $m = 2$

44. $(2, 7)$, $m = 3$

45. $(3, -10)$, $m = -2$

46. $(2, -5)$, $m = -4$

47. $(-2, 5)$, $m = \dfrac{2}{3}$

48. $(-4, 1)$, $m = \dfrac{3}{4}$

Write an equation of the line passing through each pair of points. Give the final answer in (a) slope-intercept form and (b) standard form. See Example 7.

49. $(8, 5)$ and $(9, 6)$ **50.** $(4, 10)$ and $(6, 12)$ **51.** $(-1, -7)$ and $(-8, -2)$ **52.** $(-2, -1)$ and $(3, -4)$

53. $(0, -2)$ and $(-3, 0)$ **54.** $(-4, 0)$ and $(0, 2)$ **55.** $\left(\frac{1}{2}, \frac{3}{2}\right)$ and $\left(-\frac{1}{4}, \frac{5}{4}\right)$ **56.** $\left(-\frac{2}{3}, \frac{8}{3}\right)$ and $\left(\frac{1}{3}, \frac{7}{3}\right)$

Write an equation of the line that satisfies the given conditions. See Example 8.

57. Through $(9, 5)$; slope 0   **58.** Through $(-4, -2)$; slope 0   **59.** Through $(9, 10)$; undefined slope

60. Through $(-2, 8)$; undefined slope   **61.** Through $\left(-\frac{3}{4}, -\frac{3}{2}\right)$; slope 0   **62.** Through $\left(-\frac{5}{8}, -\frac{9}{2}\right)$; slope 0

63. Through $(-7, 8)$; horizontal   **64.** Through $(2, -7)$; horizontal

65. Through $(0.5, 0.2)$; vertical   **66.** Through $(0.1, 0.4)$; vertical

Write an equation of the line satisfying the given conditions. Give the final answer in slope-intercept form. See Example 9.

67. Perpendicular to $x - 2y = 7$, y-intercept $(0, -3)$ **68.** Parallel to $5x = 2y + 10$, y-intercept $(0, 4)$ **69.** Passing through $(2, 3)$, parallel to $4x - y = -2$

70. Passing through $(4, 2)$, perpendicular to $x - 3y = 7$ **71.** Passing through $(2, -3)$, parallel to $3x = 4y + 5$ **72.** Passing through $(-1, 4)$, perpendicular to $2x + 3y = 8$

Write an equation in the form y = mx for each situation. Then give three ordered pairs associated with the equation for x-values of 0, 5, and 10. See Example 10(a).

73. *x* represents the number of hours traveling at 45 mph, and *y* represents the distance traveled (in miles).

74. *x* represents the number of t-shirts sold at $16 each, and *y* represents the total cost of the t-shirts (in dollars).

75. *x* represents the number of gallons of gasoline sold at $5.00 per gal, and *y* represents the total cost of the gasoline (in dollars).

76. *x* represents the number of days a DVD movie is rented at $2.50 per day, and *y* represents the total charge for the rental (in dollars).

For each situation, (a) write an equation in the form y = mx + b, (b) find and interpret the ordered pair associated with the equation for x = 5, and (c) answer the question. See Examples 10(b) and 10(c).

77. A membership to a health club costs $99, plus $41 per month. Let *x* represent the number of months and *y* represent the cost in dollars. How much does a one-year membership cost? (*Source:* Midwest Athletic Club.)

78. An Executive VIP/Gold membership to a health club costs $159, plus $57 per month. Let *x* represent the number of months and *y* represent the cost in dollars. How much does a one-year membership cost? (*Source:* Midwest Athletic Club.)

79. A cell phone plan includes 900 anytime minutes for $60 per month, plus a one-time activation fee of $36. Let *x* represent the number of months and *y* represent the cost in dollars. Over a two-year contract, how much will this plan cost? (We never use more than the allotted number of minutes.) (*Source:* AT&T.)

80. A cell phone plan includes 450 anytime minutes for $40 per month, plus $40 for an Acer Aspire A0722 cell phone and $36 for a one-time activation fee. Let *x* represent the number of months and *y* represent the cost in dollars. Over a two-year contract, how much will this plan cost? (We never use more than the allotted number of minutes.) (*Source:* AT&T.)

*The cost y to produce x items is, in some cases, expressed as y = mx + b. The number b gives the **fixed cost** (the cost that is the same no matter how many items are produced), and the number m gives the **variable cost** (the cost to produce an additional item). Use this information to work Exercises 81 and 82.*

81. It costs $400 to start up a business selling campaign buttons. Each button costs $0.25 to produce.

   (a) What is the fixed cost?

   (b) What is the variable cost?

   (c) Write the cost equation.

   (d) What will be the cost to produce 100 campaign buttons, based on the cost equation?

   (e) How many campaign buttons will be produced if total cost is $775?

82. It costs $2000 to purchase a copier, and each copy costs $0.02 to make.

   (a) What is the fixed cost?

   (b) What is the variable cost?

   (c) Write the cost equation.

   (d) What will be the cost to produce 10,000 copies, based on the cost equation?

   (e) How many copies will be produced if total cost is $2600?

*Solve each problem. In part (a), give equations in slope-intercept form. Round the slope
to the nearest whole number.* **See Example 11.**

83. The numbers of U.S. travelers (in thousands) to
Canada are shown in the graph, where the year 2006
corresponds to $x = 0$.

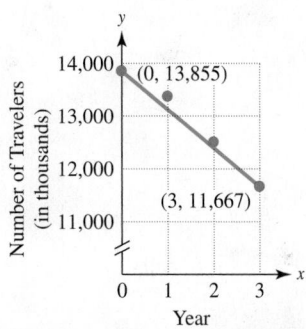

Source: U.S. Department of
Commerce.

(a) Use the ordered pairs from the graph to write
an equation that models the data. What does the
slope tell us in the context of the problem?

(b) Use the equation from part (a) to approximate the
number of U.S. travelers to Canada in 2010.

84. The numbers of international travelers (in thousands)
to the United States from South America are shown in
the graph, where the year 2006 corresponds to $x = 0$.

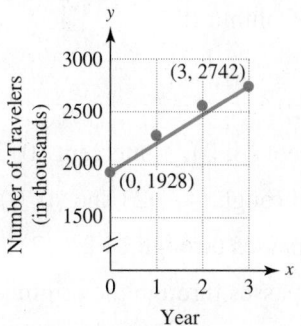

Source: U.S. Department of
Commerce.

(a) Use the ordered pairs from the graph to write
an equation that models the data. What does the
slope tell us in the context of the problem?

(b) Use the equation from part (a) to approximate
the number of travelers to the United States from
South America in 2010.

85. The table lists the average annual cost y (in dollars) of
tuition and fees at 2-year public colleges for selected
years x, where $x = 1$ represents 2007, $x = 2$
represents 2008, and so on.

(a) Write five ordered pairs for the data.

(b) Plot the ordered pairs. Do the points lie
approximately in a straight line?

Year x	Cost y (in dollars)
1	2294
2	2372
3	2558
4	2727
5	2963

Source: The College Board.

(c) Use the ordered pairs $(2, 2372)$ and $(5, 2963)$ to
find the equation of a line that approximates the
data. Write the equation in slope-intercept form.

(d) Use the equation from part (c) to estimate the
average annual cost at 2-year colleges in 2012 to
the nearest dollar. (*Hint:* What is the value of x
for 2012?)

**AVERAGE ANNUAL COSTS AT
2-YEAR COLLEGES**

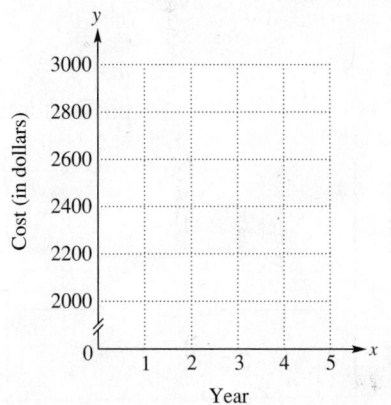

Summary Exercises *Applying Graphing and Equation-Writing Techniques for Lines*

1. CONCEPT CHECK Match the description of a line in Column I with the correct equation in Column II.

I	II
(a) Slope -0.5, $b = -2$	**A.** $y = -\frac{1}{2}x$
(b) x-intercept $(4, 0)$, y-intercept $(0, 2)$	**B.** $y = -\frac{1}{2}x - 2$
(c) Passes through $(4, -2)$ and $(0, 0)$	**C.** $x - 2y = 2$
(d) $m = \frac{1}{2}$, passes through $(-2, -2)$	**D.** $x + 2y = 4$
(e) $m = \frac{1}{2}$, passes through the origin	**E.** $y = 2x$
(f) Slope 2, $b = 0$	**F.** $x = 2y$

2. CONCEPT CHECK Which equations are equivalent to $2x + 5y = 20$?

A. $y = -\frac{2}{5}x + 4$ **B.** $y - 2 = -\frac{2}{5}(x - 5)$ **C.** $y = \frac{5}{2}x - 4$ **D.** $2x = 5y - 20$

Graph each line, using the given information or equation.

3. $m = 1$, $b = -2$

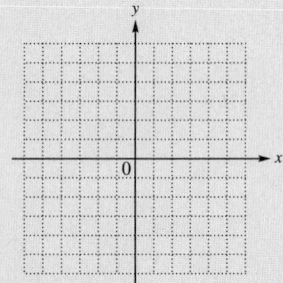

4. $m = 1$,
   y-intercept $(0, -4)$

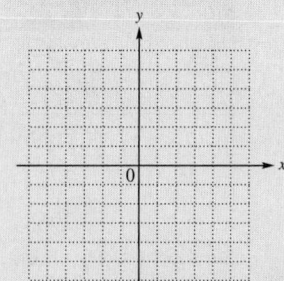

5. $y = -2x + 6$

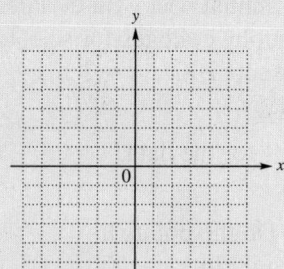

6. $x + 4 = 0$

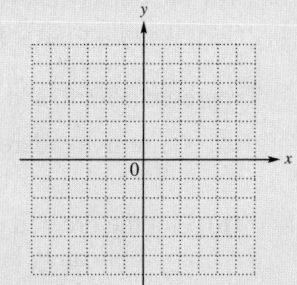

7. $m = -\frac{2}{3}$,
   passes through $(3, -4)$

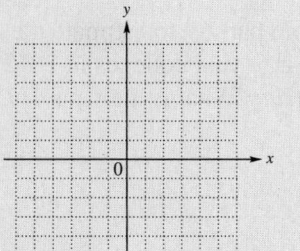

8. $y = -\frac{1}{2}x + 2$

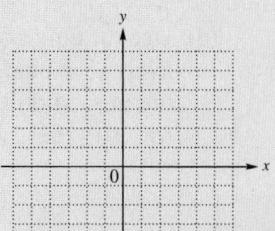

9. $y - 4 = -9$

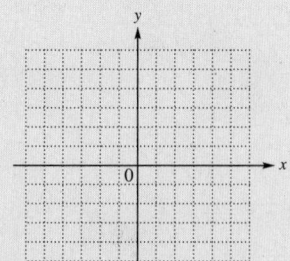

10. $m = -\frac{3}{4}$,
   passes through $(4, -4)$

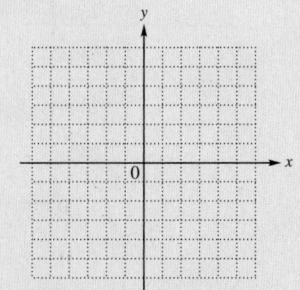

11. Undefined slope, passes through $(3.5, 0)$

12. Slope $-\frac{1}{5}$, passes through $(0, 0)$

13. $4x - 5y = 20$

14. $6x - 5y = 30$

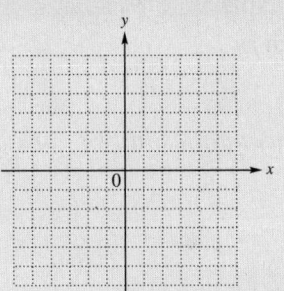

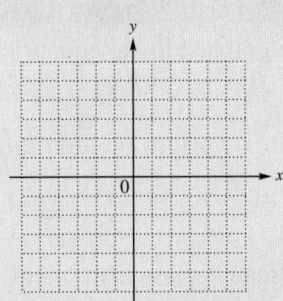

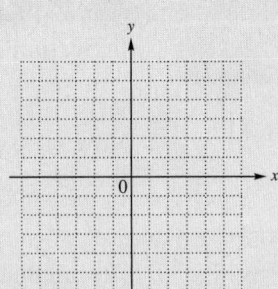

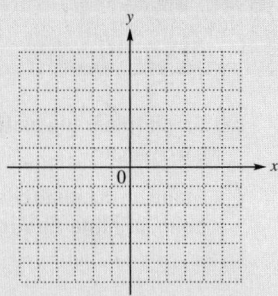

15. $x - 4y = 0$

16. $m = 0$, passes through $\left(0, \frac{3}{2}\right)$

17. $3y = 12 - 2x$

18. $8x = 6y + 24$

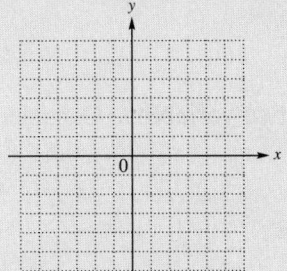

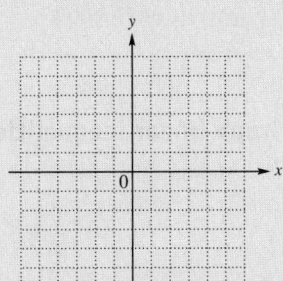

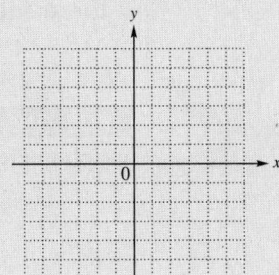

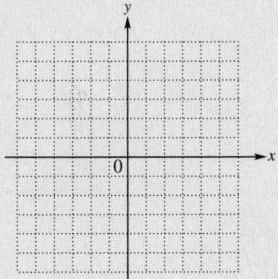

Write an equation of each line. Give the final answer in slope-intercept form if possible.

19. $m = -3, b = -6$

20. Passes through $(1, -7)$ and $(-2, 5)$

21. Passes through $(0, 0)$ and $(5, 3)$

22. Passes through $(0, 0)$, undefined slope

23. Passes through $(0, 0)$, $m = 0$

24. $m = -2$, y-intercept $(0, -4)$

25. $m = \frac{5}{3}$, through $(-3, 0)$

26. Passes through $(-2, 5)$, parallel to the graph of $3x - y = 4$

3.5 Graphing Linear Inequalities in Two Variables

OBJECTIVES

1 Graph linear inequalities in two variables.

2 Graph an inequality with a boundary line through the origin.

In **Section 3.2** we graphed linear equations, such as

$$2x + 3y = 6.$$

Now we extend this work to include *linear inequalities in two variables,* such as

$$2x + 3y \le 6.$$

(Recall that $\le$ is read "is less than or equal to.")

Linear Inequality in Two Variables

An inequality that can be written in the form

$$Ax + By < C, \ Ax + By > C, \ Ax + By \le C, \text{ or } Ax + By \ge C,$$

where A, B, and C are real numbers and A and B are not both 0, is a **linear inequality in two variables.**

OBJECTIVE ▶ 1 **Graph linear inequalities in two variables.** Consider the graph in **Figure 44.**

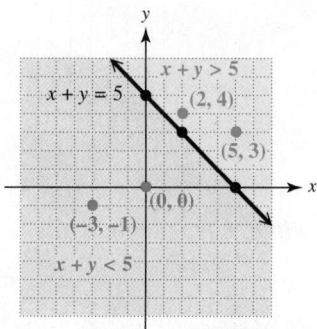

Figure 44

The graph of the line $x + y = 5$ in **Figure 44** divides the points in the rectangular coordinate system into three sets.

1. Those points that lie on the line itself and satisfy the equation $x + y = 5$ (like $(0, 5)$, $(2, 3)$, and $(5, 0)$)

2. Those points that lie in the region above the line and satisfy the inequality $x + y > 5$ (like $(5, 3)$ and $(2, 4)$)

3. Those points that lie in the region below the line and satisfy the inequality $x + y < 5$ (like $(0, 0)$ and $(-3, -1)$)

The graph of the line $x + y = 5$ is the **boundary line** for the inequalities

$$x + y > 5 \quad \text{and} \quad x + y < 5.$$

Graphs of linear inequalities in two variables are regions in the real number plane that may or may not include boundary lines.

EXAMPLE 1 **Graphing a Linear Inequality**

Graph $2x + 3y \leq 6$.

The inequality $2x + 3y \leq 6$ means that

$$2x + 3y < 6 \quad \text{or} \quad 2x + 3y = 6.$$

We begin by graphing the equation $2x + 3y = 6$, a line with intercepts $(0, 2)$ and $(3, 0)$ as shown in **Figure 45.** This boundary line divides the plane into two regions, one of which satisfies the inequality. To find the correct region, we choose a test point *not* on the boundary line and substitute it into the inequality to see whether the resulting statement is true or false. The point $(0, 0)$ is a convenient choice.

$$2x + 3y < 6$$
$$2(0) + 3(0) \overset{?}{<} 6 \quad \text{Let } x = 0 \text{ and } y = 0.$$
$$0 + 0 \overset{?}{<} 6 \quad \text{Multiply.}$$
$$0 < 6 \quad \text{True}$$

Use (0, 0) as a test point.

Since the last statement is true, we shade the region that includes the test point $(0, 0)$. See **Figure 45.** The shaded region, along with the boundary line, is the desired graph.

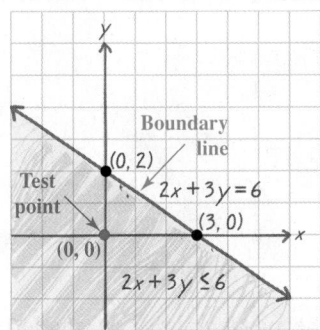

Figure 45

⋯⋯⋯⋯⋯⋯⋯⋯⋯⋯⋯⋯ **Work Problems ❶ and ❷ at the Side.** ▶

Note

Alternatively in **Example 1,** we can find the required region by solving the original inequality for y.

$$2x + 3y \leq 6 \qquad \text{Inequality from \textbf{Example 1}}$$
$$3y \leq -2x + 6 \qquad \text{Subtract } 2x.$$
$$y \leq -\frac{2}{3}x + 2 \qquad \text{Divide each term by 3.}$$

Ordered pairs in which y is equal to $-\frac{2}{3}x + 2$ are on the boundary line, so pairs in which *y is less than* $-\frac{2}{3}x + 2$ will be *below* that line. (As we move *down* vertically, the y-values *decrease*.) This gives the same region that we shaded in **Figure 45.** (Ordered pairs in which *y is greater than* $-\frac{2}{3}x + 2$ will be *above* the boundary line.)

❶ Graph $3x + 4y \leq 12$.

GS The boundary line $3x + 4y = 12$ is graphed below. Use $(0, 0)$ as a test point, and shade the appropriate region for the inequality.

$$3x + 4y < 12$$
$$3(\underline{\quad}) + 4(\underline{\quad}) \overset{?}{<} 12$$
$$\underline{\quad} < 12 \; (\textit{True} / \textit{False})$$

Shade the region of the graph that (*includes* / *does not include*) the test point $(0, 0)$.

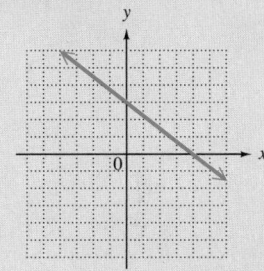

❷ Graph $4x - 5y \leq 20$.

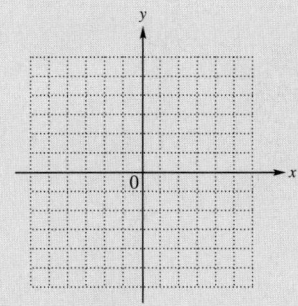

Answers

1. 0; 0; 0; True; includes

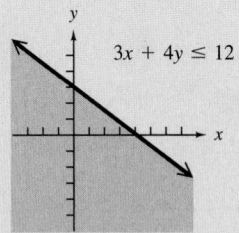

$3x + 4y \leq 12$

2.

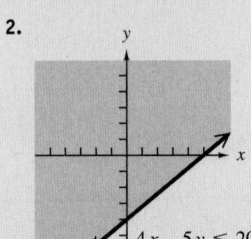

$4x - 5y \leq 20$

3 Use $(0, 0)$ as a test point and shade the appropriate region for the linear inequality.

$$3x + 5y > 15$$

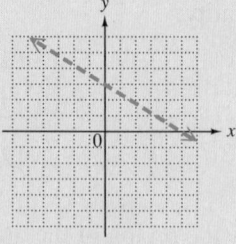

4 Graph $x + 2y > 6$.

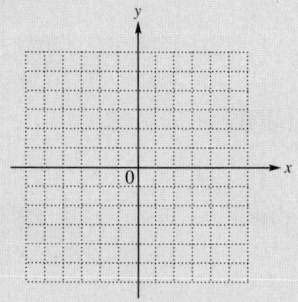

Answers

3.

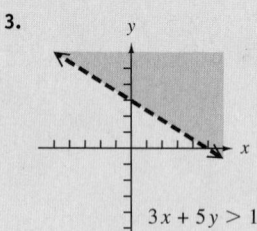

$$3x + 5y > 15$$

4.

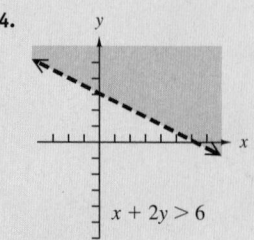

$$x + 2y > 6$$

<table>
<tr><td>**EXAMPLE 2**</td><td>**Graphing a Linear Inequality**</td></tr>
</table>

Graph $x - y > 5$.

This inequality does *not* involve equality. Therefore, the points on the line $x - y = 5$ do *not* belong to the graph. However, the line still serves as a boundary for two regions, one of which satisfies the inequality.

To graph the inequality, first graph the equation $x - y = 5$ using the intercepts $(0, -5)$ and $(5, 0)$. Use a *dashed line* to show that the points on the line are *not* solutions of the inequality $x - y > 5$. See **Figure 46.** Then choose a test point to see which region satisfies the inequality.

$$x - y > 5 \quad \text{Original inequality}$$

(0, 0) is a convenient test point.

$$0 - 0 \overset{?}{>} 5 \quad \text{Let } x = 0 \text{ and } y = 0.$$

$$0 > 5 \quad \text{False}$$

Since $0 > 5$ is false, the graph of the inequality is the region that *does not* contain $(0, 0)$. Shade the *other* region, as shown in **Figure 46.**

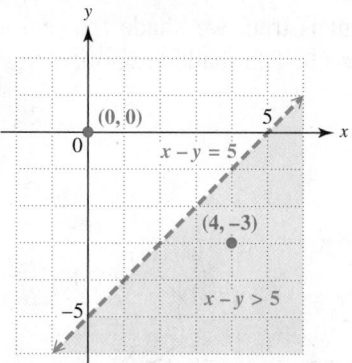

Figure 46

CHECK To check that the proper region is shaded, we test a point in the shaded region. For example, we use $(4, -3)$.

$$x - y > 5$$

$$4 - (-3) \overset{?}{>} 5 \quad \text{Let } x = 4 \text{ and } y = -3.$$

Use parentheses to avoid errors.

$$7 > 5 \checkmark \quad \text{True}$$

This verifies that the correct region is shaded in **Figure 46.**

◀ **Work Problems 3 and 4 at the Side.**

A summary of the steps used to graph a linear inequality in two variables follows.

Graphing a Linear Inequality in Two Variables

Step 1 **Graph the boundary.** Graph the line that is the boundary of the region. Use the methods of **Section 3.2.** Draw a solid line if the inequality involves $\le$ or $\ge$. Draw a dashed line if the inequality involves $<$ or $>$.

Step 2 **Shade the appropriate side.** Use any point not on the line as a test point. Substitute for x and y in the *inequality*. If a true statement results, shade the region containing the test point. If a false statement results, shade the other region.

EXAMPLE 3 Graphing a Linear Inequality with a Vertical Boundary Line

Graph $x < 3$.

First graph $x = 3$, a vertical line through the point $(3, 0)$. Use a dashed line, and choose $(0, 0)$ as a test point.

$$x < 3 \quad \text{Original inequality}$$
$$0 \overset{?}{<} 3 \quad \text{Let } x = 0.$$
$$0 < 3 \quad \text{True}$$

Because $0 < 3$ is true, we shade the region containing $(0, 0)$, as in **Figure 47.**

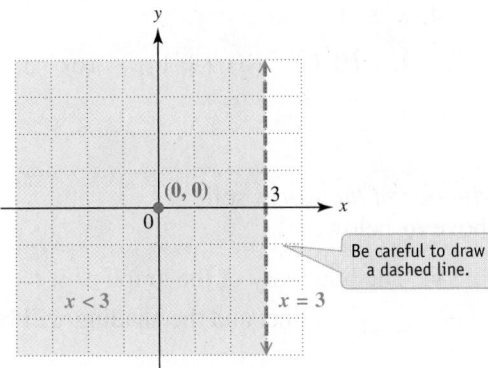

Be careful to draw a dashed line.

Figure 47

Work Problem **5** at the Side. ▶

OBJECTIVE **2** **Graph an inequality with a boundary line through the origin.** *If the graph of an inequality has a boundary line through the origin, $(0, 0)$ cannot be used as a test point.*

EXAMPLE 4 Graphing a Linear Inequality with a Boundary Line through the Origin

Graph $x \le 2y$.

We graph $x = 2y$ using a solid line through $(0, 0)$, $(6, 3)$, and $(4, 2)$. Because $(0, 0)$ is *on* the line $x = 2y$, it cannot be used as a test point. Instead, we choose a test point *off* the line, say $(1, 3)$.

$$x < 2y$$
$$1 \overset{?}{<} 2(3) \quad \text{Let } x = 1 \text{ and } y = 3.$$
$$1 < 6 \quad \text{True}$$

Because $1 < 6$ is true, we shade the side of the graph containing the test point $(1, 3)$. See **Figure 48.**

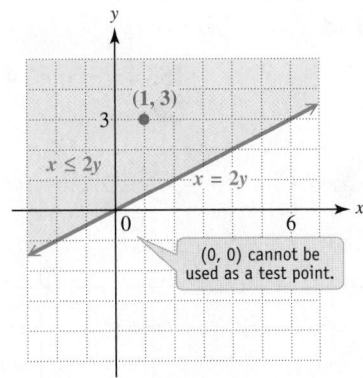

$(0, 0)$ cannot be used as a test point.

Figure 48

Work Problem **6** at the Side. ▶

5 Graph $y < 4$.

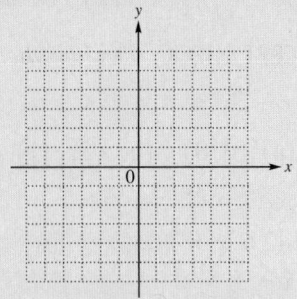

6 Graph $x \ge -3y$.

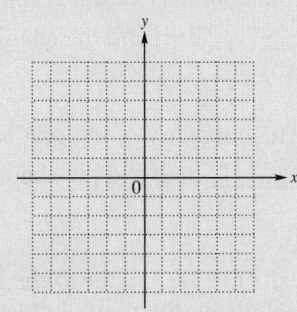

Can $(0, 0)$ be used as a test point here? (*Yes / No*)

Answers

5.

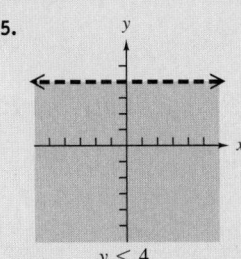

$y < 4$

6. No

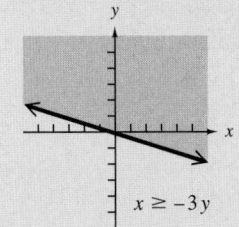

$x \ge -3y$

3.5 Exercises

  MyMathLab®

CONCEPT CHECK *Decide whether each ordered pair is a solution of the given inequality.*

1. $x - 2y \leq 4$

   (a) $(0, 0)$ **(b)** $(2, -1)$ **(c)** $(7, 1)$ **(d)** $(0, 2)$

2. $x + y > 0$

   (a) $(0, 0)$ **(b)** $(-2, 1)$ **(c)** $(2, -1)$ **(d)** $(-4, 6)$

3. $x - 5 > 0$

   (a) $(0, 0)$ **(b)** $(5, 0)$ **(c)** $(-1, 3)$ **(d)** $(6, 2)$

4. $y \leq 1$

   (a) $(0, 0)$ **(b)** $(3, 1)$ **(c)** $(2, -1)$ **(d)** $(-3, 3)$

CONCEPT CHECK *In each statement, fill in the first blank with one of the words* solid *or* dashed. *Fill in the second blank with one of the words* above *or* below.

5. The boundary of the graph of $y \leq -x + 2$ will be a _____ line, and the shading will be _____ the line.

6. The boundary of the graph of $y < -x + 2$ will be a _____ line, and the shading will be _____ the line.

7. The boundary of the graph of $y > -x + 2$ will be a _____ line, and the shading will be _____ the line.

8. The boundary of the graph of $y \geq -x + 2$ will be a _____ line, and the shading will be _____ the line.

CONCEPT CHECK *Refer to the given graph, and complete each statement with the correct inequality symbol* $<$, $\leq$, $>$, *or* $\geq$.

9. x ____ 4

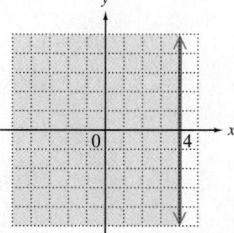

10. y ____ -3

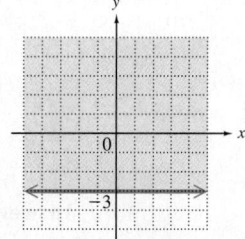

11. y ____ $3x - 2$

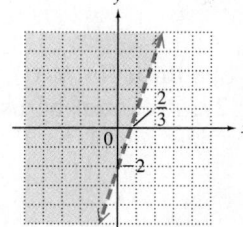

12. y ____ $-x + 3$

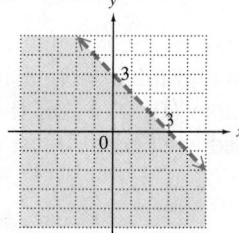

*Graph each linear inequality. **See Examples 1–4.***

13. $x + y \leq 2$

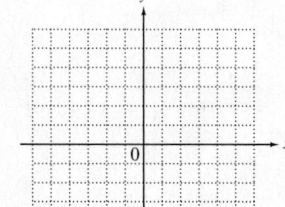

14. $x + y \leq -3$

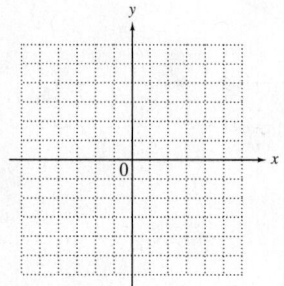

15. $4x - y < 4$

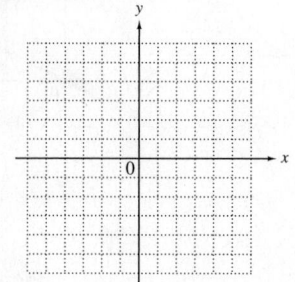

16. $3x - y < 3$

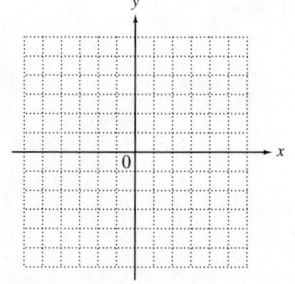

17. $x + 3y \geq -2$

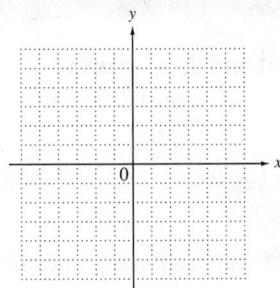

18. $x + 4y \geq -3$

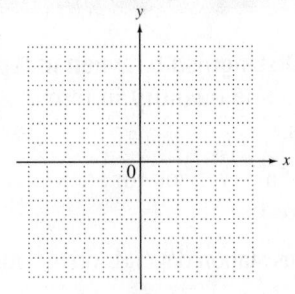

19. $y < \dfrac{1}{2}x + 3$

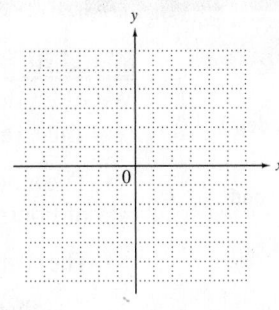

20. $y < \dfrac{1}{3}x - 2$

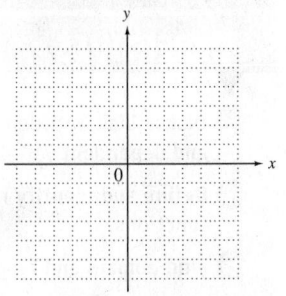

21. $y \geq -\dfrac{2}{3}x + 2$

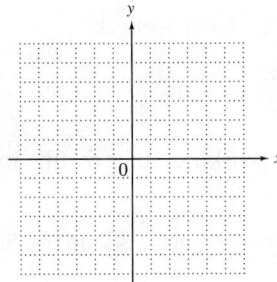

22. $y \geq -\dfrac{3}{4}x + 3$

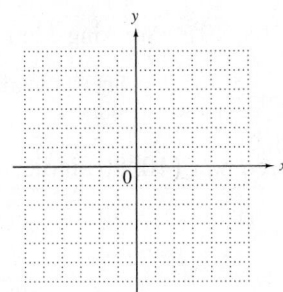

23. $x \leq -2$

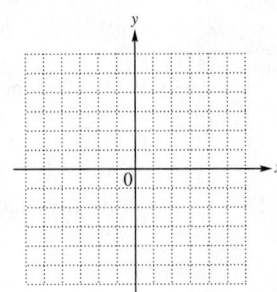

24. $x \geq 1$

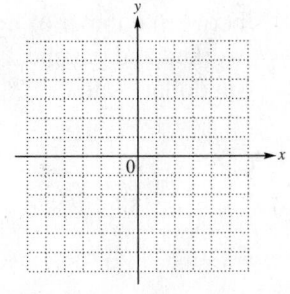

25. $y < 5$

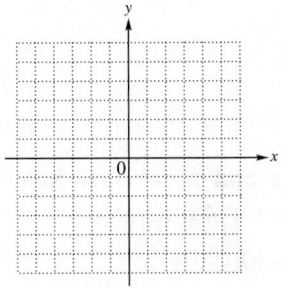

26. $y < -3$

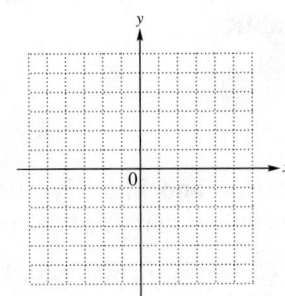

27. $x + y > 0$

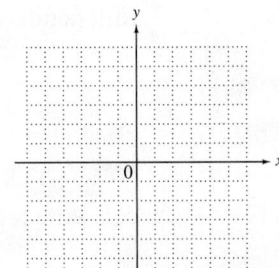

28. $x + 2y > 0$

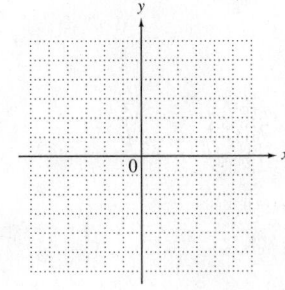

29. $x - 3y \leq 0$

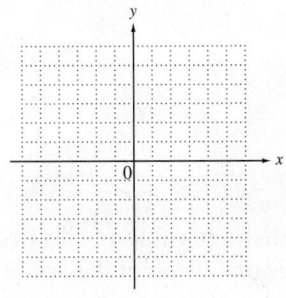

30. $x - 5y \leq 0$

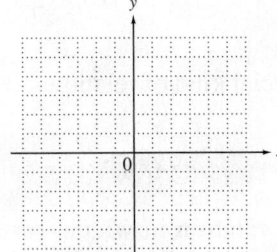

31. $y < x$

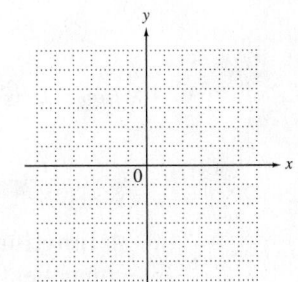

32. $y < 4x$

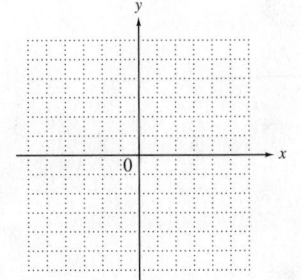

3.6 Introduction to Relations and Functions

OBJECTIVES

1. Distinguish between independent and dependent variables.
2. Define and identify relations and functions.
3. Find domain and range.
4. Identify functions defined by graphs and equations.

1 What would the ordered pair (40, 320) in the correspondence between number of hours worked and paycheck amount (in dollars) indicate?

OBJECTIVE ▶ **1** **Distinguish between independent and dependent variables.** We often describe one quantity in terms of another. Consider the following real-life examples.

- The amount of a paycheck for an hourly employee depends on the number of hours worked.
- The cost at a gas station depends on the number of gallons of gas pumped.
- The distance traveled by a car moving at a constant rate depends on the time traveled.

We can use ordered pairs to represent these corresponding quantities. For instance, we indicate the relationship between hours worked and paycheck amount as follows.

$$(\mathbf{5}, \mathbf{40}) \quad \text{Working } 5 \text{ hr results in a } \$40 \text{ paycheck.}$$

Number of hours worked ⎦　⎣ Paycheck amount in dollars

Similarly, the ordered pair (10, 80) indicates that working 10 hr results in an $80 paycheck.

◀ **Work Problem 1 at the Side.**

Since paycheck amount *depends* on number of hours worked, paycheck amount is the *dependent variable,* and number of hours worked is the *independent variable.* Generalizing, if the value of the variable y depends on the value of the variable x, then y is the **dependent variable** and x is the **independent variable.**

Independent variable ⎤ ⎡ Dependent variable
$$\downarrow\ \downarrow$$
$$(x, y)$$

OBJECTIVE ▶ **2** **Define and identify relations and functions.** Since we can write related quantities using ordered pairs, a set of ordered pairs such as

$$\{(5, 40), (10, 80), (20, 160), (40, 320)\}$$

is called a *relation.*

Relation

A **relation** is any set of ordered pairs.

A *function* is a special kind of relation.

Function

A **function** is a relation in which, for each distinct value of the first component of the ordered pairs, there is *exactly one value* of the second component.

Answer

1. It indicates that working 40 hr results in a $320 paycheck.

| **EXAMPLE 1** | **Whether Relations Are Functions** |

Determine whether each relation defines a function.

(a) $F = \{(1, 2), (-2, 4), (3, -1)\}$

For $x = 1$, there is only one value of y, **2**.

For $x = -2$, there is only one value of y, **4**.

For $x = 3$, there is only one value of y, **-1**.

Relation F is a function. For each distinct x-value, there is exactly one y-value.

(b) $G = \{(-2, -1), (-1, 0), (0, 1), (1, 2), (2, 2)\}$

Relation G is also a function. Although the last two ordered pairs have the same y-value (1 is paired with 2 and 2 is paired with 2), this does not violate the definition of a function. The first components (x-values) are distinct, and each is paired with only one second component (y-value).

(c) $H = \{(-4, 1), (-2, 1), (-2, 0)\}$

In relation H, the last two ordered pairs have the **same** x-value paired with **two different** y-values (−2 is paired with both 1 and 0). H is a relation, but *not* a function.

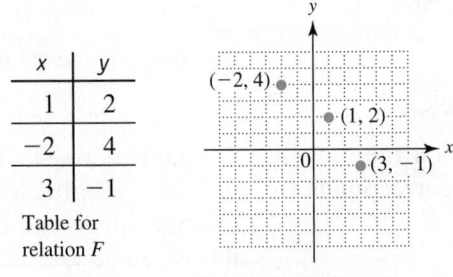

$$H = \{(-4, 1), (-2, 1), (-2, 0)\} \quad \text{Not a function}$$

Different y-values

Same x-value

In a function, no two ordered pairs can have the same first component and different second components.

················· **Work Problem 2 at the Side.** ▶

Relations and functions can be defined in several different ways, as indicated in boldface in the bulleted list that follows.

- **As a set of ordered pairs** (See Example 1.)

- **As a correspondence or** *mapping*
  See **Figure 49.** In the mapping for relation F from **Example 1(a),** 1 is mapped to 2, −2 is mapped to 4, and 3 is mapped to −1. Thus, F is a function, since each first component is paired with exactly one second component.
  In the mapping for relation H from **Example 1(c),** which is not a function, the first component −2 is paired with two different second components.

- **As a table**

- **As a graph**
  Figure 50 includes a table and graph for relation F, which is a function, from **Example 1(a).**

x	y
1	2
−2	4
3	−1

Table for relation F

Graph of relation F

Figure 50

---

2 Determine whether each relation defines a function.

GS (a) $\{(0, 3), (-1, 2), (-1, 3)\}$

The x-value ____ is paired with both ____ and ____. This relation (*is/is not*) a function.

(b) $\{(2, -2), (4, -4), (6, -6)\}$

(c) $\{(-1, 5), (0, 5)\}$

(d) $\{(1, 5), (2, 3), (1, 7), (-2, 3)\}$

Relation F

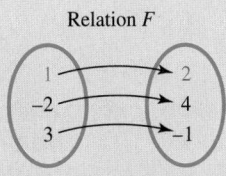

F is a function.

Relation H

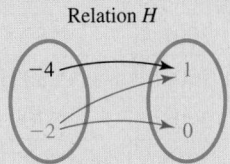

H is not a function.

Figure 49

Answers

2. (a) −1; 2; 3; is not **(b)** function
(c) function **(d)** not a function

Function machine

Another way to think of a function relationship is as an input-output (function) machine.

3 Give the domain and range of each relation. Tell whether the relation defines a function.

(a) $\{(4, 0), (4, 1), (4, 2)\}$

(b)

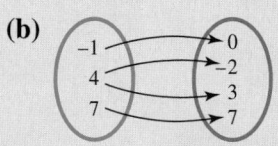

(c)

Year	Cell Phone Subscribers (in millions)
2002	141
2004	182
2006	233
2008	270
2010	303

Source: CTIA-The Wireless Association.

Answers

3. **(a)** domain: $\{4\}$; range: $\{0, 1, 2\}$;
   The relation does not define a function.
   (b) domain: $\{-1, 4, 7\}$; range: $\{0, -2, 3, 7\}$;
   The relation does not define a function.
   (c) domain: $\{2002, 2004, 2006, 2008, 2010\}$;
   range: $\{141, 182, 233, 270, 303\}$;
   The relation defines a function.

• **As an equation (or rule)**
An equation (or rule) tells how to determine the dependent variable for a specific value of the independent variable. For example, if the value of y is twice the value of any real number x, the equation is

$$\text{Dependent variable} \rightarrow y = 2x. \leftarrow \text{Independent variable}$$

The solutions of this equation define an infinite set of ordered pairs that can be represented by the graph in **Figure 51**.

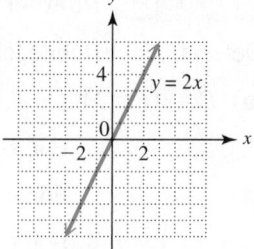

Graph of the relation defined by $y = 2x$

Figure 51

In a function, there is exactly one value of the dependent variable, the second component, for each value of the independent variable, the first component.

OBJECTIVE ▶ **3** **Find domain and range.**

Domain and Range

For every relation consisting of ordered pairs (x, y), there are two important sets of elements.

The set of all values of the independent variable (x) is the **domain.**

The set of all values of the dependent variable (y) is the **range.**

EXAMPLE 2 **Finding Domains and Ranges of Relations**

Give the domain and range of each relation. Tell whether the relation defines a function.

(a) $\{(3, -1), (4, 2), (4, 5), (6, 8)\}$
The domain, the set of x-values, is $\{3, 4, 6\}$. The range, the set of y-values, is $\{-1, 2, 5, 8\}$. This relation is not a function because the same x-value 4 is paired with two different y-values, 2 and 5.

(b)

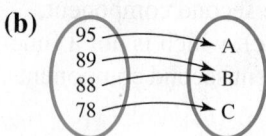

(c)

x	y
-5	2
0	2
5	2

The domain of the relation represented by this mapping is

$$\{95, 89, 88, 78\},$$

and the range is

$$\{A, B, C\}.$$

The mapping defines a function—each domain value corresponds to exactly one range value.

In this table, the domain is the set of x-values,

$$\{-5, 0, 5\}.$$

The range is the set of y-values,

$$\{2\}.$$

The table defines a function—each distinct x-value corresponds to exactly one y-value (even though it is the same y-value).

◀ **Work Problem 3** at the Side.

A graph gives a "picture" of a relation and can be used to determine its domain and range.

EXAMPLE 3 **Finding Domains and Ranges from Graphs**

Give the domain and range of each relation.

(a)

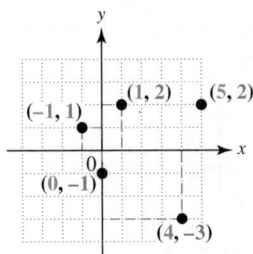

This relation includes the five ordered pairs that are graphed. The domain is the set of *x*-values.

$$\{-1, 0, 1, 4, 5\}$$

The range is the set of *y*-values.

$$\{-3, -1, 1, 2\}$$ Only list 2 once.

(b)

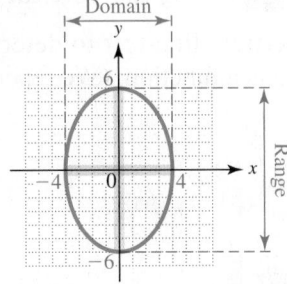

The *x*-values of the points on the graph include all numbers between −4 and 4, inclusive. The *y*-values include all numbers between −6 and 6, inclusive.

The domain is $[-4, 4]$. Use interval notation.
The range is $[-6, 6]$.

(c)

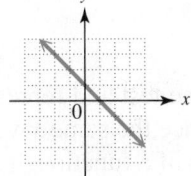

The arrowheads indicate that the line extends indefinitely left and right, as well as up and down. Therefore, both the domain and the range include all real numbers, written $(-\infty, \infty)$.

(d)

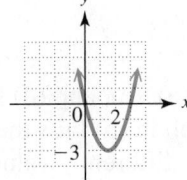

The graph extends indefinitely left and right, as well as upward. The domain is $(-\infty, \infty)$. Because there is a least *y*-value, −3, the range includes all numbers greater than or equal to −3, written $[-3, \infty)$.

················· **Work Problem** ❹ **at the Side.** ▶

OBJECTIVE ❹ **Identify functions defined by graphs and equations.**
Since each value of *x* in a function corresponds to only one value of *y*, any vertical line drawn through the graph of a function must intersect the graph in at most one point. This is the *vertical line test* for a function. **Figure 52** illustrates this test with the graphs of two relations.

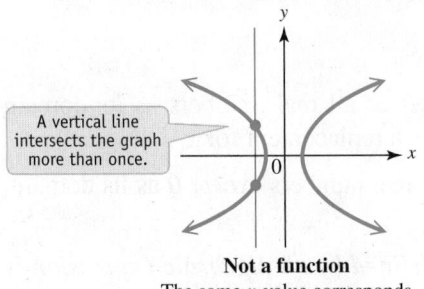

A vertical line intersects the graph more than once.

Not a function
The same *x*-value corresponds to two different *y*-values.

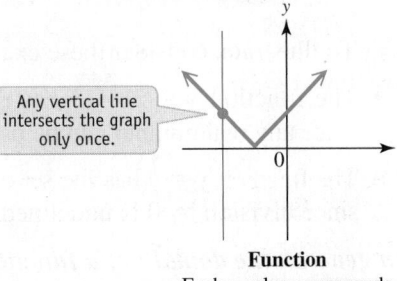

Any vertical line intersects the graph only once.

Function
Each *x*-value corresponds to only one *y*-value.

Figure 52

❹ Give the domain and range of each relation.

(a)

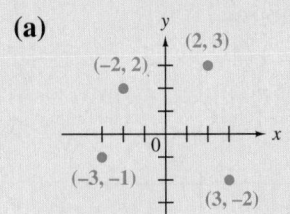

(b)

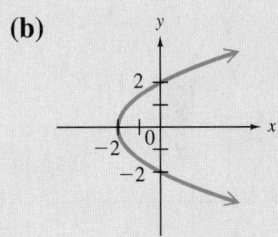

(c)

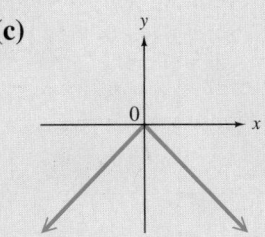

5 Use the vertical line test to decide which graphs represent functions.

A.

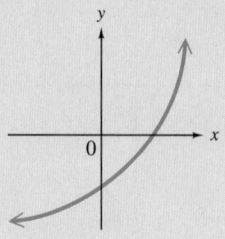

B.

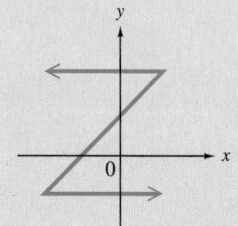

C.

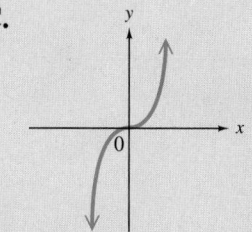

Vertical Line Test

If every vertical line intersects the graph of a relation in no more than one point, then the relation represents a function.

EXAMPLE 4 Using the Vertical Line Test

Use the vertical line test to determine whether each relation graphed in **Example 3** is a function. (We repeat the graphs here.)

(a)

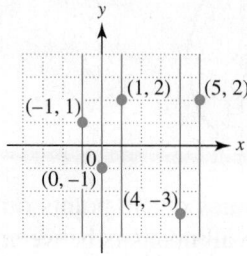

Function

(b)

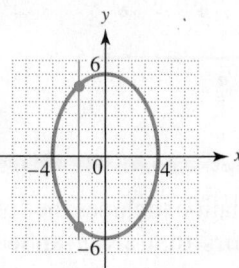

Not a function

(c)

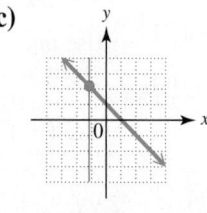

Function

(d)

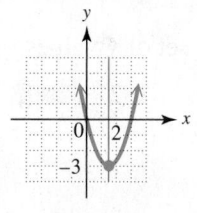

Function

The graphs in (a), (c), and (d) satisfy the vertical line test and represent functions. The graph in (b) fails the vertical line test, since the same x-value corresponds to two different y-values. It is not the graph of a function.

◀ **Work Problem 5** at the Side.

Note

Graphs that do not represent functions are still relations. *All equations and graphs represent relations, and all relations have a domain and range.*

Relations are often defined by equations, such as $y = 2x + 3$ and $y^2 = x$. We must sometimes determine the domain of a relation from its equation.

Agreement on Domain

The domain of a relation is assumed to be the set of all real numbers that produce real numbers when substituted for the independent variable.

To illustrate, consider these examples.

• The function $y = 2x + 3$ has the set of all real numbers as its domain since any real number can be used as a replacement for x.

• The function $y = \frac{1}{x}$ has the set of all real numbers *except 0* as its domain, since division by 0 is undefined.

In general, the domain of a function defined by an algebraic expression is all real numbers, except those numbers that lead to division by 0 or, as we will see in Chapter 9, an even root of a negative number.

Answer

5. A and C are graphs of functions.

EXAMPLE 5	Identifying Functions from Their Equations

Decide whether each relation defines y as a function of x. Give the domain.

(a) $y = x + 4$

In this equation, y is found by adding 4 to x. Thus, each value of x corresponds to just one value of y and the relation defines a function. Since x can be any real number, the domain is $(-\infty, \infty)$.

(b) $y^2 = x$

The ordered pairs $(16, 4)$ and $(16, -4)$ both satisfy this equation. Since one value of x, 16, corresponds to two values of y, 4 and -4, this equation does not define a function. Because x is equal to the square of y, the values of x must always be nonnegative. The domain is $[0, \infty)$.

(c) $y \leq x - 1$

By definition, y is a function of x if every value of x leads to exactly one value of y. Here a particular value of x, say 1, corresponds to many values of y. The ordered pairs $(1, 0)$, $(1, -1)$, $(1, -2)$, $(1, -3)$, and so on, all satisfy the inequality. Thus, this relation does not define a function. Any number can be used for x, so the domain is the set of real numbers, $(-\infty, \infty)$.

(d) $y = \dfrac{5}{x - 1}$

Given any value of x in the domain, we find y by subtracting 1, and then dividing the result into 5. This process produces exactly one value of y for each value in the domain, so this equation defines a function. The domain includes all real numbers *except* those that make the denominator 0.

$$x - 1 = 0 \qquad \text{Set the denominator equal to 0.}$$

$$x = 1 \qquad \text{Add 1.}$$

The domain includes all real numbers *except* 1, written

$$(-\infty, 1) \cup (1, \infty).^*$$

················· **Work Problem 6 at the Side.** ▶

In summary, we give three variations of the definition of function.

Variations of the Definition of Function

1. A **function** is a relation in which, for each distinct value of the first component of the ordered pairs, there is exactly one value of the second component.

2. A **function** is a set of ordered pairs in which no first component is repeated.

3. A **function** is an equation (rule) or correspondence (mapping) that assigns exactly one range value to each distinct domain value.

---

6 Decide whether each relation defines y as a function of x. Give the domain.

(a) $y = 6x + 12$

(b) $y \leq 4x$

(c) $y^2 = 25x$

(d) $y = \dfrac{1}{x + 2}$

Answers

6. (a) yes; $(-\infty, \infty)$ **(b)** no; $(-\infty, \infty)$
(c) no; $[0, \infty)$
(d) yes; $(-\infty, -2) \cup (-2, \infty)$

---

*The union of two sets A and B, written $A \cup B$, is the set of all elements of A together with the set of all elements of B.

3.6 Exercises

CONCEPT CHECK *Complete each statement. Choices may be used more than once.*

function	independent variable	vertical line test	relation
domain	ordered pairs	dependent variable	range

1. A _____ is any set of _____ $\{(x, y)\}$.

2. A _____ is a relation in which, for each distinct value of the first component of the _____, there is exactly one value of the second component.

3. In a relation $\{(x, y)\}$, the _____ is the set of x-values, and the _____ is the set of y-values.

4. The relation $\{(0, -2), (2, -1), (2, -4), (5, 3)\}$ (*does / does not*) define a function. The set $\{0, 2, 5\}$ is its _____, and the set $\{-2, -1, -4, 3\}$ is its _____ .

5. Consider the function $d = 50t$, where d represents distance and t represents time. The value of d depends on the value of t, so the variable t is the _____, and the variable d is the _____ .

6. The _____ is used to determine whether a graph is that of a function. It says that any vertical line can intersect the graph of a _____ in no more than (*zero / one / two*) point(s).

CONCEPT CHECK *Express each relation using a different form. (For example, if the given form is a set of ordered pairs, give a graph.) There is more than one correct way to do this.* **See Objective 2.**

7. $\{(0, 2), (2, 4), (4, 0)\}$

8.

x	y
-1	-3
0	-1
1	1
3	3

9.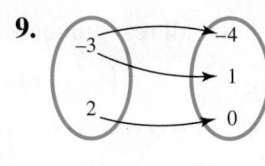

10. Does the relation given in **Exercise 9** define a function? Why or why not?

Decide whether each relation is a function, and give the domain and the range. Use the vertical line test in Exercises 25–36. **See Examples 1–4.**

11. $\{(5, 1), (3, 2), (4, 9), (7, 3)\}$

12. $\{(8, 0), (5, 4), (9, 3), (3, 9)\}$

13. $\{(2, 4), (0, 2), (2, 6)\}$

14. $\{(9, -2), (-3, 5), (9, 1)\}$

15. $\{(-3, 1), (4, 1), (-2, 7)\}$

16. $\{(-12, 5), (-10, 3), (8, 3)\}$

17. $\{(1, 1), (1, -1), (0, 0), (2, 4), (2, -4)\}$

18. $\{(2, 5), (3, 7), (4, 9), (5, 11)\}$

19.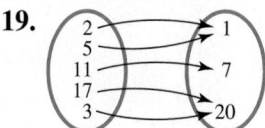

20.

21.

x	y
1	5
1	2
1	−1
1	−4

22.

x	y
−4	−4
−4	0
−4	4
−4	8

23.

x	y
4	−3
2	−3
0	−3
−2	−3

24.

x	y
−3	−6
−1	−6
1	−6
3	−6

25.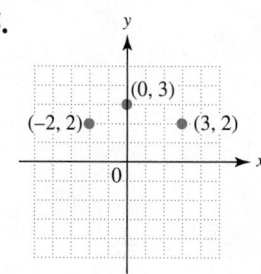

26.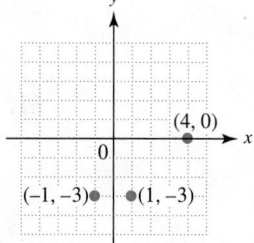

27.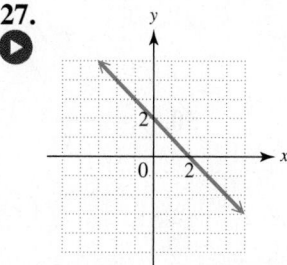

28.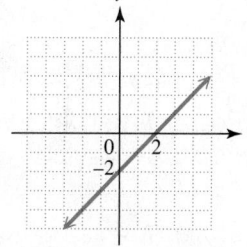

29.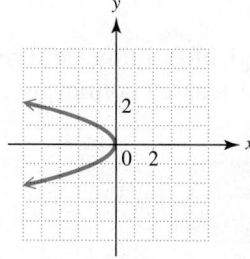

30.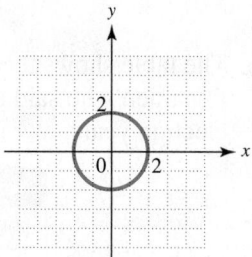

31.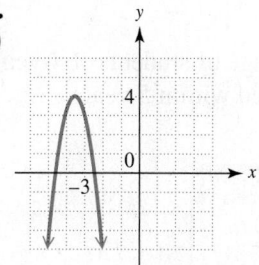

32.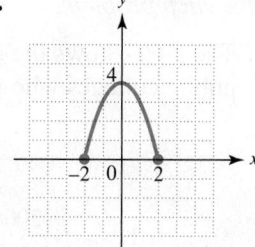

33.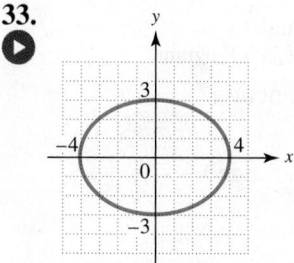

34.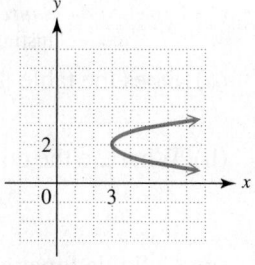

35.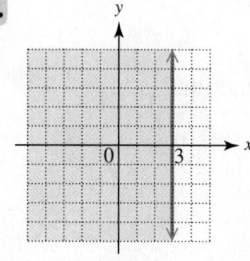

36.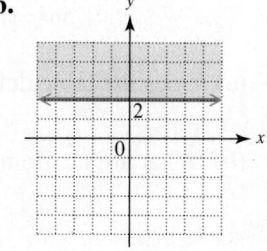

Decide whether each relation defines y as a function of x. Give the domain.
See Example 5.

37. $y = -6x$

38. $y = -9x$

39. $y = 2x - 6$
▶

40. $y = 6x + 8$

41. $y = x^2$

42. $y = x^3$

43. $x = y^6$

44. $x = y^4$

45. $x + y < 4$

46. $x - y < 3$

47. $y = \dfrac{x + 4}{5}$

48. $y = \dfrac{x - 3}{2}$

49. $y = -\dfrac{2}{x}$

50. $y = -\dfrac{6}{x}$

51. $y = \dfrac{2}{x - 4}$

52. $y = \dfrac{7}{x - 2}$

53. $xy = 1$

54. $xy = 3$

Solve each problem.

55. The table shows the percentage of students at 4-year public colleges who graduated within 5 years.

Year	Percentage
2006	39.6
2007	40.5
2008	40.3
2009	43.0
2010	39.6

Source: ACT.

(a) Does the table define a function?

(b) What are the domain and range?

(c) Call this function *f*. Give two ordered pairs that belong to *f*.

56. The table shows the percentage of full-time college freshmen who said they had discussed politics in election years.

Year	Percentage
1992	83.7
1996	73.0
2000	69.6
2004	77.4
2008	85.9

Source: Cooperative Institutional Research Program.

(a) Does the table define a function?

(b) What are the domain and range?

(c) Call this function *g*. Give two ordered pairs that belong to *g*.

3.7 Function Notation and Linear Functions

OBJECTIVE ▶ 1 Use function notation. When a function f is defined with a rule or an equation using x and y for the independent and dependent variables, we say "y *is a function of* x" to emphasize that y *depends on* x. We use the notation $y = f(x)$, called **function notation,** to express this and read $f(x)$ as "f *of* x," or "f *at* x." The letter f is a name for this particular function.

For example, if $y = 9x - 5$, we can name this function f and write

$$f(x) = 9x - 5.$$

f is the name of the function.
x is a value from the domain.
$f(x)$ is the function value (or y-value) that corresponds to x.

$f(x)$ *is just another name for the dependent variable* y.

We evaluate a function at different values of x by substituting x-values from the domain into the function.

EXAMPLE 1   Evaluating a Function

Let $f(x) = 9x - 5$. Evaluate the function f for each of the following.

(a) $x = 2$

$f(x) = 9x - 5$	Given function
$f(2) = 9 \cdot 2 - 5$	Replace x with 2.
$f(2) = 18 - 5$	Multiply.
$f(2) = 13$	Subtract.

> Read $f(2)$ as "f of 2" or "f at 2".

Thus, for $x = 2$, the corresponding function value (or y-value) is **13**. $f(2) = 13$ is an abbreviation for the statement "If $x = 2$ in the function f, then $y = 13$" and is represented by the ordered pair $(2, 13)$.

(b) $x = -3$

$f(x) = 9x - 5$	
$f(-3) = 9(-3) - 5$	Replace x with -3.
$f(-3) = -27 - 5$	Multiply.
$f(-3) = -32$	Subtract.

> Use parentheses to avoid errors.

Thus, $f(-3) = -32$ and the ordered pair $(-3, -32)$ belongs to f.

·········· **Work Problem 1 at the Side.** ▶

CAUTION

The symbol $f(x)$ *does not* indicate "f times x," but represents the y-value associated with the indicated x-value. As shown in **Example 1(a)**, $f(2)$ is the y-value that corresponds to the x-value 2 in f.

These ideas can be illustrated as follows.

Name of the function
Defining expression

$$y = f(x) = 9x - 5$$

Value of the function   Name of the independent variable

OBJECTIVES

1 Use function notation.

2 Graph linear and constant functions.

1 Let $f(x) = 6x - 2$. Evaluate the function f for each of the following.

(a) $x = -2$

$$f(x) = 6x - 2$$
$$f(\underline{\quad}) = 6(\underline{\quad}) - 2$$
$$f(-2) = \underline{\quad} - 2$$
$$f(-2) = \underline{\quad}$$

Thus, $f(-2) = \underline{\quad}$

and the ordered pair

$(\underline{\quad}, \underline{\quad})$ belongs to f.

(b) $x = 0$

(c) $x = 11$

Answers

1. (a) -2; -2; -12; -14; -14; -2; -14
   (b) -2 **(c)** 64

② Let $f(x) = -x^2 - 4x + 1$. Find the following.

(a) $f(-2)$

(b) $f(a)$

③ Find the following.

(a) Let $g(x) = 5x - 1$. Find and simplify $g(m + 2)$.

$$g(x) = 5x - 1$$
$$g(m + 2) = 5(\underline{\hspace{1cm}}) - 1$$
$$g(m + 2) = \underline{\hspace{0.7cm}} + \underline{\hspace{0.7cm}} - 1$$
$$g(m + 2) = \underline{\hspace{1cm}}$$

(b) Let $f(x) = 8x - 5$. Find and simplify $f(a - 2)$.

④ For each function, find $f(-2)$.

(a) $f = \{(0, 5), (-1, 3), (-2, 1)\}$

(b)

x	$f(x)$
-4	16
-2	4
0	0
2	4

EXAMPLE 2   **Using Function Notation**

Let $f(x) = -x^2 + 5x - 3$. Find the following.

(a) $f(4)$

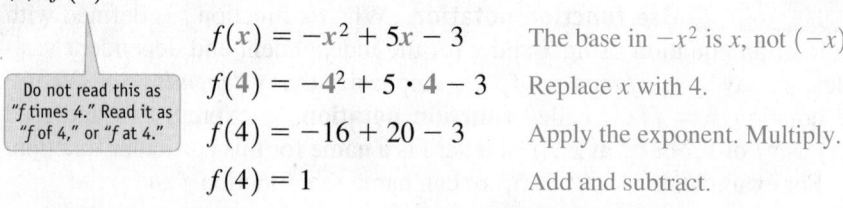

	$f(x) = -x^2 + 5x - 3$	The base in $-x^2$ is x, not $(-x)$.
Do not read this as "f times 4." Read it as "f of 4," or "f at 4."	$f(4) = -4^2 + 5 \cdot 4 - 3$	Replace x with 4.
	$f(4) = -16 + 20 - 3$	Apply the exponent. Multiply.
	$f(4) = 1$	Add and subtract.

Since $f(4) = 1$, the ordered pair $(4, 1)$ belongs to f.

(b) $f(q)$

$$f(x) = -x^2 + 5x - 3$$
$$f(q) = -q^2 + 5q - 3 \quad \text{Replace } x \text{ with } q.$$

The replacement of one variable with another is important in later courses.

◀ **Work Problem ②** at the Side.

Sometimes letters other than f, such as g, h, or capital letters F, G, and H, are used to name functions.

EXAMPLE 3   **Using Function Notation**

Let $g(x) = 2x + 3$. Find and simplify $g(a + 1)$.

$$g(x) = 2x + 3$$
$$g(a + 1) = 2(a + 1) + 3 \quad \text{Replace } x \text{ with } a + 1.$$
$$g(a + 1) = 2a + 2 + 3 \quad \text{Distributive property}$$
$$g(a + 1) = 2a + 5 \quad \text{Add.}$$

◀ **Work Problem ③** at the Side.

Functions can be evaluated in a variety of ways, as shown in **Example 4.**

EXAMPLE 4   **Evaluating Functions**

For each function, find $f(3)$.

(a) $f(x) = 3x - 7$

$$f(3) = 3(3) - 7 \quad \text{Replace } x \text{ with 3.}$$
$$f(3) = 9 - 7 \quad \text{Multiply.}$$
$$f(3) = 2 \quad \text{Subtract.}$$

(b)

x	$y = f(x)$
6	-12
3	-6 ←$f(3) = -6$
0	0
-3	6

(c) $f = \{(-3, 5), (0, 3), (3, 1), (6, -1)\}$

We want $f(3)$, the y-value of the ordered pair whose first component is 3. As indicated by the ordered pair $(3, 1)$, for $x = 3$, $y = 1$. Thus, $f(3) = 1$.

(d)

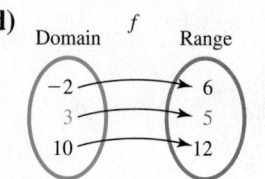

Domain f Range

The domain element 3 is paired with 5 in the range, so $f(3) = 5$.

◀ **Work Problem ④** at the Side.

2. (a) 5   **(b)** $-a^2 - 4a + 1$
3. (a) $m + 2$; $5m$; 10; $5m + 9$ **(b)** $8a - 21$
4. (a) 1   **(b)** 4

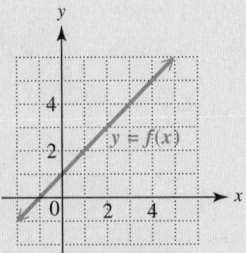

Figure 53

> **EXAMPLE 5** **Finding Function Values from a Graph**

Refer to the function f graphed in **Figure 53** at the right.

(a) Find $f(3)$.

Locate 3 on the x-axis. See **Figure 54**. Moving up to the graph of f and over to the y-axis gives 4 for the corresponding y-value. Thus, $f(3) = 4$, which corresponds to the ordered pair $(3, 4)$.

(b) Find $f(0)$.

Refer to **Figure 54** to see that $f(0) = 1$.

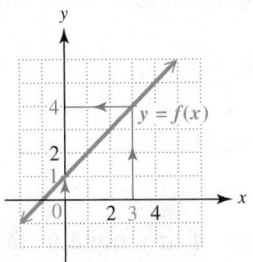

Figure 54

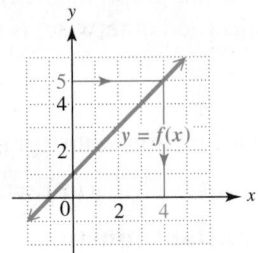

Figure 55

(c) For what value of x is $f(x) = 5$?

Since $f(x) = y$, we want the value of x that corresponds to $y = 5$. Locate 5 on the y-axis. See **Figure 55**. Moving across to the graph of f and down to the x-axis gives $x = 4$. Thus, $f(4) = 5$, which corresponds to the ordered pair $(4, 5)$.

························· **Work Problem ❺ at the Side.** ▶

⑤ Refer to the function graphed above in **Figure 53**.

(a) Find $f(2)$.

(b) Find $f(-1)$.

(c) For what value of x is $f(x) = 2$?

If a function f is defined by an equation with x and y instead of with function notation, use the following steps to find $f(x)$.

Writing an Equation Using Function Notation

Step 1 Solve the equation for y if it is not given in that form.

Step 2 Replace y with $f(x)$.

---

⑥ Write each equation defining a function f using function notation $f(x)$. Then find $f(-1)$.

(a) $3x + y = 6$

> **EXAMPLE 6** **Writing Equations Using Function Notation**

Write each equation using function notation $f(x)$. Then find $f(-2)$.

(a) $y = x^2 + 1$ ◄—— This equation is already solved for y.

$f(x) = x^2 + 1$ Replace y with $f(x)$. (Step 2)

Now find $f(-2)$. $f(-2) = (-2)^2 + 1$ Let $x = -2$.

$f(-2) = 4 + 1$ $(-2)^2 = -2(-2)$

$f(-2) = 5$ Add.

(b) $x - 4y = 5$

Step 1 $-4y = -x + 5$ Subtract x.

$y = \dfrac{1}{4}x - \dfrac{5}{4}$ Divide by -4.

Step 2 $f(x) = \dfrac{1}{4}x - \dfrac{5}{4}$,   so $f(-2) = \dfrac{1}{4}(-2) - \dfrac{5}{4} = -\dfrac{7}{4}$

(b) $2x - 5y = 4$

(c) $x^2 - 4y = 3$

Answers

5. **(a)** 3 **(b)** 0 **(c)** 1
6. **(a)** $f(x) = -3x + 6$; 9

(b) $f(x) = \dfrac{2}{5}x - \dfrac{4}{5}$; $-\dfrac{6}{5}$

(c) $f(x) = \dfrac{1}{4}x^2 - \dfrac{3}{4}$; $-\dfrac{1}{2}$

···························· **Work Problem ❻ at the Side.** ▶

❼ Graph each function. Give the domain and range.

(a) $f(x) = \dfrac{3}{4}x - 2$

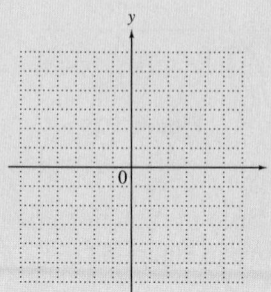

(b) $g(x) = 3$

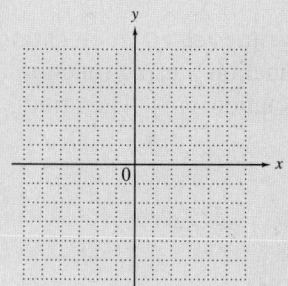

Answers

7. (a)

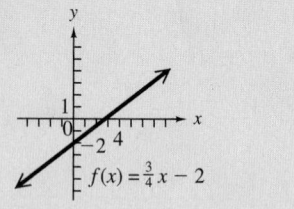

domain: $(-\infty, \infty)$; range: $(-\infty, \infty)$

(b)

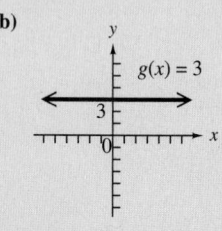

domain: $(-\infty, \infty)$; range: $\{3\}$

OBJECTIVE **2** **Graph linear and constant functions.** Linear equations (except for vertical lines with equations $x = a$) define *linear functions.*

> **Linear Function**
>
> A function f that can be written in the form
>
> $$f(x) = ax + b,$$
>
> for real numbers a and b, is a **linear function.** The value of a is the slope m of the graph of the function. The domain of a linear function, unless specified otherwise, is $(-\infty, \infty)$.

A linear function whose graph is a horizontal line has the form

$$f(x) = b \qquad \text{Constant function}$$

and is a **constant function.** While the range of any nonconstant linear function is $(-\infty, \infty)$, the range of a constant function $f(x) = b$ is $\{b\}$.

EXAMPLE 7 **Graphing Linear and Constant Functions**

Graph each function. Give the domain and range.

(a) $f(x) = \dfrac{1}{4}x - \dfrac{5}{4}$ ⟵ From **Example 6(b)**

Slope ⟶ ⟵ y-intercept is $\left(0, -\frac{5}{4}\right)$.

To graph this function, plot the y-intercept $\left(0, -\frac{5}{4}\right)$. Use the geometric definition of slope as $\frac{\text{rise}}{\text{run}}$ to find a second point on the line. Since the slope is $\frac{1}{4}$, move 1 unit up from $\left(0, -\frac{5}{4}\right)$ and 4 units to the right to the point $\left(4, -\frac{1}{4}\right)$. Draw the straight line through these points to obtain the graph shown in **Figure 56.** The domain and range are both $(-\infty, \infty)$.

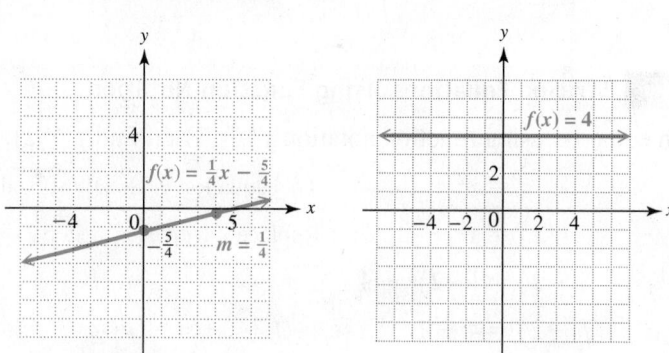

Figure 56   **Figure 57**

(b) $f(x) = 4$

The graph of this constant function is the horizontal line containing all points with y-coordinate 4. See **Figure 57.** The domain is $(-\infty, \infty)$ and the range is $\{4\}$.

◀ **Work Problem ❼ at the Side.**

3.7 Exercises

FOR EXTRA HELP

  Download the MyDashBoard App

▶ MyMathLab®

1. CONCEPT CHECK To emphasize that "y is a function of x" for a given function f, we use function notation and write $y =$ _____ . Here, f is the name of the _____ , x is a value from the _____ , and $f(x)$ is the function value (or y-value) that corresponds to _____ . We read $f(x)$ as "_____ ."

2. CONCEPT CHECK Choose the correct response: For a function f, the notation $f(3)$ means _____ .

  A. the variable f times 3 or $3f$

  B. the value of the dependent variable when the independent variable is 3

  C. the value of the independent variable when the dependent variable is 3

  D. f equals 3

Let $f(x) = -3x + 4$ and $g(x) = -x^2 + 4x + 1$. Find the following.
See Examples 1–3.

3. $f(0)$ **4.** $f(-3)$ **5.** $g(-2)$ **6.** $g(0)$ **7.** $g(10)$

8. $f(10)$ **9.** $f\left(\dfrac{1}{3}\right)$ **10.** $f\left(\dfrac{7}{3}\right)$ **11.** $g(0.5)$ **12.** $g(1.5)$

13. $f(p)$ **14.** $g(k)$ **15.** $f(-x)$ **16.** $g(-x)$ **17.** $f(x + 2)$

18. $f(x - 2)$ **19.** $f(2t + 1)$ **20.** $f(3t - 2)$ **21.** $g\left(\dfrac{p}{3}\right)$ **22.** $g\left(\dfrac{1}{x}\right)$

*For each function, find **(a)** $f(2)$ and **(b)** $f(-1)$. See Examples 4 and 5.*

23. $f = \{(-2, 2), (-1, -1), (2, -1)\}$ **24.** $f = \{(-1, -5), (0, 5), (2, -5)\}$

25. $f = \{(-1, 3), (4, 7), (0, 6), (2, 2)\}$ **26.** $f = \{(2, 5), (3, 9), (-1, 11), (5, 3)\}$

27.

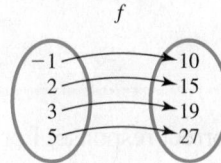

28.

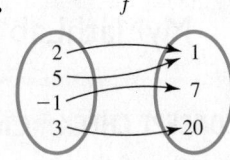

29.

x	$y = f(x)$
2	4
1	1
0	0
−1	1
−2	4

30.

x	$y = f(x)$
8	6
5	3
2	0
−1	−3
−4	−6

31.

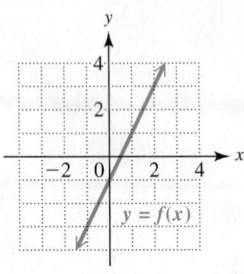

32.

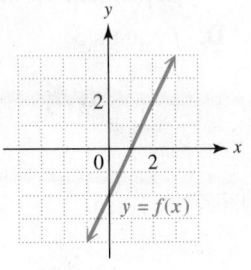

33.

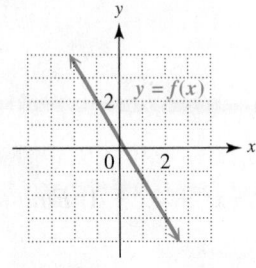

34.

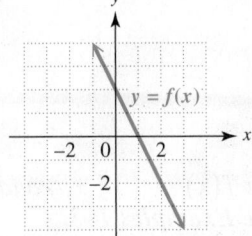

35. Refer to **Exercise 31.** Find the value of x for each value of $f(x)$. **See Example 5(c).**

(a) $f(x) = 3$

(b) $f(x) = -1$

(c) $f(x) = -3$

36. Refer to **Exercise 32.** Find the value of x for each value of $f(x)$. **See Example 5(c).**

(a) $f(x) = 4$

(b) $f(x) = -2$

(c) $f(x) = 0$

An equation that defines y as a function f of x is given. (a) Solve for y in terms of x, and replace y with the function notation $f(x)$. (b) Find $f(3)$. See Example 6.

37. $x + 3y = 12$

38. $x - 4y = 8$

39. $y + 2x^2 = 3$

40. $y - 3x^2 = 2$

41. $4x - 3y = 8$

42. $-2x + 5y = 9$

CONCEPT CHECK *Fill in each blank with the correct response.*

43. The equation $2x + y = 4$ has a straight _____ as its graph. One point that lies on the graph is $(3, \_\_\_\_)$. If we solve the equation for y and use function notation, we have a _____ function $f(x) = _____$. For this function, $f(3) = \_\_\_\_$, meaning that the point $(\_\_\_\_, \_\_\_\_)$ lies on the graph of the function.

44. A linear function f can be written in the form

$f(x) = _____$, for real numbers a and b.

The graph of a linear function is a _____. The value of a is the _____ of the line and $(0, b)$ is the _____. The domain of a linear function, unless specified otherwise, is _____.

Graph each linear or constant function. Give the domain and range. ***See Example 7.***

45. $f(x) = -2x + 5$

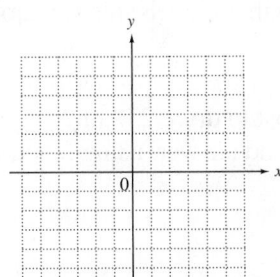

46. $g(x) = 4x - 1$

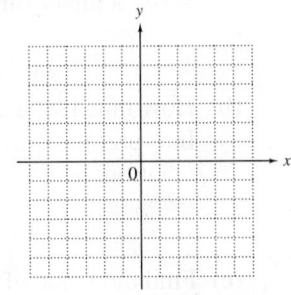

47. $h(x) = \dfrac{1}{2}x + 2$

48. $F(x) = -\dfrac{1}{4}x + 1$

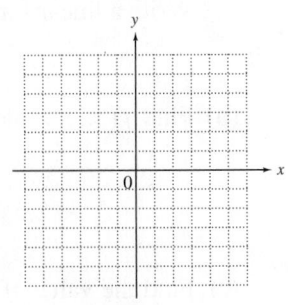

49. $g(x) = -4$

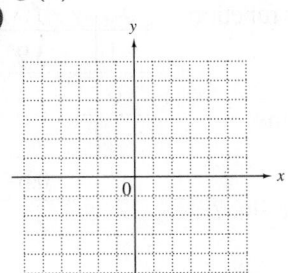

50. $f(x) = 5$

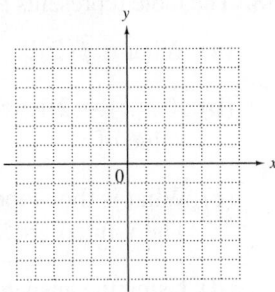

51. $f(x) = 0$

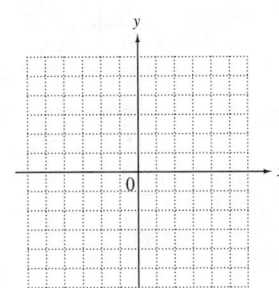

52. $f(x) = 2.5$

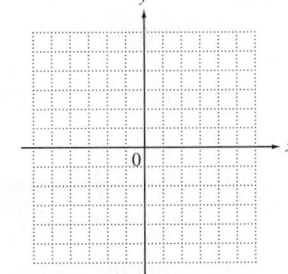

Solve each problem.

53. A taxicab driver charges $2.50 per mi.

(a) Fill in the table with the correct response for the price $f(x)$ he charges for a trip of x miles.

x	$f(x)$
0	
1	
2	
3	

(b) The linear function that gives a rule for the amount charged in dollars is $f(x) = $ _____.

(c) Graph this function for the domain $\{0, 1, 2, 3\}$.

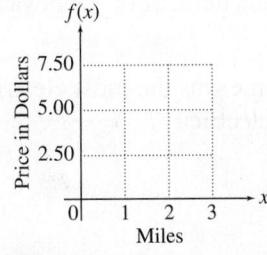

54. A package weighing x pounds costs $f(x)$ dollars to ship to a given location, where $f(x) = 3.75x$.

(a) What is the value of $f(3)$?

(b) Describe what 3 and the value $f(3)$ mean in part (a), using the terms *independent variable* and *dependent variable*.

(c) How much would it cost to mail a 5-lb package? Then write this question and its answer using function notation.

55. To print t-shirts, there is a $100 set-up fee, plus a $12 charge per t-shirt. Let x represent the number of t-shirts printed and $f(x)$ represent the total charge.

   (a) Write a linear function that models this situation.

   (b) Find $f(125)$. Interpret your answer.

   (c) Find the value of x if $f(x) = 1000$. Express this situation using function notation, and interpret it.

56. Rental on a car is $150, plus $0.20 per mile. Let x represent the number of miles the car is driven and $f(x)$ represent the total cost to rent the car.

   (a) Write a linear function that models this situation.

   (b) How much would it cost to drive 250 mi? Interpret the question and answer, using function notation.

   (c) Find the value of x if $f(x) = 230$. Interpret your answer.

57. The table represents a linear function.

   (a) What is $f(2)$?

   (b) If $f(x) = -1.3$, what is the value of x?

   (c) What is the slope of the line? The y-intercept?

   (d) Using the answers from part (c), write an equation for $f(x)$.

x	$y = f(x)$
0	3.5
1	2.3
2	1.1
3	−0.1
4	−1.3

58. The table represents a linear function.

   (a) What is $f(2)$?

   (b) If $f(x) = 2.1$, what is the value of x?

   (c) What is the slope of the line? The y-intercept?

   (d) Using the answers from part (c), write an equation for $f(x)$.

x	$y = f(x)$
−1	−3.9
0	−2.4
1	−0.9
2	0.6
3	2.1

59. Refer to the graph to answer the questions.

Gallons of Water in a Pool at Time t

   (a) What numbers are possible values of the independent variable? Dependent variable?

   (b) For how long is the water level increasing? Decreasing?

   (c) How many gallons are in the pool after 90 hr?

   (d) Call this function g. What is $g(0)$? What does it mean in this example?

60. The graph shows electricity use on a summer day.

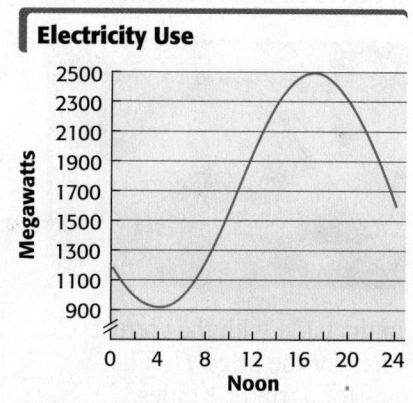

Electricity Use

   (a) Is this the graph of a function?

   (b) What is the domain?

   (c) Estimate the number of megawatts used at 8 A.M.

   (d) At what time was the most electricity used? The least electricity?

Chapter 3 Summary

Key Terms

3.1

bar graph A bar graph is a series of bars used to show comparisons between two categories of data.

line graph A line graph consists of a series of points that are connected with line segments and is used to show changes or trends in data.

linear equation in two variables An equation that can be written in the form $Ax + By = C$ is a linear equation in two variables. (A and B are real numbers that cannot both be 0.)

ordered pair A pair of numbers written between parentheses in which order is important is an ordered pair.

table of values A table showing selected ordered pairs of numbers that satisfy an equation is a table of values.

x	y
0	4
2	0
1	2

Table of values for $2x + y = 4$

x-axis The horizontal axis in a coordinate system is the x-axis.

y-axis The vertical axis in a coordinate system is the y-axis.

rectangular (Cartesian) coordinate system An x-axis and y-axis at right angles form a coordinate system.

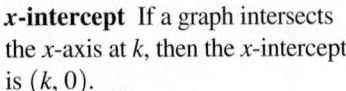

origin The point at which the x-axis and y-axis intersect is the origin.

quadrants A coordinate system divides the plane into four regions or quadrants.

plane A flat surface determined by two intersecting lines is a plane.

coordinates The numbers in an ordered pair are the coordinates of the corresponding point.

plot To plot an ordered pair is to find the corresponding point on a coordinate system.

scatter diagram A graph of ordered pairs of data is a scatter diagram.

3.2

graph The graph of an equation is the set of all points that correspond to the ordered pairs that satisfy the equation.

graphing The process of plotting the ordered pairs that satisfy a linear equation and drawing a line through them is called graphing.

y-intercept If a graph intersects the y-axis at k, then the y-intercept is $(0, k)$.

x-intercept If a graph intersects the x-axis at k, then the x-intercept is $(k, 0)$.

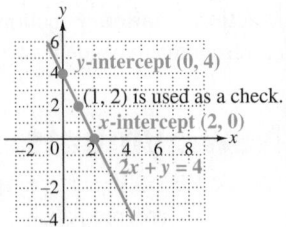

Graph of $2x + y = 4$

3.3

rise Rise is the vertical change between two different points on a line.

run Run is the horizontal change between two different points on a line.

slope The slope of a line is the ratio of the change in y compared to the change in x when moving along the line from one point to another.

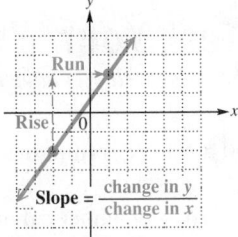

parallel lines Two lines in a plane that never intersect are parallel.

perpendicular lines Perpendicular lines intersect at a 90° angle.

3.5

linear inequality in two variables An inequality that can be written in the form $Ax + By < C$, $Ax + By > C$, $Ax + By \leq C$, or $Ax + By \geq C$ is a linear inequality in two variables.

boundary line In the graph of a linear inequality, the boundary line separates the region that satisfies the inequality from the region that does not satisfy the inequality.

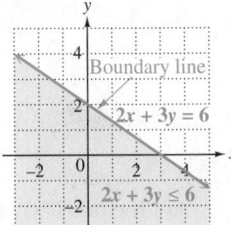

3.6

dependent variable If the quantity y depends on x, then y is the dependent variable in a relation between x and y.

independent variable If y depends on x, then x is the independent variable in a relation between x and y.

relation A relation is any set of ordered pairs.

function A function is a set of ordered pairs in which each distinct value of the first component, x, corresponds to exactly one value of the second component, y.

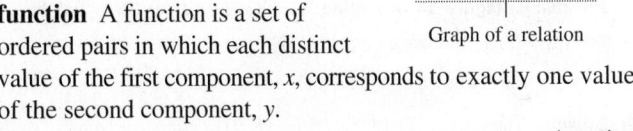

Graph of a relation

(continued)

domain The domain of a relation is the set of first components (x-values) of the ordered pairs of the relation.

range The range of a relation is the set of second components (y-values) of the ordered pairs of the relation.

linear function A function of the form $f(x) = ax + b$ is a linear function.

constant function A constant function is a linear function of the form $f(x) = b$, for a real number b.

3.7

function notation Function notation $f(x)$ is another way to represent the dependent variable y for the function f.

New Symbols

(x, y)	ordered pair		m	slope
x_1	a specific value of the variable x (read "x-sub-one")		$f(x)$	function f of x (read "f of x" or "f at x")

Test Your Word Power

See how well you have learned the vocabulary in this chapter.

1 An **ordered pair** is a pair of numbers written
   A. in numerical order between brackets
   B. between parentheses or brackets
   C. between parentheses in which order is important
   D. between parentheses in which order does not matter.

2 A **linear equation in two variables** is an equation that can be written in the form
   A. $Ax + By < C$
   B. $ax = b$
   C. $y = x^2$
   D. $Ax + By = C$.

3 An **intercept** is
   A. the point where the x-axis and y-axis intersect
   B. a pair of numbers written between parentheses in which order matters
   C. one of the four regions of a rectangular coordinate system

D. the point where a graph intersects the x-axis or the y-axis.

4 The **slope** of a line is
   A. the measure of the run over the rise of the line
   B. the distance between two points on the line
   C. the ratio of the change in y to the change in x along the line
   D. the horizontal change compared to the vertical change of two points on the line.

5 A **relation** is
   A. a set of ordered pairs
   B. the ratio of the change in y to the change in x along a line
   C. the set of all possible values of the independent variable
   D. all the second components of a set of ordered pairs.

6 A **function** is
   A. the numbers in an ordered pair

B. a set of ordered pairs in which each distinct x-value corresponds to exactly one y-value
   C. a pair of numbers written between parentheses in which order matters
   D. the set of all ordered pairs that satisfy an equation.

7 The **domain** of a function is
   A. the set of all possible values of the dependent variable y
   B. a set of ordered pairs
   C. the difference between the x-values
   D. the set of all possible values of the independent variable x.

8 The **range** of a function is
   A. the set of all possible values of the dependent variable y
   B. a set of ordered pairs
   C. the difference between the y-values
   D. the set of all possible values of the independent variable x.

Answers to Test Your Word Power

1. C; *Examples:* $(0, 3)$, $(3, 8)$, $(4, 0)$

2. D; *Examples:* $3x + 2y = 6$, $x = y - 7$, $4x = y$

3. D; *Example:* In **Figure 11** of **Section 3.2,** the x-intercept is $(2, 0)$ and the y-intercept is $(0, 4)$.

4. C; *Example:* The line through $(3, 6)$ and $(5, 4)$ has slope $\dfrac{4 - 6}{5 - 3} = \dfrac{-2}{2} = -1$.

5. A; *Example:* The set $\{(2, 0), (4, 3), (6, 6), (8, 9)\}$ defines a relation.

6. B; *Example:* The relation given in Answer 5 is a function since each distinct x-value corresponds to exactly one y-value.

7. D; *Example:* In the function in Answer 5, the domain is the set of x-values, $\{2, 4, 6, 8\}$.

8. A; *Example:* In the function in Answer 5, the range is the set of y-values, $\{0, 3, 6, 9\}$.

Quick Review

Concepts	Examples

3.1 Linear Equations in Two Variables; The Rectangular Coordinate System

An ordered pair is a solution of an equation if it makes the equation a true statement.

Is $(2, -5)$ or $(0, -6)$ a solution of $4x - 3y = 18$?

$$4(2) - 3(-5) \stackrel{?}{=} 18 \qquad\qquad 4(0) - 3(-6) \stackrel{?}{=} 18$$
$$8 + 15 \stackrel{?}{=} 18 \qquad\qquad\qquad 0 + 18 \stackrel{?}{=} 18$$
$$23 = 18 \quad \text{False} \qquad\qquad 18 = 18 \ \checkmark \ \text{True}$$

$(2, -5)$ is not a solution. $(0, -6)$ is a solution.

If a value of either variable in an equation is given, the value of the other variable can be found by substitution.

Complete the ordered pair $(0, \underline{\quad})$ for $3x = y + 4$.

$$3(0) = y + 4 \qquad \text{Let } x = 0.$$
$$0 = y + 4 \qquad \text{Multiply.}$$
$$-4 = y \qquad \text{Subtract 4.}$$

The ordered pair is $(0, -4)$.

3.2 Graphing Linear Equations in Two Variables

To graph a linear equation, follow these steps.

Step 1 Find at least two ordered pairs that are solutions of the equation. (The intercepts are good choices.)

Step 2 Plot the corresponding points.

Step 3 Draw a straight line through the points.

Graph $x - 2y = 4$.

x	y
0	-2
4	0
-2	-3

Intercepts

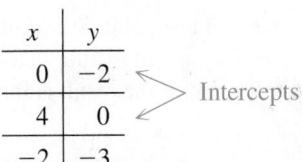

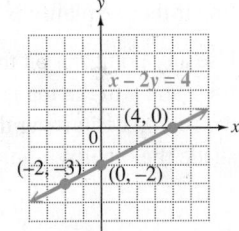

The graph of $Ax + By = 0$ passes through the origin. In this case, find and plot at least one other point that satisfies the equation. Then draw the line through these points.

The graph of $y = b$ is a horizontal line through $(0, b)$.

The graph of $x = a$ is a vertical line through $(a, 0)$.

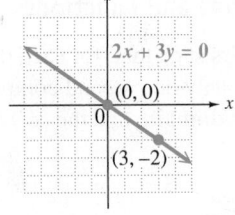

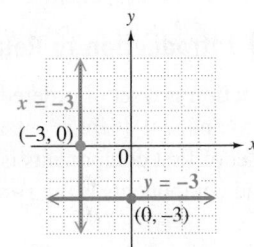

3.3 The Slope of a Line

The slope m of the line passing through the points (x_1, y_1) and (x_2, y_2) is defined as follows.

$$m = \frac{\text{change in } y}{\text{change in } x} = \frac{y_2 - y_1}{x_2 - x_1} \quad (\text{where } x_1 \neq x_2)$$

Horizontal lines have slope 0.

Vertical lines have undefined slope.

To find the slope of a line from its equation, solve for y. The slope is the coefficient of x.

The line through $(-2, 3)$ and $(4, -5)$ has slope

$$m = \frac{-5 - 3}{4 - (-2)} = \frac{-8}{6} = -\frac{4}{3}.$$

The line $y = -2$ has slope 0.

The line $x = 4$ has **undefined slope.**

Find the slope of the graph of $3x - 4y = 12$.

$$-4y = -3x + 12 \qquad \text{Add } -3x.$$
$$y = \frac{3}{4}x - 3 \qquad \text{Divide by } -4.$$

Slope ⟶

Parallel lines have the same slope.

The lines $y = 3x - 1$ and $y = 3x + 4$ are parallel because both have slope 3.

The slopes of perpendicular lines are negative reciprocals (that is, their product is -1).

The lines $y = -3x - 1$ and $y = \frac{1}{3}x + 4$ are perpendicular because their slopes are -3 and $\frac{1}{3}$, and $-3\left(\frac{1}{3}\right) = -1$.

Concepts	Examples

3.4 Writing and Graphing Equations of Lines

Slope-Intercept Form $y = mx + b$

$y = 2x + 3$ $m = 2$; y-intercept is $(0, 3)$.

Point-Slope Form $y - y_1 = m(x - x_1)$

$y - 3 = 4(x - 5)$ $(5, 3)$ is on the line; $m = 4$.

Standard Form $Ax + By = C$, where A, B, and C are real numbers, and A and B are not both 0. (We give A, B, and C integers, with $A \geq 0$.)

$2x - 5y = 8$ Standard form

Horizontal Line $y = b$

$y = 4$ Horizontal line

Vertical Line $x = a$

$x = -1$ Vertical line

3.5 Graphing Linear Inequalities in Two Variables

Graphing a Linear Inequality

Step 1 Graph the line that is the boundary of the region. Draw a solid line if the inequality is $\leq$ or $\geq$. Draw a dashed line if the inequality is $<$ or $>$.

Step 2 Use any point not on the line as a test point. Substitute for x and y in the inequality. If the result is true, shade the region containing the test point. If the result is false, shade the other region.

Graph $2x - 3y \leq 6$.
Draw the graph of $2x - 3y = 6$. Use a solid line because of the inclusion of equality in the symbol $\leq$.

Choose $(0, 0)$ as a test point.

$$2(0) - 3(0) \overset{?}{\leq} 6$$
$$0 < 6 \quad \text{True}$$

Shade the region that includes $(0, 0)$.

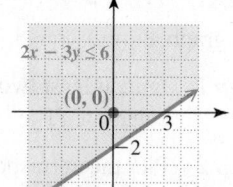

3.6 Introduction to Relations and Functions

A **function** is a set of ordered pairs such that, for each distinct first component, there is one and only one second component. The set of first components is the **domain,** and the set of second components is the **range.**

$f = \{(-1, 4), (0, 6), (1, 4)\}$ defines a function f, with domain the set of x-values $\{-1, 0, 1\}$ and range the set of y-values $\{4, 6\}$.

3.7 Function Notation and Linear Functions

To evaluate a function f using function notation $f(x)$ for a given value of x, substitute the value wherever x appears.

Let $f(x) = x^2 - 7x + 12$. Find $f(1)$.
$$f(1) = 1^2 - 7(1) + 12 \quad \text{Let } x = 1.$$
$$f(1) = 6$$

To write an equation that defines a function f in function notation, follow these steps.

Write $2x + 3y = 12$ in function notation for function f.

Step 1 Solve the equation for y if it is not given in that form.

$$3y = -2x + 12 \quad \text{Subtract } 2x.$$
$$y = -\frac{2}{3}x + 4 \quad \text{Divide by 3.}$$

Step 2 Replace y with $f(x)$.

$$f(x) = -\frac{2}{3}x + 4 \quad \text{Replace } y \text{ with } f(x).$$

Chapter 3 Review Exercises

3.1 *The percents of first-year college students at two-year public institutions who returned for a second year for the years 2006 through 2011 are shown in the graph.*

1. Write two ordered pairs of the form (year, percent) for the data shown in the graph for the years 2006 and 2011.

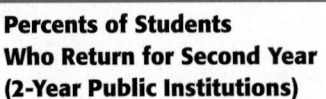
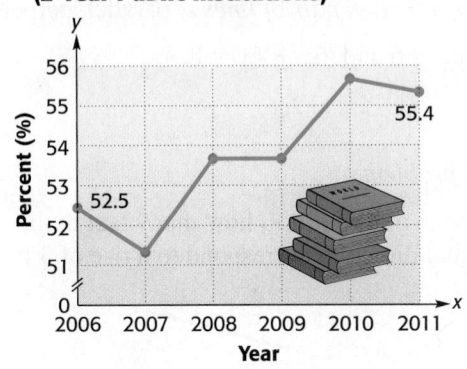

Percents of Students Who Return for Second Year (2-Year Public Institutions)

Source: ACT.

2. What does the ordered pair (2000, 52.0) mean in the context of these problems?

3. Over which two consecutive years did the percent stay the same? Estimate this percent.

4. Over which pairs of consecutive years did the percent increase? Estimate the increases.

Complete the given ordered pairs for each equation.

5. $y = 3x + 2$; $(-1, \_\_\_\_), (0, \_\_\_\_), (\_\_\_\_, 5)$

6. $4x + 3y = 6$; $(0, \_\_\_\_), (\_\_\_\_, 0), (-2, \_\_\_\_)$

Decide whether each ordered pair is a solution of the given equation.

7. $2x + y = 5$; $(-1, 3)$

8. $3x - y = 4$; $\left(\dfrac{1}{3}, -3 \right)$

3.2 *Graph each linear equation using intercepts.*

9. $2x - y = 3$

10. $x + 2y = -4$

11. $x + y = 0$

12. $y - 5 = 0$

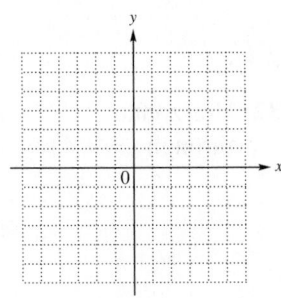
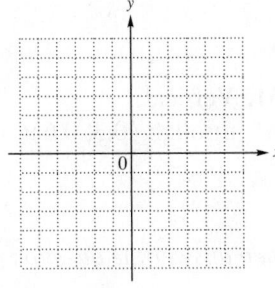
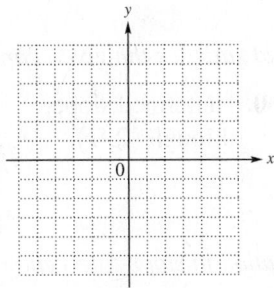
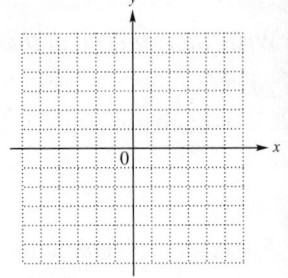

3.3 *Find the slope of each line.*

13. The line passing through $(2, 3)$ and $(-4, 6)$

14. The line passing through $(2, 5)$ and $(2, 8)$

15. $y = 4$

16. $y = 3x - 4$

17.

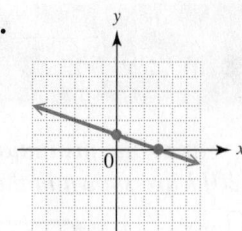

18. (a) A line parallel to the graph of $y = 2x + 3$

(b) A line perpendicular to the graph of $y = -3x + 3$

Decide whether each pair of lines is parallel, perpendicular, *or* neither.

19. $3x + 2y = 6$ and $6x + 4y = 8$

20. $x - 3y = 1$ and $3x + y = 4$

Solve each problem.

21. If the pitch of a roof is $\frac{1}{4}$, how many feet in the horizontal direction correspond to a rise of 3 ft?

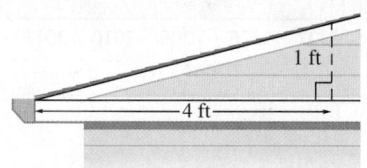

22. Family income in the United States has steadily increased for many years (primarily due to inflation). In 1980 the median family income was about $21,000 per yr. In 2008 it was about $61,500 per yr. Find the average rate of change of median family income to the nearest dollar over that period. (*Source:* U.S. Census Bureau.)

3.4 *Write an equation of each line. Give the answer in slope-intercept form. In Exercises 26–28, also give the answer in standard form.*

23. $m = -1, b = \dfrac{2}{3}$

24. The line in **Exercise 17**

25. The line passing through $(4, -3), m = 1$

26. The line passing through $(2, 1)$ and $(-2, 2)$

27. Parallel to $4x - y = 3$ and through $(6, -2)$

28. Perpendicular to $2x - 5y = 7$ and through $(0, 1)$

Write an equation of the line that satisfies the given conditions.

29. Slope 0; y-intercept $(0, 12)$

30. Undefined slope; through $(2, 7)$

31. Vertical; through $(0.3, 0.6)$

32. Horizontal; through $(-1, 4)$

For each problem, write an equation in slope-intercept form. Then answer the question posed.

33. Resident tuition at Broward College is $87.95 per credit hour. There is also a $20 health science application fee. Let x represent the number of credit hours and y represent the cost. How much does it cost for a student in health science to take 15 credit hours? (*Source:* www.broward.edu)

34. An Executive Regular/Silver membership to a health club costs $159, plus $47 per month. Let x represent the number of months and y represent the cost. How much will a one-year membership cost? (*Source:* Midwest Athletic Club.)

3.5 *Graph each linear inequality.*

35. $3x + 5y > 9$

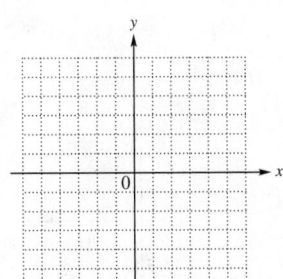

36. $2x - 3y > -6$

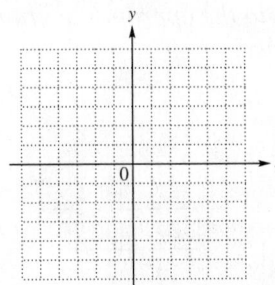

37. $x \geq -4$

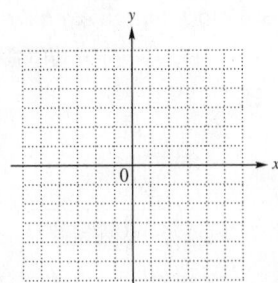

38. $y < -4x$

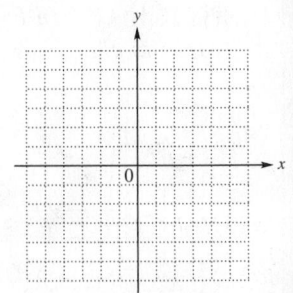

3.6 *Give the domain and range of each relation. Identify any functions.*

39. $\{(-4, 2), (-4, -2),$
$(1, 5), (1, -5)\}$

40.

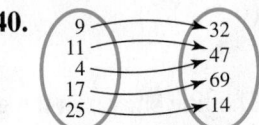

41.

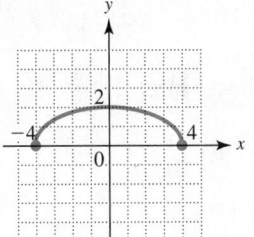

42.

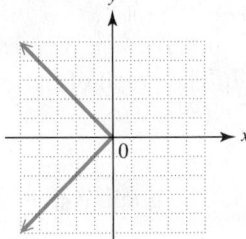

Determine whether each relation defines y as a function of x. Identify any linear functions. Give the domain in each case.

43. $y = 3x - 3$

44. $y < x + 2$

45. $x = y^2$

46. $y = \dfrac{7}{x - 36}$

3.7 *Given $f(x) = -2x^2 + 3x - 6$, find each of the following.*

47. $f(0)$

48. $f(3)$

49. $f(p)$

50. $f(-k)$

51. CONCEPT CHECK The linear equation $2x - 5y = 7$ defines y as a function of x. If $y = f(x)$, which of the following defines the same function?

A. $f(x) = -\dfrac{2}{5}x + \dfrac{7}{5}$ **B.** $f(x) = -\dfrac{2}{5}x - \dfrac{7}{5}$

C. $f(x) = \dfrac{2}{5}x - \dfrac{7}{5}$ **D.** $f(x) = \dfrac{2}{5}x + \dfrac{7}{5}$

52. CONCEPT CHECK Which of the following defines a linear function?

A. $y = \dfrac{2}{5}x - 3$ **B.** $y = \dfrac{1}{x}$

C. $y = x^2$ **D.** $y < x$

53. The equation $2x^2 - y = 0$ defines y as a function of x. Write it using $f(x)$ notation, and find $f(3)$.

54. Describe the graph of a constant function.

Mixed Review Exercises

CONCEPT CHECK *In Exercises 55–60, match each statement to the appropriate graph or graphs in choices A–D. Graphs may be used more than once.*

A.
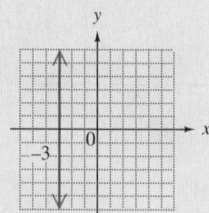

B.
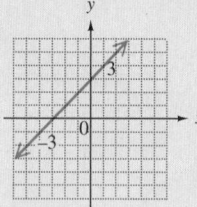

C.
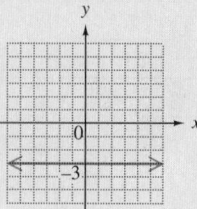

D.
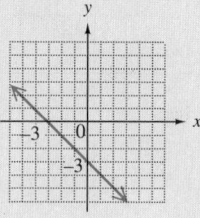

55. The line shown in the graph has undefined slope.

56. The graph of the equation has *y*-intercept $(0, -3)$.

57. The graph of the equation has *x*-intercept $(-3, 0)$.

58. The line shown in the graph has negative slope.

59. The graph is that of the equation $y = -3$.

60. The line shown in the graph has slope 1.

*Write an equation of each line. Give the final answer **(a)** in slope-intercept form and* ***(b)*** *in standard form.*

61. $m = -\dfrac{1}{4}, b = -\dfrac{5}{4}$

62. The line passing through $(8, 6), m = -3$

63. The line passing through $(3, -5)$ and $(-4, -1)$

Solve each problem.

64. CONCEPT CHECK Which equations have a graph with just one intercept?

   A. $x - 6 = 0$ **B.** $x + y = 0$

   C. $x - y = 4$ **D.** $y = -4$

65. CONCEPT CHECK Which inequality has as its graph a dashed boundary line and shading below the line?

   A. $y \geq 4x + 3$ **B.** $y > 4x + 3$

   C. $y \leq 4x + 3$ **D.** $y < 4x + 3$

66. The table shows life expectancy at birth in the United States for selected years.

   (a) Does the table define a function?

   (b) What are the domain and range?

   (c) Call this function *f*. Give two ordered pairs that belong to *f*.

   (d) Find $f(2010)$. What does it mean?   **(e)** If $f(x) = 75.4$, what does *x* equal?

Year	Life Expectancy at Birth (in years)
1950	68.2
1960	69.7
1970	70.8
1980	73.7
1990	75.4
2000	77.0
2010	78.7

Source: National Center for Health Statistics.

Chapter 3 Test

Graph each linear equation. Give the intercepts.

1. $3x + y = 6$

y-intercept: _____

x-intercept: _____

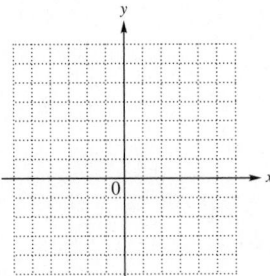

2. $y - 2x = 0$

y-intercept: _____

x-intercept: _____

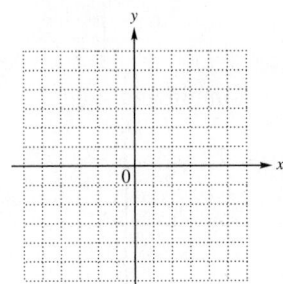

3. $x + 3 = 0$

y-intercept: _____

x-intercept: _____

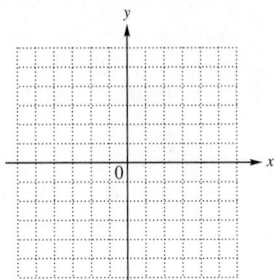

4. $y = 1$

y-intercept: _____

x-intercept: _____

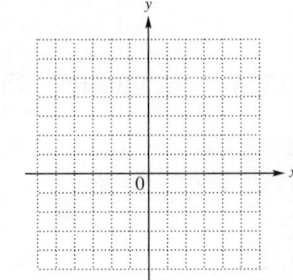

Find the slope of each line.

5. The line passing through $(-4, 6)$ and $(-1, -2)$

6. $2x + y = 10$

7. $x + 12 = 0$

8.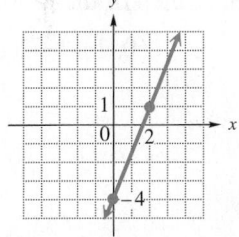

9. In 1980, there were 119,000 farms in Iowa. As of 2010, there were 92,000. Find and interpret the average rate of change in the number of farms per year. (*Source:* U.S. Department of Agriculture.)

*Write the equation of each line **(a)** in slope-intercept form and **(b)** in standard form.*

10. Through $(-2, 3)$ and $(6, -1)$

11. Through $(4, -1)$; $m = -5$

Write an equation for each line.

12. Through $(-3, 14)$; horizontal

13. Through $(5, -6)$; vertical

14. Write an equation in slope-intercept form for the line through $(-7, 2)$ and

(a) parallel to $3x + 5y = 6$.

(b) perpendicular to $y = 2x$.

Graph each linear inequality.

15. $x + y \le 3$

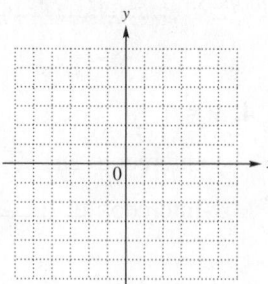

16. $3x - y > 0$

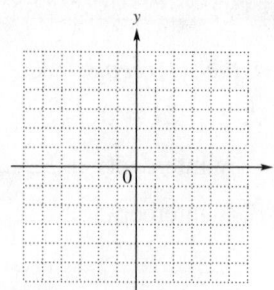

17. Which of the following is the graph of a function? Give its domain and range.

A.

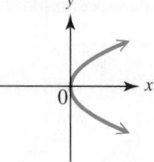

B.

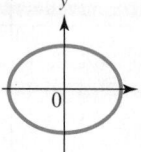

C.

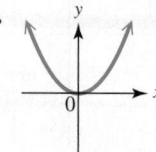

18. Which of the following does not define a function? Give its domain and range.

A. $\{(0, 1), (-2, 3), (4, 8)\}$ **B.** $y = 2x - 6$

C.

x	y
0	1
3	2
0	2
6	3

19. If $f(x) = -x^2 + 2x - 1$, find $f(1)$ and $f(a)$.

The graph shows worldwide snowmobile sales from 2000 through 2011, where 2000 corresponds to $x = 0$. Use the graph to work Exercises 20–24.

20. Is the slope of the line in the graph positive or negative? Explain.

21. Write two ordered pairs for the data points shown in the graph. Use them to find the slope of the line to the nearest tenth.

22. Use the ordered pairs and slope from **Exercise 21** to find an equation of a line that models the data. Write the equation in slope-intercept form.

Worldwide Snowmobile Sales

Source: www.snowmobile.org

23. Use the equation from **Exercise 22** to approximate worldwide snowmobile sales for 2007. How does your answer compare to the actual sales of 160.3 thousand?

24. What does the ordered pair (11, 123) mean in the context of this problem?

Chapters R–3 Cumulative Review Exercises

Perform the indicated operations.

1. $10\dfrac{5}{8} - 3\dfrac{1}{10}$

2. $\dfrac{3}{4} \div \dfrac{1}{8}$

3. $5 - (-4) + (-2)$

4. $\dfrac{(-3)^2 - (-4)(2^4)}{5(2) - (-2)^3}$

5. *True* or *false*? $\dfrac{4(3-9)}{2-6} \geq 6$

6. Find the value of $xz^3 - 5y^2$ for $x = -2$, $y = -3$, and $z = -1$.

7. What property does $3(-2 + x) = -6 + 3x$ illustrate?

8. Simplify $-4p - 6 + 3p + 8$ by combining like terms.

Solve.

9. $V = \dfrac{1}{3}\pi r^2 h$ for h

10. $6 - 3(1 + a) = 2(a + 5) - 2$

11. $-(m - 3) = 5 - 2m$

12. $\dfrac{y-2}{3} = \dfrac{2y+1}{5}$

Solve each inequality. Write the solution set in interval notation and graph it.

13. $-2.5x < 6.5$

14. $4(x + 3) - 5x < 12$

15. $\dfrac{2}{3}t - \dfrac{1}{6}t \leq -2$

Solve each problem.

16. In 2010, a person with a bachelor's degree could expect to earn $21,424 more each year than someone with a high school diploma. Together the individuals would earn $86,528. How much could a person at each level of education expect to earn? (*Source*: Bureau of Labor Statistics.)

17. Mount Mayon in the Philippines is the most perfectly shaped conical volcano in the world. Its base is a circle with circumference 80 mi, and it has a height of about 8100 ft. (One mile is 5280 ft.) Find the radius of the circular base to the nearest mile. (*Source*: www.britannica.hk)

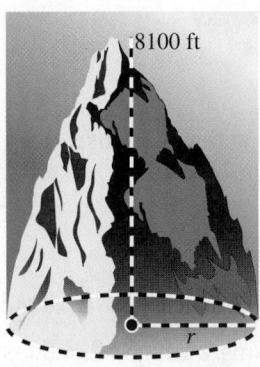

8100 ft

r

Circumference = 80 mi

18. The winning times in seconds for the women's 1000 m speed skating event in the Winter Olympics for the years 1960 through 2010 can be closely approximated by the linear equation

$$y = -0.4336x + 94.30,$$

where x is the number of years since 1960. That is, $x = 4$ represents 1964, $x = 8$ represents 1968, and so on. (*Source: World Almanac and Book of Facts.*)

(a) Use this equation to complete the table of values. Round times to the nearest hundredth of a second.

x	y
20	
38	
42	

(b) Write ordered pairs for the data given in the table.

(c) In the context of this problem, what does the ordered pair (32, 80.42) mean?

19. In 2010, approximately 60 million people from other countries visited the United States. The circle graph shows the distribution of these international visitors by their home countries or regions.

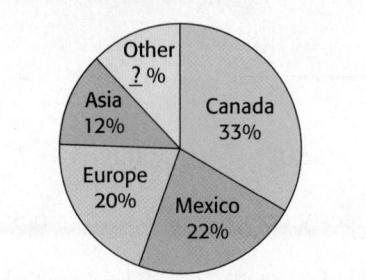

Source: U.S. Department of Commerce.

(a) About how many travelers visited the United States from Canada?

(b) About how many travelers visited the United States from Mexico?

(c) About how many travelers visited the United States from places other than Canada, Mexico, Europe, or Asia?

Consider the linear equation $-3x + 4y = 12.$ *Find the following.*

20. The x- and y-intercepts

21. The slope

22. The graph

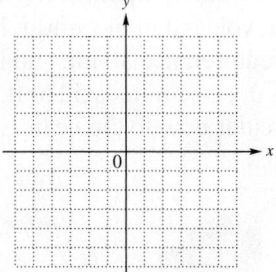

23. The y-value of the point having x-value 4

24. Are the lines with equations $x + 5y = -6$ and $y = 5x - 8$ *parallel, perpendicular,* or *neither?*

Write an equation of each line. Give the final answer in slope-intercept form if possible.

25. The line passing through $(2, -5)$ with slope 3

26. The line passing through $(0, 4)$ and $(2, 4)$

4 Systems of Linear Equations and Inequalities

The point of intersection of two lines can be found using a *system of linear equations*, the subject of this chapter.

4.1 Solving Systems of Linear Equations by Graphing

4.2 Solving Systems of Linear Equations by Substitution

4.3 Solving Systems of Linear Equations by Elimination

Summary Exercises *Applying Techniques for Solving Systems of Linear Equations*

4.4 Applications of Linear Systems

4.5 Solving Systems of Linear Inequalities

4.1 Solving Systems of Linear Equations by Graphing

1. **Decide whether a given ordered pair is a solution of a system.**

2. **Solve linear systems by graphing.**

3. **Solve special systems by graphing.**

4. **Identify special systems without graphing.**

A **system of linear equations**, also called a **linear system,** consists of two or more linear equations with the same variables.

$$2x + 3y = 4 \qquad\qquad x + 3y = 1 \qquad\qquad x - y = 1$$
$$3x - y = -5 \qquad\quad -y = 4 - 2x \qquad\quad y = 3$$

Linear systems

In the system on the right, think of $y = 3$ as an equation in two variables by writing it as $0x + y = 3$.

OBJECTIVE ▸ ① **Decide whether a given ordered pair is a solution of a system.** A **solution of a system** of linear equations is an ordered pair that makes both equations true at the same time. A solution of an equation is said to *satisfy* the equation.

---

1 Determine whether the given ordered pair is a solution of the system.

(a) $(2, 5)$

$$3x - 2y = -4$$
$$5x + y = 15$$

$$3x - 2y = -4$$
$$3 (\underline{}) - 2(\underline{}) \overset{?}{=} -4$$

$$5x + y = 15$$
$$5(2) + \underline{} \overset{?}{=} \underline{}$$

$(2, 5)$ *(is / is not)* a solution.

(b) $(1, -2)$

$$x - 3y = 7$$
$$4x + y = 5$$

$(1, -2)$ *(is / is not)* a solution.

EXAMPLE 1   **Determining Whether an Ordered Pair Is a Solution**

Determine whether the ordered pair $(4, -3)$ is a solution of each system.

(a) $x + 4y = -8$

$3x + 2y = 6$

To decide whether $(4, -3)$ is a solution of the system, substitute 4 for x and -3 for y in each equation.

$x + 4y = -8$	$3x + 2y = 6$
$4 + 4(-3) \overset{?}{=} -8$ Substitute.	$3(4) + 2(-3) \overset{?}{=} 6$ Substitute.
$4 + (-12) \overset{?}{=} -8$ Multiply.	$12 + (-6) \overset{?}{=} 6$ Multiply.
$-8 = -8$ ✓ True	$6 = 6$ ✓ True

Because $(4, -3)$ satisfies *both* equations, it is a solution of the system.

(b) $2x + 5y = -7$

$3x + 4y = 2$

Again, substitute 4 for x and -3 for y in each equation.

$2x + 5y = -7$	$3x + 4y = 2$
$2(4) + 5(-3) \overset{?}{=} -7$ Substitute.	$3(4) + 4(-3) \overset{?}{=} 2$ Substitute.
$8 + (-15) \overset{?}{=} -7$ Multiply.	$12 + (-12) \overset{?}{=} 2$ Multiply.
$-7 = -7$ ✓ True	$0 = 2$ False

The ordered pair $(4, -3)$ is *not* a solution of this system because it does not satisfy the second equation.

◂ **Work Problem** ❶ **at the Side.**

OBJECTIVE ▸ ② **Solve linear systems by graphing.** The set of all ordered pairs that are solutions of a system is its **solution set.** One way to find the solution set of a system of two linear equations is to graph both equations on the same axes. Any intersection point would be on both lines and would therefore be a solution of *both* equations. *Thus, the coordinates of any point where the lines intersect give a solution of the system.*

Answers

1. (a) 2; 5; 5; 15; is   **(b)** is not

The graph in **Figure 1** shows that the solution of the system in **Example 1(a)** is the intersection point $(4, -3)$. ***The solution (point of intersection) is always written as an ordered pair.*** Because *two different* straight lines can intersect at no more than one point, there can never be more than one solution for such a system.

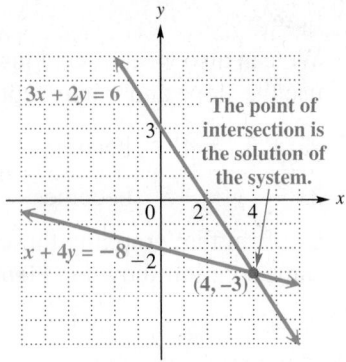

Figure 1

EXAMPLE 2 Solving a System by Graphing

Solve the system of equations by graphing both equations on the same axes.

$$2x + 3y = 4$$
$$3x - y = -5$$

We graph these two equations by plotting several points for each line. Recall from **Section 3.2** that the intercepts are often convenient choices. As review, we show finding the intercepts for $2x + 3y = 4$.

To find the y-intercept, let $x = 0$.

$$2x + 3y = 4$$
$$2(0) + 3y = 4 \quad \text{Let } x = 0.$$
$$3y = 4$$
$$y = \frac{4}{3} \quad \text{Divide by 3.}$$

To find the x-intercept, let $y = 0$.

$$2x + 3y = 4$$
$$2x + 3(0) = 4 \quad \text{Let } y = 0.$$
$$2x = 4$$
$$x = 2 \quad \text{Divide by 2.}$$

The tables show the intercepts and a check point for each graph.

$2x + 3y = 4$

x	y	
0	$\frac{4}{3}$	← y-intercept
2	0	← x-intercept
-2	$\frac{8}{3}$	

Find a third ordered pair as a check.

$3x - y = -5$

x	y	
0	5	← y-intercept
$-\frac{5}{3}$	0	← x-intercept
-2	-1	

The lines in **Figure 2** suggest that the graphs intersect at the point $(-1, 2)$. We check by substituting -1 for x and 2 for y in *both* equations.

CHECK
$$2x + 3y = 4$$
$$2(-1) + 3(2) \stackrel{?}{=} 4$$
$$4 = 4 \quad ✓ \text{ True}$$

$$3x - y = -5$$
$$3(-1) - 2 \stackrel{?}{=} -5$$
$$-5 = -5 \quad ✓ \text{ True}$$

Because $(-1, 2)$ satisfies both equations, the solution set of this system is $\{(-1, 2)\}$.

Figure 2

·········· **Work Problem ➋ at the Side.** ▶

➋ Solve each system of equations by graphing both equations on the same axes. Check each solution.

(a) $5x - 3y = 9$
$$x + 2y = 7$$

One of the lines is already graphed. Complete the table, and graph the other line.

$5x - 3y = 9$

x	y
0	-3
$\frac{9}{5}$	0
3	2

$x + 2y = 7$

x	y
0	___
___	0
-1	___

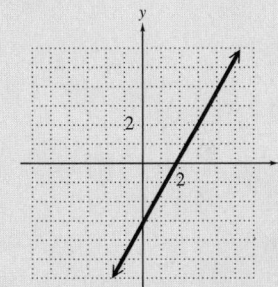

The point of intersection of the lines is the ordered pair _____, so the solution set of the system of equations is _____.

(b) $x + y = 4$
$$2x - y = -1$$

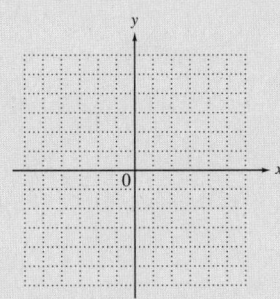

Note

We can also write each equation in a system in slope-intercept form and use the slope and y-intercept to graph each line. For **Example 2,**

$2x + 3y = 4$ becomes $y = -\frac{2}{3}x + \frac{4}{3}.$ y-intercept $\left(0, \frac{4}{3}\right)$; slope $-\frac{2}{3}$

$3x - y = -5$ becomes $y = 3x + 5.$ y-intercept $(0, 5)$; slope 3, or $\frac{3}{1}$

Confirm that graphing these equations results in the same lines and the same solution shown in **Figure 2** on the preceding page.

Solving a Linear System by Graphing

Step 1 **Graph each equation** of the system on the same coordinate axes.

Step 2 **Find the coordinates of the point of intersection** of the graphs if possible, and write it as an ordered pair. This is the solution of the system.

Step 3 **Check** the ordered-pair solution in *both* of the original equations. If it satisfies *both* equations, write the solution set.

CAUTION

With the graphing method, it may not be possible to determine from the graph the exact coordinates of the point that represents the solution, particularly if these coordinates are not integers. The graphing method does, however, show geometrically how solutions are found and is useful when approximate answers will suffice.

OBJECTIVE **3** **Solve special systems by graphing.** The graphs of the equations in a system may not intersect at all or may be the same line.

EXAMPLE 3 Solving Special Systems

Solve each system by graphing.

(a) $2x + y = 2$

$2x + y = 8$

The graphs of these lines are shown in **Figure 3.** The two lines are parallel and have no points in common. For such a system, there is no solution. We write the solution set as $\varnothing$.

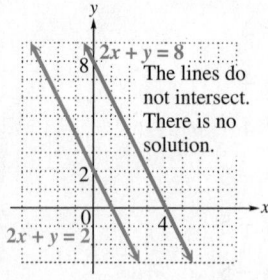

Figure 3

Continued on Next Page

(b) $2x + 5y = 1$

$6x + 15y = 3$

The graphs of these two equations are the same line. See **Figure 4.** We can obtain the second equation by multiplying each side of the first equation by 3. In this case, every point on the line is a solution of the system, and the solution set contains an infinite number of ordered pairs, each of which satisfies both equations of the system.

We write the solution set as

$$\{(x, y) \mid 2x + 5y = 1\},$$

> This is the first equation in the system. See the Note on the next page.

read "the set of ordered pairs (x, y) such that $2x + 5y = 1$." Recall from **Section 1.3** that this notation is called **set-builder notation.**

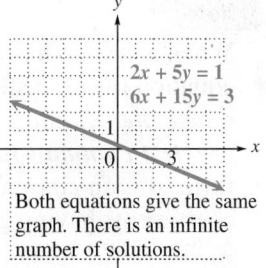

Both equations give the same graph. There is an infinite number of solutions.

Figure 4

· Work Problem **3** at the Side. ▶

The system in **Example 2** has exactly one solution. A system with at least one solution is a **consistent system.** A system of equations with no solution, such as the one in **Example 3(a)**, is an **inconsistent system.**

The equations in **Example 2** are **independent equations** with different graphs. The equations of the system in **Example 3(b)** have the same graph and are equivalent. Because they are different forms of the same equation, these equations are **dependent equations.**

Three Cases for Solutions of Linear Systems with Two Variables

Case 1 The graphs intersect at exactly one point, which gives the (single) ordered-pair solution of the system. The **system is consistent** and the **equations are independent.** See **Figure 5(a).**

Case 2 The graphs are parallel lines, so there is no solution and the solution set is ∅. The **system is inconsistent** and the **equations are independent.** See **Figure 5(b).**

Case 3 The graphs are the same line. There is an infinite number of solutions, and the solution set is written in set-builder notation as

$$\{(x, y) \mid \text{_____}\},$$

where one of the equations follows the $\mid$ symbol. The **system is consistent** and the **equations are dependent.** See **Figure 5(c).**

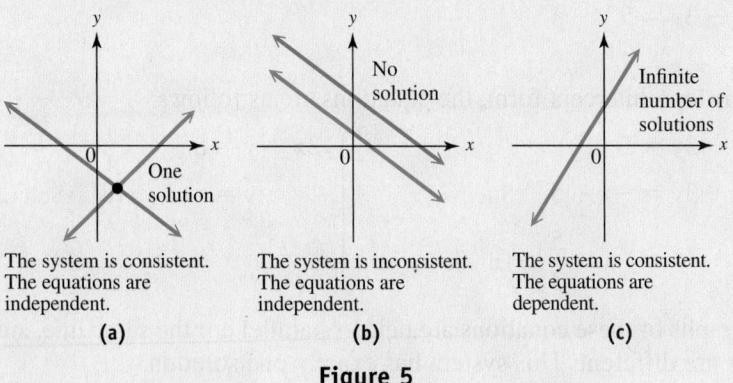

The system is consistent. The equations are independent.

(a)

No solution

The system is inconsistent. The equations are independent.

(b)

Infinite number of solutions

The system is consistent. The equations are dependent.

(c)

Figure 5

---

3 Solve each system of equations by graphing both equations on the same axes.

(GS) **(a)** $3x - y = 4$

$6x - 2y = 12$

One of the lines is already graphed.

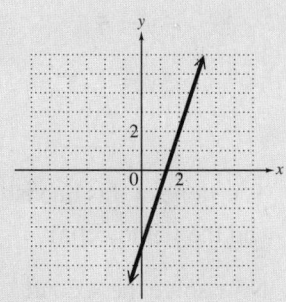

(b) $x - 3y = -2$

$2x - 6y = -4$

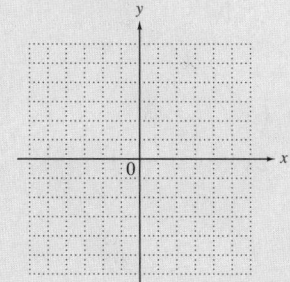

Answers

3. (a) ∅   **(b)** $\{(x, y) \mid x - 3y = -2\}$

4 Describe each system without graphing. State the number of solutions.

(a) $2x - 3y = 5$

$3y = 2x - 7$

Solve the first equation for ____.

$2x - 3y = 5$

$-3y = \underline{\hspace{1cm}} + 5$

$y = \underline{\hspace{1cm}} - \underline{\hspace{1cm}}$

The slope of this line is ____, and the y-intercept is ____.

Now solve the second equation for y, and complete the solution.

(b) $-x + 3y = 2$

$2x - 6y = -4$

(c) $6x + y = 3$

$2x - y = -11$

Answers

4. (a) $y; -2x; \dfrac{2}{3}x; \dfrac{5}{3}; \dfrac{2}{3}; \left(0, -\dfrac{5}{3}\right)$

The equations represent parallel lines. The system has no solution.

(b) The equations represent the same line. The system has an infinite number of solutions.

(c) The equations represent lines that are neither parallel nor the same line. The system has exactly one solution.

Note

When a system has an infinite number of solutions, as in Case 3, either equation of the system can be used to write the solution set. *We prefer to use the equation in standard form with integer coefficients having greatest common factor 1.* If neither of the given equations is in this form, use an *equivalent* equation that is in standard form with integer coefficients having greatest common factor 1.

OBJECTIVE ▶ **4** Identify special systems without graphing.

EXAMPLE 4 Identifying the Three Cases by Using Slopes

Describe each system without graphing. State the number of solutions.

(a) $3x + 2y = 6$

$-2y = 3x - 5$

Write each equation in slope-intercept form, $y = mx + b$.

$3x + 2y = 6$ ◁ Solve for y.

$2y = -3x + 6$ Subtract 3x.

$y = -\dfrac{3}{2}x + 3$ Divide *each* term by 2.

$-2y = 3x - 5$ ◁ Solve for y.

$y = -\dfrac{3}{2}x + \dfrac{5}{2}$ Divide *each* term by -2.

Both equations have slope $-\dfrac{3}{2}$ but they have different y-intercepts, $(0, 3)$ and $\left(0, \dfrac{5}{2}\right)$. Lines with the same slope are parallel (**Section 3.3**), so these equations have graphs that are parallel lines. Thus, the system has no solution.

(b) $2x - y = 4$

$x = \dfrac{y}{2} + 2$

Again, write each equation in slope-intercept form.

$2x - y = 4$

$-y = -2x + 4$ Subtract 2x.

$y = 2x - 4$ Multiply by -1.

$x = \dfrac{y}{2} + 2$

$\dfrac{y}{2} + 2 = x$ Interchange sides.

$y = 2x - 4$ Subtract 2. Multiply by 2.

The equations are exactly the same—their graphs are the same line. Thus, the system has an infinite number of solutions.

(c) $x - 3y = 5$

$2x + y = 8$

In slope-intercept form, the equations are as follows.

$x - 3y = 5$

$-3y = -x + 5$ Subtract x.

$y = \dfrac{1}{3}x - \dfrac{5}{3}$ Divide by -3.

$2x + y = 8$

$y = -2x + 8$ Subtract 2x.

The graphs of these equations are neither parallel nor the same line, since the slopes are different. This system has exactly one solution.

◀ **Work Problem 4 at the Side.**

4.1 Exercises

FOR EXTRA HELP

 Download the MyDashBoard App

MyMathLab®

CONCEPT CHECK *Complete each statement. The following terms may be used once, more than once, or not at all.*

consistent   system of linear equations   inconsistent   solution

ordered pair   independent   linear equation   dependent

1. A(n) _____ consists of two or more linear equations with the (*same / different*) variables.

2. A solution of a system of linear equations is a(n) _____ that makes all equations of the system (*true / false*) at the same time.

3. The equations of two parallel lines form a(n) _____ system that has (*one / no / infinitely many*) solution(s). The equations are _____ because their graphs are different.

4. If the graphs of a linear system intersect in one point, the point of intersection is the _____ of the system. The system is _____ and the equations are independent.

5. If two equations of a linear system have the same graph, the equations are _____. The system is _____ and has (*one / no / infinitely many*) solution(s).

6. If a linear system is inconsistent, the graphs of the two equations are (*intersecting / parallel / the same*) line(s). The system has no _____.

CONCEPT CHECK *Work each problem.*

7. A student determined that the ordered pair $(1, -2)$ is a solution of the following system. His reasoning was that the ordered pair satisfies the first equation $x + y = -1$, since $1 + (-2) = -1$. **What Went Wrong?**

$$x + y = -1$$
$$2x + y = 4$$

8. The following system has infinitely many solutions. Write its solution set, using set-builder notation.

$$6x - 4y = 8$$
$$3x - 2y = 4$$

9. Which ordered pair could be a solution of the system graphed? Why is it the only valid choice?

   A. $(2, 2)$
   B. $(-2, 2)$
   C. $(-2, -2)$
   D. $(2, -2)$

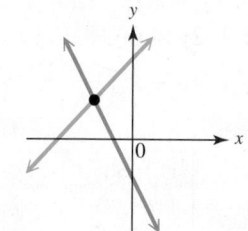

10. Which ordered pair could be a solution of the system graphed? Why is it the only valid choice?

   A. $(2, 0)$
   B. $(0, 2)$
   C. $(-2, 0)$
   D. $(0, -2)$

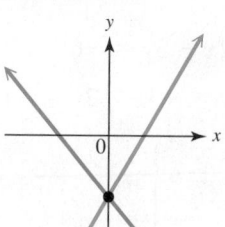

Decide whether the given ordered pair is a solution of the given system. See Example 1.

11. $(2, -3)$
$x + y = -1$
$2x + 5y = 19$

12. $(4, 3)$
$x + 2y = 10$
$3x + 5y = 3$

13. $(-1, -3)$
$3x + 5y = -18$
$4x + 2y = -10$

14. $(-9, -2)$
$2x - 5y = -8$
$3x + 6y = -39$

15. $(7, -2)$
$$4x = 26 - y$$
$$3x = 29 + 4y$$

16. $(9, 1)$
$$2x = 23 - 5y$$
$$3x = 24 + 3y$$

17. $(6, -8)$
$$-2y = x + 10$$
$$3y = 2x + 30$$

18. $(-5, 2)$
$$5y = 3x + 20$$
$$3y = -2x - 4$$

Solve each system of equations by graphing. If the system is inconsistent or the equations are dependent, say so. **See Examples 2 and 3.**

19. $x - y = 2$
$x + y = 6$

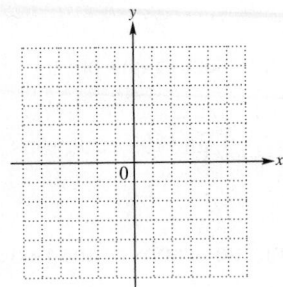

20. $x - y = 3$
$x + y = -1$

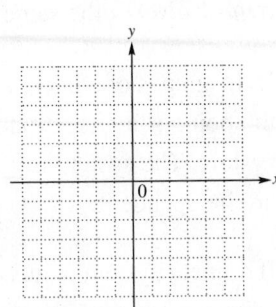

21. $x + y = 4$
$y - x = 4$

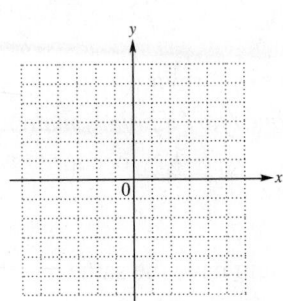

22. $x + y = -5$
$x - y = 5$

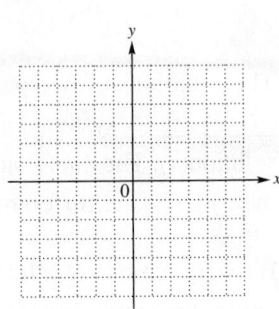

23. $x - 2y = 6$
$x + 2y = 2$

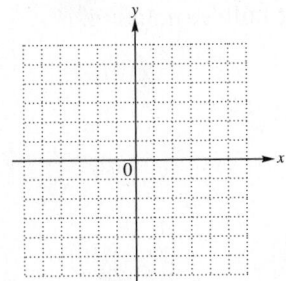

24. $2x - y = 4$
$4x + y = 2$

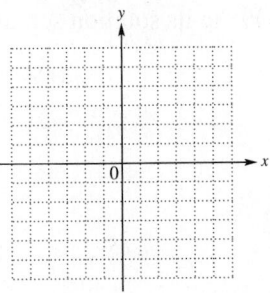

25. $3x - 2y = -3$
$-3x - y = -6$

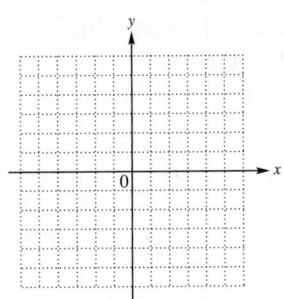

26. $2x - y = 4$
$2x + 3y = 12$

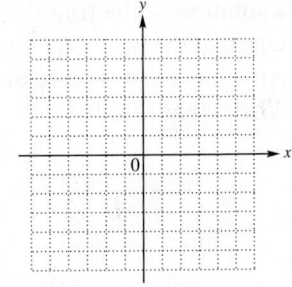

27. $2x - 3y = -6$
$y = -3x + 2$

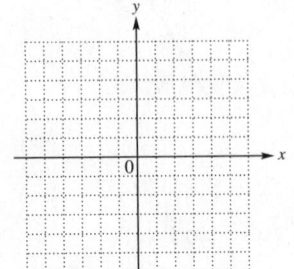

28. $-3x + y = -3$
$y = x - 3$

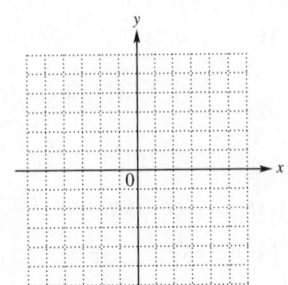

29. $x + 2y = 6$
$2x + 4y = 8$

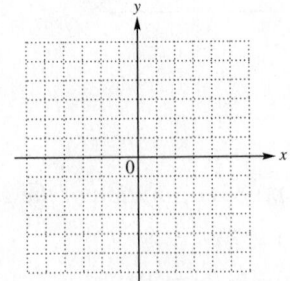

30. $2x - y = 6$
$6x - 3y = 12$

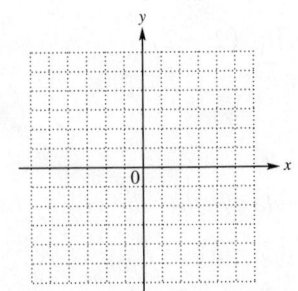

31. $5x - 3y = 2$
$\quad\quad 10x - 6y = 4$

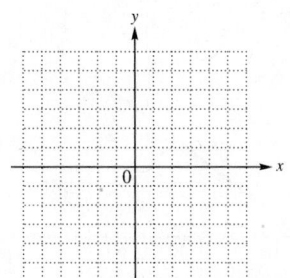

32. $2x - 5y = 8$
$\quad\quad 4x - 10y = 16$

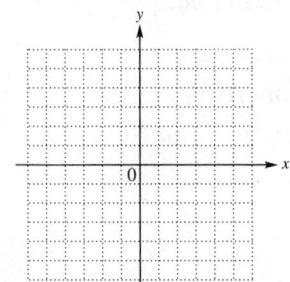

33. $3x - 4y = 24$
$\quad\quad y = -\dfrac{3}{2}x + 3$

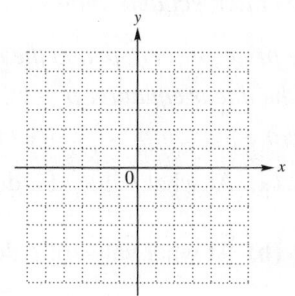

34. $3x - 2y = 12$
$\quad\quad y = -4x + 5$

35. $4x - 2y = 8$
$\quad\quad 2x = y + 4$

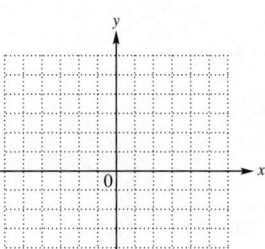

36. $3x = 5 - y$
$\quad\quad 6x + 2y = 10$

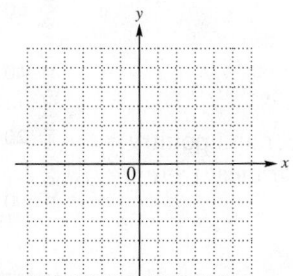

37. $3x = y + 5$
$\quad\quad 6x - 5 = 2y$

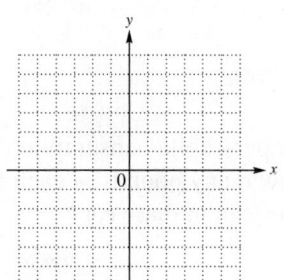

38. $2x = y - 4$
$\quad\quad 4x - 2y = -4$

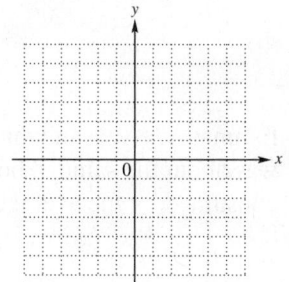

Without graphing, answer the following questions for each linear system. ***See Example 4.***

(a) Is the system inconsistent, are the equations dependent, or neither?
(b) Is the graph a pair of intersecting lines, a pair of parallel lines, or one line?
(c) Does the system have one solution, no solution, or an infinite number of solutions?

39. $y - x = -5$
$\quad\quad x + y = 1$

40. $2x + y = 6$
$\quad\quad x - 3y = -4$

41. $x + 2y = 0$
$\quad\quad 4y = -2x$

42. $4x - 6y = 10$
$\quad\quad -6x + 9y = -15$

43. $5x + 4y = 7$
$\quad\quad 10x + 8y = 4$

44. $y = 3x$
$\quad\quad y + 3 = 3x$

45. $x - 3y = 5$
$\quad\quad 2x + y = 8$

46. $2x + 3y = 12$
$\quad\quad 2x - y = 4$

*Economics deals with **supply and demand**. Typically, as the price of an item increases, the demand for the item decreases while the supply increases. If supply and demand can be described by straight-line equations, the point at which the lines intersect determines the **equilibrium supply** and **equilibrium demand**.*

The price per unit, p, and the demand, x, for a particular aluminum siding are related by the linear equation $p = 60 - \frac{3}{4}x$, while the price and the supply are related by the linear equation $p = \frac{3}{4}x$. See the figure. Use the graph to work Exercises 47 and 48.

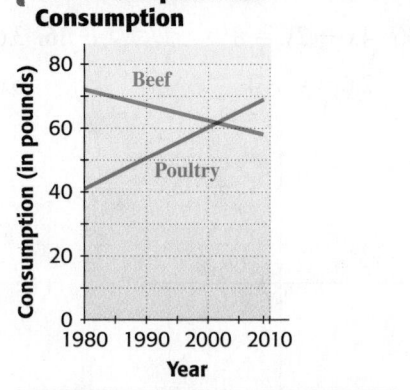

SUPPLY AND DEMAND

47. (a) At what value of x does supply equal demand?

(b) At what value of p does supply equal demand?

48. When $x > 40$, does demand exceed supply or does supply exceed demand?

The per-capita consumption of beef and poultry in the United States in selected years is shown in the graph. Use the graph to work Exercises 49 and 50.

49. For which years did per-capita consumption of beef exceed that of poultry?

50. Estimate the year in which the per-capita consumption of beef and poultry was about the same. About how many pounds of each kind of meat were consumed by the average American in that year?

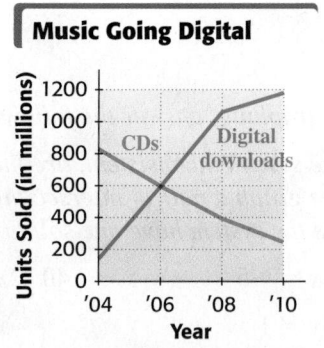

U.S. Per-Capita Meat Consumption

Source: U.S. Department of Agriculture.

The graph shows sales of music CDs and digital downloads of single songs (in millions) in the United States over selected years. Use the graph to work Exercises 51–54.

51. In what year did Americans purchase about the same number of CDs as single digital downloads? How many units was this?

52. Express the point of intersection of the two graphs as an ordered pair of the form (year, units in millions).

Music Going Digital

Source: Recording Industry Association of America.

53. Describe the trend in sales of music CDs over the years 2004 to 2010. If a straight line were used to approximate its graph, would the line have positive, negative, or zero slope? Explain.

54. If a straight line were used to approximate the graph of sales of digital downloads over the years 2004 to 2010, would the line have positive, negative, or zero slope? Explain.

4.2 Solving Systems of Linear Equations by Substitution

OBJECTIVE ▶ ① Solve linear systems by substitution.

Work Problem ① at the Side. ▶

As we see in **Margin Problem 1,** graphing to solve a system of equations has a serious drawback: It is difficult to accurately find a solution such as $\left(\frac{11}{3}, -\frac{4}{9}\right)$ from a graph.

As a result, there are algebraic methods for solving systems of equations. The **substitution method,** which gets its name from the fact that an expression in one variable is *substituted* for the other variable, is one such method.

EXAMPLE 1   **Using the Substitution Method**

Solve the system by the substitution method.

$$3x + 5y = 26 \quad (1)$$

We number the equations for reference in our discussion.

$$y = 2x \quad (2)$$

Equation (2) is already solved for y. This equation says that $y = 2x$, so we substitute $2x$ for y in equation (1).

$$3x + 5y = 26 \quad (1)$$
$$3x + 5(2x) = 26 \quad \text{Let } y = 2x.$$
$$3x + 10x = 26 \quad \text{Multiply.}$$
$$13x = 26 \quad \text{Combine like terms.}$$

Don't stop here. — $x = 2 \quad$ Divide by 13.

Now we can find the value of y by substituting 2 for x in either equation. We choose equation (2) since the substitution is easier.

$$y = 2x \quad (2)$$
$$y = 2(2) \quad \text{Let } x = 2.$$
$$y = 4 \quad \text{Multiply.}$$

We check the solution $(2, 4)$ by substituting 2 for x and 4 for y in *both* equations.

CHECK

$3x + 5y = 26 \quad (1)$	$y = 2x \quad (2)$
$3(2) + 5(4) \overset{?}{=} 26 \quad$ Substitute.	$4 \overset{?}{=} 2(2) \quad$ Substitute.
$6 + 20 \overset{?}{=} 26 \quad$ Multiply.	$4 = 4 \checkmark \quad$ True
$26 = 26 \checkmark$ True	

Since $(2, 4)$ satisfies both equations, the solution set of the system is $\{(2, 4)\}$.

···· Work Problem ② at the Side. ▶

CAUTION

A system is not completely solved until values for both x and y are found. Write the solution set as a set containing an ordered pair.

OBJECTIVES

① Solve linear systems by substitution.

② Solve special systems by substitution.

③ Solve linear systems with fractions and decimals.

---

① Solve the system by graphing.

$$2x + 3y = 6$$
$$x - 3y = 5$$

Can you determine the answer? Why or why not?

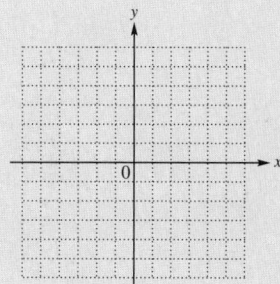

② Fill in the blanks to solve by the substitution method. Check the solution.

$$3x + 5y = 69$$
$$y = 4x$$

$$3x + 5(\underline{\quad}) = 69$$
$$3x + \underline{\quad} = 69$$
$$\underline{\quad} = 69$$
$$x = \underline{\quad}$$
$$y = 4(\underline{\quad}) = \underline{\quad}$$

The solution set is $\underline{\quad}$.

Answers

1. The answer cannot be determined from the graph because it is too difficult to read the exact coordinates.
2. $4x$; $20x$; $23x$; 3; 3; 12; $\{(3, 12)\}$

3 Solve each system by the substitution method. Check each solution.

(a) $2x + 7y = -12$

$x = 3 - 2y$

(b) $x = y - 3$

$4x + 9y = 1$

EXAMPLE 2 Using the Substitution Method

Solve the system by the substitution method.

$$2x + 5y = 7 \quad (1)$$
$$x = -1 - y \quad (2)$$

Equation (2) gives x in terms of y. We substitute $-1 - y$ for x in equation (1).

$$2x + 5y = 7 \quad (1)$$
$$2(-1 - y) + 5y = 7 \quad \text{Let } x = -1 - y.$$

> Distribute 2 to *both* −1 and −y.

$$-2 - 2y + 5y = 7 \quad \text{Distributive property}$$
$$-2 + 3y = 7 \quad \text{Combine like terms.}$$
$$3y = 9 \quad \text{Add 2.}$$
$$y = 3 \quad \text{Divide by 3.}$$

To find x, substitute 3 for y in equation (2).

$$x = -1 - y \quad (2)$$
$$x = -1 - 3 \quad \text{Let } y = 3.$$
$$x = -4 \quad \text{Subtract.}$$

> Write the
> x-coordinate first.

We check the solution $(-4, 3)$ in *both* equations.

CHECK

$$2x + 5y = 7 \quad (1) \qquad\qquad x = -1 - y \quad (2)$$
$$2(-4) + 5(3) \overset{?}{=} 7 \quad \text{Substitute.} \qquad -4 \overset{?}{=} -1 - 3 \quad \text{Substitute.}$$
$$-8 + 15 \overset{?}{=} 7 \quad \text{Multiply.} \qquad\quad -4 = -4 \checkmark \quad \text{True}$$
$$7 = 7 \checkmark \text{ True}$$

Both results are true, so the solution set of the system is $\{(-4, 3)\}$.

◀ **Work Problem 3 at the Side.**

CAUTION

Even though we found *y* first in **Example 2**, *the x-coordinate is always written first in the ordered-pair solution of a system.*

To solve a system by substitution, follow these steps.

Solving a Linear System by Substitution

Step 1 **Solve one equation for either variable.** If one of the variables has coefficient 1 or −1, choose it, since it usually makes the substitution easier.

Step 2 **Substitute** for that variable in the other equation. The result should be an equation with just one variable.

Step 3 **Solve** the equation from Step 2.

Step 4 **Substitute** the result from Step 3 into the equation from Step 1 to find the value of the other variable.

Step 5 **Check** the ordered-pair solution in both of the *original* equations. Then write the solution set.

Answers

3. (a) $\{(15, -6)\}$ (b) $\{(-2, 1)\}$

EXAMPLE 3 Using the Substitution Method

Solve the system by the substitution method.

$$2x = 4 - y \quad (1)$$

$$5x + 3y = 10 \quad (2)$$

Step 1 We must solve one of the equations for either x or y. Because the coefficient of y in equation (1) is -1, we avoid fractions by choosing this equation and solving for y.

$$2x = 4 - y \quad (1)$$

$$y + 2x = 4 \qquad \text{Add } y.$$

$$y = -2x + 4 \qquad \text{Subtract } 2x.$$

Step 2 Now substitute $-2x + 4$ for y in equation (2).

$$5x + 3y = 10 \quad (2)$$

$$5x + 3(-2x + 4) = 10 \qquad \text{Let } y = -2x + 4.$$

Step 3 Solve the equation from Step 2.

Distribute 3 to *both* $-2x$ and 4.

$$5x - 6x + 12 = 10 \qquad \text{Distributive property}$$

$$-x + 12 = 10 \qquad \text{Combine like terms.}$$

$$-x = -2 \qquad \text{Subtract 12.}$$

$$x = 2 \qquad \text{Multiply by } -1.$$

Step 4 Equation (1) solved for y is $y = -2x + 4$. Substitute 2 for x.

$$y = -2(2) + 4 \qquad \text{Substitute in (1).}$$

$$y = 0 \qquad \text{Multiply, and then add.}$$

Step 5 Check that $(2, 0)$ is the solution.

CHECK

$$2x = 4 - y \quad (1) \qquad\qquad 5x + 3y = 10 \quad (2)$$

$$2(2) \overset{?}{=} 4 - 0 \quad \text{Substitute.} \qquad 5(2) + 3(0) \overset{?}{=} 10 \quad \text{Substitute.}$$

$$4 = 4 \checkmark \quad \text{True} \qquad\qquad 10 = 10 \checkmark \quad \text{True}$$

Since both results are true, the solution set of the system is $\{(2, 0)\}$.

▶ Work Problem **4** at the Side. ▶

OBJECTIVE **2** Solve special systems by substitution.

EXAMPLE 4 Solving an Inconsistent System by Substitution

Use substitution to solve the system.

$$x = 5 - 2y \quad (1)$$

$$2x + 4y = 6 \quad (2)$$

Because equation (1) is already solved for x, we substitute $5 - 2y$ for x in equation (2).

$$2x + 4y = 6 \quad (2)$$

$$2(5 - 2y) + 4y = 6 \qquad \text{Let } x = 5 - 2y.$$

$$10 - 4y + 4y = 6 \qquad \text{Distributive property}$$

$$10 = 6 \qquad \text{False}$$

▶ **Continued on Next Page**

4 Solve each system by the substitution method. Check each solution.

GS **(a)** Fill in the blanks to solve.

$$x + 4y = -1$$

$$2x - 5y = 11$$

Solve the first equation for x.

$$x = -1 - \underline{\quad}$$

Substitute into the second equation to find y.

$$2(\underline{\quad}) - 5y = 11$$

$$-2 - 8y - 5y = 11$$

$$-2 - \underline{\quad}y = 11$$

$$\underline{\quad}y = 13$$

$$y = \underline{\quad}$$

Find x.

$$x = -1 - 4y$$

$$x = -1 - 4(\underline{\quad})$$

$$x = \underline{\quad}$$

The solution set is $\underline{\quad}$.

(b) $2x + 5y = 4$

$x + y = -1$

Answers

4. **(a)** $4y$; $-1 - 4y$; 13; -13; -1; -1; 3; $\{(3, -1)\}$

   (b) $\{(-3, 2)\}$

5 Use substitution to solve the system.

$$8x - 2y = 1$$
$$y = 4x - 8$$

A false result, here $10 = 6$, means that the equations in the system

$$x = 5 - 2y$$
$$2x + 4y = 6$$

have graphs that are parallel lines. The system is inconsistent and has no solution, so the solution set is $\emptyset$. See **Figure 6**.

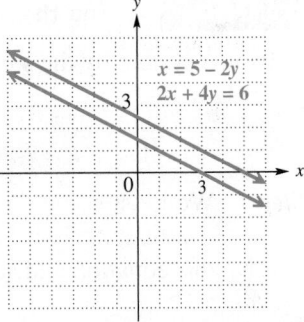

Figure 6

◄ Work Problem **5** at the Side.

CAUTION
It is a common error to give "false" as the answer to an inconsistent system. The correct response is $\emptyset$.

6 Solve each system by substitution.

(a) $7x - 6y = 10$
$-14x + 20 = -12y$

EXAMPLE 5   **Solving a System with Dependent Equations by Substitution**

Solve the system by the substitution method.

$$3x - y = 4 \qquad (1)$$
$$-9x + 3y = -12 \qquad (2)$$

Begin by solving equation (1) for y to get $y = 3x - 4$. Substitute $3x - 4$ for y in equation (2) and solve the resulting equation.

$$-9x + 3y = -12 \qquad (2)$$
$$-9x + 3(3x - 4) = -12 \qquad \text{Let } y = 3x - 4.$$
$$-9x + 9x - 12 = -12 \qquad \text{Distributive property}$$
$$0 = 0 \qquad \text{Add 12. Combine like terms.}$$

This true result means that every solution of one equation is also a solution of the other, so the system has an infinite number of solutions—all the ordered pairs corresponding to points that lie on the common graph. The solution set is

$$\left\{ (x, y) \mid 3x - y = 4 \right\}.$$

A graph of the equations of this system is shown in **Figure 7.**

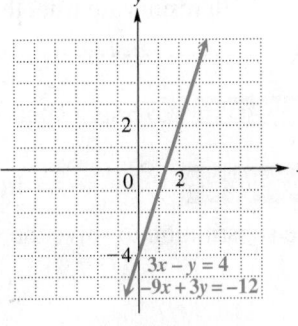

Figure 7

◄ Work Problem **6** at the Side.

(b) $8x - y = 4$
$y = 8x + 4$

CAUTION
It is a common error to give "true" as the solution of a system of dependent equations. Write the solution set in set-builder notation using the equation in the system (or an equivalent equation) that is in standard form, with integer coefficients having greatest common factor 1.

Answers
5. $\emptyset$
6. (a) $\left\{ (x, y) \mid 7x - 6y = 10 \right\}$ **(b)** $\emptyset$

OBJECTIVE ▶ ③ Solve linear systems with fractions and decimals.

EXAMPLE 6 **Using the Substitution Method with Fractions**

Solve the system by the substitution method.

$$3x + \frac{1}{4}y = 2 \qquad (1)$$

$$\frac{1}{2}x + \frac{3}{4}y = -\frac{5}{2} \qquad (2)$$

Clear equation (1) of fractions by multiplying each side by 4.

$$4\left(3x + \frac{1}{4}y\right) = 4(2) \qquad \text{Multiply by 4, the common denominator.}$$

$$4(3x) + 4\left(\frac{1}{4}y\right) = 4(2) \qquad \text{Distributive property}$$

$$12x + y = 8 \qquad (3)$$

Now clear equation (2) of fractions by multiplying each side by 4.

$$\frac{1}{2}x + \frac{3}{4}y = -\frac{5}{2} \qquad (2)$$

$$4\left(\frac{1}{2}x + \frac{3}{4}y\right) = 4\left(-\frac{5}{2}\right) \qquad \text{Multiply by 4, the common denominator.}$$

$$4\left(\frac{1}{2}x\right) + 4\left(\frac{3}{4}y\right) = 4\left(-\frac{5}{2}\right) \qquad \text{Distributive property.}$$

$$2x + 3y = -10 \qquad (4)$$

The given system of equations has been simplified to an equivalent system.

$$12x + y = 8 \qquad (3)$$

$$2x + 3y = -10 \qquad (4)$$

To solve this system by substitution, solve equation (3) for y.

$$12x + y = 8 \qquad (3)$$

$$y = -12x + 8 \qquad \text{Subtract } 12x.$$

Now substitute this result for y in equation (4).

$$2x + 3y = -10 \qquad (4)$$

$$2x + 3(-12x + 8) = -10 \qquad \text{Let } y = -12x + 8.$$

$$2x - 36x + 24 = -10 \qquad \text{Distributive property}$$

$$-34x = -34 \qquad \text{Combine like terms. Subtract 24.}$$

$$x = 1 \qquad \text{Divide by } -34.$$

> Distribute 3 to both −12x and 8.

Substitute 1 for x in $y = -12x + 8$ (equation (3) solved for y).

$$y = -12(1) + 8 = -4$$

Check $(1, -4)$ in both of the original equations. The solution set of the system is $\{(1, -4)\}$.

Work Problem ❼ at the Side. ▶

---

❼ Solve each system by the substitution method.

GS (a) $\frac{2}{3}x + \frac{1}{2}y = 6 \qquad (1)$

$$\frac{1}{2}x - \frac{3}{4}y = 0 \qquad (2)$$

First clear all fractions. The LCD for the denominators 3 and 2 in (1) is ____.

$$\underline{\quad}\left(\frac{2}{3}x + \frac{1}{2}y\right) = \underline{\quad} \quad (6)$$

$$\underline{\quad}\left(\frac{2}{3}x\right) + 6(\underline{\quad}) = 36$$

$$\underline{\quad} + 3y = 36$$

Clear fractions in (2) by multiplying both sides by ____, the LCD of denominators 2 and 4.

Complete the solution of the system. The solution set is

____ .

(b) $x + \frac{1}{2}y = \frac{1}{2}$

$$\frac{1}{6}x - \frac{1}{3}y = \frac{4}{3}$$

Answers

7. (a) $6; 6; 6; 6; \frac{1}{2}y; 4x; 4; \{(6, 4)\}$

(b) $\{(2, -3)\}$

8 Solve the system by the substitution method.

$$0.2x + 0.3y = 0.5$$

$$0.3x - 0.1y = 1.3$$

EXAMPLE 7   **Using the Substitution Method with Decimals**

Solve the system by the substitution method.

$$0.5x + 2.4y = 4.2 \quad (1)$$

$$-0.1x + 1.5y = 5.1 \quad (2)$$

Clear each equation of decimals, as in **Section 2.3,** by multiplying by 10.

$$10\,(0.5x + 2.4y) = 10\,(4.2) \qquad \text{Multiply equation (1) by 10.}$$

$$10\,(0.5x) + 10\,(2.4y) = 10\,(4.2) \qquad \text{Distributive property}$$

$$5x + 24y = 42 \qquad (3)$$

$$10\,(-0.1x + 1.5y) = 10\,(5.1) \qquad \text{Multiply equation (2) by 10.}$$

$$10\,(-0.1x) + 10\,(1.5y) = 10\,(5.1) \qquad \text{Distributive property}$$

$$\boxed{10\,(-0.1x) = -1x = -x} \longrightarrow -x + 15y = 51 \qquad (4)$$

Now solve the equivalent system of equations by substitution.

$$5x + 24y = 42 \quad (3)$$

$$-x + 15y = 51 \quad (4)$$

Equation (4) can be solved for x.

$$x = 15y - 51 \qquad \text{Equation (4) solved for } x$$

Substitute this result for x in equation (3).

$$5x + 24y = 42 \qquad (3)$$

$$5\,(15y - 51) + 24y = 42 \qquad \text{Let } x = 15y - 51.$$

$$75y - 255 + 24y = 42 \qquad \text{Distributive property}$$

$$99y = 297 \qquad \text{Combine like terms. Add 255.}$$

$$y = 3 \qquad \text{Divide by 99.}$$

Since equation (4) solved for x is $x = 15y - 51$, substitute 3 for y.

$$x = 15\,(3) - 51 = -6$$

Check $(-6, 3)$ in both of the original equations. The solution set is $\{(-6, 3)\}$.

◀ **Work Problem 8 at the Side.**

Answer

8. $\{(4, -1)\}$

4.2 Exercises

FOR EXTRA HELP

 Download the MyDashBoard App

MyMathLab®

CONCEPT CHECK *Work each problem.*

1. A student solves the following system and finds that $x = 3$, which is correct. The student gives $\{3\}$ as the solution set. **What Went Wrong?**

$$5x - y = 15$$
$$7x + y = 21$$

2. A student solves the following system and obtains the statement $0 = 0$. The student gives the solution set as $\{(0, 0)\}$. **What Went Wrong?**

$$x + y = 4$$
$$2x + 2y = 8$$

3. When we use the substitution method, how can we tell that a system has no solution?

4. When we use the substitution method, how can we tell that a system has an infinite number of solutions?

Solve each system by the substitution method. Check each solution. **See Examples 1–5.**

5. $x + y = 12$
   $y = 3x$

6. $x + 3y = -28$
   $y = -5x$

7. $3x + 2y = 27$
   $x = y + 4$

8. $4x + 3y = -5$
   $x = y - 3$

9. $3x + 5y = 14$
   $x - 2y = -10$

10. $5x + 2y = -1$
    $2x - y = -13$

11. $3x + 4 = -y$
    $2x + y = 0$

12. $2x - 5 = -y$
    $x + 3y = 0$

13. $7x + 4y = 13$
    $x + y = 1$

14. $3x - 2y = 19$
    $x + y = 8$

15. $3x - y = 5$
    $y = 3x - 5$

16. $4x - y = -3$
    $y = 4x + 3$

17. $2x + 8y = 3$
    $x = 8 - 4y$

18. $2x + 10y = 3$
    $x = 1 - 5y$

19. $2y = 4x + 24$
    $2x - y = -12$

20. $2y = 14 - 6x$

$3x + y = 7$

21. $6x - 8y = 6$

$2y = -2 + 3x$

22. $3x + 2y = 6$

$6x = 8 + 4y$

Solve each system by the substitution method. Check each solution. **See Examples 6 and 7.**

23. $\dfrac{1}{5}x + \dfrac{2}{3}y = -\dfrac{8}{5}$

$3x - y = 9$

24. $\dfrac{1}{3}x - \dfrac{1}{2}y = -\dfrac{2}{3}$

$4x + y = 6$

25. $\dfrac{1}{2}x + \dfrac{1}{3}y = -\dfrac{1}{3}$

$\dfrac{1}{2}x + 2y = -7$

26. $\dfrac{1}{6}x + \dfrac{1}{6}y = 1$

$-\dfrac{1}{2}x - \dfrac{1}{3}y = -5$

27. $\dfrac{x}{5} + 2y = \dfrac{16}{5}$

$\dfrac{3x}{5} + \dfrac{y}{2} = -\dfrac{7}{5}$

28. $\dfrac{x}{2} + \dfrac{y}{3} = \dfrac{7}{6}$

$\dfrac{x}{4} - \dfrac{3y}{2} = \dfrac{9}{4}$

29. $0.1x + 0.9y = -2$

$0.5x - 0.2y = 4.1$

30. $0.2x - 1.3y = -3.2$

$-0.1x + 2.7y = 9.8$

31. $0.8x - 0.1y = 1.3$

$2.2x + 1.5y = 8.9$

32. $0.3x - 0.1y = 2.1$

$0.6x + 0.3y = -0.3$

Relating Concepts (Exercises 33–36) For Individual or Group Work

A system of linear equations can be used to model the cost and the revenue of a business. **Work Exercises 33–36 in order.**

33. Suppose that to start a business manufacturing and selling bicycles, it costs $5000. Each bicycle will cost $400 to manufacture. Explain why the linear equation

$$y_1 = 400x + 5000 \quad (y_1 \text{ in dollars})$$

gives the *total* cost to manufacture x bicycles.

34. We decide to sell each bicycle for $600. Write an equation using y_2 (in dollars) to express the revenue when we sell x bicycles.

35. Form a system from the two equations in **Exercises 33 and 34.** Then solve the system, assuming $y_1 = y_2$, that is, cost = revenue.

36. The value of x in **Exercise 35** is the number of bicycles it takes to *break even*. Fill in the blanks: When _____ bicycles are sold, the break-even point is reached. At that point, we have spent _____ dollars and taken in _____ dollars.

4.3 Solving Systems of Linear Equations by Elimination

OBJECTIVE ▶ ① **Solve linear systems by elimination.** Recall that adding the same quantity to each side of an equation results in equal sums.

$$\text{If } A = B, \quad \text{then } A + C = B + C.$$

We can take this addition a step further. Adding *equal* quantities, rather than the *same* quantity, to each side of an equation also results in equal sums.

$$\text{If } A = B \quad \text{and} \quad C = D, \quad \text{then } A + C = B + D.$$

Using the addition property of equality to solve systems is called the **elimination method.**

OBJECTIVES

① Solve linear systems by elimination.

② Multiply when using the elimination method.

③ Use an alternative method to find the second value in a solution.

④ Solve special systems by elimination.

EXAMPLE 1 Using the Elimination Method

Use the elimination method to solve the system.

$$x + y = 5 \quad (1)$$
$$x - y = 3 \quad (2)$$

Each equation in this system is a statement of equality, so the sum of the left sides equals the sum of the right sides. Adding vertically in this way gives the following.

$$\begin{array}{ll} x + y = 5 & (1) \\ \underline{x - y = 3} & (2) \\ 2x \quad\;\; = 8 & \text{Add left sides and add right sides.} \\ x = 4 & \text{Divide by 2.} \end{array}$$

Notice that y has been eliminated. The result, $x = 4$, gives the x-value of the ordered-pair solution of the given system. To find the y-value of the solution, substitute 4 for x in either of the two equations of the system. We choose equation (1).

$$\begin{array}{ll} x + y = 5 & (1) \\ 4 + y = 5 & \text{Let } x = 4. \\ y = 1 & \text{Subtract 4.} \end{array}$$

Check the solution, $(4, 1)$, by substituting 4 for x and 1 for y in both equations of the given system.

CHECK

$$\begin{array}{llll} x + y = 5 & (1) & \quad x - y = 3 & (2) \\ 4 + 1 \overset{?}{=} 5 & \text{Substitute.} & \quad 4 - 1 \overset{?}{=} 3 & \text{Substitute.} \\ 5 = 5 \;\checkmark\; \text{True} & & \quad 3 = 3 \;\checkmark\; \text{True} \end{array}$$

Since *both* results are true, the solution set of the system is $\{(4, 1)\}$.

················· **Work Problem ①** at the Side. ▶

① Solve each system by the elimination method. Check each solution.

ⓖⓢ **(a)** Fill in the blanks to solve the following system.

$$\begin{array}{ll} x + y = 8 & (1) \\ \underline{x - y = 2} & (2) \\ \underline{\quad\;} = \underline{\quad} & \text{Add.} \\ x = \underline{\quad} \end{array}$$

Find y.

$$\begin{array}{ll} x - y = 2 & (2) \\ \underline{\quad} - y = 2 & \\ -y = \underline{\quad} & \\ y = \underline{\quad} & \end{array}$$

The solution set is ____.

(b) $3x - y = 7$
$\quad\;\; 2x + y = 3$

CAUTION

A system is not completely solved until values for both x and y are found. Do not stop after finding the value of only one variable. Remember to write the solution set as a set containing an ordered pair.

Answers
1. **(a)** $2x$; 10; 5; 5; -3; 3; $\{(5, 3)\}$
   (b) $\{(2, -1)\}$

With the elimination method, the idea is to *eliminate* one of the variables. *To do this, one pair of variable terms in the two equations must have coefficients that are opposites.*

Solving a Linear System by Elimination

Step 1   **Write both equations in standard form** $Ax + By = C$.

Step 2   **Transform so that the coefficients of one pair of variable terms are opposites.** Multiply one or both equations by appropriate numbers so that the sum of the coefficients of either the x- or y-terms is 0.

Step 3   **Add** the new equations to eliminate a variable. The sum should be an equation with just one variable.

Step 4   **Solve** the equation from Step 3 for the remaining variable.

Step 5   **Substitute** the result from Step 4 into *either* of the original equations, and solve for the other variable.

Step 6   **Check** the ordered-pair solution in *both* of the *original* equations. Then write the solution set.

It does not matter which variable is eliminated first. Usually we choose the one that is more convenient to work with.

EXAMPLE 2   **Using the Elimination Method**

Solve the system.

$$y + 11 = 2x \quad (1)$$
$$5x = y + 26 \quad (2)$$

Step 1   Write both equations in standard form $Ax + By = C$.

$$-2x + y = -11 \qquad \text{Subtract } 2x \text{ and } 11 \text{ in equation (1).}$$
$$5x - y = 26 \qquad \text{Subtract } y \text{ in equation (2).}$$

Step 2   Because the coefficients of y are 1 and -1, adding will eliminate y. It is not necessary to multiply either equation by a number.

Step 3   Add the two equations.

$$-2x + y = -11$$
$$\underline{5x - y = 26}$$
$$3x = 15 \qquad \text{Add in columns.}$$

Step 4   Solve.

$$x = 5 \qquad \text{Divide by 3.}$$

Step 5   Find the value of y by substituting 5 for x in either of the original equations.

$$y + 11 = 2x \qquad (1)$$
$$y + 11 = 2(5) \qquad \text{Let } x = 5.$$
$$y + 11 = 10 \qquad \text{Multiply.}$$
$$y = -1 \qquad \text{Subtract 11.}$$

Continued on Next Page

Step 6 Check the ordered-pair solution $(5, -1)$ by substituting $x = 5$ and $y = -1$ into both of the original equations.

CHECK

$y + 11 = 2x$ (1) $5x = y + 26$ (2)

$-1 + 11 \overset{?}{=} 2(5)$ Substitute. $5(5) \overset{?}{=} -1 + 26$ Substitute.

$10 = 10$ ✓ True $25 = 25$ ✓ True

Since $(5, -1)$ is a solution of *both* equations, the solution set is $\{(5, -1)\}$.

························ **Work Problem ② at the Side. ▶**

OBJECTIVE ▶ 2 Multiply when using the elimination method. Sometimes we need to multiply each side of one or both equations in a system by some number so that adding the equations will eliminate a variable.

EXAMPLE 3   **Using the Elimination Method**

Solve the system.

$$2x + 3y = -15 \quad (1)$$
$$5x + 2y = 1 \quad (2)$$

Adding the two equations gives $7x + 5y = -14$, which does not eliminate either variable. However, we can multiply each equation by a suitable number so that the coefficients of one of the two variables are opposites. For example, to eliminate x, multiply each side of equation (1) by 5, and each side of equation (2) by -2.

$$10x + 15y = -75 \quad \text{Multiply equation (1) by 5.}$$
$$\underline{-10x - 4y = -2} \quad \text{Multiply equation (2) by } -2.$$
$$11y = -77 \quad \text{Add.}$$
$$y = -7 \quad \text{Divide by 11.}$$

The coefficients of x are opposites.

Substituting -7 for y in either equation (1) or (2) gives $x = 3$. Check that the solution set of the system is $\{(3, -7)\}$.

························ **Work Problem ③ at the Side. ▶**

OBJECTIVE ▶ 3 Use an alternative method to find the second value in a solution. Sometimes it is easier to find the value of the second variable in a solution by using the elimination method twice.

EXAMPLE 4   **Finding the Second Value Using an Alternative Method**

Solve the system.

$$4x = 9 - 3y \quad (1)$$
$$5x - 2y = 8 \quad (2)$$

Write equation (1) in standard form by adding $3y$ to each side.

$$4x + 3y = 9 \quad (3)$$
$$5x - 2y = 8 \quad (4)$$

One way to proceed is to eliminate y by multiplying each side of equation (3) by 2 and each side of equation (2) by 3, and then adding.

························ **Continued on Next Page**

② Solve each system by the elimination method. Check each solution.

(a) $2x - y = 2$
$4x + y = 10$

(b) $8x - 5y = 32$
$4x + 5y = 4$

③ (a) Solve the system in **Example 3** by first eliminating the variable y. Check the solution.

(b) Solve the system, and check the solution.

$$6x + 7y = 4$$
$$5x + 8y = -1$$

Answers

2. (a) $\{(2, 2)\}$ **(b)** $\left\{\left(3, -\frac{8}{5}\right)\right\}$

3. (a) $\{(3, -7)\}$ **(b)** $\{(3, -2)\}$

4 Solve each system of equations.

(a) $5x = 7 + 2y$

$5y = 5 - 3x$

(b) $3y = 8 + 4x$

$6x = 9 - 2y$

5 Solve each system by the elimination method.

(a) $4x + 3y = 10$

$2x + \dfrac{3}{2}y = 12$

(b) $\quad 4x - 6y = 10$

$-10x + 15y = -25$

$$8x + 6y = 18 \quad \text{Multiply equation (3) } 4x + 3y = 9 \text{ by 2.}$$
$$\underline{15x - 6y = 24} \quad \text{Multiply equation (2) } 5x - 2y = 8 \text{ by 3.}$$
$$23x \quad = 42 \quad \text{Add.}$$

The coefficients of y are opposites.

$$x = \dfrac{42}{23} \quad \text{Divide by 23.}$$

Substituting $\dfrac{42}{23}$ for x in one of the given equations would give y, but the arithmetic would be complicated. Instead, solve for y by starting again with the original equations in standard form (equations (3) and (2)) and eliminating x.

$$20x + 15y = \quad 45 \quad \text{Multiply equation (3) } 4x + 3y = 9 \text{ by 5.}$$
$$\underline{-20x + \quad 8y = -32} \quad \text{Multiply equation (2) } 5x - 2y = 8 \text{ by } -4.$$
$$23y = \quad 13 \quad \text{Add.}$$

The coefficients of x are opposites.

$$y = \dfrac{13}{23} \quad \text{Divide by 23.}$$

Check that the solution set is $\left\{\left(\dfrac{42}{23}, \dfrac{13}{23}\right)\right\}$.

◀ **Work Problem ④ at the Side.**

Note

When the value of the first variable is a fraction, the method used in **Example 4** helps avoid arithmetic errors. This method could be used to solve any system.

OBJECTIVE ④ **Solve special systems by elimination.**

EXAMPLE 5 Solve Special Systems Using the Elimination Method

Solve each system by the elimination method.

(a) $$2x + 4y = 5 \quad (1)$$
$$4x + 8y = -9 \quad (2)$$

Multiply each side of equation (1) by -2. Then add the two equations.

$$-4x - 8y = -10 \quad \text{Multiply equation (1) by } -2.$$
$$\underline{4x + 8y = \quad -9} \quad (2)$$
$$0 = -19 \quad \text{False}$$

The false statement $0 = -19$ indicates that the system has solution set $\varnothing$.

(b) $$3x - y = 4 \quad (1)$$
$$-9x + 3y = -12 \quad (2)$$

Multiply each side of equation (1) by 3. Then add the two equations.

$$9x - 3y = \quad 12 \quad \text{Multiply equation (1) by 3.}$$
$$\underline{-9x + 3y = -12} \quad (2)$$
$$0 = 0 \quad \text{True}$$

A true statement occurs when the equations are equivalent. This indicates that every solution of one equation is also a solution of the other. The solution set is $\{(x, y) \mid 3x - y = 4\}$.

◀ **Work Problem ⑤ at the Side.**

4.3 Exercises

FOR EXTRA HELP

Download the MyDashBoard App

MyMathLab®

CONCEPT CHECK *In Exercises 1–4, answer* true *or* false *for each statement. If false, tell why.*

1. The ordered pair $(0, 0)$ *must* be a solution of a system in the following form.

$$Ax + By = 0$$
$$Cx + Dy = 0$$

2. To eliminate the y-terms in the following system, we should multiply equation (2) by 3 and then add the result to equation (1).

$$2x + 12y = 7 \quad (1)$$
$$3x + 4y = 1 \quad (2)$$

3. The following system has solution set $\emptyset$.

$$x + y = 1$$
$$x + y = 2$$

4. The ordered pair $(4, -5)$ cannot be a solution of a system that contains the following equation.

$$5x - 4y = 0$$

Solve each system by the elimination method. Check each solution. ***See Examples 1 and 2.***

5. $x + y = 2$
$2x - y = -5$

6. $3x - y = -12$
$x + y = 4$

7. $2x + y = -5$
$x - y = 2$

8. $2x + y = -15$
$-x - y = 10$

9. $3x + 2y = 0$
$-3x - y = 3$

10. $5x - y = 5$
$-5x + 2y = 0$

11. $6x - y = -1$
$5y = 17 + 6x$

12. $y = 9 - 6x$
$-6x + 3y = 15$

Solve each system by the elimination method. Check each solution. ***See Examples 3–5.***

13. $2x - y = 12$
$3x + 2y = -3$

14. $x + y = 3$
$-3x + 2y = -19$

15. $x + 3y = 19$
$2x - y = 10$

16. $4x - 3y = -19$
$2x + y = 13$

17. $x + 4y = 16$
$3x + 5y = 20$

18. $2x + y = 8$
$5x - 2y = -16$

19. $5x - 3y = -20$
$-3x + 6y = 12$

20. $4x + 3y = -28$
$5x - 6y = -35$

21. $2x - 8y = 0$
$\quad 4x + 5y = 0$

22. $3x - 15y = 0$
$\quad 6x + 10y = 0$

23. $x + y = 7$
$\quad x + y = -3$

24. $x - y = 4$
$\quad x - y = -3$

25. $-x + 3y = 4$
$\quad -2x + 6y = 8$

26. $6x - 2y = 24$
$\quad -3x + y = -12$

27. $2x + 3y = 21$
$\quad 5x - 2y = -14$

28. $5x + 4y = 12$
$\quad 3x + 5y = 15$

29. $3x - 7 = -5y$
$\quad 5x + 4y = -10$

30. $2x + 3y = 13$
$\quad 6 + 2y = -5x$

31. $2x + 3y = 0$
$\quad 4x + 12 = 9y$

32. $4x - 3y = -2$
$\quad 5x + 3 = 2y$

33. $24x + 12y = -7$
$\quad 16x - 17 = 18y$

34. $9x + 4y = -3$
$\quad 6x + 7 = -6y$

35. $5x - 2y = 3$
$\quad 10x - 4y = 5$

36. $3x - 5y = 1$
$\quad 6x - 10y = 4$

37. $6x + 3y = 0$
$\quad -18x - 9y = 0$

38. $3x - 5y = 0$
$\quad 9x - 15y = 0$

39. $3x = 3 + 2y$
$\quad -\dfrac{4}{3}x + y = \dfrac{1}{3}$

40. $3x = 27 + 2y$
$\quad x - \dfrac{7}{2}y = -25$

41. $\dfrac{1}{5}x + y = \dfrac{6}{5}$

$\dfrac{1}{10}x + \dfrac{1}{3}y = \dfrac{5}{6}$

42. $\dfrac{1}{3}x + \dfrac{1}{2}y = \dfrac{13}{6}$

$\dfrac{1}{2}x - \dfrac{1}{4}y = -\dfrac{3}{4}$

43. $2.4x + 1.7y = 7.6$

$1.2x - 0.5y = 9.2$

44. $0.5x + 3.4y = 13$

$1.5x - 2.6y = -25$

Relating Concepts (Exercises 45–48) For Individual or Group Work

The graph shows average U.S. movie theater ticket prices from 2001 through 2010. In 2001, the average ticket price was $5.66, as represented on the graph by the point P(2001, 5.66). In 2010, the average ticket price was $7.89, as represented on the graph by the point Q(2010, 7.89). **Work Exercises 45–48 in order.**

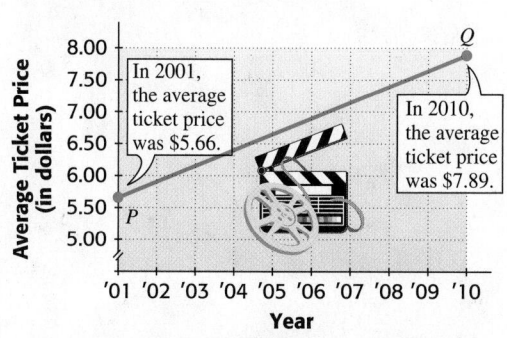

Average Movie Ticket Price

Source: Motion Picture Association of America.

45. The line segment has an equation that can be written in the form $y = ax + b$. Using the coordinates of point P with $x = 2001$ and $y = 5.66$, write an equation in the variables a and b.

46. Using the coordinates of point Q with $x = 2010$ and $y = 7.89$, write a second equation in the variables a and b.

47. Write the system of equations formed from the two equations in **Exercises 45 and 46.** Solve the system, giving the values of a and b to three decimal places. (*Hint:* Eliminate b using the elimination method.)

48. (a) What is the equation of the segment PQ?

(b) Let $x = 2008$ in the equation from part (a), and solve for y (to two decimal places). How does the result compare with the actual figure of $7.18?

Summary Exercises *Applying Techniques for Solving Systems of Linear Equations*

Use the following guidelines to help decide whether to use substitution or elimination to solve a system of linear equations.

Guidelines for Choosing a Method to Solve a System of Linear Equations

1. If one of the equations of the system is already solved for one of the variables, as indicated by the arrows in the following systems, the substitution method is the better choice.

$$3x + 4y = 9 \qquad \qquad \to x = 3y - 7$$
$$\text{and}$$
$$\to y = 2x - 6 \qquad \qquad -5x + 3y = 9$$

2. If both equations are in standard $Ax + By = C$ form and none of the variables has coefficient -1 or 1, as in the following system, the elimination method is the better choice.

$$4x - 11y = 3$$
$$-2x + 3y = 4$$

3. If one or both of the equations are in standard form and the coefficient of one of the variables is -1 or 1, as indicated by the arrows in the following systems, either method is appropriate.

$$\to 3x + y = -2 \qquad \qquad 3x - 2y = 8$$
$$\text{and}$$
$$-5x + 2y = 4 \qquad \qquad \to -x + 3y = -4$$

CONCEPT CHECK *Use the preceding guidelines to solve each problem.*

1. To minimize the amount of work required, tell whether you would use the substitution or elimination method to solve each system, and why. *Do not actually solve.*

   (a) $3x + 2y = 18$

   $y = 3x$

   (b) $3x + y = -7$

   $x - y = -5$

   (c) $3x - 2y = 0$

   $9x + 8y = 7$

2. Which system would be easier to solve using the substitution method? Why?

   $$\text{System A: } 5x - 3y = 7 \qquad \text{System B: } 7x + 2y = 4$$
   $$2x + 8y = 3 \qquad \qquad \qquad y = -3x + 1$$

*In Exercises 3 and 4, **(a)** solve the system by the elimination method, **(b)** solve the system by the substitution method, and **(c)** tell which method you prefer for that particular system and why.*

3. $4x - 3y = -8$

$x + 3y = 13$

4. $2x + 5y = 0$

$x = -3y + 1$

*Solve each system by the method of your choice. (For Exercises 5–7, see your answers for **Exercise 1**.)*

5. $3x + 2y = 18$

$y = 3x$

6. $3x + y = -7$

$x - y = -5$

7. $3x - 2y = 0$

$9x + 8y = 7$

8. $x + y = 7$

$x = -3 - y$

9. $5x - 4y = 15$

$-3x + 6y = -9$

10. $4x + 2y = 3$

$y = -x$

11. $3x = 7 - y$

$2y = 14 - 6x$

12. $3x - 5y = 7$

$2x + 3y = 30$

13. $5x = 7 + 2y$

$5y = 5 - 3x$

14. $4x + 3y = 1$

$3x + 2y = 2$

15. $2x - 3y = 7$

$-4x + 6y = 14$

16. $7x - 4y = 0$

$3x = 2y$

17. $6x + 5y = 13$

$3x + 3y = 4$

18. $x - 3y = 7$

$4x + y = 5$

Solve each system by any method. First clear all fractions or decimals.

19. $\dfrac{1}{4}x - \dfrac{1}{5}y = 9$

$y = 5x$

20. $\dfrac{1}{5}x + \dfrac{2}{3}y = -\dfrac{8}{5}$

$x - 5y = 17$

21. $\dfrac{1}{6}x + \dfrac{1}{6}y = 2$

$-\dfrac{1}{2}x - \dfrac{1}{3}y = -8$

22. $\dfrac{x}{5} + 2y = \dfrac{8}{5}$

$\dfrac{3x}{5} + \dfrac{y}{2} = -\dfrac{7}{10}$

23. $\dfrac{x}{5} + y = 6$

$\dfrac{x}{10} + \dfrac{y}{3} = \dfrac{5}{6}$

24. $\dfrac{2}{5}x + \dfrac{4}{3}y = -8$

$\dfrac{7}{10}x - \dfrac{2}{9}y = 9$

25. $0.2x + 0.3y = 1.0$

$-0.3x + 0.1y = 1.8$

26. $0.3x - 0.2y = 0.9$

$0.2x - 0.3y = 0.1$

4.4 Applications of Linear Systems

OBJECTIVES

1. Solve problems about unknown numbers.
2. Solve problems about quantities and their costs.
3. Solve problems about mixtures.
4. Solve problems about distance, rate (or speed), and time.

We modify the six-step problem-solving method from **Section 2.4** slightly to allow for two variables and two equations.

Solving an Applied Problem with Two Variables

Step 1 **Read** the problem carefully. What information is given? What is to be found?

Step 2 **Assign variables** to represent the unknown values. Use a sketch, diagram, or table, as needed. Write down what each variable represents.

Step 3 **Write two equations** using both variables.

Step 4 **Solve** the system of two equations.

Step 5 **State the answer.** Label it appropriately. Does it seem reasonable?

Step 6 **Check** the answer in the words of the *original* problem.

OBJECTIVE ▶ **1** Solve problems about unknown numbers.

EXAMPLE 1 Solving a Problem about Two Unknown Numbers

In 2011, consumer sales of sports equipment were $307 million more for snow skiing than for snowboarding. Together, total equipment sales for these two sports were $931 million. What were equipment sales for each sport? (*Source:* National Sporting Goods Association.)

Step 1 **Read** the problem carefully. We must find equipment sales (in millions of dollars) for snow skiing and for snowboarding. We know how much more equipment sales were for snow skiing than for snowboarding. Also, we know the total sales.

Step 2 **Assign variables.**

Let x = equipment sales for skiing (in millions of dollars),

and y = equipment sales for snowboarding (in millions of dollars).

Step 3 **Write two equations.**

$x = 307 + y$ Equipment sales for skiing were $307 million more than equipment sales for snowboarding.   (1)

$x + y = 931$ Total sales were $931 million.   (2)

Step 4 **Solve** the system from Step 3. We use the substitution method since the first equation is already solved for x.

$$x + y = 931 \quad (2)$$
$$(307 + y) + y = 931 \quad \text{Let } x = 307 + y.$$
$$307 + 2y = 931 \quad \text{Combine like terms.}$$
$$2y = 624 \quad \text{Subtract 307.}$$
$$y = 312 \quad \text{Divide by 2.}$$

Don't stop here. ▶

Continued on Next Page

To find the value of x, we substitute 312 for y in either equation (1) or (2).

$$x = 307 + 312 \qquad \text{Let } y = 312 \text{ in equation (1)}.$$
$$x = 619 \qquad \text{Add.}$$

Step 5 **State the answer.** Equipment sales for skiing were $619 million, and equipment sales for snowboarding were $312 million.

Step 6 **Check** the answer in the original problem. Since

$$619 = 307 + 312 \quad \text{and} \quad 619 + 312 = 931,$$

the answer satisfies the information given in the problem.

··· **Work Problem ❶ at the Side.** ▶

CAUTION
If an applied problem asks for *two* values as in **Example 1,** be sure to give both of them in your answer.

OBJECTIVE ❷ **Solve problems about quantities and their costs.**

EXAMPLE 2 Solving a Problem about Quantities and Costs

For a production of *Jersey Boys* at the August Wilson Theatre in New York, main floor tickets cost $127 and rear mezzanine tickets cost $97. The members of a club spent a total of $3150 for 30 tickets. How many tickets of each kind did they buy? (*Source:* www.broadway.com)

Step 1 **Read** the problem several times.

Step 2 **Assign variables.**

Let x = the number of main floor tickets,

and y = the number of rear mezzanine tickets.

Summarize the information given in the problem in a table. The entries in the first two rows of the Total Value column were found by multiplying Number of Tickets by Price per Ticket.

	Number of Tickets	Price per Ticket (in dollars)	Total Value (in dollars)
Main Floor	x	127	$127x$
Mezzanine	y	97	$97y$
Total	30	✕✕✕✕✕	3150

Step 3 **Write two equations.**

$$x + y = 30 \qquad \text{Total number of tickets was 30.} \qquad (1)$$
$$127x + 97y = 3150 \qquad \text{Total value of tickets was \$3150.} \qquad (2)$$

Step 4 **Solve** the system formed in Step 3 using the elimination method.

$$x + y = 30 \qquad (1)$$
$$127x + 97y = 3150 \qquad (2)$$

·· **Continued on Next Page**

---

❶ Solve each problem.

(a) In 2011, consumer sales of sports equipment were $26 million less for tennis than for archery. Together, total equipment sales for these two sports were $876 million. What were equipment sales for each sport? (*Source:* National Sporting Goods Association.)

Step 1
We must find _____.

Step 2
Let x = equipment sales for tennis (in millions of dollars), and y = equipment sales for _____ (in millions of dollars).

Step 3
Write two equations.

$$x = \text{_____} \qquad (1)$$
(See the first sentence in the problem.)

$$\text{_____} \qquad (2)$$
(See the second sentence in the problem.)

Complete Steps 4–6 to solve the problem. Give the answer.

(b) Two of the most popular movies of 2011 were *Fast Five* and *Cars 2*. Together, their domestic gross was about $401 million. *Cars 2* grossed $18 million less than *Fast Five*. How much did each movie gross? (*Source:* www.boxofficemojo.com)

Answers
1. (a) equipment sales for each sport; archery;
$y - 26$; $x + y = 876$;
equipment sales for tennis: $425 million;
equipment sales for archery: $451 million
(b) *Fast Five*: $209.5 million;
Cars 2: $191.5 million

2 For a production of *The Lion King* playing at the Minskoff Theatre in New York, orchestra tickets cost $129 and rear mezzanine tickets cost $87. If a group of 18 people attended the show and spent a total of $1776 for their tickets, how many of each kind of ticket did they buy? (*Source:* www.broadway.com)

(a) Complete the table.

	Number of Tickets	Price per Ticket (dollars)	Total Value (dollars)
Orchestra	x	____	____
Mezzanine	y	____	____
Total	____	✕✕✕	____

(b) Write a system of equations.

(c) Use the system of equations to solve the problem. Check your answer in the words of the original problem.

Answers

2. (a)

	Number of Tickets	Price per Ticket (dollars)	Total Value (dollars)
Orchestra	x	129	$129x$
Mezzanine	y	87	$87y$
Total	18	✕✕✕	1776

(b) $x + y = 18$
$129x + 87y = 1776$

(c) orchestra: 5; mezzanine: 13

$$-97x - 97y = -2910 \qquad \text{Multiply equation (1) by } -97.$$
$$\underline{127x + 97y = 3150} \qquad (2)$$
$$30x = 240 \qquad \text{Add.}$$

Main floor tickets $\longrightarrow x = 8$ Divide by 30.

Substitute 8 for x in equation (1).

$$x + y = 30 \qquad (1)$$
$$\mathbf{8} + y = 30 \qquad \text{Let } x = 8.$$

Mezzanine tickets $\longrightarrow y = 22$ Subtract 8.

Step 5 **State the answer.** The club members bought 8 main floor tickets and 22 rear mezzanine tickets.

Step 6 **Check.** The sum of 8 and 22 is 30, so the total number of tickets is correct. Since 8 tickets were purchased at $127 each and 22 at $97 each, the total spent on all the tickets is

$$\$127(8) + \$97(22) = \$3150, \qquad \text{as required.}$$

◄ **Work Problem** **2** **at the Side.**

OBJECTIVE **3** **Solve problems about mixtures.** In **Section 2.6,** we solved percent problems using one variable. Mixture problems that involve percent can be solved using a system of two equations in two variables.

EXAMPLE 3 **Solving a Mixture Problem Involving Percent**

A pharmacist needs 100 L of 50% alcohol solution. She has a 30% alcohol solution and an 80% alcohol solution, which she can mix. How many liters of each will be required to make the 100 L of a 50% alcohol solution?

Step 1 **Read** the problem. Note the percent of each solution and of the mixture.

Step 2 **Assign variables.**

Let x = the number of liters of 30% alcohol needed,

and y = the number of liters of 80% alcohol needed.

Liters of Mixture	Percent (as a decimal)	Liters of Pure Alcohol
x	0.30	$0.30x$
y	0.80	$0.80y$
100	0.50	$0.50(100)$

Summarize the information in a table. Percents are written as decimals.

Figure 8 gives an idea of what is happening in this problem.

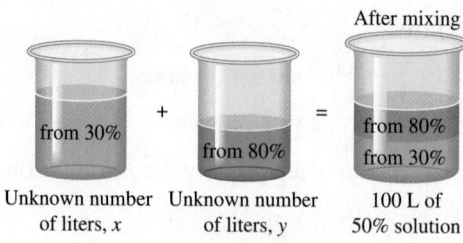

Figure 8

Continued on Next Page

Step 3 **Write two equations.** The total number of liters in the final mixture will be 100, which gives one equation.

$$x + y = 100$$

To find the amount of pure alcohol in each mixture, multiply the number of liters by the concentration. (Refer to the table.) The amount of pure alcohol in the 30% solution added to the amount of pure alcohol in the 80% solution will equal the amount of pure alcohol in the final 50% solution. This gives a second equation.

$$0.30x + 0.80y = 0.50(100)$$

These two equations form a system.

> Be sure to write two equations.

$$x + y = 100 \quad (1)$$
$$0.30x + 0.80y = 50 \quad (2) \quad 0.50(100) = 50$$

Step 4 **Solve** the system using the substitution method. Solving equation (1) for x gives $x = 100 - y$. Substitute $100 - y$ for x in equation (2).

$$0.30x + 0.80y = 50 \quad (2)$$
$$0.30(\mathbf{100 - y}) + 0.80y = 50 \quad \text{Let } x = 100 - y.$$
$$30 - 0.30y + 0.80y = 50 \quad \text{Distributive property}$$
$$30 + 0.50y = 50 \quad \text{Combine like terms.}$$
$$0.50y = 20 \quad \text{Subtract 30.}$$
Liters of 80% solution $\longrightarrow y = \mathbf{40} \quad$ Divide by 0.50.

Equation (1) solved for x is $x = 100 - y$. Substitute 40 for y to find the value of x.

$$x = 100 - \mathbf{40} \quad \text{Let } y = 40.$$
Liters of 30% solution $\longrightarrow x = \mathbf{60} \quad$ Subtract.

Step 5 **State the answer.** The pharmacist should use 60 L of the 30% solution and 40 L of the 80% solution.

Step 6 **Check** the answer in the original problem. Since

$$60 + 40 = 100 \quad \text{and} \quad 0.30(60) + 0.80(40) = 50,$$

this mixture will give the 100 L of 50% solution, as required.

············· **Work Problems ❸ and ❹ at the Side.** ▶

Note

In Example 3, we could have used the elimination method. Also, we could have cleared decimals by multiplying each side of equation (2) by 10.

OBJECTIVE ❹ **Solve problems about distance, rate (or speed), and time.** If an automobile travels at an average rate of 50 mph for 2 hr, then it travels

$$50 \times 2 = 100 \text{ mi.}$$

This is an example of the basic relationship between distance, rate, and time,

distance = rate × time, given by the formula **$d = rt$.**

Work Problem ❺ at the Side. ▶

❸ How many liters of a 25% alcohol solution must be mixed with a 12% solution to get 13 L of a 15% solution?

(a) Complete the table.

Liters	Percent (as a decimal)	Liters of Pure Alcohol
x	0.25	$0.25x$
y	0.12	_____
13	0.15	_____

(b) Write a system of equations, and use it to solve the problem.

❹ Solve the problem.
Joe needs 60 milliliters (mL) of 20% acid solution for a chemistry experiment. The lab has on hand only 10% and 25% solutions. How much of each should he mix to get the desired amount of 20% solution?

❺ Solve using the distance formula $d = rt$.
A small plane traveled from Stockholm, Sweden, to Oslo, Norway, averaging 244 km per hr. The trip took 1.7 hr. To the nearest kilometer, what is the distance between the two cities?

Answers
3. (a)

Liters	Percent (as a decimal)	Liters of Pure Alcohol
x	0.25	$0.25x$
y	0.12	$0.12y$
13	0.15	$0.15(13)$

(b) $x + y = 13$
$0.25x + 0.12y = 0.15(13)$;
3 L of 25%; 10 L of 12%
4. 20 mL of 10%; 40 mL of 25%
5. 415 km

6 Two cars that were 450 mi apart traveled toward each other. They met after 5 hr. If one car traveled twice as fast as the other, what were their rates?

(a) Complete this table.

	r	t	d
Faster Car	x	5	___
Slower Car	y	5	___

(b) Write a system of equations, and use it to solve the problem.

7 Solve the problem.
From a truck stop, two trucks travel in opposite directions on a straight highway. In 3 hr they are 405 mi apart. Find the rate of each truck if one travels 5 mph faster than the other.

EXAMPLE 4	**Solving a Problem about Distance, Rate, and Time**

Two executives in cities 400 mi apart drive to a business meeting at a location on the line between their cities. They meet after 4 hr. Find the rate (speed) of each car if one car travels 20 mph faster than the other.

Step 1 **Read** the problem carefully.

Step 2 **Assign variables.**

Let x = the rate of the faster car,

and y = the rate of the slower car.

Make a table using the formula $d = rt$, and draw a sketch. See **Figure 9.**

	r	t	d
Faster Car	x	4	$x \cdot 4$, or $4x$
Slower Car	y	4	$y \cdot 4$, or $4y$

Since each car travels for 4 hr, the time t for each car is 4. Find d, using $d = rt$ (or $rt = d$).

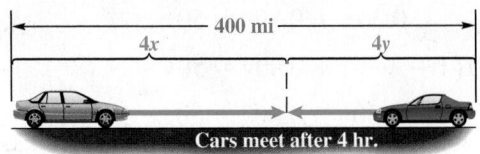

Figure 9

Step 3 **Write two equations.** The total distance traveled by both cars is 400 mi, which gives equation (1). The faster car goes 20 mph faster than the slower car, which gives equation (2).

$$4x + 4y = 400 \quad (1)$$
$$x = 20 + y \quad (2)$$

Step 4 **Solve** the system of equations by substitution. Replace x with $20 + y$ in equation (1) and then solve for y.

$4x + 4y = 400$	(1)
$4(20 + y) + 4y = 400$	Let $x = 20 + y$.
$80 + 4y + 4y = 400$	Distributive property
$80 + 8y = 400$	Combine like terms.
$8y = 320$	Subtract 80.
Slower car → $y = 40$	Divide by 8.

To find x, substitute 40 for y in equation (2), $x = 20 + y$.

$x = 20 + 40$	Let $y = 40$.
Faster car → $x = 60$	Add.

Step 5 **State the answer.** The rates of the cars are 40 mph and 60 mph.

Step 6 **Check** the answer. Since each car travels for 4 hr, total distance is

$$4(60) + 4(40) = 240 + 160 = 400 \text{ mi}, \quad \text{as required.}$$

◀ **Work Problems 6 and 7 at the Side.**

Answers

6. (a)

	r	t	d
Faster Car	x	5	5x
Slower Car	y	5	5y

(b) $5x + 5y = 450$
$x = 2y$;
faster car: 60 mph; slower car: 30 mph

7. faster truck: 70 mph; slower truck: 65 mph

> **CAUTION**
>
> Be careful. ***When you use two variables to solve a problem, you must write two equations.***

EXAMPLE 5 Solving a Problem about Distance, Rate, and Time

A plane flies 560 mi in 1.75 hr traveling with the wind. The return trip against the same wind takes the plane 2 hr. Find the rate (speed) of the plane and the wind speed.

Step 1 **Read** the problem several times.

Step 2 **Assign variables.**

$$\text{Let } x = \text{the rate of the plane,}$$
$$\text{and } y = \text{the wind speed.}$$

When the plane is traveling *with* the wind, the wind "pushes" the plane. In this case, the rate (speed) of the plane is the *sum* of the rate of the plane and the wind speed, $(x + y)$ mph. See **Figure 10.**

When the plane is traveling *against* the wind, the wind "slows" the plane down. In this case, the rate (speed) of the plane is the *difference* between the rate of the plane and the wind speed, $(x - y)$ mph. Again, see **Figure 10.**

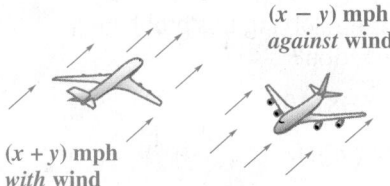

$(x - y)$ mph
against wind

$(x + y)$ mph
with wind

Figure 10

	r	t	d
With Wind	$x + y$	1.75	560
Against Wind	$x - y$	2	560

Summarize this information in a table. Use the formula $d = rt$ (or $rt = d$).

Step 3 **Write two equations.** Refer to the table to do this.

$$(x + y)\,1.75 = 560 \xrightarrow{\text{Divide by 1.75.}} x + y = 320 \quad (1)$$

$$(x - y)\,2 = 560 \xrightarrow{\text{Divide by 2.}} x - y = 280 \quad (2)$$

Step 4 **Solve** the system of equations (1) and (2) using elimination.

$$\begin{aligned} x + y &= 320 \quad (1) \\ \underline{x - y} &= \underline{280} \quad (2) \\ 2x &= 600 \quad \text{Add.} \\ x &= 300 \quad \text{Divide by 2.} \end{aligned}$$

Since $x + y = 320$ and $x = 300$, it follows that $y = 20$.

Step 5 **State the answer.** The rate of the plane is 300 mph, and the wind speed is 20 mph.

Step 6 **Check.** The answer seems reasonable, and true statements result when the values are substituted into the equations of the system.

·················· Work Problem **8** at the Side. ▶

---

8 Solve each problem.

(a) In 1 hr, Gigi can row 2 mi against the current or 10 mi with the current. Find the rate of the current and Gigi's rate in still water.

Steps 1 and 2
Let x = the rate of the current, and y = _____ in still water. Then her rate *against* the current is (_____) mph, and her rate with the current is (_____) mph.

Complete the table.

	r	t	d
Against Current	___	___	___
With Current	___	___	___

Step 3
Write two equations.

$(y - x) \cdot 1 =$ ____ (1)
(See the first row of the table.)

_____ (2)
(See the second row of the table.)

Complete Steps 4–6 to solve the problem. Give the answer.

(b) In 1 hr, a boat travels 15 mph with the current and 9 mph against the current. Find the rate of the boat and the rate of the current.

Answers

8. **(a)** Gigi's rate; $y - x$; $y + x$

	r	t	d
Against Current	$y - x$	1	2
With Current	$y + x$	1	10

2; $(y + x) \cdot 1 = 10$;
rate of the current: 4 mph;
Gigi's rate: 6 mph
(b) rate of the boat: 12 mph;
rate of the current: 3 mph

4.4 Exercises

FOR EXTRA HELP

Download the MyDashBoard App

MyMathLab®

CONCEPT CHECK   *Choose the correct response in Exercises 1–8.*

1. Which expression represents the monetary value of x 20-dollar bills?

   A. $\dfrac{x}{20}$ dollars   **B.** $\dfrac{20}{x}$ dollars   **C.** $(20 + x)$ dollars   **D.** $20x$ dollars

2. Which expression represents the cost of t pounds of candy that sells for $1.95 per lb?

   A. $1.95t$ **B.** $\dfrac{\$1.95}{t}$ **C.** $\dfrac{t}{\$1.95}$ **D.** $1.95 + t$

3. Which expression represents the amount of interest earned on d dollars invested at an interest rate of 2% for 1 yr?

   A. $2d$ dollars   **B.** $0.02d$ dollars   **C.** $0.2d$ dollars   **D.** $200d$ dollars

4. Suppose that x liters of a 40% acid solution are mixed with y liters of a 35% solution to obtain 100 L of a 38% solution. One equation in a system for solving this problem is $x + y = 100$. Which one of the following is the other equation?

   A. $0.35x + 0.40y = 0.38(100)$ **B.** $0.40x + 0.35y = 0.38(100)$

   C. $35x + 40y = 38$ **D.** $40x + 35y = 0.38(100)$

5. According to *Natural History* magazine, the speed of a cheetah is 70 mph. If a cheetah runs for x hours, how many miles does the cheetah cover?

   A. $(70 + x)$ miles   **B.** $(70 - x)$ miles   **C.** $\dfrac{70}{x}$ miles   **D.** $70x$ miles

6. How far does a car travel in 2.5 hr if it travels at an average rate of x miles per hour?

   A. $(x + 2.5)$ miles   **B.** $\dfrac{2.5}{x}$ miles   **C.** $\dfrac{x}{2.5}$ miles   **D.** $2.5x$ miles

7. What is the rate of a plane that travels at 560 mph *with* a wind of r mph?

   A. $\dfrac{r}{560}$ mph   **B.** $(560 - r)$ mph   **C.** $(560 + r)$ mph   **D.** $(r - 560)$ mph

8. What is the rate of a plane that travels at 560 mph *against* a wind of r mph?

   A. $(560 + r)$ mph   **B.** $\dfrac{560}{r}$ mph   **C.** $(560 - r)$ mph   **D.** $(r - 560)$ mph

GS *In Exercises 9 and 10, refer to the six-step problem-solving method. Fill in the blanks for Steps 2 and 3, and then complete the solution by applying Steps 4–6.*

9. The sum of two numbers is 98 and the difference between them is 48. Find the two numbers.

Step 1 **Read** the problem carefully.

Step 2 **Assign variables.**

Let x = the first number

and y = _____ .

Step 3 **Write two equations.**

First equation: $x + y = 98$

Second equation: _____

10. The sum of two numbers is 201 and the difference between them is 11. Find the two numbers.

Step 1 **Read** the problem carefully.

Step 2 **Assign variables.**

Let x = _____

and y = the second number.

Step 3 **Write two equations.**

First equation: _____

Second equation: $x - y = 11$

Solve each problem using a system of equations. **See Example 1.**

11. Two of the longest-running shows on Broadway are *The Phantom of the Opera* and *Cats.* As of August 21, 2011, there had been a total of 17,288 performances of the two shows, with 2318 more performances of *The Phantom of the Opera* than *Cats.* How many performances were there of each show? (*Source:* The Broadway League.)

12. During Broadway runs of *A Chorus Line* and *Beauty and the Beast,* there have been 676 fewer performances of *Beauty and the Beast* than of *A Chorus Line.* There were a total of 11,598 performances of the two shows. How many performances were there of each show? (*Source:* The Broadway League.)

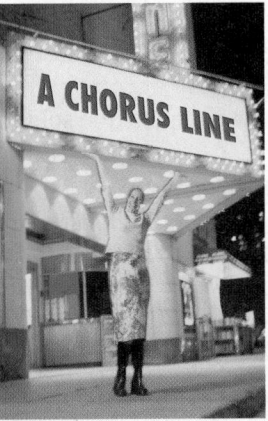

13. The two domestic top-grossing movies of 2011 were *Harry Potter and the Deathly Hallows Part 2* and *Transformers: Dark of the Moon.* The Transformers movie grossed $29 million less than the Harry Potter movie, and together the two films took in $733 million. How much did each of these movies earn? (*Source:* www.boxofficemojo.com)

14. During their opening weekends, the movies *Harry Potter and the Deathly Hallows Part 2* and *Transformers: Dark of the Moon* grossed a total of $267 million, with the Harry Potter movie grossing $71 million more than the Transformers movie. How much did each of these movies earn during their opening weekends? (*Source:* www.boxofficemojo.com)

15. The Terminal Tower in Cleveland, Ohio, is 239 ft shorter than the Key Tower, also in Cleveland. The total of the heights of the two buildings is 1655 ft. Find the heights of the buildings. (*Source: World Almanac and Book of Facts.*)

16. The total of the heights of the Chase Tower and the One America Tower, both in Indianapolis, Indiana, is 1234 ft. The Chase Tower is 168 ft taller than the One America Tower. Find the heights of the two buildings. (*Source: World Almanac and Book of Facts.*)

17. An official playing field (including end zones) for the Indoor Football League has length 38 yd longer than its width. The perimeter of the rectangular field is 188 yd. Find the length and width of the field. (*Source: Indoor Football League.*)

Steps 1 and 2 **Read** carefully, and **assign** _____.

Let x = the length (in yards),

and y = the _____ (in yards).

Step 3 **Write two equations.**

Equation (1): See the first sentence in the problem. Express the length in terms of the width.

$$x = \text{_____} \quad (1)$$

Equation (2): See the second sentence in the problem. Perimeter of a rectangle equals twice the _____ plus _____ the width.

$$2x + \text{_____} = \text{_____} \quad (2)$$

Now complete the solution.

18. Pickleball is a combination of badminton, tennis, and ping pong. The perimeter of the rectangular-shaped court is 128 ft. The width is 24 ft shorter than the length. Find the length and width of the court. (*Source: www.sportsknowhow.com*)

Steps 1 and 2 **Read** carefully, and **assign** _____.

Let x = the _____ (in feet),

and y = the width (in feet).

Step 3 **Write two equations.**

Equation (1): Perimeter of a rectangle equals _____ the length plus twice the _____.

$$\text{_____} + 2y = \text{_____} \quad (1)$$

Equation (2): See the third sentence in the problem. Express the width in terms of the length.

$$y = \text{_____} \quad (2)$$

Now complete the solution.

*Suppose that x units of a product cost C dollars to manufacture and earn revenue of R dollars. The value of x, where the expressions for C and R are equal, is the **break-even quantity**, the number of units that produce 0 profit.*

In Exercises 19 and 20, (a) find the break-even quantity, and (b) decide whether the product should be produced based on if it will earn a profit. (Profit equals revenue minus cost.)

19. $C = 85x + 900; \quad R = 105x;$
No more than 38 units can be sold.

20. $C = 105x + 6000; \quad R = 255x;$
No more than 400 units can be sold.

*For each problem, complete any tables. Then solve the problem using a system of equations. **See Example 2.***

21. A motel clerk counts his $1 and $10 bills at the end of a day. He finds that he has a total of 74 bills having a combined monetary value of $326. Find the number of bills of each denomination that he has.

Number of Bills	Denomination of Bill (in dollars)	Total Value (in dollars)
x	1	_____
y	10	_____
74	✕✕✕✕✕✕	326

22. Carly is a bank teller. At the end of a day, she has a total of 69 $5 and $10 bills. The total value of the money is $590. How many of each denomination does she have?

Number of Bills	Denomination of Bill (in dollars)	Total Value (in dollars)
x	5	$5x$
y	10	_____
_____	✕✕✕✕✕✕	_____

23. Tracy Sudak bought each of her seven nephews a DVD of *Moneyball* or a Blu-ray disc of *The Concert for George*. Each DVD cost $16.99. Each Blu-ray disc cost $22.42. Tracy spent a total of $129.79. How many of each did she buy? (*Source:* www.amazon.com)

24. Terry Wong bought each of his five nieces a DVD of *Dolphin Tale* or a Blu-ray disc of *Treasure Buddies*. Each DVD cost $14.99. Each Blu-ray disc cost $24.99. Terry spent a total of $84.95. How many of each did he buy? (*Source:* www.amazon.com)

25. Maria Lopez has twice as much money invested at 5% simple annual interest as she does at 4%. If her yearly income from these two investments is $350, how much does she have invested at each rate?

Amount Invested (in dollars)	Rate of Interest	Interest for One Year (in dollars)
x	5%, or 0.05	$0.05x$
y	_____	_____
XXXXX	XXXXX	350

Equation (1): $x =$ _____

Equation (2): $0.05x +$ _____ $=$ _____

26. Charles Miller invested in two accounts, one paying 3% simple annual interest and the other paying 2% interest. He earned a total of $880 interest. If he invested three times as much in the 3% account as in the 2% account, how much did he invest at each rate?

Amount Invested (in dollars)	Rate of Interest	Interest for One Year (in dollars)
x	_____	_____
y	_____	_____
XXXXX	XXXX	_____

Equation (1):   _____ $+$ _____ $= 880$

Equation (2): $x =$ _____

27. The two top-grossing North American concert tours in 2011 were U2 and Taylor Swift. Based on average ticket prices, it cost a total of $912 to buy six tickets for U2 and five tickets for Taylor Swift. Three tickets for U2 and four tickets for Taylor Swift cost $564. How much did an average ticket cost for each tour? (*Source:* Pollstar.)

28. Two other popular North American concert tours in 2011 were Kenny Chesney and Lady Gaga. Based on average ticket prices, it cost a total of $875 to buy eight tickets for Kenny Chesney and three tickets for Lady Gaga. Four tickets for Kenny Chesney and five tickets for Lady Gaga cost $777. How much did an average ticket cost for each tour? (*Source:* Pollstar.)

For each problem, complete any tables. Then solve the problem using a system of equations. See Example 3.

29. A 40% dye solution is to be mixed with a 70% dye solution to get 120 L of a 50% solution. How many liters of the 40% and 70% solutions will be needed?

Liters of Solution	Percent (as a decimal)	Liters of Pure Dye
x	0.40	_____
y	0.70	_____
120	0.50	_____

30. A 90% antifreeze solution is to be mixed with a 75% solution to make 120 L of a 78% solution. How many liters of the 90% and 75% solutions will be used?

Liters of Solution	Percent (as a decimal)	Liters of Pure Antifreeze
x	0.90	_____
y	0.75	_____
120	0.78	_____

31. Ahmad Hashemi wishes to mix coffee worth $6 per lb with coffee worth $3 per lb to get 90 lb of a mixture worth $4 per lb. How many pounds of the $6 and the $3 coffees will be needed?

Number of Pounds	Dollars per Pound	Cost (in dollars)
x	6	_____
y	_____	_____
90	_____	_____

32. Mariana Coanda wishes to blend candy selling for $1.20 per lb with candy selling for $1.80 per lb to get a mixture that will be sold for $1.40 per lb. How many pounds of the $1.20 and the $1.80 candies should be used to get 45 lb of the mixture?

Number of Pounds	Dollars per Pound	Cost (in dollars)
x	_____	_____
y	1.80	_____
45	_____	_____

33. How many pounds of nuts selling for $6 per lb and raisins selling for $3 per lb should Kelli Hammer combine to obtain 60 lb of a trail mix selling for $5 per lb?

34. Callie Daniels is preparing cheese trays. She uses some cheeses that sell for $8 per lb and others that sell for $12 per lb. How many pounds of each cheese should she use in order for the mixed cheeses on the trays to weigh a total of 56 lb and sell for $10.50 per lb?

For each problem, complete any tables. Then solve the problem using a system of equations. **See Examples 4 and 5.**

35. Two trains start from towns 495 mi apart and travel toward each other on parallel tracks. They pass each other 4.5 hr later. If one train travels 10 mph faster than the other, find the rate of each train.

	r	t	d
Train 1	x	_____	_____
Train 2	y	_____	_____

Equation (1): $4.5x +$ _____ $=$ _____

Equation (2): $x =$ _____ $+$ _____

36. Two trains that are 495 mi apart travel toward each other. They pass each other 5 hr later. If one train travels half as fast as the other, what are their rates?

	r	t	d
Train 1	x	_____	_____
Train 2	_____	_____	$5y$

Equation (1): _____ $+ 5y =$ _____

Equation (2): $x =$ _____y, or _____ $= y$

37. **RAGBRAI®**, the Des Moines **R**egister's **A**nnual **G**reat **B**icycle **R**ide **A**cross **I**owa, is the longest and oldest touring bicycle ride in the world. Suppose a cyclist began the 471-mi ride on July 22, 2012, in western Iowa at the same time that a car traveling toward it left eastern Iowa. If the bicycle and the car met after 7.5 hr and the car traveled 35.8 mph faster than the bicycle, find the average rate of each. (*Source:* www.ragbrai.com)

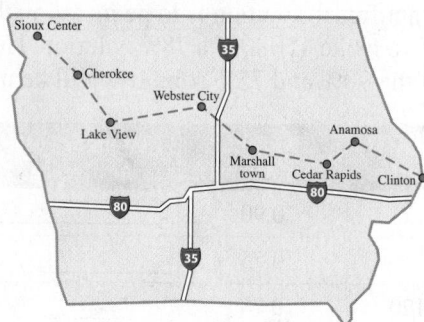

38. In 2010, Atlanta's Hartsfield Airport was the nation's busiest. Suppose two planes leave the airport at the same time, one traveling east and the other traveling west. If the planes are 2100 mi apart after 2 hr and one plane travels 50 mph faster than the other, find the rate of each plane. (*Source:* Airports Council International.)

39. Toledo and Cincinnati are 200 mi apart. A car leaves Toledo traveling toward Cincinnati, and another car leaves Cincinnati at the same time, traveling toward Toledo. The car leaving Toledo averages 15 mph faster than the other, and they meet after 1 hr, 36 min. What are the rates of the cars?

40. Kansas City and Denver are 600 mi apart. Two cars start from these cities, traveling toward each other. They meet after 6 hr. Find the rate of each car if one travels 30 mph slower than the other.

Denver ⊢————— 600 mi —————⊣ Kansas City
Rate: $y = x - 30$ Rate: x

41. At the beginning of a bicycle ride for charity, Roberto and Juana are 30 mi apart. If they leave at the same time and ride in the same direction, Roberto overtakes Juana in 6 hr. If they ride toward each other, they meet in 1 hr. What are their rates?

42. Mr. Abbot left Farmersville in a plane at noon to travel to Exeter. Mr. Baker left Exeter in his automobile at 2 P.M. to travel to Farmersville. It is 400 mi from Exeter to Farmersville. If the sum of their rates was 120 mph, and if they crossed paths at 4 P.M., find the rate of each.

43. A boat takes 3 hr to go 24 mi upstream. It can go 36 mi downstream in the same time. Find the rate of the current and the rate of the boat in still water.

Let x = the rate of the boat in still water and y = the rate of the current.

	r	t	d
Downstream	$x + y$	_____	36
Upstream	$x - y$	_____	_____

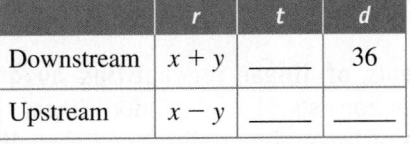

Downstream   Upstream
$x + y$ $x - y$

44. It takes a boat $1\frac{1}{2}$ hr to go 12 mi downstream, and 6 hr to return. Find the rate of the boat in still water and the rate of the current.

Let x = the rate of the boat in still water and y = the rate of the current.

	r	t	d
Downstream	$x + y$	$\frac{3}{2}$	_____
Upstream	_____	6	_____

45. If a plane can travel 440 mph against the wind and 500 mph with the wind, find the wind speed and the rate of the plane in still air.

440 mph
against wind

500 mph
with wind

46. A small plane travels 200 mph with the wind and 120 mph against it. Find the wind speed and the rate of the plane in still air.

4.5 Solving Systems of Linear Inequalities

OBJECTIVE

1 **Solve systems of linear inequalities by graphing.**

We graphed the solutions of a linear inequality in **Section 3.5.** For example, to graph the solutions of $x + 3y > 12$, we first graph the boundary line $x + 3y = 12$ by finding and plotting a few ordered pairs that satisfy the equation. (The x- and y-intercepts are good choices.) Because the $>$ symbol does not include equality, the points on the line do *not* satisfy the inequality, and we graph it using a dashed line. To decide which region includes the points that are solutions, we choose a test point not on the line.

$$x + 3y > 12 \quad \text{Original inequality}$$

(0, 0) is a convenient test point.

$$0 + 3(0) \overset{?}{>} 12 \quad \text{Let } x = 0 \text{ and } y = 0.$$

$$0 > 12 \quad \text{False}$$

This false result indicates that the solutions are those points on the side of the line that does *not* include $(0, 0)$, as shown in **Figure 11.**

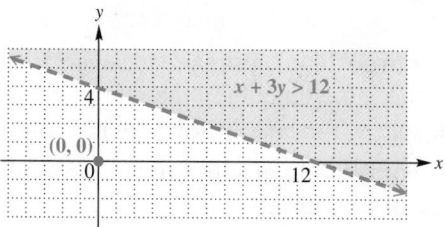

Figure 11

OBJECTIVE 1 **Solve systems of linear inequalities by graphing.** A **system of linear inequalities** consists of two or more linear inequalities. The **solution set of a system of linear inequalities** includes all points that make all inequalities of the system true at the same time.

Solving a System of Linear Inequalities

Step 1 **Graph each linear inequality.** Use the method of **Section 3.5.**

Step 2 **Choose the intersection.** Indicate the solution set of the system by shading the intersection of the graphs (the region where the graphs overlap).

EXAMPLE 1 **Solving a System of Two Linear Inequalities**

Graph the solution set of the system.

$$3x + 2y \leq 6$$
$$2x - 5y \geq 10$$

Step 1 To graph $3x + 2y \leq 6$, graph the solid boundary line $3x + 2y = 6$ using the intercepts $(0, 3)$ and $(2, 0)$. Determine the region to shade.

$$3x + 2y < 6$$
$$3(0) + 2(0) \overset{?}{<} 6 \quad \text{Use } (0, 0) \text{ as a test point.}$$
$$0 < 6 \quad \text{True}$$

Shade the region containing $(0, 0)$. See **Figure 12(a)** on the next page.

· **Continued on Next Page**

Now graph $2x - 5y \geq 10$ with solid boundary line $2x - 5y = 10$ using the intercepts $(0, -2)$ and $(5, 0)$. Determine the region to shade.

$$2x - 5y > 10$$
$$2(0) - 5(0) \overset{?}{>} 10 \qquad \text{Use } (0, 0) \text{ as a test point.}$$
$$0 > 10 \qquad \text{False}$$

Shade the region that does *not* contain $(0, 0)$. See **Figure 12(b).**

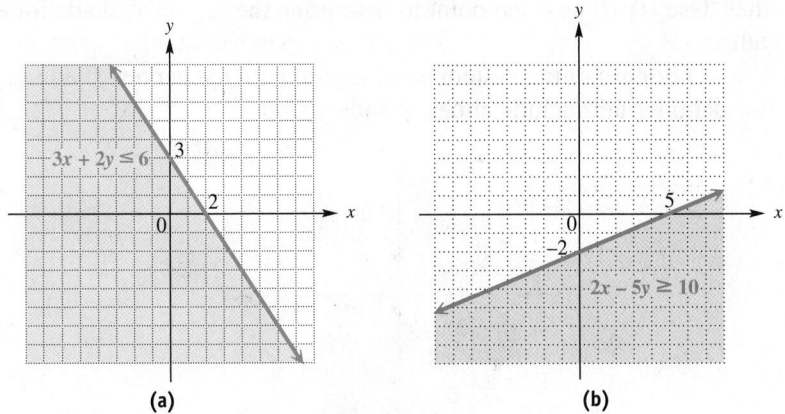

(a)   (b)

Figure 12

Step 2 The solution set of this system includes all points in the intersection (overlap) of the graphs of the two inequalities. As shown in **Figure 13,** this intersection is the gray shaded region and portions of the two boundary lines that surround it.

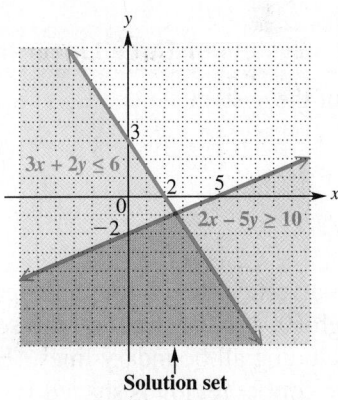

Solution set

Figure 13

CHECK To confirm that the correct region of intersection is shaded, select a test point in the gray shaded region, say $(0, -4)$, and substitute it into *both* inequalities of the given system to make sure that true statements result. Try this. (Using an ordered pair that has one coordinate 0 makes the substitution easier.) ✓

·· **Work Problem ❶ at the Side.** ▶

Note

We usually do all the work on one set of axes. In the remaining examples, only one graph is shown. Be sure that the region of the final solution set is clearly indicated.

❶ Graph the solution set of each system.

ɢs **(a)** $x - 2y \leq 8$

$3x + y \geq 6$

To get you started, the graphs of $x - 2y = 8$ and $3x + y = 6$ are shown.

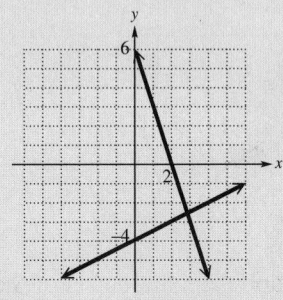

(b) $4x - 2y \leq 8$

$x + 3y \geq 3$

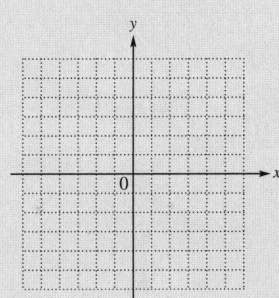

Answers

1. (a)

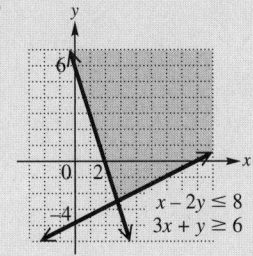

(b)

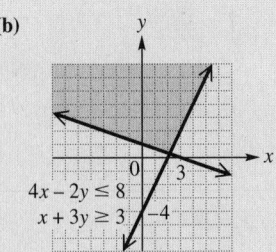

2 Graph the solution set of the system.

$$x + 2y < 0$$
$$3x - 4y < 12$$

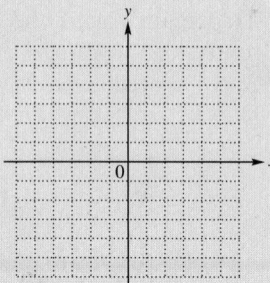

3 Graph the solution set of the system.

$$3x + 2y \le 12$$
$$x \le 2$$
$$y \le 4$$

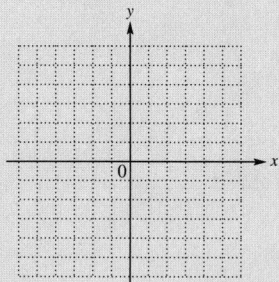

Answers

2.

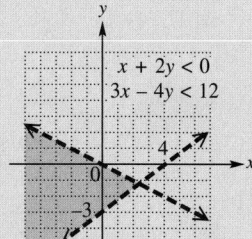

3.

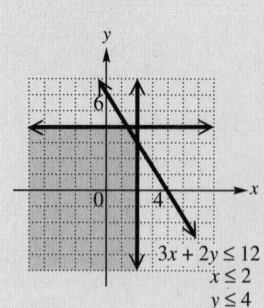

| EXAMPLE 2 | Solving a System of Two Linear Inequalities |

Graph the solution set of the system.

$$x - y > 5$$
$$2x + y < 2$$

Figure 14 shows the graphs of both $x - y > 5$ and $2x + y < 2$. Dashed lines show that the graphs of the inequalities do not include their boundary lines. Use $(0, 0)$ as a test point to determine the region to shade for each inequality.

The solution set of the system is the region with the darkest shading. The solution set does not include either boundary line.

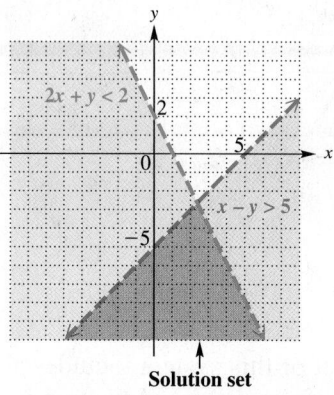

Figure 14

◀ Work Problem **2** at the Side.

| EXAMPLE 3 | Solving a System of Three Linear Inequalities |

Graph the solution set of the system.

$$4x - 3y \le 8$$
$$x \ge 2$$
$$y \le 4$$

Recall that $x = 2$ is a vertical line through the point $(2, 0)$, and $y = 4$ is a horizontal line through $(0, 4)$. The graph of the solution set is the shaded region in **Figure 15,** including all boundary lines. (Here, use $(3, 2)$ as a test point to confirm that the correct region is shaded.)

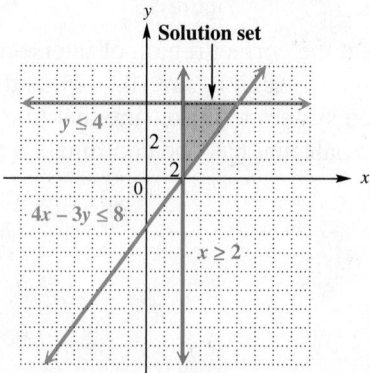

Figure 15

◀ Work Problem **3** at the Side.

4.5 Exercises

FOR EXTRA HELP

 Download the MyDashBoard App

 MyMathLab®

CONCEPT CHECK *Match each system of inequalities with the correct graph from choices A–D.*

1. $x \geq 5$

$y \leq -3$

2. $x \leq 5$

$y \geq -3$

3. $x > 5$

$y < -3$

4. $x < 5$

$y > -3$

A.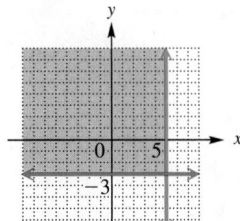

B.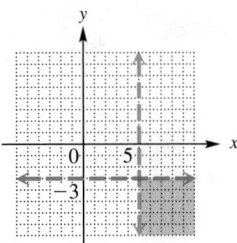

C.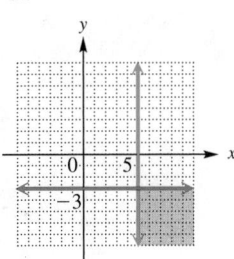

D.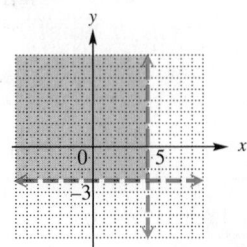

CONCEPT CHECK *In Exercises 5–8, shade the solution set of each system of inequalities. Boundary lines are already graphed.*

5. $x - 3y \leq 6$

$x \geq -4$

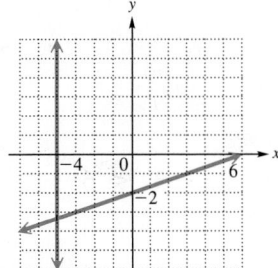

6. $x - 2y \geq 4$

$x \leq -2$

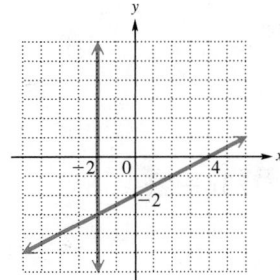

7. $x + y > 4$

$5x - 3y < 15$

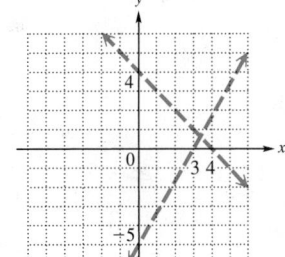

8. $3x - 2y > 12$

$4x + 3y < 12$

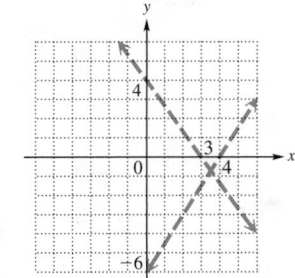

Graph the solution set of each system of linear inequalities. See Examples 1–3.

9. $x + y \leq 6$

$x - y \geq 1$

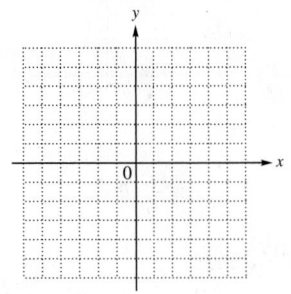

10. $x + y \leq 2$

$x - y \geq 3$

11. $4x + 5y \geq 20$

$x - 2y \leq 5$

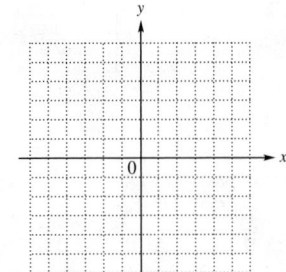

12. $x + 4y \leq 8$

$2x - y \geq 4$

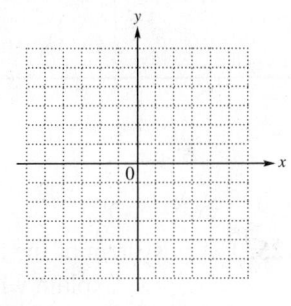

13. $2x + 3y < 6$
$x - y < 5$

14. $x + 2y < 4$
$x - y < -1$

15. $y \leq 2x - 5$
$x < 3y + 2$

16. $x \geq 2y + 6$
$y > -2x + 4$

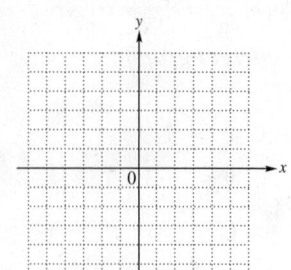

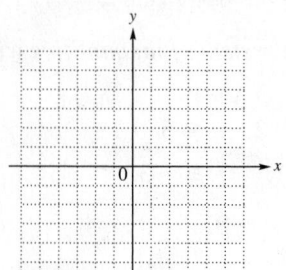

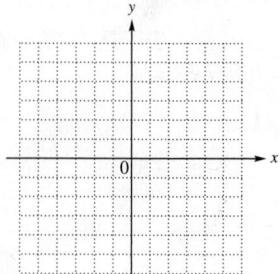

17. $4x + 3y < 6$
$x - 2y > 4$

18. $3x + y > 4$
$x + 2y < 2$

19. $x \leq 2y + 3$
$x + y < 0$

20. $x \leq 4y + 3$
$x + y > 0$

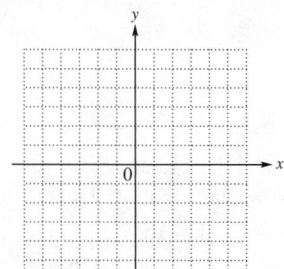

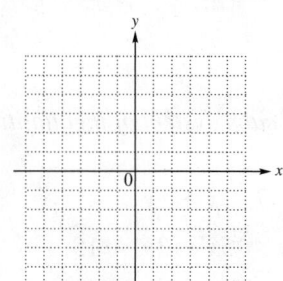

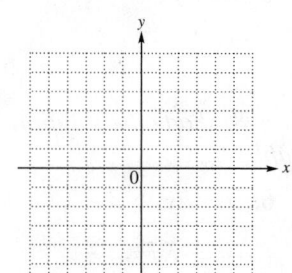

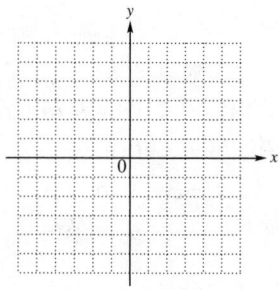

21. $4x + 5y < 8$
$y > -2$
$x > -4$

22. $x + y \geq -3$
$x - y \leq 3$
$y \leq 3$

23. $3x - 2y \geq 6$
$x + y \leq 4$
$x \geq 0$
$y \geq -4$

24. $2x - 3y < 6$
$x + y > 3$
$x < 4$
$y < 4$

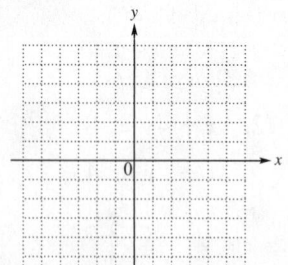

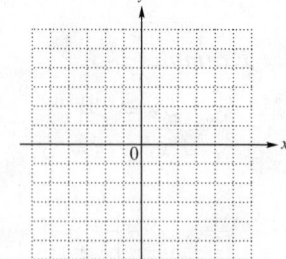

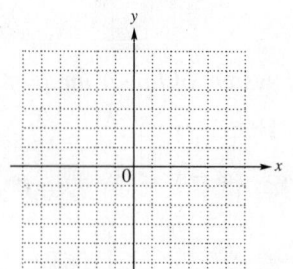

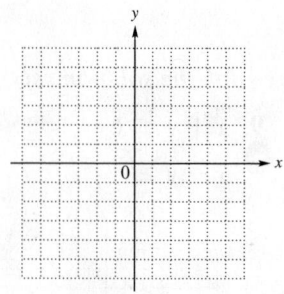

25. Every system of inequalities illustrated in the examples of this section has infinitely many solutions. Explain why this is so. Does this mean that *any* ordered pair is a solution?

Chapter 4 *Summary*

Key Terms

4.1

system of linear equations A system of linear equations (or **linear system**) consists of two or more linear equations with the same variables.

solution of a system A solution of a system of linear equations is an ordered pair that makes all the equations of the system true at the same time.

solution set of a system The set of all ordered pairs that are solutions of a system is its solution set.

consistent system A system of equations with at least one solution is a consistent system.

inconsistent system An inconsistent system is a system of equations with no solution.

independent equations Equations of a system that have different graphs are independent equations.

dependent equations Equations of a system that have the same graph (because they are different forms of the same equation) are dependent equations.

4.5

system of linear inequalities A system of linear inequalities contains two or more linear inequalities (and no other kinds of inequalities).

solution set of a system of linear inequalities The solution set of a system of linear inequalities includes all ordered pairs that make all inequalities of the system true at the same time.

Test Your Word Power

See how well you have learned the vocabulary in this chapter.

1 A **system of linear equations** consists of
   A. at least two linear equations with different variables
   B. two or more linear equations that have an infinite number of solutions
   C. two or more linear equations with the same variables
   D. two or more linear inequalities.

2 A **solution of a system** of linear equations is
   A. an ordered pair that makes one equation of the system true

   B. an ordered pair that makes all the equations of the system true at the same time
   C. any ordered pair that makes one or the other or both equations of the system true
   D. the set of values that make all the equations of the system false.

3 A **consistent system** is a system of equations
   A. with at least one solution
   B. with no solution
   C. with an infinite number of solutions
   D. that have the same graph.

4 An **inconsistent system** is a system of equations
   A. with one solution
   B. with no solution
   C. with an infinite number of solutions
   D. that have the same graph.

5 **Dependent equations**
   A. have different graphs
   B. have no solution
   C. have one solution
   D. are different forms of the same equation.

Answers to Test Your Word Power

1. C; *Example:* $2x + y = 7$
$3x - y = 3$

2. B; *Example:* The ordered pair $(2, 3)$ satisfies both equations of the system in the Answer 1 example, so it is a solution of the system.

3. A; *Example:* The system in the Answer 1 example is consistent. The graphs of the equations intersect at exactly one point, in this case the solution $(2, 3)$.

4. B; *Example:* The equations of two parallel lines make up an inconsistent system. Their graphs never intersect, so there is no solution of the system.

5. D; *Example:* The equations $4x - y = 8$ and $8x - 2y = 16$ are dependent because their graphs are the same line.

Quick Review

Concepts	Examples

4.1 Solving Systems of Linear Equations by Graphing

An ordered pair is a solution of a system if it makes all equations of the system true at the same time.

Is $(4, -1)$ a solution of the system $\begin{array}{l} x + y = 3 \\ 2x - y = 9 \end{array}$?

Because $4 + (-1) = 3$ and $2(4) - (-1) = 9$ are both true, $(4, -1)$ is a solution.

To solve a linear system by graphing, follow these steps.

Solve the system by graphing.

Step 1 Graph each equation of the system on the same axes.

$$x + y = 5$$

Step 2 Find the coordinates of the point of intersection.

$$2x - y = 4$$

Step 3 Check. Write the solution set.

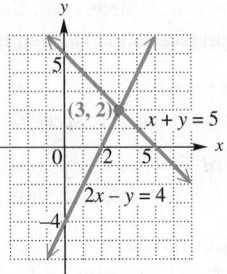

If the graphs of the equations do not intersect (that is, the lines are parallel), then the system has *no solution* and the *solution set is $\emptyset$*.

If the graphs of the equations are the same line, then the system has an *infinite number of solutions*. Use set-builder notation to write the solution set as

$$\{(x, y) \mid \underline{\hspace{2cm}} \},$$

where a form of the equation is written on the blank.

The ordered pair $(3, 2)$ satisfies both equations, so $\{(3, 2)\}$ is the solution set.

4.2 Solving Systems of Linear Equations by Substitution

Solve by substitution.

$$\begin{array}{ll} x + 2y = -5 & (1) \\ y = -2x - 1 & (2) \end{array}$$

Step 1 Solve one equation for either variable.

Equation (2) is already solved for y.

Step 2 Substitute for that variable in the other equation to get an equation in one variable.

Substitute $-2x - 1$ for y in equation (1).

Step 3 Solve the equation from Step 2.

$$\begin{array}{ll} x + 2(-2x - 1) = -5 & \text{Let } y = -2x - 1 \text{ in (1).} \\ x - 4x - 2 = -5 & \text{Distributive property} \\ -3x - 2 = -5 & \text{Combine like terms.} \\ -3x = -3 & \text{Add 2.} \\ x = 1 & \text{Divide by } -3. \end{array}$$

Step 4 Substitute the result into the equation from Step 1 to get the value of the other variable.

To find y, let $x = 1$ in equation (2).

$$\begin{array}{ll} y = -2(1) - 1 & \text{Let } x = 1. \\ y = -3 & \text{Multiply, and then subtract.} \end{array}$$

Step 5 Check. Write the solution set.

The solution $(1, -3)$ checks, so $\{(1, -3)\}$ is the solution set.

Concepts	Examples

4.3 Solving Systems of Linear Equations by Elimination

Step 1 Write both equations in standard form $Ax + By = C$.

Step 2 Multiply one or both equations by appropriate numbers as needed so that the sum of the coefficients of either the x- or y-terms is 0.

Step 3 Add the equations to get an equation with only one variable (or no variable).

Step 4 Solve the equation from Step 3.

Step 5 Substitute the solution from Step 4 into either of the original equations to find the value of the remaining variable.

Step 6 Check. Write the solution set.

Solve by elimination.

$$x + 3y = 7 \quad (1)$$
$$3x - y = 1 \quad (2)$$

Multiply equation (1) by -3 to eliminate the x-terms.

$$\begin{array}{ll} -3x - 9y = -21 & \text{Multiply (1) by } -3. \\ \underline{3x - y = \quad 1} & (2) \\ -10y = -20 & \text{Add.} \\ y = 2 & \text{Divide by } -10. \end{array}$$

Substitute to get the value of x.

$$\begin{array}{ll} x + 3(2) = 7 & \text{Let } y = 2 \text{ in (1).} \\ x + 6 = 7 & \text{Multiply.} \\ x = 1 & \text{Subtract 6.} \end{array}$$

Since $1 + 3(2) = 7$ and $3(1) - 2 = 1$, the solution $(1, 2)$ checks. The solution set is $\{(1, 2)\}$.

4.4 Applications of Linear Systems

Use the modified six-step method.

Step 1 **Read** the problem carefully.

Step 2 **Assign variables** for each unknown value. Use diagrams or tables as needed.

Step 3 **Write two equations** using both variables.

Step 4 **Solve** the system.

Step 5 **State the answer.**

Step 6 **Check** the answer in the words of the original problem.

The sum of two numbers is 30. Their difference is 6. Find the numbers.

Let x represent one number.

Let y represent the other number.

$$\begin{array}{ll} x + y = 30 & (1) \\ \underline{x - y = \; 6} & (2) \\ 2x \quad\;\; = 36 & \text{Add.} \\ x = 18 & \text{Divide by 2.} \end{array}$$

Let $x = 18$ in equation (1): $18 + y = 30$. Solve to get $y = 12$. The numbers are 18 and 12.

The sum of 18 and 12 is 30, and the difference between 18 and 12 is 6, so the answer checks.

4.5 Solving Systems of Linear Inequalities

To solve a system of linear inequalities, follow these steps.

Step 1 Graph each inequality on the same axes. (This was explained in **Section 3.5.**)

Step 2 Choose the intersection. The solution set of the system is formed by the overlap of the regions of the two graphs.

The shaded region is the solution of the following system.

$$2x + 4y \geq 5$$
$$x \geq 1$$

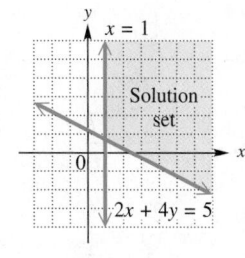

Chapter 4　*Review Exercises*

4.1 *Decide whether the given ordered pair is a solution of the given system.*

1. $(3, 4)$
$4x - 2y = 4$
$5x + y = 19$

2. $(-5, 2)$
$x - 4y = -13$
$2x + 3y = 4$

Solve each system by graphing.

3. $x + y = 4$
$2x - y = 5$

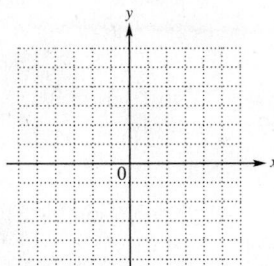

4. $x - 2y = 4$
$2x + y = -2$

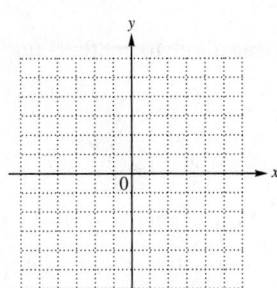

5. $x - 2 = 2y$
$2x - 4y = 4$

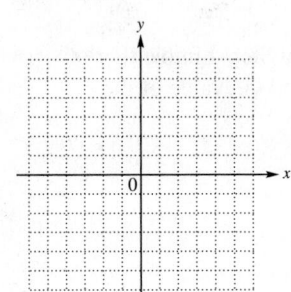

6. $2x + 4 = 2y$
$y - x = -3$

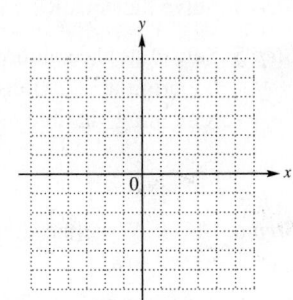

4.2 *Solve each system by the substitution method.*

7. $3x + y = 7$
$x = 2y$

8. $2x - 5y = -19$
$y = x + 2$

9. $4x + 5y = 44$
$x + 2 = 2y$

10. $5x + 15y = 3$
$x + 3y = 2$

4.3 *Solve each system by the elimination method.*

11. $2x - y = 13$
$x + y = 8$

12. $3x - y = -13$
$x - 2y = -1$

13. $-4x + 3y = 25$
$6x - 5y = -39$

14. $3x - 4y = 9$
$6x - 8y = 18$

4.1–4.3 *Solve each system by any method.*

15. $x - 2y = 5$
$y = x - 7$

16. $5x - 3y = 11$
$2y = x - 4$

17. $\dfrac{x}{2} + \dfrac{y}{3} = 7$

$\dfrac{x}{4} + \dfrac{2y}{3} = 8$

18. $\dfrac{3x}{4} - \dfrac{y}{3} = \dfrac{7}{6}$

$\dfrac{x}{2} + \dfrac{2y}{3} = \dfrac{5}{3}$

19. $0.2x + 1.2y = -1$
$0.1x + 0.3y = 0.1$

20. $0.1x + y = 1.6$
$0.6x + 0.5y = -1.4$

4.4 *For each problem, complete any tables. Then solve the problem using a system of equations.*

21. The two leading pizza chains in the United States are Pizza Hut and Domino's. In 2010, Pizza Hut had 2613 more locations than Domino's, and together the two chains had 12,471 locations. How many locations did each chain have? (*Source: PMQ Pizza Magazine.*)

22. Together, the average paid circulation for *Reader's Digest* and *People* magazines in 2011 was 9.3 million. The circulation for *People* was 2.1 million less than that of *Reader's Digest*. What were the circulation figures for each magazine? (*Source:* Audit Bureau of Circulations.)

23. Candy that sells for $1.30 per lb is to be mixed with candy selling for $0.90 per lb to get 100 lb of a mix that will sell for $1 per lb. How much of each type should be used?

Number of Pounds	Cost per Pound (in dollars)	Total Value (in dollars)
_____	1.30	1.30x
y	_____	_____
100	1.00	_____

24. A cashier has 20 bills, all of which are $10 or $20 bills. The total value of the money is $330. How many of each type does the cashier have?

Number of Bills	Denomination of Bills (in dollars)	Total Value (in dollars)
x	10	_____
_____	_____	20y
_____	✕✕✕✕✕✕	330

25. The perimeter of a rectangle is 90 m. Its length is $1\frac{1}{2}$ times its width. Find the length and width of the rectangle.

26. A certain plane flying with the wind travels 540 mi in 2 hr. Later, flying against the same wind, the plane travels 690 mi in 3 hr. Find the rate of the plane in still air and the wind speed.

27. After taxes, Ms. Cesar's game show winnings were $18,000. She invested part of it at 3% annual simple interest and the rest at 4%. Her interest income for the first year was $650. How much did she invest at each rate?

Amount Invested (in dollars)	Percent (as a decimal)	Interest (in dollars)
x	0.03	_____
y	_____	_____
18,000	✕✕✕✕✕	_____

28. A 40% antifreeze solution is to be mixed with a 70% solution to get 90 L of a 50% solution. How many liters of the 40% and 70% solutions will be needed?

Number of Liters	Percent (as a decimal)	Amount of Pure Antifreeze
x	0.40	_____
y	_____	_____
90	0.50	_____

4.5 **CONCEPT CHECK** *Answer each question.*

29. Which system of linear inequalities is graphed in the figure?

A. $x \le 3$ **B.** $x \le 3$ **C.** $x \ge 3$ **D.** $x \ge 3$
    $y \le 1$ $y \ge 1$ $y \le 1$ $y \ge 1$

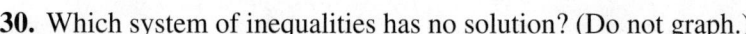

30. Which system of inequalities has no solution? (Do not graph.)

A. $x \ge 4$ **B.** $x + y > 4$ **C.** $x > 2$ **D.** $x + y < 4$
    $y \le 3$ $x + y < 3$ $y < 1$ $x - y < 3$

Graph the solution set of each system of linear inequalities.

31. $x + y \geq 2$

$x - y \leq 4$

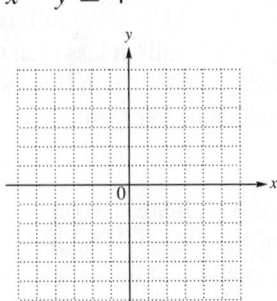

32. $y \geq 2x$

$2x + 3y \leq 6$

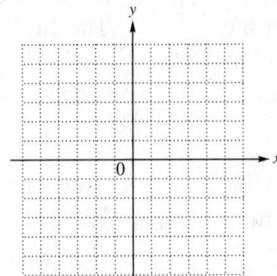

33. $x + y < 3$

$2x > y$

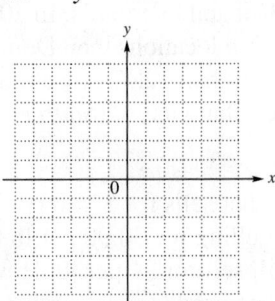

Mixed Review Exercises

Solve.

34. $3x + 4y = 6$

$4x - 5y = 8$

35. $\dfrac{3x}{2} + \dfrac{y}{5} = -3$

$4x + \dfrac{y}{3} = -11$

36. $x + 6y = 3$

$2x + 12y = 2$

37. $x + y < 5$

$x - y \geq 2$

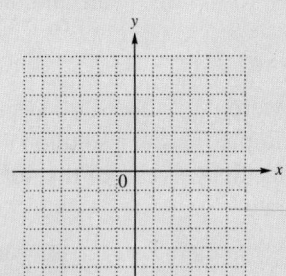

38. $y \leq 2x$

$x + 2y > 4$

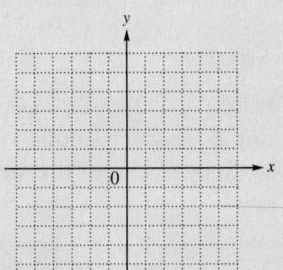

39. $y < -4x$

$y < -2$

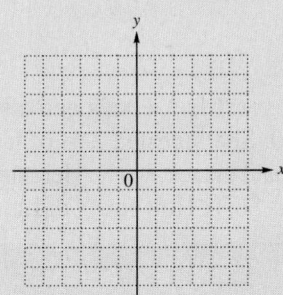

40. The perimeter of an isosceles triangle is 29 in. One side of the triangle is 5 in. longer than each of the two equal sides. Find the lengths of the sides of the triangle.

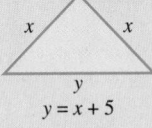

41. In Super Bowl XLVI, the New York Giants beat the New England Patriots by 4 points, and the winning score was 13 points less than twice the losing score. What was the final score of the game? (*Source:* NFL.)

42. Eboni Perkins compared the monthly payments she would incur for two types of mortgages: fixed-rate and variable-rate. Her observations led to the following graph.

(a) For which years would the monthly payment be more for the fixed-rate mortgage than for the variable-rate mortgage?

(b) In what year would the payments be the same, and what would those payments be?

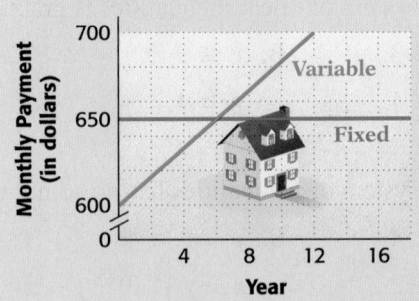

Chapter 4 **Test**  CHAPTER
Test Prep
VIDEO

The Chapter Test Prep Videos with test solutions are available
on DVD, in MyMathLab, and on YouTube—search
"LialCombinedAlg" and click on "Channels."

1. Decide whether each ordered pair is a solution of the system.

$$2x + y = -3$$
$$x - y = -9$$

(a) $(1, -5)$ (b) $(1, 10)$ (c) $(-4, 5)$

2. Solve the system by graphing.

$$2x + y = 1$$
$$3x - y = 9$$

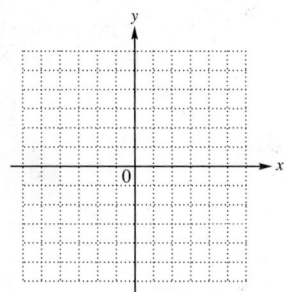

3. Suppose that the graph of a system of two linear equations consists of lines that have the same slope but different y-intercepts. How many solutions does the system have?

Solve each system by the substitution method.

4. $2x + y = -4$

$x = y + 7$

5. $4x + 3y = -35$

$x + y = 0$

Solve each system by the elimination method.

6. $2x - y = 4$

$3x + y = 21$

7. $4x + 2y = 2$

$5x + 4y = 7$

8. $3x + 4y = 9$

$2x + 5y = 13$

9. $6x - 5y = 0$

$-2x + 3y = 0$

10. $4x + 5y = 2$

$-8x - 10y = 6$

Solve each system by any method.

11. $3x = 6 + y$

  $6x - 2y = 12$

12. $\dfrac{6}{5}x - \dfrac{1}{3}y = -20$

  $-\dfrac{2}{3}x + \dfrac{1}{6}y = 11$

Solve each problem.

13. The distance between Memphis and Atlanta is 782 mi less than the distance between Minneapolis and Houston. Together, the two distances total 1570 mi. How far is it between Memphis and Atlanta? How far is it between Minneapolis and Houston? (*Source: Rand McNally Road Atlas.*)

14. In 2010, a total of 7.8 million people visited the Statue of Liberty and the National World War II Memorial, two popular tourist attractions. The Statue of Liberty had 0.2 million fewer visitors than the National World War II Memorial. How many visitors did each of these attractions have? (*Source:* National Park Service, Department of the Interior.)

15. A 15% solution of alcohol is to be mixed with a 40% solution to get 50 L of a final mixture that is 30% alcohol. How much of each of the original solutions should be used?

Liters of Solution	Percent (as a decimal)	Liters of Pure Alcohol

16. Two cars leave from Perham, Minnesota, and travel in the same direction. One car travels $1\frac{1}{3}$ times as fast as the other. After 3 hr they are 45 mi apart. What are the rates of the cars?

	r	t	d
Faster Car			
Slower Car			

Graph the solution set of each system of inequalities.

17. $2x + 7y \le 14$

  $x - y \ge 1$

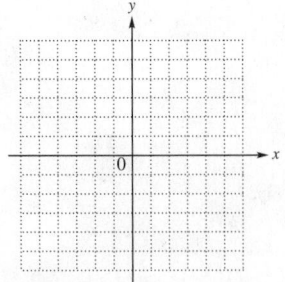

18. $2x - y > 6$

  $4y + 12 \ge -3x$

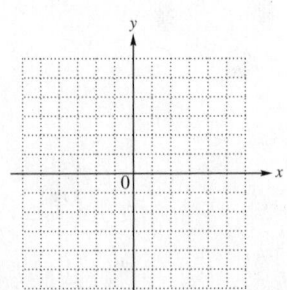

Chapters R–4 Cumulative Review Exercises

1. List all integer factors of 40.

2. Find the value of the expression for $x = 1$ and $y = 5$.

$$\frac{3x^2 + 2y^2}{10y + 3}$$

Name the property that justifies each statement.

3. $5 + (-4) = (-4) + 5$

4. $r(s - k) = rs - rk$

5. $-\dfrac{2}{3} + \dfrac{2}{3} = 0$

6. Evaluate $-2 + 6[3 - (4 - 9)]$.

7. Solve the formula $P = \dfrac{kT}{V}$ for T.

Solve each linear equation.

8. $2 - 3(6x + 2) = 4(x + 1) + 18$

9. $\dfrac{3}{2}\left(\dfrac{1}{3}x + 4\right) = 6\left(\dfrac{1}{4} + x\right)$

Solve each linear inequality. Write the solution set in interval notation.

10. $-\dfrac{5}{6}x < 15$

11. $-8 < 2x + 3$

12. The third generation iPad tablet computer, released in March 2012, has a perimeter of 33.62 in., and its width is 2.19 in. less than its length. What are its dimensions? (*Source:* www.apple.com)

Graph each linear equation.

13. $x - y = 4$

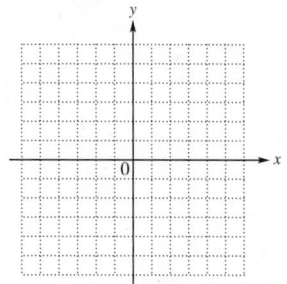

14. $3x + y = 6$

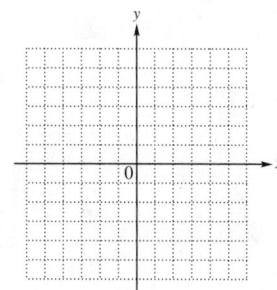

Find the slope of each line.

15. Through $(-5, 6)$ and $(1, -2)$

16. Perpendicular to the line $y = 4x - 3$

Write an equation in slope-intercept form for each line.

17. Through $(-4, 1)$ with slope $\frac{1}{2}$

18. Through the points $(1, 3)$ and $(-2, -3)$

19. **(a)** Write an equation of the vertical line through $(9, -2)$.

(b) Write an equation of the horizontal line through $(4, -1)$.

Solve each system by any method.

20. $2x - y = -8$
$x + 2y = 11$

21. $4x + 5y = -8$
$3x + 4y = -7$

22. $3x + 4y = 2$
$6x + 8y = 1$

Use a system of equations to solve each problem.

23. Admission prices at a high school football game were $6 for adults and $2 for children. The total value of the tickets sold was $2528, and 454 tickets were sold. How many adults and how many children attended the game?

Kind of Ticket	Number Sold	Cost of Each (in dollars)	Total Value (in dollars)
Adult	x	6	$6x$
Child	y	_____	_____
Total	454	XXXXXXX	_____

24. The perimeter of a triangle is 53 in. If two sides are of equal length, and the third side measures 4 in. less than each of the equal sides, what are the lengths of the three sides?

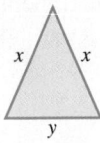

25. Graph the solution set of the system.

$x + 2y \leq 12$
$2x - y \leq 8$

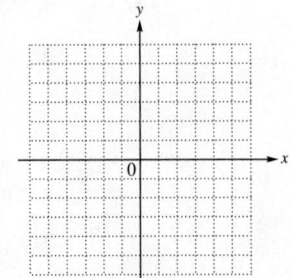

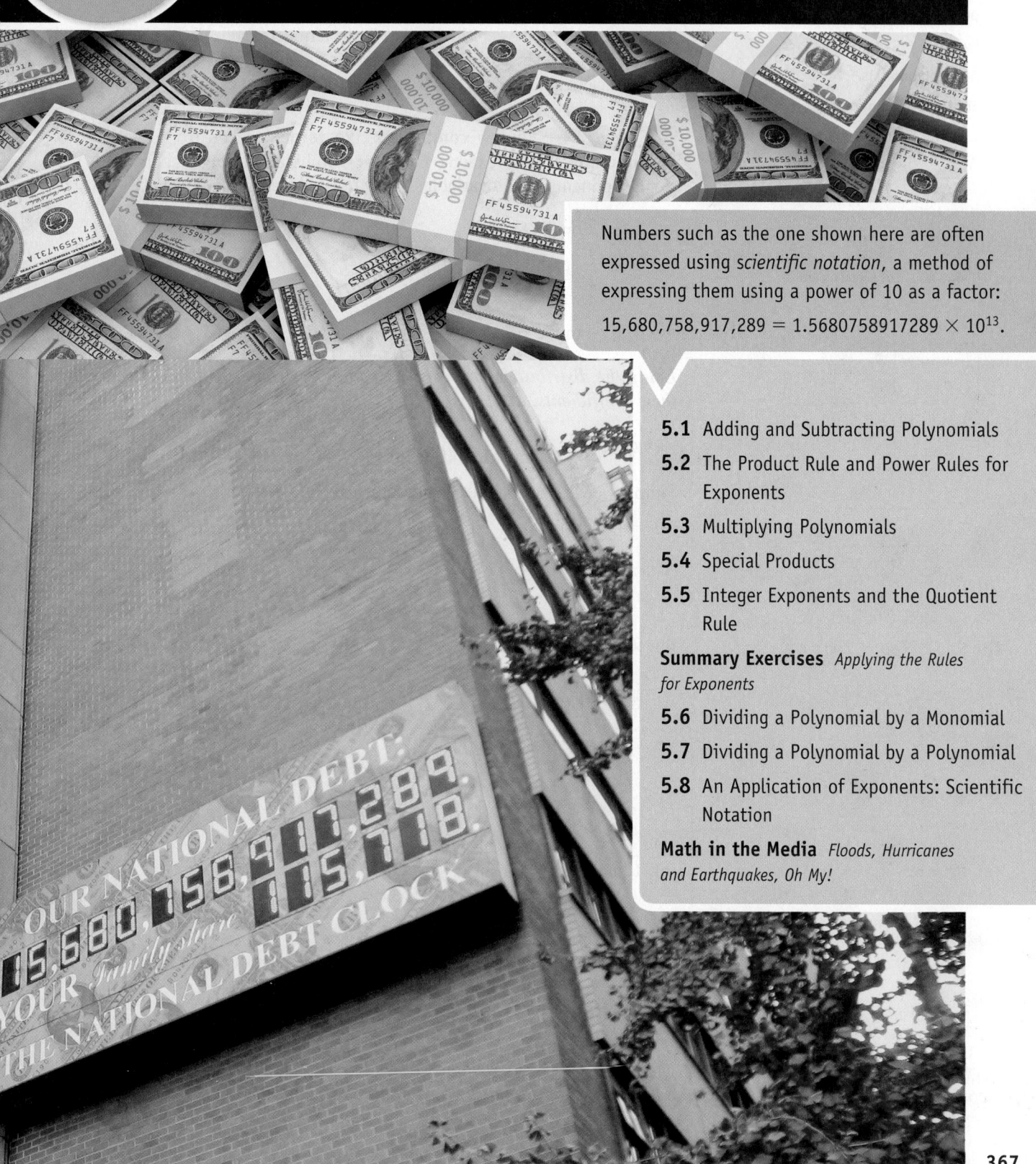

5 Exponents and Polynomials

Numbers such as the one shown here are often expressed using *scientific notation*, a method of expressing them using a power of 10 as a factor:
$$15{,}680{,}758{,}917{,}289 = 1.5680758917289 \times 10^{13}.$$

5.1 Adding and Subtracting Polynomials

5.2 The Product Rule and Power Rules for Exponents

5.3 Multiplying Polynomials

5.4 Special Products

5.5 Integer Exponents and the Quotient Rule

Summary Exercises *Applying the Rules for Exponents*

5.6 Dividing a Polynomial by a Monomial

5.7 Dividing a Polynomial by a Polynomial

5.8 An Application of Exponents: Scientific Notation

Math in the Media *Floods, Hurricanes and Earthquakes, Oh My!*

5.1 Adding and Subtracting Polynomials

OBJECTIVES

1. Review combining like terms.
2. Know the vocabulary for polynomials.
3. Evaluate polynomials.
4. Add polynomials.
5. Subtract polynomials.
6. Add and subtract polynomials with more than one variable.

Recall from **Section 1.8** that in an algebraic expression such as

$$4x^3 + 6x^2 + 5x + 8,$$

the quantities that are added, $4x^3$, $6x^2$, $5x$, and 8, are **terms.** In the term $4x^3$, the number **4** is the **numerical coefficient,** or simply the **coefficient,** of x^3. In the same way, **6** is the coefficient of x^2 in the term $6x^2$, and **5** is the coefficient of x in the term $5x$. The *constant* term 8 can be thought of as $8 \cdot 1 = 8x^0$, so 8 is the coefficient in the term 8. Other examples are given in the table at the side.

Term	Numerical Coefficient
$-7y$	-7
$34r^3$	34
$-26x^5yz^4$	-26
$-k = -1k$	-1
$r = 1r$	1
$\frac{3x}{8} = \frac{3}{8}x$	$\frac{3}{8}$
$\frac{x}{3} = \frac{1x}{3} = \frac{1}{3}x$	$\frac{1}{3}$

OBJECTIVE ➊ **Review combining like terms.** Recall from **Section 1.8** that **like terms** have exactly the same combination of variables, with the same exponents on the variables. *Only the coefficients may differ.*

$$\left.\begin{array}{l} 19m^5 \quad \text{and} \quad 14m^5 \\ -37y^9 \quad \text{and} \quad y^9 \\ 3pq \quad \text{and} \quad -2pq \\ 2xy^2 \quad \text{and} \quad -xy^2 \end{array}\right\} \begin{array}{c} \text{Examples} \\ \text{of} \\ \text{like terms} \end{array}$$

$$\left.\begin{array}{l} 7x \quad \text{and} \quad 7y \\ z^4 \quad \text{and} \quad z \\ 2pq \quad \text{and} \quad 2p \\ -4xy^2 \quad \text{and} \quad 5x^2y \end{array}\right\} \begin{array}{c} \text{Examples} \\ \text{of} \\ \text{unlike terms} \end{array}$$

Using the distributive property, we combine, or add, like terms by adding their coefficients.

EXAMPLE 1 Combining Like Terms

Simplify each expression by combining like terms.

(a) $-4x^3 + 6x^3$

$\quad = (-4 + 6)x^3$

$\quad = 2x^3$

(b) $8rs - 13rs + 9rs$

$\quad = (8 - 13 + 9)rs$

$\quad = 4rs$

(c) $9x^6 - 14x^6 + x^6$ $\boxed{x^6 = 1x^6}$

$\quad = (9 - 14 + 1)x^6$

$\quad = -4x^6$

(d) $y + \dfrac{2}{3}y$

$\quad = 1y + \dfrac{2}{3}y \qquad y = 1y$

$\quad = \left(\dfrac{3}{3} + \dfrac{2}{3}\right)y$

$\quad = \dfrac{5}{3}y$

(e) $12m^2 + 5m + 4m^2$

$\quad = (12 + 4)m^2 + 5m$

$\quad = 16m^2 + 5m$

(f) $5u + 11v$

These are unlike terms. They cannot be combined.

◀ Work Problem ➊ at the Side.

➊ Combine like terms.

(a) $5x^4 + 7x^4$

(b) $9pq + 3pq - 2pq$

(c) $r^2 + 3r + 5r^2$

(d) $x + \dfrac{1}{2}x$

(e) $8t + 6w$

(f) $3x^4 - 3x^2$

Answers

1. **(a)** $12x^4$ **(b)** $10pq$ **(c)** $6r^2 + 3r$ **(d)** $\dfrac{3}{2}x$

   (e) These are unlike terms. They cannot be combined.

   (f) These are unlike terms. They cannot be combined.

CAUTION

In **Example 1(e),** we cannot combine $16m^2$ and $5m$. These two terms are unlike because the exponents on the variables are different. *Unlike terms have different variables or different exponents on the same variables.*

OBJECTIVE **2** **Know the vocabulary for polynomials.** A **polynomial in** x is a term or the sum of a finite number of terms of the form ax^n, for any real number a and any whole number n. For example,

$$16x^8 - 7x^6 + 5x^4 - 3x^2 + 4 \qquad \text{Polynomial in } x$$

(The 4 can be written as $4x^0$.)

is a polynomial in x. It is written in **descending powers,** because the exponents on x decrease from left to right. By contrast, the expression

$$2x^3 - x^2 + \frac{4}{x} \qquad \text{Not a polynomial}$$

is not a polynomial in x. A variable appears in a denominator.

Note

We can define *polynomial* using any variable and not just x, as in **Example 1(d).** Polynomials may have terms with more than one variable, as in **Example 1(b).**

Work Problem **2** at the Side. ▶

The **degree of a term** is the sum of the exponents on the variables. The **degree of a polynomial** is the greatest degree of any nonzero term of the polynomial. The table gives several examples.

Term	Degree	Polynomial	Degree
$3x^4$	4	$3x^4 - 5x^2 + 6$	4
$5x$, or $5x^1$	1	$5x + 7$	1
-7, or $-7x^0$	0	$x^2y + xy - 5y^2$	3
$2x^2y$, or $2x^2y^1$	$2 + 1 = 3$	$x^5 + 3x^6$	6

A polynomial with only one term is a **monomial.** (*Mono-* means "one," as in *mono*rail.) A polynomial with exactly two terms is a **binomial.** (*Bi-* means "two," as in *bi*cycle.) A polynomial with exactly three terms is a **trinomial.** (*Tri-* means "three," as in *tri*angle.)

$$9m, \quad -6y^5, \quad a^2, \quad \text{and} \quad 6 \qquad \text{Monomials}$$

$$-9x^4 + 9x^3, \quad 8m^2 + 6m, \quad \text{and} \quad 3m^5 - 9m^2 \qquad \text{Binomials}$$

$$9m^3 - 4m^2 + 6, \quad \frac{19}{3}y^2 + \frac{8}{3}y + 5, \quad \text{and} \quad -3m^5 - 9m^2 + 2 \qquad \text{Trinomials}$$

EXAMPLE 2 **Classifying Polynomials**

Simplify each polynomial if possible. Then give the degree and tell whether the polynomial is a *monomial*, a *binomial*, a *trinomial*, or *none of these*.

(a) $2x^3 + 5$ We cannot simplify further. This is a binomial of degree 3.

(b) $4x - 5x + 2x$

$$= x \qquad \text{Combine like terms to simplify.}$$

The degree is 1 (since $x = x^1$). The simplified polynomial is a monomial.

Work Problem **3** at the Side. ▶

2 Choose all descriptions that apply for each of the expressions in parts (a)–(d).

A. Polynomial

B. Polynomial written in descending powers

C. Not a polynomial

(a) $3m^5 + 5m^2 - 2m + 1$

(b) $2p^4 + p^6$

(c) $\frac{1}{x} + 2x^2 + 3$

(d) $x - 3$

3 Simplify each polynomial if possible. Then give the degree and tell whether the polynomial is a *monomial*, a *binomial*, a *trinomial*, or *none of these*.

(a) $3x^2 + 2x - 4$

(b) $x^3 + 4x^3$

(c) $x^8 - x^7 + 2x^8$

Answers

2. **(a)** A and B **(b)** A **(c)** C **(d)** A and B
3. **(a)** degree 2; trinomial
   (b) degree 3; monomial (simplify to $5x^3$)
   (c) degree 8; binomial (simplify to $3x^8 - x^7$)

4 Find the value of $2x^3 + 8x - 6$ in each case.

(a) For $x = -1$

$$2x^3 + 8x - 6$$
$$= 2(\underline{})^3 + 8(\underline{}) - 6$$
$$= 2(\underline{}) + 8(-1) - 6$$
$$= \underline{} - \underline{} - 6$$
$$= \underline{}$$

(b) For $x = 4$

(c) For $x = -2$

5 Add each pair of polynomials.

(a) $4x^3 - 3x^2 + 2x$ and
$6x^3 + 2x^2 - 3x$

(b) $x^2 - 2x + 5$ and
$4x^2 - 2$

Answers

4. **(a)** -1; -1; -1; -2; 8; -16
   (b) 154   **(c)** -38
5. **(a)** $10x^3 - x^2 - x$
   (b) $5x^2 - 2x + 3$

OBJECTIVE ▶ 3 **Evaluate polynomials.** When we evaluate an expression, we find its value. A polynomial usually represents different numbers for different values of the variable.

EXAMPLE 3 **Evaluating a Polynomial**

Find the value of $3x^4 + 5x^3 - 4x - 4$ for **(a)** $x = -2$ and for **(b)** $x = 3$.

(a) $\qquad 3x^4 + 5x^3 - 4x - 4$

$= 3(-2)^4 + 5(-2)^3 - 4(-2) - 4$	Substitute -2 for x.
$= 3(16) + 5(-8) - 4(-2) - 4$	Apply the exponents.
$= 48 - 40 + 8 - 4$	Multiply.
$= 12$	Add and subtract.

Use parentheses to avoid errors.

(b) $3x^4 + 5x^3 - 4x - 4$

$= 3(3)^4 + 5(3)^3 - 4(3) - 4$	Let $x = 3$.
$= 3(81) + 5(27) - 4(3) - 4$	Apply the exponents.
$= 243 + 135 - 12 - 4$	Multiply.
$= 362$	Add and subtract.

◀ **Work Problem** **4** **at the Side.**

OBJECTIVE ▶ 4 **Add polynomials.**

Adding Polynomials

To add two polynomials, add like terms.

EXAMPLE 4 **Adding Polynomials Vertically**

(a) Add $6x^3 - 4x^2 + 3$ and $-2x^3 + 7x^2 - 5$.

$$\begin{array}{r} 6x^3 - 4x^2 + 3 \\ -2x^3 + 7x^2 - 5 \\ \hline \end{array}$$ Write like terms in columns.

Now add, column by column.

Combine the coefficients only. Do not add the exponents.

$$\begin{array}{ccc} 6x^3 & -4x^2 & 3 \\ -2x^3 & 7x^2 & -5 \\ \hline 4x^3 & 3x^2 & -2 \end{array}$$

Add the three sums together to obtain the answer.

$$4x^3 + 3x^2 + (-2) = 4x^3 + 3x^2 - 2$$

(b) Add $2x^2 - 4x + 3$ and $x^3 + 5x$.

Write like terms in columns and add column by column.

$$\begin{array}{r} 2x^2 - 4x + 3 \\ x^3 + 5x \\ \hline x^3 + 2x^2 + x + 3 \end{array}$$

Leave spaces for missing terms.

◀ **Work Problem** **5** **at the Side.**

The polynomials in **Example 4** also could be added horizontally.

EXAMPLE 5 Adding Polynomials Horizontally

Find each sum.

(a) Add $6x^3 - 4x^2 + 3$ and $-2x^3 + 7x^2 - 5$.

$(6x^3 - 4x^2 + 3) + (-2x^3 + 7x^2 - 5) = 4x^3 + 3x^2 - 2$ Same answer found in **Example 4(a)**

Combine like terms.

(b) $(2x^2 - 4x + 3) + (x^3 + 5x)$

$= x^3 + 2x^2 - 4x + 5x + 3$ Commutative property

$= x^3 + 2x^2 + x + 3$ Combine like terms.

·········· **Work Problem ⑥ at the Side.** ▶

OBJECTIVE ⑤ Subtract polynomials. In **Section 1.5**, the difference $x - y$ was defined as $x + (-y)$. (We find the difference $x - y$ by adding x and the opposite of y.)

$7 - 2$ is equivalent to $7 + (-2)$,   which equals 5.

$-8 - (-2)$ is equivalent to $-8 + 2$,   which equals -6.

A similar method is used to subtract polynomials.

Subtracting Polynomials

To subtract two polynomials, change all the signs of the subtrahend (second polynomial) and add the result to the minuend (first polynomial).

EXAMPLE 6 Subtracting Polynomials Horizontally

Perform each subtraction.

(a) $(5x - 2) - (3x - 8)$

$= (5x - 2) + [-(3x - 8)]$ Definition of subtraction

$= (5x - 2) + [-1(3x - 8)]$ $-a = -1a$

$= (5x - 2) + (-3x + 8)$ Distributive property

$= 2x + 6$ Combine like terms.

(b) Subtract $6x^3 - 4x^2 + 2$ from $11x^3 + 2x^2 - 8$.

$(11x^3 + 2x^2 - 8) - (6x^3 - 4x^2 + 2)$ *Be careful to write the problem in the correct order.*

$= (11x^3 + 2x^2 - 8) + (-6x^3 + 4x^2 - 2)$

$= 5x^3 + 6x^2 - 10$ Answer

CHECK To check a subtraction problem, use the following fact.

If $a - b = c$,   then $a = b + c$.

Here, add $6x^3 - 4x^2 + 2$ and $5x^3 + 6x^2 - 10$.

$(6x^3 - 4x^2 + 2) + (5x^3 + 6x^2 - 10)$

$= 11x^3 + 2x^2 - 8$ ✓

·········· **Work Problem ⑦ at the Side.** ▶

---

⑥ Find each sum.

(a) $(2x^4 - 6x^2 + 7)$
$+ (-3x^4 + 5x^2 + 2)$

(b) $(3x^2 + 4x + 2)$
$+ (6x^3 - 5x - 7)$

⑦ Subtract, and check your answers by addition.

GS (a) $(14y^3 - 6y^2 + 2y - 5)$
$- (2y^3 - 7y^2 - 4y + 6)$

$= ($ _____ $)$

$+ ($ _____ $)$

$=$ _____

(b) Subtract

$\left(-\dfrac{3}{2}y^2 + \dfrac{4}{3}y + 6\right)$

from $\left(\dfrac{7}{2}y^2 - \dfrac{11}{3}y + 8\right)$.

Answers

6. (a) $-x^4 - x^2 + 9$
   (b) $6x^3 + 3x^2 - x - 5$
7. (a) $14y^3 - 6y^2 + 2y - 5;$
     $-2y^3 + 7y^2 + 4y - 6;$
     $12y^3 + y^2 + 6y - 11$
   (b) $5y^2 - 5y + 2$

8 Subtract by columns.

$(4y^3 - 16y^2 + 2y)$
$- (12y^3 - 9y^2 + 16)$

9 Perform the indicated operations.

$(6p^4 - 8p^3 + 2p - 1)$
$- (-7p^4 + 6p^2 - 12)$
$+ (p^4 - 3p + 8)$

10 Add or subtract.

(a) $(3mn + 2m - 4n)$
$+ (-mn + 4m + n)$

(b) $(5p^2q^2 - 4p^2 + 2q)$
$- (2p^2q^2 - p^2 - 3q)$

Answers

8. $-8y^3 - 7y^2 + 2y - 16$
9. $14p^4 - 8p^3 - 6p^2 - p + 19$
10. (a) $2mn + 6m - 3n$
   (b) $3p^2q^2 - 3p^2 + 5q$

Subtraction also can be done in columns. We use vertical subtraction in **Section 5.7** when we study polynomial division.

EXAMPLE 7 Subtracting Polynomials Vertically

Subtract by columns: $(14y^3 - 6y^2 + 2y - 5) - (2y^3 - 7y^2 - 4y + 6)$.

$$14y^3 - 6y^2 + 2y - 5 \quad \text{Arrange like terms}$$
$$\underline{2y^3 - 7y^2 - 4y + 6} \quad \text{in columns.}$$

Change all signs in the second row, and then add.

$$14y^3 - 6y^2 + 2y - \ 5$$
$$\underline{-2y^3 + 7y^2 + 4y - \ 6} \quad \text{Change signs.}$$
$$12y^3 + \ y^2 + 6y - 11 \quad \text{Add.}$$

◀ **Work Problem 8** at the Side.

EXAMPLE 8 Adding and Subtracting More Than Two Polynomials

Perform the indicated operations to simplify the expression

$$(4 - x + 3x^2) - (2 - 3x + 5x^2) + (8 + 2x - 4x^2).$$

Rewrite, changing the subtraction to adding the opposite.

$$(4 - x + 3x^2) - (2 - 3x + 5x^2) + (8 + 2x - 4x^2)$$
$$= (4 - x + 3x^2) + (-2 + 3x - 5x^2) + (8 + 2x - 4x^2)$$
$$= (2 + 2x - 2x^2) + (8 + 2x - 4x^2) \quad \text{Combine like terms.}$$
$$= 10 + 4x - 6x^2 \quad \text{Combine like terms.}$$

◀ **Work Problem 9** at the Side.

OBJECTIVE 6 Add and subtract polynomials with more than one variable. Polynomials in more than one variable are added and subtracted by combining like terms, just as with single-variable polynomials.

EXAMPLE 9 Adding and Subtracting Multivariable Polynomials

Add or subtract as indicated.

(a) $(4a + 2ab - b) + (3a - ab + b)$
$= 4a + 2ab - b + 3a - ab + b$
$= 7a + ab \quad \text{Combine like terms.}$

(b) $(2x^2y + 3xy + y^2) - (3x^2y - xy - 2y^2)$
$= 2x^2y + 3xy + y^2 - 3x^2y + xy + 2y^2$
$= -x^2y + 4xy + 3y^2 \quad$ Be careful with signs.

◀ **Work Problem 10** at the Side.

5.1 Exercises

CONCEPT CHECK *Complete each statement.*

1. In the term $7x^5$, the coefficient is _____ and the exponent is _____ .

2. The expression $5x^3 - 4x^2$ has (*one / two / three*) term(s).

3. The degree of the term $-4x^8$ is _____ .

4. The polynomial $4x^2 - y^2$ (*is / is not*) an example of a trinomial.

5. When $x^2 + 10$ is evaluated for $x = 4$, the result is _____ .

6. _____ is an example of a monomial with coefficient 5, in the variable x, having degree 9.

For each polynomial, determine the number of terms, and give the coefficient of each term. **See Objective 2.**

7. $6x^4$

8. $-9y^5$

9. t^4

10. s^7

11. $\dfrac{x}{5}$

12. $\dfrac{z}{8}$

13. $-19r^2 - r$

14. $2y^3 - y$

15. $x - 8x^2 + \dfrac{2}{3}x^3$

16. $v - 2v^3 + \dfrac{3}{4}v^2$

In each polynomial, combine like terms whenever possible. Write the result with descending powers. **See Example 1.**

17. $-3m^5 + 5m^5$

18. $-4y^3 + 3y^3$

19. $2r^5 + (-3r^5)$

20. $-19y^2 + 9y^2$

21. $\dfrac{1}{2}x^4 + \dfrac{1}{6}x^4$

22. $\dfrac{3}{10}x^6 + \dfrac{1}{5}x^6$

23. $-0.5m^2 + 0.2m^5$

24. $-0.9y + 0.9y^2$

25. $-3x^5 + 2x^5 - 4x^5$

26. $6x^3 - 8x^3 + 9x^3$

27. $-4p^7 + 8p^7 + 5p^9$

28. $-3a^8 + 4a^8 - 3a^2$

29. $-4y^2 + 3y^2 - 2y^2 + y^2$

30. $3r^5 - 8r^5 + r^5 + 2r^5$

For each polynomial, first simplify, if possible, and write it with descending powers. Then give the degree of the resulting polynomial, and tell whether it is a monomial, *a* binomial, *a* trinomial, *or* none of these. **See Example 2.**

31. $6x^4 - 9x$

32. $7t^3 - 3t$

33. $5m^4 - 3m^2 + 6m^5 - 7m^3$

34. $6p^5 + 4p^3 - 8p^4 + 10p^2$

35. $\dfrac{5}{3}x^4 - \dfrac{2}{3}x^4 + \dfrac{1}{3}x^2 - 4$

36. $\dfrac{4}{5}r^6 + \dfrac{1}{5}r^6 - r^4 + \dfrac{2}{5}r$

37. $0.8x^4 - 0.3x^4 - 0.5x^4 + 7$

38. $1.2t^3 - 0.9t^3 - 0.3t^3 + 9$

39. $2.5x^2 + 0.5x + x^2 - x - 2x^2$

40. $8.3y - 9.2y^3 - 2.6y + 4.8y^3 - 6.7y^2$

*Find the value of each polynomial for **(a)** $x = 2$ and for **(b)** $x = -1$.* **See Example 3.**

41. $-2x + 3$

42. $5x - 4$

43. $2x^2 + 5x + 1$

44. $-3x^2 + 14x - 2$

45. $-2x^5 - 4x^4 + 5x^3 - x^2$

46. $x^4 - 6x^3 + x^2 + 1$

47. $-4x^5 + x^2$

48. $2x^6 - 4x$

Add or subtract as indicated. **See Examples 4 and 7.**

49. Add.

$$3m^2 + 5m$$
$$\underline{2m^2 - 2m}$$

50. Add.

$$4a^3 - 4a^2$$
$$\underline{6a^3 + 5a^2}$$

51. Subtract.

$$12x^4 - x^2$$
$$\underline{8x^4 + 3x^2}$$

52. Subtract.

$$13y^5 - y^3$$
$$\underline{7y^5 + 5y^3}$$

53. Add.

$$\dfrac{2}{3}x^2 + \dfrac{1}{5}x + \dfrac{1}{6}$$
$$\underline{\dfrac{1}{2}x^2 - \dfrac{1}{3}x + \dfrac{2}{3}}$$

54. Add.

$$\dfrac{4}{7}y^2 - \dfrac{1}{5}y + \dfrac{7}{9}$$
$$\underline{\dfrac{1}{3}y^2 - \dfrac{1}{3}y + \dfrac{2}{5}}$$

55. Subtract.

$$12m^3 - 8m^2 + 6m + 7$$
$$\underline{5m^2 - 4}$$

56. Subtract.

$$5a^4 - 3a^3 + 2a^2 - a + 6$$
$$\underline{-6a^4 - a^2 + a - 1}$$

57. Subtract.

$$4.3x^3 - 6.1x^2 - 3.0x - 5$$
$$\underline{1.4x^3 - 2.6x^2 - 1.5x + 4}$$

58. Subtract.

$$8.7z^3 + 4.2z^2 - 9.0z - 7$$
$$\underline{5.9z^3 - 6.6z^2 + 3.5z - 4}$$

*Perform the indicated operations. **See Examples 5, 6, and 8.***

59. $(2r^2 + 3r - 12) + (6r^2 + 2r)$

60. $(3r^2 + 5r - 6) + (2r - 5r^2)$

61. $(8m^2 - 7m) - (3m^2 + 7m - 6)$

62. $(x^2 + x) - (3x^2 + 2x - 1)$

63. $(16x^3 - x^2 + 3x) + (-12x^3 + 3x^2 + 2x)$

64. $(-2b^6 + 3b^4 - b^2) + (b^6 + 2b^4 + 2b^2)$

65. $(7y^4 + 3y^2 + 2y) - (18y^5 - 5y^3 + y)$

66. $(8t^5 + 3t^3 + 5t) - (19t^4 - 6t^2 + t)$

67. $\big[(8m^2 + 4m - 7) - (2m^3 - 5m + 2) \big]$
$- (m^2 + m)$

68. $\big[(9b^3 - 4b^2 + 3b + 2) - (-2b^3 + b) \big]$
$- (8b^3 + 6b + 4)$

69. Subtract $9x^2 - 3x + 7$ from $-2x^2 - 6x + 4$.

70. Subtract $-5w^3 + 5w^2 - 7$ from $6w^3 + 8w + 5$.

Find a polynomial that represents the perimeter of each square, rectangle, or triangle.

71. 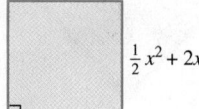 $\frac{1}{2}x^2 + 2x$

72. 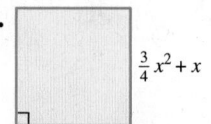 $\frac{3}{4}x^2 + x$

73. 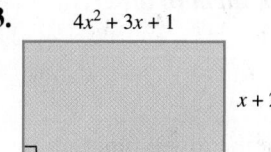 $4x^2 + 3x + 1$ $x + 2$

74. 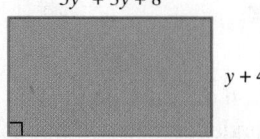 $5y^2 + 3y + 8$ $y + 4$

75. 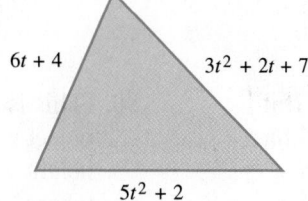 $6t + 4$ $3t^2 + 2t + 7$ $5t^2 + 2$

76. 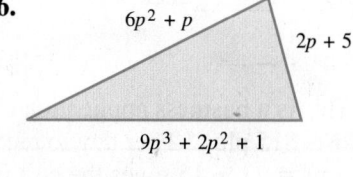 $6p^2 + p$ $2p + 5$ $9p^3 + 2p^2 + 1$

*Add or subtract as indicated. **See Example 9.***

77. $(9a^2b - 3a^2 + 2b) + (4a^2b - 4a^2 - 3b)$

78. $(4xy^3 - 3x + y) + (5xy^3 + 13x - 4y)$

79. $(2c^4d + 3c^2d^2 - 4d^2) - (c^4d + 8c^2d^2 - 5d^2)$

80. $(3k^2h^3 + 5kh + 6k^3h^2) - (2k^2h^3 - 9kh + k^3h^2)$

81. Subtract.

$$9m^3n - 5m^2n^2 + 4mn^2$$
$$-3m^3n + 6m^2n^2 + 8mn^2$$

82. Subtract.

$$12r^5t + 11r^4t^2 - 7r^3t^3$$
$$-8r^5t + 10r^4t^2 + 3r^3t^3$$

*Find (**a**) a polynomial that represents the perimeter of each triangle and (**b**) the measures of the angles of the triangle. (Hint: In part (b), the sum of the measures of the angles of any triangle is 180°.)*

83.

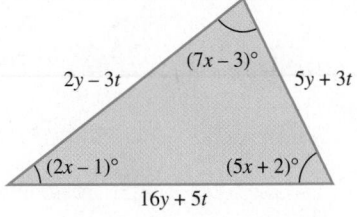

84.

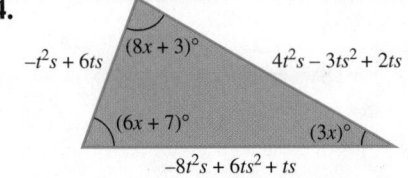

Relating Concepts (Exercises 85–88) For Individual or Group Work

*A polynomial can model the distance in feet that a car going approximately 68 mph will skid in t seconds. If we let D represent this distance, then using function notation from **Section 3.7,** we have*

$$D(t) = 100t - 13t^2.$$

*Each time we evaluate this **polynomial function** for a value of t, we get one and only one output value D(t). Exercises 85–88 illustrate this idea with this function and two others. **Work them in order.***

85. Evaluate the given function for $t = 5$. Use the result to fill in the blanks:

In _____ seconds, the car will skid

_____ feet.

86. Use the given function to find the distance the car will skid in 1 sec. Write an ordered pair of the form $(t, D(t))$.

87. (This is a business application of functions.) If it costs \$15 plus \$2 per day to rent a chain saw, then $C(x) = 2x + 15$ gives the cost in dollars to rent the chain saw for x days. Evaluate this function for $x = 6$. Use the result to fill in the blanks:

If the saw is rented for _____ days,

the cost is _____.

88. (This is a physics application of functions.) If an object is projected upward under certain conditions, its height in feet is given by $h(t) = -16t^2 + 60t + 80$, where t is in seconds. Evaluate this function for $t = 2.5$. Use the result to fill in the blanks:

If _____ seconds have elapsed,

the height of the object is _____ feet.

5.2 The Product Rule and Power Rules for Exponents

OBJECTIVES

1. Use exponents.
2. Use the product rule for exponents.
3. Use the rule $(a^m)^n = a^{mn}$.
4. Use the rule $(ab)^m = a^m b^m$.
5. Use the rule $\left(\dfrac{a}{b}\right)^m = \dfrac{a^m}{b^m}$.
6. Use combinations of the rules for exponents.
7. Use the rules for exponents in a geometry problem.

OBJECTIVE ▸ **1** **Use exponents.** In **Section 1.1,** we used exponents to write repeated products. In the expression 5^2, the number 5 is the **base** and 2 is the **exponent,** or **power**. The expression 5^2 is an **exponential expression.** Although we do not usually write the exponent when it is 1, in general,

$$a^1 = a, \quad \text{for any quantity } a.$$

EXAMPLE 1 Using Exponents

Write $3 \cdot 3 \cdot 3 \cdot 3 \cdot 3$ in exponential form and evaluate.

Since 3 occurs as a factor five times, the base is **3** and the exponent is **5**. The exponential expression is 3^5, read "3 to the fifth power," or simply "3 to the fifth."

$$\underbrace{3 \cdot 3 \cdot 3 \cdot 3 \cdot 3}_{\text{5 factors of 3}} \quad \text{means} \quad 3^5, \quad \text{which equals} \quad 243.$$

··········· Work Problem **1** at the Side. ▸

EXAMPLE 2 Evaluating Exponential Expressions

Name the base and the exponent of each expression. Then evaluate.

Expression	Base	Exponent	Value
(a) 5^4	5	4	$5 \cdot 5 \cdot 5 \cdot 5$, which equals 625
(b) -5^4	5	4	$-1 \cdot (5 \cdot 5 \cdot 5 \cdot 5)$, which equals -625
(c) $(-5)^4$	-5	4	$(-5)(-5)(-5)(-5)$, which equals 625

CAUTION

Compare **Examples 2(b) and 2(c).** In -5^4, the absence of parentheses shows that the exponent 4 applies only to the base 5, not -5. In $(-5)^4$, the parentheses show that the exponent 4 applies to the base -5. In summary, $-a^n$ and $(-a)^n$ are not necessarily the same.

Expression	Base	Exponent	Example
$-a^n$	a	n	$-3^2 = -(3 \cdot 3) = -9$
$(-a)^n$	$-a$	n	$(-3)^2 = (-3)(-3) = 9$

Work Problem **2** at the Side. ▸

OBJECTIVE ▸ **2** **Use the product rule for exponents.** To develop the product rule, we use the definition of an exponent.

$$2^4 \cdot 2^3$$

$$= \underbrace{(2 \cdot 2 \cdot 2 \cdot 2)}_{\text{4 factors}} \underbrace{(2 \cdot 2 \cdot 2)}_{\text{3 factors}}$$

$$= \underbrace{2 \cdot 2 \cdot 2 \cdot 2 \cdot 2 \cdot 2 \cdot 2}_{4 + 3 = 7 \text{ factors}}$$

$$= 2^7$$

1 Write each product in exponential form and evaluate.

(a) $2 \cdot 2 \cdot 2 \cdot 2$

(b) $(-3)(-3)(-3)$

2 Name the base and the exponent of each expression. Then evaluate.

(a) 2^5 **(b)** -2^5

(c) $(-2)^5$ **(d)** 4^2

(e) -4^2 **(f)** $(-4)^2$

Answers

1. **(a)** 2^4; 16   **(b)** $(-3)^3$; -27

2. **(a)** 2; 5; 32 **(b)** 2; 5; -32

   (c) -2; 5; -32 **(d)** 4; 2; 16

   (e) 4; 2; -16 **(f)** -4; 2; 16

3 Simplify by using the product rule, if possible.

GS **(a)** $8^2 \cdot 8^5$

$$= 8^{\underline{}+\underline{}}$$

$$= \underline{}$$

(b) $(-7)^5(-7)^3$

GS **(c)** $y^3 \cdot y$

$$= y^3 \cdot y^{\underline{}}$$

$$= y^{\underline{}+\underline{}}$$

$$= \underline{}$$

(d) $z^2 z^5 z^6$

(e) $4^2 \cdot 3^5$

(f) $6^4 + 6^2$

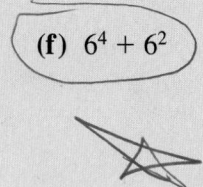

Answers

3. **(a)** $2; 5; 8^7$ **(b)** $(-7)^8$
   (c) $1; 3; 1; y^4$ **(d)** z^{13}
   (e) The product rule does not apply.
     (The product is 3888.)
   (f) The product rule does not apply.
     (The sum is 1332.)

Also,

$$6^2 \cdot 6^3$$

$$= (6 \cdot 6)(6 \cdot 6 \cdot 6)$$

$$= 6 \cdot 6 \cdot 6 \cdot 6 \cdot 6$$

$$= 6^5.$$

Generalizing from these examples, we have the following.

$2^4 \cdot 2^3$ is equal to 2^{4+3}, which equals 2^7.

$6^2 \cdot 6^3$ is equal to 6^{2+3}, which equals 6^5.

In each case, adding the exponents gives the exponent of the product, suggesting the **product rule for exponents.**

> **Product Rule for Exponents**
>
> For any positive integers m and n, $a^m \cdot a^n = a^{m+n}$.
> (Keep the same base and add the exponents.)
>
> *Example:* $6^2 \cdot 6^5 = 6^{2+5} = 6^7$

> **CAUTION**
>
> Do not multiply the bases when using the product rule. **_Keep the same base and add the exponents._** For example,
>
> $$6^2 \cdot 6^5 \text{ is equal to } 6^7, \text{ \textbf{not} } 36^7.$$

> **EXAMPLE 3** **Using the Product Rule**
>
> Use the product rule for exponents to simplify, if possible.
>
> **(a)** $6^3 \cdot 6^5$
>
> $\quad = 6^{3+5}$ Product rule
>
> $\quad = 6^8$ Add the exponents.
>
> **(b)** $(-4)^7(-4)^2$
>
> $\quad = (-4)^{7+2}$ Product rule
>
> $\quad = (-4)^9$ Add the exponents.
>
> **(c)** $x^2 \cdot x$
>
> $\quad = x^2 \cdot x^1$ $a = a^1$, for all a.
>
> $\quad = x^{2+1}$ Product rule
>
> $\quad = x^3$ Add the exponents.
>
> **(d)** $m^4 m^3 m^5$
>
> $\quad = m^{4+3+5}$ Product rule
>
> $\quad = m^{12}$ Add the exponents.
>
> **(e)** $\qquad 2^3 \cdot 3^2$ ⟵ Think: 3^2 means $3 \cdot 3$.
>
> Think: 2^3 means $2 \cdot 2 \cdot 2$.
>
> $\quad = 8 \cdot 9$ Evaluate 2^3 and 3^2.
>
> $\quad = 72$ Multiply.
>
> The product rule does not apply because the bases are different.
>
> **(f)** $2^3 + 2^4$
>
> $\quad = 8 + 16$ Evaluate 2^3 and 2^4.
>
> $\quad = 24$ Add.
>
> The product rule does not apply. This is a *sum,* not a *product.*

◀ Work Problem **3** at the Side.

EXAMPLE 4   Using the Product Rule

Multiply $2x^3$ and $3x^7$.

$2x^3 \cdot 3x^7$ ⎯ $\boxed{2x^3 = 2 \cdot x^3;\ 3x^7 = 3 \cdot x^7}$

$= (2 \cdot 3) \cdot (x^3 \cdot x^7)$ Commutative and associative properties

$= 6x^{3+7}$ Multiply; product rule

$= 6x^{10}$ Add the exponents.

·········· **Work Problem ④ at the Side.** ▶

CAUTION

Be sure that you understand the difference between *adding* and *multiplying* exponential expressions. For example,

$8x^3 + 5x^3$ means $(8 + 5)x^3$,   which equals $13x^3$,

but $(8x^3)(5x^3)$ means $(8 \cdot 5)x^{3+3}$,   which equals $40x^6$.

OBJECTIVE ③ **Use the rule $(a^m)^n = a^{mn}$.** We can simplify an expression such as $(5^2)^4$ with the product rule for exponents, as follows.

$(5^2)^4$

$= 5^2 \cdot 5^2 \cdot 5^2 \cdot 5^2$ Definition of exponent

$= 5^{2+2+2+2}$ Product rule

$= 5^8$ Add the exponents.

Observe that $2 \cdot 4 = 8$. This example suggests **power rule (a) for exponents.**

Power Rule (a) for Exponents

For any positive integers m and n, $(a^m)^n = a^{mn}$.
(Raise a power to a power by multiplying exponents.)

Example: $(3^2)^4 = 3^{2 \cdot 4} = 3^8$

EXAMPLE 5   Using Power Rule (a)

Use power rule (a) for exponents to simplify.

(a) $(2^5)^3$ **(b)** $(5^7)^2$ **(c)** $(x^2)^5$

$= 2^{5 \cdot 3}$ $= 5^{7 \cdot 2}$ $= x^{2 \cdot 5}$ Power rule (a)

$= 2^{15}$ $= 5^{14}$ $= x^{10}$ Multiply.

·········· **Work Problem ④ at the Side.** ▶

OBJECTIVE ④ **Use the rule $(ab)^m = a^m b^m$.** We can rewrite the expression $(4x)^3$ as shown below.

$(4x^3)$

$= (4x)(4x)(4x)$ Definition of exponent

$= 4 \cdot 4 \cdot 4 \cdot x \cdot x \cdot x$ Commutative and associative properties

$= 4^3 x^3$ Definition of exponent

④ Multiply.

GS (a) $5m^2 \cdot 2m^6$

$= (5 \cdot \underline{\hspace{1em}}) \cdot (m^{\underline{\hspace{0.5em}}} \cdot m^{\underline{\hspace{0.5em}}})$

$= \underline{\hspace{1em}} m^{\underline{\hspace{0.5em}}+\underline{\hspace{0.5em}}}$

$= \underline{\hspace{1em}}$

(b) $3p^5 \cdot 9p^4$

(c) $-7p^5 \cdot (3p^8)$

⑤ Simplify.

GS (a) $(5^3)^4$

$= 5^{\underline{\hspace{0.5em}} \cdot \underline{\hspace{0.5em}}}$

$= \underline{\hspace{1em}}$

(b) $(6^2)^5$

(c) $(3^2)^4$

(d) $(a^6)^5$

6 Simplify.

(GS) **(a)** $(2ab)^4$

$$= 2\text{—}a\text{—}b\text{—}$$

$$= \underline{\quad\quad}$$

(b) $5(mn)^3$

(c) $(3a^2b^4)^5$

(d) $(-5m^2)^3$

The example $(4x)^3 = 4^3x^3$ suggests **power rule (b) for exponents.**

Power Rule (b) for Exponents

For any positive integer m, $\quad (ab)^m = a^m b^m$.
(Raise a product to a power by raising each factor to the power.)

Example: $\quad (2p)^5 = 2^5 p^5$

EXAMPLE 6 **Using Power Rule (b)**

Use power rule (b) for exponents to simplify.

(a) $(3xy)^2$

$$= 3^2 x^2 y^2 \quad \text{Power rule (b)}$$
$$= 9x^2 y^2 \quad\quad 3^2 = 3 \cdot 3 = 9$$

(b) $9(pq)^2$

$$= 9(p^2 q^2) \quad \text{Power rule (b)}$$
$$= 9p^2 q^2 \quad\quad \text{Multiply.}$$

(c) $5(2m^2p^3)^4$

$$= 5\left[2^4 (m^2)^4 (p^3)^4\right] \quad \text{Power rule (b)}$$
$$= 5(2^4 m^8 p^{12}) \quad\quad \text{Power rule (a)}$$
$$= 80m^8 p^{12} \quad\quad 5 \cdot 2^4 = 5 \cdot 16 = 80$$

(d) $\quad\quad\quad (-5^6)^3$

$$= (-1 \cdot 5^6)^3 \quad -a = -1 \cdot a$$
$$= (-1)^3 (5^6)^3 \quad \text{Power rule (b)}$$

> Raise -1 to the designated power.

$$= -1 \cdot 5^{18} \quad\quad \text{Power rule (a)}$$
$$= -5^{18} \quad\quad\quad \text{Multiply.}$$

◀ **Work Problem 6 at the Side.**

CAUTION

Power rule (b) does not apply to a sum.

$$(4x)^2 = 4^2 x^2, \quad \text{but} \quad (4 + x)^2 \neq 4^2 + x^2.$$

OBJECTIVE **5** Use the rule $\left(\frac{a}{b}\right)^m = \frac{a^m}{b^m}$. Since the quotient $\frac{a}{b}$ can be written as $a \cdot \frac{1}{b}$, we can use power rule (b), together with some of the properties of real numbers, to state **power rule (c) for exponents.**

Power Rule (c) for Exponents

For any positive integer m, $\quad \left(\dfrac{a}{b}\right)^m = \dfrac{a^m}{b^m}$ (where $b \neq 0$).

(Raise a quotient to a power by raising both the numerator and the denominator to the power.)

Example: $\quad \left(\dfrac{5}{3}\right)^2 = \dfrac{5^2}{3^2}$

Answers

6. **(a)** 4; 4; 4; $16a^4b^4$ **(b)** $5m^3n^3$
 (c) $243a^{10}b^{20}$ **(d)** $-125m^6$

EXAMPLE 7 Using Power Rule (c)

Use power rule (c) for exponents to simplify.

(a) $\left(\dfrac{2}{3}\right)^5$ **(b)** $\left(\dfrac{m}{n}\right)^4$ **(c)** $\left(\dfrac{1}{5}\right)^4$

$\quad = \dfrac{2^5}{3^5}$ $= \dfrac{m^4}{n^4}$ $= \dfrac{1^4}{5^4}$ Power rule (c)

$\quad = \dfrac{32}{243}$ (where $n \neq 0$) $= \dfrac{1}{625}$ Simplify.

Note

In **Example 7(c)**, we used the fact that $1^4 = 1$ since $1 \cdot 1 \cdot 1 \cdot 1 = 1$.

     In general, $1^n = 1$,   *for any integer n.*

Work Problem **7** at the Side. ▶

The rules for exponents discussed in this section should be *memorized*.

Rules for Exponents

For positive integers m and n, the following are true.

                                          Examples

Product rule $a^m \cdot a^n = a^{m+n}$ $6^2 \cdot 6^5 = 6^{2+5} = 6^7$

Power rules   **(a)** $(a^m)^n = a^{mn}$ $(3^2)^4 = 3^{2 \cdot 4} = 3^8$

              (b) $(ab)^m = a^m b^m$ $(2p)^5 = 2^5 p^5$

              (c) $\left(\dfrac{a}{b}\right)^m = \dfrac{a^m}{b^m}$ $(b \neq 0)$ $\left(\dfrac{5}{3}\right)^2 = \dfrac{5^2}{3^2}$

OBJECTIVE ▶ **6** Use combinations of the rules for exponents.

EXAMPLE 8 Using Combinations of Rules

Simplify each expression.

(a) $\left(\dfrac{2}{3}\right)^2 \cdot 2^3$ **(b)** $(5x)^3(5x)^4$

                                       $= (5x)^7$ Product rule

$\quad = \dfrac{2^2}{3^2} \cdot \dfrac{2^3}{1}$ Power rule (c) $= 5^7 x^7$ Power rule (b)

$\quad = \dfrac{2^2 \cdot 2^3}{3^2 \cdot 1}$ Multiply fractions.

$\quad = \dfrac{2^{2+3}}{3^2}$ Product rule

$\quad = \dfrac{2^5}{3^2},$ or $\dfrac{32}{9}$

············ Continued on Next Page

---

7 Simplify. Assume that all variables represent nonzero real numbers.

GS (a) $\left(\dfrac{5}{2}\right)^4$

     $= \dfrac{5\text{—}}{2\text{—}}$

     $= \underline{\quad}$

(b) $\left(\dfrac{p}{q}\right)^2$

(c) $\left(\dfrac{r}{t}\right)^3$

(d) $\left(\dfrac{1}{3}\right)^5$

(e) $\left(\dfrac{1}{x}\right)^{10}$

Answers

7. (a) $4; 4; \dfrac{625}{16}$ **(b)** $\dfrac{p^2}{q^2}$ **(c)** $\dfrac{r^3}{t^3}$

    (d) $\dfrac{1}{243}$ **(e)** $\dfrac{1}{x^{10}}$

8 Simplify.

(a) $(2m)^5(2m)^3$

$$= (2m)\text{—}$$

$$= \underline{\quad\quad}$$

(b) $\left(\dfrac{5k^3}{3}\right)^2$

(c) $\left(\dfrac{1}{5}\right)^4(2x)^2$

(d) $(-3xy^2)^3(x^2y)^4$

9 Find a polynomial that represents the area of each figure.

(a)

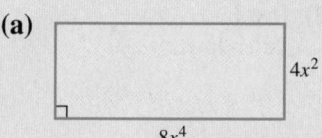

$4x^2$

$8x^4$

(b)

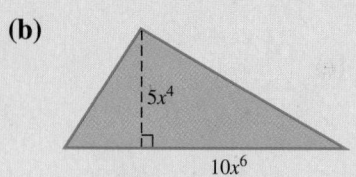

$5x^4$

$10x^6$

Answers

8. (a) $8; 2^8m^8,$ or $256m^8$ (b) $\dfrac{5^2k^6}{3^2},$ or $\dfrac{25k^6}{9}$

   (c) $\dfrac{2^2x^2}{5^4},$ or $\dfrac{4x^2}{625}$

   (d) $-3^3x^{11}y^{10},$ or $-27x^{11}y^{10}$

9. (a) $32x^6$ (b) $25x^{10}$

(c) $(2x^2y^3)^4\,(3xy^2)^3$

$$= 2^4(x^2)^4(y^3)^4 \cdot 3^3x^3(y^2)^3 \quad \text{Power rule (b)}$$

$$= 2^4x^8y^{12} \cdot 3^3x^3y^6 \quad\quad \text{Power rule (a)}$$

$$= 2^4 \cdot 3^3x^8x^3y^{12}y^6 \quad\quad \text{Commutative and associative properties}$$

$$= 16 \cdot 27x^{11}y^{18} \quad\quad\quad \text{Product rule}$$

$$= 432x^{11}y^{18} \quad\quad\quad\quad \text{Multiply.}$$

Notice that $(2x^2y^3)^4$ means $2^4x^{2\cdot4}y^{3\cdot4}$, **not** $(2\cdot4)\,x^{2\cdot4}y^{3\cdot4}$.

> Do *not* multiply the coefficient 2 and the exponent 4.

(d) $\quad\quad\quad (-x^3y)^2(-x^5y^4)^3$

> Think of the negative sign in each factor as −1.

$$= (-1x^3y)^2(-1x^5y^4)^3 \quad -a = -1 \cdot a$$

$$= (-1)^2(x^3)^2(y^2) \cdot (-1)^3(x^5)^3(y^4)^3 \quad \text{Power rule (b)}$$

$$= (-1)^2(x^6)(y^2) \cdot (-1)^3(x^{15})(y^{12}) \quad \text{Power rule (a)}$$

$$= (-1)^5(x^{6+15})(y^{2+12}) \quad \text{Product rule}$$

$$= -1x^{21}y^{14} \quad \text{Add the exponents.}$$

$$= -x^{21}y^{14} \quad -1 \cdot a = -a$$

◄ **Work Problem 8** at the Side.

OBJECTIVE **7** **Use the rules for exponents in a geometry problem.**

EXAMPLE 9   **Using Area Formulas**

Find a polynomial that represents the area in **(a) Figure 1** and **(b) Figure 2**.

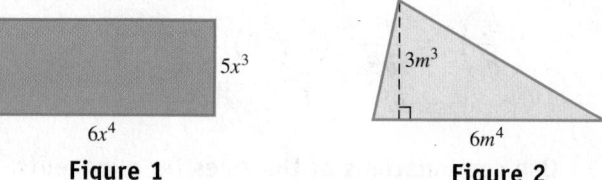

$5x^3$

$6x^4$

Figure 1

$3m^3$

$6m^4$

Figure 2

(a) For **Figure 1,** use the formula for the area of a rectangle, $A = LW$.

$$A = (6x^4)(5x^3) \quad \text{Substitute.}$$

$$A = 6 \cdot 5 \cdot x^{4+3} \quad \text{Commutative property and product rule}$$

$$A = 30x^7 \quad \text{Multiply. Add the exponents.}$$

(b) **Figure 2** is a triangle with base $6m^4$ and height $3m^3$.

$$A = \frac{1}{2}bh \quad\quad\quad \text{Area formula}$$

$$A = \frac{1}{2}(6m^4)(3m^3) \quad \text{Substitute.}$$

$$A = \frac{1}{2}(6 \cdot 3 \cdot m^{4+3}) \quad \text{Commutative property and product rule}$$

$$A = 9m^7 \quad\quad\quad \text{Multiply. Add the exponents.}$$

◄ **Work Problem 9** at the Side.

CONCEPT CHECK *Decide whether each statement is* true *or* false.

1. $3^3 = 9$

2. $(-2)^4 = 2^4$

3. $(a^2)^3 = a^5$

4. $\left(\dfrac{1}{4}\right)^2 = \dfrac{1}{4^2}$

5. CONCEPT CHECK What exponent is understood on the base x in the expression xy^2?

6. CONCEPT CHECK How are the expressions 3^2, 5^3, and 7^4 read?

Write each expression using exponents. **See Example 1.**

7. $t \cdot t \cdot t \cdot t \cdot t \cdot t \cdot t \cdot t$

8. $\dot{w} \cdot w \cdot \dot{w} \cdot w \cdot w \cdot w$

9. $\left(\dfrac{1}{2}\right)\left(\dfrac{1}{2}\right)\left(\dfrac{1}{2}\right)\left(\dfrac{1}{2}\right)\left(\dfrac{1}{2}\right)$

10. $\left(-\dfrac{1}{4}\right)\left(-\dfrac{1}{4}\right)\left(-\dfrac{1}{4}\right)\left(-\dfrac{1}{4}\right)$

11. $(-8p)(-8p)$

12. $(-7x)(-7x)(-7x)$

Identify the base and the exponent for each exponential expression. In Exercises 13–16, also evaluate the expression. **See Example 2.**

13. 3^5

14. 2^7

15. $(-3)^5$

16. $(-2)^7$

17. $(-6x)^4$

18. $(-8x)^4$

19. $-6x^4$

20. $-8x^4$

Use the product rule for exponents to simplify each expression, if possible. If it does not apply, say so. Write each answer in exponential form. **See Examples 3 and 4.**

21. $5^2 \cdot 5^6$

22. $3^6 \cdot 3^7$

23. $4^2 \cdot 4^7 \cdot 4^3$

24. $5^3 \cdot 5^8 \cdot 5^2$

25. $(-7)^3(-7)^6$

26. $(-9)^8(-9)^5$

27. $t^3 t^8 t^{13}$

28. $n^5 n^6 n^9$

29. $(-8r^4)(7r^3)$

30. $(10a^7)(-4a^3)$

31. $(-6p^5)(-7p^5)$

32. $(-5w^8)(-9w^8)$

33. $3^8 + 3^9$

34. $4^{12} + 4^5$

35. $5^8 \cdot 3^8$

36. $6^3 \cdot 8^3$

Use the power rules for exponents to simplify each expression. **See Examples 5–7.**

37. $(4^3)^2$

38. $(8^3)^6$

39. $(t^4)^5$

40. $(y^6)^5$

41. $(7r)^3$

42. $(11x)^4$

43. $(-5^2)^6$

44. $(-9^4)^8$

45. $(-8^3)^5$

46. $(-7^5)^7$

47. $(5xy)^5$

48. $(9pq)^6$

49. $8(qr)^3$

50. $4(vw)^5$

51. $\left(\dfrac{1}{2}\right)^3$

52. $\left(\dfrac{1}{3}\right)^5$

53. $\left(\dfrac{a}{b}\right)^3$ $(b \neq 0)$ **54.** $\left(\dfrac{r}{t}\right)^4$ $(t \neq 0)$ **55.** $\left(\dfrac{9}{5}\right)^8$ **56.** $\left(\dfrac{12}{7}\right)^6$

57. $(-2x^2y)^3$ **58.** $(-5m^4p^2)^3$ **59.** $(3a^3b^2)^2$ **60.** $(4x^3y^5)^4$

*Simplify each expression. **See Example 8.***

61. $\left(\dfrac{5}{2}\right)^3 \cdot \left(\dfrac{5}{2}\right)^2$ **62.** $\left(\dfrac{3}{4}\right)^5 \cdot \left(\dfrac{3}{4}\right)^6$ **63.** $\left(\dfrac{9}{8}\right)^3 \cdot 9^2$ **64.** $\left(\dfrac{8}{5}\right)^4 \cdot 8^3$

65. $(2x)^9(2x)^3$ **66.** $(6y)^5(6y)^8$ **67.** $(-6p)^4(-6p)$ **68.** $(-13q)^3(-13q)$

69. $(6x^2y^3)^5$ **70.** $(5r^5t^6)^7$ **71.** $(x^2)^3(x^3)^5$ **72.** $(y^4)^5(y^3)^5$

73. $(2w^2x^3y)^2(x^4y)^5$ **74.** $(3x^4y^2z)^3(yz^4)^5$ **75.** $(-r^4s)^2(-r^2s^3)^5$ **76.** $(-ts^6)^4(-t^3s^5)^3$

77. $\left(\dfrac{5a^2b^5}{c^6}\right)^3$ $(c \neq 0)$ **78.** $\left(\dfrac{6x^3y^9}{z^5}\right)^4$ $(z \neq 0)$ **79.** $(-5m^3p^4q)^2(p^2q)^3$

80. $(-a^4b^5)(-6a^3b^3)^2$ **81.** $(2x^2y^3z)^4(xy^2z^3)^2$ **82.** $(4q^2r^3s^5)^3(qr^2s^3)^4$

83. CONCEPT CHECK A student simplified $(10^2)^3$ as 1000^6. **What Went Wrong?**

84. CONCEPT CHECK A student simplified $(3x^2y^3)^4$ as $12x^8y^{12}$. **What Went Wrong?**

*Find a polynomial that represents the area of each figure. In Exercise 88, leave π in your answer. **See Example 9.** (If necessary, refer to the formulas inside the back cover. The ⌐ in the figures indicates 90° (right) angles.)*

85.

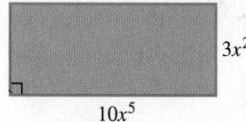

$3x^2$

$10x^5$

86.

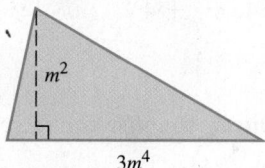

m^2

$3m^4$

87.

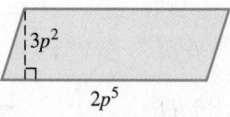

$3p^2$

$2p^5$

88.

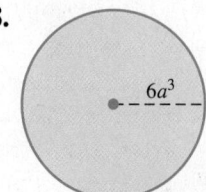

$6a^3$

5.3 Multiplying Polynomials

OBJECTIVES

1 Multiply a monomial and a polynomial.

2 Multiply two polynomials.

3 Multiply binomials by the FOIL method.

OBJECTIVE **1** **Multiply a monomial and a polynomial.** In Section 5.2, we found the product of two monomials as follows.

$$(-8m^6)(-9n^6)$$
$$= (-8)(-9)(m^6)(n^6) \quad \text{Commutative and associative properties}$$
$$= 72m^6n^6 \quad \text{Multiply.}$$

CAUTION

Do not confuse addition of terms with multiplication of terms.

$7q^5 + 2q^5$	$(7q^5)(2q^5)$
$= (7 + 2)q^5$	$= 7 \cdot 2q^{5+5}$
$= 9q^5$	$= 14q^{10}$

To find the product of a monomial and a polynomial with more than one term, we use the distributive property and multiplication of monomials.

EXAMPLE 1 **Multiplying Monomials and Polynomials**

Find each product.

(a) $4x^2(3x + 5)$
$$= 4x^2(3x) + 4x^2(5) \quad \text{Distributive property}$$
$$= 12x^3 + 20x^2 \quad \text{Multiply monomials.}$$

(b) $-8m^3(4m^3 + 3m^2 + 2m - 1)$
$$= -8m^3(4m^3) + (-8m^3)(3m^2)$$
$$\quad + (-8m^3)(2m) + (-8m^3)(-1) \quad \text{Distributive property}$$
$$= -32m^6 - 24m^5 - 16m^4 + 8m^3 \quad \text{Multiply monomials.}$$

·· **Work Problem ❶ at the Side.** ▶

OBJECTIVE **2** **Multiply two polynomials.** To find the product of the polynomials $x^2 + 3x + 5$ and $x - 4$, think of $x - 4$ as a single quantity and use the distributive property as follows.

$$(x^2 + 3x + 5)(x - 4)$$
$$= x^2(x - 4) + 3x(x - 4) + 5(x - 4) \quad \text{Distributive property}$$
$$= x^2(x) + x^2(-4) + 3x(x) + 3x(-4) + 5(x) + 5(-4)$$
$$\qquad\qquad\qquad\qquad\qquad\qquad \text{Distributive property again}$$
$$= x^3 - 4x^2 + 3x^2 - 12x + 5x - 20 \quad \text{Multiply monomials.}$$
$$= x^3 - x^2 - 7x - 20 \quad \text{Combine like terms.}$$

Multiplying Polynomials

To multiply two polynomials, multiply each term of the second polynomial by each term of the first polynomial and add the products.

---

❶ Find each product.

GS **(a)** $5m^3(2m + 7)$
$$= 5m^3(\underline{\quad}) + 5m^3(\underline{\quad})$$
$$= \underline{\qquad\qquad}$$

(b) $2x^4(3x^2 + 2x - 5)$

(c) $-4y^2(3y^3 + 2y^2 - 4y + 8)$

Answers

1. **(a)** $2m; 7; 10m^4 + 35m^3$
 (b) $6x^6 + 4x^5 - 10x^4$
 (c) $-12y^5 - 8y^4 + 16y^3 - 32y^2$

2 Multiply.

(a) $(m + 3)(m^2 - 2m + 1)$

(b) $(6p^2 + 2p - 4)(3p^2 - 5)$

3 Multiply vertically.

$$3x^2 + 4x - 5$$
$$x + 4$$

4 Use the rectangle method to find each product.

GS **(a)** $(4x + 3)(x + 2)$

	x	2
$4x$	$4x^2$	
3		

(b) $(x + 5)(x^2 + 3x + 1)$

EXAMPLE 2 **Multiplying Two Polynomials**

Multiply $(m^2 + 5)(4m^3 - 2m^2 + 4m)$.

$(m^2 + 5)(4m^3 - 2m^2 + 4m)$ Multiply each term of the second polynomial by each term of the first.

$= m^2(4m^3) + m^2(-2m^2) + m^2(4m) + 5(4m^3) + 5(-2m^2) + 5(4m)$

  Distributive property

$= 4m^5 - 2m^4 + 4m^3 + 20m^3 - 10m^2 + 20m$ Distributive property again

$= 4m^5 - 2m^4 + 24m^3 - 10m^2 + 20m$ Combine like terms.

◀ **Work Problem** **2** at the Side.

EXAMPLE 3 **Multiplying Polynomials Vertically**

Multiply $(x^3 + 2x^2 + 4x + 1)(3x + 5)$ vertically.

$$x^3 + 2x^2 + 4x + 1$$
$$3x + 5$$
  Write the polynomials vertically.

Begin by multiplying each of the terms in the top row by 5.

$$x^3 + 2x^2 + 4x + 1$$
$$3x + 5$$
$$\overline{5x^3 + 10x^2 + 20x + 5}$$ $5(x^3 + 2x^2 + 4x + 1)$

Now multiply each term in the top row by $3x$. Then add like terms.

$$x^3 + 2x^2 + 4x + 1$$
$$3x + 5$$
This process is similar to multiplication of whole numbers.

Place *like* terms in columns so they can be added.

$$\overline{5x^3 + 10x^2 + 20x + 5}$$
$$3x^4 + 6x^3 + 12x^2 + 3x$$ $3x(x^3 + 2x^2 + 4x + 1)$
$$\overline{3x^4 + 11x^3 + 22x^2 + 23x + 5}$$ Add in columns.

The product is $3x^4 + 11x^3 + 22x^2 + 23x + 5$.

◀ **Work Problem** **3** at the Side.

We can use a rectangle to model polynomial multiplication. For example, to find the product

$$(2x + 1)(3x + 2),$$

we label a rectangle with each term as shown below on the left. Then we write the product of each pair of monomials in the appropriate box as shown on the right.

	$3x$	2
$2x$		
1		

	$3x$	2
$2x$	$6x^2$	$4x$
1	$3x$	2

The product of the binomials is the sum of these four monomial products.

$$(2x + 1)(3x + 2)$$
$$= 6x^2 + 4x + 3x + 2$$
$$= 6x^2 + 7x + 2$$ Combine like terms.

◀ **Work Problem** **4** at the Side.

Answers

2. (a) $m^3 + m^2 - 5m + 3$
  (b) $18p^4 + 6p^3 - 42p^2 - 10p + 20$
3. $3x^3 + 16x^2 + 11x - 20$
4. (a)

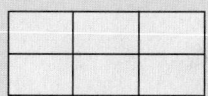

	x	2
$4x$	$4x^2$	$8x$
3	$3x$	6

  $4x^2 + 11x + 6$
  (b) $x^3 + 8x^2 + 16x + 5$

OBJECTIVE ▶ ③ **Multiply binomials by the FOIL method.** When multiplying binomials, the **FOIL method** reduces the rectangle method to a systematic approach without the rectangle. Consider this example.

$$(x + 3)(x + 5)$$
$$= (x + 3)x + (x + 3)5 \qquad \text{Distributive property}$$
$$= x(x) + 3(x) + x(5) + 3(5) \qquad \text{Distributive property again}$$
$$= x^2 + 3x + 5x + 15 \qquad \text{Multiply.}$$
$$= x^2 + 8x + 15 \qquad \text{Combine like terms.}$$

The letters of the word FOIL refer to the positions of the terms.

$(x + 3)(x + 5)$ Multiply the **First terms:** $x(x)$.   **F**

$(x + 3)(x + 5)$ Multiply the **Outer terms:** $x(5)$.   **O**
This is the **outer product.**

$(x + 3)(x + 5)$ Multiply the **Inner terms:** $3(x)$.   **I**
This is the **inner product.**

$(x + 3)(x + 5)$ Multiply the **Last terms:** $3(5)$.   **L**

We add the outer product, $5x$, and the inner product, $3x$, mentally so that the three terms of the answer can be written without extra steps.

$$(x + 3)(x + 5)$$
$$= x^2 + 8x + 15$$

Multiplying Binomials by the FOIL Method

Step 1   Multiply the two **First** terms of the binomials to get the first term of the answer.

Step 2   Find the **Outer** product and the **Inner** product and combine them (when possible) to get the middle term of the answer.

Step 3   Multiply the two **Last** terms of the binomials to get the last term of the answer.

$$\mathbf{F} = x^2 \qquad \mathbf{L} = 15$$

$$(x + 3)(x + 5) \qquad (x + 3)(x + 5)$$
$$= x^2 + 8x + 15$$

$$\mathbf{I} = 3x$$
$$\underline{\mathbf{O} = 5x}$$
$$8x \qquad \text{Add.}$$

Add the terms found in Steps 1–3.

Work Problem ⑤ at the Side.

---

⑤ For the product
$$(2p - 5)(3p + 7),$$
find and simplify the following.

(a) Product of first terms

_____ (_____)

= _____

(b) Outer product

_____ (_____)

= _____

(c) Inner product

_____ (_____)

= _____

(d) Product of last terms

_____ (_____)

= _____

(e) Complete product in simplified form

_____

Answers

5. **(a)** $2p; 3p; 6p^2$ **(b)** $2p; 7; 14p$
    (c) $-5; 3p; -15p$ **(d)** $-5; 7; -35$
    (e) $6p^2 - p - 35$

6 Use the FOIL method to find each product.

(a) $(m + 4)(m - 3)$

$$= m(\underline{\quad}) + m(\underline{\quad})$$
$$+ 4(\underline{\quad}) + 4(\underline{\quad})$$
$$= \underline{\qquad\qquad}$$

(b) $(y + 7)(y + 2)$

(c) $(r - 8)(r - 5)$

7 Find the product.

$$(4x - 3)(2y + 5)$$

8 Find each product.

(a) $(6m + 5)(m - 4)$

$$= 6m(\underline{\quad}) + 6m(\underline{\quad})$$
$$+ 5(\underline{\quad}) + 5(\underline{\quad})$$
$$= \underline{\qquad\qquad}$$

(b) $(3r + 2t)(3r + 4t)$

(c) $y^2(8y + 3)(2y + 1)$

$$= y^2(\underline{\qquad\qquad})$$
$$= \underline{\qquad\qquad}$$

Answers

6. (a) $m; -3; m; -3; m^2 + m - 12$
   (b) $y^2 + 9y + 14$
   (c) $r^2 - 13r + 40$
7. $8xy + 20x - 6y - 15$
8. (a) $m; -4; m; -4; 6m^2 - 19m - 20$
   (b) $9r^2 + 18rt + 8t^2$
   (c) $16y^2 + 14y + 3; 16y^4 + 14y^3 + 3y^2$

EXAMPLE 4 Using the FOIL Method

Use the FOIL method to find the product $(x + 8)(x - 6)$.

Step 1 **F** Multiply the **first** terms: $x(x) = x^2$.

Step 2 **O** Find the **outer** product: $x(-6) = -6x$.

      I Find the **inner** product: $8(x) = 8x$.

         Add the outer and inner products mentally: $-6x + 8x = 2x$.

Step 3 **L** Multiply the **last** terms: $8(-6) = -48$.

The product $(x + 8)(x - 6)$ is $x^2 + 2x - 48$.   Add the terms found in Steps 1–3.

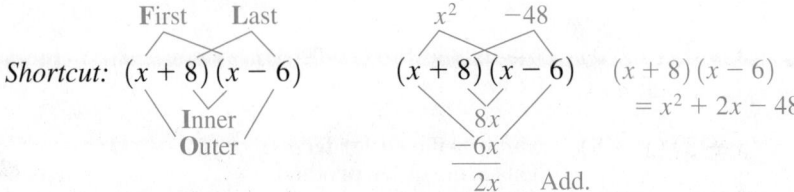

Shortcut:

◀ **Work Problem 6** at the Side.

EXAMPLE 5 Using the FOIL Method

Multiply $(9x - 2)(3y + 1)$.

First	$(9x - 2)(3y + 1)$	$27xy$
Outer	$(9x - 2)(3y + 1)$	$9x$
Inner	$(9x - 2)(3y + 1)$	$-6y$
Last	$(9x - 2)(3y + 1)$	-2

These unlike terms *cannot* be added.

          F   **O**   **I**   **L**

The product $(9x - 2)(3y + 1)$ is $27xy + 9x - 6y - 2$.

◀ **Work Problem 7** at the Side.

EXAMPLE 6 Using the FOIL Method

Find each product.

(a) $(2k + 5y)(k + 3y)$

$$= 2k(k) + 2k(3y) + 5y(k) + 5y(3y) \quad \text{FOIL method}$$
$$= 2k^2 + 6ky + 5ky + 15y^2 \quad \text{Multiply.}$$
$$= 2k^2 + 11ky + 15y^2 \quad \text{Combine like terms.}$$

(b) $(7p + 2q)(3p - q)$

$$= 21p^2 - 7pq + 6pq - 2q^2$$
$$= 21p^2 - pq - 2q^2$$

(c) $2x^2(x - 3)(3x + 4)$

$$= 2x^2(3x^2 - 5x - 12)$$
$$= 6x^4 - 10x^3 - 24x^2$$

◀ **Work Problem 8** at the Side.

Note

Alternatively, **Example 6(c)** can be simplified as follows.

$$2x^2(x - 3)(3x + 4) \quad \text{Multiply } 2x^2 \text{ and } x - 3 \text{ first.}$$
$$= (2x^3 - 6x^2)(3x + 4) \quad \text{Multiply that product and } 3x + 4.$$
$$= 6x^4 - 10x^3 - 24x^2 \quad \text{Same answer}$$

5.3 Exercises

CONCEPT CHECK *In Exercises 1 and 2, match each product in Column I with the correct polynomial in Column II.*

I	II	I	II
1. (a) $5x^3(6x^7)$	**A.** $125x^{21}$	**2. (a)** $(x-5)(x+4)$	**A.** $x^2+9x+20$
(b) $-5x^7(6x^3)$	**B.** $30x^{10}$	**(b)** $(x+5)(x+4)$	**B.** $x^2-9x+20$
(c) $(5x^7)^3$	**C.** $-216x^9$	**(c)** $(x-5)(x-4)$	**C.** x^2-x-20
(d) $(-6x^3)^3$	**D.** $-30x^{10}$	**(d)** $(x+5)(x-4)$	**D.** x^2+x-20

CONCEPT CHECK *Fill in each blank with the correct response.*

3. In multiplying a monomial by a polynomial, such as in $4x(3x^2+7x^3)=4x(3x^2)+4x(7x^3)$, the first property that is used is the _____ property.

4. The FOIL method can only be used to multiply two polynomials when both polynomials are _____.

Find each product. ***See Section 5.2.***

5. $5p(3q^2)$

6. $4a^3(3b^2)$

7. $(-6m^3)(3n^2)$

8. $(9r^3)(-2s^2)$

9. $y^5 \cdot 9y \cdot y^4$

10. $x^2 \cdot 3x^3 \cdot 2x$

11. $(4x^3)(2x^2)(-x^5)$

12. $(7t^5)(3t^4)(-t^8)$

Find each product. ***See Example 1.***

13. $-2m(3m+2)$

14. $-5p(6+3p)$

15. $\dfrac{3}{4}p(8-6p+12p^3)$

16. $\dfrac{4}{3}x(3+2x+5x^3)$

17. $2y^5(5y^4+2y+3)$

18. $2m^4(3m^2+5m+6)$

19. $2y^3(3y^3+2y+1)$

20. $2m^4(3m^2+5m+1)$

21. $-4r^3(-7r^2+8r-9)$

22. $-9a^5(-3a^6-2a^4+8a^2)$

23. $3a^2(2a^2-4ab+5b^2)$

24. $4z^3(8z^2+5zy-3y^2)$

25. $7m^3n^2(3m^2+2mn-n^3)$

$= 7m^3n^2(\underline{\quad}) + 7m^3n^2(\underline{\quad}) - 7m^3n^2(\underline{\quad})$

$= \underline{\hspace{3cm}}$

26. $2p^2q(3p^2q^2-5p+2q^2)$

$= \underline{\quad}(3p^2q^2) + \underline{\quad}(-5p) + \underline{\quad}(2q^2)$

$= \underline{\hspace{3cm}}$

Find each product. ***See Examples 2 and 3.***

27. $(6x + 1)(2x^2 + 4x + 1)$

28. $(9a + 2)(9a^2 + a + 1)$

29. $(2r - 1)(3r^2 + 4r - 4)$

30. $(9y - 2)(8y^2 - 6y + 1)$

31. $(4m + 3)(5m^3 - 4m^2 + m - 5)$

32. $(y + 4)(3y^3 - 2y^2 + y + 3)$

33. $(5x^2 + 2x + 1)(x^2 - 3x + 5)$

34. $(2m^2 + m - 3)(m^2 - 4m + 5)$

35. $(6x^4 - 4x^2 + 8x)\left(\dfrac{1}{2}x + 3\right)$

36. $(8y^6 + 4y^4 - 12y^2)\left(\dfrac{3}{4}y^2 + 2\right)$

GS *Find each product using the rectangle method shown in the text.*

37. $(x + 3)(x + 4)$

	x	4
x		
3		

Product: _____

38. $(x + 5)(x + 2)$

	___	___
x		
5		

Product: _____

39. $(2x + 1)(x^2 + 3x + 2)$

	x^2	$3x$	2
$2x$			
1			

Product: _____

40. $(x + 4)(3x^2 + 2x + 1)$

	___	$2x$	___
___			
___			

Product: _____

Find each product. Use the FOIL method. In Exercises 61–66, use the FOIL method and the distributive property. ***See Examples 4–6.***

41. $(m + 7)(m + 5)$
42. $(x + 4)(x + 7)$
43. $(n - 2)(n + 3)$
44. $(r - 6)(r + 8)$

45. $(4r + 1)(2r - 3)$
46. $(5x + 2)(2x - 7)$
47. $(3x + 2)(3x - 2)$
48. $(7x + 3)(7x - 3)$

49. $(3q + 1)(3q + 1)$
50. $(4w + 7)(4w + 7)$
51. $(5x + 7)(3y - 8)$

52. $(4x + 3)(2y - 1)$
53. $(3t + 4s)(2t + 5s)$
54. $(8v + 5w)(2v + 3w)$

55. $(-0.3t + 0.4)(t + 0.6)$

56. $(-0.5x + 0.9)(x - 0.2)$

57. $\left(x - \dfrac{2}{3}\right)\left(x + \dfrac{1}{4}\right)$

58. $\left(y + \dfrac{3}{5}\right)\left(y - \dfrac{1}{2}\right)$

59. $\left(-\dfrac{5}{4} + 2r\right)\left(-\dfrac{3}{4} - r\right)$

60. $\left(-\dfrac{8}{3} + 3k\right)\left(-\dfrac{2}{3} - k\right)$

61. $x(2x - 5)(x + 3)$

62. $m(4m - 1)(2m + 3)$

63. $3y^3(2y + 3)(y - 5)$

distribute first

64. $5t^4(t + 3)(3t - 1)$

65. $-8r^3(5r^2 + 2)(5r^2 - 2)$

66. $-5t^4(2t^4 + 1)(2t^4 - 1)$

Find polynomials that represent **(a)** *the area and* **(b)** *the perimeter of each square or rectangle. (If necessary, refer to the formulas inside the back cover.)*

67.

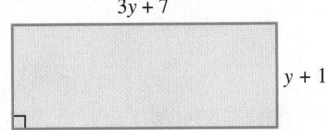

3y + 7 · y + 1

68.

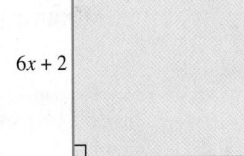

6x + 2

Relating Concepts (Exercises 69–74) For Individual or Group Work

Work Exercises 69–74 in order. *(All units are in yards.)*

69. Find a polynomial that represents the area of the rectangle.

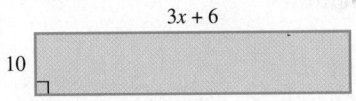

3x + 6 · 10

70. Suppose you know that the area of the rectangle is 600 yd². Use this information and the polynomial from **Exercise 69** to write an equation in x, and solve it.

71. Refer to **Exercise 70.** What are the dimensions of the rectangle?

72. Use the result of **Exercise 71** to find the perimeter of the rectangle.

73. Suppose the rectangle represents a lawn and it costs $3.50 per square yard to lay sod on the lawn. How much will it cost to sod the entire lawn?

74. Again, suppose the rectangle represents a lawn and it costs $9.00 per yard to fence the lawn. How much will it cost to fence the lawn?

5.4 Special Products

OBJECTIVES

1. Square binomials.
2. Find the product of the sum and difference of two terms.
3. Find greater powers of binomials.

---

1 Consider the binomial $x + 4$.

GS **(a)** What is the first term of the binomial? ____

Square it. ____

(b) What is the last term of the binomial? ____

Square it. ____

(c) Find twice the product of the two terms of the binomial.

$2(\underline{})(\underline{}) = \underline{}$

(d) Use the results of parts (a)–(c) to find $(x + 4)^2$.

_____

OBJECTIVE **1** **Square binomials.** The square of a binomial can be found quickly by using the method shown in **Example 1**.

EXAMPLE 1 Squaring a Binomial

Find $(m + 3)^2$.

> $(m + 3)^2$ means $(m + 3)(m + 3)$.

$$(m + 3)(m + 3)$$
$$= m^2 + 3m + 3m + 9 \qquad \text{FOIL method}$$
$$= m^2 + 6m + 9 \qquad \text{Combine like terms.}$$

This result has the squares of the first and the last terms of the binomial.

$$m^2 = m^2 \quad \text{and} \quad 3^2 = 9$$

The middle term, 6m, is twice the product of the two terms of the binomial, since the outer and inner products are $m(3)$ and $3(m)$.

$$m(3) + 3(m) = 2(m)(3) = 6m$$

◄ Work Problem **1** at the Side.

Example 1 suggests the following rules.

Square of a Binomial

The square of a binomial is a trinomial consisting of

| the square of the first term | + | twice the product of the two terms | + | the square of the last term. |

For a and b, the following are true.

$$(a + b)^2 = a^2 + 2ab + b^2$$
$$(a - b)^2 = a^2 - 2ab + b^2$$

EXAMPLE 2 Squaring Binomials

Square each binomial.

$$(a - b)^2 = a^2 \quad - 2 \cdot a \cdot b \quad + b^2$$

(a) $(5z - 1)^2 = (5z)^2 - 2(5z)(1) + (1)^2$

$$= 5^2z^2 - 10z + 1 \qquad (5z)^2 = 5^2z^2 = 25z^2 \text{ by power rule (b).}$$
$$= 25z^2 - 10z + 1$$

(b) $(3b + 5r)^2$

$$= (3b)^2 + 2(3b)(5r) + (5r)^2 \qquad (a + b)^2 = a^2 + 2ab + b^2$$
$$= 9b^2 + 30br + 25r^2 \qquad (3b)^2 = 3^2b^2 = 9b^2 \text{ and } (5r)^2 = 5^2r^2 = 25r^2$$

Answers

1. **(a)** $x; x^2$ **(b)** 4; 16 **(c)** $x; 4; 8x$
   (d) $x^2 + 8x + 16$

Continued on Next Page

(c) $(2a - 9x)^2$

$$= (2a)^2 - 2(2a)(9x) + (9x)^2 \qquad (a - b)^2 = a^2 - 2ab + b^2$$

$$= 4a^2 - 36ax + 81x^2 \qquad (2a)^2 = 2^2 a^2 = 4a^2 \text{ and}$$
$$(9x)^2 = 9^2 x^2 = 81x^2$$

(d) $\left(4m + \dfrac{1}{2}\right)^2$

$$= (4m)^2 + 2(4m)\left(\dfrac{1}{2}\right) + \left(\dfrac{1}{2}\right)^2 \qquad (a + b)^2 = a^2 + 2ab + b^2$$

$$= 16m^2 + 4m + \dfrac{1}{4} \qquad (4m)^2 = 4^2 m^2 = 16m^2 \text{ and } \left(\tfrac{1}{2}\right)^2 = \tfrac{1}{4}$$

(e) $x(4x - 3)^2$ ⟵ Remember the middle term.

$$= x(16x^2 - 24x + 9) \qquad \text{Square the binomial.}$$

$$= 16x^3 - 24x^2 + 9x \qquad \text{Distributive property}$$

In the square of a sum, all of the terms are positive, as in Examples *2(b)* and *(d)*. In the square of a difference, the middle term is negative, as in Examples *2(a)* and *(c)*.

CAUTION

A common error when squaring a binomial is to forget the middle term of the product. In general, remember the following.

$$(a + b)^2 = a^2 + 2ab + b^2, \quad \textbf{not} \quad a^2 + b^2,$$

and $\quad (a - b)^2 = a^2 - 2ab + b^2, \quad \textbf{not} \quad a^2 - b^2.$

Work Problem ❷ at the Side. ▶

OBJECTIVE ▶ ❷ Find the product of the sum and difference of two terms. In binomial products of the form $(a + b)(a - b)$, one binomial is the sum of two terms, and the other is the difference of the *same* two terms. Consider the following.

$$(x + 2)(x - 2)$$

$$= x^2 - 2x + 2x - 4 \qquad \text{FOIL method}$$

$$= x^2 - 4 \qquad \text{Combine like terms.}$$

Thus, the product of $a + b$ and $a - b$ is the **difference of two squares.**

Product of the Sum and Difference of Two Terms

$$(a + b)(a - b) = a^2 - b^2$$

Note

The expressions $a + b$ and $a - b$, the sum and difference of the *same* two terms, are **conjugates.** In the example above, $x + 2$ and $x - 2$ are conjugates.

---

❷ Square each binomial.

GS **(a)** $(t - 6)^2$

$$= (\underline{\quad})^2 - 2(\underline{\quad})(\underline{\quad})$$
$$+ (\underline{\quad})^2$$

$$= \underline{\hspace{3cm}}$$

(b) $(2m - p)^2$

GS **(c)** $(4p + 3q)^2$

$$= (\underline{\quad})^2 + \underline{\quad}(\underline{\quad})\,3q$$
$$+ (\underline{\quad})^2$$

$$= \underline{\hspace{3cm}}$$

(d) $(5r - 6s)^2$

(e) $\left(3k - \dfrac{1}{2}\right)^2$

(f) $x(2x + 7)^2$

Answers

2. **(a)** t; t; 6; 6; $t^2 - 12t + 36$
   (b) $4m^2 - 4mp + p^2$
   (c) $4p$; 2; $4p$; $3q$; $16p^2 + 24pq + 9q^2$
   (d) $25r^2 - 60rs + 36s^2$
   (e) $9k^2 - 3k + \dfrac{1}{4}$
   (f) $4x^3 + 28x^2 + 49x$

❸ Find each product.

ⓖⓢ (a) $(y + 3)(y - 3)$

$$= (\underline{})^2 - (\underline{})^2$$

$$= \underline{}$$

(b) $(10m + 7)(10m - 7)$

ⓖⓢ (c) $(7p + 2q)(7p - 2q)$

$$= (\underline{})^2 - (\underline{})^2$$

$$= \underline{}$$

(d) $\left(3r - \dfrac{1}{2}\right)\left(3r + \dfrac{1}{2}\right)$

(e) $3x(x^3 - 4)(x^3 + 4)$

EXAMPLE 3 **Finding the Product of the Sum and Difference of Two Terms**

Find each product.

(a) $(x + 4)(x - 4)$

Use the rule for the product of the sum and difference of two terms.

$$(x + 4)(x - 4)$$

$$= x^2 - 4^2 \qquad (a + b)(a - b) = a^2 - b^2$$

$$= x^2 - 16 \qquad \text{Square 4.}$$

(b) $\left(\dfrac{2}{3} - w\right)\left(\dfrac{2}{3} + w\right)$

$$= \left(\dfrac{2}{3} + w\right)\left(\dfrac{2}{3} - w\right) \qquad \text{Commutative property}$$

$$= \left(\dfrac{2}{3}\right)^2 - w^2 \qquad (a + b)(a - b) = a^2 - b^2$$

$$= \dfrac{4}{9} - w^2 \qquad \text{Square } \tfrac{2}{3}.$$

(c) $x(x + 2)(x - 2)$

$$= x(x^2 - 4) \qquad \text{Find the product of the sum and difference of two terms.}$$

$$= x^3 - 4x \qquad \text{Distributive property}$$

EXAMPLE 4 **Finding the Product of the Sum and Difference of Two Terms**

Find each product.

$$(a \;\; + \; b)\;(a \;\; - \; b)$$
$$\downarrow \quad\;\; \downarrow \quad\; \downarrow \quad\;\; \downarrow$$

(a) $(5m + 3)(5m - 3)$

Use the rule for the product of the sum and difference of two terms.

$$(5m + 3)(5m - 3)$$

$$\boxed{\text{Be careful to square } 5m \text{ correctly.}} \quad = (5m)^2 - 3^2 \qquad (a + b)(a - b) = a^2 - b^2$$

$$= 25m^2 - 9 \qquad \text{Apply the exponents.}$$

(b) $(4x + y)(4x - y)$

$$= (4x)^2 - y^2$$

$$= 16x^2 - y^2$$

$$\boxed{(4x)^2 = 4^2x^2 = 16x^2}$$

(c) $\left(z - \dfrac{1}{4}\right)\left(z + \dfrac{1}{4}\right)$

$$= z^2 - \left(\dfrac{1}{4}\right)^2$$

$$= z^2 - \dfrac{1}{16}$$

(d) $2p(p^2 + 3)(p^2 - 3)$

$$= 2p(p^4 - 9) \qquad \text{Multiply the conjugates.}$$

$$= 2p^5 - 18p \qquad \text{Distributive property}$$

◀ Work Problem **❸** at the Side.

Answers

3. (a) y; 3; $y^2 - 9$ **(b)** $100m^2 - 49$
(c) $7p$; $2q$; $49p^2 - 4q^2$ **(d)** $9r^2 - \dfrac{1}{4}$
(e) $3x^7 - 48x$

The product rules of this section will be important in **Chapters 6 and 7** and should be *memorized*.

OBJECTIVE ▶ **③ Find greater powers of binomials.** The methods used in the previous section and this section can be combined to find greater powers of binomials.

EXAMPLE 5 **Finding Greater Powers of Binomials**

Find each product.

(a) $(x + 5)^3$

$= (x + 5)^2(x + 5)$ $a^3 = a^2 \cdot a$

$= (x^2 + 10x + 25)(x + 5)$ Square the binomial.

$= x^3 + 10x^2 + 25x + 5x^2 + 50x + 125$ Multiply polynomials.

$= x^3 + 15x^2 + 75x + 125$ Combine like terms.

(b) $(2y - 3)^4$

$= (2y - 3)^2(2y - 3)^2$ $a^4 = a^2 \cdot a^2$

$= (4y^2 - 12y + 9)(4y^2 - 12y + 9)$ Square each binomial.

$= 16y^4 - 48y^3 + 36y^2 - 48y^3 + 144y^2$ Multiply polynomials.
$\quad - 108y + 36y^2 - 108y + 81$

$= 16y^4 - 96y^3 + 216y^2 - 216y + 81$ Combine like terms.

(c) $-2r(r + 2)^3$

$= -2r(r + 2)(r + 2)^2$ $a^3 = a \cdot a^2$

$= -2r(r + 2)(r^2 + 4r + 4)$ Square the binomial.

$= -2r(r^3 + 4r^2 + 4r + 2r^2 + 8r + 8)$ Multiply polynomials.

$= -2r(r^3 + 6r^2 + 12r + 8)$ Combine like terms.

$= -2r^4 - 12r^3 - 24r^2 - 16r$ Distributive property

········· **Work Problem ④ at the Side.** ▶

---

④ Find each product.

(a) $(m + 1)^3$

$= (\underline{\quad\quad})^2(\underline{\quad\quad})$

$= (\underline{\quad\quad\quad})(m + 1)$

$= \underline{\quad\quad\quad\quad}$

(b) $(3k - 2)^4$

(c) $-3x(x - 4)^3$

Answers

4. (a) $m + 1$; $m + 1$; $m^2 + 2m + 1$;
  $m^3 + 3m^2 + 3m + 1$
(b) $81k^4 - 216k^3 + 216k^2 - 96k + 16$
(c) $-3x^4 + 36x^3 - 144x^2 + 192x$

5.4 Exercises

Download the
MyDashBoard App

MyMathLab®

CONCEPT CHECK   *Fill in each blank with the correct response.*

1. The square of a binomial is a trinomial consisting of the _____ of the first term + _____ the _____ of the two terms + the _____ of the last term.

2. In the square of a sum, all of the terms are _____, while in the square of a difference, the middle term is _____.

3. The product of the sum and difference of two terms is the _____ of the _____ of the two terms.

4. The word _____ describes two expressions that represent the sum and the difference of the same two terms.

5. Consider the binomial square $(2x + 3)^2$.

   (a) What is the simplest form of the square of the first term, $(2x)^2$?

   (b) What is the simplest form of twice the product of the two terms, $2(2x)(3)$?

   (c) What is the simplest form of the square of the last term, 3^2?

   (d) Write the final product, which is a trinomial, using your results from parts (a)–(c).

6. Repeat **Exercise 5** for the binomial square $(3x - 2)^2$.

Square each binomial. See Examples 1 and 2.

7. $(p + 2)^2$

8. $(r + 5)^2$

9. $(z - 5)^2$

10. $(x - 3)^2$

11. $\left(x - \dfrac{3}{4}\right)^2$

12. $\left(y + \dfrac{5}{8}\right)^2$

13. $(v + 0.4)^2$

14. $(w - 0.9)^2$

15. $(4x - 3)^2$

16. $(9y - 4)^2$

17. $(10z + 6)^2$

18. $(5y + 2)^2$

19. $(2p + 5q)^2$

20. $(8a - 3b)^2$

21. $(0.8t + 0.7s)^2$

22. $(0.7z - 0.3w)^2$

23. $\left(5x + \dfrac{2}{5}y\right)^2$

24. $\left(6m - \dfrac{4}{5}n\right)^2$

25. $t(3t - 1)^2$

26. $x(2x + 5)^2$

27. $3t(4t + 1)^2$

28. $2x(7x - 2)^2$

29. $-(4r - 2)^2$

30. $-(3y - 8)^2$

31. Consider the product $(7x + 3y)(7x - 3y)$.

(a) What is the simplest form of the product of the first terms, $7x(7x)$?

(b) Multiply the outer terms, $7x(-3y)$. Then multiply the inner terms, $3y(7x)$. Add the results. What is this sum?

(c) What is the simplest form of the product of the last terms, $3y(-3y)$?

(d) Write the final product using your answers in parts (a) and (c). Why is the sum found in part (b) omitted here?

32. Repeat **Exercise 31** for the product $(5x + 7y)(5x - 7y)$.

Find each product. See Examples 3 and 4.

33. $(q + 2)(q - 2)$

34. $(x + 8)(x - 8)$

35. $\left(r - \dfrac{3}{4}\right)\left(r + \dfrac{3}{4}\right)$

36. $\left(q - \dfrac{7}{8}\right)\left(q + \dfrac{7}{8}\right)$

37. $(s + 2.5)(s - 2.5)$

38. $(t + 1.4)(t - 1.4)$

39. $(2w + 5)(2w - 5)$

40. $(3z + 8)(3z - 8)$

41. $(10x + 3y)(10x - 3y)$

42. $(13r + 2z)(13r - 2z)$

43. $(2x^2 - 5)(2x^2 + 5)$

44. $(9y^2 - 2)(9y^2 + 2)$

45. $\left(7x + \dfrac{3}{7}\right)\left(7x - \dfrac{3}{7}\right)$

46. $\left(9y + \dfrac{2}{3}\right)\left(9y - \dfrac{2}{3}\right)$

47. $p(3p + 7)(3p - 7)$

48. $q(5q - 1)(5q + 1)$

Find each product. See Example 5.

49. $(m - 5)^3$

50. $(p + 3)^3$

51. $(y + 2)^3$

52. $(x - 7)^3$

53. $(2a + 1)^3$

54. $(3m - 1)^3$

55. $(3r - 2t)^4$

56. $(2z + 5y)^4$

57. $3x^2(x - 3)^3$

58. $4p^3(p + 4)^3$

59. $-8x^2y(x + y)^4$

60. $-5uv^2(u - v)^4$

Determine a polynomial that represents the area of each figure. In Exercise 65, leave π in your answer. (If necessary, refer to the formulas inside the back cover.)

61.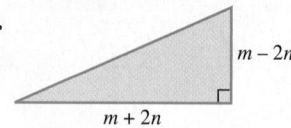

62.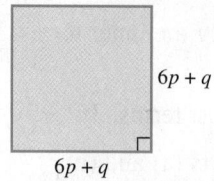

63.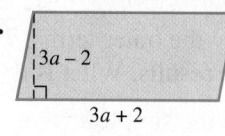

64.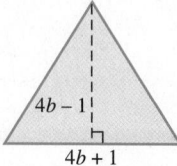

65.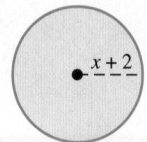

66.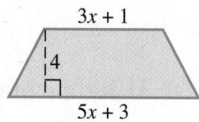

In Exercises 67 and 68, refer to the figure shown here.

67. Find a polynomial that represents the volume of the cube (in cubic units).

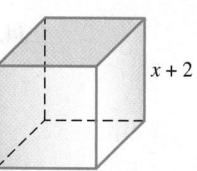

68. If the value of x is 6, what is the volume of the cube (in cubic units)?

Relating Concepts (Exercises 69–78) For Individual or Group Work

Special products can be illustrated by using areas of rectangles. Use the figure and **work Exercises 69–74 in order,** *to justify the special product* $(a + b)^2 = a^2 + 2ab + b^2$.

69. Express the area of the large square as the square of a binomial.

70. Give the monomial that represents the area of the red square.

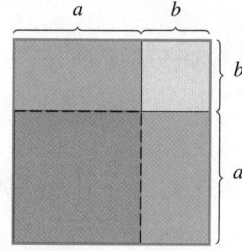

71. Give the monomial that represents the sum of the areas of the blue rectangles.

72. Give the monomial that represents the area of the yellow square.

73. What is the sum of the monomials you obtained in **Exercises 70–72?**

74. Explain why the binomial square found in **Exercise 69** must equal the polynomial found in **Exercise 73.**

To understand how the special product $(a + b)^2 = a^2 + 2ab + b^2$ *can be applied to a purely numerical problem,* **work Exercises 75–78 in order.**

75. Evaluate 35^2 using either traditional paper-and-pencil methods or a calculator.

76. The number 35 can be written as $30 + 5$. Therefore, $35^2 = (30 + 5)^2$. Use the special product for squaring a binomial with $a = 30$ and $b = 5$ to write an expression for $(30 + 5)^2$. Do not simplify at this time.

77. Use the order of operations to simplify the expression found in **Exercise 76.**

78. How do the answers in **Exercises 75 and 77** compare?

5.5 Integer Exponents and the Quotient Rule

OBJECTIVES

1. Use 0 as an exponent.
2. Use negative numbers as exponents.
3. Use the quotient rule for exponents.
4. Use combinations of rules.

Consider the following list.

$$2^4 = 16$$
$$2^3 = 8$$
$$2^2 = 4$$

Each time we decrease the exponent by 1, the value is divided by 2 (the base). Using this pattern, we can continue the list to lesser and lesser integer exponents.

$$2^1 = 2$$
$$2^0 = 1$$
$$2^{-1} = \frac{1}{2}$$

Work Problem ❶ at the Side. ▶

From the preceding list and the answers to **Margin Problem 1,** it appears that we should define 2^0 as 1 and negative exponents as reciprocals.

OBJECTIVE ❶ **Use 0 as an exponent.** We want the definitions of 0 and negative exponents to satisfy the rules for exponents from **Section 5.2.** For example, if $6^0 = 1$,

$$6^0 \cdot 6^2 = 1 \cdot 6^2 = 6^2 \quad \text{and} \quad 6^0 \cdot 6^2 = 6^{0+2} = 6^2,$$

so the product rule is satisfied. The power rules are also valid for a 0 exponent. Thus, we define a 0 exponent as follows.

Zero Exponent

For any nonzero real number a, $a^0 = 1$.

Example: $17^0 = 1$

EXAMPLE 1 Using Zero Exponents

Evaluate.

(a) $60^0 = 1$ **(b)** $(-60)^0 = 1$

(c) $-60^0 = -(1)$, or -1 **(d)** $y^0 = 1$ $(y \neq 0)$

(e) $6y^0 = 6(1)$, or 6 $(y \neq 0)$ **(f)** $(6y)^0 = 1$ $(y \neq 0)$

Work Problem ❷ at the Side. ▶

CAUTION

Look again at **Examples 1(b) and 1(c).** In $(-60)^0$, the base is -60, and since any nonzero base raised to the 0 exponent is 1, $(-60)^0 = 1$. In -60^0, which can be written $-(60)^0$, the base is 60, so $-60^0 = -1$.

❶ Continue the list of exponentials using -2, -3, and -4 as exponents.

$$2^{-2} = \underline{\hspace{1cm}}$$
$$2^{-3} = \underline{\hspace{1cm}}$$
$$2^{-4} = \underline{\hspace{1cm}}$$

❷ Evaluate.

(a) 28^0

(b) $(-16)^0$

(c) -7^0

(d) m^0 $(m \neq 0)$

(e) $-2p^0$ $(p \neq 0)$

(f) $(5r)^0$ $(r \neq 0)$

Answers

1. $2^{-2} = \frac{1}{4}; 2^{-3} = \frac{1}{8}; 2^{-4} = \frac{1}{16}$

2. **(a)** 1 **(b)** 1 **(c)** -1
   (d) 1 **(e)** -2 **(f)** 1

OBJECTIVE **2** **Use negative numbers as exponents.** Review the lists at the beginning of this section and **Margin Problem 1.** Since $2^{-2} = \frac{1}{4}$ and $2^{-3} = \frac{1}{8}$, we can think that 2^{-n} should perhaps equal $\frac{1}{2^n}$. Is the product rule valid in such cases? For example, if we multiply 6^{-2} by 6^2, we get

$$6^{-2} \cdot 6^2 = 6^{-2+2} = 6^0 = 1.$$

The expression 6^{-2} behaves as if it were the reciprocal of 6^2, because their product is 1. The reciprocal of 6^2 may be written $\frac{1}{6^2}$, leading us to define 6^{-2} as $\frac{1}{6^2}$, and generalize accordingly.

Negative Exponents

For any nonzero real number a and any integer n, $\quad a^{-n} = \dfrac{1}{a^n}.$

Example: $\quad 3^{-2} = \dfrac{1}{3^2}$

By definition, a^{-n} and a^n are reciprocals, since

$$a^n \cdot a^{-n} = a^n \cdot \frac{1}{a^n} = 1.$$

Since $1^n = 1$, the definition of a^{-n} can also be written

$$a^{-n} = \frac{1}{a^n} = \frac{1^n}{a^n} = \left(\frac{1}{a}\right)^n.$$

For example, $\quad 6^{-3} = \left(\dfrac{1}{6}\right)^3 \quad$ and $\quad \left(\dfrac{1}{3}\right)^{-2} = 3^2.$

EXAMPLE 2 **Using Negative Exponents**

Simplify by writing with positive exponents. Assume that all variables represent nonzero real numbers.

(a) $3^{-2} = \dfrac{1}{3^2}$, or $\dfrac{1}{9}$ $\quad a^{-n} = \frac{1}{a^n}$ **(b)** $5^{-3} = \dfrac{1}{5^3}$, or $\dfrac{1}{125}$ $\quad a^{-n} = \frac{1}{a^n}$

(c) $\left(\dfrac{1}{2}\right)^{-3} = 2^3$, or 8 $\quad \frac{1}{2}$ and 2 are reciprocals.

Notice that we can change the base to its reciprocal if we also change the sign of the exponent.

(d) $\left(\dfrac{2}{5}\right)^{-4}$ **(e)** $\left(\dfrac{4}{3}\right)^{-5}$

$= \left(\dfrac{5}{2}\right)^4$ $\quad \frac{2}{5}$ and $\frac{5}{2}$ are reciprocals. $\qquad = \left(\dfrac{3}{4}\right)^5$ $\quad \frac{4}{3}$ and $\frac{3}{4}$ are reciprocals.

$= \dfrac{5^4}{2^4}$ $\quad$ Power rule (c) $\qquad\qquad = \dfrac{3^5}{4^5}$ $\quad$ Power rule (c)

$= \dfrac{625}{16}$ $\quad$ Apply the exponents. $\qquad = \dfrac{243}{1024}$ $\quad$ Apply the exponents.

················· **Continued on Next Page**

(f) $4^{-1} - 2^{-1}$

$$= \frac{1}{4} - \frac{1}{2} \quad \text{Apply the exponents.}$$

$$= \frac{1}{4} - \frac{2}{4} \quad \text{Find a common denominator.}$$

$$= -\frac{1}{4} \quad \text{Subtract.}$$

(g) $3p^{-2}$

$$= \frac{3}{1} \cdot \frac{1}{p^2} \quad a^{-n} = \frac{1}{a^n}$$

$$= \frac{3}{p^2} \quad \text{Multiply.}$$

(h) $\dfrac{1}{x^{-4}}$

$$= \frac{1^{-4}}{x^{-4}} \quad \begin{array}{l} 1^n = 1, \\ \text{for any integer } n \end{array}$$

$$= \left(\frac{1}{x}\right)^{-4} \quad \text{Power rule (c)}$$

$$= x^4 \quad \frac{1}{x} \text{ and } x \text{ are reciprocals.}$$

(i) $x^3 y^{-4}$

$$= \frac{x^3}{1} \cdot \frac{1}{y^4} \quad a^{-n} = \frac{1}{a^n}$$

$$= \frac{x^3}{y^4} \quad \text{Multiply.}$$

In general, $\quad \dfrac{1}{a^{-n}} = a^n.$

CAUTION

A negative exponent does not indicate a negative number. Negative exponents lead to reciprocals.

Expression	**Example**	
a^{-n}	$3^{-2} = \dfrac{1}{3^2} = \dfrac{1}{9}$	Not negative
$-a^{-n}$	$-3^{-2} = -\dfrac{1}{3^2} = -\dfrac{1}{9}$	Negative

Work Problem ③ at the Side. ▶

Consider the following.

$$\frac{2^{-3}}{3^{-4}}$$

$$= \frac{\dfrac{1}{2^3}}{\dfrac{1}{3^4}} \quad \text{Definition of negative exponent}$$

$$= \frac{1}{2^3} \div \frac{1}{3^4} \quad \frac{a}{b} \text{ means } a \div b.$$

$$= \frac{1}{2^3} \cdot \frac{3^4}{1} \quad \begin{array}{l} \text{To divide, multiply by the} \\ \text{reciprocal of the divisor.} \end{array}$$

$$= \frac{3^4}{2^3} \quad \text{Multiply.}$$

Therefore, $\dfrac{2^{-3}}{3^{-4}} = \dfrac{3^4}{2^3}.$

---

③ Simplify by writing with positive exponents. Assume that all variables represent nonzero real numbers.

(a) 4^{-3}

(b) 6^{-2}

(c) $\left(\dfrac{1}{4}\right)^{-2}$

(d) $\left(\dfrac{2}{3}\right)^{-2}$

(e) $2^{-1} + 5^{-1}$

(f) $7m^{-5}$

(g) $\dfrac{1}{z^{-6}}$

(h) $p^2 q^{-5}$

Answers

3. (a) $\dfrac{1}{4^3}$, or $\dfrac{1}{64}$ **(b)** $\dfrac{1}{6^2}$, or $\dfrac{1}{36}$

(c) 4^2, or 16 **(d)** $\left(\dfrac{3}{2}\right)^2$, or $\dfrac{9}{4}$

(e) $\dfrac{1}{2} + \dfrac{1}{5} = \dfrac{7}{10}$ **(f)** $\dfrac{7}{m^5}$ **(g)** z^6 **(h)** $\dfrac{p^2}{q^5}$

4 Simplify. Assume that all variables represent nonzero real numbers.

(a) $\dfrac{7^{-1}}{5^{-4}}$

$= \dfrac{5\underline{\quad}}{7\underline{\quad}}$

$= \underline{\quad\quad}$

(b) $\dfrac{x^{-3}}{y^{-2}}$

(c) $\dfrac{4h^{-5}}{m^{-2}k}$

(d) $\left(\dfrac{3m}{p}\right)^{-2}$

$= \left(\dfrac{p}{3m}\right)^{\underline{\quad}}$

$= \dfrac{p\underline{\quad}}{3\underline{\quad}m\underline{\quad}}$

$= \underline{\quad\quad}$

Answers

4. (a) $4; 1; \dfrac{625}{7}$ **(b)** $\dfrac{y^2}{x^3}$ **(c)** $\dfrac{4m^2}{h^5k}$

(d) $2; 2; 2; 2; \dfrac{p^2}{9m^2}$

Changing from Negative to Positive Exponents

For any nonzero numbers a and b, and any integers m and n,

$$\frac{a^{-m}}{b^{-n}} = \frac{b^n}{a^m} \quad \text{and} \quad \left(\frac{a}{b}\right)^{-m} = \left(\frac{b}{a}\right)^m.$$

Examples: $\dfrac{3^{-5}}{2^{-4}} = \dfrac{2^4}{3^5}$ and $\left(\dfrac{4}{5}\right)^{-3} = \left(\dfrac{5}{4}\right)^3$

EXAMPLE 3 Changing from Negative to Positive Exponents

Simplify. Assume that all variables represent nonzero real numbers.

(a) $\dfrac{4^{-2}}{5^{-3}} = \dfrac{5^3}{4^2}$, or $\dfrac{125}{16}$ **(b)** $\dfrac{m^{-5}}{p^{-1}} = \dfrac{p^1}{m^5}$, or $\dfrac{p}{m^5}$

(c) $\dfrac{a^{-2}b}{3d^{-3}} = \dfrac{bd^3}{3a^2}$ Notice that b in the numerator and the coefficient 3 in the denominator are not affected.

(d) $\left(\dfrac{x}{2y}\right)^{-4}$

$= \left(\dfrac{2y}{x}\right)^4$ Negative-to-positive rule

$= \dfrac{2^4y^4}{x^4}$ Power rules (b) and (c)

$= \dfrac{16y^4}{x^4}$ Apply the exponent.

◀ **Work Problem 4** at the Side.

CAUTION

Be careful. We cannot use the rule $\dfrac{a^{-m}}{b^{-n}} = \dfrac{b^n}{a^m}$ to change negative exponents to positive exponents if the exponents occur in a *sum* or *difference* of terms. For example,

$$\frac{5^{-2} + 3^{-1}}{7 - 2^{-3}} \quad \text{would be written with positive exponents as} \quad \frac{\dfrac{1}{5^2} + \dfrac{1}{3}}{7 - \dfrac{1}{2^3}}.$$

OBJECTIVE **3** **Use the quotient rule for exponents.** Consider a quotient of two exponential expressions with the same base.

$$\frac{6^5}{6^3} = \frac{6 \cdot 6 \cdot 6 \cdot 6 \cdot 6}{6 \cdot 6 \cdot 6} = 6^2$$

The difference between the exponents, $5 - 3 = 2$, is the exponent in the quotient. Also,

$$\frac{6^2}{6^4} = \frac{6 \cdot 6}{6 \cdot 6 \cdot 6 \cdot 6} = \frac{1}{6^2} = 6^{-2}.$$

Here, $2 - 4 = -2$. These examples suggest the **quotient rule for exponents.**

Quotient Rule for Exponents

For any nonzero real number a and any integers m and n,

$$\frac{a^m}{a^n} = a^{m-n}.$$

(Keep the same base and subtract the exponents.)

Example: $\dfrac{5^8}{5^4} = 5^{8-4} = 5^4$

CAUTION

A common **error** is to write $\dfrac{5^8}{5^4} = 1^{8-4} = 1^4$. ***This is incorrect.*** By the quotient rule, the quotient must have the *same base*, which is 5 here.

$$\frac{5^8}{5^4} = 5^{8-4} = 5^4$$

We can confirm this by writing out the factors.

$$\frac{5^8}{5^4} = \frac{5 \cdot 5 \cdot 5 \cdot 5 \cdot 5 \cdot 5 \cdot 5 \cdot 5}{5 \cdot 5 \cdot 5 \cdot 5} = 5^4$$

EXAMPLE 4 Using the Quotient Rule

Simplify. Assume that all variables represent nonzero real numbers.

(a) $\dfrac{5^8}{5^6} = 5^{8-6} = 5^2 = 25$

Keep the same base.

(b) $\dfrac{4^2}{4^9} = 4^{2-9} = 4^{-7} = \dfrac{1}{4^7}$

(c) $\dfrac{5^{-3}}{5^{-7}} = 5^{-3-(-7)} = 5^4 = 625$

Be careful with signs.

(d) $\dfrac{q^5}{q^{-3}} = q^{5-(-3)} = q^8$

(e) $\dfrac{3^2 x^5}{3^4 x^3}$

$= \dfrac{3^2}{3^4} \cdot \dfrac{x^5}{x^3}$

$= 3^{2-4} \cdot x^{5-3}$ Quotient rule

$= 3^{-2} x^2$ Subtract.

$= \dfrac{x^2}{3^2}$,   or $\dfrac{x^2}{9}$

(f) $\dfrac{(m+n)^{-2}}{(m+n)^{-4}}$

$= (m+n)^{-2-(-4)}$

$= (m+n)^{-2+4}$

$= (m+n)^2$ $(m \neq -n)$

The restriction $m \neq -n$ is necessary to prevent a denominator of 0 in the original expression. Division by 0 is undefined.

(g) $\dfrac{7x^{-3}y^2}{2^{-1}x^2y^{-5}}$

$= \dfrac{7 \cdot 2^1 y^2 y^5}{x^2 x^3}$ Definition of negative exponent

$= \dfrac{14y^7}{x^5}$ Multiply; product rule

· ▶ **Work Problem ⑤ at the Side.** ▶

⑤ Simplify. Assume that all variables represent nonzero real numbers.

GS (a) $\dfrac{5^{11}}{5^8}$

$= 5^{\underline{}-\underline{}}$

$= 5^{\underline{}}$

$= \underline{}$

(b) $\dfrac{4^7}{4^{10}}$

GS (c) $\dfrac{6^{-5}}{6^{-2}}$

$= 6^{\underline{}-(\underline{})}$

$= 6^{\underline{}}$

$= \dfrac{1}{6^{\underline{}}}$

$= \underline{}$

(d) $\dfrac{8^4 m^9}{8^5 m^{10}}$

(e) $\dfrac{3^{-1}(x+y)^{-3}}{2^{-2}(x+y)^{-4}}$ $(x \neq -y)$

Answers

5. (a) $11; 8; 3; 125$ **(b)** $\dfrac{1}{64}$

(c) $-5; -2; -3; 3; \dfrac{1}{216}$

(d) $\dfrac{1}{8m}$ **(e)** $\dfrac{4}{3}(x+y)$

The definitions and rules for exponents given in this section and **Section 5.2** are summarized here.

Definitions and Rules for Exponents

For any integers m and n, the following are true.

		Examples
Product rule	$a^m \cdot a^n = a^{m+n}$	$7^4 \cdot 7^5 = 7^{4+5} = 7^9$
Zero exponent	$a^0 = 1 \quad (a \neq 0)$	$(-3)^0 = 1$
Negative exponent	$a^{-n} = \dfrac{1}{a^n} \quad (a \neq 0)$	$5^{-3} = \dfrac{1}{5^3}$
Quotient rule	$\dfrac{a^m}{a^n} = a^{m-n} \quad (a \neq 0)$	$\dfrac{2^2}{2^5} = 2^{2-5} = 2^{-3} = \dfrac{1}{2^3}$
Power rules (a)	$(a^m)^n = a^{mn}$	$(4^2)^3 = 4^{2 \cdot 3} = 4^6$
(b)	$(ab)^m = a^m b^m$	$(3k)^4 = 3^4 k^4$
(c)	$\left(\dfrac{a}{b}\right)^m = \dfrac{a^m}{b^m} \quad (b \neq 0)$	$\left(\dfrac{2}{3}\right)^2 = \dfrac{2^2}{3^2}$
Negative-to-positive rules	$\dfrac{a^{-m}}{b^{-n}} = \dfrac{b^n}{a^m} \quad (a, b \neq 0)$	$\dfrac{2^{-4}}{5^{-3}} = \dfrac{5^3}{2^4}$
	$\left(\dfrac{a}{b}\right)^{-m} = \left(\dfrac{b}{a}\right)^m$	$\left(\dfrac{4}{7}\right)^{-2} = \left(\dfrac{7}{4}\right)^2$

OBJECTIVE **4** **Use combinations of rules.** We sometimes need to use more than one rule to simplify an expression.

EXAMPLE 5 **Using a Combination of Rules**

Simplify each expression. Assume that all variables represent nonzero real numbers.

(a) $\dfrac{(4^2)^3}{4^5}$

$= \dfrac{4^6}{4^5}$ Power rule (a)

$= 4^{6-5}$ Quotient rule

$= 4^1$ Subtract the exponents.

$= 4$ $4^1 = 4$

(b) $(2x)^3(2x)^2$

$= (2x)^5$ Product rule

$= 2^5 x^5$ Power rule (b)

$= 32x^5$ $2^5 = 32$

Continued on Next Page

(c) $\left(\dfrac{2x^3}{5}\right)^{-4}$

$= \left(\dfrac{5}{2x^3}\right)^4$ Negative-to-positive rule

$= \dfrac{5^4}{2^4 x^{12}}$ Power rules (a)–(c)

$= \dfrac{625}{16x^{12}}$ Apply the exponents.

(d) $\left(\dfrac{3x^{-2}}{4^{-1}y^3}\right)^{-3}$

$= \dfrac{3^{-3}x^6}{4^3 y^{-9}}$ Power rules (a)–(c)

$= \dfrac{x^6 y^9}{4^3 \cdot 3^3}$ Negative-to-positive rule

$= \dfrac{x^6 y^9}{1728}$ $4^3 \cdot 3^3 = 64 \cdot 27 = 1728$

(e) $\dfrac{(4m)^{-3}}{(3m)^{-4}}$

$= \dfrac{4^{-3}m^{-3}}{3^{-4}m^{-4}}$ Power rule (b)

$= \dfrac{3^4 m^4}{4^3 m^3}$ Negative-to-positive rule

$= \dfrac{3^4 m^{4-3}}{4^3}$ Quotient rule

$= \dfrac{3^4 m}{4^3}$ Subtract.

$= \dfrac{81m}{64}$ Apply the exponents.

Note

Since the steps can be done in several different orders, there are many equally correct ways to simplify expressions like those in **Examples 5(c) through 5(e).**

Work Problem ❻ at the Side. ▶

❻ Simplify each expression. Assume that all variables represent nonzero real numbers.

(a) $\dfrac{(3^4)^2}{3^3}$

(b) $(4x)^2(4x)^4$

(c) $\left(\dfrac{5}{2z^4}\right)^{-3}$

(d) $\left(\dfrac{6y^{-4}}{7^{-1}z^5}\right)^{-2}$

(e) $\dfrac{(6x)^{-1}}{(3x^2)^{-2}}$

Answers

6. **(a)** 243   **(b)** $4^6 x^6$, or $4096x^6$

    (c) $\dfrac{8z^{12}}{125}$ **(d)** $\dfrac{y^8 z^{10}}{1764}$ **(e)** $\dfrac{3x^3}{2}$

5.5 Exercises

FOR EXTRA HELP

  MyMathLab®

CONCEPT CHECK *Decide whether each expression is positive, negative, or* 0.

1. $(-2)^{-3}$

2. $(-3)^{-2}$

3. -2^4

4. -3^6

5. $\left(\dfrac{1}{4}\right)^{-2}$

6. $\left(\dfrac{1}{5}\right)^{-2}$

7. $1 - 5^0$

8. $1 - 7^0$

CONCEPT CHECK *In Exercises 9 and 10, match each expression in Column I with the equivalent expression in Column II. Choices in Column II may be used once, more than once, or not at all. (In Exercise 9, $x \neq 0$.)*

I	II	I	II
9. (a) x^0	**A.** 0	**10. (a)** -2^{-4}	**A.** 8
(b) $-x^0$	**B.** 1	**(b)** $(-2)^{-4}$	**B.** 16
(c) $7x^0$	**C.** -1	**(c)** 2^{-4}	**C.** $-\dfrac{1}{16}$
(d) $(7x)^0$	**D.** 7	**(d)** $\dfrac{1}{2^{-4}}$	**D.** -8
(e) $-7x^0$	**E.** -7	**(e)** $\dfrac{1}{-2^{-4}}$	**E.** -16
(f) $(-7x)^0$	**F.** $\dfrac{1}{7}$	**(f)** $\dfrac{1}{(-2)^{-4}}$	**F.** $\dfrac{1}{16}$

Decide whether each expression is equal to 0, 1, *or* -1. ***See Example 1.***

11. 9^0

12. 5^0

13. $(-4)^0$

14. $(-10)^0$

15. -9^0

16. -5^0

17. $(-2)^0 - 2^0$

18. $(-8)^0 - 8^0$

19. $\dfrac{0^{10}}{10^0}$

20. $\dfrac{0^5}{5^0}$

Evaluate each expression. ***See Examples 1 and 2.***

21. $7^0 + 9^0$

22. $8^0 + 6^0$

23. 4^{-3}

24. 5^{-4}

25. $\left(\dfrac{1}{2}\right)^{-4}$

26. $\left(\dfrac{1}{3}\right)^{-3}$

27. $\left(\dfrac{6}{7}\right)^{-2}$

28. $\left(\dfrac{2}{3}\right)^{-3}$

29. $(-3)^{-4}$

30. $(-4)^{-3}$

31. $5^{-1} + 3^{-1}$

32. $6^{-1} + 2^{-1}$

33. $-2^{-1} + 3^{-2}$

34. $(-3)^{-2} + (-4)^{-1}$

Simplify by writing each expression with positive exponents. Assume that all variables represent nonzero real numbers. ***See Examples 2–4.***

35. $\dfrac{9^4}{9^5}$

36. $\dfrac{7^3}{7^4}$

37. $\dfrac{6^{-3}}{6^2}$

38. $\dfrac{4^{-2}}{4^3}$

39. $\dfrac{1}{6^{-3}}$

40. $\dfrac{1}{5^{-2}}$

41. $\dfrac{2}{r^{-4}}$

42. $\dfrac{3}{s^{-8}}$

43. $\dfrac{4^{-3}}{5^{-2}}$

44. $\dfrac{6^{-2}}{5^{-4}}$

45. $p^5 q^{-8}$

46. $x^{-8} y^4$

47. $\dfrac{r^5}{r^{-4}}$

48. $\dfrac{a^6}{a^{-4}}$

49. $\dfrac{6^4 x^8}{6^5 x^3}$

50. $\dfrac{3^8 y^5}{3^{10} y^2}$

51. $\dfrac{6y^3}{2y}$

52. $\dfrac{5m^2}{m}$

53. $\dfrac{3x^5}{3x^2}$

54. $\dfrac{10p^8}{2p^4}$

55. $\dfrac{x^{-3} y}{4z^{-2}}$

56. $\dfrac{p^{-5} q^{-4}}{9r^{-3}}$

57. $\dfrac{(a+b)^{-3}}{(a+b)^{-4}}$

58. $\dfrac{(x+y)^{-8}}{(x+y)^{-9}}$

Simplify by writing each expression with positive exponents. Assume that all variables represent nonzero real numbers. ***See Example 5.***

59. $\dfrac{(7^4)^3}{7^9}$

60. $\dfrac{(5^3)^2}{5^2}$

61. $x^{-3} \cdot x^5 \cdot x^{-4}$

62. $y^{-8} \cdot y^5 \cdot y^{-2}$

63. $\dfrac{(3x)^{-2}}{(4x)^{-3}}$

64. $\dfrac{(2y)^{-3}}{(5y)^{-4}}$

65. $\left(\dfrac{x^{-1} y}{z^2}\right)^{-2}$

66. $\left(\dfrac{p^{-4} q}{r^{-3}}\right)^{-3}$

67. $(6x)^4 (6x)^{-3}$

68. $(10y)^9 (10y)^{-8}$

69. $\dfrac{(m^7 n)^{-2}}{m^{-4} n^3}$

70. $\dfrac{(m^8 n^{-4})^2}{m^{-2} n^5}$

71. $\dfrac{5x^{-3}}{(4x)^2}$

72. $\dfrac{-3k^5}{(2k)^2}$

73. $\left(\dfrac{2p^{-1} q}{3^{-1} m^2}\right)^2$

74. $\left(\dfrac{4xy^2}{x^{-1} y}\right)^{-2}$

75. CONCEPT CHECK A student simplified $\frac{16^3}{2^2}$ as shown.

$$\frac{16^3}{2^2} = \left(\frac{16}{2}\right)^{3-2} = 8^1 = 8$$

What Went Wrong? Give the correct answer.

76. CONCEPT CHECK A student simplified 5^{-4} as shown.

$$5^{-4} = -5^4 = -625$$

What Went Wrong? Give the correct answer.

Summary Exercises *Applying the Rules for Exponents*

CONCEPT CHECK *Decide whether each expression is positive or negative.*

1. $(-5)^4$

2. -5^4

3. $(-2)^5$

4. -2^5

5. $\left(-\dfrac{3}{7}\right)^2$

6. $\left(-\dfrac{3}{7}\right)^3$

7. $\left(-\dfrac{3}{7}\right)^{-2}$

8. $\left(-\dfrac{3}{7}\right)^{-5}$

Simplify each expression. Assume that all variables represent nonzero real numbers.

9. $\left(\dfrac{6x^2}{5}\right)^{12}$

10. $\left(\dfrac{rs^2t^3}{3t^4}\right)^6$

11. $(10x^2y^4)^2(10xy^2)^3$

12. $(-2ab^3c)^4(-2a^2b)^3$

13. $\left(\dfrac{9wx^3}{y^4}\right)^3$

14. $(4x^{-2}y^{-3})^{-2}$

15. $\dfrac{c^{11}(c^2)^4}{(c^3)^3(c^2)^{-6}}$

16. $\left(\dfrac{k^4t^2}{k^2t^{-4}}\right)^{-2}$

17. $5^{-1} + 6^{-1}$

18. $\dfrac{(3y^{-1}z^3)^{-1}(3y^2)}{(y^3z^2)^{-3}}$

19. $\dfrac{(2xy^{-1})^3}{2^3x^{-3}y^2}$

20. $-8^0 + (-8)^0$

21. $(z^4)^{-3}(z^{-2})^{-5}$

22. $\left(\dfrac{r^2st^5}{3r}\right)^{-2}$

23. $\dfrac{(3^{-1}x^{-3}y)^{-1}(2x^2y^{-3})^2}{(5x^{-2}y^2)^{-2}}$

24. $\left(\dfrac{5x^2}{3x^{-4}}\right)^{-1}$

25. $\left(\dfrac{-2x^{-2}}{2x^2}\right)^{-2}$

26. $\dfrac{(x^{-4}y^2)^3(x^2y)^{-1}}{(xy^2)^{-3}}$

27. $\dfrac{(a^{-2}b^3)^{-4}}{(a^{-3}b^2)^{-2}(ab)^{-4}}$

28. $(2a^{-30}b^{-29})(3a^{31}b^{30})$

29. $5^{-2} + 6^{-2}$

30. $\left(\dfrac{(x^{47}y^{23})^2}{x^{-26}y^{-42}}\right)^0$

31. $\left(\dfrac{7a^2b^3}{2}\right)^3$

32. $-(-12^0)$

33. $-(-12)^0$

34. $\dfrac{0^{12}}{12^0}$

35. $\dfrac{(2xy^{-3})^{-2}}{(3x^{-2}y^4)^{-3}}$

36. $\left(\dfrac{a^2b^3c^4}{a^{-2}b^{-3}c^{-4}}\right)^{-2}$

37. $(6x^{-5}z^3)^{-3}$

38. $(2p^{-2}qr^{-3})(2p)^{-4}$

39. $\dfrac{(xy)^{-3}(xy)^5}{(xy)^{-4}}$

40. $42^0 - (-12)^0$

41. $\dfrac{(7^{-1}x^{-3})^{-2}(x^4)^{-6}}{7^{-1}x^{-3}}$

42. $\left(\dfrac{3^{-4}x^{-3}}{3^{-3}x^{-6}}\right)^{-2}$

43. $(5p^{-2}q)^{-3}(5pq^3)^4$

44. $8^{-1} + 6^{-1}$

45. $\left(\dfrac{4r^{-6}s^{-2}t}{2r^8s^{-4}t^2}\right)^{-1}$

46. $(13x^{-6}y)(13x^{-6}y)^{-1}$

47. $\dfrac{(8pq^{-2})^4}{(8p^{-2}q^{-3})^3}$

48. $\left(\dfrac{mn^{-2}p}{m^2np^4}\right)^{-2}\left(\dfrac{mn^{-2}p}{m^2np^4}\right)^3$

49. $-(-3^0)^0$

50. $5^{-1} - 8^{-1}$

5.6 Dividing a Polynomial by a Monomial

OBJECTIVE ▶ 1 **Divide a polynomial by a monomial.** We add two fractions with a common denominator as follows.

$$\frac{a}{c} + \frac{b}{c} = \frac{a+b}{c}$$

In reverse, this statement gives a rule for dividing a polynomial by a monomial.

1 Divide.

(a) $\dfrac{6p^4 + 18p^7}{3p^2}$

$= \dfrac{\overline{}}{3p^2} + \dfrac{\overline{}}{3p^2}$

$= \underline{}$

(b) $\dfrac{12m^6 + 18m^5 + 30m^4}{6m^2}$

> **Dividing a Polynomial by a Monomial**
>
> To divide a polynomial by a monomial, divide each term of the polynomial by the monomial.
>
> $$\frac{a+b}{c} = \frac{a}{c} + \frac{b}{c} \quad \text{(where } c \neq 0\text{)}$$
>
> *Examples:* $\dfrac{2+5}{3} = \dfrac{2}{3} + \dfrac{5}{3}$ and $\dfrac{x+3z}{2y} = \dfrac{x}{2y} + \dfrac{3z}{2y}$ $(y \neq 0)$

The parts of a division problem are named here.

$$\text{Dividend} \rightarrow \frac{12x^2 + 6x}{6x} = 2x + 1 \leftarrow \text{Quotient}$$
$$\text{Divisor} \rightarrow$$

EXAMPLE 1 Dividing a Polynomial by a Monomial

Divide $5m^5 - 10m^3$ by $5m^2$.

$$\frac{5m^5 - 10m^3}{5m^2}$$

$$= \frac{5m^5}{5m^2} - \frac{10m^3}{5m^2} \qquad \text{Use the preceding rule,}$$
$$\text{with + replaced by } -.$$

$$= m^3 - 2m \qquad \text{Quotient rule}$$

(c) $(18r^7 - 9r^2) \div (3r)$

CHECK Multiply. $5m^2 \cdot (m^3 - 2m) = 5m^5 - 10m^3 \leftarrow$ ✓
$\qquad\qquad\qquad\uparrow\qquad\quad\uparrow \qquad\qquad\qquad$ Original polynomial
$\qquad\qquad$ Divisor Quotient $\qquad\qquad\qquad$ (Dividend)

Because division by 0 is undefined, the quotient $\frac{5m^5 - 10m^3}{5m^2}$ is undefined if $m = 0$. From now on, we assume that no denominators are 0.

◀ **Work Problem** ❶ **at the Side.**

EXAMPLE 2 Dividing a Polynomial by a Monomial

Divide.

$$\frac{16a^5 - 12a^4 + 8a^2}{4a^3}$$

> This becomes $\frac{2}{a}$, **not** 2a.

$$= \frac{16a^5}{4a^3} - \frac{12a^4}{4a^3} + \frac{8a^2}{4a^3} \qquad \text{Divide each term by } 4a^3.$$

$$= 4a^2 - 3a + \frac{2}{a} \qquad \text{Quotient rule}$$

Answers

1. **(a)** $6p^4$; $18p^7$; $2p^2 + 6p^5$
   (b) $2m^4 + 3m^3 + 5m^2$ **(c)** $6r^6 - 3r$

Continued on Next Page

The quotient $4a^2 - 3a + \frac{2}{a}$ is *not* a polynomial because of the expression $\frac{2}{a}$, which has a variable in the denominator. While the sum, difference, and product of two polynomials are always polynomials, the quotient of two polynomials may not be.

CHECK $4a^3\left(4a^2 - 3a + \dfrac{2}{a}\right)$ Divisor × Quotient should equal Dividend.

$$= 4a^3(4a^2) + 4a^3(-3a) + 4a^3\left(\frac{2}{a}\right) \quad \text{Distributive property}$$

$$= 16a^5 - 12a^4 + 8a^2 \;\checkmark \quad \text{Dividend}$$

···· **Work Problem ② at the Side.** ▶

② Divide.

GS **(a)** $\dfrac{20x^4 - 25x^3 + 5x}{5x^2}$

$$= \underline{\quad} - \underline{\quad}$$

$$+ \underline{\quad}$$

$$= \underline{\hspace{3cm}}$$

(b) $\dfrac{50m^4 - 30m^3 + 20m}{10m^3}$

EXAMPLE 3   **Dividing a Polynomial by a Monomial with a Negative Coefficient**

Divide $-7x^3 + 12x^4 - 4x$ by $-4x$.
   Write the polynomial in descending powers as $12x^4 - 7x^3 - 4x$ before dividing.

Write in descending powers. →
$$\frac{12x^4 - 7x^3 - 4x}{-4x}$$

$$= \frac{12x^4}{-4x} - \frac{7x^3}{-4x} - \frac{4x}{-4x} \quad \begin{array}{l}\text{Divide each term}\\ \text{by } -4x.\end{array}$$

$$= -3x^3 - \frac{7x^2}{-4} - (-1) \quad \text{Quotient rule}$$

$$= -3x^3 + \frac{7}{4}x^2 + 1 \quad \boxed{\begin{array}{l}\text{Be sure to include}\\ \text{the 1 in the answer.}\end{array}}$$

CHECK $-4x\left(-3x^3 + \dfrac{7}{4}x^2 + 1\right)$ Divisor × Quotient should equal Dividend.

$$= -4x(-3x^3) - 4x\left(\frac{7}{4}x^2\right) - 4x(1) \quad \text{Distributive property}$$

$$= 12x^4 - 7x^3 - 4x \;\checkmark \quad \text{Dividend}$$

···· **Work Problem ③ at the Side.** ▶

③ Divide.

(a) $\dfrac{-9y^6 + 8y^7 - 11y - 4}{y^2}$

(b) $\dfrac{-8p^4 - 6p^3 - 12p^5}{-3p^3}$

---

④ Divide.

$$\dfrac{45x^4y^3 + 30x^3y^2 - 60x^2y}{-15x^2y}$$

EXAMPLE 4   **Dividing a Polynomial by a Monomial**

Divide $180x^4y^{10} - 150x^3y^8 + 120x^2y^6 - 90xy^4 + 100y$ by $-30xy^2$.

$$\frac{180x^4y^{10} - 150x^3y^8 + 120x^2y^6 - 90xy^4 + 100y}{-30xy^2}$$

$$= \frac{180x^4y^{10}}{-30xy^2} - \frac{150x^3y^8}{-30xy^2} + \frac{120x^2y^6}{-30xy^2} - \frac{90xy^4}{-30xy^2} + \frac{100y}{-30xy^2}$$

$$= -6x^3y^8 + 5x^2y^6 - 4xy^4 + 3y^2 - \frac{10}{3xy}$$

Check by multiplying the quotient by the divisor.

···· **Work Problem ④ at the Side.** ▶

Answers

2. (a) The complete expression is
$\dfrac{20x^4}{5x^2} - \dfrac{25x^3}{5x^2} + \dfrac{5x}{5x^2}; \; 4x^2 - 5x + \dfrac{1}{x}$

(b) $5m - 3 + \dfrac{2}{m^2}$

3. (a) $8y^5 - 9y^4 - \dfrac{11}{y} - \dfrac{4}{y^2}$

(b) $4p^2 + \dfrac{8p}{3} + 2$

4. $-3x^2y^2 - 2xy + 4$

5.6 Exercises

CONCEPT CHECK *In Exercises 1–4, complete each statement.*

1. In the statement $\dfrac{6x^2 + 8}{2} = 3x^2 + 4$, _____ is the dividend, _____ is the divisor, and _____ is the quotient.

2. The expression $\dfrac{3x + 12}{x}$ is undefined if $x =$ _____.

3. To check the division shown in **Exercise 1,** multiply _____ by _____ and show that the product is _____.

4. The expression $5x^2 - 3x + 6 + \frac{2}{x}$ *(is / is not)* a polynomial.

5. Explain why the division problem $\dfrac{16m^3 - 12m^2}{4m}$ can be performed using the method of this section, while the division problem $\dfrac{4m}{16m^3 - 12m^2}$ cannot.

6. **CONCEPT CHECK** Evaluate $\dfrac{5y + 6}{2}$ for $y = 2$.

Evaluate $5y + 3$ for $y = 2$. Is $\dfrac{5y + 6}{2}$ equivalent to $5y + 3$?

7. **CONCEPT CHECK** A polynomial in the variable x has degree 6 and is divided by a monomial in the variable x having degree 4. What is the degree of the quotient?

8. **CONCEPT CHECK** If $-60x^5 - 30x^4 + 20x^3$ is divided by $3x^2$, what is the sum of the coefficients of the third- and second-degree terms in the quotient?

Perform each division. ***See Examples 1–4.***

9. $\dfrac{12m^4 - 6m^3}{6m^2}$

10. $\dfrac{35n^5 - 5n^2}{5n}$

11.  $\dfrac{60x^4 - 20x^2 + 10x}{2x}$

12. $\dfrac{120x^6 - 60x^3 + 80x^2}{2x}$

13. $\dfrac{20m^5 - 10m^4 + 5m^2}{-5m^2}$

14. $\dfrac{12t^5 - 6t^3 + 6t^2}{-6t^2}$

15. $\dfrac{8t^5 - 4t^3 + 4t^2}{2t}$

16. $\dfrac{8r^4 - 4r^3 + 6r^2}{2r}$

17. $\dfrac{4a^5 - 4a^2 + 8}{4a}$

18. $\dfrac{5t^8 + 5t^7 + 15}{5t}$

19. $\dfrac{12x^5 - 4x^4 + 6x^3}{-6x^2}$

20. $\dfrac{24x^6 - 14x^5 + 32x^4}{-4x^2}$

21. $\dfrac{4x^2 + 20x^3 - 36x^4}{4x^2}$

22. $\dfrac{5x^2 - 30x^4 + 30x^5}{5x^2}$

23. $\dfrac{-3x^3 - 4x^4 + 2x}{-3x^2}$

24. $\dfrac{-8x + 6x^3 - 5x^4}{-3x^2}$

25. $\dfrac{27r^4 - 36r^3 - 6r^2 + 3r - 2}{3r}$

$$= \dfrac{}{3r} - \dfrac{}{3r} - \dfrac{}{3r} + \dfrac{}{3r} - \dfrac{}{3r}$$

$$= \underline{\hspace{4cm}}$$

26. $\dfrac{8k^4 - 12k^3 - 2k^2 - 2k - 3}{2k}$

$$= \dfrac{8k^4}{} - \dfrac{12k^3}{} - \dfrac{2k^2}{} - \dfrac{2k}{} - \dfrac{3}{}$$

$$= \underline{\hspace{4cm}}$$

27. $\dfrac{2m^5 - 6m^4 + 8m^2}{-2m^3}$

28. $\dfrac{6r^5 - 8r^4 + 10r^2}{-2r^4}$

29. $(120x^{11} - 60x^{10} + 140x^9 - 100x^8) \div (10x^{12})$

30. $(120x^{12} - 84x^9 + 60x^8 - 36x^7) \div (12x^9)$

31. $(20a^4b^3 - 15a^5b^2 + 25a^3b) \div (-5a^4b)$

32. $(16y^5z - 8y^2z^2 + 12yz^3) \div (-4y^2z^2)$

Use an area formula to answer each question. (If necessary, refer to the formulas inside the back cover.)

33. What expression represents the length of the rectangle?

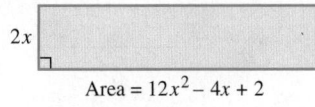

$2x$

Area $= 12x^2 - 4x + 2$

34. What expression represents the length of the base of the triangle?

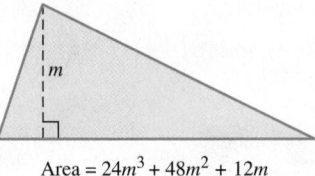

m

Area $= 24m^3 + 48m^2 + 12m$

35. CONCEPT CHECK What polynomial, when divided by $5x^3$, yields $3x^2 - 7x + 7$ as a quotient?

36. CONCEPT CHECK The quotient of a certain polynomial and $-12y^3$ is $6y^3 - 5y^2 + 2y - 3 + \frac{7}{y}$. What is this polynomial?

5.7 Dividing a Polynomial by a Polynomial

OBJECTIVE ▶ **1** **Divide a polynomial by a polynomial.** We use a method of "long division" to divide a polynomial by a polynomial (other than a monomial). *Both polynomials must be written in descending powers.*

Dividing Whole Numbers	Dividing Polynomials
Step 1	
Divide 6696 by 27.	Divide $8x^3 - 4x^2 - 14x + 15$ by $2x + 3$.
$27\overline{)6696}$	$2x + 3\overline{)8x^3 - 4x^2 - 14x + 15}$
Step 2	
66 divided by 27 = 2.	$8x^3$ divided by $2x = 4x^2$.
$2 \cdot 27 = 54$	$4x^2(2x + 3) = 8x^3 + 12x^2$
$\begin{array}{r} 2 \\ 27\overline{)6696} \\ 54 \end{array}$	$\begin{array}{r} 4x^2 \\ 2x + 3\overline{)8x^3 - 4x^2 - 14x + 15} \\ 8x^3 + 12x^2 \end{array}$
Step 3	
Subtract. Then bring down the next digit.	Subtract. Then bring down the next term.
$\begin{array}{r} 2 \\ 27\overline{)6696} \\ 54\downarrow \\ \hline 129 \end{array}$	$\begin{array}{r} 4x^2 \\ 2x + 3\overline{)8x^3 - 4x^2 - 14x + 15} \\ 8x^3 + 12x^2 \quad\downarrow \\ \hline -16x^2 - 14x \end{array}$
	(To subtract two polynomials, change the signs of the second and then add.)
Step 4	
129 divided by 27 = 4.	$-16x^2$ divided by $2x = -8x$.
$4 \cdot 27 = 108$	$-8x(2x + 3) = -16x^2 - 24x$
$\begin{array}{r} 24 \\ 27\overline{)6696} \\ 54 \\ \hline 129 \\ 108 \end{array}$	$\begin{array}{r} 4x^2 - 8x \\ 2x + 3\overline{)8x^3 - 4x^2 - 14x + 15} \\ 8x^3 + 12x^2 \\ \hline -16x^2 - 14x \\ -16x^2 - 24x \end{array}$
Step 5	
Subtract. Then bring down the next digit.	Subtract. Then bring down the next term.
$\begin{array}{r} 24 \\ 27\overline{)6696} \\ 54 \\ \hline 129 \\ 108\downarrow \\ \hline 216 \end{array}$	$\begin{array}{r} 4x^2 - 8x \\ 2x + 3\overline{)8x^3 - 4x^2 - 14x + 15} \\ 8x^3 + 12x^2 \\ \hline -16x^2 - 14x \\ -16x^2 - 24x \\ \hline 10x + 15 \end{array}$

(continued)

Step 6

216 divided by 27 = **8**.

8 · 27 = **216**

$$27\overline{\smash{)}6696}$$

the long division showing:

248 over 27)6696, 54, 129, 108, 216, → 216, Remainder → 0

6696 divided by 27 is 248.

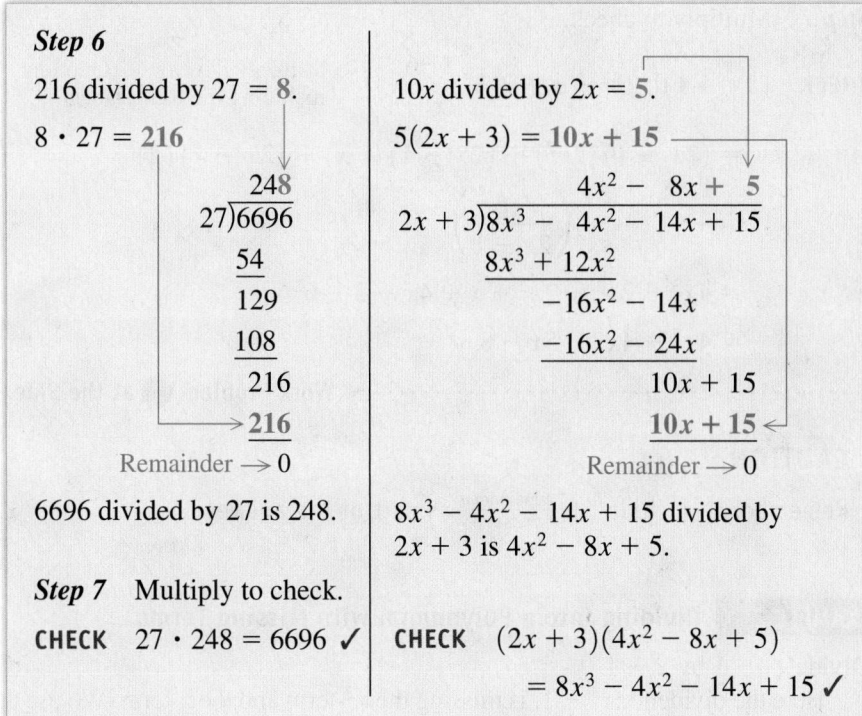

10x divided by 2x = **5**.

5(2x + 3) = 10x + 15

$$4x^2 - 8x + 5$$
$$2x + 3\overline{\smash{)}8x^3 - 4x^2 - 14x + 15}$$

$8x^3 + 12x^2$

$-16x^2 - 14x$

$-16x^2 - 24x$

$10x + 15$

$10x + 15$ ←

Remainder → 0

$8x^3 - 4x^2 - 14x + 15$ divided by $2x + 3$ is $4x^2 - 8x + 5$.

Step 7   Multiply to check.

CHECK   27 · 248 = 6696 ✓

CHECK $(2x + 3)(4x^2 - 8x + 5)$
$$= 8x^3 - 4x^2 - 14x + 15 ✓$$

EXAMPLE 1   **Dividing a Polynomial by a Polynomial**

Divide $5x + 4x^3 - 8 - 4x^2$ by $2x - 1$.

The dividend, the first polynomial here, must be written in descending powers as $4x^3 - 4x^2 + 5x - 8$. Then divide by $2x - 1$.

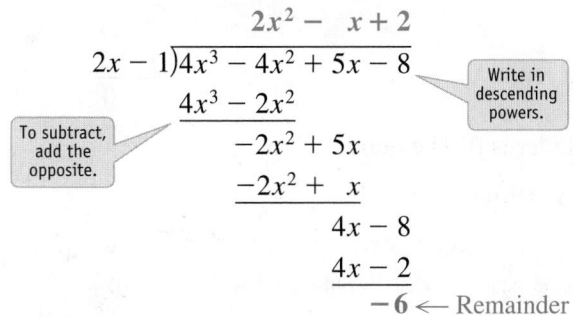

$$2x^2 - x + 2$$
$$2x - 1\overline{\smash{)}4x^3 - 4x^2 + 5x - 8}$$

> Write in descending powers.

$4x^3 - 2x^2$

> To subtract, add the opposite.

$-2x^2 + 5x$

$-2x^2 + x$

$4x - 8$

$4x - 2$

-6 ← Remainder

Step 1 $4x^3$ divided by $2x$ is $2x^2$. $2x^2(2x - 1) = 4x^3 - 2x^2$

Step 2   Subtract. Bring down the next term.

Step 3 $-2x^2$ divided by $2x$ is $-x$. $-x(2x - 1) = -2x^2 + x$

Step 4   Subtract. Bring down the next term.

Step 5 $4x$ divided by $2x$ is 2. $2(2x - 1) = 4x - 2$

Step 6   Subtract. The remainder is -6. Write the remainder as the numerator of a fraction that has $2x - 1$ as its denominator. Because of the nonzero remainder, the answer is not a polynomial.

Dividend →
Divisor →
$$\frac{4x^3 - 4x^2 + 5x - 8}{2x - 1} = \underbrace{2x^2 - x + 2}_{\substack{\text{Quotient} \\ \text{polynomial}}} + \underbrace{\frac{-6}{2x - 1}}_{\substack{\text{Fractional} \\ \text{part of} \\ \text{quotient}}}$$
← Remainder
← Divisor

Continued on Next Page

① Divide.

(a) $(x^3 + x^2 + 4x - 6)$

    $\div (x - 1)$

(b) $\dfrac{p^3 - 2p^2 - 5p + 9}{p + 2}$

② Divide.

(a) $\dfrac{r^2 - 5}{r + 4}$

(b) $(x^3 - 8) \div (x - 2)$

Answers

1. (a) $x^2 + 2x + 6$

  (b) $p^2 - 4p + 3 + \dfrac{3}{p + 2}$

2. (a) $r - 4 + \dfrac{11}{r + 4}$

  (b) $x^2 + 2x + 4$

Step 7 Multiply to check.

CHECK $(2x - 1)\left(2x^2 - x + 2 + \dfrac{-6}{2x - 1}\right)$ Multiply Divisor $\times$ (Quotient including the Remainder).

$= (2x - 1)(2x^2) + (2x - 1)(-x) + (2x - 1)(2)$

$\quad + (2x - 1)\left(\dfrac{-6}{2x - 1}\right)$

$= 4x^3 - 2x^2 - 2x^2 + x + 4x - 2 - 6$

$= 4x^3 - 4x^2 + 5x - 8$ ✓

◀ **Work Problem ① at the Side.**

> **CAUTION**
>
> Remember to include "$+ \frac{\text{remainder}}{\text{divisor}}$" as part of the answer.

EXAMPLE 2   **Dividing into a Polynomial with Missing Terms**

Divide $x^3 - 1$ by $x - 1$.

    Here the dividend, $x^3 - 1$, is missing the x^2-term and the x-term. We use 0 as the coefficient for each missing term. Thus, $x^3 - 1 = x^3 + 0x^2 + 0x - 1$.

$$
\begin{array}{r}
x^2 + x + 1 \\
x - 1 \overline{)\ x^3 + 0x^2 + 0x - 1} \\
\underline{x^3 - x^2} \\
x^2 + 0x \\
\underline{x^2 - x} \\
x - 1 \\
\underline{x - 1} \\
0
\end{array}
$$

    Insert placeholders for the missing terms.

The remainder is 0. The quotient is $x^2 + x + 1$.

CHECK $(x - 1)(x^2 + x + 1)$

$= x^3 + x^2 + x - x^2 - x - 1$

$= x^3 - 1$ ✓   Divisor $\times$ Quotient $=$ Dividend

◀ **Work Problem ② at the Side.**

EXAMPLE 3   **Dividing by a Polynomial with Missing Terms**

Divide $x^4 + 2x^3 + 2x^2 - x - 1$ by $x^2 + 1$.

    Since $x^2 + 1$ has a missing x-term, write it as $x^2 + 0x + 1$.

$$
\begin{array}{r}
x^2 + 2x + 1 \\
x^2 + 0x + 1 \overline{)\ x^4 + 2x^3 + 2x^2 - x - 1} \\
\underline{x^4 + 0x^3 + x^2} \\
2x^3 + x^2 - x \\
\underline{2x^3 + 0x^2 + 2x} \\
x^2 - 3x - 1 \\
\underline{x^2 + 0x + 1} \\
-3x - 2 \leftarrow \text{Remainder}
\end{array}
$$

Insert a placeholder for the missing term.

Continued on Next Page

When the result of subtracting ($-3x - 2$, in this case) is a constant or a polynomial of degree less than the divisor ($x^2 + 0x + 1$), that constant or polynomial is the remainder. We write the answer as follows.

$$x^2 + 2x + 1 + \frac{-3x - 2}{x^2 + 1}$$

Remember to include "$+ \frac{remainder}{divisor}$."

CHECK Show that $(x^2 + 1)\left(x^2 + 2x + 1 + \dfrac{-3x - 2}{x^2 + 1}\right)$ gives the original dividend, $x^4 + 2x^3 + 2x^2 - x - 1$. ✓

···················· Work Problem **3** at the Side. ▶

EXAMPLE 4 **Dividing a Polynomial When the Quotient Has Fractional Coefficients**

Divide $4x^3 + 2x^2 + 3x + 2$ by $4x - 4$.

$$\frac{6x^2}{4x} = \frac{3}{2}x$$
$$\frac{9x}{4x} = \frac{9}{4}$$

$$
\begin{array}{r}
x^2 + \frac{3}{2}x + \frac{9}{4} \\
4x - 4 \overline{) 4x^3 + 2x^2 + 3x + 2} \\
\underline{4x^3 - 4x^2} \\
6x^2 + 3x \\
\underline{6x^2 - 6x} \\
9x + 2 \\
\underline{9x - 9} \\
11
\end{array}
$$

The answer is $x^2 + \dfrac{3}{2}x + \dfrac{9}{4} + \dfrac{11}{4x - 4}$.

···················· Work Problem **4** at the Side. ▶

OBJECTIVE **2** **Apply division to a geometry problem.**

EXAMPLE 5 **Using an Area Formula**

The area of the rectangle in **Figure 3** is given by $(x^3 + 4x^2 + 8x + 8)$ sq. units. The width is given by $(x + 2)$ units. What is its length?

Length = ?

Width = $x + 2$

Area = $x^3 + 4x^2 + 8x + 8$

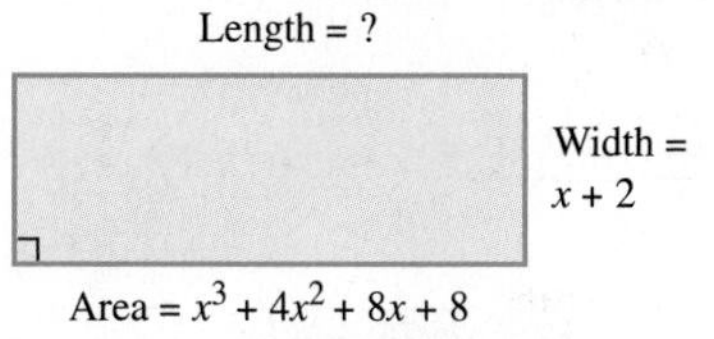

Figure 3

For a rectangle, $A = LW$. Solving for L gives $L = \frac{A}{W}$. Divide the area, $x^3 + 4x^2 + 8x + 8$, by the width, $x + 2$.

···················· Work Problem **5** at the Side. ▶

The quotient from **Margin Problem 5**, $x^2 + 2x + 4$, represents the length of the rectangle in units.

---

3 Divide.

(a) $(2x^4 + 3x^3 - x^2 + 6x + 5)$
$\div (x^2 - 1)$

(b)
$$\frac{2m^5 + m^4 + 6m^3 - 3m^2 - 18}{m^2 + 3}$$

4 Divide $3x^3 + 7x^2 + 7x + 10$ by $3x + 6$.

5 Divide $x^3 + 4x^2 + 8x + 8$ by $x + 2$.

Answers

3. (a) $2x^2 + 3x + 1 + \dfrac{9x + 6}{x^2 - 1}$

   (b) $2m^3 + m^2 - 6$

4. $x^2 + \dfrac{1}{3}x + \dfrac{5}{3}$

5. $x^2 + 2x + 4$

5.7 Exercises

FOR EXTRA HELP

Download the MyDashBoard App

MyMathLab®

CONCEPT CHECK *Answer each question.*

1. In the division problem $(4x^4 + 2x^3 - 14x^2 + 19x + 10) \div (2x + 5) = 2x^3 - 4x^2 + 3x + 2$, which polynomial is the divisor? Which is the quotient?

2. When dividing one polynomial by another, how do you know when to stop dividing?

3. In dividing $12m^2 - 20m + 3$ by $2m - 3$, what is the first step?

4. In the division in **Exercise 3,** what is the second step?

Perform each division. **See Example 1.**

5. $\dfrac{x^2 - x - 6}{x - 3}$

6. $\dfrac{m^2 - 2m - 24}{m - 6}$

7. $\dfrac{2y^2 + 9y - 35}{y + 7}$

8. $\dfrac{2y^2 + 9y + 7}{y + 1}$

9. $\dfrac{p^2 + 2p + 20}{p + 6}$

10. $\dfrac{x^2 + 11x + 16}{x + 8}$

11. $(r^2 - 8r + 15) \div (r - 3)$

12. $(t^2 + 2t - 35) \div (t - 5)$

13. $\dfrac{4a^2 - 22a + 32}{2a + 3}$

14. $\dfrac{9w^2 + 6w + 10}{3w - 2}$

15. $\dfrac{8x^3 - 10x^2 - x + 3}{2x + 1}$

16. $\dfrac{12t^3 - 11t^2 + 9t + 18}{4t + 3}$

Perform each division. **See Examples 2–4.**

17. $\dfrac{3y^3 + y^2 + 2}{y + 1}$

18. $\dfrac{2r^3 - 6r - 36}{r - 3}$

19. $\dfrac{2x^3 + x + 2}{x + 1}$

20. $\dfrac{3x^3 + x + 5}{x + 1}$

21. $\dfrac{3k^3 - 4k^2 - 6k + 10}{k^2 - 2}$

22. $\dfrac{5z^3 - z^2 + 10z + 2}{z^2 + 2}$

23. $(x^4 - x^2 - 2) \div (x^2 - 2)$

24. $(r^4 + 2r^2 - 3) \div (r^2 - 1)$

25. $\dfrac{x^4 - 1}{x^2 - 1}$

26. $\dfrac{y^3 + 1}{y + 1}$

27. $\dfrac{6p^4 - 15p^3 + 14p^2 - 5p + 10}{3p^2 + 1}$

28. $\dfrac{6r^4 - 10r^3 - r^2 + 15r - 8}{2r^2 - 3}$

29. $\dfrac{2x^5 + x^4 + 11x^3 - 8x^2 - 13x + 7}{2x^2 + x - 1}$

30. $\dfrac{4t^5 - 11t^4 - 6t^3 + 5t^2 - t + 3}{4t^2 + t - 3}$

31. $(10x^3 + 13x^2 + 4x + 1) \div (5x + 5)$

32. $(6x^3 - 19x^2 - 19x - 4) \div (2x - 8)$

*Work each problem. **See Example 5.** (If necessary, refer to the formulas inside the back cover.)*

33. Give the length of the rectangle.

$5x + 2$

The area is $(5x^3 + 7x^2 - 13x - 6)$ sq. units.

34. Find the measure of the base of the parallelogram.

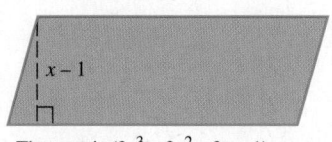

$x - 1$

The area is $(2x^3 + 2x^2 - 3x - 1)$ sq. units.

Relating Concepts (Exercises 35–38) For Individual or Group Work

We can find the value of a polynomial in x for a given value of x by substituting that number for x. Surprisingly, we can accomplish the same thing by division. For example, to find the value of $2x^2 - 4x + 3$ for $x = -3$, we would divide $2x^2 - 4x + 3$ by $x - (-3)$. The remainder will give the value of the polynomial for $x = -3$.
Work Exercises 35–38 in order.

35. Find the value of $2x^2 - 4x + 3$ for $x = -3$ by substitution.

36. Divide $2x^2 - 4x + 3$ by $x + 3$. Give the remainder.

37. Compare your answers to **Exercises 35 and 36.** What do you notice?

38. Choose another polynomial and evaluate it both ways for some value of the variable. Do the answers agree?

5.8 An Application of Exponents: Scientific Notation

OBJECTIVES

1 Express numbers in scientific notation.

2 Convert numbers in scientific notation to numbers without exponents.

3 Use scientific notation in calculations.

OBJECTIVE 1 Express numbers in scientific notation. Numbers occurring in science are often extremely large (such as the distance from Earth to the sun, 93,000,000 mi) or extremely small (the wavelength of yellow-green light, approximately 0.0000006 m). Because of the difficulty of working with many zeros, scientists often express such numbers with exponents, using a form called *scientific notation*.

> **Scientific Notation**
>
> A number is written in **scientific notation** when it is expressed in the form
>
> $$a \times 10^n,$$
>
> where $1 \le |a| < 10$ and n is an integer.

In **scientific notation,** there is always one nonzero digit to the left of the decimal point. This is shown in the following, where the first number is in scientific notation.

$3.19 \times 10^1 = 3.19 \times 10 = 31.9$ Decimal point moves 1 place to the right.

$3.19 \times 10^2 = 3.19 \times 100 = 319.$ Decimal point moves 2 places to the right.

$3.19 \times 10^3 = 3.19 \times 1000 = 3190.$ Decimal point moves 3 places to the right.

$3.19 \times 10^{-1} = 3.19 \times 0.1 = 0.319$ Decimal point moves 1 place to the left.

$3.19 \times 10^{-2} = 3.19 \times 0.01 = 0.0319$ Decimal point moves 2 places to the left.

$3.19 \times 10^{-3} = 3.19 \times 0.001 = 0.00319$ Decimal point moves 3 places to the left.

> **Note**
>
> In scientific notation, a multiplication cross (or symbol) $\times$ is commonly used.

A number in scientific notation is always written with the decimal point after the first nonzero digit and then multiplied by the appropriate power of 10. For example, 56,200 is written 5.62×10^4, since

$$56{,}200 = 5.62 \times 10{,}000 = 5.62 \times 10^4.$$

Other examples include

	42,000,000	written	4.2×10^7,
	0.000586	written	5.86×10^{-4},
and	2,000,000,000	written	2×10^9.

It is not necessary to write 2.0.

To write a positive number in scientific notation, follow the steps given on the next page. (For a negative number, follow these steps using the *absolute value* of the number. Then make the result negative.)

Writing a Positive Number in Scientific Notation

Step 1 Move the decimal point to the right of the first nonzero digit.

Step 2 Count the number of places you moved the decimal point.

Step 3 The number of places in Step 2 is the absolute value of the exponent on 10.

Step 4 The exponent on 10 is positive if the original number is greater than the number in Step 1. The exponent is negative if the original number is less than the number in Step 1. If the decimal point is not moved, the exponent is 0.

EXAMPLE 1 Using Scientific Notation

Write each number in scientific notation.

(a) 93,000,000

Move the decimal point to follow the first nonzero digit (the 9). Count the number of places the decimal point was moved.

$$93,000,000. \leftarrow \text{Decimal point}$$
$$7 \text{ places}$$

The number will be written in scientific notation as 9.3×10^n. To find the value of n, first compare the original number, 93,000,000, with 9.3. Since 93,000,000 is *greater* than 9.3, we must multiply by a *positive* power of 10 so that the product 9.3×10^n will equal the larger number.

Since the decimal point was moved 7 places, and since n is positive,

$$93,000,000 = 9.3 \times 10^7.$$

(b) $63,200,000,000 = 6.3200000000 = 6.32 \times 10^{10}$
10 places

(c) $3.021 = 3.021 \times 10^0$

(d) 0.00462

Move the decimal point to the right of the first nonzero digit and count the number of places the decimal point was moved.

$$0.00462 \quad 3 \text{ places}$$

Since 0.00462 is *less* than 4.62, the exponent must be *negative*.

$$0.00462 = 4.62 \times 10^{-3}$$

(e) $-0.0000762 = -7.62 \times 10^{-5}$
5 places   Remember the *negative* sign.

·············· **Work Problem ❶ at the Side.** ▶

Note

When writing a positive number in scientific notation, think as follows.

1. If the original number is "large," like 93,000,000, use a *positive* exponent on 10, since positive is greater than negative.

2. If the original number is "small," like 0.00462, use a *negative* exponent on 10, since negative is less than positive.

---

❶ Write each number in scientific notation.

(a) 63,000

The first nonzero digit is

_____.

The decimal point should be moved _____ places.

$$63,000 = \_\_\_ \times 10^{\_\_}$$

(b) 5,870,000

(c) 7.0065

(d) 0.0571

The first nonzero digit is

_____.

The decimal point should be moved _____ places.

$$0.0571 = \_\_\_ \times 10^{\_\_}$$

(e) −0.00062

Answers

1. (a) 6; 4; 6.3; 4 (b) 5.87×10^6
(c) 7.0065×10^0 (d) 5; 2; 5.71; −2
(e) -6.2×10^{-4}

2 Write without exponents.

GS **(a)** 4.2×10^3

Move the decimal point ____ places to the ____ .

$4.2 \times 10^3 =$ _____

(b) 8.7×10^5

GS **(c)** 6.42×10^{-3}

Move the decimal point ____ places to the ____ .

$6.42 \times 10^{-3} =$ _____

(d) -5.27×10^{-1}

3 Perform each calculation. Write answers in scientific notation and also without exponents.

(a) $(2.6 \times 10^4)(2 \times 10^{-6})$

(b) $(3 \times 10^5)(5 \times 10^{-2})$

(c) $\dfrac{4.8 \times 10^2}{2.4 \times 10^{-3}}$

Answers

2. **(a)** 3; right; 4200   **(b)** 870,000
   (c) 3; left; 0.00642   **(d)** −0.527
3. **(a)** 5.2×10^{-2}; 0.052
   (b) 1.5×10^4; 15,000
   (c) 2×10^5; 200,000

OBJECTIVE **2** **Convert numbers in scientific notation to numbers without exponents.** To convert a number written in scientific notation to a number without exponents, work in reverse.

Multiplying a positive number by a positive power of 10 will make the number greater. Multiplying by a negative power of 10 will make the number less.

EXAMPLE 2 **Writing Numbers without Exponents**

Write each number without exponents.

(a) 6.2×10^3

Since the exponent is positive, make 6.2 greater by moving the decimal point 3 places to the right. It is necessary to attach two 0s.

$$6.2 \times 10^3 = 6.200 = 6200$$

(b) $4.283 \times 10^5 = 4.28300 = 428,300$ Move 5 places to the right. Attach 0s as necessary.

(c) $-9.73 \times 10^{-2} = -09.73 = -0.0973$ Move 2 places to the left.

The exponent tells the number of places and the direction in which the decimal point is moved.

◀ **Work Problem** **2** **at the Side.**

OBJECTIVE **3** **Use scientific notation in calculations.** The next example uses scientific notation with products and quotients.

EXAMPLE 3 **Multiplying and Dividing with Scientific Notation**

Perform each calculation. Write answers in scientific notation and also without exponents.

(a)
$$(7 \times 10^3)(5 \times 10^4)$$
$$= (7 \times 5)(10^3 \times 10^4) \quad \text{Commutative and associative properties}$$
$$= 35 \times 10^7 \quad \text{Multiply and use the product rule.}$$
$$= (3.5 \times 10^1) \times 10^7 \quad \text{Write 35 in scientific notation.}$$
$$= 3.5 \times (10^1 \times 10^7) \quad \text{Associative property}$$
$$= 3.5 \times 10^8 \quad \text{Product rule}$$
$$= 350,000,000 \quad \text{Write without exponents.}$$

Don't stop. This number is *not* in scientific notation, since 35 is not between 1 and 10.

(b)
$$\frac{4 \times 10^{-5}}{2 \times 10^3}$$
$$= \frac{4}{2} \times \frac{10^{-5}}{10^3}$$
$$= 2 \times 10^{-8} \quad \text{Divide and use the quotient rule.}$$
$$= 0.00000002 \quad \text{Write without exponents.}$$

◀ **Work Problem** **3** **at the Side.**

🖩 **Calculator Tip**

Calculators usually have a key labeled EE or EXP for scientific notation. See your owner's manual for more information.

Note

Multiplying or dividing numbers written in scientific notation may produce an answer in the form $a \times 10^0$. Since $10^0 = 1$, $a \times 10^0 = a$. For example,

$$(8 \times 10^{-4})(5 \times 10^4) = 40 \times 10^0 = 40. \quad 10^0 = 1$$

Also, if $a = 1$, then $a \times 10^n = 10^n$. For example, we could write 1,000,000 as 10^6 instead of 1×10^6.

❹ The speed of light is approximately 3.0×10^5 km per sec. How far does light travel in 6.0×10^1 sec? (*Source: World Almanac and Book of Facts.*)

EXAMPLE 4 Using Scientific Notation to Solve an Application

A *nanometer* is a very small unit of measure that is equivalent to about 0.00000003937 in. About how much would 700,000 nanometers measure in inches? (*Source: World Almanac and Book of Facts.*)

Write each number in scientific notation, and then multiply.

$$700,000(0.00000003937)$$

$$= (7 \times 10^5)(3.937 \times 10^{-8}) \qquad \text{Write in scientific notation.}$$

$$= (7 \times 3.937)(10^5 \times 10^{-8}) \qquad \text{Properties of real numbers}$$

Don't stop here. → $$= 27.559 \times 10^{-3} \qquad \text{Multiply; product rule}$$

$$= (2.7559 \times 10^1) \times 10^{-3} \qquad \text{Write 27.559 in scientific notation.}$$

$$= 2.7559 \times 10^{-2} \qquad \text{Product rule}$$

$$= 0.027559 \qquad \text{Write without exponents.}$$

Thus, 700,000 nanometers would measure

$$2.7559 \times 10^{-2} \text{ in.,} \quad \text{or} \quad 0.027559 \text{ in.}$$

·············· **Work Problem ❹ at the Side.** ▶

❺ If the speed of light is approximately 3.0×10^5 km per sec, how many seconds does it take light to travel approximately 1.5×10^8 km from the Sun to Earth? (*Source: World Almanac and Book of Facts.*)

EXAMPLE 5 Using Scientific Notation to Solve an Application

In 2010, the national debt was 1.3562×10^{13} (which is more than \$13 trillion). The population of the United States was approximately 309 million that year. About how much would each person have had to contribute in order to pay off the national debt? (*Source: U.S. Department of the Treasury; U.S. Census Bureau.*)

Write the population in scientific notation. Then divide to obtain the per person contribution.

$$\frac{1.3562 \times 10^{13}}{309,000,000}$$

$$= \frac{1.3562 \times 10^{13}}{3.09 \times 10^8} \qquad \text{Write 309 million in scientific notation.}$$

$$= \frac{1.3562}{3.09} \times 10^5 \qquad \text{Quotient rule}$$

$$\approx 0.4389 \times 10^5 \qquad \text{Divide. Round to 4 decimal places.}$$

$$= 43,890 \qquad \text{Write without exponents.}$$

Each person would have to pay about \$43,890.

·············· **Work Problem ❺ at the Side.** ▶

Answers

4. 1.8×10^7 km, or 18,000,000 km
5. 5×10^2 sec, or 500 sec

CONCEPT CHECK *Match each number written in scientific notation in Column I with the correct choice from Column II. Not all choices in Column II will be used.*

I	II	I	II
1. (a) 4.6×10^{-4}	**A.** 46,000	**2. (a)** 1×10^9	**A.** 1 billion
(b) 4.6×10^4	**B.** 460,000	**(b)** 1×10^6	**B.** 100 million
(c) 4.6×10^5	**C.** 0.00046	**(c)** 1×10^8	**C.** 1 million
(d) 4.6×10^{-5}	**D.** 0.000046	**(d)** 1×10^{10}	**D.** 10 billion
	E. 4600		**E.** 100 billion

CONCEPT CHECK *Determine whether or not the given number is written in scientific notation as defined in* **Objective 1.** *If it is not, write it as such.*

3. 4.56×10^3 **4.** 7.34×10^5 **5.** 5,600,000   **6.** 34,000

7. 0.004   **8.** 0.0007   **9.** 0.8×10^2 **10.** 0.9×10^3

11. Explain what it means for a number to be written in scientific notation.

12. Explain how to multiply a number by a positive power of ten. Then explain how to multiply a number by a negative power of ten.

Write each number in scientific notation. **See Example 1.**

13. 5,876,000,000   **14.** 9,994,000,000   **15.** 82,350   **16.** 78,330

17. 0.000007   **18.** 0.0000004   **19.** −0.00203   **20.** −0.0000578

Write each number without exponents. **See Example 2.**

21. 7.5×10^5 **22.** 8.8×10^6 **23.** 5.677×10^{12} **24.** 8.766×10^9

25. 1×10^{12} **26.** 1×10^7 **27.** -6.21×10^0 **28.** -8.56×10^0

29. 7.8×10^{-4} **30.** 8.9×10^{-5} **31.** 5.134×10^{-9} **32.** 7.123×10^{-10}

Perform the indicated operations. Write the answers in scientific notation and then without exponents. **See Example 3.**

33. $(2 \times 10^8)(3 \times 10^3)$ **34.** $(3 \times 10^7)(3 \times 10^3)$ **35.** $(5 \times 10^4)(3 \times 10^2)$

36. $(8 \times 10^5)(2 \times 10^3)$

37. $(4 \times 10^{-6})(2 \times 10^3)$

38. $(3 \times 10^{-7})(2 \times 10^2)$

39. $(6 \times 10^3)(4 \times 10^{-2})$

40. $(7 \times 10^5)(3 \times 10^{-4})$

41. $(9 \times 10^4)(7 \times 10^{-7})$

42. $(6 \times 10^4)(8 \times 10^{-8})$

43. $(3.15 \times 10^{-4})(2.04 \times 10^8)$

44. $(4.92 \times 10^{-3})(2.25 \times 10^7)$

45. $\dfrac{9 \times 10^{-5}}{3 \times 10^{-1}}$

46. $\dfrac{12 \times 10^{-4}}{4 \times 10^{-3}}$

47. $\dfrac{8 \times 10^3}{2 \times 10^2}$

48. $\dfrac{15 \times 10^4}{3 \times 10^3}$

49. $\dfrac{2.6 \times 10^{-3}}{2 \times 10^2}$

50. $\dfrac{9.5 \times 10^{-1}}{5 \times 10^3}$

51. $\dfrac{4 \times 10^5}{8 \times 10^2}$

52. $\dfrac{3 \times 10^9}{6 \times 10^5}$

53. $\dfrac{2.6 \times 10^{-3} \times 7.0 \times 10^{-1}}{2 \times 10^2 \times 3.5 \times 10^{-3}}$

54. $\dfrac{9.5 \times 10^{-1} \times 2.4 \times 10^4}{5 \times 10^3 \times 1.2 \times 10^{-2}}$

55. $\dfrac{(1.65 \times 10^8)(5.24 \times 10^{-2})}{(6 \times 10^4)(2 \times 10^7)}$

56. $\dfrac{(3.25 \times 10^7)(6.48 \times 10^{-4})}{(5 \times 10^2)(4 \times 10^6)}$

Each statement taken from a book or newspaper contains a number in boldface italic type. If the number is in scientific notation, write it without exponents. If the number is not in scientific notation, write it as such. See Examples 1 and 2.

57. The muon, a close relative of the electron produced by the bombardment of cosmic rays against the upper atmosphere, has a half-life of 2 millionths of a second (***2 × 10⁻⁶*** s). (*Source:* Hewitt, Paul G., *Conceptual Physics.*)

58. There are 13 red balls and 39 black balls in a box. Mix them up and draw 13 out one at a time without returning any ball . . . the probability that the 13 drawings each will produce a red ball is . . . ***1.6 × 10⁻¹²***. (*Source:* Weaver, Warren, *Lady Luck.*)

59. An electron and a positron attract each other in two ways: the electromagnetic attraction of their opposite electric charges, and the gravitational attraction of their two masses. The electromagnetic attraction is

4,200,000,000,000,000,000,000,000,000,000,000,000,000,000

times as strong as the gravitational. (*Source:* Asimov, I., *Isaac Asimov's Book of Facts.*)

60. A googol is

10,000,000,000,000,000,000,000,000,000,000,000,000,000,000,000,000, 000,000,000,000,000,000,000,000,000,000,000,000,000.

The Web search engine Google is named after a googol. Sergey Brin, president and cofounder of Google, Inc., was a mathematics major. He chose the name Google to describe the vast reach of this search engine. (*Source: The Gazette.*)

🖩 *Use scientific notation to calculate the answer to each problem. In Exercises 63–65 and 69–70, give answers without exponents.* ***See Examples 4 and 5.***

61. The Double Helix Nebula, a conglomeration of dust ▶ and gas stretching across the center of the Milky Way galaxy, is 25,000 light-years from Earth. If one light-year is about 6,000,000,000,000 mi, about how many miles is the Double Helix Nebula from Earth? (*Source:* www.spitzer.caltech.edu)

62. Pollux, one of the brightest stars in the night sky, is 33.7 light-years from Earth. If one light-year is about 6,000,000,000,000 mi (that is, 6 trillion mi), about how many miles is Pollux from Earth? (*Source: World Almanac and Book of Facts.*)

63. In 2011, the population of the United States was about 311.6 million. To the nearest dollar, calculate how much each person in the United States would have had to contribute in order to make one person a trillionaire (that is, to give that person $1,000,000,000,000). (*Source:* U.S. Census Bureau.)

64. In 2010, the U.S. government collected about $3767 per person in individual income taxes. If the population at that time was 309,000,000, how much did the government collect in taxes for 2010? (*Source: World Almanac and Book of Facts.*)

65. In 2011, Congress raised the debt limit to 1.64×10^{13}. When this national debt limit is reached, about how much is it for every man, woman, and child in the country? Use 311 million as the population of the United States. (*Source:* www.tradingnrg.org)

66. In 2010, the state of Minnesota had about 8.1×10^4 farms with an average of 3.44×10^2 acres per farm. What was the total number of acres devoted to farmland in Minnesota that year? (*Source:* U.S. Department of Agriculture.)

67. Venus is 6.68×10^7 mi from the Sun. If light travels at a speed of 1.86×10^5 mi per sec, how long does it take light to travel from the Sun to Venus? (*Source: World Almanac and Book of Facts.*)

68. The distance to Earth from Pluto is 4.58×10^9 km. *Pioneer 10* transmitted radio signals from Pluto to Earth at the speed of light, 3.00×10^5 km per sec. How long (in seconds) did it take for the signals to reach Earth?

69. During the 2010–2011 season, Broadway shows grossed a total of 1.08×10^9 dollars. Total attendance for the season was 1.25×10^7. What was the average ticket price for a Broadway show? (*Source:* The Broadway League.)

70. In 2010, 10.6×10^9 dollars were spent to attend motion pictures in the United States and Canada. The total number of tickets sold that year totaled 1.34 billion. What was the average ticket price? (*Source:* Motion Picture Association of America.)

Chapter 5 *Summary*

Key Terms

5.1

term A term is a number, a variable, or a product or quotient of a number and one or more variables raised to powers.

like terms Terms with exactly the same variables (including the same exponents) are like terms.

polynomial A polynomial is a term or the sum of a finite number of terms with whole number exponents.

descending powers A polynomial in x is written in descending powers if the exponents on x in its terms are in decreasing order.

degree of a term The degree of a term is the sum of the exponents on the variables.

degree of a polynomial The degree of a polynomial is the greatest degree of any term of the polynomial.

monomial A monomial is a polynomial with exactly one term.

binomial A binomial is a polynomial with exactly two terms.

trinomial A trinomial is a polynomial with exactly three terms.

5.2

exponential expression A number written with an exponent is an exponential expression.

$$3^4 \leftarrow \text{Exponent} \quad \left.\right\} \text{Exponential expression}$$
$$\uparrow \text{—— Base}$$

5.3

FOIL method The FOIL method is used to find the product of two binomials. The letters of the word **FOIL** originate as follows: Multiply the **F**irst terms, multiply the **O**uter terms (to get the outer product), multiply the **I**nner terms (to get the inner product), and multiply the **L**ast terms.

outer product The outer product of $(a + b)(c + d)$ is ad.

inner product The inner product of $(a + b)(c + d)$ is bc.

5.4

conjugate The conjugate of $a + b$ is $a - b$.

5.8

scientific notation A number written as $a \times 10^n$, where $1 \le |a| < 10$ and n is an integer, is in scientific notation.

New Symbols

x^{-n} x to the negative n power

Test Your Word Power

See how well you have learned the vocabulary in this chapter.

1 A **polynomial** is an algebraic expression made up of
   A. a term or a finite product of terms with positive coefficients and exponents
   B. a term or a finite sum of terms with real coefficients and whole number exponents
   C. the product of two or more terms with positive exponents
   D. the sum of two or more terms with whole number coefficients and exponents.

2 The **degree of a term** is
   A. the number of variables in the term
   B. the product of the exponents on the variables
   C. the least exponent on the variables
   D. the sum of the exponents on the variables.

3 A **trinomial** is a polynomial with
   A. only one term
   B. exactly two terms
   C. exactly three terms
   D. more than three terms.

4 A **binomial** is a polynomial with
   A. only one term
   B. exactly two terms
   C. exactly three terms
   D. more than three terms.

5 A **monomial** is a polynomial with
   A. only one term
   B. exactly two terms
   C. exactly three terms
   D. more than three terms.

6 **FOIL** is a method for
   A. adding two binomials
   B. adding two trinomials
   C. multiplying two binomials
   D. multiplying two trinomials.

Answers to Test Your Word Power

1. B; *Example:* $5x^3 + 2x^2 - 7$

2. D; *Examples:* The term 6 has degree 0, $3x$ has degree 1, $-2x^8$ has degree 8, and $5x^2y^4$ has degree 6.

3. C; *Example:* $2a^2 - 3ab + b^2$

4. B; *Example:* $3t^3 + 5t$

5. A; *Examples:* -5 and $4xy^5$

6. C; *Example:* $(m + 4)(m - 3)$

$$\begin{array}{cccc} \text{F} & \text{O} & \text{I} & \text{L} \end{array}$$
$$= m(m) - 3m + 4m + 4(-3)$$
$$= m^2 + m - 12$$

Quick Review

Concepts	Examples
5.1 Adding and Subtracting Polynomials	
Addition Add like terms.	Add. $\begin{array}{r} 2x^2 + 5x - 3 \\ 5x^2 - 2x + 7 \\ \hline 7x^2 + 3x + 4 \end{array}$
Subtraction Change the signs of the terms in the second polynomial and add to the first polynomial.	Subtract. $(2x^2 + 5x - 3) - (5x^2 - 2x + 7)$ $= (2x^2 + 5x - 3) + (-5x^2 + 2x - 7)$ $= -3x^2 + 7x - 10$
5.2 The Product Rule and Power Rules for Exponents	
For any integers m and n, the following hold.	Perform the operations by using the rules for exponents.
Product rule $\quad a^m \cdot a^n = a^{m+n}$	$2^4 \cdot 2^5 = 2^{4+5} = 2^9$
Power rules (a) $(a^m)^n = a^{mn}$	$(3^4)^2 = 3^{4 \cdot 2} = 3^8$
(b) $(ab)^m = a^m b^m$	$(6a)^5 = 6^5 a^5$
(c) $\left(\dfrac{a}{b}\right)^m = \dfrac{a^m}{b^m}$ (where $b \neq 0$)	$\left(\dfrac{2}{3}\right)^4 = \dfrac{2^4}{3^4}$
5.3 Multiplying Polynomials	
Multiply each term of the first polynomial by each term of the second polynomial. Then add like terms.	Multiply. $\begin{array}{r} 3x^3 - 4x^2 + 2x - 7 \\ 4x + 3 \\ \hline 9x^3 - 12x^2 + 6x - 21 \\ 12x^4 - 16x^3 + 8x^2 - 28x \\ \hline 12x^4 - 7x^3 - 4x^2 - 22x - 21 \end{array}$
FOIL method	Multiply $(2x + 3)(5x - 4)$.
Step 1 Multiply the two **F**irst terms to get the first term of the answer.	$2x(5x) = 10x^2 \qquad$ F
Step 2 Find the **O**uter product and the **I**nner product and mentally add them, when possible, to get the middle term of the answer.	$2x(-4) + 3(5x) = 7x \qquad$ O, I
Step 3 Multiply the two **L**ast terms to get the last term of the answer.	$3(-4) = -12 \qquad$ L
Add the terms found in Steps 1–3.	The product is $10x^2 + 7x - 12$.
5.4 Special Products	
Square of a Binomial	Multiply.
$$(a + b)^2 = a^2 + 2ab + b^2$$ $$(a - b)^2 = a^2 - 2ab + b^2$$	$(3x + 1)^2 \qquad\qquad (2m - 5n)^2$ $= (3x)^2 + 2(3x)(1) + 1^2 \quad = (2m)^2 - 2(2m)(5n) + (5n)^2$ $= 9x^2 + 6x + 1 \qquad\qquad = 4m^2 - 20mn + 25n^2$

Concepts	Examples

5.4 Special Products *(continued)*

Product of the Sum and Difference of Two Terms
$$(a + b)(a - b) = a^2 - b^2$$

$$(4a + 3)(4a - 3)$$
$$= (4a)^2 - 3^2$$
$$= 16a^2 - 9$$

5.5 Integer Exponents and the Quotient Rule

If $a, b \neq 0$, for integers m and n, the following hold.

Zero exponent $\qquad a^0 = 1$

Negative exponent $\qquad a^{-n} = \dfrac{1}{a^n}$

Quotient rule $\qquad \dfrac{a^m}{a^n} = a^{m-n}$

Negative-to-positive rules $\qquad \dfrac{a^{-m}}{b^{-n}} = \dfrac{b^n}{a^m} \qquad \left(\dfrac{a}{b}\right)^{-m} = \left(\dfrac{b}{a}\right)^{m}$

Simplify by using the rules for exponents.

$$15^0 = 1$$

$$5^{-2} = \frac{1}{5^2} = \frac{1}{25}$$

$$\frac{4^8}{4^3} = 4^{8-3} = 4^5$$

$$\frac{6^{-2}}{7^{-3}} = \frac{7^3}{6^2} \qquad \left(\frac{5}{3}\right)^{-4} = \left(\frac{3}{5}\right)^{4}$$

5.6 Dividing a Polynomial by a Monomial

Divide each term of the polynomial by the monomial.

$$\frac{a + b}{c} = \frac{a}{c} + \frac{b}{c}$$

Divide.
$$\frac{4x^3 - 2x^2 + 6x - 8}{2x}$$

$$= \frac{4x^3}{2x} - \frac{2x^2}{2x} + \frac{6x}{2x} - \frac{8}{2x} \qquad \text{Divide each term in the dividend by } 2x, \text{ the divisor.}$$

$$= 2x^2 - x + 3 - \frac{4}{x}$$

5.7 Dividing a Polynomial by a Polynomial

Use "long division."

Divide.

$$
\begin{array}{r}
2x - 5 \\
3x + 4 \overline{\smash{)}6x^2 - 7x - 21} \\
\underline{6x^2 + 8x} \\
-15x - 21 \\
\underline{-15x - 20} \\
-1 \leftarrow \text{Remainder}
\end{array}
$$

The answer is $2x - 5 + \dfrac{-1}{3x + 4}$.

5.8 An Application of Exponents: Scientific Notation

To write a positive number in scientific notation

$$a \times 10^n,$$

move the decimal point to follow the first nonzero digit.

1. If moving the decimal point makes the number less, n is positive.
2. If it makes the number greater, n is negative.
3. If the decimal point is not moved, n is 0.

    For a negative number, follow these steps using the *absolute value* of the number. Then make the result negative.

Write in scientific notation.

$$247 = 2.47 \times 10^2$$

$$0.0051 = 5.1 \times 10^{-3}$$

Write without exponents.

$$3.25 \times 10^5 = 325{,}000$$

$$8.44 \times 10^{-6} = 0.00000844$$

Chapter 5 Review Exercises

5.1 *Combine terms where possible in each polynomial. Write the answer in descending powers of the variable. Give the degree of the answer. Identify the polynomial as a* monomial, *a* binomial, *a* trinomial, *or* none of these.

1. $9m^2 + 11m^2 + 2m^2$

2. $-4p + p^3 - p^2 + 8p + 2$

3. $12a^5 - 9a^4 + 8a^3 + 2a^2 - a + 3$

4. $-7y^5 - 8y^4 - y^5 + y^4 + 9y$

Add or subtract as indicated.

5. Add.
$$-2a^3 + 5a^2$$
$$-3a^3 - a^2$$

6. Add.
$$4r^3 - 8r^2 + 6r$$
$$-2r^3 + 5r^2 + 3r$$

7. Subtract.
$$6y^2 - 8y + 2$$
$$-5y^2 + 2y - 7$$

8. Subtract.
$$-12k^4 - 8k^2 + 7k - 5$$
$$k^4 + 7k^2 + 11k + 1$$

9. $(2m^3 - 8m^2 + 4) + (8m^3 + 2m^2 - 7)$

10. $(-5y^2 + 3y + 11) + (4y^2 - 7y + 15)$

11. $(6p^2 - p - 8) - (-4p^2 + 2p + 3)$

12. $(12r^4 - 7r^3 + 2r^2) - (5r^4 - 3r^3 + 2r^2 + 1)$

5.2 *Simplify each expression.*

13. $4^3 \cdot 4^8$

14. $(-5)^6(-5)^5$

15. $(-8x^4)(9x^3)$

16. $(2x^2)(5x^3)(x^9)$

17. $(19x)^5$

18. $(-4y)^7$

19. $5(pt)^4$

20. $\left(\dfrac{7}{5}\right)^6$

21. $(3x^2y^3)^3$

22. $(t^4)^8(t^2)^5$

23. $(6x^2z^4)^2(x^3yz^2)^4$

24. $\left(\dfrac{2m^3n}{p^2}\right)^3$

25. Find a polynomial that represents the volume of the figure. (If necessary, refer to the formulas inside the back cover.)

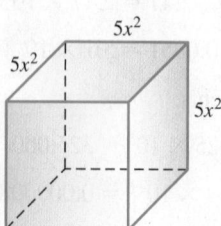

$5x^2$

$5x^2$

$5x^2$

26. Explain why the product rule for exponents does not apply to the expression
$$7^2 + 7^4.$$

5.3 *Find each product.*

27. $5x(2x + 14)$

28. $-3p^3(2p^2 - 5p)$

29. $(3r - 2)(2r^2 + 4r - 3)$

30. $(2y + 3)(4y^2 - 6y + 9)$

31. $(5p^2 + 3p)(p^3 - p^2 + 5)$

32. $(x + 6)(x - 3)$

33. $(3k - 6)(2k + 1)$

34. $(6p - 3q)(2p - 7q)$

35. $(m^2 + m - 9)(2m^2 + 3m - 1)$

5.4 *Find each product.*

36. $(a + 4)^2$

37. $(3p - 2)^2$

38. $(2r + 5s)^2$

39. $(r + 2)^3$

40. $(2x - 1)^3$

41. $(2z + 7)(2z - 7)$

42. $(6m - 5)(6m + 5)$

43. $(5a + 6b)(5a - 6b)$

44. $(2x^2 + 5)(2x^2 - 5)$

45. CONCEPT CHECK The square of a binomial leads to a polynomial with how many terms? The product of the sum and difference of two terms leads to a polynomial with how many terms?

46. Explain why $(a + b)^2$ is not equivalent to $a^2 + b^2$.

5.5 *Evaluate each expression.*

47. $5^0 + 8^0$

48. 2^{-5}

49. $\left(\dfrac{6}{5}\right)^{-2}$

50. $4^{-2} - 4^{-1}$

Simplify each expression. Assume that all variables represent nonzero numbers.

51. $\dfrac{6^{-3}}{6^{-5}}$

52. $\dfrac{x^{-7}}{x^{-9}}$

53. $\dfrac{p^{-8}}{p^4}$

54. $\dfrac{r^{-2}}{r^{-6}}$

55. $(2^4)^2$

56. $(9^3)^{-2}$

57. $(5^{-2})^{-4}$

58. $(8^{-3})^4$

59. $\dfrac{(m^2)^3}{(m^4)^2}$

60. $\dfrac{y^4 \cdot y^{-2}}{y^{-5}}$

61. $\dfrac{r^9 \cdot r^{-5}}{r^{-2} \cdot r^{-7}}$

62. $(-5m^3)^2$

63. $(2y^{-4})^{-3}$

64. $\dfrac{ab^{-3}}{a^4b^2}$

65. $\dfrac{(6r^{-1})^2 \cdot (2r^{-4})}{r^{-5}(r^2)^{-3}}$

66. $\dfrac{(2m^{-5}n^2)^3(3m^2)^{-1}}{m^{-2}n^{-4}(m^{-1})^2}$

5.6 *Perform each division.*

67. $\dfrac{-15y^4}{-9y^2}$

68. $\dfrac{-12x^3y^2}{6xy}$

69. $\dfrac{6y^4 - 12y^2 + 18y}{-6y}$

70. $\dfrac{2p^3 - 6p^2 + 5p}{2p^2}$

71. $(5x^{13} - 10x^{12} + 20x^7 - 35x^5) \div (-5x^4)$

72. $(-10m^4n^2 + 5m^3n^3 + 6m^2n^4) \div (5m^2n)$

5.7 *Perform each division.*

73. $(2r^2 + 3r - 14) \div (r - 2)$

74. $\dfrac{12m^2 - 11m - 10}{3m - 5}$

75. $\dfrac{10a^3 + 5a^2 - 14a + 9}{5a^2 - 3}$

76. $\dfrac{2k^4 + 4k^3 + 9k^2 - 8}{2k^2 + 1}$

5.8 *Write each number in scientific notation.*

77. 48,000,000

78. 28,988,000,000

79. 0.000065

80. 0.0000000824

Write each number without exponents.

81. 2.4×10^4

82. 7.83×10^7

83. 8.97×10^{-7}

84. 9.95×10^{-12}

Perform the indicated operations. Write the answers in scientific notation and then without exponents.

85. $(2 \times 10^{-3})(4 \times 10^5)$

86. $\dfrac{8 \times 10^4}{2 \times 10^{-2}}$

87. $\dfrac{12 \times 10^{-5} \times 5 \times 10^4}{4 \times 10^3 \times 6 \times 10^{-2}}$

88. $\dfrac{2.5 \times 10^5 \times 4.8 \times 10^{-4}}{7.5 \times 10^8 \times 1.6 \times 10^{-5}}$

89. A computer can perform 466,000,000 calculations per second. How many calculations can it perform per minute? Per hour?

90. In theory, there are 1×10^9 possible Social Security numbers. The population of the United States is about 3×10^8. How many Social Security numbers are available for each person? (*Source:* U.S. Census Bureau.)

Mixed Review Exercises

Perform each indicated operation. Assume that all variables represent nonzero real numbers.

91. $19^0 - 3^0$

92. $(3p)^4(3p^{-7})$

93. 7^{-2}

94. $(-7 + 2k)^2$

95. $\dfrac{2y^3 + 17y^2 + 37y + 7}{2y + 7}$

96. $\left(\dfrac{6r^2s}{5}\right)^4$

97. $-m^5(8m^2 + 10m + 6)$

98. $\left(\dfrac{1}{2}\right)^{-5}$

99. $(25x^2y^3 - 8xy^2 + 15x^3y) \div (5x)$

100. $(6r^{-2})^{-1}$

101. $(2x + y)^3$

102. $2^{-1} + 4^{-1}$

103. $(a + 2)(a^2 - 4a + 1)$

104. $(5y^3 - 8y^2 + 7) - (-3y^3 + y^2 + 2)$

105. $(2r + 5)(5r - 2)$

106. $(12a + 1)(12a - 1)$

107. What polynomial represents the area of this rectangle?   The perimeter?

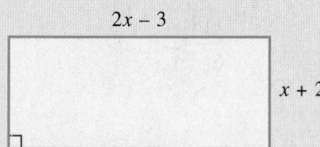

$2x - 3$

$x + 2$

108. What polynomial represents the perimeter of this square?   The area?

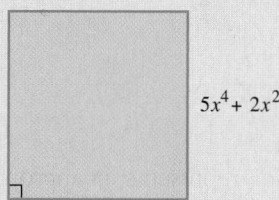

$5x^4 + 2x^2$

109. CONCEPT CHECK   One of your friends in class simplified

$$\dfrac{6x^2 - 12x}{6} \quad \text{as} \quad x^2 - 12x.$$

What Went Wrong? Give the correct answer.

110. CONCEPT CHECK   What polynomial, when multiplied by $6m^2n$, gives the following product?

$$12m^3n^2 + 18m^6n^3 - 24m^2n^2$$

Chapter 5 **Test**

The Chapter Test Prep Videos with test solutions are available on DVD, in MyMathLab, and on You Tube—search "LialCombinedAlg" and click on "Channels."

For each polynomial, combine like terms when possible and write the polynomial in descending powers of the variable. Give the degree of the simplified polynomial. Decide whether the simplified polynomial is a monomial, *a* binomial, *a* trinomial, *or* none of these.

1. $5x^2 + 8x - 12x^2$

2. $13n^3 - n^2 + n^4 + 3n^4 - 9n^2$

Perform the indicated operations.

3. $(5t^4 - 3t^2 + 7t + 3) - (t^4 - t^3 + 3t^2 + 8t + 3)$

4. $(2y^2 - 8y + 8) + (-3y^2 + 2y + 3) - (y^2 + 3y - 6)$

5. Subtract.

$9t^3 - 4t^2 + 2t + 2$
$\underline{9t^3 + 8t^2 - 3t - 6}$

6. $(-2)^3(-2)^2$

7. $\left(\dfrac{6}{m^2}\right)^3 \quad (m \neq 0)$

8. $3x^2(-9x^3 + 6x^2 - 2x + 1)$

9. $(2r - 3)(r^2 + 2r - 5)$

10. $(t - 8)(t + 3)$

11. $(4x + 3y)(2x - y)$

12. $(5x - 2y)^2$

13. $(10v + 3w)(10v - 3w)$

14. $(x + 1)^3$

15. What expression represents, in appropriate units, the perimeter of this square? The area?

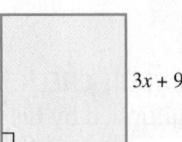

$3x + 9$

16. Determine whether each expression represents a number that is *positive, negative,* or *zero.*

(a) 3^{-4} (b) $(-3)^4$ (c) -3^4 (d) 3^0 (e) $(-3)^0 - 3^0$ (f) $(-3)^{-3}$

Evaluate each expression.

17. 5^{-4}

18. $(-3)^0 + 4^0$

19. $4^{-1} + 3^{-1}$

Perform the indicated operations. In Exercises 20 and 21, write each answer using only positive exponents. Assume that variables represent nonzero numbers.

20. $\dfrac{8^{-1} \cdot 8^4}{8^{-2}}$

21. $\dfrac{(x^{-3})^{-2}(x^{-1}y)^2}{(xy^{-2})^2}$

22. $\dfrac{8y^3 - 6y^2 + 4y + 10}{2y}$

23. $(-9x^2y^3 + 6x^4y^3 + 12xy^3) \div (3xy)$

24. $\dfrac{2x^2 + x - 36}{x - 4}$

25. $(3x^3 - x + 4) \div (x - 2)$

26. Write each number in scientific notation.

   (a) 344,000,000,000   **(b)** 0.00000557

27. Write each number without exponents.

   (a) 2.96×10^7 **(b)** 6.07×10^{-8}

28. A satellite galaxy of our own Milky Way, known as the Large Magellanic Cloud, is *1000* light-years across. A light-year is equal to *5,890,000,000,000* mi. (*Source:* "Images of Brightest Nebula Unveiled," *USA Today.*)

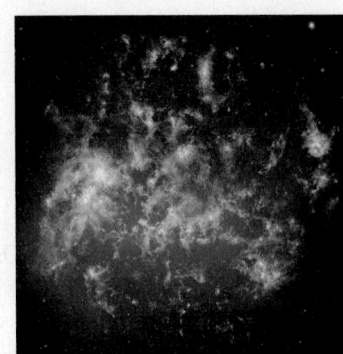

   (a) Write the two boldface italic numbers in scientific notation.

   (b) How many miles across is the Large Magellanic Cloud?

Chapters R–5   *Cumulative Review Exercises*

Perform each operation.

1. $\dfrac{2}{3} + \dfrac{1}{8}$

2. $\dfrac{7}{4} - \dfrac{9}{5}$

3. $8.32 - 4.6$

4. 7.21×8.6

5. A retailer has \$34,000 invested in her business. She finds that last year she earned 5.4% on this investment. How much did she earn?

Find the value of each expression for $x = -2$ and $y = 4$.

6. $\dfrac{4x - 2y}{x + y}$

7. $x^3 - 4xy$

Perform the indicated operations.

8. $\dfrac{(-13 + 15) - (3 + 2)}{6 - 12}$

9. $-7 - 3[2 + (5 - 8)]$

Decide what property justifies each statement.

10. $(9 + 2) + 3 = 9 + (2 + 3)$

11. $-7 + 7 = 0$

12. $6(4 + 2) = 6(4) + 6(2)$

Solve each equation.

13. $2x - 7x + 8x = 30$

14. $2 - 3(t - 5) = 4 + t$

15. $2(5h + 1) = 10h + 4$

16. $d = rt$ for r

17. $\dfrac{x}{5} = \dfrac{x - 2}{7}$

18. $\dfrac{1}{3}p - \dfrac{1}{6}p = -2$

19. $0.05x + 0.15(50 - x) = 5.50$

20. $4 - (3x + 12) = (2x - 9) - (5x - 1)$

Solve each problem.

21. A 1-oz mouse takes about 16 times as many breaths as does a 3-ton elephant. (*Source: Dinosaurs, Spitfires, and Sea Dragons*, McGowan, C., Harvard University Press.) If the two animals take a combined total of 170 breaths per minute, how many breaths does each take during that time period?

22. If a number is subtracted from 8 and this difference is tripled, the result is three times the number. Find this number to learn how many times a dolphin rests during a 24-hr period.

Solve each inequality. Write the solution set in interval notation.

23. $-8x \le -80$

24. $-2(x+4) > 3x+6$

25. $-3 \le 2x+5 < 9$

Given $2x - 3y = -6$, find the following.

26. The intercepts of the graph

27. The graph of the equation

28. The slope of the line

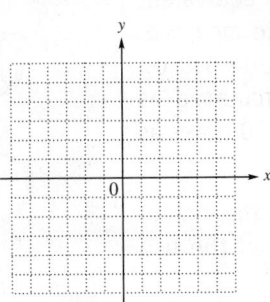

Solve the system using the method indicated.

29. $y = 2x + 5$
  $x + y = -4$ (Substitution)

30. $3x + 2y = 2$
  $2x + 3y = -7$ (Elimination)

Evaluate each expression.

31. $4^{-1} + 3^0$

32. $2^{-4} \cdot 2^5$

33. $\dfrac{8^{-5} \cdot 8^7}{8^2}$

34. Write $\dfrac{(a^{-3}b^2)^2}{(2a^{-4}b^{-3})^{-1}}$ with positive exponents only.

35. Write 34,500 in scientific notation.

Perform the indicated operations.

36. $(7x^3 - 12x^2 - 3x + 8) + (6x^2 + 4) - (-4x^3 + 8x^2 - 2x - 2)$

37. $6x^5(3x^2 - 9x + 10)$

38. $(7x + 4)(9x + 3)$

39. $(5x + 8)^2$

40. $\dfrac{y^3 - 3y^2 + 8y - 6}{y - 1}$

Math in the Media

Charles F. Richter devised a scale in 1935 to compare the intensities, or relative powers, of earthquakes. The **intensity** of an earthquake is measured relative to the intensity of a standard **zero-level** earthquake of intensity I_0. The relationship is equivalent to $I = I_0 \times 10^R$, where R is the **Richter scale** measure. For example, if an earthquake has magnitude 5.0 on the Richter scale, then its intensity is calculated as $I = I_0 \times 10^{5.0} = I_0 \times 100{,}000$, which is 100,000 times as intense as a zero-level earthquake.

To compare an earthquake that measures 8.0 on the Richter scale to one that measures 5.0, find the ratio of the two intensities.

$$\frac{\text{intensity 8.0}}{\text{intensity 5.0}} = \frac{I_0 \times 10^{8.0}}{I_0 \times 10^{5.0}} = \frac{10^8}{10^5} = 10^{8-5} = 10^3 = 1000$$

Therefore, an earthquake that measures 8.0 is 1000 times as intense as one that measures 5.0.

The Gazette

Quake rattles Iowans

Chances of 'big one'
happening here remote
UNI professor says

WIS.

IOWA   Chicago•

ILL.   IND.

Springfield   Earthquake
★   Magnitude
                5.2

MO.   West
              Salem

KY.

Area where
earthquake
was felt   TENN.

Source: WSRI; USGS.

Year	Earthquake	Richter Scale Measurement
2011	Northeast Japan	9.0
2005	Northern Sumatra, Indonesia	8.6
2004	West coast of Northern Sumatra	9.1
2003	Southeastern Iran	6.6
1998	Balleny Islands region	8.1
1906	San Francisco, CA	7.7

Source: U.S. Geological Survey.

1. Compare the intensity of the 2004 west coast of northern Sumatra earthquake to that of the 1998 Balleny Islands region earthquake.

2. Compare the intensity of the 2005 northern Sumatra, Indonesia earthquake to that of the 2003 southeastern Iran earthquake.

3. Compare the intensity of the 2011 northeast Japan earthquake to that of the 1906 San Francisco earthquake, the most powerful to strike the United States. (*Hint:* Use the exponential key of a scientific calculator to compute the required power of 10.)

4. Suppose an earthquake measures a value of x on the Richter scale. How would the intensity of a second earthquake compare if its Richter scale measure is $x + 4.0$? How would it compare if its Richter scale measure is $x - 1.0$?

Factoring and Applications

Formulas associated with the mathematicians Pythagoras (c. 380–300 B.C.) and Galileo (1564–1642) are used in applications that involve *factoring* polynomials, the subject of this chapter.

6.1 Factors; The Greatest Common Factor

6.2 Factoring Trinomials

6.3 Factoring Trinomials by Grouping

6.4 Factoring Trinomials by Using the FOIL Method

6.5 Special Factoring Techniques

6.6 A General Approach to Factoring

6.7 Solving Quadratic Equations by Factoring

6.8 Applications of Quadratic Equations

6.1 Factors; The Greatest Common Factor

OBJECTIVES

1 Find the greatest common factor of a list of numbers.

2 Find the greatest common factor of a list of variable terms.

3 Factor out the greatest common factor.

4 Factor by grouping.

To **factor** a number means to write it as the product of two or more numbers. The product is the **factored form** of the number. Consider an example.

Factors

$$12 = \overbrace{6 \cdot 2}$$

Factored form

Factoring is a process that "undoes" multiplying. We multiply $6 \cdot 2$ to get 12, but we factor 12 by writing it as $6 \cdot 2$.

OBJECTIVE 1 **Find the greatest common factor of a list of numbers.** An integer that is a factor of two or more integers is a **common factor** of those integers. For example, 6 is a common factor of 18 and 24 because 6 is a factor of both 18 and 24. Other common factors of 18 and 24 are 1, 2, and 3.

The **greatest common factor (GCF)** of a list of integers is the largest common factor of those integers. This means 6 is the greatest common factor of 18 and 24, since it is the largest of their common factors.

> **Note**
>
> *Factors* of a number are also *divisors* of the number. The *greatest common factor* is the same as the *greatest common divisor*. Here are some divisibility rules for deciding what numbers divide into a given number.
>
> | A Whole Number Divisible by | Must Have the Following Property |
> |---|---|
> | 2 | Ends in 0, 2, 4, 6, or 8 |
> | 3 | Sum of digits divisible by 3 |
> | 4 | Last two digits form a number divisible by 4 |
> | 5 | Ends in 0 or 5 |
> | 6 | Divisible by both 2 and 3 |
> | 8 | Last three digits form a number divisible by 8 |
> | 9 | Sum of digits divisible by 9 |
> | 10 | Ends in 0 |

EXAMPLE 1 **Finding the Greatest Common Factor for Numbers**

Find the greatest common factor for each list of numbers.

(a) 30, 45

$$30 = 2 \cdot 3 \cdot 5$$
$$45 = 3 \cdot 3 \cdot 5$$

Write the prime factored form of each number.

*Use each prime the **least** number of times it appears in **all** the factored forms.* There is no 2 in the prime factored form of 45, so there will be no 2 in the greatest common factor. The least number of times 3 appears in all the factored forms is 1. The least number of times 5 appears is also 1.

$$\text{GCF} = 3^1 \cdot 5^1 = 15 \quad 3^1 = 3 \text{ and } 5^1 = 5.$$

Continued on Next Page

(b) 72, 120, 432

$$72 = 2 \cdot 2 \cdot 2 \cdot 3 \cdot 3$$
$$120 = 2 \cdot 2 \cdot 2 \cdot 3 \cdot 5 \qquad \text{Write the prime factored form of each number.}$$
$$432 = 2 \cdot 2 \cdot 2 \cdot 2 \cdot 3 \cdot 3 \cdot 3$$

The least number of times 2 appears in all the factored forms is 3, and the least number of times 3 appears is 1. There is no 5 in the prime factored form of either 72 or 432.

$$\text{GCF} = 2^3 \cdot 3^1 = 24 \qquad 2^3 = 8 \text{ and } 3^1 = 3.$$

(c) 10, 11, 14

$$10 = 2 \cdot 5$$
$$11 = 11 \qquad \text{Write the prime factored form of each number.}$$
$$14 = 2 \cdot 7$$

There are no primes common to all three numbers, so the GCF is 1.

················· **Work Problem ❶ at the Side.** ▶

OBJECTIVE ▶ ❷ Find the greatest common factor of a list of variable terms. The terms $x^4, x^5, x^6,$ and x^7 have x^4 as the greatest common factor because the least exponent on the variable x in the factored forms is 4.

$$x^4 = 1 \cdot x^4, \quad x^5 = x \cdot x^4, \quad x^6 = x^2 \cdot x^4, \quad x^7 = x^3 \cdot x^4$$
$$\text{GCF} = x^4$$

Note

*The exponent on a variable in the GCF is the **least** exponent that appears on that variable in **all** the terms.*

EXAMPLE 2 Finding the Greatest Common Factor for Variable Terms

Find the greatest common factor for each list of terms.

(a) $21m^7, \quad 18m^6, \quad 45m^8, \quad 24m^5$

$$21m^7 = 3 \cdot 7 \cdot m^7$$
$$18m^6 = 2 \cdot 3 \cdot 3 \cdot m^6 \qquad \text{Here, } 3 \text{ is the greatest common factor of the coefficients 21, 18, 45, and 24.}$$
$$45m^8 = 3 \cdot 3 \cdot 5 \cdot m^8 \qquad \text{The least exponent on } m \text{ is } 5.$$
$$24m^5 = 2 \cdot 2 \cdot 2 \cdot 3 \cdot m^5 \qquad \text{GCF} = 3m^5$$

(b) $x^4y^2, \quad x^7y^5, \quad x^3y^7, \quad y^{15}$

$$x^4y^2 = x^4 \cdot y^2 \qquad \text{There is no } x \text{ in the last term, } y^{15}, \text{ so } x \text{ will not}$$
$$x^7y^5 = x^7 \cdot y^5 \qquad \text{appear in the greatest common factor. There}$$
$$x^3y^7 = x^3 \cdot y^7 \qquad \text{is a } y \text{ in each term, however, and } 2 \text{ is the least}$$
$$y^{15} = y^{15} \qquad \text{exponent on } y.$$
$$\text{GCF} = y^2$$

❶ Find the greatest common factor for each list of numbers.

(a) 30, 20, 15

$$30 = 2 \cdot 3 \cdot 5$$
$$20 = 2 \cdot \underline{\quad} \cdot \underline{\quad}$$
$$15 = 3 \cdot \underline{\quad}$$
$$\text{GCF} = \underline{\quad}$$

(b) 42, 28, 35

(c) 12, 18, 26, 32

(d) 10, 15, 21

Answers

1. (a) 2; 5; 5; 5 **(b)** 7 **(c)** 2 **(d)** 1

2 Find the greatest common factor for each list of terms.

(a) $6m^4, 9m^2, 12m^5$

$$6m^4 = 2 \cdot \underline{\quad} \cdot m^4$$

$$9m^2 = 3 \cdot \underline{\quad} \cdot \underline{\quad}$$

$$12m^5 = 2 \cdot 2 \cdot \underline{\quad} \cdot \underline{\quad}$$

$$\text{GCF} = \underline{\quad}$$

(b) $12p^5, 18q^4$

(c) y^4z^2, y^6z^8, z^9

(d) $12p^{11}, 17q^5$

Finding the Greatest Common Factor (GCF)

Step 1 **Factor.** Write each number in prime factored form.

Step 2 **List common factors.** List each prime number or each variable that is a factor of every term in the list. (If a prime does not appear in one of the prime factored forms, it *cannot* appear in the greatest common factor.)

Step 3 **Choose least exponents.** Use as exponents on the common prime factors the *least* exponents from the prime factored forms.

Step 4 **Multiply** the primes from Step 3. If there are no primes left after Step 3, the greatest common factor is 1.

◀ Work Problem **2** at the Side.

OBJECTIVE 3 Factor out the greatest common factor. Writing a polynomial (a sum) in factored form as a product is called **factoring** the polynomial. For example, the polynomial

$$3m + 12$$

has two terms, $3m$ and 12. The greatest common factor of these two terms is 3. We can write $3m + 12$ so that each term is a product with 3 as one factor.

$$3m + 12$$

$$= 3 \cdot m + 3 \cdot 4 \qquad \text{GCF} = 3$$

$$= 3(m + 4) \qquad \text{Distributive property}$$

The factored form of $3m + 12$ is $3(m + 4)$. This process is called **factoring out the greatest common factor.**

CAUTION

The polynomial $3m + 12$ is *not* in factored form when written as

$$3 \cdot m + 3 \cdot 4. \qquad \text{Not in factored form}$$

The terms are factored, but the polynomial is not. The factored form of $3m + 12$ is the *product*

$$3(m + 4). \qquad \text{In factored form}$$

EXAMPLE 3 Factoring Out the Greatest Common Factor

Write in factored form by factoring out the greatest common factor.

(a) $5y^2 + 10y$

$$= 5y(y) + 5y(2) \qquad \text{GCF} = 5y$$

$$= 5y(y + 2) \qquad \text{Distributive property}$$

CHECK Multiply the factored form.

$$5y(y + 2)$$

$$= 5y(y) + 5y(2) \qquad \text{Distributive property}$$

$$= 5y^2 + 10y \checkmark \qquad \text{Original polynomial}$$

Continued on Next Page

Answers

2. (a) 3; 3; m^2; 3; m^5; $3m^2$

(b) 6 **(c)** z^2 **(d)** 1

(b) $20m^5 + 10m^4 - 15m^3$

$\qquad = 5m^3(4m^2) + 5m^3(2m) - 5m^3(3) \qquad$ GCF $= 5m^3$

$\qquad = 5m^3(4m^2 + 2m - 3) \qquad\qquad$ Factor out $5m^3$.

CHECK $5m^3(\overset{\frown}{4m^2 + 2m} \overset{\frown}{- 3})$

$\qquad\qquad = 5m^3(4m^2) + 5m^3(2m) + 5m^3(-3) \qquad$ Distributive property

$\qquad\qquad = 20m^5 + 10m^4 - 15m^3 \checkmark \qquad$ Original polynomial

(c) $x^5 + x^3$

$\qquad = x^3(x^2) + x^3(1) \qquad$ GCF $= x^3$

$\qquad = x^3(x^2 + 1) \quad \longleftarrow$ [Don't forget the 1.]

Check mentally by distributing x^3 over each term inside the parentheses.

(d) $20m^7p^2 - 36m^3p^4$

$\qquad = 4m^3p^2(5m^4) - 4m^3p^2(9p^2) \qquad$ GCF $= 4m^3p^2$

$\qquad = 4m^3p^2(5m^4 - 9p^2) \qquad\qquad$ Factor out $4m^3p^2$.

Check mentally by distributing $4m^3p^2$ over each term inside the parentheses.

> **CAUTION**
>
> Be sure to include the **1** in a problem like **Example 3(c)**. *Check that the factored form can be multiplied out to give the original polynomial.*

Work Problem ③ at the Side. ▶

> **EXAMPLE 4**   Factoring Out a Negative Common Factor
>
> Write $-8x^4 + 16x^3 - 4x^2$ in factored form.
>
> We can factor out either $4x^2$ or $-4x^2$ here. We usually factor out $-4x^2$ so that the coefficient of the first term in the trinomial factor will be positive.
>
> $\qquad -8x^4 + 16x^3 - 4x^2$ [Be careful with signs.]
>
> $\qquad = -4x^2(2x^2) - 4x^2(-4x) - 4x^2(1) \qquad$ $-4x^2$ is a common factor.
>
> $\qquad = -4x^2(2x^2 - 4x + 1) \qquad\qquad$ Factor out $-4x^2$.
>
> **CHECK** $-4x^2(\overset{\frown}{2x^2 - 4x} \overset{\frown}{+ 1})$
>
> $\qquad\qquad = -4x^2(2x^2) - 4x^2(-4x) - 4x^2(1) \qquad$ Distributive property
>
> $\qquad\qquad = -8x^4 + 16x^3 - 4x^2 \checkmark \qquad$ Original polynomial

Work Problem ④ at the Side. ▶

> **Note**
>
> Whenever we factor a polynomial in which the coefficient of the first term is negative, we will factor out the negative common factor, even if it is just -1. However, it would also be correct to factor out $4x^2$ in **Example 4** to obtain
>
> $$4x^2(-2x^2 + 4x - 1).$$

---

③ Write in factored form by factoring out the greatest common factor.

GS (a) $4x^2 + 6x$

$\qquad = \underline{\quad}(\underline{\quad}) + 2x(\underline{\quad})$

$\qquad = \underline{\quad}(\underline{\quad} + \underline{\quad})$

(b) $10y^5 - 8y^4 + 6y^2$

GS (c) $m^7 + m^9$

$\qquad = \underline{\quad}(\underline{\quad}) + \underline{\quad}(m^2)$

$\qquad = \underline{\quad}(\underline{\quad} + \underline{\quad})$

(d) $15x^3 - 10x^2 + 5x$

(e) $8p^5q^2 + 16p^6q^3 - 12p^4q^7$

④ Write

$\qquad -14a^3b^2 - 21a^2b^3 + 7ab$

in factored form by factoring out a negative common factor.

Answers

3. (a) $2x; 2x; 3; 2x; 2x; 3$
   (b) $2y^2(5y^3 - 4y^2 + 3)$
   (c) $m^7; 1; m^7; m^7; 1; m^2$
   (d) $5x(3x^2 - 2x + 1)$
   (e) $4p^4q^2(2p + 4p^2q - 3q^5)$
4. $-7ab(2a^2b + 3ab^2 - 1)$

5 Write in factored form by factoring out the greatest common factor.

(a) $r(t-4)+5(t-4)$

$= (t-4)(\underline{\hspace{1cm}})$

(b) $x(x+2)+7(x+2)$

(c) $y^2(y+2)-3(y+2)$

(d) $x(x-1)-5(x-1)$

EXAMPLE 5 **Factoring Out a Common Binomial Factor**

Write in factored form by factoring out the greatest common factor.

Same

(a) $a(a+3)+4(a+3)$ The binomial $a+3$ is the greatest common factor.

$= (a+3)(a+4)$ Factor out $a+3$.

(b) $x^2(x+1)-5(x+1)$

$= (x+1)(x^2-5)$ Factor out $x+1$.

◀ **Work Problem 5 at the Side.**

Note

In factored forms like those in **Example 5**, the order of the factors does not matter because of the commutative property of multiplication.

$(a+3)(a+4)$ can also be written $(a+4)(a+3)$.

OBJECTIVE 4 **Factor by grouping.** *When a polynomial has four terms, common factors can sometimes be used to factor by grouping.*

EXAMPLE 6 **Factoring by Grouping**

Factor by grouping.

(a) $2x+6+ax+3a$

Group the first two terms and the last two terms, since the first two terms have a common factor of 2 and the last two terms have a common factor of a.

$2x+6+ax+3a$

$= (2x+6)+(ax+3a)$ Group the terms.

$= 2(x+3)+a(x+3)$ Factor each group.

The expression is still not in factored form because it is the *sum* of two terms. Now, however, $x+3$ is a common factor and can be factored out.

$= 2(x+3)+a(x+3)$ $x+3$ is a common factor.

$(2+a)(x+3)$ is also correct. $= (x+3)(2+a)$ Factor out $x+3$.

The final result is in factored form because it is a *product*.

CHECK $(x+3)(2+a)$

$= x(2)+x(a)+3(2)+3(a)$ Multiply using the FOIL method (**Section 5.3**).

$= 2x+ax+6+3a$ Simplify.

$= 2x+6+ax+3a$ ✓   Rearrange terms to obtain the original polynomial.

············ **Continued on Next Page**

Answers

5. (a) $r+5$ (b) $(x+2)(x+7)$
(c) $(y+2)(y^2-3)$ (d) $(x-1)(x-5)$

(b) $6ax + 24x + a + 4$

$\quad\quad = (6ax + 24x) + (a + 4)$ Group the terms.

$\quad\quad = 6x(a + 4) + 1(a + 4)$ Factor each group.

> Remember the 1.

$\quad\quad = (a + 4)(6x + 1)$ Factor out $a + 4$.

CHECK $(a + 4)(6x + 1)$

$\quad\quad = 6ax + a + 24x + 4$ FOIL method

$\quad\quad = 6ax + 24x + a + 4$ ✓   Rearrange terms to obtain the original polynomial.

(c) $2x^2 - 10x + 3xy - 15y$

$\quad\quad = (2x^2 - 10x) + (3xy - 15y)$ Group the terms.

$\quad\quad = 2x(x - 5) + 3y(x - 5)$ Factor each group.

$\quad\quad = (x - 5)(2x + 3y)$ Factor out the common factor, $x - 5$.

CHECK $(x - 5)(2x + 3y)$

$\quad\quad = 2x^2 + 3xy - 10x - 15y$ FOIL method

$\quad\quad = 2x^2 - 10x + 3xy - 15y$ ✓   Original polynomial

(d) $t^3 + 2t^2 - 3t - 6$

> Write a + sign between the groups.

$\quad\quad = (t^3 + 2t^2) + (-3t - 6)$ Group the terms.

$\quad\quad = t^2(t + 2) - 3(t + 2)$ Factor out -3 so there is a common factor, $t + 2$; $-3(t + 2) = -3t - 6$.

> Be careful with signs.

$\quad\quad = (t + 2)(t^2 - 3)$ Factor out $t + 2$.

Check by multiplying using the FOIL method.

- - -

CAUTION

Be careful with signs when grouping in a problem like **Example 6(d)**. It is wise to check the factoring in the second step, as shown in the example side comment, before continuing.

Work Problem **6** at the Side. ▶

Factoring a Polynomial with Four Terms by Grouping

Step 1   **Group terms.** Collect the terms into two groups so that each group has a common factor.

Step 2   **Factor within groups.** Factor out the greatest common factor from each group.

Step 3   **Factor the entire polynomial.** Factor a common binomial factor from the results of Step 2.

Step 4   **If necessary, rearrange terms.** If Step 2 does not result in a common binomial factor, try a different grouping.

---

6 Factor by grouping.

(a) $pq + 5q + 2p + 10$

$\quad = (\_\_ + 5q) + (\_\_ + 10)$

$\quad = \_\_(p + 5) + \_\_(p + 5)$

$\quad = (\_\_\_\_)(\_\_\_\_)$

(b) $2xy + 3y + 2x + 3$

(c) $2a^2 - 4a + 3ab - 6b$

(d) $x^3 + 3x^2 - 5x - 15$

Answers

6. **(a)** pq; $2p$; q; 2; $p + 5$; $q + 2$
     (b) $(2x + 3)(y + 1)$
     (c) $(a - 2)(2a + 3b)$
     (d) $(x + 3)(x^2 - 5)$

7 Factor by grouping.

(a) $6y^2 - 20w + 15y - 8yw$

(b) $9mn - 4 + 12m - 3n$

(c) $12p^2 - 28q - 16pq + 21p$

EXAMPLE 7   **Rearranging Terms before Factoring by Grouping**

Factor by grouping.

(a) $10x^2 - 12y + 15x - 8xy$

Factoring out the common factor 2 from the first two terms and the common factor x from the last two terms gives the following.

$$10x^2 - 12y + 15x - 8xy$$
$$= 2(5x^2 - 6y) + x(15 - 8y)$$

This does not lead to a common factor, so we try rearranging the terms.
There is usually more than one way to do this. We try the following.

$$10x^2 - 12y + 15x - 8xy$$

$$= 10x^2 - 8xy - 12y + 15x \qquad \text{Commutative property}$$

$$= (10x^2 - 8xy) + (-12y + 15x) \qquad \text{Group the terms.}$$

$$= 2x(5x - 4y) + 3(-4y + 5x) \qquad \text{Factor each group.}$$

$$= 2x(5x - 4y) + 3(5x - 4y) \qquad \text{Rewrite } -4y + 5x.$$

$$= (5x - 4y)(2x + 3) \qquad \text{Factor out } 5x - 4y.$$

CHECK $(5x - 4y)(2x + 3)$

$$= 10x^2 + 15x - 8xy - 12y \qquad \text{FOIL method}$$

$$= 10x^2 - 12y + 15x - 8xy \checkmark \qquad \text{Original polynomial}$$

(b) $2xy + 12 - 3y - 8x$

We must rearrange the terms to get two groups that each have a common factor. Trial and error suggests the following grouping.

$$2xy + 12 - 3y - 8x \qquad \text{\small Write a + sign between the two groups.}$$

$$= (2xy - 3y) + (-8x + 12) \qquad \text{Group the terms.}$$

$$= y(2x - 3) - 4(2x - 3) \qquad \begin{array}{l}\text{Factor each group;}\\ -4(2x - 3) = -8x + 12\end{array}$$

$$\qquad\qquad\qquad \text{\small Be careful with signs.}$$

$$= (2x - 3)(y - 4) \qquad \text{Factor out } 2x - 3.$$

Since the quantities in parentheses in the second step must be the same, we factored out -4 rather than 4.

CHECK $(2x - 3)(y - 4)$

$$= 2xy - 8x - 3y + 12 \qquad \text{FOIL method}$$

$$= 2xy + 12 - 3y - 8x \checkmark \qquad \text{Original polynomial}$$

CAUTION

Use negative signs carefully when grouping, as in **Example 7(b)**, or a sign error will occur. ***Always check by multiplying.***

◀ **Work Problem 7** at the Side.

Answers

7. (a) $(2y + 5)(3y - 4w)$
   (b) $(3m - 1)(3n + 4)$
   (c) $(3p - 4q)(4p + 7)$

6.1 Exercises

FOR
EXTRA
HELP

 Download the
MyDashBoard App

 MyMathLab®

CONCEPT CHECK   *Complete each statement.*

1. To factor a number or quantity means to write it as a _____. Factoring is the opposite, or inverse, process of _____.

2. An integer or variable expression that is a factor of two or more terms is a _____ _____. For example, 12 (*is / is not*) a common factor of both 36 and 72 since it _____ evenly into both integers.

Find the greatest common factor for each list of numbers. **See Example 1.**

3. 12, 16

4. 18, 24

5. 40, 20, 4

6. 50, 30, 5

7. 18, 24, 36, 48

8. 15, 30, 45, 75

9. 4, 9, 12

10. 9, 16, 24

Find the greatest common factor for each list of terms. **See Example 2.**

11. $16y$, 24

12. $18w$, 27

13. $30x^3$, $40x^6$, $50x^7$

14. $60z^4$, $70z^8$, $90z^9$

15. x^4y^3, xy^2

16. a^4b^5, a^3b

17. $42ab^3$, $36a$, $90b$, $48ab$

18. $45c^3d$, $75c$, $90d$, $105cd$

19. $12m^3n^2$, $18m^5n^4$, $36m^8n^3$

20. $25p^5r^7$, $30p^7r^8$, $50p^5r^3$

CONCEPT CHECK   *An expression is factored when it is written as a product, not a sum. Determine whether each expression is* factored *or* not factored.

21. $2k^2(5k)$

22. $2k^2(5k + 1)$

23. $2k^2 + (5k + 1)$

24. $(2k^2 + 5k) + 1$

25. CONCEPT CHECK   A student factored $18x^3y^2 + 9xy$ as $9xy(2x^2y)$. **What Went Wrong?** Factor correctly.

26. How can we check an answer when we factor a polynomial?

GS *Complete each factoring by writing each polynomial as the product of two factors.* **See Example 3.**

27. $9m^4$

$= 3m^2(\underline{\quad})$

28. $12p^5$

$= 6p^3(\underline{\quad})$

29. $-8z^9$

$= -4z^5(\underline{\quad})$

30. $-15k^{11}$

$= -5k^8(\underline{\quad})$

31. $6m^4n^5$

$= 3m^3n(\underline{\quad})$

32. $27a^3b^2$

$= 9a^2b(\underline{\quad})$

33. $12y + 24$

$= 12(\underline{\quad})$

34. $18p + 36$

$= 18(\underline{\quad})$

35. $10a^2 - 20a$

$= 10a(\underline{\quad})$

36. $15x^2 - 30x$

$= 15x(\underline{\quad})$

37. $8x^2y + 12x^3y^2$

$= 4x^2y(\underline{\quad})$

38. $18s^3t^2 + 10st$

$= 2st(\underline{\quad})$

Write in factored form by factoring out the greatest common factor (or a negative common factor if the coefficient of the term of greatest degree is negative).
See Examples 3–5.

39. $x^2 - 4x$

40. $m^2 - 7m$

41. $6t^2 + 15t$

42. $8x^2 + 6x$

43. $m^3 - m^2$

44. $p^3 - p^2$

45. $-12x^3 - 6x^2$

46. $-21b^3 + 7b^2$

47. $65y^{10} + 35y^6$

48. $100a^5 + 16a^3$

49. $11w^3 - 100$

50. $13z^5 - 80$

51. $8mn^3 + 24m^2n^3$

52. $19p^2y + 38p^2y^3$

53. $-4x^3 + 10x^2 - 6x$

54. $-9z^3 + 6z^2 - 12z$

55. $13y^8 + 26y^4 - 39y^2$

56. $5x^5 + 25x^4 - 20x^3$

57. $45q^4p^5 + 36qp^6 + 81q^2p^3$

58. $125a^3z^5 + 60a^4z^4 - 85a^5z^2$

59. $c(x + 2) + d(x + 2)$

60. $r(5 - x) + t(5 - x)$

61. $a^2(2a + b) - b(2a + b)$

62. $3x(x^2 + 5) - y(x^2 + 5)$

63. $q(p + 4) - 1(p + 4)$

64. $y^2(x - 4) + 1(x - 4)$

CONCEPT CHECK *Students often have difficulty when factoring by grouping because they are not able to tell when the polynomial is completely factored. For example,*

$$5y(2x - 3) + 8t(2x - 3) \qquad \text{Not in factored form}$$

*is not in factored form, because it is the **sum** of two terms:* $5y(2x - 3)$ *and* $8t(2x - 3)$. *However, because* $2x - 3$ *is a common factor of these two terms, the expression can now be factored.*

$$(2x - 3)(5y + 8t) \qquad \text{In factored form}$$

*The factored form is a **product** of the two factors* $2x - 3$ *and* $5y + 8t$.
    Determine whether each expression is in factored form *or is* not in factored form.
If it is not in factored form, factor it if possible.

65. $8(7t + 4) + x(7t + 4)$

66. $3r(5x - 1) + 7(5x - 1)$

67. $(8 + x)(7t + 4)$

68. $(3r + 7)(5x - 1)$

69. $18x^2(y + 4) + 7(y - 4)$

70. $12k^3(s - 3) + 7(s + 3)$

Factor by grouping. ***See Examples 6 and 7.***

71. $5m + mn + 20 + 4n$

72. $ts + 5t + 2s + 10$

73. $6xy - 21x + 8y - 28$

74. $2mn - 8n + 3m - 12$

75. $3xy + 9x + y + 3$

76. $6n + 4mn + 3 + 2m$

77. $7z^2 + 14z - az - 2a$

78. $2b^2 + 3b - 8ab - 12a$

79. $18r^2 + 12ry - 3xr - 2xy$

80. $5m^2 + 15mp - 2mr - 6pr$

81. $w^3 + w^2 + 9w + 9$

82. $y^3 + y^2 + 6y + 6$

83. $3a^3 + 6a^2 - 2a - 4$

84. $10x^3 + 15x^2 - 8x - 12$

85. $16m^3 - 4m^2p^2 - 4mp + p^3$

86. $10t^3 - 2t^2s^2 - 5ts + s^3$

87. $y^2 + 3x + 3y + xy$

88. $m^2 + 14p + 7m + 2mp$

89. $2z^2 + 6w - 4z - 3wz$

90. $2a^2 + 20b - 8a - 5ab$

91. $5m - 6p - 2mp + 15$

92. $7y - 9x - 3xy + 21$

93. $18r^2 - 2ty + 12ry - 3rt$

94. $12a^2 - 4bc + 16ac - 3ab$

Relating Concepts (Exercises 95–98) For Individual or Group Work

In many cases, the choice of which pairs of terms to group when factoring by grouping can be made in different ways. To see this for ***Example 7(b),*** *work* ***Exercises 95–98 in order.***

95. Start with the polynomial from **Example 7(b)**, $2xy + 12 - 3y - 8x$, and rearrange the terms as follows: $2xy - 8x - 3y + 12$. What property from **Section 1.7** allows this?

96. Group the first two terms and the last two terms of the rearranged polynomial in **Exercise 95.** Then factor each group.

97. Is your result from **Exercise 96** in factored form? Explain your answer.

98. If your answer to **Exercise 97** is *no*, factor the polynomial. Is the result the same as the one shown for **Example 7(b)**?

6.2 Factoring Trinomials

OBJECTIVES

1 Factor trinomials with a coefficient of 1 for the second-degree term.

2 Factor such trinomials after factoring out the greatest common factor.

1 (a) Complete the table to find pairs of positive integers whose product is 6. Then find the sum of each pair.

Factors of 6	Sums of Factors
6, ___	6 + ___ = ___
___, 2	___ + 2 = ___

(b) Which pair of factors from the table in part (a) has a sum of 5?

Using the FOIL method, we can find the product of the binomials $k - 3$ and $k + 1$.

$$(k - 3)(k + 1) = k^2 - 2k - 3 \quad \text{Multiplying}$$

Suppose instead that we are given the polynomial $k^2 - 2k - 3$ and want to rewrite it as the product $(k - 3)(k + 1)$.

$$k^2 - 2k - 3 = (k - 3)(k + 1) \quad \text{Factoring}$$

Recall from **Section 6.1** that this process is called *factoring* the polynomial. Factoring reverses or, "undoes," multiplying.

OBJECTIVE 1 **Factor trinomials with a coefficient of 1 for the second-degree term.** When factoring polynomials with integer coefficients, we use only integers in the factors. For example, we can factor $x^2 + 5x + 6$ by finding integers m and n such that

$$x^2 + 5x + 6 \quad \text{is written as} \quad (x + m)(x + n).$$

To find these integers m and n, we multiply the two binomials on the right.

$$(x + m)(x + n)$$
$$= x^2 + nx + mx + mn \quad \text{FOIL method}$$
$$= x^2 + (n + m)x + mn \quad \text{Distributive property}$$

Comparing this result with $x^2 + 5x + 6$ shows that we must find integers m and n having a sum of 5 and a product of 6.

Product of m and n is 6.

$$x^2 + \mathbf{5}x + \mathbf{6} = x^2 + (n + m)x + mn$$

Sum of m and n is 5.

Because many pairs of integers have a sum of **5**, it is best to begin by listing those pairs of integers whose product is **6**. Both 5 and 6 are positive, so we consider only pairs in which both integers are positive.

◀ **Work Problem 1 at the Side.**

From **Margin Problem 1,** we see that the numbers 6 and 1 and the numbers 3 and 2 both have a product of 6, but only the pair 3 and 2 has a sum of 5. So 3 and 2 are the required integers.

$$x^2 + 5x + 6 \quad \text{is factored as} \quad (x + 3)(x + 2).$$

Check by multiplying the binomials using the FOIL method. *Make sure that the sum of the outer and inner products produces the correct middle term.*

CHECK $\quad (x + 3)(x + 2) = x^2 + 5x + 6$ ✓ Correct

$$3x$$
$$2x$$
$$\overline{5x} \quad \text{Add.}$$

This method of factoring can be used only for trinomials that have 1 as the coefficient of the second-degree (squared) term.

Answers

1. (a) 1; 1; 7; 3; 3; 5 **(b)** 3, 2

> ### EXAMPLE 1　Factoring a Trinomial (All Positive Terms)
>
> Factor $m^2 + 9m + 14$.
>
> Look for two integers whose product is **14** and whose sum is **9**. List the pairs of integers whose products are 14, and examine the sums. Only positive integers are needed since all signs in $m^2 + 9m + 14$ are positive.
>
> | Factors of 14 | Sums of Factors |
> |---|---|
> | 14, 1 | $14 + 1 = 15$ |
> | 7, 2 | $7 + 2 = 9$ |
>
> Sum is 9.
>
> From the table, 7 and 2 are the required integers, since $7 \cdot 2 = 14$ and $7 + 2 = 9$.
>
> $$m^2 + 9m + 14 \quad \text{factors as} \quad (m + 7)(m + 2).$$
>
> $(m + 2)(m + 7)$ is also correct.
>
> **CHECK**　$(m + 7)(m + 2)$
>
> $\qquad = m^2 + 2m + 7m + 14$　FOIL method
>
> $\qquad = m^2 + 9m + 14$ ✓　Original polynomial

················ **Work Problem ❷ at the Side.** ▶

> ### Note
>
> Remember that because of the commutative property of multiplication, the order of the factors does not matter. ***Always check by multiplying.***

> ### EXAMPLE 2　Factoring a Trinomial (Negative Middle Term)
>
> Factor $x^2 - 9x + 20$.
>
> Find two integers whose product is **20** and whose sum is -9. Since the numbers we are looking for have a *positive product* and a *negative sum,* we consider only pairs of negative integers.
>
> | Factors of 20 | Sums of Factors |
> |---|---|
> | $-20, -1$ | $-20 + (-1) = -21$ |
> | $-10, -2$ | $-10 + (-2) = -12$ |
> | $-5, -4$ | $-5 + (-4) = -9$ |
>
> Sum is -9.
>
> The required integers are -5 and -4.
>
> $$x^2 - 9x + 20 \quad \text{factors as} \quad (x - 5)(x - 4).$$
>
> The order of the factors does not matter.
>
> **CHECK**　$(x - 5)(x - 4)$
>
> $\qquad = x^2 - 4x - 5x + 20$　FOIL method
>
> $\qquad = x^2 - 9x + 20$ ✓　Original polynomial

················ **Work Problem ❸ at the Side.** ▶

❷ Factor each trinomial.

(GS) (a) $y^2 + 12y + 20$

Find two integers whose product is ＿＿＿ and whose sum is ＿＿＿. Complete the table.

Factors of 20	Sums of Factors
20, 1	$20 + 1 = 21$
10, ＿＿	$10 + \_\_ = \_\_$
5, ＿＿	$5 + \_\_ = \_\_$

Which pair of factors has the required sum? ＿＿＿＿＿

Now factor the trinomial.

(b) $x^2 + 9x + 18$

❸ Factor each trinomial.

(GS) (a) $t^2 - 12t + 32$

Find two integers whose product is ＿＿＿ and whose sum is ＿＿＿. Complete the table.

Factors of 32	Sums of Factors
$-32, -1$	$-32 + (-1) = -33$
$-16, \_\_$	$-16 + (\_\_) = \_\_$
$-8, \_\_$	$-8 + (\_\_) = \_\_$

Which pair of factors has the required sum? ＿＿＿＿＿

Now factor the trinomial.

(b) $y^2 - 10y + 24$

Answers

2. **(a)** 20; 12; 2; 2; 12; 4; 4; 9; 10 and 2; $(y + 10)(y + 2)$
   (b) $(x + 3)(x + 6)$
3. **(a)** 32; -12; -2; -2; -18; -4; -4; -12; -8 and -4; $(t - 8)(t - 4)$
   (b) $(y - 6)(y - 4)$

④ Factor each trinomial.

(a) $z^2 + z - 30$

Factors of -30	Sums of Factors

(b) $x^2 + x - 42$

EXAMPLE 3 **Factoring a Trinomial (Negative Last (Constant) Term)**

Factor $x^2 + x - 6$.

We must find two integers whose product is -6 and whose sum is 1 (since the coefficient of x, or $1x$, is **1**). To get a *negative product,* the pairs of integers must have different signs.

> Once we find the required pair, we can stop listing factors.

Factors of -6	Sums of Factors	
$6, -1$	$6 + (-1) = 5$	
$-6, 1$	$-6 + 1 = -5$	
$3, -2$	$3 + (-2) = 1$	Sum is 1.

The required integers are 3 and -2.

$$x^2 + x - 6 \quad \text{factors as} \quad (x + 3)(x - 2).$$

CHECK $(x + 3)(x - 2)$

$\qquad = x^2 - 2x + 3x - 6$ FOIL method

$\qquad = x^2 + x - 6$ ✓   Original polynomial

◀ **Work Problem ④ at the Side.**

---

⑤ Factor each trinomial.

(a) $a^2 - 9a - 22$

Factors of -22	Sums of Factors

(b) $r^2 - 6r - 16$

EXAMPLE 4 **Factoring a Trinomial (Two Negative Terms)**

Factor $p^2 - 2p - 15$.

Find two integers whose product is -15 and whose sum is -2. Because the constant term, -15, is negative, we need pairs of integers with different signs.

Factors of -15	Sums of Factors	
$15, -1$	$15 + (-1) = 14$	
$-15, 1$	$-15 + 1 = -14$	
$5, -3$	$5 + (-3) = 2$	
$-5, 3$	$-5 + 3 = -2$	Sum is -2.

The required integers are -5 and 3.

$$p^2 - 2p - 15 \quad \text{factors as} \quad (p - 5)(p + 3).$$

CHECK Multiply $(p - 5)(p + 3)$ to obtain $p^2 - 2p - 15$. ✓

Note

In **Examples 1–4,** we listed factors in descending order (disregarding their signs) when we were looking for the required pair of integers. This helps avoid skipping the correct combination.

◀ **Work Problem ⑤ at the Side.**

Answers

4. (a) $(z + 6)(z - 5)$ **(b)** $(x + 7)(x - 6)$
5. (a) $(a - 11)(a + 2)$ **(b)** $(r - 8)(r + 2)$

Some trinomials cannot be factored using only integers. Such trinomials are **prime polynomials.**

Deciding Whether Polynomials Are Prime

Factor each trinomial.

(a) $x^2 - 5x + 12$

As in **Example 2,** both factors must be negative to give a positive product, 12, and a negative sum, -5. List pairs of negative integers whose product is 12, and examine the sums.

Factors of 12	Sums of Factors
$-12, -1$	$-12 + (-1) = -13$
$-6, -2$	$-6 + (-2) = -8$
$-4, -3$	$-4 + (-3) = -7$ No sum is -5.

None of the pairs of integers has a sum of -5. Therefore, the trinomial $x^2 - 5x + 12$ *cannot be factored using only integers.* It is a *prime polynomial.*

(b) $k^2 - 8k + 11$

There is no pair of integers whose product is 11 and whose sum is -8, so $k^2 - 8k + 11$ is a prime polynomial.

·· **Work Problem 6 at the Side.** ▶

Factoring $x^2 + bx + c$

Find two integers whose product is c and whose sum is b.

1. Both integers must be positive if b and c are positive. (See **Example 1.**)

2. Both integers must be negative if c is positive and b is negative. (See **Example 2.**)

3. One integer must be positive and one must be negative if c is negative. (See **Examples 3 and 4.**)

Factoring a Trinomial with Two Variables

Factor $z^2 - 2bz - 3b^2$.

Here, the coefficient of z in the middle term is $-2b$, so we need to find two expressions whose product is $-3b^2$ and whose sum is $-2b$.

Factors of $-3b^2$	Sums of Factors
$3b, -b$	$3b + (-b) = 2b$
$-3b, b$	$-3b + b = -2b$ Sum is $-2b$.

$z^2 - 2bz - 3b^2$ factors as $(z - 3b)(z + b)$.

CHECK $(z - 3b)(z + b)$

$= z^2 + zb - 3bz - 3b^2$ FOIL method

$= z^2 + 1bz - 3bz - 3b^2$ Identity and commutative properties

$= z^2 - 2bz - 3b^2$ ✓   Combine like terms.

·· **Work Problem 7 at the Side.** ▶

---

6 Factor each trinomial, if possible.

(a) $x^2 + 5x + 8$

(b) $r^2 - 3r - 4$

(c) $m^2 - 2m + 5$

7 Factor each trinomial.

GS **(a)** $b^2 - 3ab - 4a^2$

We need two expressions whose product is ____ and whose sum is ____. Complete the table.

Factors of $-4a^2$	Sums of Factors
$4a,$ ___	$4a + ($___$) = $ ___
$-4a,$ ___	$-4a + $ ___ $ = $ ___
$2a, -2a$	$2a + (-2a) = 0$

Which pair of factors has the required sum? _____

Now factor the trinomial.

(b) $r^2 - 6rs + 8s^2$

Answers

6. **(a)** prime   **(b)** $(r - 4)(r + 1)$
   (c) prime
7. **(a)** $-4a^2; -3a; -a; -a; 3a; a; a; -3a;$
   $-4a$ and $a; (b - 4a)(b + a)$
   (b) $(r - 4s)(r - 2s)$

8 Factor each trinomial completely.

(GS) **(a)** $2p^3 + 6p^2 - 8p$

$= \underline{\quad}(\underline{\quad} + 3p - \underline{\quad})$

$= \underline{\quad}(\underline{\quad\quad})(\underline{\quad\quad})$

(b) $3y^4 - 27y^3 + 60y^2$

(c) $-3x^4 + 15x^3 - 18x^2$

OBJECTIVE **2** **Factor such trinomials after factoring out the greatest common factor.** If the terms of a trinomial have a common factor, first factor it out. Then factor the remaining trinomial as in **Examples 1–6.**

EXAMPLE 7 **Factoring a Trinomial with a Common Factor**

Factor $4x^5 - 28x^4 + 40x^3$.

There is no second-degree term. Look for a common factor.

$$4x^5 - 28x^4 + 40x^3$$

$$= 4x^3(x^2 - 7x + 10) \quad \text{Factor out the greatest common factor, } 4x^3.$$

Now factor $x^2 - 7x + 10$. The integers -5 and -2 have a product of 10 and a sum of -7.

Include $4x^3$. $= 4x^3(x - 5)(x - 2)$ Completely factored form

CHECK $4x^3(x - 5)(x - 2)$

$$= 4x^3(x^2 - 2x - 5x + 10) \quad \text{FOIL method}$$

$$= 4x^3(x^2 - 7x + 10) \quad \text{Combine like terms.}$$

$$= 4x^5 - 28x^4 + 40x^3 \checkmark \quad \text{Distributive property}$$

◀ **Work Problem 8 at the Side.**

CAUTION

When factoring, always look for a common factor first. Remember to include the common factor as part of the answer. Check by multiplying out the completely factored form.

Answers

8. (a) $2p; p^2; 4; 2p; p + 4; p - 1$
 (b) $3y^2(y - 5)(y - 4)$
 (c) $-3x^2(x - 3)(x - 2)$

6.2 Exercises

FOR EXTRA HELP

Download the MyDashBoard App

▶ MyMathLab®

CONCEPT CHECK *Answer each question.*

1. When factoring a trinomial in x as $(x + a)(x + b)$, what must be true of a and b, if the coefficient of the constant term of the trinomial is negative?

2. In **Exercise 1,** what must be true of a and b if the coefficient of the constant term is positive?

3. Which one of the following is the correct factored form of $x^2 - 12x + 32$?
   A. $(x - 8)(x + 4)$ B. $(x + 8)(x - 4)$
   C. $(x - 8)(x - 4)$ D. $(x + 8)(x + 4)$

4. What would be the first step in factoring
   $$2x^3 + 8x^2 - 10x? \quad \text{(See Example 7.)}$$

5. What polynomial can be factored as
   $$(a + 9)(a + 4)?$$

6. What polynomial can be factored as
   $$(y - 7)(y + 3)?$$

7. What is meant by a *prime polynomial?*

8. How can we check our work when factoring a trinomial? Does the check ensure that the trinomial is *completely* factored?

In Exercises 9–12, list all pairs of integers with the given product. Then find the pair whose sum is given. ***See the tables in Examples 1–4.***

9. Product: 12;   Sum: 7

10. Product: 18;   Sum: 9

11. Product: -24;   Sum: -5

12. Product: -36;   Sum: -16

GS *Complete each factoring.* ***See Examples 1–4.***

13. $p^2 + 11p + 30$
    $= (p + 5)(\underline{})$

14. $x^2 + 10x + 21$
    $= (x + 7)(\underline{})$

15. $x^2 + 15x + 44$
    $= (x + 4)(\underline{})$

16. $r^2 + 15r + 56$
    $= (r + 7)(\underline{})$

17. $x^2 - 9x + 8$
    $= (x - 1)(\underline{})$

18. $t^2 - 14t + 24$
    $= (t - 2)(\underline{})$

19. $y^2 - 2y - 15$
    $= (y + 3)(\underline{})$

20. $t^2 - t - 42$
    $= (t + 6)(\underline{})$

21. $x^2 + 9x - 22$
    $= (x - 2)(\underline{})$

22. $x^2 + 6x - 27$
    $= (x - 3)(\underline{})$

23. $y^2 - 7y - 18$
    $= (y + 2)(\underline{})$

24. $y^2 - 2y - 24$
    $= (y + 4)(\underline{})$

Factor completely. If a polynomial cannot be factored, write prime.
See Examples 1–5.

25. $y^2 + 9y + 8$

26. $a^2 + 9a + 20$

27. $b^2 + 8b + 15$

28. $x^2 + 6x + 8$

29. $m^2 + m - 20$

30. $p^2 + 4p - 5$

31. $x^2 + 3x - 40$

32. $d^2 + 4d - 45$

33. $x^2 + 4x + 5$

34. $t^2 + 11t + 12$

35. $y^2 - 8y + 15$

36. $y^2 - 6y + 8$

37. $z^2 - 15z + 56$

38. $x^2 - 13x + 36$

39. $r^2 - r - 30$

40. $q^2 - q - 42$

41. $a^2 - 8a - 48$

42. $m^2 - 10m - 24$

Factor completely. **See Example 6.**

43. $r^2 + 3ra + 2a^2$

44. $x^2 + 5xa + 4a^2$

45. $x^2 + 4xy + 3y^2$

46. $p^2 + 9pq + 8q^2$

47. $t^2 - tz - 6z^2$

48. $a^2 - ab - 12b^2$

49. $v^2 - 11vw + 30w^2$

50. $v^2 - 11vx + 24x^2$

51. $a^2 + 2ab - 15b^2$

52. $m^2 + 4mn - 12n^2$

Factor completely. See **Example 7.**

53. $4x^2 + 12x - 40$

54. $5y^2 - 5y - 30$

55. $2t^3 + 8t^2 + 6t$

56. $3t^3 + 27t^2 + 24t$

57. $-2x^6 - 8x^5 + 42x^4$

58. $-4y^5 - 12y^4 + 40y^3$

59. $a^5 + 3a^4b - 4a^3b^2$

60. $z^{10} - 4z^9y - 21z^8y^2$

61. $5m^5 + 25m^4 - 40m^2$

62. $12k^5 - 6k^3 + 10k^2$

63. $m^3n - 10m^2n^2 + 24mn^3$

64. $y^3z + 3y^2z^2 - 54yz^3$

6.3 Factoring Trinomials by Grouping

OBJECTIVE ▶ 1 Factor trinomials by grouping when the coefficient of the second-degree term is not 1. We now extend our work to factor a trinomial such as $2x^2 + 7x + 6$, where the coefficient of x^2 is *not* 1.

OBJECTIVE

1 Factor trinomials by grouping when the coefficient of the second-degree term is not 1.

EXAMPLE 1 Factoring a Trinomial by Grouping (Coefficient of the Second-Degree Term Not 1)

Factor $2x^2 + 7x + 6$.

To factor this trinomial, we look for two positive integers whose product is $2 \cdot 6 = 12$ and whose sum is 7.

$$\overset{\text{Sum is 7.}}{\underset{\text{Product is } 2 \cdot 6 = 12.}{2x^2 + 7x + 6}}$$

The required integers are 3 and 4, since $3 \cdot 4 = 12$ and $3 + 4 = 7$. We use these integers to write the middle term $7x$ as $3x + 4x$.

$$2x^2 + 7x + 6$$
$$= 2x^2 + \underbrace{3x + 4x}_{7x} + 6$$
$$= (2x^2 + 3x) + (4x + 6) \qquad \text{Group the terms.}$$
$$= x(2x + 3) + 2(2x + 3) \qquad \text{Factor each group.}$$
$$\underset{\text{Must be the same}}{}$$
$$= (2x + 3)(x + 2) \qquad \text{Factor out } 2x + 3.$$

CHECK Multiply $(2x + 3)(x + 2)$ to obtain $2x^2 + 7x + 6$. ✓

························· **Work Problem ① at the Side.** ▶

1 (a) In **Example 1,** we factored
GS
$$2x^2 + 7x + 6$$
by writing $7x$ as $3x + 4x$. This trinomial can also be factored by writing $7x$ as $4x + 3x$.

Complete the following.
$$2x^2 + \mathbf{7x} + 6$$
$$= 2x^2 + \mathbf{4x} + \mathbf{3x} + 6$$
$$= (2x^2 + \underline{}) + (3x + \underline{})$$
$$= 2x(x + \underline{}) + 3(x + \underline{})$$
$$= (\underline{})(2x + 3)$$

(b) Is the answer in part (a) the same as in **Example 1?** (Remember that the order of the factors does not matter.)

(c) Factor $2z^2 + 5z + 3$ by grouping.

EXAMPLE 2 Factoring Trinomials by Grouping

Factor each trinomial.

(a) $6r^2 + r - 1$

We must find two integers with a product of $6(-1) = -6$ and a sum of 1 (since the coefficient of r, or $1r$, is 1). The integers are -2 and 3. We write the middle term r as $-2r + 3r$.

$$6r^2 + r - 1$$
$$= 6r^2 - 2r + 3r - 1 \qquad r = -2r + 3r$$
$$= (6r^2 - 2r) + (3r - 1) \qquad \text{Group the terms.}$$
$$= 2r(3r - 1) + 1(3r - 1) \qquad \text{The binomials must be the same.}$$
$$\boxed{\text{Remember the 1.}}$$
$$= (3r - 1)(2r + 1) \qquad \text{Factor out } 3r - 1.$$

CHECK Multiply $(3r - 1)(2r + 1)$ to obtain $6r^2 + r - 1$. ✓

····················· **Continued on Next Page**

Answers

1. (a) $4x$; 6; 2; 2; $x + 2$ **(b)** yes
(c) $(2z + 3)(z + 1)$

2 Factor each trinomial by grouping.

(a) $2m^2 + 7m + 3$

(b) $5p^2 - 2p - 3$

(c) $15k^2 - km - 2m^2$

3 Factor the trinomial completely.

GS $4x^2 - 2x - 30$

$$= \underline{\quad}(2x^2 - \underline{\quad} - 15)$$

$$= \underline{\quad}(2x^2 - \underline{\quad} + \underline{\quad} - 15)$$

$$= 2\big[(2x^2 - \underline{\quad}) + (5x - 15)\big]$$

$$= 2\big[\underline{\quad}(\underline{\quad\quad}) + 5(x - 3)\big]$$

$$= \underline{\quad\quad\quad\quad\quad}$$

4 Factor each trinomial completely.

(a) $18p^4 + 63p^3 + 27p^2$

(b) $6a^2 + 3ab - 18b^2$

Answers

2. (a) $(2m + 1)(m + 3)$
   (b) $(5p + 3)(p - 1)$
   (c) $(5k - 2m)(3k + m)$
3. $2; x; 2; 6x; 5x; 6x; 2x; x - 3;$
   $2(x - 3)(2x + 5)$
4. (a) $9p^2(2p + 1)(p + 3)$
   (b) $3(2a - 3b)(a + 2b)$

(b) $12z^2 - 5z - 2$

Look for two integers whose product is $12(-2) = -24$ and whose sum is -5. The required integers are 3 and -8.

$$12z^2 - 5z - 2$$

$$= 12z^2 + 3z - 8z - 2 \qquad {\scriptstyle -5z = 3z - 8z}$$

$$= (12z^2 + 3z) + (-8z - 2) \qquad \text{Group the terms.}$$

$$= 3z(4z + 1) - 2(4z + 1) \qquad \text{Factor each group.}$$

> Be careful with signs.

$$= (4z + 1)(3z - 2) \qquad \text{Factor out } 4z + 1.$$

CHECK Multiply $(4z + 1)(3z - 2)$ to obtain $12z^2 - 5z - 2.$ ✓

(c) $10m^2 + mn - 3n^2$

Two integers whose product is $10(-3) = -30$ and whose sum is 1 are -5 and 6. Rewrite the trinomial with four terms.

$$10m^2 + mn - 3n^2$$

$$= 10m^2 - 5mn + 6mn - 3n^2 \qquad {\scriptstyle mn = -5mn + 6mn}$$

$$= 5m(2m - n) + 3n(2m - n) \qquad \text{Group the terms. Factor each group.}$$

$$= (2m - n)(5m + 3n) \qquad \text{Factor out } 2m - n.$$

CHECK Multiply $(2m - n)(5m + 3n)$ to obtain $10m^2 + mn - 3n^2.$ ✓

◀ **Work Problem 2** at the Side.

EXAMPLE 3 **Factoring a Trinomial with a Common Factor by Grouping**

Factor $28x^5 - 58x^4 - 30x^3$.

$$28x^5 - 58x^4 - 30x^3$$

$$= 2x^3(14x^2 - 29x - 15) \qquad \substack{\text{Factor out the greatest} \\ \text{common factor, } 2x^3.}$$

To factor $14x^2 - 29x - 15$, find two integers whose product is $14(-15) = -210$ and whose sum is -29. Factoring 210 into prime factors helps find these integers.

$$210 = 2 \cdot 3 \cdot 5 \cdot 7$$

Combine the prime factors in pairs in different ways, using one positive factor and one negative factor to get -210. The factors 6 and -35 have the correct sum, -29.

$$28x^5 - 58x^4 - 30x^3$$

$$= 2x^3(14x^2 - 29x - 15) \qquad \text{From above}$$

> Remember the common factor.

$$= 2x^3(14x^2 + 6x - 35x - 15) \qquad {\scriptstyle -29x = 6x - 35x}$$

$$= 2x^3\big[(14x^2 + 6x) + (-35x - 15)\big] \qquad \text{Group the terms.}$$

$$= 2x^3\big[2x(7x + 3) - 5(7x + 3)\big] \qquad \text{Factor each group.}$$

$$= 2x^3\big[(7x + 3)(2x - 5)\big] \qquad \text{Factor out } 7x + 3.$$

$$= 2x^3(7x + 3)(2x - 5) \qquad \boxed{\text{Check by multiplying.}}$$

◀ **Work Problems 3 and 4** at the Side.

6.3 Exercises

 FOR EXTRA HELP

 Download the MyDashBoard App

MyMathLab®

1. **CONCEPT CHECK** Which pair of integers would be used to rewrite the middle term when factoring $12y^2 + 5y - 2$ by grouping?

   A. $-8, 3$　**B.** $8, -3$　**C.** $-6, 4$　**D.** $6, -4$

2. **CONCEPT CHECK** Which pair of integers would be used to rewrite the middle term when factoring $20b^2 - 13b + 2$ by grouping?

   A. $10, 3$　**B.** $-10, -3$　**C.** $8, 5$　**D.** $-8, -5$

GS *The middle term of each trinomial has been rewritten. Now factor by grouping.* **See Examples 1 and 2.**

3. $m^2 + 8m + 12$

   $= m^2 + 6m + 2m + 12$

4. $x^2 + 9x + 14$

   $= x^2 + 7x + 2x + 14$

5. $a^2 + 3a - 10$

   $= a^2 + 5a - 2a - 10$

6. $y^2 - 2y - 24$

   $= y^2 + 4y - 6y - 24$

7. $10t^2 + 9t + 2$

   $= 10t^2 + 5t + 4t + 2$

8. $6x^2 + 13x + 6$

   $= 6x^2 + 9x + 4x + 6$

9. $15z^2 - 19z + 6$

   $= 15z^2 - 10z - 9z + 6$

10. $12p^2 - 17p + 6$

    $= 12p^2 - 9p - 8p + 6$

11. $8s^2 + 2st - 3t^2$

    $= 8s^2 - 4st + 6st - 3t^2$

12. $3x^2 - xy - 14y^2$

    $= 3x^2 - 7xy + 6xy - 14y^2$

13. $15a^2 + 22ab + 8b^2$

    $= 15a^2 + 10ab + 12ab + 8b^2$

14. $25m^2 + 25mn + 6n^2$

    $= 25m^2 + 15mn + 10mn + 6n^2$

GS *Complete the steps to factor each trinomial by grouping.* **See Examples 1 and 2.**

15. $2m^2 + 11m + 12$

    (a) Find two integers whose product is

    　____ · ____ = ____ and whose sum is ____.

    (b) The required integers are ____ and ____.

    (c) Write the middle term $11m$ as ____ + ____.

    (d) Rewrite the given trinomial using four terms.

    (e) Factor the polynomial in part (d) by grouping.

    (f) Check by multiplying.

16. $6y^2 - 19y + 10$

    (a) Find two integers whose product is

    　____ · ____ = ____ and whose sum is ____.

    (b) The required integers are ____ and ____.

    (c) Write the middle term $-19y$ as ____ + _____.

    (d) Rewrite the given trinomial using four terms.

    (e) Factor the polynomial in part (d) by grouping.

    (f) Check by multiplying.

Factor each trinomial by grouping. **See Examples 1–3.**

17. $2x^2 + 7x + 3$

18. $3y^2 + 13y + 4$

19. $4r^2 + r - 3$

20. $4r^2 + 3r - 10$

21. $8m^2 - 10m - 3$

22. $20x^2 - 28x - 3$

23. $21m^2 + 13m + 2$

24. $38x^2 + 23x + 2$

25. $6b^2 + 7b + 2$

26. $6w^2 + 19w + 10$

27. $12y^2 - 13y + 3$

28. $15a^2 - 16a + 4$

29. $16 + 16x + 3x^2$

30. $18 + 65x + 7x^2$

31. $24x^2 - 42x + 9$

32. $48b^2 - 74b - 10$

33. $2m^3 + 2m^2 - 40m$

34. $3x^3 + 12x^2 - 36x$

35. $-32z^5 + 20z^4 + 12z^3$

36. $-18x^5 - 15x^4 + 75x^3$

37. $12p^2 + 7pq - 12q^2$

38. $6m^2 - 5mn - 6n^2$

39. $6a^2 - 7ab - 5b^2$

40. $25g^2 - 5gh - 2h^2$

41. CONCEPT CHECK On a quiz, a student factored $16x^2 - 24x + 5$ by grouping as follows.

$$16x^2 - 24x + 5$$
$$= 16x^2 - 4x - 20x + 5$$
$$= 4x(4x - 1) - 5(4x - 1) \quad \text{His answer}$$

He thought his answer was correct since it checked by multiplying. **What Went Wrong?** What is the correct factored form?

42. CONCEPT CHECK On the same quiz, another student factored $3k^3 - 12k^2 - 15k$ by first factoring out the common factor $3k$ to get $3k(k^2 - 4k - 5)$. Then she wrote the following.

$$k^2 - 4k - 5$$
$$= k^2 - 5k + k - 5$$
$$= k(k - 5) + 1(k - 5)$$
$$= (k - 5)(k + 1) \quad \text{Her answer}$$

What Went Wrong? What is the correct factored form?

6.4 Factoring Trinomials by Using the FOIL Method

Factor trinomials by using the FOIL method. This section shows an alternative method of factoring trinomials that uses trial and error.

OBJECTIVE

1 Factor trinomials by using the FOIL method.

EXAMPLE 1 **Factoring a Trinomial Using FOIL (Coefficient of the Second-Degree Term Not 1)**

Factor $2x^2 + 7x + 6$ (the trinomial in **Section 6.3, Example 1**).

We want to write $2x^2 + 7x + 6$ as the product of two binomials.

$$2x^2 + 7x + 6$$
$$= (\underline{})(\underline{})$$

We use the FOIL method in reverse.

The product of the two first terms of the binomials is $2x^2$. The possible factors of $2x^2$ are $2x$ and x, or $-2x$ and $-x$. Since all terms of the trinomial are positive, we consider only positive factors. Thus, we have the following.

$$2x^2 + 7x + 6$$
$$= (2x\underline{})(x\underline{})$$

The product of the two last terms of the binomials is 6. It can be factored as $1 \cdot 6$, $6 \cdot 1$, $2 \cdot 3$, or $3 \cdot 2$. We want the pair that gives the correct middle term, $7x$. Begin by trying 1 and 6 in $(2x\underline{})(x\underline{})$.

$$(2x + 1)(x + 6) \quad \text{Incorrect}$$

$\dfrac{x}{12x}$
$\overline{13x}$ Add.

Now try the pair 6 and 1 in $(2x\underline{})(x\underline{})$.

$$(2x + 6)(x + 1) \quad \text{Incorrect}$$

$\dfrac{6x}{2x}$
$\overline{8x}$ Add.

Since $2x + 6 = 2(x + 3)$, the terms of the binomial $2x + 6$ have a factor of 2, while the terms of $2x^2 + 7x + 6$ have no common factor other than 1. The product $(2x + 6)(x + 1)$ cannot be correct.

If the terms of the original polynomial have greatest common factor 1, then each of its factors will also have terms with GCF 1.

We try the pair 2 and 3 in $(2x\underline{})(x\underline{})$. Because of the common factor 2 in the terms of $2x + 2$, the product $(2x + 2)(x + 3)$ will not work. Finally, we try the pair 3 and 2 in $(2x\underline{})(x\underline{})$.

$$(2x + 3)(x + 2) = 2x^2 + 7x + 6 \quad \text{Correct}$$

$\dfrac{3x}{4x}$
$\overline{7x}$ Add.

Thus, $2x^2 + 7x + 6$ factors as $(2x + 3)(x + 2)$.

CHECK Multiply $(2x + 3)(x + 2)$ to obtain $2x^2 + 7x + 6$. ✓

Work Problem 1 at the Side. ▶

1 Factor each trinomial.

GS **(a)** $2p^2 + 9p + 9$

The possible factors of the first term $2p^2$ are $\underline{}$ and p.

The possible factors of the last term 9 are 9 and $\underline{}$, or $\underline{}$ and 3.

Try various combinations to find the pair of factors that gives the correct middle term, $\underline{}$.

Possible Pairs of Factors	Middle Term	Correct?
$(2p + 9)(p + 1)$	$11p$	*(Yes/No)*
$(2p + 1)(p + 9)$	$\underline{}$	*(Yes/No)*
$(2p + 3)(\underline{})$	$\underline{}$	*(Yes/No)*

$2p^2 + 9p + 9$ factors as $\underline{}$.

(b) $8y^2 + 22y + 5$

Answers

1. (a) $2p$; 1; 3; $9p$; No; $19p$; No; $p + 3$; $9p$; Yes; $(2p + 3)(p + 3)$
(b) $(4y + 1)(2y + 5)$

② Factor each trinomial.

(a) $6p^2 + 19p + 10$

(b) $8x^2 + 14x + 3$

③ Factor each trinomial.

(a) $4y^2 - 11y + 6$

(b) $9x^2 - 21x + 10$

EXAMPLE 2 **Factoring a Trinomial Using FOIL (All Positive Terms)**

Factor $8p^2 + 14p + 5$.

The number 8 has several possible pairs of factors, but 5 has only 1 and 5 or -1 and -5, so we begin by considering the factors of 5. We ignore the negative factors, since all coefficients in the trinomial are positive. If $8p^2 + 14p + 5$ can be factored, the factors will have this form.

$$(\underline{} + 5)(\underline{} + 1)$$

The possible pairs of factors of $8p^2$ are $8p$ and p, or $4p$ and $2p$. We try various combinations, checking to see if the middle term is $14p$.

$(8p + 5)(p + 1)$ Incorrect
$5p$
$8p$
$13p$ Add.

$(p + 5)(8p + 1)$ Incorrect
$40p$
p
$41p$ Add.

$(4p + 5)(2p + 1)$ Correct
$10p$
$4p$
$14p$ Add.

This last combination produces $14p$, the correct middle term.

$$8p^2 + 14p + 5 \quad \text{factors as} \quad (4p + 5)(2p + 1).$$

CHECK Multiply $(4p + 5)(2p + 1)$ to obtain $8p^2 + 14p + 5$. ✓

◀ **Work Problem ② at the Side.**

EXAMPLE 3 **Factoring a Trinomial Using FOIL (Negative Middle Term)**

Factor $6x^2 - 11x + 3$.

Since 3 has only 1 and 3 or -1 and -3 as factors, it is better here to begin by factoring 3. The last term of the trinomial $6x^2 - 11x + 3$ is positive and the middle term has a negative coefficient, so we consider only negative factors. We need two negative factors because the *product* of two negative factors is positive and their *sum* is negative, as required.

We try -3 and -1 as factors of 3.

$$(\underline{} - 3)(\underline{} - 1)$$

The factors of $6x^2$ may be either $6x$ and x, or $2x$ and $3x$.

$(6x - 3)(x - 1)$ Incorrect
$-3x$
$-6x$
$-9x$ Add.

$(2x - 3)(3x - 1)$ Correct
$-9x$
$-2x$
$-11x$ Add.

The factors $2x$ and $3x$ produce $-11x$, the correct middle term.

$$6x^2 - 11x + 3 \quad \text{factors as} \quad (2x - 3)(3x - 1).$$

Check by multiplying.

◀ **Work Problem ③ at the Side.**

Answers

2. (a) $(3p + 2)(2p + 5)$
    (b) $(4x + 1)(2x + 3)$
3. (a) $(4y - 3)(y - 2)$
    (b) $(3x - 5)(3x - 2)$

> **Note**
>
> Also in **Example 3,** our initial attempt to factor $6x^2 - 11x + 3$ as $(6x - 3)(x - 1)$ *cannot* be correct since the terms of $6x - 3$ have a common factor of 3, while those of the original polynomial do not.

4 Factor each trinomial, if possible.

(a) $6x^2 + 5x - 4$

$$= (3x + \text{\_\_\_})(\text{\_\_\_} - \text{\_\_\_})$$

(b) $6m^2 - 11m - 10$

EXAMPLE 4 Factoring a Trinomial Using FOIL (Negative Constant Term)

Factor $8x^2 + 6x - 9$.

The integer 8 has several possible pairs of factors, as does -9. Since the constant term is negative, one positive factor and one negative factor of -9 are needed. Since the coefficient of the middle term is relatively small, it is wise to avoid large factors, such as 8 or 9. We try $4x$ and $2x$ as factors of $8x^2$, and 3 and -3 as factors of -9, and check the middle term.

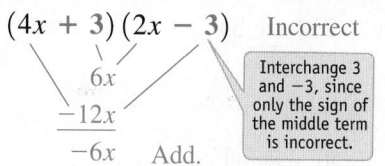

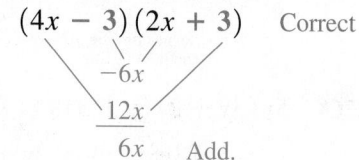

The combination on the right produces $6x$, the correct middle term.

$$8x^2 + 6x - 9 \quad \text{factors as} \quad (4x - 3)(2x + 3). \quad \boxed{\text{Check by multiplying.}}$$

(c) $4x^2 - 3x - 7$

··········· Work Problem **4** at the Side. ▶

EXAMPLE 5 Factoring a Trinomial with Two Variables

Factor $12a^2 - ab - 20b^2$.

There are several pairs of factors of $12a^2$, including

$$12a \text{ and } a, \quad 6a \text{ and } 2a, \quad \text{and} \quad 4a \text{ and } 3a.$$

There are also many possible pairs of factors of $-20b^2$, including

$$20b \text{ and } -b, \quad -20b \text{ and } b, \quad 10b \text{ and } -2b,$$
$$-10b \text{ and } 2b, \quad 4b \text{ and } -5b, \quad \text{and} \quad -4b \text{ and } 5b.$$

Once again, since the coefficient of the middle term is relatively small, avoid the larger factors. Try the factors $6a$ and $2a$, and $4b$ and $-5b$.

$$(6a + 4b)(2a - 5b) \quad \text{Incorrect}$$

This cannot be correct, since the terms of $6a + 4b$ have 2 as a common factor, while the terms of the given trinomial do not. Try $3a$ and $4a$ with $4b$ and $-5b$.

$$(3a + 4b)(4a - 5b)$$
$$= 12a^2 + ab - 20b^2 \quad \text{Incorrect}$$

Here the middle term is ab, rather than $-ab$. Interchange the signs of the last two terms in the factors.

$$(3a - 4b)(4a + 5b)$$
$$= 12a^2 - ab - 20b^2 \quad \text{Correct}$$

Thus, $12a^2 - ab - 20b^2$ factors as $(3a - 4b)(4a + 5b)$.

(d) $3y^2 + 8y - 6$

5 Factor each trinomial.

(a) $2x^2 - 5xy - 3y^2$

$$= (2x + \text{\_\_\_})(\text{\_\_\_} - \text{\_\_\_})$$

(b) $8a^2 + 2ab - 3b^2$

Answers

4. (a) 4; $2x$; 1
   (b) $(2m - 5)(3m + 2)$
   (c) $(4x - 7)(x + 1)$
   (d) prime
5. (a) y; x; $3y$
   (b) $(4a + 3b)(2a - b)$

··········· Work Problem **5** at the Side. ▶

6 Factor each trinomial.

(GS) (a) $36z^3 - 6z^2 - 72z$

$$= \underline{\quad}(\underline{\quad} - \underline{\quad} - 12)$$

$$= \underline{\quad}(\underline{\quad\quad})(\underline{\quad\quad})$$

(b) $10x^3 + 45x^2 - 90x$

(c) $-24x^3 + 32x^2 + 6x$

EXAMPLE 6   Factoring Trinomials with Common Factors

Factor each trinomial.

(a) $15y^3 + 55y^2 + 30y$

$$15y^3 + 55y^2 + 30y$$
$$= 5y(3y^2 + 11y + 6) \qquad \text{Factor out the greatest common factor, } 5y.$$

Now factor $3y^2 + 11y + 6$. Try $3y$ and y as factors of $3y^2$ and 2 and 3 as factors of 6.

$$(3y + 2)(y + 3)$$
$$= 3y^2 + 11y + 6 \qquad \text{Correct}$$

This leads to the completely factored form.

$$15y^3 + 55y^2 + 30y$$
$$= 5y(3y + 2)(y + 3)$$

> Remember the common factor.

CHECK $5y(3y + 2)(y + 3)$

$$= 5y(3y^2 + 9y + 2y + 6) \qquad \text{FOIL method}$$
$$= 5y(3y^2 + 11y + 6) \qquad \text{Combine like terms.}$$
$$= 15y^3 + 55y^2 + 30y \; \checkmark \qquad \text{Distributive property}$$

(b) $-24a^3 - 42a^2 + 45a$

The common factor could be $3a$ or $-3a$. If we factor out $-3a$, the first term of the trinomial will be positive, which makes it easier to factor the remaining trinomial.

$$-24a^3 - 42a^2 + 45a$$
$$= -3a(8a^2 + 14a - 15) \qquad \text{Factor out } -3a.$$
$$= -3a(4a - 3)(2a + 5) \qquad \text{Use trial and error.}$$

CHECK $-3a(4a - 3)(2a + 5)$

$$= -3a(8a^2 + 14a - 15) \qquad \text{FOIL method; Combine like terms.}$$
$$= -24a^3 - 42a^2 + 45a \; \checkmark \qquad \text{Distributive property}$$

CAUTION

Remember to include the common factor in the final factored form.

◀ **Work Problem 6 at the Side.**

Answers

6. (a) $6z; 6z^2; z; 6z; 3z + 4; 2z - 3$
  (b) $5x(2x - 3)(x + 6)$
  (c) $-2x(6x + 1)(2x - 3)$

6.4 Exercises

FOR EXTRA HELP

 Download the MyDashBoard App

 MyMathLab®

CONCEPT CHECK *Decide which is the correct factored form of the given polynomial.*

1. $2x^2 - x - 1$

   A. $(2x - 1)(x + 1)$ **B.** $(2x + 1)(x - 1)$

2. $3a^2 - 5a - 2$

   A. $(3a + 1)(a - 2)$ **B.** $(3a - 1)(a + 2)$

3. $4y^2 + 17y - 15$

   A. $(y + 5)(4y - 3)$ **B.** $(2y - 5)(2y + 3)$

4. $12c^2 - 7c - 12$

   A. $(6c - 2)(2c + 6)$ **B.** $(4c + 3)(3c - 4)$

5. $4k^2 + 13mk + 3m^2$

   A. $(4k + m)(k + 3m)$ **B.** $(4k + 3m)(k + m)$

6. $2x^2 + 11x + 12$

   A. $(2x + 3)(x + 4)$ **B.** $(2x + 4)(x + 3)$

GS *Complete each factoring.* **See Examples 1–6.**

7. $6a^2 + 7ab - 20b^2$

   $= (3a - 4b)(\underline{\hspace{1cm}})$

8. $9m^2 - 3mn - 2n^2$

   $= (3m + n)(\underline{\hspace{1cm}})$

9. $2x^2 + 6x - 8$

   $= 2(\underline{\hspace{2cm}})$

   $= 2(\underline{\hspace{1cm}})(\underline{\hspace{1cm}})$

10. $3x^2 - 9x - 30$

   $= 3(\underline{\hspace{2cm}})$

   $= 3(\underline{\hspace{1cm}})(\underline{\hspace{1cm}})$

11. $4z^3 - 10z^2 - 6z$

   $= 2z(\underline{\hspace{2cm}})$

   $= 2z(\underline{\hspace{1cm}})(\underline{\hspace{1cm}})$

12. $15r^3 - 39r^2 - 18r$

   $= 3r(\underline{\hspace{2cm}})$

   $= 3r(\underline{\hspace{1cm}})(\underline{\hspace{1cm}})$

13. CONCEPT CHECK A student factoring the trinomial

$$12x^2 + 7x - 12$$

wrote $(4x + 4)$ as one binomial factor. **What Went Wrong?** Factor the trinomial correctly.

14. CONCEPT CHECK Another student completely factored the trinomial

$$4x^2 + 10x - 6$$

as $(4x - 2)(x + 3)$. **What Went Wrong?** Factor the trinomial completely.

Factor each trinomial completely. **See Examples 1–6.**

15. $3a^2 + 10a + 7$

16. $7r^2 + 8r + 1$

17. $2y^2 + 7y + 6$

18. $5z^2 + 12z + 4$

19. $15m^2 + m - 2$

20. $6x^2 + x - 1$

21. $12s^2 + 11s - 5$

22. $20x^2 + 11x - 3$

23. $10m^2 - 23m + 12$

24. $6x^2 - 17x + 12$

25. $8w^2 - 14w + 3$

26. $9p^2 - 18p + 8$

27. $20y^2 - 39y - 11$

28. $10x^2 - 11x - 6$

29. $3x^2 - 15x + 16$

30. $2t^2 + 13t - 18$

31. $20x^2 + 22x + 6$

32. $36y^2 + 81y + 45$

33. $-40m^2q - mq + 6q$

34. $-15a^2b - 22ab - 8b$

35. $15n^4 - 39n^3 + 18n^2$

36. $24a^4 + 10a^3 - 4a^2$

37. $-15x^2y^2 + 7xy^2 + 4y^2$

38. $-14a^2b^3 - 15ab^3 + 9b^3$

39. $5a^2 - 7ab - 6b^2$

40. $6x^2 - 5xy - y^2$

41. $12s^2 + 11st - 5t^2$

42. $25a^2 + 25ab + 6b^2$

43. $6m^6n + 7m^5n^2 + 2m^4n^3$

44. $12k^3q^4 - 4k^2q^5 - kq^6$

If a trinomial has a negative coefficient for the second-degree term, such as $-2x^2 + 11x - 12$, it may be easier to factor by first factoring out -1.

$$-2x^2 + 11x - 12$$

$$= -1(2x^2 - 11x + 12) \quad \text{Factor out } -1.$$

$$= -1(2x - 3)(x - 4) \quad \text{Factor the trinomial.}$$

Use this method to factor each trinomial in Exercises 45–50.

45. $-x^2 - 4x + 21$

46. $-x^2 + x + 72$

47. $-3x^2 - x + 4$

48. $-5x^2 + 2x + 16$

49. $-2a^2 - 5ab - 2b^2$

50. $-3p^2 + 13pq - 4q^2$

Relating Concepts (Exercises 51–56) For Individual or Group Work

Often there are several different equivalent forms of an answer that are all correct.
Work Exercises 51–56 in order, *to see this for factoring problems.*

51. Factor the integer 35 as the product of two prime numbers.

52. Factor the integer 35 as the product of the negatives of two prime numbers.

53. Verify that $6x^2 - 11x + 4$ factors as $(3x - 4)(2x - 1)$.

54. Verify that $6x^2 - 11x + 4$ factors as $(4 - 3x)(1 - 2x)$.

55. Compare the two valid factored forms in **Exercises 53 and 54.** How do the factors in each case compare?

56. Suppose you know that the correct factored form of a particular trinomial is $(7t - 3)(2t - 5)$. From **Exercises 51–55,** what is another valid factored form?

6.5 Special Factoring Techniques

By reversing the rules for multiplication of binomials from **Section 5.4,** we obtain rules for factoring polynomials in certain forms.

OBJECTIVES

1 Factor a difference of squares.

2 Factor a perfect square trinomial.

3 Factor a difference of cubes.

4 Factor a sum of cubes.

OBJECTIVE 1 **Factor a difference of squares.** The formula for the product of the sum and difference of the same two terms is

$$(a + b)(a - b) = a^2 - b^2.$$

Reversing this rule leads to the following special factoring rule.

> **Factoring a Difference of Squares**
>
> $$a^2 - b^2 = (a + b)(a - b)$$

For example, $m^2 - 16$

$$= m^2 - 4^{2^\cdot}$$

$$= (m + 4)(m - 4).$$

The following must be true for a binomial to be a difference of squares.

1. Both terms of the binomial must be squares, such as

$$x^2, \quad 9y^2 = (3y)^2, \quad 25 = 5^2, \quad 1 = 1^2, \quad m^4 = (m^2)^2.$$

2. The terms of the binomial must have different signs (one positive and one negative).

EXAMPLE 1   **Factoring Differences of Squares**

Factor each binomial, if possible.

$$a^2 - b^2 = (a + b) (a - b)$$

(a) $x^2 - 49 = x^2 - 7^2 = (x + 7)(x - 7)$

(b) $y^2 - m^2 = (y + m)(y - m)$

(c) $x^2 - 8$

Because 8 is not the square of an integer, this binomial is not a difference of squares. It is a prime polynomial.

(d) $p^2 + 16$

Since $p^2 + 16$ is a *sum* of squares, it is not equal to $(p + 4)(p - 4)$. Also, we use the FOIL method and try the following.

$(p - 4)(p - 4)$	$(p + 4)(p + 4)$
$= p^2 - 8p + 16, \quad$ not $\quad p^2 + 16.$	$= p^2 + 8p + 16, \quad$ not $\quad p^2 + 16.$

Thus, $p^2 + 16$ is a prime polynomial.

·············· **Work Problem** 1 **at the Side.** ▶

> **CAUTION**
>
> As Example 1(d) *suggests, after any common factor is removed, a sum of squares cannot be factored.*

1 Factor each binomial, if possible.

GS **(a)** $x^2 - 81$

$$= (x + \underline{\quad})(x - \underline{\quad})$$

(b) $p^2 - 100$

(c) $t^2 - s^2$

(d) $y^2 - 10$

(e) $x^2 + 36$

(f) $x^2 + y^2$

Answers

1. **(a)** 9; 9
   (b) $(p + 10)(p - 10)$
   (c) $(t + s)(t - s)$
   (d) prime   **(e)** prime   **(f)** prime

2 Factor each difference of squares.

GS **(a)** $9m^2 - 49$

$$= (\underline{\hspace{1cm}})^2 - \underline{\hspace{1cm}}^2$$

$$= (\underline{\hspace{1cm}})(\underline{\hspace{1cm}})$$

(b) $64a^2 - 25$

(c) $25a^2 - 64b^2$

3 Factor completely.

GS **(a)** $50r^2 - 32$

$$= \underline{\hspace{0.7cm}}(\underline{\hspace{1cm}})$$

$$= \underline{\hspace{0.7cm}}[(\underline{\hspace{0.7cm}})^2 - \underline{\hspace{0.7cm}}^2]$$

$$= \underline{\hspace{0.7cm}}(\underline{\hspace{1cm}})(\underline{\hspace{1cm}})$$

(b) $27y^2 - 75$

(c) $k^4 - 49$

(d) $81r^4 - 16$

Answers

2. **(a)** $3m$; 7; $3m + 7$; $3m - 7$
   (b) $(8a + 5)(8a - 5)$
   (c) $(5a + 8b)(5a - 8b)$
3. **(a)** 2; $25r^2 - 16$; 2; $5r$; 4; 2; $5r + 4$; $5r - 4$
   (b) $3(3y + 5)(3y - 5)$
   (c) $(k^2 + 7)(k^2 - 7)$
   (d) $(9r^2 + 4)(3r + 2)(3r - 2)$

EXAMPLE 2 Factoring Differences of Squares

Factor each difference of squares.

$$a^2 - b^2 = (a + b)(a - b)$$

(a) $25m^2 - 16 = (5m)^2 - 4^2 = (5m + 4)(5m - 4)$

(b) $49z^2 - 64$

$\quad = (7z)^2 - 8^2$ Write each term as a square.

$\quad = (7z + 8)(7z - 8)$ Factor the difference of squares.

(c) $9x^2 - 4z^2$

$\quad = (3x)^2 - (2z)^2$ Write each term as a square.

$\quad = (3x + 2z)(3x - 2z)$ Factor the difference of squares.

◀ **Work Problem 2 at the Side.**

Note

Always check a factored form by multiplying.

EXAMPLE 3 Factoring More Complex Differences of Squares

Factor completely.

(a) $81y^2 - 36$

$\quad = 9(9y^2 - 4)$ Factor out the GCF, 9.

$\quad = 9[(3y)^2 - 2^2]$ Write each term as a square.

$\quad = 9(3y + 2)(3y - 2)$ Factor the difference of squares.

(b) $\qquad p^4 - 36$

Neither binomial can be factored further.

$\quad = (p^2)^2 - 6^2$ Write each term as a square.

$\quad = (p^2 + 6)(p^2 - 6)$ Factor the difference of squares.

(c) $\qquad m^4 - 16$

$\quad = (m^2)^2 - 4^2$ Write each term as a square.

$\quad = (m^2 + 4)(m^2 - 4)$ Factor the difference of squares.

Don't stop here.

$\quad = (m^2 + 4)(m + 2)(m - 2)$ Factor the difference of squares again.

CAUTION

Factor again when any of the factors is a difference of squares, as in **Example 3(c).** Check by multiplying.

◀ **Work Problem 3 at the Side.**

OBJECTIVE ▶ ② **Factor a perfect square trinomial.** The expressions 144, $4x^2$, and $81m^6$ are **perfect squares** because

$$144 = 12^2, \quad 4x^2 = (2x)^2, \quad \text{and} \quad 81m^6 = (9m^3)^2.$$

A **perfect square trinomial** is a trinomial that is the square of a binomial. For example, $x^2 + 8x + 16$ is a perfect square trinomial because it is the square of the binomial $x + 4$.

$$x^2 + 8x + 16$$
$$= (x + 4)(x + 4)$$
$$= (x + 4)^2$$

On the one hand, a necessary condition for a trinomial to be a perfect square is that *two of its terms must be perfect squares*. For this reason,

$$16x^2 + 4x + 15$$

is *not* a perfect square trinomial because only the term $16x^2$ is a perfect square.

On the other hand, even if two of the terms are perfect squares, the trinomial may *not* be a perfect square trinomial. For example,

$$x^2 + 6x + 36$$

has two perfect square terms, but it is *not* a perfect square trinomial.

We can multiply to see that the square of a binomial gives one of the following perfect square trinomials.

Factoring Perfect Square Trinomials

$$a^2 + 2ab + b^2 = (a + b)^2$$
$$a^2 - 2ab + b^2 = (a - b)^2$$

The middle term of a perfect square trinomial is always twice the product of the two terms in the squared binomial. (See Section 5.4.) Use this to check any attempt to factor a trinomial that appears to be a perfect square.

EXAMPLE 4 Factoring a Perfect Square Trinomial

Factor $x^2 + 10x + 25$.

The x^2-term is a perfect square, and so is 25. Try to factor the trinomial

$$x^2 + 10x + 25 \quad \text{as} \quad (x + 5)^2.$$

To check, take twice the product of the two terms in the squared binomial.

$$\longrightarrow 2 \cdot x \cdot 5 = 10x \longleftarrow \text{Middle term of } x^2 + 10x + 25$$

Twice First term —⤴   ⤴— Last term
of binomial   of binomial

Since $10x$ is the middle term of the given trinomial, the trinomial is a perfect square.

$$x^2 + 10x + 25 \quad \text{factors as} \quad (x + 5)^2.$$

···················· **Work Problem ④ at the Side.** ▶

---

④ Factor each trinomial.

GS **(a)** $p^2 + 14p + 49$

The term p^2 is a perfect square. Since $49 = \underline{}^2$, 49 (*is/is not*) a perfect square. Try to factor the trinomial

$$p^2 + 14p + 49$$

as $\quad (p + \underline{})^2.$

Check by taking twice the $\underline{}$ of the two terms of the squared binomial.

$$2 \cdot p \cdot \underline{} = \underline{}$$

The result (*is/is not*) the middle term of the given trinomial. The trinomial is a perfect square and factors as $\underline{}$.

(b) $m^2 + 8m + 16$

(c) $x^2 + 2x + 1$

Answers

4. **(a)** 7; is; 7; product; 7; $14p$; is; $(p + 7)^2$
    (b) $(m + 4)^2$ **(c)** $(x + 1)^2$

5 Factor each trinomial.

(a) $p^2 - 18p + 81$

(b) $16a^2 + 56a + 49$

(c) $121p^2 + 110p + 100$

(d) $64x^2 - 48x + 9$

(e) $27y^3 + 72y^2 + 48y$

EXAMPLE 5 **Factoring Perfect Square Trinomials**

Factor each trinomial.

(a) $x^2 - 22x + 121$

The first and last terms are perfect squares ($121 = 11^2$ or $(-11)^2$). Check to see whether the middle term of $x^2 - 22x + 121$ is twice the product of the first and last terms of the binomial $x - 11$.

$$2 \cdot x \cdot (-11) = -22x \leftarrow \text{Middle term of } x^2 - 22x + 121$$

Twice ⎯↑ ↑   ↳ Last
         First   term
         term

Thus, $x^2 - 22x + 121$ is a perfect square trinomial.

$$x^2 - 22x + 121 \quad \text{factors as} \quad (x - 11)^2.$$

Same sign

Notice that the sign of the second term in the squared binomial is the same as the sign of the middle term in the trinomial.

(b) $9m^2 - 24m + 16 = (3m)^2 + 2(3m)(-4) + (-4)^2 = (3m - 4)^2$

Twice ⎯↑ ↑   ↳ Last
         First   Term
         term

(c) $25y^2 + 20y + 16$

The first and last terms are perfect squares.

$$25y^2 = (5y)^2 \quad \text{and} \quad 16 = 4^2$$

Twice the product of the first and last terms of the binomial $5y + 4$ is

$$2 \cdot 5y \cdot 4 = 40y,$$

which is *not* the middle term of

$$25y^2 + 20y + 16.$$

This trinomial is not a perfect square. In fact, the trinomial cannot be factored even with the methods of the previous sections. It is a prime polynomial.

(d) $12z^3 + 60z^2 + 75z$

$$= 3z(4z^2 + 20z + 25) \qquad \text{Factor out the common factor, } 3z.$$

$$= 3z[(2z)^2 + 2(2z)(5) + 5^2] \qquad \begin{array}{l}4z^2 + 20z + 25 \text{ is a perfect square} \\ \text{trinomial.}\end{array}$$

$$= 3z(2z + 5)^2 \qquad \text{Factor.}$$

Note

1. The sign of the second term in the squared binomial is always the same as the sign of the middle term in the trinomial.

2. The first and last terms of a perfect square trinomial must be *positive*, because they are squares. For example, the polynomial $x^2 - 2x - 1$ cannot be a perfect square because the last term is negative.

3. Perfect square trinomials can also be factored using grouping or the FOIL method. Using the method of this section is often easier.

Answers

5. **(a)** $(p - 9)^2$ **(b)** $(4a + 7)^2$ **(c)** prime
   (d) $(8x - 3)^2$ **(e)** $3y(3y + 4)^2$

◄ **Work Problem 5** at the Side.

OBJECTIVE ▶ **3** **Factor a difference of cubes.** We factored a difference of squares at the beginning of this section. We can also factor a **difference of cubes.**

Factoring a Difference of Cubes

$$a^3 - b^3 = (a - b)(a^2 + ab + b^2)$$

This pattern should be memorized. To see that the pattern is correct, multiply

$$(a - b)(a^2 + ab + b^2),$$

as shown in the margin.

Notice the pattern of the terms in the factored form of $a^3 - b^3$.

- $a^3 - b^3$ factors as (a binomial factor) · (a trinomial factor).

- The binomial factor has the difference of the cube roots of the given terms.

- The terms in the trinomial factor are all positive.

- The terms in the binomial factor determine the trinomial factor.

$$
\begin{array}{c}
\text{First term} \quad\quad \text{Positive} \quad\quad \text{Second term} \\
\text{squared} \quad + \quad \text{product of} \quad + \quad \text{squared} \\
\text{the terms}
\end{array}
$$

$$a^3 - b^3 = (a - b)(\quad a^2 \quad + \quad ab \quad + \quad b^2 \quad)$$

EXAMPLE 6 | **Factoring Differences of Cubes**

Factor each difference of cubes.

(a) $m^3 - 125$ Use the pattern for a difference of cubes.

$$
\begin{array}{ccccccc}
a^3 & - & b^3 & = & (a & - & b) & (a^2 & + & ab & + & b^2) \\
\downarrow & & \downarrow & & & \downarrow & & \downarrow & & \downarrow & & \downarrow & \downarrow
\end{array}
$$

$$m^3 - 125 = m^3 - 5^3 = (m - 5)(m^2 + 5m + 5^2)$$
$$= (m - 5)(m^2 + 5m + 25)$$

(b) $8p^3 - 27$

$$= (2p)^3 - 3^3 \qquad\qquad 8p^3 = (2p)^3 \text{ and } 27 = 3^3.$$

$$= (2p - 3)[(2p)^2 + (2p)3 + 3^2] \quad \text{Let } a = 2p \text{ and } b = 3.$$

$$= (2p - 3)(4p^2 + 6p + 9) \qquad \text{Apply the exponents. Multiply.}$$

$$\boxed{(2p)^2 = 2^2p^2 = 4p^2}$$

(c) $4m^3 - 32n^3$

$$= 4(m^3 - 8n^3) \qquad\qquad\qquad \text{Factor out the common factor.}$$

$$= 4[m^3 - (2n)^3] \qquad\qquad\quad 8n^3 = (2n)^3$$

$$= 4(m - 2n)[m^2 + m(2n) + (2n)^2] \quad \text{Let } a = m \text{ and } b = 2n.$$

$$= 4(m - 2n)(m^2 + 2mn + 4n^2) \qquad \text{Apply the exponents. Multiply.}$$

Work Problem **6** **at the Side.** ▶

$$
\begin{array}{r}
a^2 + ab + b^2 \\
a - b \\
\hline
-a^2b - ab^2 - b^3 \\
a^3 + a^2b + ab^2 \\
\hline
a^3 \qquad\qquad - b^3
\end{array}
$$

Multiply vertically. (Section 5.3)

Add.

6 Factor each difference of cubes.

GS **(a)** $t^3 - 64$

$$= \underline{\quad}^3 - \underline{\quad}^3$$

$$= (\underline{\quad} - \underline{\quad}) \cdot$$

$$\left(\underline{\quad}^2 + \underline{\quad} + \underline{\quad}^2\right)$$

$$= \underline{\qquad\qquad}$$

(b) $2x^3 - 54$

(c) $8k^3 - y^3$

Answers

6. (a) $t; 4; t; 4; t; 4t; 4;$
  $(t - 4)(t^2 + 4t + 16)$
(b) $2(x - 3)(x^2 + 3x + 9)$
(c) $(2k - y)(4k^2 + 2ky + y^2)$

7 Factor each sum of cubes.

GS **(a)** $x^3 + 8$

$$= \underline{\quad}^3 + \underline{\quad}^3$$

$$= (\underline{\quad} + \underline{\quad}) \cdot$$
$$(\underline{\quad}^2 - \underline{\quad} + \underline{\quad}^2)$$

$$= \underline{\qquad\qquad}$$

(b) $64y^3 + 1$

CAUTION

A common error in factoring $a^3 - b^3 = (a - b)(a^2 + ab + b^2)$ is to try to factor $a^2 + ab + b^2$. This is usually not possible.

OBJECTIVE ▶ **4** **Factor a sum of cubes.** A sum of squares, such as $m^2 + 25$, cannot be factored using real numbers, but a **sum of cubes** can.

Factoring a Sum of Cubes

$$a^3 + b^3 = (a + b)(a^2 - ab + b^2)$$

Observe the positive and negative signs in these factoring patterns.

Positive

$$a^3 - b^3 = (a - b)(a^2 + ab + b^2) \qquad \text{Difference of cubes}$$

Same sign   Opposite sign

The only difference between the patterns is the positive and negative signs.

Positive

$$a^3 + b^3 = (a + b)(a^2 - ab + b^2) \qquad \text{Sum of cubes}$$

Same sign   Opposite sign

(c) $27m^3 + 343n^3$

EXAMPLE 7 **Factoring Sums of Cubes**

Factor each sum of cubes.

(a) $k^3 + 27$

$$= k^3 + 3^3 \qquad\qquad 27 = 3^3$$

$$= (k + 3)(k^2 - 3k + 3^2) \quad \text{Let } a = k \text{ and } b = 3.$$

$$= (k + 3)(k^2 - 3k + 9) \quad \text{Apply the exponent.}$$

(b) $8m^3 + 125p^3$

$$= (2m)^3 + (5p)^3 \quad 8m^3 = (2m)^3; 125p^3 = (5p)^3$$

$$= (2m + 5p)[(2m)^2 - (2m)(5p) + (5p)^2] \quad \text{Let } a = 2m \text{ and } b = 5p.$$

$$= (2m + 5p)(4m^2 - 10mp + 25p^2) \quad \begin{array}{l}\text{Apply the exponents.}\\ \text{Multiply.}\end{array}$$

◀ Work Problem **7** at the Side.

Answers

7. (a) $x; 2; x; 2; x; 2x; 2;$
    $(x + 2)(x^2 - 2x + 4)$
(b) $(4y + 1)(16y^2 - 4y + 1)$
(c) $(3m + 7n)(9m^2 - 21mn + 49n^2)$

6.5 Exercises

Download the MyDashBoard App

MyMathLab®

CONCEPT CHECK   *Work each problem.*

1. To help factor a difference of squares, complete the following list of squares.

$1^2 =$ _____ $2^2 =$ _____ $3^2 =$ _____ $4^2 =$ _____ $5^2 =$ _____

$6^2 =$ _____ $7^2 =$ _____ $8^2 =$ _____ $9^2 =$ _____ $10^2 =$ _____

$11^2 =$ _____ $12^2 =$ _____ $13^2 =$ _____ $14^2 =$ _____ $15^2 =$ _____

$16^2 =$ _____ $17^2 =$ _____ $18^2 =$ _____ $19^2 =$ _____ $20^2 =$ _____

2. To use the factoring techniques described in this section, it is helpful to recognize fourth powers of integers. Complete the following list of fourth powers.

$1^4 =$ _____ $2^4 =$ _____ $3^4 =$ _____ $4^4 =$ _____ $5^4 =$ _____

3. Which of the following are differences of squares?

A. $x^2 - 4$ **B.** $y^2 + 9$ **C.** $2a^2 - 25$ **D.** $9m^2 - 1$

4. On a quiz, a student indicated *prime* when asked to factor $4x^2 + 16$, since she said that a sum of squares cannot be factored. **What Went Wrong?**

Factor each binomial completely. **See Examples 1–3.**

5. $y^2 - 25$

6. $t^2 - 16$

7. $x^2 - 144$

8. $x^2 - 400$

9. $m^2 - 12$

10. $k^2 - 18$

11. $m^2 + 64$

12. $k^2 + 49$

13. $9r^2 - 4$

14. $4x^2 - 9$

15. $36x^2 - 16$

16. $32a^2 - 8$

17. $196p^2 - 225$

18. $361q^2 - 400$

19. $16r^2 - 25a^2$

20. $49m^2 - 100p^2$

21. $100x^2 + 49$

22. $81w^2 + 16$

23. $p^4 - 49$

24. $r^4 - 25$

25. $x^4 - 1$

26. $y^4 - 10,000$

27. $p^4 - 256$

28. $x^4 - 625$

29. CONCEPT CHECK Which of the following are perfect square trinomials?

    A. $y^2 - 13y + 36$ **B.** $x^2 + 6x + 9$ **C.** $4z^2 - 4z + 1$ **D.** $16m^2 + 10m + 1$

30. In the polynomial $9y^2 + 14y + 25$, the first and last terms are perfect squares. Can the polynomial be factored? If it can, factor it. If it cannot, explain why it is not a perfect square trinomial.

Factor each trinomial completely. It may be necessary to factor out the greatest common factor first. ***See Examples 4 and 5.***

31. $w^2 + 2w + 1$ **32.** $p^2 + 4p + 4$ **33.** $x^2 - 8x + 16$

34. $x^2 - 10x + 25$ **35.** $x^2 - 10x + 100$ **36.** $x^2 - 18x + 36$

37. $2x^2 + 24x + 72$ **38.** $3y^2 + 48y + 192$ **39.** $4x^2 + 12x + 9$

40. $25x^2 + 10x + 1$ **41.** $16x^2 - 40x + 25$ **42.** $36y^2 - 60y + 25$

43. $49x^2 - 28xy + 4y^2$ **44.** $4z^2 - 12zw + 9w^2$ **45.** $64x^2 + 48xy + 9y^2$

46. $9t^2 + 24tr + 16r^2$ **47.** $-50h^3 + 40h^2y - 8hy^2$ **48.** $-18x^3 - 48x^2y - 32xy^2$

Although we usually factor polynomials using integers, we can apply the same concepts to factoring using fractions and decimals.

$$z^2 - \frac{9}{16}$$

$$= z^2 - \left(\frac{3}{4}\right)^2 \qquad \frac{9}{16} = \left(\frac{3}{4}\right)^2$$

$$= \left(z + \frac{3}{4}\right)\left(z - \frac{3}{4}\right) \qquad \text{Factor the difference of squares.}$$

Factor each binomial or trinomial.

49. $p^2 - \dfrac{1}{9}$ **50.** $q^2 - \dfrac{1}{4}$ **51.** $4m^2 - \dfrac{9}{25}$ **52.** $100b^2 - \dfrac{4}{49}$

53. $x^2 - 0.64$ **54.** $y^2 - 0.36$ **55.** $t^2 + t + \dfrac{1}{4}$

56. $m^2 + \dfrac{2}{3}m + \dfrac{1}{9}$ **57.** $x^2 - 1.0x + 0.25$ **58.** $y^2 - 1.4y + 0.49$

CONCEPT CHECK *Work each problem.*

59. To help factor the sum or difference of cubes, complete the following list of cubes.

$1^3 =$ ____ $2^3 =$ ____ $3^3 =$ ____ $4^3 =$ ____ $5^3 =$ ____

$6^3 =$ ____ $7^3 =$ ____ $8^3 =$ ____ $9^3 =$ ____ $10^3 =$ ____

60. The following powers of x are all perfect cubes: $x^3, x^6, x^9, x^{12}, x^{15}$. Based on this observation, we may make a conjecture that if the power of a variable is divisible by ____ (with 0 remainder), then we have a perfect cube.

61. Which of the following are differences of cubes?

A. $9x^3 - 125$ **B.** $x^3 - 16$ **C.** $x^3 - 1$ **D.** $8x^3 - 27y^3$

62. Which of the following are sums of cubes?

A. $x^3 + 1$ **B.** $x^3 + 36$ **C.** $12x^3 + 27$ **D.** $64x^3 + 216y^3$

*Factor. Use your answers in **Exercises 59 and 60** as necessary. **See Examples 6 and 7.***

63. $x^3 + 1$ **64.** $m^3 + 8$ **65.** $x^3 - 1$

66. $m^3 - 8$ **67.** $p^3 + q^3$ **68.** $w^3 + z^3$

69. $y^3 - 216$ **70.** $x^3 - 343$ **71.** $k^3 + 1000$

72. $p^3 + 512$ **73.** $27x^3 - 1$ **74.** $64y^3 - 27$

75. $125x^3 + 8$ **76.** $216x^3 + 125$ **77.** $y^3 - 8x^3$

78. $w^3 - 216z^3$ **79.** $27x^3 - 64y^3$ **80.** $125m^3 - 8n^3$

81. $8p^3 + 729q^3$ **82.** $27x^3 + 1000y^3$

83. $16t^3 - 2$ **84.** $3p^3 - 81$

85. $40w^3 + 135$

86. $32z^3 + 500$

87. $x^3 + y^6$

88. $p^9 + q^3$

89. $125k^3 - 8m^9$

90. $125c^6 - 216d^3$

Relating Concepts (Exercises 91–98) For Individual or Group Work

A binomial may be both a difference of squares and a difference of cubes. One example of such a binomial is $x^6 - 1$. Using the techniques of this section, one factoring method will give the completely factored form, while the other will not. **Work Exercises 91–98 in order,** *to determine the method to use.*

91. Factor $x^6 - 1$ as the difference of two squares.

92. The factored form obtained in **Exercise 91** consists of a difference of cubes multiplied by a sum of cubes. Factor each binomial further.

93. Now start over and factor $x^6 - 1$ as a difference of cubes.

94. The factored form obtained in **Exercise 93** consists of a binomial that is a difference of squares and a trinomial. Factor the binomial further.

95. Compare the results in **Exercises 92 and 94.** Which one of these is the completely factored form?

96. Verify that the trinomial in the factored form in **Exercise 94** is the product of the two trinomials in the factored form in **Exercise 92.**

97. Use the results of **Exercises 91–96** to complete the following statement.

In general, if I must choose between factoring first using the method for a difference of squares or the method for a difference of cubes, I should choose the _____ method to eventually obtain the completely factored form.

98. Find the *completely* factored form of $x^6 - 729$ using the knowledge gained in **Exercises 91–97.**

6.6 A General Approach to Factoring

A polynomial is *completely factored* when it is written as a product of prime polynomials with integer coefficients.

OBJECTIVES

1. Factor out any common factor.
2. Factor binomials.
3. Factor trinomials.
4. Factor polynomials of more than three terms.

Factoring a Polynomial

Step 1   **Factor out any common factor.**

Step 2   **If the polynomial is a binomial,** check to see if it is the difference of squares, the difference of cubes, or the sum of cubes.

        If the polynomial is a trinomial, check to see if it is a perfect square trinomial. If it is not, factor as in **Sections 6.2–6.4.**

        If the polynomial has more than three terms, try to factor by grouping as in **Section 6.1.**

Step 3   **If any of the factors can be factored further, do so.**

Step 4   **Check the factored form by multiplying.**

OBJECTIVE ▸ 1 **Factor out any common factor.** *This step is always the same, regardless of the number of terms in the polynomial.*

EXAMPLE 1   **Factoring Out a Common Factor**

Factor each polynomial.

(a) $8m^2p^2 + 4mp$

    $= 4mp(2mp + 1)$
       GCF $= 4mp$

(b) $5x(a + b) - y(a + b)$

    $= (a + b)(5x - y)$
       Factor out $a + b$.

Work Problem ❶ **at the Side.** ▶

OBJECTIVE ▸ 2 **Factor binomials.** Check for the following patterns.

Factoring a Binomial (Two Terms)

Difference of squares $x^2 - y^2 = (x + y)(x - y)$

Difference of cubes $x^3 - y^3 = (x - y)(x^2 + xy + y^2)$

Sum of cubes $x^3 + y^3 = (x + y)(x^2 - xy + y^2)$

EXAMPLE 2   **Factoring Binomials**

Factor each binomial if possible.

(a) $64m^2 - 9n^2$

    $= (8m)^2 - (3n)^2$
       Difference of squares
    $= (8m + 3n)(8m - 3n)$

(b) $8p^3 - 27$

    $= (2p)^3 - 3^3$ Difference of cubes
    $= (2p - 3)[(2p)^2 + (2p)(3) + 3^2]$
    $= (2p - 3)(4p^2 + 6p + 9)$

(c) $1000m^3 + 1$

    $= (10m)^3 + 1^3$ Sum of cubes
    $= (10m + 1)[(10m)^2 - (10m)(1) + 1^2]$
    $= (10m + 1)(100m^2 - 10m + 1)$

(d) $25m^2 + 121$ is prime.
It is the *sum* of squares.
There is no common factor.

Work Problem ❷ **at the Side.** ▶

---

❶ Factor each polynomial.

  (a) $8x - 80$

  (b) $2x^3 + 10x^2 - 2x$

  (c) $12m(p - q) - 7n(p - q)$

❷ Factor each binomial if possible.

  (a) $36x^2 - y^2$

  (b) $4t^2 + 1$

  (c) $125x^3 - 27y^3$

  (d) $x^3 + 343y^3$

Answers

1. (a) $8(x - 10)$
  (b) $2x(x^2 + 5x - 1)$
  (c) $(p - q)(12m - 7n)$

2. (a) $(6x + y)(6x - y)$
  (b) prime
  (c) $(5x - 3y)(25x^2 + 15xy + 9y^2)$
  (d) $(x + 7y)(x^2 - 7xy + 49y^2)$

3 Factor each trinomial.

(a) $16m^2 + 56m + 49$

(b) $r^2 + 18r + 72$

(c) $8t^2 - 13t + 5$

(d) $6x^2 - 3x - 63$

4 Factor each polynomial.

(a) $20 - 5m - 12n + 3mn$

(b) $p^3 - 2pq^2 + p^2q - 2q^3$

(c) $9x^2 + 24x + 16 - y^2$

Answers

3. (a) $(4m + 7)^2$
   (b) $(r + 6)(r + 12)$
   (c) $(8t - 5)(t - 1)$
   (d) $3(2x - 7)(x + 3)$
4. (a) $(4 - m)(5 - 3n)$
   (b) $(p + q)(p^2 - 2q^2)$
   (c) $(3x + 4 + y)(3x + 4 - y)$

The binomial $25m^2 + 625$ is a sum of squares. It *can* be factored as $25(m^2 + 25)$ because its terms have greatest common factor 25.

OBJECTIVE ▶ 3 Factor trinomials. Consider the following.

> **Factoring a Trinomial (Three Terms)**
>
> For a **trinomial,** decide if it is a perfect square trinomial.
>
> $$x^2 + 2xy + y^2 = (x + y)^2 \quad \text{or} \quad x^2 - 2xy + y^2 = (x - y)^2$$
>
> If not, use the methods of **Sections 6.2–6.4.**

EXAMPLE 3 Factoring Trinomials

Factor each trinomial.

(a) $p^2 + 10p + 25$

$= (p + 5)^2$ Perfect square trinomial

(b) $49z^2 - 42z + 9$

$= (7z - 3)^2$ Perfect square trinomial

(c) $y^2 - 5y - 6$ The numbers -6 and 1 have a product of -6 and a sum of -5.

$= (y - 6)(y + 1)$

(d) $2k^2 - k - 6$

$= (2k + 3)(k - 2)$

(e) $28z^2 + 6z - 10$

$= 2(14z^2 + 3z - 5)$

$= 2(7z + 5)(2z - 1)$

◀ **Work Problem 3 at the Side.**

OBJECTIVE ▶ 4 Factor polynomials of more than three terms. Consider factoring by grouping as in **Section 6.1.**

EXAMPLE 4 Factoring Polynomials of More than Three Terms

Factor each polynomial.

(a) $4 - 2q - 6p + 3pq$

$= (4 - 2q) + (-6p + 3pq)$ Group the terms.

$= 2(2 - q) - 3p(2 - q)$ Factor each group.

$= (2 - q)(2 - 3p)$ Factor out $2 - q$.

(b) $20k^3 + 4k^2 - 45k - 9$ Be careful with signs.

$= (20k^3 + 4k^2) + (-45k - 9)$ Group the terms.

$= 4k^2(5k + 1) - 9(5k + 1)$ Factor each group.

$= (5k + 1)(4k^2 - 9)$ $5k + 1$ is a common factor.

$= (5k + 1)(2k + 3)(2k - 3)$ Difference of squares

(c) $4a^2 + 4a + 1 - b^2$

$= (4a^2 + 4a + 1) - b^2$ Group the first three terms.

$= (2a + 1)^2 - b^2$ Perfect square trinomial

$= (2a + 1 + b)(2a + 1 - b)$ Difference of squares

◀ **Work Problem 4 at the Side.**

6.6 Exercises

CONCEPT CHECK *In Exercises 1 and 2, match each polynomial in Column I with the method or methods for factoring it in Column II. The choices in Column II may be used once, more than once, or not at all.*

I	II
1. (a) $49x^2 - 81y^2$	**A.** Factor out the GCF.
(b) $125z^3 + 1$	**B.** Factor a difference of squares.
(c) $88r^2 - 55s^2$	**C.** Factor a difference of cubes.
(d) $8a^3 - b^6$	**D.** Factor a sum of cubes.
(e) $50x^2 - 128y^4$	**E.** The polynomial is prime.

I	II
2. (a) $ab - 5a + 3b - 15$	**A.** Factor out the GCF.
(b) $z^2 - 3z + 6$	**B.** Factor a perfect square trinomial.
(c) $x^2 - 12x + 36 - 4p^2$	**C.** Factor by grouping.
(d) $r^2 - 24r + 144$	**D.** Factor into two distinct binomials.
(e) $2y^2 + 36y + 162$	**E.** The polynomial is prime.

The following exercises are of mixed variety. Factor each polynomial.
See Examples 1–4.

3. $32m^9 + 16m^5 + 24m^3$

4. $2m^2 - 10m - 48$

5. $14k^3 + 7k^2 - 70k$

6. $9z^2 + 64$

7. $m^2 - 3mn - 4n^2$

8. $49z^2 - 16y^2$

9. $100n^2r^2 + 30nr^3 - 50n^2r$

10. $16x^2 + 20x$

11. $20 + 5m + 12n + 3mn$

12. $x^3 + 1000$

13. $y^4 - 81$

14. $m^2 + 2m - 15$

15. $8p^3 - 125$

16. $32z^3 + 56z^2 - 16z$

17. $p^2 - 24p + 144$

18. $9m^2 - 64$

19. $16z^2 - 8z + 1$

20. $6y^2 - 6y - 12$

21. $x^2 + 2x + 16$

22. $p^2 - 17p + 66$

23. $p^3 + 64$

24. $k^2 + 100$

25. $2a^3 + a^2 - 14a - 7$

26. $a^2 - 3ab - 28b^2$

27. $16r^2 + 24rm + 9m^2$

28. $3k^2 + 4k - 4$

29. $n^2 - 12n - 35$

30. $a^4 - 625$

31. $16k^2 - 48k + 36$

32. $36y^6 - 42y^5 - 120y^4$

33. $8k^2 - 2kh - 3h^2$

34. $54m^2 - 24z^2$

35. $10z^2 - 7z - 6$

36. $125m^3 - 216n^3$

37. $9y^2 + 12y - 5$

38. $9u^2 + 66uv + 121v^2$

39. $36x^2 + 32x + 9$

40. $10m^2 + 25m - 60$

41. $4 - 2q - 6p + 3pq$

42. $k^2 - 121$

43. $1000z^3 + 512$

44. $m^3 + 4m^2 - 6m - 24$

45. $100a^2 - 81y^2$

46. $8a^2 + 23ab - 3b^2$

47. $a^2 + 8a + 16$

48. $4y^2 - 25$

49. $2x^2 + 5x + 6$

50. $-3x^3 + 12xy^2$

51. $25a^2 - 70ab + 49b^2$

52. $-4x^2 + 24xy - 36y^2$

53. $2m^2 - 15n - 5mn + 6m$

54. $y^6 + 5y^4 - 3y^2 - 15$

55. $54m^3 - 2000$

56. $12p^3 - 54p^2 - 30p$

6.7 Solving Quadratic Equations by Factoring

OBJECTIVES

1. Solve quadratic equations by factoring.

2. Solve other equations by factoring.

Galileo Galilei developed theories to explain physical phenomena. According to legend, Galileo dropped objects of different weights from the Leaning Tower of Pisa to disprove the belief that heavier objects fall faster than lighter objects. He developed a formula for freely falling objects described by

$$d = 16t^2,$$

where d is the distance in feet that an object falls (disregarding air resistance) in t seconds, regardless of weight.

The equation $d = 16t^2$ is a *quadratic equation*. A quadratic equation contains a second-degree (squared) term and no terms of greater degree.

Quadratic Equation

A **quadratic equation** is an equation that can be written in the form

$$ax^2 + bx + c = 0,$$

where a, b, and c are real numbers, with $a \neq 0$. The given form is called **standard form.**

$$x^2 + 5x + 6 = 0, \quad 2t^2 - 5t = 3, \quad y^2 = 4 \qquad \text{Quadratic equations}$$

In these examples, only $x^2 + 5x + 6 = 0$ is in standard form.

Work Problems ❶ and ❷ at the Side. ▶

We have factored many quadratic *expressions* of the form $ax^2 + bx + c$. We use factored quadratic expressions to solve quadratic *equations*.

OBJECTIVE ▶ ❶ Solve quadratic equations by factoring. We do this using the **zero-factor property.**

Zero-Factor Property

If a and b are real numbers and $ab = 0$, then $a = 0$ or $b = 0$.

In words, if the product of two numbers is 0, then at least one of the numbers must be 0. One number *must* be 0, but both *may* be 0.

EXAMPLE 1 Using the Zero-Factor Property

Solve each equation.

(a) $(x + 3)(2x - 1) = 0$

The product $(x + 3)(2x - 1)$ is equal to 0. By the zero-factor property, the only way that the product of these two factors can be 0 is if at least one of the factors equals 0. Therefore, either $x + 3 = 0$ or $2x - 1 = 0$.

$$x + 3 = 0 \quad \text{or} \quad 2x - 1 = 0 \qquad \text{Zero-factor property}$$

$$x = -3 \quad \text{or} \qquad 2x = 1 \qquad \text{Solve each equation.}$$

$$x = \frac{1}{2} \qquad \text{Divide each side by 2.}$$

Continued on Next Page

Galileo Galilei (1564–1642)

❶ Which of the following equations are quadratic equations?

A. $y^2 - 4y - 5 = 0$

B. $x^3 - x^2 + 16 = 0$

C. $2z^2 + 7z = -3$

D. $x + 2y = -4$

❷ Write each quadratic equation in standard form.

(a) $x^2 - 3x = 4$

(b) $y^2 = 9y - 8$

Answers

1. A, C
2. (a) $x^2 - 3x - 4 = 0$
   (b) $y^2 - 9y + 8 = 0$

3 Solve each equation. Check your solutions.

(GS) **(a)** $(x - 5)(x + 2) = 0$

To solve, use the zero-factor property.

_____ = 0 or _____ = 0

Solve these two equations, obtaining

$x = $ _____ or $x = $ _____.

CHECK Let $x = 5$.

$$(x - 5)(x + 2) = 0$$
$$(\underline{\quad} - 5)(\underline{\quad} + 2) \overset{?}{=} 0$$
$$\underline{\quad} (7) = 0$$
$$(\textit{True / False})$$

To check the other solution, substitute _____ for x in the original equation. A (*true / false*) statement results.

The solution set is $\{\underline{\quad}, \underline{\quad}\}$.

(b) $(3x - 2)(x + 6) = 0$

(c) $z(2z + 5) = 0$

The original equation, $(x + 3)(2x - 1) = 0$, has two solutions, -3 and $\frac{1}{2}$. *Check* these solutions by substituting -3 for x in this equation. ***Then start over*** and substitute $\frac{1}{2}$ for x.

CHECK Let $x = -3$.

$$(x + 3)(2x - 1) = 0$$
$$(-3 + 3)[2(-3) - 1] \overset{?}{=} 0$$
$$0(-7) \overset{?}{=} 0$$
$$0 = 0 \checkmark \text{ True}$$

Let $x = \frac{1}{2}$.

$$(x + 3)(2x - 1) = 0$$
$$\left(\frac{1}{2} + 3\right)\left(2 \cdot \frac{1}{2} - 1\right) \overset{?}{=} 0$$
$$\frac{7}{2}(1 - 1) \overset{?}{=} 0$$
$$0 = 0 \checkmark \text{ True}$$

Both -3 and $\frac{1}{2}$ result in true equations, so the solution set is $\left\{-3, \frac{1}{2}\right\}$.

(b) $\qquad y(3y - 4) = 0$

$\quad y = 0 \quad$ or $\quad 3y - 4 = 0 \quad$ Zero-factor property

Don't forget that 0 is a solution.

$\qquad\qquad 3y = 4 \qquad$ Add 4.

$\qquad\qquad y = \frac{4}{3} \qquad$ Divide by 3.

Check these solutions by substituting each one in the original equation. The solution set is $\left\{0, \frac{4}{3}\right\}$.

◀ **Work Problem 3** at the Side.

Note

The word *or* as used in **Example 1** means "one or the other or both."

If the polynomial in an equation is not already factored, first make sure that the equation is in standard form. Then factor and solve.

EXAMPLE 2 Solving Quadratic Equations

Solve each equation.

(a) $x^2 - 5x = -6$

First, write the equation in standard form by adding 6 to each side.

Don't factor x out at this step.

$$x^2 - 5x = -6$$
$$x^2 - 5x + 6 = 0 \qquad \text{Add 6.}$$

Now factor $x^2 - 5x + 6$. Find two numbers whose product is 6 and whose sum is -5. These two numbers are -2 and -3, so we factor as follows.

$$(x - 2)(x - 3) = 0 \qquad \text{Factor.}$$
$$x - 2 = 0 \quad \text{or} \quad x - 3 = 0 \qquad \text{Zero-factor property}$$
$$x = 2 \quad \text{or} \qquad x = 3 \qquad \text{Solve each equation.}$$

Continued on Next Page

CHECK Let $x = 2$. | Let $x = 3$.

$$x^2 - 5x = -6$$
$$2^2 - 5(2) \stackrel{?}{=} -6$$
$$4 - 10 \stackrel{?}{=} -6$$
$$-6 = -6 \checkmark \text{ True}$$

$$x^2 - 5x = -6$$
$$3^2 - 5(3) \stackrel{?}{=} -6$$
$$9 - 15 \stackrel{?}{=} -6$$
$$-6 = -6 \checkmark \text{ True}$$

Both solutions check, so the solution set is $\{2, 3\}$.

(b) $$y^2 = y + 20 \quad \boxed{\text{Write this equation in standard form.}}$$

Standard form $\longrightarrow$ $y^2 - y - 20 = 0$ Subtract y and 20.

$$(y - 5)(y + 4) = 0 \quad \text{Factor.}$$

$$y - 5 = 0 \quad \text{or} \quad y + 4 = 0 \quad \text{Zero-factor property}$$

$$y = 5 \quad \text{or} \quad y = -4 \quad \text{Solve each equation.}$$

Check by substituting in the original equation. The solution set is $\{-4, 5\}$.

···················· **Work Problem ④ at the Side.** ▶

Solving a Quadratic Equation by Factoring

Step 1   **Write the equation in standard form,** that is, with all terms on one side of the equality symbol in descending powers of the variable and 0 on the other side.

Step 2   **Factor** completely.

Step 3   **Use the zero-factor property** to set each factor with a variable equal to 0.

Step 4   **Solve** the resulting equations.

Step 5   **Check** each solution in the original equation.

EXAMPLE 3   **Solving a Quadratic Equation (Common Factor)**

Solve $4p^2 + 40 = 26p$.

$$4p^2 + 40 = 26p$$

$$4p^2 - 26p + 40 = 0 \quad \text{Standard form}$$

$\boxed{\text{This 2 is } \textit{not} \text{ a solution of the equation.}}$ $2(2p^2 - 13p + 20) = 0 \quad \text{Factor out 2.}$

$$2p^2 - 13p + 20 = 0 \quad \text{Divide each side by 2.}$$

$$(2p - 5)(p - 4) = 0 \quad \text{Factor.}$$

$$2p - 5 = 0 \quad \text{or} \quad p - 4 = 0 \quad \text{Zero-factor property}$$

$$p = \frac{5}{2} \quad \text{or} \quad p = 4 \quad \text{Solve each equation.}$$

Check that the solution set is $\left\{\frac{5}{2}, 4\right\}$ by substituting in the original equation.

···················· **Work Problem ⑤ at the Side.** ▶

CAUTION

A common error is to include the common factor **2** as a solution in **Example 3.** *Only factors containing variables lead to solutions.*

---

④ Solve each equation. Check your solutions.

(a) $m^2 - 3m - 10 = 0$

(b) $r^2 + 2r = 8$

⑤ Solve each equation. Check your solutions.

GS **(a)** $10a^2 - 5a - 15 = 0$

$$\underline{\quad}(\underline{\quad} - a - 3) = 0$$

Divide each side by $\underline{\quad}$.

$$2a^2 - a - 3 = 0$$

$$(2a - 3)(\underline{\quad}) = 0$$

$$2a - 3 = \underline{\quad} \quad \text{or} \quad \underline{\quad} = 0$$

$$a = \underline{\quad} \quad \text{or} \quad a = \underline{\quad}$$

Check by substituting the solutions in the original equation.

The solution set is $\underline{\quad\quad}$.

(b) $4x^2 - 2x = 42$

Answers

4. **(a)** $\{-2, 5\}$ **(b)** $\{-4, 2\}$

5. **(a)** $5; 2a^2; 5; a + 1; 0; a + 1; \dfrac{3}{2}; -1;$

$$\left\{-1, \frac{3}{2}\right\}, \text{ or } \left\{\frac{3}{2}, -1\right\}$$

(b) $\left\{-3, \dfrac{7}{2}\right\}$

6 Solve each equation. Check your solutions.

(a) $49m^2 - 9 = 0$

(b) $m^2 = 3m$

(c) $p(4p + 7) = 2$

EXAMPLE 4 Solving Quadratic Equations

Solve each equation.

(a)
$$16m^2 - 25 = 0$$
$$(4m + 5)(4m - 5) = 0$$ Factor the difference of squares. (**Section 6.5**)

$4m + 5 = 0$ or $4m - 5 = 0$ Zero-factor property

$4m = -5$ or $4m = 5$ Solve each equation.

$m = -\dfrac{5}{4}$ or $m = \dfrac{5}{4}$ Divide by 4.

Check the solutions $-\frac{5}{4}$ and $\frac{5}{4}$ in the original equation. The solution set is $\left\{-\frac{5}{4}, \frac{5}{4}\right\}$.

(b)
$$y^2 = 2y$$
$$y^2 - 2y = 0$$ Standard form

> Don't forget to set the variable factor y equal to 0.

$$y(y - 2) = 0$$ Factor.

$y = 0$ or $y - 2 = 0$ Zero-factor property

$y = 2$ Solve.

The solution set is $\{0, 2\}$.

(c)
$$k(2k + 5) = 3$$ ◁ To be in standard form, 0 must be on one side.

$$2k^2 + 5k = 3$$ Multiply.

Standard form ⟶ $2k^2 + 5k - 3 = 0$ Subtract 3.

$$(2k - 1)(k + 3) = 0$$ Factor.

$2k - 1 = 0$ or $k + 3 = 0$ Zero-factor property

$2k = 1$ or $k = -3$ Solve each equation.

$k = \dfrac{1}{2}$

The solution set is $\left\{-3, \frac{1}{2}\right\}$.

CAUTION

In **Example 4(b)**, it is tempting to begin by dividing both sides of
$$y^2 = 2y$$
by y to get $y = 2$. We do not get the other solution, 0, if we divide by a variable. (We *may* divide each side of an equation by a *nonzero* real number, however. In **Example 3** we divided each side by 2.)

In **Example 4(c)**, we cannot use the zero-factor property to solve
$$k(2k + 5) = 3$$
in its given form because of the 3 on the right side of the equation. *The zero-factor property applies only to a product that equals 0.*

Answers

6. (a) $\left\{-\frac{3}{7}, \frac{3}{7}\right\}$ (b) $\{0, 3\}$ (c) $\left\{-2, \frac{1}{4}\right\}$

◀ **Work Problem 6** at the Side.

EXAMPLE 5 Solving Quadratic Equations (Double Solutions)

Solve each equation.

(a)
$$z^2 - 22z + 121 = 0$$
$$(z - 11)^2 = 0 \qquad \text{Factor the perfect square trinomial.}$$
$$(z - 11)(z - 11) = 0 \qquad a^2 = a \cdot a$$
$$z - 11 = 0 \quad \text{or} \quad z - 11 = 0 \qquad \text{Zero-factor property}$$

Because the two factors are identical, they both lead to the same solution, called a **double solution.**

$$z = 11 \qquad \text{Add 11.}$$

CHECK
$$z^2 - 22z + 121 = 0 \qquad \text{Original equation}$$
$$11^2 - 22(11) + 121 \overset{?}{=} 0 \qquad \text{Let } z = 11.$$
$$121 - 242 + 121 \overset{?}{=} 0 \qquad \text{Simplify.}$$
$$0 = 0 \checkmark \qquad \text{True}$$

The solution set is $\{11\}$.

(b)
$$9t^2 - 30t = -25$$
$$9t^2 - 30t + 25 = 0 \qquad \text{Standard form}$$
$$(3t - 5)^2 = 0 \qquad \text{Factor the perfect square trinomial.}$$
$$3t - 5 = 0 \quad \text{or} \quad 3t - 5 = 0 \qquad \text{Zero-factor property}$$
$$3t = 5 \qquad \text{Add 5.}$$
$$t = \frac{5}{3} \qquad \tfrac{5}{3} \text{ is a double solution.}$$

CHECK
$$9t^2 - 30t = -25 \qquad \text{Original equation}$$
$$9\left(\frac{5}{3}\right)^2 - 30\left(\frac{5}{3}\right) \overset{?}{=} -25 \qquad \text{Let } t = \tfrac{5}{3}.$$
$$9\left(\frac{25}{9}\right) - 30\left(\frac{5}{3}\right) \overset{?}{=} -25 \qquad \text{Apply the exponent.}$$
$$25 - 50 \overset{?}{=} -25 \qquad \text{Multiply.}$$
$$-25 = -25 \checkmark \qquad \text{True}$$

The solution set is $\left\{\frac{5}{3}\right\}$.

·········· **Work Problem 7 at the Side.** ▶

CAUTION

Each equation in **Example 5** has only *one* distinct solution. **We will write a double solution only once in a solution set.**

7 Solve each equation. Check your solutions.

(a)
$$x^2 + 16x = -64$$
$$x^2 + 16x + \underline{\quad} = 0$$
$$(\underline{\quad})^2 = 0$$
$$\underline{\quad} = 0 \quad \text{or} \quad x + 8 = \underline{\quad}$$

Solve to obtain $x = \underline{\quad}$, so -8 is a $\underline{\quad}$ solution. The solution set is $\underline{\quad}$.

(b) $4x^2 - 4x + 1 = 0$

(c) $4z^2 + 20z = -25$

Answers

7. (a) 64; $x + 8$; $x + 8$; 0; -8; double; $\{-8\}$

(b) $\left\{\frac{1}{2}\right\}$ **(c)** $\left\{-\frac{5}{2}\right\}$

8 Solve each equation. Check your solutions.

(a) $r^3 - 16r = 0$

(b) $x^3 - 3x^2 - 18x = 0$

9 Solve each equation. Check your solutions.

(a) $(m + 3)(m^2 - 11m + 10) = 0$

(b) $(2x + 5)(4x^2 - 9) = 0$

Note

Not all quadratic equations can be solved by factoring. A more general method for solving such equations is given in **Chapter 10.**

OBJECTIVE ▸ **2** **Solve other equations by factoring.** We can extend the zero-factor property to solve equations that involve more than two factors with variables. (These equations will have at least one term greater than second degree. They are *not* quadratic equations.)

EXAMPLE 6 Solving an Equation with More Than Two Variable Factors

Solve $6z^3 - 6z = 0$.

$$6z^3 - 6z = 0$$
$$6z(z^2 - 1) = 0 \quad \text{Factor out } 6z.$$
$$6z(z + 1)(z - 1) = 0 \quad \text{Factor } z^2 - 1.$$

By an extension of the zero-factor property, this product can equal 0 only if at least one of the factors equals 0. Write and solve three equations, one for each factor with a variable.

$$6z = 0 \quad \text{or} \quad z + 1 = 0 \quad \text{or} \quad z - 1 = 0 \quad \text{Zero-factor property}$$
$$z = 0 \quad \text{or} \quad z = -1 \quad \text{or} \quad z = 1 \quad \text{Solve each equation.}$$

Check by substituting, in turn, 0, -1, and 1 in the original equation. The solution set is $\{-1, 0, 1\}$.

◀ **Work Problem 8** at the Side.

EXAMPLE 7 Solving an Equation with a Quadratic Factor

Solve $(2x - 1)(x^2 - 9x + 20) = 0$.

$$(2x - 1)(x^2 - 9x + 20) = 0$$
$$(2x - 1)(x - 5)(x - 4) = 0 \quad \text{Factor } x^2 - 9x + 20.$$
$$2x - 1 = 0 \quad \text{or} \quad x - 5 = 0 \quad \text{or} \quad x - 4 = 0 \quad \text{Zero-factor property}$$
$$x = \frac{1}{2} \quad \text{or} \quad x = 5 \quad \text{or} \quad x = 4 \quad \text{Solve each equation.}$$

Check to verify that the solution set is $\left\{\frac{1}{2}, 4, 5\right\}$.

◀ **Work Problem 9** at the Side.

CAUTION

In **Example 7,** it would be unproductive to begin by multiplying the two factors together. The zero-factor property requires the *product* of two or more factors to equal 0. *Always consider first whether an equation is given in the appropriate form to apply the zero-factor property.*

Answers

8. (a) $\{-4, 0, 4\}$ (b) $\{-3, 0, 6\}$

9. (a) $\{-3, 1, 10\}$ (b) $\left\{-\frac{5}{2}, -\frac{3}{2}, \frac{3}{2}\right\}$

6.7 Exercises

FOR EXTRA HELP

 Download the MyDashBoard App

MyMathLab®

CONCEPT CHECK *In Exercises 1–4, fill in each blank with the correct response.*

1. A quadratic equation in x is an equation that can be put into the form _____ = 0.

2. The form $ax^2 + bx + c = 0$ is called _____ form.

3. If a quadratic equation is in standard form, to solve the equation we should begin by attempting to _____ the polynomial.

4. If the product of two numbers is 0, then at least one of the numbers is _____. This is the _____-_____ property.

CONCEPT CHECK *Work each problem.*

5. Identify each equation as *linear* or *quadratic*.

   (a) $2x - 5 = 6$ (b) $x^2 - 5 = -4$

   (c) $x^2 + 2x - 1 = 2x^2$ (d) $5^2x + 2 = 0$

6. The number 9 is a *double solution* of the equation
   $$(x - 9)^2 = 0.$$
   Why is this so?

7. Look at this "solution." **What Went Wrong?**

   $$2x(3x - 4) = 0$$
   $$x = 2 \quad \text{or} \quad x = 0 \quad \text{or} \quad 3x - 4 = 0$$
   $$x = \frac{4}{3}$$

   The solution set is $\left\{2, 0, \frac{4}{3}\right\}$.

8. Look at this "solution." **What Went Wrong?**

   $$x(7x - 1) = 0$$
   $$7x - 1 = 0 \quad \text{Zero-factor property}$$
   $$x = \frac{1}{7}$$

   The solution set is $\left\{\frac{1}{7}\right\}$.

Solve each equation, and check your solutions. **See Example 1.**

9. $(x + 5)(x - 2) = 0$

10. $(x - 1)(x + 8) = 0$

11. $(2m - 7)(m - 3) = 0$

12. $(6k + 5)(k + 4) = 0$

13. $(2x + 1)(6x - 1) = 0$

14. $(3x + 2)(10x - 1) = 0$

15. $t(6t + 5) = 0$

16. $w(4w + 1) = 0$

17. $2x(3x - 4) = 0$

18. $6x(4x + 9) = 0$

19. $(x - 9)(x - 9) = 0$

20. $(2x + 1)(2x + 1) = 0$

Solve each equation, and check your solutions. ***See Examples 2–7.***

21. $y^2 + 3y + 2 = 0$

22. $p^2 + 8p + 7 = 0$

23. $y^2 - 3y + 2 = 0$

24. $r^2 - 4r + 3 = 0$

25. $x^2 = 24 - 5x$

26. $t^2 = 2t + 15$

27. $x^2 = 3 + 2x$

28. $m^2 = 4 + 3m$

29. $z^2 + 3z = -2$

30. $p^2 - 2p = 3$

31. $m^2 + 8m + 16 = 0$

32. $b^2 - 6b + 9 = 0$

33. $3x^2 + 5x - 2 = 0$

34. $6r^2 - r - 2 = 0$

35. $12p^2 = 8 - 10p$

36. $18x^2 = 12 + 15x$

37. $9s^2 + 12s = -4$

38. $36x^2 + 60x = -25$

39. $y^2 - 9 = 0$

40. $m^2 - 100 = 0$

41. $16k^2 - 49 = 0$

42. $4w^2 - 9 = 0$

43. $n^2 = 121$

44. $x^2 = 400$

45. $x^2 = 7x$

46. $t^2 = 9t$

47. $6r^2 = 3r$

48. $10y^2 = -5y$

49. $g(g - 7) = -10$

50. $r(r - 5) = -6$

51. $z(2z + 7) = 4$

52. $b(2b + 3) = 9$

53. $2(y^2 - 66) = -13y$

54. $3(t^2 + 4) = 20t$

55. $5x^3 - 20x = 0$

56. $3x^3 - 48x = 0$

57. $9y^3 - 49y = 0$

58. $16r^3 - 9r = 0$

59. $(2r + 5)(3r^2 - 16r + 5) = 0$

60. $(3m + 4)(6m^2 + m - 2) = 0$

61. $(2x + 7)(x^2 + 2x - 3) = 0$

62. $(x + 1)(6x^2 + x - 12) = 0$

63. $x^3 + x^2 - 20x = 0$

64. $y^3 - 6y^2 + 8y = 0$

65. $r^4 = 2r^3 + 15r^2$

66. $x^3 = 3x + 2x^2$

67. $3x(x + 1) = (2x + 3)(x + 1)$

68. $2x(x + 3) = (3x + 1)(x + 3)$

69. Galileo's formula describing the motion of freely falling objects is

$$d = 16t^2.$$

The distance d in feet an object falls depends on the time t elapsed, in seconds. (This is an example of a **function,** introduced in **Section 3.6.**)

(a) Use Galileo's formula and complete the following table. (*Hint:* Substitute each given value into the formula and solve for the unknown value.)

t in seconds	0	1	2	3	__	__
d in feet	0	16	__	__	256	576

(b) When $t = 0$, $d = 0$. Explain this in the context of the problem.

70. In **Exercise 69,** when you substituted 256 for d and solved for t, you should have found two solutions: 4 and -4. Why doesn't -4 make sense as an answer?

6.8 Applications of Quadratic Equations

OBJECTIVES

1. Solve problems involving geometric figures.
2. Solve problems involving consecutive integers.
3. Solve problems by applying the Pythagorean theorem.
4. Solve problems by using given quadratic models.

We can use factoring to solve quadratic equations that arise in applications by following the same six problem-solving steps given in **Section 2.4.**

Solving an Applied Problem

Step 1 **Read** the problem carefully. What information is given? What is to be found?

Step 2 **Assign a variable** to represent the unknown value. Use a sketch, diagram, or table, as needed. Express any other unknown values in terms of the variable.

Step 3 **Write an equation** using the variable expression(s).

Step 4 **Solve** the equation.

Step 5 **State the answer.** Label it appropriately. Does it seem reasonable?

Step 6 **Check** the answer in the words of the *original* problem.

OBJECTIVE ▶ **1** **Solve problems involving geometric figures.** Refer to the formulas inside the back cover of the text, if necessary.

EXAMPLE 1   Solving an Area Problem

The Monroes want to plant a rectangular garden in their yard. The width of the garden will be 4 ft less than its length, and they want it to have an area of 96 ft^2. (ft^2 means square feet.) Find the length and width of the garden.

Step 1 **Read** the problem carefully. We need to find the dimensions of a garden with area 96 ft^2.

Step 2 **Assign a variable.**

Let x = the length of the garden.

Then $x - 4$ = the width. (The width is 4 ft less than the length.)

See **Figure 1.**

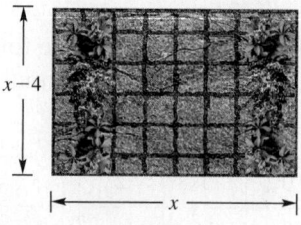

Figure 1

Step 3 **Write an equation.** The area of a rectangle is given by

Area = Length × Width.   Area formula

Substitute 96 for area, x for length, and $x - 4$ for width.

$$A = LW$$

$$96 = x(x - 4) \quad \text{Let } A = 96, L = x, W = x - 4.$$

Continued on Next Page

Step 4 **Solve.** $96 = x(x - 4)$ Equation from Step 3

$96 = x^2 - 4x$ Distributive property

$x^2 - 4x - 96 = 0$ Standard form

$(x - 12)(x + 8) = 0$ Factor.

$x - 12 = 0$ or $x + 8 = 0$ Zero-factor property

$x = \mathbf{12}$ or $x = \mathbf{-8}$ Solve each equation.

Step 5 **State the answer.** The solutions are 12 and −8. A rectangle cannot have a side of negative length, so we discard −**8**. The length of the garden will be **12** ft. The width will be $12 - 4 = 8$ ft.

Step 6 **Check.** The width of the garden is 4 ft less than the length, and the area is $12 \cdot 8 = 96$ ft^2, as required.

.... Work Problem ❶ at the Side. ▶

> **Problem-Solving Hint**
>
> *When solving applied problems, always check solutions against physical facts* and discard any answers that are not appropriate.

OBJECTIVE ▶ ❷ Solve problems involving consecutive integers. Recall from **Section 2.4** that **consecutive integers** are integers that are next to each other on a number line, such as 1 and 2, or −11 and −10. See **Figure 2.**

  Consecutive odd integers are *odd* integers that are next to each other, such as 1 and 3, or −13 and −11. **Consecutive even integers** are defined similarly, so 2 and 4 are consecutive even integers, as are −10 and −8. See **Figure 3.** (We list consecutive integers in increasing order from left to right.)

> **Problem-Solving Hint**
>
> If x represents the lesser integer, then the following apply.
>
> For two consecutive integers, use $x,\ \ x + 1.$
>
> For three consecutive integers, use $x,\ \ x + 1,\ \ x + 2.$
>
> For two consecutive even or odd integers, use $x,\ \ x + 2.$
>
> For three consecutive even or odd integers, use $x,\ \ x + 2,\ \ x + 4.$

EXAMPLE 2   **Solving a Consecutive Integer Problem**

The product of the numbers on two consecutive post-office boxes is 210. Find the box numbers.

Step 1 **Read** the problem. Note that the boxes are numbered consecutively.

Step 2 **Assign a variable.** See **Figure 4.**

Let $x =$ the first box number.

Then $x + 1 =$ the next consecutive box number.

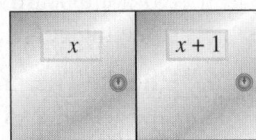

Figure 4

.... **Continued on Next Page**

❶ Solve each problem.

(a) The length of a rectangular room is 2 m more than the width. The area of the floor is 48 m^2. Find the length and width of the room.

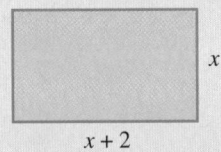

Give the equation using x as the variable, and give the answer.

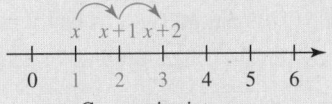

Consecutive integers

Figure 2

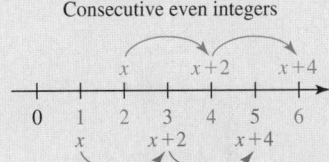

Consecutive even integers

Consecutive odd integers

Figure 3

(b) The length of each side of a square is increased by 4 in. The sum of the areas of the original square and the larger square is 106 in^2. What is the length of a side of the original square?

Answers

1. (a) $48 = (x + 2)x$; length: 8 m; width: 6 m
  (b) 5 in.

2 Solve the problem.

The product of the numbers on two consecutive lockers at a health club is 132. Find the locker numbers.

Step 3 **Write an equation.** The product of the box numbers is 210.

$$x(x + 1) = 210$$

Step 4 **Solve.**

$x^2 + x = 210$	Distributive property
$x^2 + x - 210 = 0$	Standard form
$(x + 15)(x - 14) = 0$	Factor.
$x + 15 = 0$ or $x - 14 = 0$	Zero-factor property
$x = -15$ or $x = 14$	Solve each equation.

Step 5 **State the answer.** The solutions are -15 and 14. We discard -15 since a box number cannot be negative. If $x = 14$, then $x + 1 = 15$, so the boxes have numbers 14 and 15.

Step 6 **Check.** The numbers 14 and 15 are consecutive and their product is $14 \cdot 15 = 210$, as required.

◀ **Work Problem 2** at the Side.

3 Solve each problem. Give the equation using x to represent the least integer, and give the answer.

(a) The product of two consecutive even integers is 4 more than two times their sum. Find the integers.

> **EXAMPLE 3** Solving a Consecutive Integer Problem

The product of two consecutive odd integers is 1 less than five times their sum. Find the integers.

Step 1 **Read** carefully. This problem is a little more complicated.

Step 2 **Assign a variable.** We must find two consecutive *odd* integers.

Let $x =$ the lesser integer.

Then $x + 2 =$ the next greater odd integer.

Step 3 **Write an equation.**

The product	is	five times the sum	less 1.
↓	↓	↓	↓
$x(x + 2)$	$=$	$5(x + x + 2)$	$- 1$

Step 4 **Solve.**

$x^2 + 2x = 5x + 5x + 10 - 1$	Distributive property
$x^2 + 2x = 10x + 9$	Combine like terms.
$x^2 - 8x - 9 = 0$	Standard form
$(x - 9)(x + 1) = 0$	Factor.
$x - 9 = 0$ or $x + 1 = 0$	Zero-factor property
$x = 9$ or $x = -1$	Solve each equation.

(b) Find three consecutive odd integers such that the product of the least and greatest is 16 more than the middle integer.

Step 5 **State the answer.** We need to find two consecutive odd integers.

If $x = 9$ is the lesser, then $x + 2 = 9 + 2 = 11$ is the greater.

If $x = -1$ is the lesser, then $x + 2 = -1 + 2 = 1$ is the greater.

Do not discard the solution -1 here. There are two sets of answers since integers can be positive *or* negative.

Step 6 **Check.** The product of the first pair of integers is $9 \cdot 11 = 99$. One less than five times their sum is $5(9 + 11) - 1 = 99$. Thus, 9 and 11 satisfy the problem. Repeat the check with -1 and 1.

◀ **Work Problem 3** at the Side.

Answers

2. 11 and 12
3. (a) $x(x + 2) = 4 + 2(x + x + 2)$; 4 and 6 or -2 and 0
   (b) $x(x + 4) = 16 + (x + 2)$; 3, 5, 7

CAUTION

Do *not* use $x, x + 1, x + 3$ to represent consecutive odd integers. To see why, let $x = 3$. Then $x + 1 = 3 + 1$ or 4, and $x + 3 = 3 + 3$ or 6. The numbers 3, 4, and 6 are not consecutive odd integers.

OBJECTIVE 3 Solve problems by applying the Pythagorean theorem.
The next example requires the Pythagorean theorem from geometry.

Pythagorean Theorem

If a right triangle (a triangle with a 90° angle) has longest side of length c and two other sides of lengths a and b, then

$$a^2 + b^2 = c^2.$$

The longest side, the **hypotenuse,** is opposite the right angle. The two shorter sides are the **legs** of the triangle.

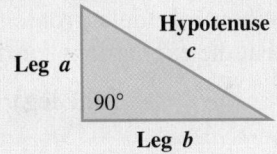

Pythagoras (c. 380–300 B.C.)

EXAMPLE 4 **Applying the Pythagorean Theorem**

Amy and Kevin leave their office, with Amy traveling north and Kevin traveling east. When Kevin is 1 mi farther than Amy from the office, the distance between them is 2 mi more than Amy's distance from the office. Find their distances from the office and the distance between them.

Step 1 **Read** the problem again. We must find three distances.

Step 2 **Assign a variable.**

Let $x =$ Amy's distance from the office.

Then $x + 1 =$ Kevin's distance from the office,

and $x + 2 =$ the distance between them.

Place these expressions on a right triangle, as in **Figure 5.**

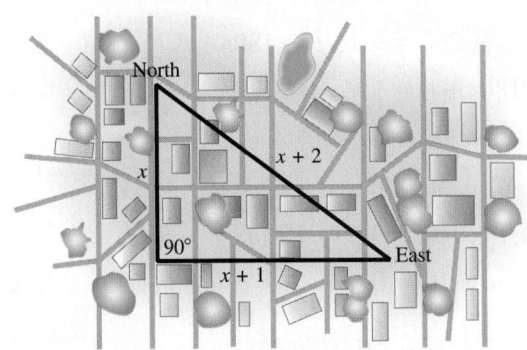

Figure 5

Step 3 **Write an equation.** Substitute into the Pythagorean theorem.

$$a^2 + b^2 = c^2$$

$$x^2 + (x + 1)^2 = (x + 2)^2$$ ◁ Be careful to substitute properly.

Continued on Next Page

④ Solve the problem.

The hypotenuse of a right triangle is 3 in. longer than the longer leg. The shorter leg is 3 in. shorter than the longer leg. Find the lengths of the sides of the triangle.

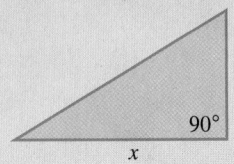

⑤ Solve each problem.

(a) Refer to **Example 5.** How long will it take for the ball to reach a height of 50 ft?

(b) The number y of impulses fired after a nerve has been stimulated is modeled by

$$y = -x^2 + 2x + 60,$$

where x is in milliseconds (ms) after the stimulation. When will 45 impulses occur? Do we get two solutions? Why is only one answer acceptable?

Step 4   Solve.
$$x^2 + x^2 + 2x + 1 = x^2 + 4x + 4$$

$$x^2 - 2x - 3 = 0 \qquad \text{Standard form}$$

$$(x - 3)(x + 1) = 0 \qquad \text{Factor.}$$

$$x - 3 = 0 \quad \text{or} \quad x + 1 = 0 \qquad \text{Zero-factor property}$$

$$x = 3 \quad \text{or} \qquad x = -1 \quad \text{Solve each equation.}$$

Step 5   State the answer. Since -1 cannot represent a distance, 3 is the only possible answer. Amy's distance is 3 mi, Kevin's distance is $3 + 1 = 4$ mi, and the distance between them is $3 + 2 = 5$ mi.

Step 6   Check. Since $3^2 + 4^2 = 5^2$ is true, the answer is correct.

◀ **Work Problem ④ at the Side.**

> **Problem-Solving Hint**
>
> When solving a problem involving the Pythagorean theorem, be sure that the expressions for the sides of the triangle are properly placed.
>
> $$(\textbf{one leg})^2 + (\textbf{other leg})^2 = \textbf{hypotenuse}^2$$

OBJECTIVE ④ Solve problems by using given quadratic models. In **Examples 1–4,** we wrote quadratic equations to model, or mathematically describe, various situations and then solved the equations. Now we are given the quadratic models and must use them to determine data.

EXAMPLE 5   Finding the Height of a Ball

A tennis player can hit a ball 180 ft per sec (123 mph). If she hits a ball directly upward, the height h of the ball in feet at time t in seconds is modeled by the quadratic equation

$$h = -16t^2 + 180t + 6.$$

How long will it take for the ball to reach a height of 206 ft?

A height of 206 ft means $h = 206$, so we substitute 206 for h in the equation and then solve for t.

$$h = -16t^2 + 180t + 6$$

$$\mathbf{206} = -16t^2 + 180t + 6 \qquad \text{Let } h = 206.$$

$$-16t^2 + 180t + 6 = 206 \qquad \text{Interchange sides.}$$

$$-16t^2 + 180t - 200 = 0 \qquad \text{Standard form}$$

$$4t^2 - 45t + 50 = 0 \qquad \text{Divide by } -4.$$

$$(4t - 5)(t - 10) = 0 \qquad \text{Factor.}$$

$$4t - 5 = 0 \quad \text{or} \quad t - 10 = 0 \qquad \text{Zero-factor property}$$

$$t = \frac{5}{4} \quad \text{or} \qquad t = 10 \quad \text{Solve each equation.}$$

Since we found two acceptable answers, the ball will be 206 ft above the ground twice (once on its way up and once on its way down)—at $\frac{5}{4}$ sec and at 10 sec after it is hit. See **Figure 6.**

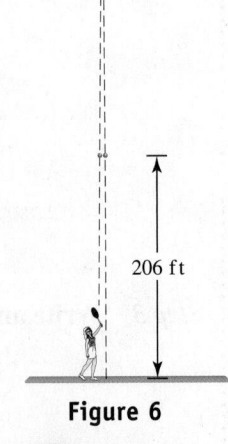

206 ft

Figure 6

◀ **Work Problem ⑤ at the Side.**

EXAMPLE 6   Modeling the Foreign-Born Population of the United States

The foreign-born population of the United States over the years 1930–2010 can be modeled by the quadratic equation

$$y = 0.009665x^2 - 0.4942x + 15.12,$$

where $x = 0$ represents 1930, $x = 10$ represents 1940, and so on, and y is the number of people in millions. (*Source:* U.S. Census Bureau.)

(a) Use the model to find the foreign-born population in 1980 to the nearest tenth of a million.

Since $x = 0$ represents 1930, $x = 50$ represents 1980. Substitute 50 for x in the equation.

$y = 0.009665\,(50)^2 - 0.4942\,(50) + 15.12$ Let $x = 50$.

$y = 14.6$ Round to the nearest tenth.

In 1980, the foreign-born population of the United States was about 14.6 million.

(b) Repeat part (a) for 2010.

$y = 0.009665\,(80)^2 - 0.4942\,(80) + 15.12$ For 2010, let $x = 80$.

$y = 37.4$ Round to the nearest tenth.

In 2010, the foreign-born population of the United States was about 37.4 million.

(c) The model used in parts (a) and (b) was developed using the data in the table below. How do the results in parts (a) and (b) compare to the actual data from the table?

Year	Foreign-Born Population (millions)
1930	14.2
1940	11.6
1950	10.3
1960	9.7
1970	9.6
1980	14.1
1990	19.8
2000	28.4
2010	37.6

From the table, the actual value for 1980 is 14.1 million. By comparison, our answer in part (a), 14.6 million, is slightly high. For 2010, the actual value is 37.6 million, so our answer of 37.4 million in part (b) is slightly low, but a good estimate.

· **Work Problem** ❻ **at the Side.** ▶

⬤ Use the model in **Example 6** to find the foreign-born population of the United States in 1990. Give your answer to the nearest tenth of a million. How does it compare to the actual value from the table?

Answer

6. 20.3 million; The actual value is 19.8 million, so our answer using the model is somewhat high.

6.8 Exercises

1. **CONCEPT CHECK** To review the six problem-solving steps first introduced in **Section 2.4,** complete each statement.

Step 1: _____ the problem carefully.

Step 2: Assign a _____ to represent the unknown value.

Step 3: Write a(n) _____ using the variable expression(s).

Step 4: _____ the equation.

Step 5: State the _____.

Step 6: _____ the answer in the words of the _____ problem.

2. **CONCEPT CHECK** A student solves an applied problem and gets 6 or -3 for the length of the side of a square. Which of these answers is reasonable? Why?

GS *In Exercises 3–6, a figure and a corresponding geometric formula are given. Using x as the variable, complete Steps 3–6 for each problem. (Refer to the steps in **Exercise 1** as needed.)*

3.

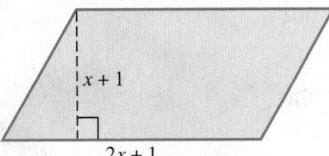

$x + 1$

$2x + 1$

Area of a parallelogram: $A = bh$

The area of this parallelogram is 45 sq. units. Find its base and height.

Step 3: $45 =$ _____

Step 4: $x =$ _____ or $x =$ _____

Step 5: base: _____ units; height: _____ units

Step 6: _____ = 45

4.

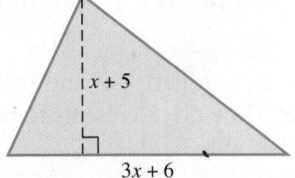

$x + 5$

$3x + 6$

Area of a triangle: $A = \frac{1}{2}bh$

The area of this triangle is 60 sq. units. Find its base and height.

Step 3: $60 =$ _____

Step 4: $x =$ _____ or $x =$ _____

Step 5: base: _____ units; height: _____ units

Step 6: _____ = 60

5.

4

$x + 2$ x

Volume of a rectangular jewelry box: $V = LWH$

The volume of this box is 192 cu. units. Find its length and width.

Step 3: _____ = _____ $(x + 2)$

Step 4: $x =$ _____ or $x =$ _____

Step 5: length: _____ units; width: _____ units

Step 6: _____ · 4 = _____

6.

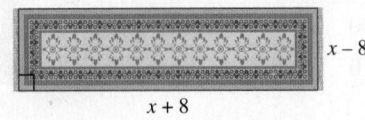

$x - 8$

$x + 8$

Area of a rectangular rug: $A = LW$

The area of this rug is 80 sq. units. Find its length and width.

Step 3: _____ = $(x + 8)$ _____

Step 4: $x =$ _____ or $x =$ _____

Step 5: length: _____ units; width: _____ units

Step 6: _____ = 80

Solve each problem. Check answers to be sure they are reasonable. Refer to the formulas inside the back cover. **See Example 1.**

7. The length of a standard jewel case is 2 cm more than its width. The area of the rectangular top of the case is 168 cm². Find the length and width of the jewel case.

8. A standard DVD case is 6 cm longer than it is wide. The area of the rectangular top of the case is 247 cm². Find the length and width of the case.

9. The area of a triangle is 30 in.². The base of the triangle measures 2 in. more than twice the height of the triangle. Find the measures of the base and the height.

10. A certain triangle has its base equal in measure to its height. The area of the triangle is 72 m². Find the equal base and height measure.

11. The dimensions of a flat-panel monitor are such that its length is 3 in. more than its width. If the length were doubled and if the width were decreased by 1 in., the area would be increased by 150 in.². What are the length and width of the flat panel?

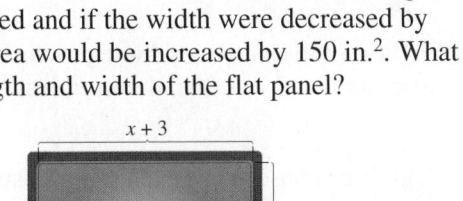

12. A computer keyboard is 11 in. longer than it is wide. If the length were doubled and if 2 in. were added to the width, the area would be increased by 198 in.². What are the length and width of the keyboard?

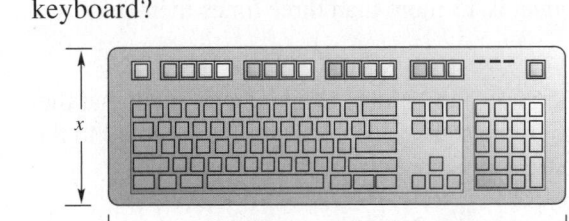

13. A 10-gal aquarium is 3 in. higher than it is wide. Its length is 21 in., and its volume is 2730 in.³. What are the height and width of the aquarium?

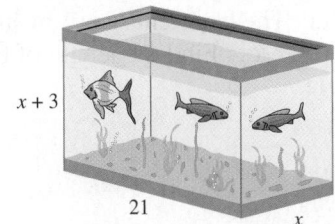

14. A toolbox is 2 ft high, and its width is 3 ft less than its length. If its volume is 80 ft³, find the length and width of the box.

15. A square mirror has sides measuring 2 ft less than the sides of a square painting. If the difference between their areas is 32 ft², find the lengths of the sides of the mirror and the painting.

16. The sides of one square have length 3 m more than the sides of a second square. If the area of the larger square is subtracted from 4 times the area of the smaller square, the result is 36 m². What are the lengths of the sides of each square?

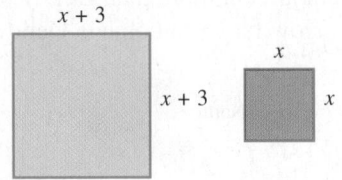

*Solve each problem about consecutive integers. **See Examples 2 and 3.***

17. The product of the numbers on two consecutive volumes of research data is 420. Find the volume numbers.

18. The product of the page numbers on two facing pages of a book is 600. Find the page numbers.

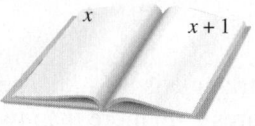

19. The product of two consecutive integers is 11 more than their sum. Find the integers.

20. The product of two consecutive integers is 4 less than four times their sum. Find the integers.

21. Find two consecutive odd integers such that their product is 15 more than three times their sum.

22. Find two consecutive odd integers such that five times their sum is 23 less than their product.

23. Find three consecutive even integers such that the sum of the squares of the lesser two is equal to the square of the greatest.

24. Find three consecutive even integers such that the square of the sum of the lesser two is equal to twice the greatest.

25. Find three consecutive odd integers such that 3 times the sum of all three is 18 more than the product of the first and second integers.

26. Find three consecutive odd integers such that the sum of all three is 42 less than the product of the second and third integers.

*Use the Pythagorean theorem to solve each problem. **See Example 4.***

27. The hypotenuse of a right triangle is 1 cm longer than the longer leg. The shorter leg is 7 cm shorter than the longer leg. Find the length of the longer leg of the triangle.

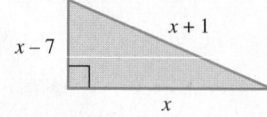

28. The longer leg of a right triangle is 1 m longer than the shorter leg. The hypotenuse is 1 m shorter than twice the shorter leg. Find the length of the shorter leg of the triangle.

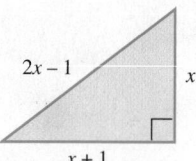

29. Terri works due north of home. Her husband Denny works due east. They leave for work at the same time. By the time Terri is 5 mi from home, the distance between them is 1 mi more than Denny's distance from home. How far from home is Denny?

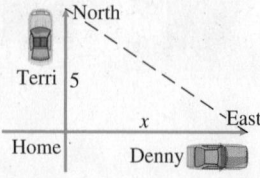

30. Two cars left an intersection at the same time. One traveled north. The other traveled 14 mi farther, but to the east. How far apart were they then, if the distance between them was 4 mi more than the distance traveled east?

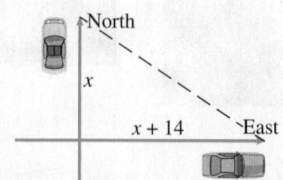

31. A ladder is leaning against a building. The distance from the bottom of the ladder to the building is 4 ft less than the length of the ladder. How high up the side of the building is the top of the ladder if that distance is 2 ft less than the length of the ladder?

32. A lot has the shape of a right triangle with one leg 2 m longer than the other. The hypotenuse is 2 m less than twice the length of the shorter leg. Find the length of the shorter leg.

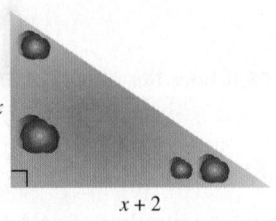

Solve each problem. See Example 5.

33. An object projected from a height of 48 ft with an initial velocity of 32 ft per sec after t seconds has height

$$h = -16t^2 + 32t + 48.$$

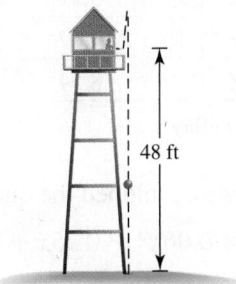

(a) After how many seconds is the height 64 ft? (*Hint:* Let $h = 64$ and solve.)

(b) After how many seconds is the height 60 ft?

(c) After how many seconds does the object hit the ground? (*Hint:* When the object hits the ground, $h = 0$.)

(d) The quadratic equation from part (c) has two solutions, yet only one of them is appropriate for answering the question. Why is this so?

34. If an object is projected upward from ground level with an initial velocity of 64 ft per sec, its height h in feet t seconds later is

$$h = -16t^2 + 64t.$$

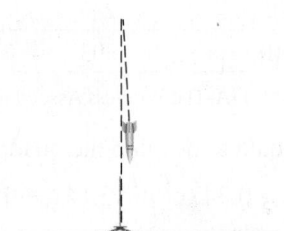

(a) After how many seconds is the height 48 ft? (*Hint:* Let $h = 48$ and solve.)

(b) The object reaches its maximum height 2 sec after it is projected. What is this maximum height?

(c) After how many seconds does the object hit the ground? (*Hint:* When the object hits the ground, $h = 0$.)

(d) The quadratic equation from part (c) has two solutions, yet only one of them is appropriate for answering the question. Why is this so?

*If an object is projected upward with an initial velocity of 128 ft per sec, its height h
in feet after t seconds is*

$$h = 128t - 16t^2.$$

Find the height of the object after each period of time. **See Example 5.**

35. 1 sec **36.** 2 sec **37.** 4 sec

38. How long does it take the object just described to return to the ground?

🖩 *Solve each problem.* **See Example 6.**

39. The table shows the number of cellular phone
subscribers (in millions) in the United States.

Year	Subscribers (in millions)
1994	24
1996	44
1998	69
2000	109
2002	141
2004	182
2006	233
2008	286
2010	303

Source: CTIA-The Wireless Association.

We used the data to develop the quadratic equation

$$y = 0.347x^2 + 13.14x + 17.98,$$

which models the number of cellular phone subscribers y (in millions) in the year x, where $x = 0$ represents
1994, $x = 2$ represents 1996, and so on.

(a) Use the model to find the number of subscribers
in 2000 to the nearest million. How does the
result compare to the actual data in the table?

(b) What value of x corresponds to 2008?

(c) Use the model to find the number of subscribers
in 2008 to the nearest million. How does the
result compare to the actual data in the table?

(d) Assuming that the trend in the data continues, use
the quadratic equation to estimate the number of
subscribers in 2013 to the nearest million.

40. Annual revenue in billions of dollars for eBay is
shown in the table.

Year	Annual Revenue (in billions of dollars)
2001	0.749
2002	1.21
2003	2.17
2004	3.27
2005	4.55
2006	5.97
2007	7.67
2008	8.54

Source: eBay.

Using the data, we developed the quadratic equation

$$y = 0.06x^2 + 0.75x + 0.55$$

to model eBay revenues y in year x, where $x = 0$
represents 2001, $x = 1$ represents 2002, and so on.

(a) Use the model to find annual revenue for eBay in
2005 and 2008. How do the results compare with
the actual data in the table?

(b) Use the model to estimate annual revenue for
eBay in 2009, to the nearest tenth.

(c) Actual revenue in 2009 was $8.73 billion.
How does the result from part (b) compare with
the actual revenue in 2009?

(d) Should the quadratic equation be used to estimate
eBay revenue for years after 2008? Explain.

Chapter 6 *Summary*

Key Terms

6.1

factor For integers a and b, if $a \cdot b = c$, then a and b are factors of c.

factored form An expression is in factored form when it is written as a product.

greatest common factor (GCF) The greatest common factor is the largest quantity that is a factor of each of a group of quantities.

factoring The process of writing a polynomial as a product is called factoring.

6.2

prime polynomial A prime polynomial is a polynomial that cannot be factored using only integers.

6.5

perfect square trinomial A perfect square trinomial is a trinomial that can be factored as the square of a binomial.

6.7

quadratic equation A quadratic equation is an equation that can be written in the form $ax^2 + bx + c = 0$, with $a \neq 0$.

standard form The form $ax^2 + bx + c = 0$ is the standard form of a quadratic equation.

6.8

hypotenuse The longest side of a right triangle, opposite the right angle, is the hypotenuse.

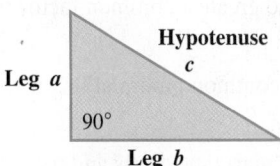

legs The two shorter sides of a right triangle are the legs.

Test Your Word Power

See how well you have learned the vocabulary in this chapter.

1 **Factoring** is
   A. a method of multiplying polynomials
   B. the process of writing a polynomial as a product
   C. the answer in a multiplication problem
   D. a way to add the terms of a polynomial.

2 A polynomial is in **factored form** when
   A. it is prime
   B. it is written as a sum
   C. the second-degree term has a coefficient of 1
   D. it is written as a product.

3 The **greatest common factor** of a polynomial is
   A. the least integer that divides evenly into all its terms
   B. the least expression that is a factor of all its terms
   C. the greatest expression that is a factor of all its terms
   D. the variable that is common to all its terms.

4 A **perfect square trinomial** is a trinomial
   A. that can be factored as the square of a binomial
   B. that cannot be factored
   C. that is multiplied by a binomial

   D. where all terms are perfect squares.

5 A **quadratic equation** is an equation that can be written in the form
   A. $y = mx + b$
   B. $ax^2 + bx + c = 0 \ (a \neq 0)$
   C. $Ax + By = C$
   D. $x = k$.

6 A **hypotenuse** is
   A. either of the two shorter sides of a triangle
   B. the shortest side of a right triangle
   C. the side opposite the right angle in a right triangle
   D. the longest side in any triangle.

Answers to Test Your Word Power

1. B; *Example:* $x^2 - 5x - 14$ factors as $(x - 7)(x + 2)$.

2. D; *Example:* The factored form of $x^2 - 5x - 14$ is $(x - 7)(x + 2)$.

3. C; *Example:* The greatest common factor of $8x^2$, $22xy$, and $16x^3y^2$ is $2x$.

4. A; *Example:* $a^2 + 2a + 1$ is a perfect square trinomial. Its factored form is $(a + 1)^2$.

5. B; *Examples:* $y^2 - 3y + 2 = 0$, $x^2 - 9 = 0$, $2m^2 = 6m + 8$

6. C; *Example:* See the triangle included in the Key Terms above.

Quick Review

Concepts	Examples

6.1 Factors; The Greatest Common Factor

Finding the Greatest Common Factor (GCF)

Step 1 Write each number in prime factored form.

Step 2 List each prime number or each variable that is a factor of every term in the list.

Step 3 Use as exponents on the common prime factors the *least* exponents from the prime factored forms.

Step 4 Multiply the primes from Step 3.

Find the greatest common factor of $4x^2y$, $6x^2y^3$, and $2xy^2$.

$$4x^2y = 2 \cdot 2 \cdot x^2 \cdot y$$
$$6x^2y^3 = 2 \cdot 3 \cdot x^2 \cdot y^3$$
$$2xy^2 = 2 \cdot x \cdot y^2$$

The greatest common factor is $2xy$.

Factoring by Grouping

Step 1 Group the terms.

Step 2 Factor out the greatest common factor from each group.

Step 3 Factor out a common binomial factor from the results of Step 2.

Step 4 If necessary, rearrange terms and try a different grouping.

Factor by grouping.

$$2a^2 + 2ab + a + b$$

$$= (2a^2 + 2ab) + (a + b) \qquad \text{Group the terms.}$$

$$= 2a(a + b) + 1(a + b) \qquad \text{Factor each group.}$$

$$= (a + b)(2a + 1) \qquad \text{Factor out } a + b.$$

6.2 Factoring Trinomials

To factor $x^2 + bx + c$, find m and n such that $mn = c$ and $m + n = b$.

$$\overset{\displaystyle mn = c}{\underset{\displaystyle m + n = b}{x^2 + bx + c}}$$

Then $x^2 + bx + c$ factors as $(x + m)(x + n)$.

Check by multiplying.

Factor $x^2 + 6x + 8$.

$$\overset{\displaystyle mn = 8}{\underset{\displaystyle m + n = 6}{x^2 + 6x + 8}} \quad \text{Here, } m = 2 \text{ and } n = 4.$$

$x^2 + 6x + 8$ factors as $(x + 2)(x + 4)$.

CHECK $(x + 2)(x + 4)$

$$= x^2 + 4x + 2x + 8 \qquad \text{FOIL method}$$

$$= x^2 + 6x + 8 ✓ \qquad \text{Combine like terms.}$$

6.3 Factoring Trinomials by Grouping

To factor $ax^2 + bx + c$, find m and n such that $mn = ac$ and $m + n = b$.

$$\overset{\displaystyle m + n = b}{\underset{\displaystyle mn = ac}{ax^2 + bx + c}}$$

Then factor $ax^2 + mx + nx + b$ by grouping.

Factor $3x^2 + 14x - 5$.

$$3x^2 + 14x - 5$$
$$\underset{-15}{}$$

Find two integers with a product of $3(-5) = -15$ and a sum of 14. The integers are -1 and 15.

$$3x^2 + 14x - 5$$

$$= 3x^2 - x + 15x - 5 \qquad 14x = -x + 15x$$

$$= (3x^2 - x) + (15x - 5) \qquad \text{Group the terms.}$$

$$= x(3x - 1) + 5(3x - 1) \qquad \text{Factor each group.}$$

$$= (3x - 1)(x + 5) \qquad \text{Factor out } 3x - 1.$$

Concepts	Examples

6.4 Factoring Trinomials by Using the FOIL Method

To factor $ax^2 + bx + c$ by trial and error, use the FOIL method in reverse.

Factor $3x^2 + 14x - 5$.

The only positive factors of 3 are 3 and 1, and -5 has possible factors of 1 and -5, or -1 and 5. Using trial and error,

$$3x^2 + 14x - 5 \quad \text{factors as} \quad (3x - 1)(x + 5).$$

6.5 Special Factoring Techniques

Difference of Squares
$$a^2 - b^2 = (a + b)(a - b)$$

Perfect Square Trinomials
$$a^2 + 2ab + b^2 = (a + b)^2$$
$$a^2 - 2ab + b^2 = (a - b)^2$$

Difference of Cubes
$$x^3 - y^3 = (x - y)(x^2 + xy + y^2)$$

Sum of Cubes
$$x^3 + y^3 = (x + y)(x^2 - xy + y^2)$$

Factor.

$$4x^2 - 9$$
$$= (2x + 3)(2x - 3)$$

$$9x^2 + 6x + 1 \qquad\qquad 4x^2 - 20x + 25$$
$$= (3x + 1)^2 \qquad\qquad = (2x - 5)^2$$

$$8 - 27a^3$$
$$= (2 - 3a)(4 + 6a + 9a^2)$$

$$64z^3 + 1$$
$$= (4z + 1)(16z^2 - 4z + 1)$$

6.7 Solving Quadratic Equations by Factoring

Zero-Factor Property

If a and b are real numbers and $ab = 0$, then $a = 0$ or $b = 0$.

If $(x - 2)(x + 3) = 0$, then $x - 2 = 0$ or $x + 3 = 0$.

Solving a Quadratic Equation by Factoring

Step 1 Write the equation in standard form.

Step 2 Factor.

Step 3 Use the zero-factor property.

Step 4 Solve the resulting equations.

Solve $2x^2 = 7x + 15$.

$2x^2 - 7x - 15 = 0$	Standard form
$(2x + 3)(x - 5) = 0$	Factor.
$2x + 3 = 0 \quad$ or $\quad x - 5 = 0$	Zero-factor property
$2x = -3$ or $\qquad x = 5$	Solve each equation.

$$x = -\frac{3}{2}$$

Step 5 Check.

The solutions $-\frac{3}{2}$ and 5 satisfy the original equation. The solution set is $\left\{-\frac{3}{2}, 5\right\}$.

6.8 Applications of Quadratic Equations

Pythagorean Theorem
In a right triangle, the square of the hypotenuse equals the sum of the squares of the legs.

$$a^2 + b^2 = c^2$$

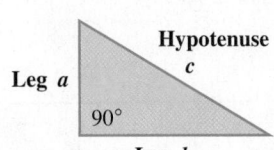

In a right triangle, one leg measures 2 ft longer than the other. The hypotenuse measures 4 ft longer than the shorter leg. Find the lengths of the three sides of the triangle.

Let $x =$ the length of the shorter leg. Then

$$x^2 + (x + 2)^2 = (x + 4)^2.$$

Solve this equation to get $x = 6$ or $x = -2$. Discard -2 as a solution. Check that the sides measure

$$6 \text{ ft}, \quad 6 + 2 = 8 \text{ ft}, \quad \text{and} \quad 6 + 4 = 10 \text{ ft}.$$

Chapter 6 *Review Exercises*

6.1 *Factor out the greatest common factor or factor by grouping.*

1. $15t + 45$

2. $60z^3 + 30z$

3. $44x^3 + 55x^2$

4. $100m^2n^3 - 50m^3n^4 + 150m^2n^2$

5. $2xy - 8y + 3x - 12$

6. $6y^2 + 9y + 4xy + 6x$

6.2 *Factor completely.*

7. $x^2 + 10x + 21$

8. $y^2 - 13y + 40$

9. $q^2 + 6q - 27$

10. $r^2 - r - 56$

11. $x^2 + x + 1$

12. $3x^2 + 6x + 6$

13. $r^2 - 4rs - 96s^2$

14. $p^2 + 2pq - 120q^2$

15. $-8p^3 + 24p^2 + 80p$

16. $3x^4 + 30x^3 + 48x^2$

17. $m^2 - 3mn - 18n^2$

18. $y^2 - 8yz + 15z^2$

19. $p^7 - p^6q - 2p^5q^2$

20. $-3r^5 + 6r^4s + 45r^3s^2$

6.3–6.4

21. CONCEPT CHECK To begin factoring $6r^2 - 5r - 6$, what are the possible first terms of the two binomial factors, if we consider only positive integer coefficients?

22. CONCEPT CHECK What is the first step you would use to factor $2z^3 + 9z^2 - 5z$?

Factor completely.

23. $2k^2 - 5k + 2$

24. $3r^2 + 11r - 4$

25. $6r^2 - 5r - 6$

26. $10z^2 - 3z - 1$

27. $5t^2 - 11t + 12$

28. $24x^5 - 20x^4 + 4x^3$

29. $-6x^2 + 3x + 30$

30. $10r^3s + 17r^2s^2 + 6rs^3$

31. $-30y^3 - 5y^2 + 10y$

32. $4z^2 - 5z + 7$

33. $-3m^3n + 19m^2n + 40mn$

34. $14a^2 - 27ab - 20b^2$

6.5

35. CONCEPT CHECK Which one of the following is a difference of squares?

A. $32x^2 - 1$ **B.** $4x^2y^2 - 25z^2$

C. $x^2 + 36$ **D.** $25y^3 - 1$

36. CONCEPT CHECK Which one of the following is a perfect square trinomial?

A. $x^2 + x + 1$ **B.** $y^2 - 4y + 9$

C. $4x^2 + 10x + 25$ **D.** $x^2 - 20x + 100$

Factor completely.

37. $n^2 - 64$

38. $25b^2 - 121$

39. $49y^2 - 25w^2$

40. $144p^2 - 36q^2$

41. $x^2 + 100$

42. $z^2 + 10z + 25$

43. $9t^2 - 42t + 49$

44. $16m^2 + 40mn + 25n^2$

45. $125x^3 - 1$

46. $1000p^3 + 27$

6.7

Solve each equation, and check the solutions.

47. $(4t + 3)(t - 1) = 0$

48. $(x + 7)(x - 4)(x + 3) = 0$

49. $x(2x - 5) = 0$

50. $z^2 + 4z + 3 = 0$

51. $m^2 - 5m + 4 = 0$

52. $x^2 = -15 + 8x$

53. $3z^2 - 11z - 20 = 0$

54. $81t^2 - 64 = 0$

55. $y^2 = 8y$

56. $n(n - 5) = 6$

57. $t^2 - 14t + 49 = 0$

58. $t^2 = 12(t - 3)$

59. $(5z + 2)(z^2 + 3z + 2) = 0$

60. $x^2 = 9$

61. $64x^3 - 9x = 0$

62. $(2r + 1)(12r^2 + 5r - 3) = 0$

63. $25w^2 - 90w + 81 = 0$

64. $r(r - 7) = 30$

6.8

Solve each problem.

65. The length of a rug is 6 ft more than the width. The area is 40 ft². Find the length and width of the rug.

66. The surface area S of a box is given by

$$S = 2WH + 2WL + 2LH.$$

A treasure chest from a sunken galleon has dimensions (in feet) as shown in the figure. Its surface area is 650 ft². Find its width.

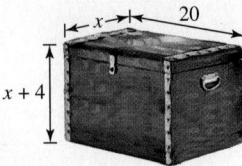

67. The product of two consecutive integers is 29 more than their sum. What are the integers?

68. The product of the lesser two of three consecutive integers is equal to 23 plus the greatest. Find the integers.

69. Two cars left an intersection at the same time. One traveled west, and the other traveled 14 mi less, but to the south. How far apart were they then, if the distance between them was 16 mi more than the distance traveled south?

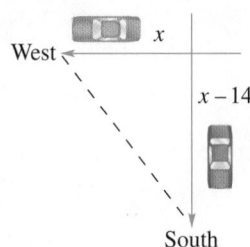

70. The triangular sail of a schooner has an area of 30 m². The height of the sail is 4 m more than the base. Find the length of the base of the sail.

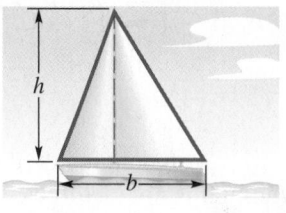

71. The floor plan for a house is a rectangle with length 7 m more than its width. The area is 170 m². Find the width and length of the house.

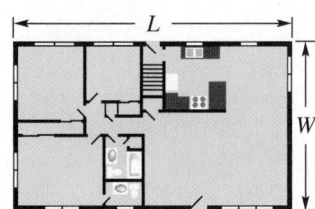

72. If an object is dropped, the distance d in feet it falls in t seconds (disregarding air resistance) is given by the quadratic equation

$$d = 16t^2.$$

Find the distance an object would fall in each of the following times.

(a) 4 sec   **(b)** 8 sec

The numbers of alternative-fueled vehicles, in thousands, in use in the United States for selected years over the period 2001–2009 are given in the table.

Year	Number (in thousands)
2001	425
2003	534
2005	592
2007	696
2009	826

Source: Energy Information Administration.

To model the number of vehicles y in year x, we developed the quadratic equation

$$y = 1.57x^2 + 32.5x + 400.$$

Here we used $x = 1$ for 2001, $x = 2$ for 2002, and so on.

73. Use the model to find the number of alternative-fueled vehicles in 2005, to the nearest thousand. How does the result compare with the actual data in the table?

74. What estimate for 2010 is given by the model? Why might the estimate for 2010 be unreliable?

Mixed Review Exercises

75. CONCEPT CHECK   Which of the following is *not* factored completely?

    A. $3(7t)$ **B.** $3x(7t + 4)$ **C.** $(3 + x)(7t + 4)$ **D.** $3(7t + 4) + x(7t + 4)$

76. CONCEPT CHECK   A student factored $6x^2 + 16x - 32$ as $(2x + 8)(3x - 4)$. Tell why the polynomial is not factored completely, and give the completely factored form.

Factor completely.

77. $15m^2 + 20mp - 12m - 16p$

78. $24ab^3c^2 - 56a^2bc^3 + 72a^2b^2c$

79. $z^2 - 11zx + 10x^2$

80. $3k^2 + 11k + 10$

81. $y^4 - 625$

82. $6m^3 - 21m^2 - 45m$

83. $25a^2 + 15ab + 9b^2$

84. $8z^3 + 64y^3$

85. $-12r^2 - 8rq + 15q^2$

86. $100a^2 - 9$

87. $49t^2 + 56t + 16$

Solve each equation, and check the solutions.

88. $t(t - 7) = 0$

89. $x^2 + 3x = 10$

90. $25x^2 + 20x + 4 = 0$

Solve each problem.

91. A lot is in the shape of a right triangle. The hypotenuse is 3 m longer than the longer leg. The longer leg is 6 m longer than twice the length of the shorter leg. Find the lengths of the sides of the lot.

92. A pyramid has a rectangular base with a length that is 2 m more than the width. The height of the pyramid is 6 m, and its volume is 48 m^3. Find the length and width of the base.

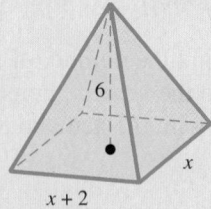

Chapter 6 *Test*

1. Which one of the following is the correct, completely factored form of $2x^2 - 2x - 24$?

   A. $(2x + 6)(x - 4)$ **B.** $(x + 3)(2x - 8)$

   C. $2(x + 4)(x - 3)$ **D.** $2(x + 3)(x - 4)$

Factor each polynomial completely. If the polynomial is prime, *say so.*

2. $12x^2 - 30x$

3. $2m^3n^2 + 3m^3n - 5m^2n^2$

4. $2ax - 2bx + ay - by$

5. $x^2 - 9x + 14$

6. $2x^2 + x - 3$

7. $6x^2 - 19x - 7$

8. $3x^2 - 12x - 15$

9. $10z^2 - 17z + 3$

10. $t^2 + 6t + 10$

11. $x^2 + 36$

12. $y^2 - 49$

13. $81a^2 - 121b^2$

14. $x^2 + 16x + 64$

15. $4x^2 - 28xy + 49y^2$

16. $-2x^2 - 4x - 2$

17. $4t^3 + 32t^2 + 64t$

18. $x^4 - 81$

19. $x^3 - 512$

20. $8k^3 + 64$

Solve each equation.

21. $(x + 3)(x - 9) = 0$

22. $2r^2 - 13r + 6 = 0$

23. $25x^2 - 4 = 0$

24. $x(x - 20) = -100$

25. $t^2 = 3t$

26. $(s + 8)(6s^2 + 13s - 5) = 0$

Solve each problem.

27. The length of a rectangular flower bed is 3 ft less than twice its width. The area of the bed is 54 ft^2. Find the dimensions of the flower bed.

28. Find two consecutive integers such that the square of the sum of the two integers is 11 more than the lesser integer.

29. A carpenter needs to cut a brace to support a wall stud, as shown in the figure. The brace should be 7 ft less than three times the length of the stud. If the brace will be anchored on the floor 15 ft away from the stud, how long should the brace be?

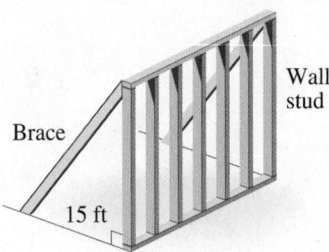

30. The number of pieces of first-class mail sent through the U.S. Postal Service (USPS) from 2001 through 2010 can be approximated by the quadratic equation

$$y = -0.3470x^2 + 1.374x + 100.7,$$

where $x = 1$ represents 2001, $x = 2$ represents 2002, and so on, and y is the number of pieces of mail in billions. Use the model to estimate the number of pieces of first-class mail sent through the USPS in 2009. Round your answer to the nearest billion. (*Source:* U.S. Postal Service.)

Chapters R–6 Cumulative Review Exercises

Solve each equation.

1. $3x + 2(x - 4) = 4(x - 2)$

2. $0.3x + 0.9x = 0.06$

3. $\dfrac{2}{3}n - \dfrac{1}{2}(n - 4) = 3$

4. Solve for t: $A = P + Prt$

5. From a list of "technology-related items," adults were recently surveyed as to those items they couldn't live without. Complete the results shown in the table if 500 adults were surveyed.

Item	Percent That Couldn't Live Without	Number That Couldn't Live Without
Personal computer	46%	____
Cell phone	41%	____
High-speed Internet	____	190
MP3 player	____	60

(Other items included digital cable, HDTV, and electronic gaming console.)
Source: Ipsos for AP.

Solve each problem.

6. At the 2010 Winter Olympics in Vancouver, Canada, the top medal winner was the United States, which won a total of 37 medals. The United States won 2 more silver medals than bronze and 4 fewer gold medals than bronze. Find the number of each type of medal won. (*Source: World Almanac and Book of Facts.*)

7. In 2010, American women working full time earned, on average, 81.2 cents for every dollar earned by men working full time. The median weekly earnings for full-time male workers were $824. To the nearest dollar, what were the median weekly earnings for full-time female workers? (*Source*: U.S. Bureau of Labor Statistics.)

8. Find the measures of the marked angles.

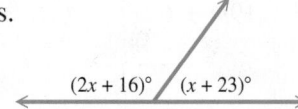

$(2x + 16)°$ $(x + 23)°$

9. Fill in each blank with *positive* or *negative*. The point with coordinates (a, b) is in

   (a) quadrant II if a is _____ and b is _____ .

   (b) quadrant III if a is _____ and b is _____ .

Consider the equation $y = 4x + 3$. Find the following.

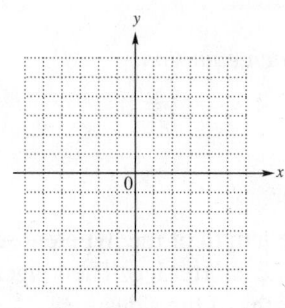

10. The x- and y-intercepts

11. The slope

12. The graph

13. The points on the graph show the total retail sales of prescription drugs in the United States in the years 2004–2010, along with a graph of a linear equation that models the data.

   (a) Use the ordered pairs shown on the graph to find the slope of the line to the nearest whole number. Interpret the slope.

   (b) Use the graph to estimate sales in the year 2008. Write your answer as an ordered pair of the form (year, sales in billions of dollars).

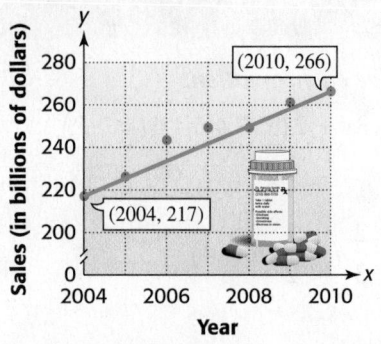

Retail Prescription Drug Sales

Source: National Association of Chain Drug Stores.

Solve each system of equations.

14. $4x - y = -6$
$2x + 3y = 4$

15. $5x + 3y = 10$
$2x + \dfrac{6}{5}y = 5$

Evaluate each expression.

16. $2^{-3} \cdot 2^5$

17. $\left(\dfrac{3}{4}\right)^{-2}$

18. $\dfrac{6^5 \cdot 6^{-2}}{6^3}$

19. $\left(\dfrac{4^{-3} \cdot 4^4}{4^5}\right)^{-1}$

Simplify each expression and write the answer using only positive exponents. Assume no denominators are 0.

20. $\dfrac{(p^2)^3 p^{-4}}{(p^{-3})^{-1} p}$

21. $\dfrac{(m^{-2})^3 m}{m^5 m^{-4}}$

Perform the indicated operations.

22. $(2k^2 + 4k) - (5k^2 - 2) - (k^2 + 8k - 6)$

23. $(9x + 6)(5x - 3)$

24. $(3p + 2)^2$

25. $\dfrac{8x^4 + 12x^3 - 6x^2 + 20x}{2x}$

Factor completely.

26. $2a^2 + 7a - 4$

27. $10m^2 + 19m + 6$

28. $8t^2 + 10tv + 3v^2$

29. $4p^2 - 12p + 9$

30. $25r^2 - 81t^2$

31. $2pq + 6p^3q + 8p^2q$

Solve each equation.

32. $6m^2 + m - 2 = 0$

33. $8x^2 = 64x$

34. $49x^2 - 56x + 16 = 0$

35. The length of the hypotenuse of a right triangle is 3 m more than twice the length of the shorter leg. The longer leg is 7 m longer than the shorter leg. Find the lengths of the sides.

7 Rational Expressions and Functions

Ratios and *proportions,* topics covered in this chapter and used in comparisons of quantities, involve *rational expressions.*

7.1 Rational Expressions and Functions; Multiplying and Dividing

7.2 Adding and Subtracting Rational Expressions

7.3 Complex Fractions

7.4 Equations with Rational Expressions and Graphs

Summary Exercises *Simplifying Rational Expressions vs. Solving Rational Equations*

7.5 Applications of Rational Expressions

7.6 Variation

7.1 Rational Expressions and Functions; Multiplying and Dividing

OBJECTIVES

1. Define rational expressions.
2. Define rational functions and describe their domains.
3. Write rational expressions in lowest terms.
4. Multiply rational expressions.
5. Find reciprocals for rational expressions.
6. Divide rational expressions.

OBJECTIVE 1 Define rational expressions. In arithmetic, a rational number is the quotient of two integers, with the denominator not 0. In algebra, a **rational expression,** or *algebraic fraction,* is the quotient of two polynomials, again with the denominator not 0.

$$\frac{x}{y},\ \frac{-a}{4},\ \frac{m+4}{m-2},\ \frac{8x^2-2x+5}{4x^2+5x},\ \text{ and }\ x^5\left(\text{or }\frac{x^5}{1}\right) \quad \text{Rational expressions}$$

Rational expressions are elements of the set

$$\left\{\frac{P}{Q}\ \middle|\ P \text{ and } Q \text{ are polynomials, with } Q \neq 0\right\}.$$

OBJECTIVE 2 Define rational functions and describe their domains.

> **Rational Function**
>
> A function that is defined by a quotient of polynomials is a **rational function.** It has the form
>
> $$f(x) = \frac{P(x)}{Q(x)}, \quad \text{where } Q(x) \neq 0.$$
>
> The domain of a rational function includes all real numbers except those that make $Q(x)$, that is, the denominator, equal to 0.

For example, the domain of the rational function

$$f(x) = \frac{2}{x - 5} \leftarrow \text{Cannot equal 0}$$

includes all real numbers except 5, because 5 would make the denominator equal to 0. **Figure 1** shows a graph of this function. Notice that the graph does not exist when $x = 5$. (It does not intersect the dashed vertical line whose equation is $x = 5$. This line is an **asymptote.**)

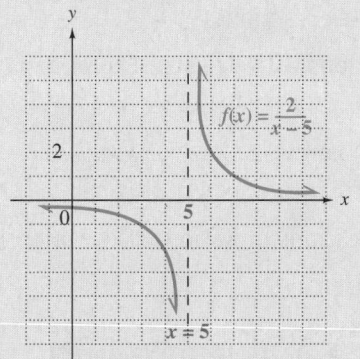

Figure 1

> **EXAMPLE 1   Finding Domains of Rational Functions**
>
> For each rational function, find all numbers that are not in the domain. Then give the domain, using set-builder notation.
>
> **(a)** $f(x) = \dfrac{3}{7x - 14}$
> Values that make the denominator 0 cannot be used. To find these values, set the denominator equal to 0 and solve.
>
> $$7x - 14 = 0$$
> $$7x = 14 \quad \text{Add 14.}$$
> $$x = 2 \quad \text{Divide by 7.}$$
>
> The number 2 cannot be used as a replacement for x. The domain of f includes all real numbers except 2, written using set-builder notation as $\{x \mid x \neq 2\}$.

···· **Continued on Next Page**

(b) $g(x) = \dfrac{3 + x}{x^2 - 4x + 3}$ ← Values that make the denominator 0 must be excluded.

$$x^2 - 4x + 3 = 0 \quad \text{Set the denominator equal to 0.}$$
$$(x - 1)(x - 3) = 0 \quad \text{Factor.}$$
$$x - 1 = 0 \quad \text{or} \quad x - 3 = 0 \quad \text{Zero-factor property}$$
$$x = 1 \quad \text{or} \quad x = 3 \quad \text{Solve each equation.}$$

The domain of g includes all real numbers except 1 and 3, written $\{x \mid x \neq 1, 3\}$.

(c) $h(x) = \dfrac{8x + 2}{3}$

The denominator, 3, can never be 0, so the domain of h includes all real numbers, written in set-builder notation as $\{x \mid x \text{ is a real number}\}$.

(d) $f(x) = \dfrac{2}{x^2 + 4}$

Setting $x^2 + 4$ equal to 0 leads to $x^2 = -4$. There is no real number whose square is -4. Therefore, any real number can be used as a replacement for x. The domain of f is $\{x \mid x \text{ is a real number}\}$.

············· **Work Problem** ❶ **at the Side.** ▶

OBJECTIVE ❸ **Write rational expressions in lowest terms.** In arithmetic, we write the fraction $\frac{15}{20}$ in lowest terms by dividing the numerator and denominator by 5 to get $\frac{3}{4}$. We write rational expressions in lowest terms in a similar way, using the **fundamental property of rational numbers.**

Fundamental Property of Rational Numbers

If $\frac{a}{b}$ is a rational number and if c is any nonzero real number, then

$$\frac{a}{b} = \frac{ac}{bc}.$$

In words, the numerator and denominator of a rational number may either be multiplied or divided by the same *nonzero number* without changing the value of the rational number.

Because $\frac{c}{c}$ is equivalent to 1, the fundamental property is based on the identity property for multiplication.

Note

A rational expression is a quotient of two polynomials. Since the value of a polynomial is a real number for every value of the variable for which it is defined, any statement that applies to rational numbers will also apply to rational expressions.

We use the following steps to write rational expressions in lowest terms.

Writing a Rational Expression in Lowest Terms

Step 1 **Factor** both numerator and denominator to find their greatest common factor (GCF).

Step 2 **Apply the fundamental property.** Divide out common factors.

---

❶ For each rational function, find all numbers that are not in the domain. Then give the domain, using set-builder notation.

(a) $f(x) = \dfrac{x + 4}{x - 6}$

(b) $f(x) = \dfrac{x + 6}{x^2 - x - 6}$

(c) $f(x) = \dfrac{3 + 2x}{5}$

(d) $f(x) = \dfrac{2}{x^2 + 1}$

Answers
1. **(a)** 6; $\{x \mid x \neq 6\}$
   (b) $-2, 3$; $\{x \mid x \neq -2, 3\}$
   (c) none; $\{x \mid x \text{ is a real number}\}$
   (d) none; $\{x \mid x \text{ is a real number}\}$

EXAMPLE 2　Writing Rational Expressions in Lowest Terms

Write each rational expression in lowest terms.

(a) $\dfrac{8k}{16}$

$= \dfrac{k \cdot 8}{2 \cdot 8}$　　Factor.

$= \dfrac{k}{2} \cdot 1$　　Apply the fundamental property.

$= \dfrac{k}{2}$　　Lowest terms

(b) $\dfrac{8 + k}{16}$　Be careful. The numerator cannot be factored.　This expression cannot be simplified further and is in lowest terms.

(c) $\dfrac{a^2 - a - 6}{a^2 + 5a + 6}$

$= \dfrac{(a - 3)(a + 2)}{(a + 3)(a + 2)}$　　Factor the numerator. Factor the denominator.

$= \dfrac{a - 3}{a + 3}$　　$\dfrac{a + 2}{a + 2} = 1$; Fundamental property

(d) $\dfrac{y^2 - 4}{2y + 4}$

$= \dfrac{(y + 2)(y - 2)}{2(y + 2)}$　　Factor the difference of squares in the numerator. Factor the denominator.

$= \dfrac{y - 2}{2}$　　Lowest terms

(e) $\dfrac{x^3 - 27}{x - 3}$

$= \dfrac{(x - 3)(x^2 + 3x + 9)}{x - 3}$　　Factor the difference of cubes in the numerator.

$= x^2 + 3x + 9$　　Lowest terms

(f) $\dfrac{pr + qr + ps + qs}{pr + qr - ps - qs}$

$= \dfrac{(pr + qr) + (ps + qs)}{(pr + qr) - (ps + qs)}$　　Group the terms.

Be careful with signs.

$= \dfrac{r(p + q) + s(p + q)}{r(p + q) - s(p + q)}$　　Factor within the groups.

$= \dfrac{(p + q)(r + s)}{(p + q)(r - s)}$　　Factor by grouping.

$= \dfrac{r + s}{r - s}$　　Lowest terms

CAUTION

When using the fundamental property of rational numbers, only common factors may be divided out. For example,

$$\frac{y-2}{2} \neq y \quad \text{and} \quad \frac{y-2}{2} \neq y-1.$$ The 2 in $y-2$ is **not** a *factor* of the numerator.

The expression $\frac{y-2}{2}$ indicates that the *entire* numerator is being divided by 2, not just certain terms. It is already in lowest terms, although it could be written in the equivalent form $\frac{y}{2} - \frac{2}{2}$, or $\frac{y}{2} - 1$.
We must factor before writing a fraction in lowest terms.

Work Problem **2** at the Side. ▶

Look again at the rational expression from **Example 2(c).**

$$\frac{a^2 - a - 6}{a^2 + 5a + 6}, \quad \text{or} \quad \frac{(a-3)(a+2)}{(a+3)(a+2)}$$

In this expression, a can take any value except -3 or -2, since these values make the denominator 0. In the simplified rational expression

$$\frac{a-3}{a+3}, \quad a \text{ cannot equal } -3.$$

Because of this,

$$\frac{a^2 - a - 6}{a^2 + 5a + 6} = \frac{a-3}{a+3}, \quad \text{for all values of } a \text{ except } -3 \text{ or } -2.$$

From now on such statements of equality will be made with the understanding that they apply only for those real numbers that make neither denominator equal 0. We will no longer state such restrictions.

EXAMPLE 3   Writing Rational Expressions in Lowest Terms

Write each rational expression in lowest terms.

(a) $\frac{m-3}{3-m}$ — Here, the numerator and denominator are opposites.

To write this expression in lowest terms, we write the denominator as $-1(m-3)$.

$$\frac{m-3}{3-m} = \frac{m-3}{-1(m-3)} = \frac{1}{-1} = -1$$ Factor out -1 in the denominator.

Alternatively, we could have written the numerator as $-1(3-m)$ and obtained the same result.

(b)
$$\frac{r^2 - 16}{4 - r}$$

$$= \frac{(r+4)(r-4)}{4-r}$$ Factor the difference of squares in the numerator.

$$= \frac{(r+4)(r-4)}{-1(r-4)}$$ Factor out -1 in the denominator to write $4-r$ as $-1(r-4)$.

Distribute to check.
$-1(r-4)$
$= -1 \cdot r - 1(-4)$
$= -r + 4,$ or $4 - r$

$$= \frac{r+4}{-1}$$ Fundamental property

$$= -(r+4), \quad \text{or} \quad -r-4$$ Lowest terms

---

2 Write each rational expression in lowest terms.

(a) $\dfrac{y^2 + 2y - 3}{y^2 - 3y + 2}$

(b) $\dfrac{3y + 9}{y^2 - 9}$

(c) $\dfrac{x - 9}{12}$

(d) $\dfrac{y + 2}{y^2 + 4}$

(e) $\dfrac{1 + p^3}{1 + p}$

(f) $\dfrac{3x + 3y + rx + ry}{5x + 5y - rx - ry}$

Answers

2. (a) $\dfrac{y+3}{y-2}$ **(b)** $\dfrac{3}{y-3}$
   (c) already in lowest terms
   (d) already in lowest terms
   (e) $1 - p + p^2$ **(f)** $\dfrac{3+r}{5-r}$

3 Write each rational expression in lowest terms.

(GS) **(a)** $\dfrac{10 - a}{a - 10}$

Are the numerator and denominator opposites? (*Yes / No*)

$$\dfrac{10 - a}{a - 10}$$

$$= \dfrac{-1(\underline{})}{a - 10}$$

$$= \underline{}$$

(b) $\dfrac{y - 2}{2 - y}$

(c) $\dfrac{8 - b}{8 + b}$

(d) $\dfrac{p - 2}{4 - p^2}$

Answers

3. **(a)** Yes; $a - 10$; -1 **(b)** -1
    (c) already in lowest terms   **(d)** $\dfrac{-1}{2 + p}$

As shown in **Example 3,** the quotient $\frac{a}{-a}$ ($a \neq 0$) can be simplified.

$$\frac{a}{-a} = \frac{a}{-1(a)} = \frac{1}{-1} = -1$$

Quotient of Opposites

In general, if the numerator and the denominator of a rational expression are opposites, then the expression equals -1.

Based on this result, the following are true statements.

$$\frac{q - 7}{7 - q} = -1 \quad \text{and} \quad \frac{-5a + 2b}{5a - 2b} = -1$$

Numerator and denominator in each expression are opposites.

However, the following expression *cannot* be simplified further.

$$\frac{r - 2}{r + 2} \overset{\ulcorner}{\underset{\llcorner}{}} \begin{array}{l}\text{Numerator and denominator} \\ \text{are } not \text{ opposites.}\end{array}$$

◀ **Work Problem 3** at the Side.

OBJECTIVE ▶ 4 **Multiply rational expressions.** To multiply rational expressions, follow these steps. (In practice, we usually simplify before multiplying.)

Multiplying Rational Expressions

Step 1 **Factor** all numerators and denominators as completely as possible.

Step 2 **Apply the fundamental property.**

Step 3 **Multiply** remaining factors in the numerator and remaining factors in the denominator. Leave the denominator in factored form.

Step 4 **Check** to be sure the product is in lowest terms.

EXAMPLE 4 **Multiplying Rational Expressions**

Multiply.

(a) $\dfrac{5p - 5}{p} \cdot \dfrac{3p^2}{10p - 10}$

$$= \dfrac{5(p - 1)}{p} \cdot \dfrac{3p \cdot p}{2 \cdot 5(p - 1)} \qquad \text{Factor.}$$

$$= \dfrac{5(p - 1)}{5(p - 1)} \cdot \dfrac{p}{p} \cdot \dfrac{3p}{2} \qquad \text{Commutative property}$$

$$= \dfrac{1}{1} \cdot \dfrac{1}{1} \cdot \dfrac{1}{1} \cdot \dfrac{3p}{2} \qquad \text{Fundamental property}$$

> In practice, this step is usually done mentally.

$$= \dfrac{3p}{2} \qquad \text{Lowest terms}$$

······ **Continued on Next Page**

(b) $\dfrac{k^2 + 2k - 15}{k^2 - 4k + 3} \cdot \dfrac{k^2 - k}{k^2 + k - 20}$

$= \dfrac{(k + 5)(k - 3)}{(k - 3)(k - 1)} \cdot \dfrac{k(k - 1)}{(k + 5)(k - 4)}$ Factor.

$= \dfrac{k}{k - 4}$ Fundamental property; Multiply.

(c) $(p - 4) \cdot \dfrac{3}{5p - 20}$

$= \dfrac{p - 4}{1} \cdot \dfrac{3}{5p - 20}$ Write $p - 4$ as $\frac{p - 4}{1}$.

$= \dfrac{p - 4}{1} \cdot \dfrac{3}{5(p - 4)}$ Factor.

$= \dfrac{3}{5}$ Fundamental property; Multiply.

(d) $\dfrac{x^2 + 2x}{x + 1} \cdot \dfrac{x^2 - 1}{x^3 + x^2}$

$= \dfrac{x(x + 2)}{x + 1} \cdot \dfrac{(x + 1)(x - 1)}{x^2(x + 1)}$ Factor.

$= \dfrac{(x + 2)(x - 1)}{x(x + 1)}$ Fundamental property; Multiply.

(e) $\dfrac{x - 6}{x^2 - 12x + 36} \cdot \dfrac{x^2 - 3x - 18}{x^2 + 7x + 12}$

$= \dfrac{x - 6}{(x - 6)^2} \cdot \dfrac{(x + 3)(x - 6)}{(x + 3)(x + 4)}$ Factor.

$= \dfrac{1}{x + 4}$ *Remember to include 1 in the numerator when all other factors are eliminated.*   Fundamental property; Multiply.

················· **Work Problem ④ at the Side.** ▶

OBJECTIVE ▶ ⑤ Find reciprocals for rational expressions. The rational numbers $\frac{a}{b}$ and $\frac{c}{d}$ are reciprocals of each other if they have a product of 1. The **reciprocal** of a rational expression is defined in the same way. *Two rational expressions are reciprocals of each other if they have a product of 1.*

    The table shows several rational expressions and their reciprocals.

Rational Expression	Reciprocal
3, or $\dfrac{3}{1}$	$\dfrac{1}{3}$
$\dfrac{5}{k}$	$\dfrac{k}{5}$
$\dfrac{m^2 - 9m}{2}$	$\dfrac{2}{m^2 - 9m}$
$\dfrac{0}{4}$	undefined

Reciprocals have a product of 1.

Recall that 0 has no reciprocal.

---

④ Multiply.

GS (a) $\dfrac{2r + 4}{5r} \cdot \dfrac{3r}{5r + 10}$

$= \dfrac{2(\underline{\quad\quad})}{5r} \cdot \dfrac{3r}{5(\underline{\quad\quad})}$

Now divide out all common factors.

$= \dfrac{2 \cdot \underline{\quad}}{5 \cdot \underline{\quad}}$

$= \underline{\quad\quad}$

(b) $\dfrac{c^2 + 2c}{c^2 - 4} \cdot \dfrac{c^2 - 4c + 4}{c^2 - c}$

(c) $\dfrac{m^2 - 16}{m + 2} \cdot \dfrac{1}{m + 4}$

(d)

$\dfrac{x - 3}{x^2 + 2x - 15} \cdot \dfrac{x^2 - 25}{x^2 + 3x - 40}$

Answers

4. (a) $r + 2$; $r + 2$; 3; 5; $\dfrac{6}{25}$

  (b) $\dfrac{c - 2}{c - 1}$ **(c)** $\dfrac{m - 4}{m + 2}$ **(d)** $\dfrac{1}{x + 8}$

5 Find each reciprocal.

(a) $\dfrac{-3}{r}$

(b) $\dfrac{7}{y+8}$

(c) $\dfrac{a^2 + 7a}{2a - 1}$

(d) $\dfrac{0}{-5}$

6 Divide.

(a) $\dfrac{16k^2}{5} \div \dfrac{3k}{10}$

(b) $\dfrac{5p + 2}{6} \div \dfrac{15p + 6}{5}$

(c) $\dfrac{y^2 - 2y - 3}{y^2 + 4y + 4} \div \dfrac{y^2 - 1}{y^2 + y - 2}$

Answers

5. (a) $\dfrac{r}{-3}$ (b) $\dfrac{y+8}{7}$ (c) $\dfrac{2a-1}{a^2+7a}$
   (d) There is no reciprocal.
6. (a) $\dfrac{32k}{3}$ (b) $\dfrac{5}{18}$ (c) $\dfrac{y-3}{y+2}$

The examples in the table on the previous page suggest the following.

> **Finding the Reciprocal**
>
> To find the reciprocal of a nonzero rational expression, interchange the numerator and denominator of the expression.

◀ **Work Problem** **5** at the Side.

OBJECTIVE **6** **Divide rational expressions.** Dividing rational expressions is like dividing rational numbers.

> **Dividing Rational Expressions**
>
> To divide two rational expressions, multiply the first expression (the dividend) by the reciprocal of the second expression (the divisor).

EXAMPLE 5 **Dividing Rational Expressions**

Divide.

(a) $\dfrac{2z}{9} \div \dfrac{5z^2}{18}$

$= \dfrac{2z}{9} \cdot \dfrac{18}{5z^2}$ Multiply by the reciprocal of the divisor.

$= \dfrac{2z}{9} \cdot \dfrac{2 \cdot 9}{5 \cdot z \cdot z}$ Factor.

$= \dfrac{4}{5z}$ Fundamental property; Multiply.

(b) $\dfrac{8k - 16}{3k} \div \dfrac{3k - 6}{4k^2}$

$= \dfrac{8k - 16}{3k} \cdot \dfrac{4k^2}{3k - 6}$ Multiply by the reciprocal.

$= \dfrac{8(k - 2)}{3k} \cdot \dfrac{4k \cdot k}{3(k - 2)}$ Factor.

$= \dfrac{32k}{9}$ Fundamental property; Multiply.

(c) $\dfrac{5m^2 + 17m - 12}{3m^2 + 7m - 20} \div \dfrac{5m^2 + 2m - 3}{15m^2 - 34m + 15}$

$= \dfrac{5m^2 + 17m - 12}{3m^2 + 7m - 20} \cdot \dfrac{15m^2 - 34m + 15}{5m^2 + 2m - 3}$ Definition of division

$= \dfrac{(5m - 3)(m + 4)}{(m + 4)(3m - 5)} \cdot \dfrac{(3m - 5)(5m - 3)}{(5m - 3)(m + 1)}$ Factor.

$= \dfrac{5m - 3}{m + 1}$ Fundamental property; Multiply.

◀ **Work Problem** **6** at the Side.

7.1 Exercises

  ▶ MyMathLab®

CONCEPT CHECK *As review, multiply or divide the rational numbers as indicated. Write answers in lowest terms.*

1. $\dfrac{4}{21} \cdot \dfrac{7}{10}$

2. $\dfrac{5}{9} \cdot \dfrac{12}{25}$

3. $\dfrac{3}{8} \div \dfrac{5}{12}$

4. $\dfrac{5}{6} \div \dfrac{14}{15}$

5. $\dfrac{2}{3} \div \dfrac{8}{9}$

6. $\dfrac{3}{8} \div \dfrac{9}{14}$

7. CONCEPT CHECK Rational expressions can often be written in lowest terms in seemingly different ways. For example,

$$\frac{y-3}{-5} \quad \text{and} \quad \frac{-y+3}{5}$$

look different, but we obtain the second expression by multiplying the first by -1 in both the numerator and denominator. To practice recognizing equivalent rational expressions, match the expressions in parts (a)–(f) with their equivalents in choices A–F.

(a) $\dfrac{x-3}{x+4}$ **(b)** $\dfrac{x+3}{x-4}$ **(c)** $\dfrac{x-3}{x-4}$ **(d)** $\dfrac{x+3}{x+4}$ **(e)** $\dfrac{3-x}{x+4}$ **(f)** $\dfrac{x+3}{4-x}$

A. $\dfrac{-x-3}{4-x}$ **B.** $\dfrac{-x-3}{-x-4}$ **C.** $\dfrac{3-x}{-x-4}$ **D.** $\dfrac{-x+3}{-x+4}$ **E.** $\dfrac{x-3}{-x-4}$ **F.** $\dfrac{-x-3}{x-4}$

8. CONCEPT CHECK Which rational expressions are equivalent to $-\dfrac{x}{y}$?

A. $\dfrac{-x}{-y}$ **B.** $\dfrac{x}{-y}$ **C.** $\dfrac{x}{y}$ **D.** $-\dfrac{x}{-y}$ **E.** $\dfrac{-x}{y}$ **F.** $-\dfrac{-x}{-y}$

For each rational function, find all numbers that are not in the domain. Then give the domain, using set-builder notation. **See Example 1.**

9. $f(x) = \dfrac{x}{x-7}$

10. $f(x) = \dfrac{x}{x+3}$

11. $f(x) = \dfrac{6x-5}{7x+1}$

12. $f(x) = \dfrac{8x-3}{2x+7}$

13. $f(x) = \dfrac{12x+3}{x}$

14. $f(x) = \dfrac{9x+8}{x}$

15. $f(x) = \dfrac{3x+1}{2x^2+x-6}$

16. $f(x) = \dfrac{2x+4}{3x^2+11x-42}$

17. $f(x) = \dfrac{x+2}{14}$

18. $f(x) = \dfrac{x-9}{26}$

19. $f(x) = \dfrac{2x^2-3x+4}{3x^2+8}$

20. $f(x) = \dfrac{9x^2-8x+3}{4x^2+1}$

21. CONCEPT CHECK Identify the two *terms* in the numerator and the two *terms* in the denominator of the following rational expression, and then write it in lowest terms.

$$\frac{x^2 + 4x}{x + 4}$$

22. CONCEPT CHECK Which rational expression can be simplified?

A. $\dfrac{x^2 + 2}{x^2}$ **B.** $\dfrac{x^2 + 2}{2}$

C. $\dfrac{x^2 + y^2}{y^2}$ **D.** $\dfrac{x^2 - 5x}{x}$

23. CONCEPT CHECK Which rational expression is *not* equivalent to $\frac{x-3}{4-x}$?

A. $\dfrac{3-x}{x-4}$ **B.** $\dfrac{x+3}{4+x}$

C. $-\dfrac{3-x}{4-x}$ **D.** $-\dfrac{x-3}{x-4}$

24. CONCEPT CHECK Which rational expressions are equivalent to -1?

A. $\dfrac{2x+3}{2x-3}$ **B.** $\dfrac{2x-3}{3-2x}$

C. $\dfrac{2x+3}{3+2x}$ **D.** $\dfrac{2x+3}{-2x-3}$

Write each rational expression in lowest terms. ***See Example 2.***

25. $\dfrac{x^2(x+1)}{x(x+1)}$

26. $\dfrac{y^3(y-4)}{y^2(y-4)}$

27. $\dfrac{(x+4)(x-3)}{(x+5)(x+4)}$

28. $\dfrac{(2x+7)(x-1)}{(2x+3)(2x+7)}$

29. $\dfrac{4x(x+3)}{8x^2(x-3)}$

30. $\dfrac{5y^2(y+8)}{15y(y-8)}$

31. $\dfrac{3x+7}{3}$

32. $\dfrac{4x-9}{4}$

33. $\dfrac{6m+18}{7m+21}$

34. $\dfrac{5r-20}{3r-12}$

35. $\dfrac{3z^2+z}{18z+6}$

36. $\dfrac{2x^2-5x}{16x-40}$

37. $\dfrac{2t+6}{t^2-9}$

38. $\dfrac{5s-25}{s^2-25}$

39. $\dfrac{x^2+2x-15}{x^2+6x+5}$

40. $\dfrac{y^2 - 5y - 14}{y^2 + y - 2}$

41. $\dfrac{8x^2 - 10x - 3}{8x^2 - 6x - 9}$

42. $\dfrac{12x^2 - 4x - 5}{8x^2 - 6x - 5}$

43. $\dfrac{a^3 + b^3}{a + b}$

44. $\dfrac{r^3 - s^3}{r - s}$

45. $\dfrac{2c^2 + 2cd - 60d^2}{2c^2 - 12cd + 10d^2}$

46. $\dfrac{3s^2 - 9st - 54t^2}{3s^2 - 6st - 72t^2}$

47. $\dfrac{ac - ad + bc - bd}{ac - ad - bc + bd}$

48. $\dfrac{2xy + 2xw + y + w}{2xy + y - 2xw - w}$

Write each rational expression in lowest terms. ***See Example 3.***

49. $\dfrac{7 - b}{b - 7}$

50. $\dfrac{r - 13}{13 - r}$

51. $\dfrac{x^2 - 4}{2 - x}$

52. $\dfrac{x^2 - 81}{9 - x}$

53. $\dfrac{x^2 - y^2}{y - x}$

54. $\dfrac{m^2 - n^2}{n - m}$

55. $\dfrac{(a - 3)(x + y)}{(3 - a)(x - y)}$

56. $\dfrac{(8 - p)(x + 2)}{(p - 8)(x - 2)}$

57. $\dfrac{5k - 10}{20 - 10k}$

58. $\dfrac{7x - 21}{63 - 21x}$

59. $\dfrac{a^2 - b^2}{a^2 + b^2}$

60. $\dfrac{p^2 + q^2}{p^2 - q^2}$

Multiply or divide as indicated. ***See Examples 4 and 5.***

61. $\dfrac{(x + 2)(x + 1)}{(x + 3)(x - 2)} \cdot \dfrac{(x + 3)(x + 4)}{(x + 2)(x + 1)}$

62. $\dfrac{(x + 3)(x - 6)}{(x - 4)(x + 2)} \cdot \dfrac{(x + 5)(x - 4)}{(x + 3)(x - 6)}$

63. $\dfrac{(2x + 3)(x - 4)}{(x + 8)(x - 4)} \div \dfrac{(x - 4)(x + 2)}{(x - 4)(x + 8)}$

64. $\dfrac{(6x + 5)(x - 1)}{(x + 9)(x - 1)} \div \dfrac{(x - 3)(2x + 7)}{(x - 3)(x + 9)}$

65. $\dfrac{7t + 7}{-6} \div \dfrac{4t + 4}{15}$

66. $\dfrac{8z - 16}{-20} \div \dfrac{3z - 6}{40}$

67. $\dfrac{4x}{8x + 4} \cdot \dfrac{14x + 7}{6}$

68. $\dfrac{12x - 20}{5x} \cdot \dfrac{6}{9x - 15}$

69. $\dfrac{p^2 - 25}{4p} \cdot \dfrac{2}{5 - p}$

70. $\dfrac{a^2 - 1}{4a} \cdot \dfrac{2}{1 - a}$

71. $\dfrac{m^2 - 49}{m + 1} \div \dfrac{7 - m}{m}$

72. $\dfrac{k^2 - 4}{3k^2} \div \dfrac{2 - k}{11k}$

73. $\dfrac{12x - 10y}{3x + 2y} \cdot \dfrac{6x + 4y}{10y - 12x}$

74. $\dfrac{9s - 12t}{2s + 2t} \cdot \dfrac{3s + 3t}{4t - 3s}$

75. $\dfrac{x^2 - 25}{x^2 + x - 20} \cdot \dfrac{x^2 + 7x + 12}{x^2 - 2x - 15}$

76. $\dfrac{t^2 - 49}{t^2 + 4t - 21} \cdot \dfrac{t^2 + 8t + 15}{t^2 - 2t - 35}$

77. $\dfrac{8x^3 - 27}{2x^2 - 18} \cdot \dfrac{2x + 6}{8x^2 + 12x + 18}$

78. $\dfrac{64x^3 + 1}{4x^2 - 100} \cdot \dfrac{4x + 20}{64x^2 - 16x + 4}$

79. $\dfrac{6x^2 + 5xy - 6y^2}{12x^2 - 11xy + 2y^2} \div \dfrac{4x^2 - 12xy + 9y^2}{8x^2 - 14xy + 3y^2}$

80. $\dfrac{8a^2 - 6ab - 9b^2}{6a^2 - 5ab - 6b^2} \div \dfrac{4a^2 + 11ab + 6b^2}{9a^2 + 12ab + 4b^2}$

81. $\dfrac{3k^2 + 17kp + 10p^2}{6k^2 + 13kp - 5p^2} \div \dfrac{6k^2 + kp - 2p^2}{6k^2 - 5kp + p^2}$

82. $\dfrac{16c^2 + 24cd + 9d^2}{16c^2 - 16cd + 3d^2} \div \dfrac{16c^2 - 9d^2}{16c^2 - 24cd + 9d^2}$

83. $\left(\dfrac{6k^2 - 13k - 5}{k^2 + 7k} \div \dfrac{2k - 5}{k^3 + 6k^2 - 7k} \right)$
$\cdot \dfrac{k^2 - 5k + 6}{3k^2 - 8k - 3}$

84. $\left(\dfrac{2x^3 + 3x^2 - 2x}{3x - 15} \div \dfrac{2x^3 - x^2}{x^2 - 3x - 10} \right)$
$\cdot \dfrac{5x^2 - 10x}{3x^2 + 12x + 12}$

7.2 Adding and Subtracting Rational Expressions

OBJECTIVE ➊ **Add and subtract rational expressions with the same denominator.** We do this as we would with rational numbers.

OBJECTIVES

➊ **Add and subtract rational expressions with the same denominator.**

➋ **Find a least common denominator.**

➌ **Add and subtract rational expressions with different denominators.**

Adding or Subtracting Rational Expressions

Step 1 **If the denominators are the same,** add or subtract the numerators. Place the result over the common denominator.

If the denominators are different, first find the least common denominator (LCD). Write all rational expressions with this LCD, and then add or subtract the numerators. Place the result over the common denominator.

Step 2 **Simplify.** Write all answers in lowest terms.

EXAMPLE 1   **Adding and Subtracting Rational Expressions (Same Denominators)**

Add or subtract as indicated.

(a) $\dfrac{3y}{5} + \dfrac{x}{5} = \dfrac{3y + x}{5}$ ← Add the numerators.
    ← Keep the common denominator.

(b) $\dfrac{7}{2r^2} - \dfrac{11}{2r^2}$

$ = \dfrac{7 - 11}{2r^2}$ Subtract the numerators.
    Keep the common denominator.

$ = \dfrac{-4}{2r^2},$ or $-\dfrac{2}{r^2}$ Write in lowest terms.

(c) $\dfrac{m}{m^2 - p^2} + \dfrac{p}{m^2 - p^2}$

$ = \dfrac{m + p}{m^2 - p^2}$ Add the numerators.
    Keep the common denominator.

$ = \dfrac{m + p}{(m + p)(m - p)}$ Factor.

$ = \dfrac{1}{m - p}$ Remember to write 1 in the numerator.   Write in lowest terms.

(d) $\dfrac{4}{x^2 + 2x - 8} + \dfrac{x}{x^2 + 2x - 8}$

$ = \dfrac{4 + x}{x^2 + 2x - 8}$ Add.

$ = \dfrac{4 + x}{(x - 2)(x + 4)}$ Factor.

$ = \dfrac{1}{x - 2}$ Write in lowest terms.

·········· **Work Problem** ➊ **at the Side.** ▶

➊ Add or subtract.

(a) $\dfrac{3m}{8} + \dfrac{5n}{8}$

(b) $\dfrac{7}{3a} + \dfrac{10}{3a}$

(c) $\dfrac{2}{y^2} - \dfrac{5}{y^2}$

(d) $\dfrac{a}{a + b} + \dfrac{b}{a + b}$

(e) $\dfrac{2y - 1}{y^2 + y - 2} - \dfrac{y}{y^2 + y - 2}$

Answers

1. (a) $\dfrac{3m + 5n}{8}$ (b) $\dfrac{17}{3a}$

  (c) $-\dfrac{3}{y^2}$ (d) 1 (e) $\dfrac{1}{y + 2}$

2 Find the LCD for each group of denominators.

(GS) **(a)** $5k^3s$, $10ks^4$

$$5k^3s = 5 \cdot k^3 \cdot s$$

$$10ks^4 = 2 \cdot \underline{} \cdot k \cdot s^4$$

For the LCD, include each factor that appears in either of the two monomials raised to the (*least* / *greatest*) power.

$$LCD = 2 \cdot \underline{} \cdot k\text{---} \cdot s\text{---}$$

$$= \underline{}$$

(b) $3 - x$, $9 - x^2$

(c) z, $z + 6$

(d) $2y^2 - 3y - 2$,
$2y^2 + 3y + 1$

(e) $x^2 - 2x + 1$,
$x^2 - 4x + 3$,
$4x - 4$

Answers

2. **(a)** 5; greatest; 5; 3; 4; $10k^3s^4$
   (b) $(3 + x)(3 - x)$
   (c) $z(z + 6)$
   (d) $(y - 2)(2y + 1)(y + 1)$
   (e) $4(x - 3)(x - 1)^2$

OBJECTIVE 2 **Find a least common denominator.** We add or subtract rational expressions with different denominators by first writing them with a common denominator, usually the **least common denominator (LCD).**

Finding the Least Common Denominator

Step 1 **Factor** each denominator.

Step 2 **Find the least common denominator.** The LCD is the product of all different factors from each denominator, with each factor raised to the *greatest* power that occurs in any denominator.

EXAMPLE 2   **Finding Least Common Denominators**

Suppose that the given expressions are denominators of fractions. Find the LCD for each group of denominators.

(a) $5xy^2$, $2x^3y$

Each denominator is already factored.

$$5xy^2 = 5 \cdot x \cdot y^2$$

$$2x^3y = 2 \cdot x^3 \cdot y$$

$$LCD = 5 \cdot 2 \cdot x^3 \cdot y^2 \quad \substack{\text{Greatest exponent on } x \text{ is 3.} \\ \leftarrow \text{Greatest exponent on } y \text{ is 2.}}$$

$$= 10x^3y^2$$

(b) $k - 3$, k

Each denominator is already factored. The LCD must be divisible by *both* $k - 3$ and k.

$$\boxed{\substack{\text{Don't forget} \\ \text{the factor } k.}} \longrightarrow k(k - 3)$$

It is usually best to leave a least common denominator in factored form.

(c) $y^2 - 2y - 8$, $y^2 + 3y + 2$

Factor the denominators.

$$\left.\begin{array}{l} y^2 - 2y - 8 = (y - 4)(y + 2) \\ y^2 + 3y + 2 = (y + 2)(y + 1) \end{array}\right\} \text{Factor.}$$

$$LCD = (y - 4)(y + 2)(y + 1)$$

(d) $8z - 24$, $5z^2 - 15z$

$$\left.\begin{array}{l} 8z - 24 = 8(z - 3) \\ 5z^2 - 15z = 5z(z - 3) \end{array}\right\} \text{Factor.}$$

$$LCD = 8 \cdot 5z \cdot (z - 3), \quad \text{or} \quad 40z(z - 3)$$

(e) $m^2 + 5m + 6$, $m^2 + 4m + 4$, $2m + 6$

$$\left.\begin{array}{l} m^2 + 5m + 6 = (m + 3)(m + 2) \\ m^2 + 4m + 4 = (m + 2)^2 \\ 2m + 6 = 2(m + 3) \end{array}\right\} \text{Factor.}$$

$$LCD = 2(m + 3)(m + 2)^2$$

◀ **Work Problem 2** at the Side.

OBJECTIVE **3** **Add and subtract rational expressions with different denominators.** Before adding or subtracting such rational expressions, we must write each expression with the least common denominator by multiplying its numerator and denominator by the factors needed to get the LCD. This procedure is valid because we are multiplying each rational expression by a form of 1, the identity element for multiplication.

Consider the sum $\frac{7}{15} + \frac{5}{12}$.

$$\frac{7}{15} + \frac{5}{12}$$ The LCD for 15 and 12 is 60.

$\frac{4}{4}$ and $\frac{5}{5}$ are forms of 1.

$$= \frac{7 \cdot 4}{15 \cdot 4} + \frac{5 \cdot 5}{12 \cdot 5}$$ Fundamental property

$$= \frac{28}{60} + \frac{25}{60}$$ Write each fraction with the common denominator.

$$= \frac{28 + 25}{60}$$ Add the numerators. Keep the common denominator.

$$= \frac{53}{60}$$

EXAMPLE 3 **Adding and Subtracting Rational Expressions (Different Denominators)**

Add or subtract as indicated.

(a) $\frac{5}{2p} + \frac{3}{8p}$ The LCD for $2p$ and $8p$ is $8p$.

$$= \frac{5 \cdot 4}{2p \cdot 4} + \frac{3}{8p}$$ Fundamental property

$$= \frac{20}{8p} + \frac{3}{8p}$$ Write the first fraction with the common denominator.

$$= \frac{20 + 3}{8p}$$ Add the numerators. Keep the common denominator.

$$= \frac{23}{8p}$$

(b) $\frac{6}{r} - \frac{5}{r-3}$ The LCD is $r(r-3)$.

$$= \frac{6(r-3)}{r(r-3)} - \frac{r \cdot 5}{r(r-3)}$$ Fundamental property

$$= \frac{6r - 18}{r(r-3)} - \frac{5r}{r(r-3)}$$ Distributive and commutative properties

$$= \frac{6r - 18 - 5r}{r(r-3)}$$ Subtract the numerators. Keep the common denominator.

$$= \frac{r - 18}{r(r-3)}$$ Combine like terms in the numerator.

Work Problem **3** **at the Side.** ▶

3 Add or subtract.

(a) $\frac{6}{7} + \frac{1}{5}$

(b) $\frac{8}{3k} - \frac{2}{9k}$

(c) $\frac{2}{y} - \frac{1}{y+4}$

Answers

3. **(a)** $\frac{37}{35}$ **(b)** $\frac{22}{9k}$ **(c)** $\frac{y+8}{y(y+4)}$

4 Subtract.

GS **(a)** $\dfrac{5x+7}{2x+7} - \dfrac{-x-14}{2x+7}$

The denominators here are (*the same / different*), so we subtract the numerators and keep the _____.

$$\dfrac{5x+7}{2x+7} - \dfrac{-x-14}{2x+7}$$

$$= \dfrac{5x+7-(\underline{})}{\underline{}}$$

$$= \dfrac{5x+7+\underline{}}{2x+7}$$

$$= \dfrac{6x+\underline{}}{2x+7}$$

$$= \dfrac{\underline{}(2x+7)}{(2x+7)}$$

$$= \underline{}$$

(b) $\dfrac{18x-7}{4x-5} - \dfrac{2x+13}{4x-5}$

(c) $\dfrac{2}{r-2} - \dfrac{r}{r-1}$

CAUTION

Sign errors occur easily when a rational expression with two or more terms in the numerator is being subtracted. *In this case, the subtraction sign must be distributed to every term in the numerator of the fraction that follows it.* Study **Example 4** carefully to see how this is done.

EXAMPLE 4 **Subtracting Rational Expressions**

Subtract.

(a) $\dfrac{7x}{3x+1} - \dfrac{x-2}{3x+1}$

The denominators are the same for both rational expressions. *The subtraction sign must be applied to both terms in the numerator of the second rational expression.*

$$\dfrac{7x}{3x+1} - \dfrac{x-2}{3x+1}$$ Use parentheses to avoid errors.

$$= \dfrac{7x-(x-2)}{3x+1}$$ Subtract the numerators. Keep the common denominator.

$$= \dfrac{7x-x+2}{3x+1}$$ Be careful with signs. Distributive property

$$= \dfrac{6x+2}{3x+1}$$ Combine like terms in the numerator.

$$= \dfrac{2(3x+1)}{3x+1}$$ Factor the numerator.

$$= 2$$ Write in lowest terms.

(b) $\dfrac{1}{q-1} - \dfrac{1}{q+1}$ The LCD is $(q-1)(q+1)$.

$$= \dfrac{1(q+1)}{(q-1)(q+1)} - \dfrac{1(q-1)}{(q+1)(q-1)}$$ Fundamental property

$$= \dfrac{(q+1)-(q-1)}{(q-1)(q+1)}$$ Subtract the numerators. Keep the common denominator.

$$= \dfrac{q+1-q+1}{(q-1)(q+1)}$$ Be careful with signs. Distributive property

$$= \dfrac{2}{(q-1)(q+1)}$$ Combine like terms in the numerator.

◀ **Work Problem 4** at the Side.

In some problems, rational expressions to be added or subtracted have denominators that are opposites of each other, such as

$$\dfrac{y}{y-2} + \dfrac{8}{2-y}.$$ Denominators are opposites.

The next example illustrates how to proceed in such a problem.

EXAMPLE 5 Adding Rational Expressions (Denominators Are Opposites)

Add.

$$\frac{y}{y-2} + \frac{8}{2-y} \longleftarrow \text{Denominators are opposites.}$$

$$= \frac{y}{y-2} + \frac{8(-1)}{(2-y)(-1)} \qquad \text{Multiply the second expression by } \frac{-1}{-1}.$$

$$= \frac{y}{y-2} + \frac{-8}{y-2} \longleftarrow \qquad \text{The LCD is } y-2.$$

$$= \frac{y-8}{y-2} \qquad \text{Add the numerators.}$$

Alternatively, we could use $2 - y$ as the common denominator and rewrite the first expression.

$$\frac{y}{y-2} + \frac{8}{2-y}$$

$$= \frac{y(-1)}{(y-2)(-1)} + \frac{8}{2-y} \qquad \text{Multiply the first expression by } \frac{-1}{-1}.$$

$$= \frac{-y+8}{2-y} \qquad \text{The LCD is } 2-y.$$

$$= \frac{8-y}{2-y} \qquad \text{This is an equivalent form of the answer given in color above.}$$

···· **Work Problem ⑤ at the Side.** ▶

EXAMPLE 6 Adding and Subtracting Three Rational Expressions

Add and subtract as indicated.

$$\frac{3}{x-2} + \frac{5}{x} - \frac{6}{x^2-2x}$$

$$= \frac{3}{x-2} + \frac{5}{x} - \frac{6}{x(x-2)} \qquad \text{Factor the third denominator.}$$

$$= \frac{3x}{x(x-2)} + \frac{5(x-2)}{x(x-2)} - \frac{6}{x(x-2)} \qquad \begin{array}{l}\text{The LCD is } x(x-2);\\ \text{fundamental property}\end{array}$$

$$= \frac{3x + 5(x-2) - 6}{x(x-2)} \qquad \text{Add and subtract the numerators.}$$

$$= \frac{3x + 5x - 10 - 6}{x(x-2)} \qquad \text{Distributive property}$$

$$= \frac{8x - 16}{x(x-2)} \qquad \text{Combine like terms in the numerator.}$$

$$= \frac{8(x-2)}{x(x-2)} \qquad \text{Factor the numerator.}$$

$$= \frac{8}{x} \qquad \text{Lowest terms}$$

···· **Work Problem ⑥ at the Side.** ▶

⑤ Add or subtract as indicated.

(a) $\dfrac{8}{x-4} + \dfrac{2}{4-x}$

(b) $\dfrac{9}{2x-9} - \dfrac{4}{9-2x}$

⑥ Add and subtract as indicated.

$$\frac{4}{x-5} + \frac{-2}{x} - \frac{10}{x^2-5x}$$

Answers

5. (a) $\dfrac{6}{x-4}$, or $\dfrac{-6}{4-x}$ (b) $\dfrac{13}{2x-9}$, or $\dfrac{-13}{9-2x}$

6. $\dfrac{2}{x-5}$

7 Subtract.

$$\frac{-a}{a^2 + 3a - 4} - \frac{4a}{a^2 + 7a + 12}$$

EXAMPLE 7 **Subtracting Rational Expressions**

Subtract.

$$\frac{m + 4}{m^2 - 2m - 3} - \frac{2m - 3}{m^2 - 5m + 6}$$

$$= \frac{m + 4}{(m - 3)(m + 1)} - \frac{2m - 3}{(m - 3)(m - 2)} \quad \text{Factor each denominator.}$$

$$= \frac{(m + 4)(m - 2)}{(m - 3)(m + 1)(m - 2)} - \frac{(2m - 3)(m + 1)}{(m - 3)(m - 2)(m + 1)} \quad \begin{array}{l}\text{Fundamental}\\\text{property}\end{array}$$

The LCD is $(m - 3)(m + 1)(m - 2)$.

$$= \frac{(m + 4)(m - 2) - (2m - 3)(m + 1)}{(m - 3)(m + 1)(m - 2)} \quad \text{Subtract the numerators.}$$

> Note the careful use of parentheses.

$$= \frac{m^2 + 2m - 8 - (2m^2 - m - 3)}{(m - 3)(m + 1)(m - 2)} \quad \text{Multiply in the numerator.}$$

> Be careful with signs.

$$= \frac{m^2 + 2m - 8 - 2m^2 + m + 3}{(m - 3)(m + 1)(m - 2)} \quad \text{Distributive property}$$

$$= \frac{-m^2 + 3m - 5}{(m - 3)(m + 1)(m - 2)} \quad \begin{array}{l}\text{Combine like terms in the}\\\text{numerator.}\end{array}$$

If we try to factor the numerator, we find that this rational expression is in lowest terms.

◀ Work Problem **7** at the Side.

8 Add.

$$\frac{4}{p^2 - 6p + 9} + \frac{1}{p^2 + 2p - 15}$$

EXAMPLE 8 **Adding Rational Expressions**

Add.

$$\frac{5}{x^2 + 10x + 25} + \frac{2}{x^2 + 7x + 10}$$

$$= \frac{5}{(x + 5)^2} + \frac{2}{(x + 5)(x + 2)} \quad \text{Factor each denominator.}$$

$$= \frac{5(x + 2)}{(x + 5)^2(x + 2)} + \frac{2(x + 5)}{(x + 5)(x + 2)(x + 5)} \quad \begin{array}{l}\text{The LCD is}\\(x + 5)^2(x + 2);\\\text{fundamental property}\end{array}$$

$$= \frac{5(x + 2) + 2(x + 5)}{(x + 5)^2(x + 2)} \quad \text{Add the numerators.}$$

$$= \frac{5x + 10 + 2x + 10}{(x + 5)^2(x + 2)} \quad \text{Distributive property}$$

$$= \frac{7x + 20}{(x + 5)^2(x + 2)} \quad \begin{array}{l}\text{Combine like terms in the}\\\text{numerator.}\end{array}$$

◀ Work Problem **8** at the Side.

Answers

7. $\dfrac{-5a^2 + a}{(a + 4)(a - 1)(a + 3)}$

8. $\dfrac{5p + 17}{(p - 3)^2(p + 5)}$

7.2 Exercises

CONCEPT CHECK *As review, add or subtract the rational numbers as indicated. Write answers in lowest terms.*

1. $\dfrac{8}{15} + \dfrac{4}{15}$

2. $\dfrac{5}{16} + \dfrac{9}{16}$

3. $\dfrac{5}{6} - \dfrac{8}{9}$

4. $\dfrac{3}{4} - \dfrac{5}{6}$

5. $\dfrac{5}{18} + \dfrac{7}{12}$

6. $\dfrac{3}{10} + \dfrac{7}{15}$

Add or subtract as indicated. Write all answers in lowest terms. **See Example 1.**

7. $\dfrac{7}{t} + \dfrac{2}{t}$

8. $\dfrac{5}{r} + \dfrac{9}{r}$

9. $\dfrac{11}{5x} - \dfrac{1}{5x}$

10. $\dfrac{7}{4y} - \dfrac{3}{4y}$

11. $\dfrac{5x + 4}{6x + 5} + \dfrac{x + 1}{6x + 5}$

12. $\dfrac{6y + 12}{4y + 3} + \dfrac{2y - 6}{4y + 3}$

13. $\dfrac{x^2}{x + 5} - \dfrac{25}{x + 5}$

14. $\dfrac{y^2}{y + 6} - \dfrac{36}{y + 6}$

15. $\dfrac{4}{p^2 + 7p + 12} + \dfrac{p}{p^2 + 7p + 12}$

16. $\dfrac{5}{x^2 + x - 20} + \dfrac{x}{x^2 + x - 20}$

17. $\dfrac{a^3}{a^2 + ab + b^2} - \dfrac{b^3}{a^2 + ab + b^2}$

18. $\dfrac{p^3}{p^2 - pq + q^2} + \dfrac{q^3}{p^2 - pq + q^2}$

Suppose that the expressions given are denominators of fractions. Find the least common denominator (LCD) for each group. **See Example 2.**

19. $18x^2y^3, \quad 24x^4y^5$

20. $24a^3b^4, \quad 18a^5b^2$

21. $z - 2, \quad z$

22. $k + 3, \quad k$

23. $2y + 8, \quad y + 4$

24. $3r - 21, \quad r - 7$

25. $x^2 - 81, \quad x^2 + 18x + 81$

26. $y^2 - 16, \quad y^2 - 8y + 16$

27. $m + n, \quad m - n, \quad m^2 - n^2$

28. $r + s, \quad r - s, \quad r^2 - s^2$

29. $x^2 - 3x - 4, \quad x + x^2$

30. $y^2 - 8y + 12, \quad y^2 - 6y$

31. $2t^2 + 7t - 15, \quad t^2 + 3t - 10$

32. $s^2 - 3s - 4, \quad 3s^2 + s - 2$

33. $2y + 6, \quad y^2 - 9, \quad y$

34. $9x + 18, \quad x^2 - 4, \quad x$

35. CONCEPT CHECK Consider the following incorrect work. **What Went Wrong?**

$$\frac{x}{x+2} - \frac{4x-1}{x+2}$$

$$= \frac{x - 4x - 1}{x+2}$$

$$= \frac{-3x - 1}{x+2}$$

36. One student added two rational expressions and obtained the answer

$$\frac{3}{5-y}.$$

Another student obtained the answer

$$\frac{-3}{y-5}$$

for the same problem. Is it possible that both answers are correct? Explain.

Add or subtract as indicated. Write all answers in lowest terms. ***See Examples 3–8.***

37. $\dfrac{8}{t} + \dfrac{7}{3t}$

38. $\dfrac{5}{x} + \dfrac{9}{4x}$

39. $\dfrac{5}{12x^2y} - \dfrac{11}{6xy}$

40. $\dfrac{7}{18a^3b^2} - \dfrac{2}{9ab}$

41. $\dfrac{1}{x-1} - \dfrac{1}{x}$

42. $\dfrac{3}{x-3} - \dfrac{1}{x}$

43. $\dfrac{3a}{a+1} + \dfrac{2a}{a-3}$

44. $\dfrac{2x}{x+4} + \dfrac{3x}{x-7}$

45. $\dfrac{17y+3}{9y+7} - \dfrac{-10y-18}{9y+7}$

46. $\dfrac{7x+8}{3x+2} - \dfrac{x+4}{3x+2}$

47. $\dfrac{2}{4-x} + \dfrac{5}{x-4}$

48. $\dfrac{3}{2-t} + \dfrac{1}{t-2}$

49. $\dfrac{w}{w-z} - \dfrac{z}{z-w}$

50. $\dfrac{a}{a-b} - \dfrac{b}{b-a}$

51. $\dfrac{5}{12+4x} - \dfrac{7}{9+3x}$

52. $\dfrac{3}{10x+15} - \dfrac{8}{12x+18}$

53. $\dfrac{4x}{x-1} - \dfrac{2}{x+1} - \dfrac{4}{x^2-1}$

54. $\dfrac{4}{x+3} - \dfrac{x}{x-3} - \dfrac{18}{x^2-9}$

55. $\dfrac{15}{y^2+3y} + \dfrac{2}{y} + \dfrac{5}{y+3}$

56. $\dfrac{7}{t-2} - \dfrac{6}{t^2-2t} - \dfrac{3}{t}$

57. $\dfrac{5}{x-2} + \dfrac{1}{x} + \dfrac{2}{x^2-2x}$

58. $\dfrac{5x}{x-3} + \dfrac{2}{x} + \dfrac{6}{x^2-3x}$

59. $\dfrac{3x}{x+1} + \dfrac{4}{x-1} - \dfrac{6}{x^2-1}$

60. $\dfrac{5x}{x+3} + \dfrac{x+2}{x} - \dfrac{6}{x^2+3x}$

61. $\dfrac{4}{x+1} + \dfrac{1}{x^2-x+1} - \dfrac{12}{x^3+1}$

62. $\dfrac{5}{x+2} + \dfrac{2}{x^2-2x+4} - \dfrac{60}{x^3+8}$

63. $\dfrac{2x+4}{x+3} + \dfrac{3}{x} - \dfrac{6}{x^2+3x}$

64. $\dfrac{4x+1}{x+5} - \dfrac{2}{x} + \dfrac{10}{x^2+5x}$

65. $\dfrac{3}{x^2-5x+6} - \dfrac{2}{x^2-4x+4}$

66. $\dfrac{2}{m^2-4m+4} + \dfrac{3}{m^2+m-6}$

67. $\dfrac{3}{x^2+4x+4} + \dfrac{7}{x^2+5x+6}$

68. $\dfrac{5}{x^2+6x+9} - \dfrac{2}{x^2+4x+3}$

69. $\dfrac{5x}{x^2+xy-2y^2} - \dfrac{3x}{x^2+5xy-6y^2}$

70. $\dfrac{6x}{6x^2+5xy-4y^2} - \dfrac{2y}{9x^2-16y^2}$

*A **concours d'elegance** is a competition in which a maximum of* 100 *points is awarded to a car based on its general attractiveness. The function defined by the rational expression*

$$c(x) = \dfrac{1010}{49(101-x)} - \dfrac{10}{49}$$

approximates the cost, in thousands of dollars, of restoring a car so that it will win x points.
  Use this information to work Exercises 71 and 72.

71. Simplify the expression for $c(x)$ by performing the indicated subtraction.

72. Use the simplified expression to determine how much it would cost to win 95 points.

7.3 Complex Fractions

OBJECTIVES

1. Simplify complex fractions by simplifying the numerator and denominator (Method 1).
2. Simplify complex fractions by multiplying by a common denominator (Method 2).
3. Compare the two methods of simplifying complex fractions.
4. Simplify rational expressions with negative exponents.

A **complex fraction** is a quotient that has a fraction in the numerator, denominator, or both.

$$\frac{1 + \dfrac{1}{x}}{2}, \quad \frac{\dfrac{4}{y}}{6 - \dfrac{3}{y}}, \quad \text{and} \quad \frac{\dfrac{m^2 - 9}{m + 1}}{\dfrac{m + 3}{m^2 - 1}} \qquad \text{Complex fractions}$$

OBJECTIVE 1 Simplify complex fractions by simplifying the numerator and denominator (Method 1). There are two different methods for simplifying complex fractions.

Simplifying a Complex Fraction (Method 1)

Step 1 Simplify the numerator and denominator separately.

Step 2 Divide by multiplying the numerator by the reciprocal of the denominator.

Step 3 Simplify the resulting fraction, if possible.

In Step 2, we are treating the complex fraction as a quotient of two rational expressions and dividing. ***Before performing this step, be sure that both the numerator and denominator are single fractions.***

$$\frac{q - 5}{8} \leftarrow \text{Single fraction} \qquad \frac{\dfrac{1}{x} + x}{\dfrac{x^2 + 1}{8}} \leftarrow \text{Not a single fraction} \qquad \frac{6 + \dfrac{3}{x}}{\dfrac{x}{4} + \dfrac{7}{8}} \leftarrow \text{Not a single fraction}$$

$$\frac{q + 5}{3} \leftarrow \text{Single fraction} \qquad \qquad \leftarrow \text{Single fraction} \qquad \qquad \leftarrow \text{Not a single fraction}$$

EXAMPLE 1 Simplifying Complex Fractions (Method 1)

Use Method 1 to simplify each complex fraction.

(a) $\dfrac{\dfrac{x + 1}{x}}{\dfrac{x - 1}{2x}}$

Both the numerator and the denominator are single fractions. Each is already simplified. (Step 1)

$$= \frac{x + 1}{x} \div \frac{x - 1}{2x} \qquad \text{Write as a division problem.}$$

$$= \frac{x + 1}{x} \cdot \frac{2x}{x - 1} \qquad \text{Multiply by the reciprocal of } \tfrac{x-1}{2x}. \text{ (Step 2)}$$

$$= \frac{2x(x + 1)}{x(x - 1)} \qquad \text{Multiply.}$$

$$= \frac{2(x + 1)}{x - 1} \qquad \text{Simplify. (Step 3)}$$

Continued on Next Page

(b)
$$\frac{2+\dfrac{1}{y}}{3-\dfrac{2}{y}}$$

The numerator and denominator are *not* single fractions. Simplify them separately. (Step 1)

$$=\frac{\dfrac{2y}{y}+\dfrac{1}{y}}{\dfrac{3y}{y}-\dfrac{2}{y}}$$

Prepare to write the numerator and denominator as single fractions.

The numerator and denominator are now single fractions.

$$=\frac{\dfrac{2y+1}{y}}{\dfrac{3y-2}{y}}$$

$\dfrac{\frac{2y+1}{y}}{\frac{3y-2}{y}}$ means $\dfrac{2y+1}{y}\div\dfrac{3y-2}{y}$.

$$=\frac{2y+1}{y}\cdot\frac{y}{3y-2}$$

Multiply by the reciprocal of $\frac{3y-2}{y}$. (Step 2)

$$=\frac{2y+1}{3y-2}$$

Multiply and simplify. (Step 3)

························· **Work Problem ❶ at the Side.** ▶

OBJECTIVE ❷ Simplify complex fractions by multiplying by a common denominator (Method 2). This method uses the identity property for multiplication.

Simplifying a Complex Fraction (Method 2)

Step 1 Multiply numerator and denominator of the complex fraction by the least common denominator of the fractions in the numerator and the fractions in the denominator of the complex fraction.

Step 2 Simplify the resulting fraction, if possible.

EXAMPLE 2 Simplifying Complex Fractions (Method 2)

Use Method 2 to simplify each complex fraction.

(a) $\dfrac{2+\dfrac{1}{y}}{3-\dfrac{2}{y}}$ This is the same fraction as in **Example 1(b)** above. Compare the solution methods.

$$=\frac{\left(2+\dfrac{1}{y}\right)\cdot y}{\left(3-\dfrac{2}{y}\right)\cdot y}$$

The LCD of all the fractions is y. Multiply the numerator and denominator by y, since $\frac{y}{y}=1$. (Step 1)

$$=\frac{2\cdot y+\dfrac{1}{y}\cdot y}{3\cdot y-\dfrac{2}{y}\cdot y}$$

Distributive property (Step 2)

$$=\frac{2y+1}{3y-2}$$

Multiply.

···················· **Continued on Next Page**

❶ Use Method 1 to simplify each complex fraction.

(a) $\dfrac{\dfrac{a+2}{5a}}{\dfrac{a-3}{7a}}$

(b) $\dfrac{2+\dfrac{1}{k}}{2-\dfrac{1}{k}}$

(c) $\dfrac{\dfrac{r^2-4}{4}}{1+\dfrac{2}{r}}$

Answers

1. **(a)** $\dfrac{7(a+2)}{5(a-3)}$ **(b)** $\dfrac{2k+1}{2k-1}$ **(c)** $\dfrac{r(r-2)}{4}$

2 Use Method 2 to simplify each complex fraction.

(a) $\dfrac{\dfrac{5}{y} + 6}{\dfrac{8}{3y} - 1}$

(b) $\dfrac{\dfrac{1}{y} + \dfrac{1}{y-1}}{\dfrac{1}{y} - \dfrac{2}{y-1}}$

(b) $\dfrac{2p + \dfrac{5}{p-1}}{3p - \dfrac{2}{p}}$

$= \dfrac{\left(2p + \dfrac{5}{p-1}\right) \cdot p(p-1)}{\left(3p - \dfrac{2}{p}\right) \cdot p(p-1)}$ 　Multiply the numerator and denominator by the LCD, $p(p-1)$. (Step 1)

$= \dfrac{2p[p(p-1)] + \dfrac{5}{p-1} \cdot p(p-1)}{3p[p(p-1)] - \dfrac{2}{p} \cdot p(p-1)}$ 　Distributive property (Step 2)

$= \dfrac{2p[p(p-1)] + 5p}{3p[p(p-1)] - 2(p-1)}$ 　Multiply.

$= \dfrac{2p[p^2 - p] + 5p}{3p[p^2 - p] - 2p + 2}$ 　Distributive property

$= \dfrac{2p^3 - 2p^2 + 5p}{3p^3 - 3p^2 - 2p + 2}$ 　Distributive property again

◀ **Work Problem ❷ at the Side.**

OBJECTIVE 3 **Compare the two methods of simplifying complex fractions.** Some students prefer one method over the other, while other students feel comfortable with both methods and rely on practice with many examples to determine which method to use on a particular problem.

EXAMPLE 3 **Simplifying Complex Fractions (Both Methods)**

Use both Method 1 and Method 2 to simplify each complex fraction.

Method 1	**Method 2**
(a) $\dfrac{\dfrac{2}{x-3}}{\dfrac{5}{x^2-9}}$	(a) $\dfrac{\dfrac{2}{x-3}}{\dfrac{5}{x^2-9}}$
$= \dfrac{\dfrac{2}{x-3}}{\dfrac{5}{(x-3)(x+3)}}$	$= \dfrac{\dfrac{2}{x-3}}{\dfrac{5}{(x-3)(x+3)}}$
$= \dfrac{2}{x-3} \div \dfrac{5}{(x-3)(x+3)}$	$= \dfrac{\dfrac{2}{x-3} \cdot (x-3)(x+3)}{\dfrac{5}{(x-3)(x+3)} \cdot (x-3)(x+3)}$
$= \dfrac{2}{x-3} \cdot \dfrac{(x-3)(x+3)}{5}$	
$= \dfrac{2(x+3)}{5}$	$= \dfrac{2(x+3)}{5}$

Answers

2. **(a)** $\dfrac{15 + 18y}{8 - 3y}$ 　**(b)** $\dfrac{2y-1}{-y-1}$, or $\dfrac{1-2y}{y+1}$

Continued on Next Page

Method 1

(b) $\dfrac{\dfrac{1}{x} + \dfrac{1}{y}}{\dfrac{1}{x^2} - \dfrac{1}{y^2}}$

$= \dfrac{\dfrac{y}{xy} + \dfrac{x}{xy}}{\dfrac{y^2}{x^2y^2} - \dfrac{x^2}{x^2y^2}}$

$= \dfrac{\dfrac{y+x}{xy}}{\dfrac{y^2-x^2}{x^2y^2}}$

$= \dfrac{y+x}{xy} \div \dfrac{y^2-x^2}{x^2y^2}$

$= \dfrac{y+x}{xy} \cdot \dfrac{x^2y^2}{(y-x)(y+x)}$

$= \dfrac{xy}{y-x}$

Method 2

(b) $\dfrac{\dfrac{1}{x} + \dfrac{1}{y}}{\dfrac{1}{x^2} - \dfrac{1}{y^2}}$

$= \dfrac{\left(\dfrac{1}{x} + \dfrac{1}{y}\right) \cdot x^2y^2}{\left(\dfrac{1}{x^2} - \dfrac{1}{y^2}\right) \cdot x^2y^2}$

$= \dfrac{\left(\dfrac{1}{x}\right)x^2y^2 + \left(\dfrac{1}{y}\right)x^2y^2}{\left(\dfrac{1}{x^2}\right)x^2y^2 - \left(\dfrac{1}{y^2}\right)x^2y^2}$

$= \dfrac{xy^2 + x^2y}{y^2 - x^2}$

$= \dfrac{xy(y+x)}{(y+x)(y-x)}$

$= \dfrac{xy}{y-x}$

·····Work Problem **3** at the Side. ▶

OBJECTIVE ▶ **4** **Simplify rational expressions with negative exponents.**
We begin by rewriting these expressions with only positive exponents.

EXAMPLE 4 **Simplifying a Rational Expression with Negative Exponents**

Simplify, using only positive exponents in the answer.

$\dfrac{m^{-1} + p^{-2}}{2m^{-2} - p^{-1}}$ $a^{-n} = \dfrac{1}{a^n}$ (Section 5.5)

$= \dfrac{\dfrac{1}{m} + \dfrac{1}{p^2}}{\dfrac{2}{m^2} - \dfrac{1}{p}}$

Write with positive exponents.
$2m^{-2} = 2 \cdot m^{-2} = \dfrac{2}{1} \cdot \dfrac{1}{m^2} = \dfrac{2}{m^2}$

> The base of $2m^{-2}$ is m, not $2m$:
> $2m^{-2} = \dfrac{2}{m^2}$.

$= \dfrac{m^2p^2\left(\dfrac{1}{m} + \dfrac{1}{p^2}\right)}{m^2p^2\left(\dfrac{2}{m^2} - \dfrac{1}{p}\right)}$

Simplify by Method 2.
Multiply the numerator and denominator by the LCD, m^2p^2.

$= \dfrac{m^2p^2 \cdot \dfrac{1}{m} + m^2p^2 \cdot \dfrac{1}{p^2}}{m^2p^2 \cdot \dfrac{2}{m^2} - m^2p^2 \cdot \dfrac{1}{p}}$

Distributive property

$= \dfrac{mp^2 + m^2}{2p^2 - m^2p}$

Write in lowest terms.

·····Work Problem **4** at the Side. ▶

3 Use both methods to simplify each complex fraction.

(a) $\dfrac{\dfrac{5}{y+2}}{\dfrac{-3}{y^2-4}}$

(b) $\dfrac{\dfrac{1}{a} - \dfrac{1}{b}}{\dfrac{1}{a^2} - \dfrac{1}{b^2}}$

4 Simplify each expression, using only positive exponents in the answer.

(a) $\dfrac{r^{-2} - s^{-1}}{4r^{-1} + s^{-2}}$

(b) $\dfrac{x^{-2} - 2y^{-1}}{y - 2x^2}$

Answers

3. (Both methods give the same answers.)
(a) $\dfrac{5(y-2)}{-3}$ (b) $\dfrac{ab}{b+a}$
4. (a) $\dfrac{s^2 - r^2s}{4rs^2 + r^2}$ (b) $\dfrac{1}{x^2y}$

7.3 Exercises

  MyMathLab®

CONCEPT CHECK *Fill in each blank with the correct response.*

1. A(n) _____ fraction is a quotient that has a fraction in the _____, denominator, or _____ .

2. To use Method 1 for simplifying a complex fraction, both the numerator and denominator must be single fractions. Divide by multiplying the numerator by the _____ of the _____ , and then simplify.

3. To use Method 2 for simplifying a complex fraction, multiply both the numerator and denominator by the _____ of all the fractions in the numerator and denominator. This is an application of the _____ property for multiplication.

4. **CONCEPT CHECK** Find the slope of the line that passes through each pair of points. (*Hint:* This will involve simplifying complex fractions. Recall that slope $m = \frac{y_2 - y_1}{x_2 - x_1}$.)

    (a) $\left(-\frac{5}{2}, \frac{1}{6}\right)$ and $\left(\frac{5}{3}, \frac{3}{8}\right)$ **(b)** $\left(-\frac{5}{6}, -\frac{1}{2}\right)$ and $\left(-\frac{1}{3}, -\frac{3}{2}\right)$

CONCEPT CHECK *Simplify. Write answers in lowest terms.*

5. $\dfrac{\dfrac{2}{3}}{4}$

6. $\dfrac{\dfrac{3}{4}}{\dfrac{5}{12}}$

7. $\dfrac{\dfrac{5}{9} - \dfrac{1}{3}}{\dfrac{2}{3} + \dfrac{1}{6}}$

8. $\dfrac{\dfrac{4}{3} - 2}{1 - \dfrac{3}{8}}$

Use either method to simplify each complex fraction. See Examples 1–3.

9. $\dfrac{\dfrac{12}{x - 1}}{\dfrac{6}{x}}$

10. $\dfrac{\dfrac{24}{t + 4}}{\dfrac{6}{t}}$

11. $\dfrac{\dfrac{k + 1}{2k}}{\dfrac{3k - 1}{4k}}$

12. $\dfrac{\dfrac{1 - r}{4r}}{\dfrac{-1 - r}{8r}}$

13. $\dfrac{\dfrac{4z^2x^4}{9}}{\dfrac{12x^2z^5}{15}}$

14. $\dfrac{\dfrac{3y^2x^3}{8}}{\dfrac{9y^3x^4}{16}}$

15. $\dfrac{\dfrac{1}{x} + 1}{-\dfrac{1}{x} + 1}$

16. $\dfrac{\dfrac{2}{k} - 1}{\dfrac{2}{k} + 1}$

17. $\dfrac{6 + \dfrac{1}{x}}{7 - \dfrac{3}{x}}$

18. $\dfrac{4 - \dfrac{1}{p}}{9 + \dfrac{5}{p}}$

19. $\dfrac{\dfrac{3}{x} + \dfrac{3}{y}}{\dfrac{3}{x} - \dfrac{3}{y}}$

20. $\dfrac{\dfrac{4}{t} - \dfrac{4}{s}}{\dfrac{4}{t} + \dfrac{4}{s}}$

21. $\dfrac{\dfrac{8x - 24y}{10}}{\dfrac{x - 3y}{5x}}$

22. $\dfrac{\dfrac{10x - 5y}{12}}{\dfrac{2x - y}{6y}}$

23. $\dfrac{\dfrac{6}{y - 4}}{\dfrac{12}{y^2 - 16}}$

24. $\dfrac{\dfrac{8}{t + 7}}{\dfrac{24}{t^2 - 49}}$

25. $\dfrac{\dfrac{x^2 - 16y^2}{xy}}{\dfrac{1}{y} - \dfrac{4}{x}}$

26. $\dfrac{\dfrac{4t^2 - 9s^2}{st}}{\dfrac{2}{s} - \dfrac{3}{t}}$

27. $\dfrac{\dfrac{1}{b^2} - \dfrac{1}{a^2}}{\dfrac{1}{b} - \dfrac{1}{a}}$

28. $\dfrac{\dfrac{1}{x^2} - \dfrac{1}{y^2}}{\dfrac{1}{x} + \dfrac{1}{y}}$

29. $\dfrac{x + y}{\dfrac{1}{y} + \dfrac{1}{x}}$

30. $\dfrac{s - r}{\dfrac{1}{r} - \dfrac{1}{s}}$

31. $\dfrac{y - \dfrac{y - 3}{3}}{\dfrac{4}{9} + \dfrac{2}{3y}}$

32. $\dfrac{p - \dfrac{p + 2}{4}}{\dfrac{3}{4} - \dfrac{5}{2p}}$

33. $\dfrac{\dfrac{x + 2}{x} + \dfrac{1}{x + 2}}{\dfrac{5}{x} + \dfrac{x}{x + 2}}$

34. $\dfrac{\dfrac{y + 3}{y} - \dfrac{4}{y - 1}}{\dfrac{y}{y - 1} + \dfrac{1}{y}}$

Simplify each expression, using only positive exponents in the answer. ***See Example 4.***

35. $\dfrac{1}{x^{-2} + y^{-2}}$

36. $\dfrac{1}{p^{-2} - q^{-2}}$

37. $\dfrac{x^{-2} + y^{-2}}{x^{-1} + y^{-1}}$

38. $\dfrac{x^{-1} - y^{-1}}{x^{-2} - y^{-2}}$

39. $\dfrac{2y^{-1} - 3y^{-2}}{y^{-2} + 3x^{-1}}$

40. $\dfrac{k^{-1} + p^{-2}}{k^{-1} - 3p^{-2}}$

41. $\dfrac{x^{-1} + 2y^{-1}}{2y + 4x}$

42. $\dfrac{a^{-2} - 4b^{-2}}{3b - 6a}$

7.4 Equations with Rational Expressions and Graphs

OBJECTIVES

1. Determine the domain of the variable in a rational equation.
2. Solve rational equations.
3. Recognize the graph of a rational function.

1 Find the domain of the variable in each equation.

(a) $\dfrac{3}{x} + \dfrac{1}{2} = \dfrac{5}{6x}$

(b) $\dfrac{1}{x-6} - \dfrac{1}{x+2} = 0$

(c) $\dfrac{4}{x-5} - \dfrac{2}{x+5} = \dfrac{1}{x^2-25}$

In **Section 7.1** we defined the domain of a rational function as the set of all possible values of the variable. (We also refer to this as "the domain of the variable.") Any value that makes the denominator 0 is excluded.

OBJECTIVE ▸ 1 **Determine the domain of the variable in a rational equation.** The **domain of the variable in a rational equation** is the intersection (overlap) of the domains of the rational expressions in the equation.

EXAMPLE 1   Determining Domains of Variables

Find the domain of the variable in each equation.

(a) $\dfrac{2}{x} - \dfrac{3}{2} = \dfrac{7}{2x}$

The domains of the three expressions $\frac{2}{x}$, $\frac{3}{2}$, and $\frac{7}{2x}$ in the equation are,

$$\{x \mid x \neq 0\}, \quad \{x \mid x \text{ is a real number}\}, \quad \text{and} \quad \{x \mid x \neq 0\}.$$

The intersection of these three domains is all real numbers except 0, written using set-builder notation as $\{x \mid x \neq 0\}$.

(b) $\dfrac{2}{x-3} - \dfrac{3}{x+3} = \dfrac{12}{x^2-9}$

The domains of the three expressions are, respectively,

$$\{x \mid x \neq 3\}, \quad \{x \mid x \neq -3\}, \quad \text{and} \quad \{x \mid x \neq \pm 3\}.$$

> $\pm$ is read "positive or negative," or "plus or minus."

The domain of the variable is the intersection of the three domains, all real numbers except 3 and -3, written $\{x \mid x \neq \pm 3\}$.

◀ **Work Problem ❶ at the Side.**

OBJECTIVE ▸ 2 **Solve rational equations.** To solve rational equations, we usually multiply all terms in the equation by the least common denominator to clear the fractions. *We can do this only with equations, not expressions.*

Solving an Equation with Rational Expressions

Step 1 **Determine the domain of the variable.**

Step 2 **Multiply each side of the equation by the LCD** to clear the fractions.

Step 3 **Solve** the resulting equation.

Step 4 **Check** that each proposed solution is in the domain, and discard any values that are not. Check the remaining proposed solution(s) in the original equation.

CAUTION

When each side of an equation is multiplied by a *variable* expression, the resulting "solutions" may not satisfy the original equation. *We must either determine and observe the domain or check all proposed solutions in the original equation. It is wise to do both.*

Answers

1. (a) $\{x \mid x \neq 0\}$ (b) $\{x \mid x \neq -2, 6\}$
   (c) $\{x \mid x \neq \pm 5\}$

EXAMPLE 2 Solving a Rational Equation

Solve $\dfrac{2}{x} - \dfrac{3}{2} = \dfrac{7}{2x}$.

Step 1 The domain, which excludes 0, was found in **Example 1(a).**

Step 2 $\qquad 2x\left(\dfrac{2}{x} - \dfrac{3}{2}\right) = 2x\left(\dfrac{7}{2x}\right)$ Multiply by the LCD, $2x$.

Step 3 $\quad 2x\left(\dfrac{2}{x}\right) - 2x\left(\dfrac{3}{2}\right) = 2x\left(\dfrac{7}{2x}\right)$ Distributive property

$\qquad\qquad\qquad 4 - 3x = 7$ Multiply.

$\qquad\qquad\qquad -3x = 3$ Subtract 4.

> This proposed solution is in the domain.

$\qquad\qquad\qquad x = -1$ Divide by -3.

Step 4 **CHECK** $\qquad \dfrac{2}{x} - \dfrac{3}{2} = \dfrac{7}{2x}$ Original equation

> Don't forget this step.

$\qquad\qquad \dfrac{2}{-1} - \dfrac{3}{2} \overset{?}{=} \dfrac{7}{2(-1)}$ Let $x = -1$.

$\qquad\qquad\qquad -\dfrac{7}{2} = -\dfrac{7}{2}$ ✓ True

A true statement results, so the solution set is $\{-1\}$.

················· **Work Problem 2** at the Side. ▶

EXAMPLE 3 Solving a Rational Equation with No Solution

Solve $\dfrac{2}{x-3} - \dfrac{3}{x+3} = \dfrac{12}{x^2 - 9}$.

Step 1 The domain, which excludes ± 3, was found in **Example 1(b).**

Step 2 Factor $x^2 - 9$. Then multiply each side by the LCD, $(x+3)(x-3)$.

$(x+3)(x-3)\left(\dfrac{2}{x-3} - \dfrac{3}{x+3}\right) = (x+3)(x-3)\left[\dfrac{12}{(x+3)(x-3)}\right]$

Step 3 $(x+3)(x-3)\left(\dfrac{2}{x-3}\right) - (x+3)(x-3)\left(\dfrac{3}{x+3}\right)$

$\qquad\qquad\qquad = (x+3)(x-3)\left[\dfrac{12}{(x+3)(x-3)}\right]$

Distributive property

$\qquad 2(x+3) - 3(x-3) = 12$ Multiply.

$\qquad 2x + 6 - 3x + 9 = 12$ Distributive property

$\qquad\qquad -x + 15 = 12$ Combine like terms.

$\qquad\qquad\qquad -x = -3$ Subtract 15.

Proposed solution $\rightarrow x = 3$ Multiply by -1.

················· **Continued on Next Page**

2 Solve each equation.

(a) $\dfrac{1}{3x} - \dfrac{3}{4x} = \dfrac{1}{3}$

The domain excludes ____.

Multiply by the LCD, ____.

$\underline{\quad}\left(\dfrac{1}{3x} - \dfrac{3}{4x}\right) = 12x\left(\dfrac{1}{3}\right)$

$\underline{\quad}\left(\dfrac{1}{3x}\right) - 12x\left(\dfrac{3}{4x}\right) = 4x$

$4 - \underline{\quad} = 4x$

$-5 = 4x$

$\underline{\quad} = x$

The proposed solution is ____.

Does it check in the original equation? (*Yes / No*)

The solution set is ____.

(b) $-\dfrac{3}{20} + \dfrac{2}{x} = \dfrac{5}{4x}$

❸ Solve each equation.

(a) $\dfrac{3}{x+1} = \dfrac{1}{x-1} - \dfrac{2}{x^2-1}$

(b) $\dfrac{1}{x-3} + \dfrac{1}{x+3} = \dfrac{6}{x^2-9}$

❹ Solve.

$$\dfrac{4}{x^2+x-6} - \dfrac{1}{x^2-4}$$

$$= \dfrac{2}{x^2+5x+6}$$

Step 4 Since the proposed solution, 3, is not in the domain, it cannot be a solution of the equation. Substituting 3 into the original equation shows why.

CHECK $\dfrac{2}{x-3} - \dfrac{3}{x+3} = \dfrac{12}{x^2-9}$ Original equation

$\dfrac{2}{3-3} - \dfrac{3}{3+3} \overset{?}{=} \dfrac{12}{3^2-9}$ Let $x = 3$.

$\dfrac{2}{0} - \dfrac{3}{6} \overset{?}{=} \dfrac{12}{0}$ Division by 0 is undefined.

The equation has no solution. The solution set is $\varnothing$.

◀ **Work Problem ❸ at the Side.**

EXAMPLE 4 **Solving a Rational Equation**

Solve $\dfrac{3}{p^2+p-2} - \dfrac{1}{p^2-1} = \dfrac{7}{2(p^2+3p+2)}$.

Factor each denominator to find the domain and the LCD.

$\dfrac{3}{(p-1)(p+2)} - \dfrac{1}{(p+1)(p-1)}$

$= \dfrac{7}{2(p+2)(p+1)}$ Factor the denominators.

The domain excludes 1, −2, and −1. Multiply each side of the equation by the LCD, $2(p-1)(p+2)(p+1)$.

$$2(p-1)(p+2)(p+1)\left(\dfrac{3}{(p-1)(p+2)} - \dfrac{1}{(p+1)(p-1)}\right)$$

$$= 2(p-1)(p+2)(p+1)\left(\dfrac{7}{2(p+2)(p+1)}\right)$$

$2 \cdot 3(p+1) - 2(p+2) = 7(p-1)$ Distributive property

$6p+6-2p-4 = 7p-7$ $2 \cdot 3 = 6$; Distributive property

$4p+2 = 7p-7$ Combine like terms.

$9 = 3p$ Subtract $4p$. Add 7.

Proposed solution → $3 = p$ Divide by 3.

CHECK 3 is in the domain. Substitute it in the original equation to check.

$\dfrac{3}{p^2+p-2} - \dfrac{1}{p^2-1} = \dfrac{7}{2(p^2+3p+2)}$ Original equation

$\dfrac{3}{3^2+3-2} - \dfrac{1}{3^2-1} \overset{?}{=} \dfrac{7}{2(3^2+3\cdot 3+2)}$ Let $p = 3$.

$\dfrac{3}{10} - \dfrac{1}{8} \overset{?}{=} \dfrac{7}{40}$ Work in the denominators.

$\begin{aligned}\tfrac{3}{10} - \tfrac{1}{8} \\ = \tfrac{12}{40} - \tfrac{5}{40}\end{aligned}$

$\dfrac{7}{40} = \dfrac{7}{40}$ ✓ True

The solution set is $\{3\}$.

◀ **Work Problem ❹ at the Side.**

EXAMPLE 5   Solving a Rational Equation

Solve $\dfrac{2}{3x + 1} = \dfrac{1}{x} - \dfrac{6x}{3x + 1}$.

Since the denominator $3x + 1$ cannot equal 0, $-\frac{1}{3}$ is excluded from the domain, as is 0.

$$x(3x + 1)\left(\frac{2}{3x + 1}\right) = x(3x + 1)\left(\frac{1}{x} - \frac{6x}{3x + 1}\right) \quad \begin{array}{l} \text{Multiply by the} \\ \text{LCD, } x(3x + 1). \end{array}$$

$$x(3x + 1)\left(\frac{2}{3x + 1}\right) = x(3x + 1)\left(\frac{1}{x}\right) - x(3x + 1)\left(\frac{6x}{3x + 1}\right)$$

Distributive property

$\boxed{\text{This is a quadratic equation.}}$ $\quad 2x = 3x + 1 - 6x^2$ Multiply.

$6x^2 - x - 1 = 0$ Standard form

$(3x + 1)(2x - 1) = 0$ Factor.

$3x + 1 = 0 \quad$ or $\quad 2x - 1 = 0$ Zero-factor property

$\begin{array}{l}\text{Proposed} \\ \text{solutions}\end{array} \rightarrow x = -\dfrac{1}{3} \quad$ or $\quad x = \dfrac{1}{2}$ Solve each equation.

Because $-\frac{1}{3}$ is not in the domain, it is not a solution. Check that the solution set is $\left\{\frac{1}{2}\right\}$.

········· **Work Problem ❺ at the Side.** ▶

OBJECTIVE ▶ ❸ Recognize the graph of a rational function. Recall that a function defined by a quotient of polynomials is a **rational function.** Because one or more values of the variable may be excluded from the domain of a rational function, its graph is often **discontinuous**—that is, there will be one or more breaks in the graph.

One simple rational function is the **reciprocal function** $f(x) = \frac{1}{x}$. The domain of this function includes all real numbers except 0. Thus, this function pairs every real number except 0 with its reciprocal.

| The closer negative values of x are to 0, the less ("more negative") y is. $\longrightarrow$ | | | | | The closer positive values of x are to 0, the greater y is. $\longleftarrow$ | | | | | | |

x	-3	-2	-1	-0.5	-0.25	-0.1	0.1	0.25	0.5	1	2	3
y	$-\frac{1}{3}$	$-\frac{1}{2}$	-1	-2	-4	-10	10	4	2	1	$\frac{1}{2}$	$\frac{1}{3}$

Plotting the points from the table, we obtain the graph in **Figure 2.**

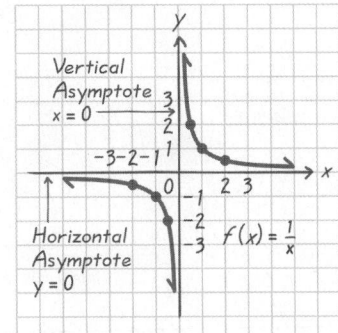

Figure 2

Reciprocal function

$$f(x) = \frac{1}{x}$$

Domain: $\{x \mid x \neq 0\}$

Range: $\{y \mid y \neq 0\}$

❺ Solve each equation.

(a) $\dfrac{2x}{x - 2} = \dfrac{-3}{x} + \dfrac{4}{x - 2}$

(b) $\dfrac{1}{x + 4} + \dfrac{x}{x - 4} = \dfrac{-8}{x^2 - 16}$

Answers

5. (a) $\left\{-\dfrac{3}{2}\right\}$ (b) $\{-1\}$

6 Graph each rational function, and give the equations of the vertical and horizontal asymptotes.

(a) $f(x) = -\dfrac{1}{x}$

(b) $f(x) = \dfrac{2}{x+3}$

Answers

6. **(a)** vertical asymptote: $x = 0$;
horizontal asymptote: $y = 0$

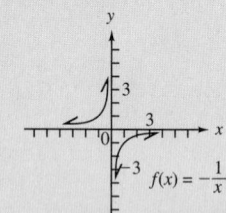

(b) vertical asymptote: $x = -3$;
horizontal asymptote: $y = 0$

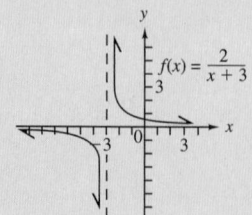

Since the domain of the function $f(x) = \frac{1}{x}$ includes all real numbers except 0, there is no point on the graph with $x = 0$. The vertical line with equation $x = 0$ is a **vertical asymptote** of the graph. Also, the horizontal line with equation $y = 0$ is a **horizontal asymptote**. The graph approaches these asymptotes but does not cross them.

> **Note**
>
> In general, the following are true of a rational function.
>
> 1. If the y-values approach ∞ or $-\infty$ as the x-values approach a real number a, the **vertical line $x = a$ is a vertical asymptote** of the graph.
>
> 2. If the y-values approach a real number b as $|x|$ increases without bound, the **horizontal line $y = b$ is a horizontal asymptote** of the graph.

EXAMPLE 6 **Graphing a Rational Function**

Graph the rational function, and give the equations of the vertical and horizontal asymptotes.

$$g(x) = \dfrac{-2}{x-3}$$

Some ordered pairs that belong to the function are listed in the table.

x	-2	-1	0	1	2	2.5	2.75	3.25	3.5	4	5	6
y	$\frac{2}{5}$	$\frac{1}{2}$	$\frac{2}{3}$	1	2	4	8	-8	-4	-2	-1	$-\frac{2}{3}$

There is no point on the graph, shown in **Figure 3**, for $x = 3$ because 3 is excluded from the domain. The dashed line $x = 3$ represents the vertical asymptote and is not part of the graph. The graph gets closer to the vertical asymptote as the x-values get closer to 3. Again, $y = 0$ is a horizontal asymptote.

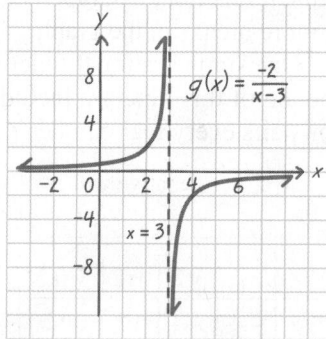

Figure 3

◀ **Work Problem 6 at the Side.**

7.4 Exercises

FOR EXTRA HELP

 Download the MyDashBoard App

 ▶ MyMathLab®

1. **CONCEPT CHECK** *Decide whether each of the following is an expression or an equation.*

(a) $\dfrac{2}{x} = \dfrac{4}{3x} + \dfrac{1}{3}$ (b) $\dfrac{4}{3x} + \dfrac{1}{3}$

(c) $\dfrac{5}{x+1} - \dfrac{2}{x-1}$ (d) $\dfrac{5}{x+1} - \dfrac{2}{x-1} = \dfrac{4}{x^2-1}$

2. What is wrong with the following problem?

"Solve $\dfrac{2x+1}{3x-4} + \dfrac{1}{2x+3}$."

As explained in this section, any values that would cause a denominator to equal 0 must be excluded from the domain and, consequently, as solutions of an equation that has variable expressions in the denominators.

(a) Without actually solving each equation, list all possible values that would have to be rejected if they appeared as proposed solutions.

(b) Then give the domain, using set-builder notation.

See Example 1.

3. ▶ $\dfrac{1}{3x} + \dfrac{1}{2x} = \dfrac{x}{3}$

4. $\dfrac{5}{6x} - \dfrac{8}{2x} = \dfrac{x}{4}$

5. $\dfrac{1}{x+1} - \dfrac{1}{x-2} = 0$

6. $\dfrac{3}{x+4} - \dfrac{2}{x-9} = 0$

7. $\dfrac{5}{3x+5} - \dfrac{1}{x} = \dfrac{1}{2x+3}$

8. $\dfrac{6}{4x+7} - \dfrac{3}{x} = \dfrac{5}{6x-13}$

9. $\dfrac{3x+1}{x-4} = \dfrac{6x+5}{2x-7}$

10. $\dfrac{4x-1}{2x+3} = \dfrac{12x-25}{6x-2}$

11. $\dfrac{2}{x^2-x} + \dfrac{1}{x+3} = \dfrac{4}{x-2}$

12. $\dfrac{3}{x^2+x} - \dfrac{1}{x+5} = \dfrac{2}{x-7}$

Solve each equation. See Examples 2–5.

13. $\dfrac{-5}{2x} + \dfrac{3}{4x} = \dfrac{-7}{4}$

14. $\dfrac{6}{5x} - \dfrac{2}{3x} = \dfrac{-8}{45}$

15. ▶ $x - \dfrac{24}{x} = -2$

16. $p + \dfrac{15}{p} = -8$

17. $\dfrac{x}{4} - \dfrac{21}{4x} = -1$

18. $\dfrac{x}{2} - \dfrac{12}{x} = 1$

19. $\dfrac{x-4}{x+6} = \dfrac{2x+3}{2x-1}$

20. $\dfrac{5x-8}{x+2} = \dfrac{5x-1}{x+3}$

21. $\dfrac{3x + 1}{x - 4} = \dfrac{6x + 5}{2x - 7}$

22. $\dfrac{4x - 1}{2x + 3} = \dfrac{12x - 25}{6x - 2}$

23. $\dfrac{1}{y - 1} + \dfrac{5}{12} = \dfrac{-2}{3y - 3}$

24. $\dfrac{4}{m + 2} - \dfrac{11}{9} = \dfrac{1}{3m + 6}$

25. $\dfrac{-2}{3t - 6} - \dfrac{1}{36} = \dfrac{-3}{4t - 8}$

26. $\dfrac{6}{t + 1} - \dfrac{34}{15} = \dfrac{-4}{5t + 5}$

27. $\dfrac{7}{6x + 3} - \dfrac{1}{3} = \dfrac{2}{2x + 1}$

28. $\dfrac{3}{4m + 2} = \dfrac{17}{2} - \dfrac{7}{2m + 1}$

29. $\dfrac{3}{k + 2} - \dfrac{2}{k^2 - 4} = \dfrac{1}{k - 2}$

30. $\dfrac{3}{x - 2} + \dfrac{21}{x^2 - 4} = \dfrac{14}{x + 2}$

31. $\dfrac{9}{x} + \dfrac{4}{6x - 3} = \dfrac{2}{6x - 3}$

32. $\dfrac{5}{n} + \dfrac{4}{6 - 3n} = \dfrac{2n}{6 - 3n}$

33. $\dfrac{x}{x - 3} + \dfrac{4}{x + 3} = \dfrac{18}{x^2 - 9}$

34. $\dfrac{2x}{x - 3} + \dfrac{4}{x + 3} = \dfrac{-24}{x^2 - 9}$

35. $\dfrac{6}{x - 4} + \dfrac{5}{x} = \dfrac{-20}{x^2 - 4x}$

36. $\dfrac{7}{x - 4} + \dfrac{3}{x} = \dfrac{-12}{x^2 - 4x}$

37. $\dfrac{2}{4x + 7} + \dfrac{x}{3} = \dfrac{6}{12x + 21}$

38. $\dfrac{3}{2x + 5} + \dfrac{x}{2} = \dfrac{9}{6x + 15}$

39. $\dfrac{1}{x - 2} + \dfrac{1}{4} = \dfrac{1}{4\left(x^2 - 4\right)}$

40. $\dfrac{1}{x + 4} + \dfrac{1}{3} = \dfrac{-10}{3\left(x^2 - 16\right)}$

41. $\dfrac{1}{y + 2} + \dfrac{3}{y + 7} = \dfrac{5}{y^2 + 9y + 14}$

42. $\dfrac{1}{t + 3} + \dfrac{4}{t + 5} = \dfrac{2}{t^2 + 8t + 15}$

43. $\dfrac{6}{w + 3} + \dfrac{-7}{w - 5} = \dfrac{-48}{w^2 - 2w - 15}$

44. $\dfrac{2}{r - 5} + \dfrac{3}{2r + 1} = \dfrac{22}{2r^2 - 9r - 5}$

45. $\dfrac{4x - 7}{4x^2 - 9} = \dfrac{-2x^2 + 5x - 4}{4x^2 - 9} + \dfrac{x + 1}{2x + 3}$

46. $\dfrac{5x + 14}{x^2 - 9} = \dfrac{-2x^2 - 5x + 2}{x^2 - 9} + \dfrac{2x + 4}{x - 3}$

Graph each rational function. Give the equations of the vertical and horizontal asymptotes. See Example 6.

47. $f(x) = \dfrac{2}{x}$

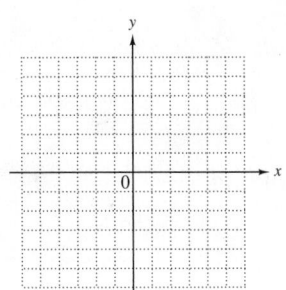

48. $f(x) = \dfrac{3}{x}$

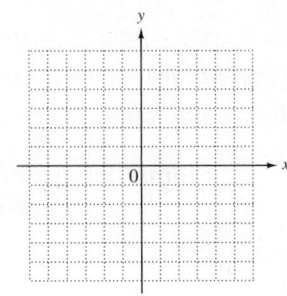

49. $g(x) = -\dfrac{3}{x}$

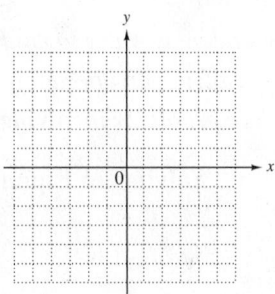

50. $g(x) = -\dfrac{2}{x}$

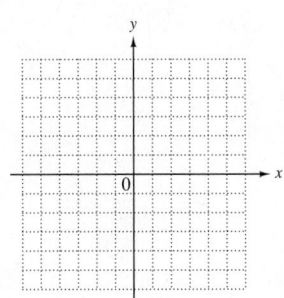

51. $f(x) = \dfrac{1}{x - 2}$

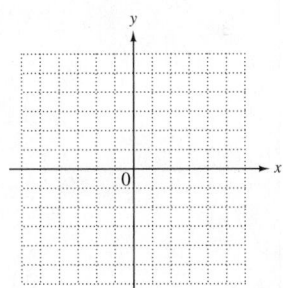

52. $f(x) = \dfrac{1}{x + 2}$

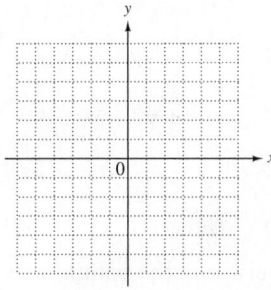

Solve each problem.

53. The average number of vehicles waiting in line to enter a parking area is modeled by the rational function

$$w(x) = \dfrac{x^2}{2(1 - x)},$$

where x is a quantity between 0 and 1 known as the **traffic intensity.** (*Source*: Mannering, F. and W. Kilareski, *Principles of Highway Engineering and Traffic Control,* John Wiley and Sons.) To the nearest tenth, find the average number of vehicles waiting for each traffic intensity.

(a) 0.1

(b) 0.8

(c) 0.9

(d) What happens to waiting time as traffic intensity increases?

54. The force required to keep a 2000-lb car going 30 mph from skidding on a curve, where r is the radius of the curve in feet, is given by

$$F(r) = \dfrac{225{,}000}{r}.$$

(a) What radius must a curve have if a force of 450 lb is needed to keep the car from skidding?

(b) As the radius of the curve is lengthened, how is the force affected?

Summary Exercises *Simplifying Rational Expressions vs. Solving Rational Equations*

A common student error is to confuse an equation, such as $\frac{x}{2} + \frac{x}{3} = -5$, with an expression involving an operation, such as $\frac{x}{2} + \frac{x}{3}$. **Equations are solved for a numerical answer, while problems involving operations result in simplified expressions.**

Solving an Equation	Simplifying an Expression Involving an Operation

Solve: $\dfrac{x}{2} + \dfrac{x}{3} = -5$ Look for the equality symbol.

Multiply each side by the LCD, 6.

$$6\left(\frac{x}{2} + \frac{x}{3}\right) = 6(-5)$$

$$6\left(\frac{x}{2}\right) + 6\left(\frac{x}{3}\right) = 6(-5)$$

$$3x + 2x = -30$$

$$5x = -30$$

$$x = -6$$

Check that the solution set is $\{-6\}$.

Add: $\dfrac{x}{2} + \dfrac{x}{3}$

Write both fractions with the LCD, 6.

$$\frac{x}{2} + \frac{x}{3}$$

$$= \frac{x \cdot 3}{2 \cdot 3} + \frac{x \cdot 2}{3 \cdot 2}$$

$$= \frac{3x}{6} + \frac{2x}{6}$$

$$= \frac{3x + 2x}{6}$$

$$= \frac{5x}{6}$$

Identify each exercise as an **expression** *or an* **equation.** *Then simplify the expression by performing the indicated operation, or solve the equation, as appropriate.*

1. $\dfrac{x}{2} - \dfrac{x}{4} = 5$

2. $\dfrac{4x - 20}{x^2 - 25} \cdot \dfrac{(x + 5)^2}{10}$

3. $\dfrac{6}{7x} - \dfrac{4}{x}$

4. $\dfrac{\dfrac{1}{x} + \dfrac{1}{y}}{\dfrac{1}{x} - \dfrac{1}{y}}$

5. $\dfrac{5}{7t} = \dfrac{52}{7} - \dfrac{3}{t}$

6. $\dfrac{x - 5}{3} + \dfrac{1}{3} = \dfrac{x - 2}{5}$

7. $\dfrac{7}{6x} + \dfrac{5}{8x}$

8. $\dfrac{4}{x} - \dfrac{8}{x + 1} = 0$

9. $\dfrac{\dfrac{6}{x + 1} - \dfrac{1}{x}}{\dfrac{2}{x} - \dfrac{4}{x + 1}}$

10. $\dfrac{8}{r+2} - \dfrac{7}{4r+8}$

11. $\dfrac{x}{x+y} + \dfrac{2y}{x-y}$

12. $\dfrac{3p^2-6p}{p+5} \div \dfrac{p^2-4}{8p+40}$

13. $\dfrac{x-2}{9} \cdot \dfrac{5}{8-4x}$

14. $\dfrac{a-4}{3} + \dfrac{11}{6} = \dfrac{a+1}{2}$

15. $\dfrac{b^2+b-6}{b^2+2b-8} \cdot \dfrac{b^2+8b+16}{3b+12}$

16. $\dfrac{10z^2-5z}{3z^3-6z^2} \div \dfrac{2z^2+5z-3}{z^2+z-6}$

17. $\dfrac{5}{x^2-2x} - \dfrac{3}{x^2-4}$

18. $\dfrac{3}{t-1} + \dfrac{1}{t} = \dfrac{7}{2}$

19. $\dfrac{-1}{3-x} - \dfrac{2}{x-3}$

20. $\dfrac{\dfrac{t}{4}-\dfrac{1}{t}}{1+\dfrac{t+4}{t}}$

21. $\dfrac{3r}{r-2} = 1 + \dfrac{6}{r-2}$

22. $\dfrac{7}{2x^2-8x} + \dfrac{3}{x^2-16}$

23. $\dfrac{\dfrac{5}{x}-\dfrac{3}{y}}{\dfrac{9x^2-25y^2}{x^2y}}$

24. $\dfrac{2k+\dfrac{5}{k-1}}{3k-\dfrac{2}{k}}$

25. $\dfrac{2}{y+1} - \dfrac{3}{y^2-y-2} = \dfrac{3}{y-2}$

26. $\dfrac{-2}{a^2+2a-3} - \dfrac{5}{3-3a} = \dfrac{4}{3a+9}$

27. $\dfrac{4y^2-13y+3}{2y^2-9y+9} \div \dfrac{4y^2+11y-3}{6y^2-5y-6}$

28. $\dfrac{8}{3k+9} - \dfrac{8}{15} = \dfrac{2}{5k+15}$

29. $\dfrac{3}{y-3} - \dfrac{3}{y^2-5y+6} = \dfrac{2}{y-2}$

30. $\dfrac{6z^2-5z-6}{6z^2+5z-6} \cdot \dfrac{12z^2-17z+6}{12z^2-z-6}$

7.5 Applications of Rational Expressions

OBJECTIVES

1. Find the value of an unknown variable in a formula.
2. Solve a formula for a specified variable.
3. Solve applications using proportions.
4. Solve applications about distance, rate, and time.
5. Solve applications about work rates.

❶ Use the formula given in **Example 1** to answer each part.

(a) Find p if $f = 15$ and $q = 25$.

(b) Find f if $p = 6$ and $q = 9$.

(c) Find q if $f = 12$ and $p = 16$.

❷ Solve $\dfrac{3}{p} + \dfrac{3}{q} = \dfrac{5}{r}$ for q.

OBJECTIVE ❶ **Find the value of an unknown variable in a formula.** Formulas may contain rational expressions, such as $t = \frac{d}{r}$ and $\frac{1}{f} = \frac{1}{p} + \frac{1}{q}$.

EXAMPLE 1 Finding the Value of a Variable in a Formula

In physics, the focal length, f, of a lens is given by the formula

$$\frac{1}{f} = \frac{1}{p} + \frac{1}{q},$$

where p is the distance from the object to the lens and q is the distance from the lens to the image. See **Figure 4.** Find q if $p = 20$ cm and $f = 10$ cm.

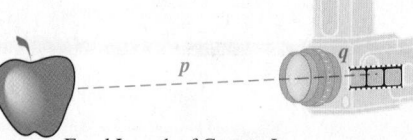

Focal Length of Camera Lens

Figure 4

$$\frac{1}{f} = \frac{1}{p} + \frac{1}{q} \qquad \text{Solve this equation for } q.$$

$$\frac{1}{10} = \frac{1}{20} + \frac{1}{q} \qquad \text{Let } f = 10, p = 20.$$

$$20q \cdot \frac{1}{10} = 20q\left(\frac{1}{20} + \frac{1}{q}\right) \qquad \text{Multiply by the LCD, } 20q.$$

$$20q \cdot \frac{1}{10} = 20q\left(\frac{1}{20}\right) + 20q\left(\frac{1}{q}\right) \qquad \text{Distributive property}$$

$$2q = q + 20 \qquad \text{Multiply.}$$

$$q = 20 \qquad \text{Subtract } q.$$

The distance from the lens to the image is 20 cm.

◀ **Work Problem** ❶ **at the Side.**

OBJECTIVE ❷ **Solve a formula for a specified variable.** *The goal is to isolate the specified variable on one side of the equality symbol.*

EXAMPLE 2 Solving a Formula for a Specified Variable

Solve $\dfrac{1}{f} = \dfrac{1}{p} + \dfrac{1}{q}$ for p.

$$\frac{1}{f} = \frac{1}{p} + \frac{1}{q}$$

$$fpq \cdot \frac{1}{f} = fpq\left(\frac{1}{p} + \frac{1}{q}\right) \qquad \text{To clear the fractions, multiply by the LCD, } fpq.$$

We want the terms with p on the same side.

$$pq = fq + fp \qquad \text{Distributive property}$$

$$pq - fp = fq \qquad \text{Subtract } fp.$$

This is a key step.

$$p(q - f) = fq \qquad \text{Factor out } p.$$

$$p = \frac{fq}{q - f} \qquad \text{Divide by } q - f.$$

◀ **Work Problem** ❷ **at the Side.**

Answers

1. (a) $\dfrac{75}{2}$ (b) $\dfrac{18}{5}$ (c) 48

2. $q = \dfrac{3rp}{5p - 3r}$, or $q = \dfrac{-3rp}{3r - 5p}$

EXAMPLE 3 Solving a Formula for a Specified Variable

Solve $I = \dfrac{nE}{R + nr}$ for n.

$$I = \frac{nE}{R + nr}$$

$$(R + nr)I = (R + nr)\frac{nE}{R + nr} \qquad \text{To clear the fraction, multiply by } R + nr.$$

$$RI + nrI = nE \qquad \text{Distributive property on the left}$$

> Write the *n*-terms on the **same** side in preparation for factoring.

$$RI = nE - nrI \qquad \text{Subtract } nrI.$$

$$RI = n(E - rI) \qquad \text{Factor out } n.$$

$$\frac{RI}{E - rI} = n, \quad \text{or} \quad n = \frac{RI}{E - rI} \qquad \text{Divide by } E - rI.$$

··· **Work Problem ❸ at the Side.** ▶

CAUTION

Refer to the steps in **Examples 2 and 3** that factor out the desired variable. *This variable must be a factor on only one side of the equation,* so that each side can be divided by the remaining factor in the last step.

OBJECTIVE ▶ ❸ Solve applications using proportions. Recall from **Section 2.6** that a **ratio** is a comparison of two quantities. The ratio of a to b may be written in any of the following ways.

$$a \text{ to } b, \quad a : b, \quad \text{or} \quad \frac{a}{b} \qquad \text{Ratio of } a \text{ to } b$$

Ratios are usually written as quotients in algebra. A **proportion** is a statement that two ratios are equal.

$$\frac{a}{b} = \frac{c}{d} \qquad \text{Proportion}$$

EXAMPLE 4 Solving a Proportion

In 2010, about 16 of every 100 Americans had no health insurance coverage. The population at that time was about 309 million. How many million Americans had no health insurance? (*Source*: U.S. Census Bureau.)

Step 1 **Read** the problem.

Step 2 **Assign a variable.**

Let x = the number (in millions) who had no health insurance.

Step 3 **Write an equation.** We set up a proportion. The ratio 16 to 100 should equal the ratio x to 309.

$$\frac{16}{100} = \frac{x}{309} \qquad \text{Write a proportion.}$$

··· **Continued on Next Page**

❸ Solve each formula for the specified variable.

(a) $A = \dfrac{Rr}{R + r}$ for R

ᴳˢ (b) $R = \dfrac{M - P}{PT}$ for P

Our goal is to isolate the variable ____.

$$R = \frac{M - P}{PT}$$

$$(\underline{\quad})R = (\underline{\quad})\frac{M - P}{PT}$$

$$PTR = M - P$$

$$PTR + \underline{\quad} = M$$

$$P(\underline{\quad}) = M$$

$$P = \underline{\quad}$$

Answers

3. (a) $R = \dfrac{-Ar}{A - r}$, or $R = \dfrac{Ar}{r - A}$

(b) P; PT; PT; P; $TR + 1$; $\dfrac{M}{TR + 1}$

④ Solve the problem.

In 2010, approximately 9.8% (that is, 9.8 of every 100) of the 74,165,000 children under 18 yr of age in the United States had no health insurance. How many such children were uninsured? (*Source*: U.S. Census Bureau.)

Step 4 **Solve.**

$$\frac{16}{100} = \frac{x}{309}$$ Proportion from Step 3

$$30{,}900\left(\frac{16}{100}\right) = 30{,}900\left(\frac{x}{309}\right)$$ Multiply by a common denominator. Here, $100 \cdot 309 = 30{,}900$.

$$4944 = 100x$$ Simplify.

$$49.44 = x$$ Divide by 100.

Step 5 **State the answer.** There were about 49.44 million Americans with no health insurance in 2010.

Step 6 **Check** that the ratio of 49.44 million to 309 million equals $\frac{16}{100}$.

$$\frac{49.44}{309} = \frac{16}{100}$$ Use a calculator to divide 49.44 by 309. A true statement results.

◀ **Work Problem ④ at the Side.**

EXAMPLE 5 **Solving a Proportion Involving Rates**

Marissa's car uses 10 gal of gas to travel 210 mi. She has 5 gal of gas in the car, and she still needs to drive 640 mi. If we assume the car continues to use gas at the same rate, how many more gallons will she need?

Step 1 **Read** the problem.

Step 2 **Assign a variable.**

Let x = the additional number of gallons of gas.

Step 3 **Write an equation.** We set up a proportion.

$$\frac{\text{gallons} \longrightarrow}{\text{miles} \longrightarrow} \frac{10}{210} = \frac{5 + x}{640} \frac{\longleftarrow \text{gallons}}{\longleftarrow \text{miles}}$$

⑤ Solve the problem.

Lauren's car uses 15 gal of gasoline to drive 495 mi. She has 6 gal of gasoline in the car, and she wants to know how much more gasoline she will need to drive 600 mi. If we assume that the car continues to use gasoline at the same rate, how many more gallons will she need? (Round your answer to the nearest tenth.)

Step 4 **Solve.** We could multiply by the LCD $10 \cdot 21 \cdot 64$. Instead we use an alternative method that involves **cross products**.

For $\frac{a}{b} = \frac{c}{d}$ to be true, the cross products ad and bc must be equal.

$$a \cdot d = b \cdot c$$ Cross products **(Section 2.6)**

$$10 \cdot 640 = 210\,(5 + x)$$ If $\frac{a}{b} = \frac{c}{d}$, then $ad = bc$.

$$6400 = 1050 + 210x$$ Multiply; distributive property

$$5350 = 210x$$ Subtract 1050.

$$25.5 \approx x$$ Divide by 210. Round to the nearest tenth.

Step 5 **State the answer.** Marissa will need about 25.5 more gallons of gas.

Step 6 **Check.** The **25.5** additional gallons plus the 5 gal she has equals 30.5 gal. From Step 3, the ratio of 10 gal to 210 mi must equal the ratio of 30.5 gal to 640 mi.

$$\frac{10}{210} \approx 0.048 \quad \text{and} \quad \frac{30.5}{640} \approx 0.048$$ Use a calculator.

Since the ratios are approximately equal, the answer is correct.

◀ **Work Problem ⑤ at the Side.**

Answers

4. 7,268,170

5. 12.2 more gallons

OBJECTIVE ④ **Solve applications about distance, rate, and time.** If an automobile travels at an average rate of 65 mph for 2 hr, then it travels

$$65 \times 2 = 130 \text{ mi.} \quad rt = d, \text{ or } d = rt \text{ (Relationship between distance, rate, and time from Section 4.4)}$$

By solving, in turn, for r and t in the formula $d = rt$, we obtain two other equivalent forms of the formula. The three forms are given below.

Distance, Rate, and Time Relationship

$$d = rt \qquad r = \frac{d}{t} \qquad t = \frac{d}{r} \qquad \text{The last two forms are rational expressions.}$$

EXAMPLE 6 Finding Distance, Rate, or Time

Solve each problem using a form of the distance formula.

(a) The speed (rate) of sound is 1088 ft per sec at sea level at 32°F. Find the distance sound travels in 5 sec under these conditions.
We must find distance, given rate and time, using $d = rt$ (or $rt = d$).

$$\underset{\text{Rate}}{1088} \times \underset{\text{Time}}{5} = \underset{\text{Distance}}{5440 \text{ ft}}$$

(b) The winner of the first Indianapolis 500 race (in 1911) was Ray Harroun, driving a Marmon Wasp at an average rate of 74.60 mph. (*Source: World Almanac and Book of Facts.*) How long did it take him to complete the 500 mi?
We must find time, given rate and distance, using $t = \frac{d}{r}$ (or $\frac{d}{r} = t$).

$$\underset{\text{Rate} \to}{\overset{\text{Distance} \to}{\frac{500}{74.60}}} = 6.70 \text{ hr (rounded)} \leftarrow \text{Time}$$

To convert 0.70 hr to minutes, we multiply by 60 to get $0.70(60) = 42$. It took Harroun about 6 hr, 42 min, to complete the race.

(c) At the 2008 Olympic Games, Australian swimmer Leisel Jones set an Olympic record of 65.17 sec in the women's 100-m breaststroke swimming event. (*Source: World Almanac and Book of Facts.*) Find her rate.
We must find rate, given distance and time, using $r = \frac{d}{t}$ (or $\frac{d}{t} = r$).

$$\underset{\text{Time} \to}{\overset{\text{Distance} \to}{\frac{100}{65.17}}} = 1.53 \text{ m per sec (rounded)} \leftarrow \text{Rate}$$

······ **Work Problem ⑥ at the Side.** ▶

Problem-Solving Hint

Many applied problems use the formulas just discussed. The next two examples show how to solve typical applications of the formula $d = rt$.

A helpful strategy for solving such problems is to *first make a sketch* showing what is happening in the problem. *Then make a table* using the information given, along with the unknown quantities. The table will help organize the information, and the sketch will help set up the equation.

⑥ Solve each problem.

(a) A new world record in the men's 100-m dash was set by Usain Bolt of Jamaica in 2009. He ran it in 9.58 sec. What was his rate in meters per second, to the nearest hundredth? (*Source: World Almanac and Book of Facts.*)

(b) A new world record for the women's 3000-m steeplechase was set by Gulnara Samitova of Russia in 2008. Her rate was 5.568 m per sec. To the nearest second, what was her time? (*Source: World Almanac and Book of Facts.*)

(c) A small plane flew from Chicago to St. Louis averaging 145 mph. The trip took 2 hr. What is the distance between Chicago and St. Louis?

Answers

6. (a) 10.44 m per sec
(b) 539 sec, or 8 min, 59 sec
(c) 290 mi

| EXAMPLE 7 | **Solving a Problem about Distance, Rate, and Time** |

A paddle wheeler travels 10 mi against the current in a river in the same time that it travels 15 mi with the current. If the rate of the current is 3 mph, find the rate of the boat in still water.

Step 1   **Read** the problem. We must find the rate of the boat in still water.

Step 2   **Assign a variable.**

Let x = the rate of the boat in still water.

Traveling **against** the current slows the boat down, so the rate of the boat is the **difference** between its rate in still water and the rate of the current—that is, $(x - 3)$ mph.

Traveling **with** the current speeds the boat up, so the rate of the boat is the **sum** of its rate in still water and the rate of the current—that is, $(x + 3)$ mph.

Thus, $x - 3$ = the rate of the boat *against* the current,

and $x + 3$ = the rate of the boat *with* the current.

Because the time is the *same* going against the current as with the current, find time in terms of distance and rate for each situation. Against the current, the distance is 10 mi and the rate is $(x - 3)$ mph.

$$t = \frac{d}{r} = \frac{10}{x - 3} \qquad \text{Time \textit{against} the current}$$

With the current, the distance is 15 mi and the rate is $(x + 3)$ mph.

$$t = \frac{d}{r} = \frac{15}{x + 3} \qquad \text{Time \textit{with} the current}$$

This information is summarized in the following table.

	Distance	Rate	Time
Against Current	10	$x - 3$	$\frac{10}{x - 3}$
With Current	15	$x + 3$	$\frac{15}{x + 3}$

The times are equal.

Step 3   **Write an equation,** using the fact that the times are equal.

$$\frac{10}{x - 3} = \frac{15}{x + 3}$$

························· **Continued on Next Page**

Step 4 **Solve.** $\dfrac{10}{x-3} = \dfrac{15}{x+3}$

> To solve, we can multiply by the LCD or use cross products.

$(x+3)(x-3)\left(\dfrac{10}{x-3}\right) = (x+3)(x-3)\left(\dfrac{15}{x+3}\right)$ Multiply by the LCD, $(x+3)(x-3)$.

$10(x+3) = 15(x-3)$ Multiply.

$10x + 30 = 15x - 45$ Distributive property

$30 = 5x - 45$ Subtract $10x$.

$75 = 5x$ Add 45.

$15 = x$ Divide by 5.

Step 5 **State the answer.** The rate of the boat in still water is 15 mph.

Step 6 **Check** the answer: $\dfrac{10}{15-3} = \dfrac{15}{15+3}$ is true.

································· **Work Problem ⑦ at the Side.** ▶

EXAMPLE 8 **Solving a Problem about Distance, Rate, and Time**

At O'Hare International Airport in Chicago, Cheryl and Bill are walking to the gate at the same rate to catch their flight to Denver. Bill wants a window seat, so he steps onto the moving sidewalk and continues to walk while Cheryl uses the stationary sidewalk. If the sidewalk moves at 1 m per sec and Bill saves 50 sec covering the 300-m distance, what is their walking rate?

Step 1 **Read** the problem. We must find their walking rate.

Step 2 **Assign a variable.**

Let $x =$ their walking rate in meters per second.

Thus, Cheryl travels at x meters per second and Bill travels at $(x+1)$ meters per second. Express their times in terms of the known distances and the variable rates, as in **Example 7.** Cheryl travels 300 m at a rate of x meters per second.

$$t = \frac{d}{r} = \frac{300}{x} \quad \text{Cheryl's time}$$

Bill travels 300 m at a rate of $(x+1)$ meters per second.

$$t = \frac{d}{r} = \frac{300}{x+1} \quad \text{Bill's time}$$

	Distance	Rate	Time
Cheryl	300	x	$\frac{300}{x}$
Bill	300	$x+1$	$\frac{300}{x+1}$

Step 3 **Write an equation,** using the times from the table.

$$\underbrace{\frac{300}{x+1}}_{\substack{\text{Bill's} \\ \text{time}}} \quad \underset{\text{is}}{=} \quad \underbrace{\frac{300}{x}}_{\substack{\text{Cheryl's} \\ \text{time}}} \quad \underbrace{- \quad 50}_{\substack{\text{less 50} \\ \text{seconds.}}}$$

···················· **Continued on Next Page**

⑦ Solve each problem.

GS **(a)** A plane travels 100 mi against the wind in the same time that it takes to travel 120 mi with the wind. The wind speed is 20 mph. Find the rate of the plane in still air.

Let $x =$ _____.

Complete the table.

	d	r	t
Against Wind	100	$x - 20$	_____
With Wind	120	$x + 20$	_____

Write an equation, and complete the solution.

(b) A small fishing boat travels 36 mi against the current in a river in the same time that it travels 44 mi with the current. If the rate of the boat in still water is 20 mph, find the rate of the current.

Answers

7. (a) the rate of the plane in still air;
$\dfrac{100}{x-20}; \dfrac{120}{x+20}; \dfrac{100}{x-20} = \dfrac{120}{x+20};$
220 mph

(b) 2 mph

8 Solve each problem.

(a) Kathy Manley drove 300 mi north from San Antonio, mostly on the freeway. She usually averaged 55 mph, but an accident slowed her speed through Dallas to 15 mph. If her trip took 6 hr, how many miles did she drive at the reduced rate?

Let $x =$ _____.

Complete the table.

	d	r	t
Normal Speed	$300 - x$	55	___
Reduced Speed	x	15	___

Write an equation, and complete the solution.

(b) James and Pat are driving from Atlanta to Jacksonville, a distance of 310 mi. James, whose average rate is 5 mph faster than Pat's, will drive the first 130 mi and then Pat will drive the rest of the way to their destination. If the total driving time is 5 hr, determine the average rate of each driver.

Answers

8. (a) the distance driven at the reduced rate;
$\dfrac{300-x}{55}$; $\dfrac{x}{15}$; $\dfrac{300-x}{55} + \dfrac{x}{15} = 6$; $11\frac{1}{4}$ mi
(b) James: 65 mph; Pat: 60 mph

Step 4 Solve.

$$\frac{300}{x+1} = \frac{300}{x} - 50$$

> We **cannot** solve using cross products because there are two terms on the right.

$$x(x+1)\left(\frac{300}{x+1}\right) = x(x+1)\left(\frac{300}{x} - 50\right)$$

Multiply by the LCD, $x(x+1)$.

$$x(x+1)\left(\frac{300}{x+1}\right) = x(x+1)\left(\frac{300}{x}\right) - x(x+1)(50)$$

Distributive property

$$300x = 300(x+1) - 50x(x+1) \quad \text{Multiply.}$$

$$300x = 300x + 300 - 50x^2 - 50x \quad \text{Distributive property}$$

$$50x^2 + 50x - 300 = 0 \quad \text{Standard form}$$

$$x^2 + x - 6 = 0 \quad \text{Divide by 50.}$$

$$(x+3)(x-2) = 0 \quad \text{Factor.}$$

$$x + 3 = 0 \quad \text{or} \quad x - 2 = 0 \quad \text{Zero-factor property}$$

$$x = -3 \quad \text{or} \quad x = 2 \quad \text{Solve each equation.}$$

Discard the negative answer, since rate (speed) cannot be negative.

Step 5 State the answer. Their walking rate is 2 m per sec.

Step 6 Check the answer in the words of the original problem.

◄ **Work Problem 8** at the Side.

OBJECTIVE 5 Solve applications about work rates. Suppose that we can mow a lawn in 4 hr. Then after 1 hr, we will have mowed $\frac{1}{4}$ of the lawn. After 2 hr, we will have mowed $\frac{2}{4}$, or $\frac{1}{2}$, of the lawn, and so on. This idea is generalized as follows.

Rate of Work

If a job can be completed in t units of time, then the rate of work is

$$\frac{1}{t} \text{ job per unit of time.}$$

Problem-Solving Hint

The formula $d = rt$ says that distance traveled is equal to rate of travel multiplied by time traveled. Similarly, the fractional part of a job accomplished is equal to the rate of work multiplied by the time worked.

In the lawn mowing example, after 3 hr, the fractional part of the job done is found as follows.

$$\underset{\substack{\text{Rate of}\\\text{work}}}{\frac{1}{4}} \cdot \underset{\substack{\text{Time}\\\text{worked}}}{3} = \underset{\substack{\text{Fractional part}\\\text{of job done}}}{\frac{3}{4}}$$

After 4 hr, $\frac{1}{4}(4) = 1$ whole job has been done.

EXAMPLE 9 Solving a Problem about Work Rates

With spraying equipment, Mateo can paint the woodwork in a small house in 8 hr. His assistant, Chet, needs 14 hr to complete the same job painting by hand. If both Mateo and Chet work together, how long will it take them to paint the woodwork?

Step 1 **Read** the problem again. We are looking for time working together.

Step 2 **Assign a variable.**

Let x = the number of hours it will take for Mateo and Chet to paint the woodwork, working together.

Begin by making a table. Based on the previous discussion, Mateo's rate alone is $\frac{1}{8}$ job per hour, and Chet's rate is $\frac{1}{14}$ job per hour.

	Rate	Time Working Together	Fractional Part of the Job Done When Working Together
Mateo	$\frac{1}{8}$	x	$\frac{1}{8}x$
Chet	$\frac{1}{14}$	x	$\frac{1}{14}x$

Sum is 1 whole job.

Step 3 **Write an equation.**

$$\underbrace{\text{Fractional part}}_{\text{done by Mateo}} + \underbrace{\text{Fractional part}}_{\text{done by Chet}} = \underbrace{1 \text{ whole job}}$$

$$\frac{1}{8}x \quad + \quad \frac{1}{14}x \quad = \quad 1$$

Together, Mateo and Chet complete 1 whole job. Add the fractional parts and set the sum equal to 1.

Step 4 **Solve.** $56\left(\frac{1}{8}x + \frac{1}{14}x\right) = 56\,(1)$ Multiply by the LCD, 56.

$$56\left(\frac{1}{8}x\right) + 56\left(\frac{1}{14}x\right) = 56\,(1)$$ Distributive property

$$7x + 4x = 56$$ Multiply.

$$11x = 56$$ Combine like terms.

$$x = \frac{56}{11}$$ Divide by 11.

Step 5 **State the answer.** Working together, Mateo and Chet can paint the woodwork in $\frac{56}{11}$ hr, or $5\frac{1}{11}$ hr.

Step 6 **Check.** Substitute $\frac{56}{11}$ for x in the equation from Step 3.

$$\frac{1}{8}x + \frac{1}{14}x = 1$$ Equation from Step 3

$$\frac{1}{8}\left(\frac{56}{11}\right) + \frac{1}{14}\left(\frac{56}{11}\right) \stackrel{?}{=} 1$$ Let $x = \frac{56}{11}$.

$$\frac{7}{11} + \frac{4}{11} = 1 \quad \checkmark \text{ True}$$

Our answer, $\frac{56}{11}$ hr, or $5\frac{1}{11}$ hr, is correct.

9 Solve each problem.

(a) Michael can paint a room, working alone, in 8 hr. Lindsay can paint the same room, working alone, in 6 hr. How long will it take them if they work together?

Let $x =$ the number of

_____ it will take for Michael and Lindsay to paint the room, working _____.

Complete the table.

	Rate	Time Working Together	Fractional Part of the Job Done When Working Together
Michael	_____	x	_____
Lindsay	_____	x	_____

Together, Michael and Lindsay complete _____ whole job(s). Write an equation, and complete the solution.

(b) Roberto can detail his Camaro in 2 hr working alone. His brother Marco can do the job in 3 hr working alone. How long would it take them if they worked together?

Problem-Solving Hint

A common error students make when solving a work problem like that in **Example 9** is to add the two times.

$$8 \text{ hr} + 14 \text{ hr} = 22 \text{ hr} \leftarrow \text{Incorrect answer}$$

The answer 22 hr is unreasonable, because the slower worker (Chet) can do the job *alone* in 14 hr. The correct answer *must* be less than 14 hr.

Keep in mind that the correct time for the answer must also be less than that of the *faster* worker (Mateo, at 8 hr) because he is getting help. Based on this reasoning, does our answer of $5\frac{1}{11}$ hr in **Example 9** make sense?

◄ **Work Problem 9** at the Side.

An alternative approach is to consider the part of the job that can be done in 1 hr. For instance, in **Example 9** Mateo can do the entire job in 8 hr, and Chet can do it in 14 hr. Thus, their work rates, as we saw in **Example 9,** are $\frac{1}{8}$ and $\frac{1}{14}$, respectively. Since it takes them x hours to complete the job when working together, in 1 hr they can paint $\frac{1}{x}$ of the woodwork.

The amount painted by Mateo in 1 hr plus the amount painted by Chet in 1 hr must equal the amount they can do together. This leads to the following alternative equation.

$$\overset{\text{Amount by Chet}}{\underset{\text{Amount by Mateo} \rightarrow}{\frac{1}{8} + \frac{1}{14}}} = \underset{}{\frac{1}{x}} \leftarrow \text{Amount together}$$

Compare this with the equation in Step 3 of **Example 9.** Multiplying each side by $56x$ leads to the same equation found in the third line of Step 4 in the example.

$$7x + 4x = 56$$

The same solution results.

Answers

9. **(a)** hours; together

	Rate	Time Working Together	Fractional Part of the Job Done When Working Together
Michael	$\frac{1}{8}$	x	$\frac{1}{8}x$
Lindsay	$\frac{1}{6}$	x	$\frac{1}{6}x$

$1; \frac{1}{8}x + \frac{1}{6}x = 1; \frac{24}{7}$ hr, or $3\frac{3}{7}$ hr

(b) $\frac{6}{5}$ hr, or $1\frac{1}{5}$ hr

7.5 Exercises

 MyMathLab®

CONCEPT CHECK *In Exercises 1–4, a formula is given. Give the letter of the choice that is an equivalent form of the given formula.*

1. $p = br$ (percent)

  A. $b = \dfrac{p}{r}$ **B.** $r = \dfrac{b}{p}$

  C. $b = \dfrac{r}{p}$ **D.** $p = \dfrac{r}{b}$

2. $V = LWH$ (geometry)

  A. $H = \dfrac{LW}{V}$ **B.** $L = \dfrac{V}{WH}$

  C. $L = \dfrac{WH}{V}$ **D.** $W = \dfrac{H}{VL}$

3. $m = \dfrac{F}{a}$ (physics)

  A. $a = mF$ **B.** $F = \dfrac{m}{a}$

  C. $F = \dfrac{a}{m}$ **D.** $F = ma$

4. $I = \dfrac{E}{R}$ (electricity)

  A. $R = \dfrac{I}{E}$ **B.** $R = IE$

  C. $E = \dfrac{I}{R}$ **D.** $E = RI$

Solve each problem. See Example 1.

5. A gas law in chemistry says that
$$\frac{PV}{T} = \frac{pv}{t}.$$
Suppose that $T = 300$, $t = 350$, $V = 9$, $P = 50$, and $v = 8$. Find p.

6. In work with electric circuits, the formula
$$\frac{1}{a} = \frac{1}{b} + \frac{1}{c}$$
occurs. Find b if $a = 8$ and $c = 12$.

7. A formula from anthropology says that
$$c = \frac{100b}{L}.$$
Find L if $c = 80$ and $b = 5$.

8. The gravitational force between two masses is given by
$$F = \frac{GMm}{d^2}.$$
Find M to the nearest thousandth if $F = 10$, $G = 6.67 \times 10^{-11}$, $m = 1$, and $d = 3 \times 10^{-6}$.

9. **CONCEPT CHECK** To solve the equation
$$m = \frac{ab}{a - b}$$
for a, what is the first step?

10. **CONCEPT CHECK** To solve the equation
$$rp - rq = p + q$$
for r, what is the first step?

Solve each formula for the specified variable. See Examples 2 and 3.

11. $F = \dfrac{GMm}{d^2}$ for G (physics)   **12.** $F = \dfrac{GMm}{d^2}$ for M (physics)   **13.** $\dfrac{1}{a} = \dfrac{1}{b} + \dfrac{1}{c}$ for a (electricity)

14. $\dfrac{1}{a} = \dfrac{1}{b} + \dfrac{1}{c}$ for b (electricity)

15. $\dfrac{PV}{T} = \dfrac{pv}{t}$ for v (chemistry)

16. $\dfrac{PV}{T} = \dfrac{pv}{t}$ for T (chemistry)

17. $I = \dfrac{nE}{R + nr}$ for r
(engineering)

18. $a = \dfrac{V - v}{t}$ for V (physics)

19. $A = \dfrac{1}{2}h(b + B)$ for b
(mathematics)

20. $S = \dfrac{n}{2}(a + \ell)d$ for n
(mathematics)

21. $\dfrac{E}{e} = \dfrac{R + r}{r}$ for r
(engineering)

22. $y = \dfrac{x + z}{a - x}$ for x

23. $D = \dfrac{R}{1 + RT}$ for R (banking)

24. $R = \dfrac{D}{1 - DT}$ for D (banking)

CONCEPT CHECK *Use proportions to solve each problem mentally.*

25. In a mathematics class, 3 of every 4 students are girls. If there are 28 students in the class, how many are girls? How many are boys?

26. In a small town in Louisiana, sales tax on a purchase of $1.50 is $0.12. What is the sales tax on a purchase of $9.00?

The water content of snow is affected by the temperature, wind speed, and other factors present when the snow is falling. The average snow-to-liquid ratio is 10 in. of snow to 1 in. of liquid precipitation. This means that if 10 in. of snow fell and was melted, it would produce 1 in. of liquid precipitation in a rain gauge. (Source: www.theweatherprediction.com)
*Use a proportion to solve each problem. **See Examples 4 and 5.***

27. A dry snow might have a snow-to-liquid ratio of 18 to 1. Using this ratio, how much liquid precipitation would be produced by 31.5 in. of snow?

28. A wet, sticky snow good for making a snow man might have a snow-to-liquid ratio of 5 to 1. How many inches of fresh snow would produce 3.25 in. of liquid precipitation using this ratio?

Solve each problem. Give answers to the nearest tenth if an approximation is needed.
See Examples 4 and 5.

29. On a map of the United States, the distance between Seattle and Durango is 4.125 in. The two cities are actually 1238 miles apart. On this same map, what would be the distance between Chicago and El Paso, two cities that are actually 1606 mi apart? (*Source:* Universal Map Atlas.)

30. On a map of the United States, the distance between Reno and Phoenix is 2.5 in. The two cities are actually 768 miles apart. On this same map, what would be the distance between St. Louis and Jacksonville, two cities that are actually 919 mi apart? (*Source:* Universal Map Atlas.)

31. On June 3, 2012, the Chicago White Sox were in first place in the Central Division of the American League, having won 31 of their first 54 regular season games. If the team continued to win the same fraction of its games, how many games would the White Sox win for the complete 162-game season? (*Source:* www.mlb.com)

32. During 2010–2011, the ratio of teachers to students in public elementary and secondary schools was approximately 1 to 15. If a public school had 846 students, how many teachers would be at the school according to this ratio? Round the answer to the nearest whole number. (*Source*: U.S. National Center for Education Statistics.)

33. Biologists tagged 500 fish in a lake on January 1. On February 1 they returned and collected a random sample of 400 fish, 8 of which had been previously tagged. How many fish does the lake have based on this experiment?

34. Suppose that in the experiment of **Exercise 33,** 10 of the previously tagged fish were collected on February 1. What would be the estimate of the fish population?

35. Bruce Johnston's Shelby Cobra uses 5 gal of gasoline to drive 156 mi. He has 3 gal of gasoline in the car, and he wants to know how much more gasoline he will need to drive 300 mi. If we assume that the car continues to use gasoline at the same rate, how many more gallons will he need?

36. Mike Love's T-bird uses 6 gal of gasoline to drive 141 miles. He has 4 gal of gasoline in the car, and he wants to know how much more gasoline he will need to drive 275 mi. If we assume that the car continues to use gasoline at the same rate, how many more gallons will he need?

Nurses use proportions to determine the amount of a drug to administer when the dose of the drug is measured in milligrams but the drug is packaged in a diluted form in milliliters. (Source: Hoyles, Celia, Richard Noss, and Stefano Pozzi, "Proportional Reasoning in Nursing Practice," Journal for Research in Mathematics Education.) For example, to find the number of milliliters of fluid needed to administer 300 mg of a drug that comes packaged as 120 mg in 2 mL of fluid, a nurse sets up the proportion

$$\frac{120 \text{ mg}}{2 \text{ mL}} = \frac{300 \text{ mg}}{x \text{ mL}},$$

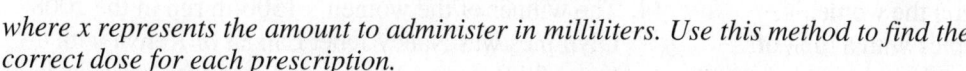

where x represents the amount to administer in milliliters. Use this method to find the correct dose for each prescription.

37. 120 mg of Amakacine packaged as 100 mg in 2-mL vials

38. 1.5 mg of morphine packaged as 20 mg ampules diluted in 10 mL of fluid

*In geometry, it is shown that two triangles with corresponding angle measures equal, called **similar triangles,** have corresponding sides proportional.*

For example, in the figure, angle A = angle D, angle B = angle E, and angle C = angle F, so the triangles are similar. Then the following ratios of corresponding sides are equal.

$$\frac{4}{6} = \frac{6}{9} = \frac{2x + 1}{2x + 5}$$

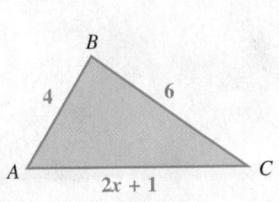
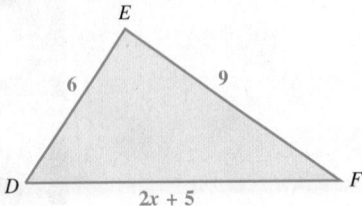

39. Solve for *x* using the given proportion to find the lengths of the third sides of the triangles.

40. Suppose the following triangles are similar. Find *y* and the lengths of the two longest sides of each triangle.

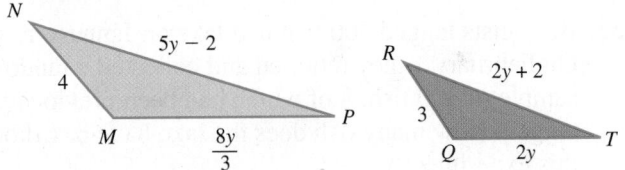

*Solve each problem. **See Example 6.***

41. In 2007, British explorer and endurance swimmer Lewis Gordon Pugh became the first person to swim at the North Pole. He swam 0.6 mi at 0.0319 mi per min in waters created by melted sea ice. What was his time (to three decimal places)? (*Source: The Gazette.*)

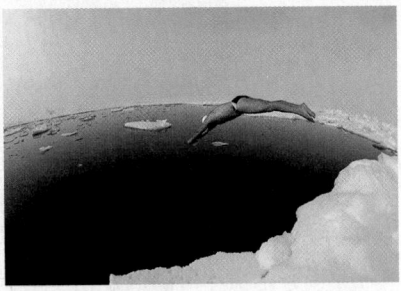

42. In the 2008 Summer Olympics, Britta Steffen of Germany won the women's 100-m freestyle swimming event. Her rate was 1.8825 m per sec. What was her time (to two decimal places)? (*Source: World Almanac and Book of Facts.*)

43. Tirunesh Dibaba of Ethiopia won the women's 5000-m race in the 2008 Olympics with a time of 15.911 min. What was her rate (to three decimal places)? (*Source: World Almanac and Book of Facts.*)

44. The winner of the women's 1500-m run in the 2008 Olympics was Nancy Jebet Langat of Kenya with a time of 4.004 min. What was her rate (to three decimal places)? (*Source: World Almanac and Book of Facts.*)

Complete any tables. Then solve each problem. See Examples 7 and 8.

45. Kellen's boat travels 12 mph. Find the rate of the current of the river if she can travel 6 mi upstream in the same amount of time she can travel 10 mi downstream.

	Distance	Rate	Time
Downstream	10	$12 + x$	_____
Upstream	6	$12 - x$	_____

46. Kasey can travel 8 mi upstream in the same time it takes her to travel 12 mi downstream. Her boat travels 15 mph in still water. What is the rate of the current?

	Distance	Rate	Time
Downstream	_____	_____	_____
Upstream	_____	$15 - x$	_____

47. In his boat, Sheldon can travel 30 mi downstream in the same time that it takes to travel 10 mi upstream. If the rate of the current is 5 mph, find the rate of the boat in still water.

48. In his boat, Leonard can travel 24 mi upstream in the same time that it takes to travel 36 mi downstream. If the rate of the current is 2 mph, find the rate of the boat in still water.

49. On his drive from Montpelier, Vermont, to Columbia, South Carolina, Dylan Davis averaged 51 mph. If he had been able to average 60 mph, he would have reached his destination 3 hr earlier. What is the driving distance between Montpelier and Columbia?

50. Leah drove from her apartment to her parents' house for the weekend. Driving to their house, she was able to average 60 mph. However, returning she was able to average only 45 mph on the same route, because traffic was heavy. The return drive took her 1.5 hr longer than the drive on Saturday. What is the distance between Leah's apartment and her parents' house?

51. A private plane traveled from San Francisco to a secret rendezvous. It averaged 200 mph. On the return trip, the average speed was 300 mph. If the total traveling time was 4 hr, how far from San Francisco was the secret rendezvous?

52. Johnny averages 30 mph when he drives on the old highway to his favorite fishing hole, and he averages 50 mph when most of his route is on the interstate. If both routes are the same length, and he saves 2 hr by traveling on the interstate, how far away is the fishing hole?

53. On the first part of a trip to Carmel traveling on the freeway, Marge averaged 60 mph. On the rest of the trip, which was 10 mi longer than the first part, she averaged 50 mph. Find the total distance to Carmel if the second part of the trip took 30 min more than the first part.

54. During the first part of a trip on the highway, Jim and Annie averaged 60 mph. When they got to Houston, traffic caused them to average only 30 mph. The distance they drove in Houston was 100 mi less than their distance on the highway. What was their total driving distance if they spent 50 min more on the highway than they did in Houston?

*Complete any tables. Then solve each problem. **See Example 9.***

55. Butch and Peggy want to pick up the mess that their grandson, Grant, has made in his playroom. Butch could do it in 15 min working alone. Peggy, working alone, could clean it in 12 min. How long will it take them if they work together?

	Rate	Time Working Together	Fractional Part of the Job Done
Butch	$\frac{1}{15}$	x	_____
Peggy	$\frac{1}{12}$	x	_____

56. Lou can groom Jay Beckenstein's dogs in 8 hr, but it takes his business partner, Janet, only 5 hr to groom the same dogs. How long will it take them to groom Jay's dogs if they work together?

	Rate	Time Working Together	Fractional Part of the Job Done
Lou	$\frac{1}{8}$	x	_____
Janet	$\frac{1}{5}$	x	_____

57. Jerry and Kuba are laying a hardwood floor. Working alone, Jerry can do the job in 20 hr. If the two of them work together, they can complete the job in 12 hr. How long would it take Kuba to lay the floor working alone? (Let x = the time it would take Kuba working alone.)

	Rate	Time Working Together	Fractional Part of the Job Done
Jerry	_____	12	_____
Kuba	_____	12	_____

58. Mrs. Disher is a mathematics teacher. She can grade a set of chapter tests in 5 hr working alone. If her student teacher Mr. Howes helps her, it will take 3 hr to grade the tests. How long would it take Mr. Howes to grade the tests if he worked alone? (Let x = the time it would take Mr. Howes working alone.)

	Rate	Time Working Together	Fractional Part of the Job Done
Mrs. Disher	_____	3	_____
Mr. Howes	_____	3	_____

59. If a vat of acid can be filled by an inlet pipe in 10 hr and emptied by an outlet pipe in 20 hr, how long will it take to fill the vat if both pipes are open?

60. A winery has a vat to hold Chardonnay. An inlet pipe can fill the vat in 9 hr, while an outlet pipe can empty it in 12 hr. How long will it take to fill the vat if both the outlet and the inlet pipes are open?

61. Suppose that Hortense and Mort can clean their entire house in 7 hr, while their toddler, Mimi, just by being around, can completely mess it up in only 2 hr. If Hortense and Mort clean the house while Mimi is at her grandma's, and then start cleaning up after Mimi the minute she gets home, how long does it take from the time Mimi gets home until the whole place is a shambles?

62. An inlet pipe can fill an artificial lily pond in 60 min, while an outlet pipe can empty it in 80 min. Through an error, both pipes are left open. How long will it take for the pond to fill?

7.6 Variation

Functions in which *y depends on a multiple of x* or *y depends on a number divided by x* are common in business and the physical sciences.

OBJECTIVE ▶ ① Write an equation expressing direct variation. The circumference of a circle is given by the formula $C = 2\pi r$, where r is the radius of the circle. See **Figure 5.** Circumference is always a constant multiple of the radius—that is, C is always found by multiplying r by the constant 2π.

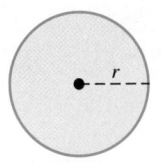

$C = 2\pi r$

Figure 5

As the **radius increases,** the **circumference increases.**

As the **radius decreases,** the **circumference decreases.**

As a result, the circumference is said to *vary directly* as the radius.

Direct Variation

y varies directly as x if there exists a real number *k* such that

$$y = kx.$$

y is **proportional to** *x*. The number *k* is the **constant of variation.** *In direct variation, for k > 0, as the value of x increases, the value of y also increases. Similarly, as x decreases, y decreases.*

OBJECTIVE ▶ ② Find the constant of variation, and solve direct variation problems. *The direct variation equation* $y = kx$ *defines a linear function, where the constant of variation k is the slope of the line.* For example, the following equation describes the cost *y* to buy *x* gallons of gasoline.

$$y = 4.50x$$

The cost varies directly as the number of gallons of gasoline purchased.

As the *number* of gallons of gasoline *increases,* the *cost increases.*

As the *number* of gallons of gasoline *decreases,* the *cost decreases.*

The constant of variation *k* is **4.50,** the cost of 1 gallon of gasoline.

EXAMPLE 1   Solving a Direct Variation Problem

Stella Frolick is paid an hourly wage. One week she worked 43 hr and was paid $795.50. How much does she earn per hour?

Let *h* represent the number of hours she works and *P* represent her corresponding pay. Write a variation equation.

k represents Stella's hourly wage. → $P = kh$ P varies directly as h.

$795.50 = k \cdot 43$ Let $P = 795.50$ and $h = 43$.

This is the constant of variation. → $18.50 = k$ Divide by 43. Use a calculator.

Thus, her hourly wage is $18.50, and *P* and *h* are related by $P = 18.50h$.

········· **Work Problem ① at the Side.** ▶

OBJECTIVES

① Write an equation expressing direct variation.

② Find the constant of variation, and solve direct variation problems.

③ Solve inverse variation problems.

④ Solve joint variation problems.

⑤ Solve combined variation problems.

---

① Find the constant of variation, and write a direct variation equation.

GS **(a)** Ginny Michaud is paid a daily wage. One month she worked 17 days and earned $1334.50.

Let d = days she worked and E = her corresponding _____.

Write a variation equation using these variables.

____ = $k \cdot$ ____

Substitute ____ for E and ____ for d.

$1334.50 = k \cdot$ ____

____ = k

Thus, the constant of variation is $k =$ ____, and E and d are related by the equation

$E =$ ____.

(b) Distance varies directly as time (at a constant rate). A car travels 100 mi at a constant rate in 2 hr.

Answers

1. **(a)** earnings; E; d; 1334.50; 17; 17; 78.50; 78.50; 78.50d
   (b) $k = 50$; Let d represent the distance traveled in h hours. Then $d = 50h$.

❷ Solve the problem.

The charge (in dollars) to customers for electricity (in kilowatt-hours) varies directly as the number of kilowatt-hours used. It costs $52 to use 800 kilowatt-hours. Find the cost to use 1000 kilowatt-hours.

EXAMPLE 2   Solving a Direct Variation Problem

Hooke's law for an elastic spring states that the distance a spring stretches is proportional to the force applied. If a force of 150 newtons* stretches a certain spring 8 cm, how much will a force of 400 newtons stretch the spring?

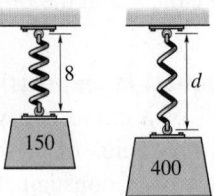

Figure 6

See **Figure 6.** If d is the distance the spring stretches and f is the force applied, then $d = kf$ for some constant k. Since a force of 150 newtons stretches the spring 8 cm, we use these values to find k.

$$d = kf \qquad \text{Variation equation}$$

$$8 = k \cdot 150 \qquad \text{Let } d = 8 \text{ and } f = 150.$$

$$k = \frac{8}{150}, \quad \text{or} \quad \frac{4}{75} \qquad \begin{array}{l}\text{Solve for } k.\\ \text{Write in lowest terms.}\end{array}$$

Substitute $\frac{4}{75}$ for k in the variation equation $d = kf$.

$$d = \frac{4}{75}f \qquad \text{Here, } k = \frac{4}{75}.$$

For a force of 400 newtons, substitute 400 for f.

$$d = \frac{4}{75}(400) = \frac{64}{3} \qquad \text{Let } f = 400.$$

The spring will stretch $\frac{64}{3}$ cm, or $21\frac{1}{3}$ cm, if a force of 400 newtons is applied.

··◀ **Work Problem ❷ at the Side.**

Solving a Variation Problem

Step 1 Write the variation equation.

Step 2 Substitute the initial values and solve for k.

Step 3 Rewrite the variation equation with the value of k from Step 2.

Step 4 Substitute the remaining values, solve for the unknown, and find the required answer.

One variable can be proportional to a power of another variable.

Direct Variation as a Power

y varies directly as the nth power of x if there exists a real number k such that

$$y = kx^n.$$

Answer

2. $65

*A newton is a unit of measure of force used in physics.

An example of direct variation as a power is the formula for the area of a circle, $A = \pi r^2$. Here, π is the constant of variation, and the area varies directly as the square of the radius.

EXAMPLE 3   Solving a Direct Variation Problem

The distance a body falls from rest varies directly as the square of the time it falls (disregarding air resistance). If a skydiver falls 64 ft in 2 sec, how far will she fall in 8 sec?

Step 1   If d represents the distance the skydiver falls and t the time it takes to fall, then d is a function of t for some constant k.

$$d = kt^2$$

Step 2   To find the value of k, use the fact that the skydiver falls 64 ft in 2 sec.

$$d = kt^2 \qquad \text{Variation equation}$$
$$64 = k\,(2)^2 \qquad \text{Let } d = 64 \text{ and } t = 2.$$
$$k = 16 \qquad \text{Find } k.$$

Step 3   Now we rewrite the variation equation $d = kt^2$ using 16 for k.

$$d = 16t^2 \qquad \text{Here, } k = 16.$$

Step 4   Let $t = 8$ to find the number of feet the skydiver will fall in 8 sec.

$$d = 16\,(8)^2 = 1024 \qquad \text{Let } t = 8.$$

The skydiver will fall 1024 ft in 8 sec.

···················· **Work Problem ❸ at the Side.** ▶

OBJECTIVE ▶ ❸ Solve inverse variation problems. *With inverse variation, where $k > 0$, as one variable increases, the other variable decreases.*

For example, in a closed space, volume decreases as pressure increases, as illustrated by a trash compactor. See **Figure 7.** As the compactor presses down, the pressure on the trash increases, and in turn, the trash occupies a smaller space.

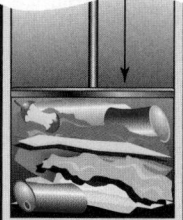

As pressure on trash increases, volume of trash decreases.

Figure 7

Inverse Variation

y **varies inversely as** x if there exists a real number k such that

$$y = \frac{k}{x}.$$

Also, y **varies inversely as the nth power of** x if there exists a real number k such that

$$y = \frac{k}{x^n}.$$

The inverse variation equation defines a rational function.

---

❸ The area of a circle varies directly as the square of its radius. A circle with radius 3 in. has area 28.278 in.^{2}.

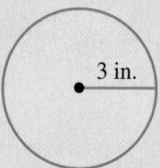

3 in.

(a) Write a variation equation and give the value of k.

(b) What is the area of a circle with radius 4.1 in.? (Use the answers from part (a).)

4 Solve each problem.

(a) For a constant area, the height of a triangle varies inversely as the base. If the height is 7 cm when the base is 8 cm, find the height when the base is 14 cm.

Another example of inverse variation comes from the distance formula.

$$d = rt \quad \text{Distance formula}$$

$$t = \frac{d}{r} \quad \text{Divide each side by } r.$$

Here, t (time) varies inversely as r (rate or speed), with d (distance) serving as the constant of variation. For example, if the distance between Chicago and Des Moines is 300 mi, then

$$t = \frac{300}{r}.$$

The values of r and t might be any of the following.

$$\left. \begin{array}{l} r = 50, t = 6 \\ r = 60, t = 5 \\ r = 75, t = 4 \end{array} \right\} \begin{array}{l} \text{As } r \text{ increases,} \\ t \text{ decreases.} \end{array} \qquad \left. \begin{array}{l} r = 30, t = 10 \\ r = 25, t = 12 \\ r = 20, t = 15 \end{array} \right\} \begin{array}{l} \text{As } r \text{ decreases,} \\ t \text{ increases.} \end{array}$$

If we *increase* the rate (speed) at which we drive, time *decreases*. If we *decrease* the rate (speed) at which we drive, time *increases*.

EXAMPLE 4 **Solving an Inverse Variation Problem**

In the manufacture of a certain medical syringe, the cost of producing the syringe varies inversely as the number produced. If 10,000 syringes are produced, the cost is $2 per syringe. Find the cost per syringe of producing 25,000 syringes.

$$\text{Let} \quad x = \text{the number of syringes produced,}$$

$$\text{and} \quad c = \text{the cost per syringe.}$$

Here, as production increases, cost decreases, and as production decreases, cost increases. We write a variation equation using the variables c and x and the constant k.

$$c = \frac{k}{x} \quad c \text{ varies inversely as } x.$$

To find k, we replace c with 2 and x with 10,000.

$$2 = \frac{k}{10,000} \quad \text{Substitute in the variation equation.}$$

$$\mathbf{20{,}000 = k} \quad \text{Multiply by 10,000.}$$

Thus, $c = \frac{k}{x}$ becomes $c = \frac{20,000}{x}$. When $x = 25,000$,

$$c = \frac{20,000}{25,000} = 0.80. \quad \text{Let } x = 25,000.$$

The cost per syringe to make 25,000 syringes is $0.80.

(b) The current in a simple electrical circuit varies inversely as the resistance. If the current is 80 amps when the resistance is 10 ohms, find the current when the resistance is 16 ohms.

◀ **Work Problem** **4** at the Side.

Answers

4. (a) 4 cm **(b)** 50 amps

EXAMPLE 5 Solving an Inverse Variation Problem

The weight of an object above Earth varies inversely as the square of its distance from the center of Earth. A space shuttle in an elliptical orbit has a maximum distance from the center of Earth (**apogee**) of 6700 mi. Its minimum distance from the center of Earth (**perigee**) is 4090 mi. See **Figure 8**. If an astronaut in the shuttle weighs 57 lb at its apogee, what does the astronaut weigh at its perigee?

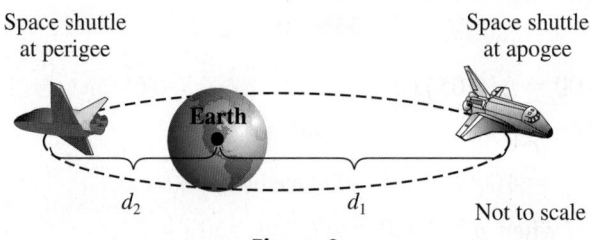

Space shuttle at perigee Space shuttle at apogee

d_2 d_1 Not to scale

Figure 8

Let w = the weight and d = the distance from the center of Earth, for some constant k.

$$w = \frac{k}{d^2} \qquad w \text{ varies inversely as the square of } d.$$

At the apogee, the astronaut weighs 57 lb, and the distance from the center of Earth is 6700 mi. Use these values to find k.

$$57 = \frac{k}{(6700)^2} \qquad \text{Let } w = 57 \text{ and } d = 6700.$$

$$k = 57(6700)^2 \qquad \text{Solve for } k.$$

Substitute $k = 57(6700)^2$ and $d = 4090$ to find the weight at the perigee.

$$w = \frac{57(6700)^2}{(4090)^2} \approx 153 \text{ lb} \qquad \text{Use a calculator.}$$

···· **Work Problem ❺ at the Side.** ▶

OBJECTIVE ④ Solve joint variation problems. If one variable varies directly as the *product* of several other variables (perhaps raised to powers), the first variable is said to *vary jointly* as the others.

Joint Variation

y **varies jointly as *x* and *z*** if there exists a real number *k* such that

$$y = kxz.$$

An example of joint variation is the formula for the area of a triangle, $A = \frac{1}{2}bh$. Here, $\frac{1}{2}$ is the constant of variation, and the area varies jointly as the length of its base and its height.

CAUTION

Note that *and* in the expression "*y* varies jointly as *x* and *z*" translates as a product in

$$y = kxz.$$

The word *and* does not indicate addition here.

---

❺ If the temperature is constant, the volume of a gas varies inversely as the pressure. For a certain gas, the volume is 10 cm³ when the pressure is 6 kg per cm².

(a) Find the variation equation.

(b) Find the volume when the pressure is 12 kg per cm².

Answers

5. (a) $V = \dfrac{60}{P}$ **(b)** 5 cm³

6 Solve the problem.

The volume of a rectangular box of a given height is proportional to its width and length. A box with width 2 ft and length 4 ft has volume 12 ft^3. Find the volume of a box with the same height that is 3 ft wide and 5 ft long.

7 Solve the problem.

The maximum load that a cylindrical column with a circular cross section can hold varies directly as the fourth power of the diameter of the cross section and inversely as the square of the height. A 9-m column 1 m in diameter will support 8 metric tons. How many metric tons can be supported by a column 12 m high and $\frac{2}{3}$ m in diameter?

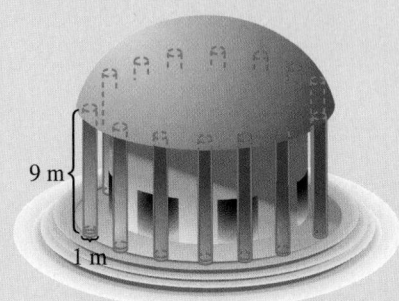

9 m

1 m

Load = 8 metric tons

Answers

6. 22.5 ft^3

7. $\frac{8}{9}$ metric ton

| EXAMPLE 6 | Solving a Joint Variation Problem |

The interest on a loan or an investment is given by the formula $I = prt$. Here, for a given principal p, the interest earned I varies jointly as the interest rate r and the time t that the principal is left earning interest. If an investment earns $100 interest at 5% for 2 yr, how much interest will the same principal earn at 4.5% for 3 yr?

We use the formula $I = prt$, where p is the constant of variation because it is the same for both investments.

$$I = prt$$

$$100 = p(0.05)(2) \qquad \text{Let } I = 100, r = 0.05, \text{ and } t = 2.$$

$$100 = 0.1p \qquad \text{Multiply.}$$

$$p = 1000 \qquad \text{Divide by 0.1. Rewrite.}$$

Now we find I when $p = 1000$, $r = 0.045$, and $t = 3$.

$$I = 1000(0.045)(3) = 135 \qquad \text{Let } p = 1000, r = 0.045, \text{ and } t = 3.$$

The interest will be $135.

◄ **Work Problem 6** at the Side.

OBJECTIVE ▶ **5** **Solve combined variation problems.** There are combinations of direct and inverse variation, called **combined variation.**

| EXAMPLE 7 | Solving a Combined Variation Problem |

Body mass index, or BMI, is used to assess a person's level of fatness. A BMI from 19 through 25 is considered desirable. BMI varies directly as an individual's weight in pounds and inversely as the square of the individual's height in inches. (*Source: Washington Post.*)

A woman who weighs 118 lb and is 64 in. tall has a BMI of 20. (The BMI is rounded to the nearest whole number.) Find the BMI of a woman who weighs 165 lb and is 70 in. tall.

Let B represent the BMI, w the weight, and h the height.

$$B = \frac{kw}{h^2} \quad \begin{array}{l} \longleftarrow \text{ BMI varies directly as the weight.} \\ \longleftarrow \text{ BMI varies inversely as the square of the height.} \end{array}$$

To find k, let $B = 20$, $w = 118$, and $h = 64$.

$$20 = \frac{k(118)}{64^2} \qquad B = \frac{kw}{h^2}$$

$$k = \frac{20(64^2)}{118} \qquad \begin{array}{l} \text{Multiply by } 64^2. \\ \text{Divide by 118.} \end{array}$$

$$k \approx 694 \qquad \text{Use a calculator.}$$

Now find B when $k = 694$, $w = 165$, and $h = 70$.

$$B = \frac{694(165)}{70^2} \approx 23 \qquad \begin{array}{l} \text{Nearest whole} \\ \text{number} \end{array}$$

The woman's BMI is 23.

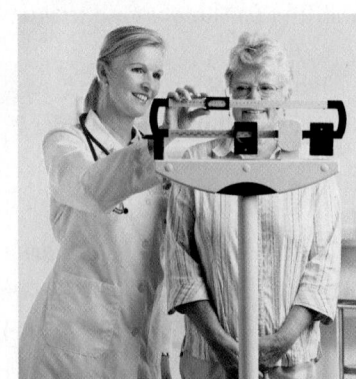

◄ **Work Problem 7** at the Side.

7.6 Exercises

Download the MyDashBoard App

MyMathLab®

CONCEPT CHECK *Use personal experience or intuition to determine whether the situation suggests* direct *or* inverse *variation.**

1. The rate and the distance traveled by a pickup truck in 3 hr

2. The number of different lottery tickets you buy and your probability of winning that lottery

3. Your age and the probability that you believe in Santa Claus

4. The amount of gasoline that you pump and the amount of empty space left in your tank

5. The amount of pressure put on the accelerator of a car and the rate of the car

6. The number of days until the end of the baseball season and the number of home runs that Jose Bautista has hit during the season

7. CONCEPT CHECK Fill in each blank with the correct response.

(a) For $k > 0$, if y varies directly as x, when x increases, y _____, and when x decreases, y _____.

(b) For $k > 0$, if y varies inversely as x, when x increases, y _____, and when x decreases, y _____.

8. Bill Veeck was the owner of several major league baseball teams in the 1950s and 1960s. He was known to often sit in the stands and enjoy games with his paying customers. Here is a quote attributed to him:

"I have discovered in 20 years of moving around a ballpark, that the knowledge of the game is usually in inverse proportion to the price of the seats."

Explain in your own words the meaning of this statement. (To prove his point, Veeck once allowed the fans to vote on managerial decisions.)

CONCEPT CHECK *Determine whether each equation represents* direct, inverse, joint, *or* combined *variation.*

9. $y = \dfrac{3}{x}$

10. $y = \dfrac{8}{x}$

11. $y = 10x^2$

12. $y = 2x^3$

13. $y = 3xz^4$

14. $y = 6x^3z^2$

15. $y = \dfrac{4x}{wz}$

16. $y = \dfrac{6x}{st}$

---

*The authors thank Linda Kodama for suggesting these exercises.

CONCEPT CHECK *Write each formula using the "language" of variation. For example, the formula for the circumference of a circle, $C = 2\pi r$, can be written as*

> *"The circumference of a circle varies directly as the length of its radius."*

17. $P = 4s$, where P is the perimeter of a square with side of length s

18. $d = 2r$, where d is the diameter of a circle with radius r

19. $S = 4\pi r^2$, where S is the surface area of a sphere with radius r

20. $V = \frac{4}{3}\pi r^3$, where V is the volume of a sphere with radius r

21. $A = \frac{1}{2}bh$, where A is the area of a triangle with base b and height h

22. $V = \frac{1}{3}\pi r^2 h$, where V is the volume of a cone with radius r and height h

Solve each problem. ***See Examples 1–6.***

23. If x varies directly as y, and $x = 9$ when $y = 3$, find x when $y = 12$.

24. If x varies directly as y, and $x = 10$ when $y = 7$, find y when $x = 50$.

25. If a varies directly as the square of b, and $a = 4$ when $b = 3$, find a when $b = 2$.

26. If h varies directly as the square of m, and $h = 15$ when $m = 5$, find h when $m = 7$.

27. If z varies inversely as w, and $z = 10$ when $w = 0.5$, find z when $w = 8$.

28. If t varies inversely as s, and $t = 3$ when $s = 5$, find s when $t = 5$.

29. If m varies inversely as the square of p, and $m = 20$ when $p = 2$, find m when $p = 5$.

30. If a varies inversely as the square of b, and $a = 48$ when $b = 4$, find a when $b = 7$.

31. p varies jointly as q and the square of r, and $p = 200$ when $q = 2$ and $r = 3$. Find p when $q = 5$ and $r = 2$.

32. f varies jointly as h and the square of g, and $f = 50$ when $h = 2$ and $g = 4$. Find f when $h = 6$ and $g = 3$.

🔲 *Solve each problem involving variation. See Examples 1–7.*

33. Matt bought 8 gal of gasoline and paid $36.79. To the nearest tenth of a cent, what is the price of gasoline per gallon?

34. Nora gives horseback rides at Shadow Mountain Ranch. A 2.5-hr ride costs $50.00. What is the price per hour?

35. The weight of an object on Earth is directly proportional to the weight of that same object on the moon. A 200-lb astronaut would weigh 32 lb on the moon. How much would a 50-lb dog weigh on the moon?

36. The pressure exerted by a certain liquid at a given point is directly proportional to the depth of the point beneath the surface of the liquid. The pressure at 30 m is 80 newtons. What pressure is exerted at 50 m?

37. The volume of a can of tomatoes is directly proportional to the height of the can. If the volume of the can is 300 cm³ when its height is 10.62 cm, find the volume of a can with height 15.92 cm.

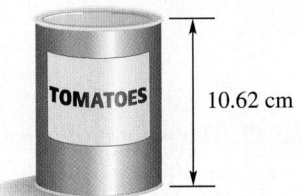

38. The force required to compress a spring is directly proportional to the change in length of the spring. If a force of 20 newtons is required to compress a certain spring 2 cm, how much force is required to compress the spring from 20 cm to 8 cm?

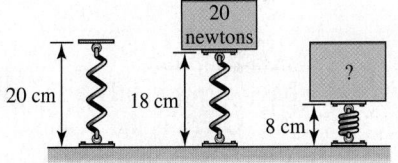

39. For a body falling freely from rest (disregarding air resistance), the distance the body falls varies directly as the square of the time. If an object is dropped from the top of a tower 576 ft high and hits the ground in 6 sec, how far did it fall in the first 4 sec?

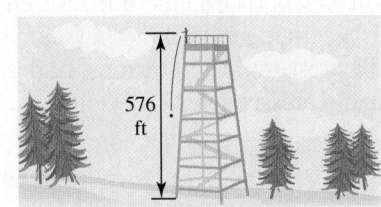

40. The amount of water emptied by a pipe varies directly as the square of the diameter of the pipe. For a certain constant water flow, a pipe emptying into a canal will allow 200 gal of water to escape in an hour. The diameter of the pipe is 6 in. How much water would a 12-in. pipe empty into the canal in an hour, assuming the same water flow?

41. The current in a simple electrical circuit is inversely proportional to the resistance. If the current is 20 amperes (an **ampere** is a unit for measuring current) when the resistance is 5 ohms, find the current when the resistance is 7.5 ohms.

42. The frequency (number of vibrations per second) of a vibrating string varies inversely as its length. That is, a longer string vibrates fewer times in a second than a shorter string. Suppose a piano string 2 ft long vibrates 250 cycles per sec. What frequency would a string 5 ft long have?

43. The amount of light (measured in foot-candles) produced by a light source varies inversely as the square of the distance from the source. If the illumination produced 1 m from a light source is 768 foot-candles, find the illumination produced 6 m from the same source.

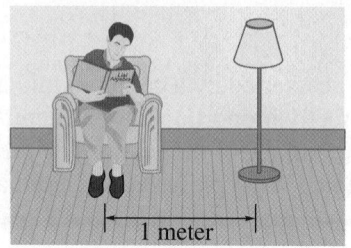

1 meter

44. The force with which Earth attracts an object above Earth's surface varies inversely with the square of the distance of the object from the center of Earth. If an object 4000 mi from the center of Earth is attracted with a force of 160 lb, find the force of attraction if the object were 6000 mi from the center of Earth.

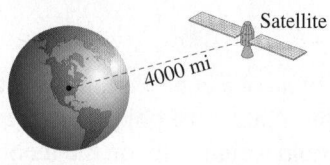

Satellite

4000 mi

45. For a given interest rate, simple interest varies jointly as principal and time. If $2000 left in an account for 4 yr earned interest of $280, how much interest would be earned in 6 yr?

46. The collision impact of an automobile varies jointly as its weight and the square of its speed. Suppose a 2000-lb car traveling at 55 mph has a collision impact of 6.1. What is the collision impact of the same car at 65 mph?

47. The weight of a bass varies jointly as its girth and the square of its length. (**Girth** is the distance around the body of a fish.) A prize-winning bass weighed in at 22.7 lb and measured 36 in. long with a 21-in. girth. How much would a bass 28 in. long with an 18-in. girth weigh (to the nearest tenth)? (*Source: Sacramento Bee.*)

48. See **Exercise 47.** The weight of a trout varies jointly as its length and the square of its girth. One angler caught a trout that weighed 10.5 lb and measured 26 in. long with an 18-in. girth. Find the weight of a trout that is 22 in. long with a 15-in. girth (to the nearest tenth). (*Source: Sacramento Bee.*)

49. The force needed to keep a car from skidding on a curve varies inversely as the radius of the curve and jointly as the weight of the car and the square of the speed. If 242 lb of force keep a 2000-lb car from skidding on a curve of radius 500 ft at 30 mph, what force (to the nearest tenth) would keep the same car from skidding on a curve of radius 750 ft at 50 mph?

50. The volume of gas varies inversely as the pressure and directly as the temperature. (Temperature must be measured in *Kelvin* (K), a unit of measurement used in physics.) If a certain gas occupies a volume of 1.3 L at 300 K and a pressure of 18 newtons, find the volume at 340 K and a pressure of 24 newtons.

51. The number of long-distance phone calls between two cities in a certain time period varies jointly as the populations of the cities, p_1 and p_2, and inversely as the distance between them. If 80,000 calls are made between two cities 400 mi apart, with populations of 70,000 and 100,000, how many calls are made between cities with populations of 50,000 and 75,000 that are 250 mi apart?

52. The maximum load of a horizontal beam that is supported at both ends varies jointly as the width and the square of the height and inversely as the length between the supports. A beam 6 m long, 0.1 m wide, and 0.06 m high supports a load of 360 kg. What is the maximum load supported by a beam 16 m long, 0.2 m wide, and 0.08 m high?

Chapter 7 Summary

Key Terms

7.1

rational expression A rational expression is the quotient of two polynomials with denominator not 0.

rational function A rational function is a function that is defined by a quotient of polynomials in the form $f(x) = \frac{P(x)}{Q(x)}$, where $Q(x) \neq 0$.

7.2

least common denominator (LCD) The least common denominator in a group of denominators is the product of all different factors from each denominator, with each factor raised to the greatest power that occurs in any denominator.

7.3

complex fraction A complex fraction is a quotient having a fraction in the numerator, denominator, or both.

7.4

domain of the variable in a rational equation The domain of the variable in a rational equation is the intersection (overlap) of the domains of the rational expressions in the equation.

discontinuous A graph of a function is discontinuous if there are one or more breaks in the graph.

vertical asymptote A rational function in simplest form $f(x) = \frac{P(x)}{x - a}$ has the line $x = a$ as a vertical asymptote. The graph approaches the line on each side but does not intersect it.

horizontal asymptote A horizontal line that a graph approaches as $|x|$ gets larger and larger without bound is a horizontal asymptote.

7.5

ratio A ratio is a comparison of two quantities using a quotient.

proportion A proportion is a statement that two ratios are equal.

7.6

varies directly y varies directly as x if there exists a real number k such that $y = kx$.

varies inversely y varies inversely as x if there exists a real number k such that $y = \frac{k}{x}$.

constant of variation In the equations for direct and inverse variation, k is the constant of variation.

joint variation y varies jointly as x and z if there exists a real number k such that $y = kxz$.

combined variation Combined variation occurs when both direct and inverse variation are involved in the same equation.

Test Your Word Power

See how well you have learned the vocabulary in this chapter.

1 A **rational expression** is
  A. an algebraic expression made up of a term or the sum of a finite number of terms with real coefficients and integer exponents
  B. a polynomial equation of degree 2
  C. a quotient with one or more fractions in the numerator, denominator, or both
  D. the quotient of two polynomials with denominator not zero.

2 In a given set of fractions, the **least common denominator** is
  A. the smallest denominator of all the denominators

  B. the smallest expression that is divisible by all the denominators
  C. the largest integer that evenly divides the numerator and denominator of all the fractions
  D. the largest denominator of all the denominators.

3 A **complex fraction** is
  A. an algebraic expression made up of a term or the sum of a finite number of terms with real coefficients and integer exponents
  B. a polynomial equation of degree 2
  C. a quotient with one or more fractions in the numerator, denominator, or both

  D. the quotient of two polynomials with denominator not zero.

4 A **ratio**
  A. compares two quantities using a quotient
  B. says that two quotients are equal
  C. is a product of two quantities
  D. is a difference between two quantities.

5 A **proportion**
  A. compares two quantities using a quotient
  B. says that two ratios are equal
  C. is a product of two quantities
  D. is a difference between two quantities.

Answers To Test Your Word Power

1. D; *Examples:* $-\dfrac{3}{4y^2}, \dfrac{5x^3}{x+2}, \dfrac{a+3}{a^2-4a-5}$

2. B; *Example:* The LCD of $\dfrac{1}{x}, \dfrac{2}{3}$, and $\dfrac{5}{x+1}$ is $3x(x+1)$.

3. C; *Examples:* $\dfrac{\frac{2}{3}}{\frac{4}{7}}, \dfrac{x-\frac{1}{x}}{x+\frac{1}{y}}, \dfrac{\frac{2}{a+1}}{a^2-1}$

4. A; *Example:* $\dfrac{7\,\text{in.}}{12\,\text{in.}}$ compares two quantities.

5. B; *Example:* The proportion $\dfrac{2}{3} = \dfrac{8}{12}$ states that the two ratios are equal.

Quick Review

Concepts		Examples

7.1 Rational Expressions and Functions; Multiplying and Dividing

Rational Function

A function of the form

$$f(x) = \frac{P(x)}{Q(x)}, \quad \text{where } Q(x) \neq 0,$$

is a rational function. Its domain consists of all real numbers except those that make $Q(x) = 0$.

Find the domain.

$$f(x) = \frac{2x+1}{3x+6}$$

Solve $3x + 6 = 0$ to find $x = -2$. This is the only real number excluded from the domain, written $\{x \mid x \neq -2\}$.

Fundamental Property of Rational Numbers

If $\frac{a}{b}$ is a rational number and if c is any nonzero real number, then

$$\frac{a}{b} = \frac{ac}{bc}.$$

$$\frac{3}{4} = \frac{3 \cdot 5}{4 \cdot 5} = \frac{15}{20} \qquad \tfrac{3}{4} \text{ and } \tfrac{15}{20} \text{ are equivalent.}$$

Writing a Rational Expression in Lowest Terms

Step 1 Factor the numerator and the denominator completely.

Step 2 Apply the fundamental property.

Write in lowest terms.

$$\frac{2x+8}{x^2-16} = \frac{2(x+4)}{(x-4)(x+4)} = \frac{2}{x-4}$$

Multiplying Rational Expressions

Step 1 Factor numerators and denominators.

Step 2 Apply the fundamental property.

Step 3 Multiply the remaining factors in the numerator and in the denominator.

Step 4 Check that the product is in lowest terms.

Multiply.

$$\frac{x^2+2x+1}{x^2-1} \cdot \frac{5}{3x+3}$$

$$= \frac{(x+1)^2}{(x-1)(x+1)} \cdot \frac{5}{3(x+1)} \qquad \text{Factor.}$$

$$= \frac{5}{3(x-1)} \qquad \text{Multiply; fundamental property}$$

Dividing Rational Expressions

Multiply the first rational expression (the dividend) by the reciprocal of the second (the divisor).

Divide.

$$\frac{2x+5}{x-3} \div \frac{2x^2+3x-5}{x^2-9}$$

$$= \frac{2x+5}{x-3} \cdot \frac{x^2-9}{2x^2+3x-5} \qquad \text{Multiply by the reciprocal.}$$

$$= \frac{2x+5}{x-3} \cdot \frac{(x+3)(x-3)}{(2x+5)(x-1)} \qquad \text{Factor.}$$

$$= \frac{x+3}{x-1} \qquad \text{Multiply; fundamental property}$$

Concepts	Examples

7.2 Adding and Subtracting Rational Expressions

Adding or Subtracting Rational Expressions

Step 1 If the denominators are the same, add or subtract the numerators. Place the result over the common denominator.

If the denominators are different, write all rational expressions with the LCD. Then add or subtract the numerators, and place the result over the common denominator.

Step 2 Make sure that the answer is in lowest terms.

Subtract.

$$\frac{1}{x+6} - \frac{3}{x+2} \qquad \text{The LCD is } (x+6)(x+2).$$

$$= \frac{x+2}{(x+6)(x+2)} - \frac{3(x+6)}{(x+6)(x+2)}$$

$$= \frac{x+2 - 3(x+6)}{(x+6)(x+2)} \qquad \text{Subtract numerators.}$$

$$= \frac{x+2 - 3x - 18}{(x+6)(x+2)} \qquad \text{Distributive property}$$

$$= \frac{-2x - 16}{(x+6)(x+2)} \qquad \text{Combine like terms.}$$

7.3 Complex Fractions

Simplifying a Complex Fraction

Method 1

Step 1 Simplify the numerator and denominator separately, as much as possible.

Step 2 Multiply the numerator by the reciprocal of the denominator.

Step 3 Then simplify the result.

Method 2

Step 1 Multiply the numerator and denominator of the complex fraction by the least common denominator of all fractions appearing in the complex fraction.

Step 2 Then simplify the result.

Method 1

$$\frac{\dfrac{1}{x^2} - \dfrac{1}{y^2}}{\dfrac{1}{x} + \dfrac{1}{y}}$$

$$= \frac{\dfrac{y^2}{x^2y^2} - \dfrac{x^2}{x^2y^2}}{\dfrac{y}{xy} + \dfrac{x}{xy}}$$

$$= \frac{\dfrac{y^2 - x^2}{x^2y^2}}{\dfrac{y+x}{xy}}$$

$$= \frac{y^2 - x^2}{x^2y^2} \div \frac{y+x}{xy}$$

$$= \frac{(y+x)(y-x)}{x^2y^2} \cdot \frac{xy}{y+x}$$

$$= \frac{y-x}{xy}$$

Method 2

$$\frac{\dfrac{1}{x^2} - \dfrac{1}{y^2}}{\dfrac{1}{x} + \dfrac{1}{y}}$$

$$= \frac{x^2y^2\left(\dfrac{1}{x^2} - \dfrac{1}{y^2}\right)}{x^2y^2\left(\dfrac{1}{x} + \dfrac{1}{y}\right)}$$

$$= \frac{y^2 - x^2}{xy^2 + x^2y}$$

$$= \frac{(y-x)(y+x)}{xy(y+x)}$$

$$= \frac{y-x}{xy}$$

7.4 Equations with Rational Expressions and Graphs

Solving an Equation with Rational Expressions

Step 1 Determine the domain of the variable.

Step 2 Multiply each side of the equation by the least common denominator.

Step 3 Solve the resulting equation.

Step 4 Check that each proposed solution is in the domain, and discard any values that are not. Check the remaining proposed solutions in the original equation.

Solve.

$$\frac{1}{x} + x = \frac{26}{5} \qquad \begin{array}{l}\text{Note that 0 is excluded} \\ \text{from the domain.}\end{array}$$

$$5 + 5x^2 = 26x \qquad \text{Multiply by } 5x.$$

$$5x^2 - 26x + 5 = 0 \qquad \text{Subtract } 26x.$$

$$(5x - 1)(x - 5) = 0 \qquad \text{Factor.}$$

$$5x - 1 = 0 \quad \text{or} \quad x - 5 = 0 \qquad \text{Zero-factor property}$$

$$x = \frac{1}{5} \quad \text{or} \qquad x = 5 \qquad \text{Solve each equation.}$$

Both proposed solutions check. The solution set is $\left\{\frac{1}{5}, 5\right\}$.

(continued)

Concepts	Examples

7.4 Equations with Rational Expressions and Graphs *(continued)*

Graphing a Rational Function

The graph of a rational function of the type covered in this section may have one or more breaks. At such points, the graph will approach an asymptote.

Graph $f(x) = \dfrac{1}{x+2}$.

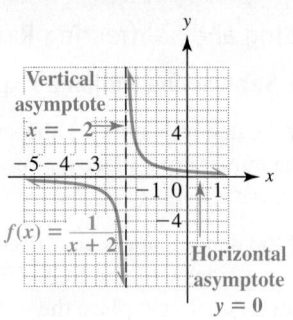

7.5 Applications of Rational Expressions

To solve a motion problem, use the formula

$$d = rt$$

or one of its equivalents,

$$t = \frac{d}{r} \quad \text{or} \quad r = \frac{d}{t}.$$

To solve a work problem, use the fact that if a complete job is done in t units of time, the rate of work is $\frac{1}{t}$ job per unit of time.

Solve.

A canal has a current of 2 mph. Find the rate of Amy's boat in still water if it travels 11 mi downstream in the same time that it travels 8 mi upstream.

Let x = the rate of the boat in still water.

	Distance	Rate	Time	
Downstream	11	$x + 2$	$\dfrac{11}{x+2}$	⎤ The times
Upstream	8	$x - 2$	$\dfrac{8}{x-2}$	⎦ are equal.

$$\frac{11}{x+2} = \frac{8}{x-2} \qquad \text{Use } t = \frac{d}{r}.$$

$$11\,(x-2) = 8\,(x+2) \qquad \begin{array}{l}\text{Multiply by the LCD,}\\ (x+2)\,(x-2).\end{array}$$

$$11x - 22 = 8x + 16 \qquad \text{Distributive property}$$

$$3x = 38 \qquad \text{Subtract } 8x. \text{ Add 22.}$$

$$x = \frac{38}{3}, \quad \text{or} \quad 12\frac{2}{3} \qquad \text{Divide by 3.}$$

The rate in still water is $12\frac{2}{3}$ mph.

7.6 Variation

Let k be a real number.

If $\quad y = kx^n,\quad$ then y varies directly as x^n.

If $\quad y = \dfrac{k}{x^n},\quad$ then y varies inversely as x^n.

If $\quad y = kxz,\quad$ then y varies jointly as x and z.

The area of a circle **varies directly as** the square of the radius.

$$A = kr^2 \qquad \text{Here, } k = \pi.$$

Pressure **varies inversely as** volume.

$$p = \frac{k}{V}$$

For a given principal, interest **varies jointly as** interest rate and time.

$$I = krt \qquad k \text{ is the given principal.}$$

Chapter 7 *Review Exercises*

7.1 *For each rational function, find all numbers that are not in the domain. Then give the domain, using set-builder notation.*

1. $f(x) = \dfrac{-7}{3x + 18}$

2. $f(x) = \dfrac{5x + 17}{x^2 - 7x + 10}$

3. $f(x) = \dfrac{9}{x^2 - 18x + 81}$

Write in lowest terms.

4. $\dfrac{12x^2 + 6x}{24x + 12}$

5. $\dfrac{25m^2 - n^2}{25m^2 - 10mn + n^2}$

6. $\dfrac{r - 2}{4 - r^2}$

Multiply or divide. Write the answer in lowest terms.

7. $\dfrac{(2y + 3)^2}{5y} \cdot \dfrac{15y^3}{4y^2 - 9}$

8. $\dfrac{w^2 - 16}{w} \cdot \dfrac{3}{4 - w}$

9. $\dfrac{z^2 - z - 6}{z - 6} \div \dfrac{z^2 + 2z - 15}{z^2 - 6z}$

10. $\dfrac{m^3 - n^3}{m^2 - n^2} \div \dfrac{m^2 + mn + n^2}{m + n}$

7.2 *Assume that each expression is the denominator of a rational expression. Find the least common denominator for each group.*

11. $32b^3, \quad 24b^5$

12. $9r^2, \quad 3r + 1$

13. $6x^2 + 13x - 5, \quad 9x^2 + 9x - 4$

14. $3x - 12, \quad x^2 - 2x - 8, \quad x^2 - 8x + 16$

Add or subtract as indicated.

15. $\dfrac{8}{z} - \dfrac{3}{2z^2}$

16. $\dfrac{5y + 13}{y + 1} - \dfrac{1 - 7y}{y + 1}$

17. $\dfrac{6}{5a + 10} + \dfrac{7}{6a + 12}$

18. $\dfrac{3r}{10r^2 - 3rs - s^2} + \dfrac{2r}{2r^2 + rs - s^2}$

7.3 *Simplify each expression.*

19. $\dfrac{\dfrac{3}{t} + 2}{\dfrac{4}{t} - 7}$

20. $\dfrac{\dfrac{2}{m - 3n}}{\dfrac{1}{3n - m}}$

21. $\dfrac{\dfrac{3}{p} - \dfrac{2}{q}}{\dfrac{9q^2 - 4p^2}{qp}}$

22. $\dfrac{x^{-2} - y^{-2}}{x^{-1} - y^{-1}}$

7.4 *Solve each equation.*

23. $\dfrac{1}{t + 4} + \dfrac{1}{2} = \dfrac{3}{2t + 8}$

24. $\dfrac{-5m}{m + 1} + \dfrac{m}{3m + 3} = \dfrac{56}{6m + 6}$

25. $\dfrac{2}{k - 1} - \dfrac{4k + 1}{k^2 - 1} = \dfrac{-1}{k + 1}$

26. $\dfrac{5}{x + 2} + \dfrac{3}{x + 3} = \dfrac{x}{x^2 + 5x + 6}$

27. Decide whether each of the following is an *expression* or an *equation*. Simplify the one that is an expression, and solve the one that is an equation.

   (a) $\dfrac{4}{x} + \dfrac{1}{2} = \dfrac{1}{3}$ **(b)** $\dfrac{4}{x} + \dfrac{1}{2} - \dfrac{1}{3}$

28. After solving the equation

$$\dfrac{3}{x - 3} - \dfrac{2}{x - 2} = \dfrac{3}{x^2 - 5x + 6},$$

a student got $x = 3$ as her final step. The answer in the back of the book was "$\varnothing$." She was sure that all her algebraic work was correct. Was she wrong or was the answer in the back of the book wrong? Explain.

29. Which is the graph of a rational function? Give the equations of its vertical and horizontal asymptotes.

  A.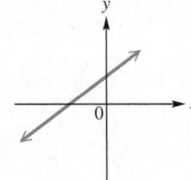

  B.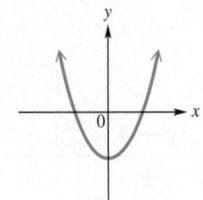

  C.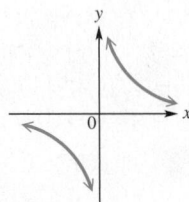

  D.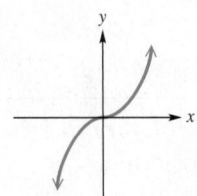

30. Graph the rational function

$$f(x) = \dfrac{2}{x + 1}.$$

Give the equations of its vertical and horizontal asymptotes.

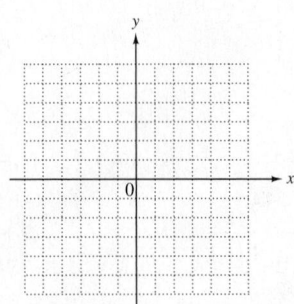

7.5 *Solve each problem.*

31. According to a law from physics,

$$\frac{1}{A} = \frac{1}{B} + \frac{1}{C}.$$

Find A if $B = 30$ and $C = 10$.

32. In banking and finance, the formula

$$P = \frac{M}{1 + RT}$$

gives present value at simple interest. If $P = 600$, $M = 750$, and $R = 0.125$, find T.

Solve each formula for the specified variable.

33. $\dfrac{1}{a} = \dfrac{1}{b} + \dfrac{1}{c}$ for c (electricity)

34. $\mu = \dfrac{Mv}{M + m}$ for M (electronics)

Complete any tables. Then solve each problem.

35. To estimate the deer population of a forest preserve, wildlife biologists caught, tagged, and then released 42 deer. A month later, they returned and caught a sample of 75 deer and found that 15 of them were tagged. Based on this experiment, approximately how many deer lived in the forest preserve?

36. Clayton's SUV uses 28 gal of gasoline to drive 500 mi. He has 10 gal of gasoline in the SUV, and wants to know how much more gasoline he will need to drive 400 mi. If we assume the car continues to use gasoline at the same rate, how many more gallons will he need?

37. A river has a current of 4 km per hr. Find the rate of Lynn McTernan's boat in still water if it travels 40 km downstream in the same time that it takes to travel 24 km upstream.

	d	r	t
Upstream	24	$x - 4$	____
Downstream	40	____	____

38. A sink can be filled by a cold-water tap in 8 min, and filled by the hot-water tap in 12 min. How long would it take to fill the sink with both taps open?

	Rate	Time Working Together	Fractional Part of the Job Done
Cold	____	x	_____
Hot	____	x	_____

7.6 *Solve each variation problem.*

39. In which one of the following does y vary inversely as x?

A. $y = 2x$ **B.** $y = \dfrac{x}{3}$ **C.** $y = \dfrac{3}{x}$ **D.** $y = x^2$

40. If m varies inversely as r, and $m = 12$ when $r = 8$, find m when $r = 16$.

41. For a particular camera, the viewing distance varies directly as the amount of enlargement. A picture taken with this camera that is enlarged 5 times should be viewed from a distance of 250 mm. Suppose a print 8.6 times the size of the negative is made. From what distance should it be viewed?

42. The volume of a rectangular box of a given height is proportional to its width and length. A box with width 4 ft and length 8 ft has volume 64 ft^3. Find the volume of a box with the same height that is 3 ft wide and 6 ft long.

Mixed Review Exercises

Write in lowest terms.

43. $\dfrac{x + 2y}{x^2 - 4y^2}$

44. $\dfrac{x^2 + 2x - 15}{x^2 - x - 6}$

Perform the indicated operations.

45. $\dfrac{2}{m} + \dfrac{5}{3m^2}$

46. $\dfrac{9}{3 - x} - \dfrac{2}{x - 3}$

47. $\dfrac{\dfrac{-3}{x} + \dfrac{x}{2}}{1 + \dfrac{x + 1}{x}}$

48. $\dfrac{t^{-2} + s^{-2}}{t^{-1} - s^{-1}}$

49. $\dfrac{\dfrac{3}{x} - 5}{6 + \dfrac{1}{x}}$

50. $\dfrac{a}{b} + \dfrac{b}{c} + \dfrac{c}{d}$

51. $\dfrac{4y + 16}{30} \div \dfrac{2y + 8}{5}$

52. $\dfrac{k^2 - 6k + 9}{1 - 216k^3} \cdot \dfrac{6k^2 + 17k - 3}{9 - k^2}$

53. $\dfrac{4a}{a^2 - ab - 2b^2} - \dfrac{6b - a}{a^2 + 4ab + 3b^2}$

54. $\dfrac{9x^2 + 46x + 5}{3x^2 - 2x - 1} \div \dfrac{x^2 + 11x + 30}{x^3 + 5x^2 - 6x}$

Solve each equation.

55. $\dfrac{x + 3}{x^2 - 5x + 4} - \dfrac{1}{x} = \dfrac{2}{x^2 - 4x}$

56. $A = \dfrac{Rr}{R + r}$ for r

57. $1 - \dfrac{5}{r} = \dfrac{-4}{r^2}$

58. $\dfrac{3x}{x - 4} + \dfrac{2}{x} = \dfrac{48}{x^2 - 4x}$

Solve each problem.

59. Anna and Matthew Sudak need to sort a pile of bottles at the recycling center. Working alone, Anna could do the entire job in 9 hr, while Matthew could do the entire job in 6 hr. How long will it take them if they work together?

60. A college student rides her bike from home to campus some days, while other days she walks. When she rides her bike, she gets to her first class 36 min faster than when she walks. If her average walking rate is 3 mph and her average biking rate is 12 mph, how far is it from her home to her first class?

61. The frequency (number of vibrations per second) of a vibrating guitar string varies inversely as its length. That is, a longer string vibrates fewer times in a second than a shorter string. Suppose a guitar string 0.65 m long vibrates 4.3 times per sec. What frequency would a string 0.5 m long have?

62. The area of a triangle varies jointly as the lengths of the base and height. A triangle with base 10 ft and height 4 ft has area 20 ft^2. Find the area of a triangle with base 3 ft and height 8 ft.

Chapter 7 *Test*

The Chapter Test Prep Videos with test solutions are available on DVD, in **MyMathLab**, and on **You**Tube—*search "Lial CombinedAlg" and click on "Channels."*

1. Find all real numbers excluded from the domain of

$$f(x) = \frac{x + 3}{3x^2 + 2x - 8}.$$

Then give the domain using set-builder notation.

2. Write in lowest terms.

$$\frac{6x^2 - 13x - 5}{9x^3 - x}$$

Multiply or divide.

3. $\dfrac{(x + 3)^2}{4} \cdot \dfrac{6}{2x + 6}$

4. $\dfrac{y^2 - 16}{y^2 - 25} \cdot \dfrac{y^2 + 2y - 15}{y^2 - 7y + 12}$

5. $\dfrac{3 - t}{5} \div \dfrac{t - 3}{10}$

6. $\dfrac{x^2 - 9}{x^3 + 3x^2} \div \dfrac{x^2 + x - 12}{x^3 + 9x^2 + 20x}$

7. Find the least common denominator for the following group of denominators: $t^2 + t - 6, \ t^2 + 3t, \ t^2$.

Add or subtract as indicated.

8. $\dfrac{7}{6t^2} - \dfrac{1}{3t}$

9. $\dfrac{9}{x - 7} + \dfrac{4}{x + 7}$

10. $\dfrac{6}{x + 4} + \dfrac{1}{x + 2} - \dfrac{3x}{x^2 + 6x + 8}$

11. $\dfrac{9}{x^2 - 6x + 9} + \dfrac{2}{x^2 - 9}$

Simplify each expression.

12. $\dfrac{\dfrac{12}{r + 4}}{\dfrac{11}{6r + 24}}$

13. $\dfrac{\dfrac{1}{a} - \dfrac{1}{b}}{\dfrac{a}{b} - \dfrac{b}{a}}$

14. $\dfrac{2x^{-2} + y^{-2}}{x^{-1} - y^{-1}}$

15. Decide whether each of the following is an *expression* or an *equation*. Simplify the one that is an expression, and solve the one that is an equation.

(a) $\dfrac{2x}{3} + \dfrac{x}{4} - \dfrac{11}{2}$

(b) $\dfrac{2x}{3} + \dfrac{x}{4} = \dfrac{11}{2}$

Solve each equation.

16. $\dfrac{1}{x} - \dfrac{4}{3x} = \dfrac{1}{x-2}$

17. $\dfrac{y}{y+2} - \dfrac{1}{y-2} = \dfrac{8}{y^2-4}$

18. Solve for the variable ℓ in this formula from mathematics: $\ \ S = \dfrac{n}{2}(a + \ell)$.

19. Sketch the graph of the function

$$f(x) = \dfrac{-2}{x+1}.$$

Give the equations of its vertical and horizontal asymptotes.

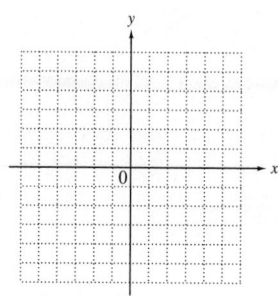

Solve each problem.

20. Wayne can do a job in 9 hr, while Sandra can do the same job in 5 hr. How long would it take them to do the job if they worked together?

21. The rate of the current in a stream is 3 mph. Danielle Lalezhar's boat can travel 36 mi downstream in the same time that it takes to travel 24 mi upstream. Find the rate of her boat in still water.

22. Biologists collected a sample of 600 fish from West Lake Okoboji on May 1 and tagged each of them. When they returned on June 1, a new sample of 800 fish was collected, and 10 of these had been previously tagged. Use this experiment to determine the approximate fish population of West Lake Okoboji.

23. In biology, the function

$$g(x) = \dfrac{5x}{2+x}$$

gives the growth rate g of a population for x units of available food. (*Source:* Smith, J. Maynard, *Models in Ecology,* Cambridge University Press.)

(a) What amount of food (in appropriate units) would produce a growth rate of 3 units of growth per unit of food?

(b) What is the growth rate if no food is available?

24. The current in a simple electrical circuit is inversely proportional to the resistance. If the current is 80 amps when the resistance is 30 ohms, find the current when the resistance is 12 ohms.

25. The force of the wind blowing on a vertical surface varies jointly as the area of the surface and the square of the velocity. If a wind blowing at 40 mph exerts a force of 50 lb on a surface of 500 ft^2, how much force will a wind of 80 mph place on a surface of 2 ft^2?

Chapters R–7 *Cumulative Review Exercises*

Solve each equation or inequality.

1. $7(2x + 3) - 4(2x + 1)$
    $= 2(x + 1)$

2. $0.04x + 0.06(x - 1) = 1.04$

3. $\dfrac{2}{3}x + \dfrac{5}{12}x \le 20$

Solve each problem.

4. Otis Taylor invested some money at 4% interest and twice as much at 3% interest. His interest for the first year was $400. How much did he invest at each rate?

5. A triangle has an area of 42 m². The base is 14 m long. Find the height of the triangle.

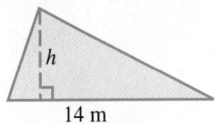

6. Find the slope of each line.

 (a) Through $(-5, 8)$ and $(-1, 2)$

 (b) Perpendicular to $4x + 3y = 12$, through $(5, 2)$

7. Write an equation of each line in **Exercise 6** in the form $y = mx + b$.

Graph.

8. $-4x + 2y = 8$

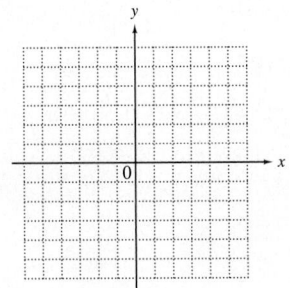

9. $2x + 5y > 10$

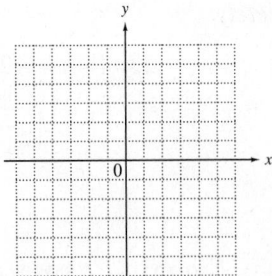

10. Consider the equation $5x - 3y = 8$.

 (a) Write y as a function f of x, using function notation $f(x)$.

 (b) Find $f(1)$.

11. Consider the relation $y = -x + 2$.

 (a) Does it define a function?

 (b) Give its domain and range.

Solve each system.

12. $4x - y = -7$
$5x + 2y = 1$

13. $3x + 4y = 5$
$6x + 7y = 8$

14. $2x - y = 6$
$3y = 6x - 18$

Simplify. Write each answer with only positive exponents. Assume that all variables represent nonzero real numbers.

15. $\left(\dfrac{a^{-3}b^4}{a^2b^{-1}}\right)^{-2}$

16. $\left(\dfrac{m^{-4}n^2}{m^2n^{-3}}\right) \cdot \left(\dfrac{m^5n^{-1}}{m^{-2}n^5}\right)$

Perform the indicated operations.

17. $(3y^2 - 2y + 6) - (-y^2 + 5y + 12)$

18. $(3x^3 + 13x^2 - 17x - 7) \div (3x + 1)$

19. $(4x + 3)(3x - 1)$

20. $(7t^3 + 8)(7t^3 - 8)$

21. $(4x + 5)^2$

Factor each polynomial completely.

22. $2x^2 - 13x - 45$

23. $100t^2 - 25$

24. $8p^3 + 125$

Perform the indicated operations. Express the answer in lowest terms.

25. $\dfrac{2a^2}{a + b} \cdot \dfrac{a - b}{4a}$

26. $\dfrac{x + 4}{x - 2} + \dfrac{2x - 10}{x - 2}$

27. $\dfrac{2x}{2x - 1} + \dfrac{4}{2x + 1} + \dfrac{8}{4x^2 - 1}$

Solve.

28. $3x^2 + 4x = 7$

29. $\dfrac{-3x}{x + 1} + \dfrac{4x + 1}{x} = \dfrac{-3}{x^2 + x}$

30. $\dfrac{1}{f} = \dfrac{1}{p} + \dfrac{1}{q}$ for q

8 Equations, Inequalities, and Systems Revisited

A centered bubble in a carpenter's level indicates a level surface (suggesting *equality*), while a surface that is not level causes the bubble to be off center (suggesting *inequality*).

8.1 Review of Solving Linear Equations and Inequalities

8.2 Set Operations and Compound Inequalities

8.3 Absolute Value Equations and Inequalities

Summary Exercises *Solving Linear and Absolute Value Equations and Inequalities*

8.4 Review of Systems of Linear Equations in Two Variables

8.5 Systems of Linear Equations in Three Variables; Applications

Math in the Media *To Pass or to Fail? That Is the Question*

8.1 Review of Solving Linear Equations and Inequalities

OBJECTIVES

1. Distinguish between expressions and equations.
2. Solve linear equations.
3. Solve linear inequalities.
4. Solve three-part inequalities.

---

1 Decide whether each of the following is an *equation* or an *expression*.

(a) $9x = 10$

(b) $9x + 10$

(c) $3 + 5x - 8x + 9$

(d) $3 + 5x = -8x + 9$

OBJECTIVE ▶ 1 **Distinguish between expressions and equations.** Recall that an *algebraic expression* is a collection of constants, variables, operation symbols, and grouping symbols.

$$8x + 9, \quad y - 4, \quad \text{and} \quad \frac{x^3 y^3}{z} \qquad \text{Algebraic expressions}$$

An *equation* is a statement that two algebraic expressions are equal. *An equation always contains an equality symbol, while an expression does not.*

EXAMPLE 1   Distinguishing between Expressions and Equations

Decide whether each of the following is an *expression* or an *equation*.

(a) $3x - 7 = 2$ 　　　　　　　　(b) $3x - 7$

In part (a) we have an equation, because there is an equality symbol. In part (b), there is no equality symbol, so it is an expression.

Equation

$$\underbrace{3x - 7}_{\text{Left side}} \;=\; \underbrace{2}_{\text{Right side}}$$

An equation can be *solved*.

Expression

$$3x - 7$$

An expression **cannot** be solved. It can often be *evaluated* or *simplified*.

◀ **Work Problem 1 at the Side.**

OBJECTIVE ▶ 2 **Solve linear equations.** A *linear equation in one variable* involves only real numbers and one variable raised to the first power.

> **Linear Equation in One Variable**
>
> A **linear equation in one variable** can be written in the form
> $$Ax + B = C,$$
> where A, B, and C are real numbers, with $A \neq 0$.

A linear equation is a **first-degree equation** since the greatest power on the variable is one.

$$x + 1 = -2, \quad x - \frac{3}{4} = 5, \quad \text{and} \quad 2k + 5 = 10 \qquad \text{Linear equations}$$

$$x^2 + 3y = 5, \quad \frac{8}{x} = -22, \quad \text{and} \quad \sqrt{x} = 6 \qquad \textit{Non}\text{linear equations}$$

If the variable in an equation can be replaced by a real number that makes the statement true, then that number is a **solution** of the equation. For example, **8** is a solution of $x - 3 = 5$, since replacing x with 8 gives a true statement, $8 - 3 = 5$. An equation is *solved* by finding its **solution set,** the set of all solutions. The solution set of $x - 3 = 5$ is $\{8\}$.

Equivalent equations are related equations that have the same solution set. To solve an equation, we usually start with the given equation and replace it with a series of simpler equivalent equations. The following are all equivalent since each has solution set $\{3\}$.

$$5x + 2 = 17, \quad 5x = 15, \quad \text{and} \quad x = 3 \qquad \text{Equivalent equations}$$

Answers

1. (a) equation (b) expression
   (c) expression (d) equation

We use two important properties of equality to produce equivalent equations.

Addition and Multiplication Properties of Equality

Addition Property of Equality
For all real numbers A, B, and C, the equations

$$A = B \quad \text{and} \quad A + C = B + C \quad \text{are equivalent.}$$

In words, *the same number may be added to each side of an equation without changing the solution set.*

Multiplication Property of Equality
For all real numbers A and B, and for $C \neq 0$, the equations

$$A = B \quad \text{and} \quad AC = BC \quad \text{are equivalent.}$$

In words, *each side of an equation may be multiplied by the same nonzero number without changing the solution set.*

Because subtraction and division are defined in terms of addition and multiplication, respectively, these properties can be extended:

The same number may be subtracted from each side of an equation, and each side of an equation may be divided by the same nonzero number, without changing the solution set.

EXAMPLE 2 Solving a Linear Equation

Solve $4x - 2x - 5 = 4 + 6x + 3$.

The goal is to isolate x on one side of the equation.

$$4x - 2x - 5 = 4 + 6x + 3$$
$$2x - 5 = 7 + 6x \qquad \text{Combine like terms.}$$
$$2x - 5 - 6x = 7 + 6x - 6x \qquad \text{Subtract } 6x \text{ from each side.}$$
$$-4x - 5 = 7 \qquad \text{Combine like terms.}$$
$$-4x - 5 + 5 = 7 + 5 \qquad \text{Add 5 to each side.}$$
$$-4x = 12 \qquad \text{Combine like terms.}$$
$$\frac{-4x}{-4} = \frac{12}{-4} \qquad \text{Divide each side by } -4.$$
$$x = -3$$

CHECK Substitute -3 for x in the *original* equation.

$$4x - 2x - 5 = 4 + 6x + 3 \qquad \text{Original equation}$$
$$4(-3) - 2(-3) - 5 \stackrel{?}{=} 4 + 6(-3) + 3 \qquad \text{Let } x = -3.$$
$$-12 + 6 - 5 \stackrel{?}{=} 4 - 18 + 3 \qquad \text{Multiply.}$$
$$-11 = -11 ✓ \qquad \text{True}$$

Use parentheses around substituted values to avoid errors.

This is *not* the solution.

The true statement indicates that $\{-3\}$ is the solution set.

Work Problem ② at the Side. ▶

② Solve and check.

(a) $3p + 2p + 1 = -24$
$$\underline{\quad} + 1 = -24$$
$$5p + 1 - \underline{\quad} = -24 - \underline{\quad}$$
$$5p = -25$$
$$\frac{5p}{\underline{\quad}} = \frac{-25}{\underline{\quad}}$$
$$p = \underline{\quad}$$

CHECK Substitute $\underline{\quad}$ in the original equation. Does the solution check? (*Yes / No*)
The solution set is $\{\underline{\quad}\}$.

(b) $3p = 2p + 4p + 5$

(c) $4x + 8x = 17x - 9 - 1$

(d) $-7 + 3t - 9t = 12t - 5$

Answers

2. **(a)** $5p$; 1; 1; 5; 5; -5; -5; Yes; -5
(b) $\left\{-\frac{5}{3}\right\}$ **(c)** $\{2\}$ **(d)** $\left\{-\frac{1}{9}\right\}$

3 Solve and check.

(a) $5p + 4(3 - 2p)$
$= 2 + p - 10$

(b) $3(z - 2) + 5z = 2$

(c) $-2 + 3(x + 4) = 8x$

Solving a Linear Equation in One Variable

Step 1 **Simplify each side separately.** Clear (eliminate) parentheses, fractions, and decimals using the distributive property and combine like terms, as needed.

Step 2 **Isolate the variable terms on one side.** Use the addition property to get all terms with variables on one side of the equation and all constants (numbers) on the other.

Step 3 **Isolate the variable.** Use the multiplication property to get an equation with just the variable (with coefficient 1) on one side.

Step 4 **Check.** Substitute the proposed solution into the *original* equation to see if a true statement results. If not, rework the problem.

EXAMPLE 3 Solving a Linear Equation

Solve $2(k - 5) + 3k = k + 6$.

Step 1 Clear parentheses on the left. Then combine like terms.

Be sure to distribute over *all* terms within parentheses.

$$2(k - 5) + 3k = k + 6$$
$$2k + 2(-5) + 3k = k + 6 \quad \text{Distributive property}$$
$$2k - 10 + 3k = k + 6 \quad \text{Multiply.}$$
$$5k - 10 = k + 6 \quad \text{Combine like terms.}$$

Step 2 Next, use the addition property of equality.

$$5k - 10 - k = k + 6 - k \quad \text{Subtract } k.$$
$$4k - 10 = 6 \quad \text{Combine like terms.}$$
$$4k - 10 + 10 = 6 + 10 \quad \text{Add 10.}$$
$$4k = 16 \quad \text{Combine like terms.}$$

Step 3 Use the multiplication property of equality to isolate k on the left.

$$\frac{4k}{4} = \frac{16}{4} \quad \text{Divide by 4.}$$
$$k = 4$$

Step 4 Check by substituting 4 for k in the original equation.

CHECK
$$2(k - 5) + 3k = k + 6 \quad \text{Original equation}$$
$$2(4 - 5) + 3(4) \stackrel{?}{=} 4 + 6 \quad \text{Let } k = 4.$$
$$2(-1) + 12 \stackrel{?}{=} 10 \quad \text{Perform operations.}$$
$$10 = 10 \checkmark \quad \text{True}$$

Always check your work.

The proposed solution checks, so the solution set is $\{4\}$.

◀ **Work Problem 3 at the Side.**

Answers

3. (a) $\{5\}$ (b) $\{1\}$ (c) $\{2\}$

EXAMPLE 4 Solving a Linear Equation

Solve $4(3x - 2) = 38 - 2(2x - 1)$.

$$4(3x - 2) = 38 - 2(2x - 1)$$

Step 1 $\quad 4(3x) + 4(-2) = 38 - 2(2x) - 2(-1)$ — Distributive property

> Be careful with signs when distributing.

$12x - 8 = 38 - 4x + 2$ — Multiply.

$12x - 8 = 40 - 4x$ — Combine like terms.

Step 2 $\quad 12x - 8 + 4x = 40 - 4x + 4x$ — Add $4x$.

$16x - 8 = 40$ — Combine like terms.

$16x - 8 + 8 = 40 + 8$ — Add 8.

$16x = 48$ — Combine like terms.

Step 3 $\quad \dfrac{16x}{16} = \dfrac{48}{16}$ — Divide by 16.

$x = 3$

Step 4 CHECK $\quad 4(3x - 2) = 38 - 2(2x - 1)$ — Original equation

$4[3(3) - 2] \stackrel{?}{=} 38 - 2[2(3) - 1]$ — Let $x = 3$.

$4[7] \stackrel{?}{=} 38 - 2[5]$ — Work inside the brackets.

$28 = 28 \checkmark$ — True

The proposed solution checks, so the solution set is $\{3\}$.

············· **Work Problem ④ at the Side.** ▶

EXAMPLE 5 Solving a Linear Equation with Fractions

Solve $\dfrac{x + 7}{6} + \dfrac{2x - 8}{2} = -4$.

Step 1 $\quad 6\left(\dfrac{x + 7}{6} + \dfrac{2x - 8}{2}\right) = 6(-4)$ — Eliminate the fractions. Multiply each side by the LCD, 6.

$6\left(\dfrac{x + 7}{6}\right) + 6\left(\dfrac{2x - 8}{2}\right) = 6(-4)$ — Distributive property

> This equivalent equation has integer coefficients.

$(x + 7) + 3(2x - 8) = -24$ — Multiply.

$x + 7 + 3(2x) + 3(-8) = -24$ — Distributive property

$x + 7 + 6x - 24 = -24$ — Multiply.

$7x - 17 = -24$ — Combine like terms.

Step 2 $\quad 7x - 17 + 17 = -24 + 17$ — Add 17.

$7x = -7$ — Combine like terms.

Step 3 $\quad \dfrac{7x}{7} = \dfrac{-7}{7}$ — Divide by 7.

$x = -1$

Step 4 Check that the solution set is $\{-1\}$.

············· **Work Problem ⑤ at the Side.** ▶

④ Solve and check.

(a) $2(2x + 1) - 3(2x - 1) = 9$

(b) $2 - 3(2 + 6x)$
$= 4(x + 1) + 18$

⑤ Solve and check.

GS (a) $\dfrac{2p}{7} - \dfrac{p}{2} = -3$

Step 1
What is the LCD of all the fractions in the equation? ____
Multiply by this LCD.

$$\underline{\quad}\left(\dfrac{2p}{7} - \dfrac{p}{2}\right) = 14(-3)$$

Step 2
Apply the ____ property.

$$\underline{\quad}\left(\dfrac{2p}{7}\right) + 14(\underline{\quad}) = 14(-3)$$

$$\underline{\quad}(2p) - \underline{\quad} = -42$$

$$4p - 7p = -42$$

$$\underline{\quad} = -42$$

Now complete the solution. Give the solution set.

(b) $\dfrac{k + 1}{2} + \dfrac{k + 3}{4} = \dfrac{1}{2}$

Answers

4. (a) $\{-2\}$ **(b)** $\left\{-\dfrac{13}{11}\right\}$

5. (a) 14; 14; distributive; 14; $-\dfrac{p}{2}$; 2; 7p; -3p; $\{14\}$

(b) $\{-1\}$

6 Solve and check.

(a) $0.04x + 0.06(20 - x)$
$= 0.05(50)$

(b) $0.10(x - 6) + 0.05x$
$= 0.06(50)$

EXAMPLE 6 **Solving a Linear Equation with Decimals**

Solve $0.06x + 0.09(15 - x) = 0.07(15)$.

$$0.06x + 0.09(15 - x) = 0.07(15)$$

      Multiply by 100 to clear the decimals.

$$\mathbf{0.06}x + \mathbf{0.09}(15 - x) = \mathbf{0.07}(15)$$

Move decimal points 2 places to the right.

$$6x + 9(15 - x) = 7(15)$$

This is an equivalent equation without decimals.

$$6x + 9(15) - 9(x) = 7(15)$$ Distributive property

$$6x + 135 - 9x = 105$$ Multiply.

$$-3x + 135 = 105$$ Combine like terms.

$$-3x + 135 - \mathbf{135} = 105 - \mathbf{135}$$ Subtract 135.

$$-3x = -30$$ Combine like terms.

$$\frac{-3x}{-3} = \frac{-30}{-3}$$ Divide by -3.

$$x = 10$$

Check that the solution set is $\{10\}$.

◀ **Work Problem 6** at the Side.

Note

Some students prefer to solve an equation with decimal coefficients without clearing the decimals.

$$0.06x + 0.09(15 - x) = 0.07(15)$$ Equation from **Example 6**

$$0.06x + 1.35 - 0.09x = 1.05$$ Distributive property

Be careful with decimal points.

$$-0.03x + 1.35 = 1.05$$ Combine like terms.

$$-0.03x + 1.35 - \mathbf{1.35} = 1.05 - \mathbf{1.35}$$ Subtract 1.35.

$$-0.03x = -0.3$$ Combine like terms.

$$\frac{-0.03x}{-0.03} = \frac{-0.3}{-0.03}$$ Divide by -0.03.

$$x = 10$$

The same solution results.

    All of the preceding equations had solution sets containing *one* element, such as $\{10\}$ in **Example 6.** Some equations, however have no solution, while others have an infinite number of solutions. The table below summarizes these types of equations.

Type of Linear Equation	Number of Solutions	Indication When Solving
Conditional	One solution	Final line is $x = $ a number. (See **Example 7(a).**)
Identity	Infinite number of solutions; solution set {all real numbers}	Final line is true, such as $0 = 0$. (See **Example 7(b).**)
Contradiction	No solution; solution set $\emptyset$	Final line is false, such as $-15 = -20$. (See **Example 7(c).**)

Answers

6. (a) $\{-65\}$ (b) $\{24\}$

EXAMPLE 7   Recognizing Conditional Equations, Identities, and Contradictions

Solve each equation. Decide whether it is a *conditional equation*, an *identity*, or a *contradiction*.

(a) $5(2x + 6) - 2 = 7(x + 4)$

$10x + 30 - 2 = 7x + 28$ Distributive property

$10x + 28 = 7x + 28$ Combine like terms.

$10x + 28 - 7x - 28 = 7x + 28 - 7x - 28$ Subtract $7x$. Subtract 28.

$3x = 0$ Combine like terms.

> The last line has a variable. The number following "=" is a solution.

$\dfrac{3x}{3} = \dfrac{0}{3}$ Divide by 3.

$x = 0$ $\dfrac{0}{a} = 0$ (See **Section 1.6.**)

CHECK $5(2x + 6) - 2 = 7(x + 4)$ Original equation

$5[2(0) + 6] - 2 \stackrel{?}{=} 7(0 + 4)$ Let $x = 0$.

$5(6) - 2 \stackrel{?}{=} 7(4)$ Multiply and then add.

$28 = 28$ ✓   True

The solution 0 checks, so the solution set is $\{0\}$. Since the solution set has only one element, $5(2x + 6) - 2 = 7(x + 4)$ is a conditional equation.

(b) $5x - 15 = 5(x - 3)$

$5x - 15 = 5x - 15$ Distributive property

$5x - 15 - 5x + 15 = 5x - 15 - 5x + 15$ Subtract $5x$. Add 15.

> The variable has "disappeared."

$0 = 0$ True

Here, the final line, the *true* statement $0 = 0$, indicates that the solution set is $\{$all real numbers$\}$. The equation $5x - 15 = 5(x - 3)$ is an identity. Notice that the first step yielded

$5x - 15 = 5x - 15$,   which is true for *all* values of x.

We could have identified the equation as an identity at that point.

(c) $5x - 15 = 5(x - 4)$

$5x - 15 = 5x - 20$ Distributive property

$5x - 15 - 5x = 5x - 20 - 5x$ Subtract $5x$.

> The variable has "disappeared."

$-15 = -20$ False

Since the result, $-15 = -20$, is *false*, the equation has no solution. The solution set is $\emptyset$, so the equation $5x - 15 = 5(x - 4)$ is a contradiction.

························· **Work Problem ➐ at the Side.** ▶

CAUTION

A common error in solving an equation like that in **Example 7(a)** is to think that the equation has no solution and write the solution set as $\emptyset$. This equation has one solution, the number 0, so it is a conditional equation with solution set $\{0\}$.

➐ Solve each equation. Decide whether it is a *conditional equation*, an *identity*, or a *contradiction*.

(a) $5(x + 2) - 2(x + 1)$
    $= 3x + 1$

(b) $9x - 3(x + 4) = 6(x - 2)$

(c) $5(3x + 1) = x + 5$

(d) $3(2x - 4) = 20 - 2x$

Answers

7. (a) $\emptyset$; contradiction
 (b) $\{$all real numbers$\}$; identity
 (c) $\{0\}$; conditional
 (d) $\{4\}$; conditional

8 Solve each inequality, check your solutions, and graph the solution set.

(a) $x - 3 < -9$

(b) $p + 6 < 8$

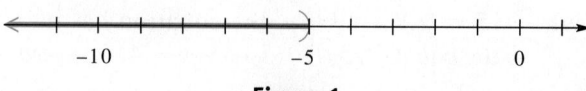

OBJECTIVE **3** **Solve linear inequalities.** Recall from **Section 2.7** that an **inequality** consists of algebraic expressions related by one of the following symbols.

$<$	"is less than"	$\leq$	"is less than or equal to"
$>$	"is greater than"	$\geq$	"is greater than or equal to"

Linear Inequality in One Variable

A **linear inequality in one variable** can be written in the form

$$Ax + B < C, \quad Ax + B \leq C, \quad Ax + B > C, \quad \text{or} \quad Ax + B \geq C,$$

where A, B, and C are real numbers, with $A \neq 0$.

$$x + 5 < 2, \quad x - 3 \geq 5, \quad \text{and} \quad 2k + 5 \leq 10 \qquad \text{Linear inequalities}$$

We solve an inequality by finding all numbers that make the inequality true. Usually, an inequality has an infinite number of solutions. These solutions, like solutions of equations, are found by producing a series of simpler equivalent inequalities. **Equivalent inequalities** are inequalities with the same solution set.

We use two important properties to produce equivalent inequalities.

Addition Property of Inequality

For all real numbers A, B, and C, the inequalities

$$A < B \quad \text{and} \quad A + C < B + C$$

are equivalent.

In words, adding the same number to each side of an inequality does not change the solution set.

EXAMPLE 8 **Using the Addition Property of Inequality**

Solve $x - 7 < -12$, and graph the solution set.

$$x - 7 < -12$$
$$x - 7 + 7 < -12 + 7 \qquad \text{Add 7.}$$
$$x < -5 \qquad \text{Combine like terms.}$$

A graph of the solution set, written $(-\infty, -5)$ using interval notation (introduced in **Section 2.7**), is shown in **Figure 1.**

Figure 1

As with equations, the addition property of inequality can be used to *subtract* the same number from each side of an inequality.

◀ **Work Problem** **8** at the Side.

Answers

8. (a) $(-\infty, -6)$

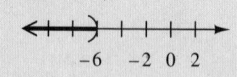

(b) $(-\infty, 2)$

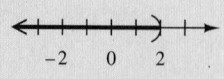

The multiplication property of inequality has two parts.

Multiplication Property of Inequality

For all real numbers A, B, and C, with $C \neq 0$, the following hold.

(a) The inequalities

$$A < B \quad \text{and} \quad AC < BC \quad \text{are equivalent} \quad \text{if } C > 0.$$

(b) The inequalities

$$A < B \quad \text{and} \quad AC > BC \quad \text{are equivalent} \quad \text{if } C < 0.$$

In words, each side of an inequality may be multiplied (or divided) by a *positive* number without changing the direction of the inequality symbol. *Multiplying (or dividing) by a negative number requires that we reverse the direction of the inequality symbol.*

EXAMPLE 9 **Using the Multiplication Property of Inequality**

Solve each inequality, and graph the solution set.

(a) $5m \leq -30$

Divide each side by 5. *Since $5 > 0$, do not reverse the direction of the inequality symbol.*

$$5m \leq -30$$

$$\frac{5m}{5} \leq \frac{-30}{5} \quad \text{Divide by 5.}$$

$$m \leq -6$$

Check that the solution set is the interval $(-\infty, -6\,]$, graphed in **Figure 2**.

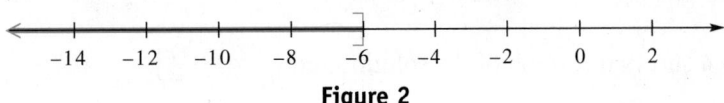

Figure 2

(b) $-4k \leq 32$

Divide each side by -4. *Since $-4 < 0$, reverse the direction of the inequality symbol.*

$$-4k \leq 32$$

$$\frac{-4k}{-4} \geq \frac{32}{-4} \quad \begin{array}{l}\text{Divide by } -4. \text{ Reverse the} \\ \text{direction of the symbol.}\end{array}$$

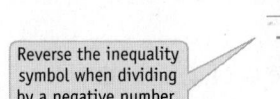
Reverse the inequality symbol when dividing by a negative number.

$$k \geq -8$$

Check the solution set. **Figure 3** shows the graph of the solution set, $[-8, \infty)$.

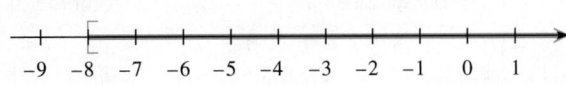

Figure 3

··· Work Problem **9** at the Side. ▶

CAUTION

Reverse the direction of the inequality symbol only when multiplying or dividing each side of an inequality by a negative number.

---

9 Solve, check, and graph the solution set of each inequality.

(a) $2x < -10$

_____⟶

GS (b) $-7k \geq 8$

Divide each side by ____.
Since $-7 \,(< \,/\, >)\, 0$, (*reverse / do not reverse*) the direction of the inequality symbol.

$$k \,(\leq \,/\, \geq)\, \underline{\quad}$$

The solution set is ____.

_____⟶

(c) $-9m < -81$

_____⟶

Answers

9. (a) $(-\infty, -5)$

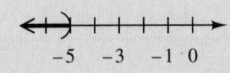

(b) -7; $<$; reverse; $\leq$; $-\dfrac{8}{7}$; $\left(-\infty, -\dfrac{8}{7}\right]$

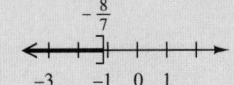

(c) $(9, \infty)$

10 Solve, check, and graph the solution set of each inequality.

GS **(a)** $x + 4(2x - 1) \geq x + 2$

Step 1
Use the distributive property to clear parentheses.

———————

Combine like terms.

———————

Follow Steps 2 and 3 to complete the solution.

———————————→

$x\ (\leq / \geq)\ \_\_\_\_$

The solution set is ____.

———————————→

(b) $m - 2(m - 4) \leq 3m$

———————————→

Solving a Linear Inequality in One Variable

Step 1 **Simplify each side separately.** Clear parentheses, fractions, and decimals using the distributive property as needed, and combine like terms.

Step 2 **Isolate the variable terms on one side.** Use the addition property of inequality to get all terms with variables on one side of the inequality and all constants (numbers) on the other side.

Step 3 **Isolate the variable.** Use the multiplication property of inequality to change the inequality to the form $x < k$ or $x > k$.

EXAMPLE 10 **Solving a Linear Inequality**

Solve $-3(x + 4) + 2 \geq 7 - x$, and graph the solution set.

Step 1	$-3(x + 4) + 2 \geq 7 - x$	
	$-3x - 12 + 2 \geq 7 - x$	Distributive property
	$-3x - 10 \geq 7 - x$	Combine like terms.
Step 2	$-3x - 10 + x \geq 7 - x + x$	Add x.
	$-2x - 10 \geq 7$	Combine like terms.
	$-2x - 10 + 10 \geq 7 + 10$	Add 10.
	$-2x \geq 17$	Combine like terms.
Step 3	$\dfrac{-2x}{-2} \leq \dfrac{17}{-2}$	Divide by -2. Change $\geq$ to $\leq$.
	$x \leq -\dfrac{17}{2}$	

Be sure to reverse the direction of the inequality symbol.

Figure 4 shows the graph of the solution set, $\left(-\infty, -\frac{17}{2}\right]$.

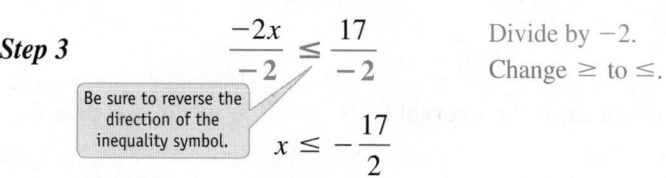

Figure 4

◀ Work Problem **10** at the Side.

Note

In Step 2 of **Example 10**, we could add $3x$ (instead of x) to both sides.

$-3x - 10 + 3x \geq 7 - x + 3x$	Add $3x$.
$-10 \geq 2x + 7$	Combine like terms.
$-10 - 7 \geq 2x + 7 - 7$	Subtract 7.
$-17 \geq 2x$	Combine like terms.
$-\dfrac{17}{2} \geq x, \quad$ or $\quad x \leq -\dfrac{17}{2}$	Divide by 2. Rewrite. The same solution results.

The symbol points to x in each case.

Answers

10. **(a)** $x + 8x - 4 \geq x + 2;\ 9x - 4 \geq x + 2;$
$\geq \dfrac{3}{4};\ \left[\dfrac{3}{4}, \infty\right)$

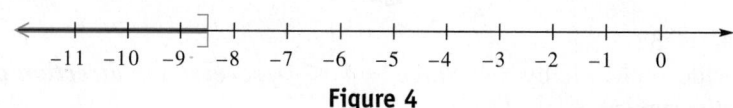

(b) $[2, \infty)$

OBJECTIVE 4 **Solve three-part inequalities.** An inequality that says that one number is *between* two other numbers is a *three-part inequality*. (See **Section 2.7**.)

$$-1 < t < 8 \quad \text{and} \quad 4 \le 3x - 5 < 6 \quad \text{Three-part inequalities}$$

To solve a three-part inequality, we work with all three parts at the same time.

EXAMPLE 11   **Solving a Three-Part Inequality**

Solve $-2 \le -3k - 1 \le 5$, and graph the solution set.

$$-2 \le \quad -3k - 1 \quad \le 5$$

$$-2 + 1 \le -3k - 1 + 1 \le 5 + 1 \quad \text{Add 1 to each part.}$$

$$-1 \le \quad -3k \quad \le 6$$

$$\frac{-1}{-3} \ge \quad \frac{-3k}{-3} \quad \ge \frac{6}{-3} \quad \begin{array}{l}\text{Divide each part by } -3.\\ \text{Reverse the direction of the}\\ \text{inequality symbols.}\end{array}$$

$$\frac{1}{3} \ge \quad k \quad \ge -2$$

$$-2 \le \quad k \quad \le \frac{1}{3} \quad \boxed{\begin{array}{l}\text{Rewrite in the}\\ \text{order on the}\\ \text{number line.}\end{array}}$$

Check that the solution set is the closed interval $\left[-2, \frac{1}{3}\right]$. See **Figure 5**.

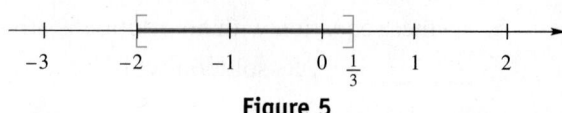

Figure 5

·········· **Work Problem 11 at the Side. ▶**

Types of solution sets for linear equations and linear inequalities are reviewed here.

SOLUTIONS OF LINEAR EQUATIONS AND INEQUALITIES*

Equation or Inequality	Typical Solution Set	Graph of Solution Set
Linear equation $5x + 4 = 14$	$\{2\}$	![graph] 2
Linear inequality $5x + 4 < 14$	$(-\infty, 2)$	![graph] 2
$5x + 4 > 14$	$(2, \infty)$	![graph] 2
Three-part inequality $-1 \le 5x + 4 \le 14$	$[-1, 2]$	![graph] -1 ... 2

---

*In the rest of this chapter, we use set notation for equations that have all real numbers as solutions—that is, {all real numbers}—and interval notation for similar inequalities—that is, $(-\infty, \infty)$.

11 Solve, check, and graph the solution set of each inequality.

(a) $5 < 3x - 4 < 9$

(b) $-2 < -4x - 5 \le 7$

Answers

11. (a) $\left(3, \dfrac{13}{3}\right)$

$$\text{--+++++}(-)\text{+--}$$
$$0\ 1\ 2\ 3\ 4\ 5$$

(b) $\left[-3, -\dfrac{3}{4}\right)$

$$\text{--++}[-)\text{++--}$$
$$-4\ -3\ -2\ -1\ \ 0\ \ 1$$

8.1 Exercises

FOR EXTRA HELP

  Download the MyDashBoard App

MyMathLab®

CONCEPT CHECK *Complete each statement. The following key terms may be used once, more than once, or not at all.*

linear equation	solution	algebraic expression	contradiction	all real numbers
solution set	identity	conditional equation	first-degree equation	empty set $\emptyset$

1. A collection of numbers, variables, operation symbols, and grouping symbols, such as $2(8x - 15)$, is a(n) _____. While an equation (*does / does not*) include an equality symbol, there (*is / is not*) an equality symbol in an algebraic expression.

2. A(n) _____ in one variable can be written in the form $Ax + B (= / > / <) C$, with $A \neq 0$. Another name for a linear equation is a(n) _____, since the greatest power on the variable is (*one / two / three*).

3. If we let $x = 2$ in the linear equation $2x + 5 = 9$, a (*true / false*) statement results. The number 2 is a(n) _____ of the equation, and $\{2\}$ is the _____.

4. A linear equation with one solution in its _____, such as the equation in **Exercise 3**, is a(n) _____.

5. A linear equation with an infinite number of solutions is a(n) _____. Its solution set is $\{$_____$\}$.

6. A linear equation with no solution is a(n) _____. Its solution set is the _____.

7. **CONCEPT CHECK** Which equations are linear equations in x?

   A. $3x + x - 2 = 0$ **B.** $12 = x^2$

   C. $9x - 4 = 9$ **D.** $\dfrac{1}{8}x - \dfrac{1}{x} = 0$

8. Which of the equations in **Exercise 7** are nonlinear equations in x? Explain why.

Decide whether each of the following is an expression *or an* equation. *See Example 1.*

9. $-3x + 2 - 4 = x$

10. $-3x + 2 - 4 - x = 4$

11. $4(x + 3) - 2(x + 1) - 10$

12. $4(x + 3) - 2(x + 1) + 10$

13. $-10x + 12 - 4x = -3$

14. $-10x + 12 - 4x + 3 = 0$

Solve each equation, and check your solution. See Examples 2–4.

15. $9x + 10 = 1$

16. $7x - 4 = 31$

17. $5x + 2 = 3x - 6$

18. $9p + 1 = 7p - 9$

19. $7x - 5x + 15 = x + 8$

20. $2x + 4 - x = 4x - 5$

21. $12w + 15w - 9 + 5 = -3w + 5 - 9$

22. $-4t + 5t - 8 + 4 = 6t - 4$

23. $-5(x + 1) + 3x + 2 = 6x + 4$

24. $5(x + 3) + 4x - 5 = 4 - 2x$

25. $3(2w + 1) - 2(w - 2) = 5$

26. $4(x + 2) - 2(x + 3) = 5$

27. $6p - 4(3 - 2p) = 5(p - 4) - 10$

28. $-2k - 3(4 - 2k) = 2(k - 3) + 2$

29. $2[w - (2w + 4) + 3] = 2(w + 1)$

30. $4[2t - (3 - t) + 5] = -(2 + 7t)$

*Solve each equation, and check your solution. **See Examples 5 and 6.***

31. $\dfrac{3}{4}x + \dfrac{5}{2}x = 13$

32. $\dfrac{8}{3}x - \dfrac{1}{2}x = -13$

33. $\dfrac{x - 8}{5} + \dfrac{8}{5} = -\dfrac{x}{3}$

34. $\dfrac{2r - 3}{7} + \dfrac{3}{7} = -\dfrac{r}{3}$

35. $\dfrac{4x - 2}{5} - \dfrac{3x - 1}{2} = -2$

36. $\dfrac{3x + 2}{7} - \dfrac{x + 4}{5} = 2$

37. $0.05x + 0.12(x + 5000) = 940$

38. $0.09k + 0.13(k + 300) = 61$

39. $0.02(50) + 0.08r = 0.04(50 + r)$

40. $0.20(14{,}000) + 0.14t = 0.18(14{,}000 + t)$

41. $0.006(x + 2) = 0.007x + 0.009$

42. $0.004x + 0.006(50 - x) = 0.004(68)$

*Solve each equation. Decide whether it is a conditional equation, an identity, or a contradiction. **See Example 7.***

43. $-x + 4x - 9 = 3(x - 4) - 5$

44. $-12x + 2x - 11 = -2(5x - 3) + 4$

45. $-11x + 4(x - 3) + 6x = 4x - 12$

46. $3x - 5(x + 4) + 9 = -11 + 15x$

47. $-2(t + 3) - t - 4 = -3(t + 4) + 2$

48. $4(2d + 7) = 2d + 25 + 3(2d + 1)$

CONCEPT CHECK *Match each inequality in Column I with the correct graph or interval in Column II.*

I		**II**

49. $x \leq 3$ **50.** $x > 3$

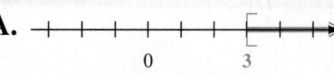

A.

51. $x < 3$ **52.** $x \geq 3$

B.

C. $(3, \infty)$ **D.** $(-\infty, 3]$

53. $-3 \leq x \leq 3$ **54.** $-3 < x < 3$

E. $(-3, 3)$ **F.** $[-3, 3]$

Solve each inequality, giving solution sets in both interval and graph forms. Check your answers. **See Examples 8–10.**

55. $x - 4 \leq 3$

56. $t - 3 \leq 1$

57. $4x + 1 \geq 21$

58. $5t + 2 \geq 52$

59. $5x > -25$

60. $7x < -28$

61. $-4x < 16$

62. $-2m > 10$

63. $-\dfrac{3}{4}r \geq 30$

64. $-\dfrac{2}{3}x \leq 12$

65. $\dfrac{3k - 1}{4} > 5$

66. $\dfrac{5z - 6}{8} < 8$

67. $\dfrac{2k - 5}{-4} > 5$

68. $\dfrac{3z - 2}{-5} < 6$

69. $3k + 1 < -20$

70. $5z + 6 > -29$

71. $x + 4(2x - 1) \geq x$

72. $m - 2(m - 4) \leq 3m$

73. $-(4 + r) + 2 - 3r < -14$

74. $-(9 + k) - 5 + 4k \geq 4$

Solve each inequality, giving solution sets in both interval and graph forms. Check your answers. **See Example 11.**

75. $-4 < x - 5 < 6$

76. $-1 < x + 1 < 8$

77. $-6 \leq 2(z + 2) \leq 16$

78. $-15 < 3(p + 2) < 24$

79. $-16 < 3t + 2 < -10$

80. $-19 < 3x - 5 \leq 1$

81. $4 < -9x + 5 \leq 8$

82. $4 < -2x + 3 \leq 8$

8.2 Set Operations and Compound Inequalities

OBJECTIVES

1 Find the intersection of two sets.

2 Solve compound inequalities with the word *and*.

3 Find the union of two sets.

4 Solve compound inequalities with the word *or*.

Consider the two sets A and B, defined as follows.

$$A = \{1, 2, 3\}, \qquad B = \{2, 3, 4\}$$

The set of all elements that belong to both A **and** B, called their *intersection* and symbolized $A \cap B$, is given by

$$A \cap B = \{2, 3\}. \qquad \text{Intersection}$$

The set of all elements that belong to either A **or** B, or both, called their *union* and symbolized $A \cup B$, is given by

$$A \cup B = \{1, 2, 3, 4\}. \qquad \text{Union}$$

We discuss the use of the words *and* and *or* as they relate to sets and inequalities.

① List the elements in each set.

(a) $A \cap B$, if $A = \{3, 4, 5, 6\}$
and $B = \{5, 6, 7\}$

OBJECTIVE ▶ ① Find the intersection of two sets.

> **Intersection of Sets**
>
> For any two sets A and B, the **intersection** of A and B, symbolized $A \cap B$, is defined as follows.
>
> $$A \cap B = \{x \mid x \text{ is an element of } A \text{ and } x \text{ is an element of } B\}$$
>
>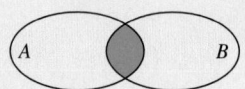

(b) $R \cap S$, if $R = \{1, 3, 5\}$
and $S = \{2, 4, 6\}$

> **EXAMPLE 1   Finding the Intersection of Two Sets**
>
> Let $A = \{1, 2, 3, 4\}$ and $B = \{2, 4, 6\}$. Find $A \cap B$.
> The set $A \cap B$ contains those elements that belong to both A *and* B.
>
> $$A \cap B = \{1, 2, 3, 4\} \cap \{2, 4, 6\}$$
> $$= \{2, 4\}$$

◀ **Work Problem ①** at the Side.

A **compound inequality** consists of two inequalities linked by a connective word.

$$x + 1 \le 9 \quad \text{and} \quad x - 2 \ge 3$$
$$2x > 4 \quad \text{or} \quad 3x - 6 < 5$$

Compound inequalities

OBJECTIVE ▶ ② Solve compound inequalities with the word *and*.

> **Solving a Compound Inequality with *and***
>
> *Step 1*   Solve each inequality individually.
>
> *Step 2*   Since the inequalities are joined with *and*, the solution set of the compound inequality includes all numbers that satisfy both inequalities in Step 1 (the *intersection* of the solution sets).

Answers

1. (a) $\{5, 6\}$ (b) $\varnothing$

EXAMPLE 2 Solving a Compound Inequality with *and*

Solve the compound inequality, and graph the solution set.

$$x + 1 \leq 9 \quad \text{and} \quad x - 2 \geq 3$$

Step 1 Solve each inequality individually.

$$x + 1 \leq 9 \qquad \text{and} \qquad x - 2 \geq 3$$
$$x + 1 - 1 \leq 9 - 1 \quad \text{and} \quad x - 2 + 2 \geq 3 + 2$$
$$x \leq 8 \qquad \text{and} \qquad x \geq 5$$

Step 2 The solution set includes all numbers that satisfy *both* inequalities in Step 1 at the same time. The compound inequality is true whenever $x \leq 8$ and $x \geq 5$ are both true. See the graphs in **Figure 6.**

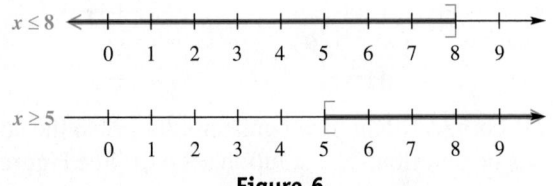

The set of points where the graphs "overlap" represents the intersection.

Figure 6

The intersection of the two graphs in **Figure 6** is the solution set. **Figure 7** shows this solution set, $[5, 8]$.

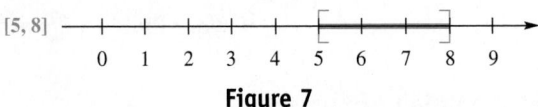

Figure 7

···················· Work Problem **2** at the Side. ▶

EXAMPLE 3 Solving a Compound Inequality with *and*

Solve the compound inequality, and graph the solution set.

$$-3x - 2 > 5 \quad \text{and} \quad 5x - 1 \leq -21$$

Step 1 Solve each inequality individually.

$$-3x - 2 > 5 \qquad \text{and} \qquad 5x - 1 \leq -21$$
$$-3x > 7 \qquad \text{and} \qquad 5x \leq -20$$
$$x < -\frac{7}{3} \qquad \text{and} \qquad x \leq -4$$

> Remember to reverse the direction of the inequality symbol.

The graphs of $x < -\frac{7}{3}$ and $x \leq -4$ are shown in **Figure 8.**

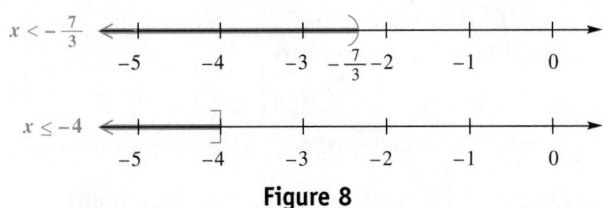

Figure 8

Step 2 Now find all values of x that are less than $-\frac{7}{3}$ and also less than or equal to -4. As shown in **Figure 9,** the solution set is $(-\infty, -4]$.

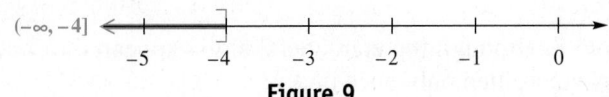

Figure 9

···················· Work Problem **3** at the Side. ▶

2 Solve each compound inequality, and graph the solution set.

(a) $x < 10$ and $x > 2$

_____

(b) $x + 3 \leq 1$ and
$x - 4 \geq -12$

_____

3 Solve and graph the solution set.

$2x \geq x - 1$ and $3x \geq 3 + 2x$

Step 1
Solve each inequality individually.

$2x \geq x - 1$ and $3x \geq 3 + 2x$

$x \geq$ _____ and $x \geq$ _____

Step 2
The solution set of the compound inequality includes all numbers that satisfy (*one* / *none* / *both*) of the inequalities in Step 1.

This solution set in interval form is _____.

_____

Answers

2. (a) $(2, 10)$

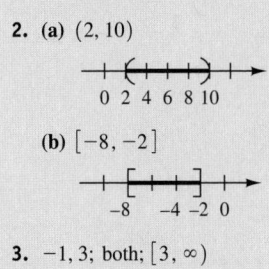

(b) $[-8, -2]$

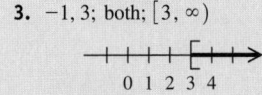

3. $-1, 3;$ both; $[3, \infty)$

4 Solve.

(a) $x < 5$ and $x > 5$

(GS) (b) $x + 2 > 3$ and
$2x + 1 < -3$

Step 1
Solve each inequality
individually.

$x + 2 > 3$ and $2x + 1 < -3$

$x > \underline{\quad}$ and $x < \underline{\quad}$

Step 2
There is no number that is
both greater than _____ and
less than _____ at the same
time.

The solution set is _____ .

5 List the elements in each set.

(a) $A \cup B$, if $A = \{3, 4, 5, 6\}$
and $B = \{5, 6, 7\}$

(b) $R \cup S$, if $R = \{1, 3, 5\}$
and $S = \{2, 4, 6\}$

EXAMPLE 4 **Solving a Compound Inequality with *and***

Solve the compound inequality, and graph the solution set.

$$x + 2 < 5 \quad \text{and} \quad x - 10 > 2$$

Step 1 Solve each inequality individually.

$$x + 2 < 5 \quad \text{and} \quad x - 10 > 2$$
$$x < 3 \quad \text{and} \quad x > 12$$

The graphs of $x < 3$ and $x > 12$ are shown in **Figure 10**.

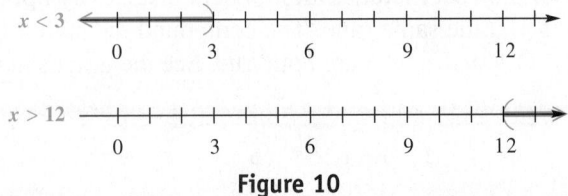

Figure 10

Step 2 No number is both less than 3 *and* greater than 12, so the compound inequality has no solution. The solution set is $\emptyset$. See **Figure 11.**

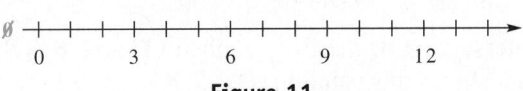

Figure 11

◀ **Work Problem** **4** at the Side.

OBJECTIVE **3** **Find the union of two sets.**

Union of Sets

For any two sets A and B, the **union** of A and B, symbolized $A \cup B$, is defined as follows.

$$A \cup B = \{x \mid x \text{ is an element of } A \text{ or } x \text{ is an element of } B\}$$

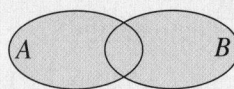

EXAMPLE 5 **Finding the Union of Two Sets**

Let $A = \{1, 2, 3, 4\}$ and $B = \{2, 4, 6\}$. Find $A \cup B$.

Begin by listing all the elements of set A: 1, 2, 3, 4. Then list any additional elements from set B. In this case the elements 2 and 4 are already listed, so the only additional element is 6.

$$A \cup B = \{1, 2, 3, 4\} \cup \{2, 4, 6\}$$
$$= \{1, 2, 3, 4, 6\}$$

The union consists of all elements in either A *or* B (or both).

◀ **Work Problem** **5** at the Side.

Note

In **Example 5,** although the elements 2 and 4 appeared in both sets A and B, they are written only once in $A \cup B$.

Answers

4. (a) $\emptyset$ (b) 1; −2; 1; −2; $\emptyset$

5. (a) $\{3, 4, 5, 6, 7\}$ (b) $\{1, 2, 3, 4, 5, 6\}$

OBJECTIVE ▶ ④ **Solve compound inequalities with the word *or*.**

> **Solving a Compound Inequality with *or***
>
> **Step 1** Solve each inequality individually.
>
> **Step 2** Since the inequalities are joined with *or,* the solution set of the compound inequality includes all numbers that satisfy either one of the two inequalities in Step 1 (the *union* of the solution sets).

EXAMPLE 6 Solving a Compound Inequality with *or*

Solve the compound inequality, and graph the solution set.
$$6x - 4 < 2x \quad \text{or} \quad -3x \le -9$$

Step 1 Solve each inequality individually.

$$6x - 4 < 2x \quad \text{or} \quad -3x \le -9$$
$$4x < 4$$

> Remember to reverse the inequality symbol.

$$x < 1 \quad \text{or} \quad x \ge 3$$

The graphs of these two inequalities are shown in **Figure 12**.

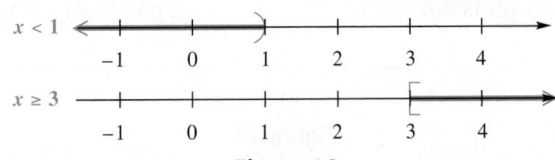

Figure 12

Step 2 Since the inequalities are joined with *or,* find the union of the two solution sets. The union is shown in **Figure 13**.

$$(-\infty, 1) \cup [3, \infty)$$

> Keep the numbers 1 and 3 in their order on the number line.

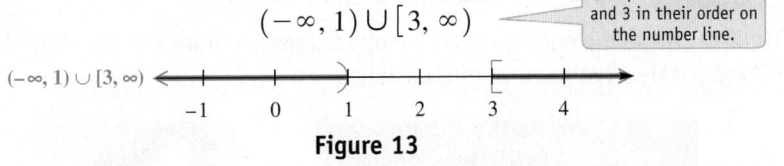

Figure 13

·········· **Work Problem ⑥ at the Side.** ▶

EXAMPLE 7 Solving a Compound Inequality with *or*

Solve the compound inequality, and graph the solution set.

$$-4x + 1 \ge 9 \quad \text{or} \quad 5x + 3 \le -12$$
$$-4x \ge 8 \quad \text{or} \quad 5x \le -15$$
$$x \le -2 \quad \text{or} \quad x \le -3$$

(Step 1) Solve each inequality individually.

The graphs of these two inequalities are shown in **Figure 14**.

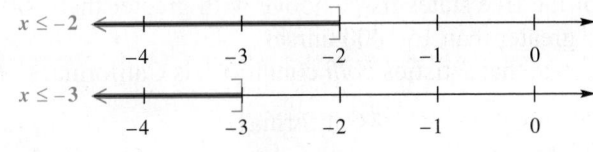

Figure 14

We take the union to obtain $(-\infty, -2]$. See **Figure 15**.   *(Step 2)*

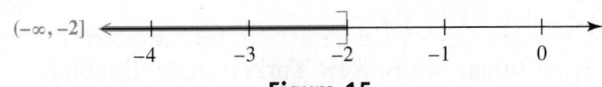

Figure 15

·········· **Work Problem ⑦ at the Side.** ▶

⑥ Solve. Give each solution set in both interval and graph forms.

(a) $x + 2 > 3$ or
$2x + 1 < -3$

_____

(b) $x - 1 > 2$ or
$3x + 5 < 2x + 6$

_____

⑦ Solve. Give each solution set in both interval and graph forms.

(a) $2x + 1 \le 9$ or $2x + 3 \le 5$

_____

(b) $3x - 4 > 2$ or
$-2x + 5 < 3$

_____

Answers

6. (a) $(-\infty, -2) \cup (1, \infty)$

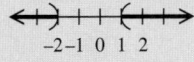

(b) $(-\infty, 1) \cup (3, \infty)$

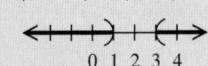

7. (a) $(-\infty, 4]$

(b) $(1, \infty)$

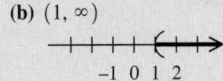

8 Solve.

$$3x - 2 \leq 13 \text{ or } x + 5 \geq 7$$

Give the solution set in both interval and graph forms.

Step 1
Solve each inequality individually.

$$3x - 2 \leq 13 \text{ or } x + 5 \geq 7$$

$$3x \leq 15 \text{ or } \quad x \geq \underline{\quad}$$

$$x \leq \underline{\quad}$$

Step 2
By taking the (*intersection / union*), we obtain every _____ as a solution.

The solution set is ____.

---

9 Refer to **Example 9.** List the elements that satisfy each set.

(a) The set of the five states listed with greater than 50,000 physicians and less than 150,000 nurses.

(b) The set of the five states listed with greater than 50,000 physicians or less than 150,000 nurses.

Answers

8. 5; 2; union; real number; $(-\infty, \infty)$

9. (a) $\emptyset$
   (b) {California, Texas, New York, Illinois}

EXAMPLE 8   **Solving a Compound Inequality with *or***

Solve the compound inequality, and graph the solution set.

$$-2x + 5 \geq 11 \quad \text{or} \quad 4x - 7 \geq -27$$

Step 1   Solve each inequality individually.

$$-2x + 5 \geq 11 \quad \text{or} \quad 4x - 7 \geq -27$$

$$-2x \geq 6 \quad \text{or} \quad 4x \geq -20$$

$$x \leq -3 \quad \text{or} \quad x \geq -5$$

The graphs of these two inequalities are shown in **Figure 16.**

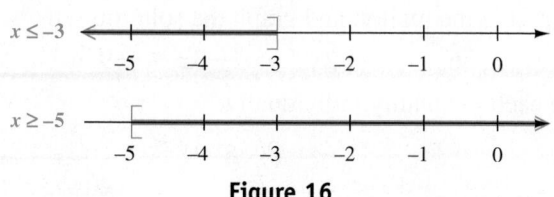

Figure 16

Step 2   By taking the union, we obtain every real number as a solution, since every real number satisfies at least one of the two inequalities. The solution set is written $(-\infty, \infty)$ and graphed as in **Figure 17.**

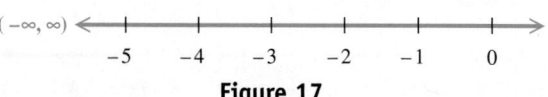

Figure 17

.... ◀ **Work Problem 8** at the Side.

EXAMPLE 9   **Applying Intersection and Union**

The table shows the number of active physicians and nurses in the United States in 2009 for the five most populous states.

State	Number of Physicians	Number of Nurses
California	100,131	233,030
Texas	53,546	168,020
New York	77,042	165,730
Florida	46,645	150,940
Illinois	36,528	116,340

Source: U.S. Bureau of Labor Statistics.

List the elements that satisfy each set.

(a) The set of the five states listed above with greater than 100,000 physicians *and* greater than 150,000 nurses
   The only state that satisfies *both* conditions is California, so the set is

$$\{California\}.$$

(b) The set of the five states listed above with less than 100,000 physicians *or* greater than 100,000 nurses
   Here, a state that satisfies *at least one* of the conditions is in the set. This includes all five states, so the set is

$$\{California, Texas, New York, Florida, Illinois\}.$$

.... ◀ **Work Problem 9** at the Side.

8.2 Exercises

FOR EXTRA HELP

 Download the MyDashBoard App

▶ MyMathLab®

CONCEPT CHECK *Decide whether each statement is* true *or* false. *If it is false, explain why.*

1. The union of the solution sets of $2x + 1 = 3$, $2x + 1 > 3$, and $2x + 1 < 3$ is $(-\infty, \infty)$.

2. The intersection of the sets $\{x \mid x \geq 5\}$ and $\{x \mid x \leq 5\}$ is $\emptyset$.

3. The union of the sets $(-\infty, 6)$ and $(6, \infty)$ is $\{6\}$.

4. The intersection of the sets $[6, \infty)$ and $(-\infty, 6]$ is $\{6\}$.

Let $A = \{1, 2, 3, 4, 5, 6\}$, $B = \{1, 3, 5\}$, $C = \{1, 6\}$, and $D = \{4\}$. Specify each set.
See Examples 1 and 5.

5. $A \cap D$

6. $B \cap C$

7. $B \cap \emptyset$

8. $A \cap \emptyset$

9. $A \cup B$

10. $B \cup D$

11. $B \cup C$

12. $C \cup B$

CONCEPT CHECK *Two sets are specified by graphs. Graph the intersection of the two sets.*

13.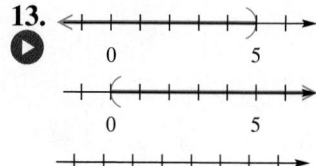

14.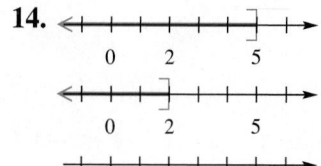

15.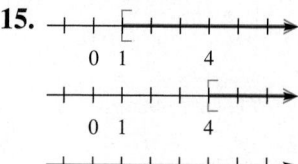

CONCEPT CHECK *Two sets are specified by graphs. Graph the union of the two sets.*

16.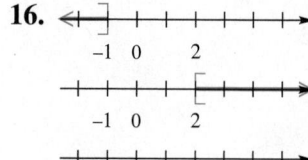

17.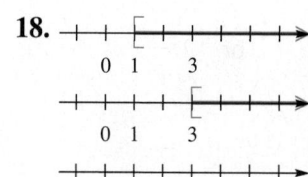

18.

19. Give an example of intersection applied to a real-life situation.

20. A compound inequality uses one of the words *and* or *or*. Explain how you will determine whether to use *intersection* or *union* when graphing the solution set.

For each compound inequality, give the solution set in both interval and graph forms.
See Examples 2–4.

21. $x < 2$ and $x > -3$

22. $x < 5$ and $x > 0$

23. $x \leq 2$ and $x \leq 5$

24. $x \geq 3$ and $x \geq 6$

25. $x \leq 3$ and $x \geq 6$

26. $x \leq -1$ and $x \geq 3$

27. $x - 3 \leq 6$ and $x + 2 \geq 7$

28. $x + 5 \leq 11$ and $x - 3 \geq -1$

29. $3x - 4 \leq 8$ and $4x - 1 \leq 15$

30. $7x + 6 \leq 48$ and $-4x \geq -24$

For each compound inequality, give the solution set in both interval and graph forms.
See Examples 6–8.

31. $x \leq 1$ or $x \leq 8$

32. $x \geq 1$ or $x \geq 8$

33. $x \geq -2$ or $x \geq 5$

34. $x \leq -2$ or $x \leq 6$

35. $x + 3 \geq 1$ or $x - 8 \leq -4$

36. $x + 6 \geq 11$ or $x - 4 \leq 3$

37. $x + 2 > 7$ or $1 - x > 6$

38. $x + 1 > 3$ or $x + 4 < 2$

39. $x + 1 > 3$ or $-4x + 1 \geq 5$

40. $3x < x + 12$ or $x + 1 > 10$

41. $4x - 8 > 0$ or $4x - 1 < 7$

42. $3x < x + 12$ or $3x - 8 > 10$

Express each set in the simplest interval form.

43. $(-\infty, -1] \cap [-4, \infty)$

44. $[-1, \infty) \cap (-\infty, 9]$

45. $(-\infty, -6] \cap [-9, \infty)$

46. $(5, 11] \cap [6, \infty)$

47. $(-\infty, 3) \cup (-\infty, -2)$

48. $[-9, 1] \cup (-\infty, -3)$

49. $[3, 6] \cup (4, 9)$

50. $[-1, 2] \cup (0, 5)$

For each compound inequality, state whether intersection or union should be used. Then give the solution set in both interval and graph forms. **See Examples 2–4 and 6–8.**

51. $x < -1$ and $x > -5$

52. $x > -1$ and $x < 7$

53. $x < 4$ or $x < -2$

54. $x < 5$ or $x < -3$

55. $x + 1 \geq 5$ and $x - 2 \leq 10$

56. $2x - 6 \leq -18$ and $2x \geq -18$

57. $-3x \leq -6$ or $-3x \geq 0$

58. $-8x \leq -24$ or $-5x \geq 15$

Average expenses for full-time college students at 4-year institutions in the United States during the 2009–2010 academic year are shown in the table.

COLLEGE EXPENSES (IN DOLLARS), 4-YEAR INSTITUTIONS

Type of Expense	Public Schools	Private Schools
Tuition and fees	6695	23,210
Board rates	3754	4331
Dormitory charges	4565	5249

Source: National Center for Education Statistics.

Use the table to list the elements of each set. **See Example 9.**

59. The set of expenses that are less than $7500 for public schools *and* are greater than $10,000 for private schools

60. The set of expenses that are greater than $3500 for public schools *and* are less than $4500 for private schools

61. The set of expenses that are less than $5000 for public schools *or* are greater than $10,000 for private schools

62. The set of expenses that are greater than $23,000 *or* are less than $3700

Relating Concepts (Exercises 63–68) For Individual or Group Work

The figures represent the backyards of neighbors Luigi, Mario, Than, and Joe. Find the area and the perimeter of each yard. Suppose that each resident has 150 ft of fencing and enough sod to cover 1400 ft² of lawn.

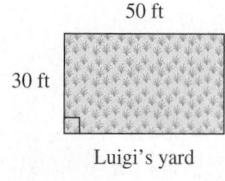

50 ft
30 ft
Luigi's yard

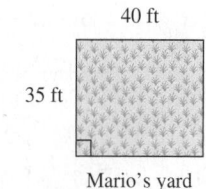

40 ft
35 ft
Mario's yard

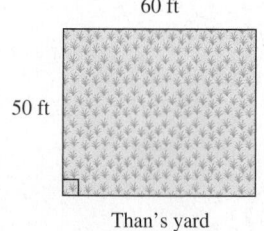

60 ft
50 ft
Than's yard

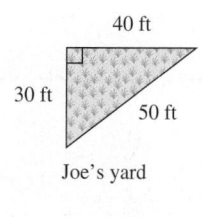

40 ft
30 ft
50 ft
Joe's yard

Give the name or names of the residents whose yards satisfy each description.

63. The yard can be fenced *and* the yard can be sodded.

64. The yard can be fenced *and* the yard cannot be sodded.

65. The yard cannot be fenced *and* the yard can be sodded.

66. The yard cannot be fenced *and* the yard cannot be sodded.

67. The yard can be fenced *or* the yard can be sodded.

68. The yard cannot be fenced *or* the yard can be sodded.

8.3 Absolute Value Equations and Inequalities

OBJECTIVES

1. Use the distance definition of absolute value.

2. Solve equations of the form $|ax + b| = k$, for $k > 0$.

3. Solve inequalities of the form $|ax + b| < k$ and of the form $|ax + b| > k$, for $k > 0$.

4. Solve absolute value equations that involve rewriting.

5. Solve equations of the form $|ax + b| = |cx + d|$.

6. Solve special cases of absolute value equations and inequalities.

Suppose a government will impose a restriction on greenhouse gas emissions *within* 3 years of 2020. This means that the *difference* between the year it will comply and 2020 is less than 3, *without regard to sign*. We state this mathematically as follows, where x represents the year in which it complies.

$$|x - 2020| < 3 \qquad \text{Absolute value inequality}$$

We can intuitively reason that the year must be between 2017 and 2023, and thus $2017 < x < 2023$ makes this inequality true.

OBJECTIVE ▶ 1 Use the distance definition of absolute value. Recall that the absolute value of a number x, written $|x|$, represents the distance from x to 0 on a number line. For example, the solutions of $|x| = 4$ are 4 and -4, as shown in **Figure 18**.

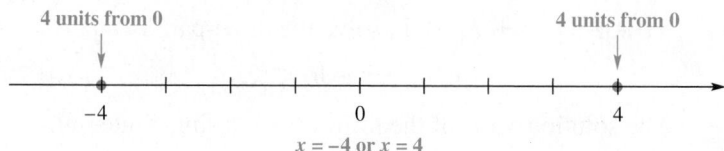

$$x = -4 \text{ or } x = 4$$

Figure 18

The solution set of $|x| > 4$ consists of all numbers that are *more* than 4 units from 0. The set $(-\infty, -4) \cup (4, \infty)$ fits this description. The graph consists of two separate intervals, which means $x < -4$ *or* $x > 4$, as shown in **Figure 19**.

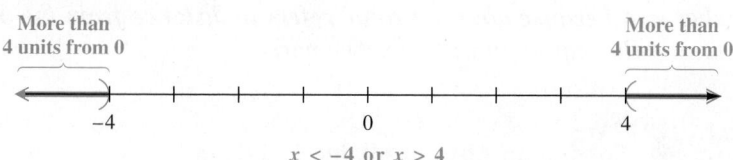

$$x < -4 \text{ or } x > 4$$

Figure 19

The solution set of $|x| < 4$ consists of all numbers that are *less* than 4 units from 0 on a number line. This is represented by all numbers *between* -4 and 4, which is given by $(-4, 4)$, as shown in **Figure 20**. Here, $-4 < x < 4$, which means $x > -4$ *and* $x < 4$.

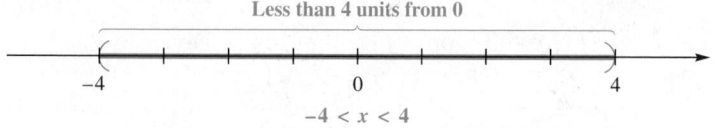

$$-4 < x < 4$$

Figure 20

Work Problem ❶ at the Side. ▶

Absolute value equations and inequalities involve the absolute value of a variable expression and generally take the form

$$|ax + b| = k, \quad |ax + b| > k, \quad \text{or} \quad |ax + b| < k,$$

where k is a positive number. From **Figures 18–20**, we see that

$|x| = 4$ has the same solution set as $x = -4$ or $x = 4$,

$|x| > 4$ has the same solution set as $x < -4$ or $x > 4$,

$|x| < 4$ has the same solution set as $x > -4$ and $x < 4$.

❶ Graph the solution set of each equation or inequality.

(a) $|x| = 3$

_____

(b) $|x| > 3$

_____

(c) $|x| < 3$

_____

Answers

1. (a)
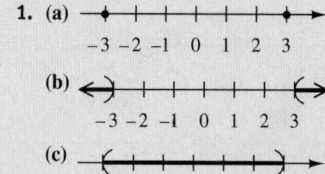

(b)

(c)

2 Solve, check, and graph the solution set of each equation.

(GS) (a) $|x + 2| = 3$

For $|x + 2|$ to equal 3, $x + 2$ must be ____ units from 0 on the number line.

$x + 2 =$ ____ or $x + 2 =$ ____

$x =$ ____ or $x =$ ____

The solution set is ____.

(b) $|3x - 4| = 11$

Solving Absolute Value Equations and Inequalities

Let k be a positive real number, and p and q be real numbers.

Case 1 To solve $|ax + b| = k$, solve the compound equation

$$ax + b = k \quad \text{or} \quad ax + b = -k.$$

The solution set is usually of the form $\{p, q\}$.

Case 2 To solve $|ax + b| > k$, solve the compound inequality

$$ax + b > k \quad \text{or} \quad ax + b < -k.$$

The solution set is of the form $(-\infty, p) \cup (q, \infty)$, which is a disjoint interval.

Case 3 To solve $|ax + b| < k$, solve the three-part inequality

$$-k < ax + b < k.$$

The solution set is of the form (p, q), a single interval.

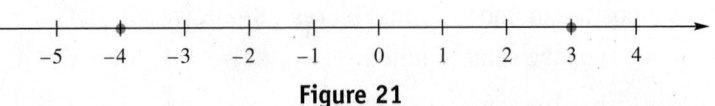

OBJECTIVE ▸ 2 Solve equations of the form $|ax + b| = k$, for $k > 0$.
Remember that because absolute value refers to distance from the origin, an absolute value equation will have two parts.

EXAMPLE 1   Solving an Absolute Value Equation

Solve $|2x + 1| = 7$. Graph the solution set.

For $|2x + 1|$ to equal 7, $2x + 1$ must be 7 units from 0 on a number line. This happens only when $2x + 1 = 7$ or $2x + 1 = -7$. This is Case 1.

$$2x + 1 = 7 \quad \text{or} \quad 2x + 1 = -7$$

$$2x = 6 \quad \text{or} \quad 2x = -8 \qquad \text{Subtract 1.}$$

$$x = 3 \quad \text{or} \quad x = -4 \qquad \text{Divide by 2.}$$

CHECK $|2x + 1| = 7$

$|2(3) + 1| \overset{?}{=} 7$ Let $x = 3$.	$|2(-4) + 1| \overset{?}{=} 7$ Let $x = -4$.
$|6 + 1| \overset{?}{=} 7$	$|-8 + 1| \overset{?}{=} 7$
$|7| \overset{?}{=} 7$	$|-7| \overset{?}{=} 7$
$7 = 7$ ✓ True	$7 = 7$ ✓ True

The solution set is $\{-4, 3\}$. The graph is shown in **Figure 21**.

Answers

2. (a) 3; 3; −3; 1; −5; $\{-5, 1\}$

−5 −4 −3 −2 −1 0 1

(b) $\left\{-\dfrac{7}{3}, 5\right\}$

−$\dfrac{7}{3}$ 0   2   4 5

Figure 21

◀ **Work Problem 2** at the Side.

Note

It is also acceptable to write the compound statements in Cases 1 and 2 of the box on the previous page as the following equivalent forms.

$$ax + b = k \quad \text{or} \quad -(ax + b) = k$$

and $\qquad ax + b > k \quad \text{or} \quad -(ax + b) > k$

These forms produce the same results.

OBJECTIVE ▶ **3** **Solve inequalities of the form $|ax + b| < k$ and of the form $|ax + b| > k$, for $k > 0$.**

EXAMPLE 2 Solving an Absolute Value Inequality Involving >

Solve $|2x + 1| > 7$. Graph the solution set.

By Case 2 in the box on the preceding page, this absolute value inequality is rewritten as

$$2x + 1 > 7 \quad \text{or} \quad 2x + 1 < -7,$$

because $2x + 1$ must represent a number that is *more* than 7 units from 0 on either side of the number line. Now, solve the compound inequality.

$2x + 1 > 7$	or	$2x + 1 < -7$
$2x > 6$	or	$2x < -8$ Subtract 1.
$x > 3$	or	$x < -4$ Divide by 2.

Check these solutions. The solution set is $(-\infty, -4) \cup (3, \infty)$. See **Figure 22.** The graph consists of disjoint intervals.

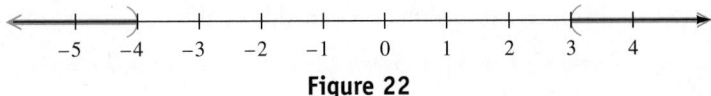

Figure 22

············· Work Problem **3** at the Side. ▶

EXAMPLE 3 Solving an Absolute Value Inequality Involving <

Solve $|2x + 1| < 7$. Graph the solution set.

The expression $2x + 1$ must represent a number that is less than 7 units from 0 on either side of the number line. Another way of thinking of this is to realize that $2x + 1$ must be between -7 and 7. As Case 3 in the box on the preceding page shows, this is written as a three-part inequality.

$$-7 < 2x + 1 < 7$$
$$-8 < \quad 2x \quad < 6 \qquad \text{Subtract 1 from each part.}$$
$$-4 < \quad x \quad < 3 \qquad \text{Divide each part by 2.}$$

Check that the solution set is $(-4, 3)$. The graph consists of the single interval shown in **Figure 23.**

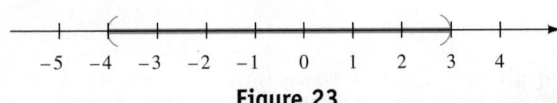

Figure 23

············· Work Problem **4** at the Side. ▶

---

3 Solve, check, and graph the solution set of each inequality.

(a) $|x + 2| > 3$

_____

(b) $|3x - 4| \geq 11$

_____

4 Solve, check, and graph the solution set of each inequality.

(a) $|x + 2| < 3$

_____

(b) $|3x - 4| \leq 11$

_____

Answers

3. (a) $(-\infty, -5) \cup (1, \infty)$

(b) $\left(-\infty, -\dfrac{7}{3}\right] \cup [5, \infty)$

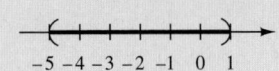

4. (a) $(-5, 1)$

(b) $\left[-\dfrac{7}{3}, 5\right]$

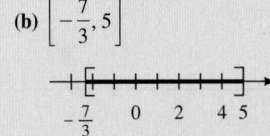

5 Solve each equation, and check your solutions.

GS **(a)** $|5x + 2| - 9 = -7$

Isolate the absolute value on one side of the equality symbol.

$$|5x + 2| = \underline{\quad}$$

$$5x + 2 = 2 \quad \text{or} \quad 5x + 2 = \underline{\quad}$$

$$5x = \underline{\quad} \quad \text{or} \quad 5x = \underline{\quad}$$

$$x = \underline{\quad} \quad \text{or} \quad x = \underline{\quad}$$

Check these solutions in the original equation.

The solution set is ____.

(b) $|10x - 2| - 2 = 12$

Look back at **Figures 21, 22, and 23,** with the graphs of

$$|2x + 1| = 7, \quad |2x + 1| > 7, \quad \text{and} \quad |2x + 1| < 7.$$

If we find the union of the three sets, we get the set of all real numbers. This is because, for any value of x, $|2x + 1|$ will satisfy one and only one of the following: It is either equal to 7, greater than 7, or less than 7.

CAUTION

When solving absolute value equations and inequalities of the types in **Examples 1, 2, and 3,** remember the following.

1. The methods described apply when the constant is alone on one side of the equation or inequality and is *positive*.

2. Absolute value equations and absolute value inequalities of the form $|ax + b| > k$ translate into "or" compound statements.

3. Absolute value inequalities of the form $|ax + b| < k$ translate into "and" compound statements, which may be written as three-part inequalities.

4. An "or" statement *cannot* be written in three parts. It would be incorrect to use $-7 > 2x + 1 > 7$ in **Example 2,** because this would imply that $-7 > 7$, which is *false*.

OBJECTIVE ▶ **4** **Solve absolute value equations that involve rewriting.**

EXAMPLE 4 **Solving an Absolute Value Equation That Involves Rewriting**

Solve $|x + 3| + 5 = 12$.

Isolate the absolute value alone on one side of the equality symbol.

$$|x + 3| + 5 = 12$$
$$|x + 3| + 5 - 5 = 12 - 5 \quad \text{Subtract 5.}$$
$$|x + 3| = 7 \quad \text{Combine like terms.}$$

Now use the method shown in **Example 1** to solve $|x + 3| = 7$.

$$x + 3 = 7 \quad \text{or} \quad x + 3 = -7$$
$$x = 4 \quad \text{or} \quad x = -10 \quad \text{Subtract 3.}$$

Check these solutions by substituting each one in the original equation.

CHECK $\quad |x + 3| + 5 = 12$

$\|4 + 3\| + 5 \overset{?}{=} 12$ Let $x = 4$.	$\|-10 + 3\| + 5 \overset{?}{=} 12$ Let $x = -10$.
$\|7\| + 5 \overset{?}{=} 12$	$\|-7\| + 5 \overset{?}{=} 12$
$12 = 12$ ✓ True	$12 = 12$ ✓ True

The check confirms that the solution set is $\{-10, 4\}$.

◀ **Work Problem** **5** **at the Side.**

Answers

5. (a) $2; -2; 0; -4; 0; -\dfrac{4}{5}; \left\{-\dfrac{4}{5}, 0\right\}$

(b) $\left\{-\dfrac{6}{5}, \dfrac{8}{5}\right\}$

CAUTION

When solving an equation like the one in **Example 4,** do *not* simply drop the absolute value bars.

EXAMPLE 5 Solving Absolute Value Inequalities That Involve Rewriting

Solve each inequality.

(a)
$$|x + 3| + 5 \geq 12$$
$$|x + 3| \geq 7$$
$$x + 3 \geq 7 \quad \text{or} \quad x + 3 \leq -7$$
$$x \geq 4 \quad \text{or} \quad x \leq -10$$

The solution set is $(-\infty, -10] \cup [4, \infty)$.

(b)
$$|x + 3| + 5 \leq 12$$
$$|x + 3| \leq 7$$
$$-7 \leq x + 3 \leq 7$$
$$-10 \leq \quad x \quad \leq 4$$

The solution set is $[-10, 4]$.

········· Work Problem **6** at the Side. ▶

OBJECTIVE ▶ **5** Solve equations of the form $|ax + b| = |cx + d|$. *If two expressions have the same absolute value, they must either be equal or be negatives of each other.*

Solving $|ax + b| = |cx + d|$

To solve an absolute value equation of the form

$$|ax + b| = |cx + d|,$$

solve the following compound equation.

$$ax + b = cx + d \quad \text{or} \quad ax + b = -(cx + d)$$

EXAMPLE 6 Solving an Equation with Two Absolute Values

Solve $|z + 6| = |2z - 3|$.

This equation is satisfied either if $z + 6$ and $2z - 3$ are equal to each other, or if $z + 6$ and $2z - 3$ are negatives of each other.

$$z + 6 = 2z - 3 \quad \text{or} \quad z + 6 = -(2z - 3)$$
$$z + 9 = 2z \quad \text{or} \quad z + 6 = -2z + 3$$
$$9 = z \quad \text{or} \quad 3z = -3$$
$$z = 9 \quad \text{or} \quad z = -1$$

CHECK $\qquad |z + 6| = |2z - 3|$

$|9 + 6| \overset{?}{=} |2(9) - 3|$ Let $z = 9$. $\quad |-1 + 6| \overset{?}{=} |2(-1) - 3|$ Let $z = -1$.

$\quad |15| \overset{?}{=} |18 - 3| \qquad\qquad\qquad |5| \overset{?}{=} |-2 - 3|$

$\quad |15| \overset{?}{=} |15| \qquad\qquad\qquad\qquad |5| \overset{?}{=} |-5|$

$\qquad 15 = 15 ✓ \qquad$ True $\qquad\qquad 5 = 5 ✓ \qquad$ True

The check confirms that the solution set is $\{9, -1\}$.

········· Work Problem **7** at the Side. ▶

---

6 Solve each inequality, and graph the solution set.

(a) $|x + 2| - 3 > 2$

(b) $|3x + 2| + 4 \leq 15$

7 Solve each equation, and check your solutions.

(a) $|k - 1| = |5k + 7|$

(b) $|4r - 1| = |3r + 5|$

Answers

6. (a) $(-\infty, -7) \cup (3, \infty)$

(b) $\left[-\dfrac{13}{3}, 3\right]$

7. (a) $\{-1, -2\}$ **(b)** $\left\{-\dfrac{4}{7}, 6\right\}$

8 Solve each equation.

(a) $|6x + 7| = -5$

(b) $\left|\dfrac{1}{4}x - 3\right| = 0$

9 Solve each inequality.

(a) $|x| > -1$

(b) $|x| < -5$

(c) $|x + 2| \leq 0$

(d) $|t - 10| - 2 \leq -3$

OBJECTIVE ▶ **6** **Solve special cases of absolute value equations and inequalities.** When a typical absolute value equation or inequality involves a *negative constant or 0* alone on one side, we use the properties of absolute value to solve the equation or inequality.

> **Special Cases of Absolute Value**
>
> ***Case 1*** The absolute value of an expression can never be negative—that is, $|a| \geq 0$ for all real numbers a.
>
> ***Case 2*** The absolute value of an expression equals 0 only when the expression is equal to 0.

EXAMPLE 7 **Solving Special Cases of Absolute Value Equations**

Solve each equation.

(a) $|5r - 3| = -4$

See Case 1 in the box. ***The absolute value of an expression can never be negative,*** so there are no solutions for this equation. The solution set is $\emptyset$.

(b) $|7x - 3| = 0$

See Case 2 in the box. The expression $|7x - 3|$ will equal 0 *only* if $7x - 3$ equals 0.

$$7x - 3 = 0 \quad |a| = 0 \text{ implies } a = 0.$$
$$7x = 3 \quad \text{Add 3.}$$
$$x = \frac{3}{7} \quad \text{Divide by 7.}$$

The solution set of the original equation is $\left\{\frac{3}{7}\right\}$, having just one element. Check this solution by substituting it in the original equation.

◀ **Work Problem 8** at the Side.

EXAMPLE 8 **Solving Special Cases of Absolute Value Inequalities**

Solve each inequality.

(a) $|x| \geq -4$

The absolute value of a number is always greater than or equal to 0. Thus, $|x| \geq -4$ is true for *all* real numbers. The solution set is $(-\infty, \infty)$.

(b)
$$|x + 6| - 3 < -5$$
$$|x + 6| < -2 \quad \text{Add 3 to each side.}$$

There is no number whose absolute value is less than -2, so this inequality has no solution. The solution set is $\emptyset$.

(c)
$$|x - 7| + 4 \leq 4$$
$$|x - 7| \leq 0 \quad \text{Subtract 4 from each side.}$$

The value of $|x - 7|$ will never be less than 0. However, $|x - 7|$ will *equal* 0 when $x = 7$. Therefore, the solution set is $\{7\}$.

◀ **Work Problem 9** at the Side.

8.3 Exercises

 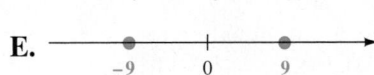

CONCEPT CHECK *Match each absolute value equation or inequality in Column I with the graph of its solution set in Column II.*

I   **II**   **I**   **II**

1. $|x| = 5$ **A.**   **2.** $|x| = 9$ **A.**

     $|x| < 5$ **B.** $|x| > 9$ **B.**

     $|x| > 5$ **C.** $|x| \geq 9$ **C.**

     $|x| \leq 5$ **D.** $|x| < 9$ **D.**

     $|x| \geq 5$ **E.** $|x| \leq 9$ **E.**

3. CONCEPT CHECK How many solutions will $|ax + b| = k$ have for each of the following conditions?

   (a) $k = 0$ **(b)** $k > 0$ **(c)** $k < 0$

4. Explain when to use *and* and when to use *or* if you are solving an absolute value equation or inequality of the form $|ax + b| = k$, $|ax + b| < k$, or $|ax + b| > k$, where k is a positive number.

Solve each equation. See Example 1.

5. $|x| = 12$ **6.** $|x| = 14$ **7.** $|4x| = 20$ **8.** $|5x| = 30$

9. $|x - 3| = 9$ **10.** $|p - 5| = 13$ **11.** $|2x + 1| = 9$ **12.** $|2x + 3| = 19$

13. $|4r - 5| = 17$ **14.** $|5t - 1| = 21$ **15.** $|2x + 5| = 14$ **16.** $|2x - 9| = 18$

17. $\left| \dfrac{1}{2}x + 3 \right| = 2$ **18.** $\left| \dfrac{2}{3}q - 1 \right| = 5$ **19.** $\left| 1 - \dfrac{3}{4}k \right| = 7$ **20.** $\left| 2 - \dfrac{5}{2}m \right| = 14$

*Solve each inequality, and graph the solution set. **See Example 2.***

21. $|x| > 3$

22. $|x| > 2$

23. $|k| \geq 4$

24. $|r| \geq 1$

25. $|t + 2| > 8$

26. $|r + 5| > 20$

27. $|3x - 1| \geq 8$

28. $|4x + 1| \geq 21$

29. $|3 - x| > 5$

30. $|5 - x| > 3$

31. CONCEPT CHECK The graph of the solution set of $|2x + 1| = 9$ is given here.

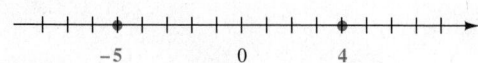

Without actually doing the algebraic work, graph the solution set of each inequality, referring to the graph above.

(a) $|2x + 1| < 9$

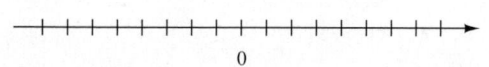

(b) $|2x + 1| > 9$

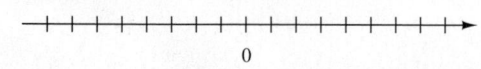

32. CONCEPT CHECK The graph of the solution set of $|3x - 4| < 5$ is given here.

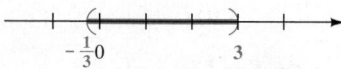

Without actually doing the algebraic work, graph the solution set of the equation and the inequality, referring to the graph above.

(a) $|3x - 4| = 5$

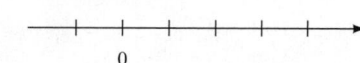

(b) $|3x - 4| > 5$

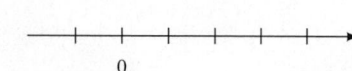

*Solve each inequality, and graph the solution set. **See Example 3.** (Hint: Compare your answers to those in **Exercises 21–30**.)*

33. $|x| \leq 3$

34. $|x| \leq 2$

35. $|k| < 4$

36. $|r| < 1$

37. $|t + 2| \leq 8$

38. $|r + 5| \leq 20$

39. $|3x - 1| < 8$

40. $|4x + 1| < 21$

41. $|3 - x| \le 5$

42. $|5 - x| \le 3$

Exercises 43–50 *represent a sampling of the various types of absolute value equations and inequalities. Decide which method of solution applies, find the solution set, and graph.* **See Examples 1–3.**

43. $|-4 + k| > 6$

44. $|-3 + t| > 5$

45. $|7 + 2z| = 5$

46. $|9 - 3p| = 3$

47. $|3r - 1| \le 11$

48. $|2s - 6| \le 6$

49. $|-3x - 8| \le 4$

50. $|-2x - 6| \le 5$

Solve each equation or inequality. Give the solution set using set notation for equations and interval notation for inequalities. **See Examples 4 and 5.**

51. $|x| - 1 = 4$

52. $|x| + 3 = 10$

53. $|x + 4| + 1 = 2$

54. $|x + 5| - 2 = 12$

55. $|2x + 1| + 3 > 8$

56. $|6x - 1| - 2 > 6$

57. $|x + 5| - 6 \le -1$

58. $|r - 2| - 3 \le 4$

Solve each equation. ***See Example 6.***

59. $|3x + 1| = |2x + 4|$

60. $|7x + 12| = |x - 8|$

61. $\left| m - \dfrac{1}{2} \right| = \left| \dfrac{1}{2}m - 2 \right|$

62. $\left| \dfrac{2}{3}r - 2 \right| = \left| \dfrac{1}{3}r + 3 \right|$

63. $|6x| = |9x + 1|$

64. $|13x| = |2x + 1|$

65. $|2p - 6| = |2p + 11|$

66. $|3x - 1| = |3x + 9|$

Solve each equation or inequality. ***See Examples 7 and 8.***

67. $|x| \geq -10$

68. $|x| \geq -15$

69. $|12t - 3| = -8$

70. $|13w + 1| = -3$

71. $|4x + 1| = 0$

72. $|6r - 2| = 0$

73. $|2q - 1| < -6$

74. $|8n + 4| < -4$

75. $|x + 5| > -9$

76. $|x + 9| > -3$

77. $|7x + 3| \leq 0$

78. $|4x - 1| \leq 0$

79. $|5x - 2| \geq 0$

80. $|4 + 7x| \geq 0$

81. $|10z + 7| > 0$

82. $|4x + 1| > 0$

83. $|x - 2| + 3 \geq 2$

84. $|k - 4| + 5 \geq 4$

Solve each problem.

85. The recommended daily intake (RDI) of calcium for females aged 19–50 is 1000 mg/day. Actual vitamin needs vary from person to person. Write an absolute value inequality in x to express the RDI plus or minus 100 mg and solve it. (*Source:* National Academy of Sciences Institute of Medicine.)

86. The average clotting time of blood is 7.45 sec with a variation of plus or minus 3.6 sec. Write this statement as an absolute value inequality in x and solve it.

The 10 tallest buildings in Houston, Texas, as of 2011 are listed, along with their heights.

Building	Height (in feet)
JPMorgan Chase Tower	1002
Wells Fargo Plaza	992
Williams Tower	901
Bank of America Center	780
Texaco Heritage Plaza	762
Enterprise Plaza	756
Centerpoint Energy Plaza	741
Continental Center 1	732
Fulbright Tower	725
One Shell Plaza	714

Source: World Almanac and Book of Facts.

Use this information to work Exercises 87–90.

87. To find the average of a group of numbers, we add the numbers and then divide by the number of numbers added. Find the average of the heights.

88. Let k represent the average height of these buildings. If a height x satisfies the inequality

$$|x - k| < t,$$

then the height is said to be within t feet of the average. Using the result from **Exercise 87,** list the buildings that are within 50 ft of the average.

89. Repeat **Exercise 88,** but list the buildings that are within 95 ft of the average.

90. Answer each of the following.

(a) Write an absolute value inequality that describes the height of a building that is *not* within 95 ft of the average.

(b) Solve the inequality from part (a).

(c) Use the result of part (b) to list the buildings that are not within 95 ft of the average.

(d) Confirm that the answer to part (c) makes sense by comparing it with the answer to **Exercise 89.**

Relating Concepts (Exercises 91–104)   For Individual or Group Work

As discussed in **Section 8.2,** a pair of inequalities joined with the word *and* is interpreted as the intersection of the solution sets of the inequalities. *The graph of the intersection of two or more inequalities in two variables is the region of the plane where all points satisfy all of the inequalities at the same time.*

For example, to graph

$$2x + 4y \geq 5 \quad \text{and} \quad x \geq 1,$$

we graph each of the two inequalities $2x + 4y \geq 5$ and $x \geq 1$ separately, as shown in **Figures A and B.** Then we use heavy shading to identify the intersection of the graphs, as shown in **Figure C.**

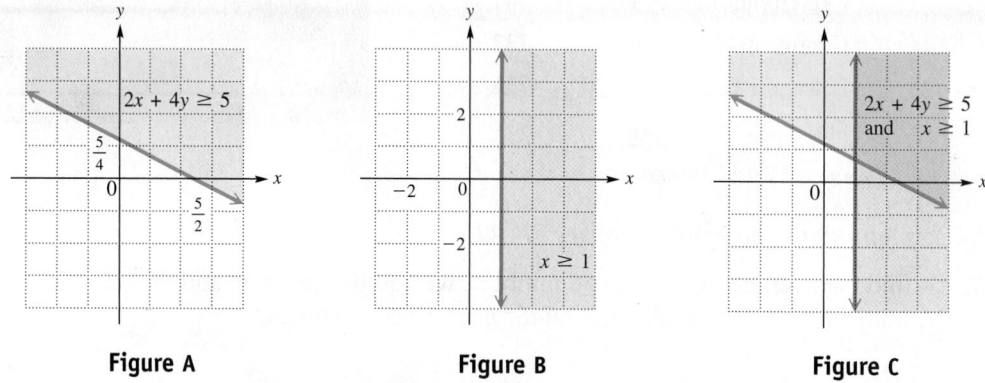

Figure A   Figure B   Figure C

In practice, the two graphs shown in **Figures A and B** are graphed on the same set of axes.

CHECK   Using **Figure C,** we can choose a test point from each of the four regions formed by the intersection of the boundary lines.

$$(2, 1), \qquad (0, 2), \qquad (0, 0), \qquad \text{and} \qquad (2, -1) \qquad \text{Possible test points}$$

| Heavily shaded region | Blue shaded region | Unshaded region | Red shaded region |

Verify that only ordered pairs in the heavily shaded region satisfy *both* inequalities. Ordered pairs in the other regions satisfy only one of the inequalities or neither of them.

When two inequalities are joined by the word *or,* we must find the union of the graphs of the inequalities. *The graph of the union of two inequalities includes all of the points that satisfy either inequality.*

For example, to graph

$$2x + 4y \geq 5 \quad \text{or} \quad x \geq 1,$$

the graphs of the two inequalities are shown in **Figures A and B** above. The graph of the union includes all points in *either* inequality, as shown in **Figure D.**

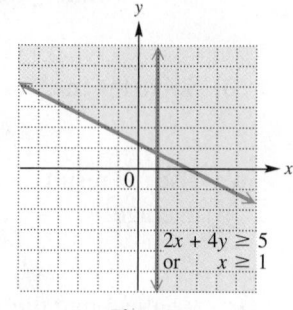

Figure D

Graph the intersection of each pair of inequalities.

91. $x + y \leq 1$ and $x \geq 1$

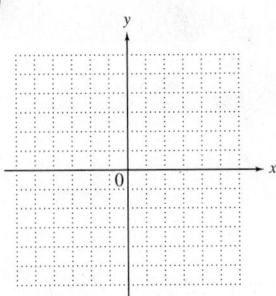

92. $x - y \geq 2$ and $x \geq 3$

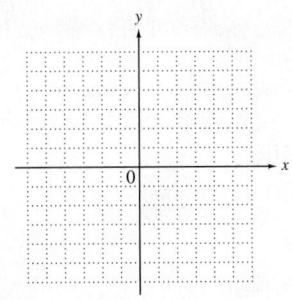

93. $2x - y \geq 2$ and $y < 4$

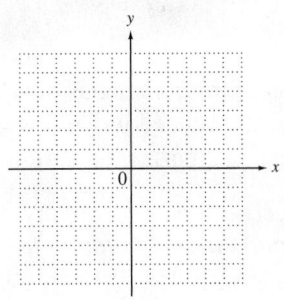

94. $3x - y \geq 3$ and $y < 3$

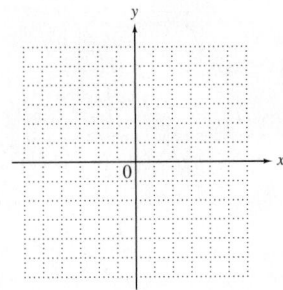

95. $x + y > -5$ and $y < -2$

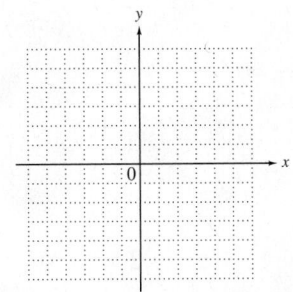

96. $6x - 4y < 10$ and $y > 2$

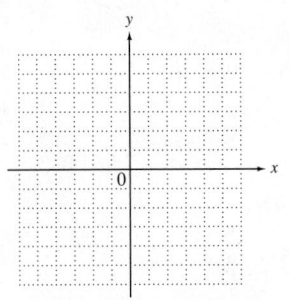

*Use the method described in **Section 8.3** to write each inequality as a compound inequality, and graph its solution set in the rectangular coordinate plane.*

97. $|x| \geq 3$

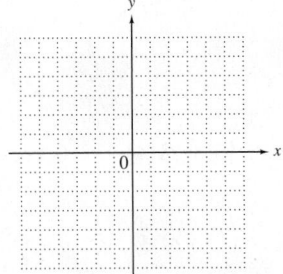

98. $|y| < 5$

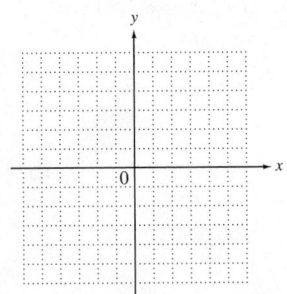

99. $|y + 1| < 2$

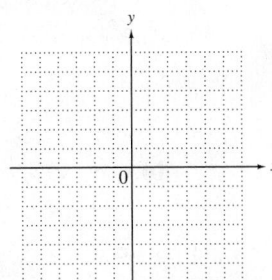

100. $|x - 2| \geq 1$

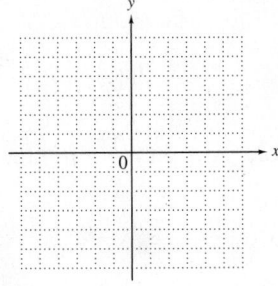

Graph the union of each pair of inequalities.

101. $x - y \geq 1$
   or $y \geq 2$

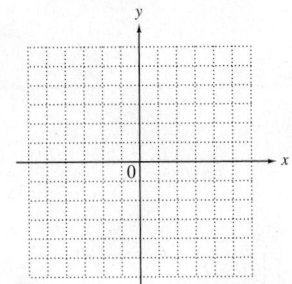

102. $x - 2 > y$
   or $x < 1$

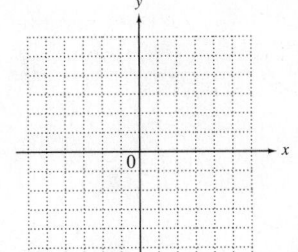

103. $3x + 2y < 6$ or
   $x - 2y > 2$

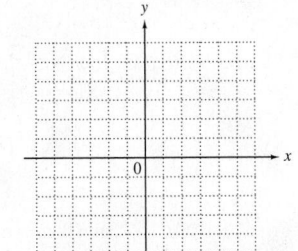

104. $x - y \geq 1$ or
   $x + y \leq 4$

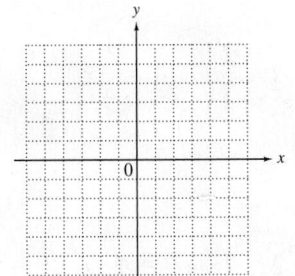

Summary Exercises *Solving Linear and Absolute Value Equations and Inequalities*

This section of miscellaneous equations and inequalities provides practice in solving all the types introduced in **Chapters 2 and 8.** As needed, refer to the boxes in these chapters that summarize the various methods of solution.

Solve each equation or inequality. Give the solution set using set notation for equations and interval notation for inequalities.

1. $4z + 1 = 49$

2. $|m - 1| = 6$

3. $6q - 9 = 12 + 3q$

4. $3p + 7 = 9 + 8p$

5. $|a + 3| = -4$

6. $2m + 1 \leq m$

7. $8r + 2 \geq 5r$

8. $4(a - 11) + 3a = 20a - 31$

9. $2q - 1 = -7$

10. $|3q - 7| - 4 = 0$

11. $6z - 5 \leq 3z + 10$

12. $|5z - 8| + 9 \geq 7$

13. $9x - 3(x + 1) = 8x - 7$

14. $|x| \geq 8$

15. $9x - 5 \geq 9x + 3$

16. $13p - 5 > 13p - 8$

17. $|q| < 5.5$

18. $4z - 1 = 12 + z$

19. $\dfrac{2}{3}x + 8 = \dfrac{1}{4}x$

20. $-\dfrac{5}{8}x \geq -20$

21. $\dfrac{1}{4}p < -6$

22. $7z - 3 + 2z = 9z - 8z$

23. $\dfrac{3}{5}q - \dfrac{1}{10} = 2$

24. $|r - 1| < 7$

25. $r + 9 + 7r = 4(3 + 2r) - 3$

26. $6 - 3(2 - p) < 2(1 + p) + 3$

27. $|2p - 3| > 11$

28. $\dfrac{x}{4} - \dfrac{2x}{3} = -10$

29. $|5a + 1| \le 0$

30. $5z - (3 + z) \ge 2(3z + 1)$

31. $-2 \le 3x - 1 \le 8$

32. $-1 \le 6 - x \le 5$

33. $|7z - 1| = |5z + 3|$

34. $|p + 2| = |p + 4|$

35. $|1 - 3x| \ge 4$

36. $\dfrac{1}{2} \le \dfrac{2}{3}r \le \dfrac{5}{4}$

37. $-(m + 4) + 2 = 3m + 8$

38. $\dfrac{p}{6} - \dfrac{3p}{5} = p - 86$

39. $-6 \le \dfrac{3}{2} - x \le 6$

40. $|5 - x| < 4$

41. $|x - 1| \ge -6$

42. $|2r - 5| = |r + 4|$

43. $8q - (1 - q) = 3(1 + 3q) - 4$

44. $8x - (x + 3) = -(2x + 1) - 12$

45. $|r - 5| = |r + 9|$

46. $|r + 2| < -3$

47. $2x + 1 > 5$ or $3x + 4 < 1$

48. $1 - 2x \ge 5$ and $7 + 3x \ge -2$

8.4 Review of Systems of Linear Equations in Two Variables

OBJECTIVES

1. Solve linear systems with two equations and two variables.
2. Solve special systems.

OBJECTIVE ▶ ① **Solve linear systems with two equations and two variables.** Recall from **Section 4.1** that a **system of linear equations** (often called a **linear system**) consists of two or more linear equations with the same variables.

$$x + y = 5$$
$$2x - y = 4$$

Linear system of equations

The **solution set of a system of linear equations** contains all ordered pairs that satisfy all equations of the system *at the same time.*

We introduced three methods for solving linear systems in **Chapter 4:** the *graphing method,* the *substitution method,* and the *elimination method.* The **graphing method** involves graphing each equation of a system on the same set of axes and finding the point where the graphs intersect.

EXAMPLE 1 Solving a System by Graphing

Solve the system of equations by graphing.

$$x + y = 5 \qquad (1)$$
$$2x - y = 4 \qquad (2)$$

To graph these linear equations, we plot several points for each line.

$$x + y = 5 \qquad\qquad 2x - y = 4$$

The intercepts are a convenient choice.

x	y
0	5
5	0
2	3

x	y
0	−4
2	0
4	4

Find a third ordered pair as a check.

As shown in **Figure 24,** the graph suggests that the point of intersection is the ordered pair $(3, 2)$.

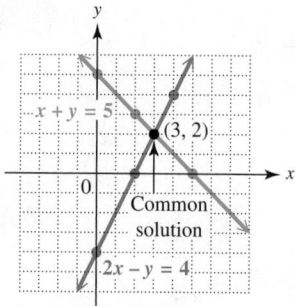

Figure 24

To be sure that $(3, 2)$ is a solution of *both* equations, we check by substituting 3 for x and 2 for y in each equation.

······Continued on Next Page

CHECK

$x + y = 5$ (1)	$2x - y = 4$ (2)
$3 + 2 \overset{?}{=} 5$	$2(3) - 2 \overset{?}{=} 4$
$5 = 5$ ✓ True	$6 - 2 \overset{?}{=} 4$
	$4 = 4$ ✓ True

Since $(3, 2)$ makes both equations true, $\{(3, 2)\}$ is the solution set of the system.

·············· **Work Problem 1 at the Side.** ▶

> 🖩 **Calculator Tip**
>
> A graphing calculator can be used to solve a system. Each equation must be solved for y and entered in the calculator. The point of intersection of the graphs, which is the solution of the system, can be displayed.

There are three possibilities for the solution set of a linear system in two variables. **Example 1** illustrates Case 1.

Graphs of Linear Systems in Two Variables

Case 1 **The two graphs intersect in a single point.** The coordinates of this point give the only solution of the system. Since the system has a solution, it is **consistent**. The equations are *not* equivalent, so they are **independent**. See **Figure 25(a)**.

Case 2 **The graphs are parallel lines.** There is no solution common to both equations, so the solution set is ∅ and the system is **inconsistent**. Since the equations are *not* equivalent, they are **independent**. See **Figure 25(b)**.

Case 3 **The graphs are the same line.** Since any solution of one equation of the system is a solution of the other, the solution set is an infinite set of ordered pairs representing the points on the line. This type of system is **consistent** because there is a solution. The equations are equivalent, so they are **dependent**. See **Figure 25(c)**.

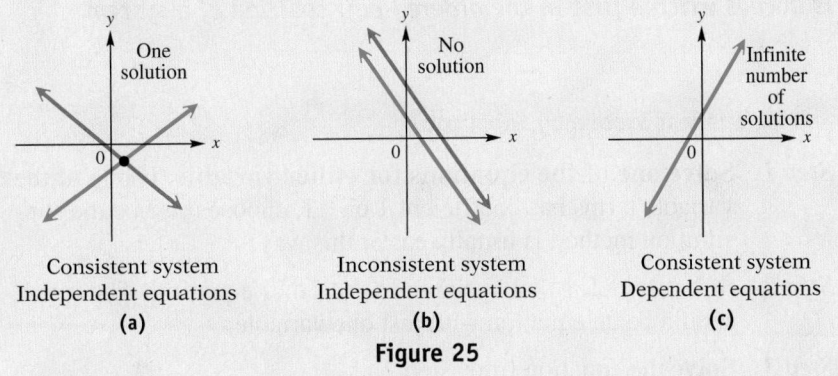

Consistent system	Inconsistent system	Consistent system
Independent equations	Independent equations	Dependent equations
(a)	**(b)**	**(c)**

Figure 25

Since it can be difficult to read exact coordinates from a graph, especially if they are not integers, we usually use algebraic methods to solve systems. The **substitution method** is most useful for solving linear systems in which one equation is solved or can be easily solved for one variable in terms of the other.

1 Solve each system of equations by graphing.

(a) $x - y = 3$ (1)
   $2x - y = 4$ (2)

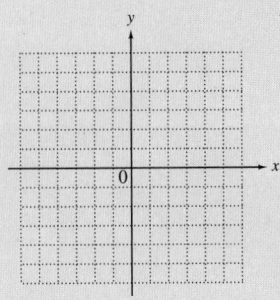

(b) $2x + y = -5$ (1)
   $-x + 3y = 6$ (2)

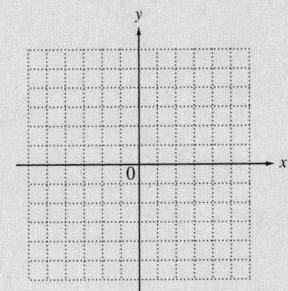

Answers

1. (a) $\{(1, -2)\}$

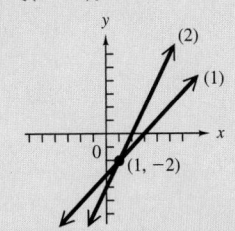

(b) $\{(-3, 1)\}$

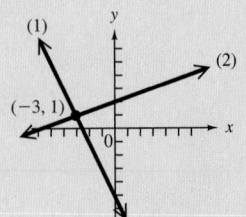

❷ Solve by substitution.

(a) $7x - 2y = -2$

$y = 3x$

(b) $5x - 3y = -6$

$x = 2 - y$

EXAMPLE 2   **Solving a System by Substitution**

Solve the system.

$$2x - y = 6 \quad (1)$$
$$x = y + 2 \quad (2)$$

Since equation (2) is solved for x, substitute $y + 2$ for x in equation (1).

$$2x - y = 6 \quad (1)$$
$$2(y + 2) - y = 6 \qquad \text{Let } x = y + 2.$$

> Be sure to use parentheses here.

$$2y + 4 - y = 6 \qquad \text{Distributive property}$$
$$y + 4 = 6 \qquad \text{Combine like terms.}$$
$$y = 2 \qquad \text{Subtract 4.}$$

We found y. Now we find x by substituting **2** for y in equation (2).

$$x = y + 2 \quad (2)$$
$$x = 2 + 2 \qquad \text{Let } y = 2.$$
$$x = 4 \qquad \text{Add.}$$

> Write the x-value first in the ordered pair.

Thus, $x = 4$ and $y = 2$, giving the ordered pair $(\mathbf{4}, \mathbf{2})$. Check this solution in both equations of the original system.

CHECK

$$2x - y = 6 \quad (1) \qquad\qquad x = y + 2 \quad (2)$$
$$2(4) - 2 \overset{?}{=} 6 \qquad\qquad\qquad 4 \overset{?}{=} 2 + 2$$
$$8 - 2 \overset{?}{=} 6 \qquad\qquad\qquad 4 = 4 \ \checkmark \quad \text{True}$$
$$6 = 6 \ \checkmark \quad \text{True}$$

Since $(4, 2)$ makes both equations true, the solution set is $\{(4, 2)\}$.

◀ **Work Problem ❷ at the Side.**

CAUTION

Be careful. Even though we found y first in **Example 2,** *the x-coordinate is always written first in the ordered-pair solution of a system.*

Solving a Linear System by Substitution

Step 1   **Solve one of the equations for either variable.** If one of the variable terms has coefficient 1 or −1, choose it, since the substitution method is usually easier this way.

Step 2   **Substitute** for that variable in the other equation. The result should be an equation with just one variable.

Step 3   **Solve** the equation from Step 2.

Step 4   **Find the other value.** Substitute the result from Step 3 into the equation from Step 1 to find the value of the other variable.

Step 5   **Check** the ordered-pair solution in *both* of the *original* equations. Then write the solution set.

Answers

2. (a) $\{(-2, -6)\}$ **(b)** $\{(0, 2)\}$

| EXAMPLE 3 | Solving a System by Substitution |

Solve the system.

$$3x + 2y = 13 \quad (1)$$
$$4x - y = -1 \quad (2)$$

Step 1 First solve one of the equations for x or y. Since the coefficient of y in equation (2) is -1, it is easiest to solve for y in equation (2).

$$4x - y = -1 \quad (2)$$
$$-y = -1 - 4x \quad \text{Subtract } 4x.$$
$$y = 1 + 4x \quad \text{Multiply by } -1.$$

Step 2 Substitute $1 + 4x$ for y in equation (1).

$$3x + 2y = 13 \quad (1)$$
$$3x + 2(1 + 4x) = 13 \quad \text{Let } y = 1 + 4x.$$

Step 3 Solve for x.

$$3x + 2 + 8x = 13 \quad \text{Distributive property}$$
$$11x = 11 \quad \text{Combine like terms. Subtract 2.}$$
$$x = 1 \quad \text{Divide by 11.}$$

Step 4 Now find y. From Step 1, $y = 1 + 4x$, so if $x = 1$, then

$$y = 1 + 4(1) = 5. \quad \text{Let } x = 1.$$

Step 5 Check the solution $(1, 5)$ in both equations (1) and (2).

CHECK

$$3x + 2y = 13 \quad (1) \qquad 4x - y = -1 \quad (2)$$
$$3(1) + 2(5) \stackrel{?}{=} 13 \qquad 4(1) - 5 \stackrel{?}{=} -1$$
$$3 + 10 \stackrel{?}{=} 13 \qquad 4 - 5 \stackrel{?}{=} -1$$
$$13 = 13 \ \checkmark \ \text{True} \qquad -1 = -1 \ \checkmark \ \text{True}$$

The solution set is $\{(1, 5)\}$.

⸺⸺⸺⸺⸺⸺ **Work Problem 3 at the Side.** ▶

The **elimination method** involves combining the two equations in a system so that one variable is eliminated. This is done using the following logic.

$$\text{If } a = b \text{ and } c = d, \quad \text{then} \quad a + c = b + d.$$

| EXAMPLE 4 | Solving a System by Elimination |

Solve the system.

$$2x + 3y = -6 \quad (1)$$
$$4x - 3y = 6 \quad (2)$$

Notice that adding the equations together will eliminate the variable y.

$$\begin{array}{ll} 2x + 3y = -6 & (1) \\ \underline{4x - 3y = 6} & (2) \\ 6x = 0 & \text{Add.} \\ x = 0 & \text{Solve for } x. \end{array}$$

⸺⸺⸺⸺ **Continued on Next Page**

3 Solve by substitution.

(a) $3x - y = 10$
$2x + 5y = 1$

(b) $4x - 5y = -11 \quad (1)$
$x + 2y = 7 \quad (2)$

Step 1
Solve equation (2) for x.

$$x = \underline{}$$

Step 2
Substitute $\underline{}$ for x in equation (1).

$$4(\underline{}) - 5y = -11$$

Step 3
Solve for y.

$$y = \underline{}$$

Step 4
Now find x.

$$x = 7 - 2y$$
$$x = 7 - 2(\underline{})$$
$$x = \underline{}$$

Step 5
Check the solution $\underline{}$ in both equations (1) and (2). The solution set is $\underline{}$.

Answers

3. **(a)** $\{(3, -1)\}$
   (b) $7 - 2y; 7 - 2y; 7 - 2y; 3; 3; 1; (1, 3);$ $\{(1, 3)\}$

4 Solve by elimination.

(a) $3x - y = -7$

$2x + y = -3$

To find y, substitute 0 for x in either equation (1) or equation (2).

$$2x + 3y = -6 \quad (1)$$
$$2(0) + 3y = -6 \quad \text{Let } x = 0.$$
$$0 + 3y = -6 \quad \text{Multiply.}$$
$$3y = -6 \quad \text{Add.}$$
$$y = -2 \quad \text{Divide by 3.}$$

Check by substituting 0 for x and -2 for y in both equations of the original system. The solution set is $\{(0, -2)\}$.

◀ **Work Problem 4 at the Side.**

By adding the equations in **Example 4,** we eliminated the variable y because the coefficients of the y-terms were opposites. In many cases the coefficients will *not* be opposites, and we must transform one or both equations so that the coefficients of one pair of variable terms are opposites.

Solving a Linear System by Elimination

Step 1 **Write both equations in standard form $Ax + By = C$.**

Step 2 **Make the coefficients of one pair of variable terms opposites.** Multiply one or both equations by appropriate number(s) so that the sum of the coefficients of either the x- or y-terms is 0.

Step 3 **Add the new equations to eliminate a variable. The sum should be an equation with just one variable.**

Step 4 **Solve the equation from Step 3 for the remaining variable.**

Step 5 **Find the other value. Substitute the result from Step 4 into either of the original equations and solve for the other variable.**

Step 6 **Check the ordered-pair solution in *both* of the *original* equations. Then write the solution set.**

GS **(b)** $-2x + 3y = -10 \quad (1)$

$2x + 2y = 5 \quad (2)$

Add the equations to eliminate the variable _____.

$$-2x + 3y = -10 \quad (1)$$
$$\underline{2x + 2y = 5} \quad (2)$$
$$\underline{} y = \underline{}$$
$$y = \underline{}$$

To find x, substitute -1 for y in equation (2).

$2x + 2(\underline{}) = 5$

$2x = \underline{}$

$x = \underline{}$

The solution is _____.

Check and write the solution set, _____.

EXAMPLE 5 **Solving a System by Elimination**

Solve the system.

$$5x - 2y = 4 \quad (1)$$
$$2x + 3y = 13 \quad (2)$$

Step 1 Both equations are in standard form.

Step 2 Suppose that you wish to eliminate the variable x. One way to do this is to multiply equation (1) by 2 and equation (2) by -5.

The goal is to have *opposite* coefficients.

$$10x - 4y = 8 \quad \text{2 times each side of equation (1)}$$
$$-10x - 15y = -65 \quad -5 \text{ times each side of equation (2)}$$

Step 3 Now add.

$$10x - 4y = 8$$
$$\underline{-10x - 15y = -65}$$
$$-19y = -57 \quad \text{Add.}$$

Step 4 Solve for y. $\qquad y = 3 \qquad$ Divide by -19.

Continued on Next Page

Answers

4. (a) $\{(-2, 1)\}$

(b) $x; 5; -5; -1; -1; 7; \dfrac{7}{2}; \left(\dfrac{7}{2}, -1\right);$

$$\left\{\left(\dfrac{7}{2}, -1\right)\right\}$$

Step 5 To find x, substitute 3 for y in either equation (1) or (2).

$$2x + 3y = 13 \quad (2)$$
$$2x + 3(3) = 13 \quad \text{Let } y = 3.$$
$$2x + 9 = 13 \quad \text{Multiply.}$$
$$2x = 4 \quad \text{Subtract 9.}$$
$$x = 2 \quad \text{Divide by 2.}$$

Step 6 To check, substitute 2 for x and 3 for y in equations (1) and (2).

CHECK

$$5x - 2y = 4 \quad (1) \qquad\qquad 2x + 3y = 13 \quad (2)$$
$$5(2) - 2(3) \overset{?}{=} 4 \qquad\qquad 2(2) + 3(3) \overset{?}{=} 13$$
$$4 = 4 \ \checkmark \ \text{True} \qquad\qquad 13 = 13 \ \checkmark \ \text{True}$$

The solution set is $\{(2, 3)\}$.

·· **Work Problem ❺ at the Side. ▶**

OBJECTIVE ▶ 2 Solve special systems. **Examples 6 and 7** illustrate Cases 2 and 3 for the solution set of a linear system in two variables as given earlier in this section.

EXAMPLE 6 | **Solving a System of Dependent Equations**

Solve the system.

$$2x - y = 3 \quad (1)$$
$$6x - 3y = 9 \quad (2)$$

We multiply equation (1) by -3, and then add the result to equation (2).

$$-6x + 3y = -9 \quad \text{-3 times each side of equation (1)}$$
$$\underline{6x - 3y = 9 \quad (2)}$$
$$0 = 0 \quad \text{True}$$

Adding gives the true statement $0 = 0$. In the original system, we could get equation (2) from equation (1) by multiplying equation (1) by 3. Equations (1) and (2) are equivalent and have the same graph, as shown in **Figure 26.** The equations are dependent.

The solution set is the set of all points on the line with equation $2x - y = 3$, written in set-builder notation **(Section 1.3)** as

$$\{(x, y) \mid 2x - y = 3\}$$

and read "the set of all ordered pairs (x, y), such that $2x - y = 3$."

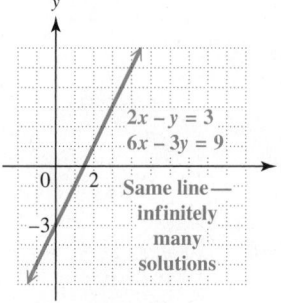

Figure 26

·· **Work Problem ❻ at the Side. ▶**

Note

To write the solution set in **Example 6,** *we use the equation in standard form with coefficients that are integers having greatest common factor 1 and positive coefficient of x.*

---

❺ Solve by elimination.

(a) $x + 3y = 8$
  $2x - 5y = -17$

(b) $6x - 2y = -21$
  $-3x + 4y = 36$

(c) $2x + 3y = 19$
  $3x - 7y = -6$

❻ Solve the system. Then graph both equations.

$$2x + y = 6 \quad (1)$$
$$-8x - 4y = -24 \quad (2)$$

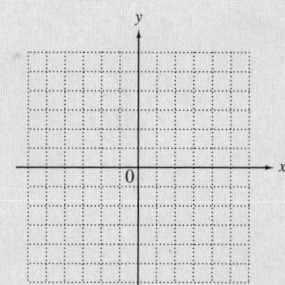

Answers

5. (a) $\{(-1, 3)\}$ (b) $\left\{\left(-\dfrac{2}{3}, \dfrac{17}{2}\right)\right\}$
  (c) $\{(5, 3)\}$
6. $\{(x, y) \mid 2x + y = 6\}$

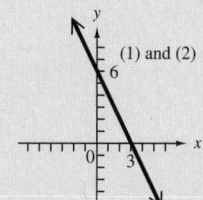

7 Solve the system. Then graph both equations.

$$2x - y = 4$$
$$-6x + 3y = 0$$

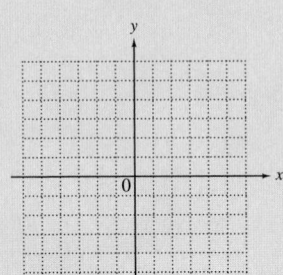

EXAMPLE 7 **Solving an Inconsistent System**

Solve the system.

$$x + 3y = 4 \quad (1)$$
$$-2x - 6y = 3 \quad (2)$$

Multiply equation (1) by 2, and then add the result to equation (2).

$$2x + 6y = 8 \quad \text{Equation (1) multiplied by 2}$$
$$\underline{-2x - 6y = 3} \quad (2)$$
$$0 = 11 \quad \text{False}$$

The result of the addition step is a false statement, $0 = 11$, which indicates that the system is inconsistent. As shown in **Figure 27,** the graphs of the equations are parallel lines.

    There are no ordered pairs that satisfy both equations, so there is no solution for the system. The solution set is $\emptyset$.

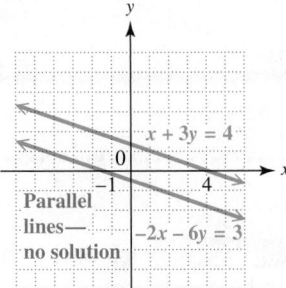

Figure 27

◀ **Work Problem 7** at the Side.

Answer

7. $\emptyset$

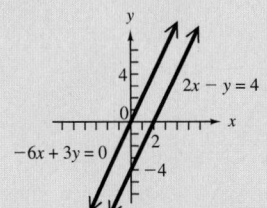

8.4 Exercises

  Download the MyDashBoard App MyMathLab®

1. CONCEPT CHECK Which ordered pair could possibly be a solution of the graphed system of equations? Why?

A. $(3, 3)$
B. $(-3, 3)$
C. $(-3, -3)$
D. $(3, -3)$

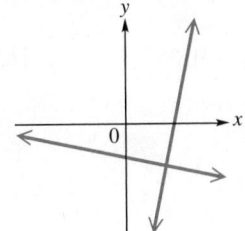

2. CONCEPT CHECK Which ordered pair could possibly be a solution of the graphed system of equations? Why?

A. $(3, 0)$
B. $(-3, 0)$
C. $(0, 3)$
D. $(0, -3)$

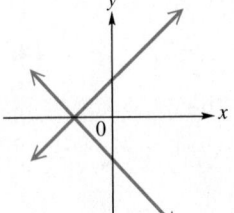

3. CONCEPT CHECK Match each system with the correct graph.

(a) $x + y = 6$
　　$x - y = 0$

(b) $x + y = -6$
　　$x - y = 0$

(c) $x + y = 0$
　　$x - y = -6$

(d) $x + y = 0$
　　$x - y = 6$

A.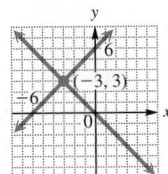

B.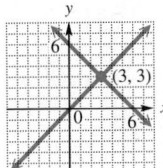

C.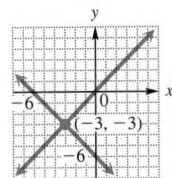

D.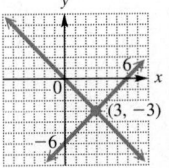

4. CONCEPT CHECK To minimize the amount of work required, tell whether you would use the substitution or elimination method to solve each system. *Do not actually solve.*

(a) $6x - y = 5$
　　$y = 11x$

(b) $3x + y = -7$
　　$x - y = -5$

(c) $3x - 2y = 0$
　　$9x + 8y = 7$

*Solve each system by graphing. **See Example 1.***

5. $x + y = -5$
　$-2x + y = 1$

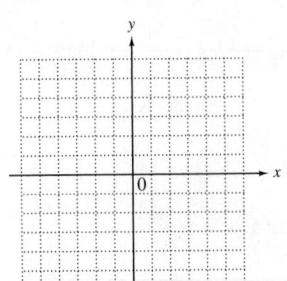

6. $x + y = 4$
　$2x - y = 2$

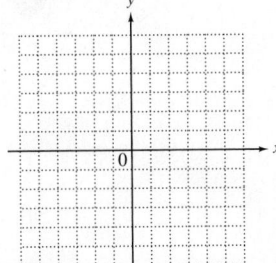

7. $x - 4y = -4$
　$3x + y = 1$

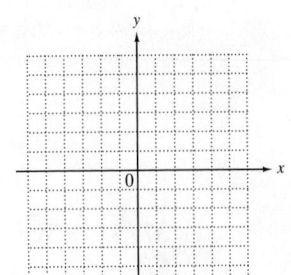

8. $2x + 3y = -6$
　$x - 3y = -3$

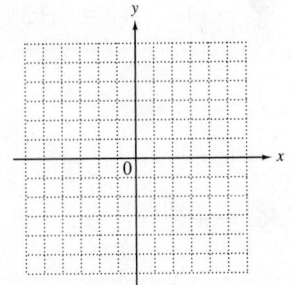

Solve each system by substitution. If the system is inconsistent or has dependent equations, say so. **See Examples 2, 3, 6, and 7.**

9. $4x + y = 6$
$y = 2x$

10. $2x - y = 6$
$y = 5x$

11. $-x - 4y = -14$
$y = 2x - 1$

12. $-3x - 5y = -17$
$y = 4x + 8$

13. $3x - 4y = -22$
$-3x + y = 0$

14. $-3x + y = -5$
$x + 2y = 0$

15. $5x - 4y = 9$
$3 - 2y = -x$

16. $6x - y = -9$
$4 + 7x = -y$

17. $x = 3y + 5$
$x = \dfrac{3}{2}y$

18. $x = 6y - 2$
$x = \dfrac{3}{4}y$

19. $\dfrac{1}{2}x + \dfrac{1}{3}y = 3$
$-3x + y = 0$

20. $\dfrac{1}{4}x - \dfrac{1}{5}y = 9$
$5x - y = 0$

21. $y = 2x$
$4x - 2y = 0$

22. $x = 3y$
$3x - 9y = 0$

23. $5x - 25y = 5$
$x = 5y$

24. $8x + 2y = 4$
$y = -4x$

Solve each system by elimination. If the system is inconsistent or has dependent equations, say so. **See Examples 4–7.**

25. $-2x + 3y = -16$
$2x - 5y = 24$

26. $6x + 5y = -7$
$-6x - 11y = 1$

27. $2x - 5y = 11$
$3x + y = 8$

28. $-2x + 3y = 1$
$-4x + y = -3$

29. $3x + 4y = -6$
$5x + 3y = 1$

30. $4x + 3y = 1$
$3x + 2y = 2$

31. $3x + 3y = 0$
$4x + 2y = 3$

32. $8x + 4y = 0$
$4x - 2y = 2$

33. $7x + 2y = 6$
$-14x - 4y = -12$

34. $x - 4y = 2$
$4x - 16y = 8$

35. $\dfrac{x}{2} + \dfrac{y}{3} = -\dfrac{1}{3}$

$\dfrac{x}{2} + 2y = -7$

36. $\dfrac{x}{4} + \dfrac{y}{3} = -\dfrac{1}{3}$

$\dfrac{x}{3} - \dfrac{y}{4} = -6$

37. $5x - 5y = 3$
$x - y = 12$

38. $2x - 3y = 7$
$-4x + 6y = 14$

8.5 Systems of Linear Equations in Three Variables; Applications

OBJECTIVES

1. Understand the geometry of systems of three equations in three variables.

2. Solve linear systems (with three equations and three variables) by elimination.

3. Solve linear systems (with three equations and three variables) in which some of the equations have missing terms.

4. Solve special systems.

5. Solve application problems with three variables using a system of three equations.

A solution of an equation in three variables, such as

$$2x + 3y - z = 4, \quad \text{Linear equation in three variables}$$

is an **ordered triple** and is written (x, y, z). For example, the ordered triple $(0, 1, -1)$ is a solution of the equation, because

$$2(0) + 3(1) - (-1) = 4$$

is a true statement. Verify that another solution of this equation is $(10, -3, 7)$.

We now extend the term *linear equation* to equations of the form

$$Ax + By + Cz + \ldots + Dw = K,$$

where not all the coefficients $A, B, C, \ldots, D$ equal 0. For example,

$$2x + 3y - 5z = 7 \quad \text{and} \quad x - 2y - z + 3u - 2w = 8$$

are linear equations, the first with three variables and the second with five.

OBJECTIVE ▶ 1 Understand the geometry of systems of three equations in three variables. Consider the solution of a system such as the following.

$$4x + 8y + z = 2$$
$$x + 7y - 3z = -14 \quad \begin{array}{l}\text{System of linear equations}\\\text{in three variables}\end{array}$$
$$2x - 3y + 2z = 3$$

Theoretically, a system of this type can be solved by graphing. However, the graph of a linear equation with three variables is a *plane*, not a line. Since visualizing a plane requires three-dimensional graphing, the graphing method is not practical with these systems. However, it does illustrate the number of solutions possible for such systems, as shown in **Figure 28**.

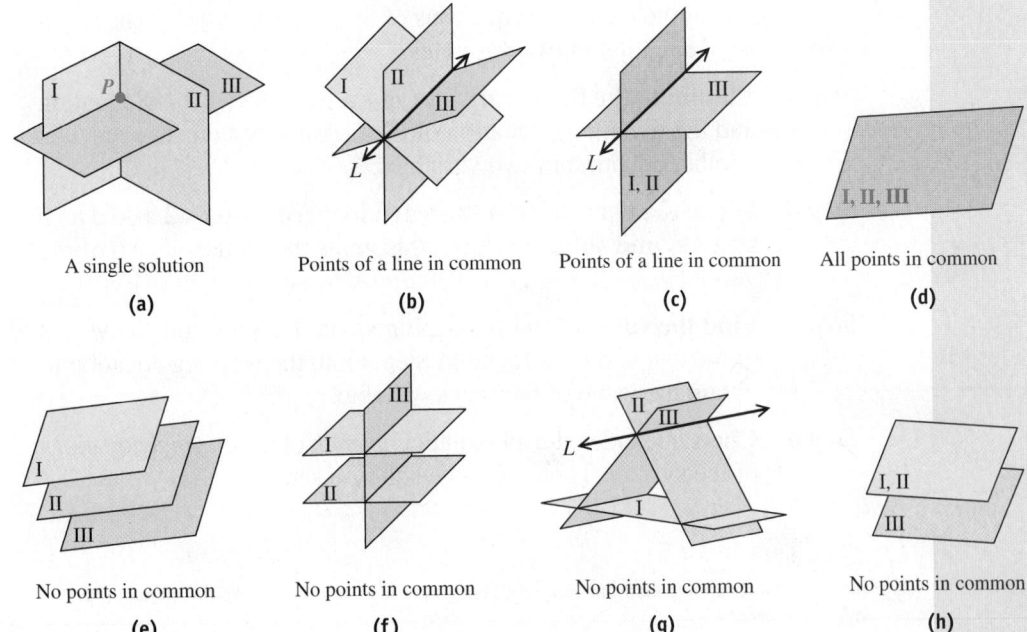

A single solution	Points of a line in common	Points of a line in common	All points in common
(a)	(b)	(c)	(d)

No points in common	No points in common	No points in common	No points in common
(e)	(f)	(g)	(h)

Figure 28

Figure 28 on the preceding page illustrates the following cases.

Graphs of Linear Systems in Three Variables

Case 1 **The three planes may meet at a single, common point** that forms the solution set of the system. See **Figure 28(a).**

Case 2 **The three planes may have the points of a line in common,** so that the infinite set of points that satisfy the equation of the line is the solution of the system. See **Figures 28(b) and (c).**

Case 3 **The three planes may coincide,** so that the solution set of the system is the set of all points on a plane. See **Figure 28(d).**

Case 4 **The planes may have no points common to all three,** so that there is no solution of the system. See **Figures 28(e)–(h).**

OBJECTIVE ▶ 2 **Solve linear systems (with three equations and three variables) by elimination.** Since graphing to find the solution set of a system of three equations in three variables is impractical, these systems are solved with an extension of the elimination method reviewed in **Section 8.4.**

In the steps that follow, we use the term **focus variable** to identify the first variable to be eliminated in the process. The focus variable will always be present in the **working equation,** which will be used twice to eliminate this variable.

Solving a Linear System in Three Variables*

Step 1 **Select a variable and an equation.** A good choice for the variable, which we call the *focus variable,* is one that has coefficient 1 or −1. Then select an equation, usually the one that contains the focus variable, as the *working equation.*

Step 2 **Eliminate the focus variable.** Use the working equation and one of the other two equations of the original system. The result is an equation in two variables.

Step 3 **Eliminate the focus variable again.** Use the working equation and the remaining equation of the original system. The result is another equation in two variables.

Step 4 **Write the equations in two variables from Steps 2 and 3 as a system, and solve it.** Doing this gives the values of two of the variables.

Step 5 **Find the value of the remaining variable.** Substitute the values of the two variables found in Step 4 into the working equation to obtain the value of the focus variable.

Step 6 **Check** the ordered-triple solution in *each* of the *original* equations of the system. Then write the solution set.

---

*The authors wish to thank Christine Heinecke Lehmann of Purdue University North Central for her suggestions here.

EXAMPLE 1 **Solving a System in Three Variables**

Solve the system.

$$4x + 8y + z = 2 \qquad (1)$$
$$x + 7y - 3z = -14 \qquad (2)$$
$$2x - 3y + 2z = 3 \qquad (3)$$

Step 1 Since z in equation (1) has coefficient 1, we choose z as the focus variable and (1) as the working equation. (Another option would be to choose x as the focus variable, since it also has coefficient 1, and use (2) as the working equation.)

$$\overset{\text{Focus variable}}{4x + 8y + z = 2} \qquad (1) \leftarrow \text{Working equation}$$

Step 2 Multiply working equation (1) by 3 and add the result to equation (2).

$$12x + 24y + 3z = 6 \qquad \text{Multiply each side of (1) by 3.}$$

| Focus variable z was eliminated. |

$$\underline{x + 7y - 3z = -14} \qquad (2)$$
$$13x + 31y = -8 \qquad \text{Add.} \quad (4)$$

Step 3 Multiply working equation (1) by -2 and add the result to remaining equation (3) to again eliminate focus variable z.

$$-8x - 16y - 2z = -4 \qquad \text{Multiply each side of (1) by } -2.$$

| Focus variable z was eliminated. |

$$\underline{2x - 3y + 2z = 3} \qquad (3)$$
$$-6x - 19y = -1 \qquad \text{Add.} \quad (5)$$

Step 4 Write the equations that result in Steps 2 and 3 as a system.

| Make sure these equations have the same two variables. |

$$13x + 31y = -8 \qquad (4) \qquad \text{The result from Step 2}$$
$$-6x - 19y = -1 \qquad (5) \qquad \text{The result from Step 3}$$

Now solve this system. We choose to eliminate x.

$$78x + 186y = -48 \qquad \text{Multiply each side of (4) by 6.}$$
$$\underline{-78x - 247y = -13} \qquad \text{Multiply each side of (5) by 13.}$$
$$-61y = -61 \qquad \text{Add.}$$
$$y = 1 \qquad \text{Divide by } -61.$$

Substitute 1 for y in either equation (4) or (5) to find x.

$$-6x - 19y = -1 \qquad (5)$$
$$-6x - 19(1) = -1 \qquad \text{Let } y = 1.$$
$$-6x - 19 = -1 \qquad \text{Multiply.}$$
$$-6x = 18 \qquad \text{Add 19.}$$
$$x = -3 \qquad \text{Divide by } -6.$$

Step 5 Now substitute the two values we found in Step 4 in working equation (1) to find the value of the remaining variable, focus variable z.

$$4x + 8y + z = 2 \qquad (1)$$
$$4(-3) + 8(1) + z = 2 \qquad \text{Let } x = -3 \text{ and } y = 1.$$
$$-4 + z = 2 \qquad \text{Multiply and then add.}$$
$$z = 6 \qquad \text{Add 4.}$$

Continued on Next Page

1 Check that the solution $(-3, 1, 6)$ also satisfies both equations (2) and (3) of **Example 1.**

(a) $x + 7y - 3z = -14$ (2)

Does the solution satisfy equation (2)? (*Yes / No*)

(b) $2x - 3y + 2z = 3$ (3)

Does the solution satisfy equation (3)? (*Yes / No*)

2 Solve each system.

(a)
$$x + y + z = 2$$
$$x - y + 2z = 2$$
$$-x + 2y - z = 1$$

(b)
$$2x + y + z = 9$$
$$-x - y + z = 1$$
$$3x - y + z = 9$$

Answers

1. (a) Yes (b) Yes
2. (a) $\{(-1, 1, 2)\}$ (b) $\{(2, 1, 4)\}$

> Write the values of x, y, and z in the correct order

Step 6 It appears that the ordered triple $(-3, 1, 6)$ is the only solution of the system. We must check that the solution satisfies all three original equations of the system. We begin with equation (1).

CHECK
$$4x + 8y + z = 2 \quad \text{(1)}$$
$$4(-3) + 8(1) + 6 \overset{?}{=} 2 \qquad \text{Substitute.}$$
$$-12 + 8 + 6 \overset{?}{=} 2 \qquad \text{Multiply.}$$
$$2 = 2 \ \checkmark \quad \text{True}$$

◀ **Work Problem ❶ at the Side.**

Because $(-3, 1, 6)$ also satisfies equations (2) and (3), the solution set is $\{(-3, 1, 6)\}$. This is Case 1 as illustrated in **Figure 28(a)** at the beginning of this section.

◀ **Work Problem ❷ at the Side.**

OBJECTIVE ▶ ❸ Solve linear systems (with three equations and three variables) in which some of the equations have missing terms. If a linear system has an equation missing a term or terms, one elimination step can be omitted.

EXAMPLE 2 **Solving a System with Missing Terms**

Solve the system.

$$6x - 12y = -5 \quad \text{(1)} \qquad \text{Missing } z$$
$$8y + z = 0 \quad \text{(2)} \qquad \text{Missing } x$$
$$9x - z = 12 \quad \text{(3)} \qquad \text{Missing } y$$

Since equation (3) is missing the variable y, one way to begin is to eliminate y again using equations (1) and (2).

> Leave space for the missing terms.

$$
\begin{array}{ll}
12x - 24y = -10 & \text{Multiply each side of (1) by 2.} \\
\underline{ 24y + 3z = 0} & \text{Multiply each side of (2) by 3.} \\
12x + 3z = -10 & \text{Add.} \quad \text{(4)}
\end{array}
$$

Use the resulting equation (4) in x and z, together with equation (3), $9x - z = 12$, to eliminate z. Multiply equation (3) by 3.

$$
\begin{array}{ll}
27x - 3z = 36 & \text{Multiply each side of (3) by 3.} \\
\underline{12x + 3z = -10} & \text{(4)} \\
39x = 26 & \text{Add.}
\end{array}
$$

$$x = \frac{26}{39}, \text{ or } \frac{2}{3} \qquad \begin{array}{l} \text{Divide by 39.} \\ \text{Write in lowest terms.} \end{array}$$

We can find z by substituting this value for x into equation (3).

$$9x - z = 12 \quad \text{(3)}$$
$$9\left(\frac{2}{3}\right) - z = 12 \qquad \text{Let } x = \tfrac{2}{3}.$$
$$6 - z = 12 \qquad \text{Multiply.}$$
$$z = -6 \qquad \text{Subtract 6. Multiply by } -1.$$

•••••••••• **Continued on Next Page**

We can find y by substituting -6 for z in equation (2).

$$8y + z = 0 \qquad (2)$$

$$8y - 6 = 0 \qquad \text{Let } z = -6.$$

$$8y = 6 \qquad \text{Add 6.}$$

$$y = \frac{6}{8}, \quad \text{or} \quad \frac{3}{4} \qquad \text{Divide by 8. Write in lowest terms.}$$

Thus, $x = \frac{2}{3}$, $y = \frac{3}{4}$, and $z = -6$. Check to verify that the solution set is $\left\{ \left(\frac{2}{3}, \frac{3}{4}, -6 \right) \right\}$. This is also an example of Case 1.

Note

Another way to solve the system in **Example 2** is to begin by eliminating the variable z from equations (2) and (3). The resulting equation together with equation (1) forms a system of two equations in the variables x and y. Try working **Example 2** this way to see that the same solution results.

There are often multiple ways to solve a system of equations. Some ways may involve more work than others.

Work Problem ❸ at the Side. ▶

OBJECTIVE ▶ ❹ Solve special systems. Linear systems with three variables may include dependent equations or may be inconsistent.

EXAMPLE 3 Solving a System of Dependent Equations

Solve the system.

$$2x - 3y + 4z = 8 \qquad (1)$$

$$-x + \frac{3}{2}y - 2z = -4 \qquad (2)$$

$$6x - 9y + 12z = 24 \qquad (3)$$

Multiplying each side of equation (1) by 3 gives equation (3). Multiplying each side of equation (2) by -6 also gives equation (3). Because of this, the equations are dependent. All three equations have the same graph, as illustrated in **Figure 28(d)** at the beginning of this section. This is Case 3. The solution set is written as follows.

$$\{(x, y, z) \mid 2x - 3y + 4z = 8\} \qquad \text{Set-builder notation}$$

Although any one of the three equations could be used to write the solution set, we use the equation with coefficients that are integers with greatest common factor 1, as we did in **Section 8.4.**

Work Problem ❹ at the Side. ▶

---

❸ Solve each system.

(a) $\begin{aligned} x - y &= 6 \\ 2y + 5z &= 1 \\ 3x - 4z &= 8 \end{aligned}$

(b) $\begin{aligned} 5x - y &= 26 \\ 4y + 3z &= -4 \\ x + z &= 5 \end{aligned}$

❹ Solve each system.

⑤ (a) $\begin{aligned} x - y + z &= 4 \qquad (1) \\ -3x + 3y - 3z &= -12 \qquad (2) \\ 2x - 2y + 2z &= 8 \qquad (3) \end{aligned}$

Multiplying each side of equation (1) by -3 gives equation ____. Multiplying each side of equation (1) by 2 gives equation ____. Therefore, the equations are ____, and the graph of all three equations is the same ____. The solution set is written

_____.

(b) $\begin{aligned} x - 3y + 2z &= 10 \\ -2x + 6y - 4z &= -20 \\ \frac{1}{2}x - \frac{3}{2}y + z &= 5 \end{aligned}$

Answers

3. (a) $\{(4, -2, 1)\}$ **(b)** $\{(5, -1, 0)\}$

4. (a) (2); (3); dependent; plane; $\{(x, y, z) \mid x - y + z = 4\}$

(b) $\{(x, y, z) \mid x - 3y + 2z = 10\}$

5 Solve each system.

(a)
$$3x - 5y + 2z = 1$$
$$5x + 8y - z = 4$$
$$-6x + 10y - 4z = 5$$

(b)
$$7x - 9y + 2z = 0$$
$$y + z = 0$$
$$8x - z = 0$$

6 Solve each system.

(GS) (a)
$$2x + 3y - z = 8 \quad (1)$$
$$\frac{1}{2}x + \frac{3}{4}y - \frac{1}{4}z = 2 \quad (2)$$
$$x + \frac{3}{2}y - \frac{1}{2}z = -6 \quad (3)$$

Multiplying each side of equation (2) by _____ gives equation (1).

Multiplying each side of equation (3) by 2 gives the equation

_____,

which (*is / is not*) equivalent to equation (1).

Therefore, the system is _____ and the solution set is _____.

(b)
$$x - 3y + 2z = 4$$
$$\frac{1}{3}x - y + \frac{2}{3}z = 7$$
$$\frac{1}{2}x - \frac{3}{2}y + z = 2$$

Answers

5. (a) $\emptyset$ (b) $\{(0, 0, 0)\}$
6. (a) 4; $2x + 3y - z = -12$; is not; inconsistent; $\emptyset$
   (b) $\emptyset$

EXAMPLE 4 Solving an Inconsistent System

Solve the system.

$$2x - 4y + 6z = 5 \quad (1)$$
$$-x + 3y - 2z = -1 \quad (2)$$
$$x - 2y + 3z = 1 \quad (3)$$ ◁ Use as the working equation, with focus variable x.

Eliminate the focus variable, x, using equations (1) and (3).

$$-2x + 4y - 6z = -2 \quad \text{Multiply each side of (3) by } -2.$$
$$\underline{2x - 4y + 6z = 5 \quad (1)}$$
$$0 = 3 \quad \text{False}$$

The resulting false statement indicates that equations (1) and (3) have no common solution. Thus, the system is inconsistent and the solution set is $\emptyset$. The graph of this system would show these two planes parallel to one another, as illustrated in **Figure 28(f)** at the beginning of this section. This is Case 4.

Note

If a false statement results when adding as in **Example 4,** it is not necessary to go any further with the solution. Since two of the three planes are parallel, it is not possible for the three planes to have any points in common.

◁ **Work Problem 5** at the Side.

EXAMPLE 5 Solving Another Special System

Solve the system.

$$2x - y + 3z = 6 \quad (1)$$
$$x - \frac{1}{2}y + \frac{3}{2}z = 3 \quad (2)$$
$$4x - 2y + 6z = 1 \quad (3)$$

Multiplying each side of equation (2) by 2 gives equation (1), so these two equations are dependent. Equations (1) and (3) are not equivalent, however. Multiplying equation (3) by $\frac{1}{2}$ does not give equation (1). Instead, we obtain two equations with the same coefficients, but with different constant terms.

The graphs of equations (1) and (3) have no points in common (that is, the planes are parallel). Thus, the system is inconsistent and the solution set is $\emptyset$, as illustrated in **Figure 28(h).** This is also an example of Case 4.

◁ **Work Problem 6** at the Side.

OBJECTIVE 5 Solve application problems with three variables using a system of three equations. We extend the method from **Section 4.4.**

EXAMPLE 6 Solving a Problem Involving Prices

At Panera Bread, a loaf of honey wheat bread costs $2.95, a loaf of sunflower bread costs $2.99, and a loaf of French bread costs $5.79. On a recent day, three times as many loaves of honey wheat bread were sold as sunflower bread. The number of loaves of French bread sold was 5 less than the number of loaves of honey wheat bread sold. Total receipts for these breads were $87.89. How many loaves of each type of bread were sold? (*Source:* Panera Bread.)

Step 1 **Read** the problem again. There are three unknowns.

Step 2 **Assign variables** to represent the three unknowns.

Let x = the number of loaves of honey wheat bread,

y = the number of loaves of sunflower bread,

and z = the number of loaves of French bread.

Step 3 **Write a system of three equations.** Three times as many loaves of honey wheat bread were sold as sunflower bread.

$$x = 3y, \quad \text{or} \quad x - 3y = 0 \quad \text{Subtract } 3y. \quad (1)$$

Also, we have the information needed for another equation.

Number of loaves of French	equals	5 less than the number of loaves of honey wheat.
↓	↓	↓
z	$=$	$x - 5$

$$-x + z = -5 \quad \text{Subtract } x.$$
$$x - z = 5 \quad \text{Multiply by } -1. \quad (2)$$

Multiplying the cost of a loaf of each kind of bread by the number of loaves of that kind sold and adding gives the total receipts.

$$2.95x + 2.99y + 5.79z = 87.89$$
$$295x + 299y + 579z = 8789 \quad \begin{array}{l}\text{Multiply by 100 to clear}\\ \text{decimals.} \quad (3)\end{array}$$

Step 4 **Solve** the system of three equations.

$$x - 3y = 0 \qquad (1)$$
$$x - z = 5 \qquad (2)$$
$$295x + 299y + 579z = 8789 \qquad (3)$$

Work Problem 7 at the Side. ▶

We find that $x = 12$, $y = 4$, and $z = 7$.

Step 5 **State the answer.** The solution set is $\{(12, 4, 7)\}$, meaning that 12 loaves of honey wheat bread, 4 loaves of sunflower bread, and 7 loaves of French bread were sold.

Step 6 **Check.** Since $12 = 3 \cdot 4$, the number of loaves of honey wheat bread is three times the number of loaves of sunflower bread. Also, $12 - 7 = 5$, so the number of loaves of French bread is 5 less than the number of loaves of honey wheat bread. Multiply the appropriate cost per loaf by the number of loaves sold and add the results to check that total receipts were $87.89. The answer checks.

Work Problem 8 at the Side. ▶

7 Solve the system of equations from Step 4 of **Example 6.**

$$x - 3y = 0 \qquad (1)$$
$$x - z = 5 \qquad (2)$$
$$295x + 299y + 579z = 8789 \qquad (3)$$

8 Solve the problem.

A department store display features three kinds of perfume: Felice, Vivid, and Joy. There are 10 more bottles of Felice than Vivid, and 3 fewer bottles of Joy than Vivid. Each bottle of Felice costs $8, Vivid costs $15, and Joy costs $32. The total value of all the perfume is $589. How many bottles of each are there?

Answers

7. $\{(12, 4, 7)\}$
8. 21 bottles of Felice; 11 of Vivid; 8 of Joy

9 Solve the problem.
A paper mill makes newsprint, bond, and copy machine paper.

- Each ton of newsprint requires 3 tons of recycled paper and 1 ton of wood pulp.

- Each ton of bond requires 2 tons of recycled paper, 4 tons of wood pulp, and 3 tons of rags.

- Each ton of copy machine paper requires 2 tons of recycled paper, 3 tons of wood pulp, and 2 tons of rags.

The mill has 4200 tons of recycled paper, 5800 tons of wood pulp, and 3900 tons of rags. How much of each kind of paper can be made from these supplies?

EXAMPLE 7   **Solving a Business Production Problem**

A company produces three sets: models X, Y, and Z.

- Each model X set requires 2 hr of electronics work, 2 hr of assembly time, and 1 hr of finishing time.

- Each model Y requires 1 hr of electronics work, 3 hr of assembly time, and 1 hr of finishing time.

- Each model Z requires 3 hr of electronics work, 2 hr of assembly time, and 2 hr of finishing time.

There are 100 hr available for electronics, 100 hr available for assembly, and 65 hr available for finishing per week. How many of each model should be produced each week if all available time must be used?

Step 1   **Read** the problem again. There are three unknowns.

Step 2   **Assign variables.** Then organize the problem information in a table.

Let x = the number of model X produced per week,

y = the number of model Y produced per week,

and z = the number of model Z produced per week.

	Each Model X	Each Model Y	Each Model Z	Totals	
Hours of Electronics Work	2	1	3	100	Gives equation (1)
Hours of Assembly Time	2	3	2	100	Gives equation (2)
Hours of Finishing Time	1	1	2	65	Gives equation (3)

Step 3   **Write a system of three equations.** The x model X sets require $2x$ hours of electronics, the y model Y sets require $1y$ (or y) hours of electronics, and the z model Z sets require $3z$ hours of electronics. There are 100 hr available for electronics. Also, there are 100 hr available for assembly. There are 65 hr available for finishing.

$$2x + y + 3z = 100 \quad \text{Electronics} \quad (1)$$
$$2x + 3y + 2z = 100 \quad \text{Assembly} \quad (2)$$
$$x + y + 2z = 65 \quad \text{Finishing} \quad (3)$$

Notice that by reading *across* the table, we can quickly determine the coefficients and constants in the equations of the system.

Step 4   **Solve** the system of equations (1), (2), and (3). We find that

$$x = 15, \quad y = 10, \quad \text{and} \quad z = 20.$$

Step 5   **State the answer.** The company should produce 15 model X, 10 model Y, and 20 model Z sets per week.

Step 6   **Check** that these values satisfy the conditions of the problem.

◀ **Work Problem 9** at the Side.

Answer

9. newsprint: 400 tons; bond: 900 tons; copy machine paper: 600 tons

8.5 Exercises

Download the MyDashBoard App   ▶ MyMathLab®

1. CONCEPT CHECK Using your immediate surroundings, give an example of three planes that satisfy the condition.

  (a) They intersect in a single point.

  (b) They do not intersect.

  (c) They intersect in infinitely many points.

2. CONCEPT CHECK Suppose that a system has infinitely many ordered triple solutions of the form (x, y, z) such that

$$x + y + 2z = 1.$$

Give three specific ordered triples that are solutions of the system.

3. Explain what the following statement means. "The solution set of the system

$$2x + y + z = 3$$
$$3x - y + z = -2$$
$$4x - y + 2z = 0$$

is $\{(-1, 2, 3)\}$."

4. CONCEPT CHECK The two equations

$$x + y + z = 6$$
$$2x - y + z = 3$$

have a common solution of $(1, 2, 3)$. Which equation would complete a system of three linear equations in three variables having solution set $\{(1, 2, 3)\}$?

  A. $3x + 2y - z = 1$ **B.** $3x + 2y - z = 4$

  C. $3x + 2y - z = 5$ **D.** $3x + 2y - z = 6$

Solve each system of equations. ***See Example 1.***

5. $2x - 5y + 3z = -1$
$x + 4y - 2z = 9$
$x - 2y - 4z = -5$

6. $x + 3y - 6z = 7$
$2x - y + z = 1$
$x + 2y + 2z = -1$

7. $3x + 2y + z = 8$
$2x - 3y + 2z = -16$
$x + 4y - z = 20$

8. $-3x + y - z = -10$
$-4x + 2y + 3z = -1$
$2x + 3y - 2z = -5$

9. $x + 2y + z = 4$
$2x + y - z = -1$
$x - y - z = -2$

10. $x - 2y + 5z = -7$
$-2x - 3y + 4z = -14$
$-3x + 5y - z = -7$

11. $-x + 2y + 6z = 2$
$3x + 2y + 6z = 6$
$x + 4y - 3z = 1$

12. $2x + y + 2z = 1$
$x + 2y + z = 2$
$x - y - z = 0$

13. $2x + 5y + 2z = 0$
$4x - 7y - 3z = 1$
$3x - 8y - 2z = -6$

14. $5x - 2y + 3z = -9$
$4x + 3y + 5z = 4$
$2x + 4y - 2z = 14$

15. $x + 2y + 3z = 1$
$-x - y + 3z = 2$
$-6x + y + z = -2$

16. $x + y - z = -2$
$2x - y + z = -5$
$-x + 2y - 3z = -4$

Solve each system of equations. **See Example 2.**

17. $2x - 3y + 2z = -1$
$x + 2y + z = 17$
$2y - z = 7$

18. $2x - y + 3z = 6$
$x + 2y - z = 8$
$2y + z = 1$

19. $4x + 2y - 3z = 6$
$x - 4y + z = -4$
$-x + 2z = 2$

20. $2x + 3y - 4z = 4$
$x - 6y + z = -16$
$-x + 3z = 8$

21. $-5x + 2y + z = 5$
$-3x - 2y - z = 3$
$-x + 6y = 1$

22. $x + y - z = 0$
$2y - z = 1$
$2x + 3y - 4z = -4$

23. $2x + y = 6$
$3y - 2z = -4$
$3x - 5z = -7$

24. $4x - 8y = -7$
$4y + z = 7$
$-8x + z = -4$

Solve each system of equations. If the system is inconsistent or has dependent equations, say so. **See Examples 1, 3, 4, and 5.**

25. $2x + 2y - 6z = 5$
$-3x + y - z = -2$
$-x - y + 3z = 4$

26. $-2x + 5y + z = -3$
$5x + 14y - z = -11$
$7x + 9y - 2z = -5$

27. $-5x + 5y - 20z = -40$
$x - y + 4z = 8$
$3x - 3y + 12z = 24$

28. $x + 4y - z = 3$
$-2x - 8y + 2z = -6$
$3x + 12y - 3z = 9$

29. $2x + y - z = 6$
$4x + 2y - 2z = 12$
$-x - \dfrac{1}{2}y + \dfrac{1}{2}z = -3$

30. $2x - 8y + 2z = -10$
$-x + 4y - z = 5$
$\dfrac{1}{8}x - \dfrac{1}{2}y + \dfrac{1}{8}z = -\dfrac{5}{8}$

31. $x + y - 2z = 0$
$3x - y + z = 0$
$4x + 2y - z = 0$

32. $2x + 3y - z = 0$
$x - 4y + 2z = 0$
$3x - 5y - z = 0$

33. $x - 2y + \dfrac{1}{3}z = 4$
$3x - 6y + z = 12$
$-6x + 12y - 2z = -3$

34. $4x + y - 2z = 3$
$x + \dfrac{1}{4}y - \dfrac{1}{2}z = \dfrac{3}{4}$
$2x + \dfrac{1}{2}y - z = 1$

35. $x + 5y - 2z = -1$
$-2x + 8y + z = -4$
$3x - y + 5z = 19$

36. $x + 3y + z = 2$
$4x + y + 2z = -4$
$5x + 2y + 3z = -2$

Solve each problem involving three unknowns. **See Examples 6 and 7.** *(In Exercises 37–40, the sum of the measures of the angles of a triangle is* 180°.)

37. In the figure, $z = x + 10$ and $x + y = 100$. Determine a third equation involving x, y, and z, and then find the measures of the three angles.

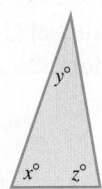

38. In the figure, x is 10 less than y and x is 20 less than z. Write a system of three equations and find the measures of the three angles.

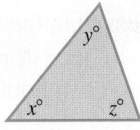

39. In a certain triangle, the measure of the second angle is 10° more than three times the first. The third angle measure is equal to the sum of the measures of the other two. Find the measures of the three angles.

40. The measure of the largest angle of a triangle is 12° less than the sum of the measures of the other two. The smallest angle measures 58° less than the largest. Find the measures of the angles.

41. The perimeter of a triangle is 70 cm. The longest side is 4 cm less than the sum of the other two sides. Twice the shortest side is 9 cm less than the longest side. Find the length of each side of the triangle.

42. The perimeter of a triangle is 56 in. The longest side measures 4 in. less than the sum of the other two sides. Three times the shortest side is 4 in. more than the longest side. Find the lengths of the three sides.

43. In a random sample of Americans of voting age, 8% more people identified themselves as Independents than as Republicans, while 6% fewer people identified themselves as Republicans than as Democrats. Of those sampled, 2% did not identify with any of the three categories. What percent identified themselves with each of the three political affiliations? *(Source:* Gallup, Inc.)

44. In the 2012 Summer Olympics, Russia earned 8 fewer gold medals than bronze. The number of silver medals earned was 38 less than twice the number of bronze medals. Russia earned a total of 82 medals. How many of each kind of medal did Russia earn? *(Source:* www.london2012.com)

45. Tickets for a Harlem Globetrotters show cost $16, $23, or, for VIP seats, $40. If nine times as many $16 tickets were sold as VIP tickets, and the number of $16 tickets sold was 55 more than the sum of the number of $23 tickets and VIP tickets, sales of all three kinds of tickets would total $46,575. How many of each kind of ticket would have been sold? *(Source:* MSU Breslin Student Events Center.)

46. Three kinds of tickets are available for a *Cowboy Mouth* concert: "up close," "in the middle," and "far out." "Up close" tickets cost $10 more than "in the middle" tickets, while "in the middle" tickets cost $10 more than "far out" tickets. Twice the cost of an "up close" ticket is $20 more than 3 times the cost of a "far out" ticket. Find the price of each kind of ticket.

47. A wholesaler supplies college t-shirts to three college bookstores: A, B, and C. The wholesaler recently shipped a total of 800 t-shirts to the three bookstores. Twice as many t-shirts were shipped to bookstore B as to bookstore A, and the number shipped to bookstore C was 40 less than the sum of the numbers shipped to the other two bookstores. How many t-shirts were shipped to each bookstore?

48. An office supply store sells three models of computer desks: A, B, and C. In January, the store sold a total of 85 computer desks. The number of model B desks was five more than the number of model C desks, and the number of model A desks was four more than twice the number of model C desks. How many of each model did the store sell in January?

In the National Hockey League, a point system is used to determine team standings. A team is awarded 2 points for a win (W), 0 points for a loss in regulation play (L), and 1 point for an overtime loss (OTL). Use this information in Exercises 49 and 50.

49. During the 2011–2012 NHL regular season, the Boston Bruins played 82 games. Their wins and overtime losses resulted in a total of 102 points. They had 25 more losses in regulation play than overtime losses. How many wins, losses, and overtime losses did they have that year?

50. During the 2011–2012 NHL regular season, the Columbus Blue Jackets played 82 games. Their wins and overtime losses resulted in a total of 65 points. They had 24 more total losses (in regulation play and overtime) than wins. How many wins, losses, and overtime losses did they have that year?

EASTERN CONFERENCE, NORTHEAST DIVISION

Team	GP	W	L	OTL	Points
Boston	82	____	____	____	102
Ottawa	82	41	31	10	92
Buffalo	82	39	32	11	89
Toronto	82	35	37	10	80
Montreal	82	31	35	16	78

Source: www.nhl.com

WESTERN CONFERENCE, CENTRAL DIVISION

Team	GP	W	L	OTL	Points
St. Louis	82	49	22	11	109
Nashville	82	48	26	8	104
Detroit	82	48	28	6	102
Chicago	82	45	26	11	101
Columbus	82	____	____	____	65

Source: www.nhl.com

Chapter 8 *Summary*

Key Terms

8.1

linear (first-degree) equation in one variable A linear equation in one variable can be written in the form $Ax + B = C$, where A, B, and C are real numbers, with $A \neq 0$.

solution A solution of an equation is a number that makes the equation true when substituted for the variable.

solution set The solution set of an equation is the set of all its solutions.

equivalent equations Equivalent equations are equations that have the same solution set.

conditional equation An equation that is true only for certain value(s) of the variable is a conditional equation.

contradiction An equation that has no solution (that is, its solution set is $\emptyset$) is a contradiction.

identity An equation that is satisfied by every valid replacement of the variable is an identity.

inequality An inequality consists of algebraic expressions related by $<$, $>$, $\leq$, or $\geq$.

linear inequality in one variable A linear inequality in one variable can be written in the form $Ax + B < C$, $Ax + B \leq C$, $Ax + B > C$, or $Ax + B \geq C$, where A, B, and C are real numbers, with $A \neq 0$.

equivalent inequalities Equivalent inequalities are inequalities that have the same solution set.

8.2

intersection The intersection of two sets A and B is the set of elements that belong to both A and B.

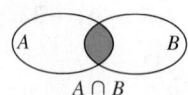

$A \cap B$

compound inequality A compound inequality is formed by joining two inequalities with a connective word such as *and* or *or*.

union The union of two sets A and B is the set of elements that belong to either A or B (or both).

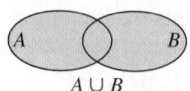

$A \cup B$

8.3

absolute value equation; absolute value inequality Absolute value equations and inequalities are equations and inequalities that involve the absolute value of a variable expression.

8.4

system of equations Two or more equations that are to be solved at the same time form a system of equations.

linear system A linear system is a system of equations that contains only linear equations.

solution set of a system All ordered pairs that satisfy all the equations of a system at the same time make up the solution set of the system.

consistent system A system is consistent if it has a solution.

independent equations Independent equations are equations whose graphs are different lines.

inconsistent system A system is inconsistent if it has no solution.

dependent equations Dependent equations are equations whose graphs are the same line.

New Symbols

$\cap$	set intersection	$\cup$	set union	(x, y, z)	ordered triple

Test Your Word Power

See how well you have learned the vocabulary in this chapter.

1 An **equation** is
  A. an algebraic expression
  B. an expression that contains fractions
  C. a statement that two algebraic expressions are equal

  D. an expression that uses any of the four basic operations or the operations of raising to powers or taking roots on any collection of variables and numbers.

2 A **solution set** is the set of numbers that
  A. make an expression undefined
  B. make an equation false
  C. make an equation true
  D. make an expression equal to 0.

(continued)

Test Your Word Power *(continued)*

3 An **inequality** is
  A. a statement that two algebraic expressions are equal
  B. a point on a number line
  C. an equation with no solutions
  D. a statement consisting of algebraic expressions related by $<$, $\leq$, $>$, or $\geq$.

4 The **intersection** of two sets A and B is the set of elements that belong
  A. to both A and B
  B. to either A or B, or both
  C. to either A or B, but not both
  D. to just A.

5 The **union** of two sets A and B is the set of elements that belong
  A. to both A and B
  B. to either A or B, or both
  C. to either A or B, but not both
  D. to just B.

6 A **system of equations** consists of
  A. at least two equations with different variables
  B. two or more equations that have an infinite number of solutions
  C. two or more equations that are to be solved at the same time

  D. two or more inequalities that are to be solved.

7 An **inconsistent system** is a system of equations
  A. with one solution
  B. with no solution
  C. with an infinite number of solutions
  D. that have the same graph.

8 A **consistent system** has
  A. no solution
  B. a solution
  C. exactly two solutions
  D. exactly three solutions.

Answers to Test Your Word Power

1. C; *Examples:* $2a + 3 = 7$, $3y = -8$, $x^2 = 4$

2. C; *Example:* $\{8\}$ is the solution set of $2x + 5 = 21$.

3. D; *Examples:* $x < 5$, $7 + 2k \geq 11$, $-5 < 2z - 1 \leq 3$

4. A; *Example:* If $A = \{2, 4, 6, 8\}$ and $B = \{1, 2, 3\}$, $A \cap B = \{2\}$.

5. B; *Example:* Using the preceding sets A and B, $A \cup B = \{1, 2, 3, 4, 6, 8\}$.

6. C; *Example:* $\begin{array}{l} 3x - y = 3 \\ 2x + y = 7 \end{array}$

7. B; *Example:* The equations of two parallel lines form an inconsistent system. Their graphs never intersect, so the system has no solution.

8. B; *Example:* The system in Answer 6 is a consistent system.

Quick Review

Concepts	Examples

8.1 Review of Solving Linear Equations and Inequalities

Solving a Linear Equation in One Variable

Step 1 Simplify each side separately.

Solve the equation.

$$4(8 - 3t) = 32 - 8(t + 2)$$

$$32 - 12t = 32 - 8t - 16 \qquad \text{Distributive property}$$

$$32 - 12t = 16 - 8t \qquad \text{Combine like terms.}$$

Step 2 Isolate the variable terms on one side.

$$32 - 12t + 12t = 16 - 8t + 12t \qquad \text{Add } 12t.$$

$$32 = 16 + 4t \qquad \text{Combine like terms.}$$

Step 3 Isolate the variable.

$$32 - 16 = 16 + 4t - 16 \qquad \text{Subtract 16.}$$

$$16 = 4t \qquad \text{Combine like terms.}$$

$$\frac{16}{4} = \frac{4t}{4} \qquad \text{Divide by 4.}$$

$$4 = t$$

Step 4 Check.

To check, substitute 4 for t in the original equation. The solution set is $\{4\}$.

Concepts	**Examples**
Solving a Linear Inequality in One Variable	Solve $3(x + 2) - 5x \leq 12$.

Step 1 Simplify each side of the inequality by clearing parentheses, fractions, and decimals, as needed, and combining like terms.

Step 2 Use the addition property of inequality to get all terms with variables on one side and all terms without variables on the other side.

Step 3 Use the multiplication property of inequality to write the inequality in the form $x < k$ or $x > k$.

$$3x + 6 - 5x \leq 12 \qquad \text{Distributive property}$$
$$-2x + 6 \leq 12 \qquad \text{Combine like terms.}$$
$$-2x + 6 - 6 \leq 12 - 6 \qquad \text{Subtract 6.}$$
$$-2x \leq 6 \qquad \text{Combine like terms.}$$
$$\frac{-2x}{-2} \geq \frac{6}{-2} \qquad \begin{array}{l}\text{Divide by } -2.\\ \text{Change } \leq \text{ to } \geq.\end{array}$$
$$x \geq -3$$

If an inequality is multiplied or divided by a negative number, the direction of the inequality symbol must be reversed.

The solution set $[-3, \infty)$ is graphed below.

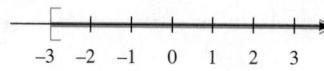

8.2 Set Operations and Compound Inequalities

Solving a Compound Inequality

Step 1 Solve each inequality in the compound inequality individually.

Step 2 If the inequalities are joined with *and*, the solution set is the intersection of the two individual solution sets.

If the inequalities are joined with *or*, the solution set is the union of the two individual solution sets.

Solve $x + 1 > 2$ and $2x < 6$.

$$x + 1 > 2 \quad \text{and} \quad 2x < 6$$
$$x > 1 \quad \text{and} \quad x < 3$$

The solution set is $(1, 3)$.

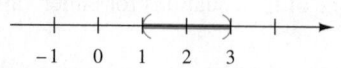

Solve $x \geq 4$ or $x \leq 0$.
The solution set is $(-\infty, 0] \cup [4, \infty)$.

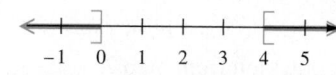

8.3 Absolute Value Equations and Inequalities

Let k be a positive number.
To solve $|ax + b| = k$, solve the compound equation

$$ax + b = k \quad \text{or} \quad ax + b = -k.$$

Solve $|x - 7| = 3$.

$$x - 7 = 3 \quad \text{or} \quad x - 7 = -3$$
$$x = 10 \quad \text{or} \quad x = 4$$

The solution set is $\{4, 10\}$.

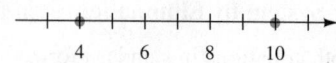

To solve $|ax + b| > k$, solve the compound inequality

$$ax + b > k \quad \text{or} \quad ax + b < -k.$$

Solve $|x - 7| > 3$.

$$x - 7 > 3 \quad \text{or} \quad x - 7 < -3$$
$$x > 10 \quad \text{or} \quad x < 4$$

The solution set is $(-\infty, 4) \cup (10, \infty)$.

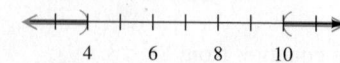

(continued)

Concepts	Examples

8.3 **Absolute Value Equations and Inequalities (continued)**

To solve $|ax + b| < k$, solve the compound inequality

$$-k < ax + b < k.$$

Solve $|x - 7| < 3$.

$$-3 < x - 7 < 3$$

$$4 < \quad x \quad < 10 \quad \text{Add 7 to each part.}$$

The solution set is $(4, 10)$.

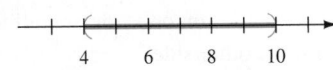

To solve an absolute value equation of the form

$$|ax + b| = |cx + d|,$$

solve the compound equation

$$ax + b = cx + d \quad \text{or} \quad ax + b = -(cx + d).$$

Solve $|x + 2| = |2x - 6|$.

$$x + 2 = 2x - 6 \quad \text{or} \quad x + 2 = -(2x - 6)$$

$$x = 8 \qquad \text{or} \quad x + 2 = -2x + 6$$

$$3x = 4$$

$$x = \frac{4}{3}$$

The solution set is $\left\{\frac{4}{3}, 8\right\}$.

8.4 **Review of Systems of Linear Equations in Two Variables**

Solving a Linear System by Substitution

Step 1 Solve one of the equations for either variable.

Step 2 Substitute for that variable in the other equation. The result should be an equation with just one variable.

Step 3 Solve the equation from Step 2.

Step 4 Find the value of the other variable by substituting the result from Step 3 into the equation from Step 1.

Step 5 Check the ordered-pair solution in *both* of the *original* equations. Then write the solution set.

Solving a Linear System by Elimination

Step 1 Write both equations in standard form.

Step 2 Make the coefficients of one pair of variable terms opposites.

Step 3 Add the new equations. The sum should be an equation with just one variable.

Step 4 Solve the equation from Step 3.

Solve by substitution.

$$4x - \quad y = 7 \quad (1)$$
$$3x + 2y = 30 \quad (2)$$

Solve for y in equation (1).

$$y = 4x - 7$$

Substitute $4x - 7$ for y in equation (2), and solve for x.

$$3x + 2y = 30 \quad (2)$$
$$3x + 2(4x - 7) = 30 \quad \text{Let } y = 4x - 7.$$
$$3x + 8x - 14 = 30 \quad \text{Distributive property}$$
$$11x - 14 = 30 \quad \text{Combine like terms.}$$
$$11x = 44 \quad \text{Add 14.}$$
$$x = 4 \quad \text{Divide by 11.}$$

Substitute 4 for x in the equation $y = 4x - 7$ to find that $y = 9$.

Check to verify that $\{(4, 9)\}$ is the solution set.

Solve by elimination.

$$5x + \quad y = 2 \quad (1)$$
$$2x - 3y = 11 \quad (2)$$

To eliminate y, multiply equation (1) by 3, and add the result to equation (2).

$$15x + 3y = \quad 6 \quad \text{3 times equation (1)}$$
$$\underline{2x - 3y = 11} \quad (2)$$
$$17x \qquad = 17 \quad \text{Add.}$$
$$x = 1 \quad \text{Divide by 17.}$$

Concepts	Examples

Step 5 Find the value of the other variable by substituting the result from Step 4 into either of the original equations.

If the result of the addition step (Step 3) is a false statement, such as $0 = 4$, the graphs are parallel lines and *there is no solution. The solution set is* $\emptyset$.

If the result is a true statement, such as $0 = 0$, the graphs are the same line, and an *infinite number of ordered pairs are solutions. The solution set is written in set-builder notation as*

$$\{(x, y) \mid \underline{\hspace{2cm}}\},$$

where a form of the equation is written in the blank.

Let $x = 1$ in equation (1), and solve for y.

$$5(1) + y = 2$$
$$y = -3$$

Check to verify that $\{(1, -3)\}$ is the solution set.

$$\begin{array}{r} x - 2y = 6 \\ -x + 2y = -2 \\ \hline 0 = 4 \end{array} \quad \text{Solution set: } \emptyset$$

$$\begin{array}{r} x - 2y = 6 \\ -x + 2y = -6 \\ \hline 0 = 0 \end{array} \quad \text{Solution set: } \{(x, y) \mid x - 2y = 6\}$$

Step 6 Check the ordered-pair solution in *both* of the *original* equations. Then write the solution set.

8.5 Systems of Linear Equations in Three Variables; Applications

Solving a Linear System in Three Variables

Step 1 Select a focus variable, preferably one with coefficient 1 or -1, and a working equation.

Step 2 Eliminate the focus variable, using the working equation and one of the equations of the system.

Step 3 Eliminate the focus variable again, using the working equation and the remaining equation of the system.

Step 4 Solve the system of two equations in two variables formed by the equations from Steps 2 and 3.

Step 5 Find the value of the remaining variable.

Step 6 Check the ordered-triple solution in each of the original equations of the system. Then write the solution set.

Solve the system.

$$\begin{array}{rl} x + 2y - z = 6 & (1) \\ x + y + z = 6 & (2) \\ 2x + y - z = 7 & (3) \end{array}$$

We choose z as the focus variable and (2) as the working equation.

Add equations (1) and (2).

$$2x + 3y = 12 \quad (4)$$

Add equations (2) and (3).

$$3x + 2y = 13 \quad (5)$$

Use equations (4) and (5) to eliminate x.

$$\begin{array}{rl} -6x - 9y = -36 & \text{Multiply (4) by } -3. \\ 6x + 4y = 26 & \text{Multiply (5) by 2.} \\ \hline -5y = -10 & \text{Add.} \\ y = 2 & \text{Divide by } -5. \end{array}$$

To find x, substitute 2 for y in equation (4).

$$\begin{array}{rl} 2x + 3(2) = 12 & \text{Let } y = 2 \text{ in (4).} \\ 2x + 6 = 12 & \text{Multiply.} \\ 2x = 6 & \text{Subtract 6.} \\ x = 3 & \text{Divide by 2.} \end{array}$$

Substitute 3 for x and 2 for y in working equation (2).

$$\begin{array}{rl} x + y + z = 6 & (2) \\ 3 + 2 + z = 6 & \text{Substitute.} \\ z = 1 & \text{Subtract 5.} \end{array}$$

A check of the solution $(3, 2, 1)$ confirms that the solution set is $\{(3, 2, 1)\}$.

Chapter 8 Review Exercises

8.1 *Solve each equation.*

1. $-(8 + 3x) + 5 = 2x + 6$

2. $\dfrac{m-2}{4} + \dfrac{m+2}{2} = 8$

3. $-(r + 5) - (2 + 7r) + 8r = 3r - 8$

4. $0.05x + 0.03(1200 - x) = 42$

Solve each equation. Decide whether it is a conditional equation, *an* identity, *or a* contradiction.

5. $7r - 3(2r - 5) + 5 + 3r = 4r + 20$

6. $8p - 4p - (p - 7) + 9p + 13 = 12p$

7. $-2r + 6(r - 1) + 3r - (4 - r) = -(r + 5) - 5$

8. $\dfrac{2}{3}x + \dfrac{5}{8}x = \dfrac{31}{24}x$

Solve each inequality. Give the solution set in both interval and graph forms.

9. $-\dfrac{2}{3}x < 6$

10. $-5x - 4 \geq 11$

11. $5 - (6 - 4t) \geq 2t - 7$

12. $-6 \leq 2k \leq 24$

13. $8 \leq 3x - 1 < 14$

14. $-4 < 3 - 2z < 9$

8.2 *Let* $A = \{a, b, c, d\}$, $B = \{a, c, e, f\}$, *and* $C = \{a, e, f, g\}$. *Find each set.*

15. $A \cap B$

16. $A \cap C$

17. $B \cup C$

18. $A \cup C$

Solve each compound inequality. Give the solution set in both interval and graph forms.

19. $x > 4$ and $x < 7$

20. $x + 4 > 12$ and $x - 2 < 12$

21. $x > 5$ or $x \le -3$

22. $x \ge -2$ or $x < 2$

23. $x - 4 > 6$ and $x + 3 \le 10$

24. $-5x + 1 \ge 11$ or $3x + 5 \ge 26$

Express each union or intersection in simplest interval form.

25. $(-3, \infty) \cap (-\infty, 4)$ **26.** $(-\infty, 6) \cap (-\infty, 2)$ **27.** $(4, \infty) \cup (9, \infty)$ **28.** $(1, 2) \cup (1, \infty)$

8.3 *Solve each absolute value equation.*

29. $|x| = 7$ **30.** $|x + 2| = 9$ **31.** $|3k - 7| = 8$ **32.** $|z - 4| = -12$

33. $|2k - 7| + 4 = 11$ **34.** $|4a + 2| - 7 = -3$ **35.** $|3p + 1| = |p + 2|$ **36.** $|5x + 8| = 0$

Solve each absolute value inequality. Give the solution set in both interval and graph forms.

37. $|x| < 12$

38. $|-x + 6| \le 7$

39. $|2p + 5| \le 1$

40. $|x + 1| \ge -3$

41. $|5r - 1| > 9$

42. $|3x + 6| \ge 0$

8.4 **43.** Solve by graphing: $x + 3y = 8$
$$2x - y = 2.$$

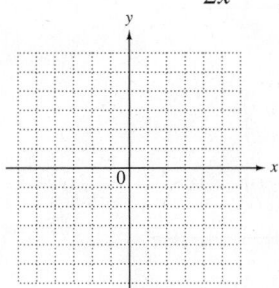

44. CONCEPT CHECK Which ordered pair is a solution of the following system?
$$3x + 2y = 6$$
$$2x - y = 11$$

A. $(2, 0)$ **B.** $(0, -11)$

C. $(4, -3)$ **D.** $(3, -2)$

Solve each system using the substitution method.

45. $3x + y = -4$

$$x = \frac{2}{3}y$$

46. $9x - y = -4$

$$y = x + 4$$

47. $-5x + 2y = -2$

$$x + 6y = 26$$

Solve each system using the elimination method. If a system is inconsistent or has dependent equations, say so.

48. $6x + 5y = 4$

   $-4x + 2y = 8$

49. $\dfrac{x}{6} + \dfrac{y}{6} = -\dfrac{1}{2}$

   $x - y = -9$

50. $4x + 5y = 9$

   $3x + 7y = -1$

51. $-3x + y = 6$

   $2y = 12 + 6x$

52. $5x - 4y = 2$

   $-10x + 8y = 7$

53. $3x + 3y = 0$

   $-2x - y = 0$

8.5 *Solve each system of equations. If a system is inconsistent or has dependent equations, say so.*

54. $2x + 3y - z = -16$

   $x + 2y + 2z = -3$

   $-3x + y + z = -5$

55. $3x - y - z = -8$

   $4x + 2y + 3z = 15$

   $-6x + 2y + 2z = 10$

56. $4x - y = 2$

   $3y + z = 9$

   $x + 2z = 7$

Solve each problem using a system of equations.

57. The sum of the measures of the angles of a triangle is 180°. The largest angle measures 10° less than the sum of the other two. The measure of the middle-sized angle is the average of the other two. Find the measures of the three angles.

58. In the great baseball year of 1961, Yankee teammates Mickey Mantle, Roger Maris, and Yogi Berra combined for 137 home runs. Mantle hit 7 fewer than Maris. Maris hit 39 more than Berra. What were the home run totals for each player? (*Source:* Neft, David S. and Richard M. Cohen, *The Sports Encyclopedia: Baseball.*)

Mixed Review Exercises

Solve each equation or inequality.

59. $(7 - 2k) + 3(5 - 3k) = k + 8$ **60.** $-5(6p + 4) - 2p = -32p + 14$ **61.** $x < 5$ and $x \geq -4$

62. $-5r \geq -10$ **63.** $|7x - 2| > 9$ **64.** $|2x - 10| = 20$

65. $|m + 3| \leq 13$ **66.** $x \geq -2$ or $x < 4$ **67.** $|m - 1| = |2m + 3|$ **68.** $-6 \leq 3x - 5 \leq 8$

In Exercises 69 and 70, sketch the graph of each solution set.

69. $x > 6$ and $x < 8$ **70.** $-5x + 1 \geq 6$ or $3x + 5 \geq 26$

Solve by any method.

71. $2x - 5y = 8$ **72.** $x = 7y + 10$ **73.** $x + 4y = 17$
    $3x + 4y = 10$ $2x + 3y = 3$ $-3x + 2y = -9$

74. $-7x + 3y = 12$ **75.** $3x - 4y + z = 8$ **76.** $2x - y + 3z = 0$
    $5x + 2y = 8$ $-6x + 8y - 2z = -16$ $5x + y - z = 0$
                                       $\dfrac{3}{2}x - 2y + \dfrac{1}{2}z = 4$ $-2x + 3y + 4z = 0$

77. In the 2010 Vancouver Winter Olympics, Germany, the U.S.A., and Canada won a combined total of 93 medals. Germany won seven fewer medals than the U.S.A. Canada won 11 fewer medals than the U.S.A. How many medals did each country win? (*Source: World Almanac and Book of Facts.*)

78. CONCEPT CHECK If $k < 0$, find the solution set for each of the following.

   (a) $|5x + 3| < k$ **(b)** $|5x + 3| > k$

   (c) $|5x + 3| = k$

Chapter 8 *Test*  CHAPTER **Test Prep** VIDEO

The Chapter Test Prep Videos with test solutions are available on DVD, in MyMathLab, *and on* You Tube—*search "LialCombinedAlgebra" and click on "Channels."*

Solve each equation. In Problems 4–6, decide whether the equation is a conditional equation, *an* identity, *or a* contradiction.

1. $3(2x - 2) - 4(x + 6) = 4x + 8$

2. $0.08x + 0.06(x + 9) = 1.24$

3. $\dfrac{x + 6}{10} + \dfrac{x - 4}{15} = 1$

4. $3x - (2 - x) + 4x + 2 = 8x + 3$

5. $\dfrac{x}{3} + 7 = \dfrac{5x}{6} - 2 - \dfrac{x}{2} + 9$

6. $-4(2x - 6) = 5x + 24 - 7x$

Solve each inequality. Give the solution set in both interval and graph forms.

7. $2 + 4x \le 5x$

8. $4 - 6(x + 3) \le -2 - 3(x + 6) + 3x$

9. $-\dfrac{4}{7}x > -16$

10. $-6 \le \dfrac{4}{3}x - 2 \le 2$

Let $A = \{1, 2, 5, 7\}$ *and* $B = \{1, 5, 9, 12\}$. *Find each of the following.*

11. $A \cap B$

12. $A \cup B$

Solve each compound or absolute value inequality. Give the solution set in both interval and graph forms.

13. $3k \ge 6$ and $k - 4 < 5$

14. $-4x \le -24$ or $4x - 2 < 10$

15. $|4x + 3| \le 7$

16. $|5 - 6x| > 12$

17. $|-3x + 4| - 4 < -1$

18. $|7 - x| \le -1$

Solve each absolute value equation.

19. $|3k - 2| + 1 = 8$

20. $|3 - 5x| = |2x + 8|$

21. $|4x + 3| + 5 = 4$

22. If $k < 0$, find the solution set for each of the following.

(a) $|8x - 5| < k$

(b) $|8x - 5| > k$

(c) $|8x - 5| = k$

23. Use a graph to solve the system.

$$x + y = 7$$
$$x - y = 5$$

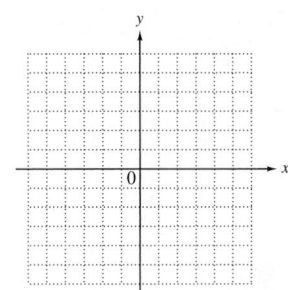

Solve each system by substitution or elimination. If a system is inconsistent or has dependent equations, say so.

24. $3x - y = -8$

$2x + 6y = 3$

25. $12x - 5y = 8$

$3x = \dfrac{5}{4}y + 2$

26. $-5x + 2y = -4$

$6x + 3y = -6$

27. $3x + 4y = 8$

$8y = 7 - 6x$

28. $3x + 5y + 3z = 2$

$6x + 5y + z = 0$

$3x + 10y - 2z = 6$

29. $4x + y + z = 11$

$x - y - z = 4$

$y + 2z = 0$

30. The owner of a tea shop wants to mix three kinds of tea to make 100 oz of a mixture that will sell for $0.83 per oz. He uses Orange Pekoe, which sells for $0.80 per oz, Irish Breakfast, for $0.85 per oz, and Earl Grey, for $0.95 per oz. If he wants to use twice as much Orange Pekoe as Irish Breakfast, how much of each kind of tea should he use?

Chapters R–8 Cumulative Review Exercises

Simplify each expression.

1. $-2(m-3)$

2. $-(-4m+3)$

3. $3x^2 - 4x + 4 + 9x - x^2$

Evaluate for $p = -4$, $q = -2$, and $r = 5$.

4. $-3(2q - 3p)$

5. $8r^2 + q^2$

6. $\dfrac{\sqrt{r}}{-p + 2q}$

7. $\dfrac{rp + 6r^2}{p^2 + q - 1}$

Solve.

8. $2z - 5 + 3z = 4 - (z + 2)$

9. $\dfrac{3a - 1}{5} + \dfrac{a + 2}{2} = -\dfrac{3}{10}$

10. $-\dfrac{4}{3}d \geq -5$

11. $3 - 2(m + 3) < 4m$

12. $2k + 4 < 10$ and $3k - 1 > 5$

13. $2k + 4 > 10$ or $3k - 1 < 5$

14. $\left|5x + 3\right| - 10 = 3$

15. $\left|x + 2\right| < 9$

16. $\left|2y - 5\right| \geq 9$

17. $V = lwh$ for h

18. Two planes leave the Dallas-Fort Worth airport at the same time. One travels east at 550 mph, and the other travels west at 500 mph. Assuming no wind, how long will it take for the planes to be 2100 mi apart?

	r	t	d
Eastbound plane	550	x	___
Westbound plane	500	x	___

19. Graph $4x + 2y = -8$.

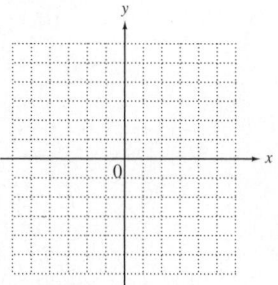

20. Find the slope of the line through the points $(-4, 8)$ and $(-2, 6)$.

21. What is the slope of the line shown here?

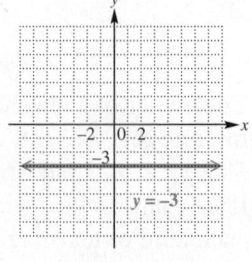

Use the function $f(x) = 2x + 7$ *to find the following.*

22. $f(-4)$ **23.** The x-intercept of its graph **24.** The y-intercept of its graph

Solve each system.

25. $3x - 2y = -7$
   $2x + 3y = 17$

26. $2x + 3y - 6z = 5$
   $8x - y + 3z = 7$
   $3x + 4y - 3z = 7$

Perform the indicated operations. Assume that variables represent nonzero real numbers.

27. $(3x^2y^{-1})^{-2}(2x^{-3}y)^{-1}$

28. $\dfrac{5m^{-2}y^3}{3m^{-3}y^{-1}}$

Perform the indicated operations.

29. $(3x^3 + 4x^2 - 7) - (2x^3 - 8x^2 + 3x)$ **30.** $(7x + 3y)^2$ **31.** $(2p + 3)(5p^2 - 4p - 8)$

Factor.

32. $16w^2 + 50wz - 21z^2$ **33.** $4y^2 - 36y + 81$ **34.** $8p^3 + 27$

Solve.

35. $(p + 4)(2p + 3)(p - 1) = 0$ **36.** $9q^2 = 6q - 1$ **37.** $\dfrac{1}{x} = \dfrac{1}{x + 1} + \dfrac{1}{2}$

Perform each operation, and write the answer in lowest terms.

38. $\dfrac{5}{q} - \dfrac{1}{q}$

39. $\dfrac{3}{7} + \dfrac{4}{r}$

40. $\dfrac{4}{5q - 20} - \dfrac{1}{3q - 12}$

41. $\dfrac{7z^2 + 49z + 70}{16z^2 + 72z - 40} \div \dfrac{3z + 6}{4z^2 - 1}$

42. Simplify the complex fraction $\dfrac{\dfrac{4}{a} + \dfrac{5}{2a}}{\dfrac{7}{6a} - \dfrac{1}{5a}}$.

Math in the Media

TO PASS OR TO FAIL? THAT IS THE QUESTION

A recent issue of *USA Today* included an article entitled "Great Education Debate: Reforming the Grade System." Pros and cons of a new movement to make 50% (rather than 0%) the minimum score for teachers to assign on graded assignments were discussed. Here is one typical example of a grading procedure.

An intermediate algebra teacher bases final grades on points earned for activities as given in the Graded Classwork table on the left. To determine final grades, the teacher strictly adheres to the point ranges given in the Grade Distribution table on the right.

GRADED CLASSWORK

Activity	Points Available
Homework and vocabulary	45
Daily activities (scaled)	55
Lab participation and completion	100
Major exams (3 at 100 points)	300
Final Exam	150
Total points	650

GRADE DISTRIBUTION

Grade	Points Required
A	585–650
B	520–584
C	455–519
IP*	< 455 and active
F	< 455 and inactive

*In Progress

Exams account for 450 of the possible 650 points.

Assumption: You earn a "baseline" number of points based on three criteria:

(1) You earn *all* of the homework and vocabulary points.
(2) You earn a minimum of 50 points based on daily activities.
(3) You earn a minimum of 90 lab participation and completion points.

1. Assume that you earn the baseline number of points. Let x = the test points to be earned. Write and solve linear inequalities to find the minimum number of points that you need in test scores to earn grades no lower than A, B, and C. What "test average" is each minimum score? Round *up* to the nearest whole percent.

2. Write and solve a compound inequality to find the range of points that you need in test scores to earn a B average. What range of "test averages" are those minimum scores? Round *up* to the nearest whole percent.

3. Suppose that Mark earns only 15 points in homework and vocabulary, 40 points in daily activities, and 50 points in lab participation. Write and solve linear inequalities to find the minimum number of points that Mark needs in test scores to earn grades no lower than A, B, and C. What "test average" is each minimum score? Round *up* to the nearest whole percent.

9 Roots, Radicals, and Root Functions

The formula for calculating the distance one can see to the horizon from the top of a tall building involves a *square root radical,* one of the topics covered in this chapter.

9.1 Radical Expressions and Graphs

9.2 Rational Exponents

9.3 Simplifying Radical Expressions

9.4 Adding and Subtracting Radical Expressions

9.5 Multiplying and Dividing Radical Expressions

Summary Exercises *Performing Operations with Radicals and Rational Exponents*

9.6 Solving Equations with Radicals

9.7 Complex Numbers

9.1 Radical Expressions and Graphs

OBJECTIVES

1. Find square roots.
2. Decide whether a given root is rational, irrational, or not a real number.
3. Find cube, fourth, and other roots.
4. Graph functions defined by radical expressions.
5. Find nth roots of nth powers.
6. Use a calculator to find roots.

OBJECTIVE ▶ 1 **Find square roots.** Recall that *squaring* a number means multiplying the number by itself.

7^2 means $7 \cdot 7$, which equals 49. The square of 7 is 49.

The opposite (inverse) of squaring a number is taking its *square root*. This is equivalent to asking

"What number when multiplied by itself equals 49?"

For the example above, one answer is 7, since $7 \cdot 7 = 49$.
This discussion can be generalized.

Square Root

A number b is a **square root** of a if $b^2 = a$.

EXAMPLE 1 Finding All Square Roots of a Number

Find all square roots of 49.
　　We ask, "What number when multiplied by itself equals 49?" As mentioned above, one square root is 7. Another square root of 49 is -7, because

$$(-7)(-7) = 49.$$

Thus, the number 49 has *two* square roots: 7 and -7. One square root is positive, and one is negative.

◀ Work Problem ❶ at the Side.

❶ Find all square roots of the given number.

(a) 100
Ask "What number when multiplied by itself equals 100?" There are two answers.

(b) 25　**(c)** 36　**(d)** $\dfrac{25}{36}$

The **positive** or **principal square root** of a number is written with the symbol $\sqrt{}$. For example, the positive square root of 121 is 11.

$$\sqrt{121} = 11 \qquad 11^2 = 121$$

The symbol $-\sqrt{}$ is used for the **negative square root** of a number. For example, the negative square root of 121 is -11.

$$-\sqrt{121} = -11 \qquad (-11)^2 = 121$$

The **radical symbol** $\sqrt{}$ always represents the positive square root (except that $\sqrt{0} = 0$). The number inside the radical symbol is the **radicand,** and the entire expression—radical symbol and radicand—is a **radical.**

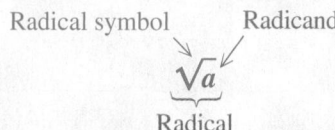

Radical symbol　　　Radicand
$$\sqrt{a}$$
Radical

An algebraic expression containing a radical is a **radical expression.**
　　The radical symbol $\sqrt{}$ has been used since sixteenth-century Germany and was probably derived from the letter R. The radical symbol at the right comes from the Latin word for root, *radix*. It was first used by Leonardo of Pisa (Fibonacci) in 1220.

**Leonardo of Pisa
(c. 1170–1250)**

Early radical symbol

Answers
1. **(a)** 10, -10　**(b)** 5, -5
　(c) 6, -6　**(d)** $\dfrac{5}{6}, -\dfrac{5}{6}$

We summarize our discussion of square roots as follows.

Square Roots of *a*

Let *a* be a positive real number.

$\quad\quad \sqrt{a}$ is the positive or principal square root of *a*.

$\quad\quad -\sqrt{a}$ is the negative square root of *a*.

For nonnegative *a*, the following hold.

$$\sqrt{a} \cdot \sqrt{a} = (\sqrt{a})^2 = a \quad \text{and} \quad -\sqrt{a} \cdot (-\sqrt{a}) = (-\sqrt{a})^2 = a$$

Also, $\sqrt{0} = 0$.

🖩 **Calculator Tip**

Most calculators have a square root key, usually labeled $\boxed{\sqrt{x}}$, for finding the square root of a number. On some models, the square root key must be used in conjunction with the key marked $\boxed{\text{INV}}$ or $\boxed{\text{2nd}}$.

EXAMPLE 2 **Finding Square Roots**

Find each square root.

(a) $\sqrt{144}$

The radical $\sqrt{144}$ represents the positive or principal square root of 144. Think of a positive number whose square is 144.

$$12^2 = 144, \quad \text{so} \quad \sqrt{144} = 12.$$

(b) $-\sqrt{1024}$

This symbol represents the negative square root of 1024. A calculator with a square root key can be used to find $\sqrt{1024} = 32$. Then, $-\sqrt{1024} = -32$.

(c) $\sqrt{\dfrac{4}{9}} = \dfrac{2}{3}$ 　　　**(d)** $-\sqrt{\dfrac{16}{49}} = -\dfrac{4}{7}$ 　　　**(e)** $\sqrt{0.81} = 0.9$

········· **Work Problem ❷ at the Side.** ▶

As noted above, when the square root of a positive real number is squared, the result is that positive real number. $\left(\text{Also, } (\sqrt{0})^2 = 0.\right)$

EXAMPLE 3 **Squaring Radical Expressions**

Find the *square* of each radical expression.

(a) $\sqrt{13}$　The square of $\sqrt{13}$ is $(\sqrt{13})^2 = 13$.　*Definition of square root*

(b) $-\sqrt{29}$

$(-\sqrt{29})^2 = 29$

The square of a *negative* number is positive.

(c) $\sqrt{p^2 + 1}$

$(\sqrt{p^2 + 1})^2 = p^2 + 1$

········· **Work Problem ❸ at the Side.** ▶

---

❷ Find each square root.

GS **(a)** $\sqrt{16}$

$$4^2 = \underline{\quad\quad}, \text{ so}$$
$$\sqrt{16} = \underline{\quad\quad}.$$

GS **(b)** $-\sqrt{169}$

$$13^2 = \underline{\quad\quad}, \text{ so}$$
$$-\sqrt{169} = \underline{\quad\quad}.$$

(c) $-\sqrt{225}$ 　　　**(d)** $\sqrt{729}$

(e) $-\sqrt{\dfrac{36}{25}}$ 　　　**(f)** $\sqrt{0.49}$

❸ Find the *square* of each radical expression.

GS **(a)** $\sqrt{41}$

$$(\sqrt{41})^2 = \underline{\quad\quad}$$

GS **(b)** $-\sqrt{39}$

$$(-\sqrt{39})^2 = \underline{\quad\quad}$$

(c) $\sqrt{120}$

(d) $\sqrt{2x^2 + 3}$

Answers

2. **(a)** 16; 4　**(b)** 169; −13　**(c)** −15
　(d) 27　**(e)** $-\dfrac{6}{5}$　**(f)** 0.7

3. **(a)** 41　**(b)** 39　**(c)** 120　**(d)** $2x^2 + 3$

4 Tell whether each square root is *rational, irrational,* or *not a real number.*

(a) $\sqrt{9}$

(b) $\sqrt{7}$

(c) $\sqrt{\dfrac{9}{16}}$

(d) $\sqrt{72}$

(e) $\sqrt{-43}$

Answers

4. (a) rational (b) irrational (c) rational
   (d) irrational (e) not a real number

OBJECTIVE **2** **Decide whether a given root is rational, irrational, or not a real number.** Numbers with rational square roots are **perfect squares.**

$$\left.\text{Perfect squares}\left\{\begin{array}{l} 25 \\[4pt] 144 \quad \text{are perfect squares since} \\[4pt] \dfrac{4}{9} \end{array}\right.\quad\begin{array}{l}\sqrt{25}=5 \\[4pt] \sqrt{144}=12 \\[4pt] \sqrt{\dfrac{4}{9}}=\dfrac{2}{3}\end{array}\right\}\begin{array}{l}\text{Rational}\\\text{square}\\\text{roots}\end{array}$$

A number that is not a perfect square has a square root that is not a rational number. For example, $\sqrt{5}$ is not a rational number because it cannot be written as the ratio of two integers. Its decimal equivalent (or approximation) neither terminates nor repeats. However, $\sqrt{5}$ is a real number and corresponds to a point on the number line.

A real number that is not rational is an **irrational number.** The number $\sqrt{5}$ is irrational. *Many square roots of integers are irrational.*

> If a is a *positive* real number that is *not* a perfect square, then
> $$\sqrt{a} \text{ is irrational.}$$

Not every number has a real number square root. For example, there is no real number that can be squared to obtain -36. (The square of a real number can never be negative.) Because of this, $\sqrt{-36}$ *is not a real number.*

> If a is a *negative* real number, then $\sqrt{a}$ is *not* a real number.

CAUTION

Do not confuse $\sqrt{-36}$ and $-\sqrt{36}$. $\sqrt{-36}$ is not a real number since there is no real number that can be squared to obtain -36. However, $-\sqrt{36}$ is the negative square root of 36, which is -6.

EXAMPLE 4 **Identifying Types of Square Roots**

Tell whether each square root is *rational, irrational,* or *not a real number.*

(a) $\sqrt{17}$ Because 17 is not a perfect square, $\sqrt{17}$ is irrational.

(b) $\sqrt{64}$ 64 is a perfect square, 8^2, so $\sqrt{64}=8$ is a rational number.

(c) $\sqrt{-25}$ There is no real number whose square is -25. Therefore, $\sqrt{-25}$ is not a real number.

◀ **Work Problem** **4** **at the Side.**

Note

Not all irrational numbers are square roots of integers. For example, π (approximately 3.14159) is an irrational number that is not a square root of any integer.

OBJECTIVE ③ **Find cube, fourth, and other roots.** Finding the square root of a number is the inverse (opposite) of squaring a number. There are inverses to finding the cube of a number and to finding the fourth or greater power of a number. These inverses are, respectively, the **cube root** $\sqrt[3]{a}$, and the **fourth root** $\sqrt[4]{a}$. Similar symbols are used for other roots.

$\sqrt[n]{a}$

The *n*th root of *a*, written $\sqrt[n]{a}$, is a number whose *n*th power equals *a*. That is,

$$\sqrt[n]{a} = b \quad \text{means} \quad b^n = a.$$

In $\sqrt[n]{a}$, the number *n* is the **index,** or **order,** of the radical.

We could write $\sqrt[2]{a}$ instead of $\sqrt{a}$, but the simpler symbol $\sqrt{a}$ is customary since the square root is the most commonly used root.

🖩 **Calculator Tip**

A calculator that has a key marked $\boxed{\sqrt[x]{y}}$, $\boxed{x^y}$, or $\boxed{y^x}$ (again perhaps in conjunction with the $\boxed{\text{INV}}$ or $\boxed{\text{2nd}}$ key) can be used to find other roots.

When working with cube roots or fourth roots, it is helpful to memorize the first few **perfect cubes** ($1^3 = 1$, $2^3 = 8$, $3^3 = 27$, and so on) and the first few **perfect fourth powers** ($1^4 = 1$, $2^4 = 16$, $3^4 = 81$, and so on).

Work Problem ⑤ at the Side. ▶

EXAMPLE 5   Finding Cube Roots

Find each cube root.

(a) $\sqrt[3]{8}$
   What number can be cubed to give 8? Because $2^3 = 8$, $\sqrt[3]{8} = 2$.

(b) $\sqrt[3]{-8} = -2$, because $(-2)^3 = -8$.

(c) $\sqrt[3]{216} = 6$, because $6^3 = 216$.

Work Problem ⑥ at the Side. ▶

Notice in **Example 5(b)** that we can find the cube root of a negative number. (Contrast this with the square root of a negative number, which is not real.) In fact, the cube root of a positive number is positive, and the cube root of a negative number is negative. ***There is only one real number cube root for each real number.***

When a radical has an *even index* (square root, fourth root, and so on), ***the radicand must be nonnegative*** to yield a real number root. Also, for $a > 0$,

$$\sqrt{a}, \ \sqrt[4]{a}, \ \sqrt[6]{a}, \text{ and so on are positive (principal) roots.}$$
$$-\sqrt{a}, \ -\sqrt[4]{a}, \ -\sqrt[6]{a}, \text{ and so on are negative roots.}$$

---

⑤ Complete the following list of perfect cubes and perfect fourth powers.

Perfect Cubes	Perfect Fourth Powers
$1^3 = 1$	$1^4 = 1$
$2^3 = 8$	$2^4 = 16$
$3^3 = 27$	$3^4 = 81$
$4^3 = \underline{\quad}$	$4^4 = \underline{\quad}$
$5^3 = \underline{\quad}$	$5^4 = \underline{\quad}$
$6^3 = \underline{\quad}$	$6^4 = \underline{\quad}$
$7^3 = \underline{\quad}$	$7^4 = \underline{\quad}$
$8^3 = \underline{\quad}$	$8^4 = \underline{\quad}$
$9^3 = \underline{\quad}$	$9^4 = \underline{\quad}$
$10^3 = \underline{\quad}$	$10^4 = \underline{\quad}$

⑥ Find each cube root.

ⒼⓈ **(a)** $\sqrt[3]{27}$

$$\underline{\quad}^3 = 27, \text{ so}$$
$$\sqrt[3]{27} = \underline{\quad}.$$

(b) $\sqrt[3]{1}$

(c) $\sqrt[3]{-125}$

Answers

5. Perfect cubes: 64; 125; 216; 343; 512; 729; 1000
   Perfect fourth powers: 256; 625; 1296; 2401; 4096; 6561; 10,000
6. **(a)** 3; 3   **(b)** 1   **(c)** −5

7 Find each root.

(a) $\sqrt[4]{81}$

(b) $\sqrt[4]{-81}$

(c) $-\sqrt[4]{81}$

(d) $\sqrt[5]{243}$

(e) $\sqrt[5]{-243}$

Answers

7. **(a)** 3 **(b)** not a real number
   (c) −3 **(d)** 3 **(e)** −3

EXAMPLE 6 Finding Other Roots

Find each root.

(a) $\sqrt[4]{16} = 2$, because 2 is positive and $2^4 = 16$.

(b) $-\sqrt[4]{16}$

From part (a), $\sqrt[4]{16} = 2$, so the negative root is

$$-\sqrt[4]{16} = -2.$$

(c) $\sqrt[4]{-16}$

For a real number fourth root, the radicand must be nonnegative. There is no real number that equals $\sqrt[4]{-16}$.

(d) $-\sqrt[5]{32}$

First find $\sqrt[5]{32}$. Because 2 is the number whose fifth power is 32, $\sqrt[5]{32} = 2$. Since $\sqrt[5]{32} = 2$, it follows that

$$-\sqrt[5]{32} = -2.$$

(e) $\sqrt[5]{-32} = -2$, because $(-2)^5 = -32$.

◀ **Work Problem 7 at the Side.**

OBJECTIVE 4 Graph functions defined by radical expressions. A **radical expression** is an algebraic expression that contains radicals.

$$3 - \sqrt{x}, \quad \sqrt[3]{x}, \quad \text{and} \quad \sqrt{2x - 1} \qquad \text{Radical expressions}$$

In earlier chapters we graphed functions defined by polynomial and rational expressions. Now we examine the graphs of functions defined by the radical expressions

$$f(x) = \sqrt{x} \quad \text{and} \quad f(x) = \sqrt[3]{x}.$$

Figure 1 shows the graph of the **square root function**

$$f(x) = \sqrt{x},$$

together with a table of selected points. Only nonnegative values can be used for x, so the domain is $[0, \infty)$. Because $\sqrt{x}$ is the principal square root of x, it always has a nonnegative value, so the range is also $[0, \infty)$.

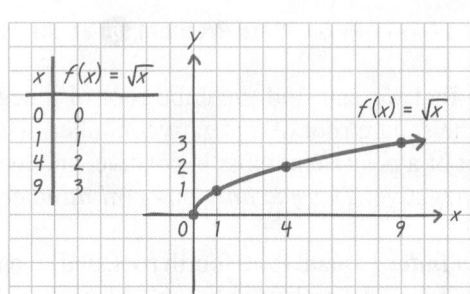

Square root function

$$f(x) = \sqrt{x}$$

Domain: $[0, \infty)$

Range: $[0, \infty)$

Figure 1

Figure 2 shows the graph of the **cube root function.** Since any real number (positive, negative, or 0) can be used for x in the cube root function, $\sqrt[3]{x}$ can be positive, negative, or 0. Thus, both the domain and the range of the cube root function are $(-\infty, \infty)$.

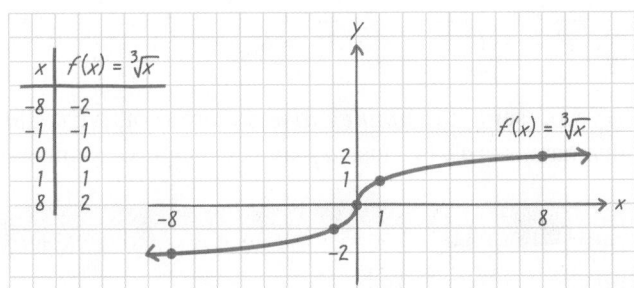

Figure 2

Cube root function

$$f(x) = \sqrt[3]{x}$$

Domain: $(-\infty, \infty)$

Range: $(-\infty, \infty)$

EXAMPLE 7 **Graphing Functions Defined with Radicals**

Graph each function by creating a table of values. Give the domain and the range.

(a) $f(x) = \sqrt{x - 3}$

A table of values is given with the graph in **Figure 3.** The x-values were chosen so that the function values are all integers. For the radicand to be nonnegative, we must have

$$x - 3 \geq 0, \quad \text{or} \quad x \geq 3.$$

Therefore, the domain is $[3, \infty)$. Function values are positive or 0, so the range is $[0, \infty)$.

x	$f(x) = \sqrt{x - 3}$
3	$\sqrt{3 - 3} = 0$
4	$\sqrt{4 - 3} = 1$
7	$\sqrt{7 - 3} = 2$

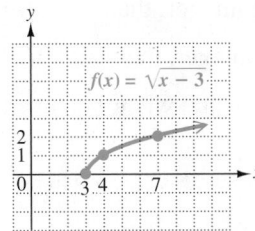

Figure 3

This graph is shifted 3 units to the right compared to the graph of $y = \sqrt{x}$.

(b) $f(x) = \sqrt[3]{x} + 2$

See **Figure 4.** Both the domain and the range are $(-\infty, \infty)$.

x	$f(x) = \sqrt[3]{x} + 2$
-8	$\sqrt[3]{-8} + 2 = 0$
-1	$\sqrt[3]{-1} + 2 = 1$
0	$\sqrt[3]{0} + 2 = 2$
1	$\sqrt[3]{1} + 2 = 3$
8	$\sqrt[3]{8} + 2 = 4$

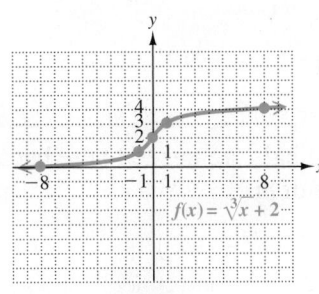

Figure 4

This graph is shifted 2 units up compared to the graph of $y = \sqrt[3]{x}$.

▸ **Work Problem** ❽ **at the Side.** ▶

---

❽ Graph each function by creating a table of values. Give the domain and the range.

(a) $f(x) = \sqrt{x} + 2$

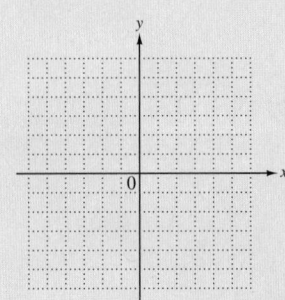

(b) $f(x) = \sqrt[3]{x} - 1$

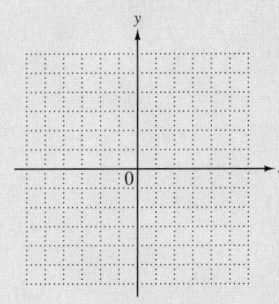

Answers

8. (a) domain: $[0, \infty)$; range: $[2, \infty)$

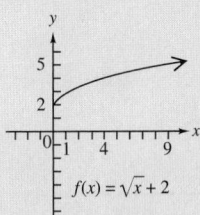

(b) domain: $(-\infty, \infty)$; range: $(-\infty, \infty)$

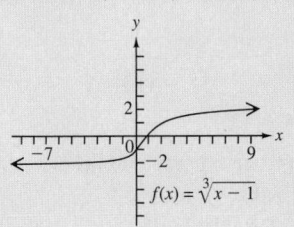

9 Find each square root. In parts (c) and (d), r is a real number.

(a) $\sqrt{15^2}$

(b) $\sqrt{(-12)^2}$

(c) $\sqrt{r^2}$

(d) $\sqrt{(-r)^2}$

10 Simplify each root.

(a) $\sqrt[4]{(-5)^4}$

(b) $\sqrt[5]{(-7)^5}$

(c) $-\sqrt[6]{(-3)^6}$

(d) $-\sqrt[4]{m^8}$

(e) $\sqrt[3]{x^{24}}$

(f) $\sqrt[6]{y^{18}}$

OBJECTIVE **5** **Find nth roots of nth powers.** Consider the expression $\sqrt{a^2}$. At first glance, we may think that it is equivalent to a. However, this is not necessarily true. For example, consider the following.

If $a = 6$,   then $\sqrt{a^2} = \sqrt{6^2} = \sqrt{36} = 6$.

If $a = -6$,   then $\sqrt{a^2} = \sqrt{(-6)^2} = \sqrt{36} = 6$. ← Instead of -6, we get 6, the *absolute value* of -6.

Since the symbol $\sqrt{a^2}$ represents the *nonnegative* square root, we write $\sqrt{a^2}$ with absolute value bars, as $|a|$, because a may be a negative number.

> **Meaning of $\sqrt{a^2}$**
>
> For any real number a, $\sqrt{a^2} = |a|$.
>
> In words, the principal square root of a^2 is the absolute value of a.

EXAMPLE 8   **Simplifying Square Roots by Using Absolute Value**

Find each square root. In parts (c) and (d), k is a real number.

(a) $\sqrt{7^2} = |7| = 7$ **(b)** $\sqrt{(-7)^2} = |-7| = 7$

(c) $\sqrt{k^2} = |k|$ **(d)** $\sqrt{(-k)^2} = |-k| = |k|$

◀ **Work Problem** **9** **at the Side.**

We can generalize this idea to any *n*th root.

> **Meaning of $\sqrt[n]{a^n}$**
>
> If n is an *even* positive integer, then $\sqrt[n]{a^n} = |a|$.
>
> If n is an *odd* positive integer, then $\sqrt[n]{a^n} = a$.
>
> In words, use absolute value when n is even. Absolute value is not necessary when n is odd.

EXAMPLE 9   **Simplifying Higher Roots by Using Absolute Value**

Simplify each root.

(a) $\sqrt[6]{(-3)^6} = |-3| = 3$ n is even. Use absolute value.

(b) $\sqrt[5]{(-4)^5} = -4$ n is odd.

(c) $-\sqrt[4]{(-9)^4} = -|-9| = -9$ n is even. Use absolute value.

(d) $-\sqrt{m^4} = -|m^2| = -m^2$ For all m, $|m^2| = m^2$.

No absolute value bars are needed here because m^2 is nonnegative for any real number value of m.

(e) $\sqrt[3]{a^{12}} = a^4$, because $a^{12} = (a^4)^3$.

(f) $\sqrt[4]{x^{12}} = |x^3|$

Absolute value bars guarantee that the result is not negative (because x^3 is negative when x is negative). Also, $|x^3|$ can be written as $x^2 \cdot |x|$.

◀ **Work Problem** **10** **at the Side.**

Answers

9. (a) 15 **(b)** 12 **(c)** $|r|$ **(d)** $|r|$

10. (a) 5 **(b)** -7 **(c)** -3

    (d) $-m^2$ **(e)** x^8 **(f)** $|y^3|$

OBJECTIVE ▶ ⑥ **Use a calculator to find roots.** Radical expressions often represent irrational numbers. To find approximations of such radicals, we usually use a calculator. For example,

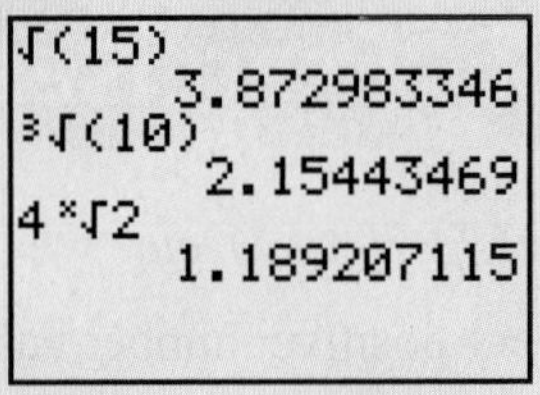

$$\sqrt{15} \approx 3.872983346, \quad \sqrt[3]{10} \approx 2.15443469, \quad \text{and} \quad \sqrt[4]{2} \approx 1.189207115,$$

where the symbol $\approx$ means **"is approximately equal to."** In this book, we often give approximations rounded to three decimal places. Thus,

$$\sqrt{15} \approx 3.873, \quad \sqrt[3]{10} \approx 2.154, \quad \text{and} \quad \sqrt[4]{2} \approx 1.189.$$

> ▦ **Calculator Tip**
>
> *When finding approximations, always consult your owner's manual for keystroke instructions.* Graphing calculators often differ from scientific calculators in the order in which keystrokes are made.

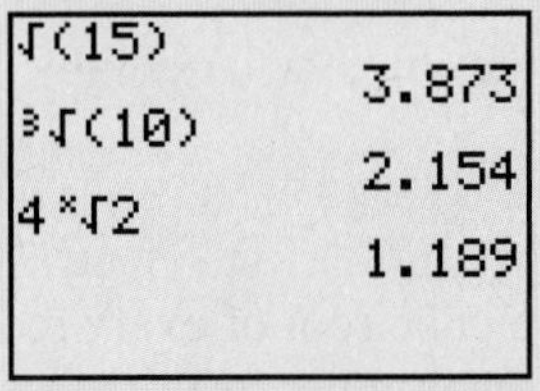

Figure 5 shows how the preceding approximations are displayed on a TI-83/84 Plus graphing calculator. In **Figure 5(a),** eight or nine decimal places are shown, while in **Figure 5(b),** the number of decimal places is fixed at three.

There is a simple way to check that a calculator approximation is "in the ballpark." For example, because 16 is slightly larger than 15, $\sqrt{16} = 4$ should be slightly larger than $\sqrt{15}$. Thus, 3.873 is a reasonable approximation for $\sqrt{15}$.

EXAMPLE 10   Finding Approximations for Roots

Use a calculator to verify that each approximation is correct.

(a) $\sqrt{39} \approx 6.245$ **(b)** $-\sqrt{72} \approx -8.485$

(c) $\sqrt[3]{93} \approx 4.531$ **(d)** $\sqrt[4]{39} \approx 2.499$

·················· **Work Problem ⑪ at the Side.** ▶

EXAMPLE 11   Using Roots to Calculate Resonant Frequency

In electronics, the resonant frequency f of a circuit may be found by the formula

$$f = \frac{1}{2\pi\sqrt{LC}},$$

where f is in cycles per second, L is in henrys, and C is in farads. (Henrys and farads are units of measure in electronics.) Find the resonant frequency f if $L = 5 \times 10^{-4}$ and $C = 3 \times 10^{-10}$. Give the answer to the nearest thousand.

Find the value of f when $L = 5 \times 10^{-4}$ and $C = 3 \times 10^{-10}$.

$$f = \frac{1}{2\pi\sqrt{LC}} \qquad \text{Given formula}$$

$$f = \frac{1}{2\pi\sqrt{(5 \times 10^{-4})(3 \times 10^{-10})}} \qquad \text{Substitute for } L \text{ and } C.$$

$$f \approx 411{,}000 \qquad \text{Use a calculator.}$$

The resonant frequency f is approximately 411,000 cycles per sec.

·················· **Work Problem ⑫ at the Side.** ▶

(a)

(b)
Figure 5

⑪ Use a calculator to approximate each radical to three decimal places.

(a) $\sqrt{17}$

(b) $-\sqrt{362}$

(c) $\sqrt[3]{9482}$

(d) $\sqrt[4]{6825}$

⑫ Use the formula in **Example 11** to approximate f to the nearest thousand if

$$L = 6 \times 10^{-5}$$

and $C = 4 \times 10^{-9}$.

Answers

11. (a) 4.123   **(b)** -19.026
    (c) 21.166   **(d)** 9.089
12. 325,000 cycles per sec

9.1 Exercises

CONCEPT CHECK *Decide whether each statement is* true *or* false. *If false, tell why.*

1. Every positive number has two real square roots.

2. A negative number has negative square roots.

3. Every nonnegative number has two real square roots.

4. The positive square root of a positive number is its principal square root.

5. The cube root of every real number has the same sign as the number itself.

6. Every positive number has three real cube roots.

CONCEPT CHECK *Match each expression from Column I with the equivalent choice from Column II. Answers may be used once, more than once, or not at all.*

I		II	
7. $-\sqrt{16}$	**8.** $\sqrt{-16}$	**A.** 3	**B.** -2
9. $\sqrt[3]{-27}$	**10.** $\sqrt[5]{-32}$	**C.** 2	**D.** -3
11. $\sqrt[4]{16}$	**12.** $-\sqrt[3]{64}$	**E.** -4	**F.** Not a real number

CONCEPT CHECK *Choose the closest approximation of each square root. Do not use a calculator.*

13. $\sqrt{123.5}$

   A. 9   **B.** 10   **C.** 11   **D.** 12

14. $\sqrt{67.8}$

   A. 7   **B.** 8   **C.** 9   **D.** 10

CONCEPT CHECK *Refer to the figure to answer the questions in Exercises 15 and 16.*

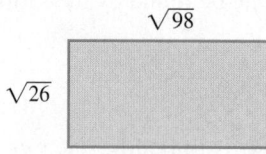

15. Which one of the following is the best estimate of its area?

   A. 2500   **B.** 250   **C.** 50   **D.** 100

16. Which one of the following is the best estimate of its perimeter?

   A. 15   **B.** 250   **C.** 100   **D.** 30

17. CONCEPT CHECK Consider the expression $-\sqrt{-a}$. Decide whether it is positive, negative, 0, or not a real number in each case.

   (a) $a > 0$ **(b)** $a < 0$ **(c)** $a = 0$

18. CONCEPT CHECK If n is odd, under what conditions is $\sqrt[n]{a}$ the following?

   (a) positive   **(b)** negative   **(c)** 0

Find all square roots of each number. **See Example 1.**

19. 9   **20.** 16   **21.** 64   **22.** 100   **23.** 169

24. 225   **25.** $\dfrac{25}{196}$ **26.** $\dfrac{81}{400}$ **27.** 900   **28.** 1600

Find each square root. **See Examples 2 and 4(c).**

29. $\sqrt{64}$
  $8^2 = \underline{\quad}$, so
  $\sqrt{64} = \underline{\quad}$.

30. $\sqrt{9}$
  $3^2 = \underline{\quad}$, so
  $\sqrt{9} = \underline{\quad}$.

31. $\sqrt{1}$

32. $\sqrt{4}$

33. $\sqrt{49}$ **34.** $\sqrt{81}$ **35.** $-\sqrt{256}$ **36.** $-\sqrt{196}$

37. $-\sqrt{\dfrac{144}{121}}$ **38.** $-\sqrt{\dfrac{49}{36}}$ **39.** $\sqrt{0.64}$ **40.** $\sqrt{0.16}$

41. $\sqrt{-121}$ **42.** $\sqrt{-64}$ **43.** $-\sqrt{-49}$ **44.** $-\sqrt{-100}$

Find the square of each radical expression. **See Example 3.**

45. $\sqrt{100}$
  $\left(\sqrt{100}\right)^2 = \underline{\quad}$

46. $\sqrt{36}$
  $\left(\sqrt{36}\right)^2 = \underline{\quad}$

47. $-\sqrt{19}$

48. $-\sqrt{99}$

49. $\sqrt{\dfrac{2}{3}}$ **50.** $\sqrt{\dfrac{5}{7}}$ **51.** $\sqrt{3x^2 + 4}$ **52.** $\sqrt{9y^2 + 3}$

Determine whether each number is rational, irrational, *or* not a real number. *If a number is rational, give its exact value. If a number is irrational, give a decimal approximation to the nearest thousandth. Use a calculator as necessary.* **See Examples 4 and 10.**

53. $\sqrt{25}$ **54.** $\sqrt{169}$ **55.** $\sqrt{29}$ **56.** $\sqrt{33}$

57. $-\sqrt{64}$ **58.** $-\sqrt{81}$ **59.** $-\sqrt{300}$ **60.** $-\sqrt{500}$

61. $\sqrt{-29}$ **62.** $\sqrt{-47}$ **63.** $\sqrt{1200}$ **64.** $\sqrt{1500}$

Find each root that is a real number. Use a calculator as necessary.
See Examples 5 and 6.

65. $\sqrt[3]{216}$

66. $\sqrt[3]{343}$

67. $\sqrt[3]{-64}$

68. $\sqrt[3]{-125}$

69. $-\sqrt[3]{512}$

70. $-\sqrt[3]{1000}$

71. $\sqrt[4]{1296}$

72. $\sqrt[4]{625}$

73. $-\sqrt[4]{16}$

74. $-\sqrt[4]{256}$

75. $\sqrt[4]{-625}$

76. $\sqrt[4]{-256}$

77. $\sqrt[6]{729}$

78. $\sqrt[6]{64}$

79. $\sqrt[6]{-64}$

80. $\sqrt[6]{-1}$

81. $\sqrt[3]{\dfrac{64}{27}}$

82. $\sqrt[4]{\dfrac{81}{16}}$

83. $-\sqrt[6]{\dfrac{1}{64}}$

84. $-\sqrt[5]{\dfrac{1}{32}}$

85. $\sqrt[3]{0.001}$

86. $\sqrt[3]{0.125}$

Graph each function and give its domain and range. ***See Example 7.***

87. $f(x) = \sqrt{x + 3}$

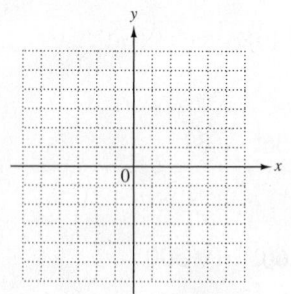

88. $f(x) = \sqrt{x - 2}$

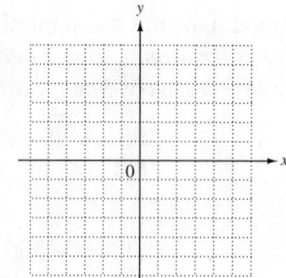

89. $f(x) = \sqrt{x} - 2$

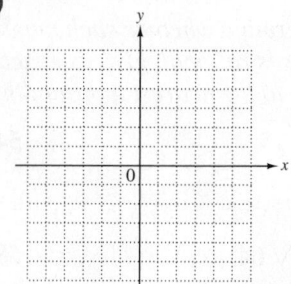

90. $f(x) = \sqrt{x} + 2$

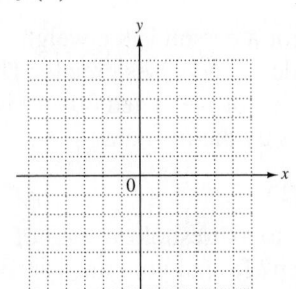

91. $f(x) = \sqrt[3]{x} - 3$

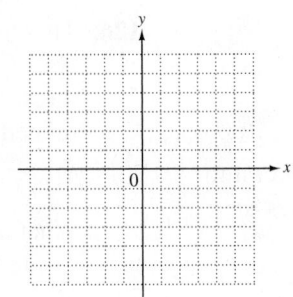

92. $f(x) = \sqrt[3]{x} + 1$

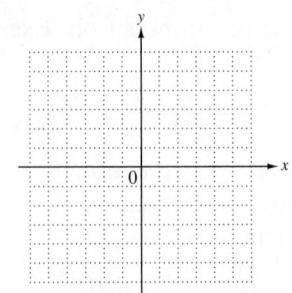

Simplify each root. ***See Examples 8 and 9.***

93. $\sqrt{12^2}$

94. $\sqrt{19^2}$

95. $\sqrt{(-10)^2}$

96. $\sqrt{(-13)^2}$

97. $\sqrt[6]{(-2)^6}$

98. $\sqrt[6]{(-4)^6}$

99. $\sqrt[5]{(-9)^5}$

100. $\sqrt[5]{(-8)^5}$

101. $-\sqrt[6]{(-5)^6}$

102. $-\sqrt[6]{(-7)^6}$

103. $\sqrt{x^2}$

104. $-\sqrt{x^2}$

105. $\sqrt{(-z)^2}$

106. $\sqrt{(-q)^2}$

107. $\sqrt[3]{x^3}$

108. $-\sqrt[3]{x^3}$

109. $\sqrt[3]{x^{15}}$

110. $\sqrt[3]{m^9}$

111. $\sqrt[6]{x^{30}}$

112. $\sqrt[4]{k^{20}}$

▦ *Find a decimal approximation for each radical. Round answers to three decimal places.* ***See Example 10.***

113. $\sqrt{9483}$

114. $\sqrt{6825}$

115. $\sqrt{284.361}$

116. $\sqrt{846.104}$

117. $-\sqrt{82}$

118. $-\sqrt{91}$

119. $\sqrt[3]{423}$

120. $\sqrt[3]{555}$

121. $\sqrt[4]{100}$

122. $\sqrt[4]{250}$

123. $\sqrt[5]{23.8}$

124. $\sqrt[5]{98.4}$

Solve each problem. See Example 11.

125. Use the formula from **Example 11**

$$f = \frac{1}{2\pi\sqrt{LC}}$$

to calculate the resonant frequency f of a circuit to the nearest thousand for the following values of L and C.

(a) $L = 7.237 \times 10^{-5}$ and $C = 2.5 \times 10^{-10}$

(b) $L = 5.582 \times 10^{-4}$ and $C = 3.245 \times 10^{-9}$

126. The threshold weight T for a person is the weight above which the risk of death increases greatly. The threshold weight in pounds for men aged 40–49 is related to height in inches by the formula

$$h = 12.3\sqrt[3]{T}.$$

What height corresponds to a threshold weight of 216 lb for a 43-yr-old man? Round your answer to the nearest inch, and then to the nearest tenth of a foot.

127. According to an article in *The World Scanner Report,* the distance D, in miles, to the horizon from an observer's point of view over water or "flat" earth is given by

$$D = \sqrt{2H},$$

where H is the height of the point of view, in feet. If a person whose eyes are 6 ft above ground level is standing at the top of a hill 44 ft above "flat" earth, approximately how far to the horizon will she be able to see?

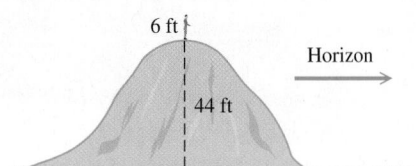

128. The time for one complete swing of a simple pendulum is given by

$$t = 2\pi\sqrt{\frac{L}{g}},$$

where t is time in seconds, L is the length of the pendulum in feet, and g, the force due to gravity, is about 32 ft per sec^2. Find the time of a complete swing of a 2-ft pendulum to the nearest tenth of a second.

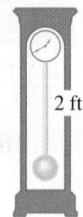

129. **Heron's formula** gives a method of finding the area of a triangle if the lengths of its sides are known. Suppose that a, b, and c are the lengths of the sides. Let s denote one-half of the perimeter of the triangle (called the **semiperimeter**)—that is,

$$s = \frac{1}{2}(a + b + c).$$

Then the area of the triangle is

$$A = \sqrt{s(s - a)(s - b)(s - c)}.$$

Find the area of the Bermuda Triangle, to the nearest thousand square miles, if the "sides" of this triangle measure approximately 850 mi, 925 mi, and 1300 mi.

130. The Vietnam Veterans' Memorial in Washington, D.C., is in the shape of an unenclosed isosceles triangle with equal sides of length 246.75 ft. If the triangle were enclosed, the third side would have length 438.14 ft. Use Heron's formula from **Exercise 129** to find the area of this enclosure to the nearest hundred square feet. (*Source:* Information pamphlet obtained at the Vietnam Veterans' Memorial.)

Not to scale

9.2 Rational Exponents

OBJECTIVE ▶ 1 **Use exponential notation for *n*th roots.** Consider the product $(3^{1/2})^2 = 3^{1/2} \cdot 3^{1/2}$. Using the rules of exponents from **Section 5.2**, extended to rational exponents, we simplify as follows.

$$(3^{1/2})^2 = 3^{1/2} \cdot 3^{1/2} \qquad a^2 = a \cdot a$$
$$= 3^{1/2+1/2} \qquad \text{Product rule: } a^m \cdot a^n = a^{m+n}$$
$$= 3^1 \qquad \text{Add exponents.}$$
$$= 3 \qquad a^1 = a$$

Also, by definition,

$$\left(\sqrt{3}\right)^2 = \sqrt{3} \cdot \sqrt{3} = 3.$$

Since both $(3^{1/2})^2$ and $\left(\sqrt{3}\right)^2$ equal 3, it seems reasonable to define

$$3^{1/2} = \sqrt{3}.$$

This suggests the following generalization.

Meaning of $a^{1/n}$

If $\sqrt[n]{a}$ is a real number, then $\qquad a^{1/n} = \sqrt[n]{a}.$

$$4^{1/2} = \sqrt{4}, \quad 8^{1/3} = \sqrt[3]{8}, \quad \text{and} \quad 16^{1/4} = \sqrt[4]{16} \qquad \text{Examples of } a^{1/n}$$

The denominator of the rational exponent is the index of the radical.

EXAMPLE 1 **Evaluating Exponentials of the Form $a^{1/n}$**

Evaluate each exponential.

The denominator is the index, or root.

The denominator is the index, or root. $\sqrt{}$ means $\sqrt[2]{}$.

(a) $64^{1/3} = \sqrt[3]{64} = 4$ **(b)** $100^{1/2} = \sqrt{100} = 10$

(c) $-256^{1/4} = -\sqrt[4]{256} = -4$

(d) $(-256)^{1/4} = \sqrt[4]{-256}$ is not a real number, because the radicand, -256, is negative and the index is even.

(e) $(-32)^{1/5} = \sqrt[5]{-32} = -2$ **(f)** $\left(\dfrac{1}{8}\right)^{1/3} = \sqrt[3]{\dfrac{1}{8}} = \dfrac{1}{2}$

···· **Work Problem** ❶ **at the Side.** ▶

CAUTION

Notice the distinction between **Examples 1(c) and (d).** The radical in part (c) is the ***negative fourth root of a positive number,*** while the radical in part (d) is the ***principal fourth root of a negative number,*** ***which is not a real number.***

OBJECTIVES

1. Use exponential notation for *n*th roots.

2. Define and use expressions of the form $a^{m/n}$.

3. Convert between radicals and rational exponents.

4. Use the rules for exponents with rational exponents.

---

❶ Evaluate each exponential.

(a) $8^{1/3}$

(b) $9^{1/2}$

(c) $-81^{1/4}$

(d) $(-81)^{1/4}$

(e) $(-64)^{1/3}$

(f) $\left(\dfrac{1}{32}\right)^{1/5}$

Answers

1. **(a)** 2   **(b)** 3   **(c)** -3
    (d) not a real number
    (e) -4 **(f)** $\dfrac{1}{2}$

❷ Evaluate each exponential.

GS **(a)** $25^{3/2}$

Use the definitions

$a^{1/n} = \sqrt[n]{a}$ and

$a^{m/n} = (a^{1/n})^m$.

$25^{3/2}$

$= (25^{1/\text{—}})\text{—}$

$= \underline{\hphantom{xx}} \text{—}$

(because $25^{1/2} = \sqrt{25} = \underline{\hphantom{xx}}$)

$= \underline{\hphantom{xx}}$

(b) $27^{2/3}$

(c) $-16^{3/2}$

(d) $(-64)^{2/3}$

(e) $(-36)^{3/2}$

OBJECTIVE ❷ **Define and use expressions of the form $a^{m/n}$.** We know that $8^{1/3} = \sqrt[3]{8}$. Now we define a number like $8^{2/3}$, where the numerator of the exponent is *not* 1. For past rules of exponents to be valid,

$$8^{2/3} = 8^{(1/3)2} = (8^{1/3})^2.$$

Since $8^{1/3} = \sqrt[3]{8}$,

$$8^{2/3} = \left(\sqrt[3]{8}\right)^2 = 2^2 = 4.$$

Generalizing from this example, we define $a^{m/n}$ as follows.

Meaning of $a^{m/n}$

If m and n are positive integers with m/n in lowest terms, then

$$a^{m/n} = (a^{1/n})^m,$$

provided that $a^{1/n}$ is a real number. If $a^{1/n}$ is not a real number, then $a^{m/n}$ is not a real number.

EXAMPLE 2 Evaluating Exponentials of the Form $a^{m/n}$

Evaluate each exponential.

> Think:
> $36^{1/2} = \sqrt{36} = 6$

> Think:
> $125^{1/3} = \sqrt[3]{125} = 5$

(a) $36^{3/2} = (36^{1/2})^3 = 6^3 = 216$ **(b)** $125^{2/3} = (125^{1/3})^2 = 5^2 = 25$

> Be careful.
> The base is 4.

(c) $-4^{5/2} = -(4^{5/2}) = -(4^{1/2})^5 = -(2)^5 = -32$
Because the base is 4, the negative sign is *not* affected by the exponent.

(d) $(-27)^{2/3} = \left[(-27)^{1/3}\right]^2 = (-3)^2 = 9$

Notice in part (c) that we first evaluate the exponential and then find its negative. In part (d), the $-$ sign is part of the base, -27.

(e) $(-100)^{3/2} = \left[(-100)^{1/2}\right]^3$, which is *not* a real number, because

$$(-100)^{1/2}, \quad \text{or} \quad \sqrt{-100}, \quad \text{is not a real number.}$$

◀ **Work Problem** ❷ **at the Side.**

Recall from **Section 5.5** that for any natural number n,

$$a^{-n} = \frac{1}{a^n} \quad (\text{where } a \neq 0).$$

When a rational exponent is negative, we apply this interpretation of negative exponents.

Meaning of $a^{-m/n}$

If $a^{m/n}$ is a real number, then

$$a^{-m/n} = \frac{1}{a^{m/n}} \quad (\text{where } a \neq 0).$$

Answers

2. (a) 2; 3; 5; 3; 5; 125
  (b) 9 **(c)** -64 **(d)** 16
  (e) not a real number

EXAMPLE 3 Evaluating Exponentials of the Form $a^{-m/n}$

Evaluate each exponential.

(a) $16^{-3/4} = \dfrac{1}{16^{3/4}} = \dfrac{1}{(16^{1/4})^3} = \dfrac{1}{(\sqrt[4]{16})^3} = \dfrac{1}{2^3} = \dfrac{1}{8}$

> The denominator of 3/4 is the index and the numerator is the exponent.

(b) $25^{-3/2} = \dfrac{1}{25^{3/2}} = \dfrac{1}{(25^{1/2})^3} = \dfrac{1}{(\sqrt{25})^3} = \dfrac{1}{5^3} = \dfrac{1}{125}$

(c) $\left(\dfrac{8}{27}\right)^{-2/3} = \dfrac{1}{\left(\dfrac{8}{27}\right)^{2/3}} = \dfrac{1}{\left(\sqrt[3]{\dfrac{8}{27}}\right)^2} = \dfrac{1}{\left(\dfrac{2}{3}\right)^2} = \dfrac{1}{\dfrac{4}{9}} = \dfrac{9}{4}$

> $\dfrac{1}{\frac{4}{9}} = 1 \div \dfrac{4}{9} = 1 \cdot \dfrac{9}{4}$

We could also use the rule $\left(\dfrac{b}{a}\right)^{-m} = \left(\dfrac{a}{b}\right)^m$ here, as follows.

$\left(\dfrac{8}{27}\right)^{-2/3} = \left(\dfrac{27}{8}\right)^{2/3} = \left(\sqrt[3]{\dfrac{27}{8}}\right)^2 = \left(\dfrac{3}{2}\right)^2 = \dfrac{9}{4}$

> Take the reciprocal only of the base, *not* the exponent.

· · · · · · · · · · · · · **Work Problem 3** at the Side. ▶

CAUTION

Be careful to distinguish between exponential expressions like the following.

$16^{-1/4}$, which equals $\dfrac{1}{2}$, $-16^{1/4}$, which equals -2, and

$-16^{-1/4}$, which equals $-\dfrac{1}{2}$

A negative exponent does not necessarily lead to a negative result. Negative exponents lead to reciprocals, which may be positive.

We obtain an alternative meaning of $a^{m/n}$ by applying the power rule a little differently than in the earlier definition.

Alternative Meaning of $a^{m/n}$

If all indicated roots are real numbers, then

$$a^{m/n} = (a^{1/n})^m = (a^m)^{1/n}.$$

As a result, we can evaluate an expression such as $27^{2/3}$ in two ways.

$27^{2/3} = (27^{1/3})^2 = 3^2 = 9$

or $\qquad 27^{2/3} = (27^2)^{1/3} = 729^{1/3} = 9$

> The result is the same.

In most cases, it is easier to use $(a^{1/n})^m$.

3 Evaluate each exponential.

(a) $36^{-3/2}$

Use the definition of a negative exponent $a^{-n} = \dfrac{1}{a^n}$.

$36^{-3/2}$

$= \dfrac{1}{36^{\underline{}}}$

$= \dfrac{1}{(36^{\underline{}})^{\underline{}}}$

$= \dfrac{1}{(\sqrt{36})^{\underline{}}}$

$= \dfrac{1}{\underline{}^{\underline{}}}$

$= \underline{}$

(b) $32^{-4/5}$

(c) $\left(\dfrac{4}{9}\right)^{-5/2}$

4 Write each exponential as a radical. Assume that all variables represent positive real numbers. Use the definition that takes the root first.

(a) $19^{1/2}$

(b) $5^{2/3}$

(c) $4k^{3/5}$

(d) $5x^{3/5} - (2x)^{3/5}$

(e) $x^{-5/7}$

(f) $(m^3 + n^3)^{1/3}$

5 Write each radical as an exponential and simplify. Assume that all variables represent positive real numbers.

(a) $\sqrt{37}$

(b) $\sqrt[4]{9^8}$

(c) $\sqrt[4]{t^4}$

Answers

4. **(a)** $\sqrt{19}$ **(b)** $\left(\sqrt[3]{5}\right)^2$ **(c)** $4\left(\sqrt[5]{k}\right)^3$
  (d) $5\left(\sqrt[5]{x}\right)^3 - \left(\sqrt[5]{2x}\right)^3$
  (e) $\dfrac{1}{\left(\sqrt[7]{x}\right)^5}$ **(f)** $\sqrt[3]{m^3 + n^3}$

5. **(a)** $37^{1/2}$ **(b)** 9^2, or 81 **(c)** t

> **Radical Form of $a^{m/n}$**
>
> If all indicated roots are real numbers, then
> $$a^{m/n} = \sqrt[n]{a^m} = \left(\sqrt[n]{a}\right)^m.$$
>
> In words, raise a to the mth power and then take the nth root, or take the nth root of a and then raise to the mth power.

For example,
$$8^{2/3} = \sqrt[3]{8^2} = \sqrt[3]{64} = 4, \quad \text{and} \quad 8^{2/3} = \left(\sqrt[3]{8}\right)^2 = 2^2 = 4,$$
so
$$8^{2/3} = \sqrt[3]{8^2} = \left(\sqrt[3]{8}\right)^2.$$

OBJECTIVE **3** **Convert between radicals and rational exponents.** Using the definition of rational exponents, we can simplify many problems involving radicals by converting the radicals to numbers with rational exponents. After simplifying, we convert the answer back to radical form if required.

EXAMPLE 4 **Converting between Rational Exponents and Radicals**

In parts (a)–(f), write each exponential as a radical. Assume that all variables represent positive real numbers. Use the definition that takes the root first.

(a) $13^{1/2} = \sqrt{13}$ **(b)** $6^{3/4} = \left(\sqrt[4]{6}\right)^3$ **(c)** $9m^{5/8} = 9\left(\sqrt[8]{m}\right)^5$

(d) $6x^{2/3} - (4x)^{3/5} = 6\left(\sqrt[3]{x}\right)^2 - \left(\sqrt[5]{4x}\right)^3$

(e) $r^{-2/3} = \dfrac{1}{r^{2/3}} = \dfrac{1}{\left(\sqrt[3]{r}\right)^2}$

(f) $(a^2 + b^2)^{1/2} = \sqrt{a^2 + b^2}$ ◁ $\sqrt{a^2 + b^2} \neq a + b$

In parts (g)–(i), write each radical as an exponential and simplify. Assume that all variables represent positive real numbers.

(g) $\sqrt{10} = 10^{1/2}$

(h) $\sqrt[4]{3^8} = 3^{8/4} = 3^2 = 9$

(i) $\sqrt[6]{z^6} = z^{6/6} = z^1 = z$, since z is positive.

◀ **Work Problem** **4** **at the Side.**

> **Note**
>
> In **Example 4(i),** it was not necessary to use absolute value bars, since the directions specifically stated that the variable represents a positive real number. Because the absolute value of the positive real number z is z itself, the answer is simply z.

◀ **Work Problem** **5** **at the Side.**

OBJECTIVE ▶ 4 Use the rules for exponents with rational exponents.
The definition of rational exponents allows us to apply the rules for exponents from **Chapter 5.**

> **Rules for Rational Exponents**
>
> Let r and s be rational numbers. For all real numbers a and b for which the indicated expressions exist, the following are true.
>
> $$a^r \cdot a^s = a^{r+s} \qquad a^{-r} = \frac{1}{a^r} \qquad \frac{a^r}{a^s} = a^{r-s} \qquad \left(\frac{a}{b}\right)^{-r} = \frac{b^r}{a^r}$$
>
> $$(a^r)^s = a^{rs} \qquad (ab)^r = a^r b^r \qquad \left(\frac{a}{b}\right)^r = \frac{a^r}{b^r} \qquad a^{-r} = \left(\frac{1}{a}\right)^r$$

EXAMPLE 5   Applying Rules for Rational Exponents

Write with only positive exponents. Assume that all variables represent positive real numbers.

(a) $2^{1/2} \cdot 2^{1/4}$

$= 2^{1/2 + 1/4}$ Product rule

$= 2^{3/4}$ Add exponents.

(b) $\dfrac{5^{2/3}}{5^{7/3}}$

$= 5^{2/3 - 7/3}$ Quotient rule

$= 5^{-5/3}$ Subtract exponents.

$= \dfrac{1}{5^{5/3}}$ $a^{-r} = \frac{1}{a^r}$

(c) $\dfrac{(x^{1/2}y^{2/3})^4}{y}$

$= \dfrac{(x^{1/2})^4 (y^{2/3})^4}{y}$ Power rule

$= \dfrac{x^2 y^{8/3}}{y^1}$ Power rule; $y = y^1$

$= x^2 y^{8/3 - 1}$ Quotient rule

$= x^2 y^{5/3}$ $\frac{8}{3} - 1 = \frac{8}{3} - \frac{3}{3} = \frac{5}{3}$

(d) $\left(\dfrac{x^4 y^{-6}}{x^{-2} y^{1/3}}\right)^{-2/3}$

$= \dfrac{(x^4)^{-2/3} (y^{-6})^{-2/3}}{(x^{-2})^{-2/3} (y^{1/3})^{-2/3}}$ Power rule

$= \dfrac{x^{-8/3} y^4}{x^{4/3} y^{-2/9}}$ Power rule

$= x^{-8/3 - 4/3} y^{4 - (-2/9)}$ Quotient rule

$= x^{-4} y^{38/9}$ [Use parentheses to avoid errors.] $4 - \left(-\frac{2}{9}\right) = \frac{36}{9} + \frac{2}{9} = \frac{38}{9}$

$= \dfrac{y^{38/9}}{x^4}$ Definition of negative exponent

Continued on Next Page

6 Write with only positive exponents. Assume that all variables represent positive real numbers.

(a) $11^{3/4} \cdot 11^{5/4}$

(b) $\dfrac{7^{3/4}}{7^{7/4}}$

(c) $\dfrac{9^{2/3}(x^{1/3})^4}{9^{-1/3}}$

(d) $\left(\dfrac{a^3 b^{-4}}{a^{-2} b^{1/5}}\right)^{-1/2}$

(e) $a^{2/3}(a^{7/3} + a^{1/3})$

7 Write all radicals as exponentials, and then apply the rules for rational exponents. Leave answers in exponential form. Assume that all variables represent positive real numbers.

(a) $\sqrt[5]{m^3} \cdot \sqrt{m}$

(b) $\dfrac{\sqrt[3]{p^5}}{\sqrt{p^3}}$

(c) $\sqrt[4]{\sqrt[3]{x}}$

Answers

6. **(a)** 11^2, or 121 **(b)** $\dfrac{1}{7}$ **(c)** $9x^{4/3}$

   (d) $\dfrac{b^{21/10}}{a^{5/2}}$ **(e)** $a^3 + a$

7. **(a)** $m^{11/10}$ **(b)** $p^{1/6}$ **(c)** $x^{1/12}$

The same result is obtained if we simplify within the parentheses first.

$$\left(\dfrac{x^4 y^{-6}}{x^{-2} y^{1/3}}\right)^{-2/3}$$

$$= (x^{4-(-2)} y^{-6-1/3})^{-2/3} \qquad \text{Quotient rule}$$

$$= (x^6 y^{-19/3})^{-2/3} \qquad -6 - \tfrac{1}{3} = -\tfrac{18}{3} - \tfrac{1}{3} = -\tfrac{19}{3}$$

$$= (x^6)^{-2/3}(y^{-19/3})^{-2/3} \qquad \text{Power rule}$$

$$= x^{-4} y^{38/9} \qquad \text{Power rule}$$

$$= \dfrac{y^{38/9}}{x^4} \qquad \text{Definition of negative exponent}$$

(e) $\qquad m^{3/4}(m^{5/4} - m^{1/4})$

$$= m^{3/4}(m^{5/4}) - m^{3/4}(m^{1/4}) \qquad \text{Distributive property}$$

> Do not make the common mistake of multiplying exponents in the first step.

$$= m^{3/4+5/4} - m^{3/4+1/4} \qquad \text{Product rule}$$

$$= m^{8/4} - m^{4/4} \qquad \text{Add in the exponents.}$$

$$= m^2 - m \qquad \text{Write the exponents in lowest terms.}$$

◄ **Work Problem 6 at the Side.**

CAUTION

Use the rules of exponents in problems like those in **Example 5**. Do not convert the expressions to radical form.

EXAMPLE 6 **Applying Rules for Rational Exponents**

Write all radicals as exponentials, and then apply the rules for rational exponents. Leave answers in exponential form. Assume that all variables represent positive real numbers.

(a) $\sqrt[3]{x^2} \cdot \sqrt[4]{x}$

$$= x^{2/3} \cdot x^{1/4} \qquad \text{Convert to rational exponents.}$$

$$= x^{2/3+1/4} \qquad \text{Product rule}$$

$$= x^{8/12+3/12} \qquad \text{Write exponents with a common denominator.}$$

$$= x^{11/12} \qquad \text{Add in the exponent.}$$

(b) $\dfrac{\sqrt{x^3}}{\sqrt[3]{x^2}}$

$$= \dfrac{x^{3/2}}{x^{2/3}} \qquad \text{Convert to rational exponents.}$$

$$= x^{3/2-2/3} \qquad \text{Quotient rule}$$

$$= x^{5/6} \qquad \tfrac{3}{2} - \tfrac{2}{3} = \tfrac{9}{6} - \tfrac{4}{6} = \tfrac{5}{6}$$

(c) $\sqrt{\sqrt[4]{z}}$

$$= \sqrt{z^{1/4}} \qquad \text{Convert the inside radical to rational exponents.}$$

$$= (z^{1/4})^{1/2} \qquad \text{Convert to rational exponents.}$$

$$= z^{1/8} \qquad \text{Power rule}$$

◄ **Work Problem 7 at the Side.**

9.2 Exercises

Download the
MyDashBoard App

MyMathLab®

CONCEPT CHECK   *Match each expression from Column I with the equivalent choice from Column II.*

	I				II	
1. $2^{1/2}$		**2.** $(-27)^{1/3}$		**A.** -4		**B.** 8

1. $2^{1/2}$ **2.** $(-27)^{1/3}$ **A.** -4 **B.** 8

3. $-16^{1/2}$ **4.** $(-16)^{1/2}$ **C.** $\sqrt{2}$ **D.** $-\sqrt{6}$

5. $(-32)^{1/5}$ **6.** $(-32)^{2/5}$ **E.** -3 **F.** $\sqrt{6}$

7. $4^{3/2}$ **8.** $6^{2/4}$ **G.** 4 **H.** -2

9. $-6^{2/4}$ **10.** $36^{0.5}$ **I.** 6 **J.** Not a real number

Evaluate each exponential. See **Examples 1–3.**

11. $169^{1/2}$ **12.** $121^{1/2}$ **13.** $729^{1/3}$ **14.** $512^{1/3}$ **15.** $16^{1/4}$

16. $625^{1/4}$ **17.** $\left(\dfrac{64}{81}\right)^{1/2}$ **18.** $\left(\dfrac{8}{27}\right)^{1/3}$ **19.** $(-27)^{1/3}$ **20.** $(-32)^{1/5}$

21. $(-144)^{1/2}$ **22.** $(-36)^{1/2}$ **23.** $100^{3/2}$ **24.** $64^{3/2}$

25. $81^{3/4}$ **26.** $216^{2/3}$ **27.** $-16^{5/2}$ **28.** $-32^{3/5}$

29. $(-8)^{4/3}$ **30.** $(-243)^{2/5}$ **31.** $32^{-3/5}$ **32.** $27^{-4/3}$

33. $64^{-3/2}$ **34.** $81^{-3/2}$ **35.** $\left(\dfrac{125}{27}\right)^{-2/3}$ **36.** $\left(\dfrac{64}{125}\right)^{-2/3}$

Write with radicals. Assume that all variables represent positive real numbers.
*Use the definition that takes the root first. **See Example 4.***

37. $12^{1/2}$

38. $3^{1/2}$

39. $8^{3/4}$

40. $7^{2/3}$

41. $(9q)^{5/8} - (2x)^{2/3}$

42. $(3p)^{3/4} + (4x)^{1/3}$

43. $(2m)^{-3/2}$

44. $(5y)^{-3/5}$

45. $(2y + x)^{2/3}$

46. $(r + 2z)^{3/2}$

47. $(3m^4 + 2k^2)^{-2/3}$

48. $(5x^2 + 3z^3)^{-5/6}$

49. CONCEPT CHECK Show that, in general, $\sqrt{a^2 + b^2} \neq a + b$ by replacing a with 3 and b with 4.

50. Suppose someone claims that $\sqrt[n]{a^n + b^n}$ must equal $a + b$, since when $a = 1$ and $b = 0$, a true statement results:

$$\sqrt[n]{a^n + b^n} = \sqrt[n]{1^n + 0^n} = \sqrt[n]{1^n} = 1 = 1 + 0 = a + b.$$

Explain why this is faulty reasoning.

*Simplify by first converting to rational exponents. Assume that all variables represent positive real numbers. **See Example 4.***

51. $\sqrt{2^{12}}$

52. $\sqrt{5^{10}}$

53. $\sqrt[3]{4^9}$

54. $\sqrt[4]{6^8}$

55. $\sqrt{x^{20}}$

56. $\sqrt{r^{50}}$

57. $\sqrt[3]{x} \cdot \sqrt{x}$

58. $\sqrt[4]{y} \cdot \sqrt[5]{y^2}$

59. $\dfrac{\sqrt[3]{t^4}}{\sqrt[5]{t^4}}$

60. $\dfrac{\sqrt[4]{w^3}}{\sqrt[6]{w}}$

*Simplify each expression. Write all answers with positive exponents. Assume that all variables represent positive real numbers. **See Example 5.***

61. $3^{1/2} \cdot 3^{3/2}$

62. $6^{4/3} \cdot 6^{2/3}$

63. $\dfrac{64^{5/3}}{64^{4/3}}$

64. $\dfrac{125^{7/3}}{125^{5/3}}$

65. $y^{7/3} \cdot y^{-4/3}$

66. $r^{-8/9} \cdot r^{17/9}$

67. $x^{2/3} \cdot x^{-1/4}$

68. $x^{2/5} \cdot x^{-1/3}$

69. $\dfrac{k^{1/3}}{k^{2/3} \cdot k^{-1}}$

70. $\dfrac{z^{3/4}}{z^{5/4} \cdot z^{-2}}$

71. $\dfrac{(x^{1/4}y^{2/5})^{20}}{x^2}$

72. $\dfrac{(r^{1/5}s^{2/3})^{15}}{r^2}$

73. $\dfrac{(x^{2/3})^2}{(x^2)^{7/3}}$

74. $\dfrac{(p^3)^{1/4}}{(p^{5/4})^2}$

75. $\dfrac{m^{3/4}n^{-1/4}}{(m^2n)^{1/2}}$

76. $\dfrac{(a^2b^5)^{-1/4}}{(a^{-3}b^2)^{1/6}}$

77. $\dfrac{p^{1/5}p^{7/10}p^{1/2}}{(p^3)^{-1/5}}$

78. $\dfrac{z^{1/3}z^{-2/3}z^{1/6}}{(z^{-1/6})^3}$

79. $\left(\dfrac{b^{-3/2}}{c^{-5/3}}\right)^2 (b^{-1/4}c^{-1/3})^{-1}$

80. $\left(\dfrac{m^{-2/3}}{a^{-3/4}}\right)^4 (m^{-3/8}a^{1/4})^{-2}$

81. $\left(\dfrac{p^{-1/4}q^{-3/2}}{3^{-1}p^{-2}q^{-2/3}}\right)^{-2}$

82. $\left(\dfrac{2^{-2}w^{-3/4}x^{-5/8}}{w^{3/4}x^{-1/2}}\right)^{-3}$

83. $p^{2/3}(p^{1/3} + 2p^{4/3})$

84. $z^{5/8}(3z^{5/8} + 5z^{11/8})$

85. $k^{1/4}(k^{3/2} - k^{1/2})$

86. $r^{3/5}(r^{1/2} + r^{3/4})$

87. $6a^{7/4}(a^{-7/4} + 3a^{-3/4})$

88. $4m^{5/3}(m^{-2/3} - 4m^{-5/3})$

89. $5m^{-2/3}(m^{2/3} + m^{-7/3})$

90. $7z^{-4/5}(z^{4/5} - z^{-6/5})$

Write radicals as exponentials, and then apply the rules for rational exponents. Give answers in exponential form. Assume that all radicands represent positive real numbers. ***See Example 6.***

91. $\sqrt[6]{y^5} \cdot \sqrt[3]{y^2}$

92. $\sqrt[5]{x^3} \cdot \sqrt[4]{x}$

93. $\dfrac{\sqrt[3]{k^5}}{\sqrt[3]{k^7}}$

94. $\dfrac{\sqrt{x^5}}{\sqrt{x^8}}$

95. $\sqrt[3]{xz} \cdot \sqrt{z}$

96. $\sqrt{y} \cdot \sqrt[3]{yz}$

97. $\sqrt[3]{\sqrt{k}}$

98. $\sqrt[4]{\sqrt[3]{m}}$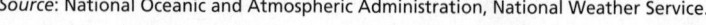

99. $\sqrt[3]{\sqrt[5]{\sqrt{y}}}$

100. $\sqrt{\sqrt[3]{\sqrt[4]{x}}}$

101. $\sqrt[3]{x^{5/9}}$

102. $\sqrt{y^{5/4}}$

Solve each problem.

103. The threshold weight T, in pounds, for a person is the weight above which the risk of death increases greatly. The threshold weight in pounds for men aged 40–49 is related to height in inches by the function

$$h(T) = (1860.867T)^{1/3}.$$

What height corresponds to a threshold weight of 200 lb for a 46-yr-old man? Round the answer to the nearest inch, and then to the nearest tenth of a foot.

104. Meteorologists can determine the duration of a storm by using the function

$$T(D) = 0.07D^{3/2},$$

where D is the diameter of the storm in miles and T is the time in hours. Find the duration of a storm with a diameter of 16 mi. Round the answer to the nearest tenth of an hour.

*The **windchill factor** is a measure of the cooling effect that the wind has on a person's skin. It calculates the equivalent cooling temperature if there were no wind. The National Weather Service uses the formula*

$$\text{Windchill temperature} = 35.74 + 0.6215T - 35.75V^{4/25} + 0.4275TV^{4/25},$$

where T is the temperature in °F and V is the wind speed in miles per hour, to calculate windchill. The table gives the windchill factor for various wind speeds and temperatures at which frostbite is a risk, and how quickly it may occur.

				Temperature (°F)					
Calm	40	30	20	10	0	−10	−20	−30	−40
5	36	25	13	1	−11	−22	−34	−46	−57
10	34	21	9	−4	−16	−28	−41	−53	−66
15	32	19	6	−7	−19	−32	−45	−58	−71
20	30	17	4	−9	−22	−35	−48	−61	−74
25	29	16	3	−11	−24	−37	−51	−64	−78
30	28	15	1	−12	−26	−39	−53	−67	−80
35	28	14	0	−14	−27	−41	−55	−69	−82
40	27	13	−1	−15	−29	−43	−57	−71	−84

Wind speed (mph)

Frostbites times: ☐ 30 minutes   ☐ 10 minutes   ☐ 5 minutes

Source: National Oceanic and Atmospheric Administration, National Weather Service.

Use the formula to determine the windchill to the nearest tenth of a degree, given the following conditions. Compare answers with the appropriate entries in the table.

105. 10°F, 30-mph wind

106. 30°F, 15-mph wind

107. 20°F, 20-mph wind

108. 40°F, 10-mph wind

9.3 Simplifying Radical Expressions

OBJECTIVE ▶ 1 Use the product rule for radicals. Consider the expressions $\sqrt{36 \cdot 4}$ and $\sqrt{36} \cdot \sqrt{4}$.

$$\sqrt{36 \cdot 4} = \sqrt{144} = 12$$

The result is the same.

$$\sqrt{36} \cdot \sqrt{4} = 6 \cdot 2 = 12$$

This is an example of the **product rule for radicals.**

Product Rule for Radicals

If $\sqrt[n]{a}$ and $\sqrt[n]{b}$ are real numbers and n is a natural number, then

$$\sqrt[n]{a} \cdot \sqrt[n]{b} = \sqrt[n]{ab}.$$

In words, the product of two nth roots is the nth root of the product.

We justify the product rule using the rules for rational exponents. Since $\sqrt[n]{a} = a^{1/n}$ and $\sqrt[n]{b} = b^{1/n}$,

$$\sqrt[n]{a} \cdot \sqrt[n]{b} = a^{1/n} \cdot b^{1/n} = (ab)^{1/n} = \sqrt[n]{ab}.$$

CAUTION

Use the product rule only when the radicals have the same index.

EXAMPLE 1 **Using the Product Rule**

Multiply. Assume that all variables represent positive real numbers.

(a) $\sqrt{5} \cdot \sqrt{7}$

$\quad = \sqrt{5 \cdot 7}$

$\quad = \sqrt{35}$

(b) $\sqrt{11} \cdot \sqrt{p}$

$\quad = \sqrt{11p}$

(c) $\sqrt{7} \cdot \sqrt{11xyz}$

$\quad = \sqrt{77xyz}$

···· **Work Problem ❶ at the Side.** ▶

EXAMPLE 2 **Using the Product Rule**

Multiply. Assume that all variables represent positive real numbers.

(a) $\sqrt[3]{3} \cdot \sqrt[3]{12}$

$\quad = \sqrt[3]{3 \cdot 12}$

$\quad = \sqrt[3]{36}$ ◁ Remember to write the index.

(b) $\sqrt[4]{8y} \cdot \sqrt[4]{3r^2}$

$\quad = \sqrt[4]{24yr^2}$

(c) $\sqrt[6]{10m^4} \cdot \sqrt[6]{5m}$

$\quad = \sqrt[6]{50m^5}$

(d) $\sqrt[4]{2} \cdot \sqrt[5]{2}$ cannot be multiplied using the product rule for radicals, because the indexes (4 and 5) are different.

···· **Work Problem ❷ at the Side.** ▶

OBJECTIVES

① Use the product rule for radicals.

② Use the quotient rule for radicals.

③ Simplify radicals.

④ Simplify products and quotients of radicals with different indexes.

⑤ Use the Pythagorean theorem.

⑥ Use the distance formula.

---

❶ Multiply. Assume that all variables represent positive real numbers.

(a) $\sqrt{5} \cdot \sqrt{13}$

(b) $\sqrt{10y} \cdot \sqrt{3k}$

---

❷ Multiply. Assume that all variables represent positive real numbers.

(a) $\sqrt[3]{2} \cdot \sqrt[3]{7}$

(b) $\sqrt[6]{8r^2} \cdot \sqrt[6]{2r^3}$

(c) $\sqrt[5]{9y^2x} \cdot \sqrt[5]{8xy^2}$

(d) $\sqrt{7} \cdot \sqrt[3]{5}$

Answers

1. (a) $\sqrt{65}$ (b) $\sqrt{30yk}$

2. (a) $\sqrt[3]{14}$ (b) $\sqrt[6]{16r^5}$ (c) $\sqrt[5]{72y^4x^2}$
      (d) cannot be multiplied using the product rule

3 Simplify. Assume that all variables represent positive real numbers.

(a) $\sqrt{\dfrac{100}{81}}$

(b) $\sqrt{\dfrac{11}{25}}$

(c) $\sqrt[3]{-\dfrac{125}{216}}$

(d) $\sqrt{\dfrac{y^8}{16}}$

(e) $-\sqrt[3]{\dfrac{x^2}{r^{12}}}$

OBJECTIVE ▶ 2 **Use the quotient rule for radicals.** The **quotient rule for radicals** is similar to the product rule.

> **Quotient Rule for Radicals**
>
> If $\sqrt[n]{a}$ and $\sqrt[n]{b}$ are real numbers, $b \neq 0$, and n is a natural number, then
> $$\sqrt[n]{\dfrac{a}{b}} = \dfrac{\sqrt[n]{a}}{\sqrt[n]{b}}.$$
>
> In words, the nth root of a quotient is the quotient of the nth roots.

EXAMPLE 3 **Using the Quotient Rule**

Simplify. Assume that all variables represent positive real numbers.

(a) $\sqrt{\dfrac{16}{25}} = \dfrac{\sqrt{16}}{\sqrt{25}} = \dfrac{4}{5}$

(b) $\sqrt{\dfrac{7}{36}} = \dfrac{\sqrt{7}}{\sqrt{36}} = \dfrac{\sqrt{7}}{6}$

(c) $\sqrt[3]{-\dfrac{8}{125}} = \sqrt[3]{\dfrac{-8}{125}} = \dfrac{\sqrt[3]{-8}}{\sqrt[3]{125}} = \dfrac{-2}{5} = -\dfrac{2}{5}$ $\dfrac{-a}{b} = -\dfrac{a}{b}$

(d) $\sqrt[3]{\dfrac{7}{216}} = \dfrac{\sqrt[3]{7}}{\sqrt[3]{216}} = \dfrac{\sqrt[3]{7}}{6}$

(e) $\sqrt[5]{\dfrac{x}{32}} = \dfrac{\sqrt[5]{x}}{\sqrt[5]{32}} = \dfrac{\sqrt[5]{x}}{2}$

(f) $-\sqrt[3]{\dfrac{m^6}{125}} = -\dfrac{\sqrt[3]{m^6}}{\sqrt[3]{125}} = -\dfrac{m^2}{5}$ Think: $\sqrt[3]{m^6} = m^{6/3} = m^2$

◀ **Work Problem 3 at the Side.**

OBJECTIVE ▶ 3 **Simplify radicals.** We use the product and quotient rules to simplify radicals. A radical is **simplified** if the following four conditions are met.

> **Conditions for a Simplified Radical**
>
> **1.** The radicand has no factor raised to a power greater than or equal to the index.
>
> **2.** The radicand has no fractions.
>
> **3.** No denominator contains a radical.
>
> **4.** Exponents in the radicand and the index of the radical have greatest common factor 1.

Consider the following examples.

$\sqrt{22}, \quad \sqrt{15xy}, \quad \sqrt[3]{18}, \quad \dfrac{\sqrt[4]{m^3}}{m}$ These radicals are simplified.

$\sqrt{28}, \quad \sqrt[3]{\dfrac{3}{5}}, \quad \dfrac{7}{\sqrt{7}}, \quad \sqrt[3]{r^{12}}$ These radicals are not simplified. Each violates one of the above conditions.

Answers

3. (a) $\dfrac{10}{9}$ (b) $\dfrac{\sqrt{11}}{5}$ (c) $-\dfrac{5}{6}$

(d) $\dfrac{y^4}{4}$ (e) $-\dfrac{\sqrt[3]{x^2}}{r^4}$

EXAMPLE 4 **Simplifying Roots of Numbers**

Simplify.

(a) $\sqrt{24}$

Check to see whether 24 is divisible by a perfect square (the square of a natural number) such as 4, 9, 16, …. The greatest perfect square that divides into 24 is 4.

$$\sqrt{24}$$
$$= \sqrt{4 \cdot 6} \qquad \text{Factor. 4 is a perfect square.}$$
$$= \sqrt{4} \cdot \sqrt{6} \qquad \text{Product rule}$$
$$= 2\sqrt{6} \qquad \sqrt{4} = 2$$

(b) $\sqrt{108}$

As shown on the left below, the number 108 is divisible by the perfect square 36. If this perfect square is not immediately clear, try factoring 108 into its prime factors, as shown on the right.

$$\sqrt{108} \qquad \qquad \sqrt{108}$$
$$= \sqrt{36 \cdot 3} \qquad \quad = \sqrt{2^2 \cdot 3^3} \qquad \text{Factor into prime factors.}$$
$$= \sqrt{36} \cdot \sqrt{3} \qquad = \sqrt{2^2 \cdot 3^2 \cdot 3} \qquad a^3 = a^2 \cdot a$$
$$= 6\sqrt{3} \qquad \qquad = \sqrt{2^2} \cdot \sqrt{3^2} \cdot \sqrt{3} \qquad \text{Product rule}$$
$$\qquad \qquad = 2 \cdot 3 \cdot \sqrt{3} \qquad \sqrt{2^2} = 2, \sqrt{3^2} = 3$$
$$\qquad \qquad = 6\sqrt{3} \qquad \text{Multiply.}$$

(c) $\sqrt{10}$ No perfect square (other than 1) divides into 10, so $\sqrt{10}$ cannot be simplified further.

(d) $\sqrt[3]{16}$

The greatest perfect *cube* that divides into 16 is 8, so factor 16 as 8 · 2.

$$\sqrt[3]{16}$$

Remember to write the index.

$$= \sqrt[3]{8 \cdot 2} \qquad \text{8 is a perfect cube.}$$
$$= \sqrt[3]{8} \cdot \sqrt[3]{2} \qquad \text{Product rule}$$
$$= 2\sqrt[3]{2} \qquad \sqrt[3]{8} = 2$$

(e) $-\sqrt[4]{162}$

$$= -\sqrt[4]{81 \cdot 2} \qquad \text{81 is a perfect 4th power.}$$

Remember the negative sign in each line.

$$= -\sqrt[4]{81} \cdot \sqrt[4]{2} \qquad \text{Product rule}$$
$$= -3\sqrt[4]{2} \qquad \sqrt[4]{81} = 3$$

· **Work Problem 4 at the Side.** ▶

CAUTION

Be careful with which factors belong outside the radical symbol and which belong inside. In **Example 4(b),** the 2 · 3 is written outside because $\sqrt{2^2} = 2$ and $\sqrt{3^2} = 3$. The remaining 3 is left inside.

---

4 Simplify.

(a) $\sqrt{32}$

The greatest perfect square that divides into 32 is ____.

$$\sqrt{32}$$
$$= \sqrt{\underline{\quad} \cdot 2}$$
$$= \sqrt{\underline{\quad}} \cdot \sqrt{\underline{\quad}}$$
$$= \underline{\quad}$$

(b) $\sqrt{45}$

(c) $\sqrt{300}$

(d) $\sqrt{35}$

(e) $-\sqrt[3]{54}$

(f) $\sqrt[4]{243}$

Answers

4. **(a)** 16; 16; 16; 2; $4\sqrt{2}$
  (b) $3\sqrt{5}$ **(c)** $10\sqrt{3}$
  (d) cannot be simplified further
  (e) $-3\sqrt[3]{2}$ **(f)** $3\sqrt[4]{3}$

5 Simplify. Assume that all variables represent positive real numbers.

(a) $\sqrt{25p^7}$

(b) $\sqrt{72xy^3}$

(c) $\sqrt[3]{-27x^5y^7z^6}$

(d) $-\sqrt[4]{32a^5b^7}$

EXAMPLE 5 **Simplifying Radicals Involving Variables**

Simplify. Assume that all variables represent positive real numbers.

(a) $\sqrt{16m^3}$

$$= \sqrt{16m^2 \cdot m} \qquad \text{Factor.}$$
$$= \sqrt{16m^2} \cdot \sqrt{m} \qquad \text{Product rule}$$
$$= 4m\sqrt{m} \qquad \text{Take the square root.}$$

Absolute value bars are not needed around the m in color because of the assumption that all the variables represent *positive* real numbers.

(b) $\sqrt{200k^7q^8}$

$$= \sqrt{10^2 \cdot 2 \cdot (k^3)^2 \cdot k \cdot (q^4)^2} \qquad \text{Factor.}$$
$$= 10k^3q^4\sqrt{2k} \qquad \text{Remove perfect square factors.}$$

(c) $\sqrt[3]{-8x^4y^5}$

$$= \sqrt[3]{(-8x^3y^3)(xy^2)} \qquad \begin{array}{l}\text{Choose } -8x^3y^3 \text{ as the perfect cube}\\\text{that divides into } -8x^4y^5.\end{array}$$
$$= \sqrt[3]{-8x^3y^3} \cdot \sqrt[3]{xy^2} \qquad \text{Product rule}$$
$$= -2xy\sqrt[3]{xy^2} \qquad \text{Take the cube root.}$$

(d) $-\sqrt[4]{32y^9}$

$$= -\sqrt[4]{(16y^8)(2y)} \qquad \begin{array}{l}16y^8 \text{ is the greatest 4th power that}\\\text{divides into } 32y^9.\end{array}$$
$$= -\sqrt[4]{16y^8} \cdot \sqrt[4]{2y} \qquad \text{Product rule}$$
$$= -2y^2\sqrt[4]{2y} \qquad \text{Take the fourth root.}$$

◀ **Work Problem 5 at the Side.**

Note

From **Example 5** we see that if a variable is raised to a power with an exponent divisible by 2, it is a perfect square. If it is raised to a power with an exponent divisible by 3, it is a perfect cube. *In general, if it is raised to a power with an exponent divisible by n, it is a perfect nth power.*

The conditions for a simplified radical given earlier state that an exponent in the radicand and the index of the radical should have greatest common factor 1.

EXAMPLE 6 **Simplifying Radicals by Using Lesser Indexes**

Simplify. Assume that all variables represent positive real numbers.

(a) $\sqrt[9]{5^6}$

We can write this radical using rational exponents and then write the exponent in lowest terms. We then express the answer as a radical.

$$\sqrt[9]{5^6} = (5^6)^{1/9} = 5^{6/9} = 5^{2/3} = \sqrt[3]{5^2}, \quad \text{or} \quad \sqrt[3]{25}$$

(b) $\sqrt[4]{p^2} = p^{2/4} = p^{1/2} = \sqrt{p}$ (Recall the assumption that $p > 0$.)

Answers

5. **(a)** $5p^3\sqrt{p}$ **(b)** $6y\sqrt{2xy}$
   (c) $-3xy^2z^2\sqrt[3]{x^2y}$ **(d)** $-2ab\sqrt[4]{2ab^3}$

These examples suggest the following rule.

Meaning of $\sqrt[kn]{a^{km}}$

If m is an integer, n and k are natural numbers, and all indicated roots exist, then

$$\sqrt[kn]{a^{km}} = \sqrt[n]{a^m}.$$

Work Problem 6 at the Side. ▶

OBJECTIVE ▶ 4 Simplify products and quotients of radicals with different indexes. We multiply and divide radicals with different indexes by using rational exponents.

EXAMPLE 7 Multiplying Radicals with Different Indexes

Simplify $\sqrt{7} \cdot \sqrt[3]{2}$.

Because the different indexes, 2 and 3, have least common multiple index of 6, we use rational exponents to write each radical as a sixth root.

$$\sqrt{7} = 7^{1/2} = 7^{3/6} = \sqrt[6]{7^3} = \sqrt[6]{343}$$

$$\sqrt[3]{2} = 2^{1/3} = 2^{2/6} = \sqrt[6]{2^2} = \sqrt[6]{4}$$

Now we multiply.

$$\sqrt{7} \cdot \sqrt[3]{2}$$

$$= \sqrt[6]{343} \cdot \sqrt[6]{4} \qquad \sqrt{7} = \sqrt[6]{343}, \ \sqrt[3]{2} = \sqrt[6]{4}$$

$$= \sqrt[6]{1372} \qquad \text{Product rule}$$

Work Problem 7 at the Side. ▶

OBJECTIVE ▶ 5 Use the Pythagorean theorem. The **Pythagorean theorem** relates the lengths of the three sides of a right triangle.

Pythagorean Theorem

If a and b are the lengths of the shorter sides of a right triangle and c is the length of the longest side, then

$$a^2 + b^2 = c^2.$$

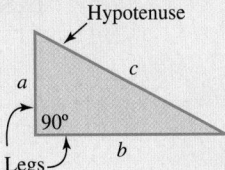

The two shorter sides are the **legs** of the triangle, and the longest side is the **hypotenuse.** The hypotenuse is the side opposite the right angle. Thus,

$$\textbf{leg}^2 + \textbf{leg}^2 = \textbf{hypotenuse}^2.$$

In **Section 10.1,** we will see that an equation such as $x^2 = 7$ has two solutions: $\sqrt{7}$ (the principal, or positive, square root of 7) and $-\sqrt{7}$. Similarly, $c^2 = 52$ has two solutions, $\pm\sqrt{52}$, or $\pm 2\sqrt{13}$. In applications we often choose only the positive square root.

6 Simplify. Assume that all variables represent positive real numbers.

(a) $\sqrt[12]{2^3}$

(b) $\sqrt[6]{t^2}$

7 Simplify.

GS (a) $\sqrt{5} \cdot \sqrt[3]{4}$

The indexes of $\sqrt{5}$ and $\sqrt[3]{4}$ are _____ and 3, which have least common multiple index of _____. Thus, we use rational exponents to write each radical as a _____ root.

$$\sqrt{5} = 5^{1/2} = 5^{\underline{\ \ }/6} = \sqrt[6]{5^{\underline{\ \ }}} = \sqrt[6]{\underline{\ \ }}$$

$$\sqrt[3]{4} = 4^{1/3} = 4^{\underline{\ \ }/6} = \sqrt[6]{4^{\underline{\ \ }}} = \sqrt[6]{16}$$

$$\sqrt{5} \cdot \sqrt[3]{4}$$

$$= \sqrt[6]{\underline{\ \ }} \cdot \sqrt[6]{16}$$

$$= \underline{\ \ }$$

(b) $\sqrt[3]{3} \cdot \sqrt{6}$

Answers

6. (a) $\sqrt[4]{2}$ (b) $\sqrt[3]{t}$

7. (a) 2; 6; sixth; 3; 3; 125; 2; 2; 125; $\sqrt[6]{2000}$
(b) $\sqrt[6]{1944}$

8 Find the length of the unknown side in each triangle.

(a)

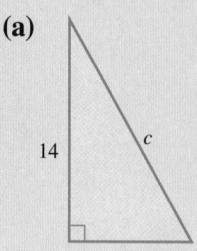

EXAMPLE 8   **Using the Pythagorean Theorem**

Use the Pythagorean theorem to find the length of the hypotenuse in the triangle in **Figure 6**.

To find the length of the hypotenuse c, let $a = 4$ and $b = 6$.

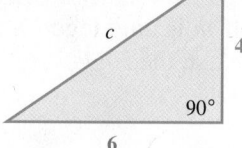

Figure 6

$$a^2 + b^2 = c^2 \qquad \text{Pythagorean theorem}$$

Substitute carefully. → $$4^2 + 6^2 = c^2 \qquad \text{Let } a = 4 \text{ and } b = 6.$$

$$16 + 36 = c^2 \qquad \text{Apply the exponents.}$$

$$c^2 = 52 \qquad \text{Add. Interchange sides.}$$

$$c = \sqrt{52} \qquad \text{Choose the principal root.}$$

$$c = \sqrt{4 \cdot 13} \qquad \text{Factor.}$$

$$c = \sqrt{4} \cdot \sqrt{13} \qquad \text{Product rule}$$

$$c = 2\sqrt{13} \qquad \text{Simplify.}$$

The length of the hypotenuse is $2\sqrt{13}$.

◀ **Work Problem** **8** at the Side.

OBJECTIVE **6** **Use the distance formula.** The *distance formula* allows us to find the distance between two points in the coordinate plane, or the length of the line segment joining those two points.

Figure 7 shows the points $(3, -4)$ and $(-5, 3)$. The vertical line through $(-5, 3)$ and the horizontal line through $(3, -4)$ intersect at the point $(-5, -4)$. Thus, the point $(-5, -4)$ becomes the vertex of the right angle in a right triangle.

By the Pythagorean theorem, the sum of the squares of the lengths of the two legs a and b of the right triangle in **Figure 7** is equal to the square of the length of the hypotenuse, c.

$$a^2 + b^2 = c^2, \quad \text{or} \quad c^2 = a^2 + b^2$$

(b)

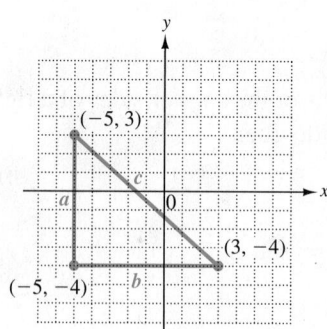

Figure 7

The length a is the difference between the y-coordinates of the endpoints. Since the x-coordinate of both of these points in **Figure 7** is -5, the side is vertical, and we can find a by finding the difference between the y-coordinates. We subtract -4 from 3 to get a positive value for a.

$$a = 3 - (-4) = 7$$

Similarly, we find b by subtracting -5 from 3.

$$b = 3 - (-5) = 8$$

Answers

8. **(a)** $2\sqrt{65}$ **(b)** $2\sqrt{5}$

Now substitute these values for a and b into the Pythagorean theorem.

$$c^2 = a^2 + b^2$$

$$c^2 = 7^2 + 8^2 \qquad \text{Let } a = 7 \text{ and } b = 8.$$

$$c^2 = 49 + 64 \qquad \text{Apply the exponents.}$$

$$c^2 = 113 \qquad \text{Add.}$$

$$c = \sqrt{113} \qquad \text{Choose the principal root.}$$

We choose the principal root since distance cannot be negative. Therefore, the distance between $(-5, 3)$ and $(3, -4)$ is $\sqrt{113}$.

> **Note**
>
> It is customary to leave the distance in radical form. Do not use a calculator to get an approximation unless specifically directed to do so.

This work can be generalized. **Figure 8** shows the two points (x_1, y_1) and (x_2, y_2). The distance a between (x_1, y_1) and (x_2, y_1) is given by

$$a = |x_2 - x_1|.$$

The distance b between (x_2, y_2) and (x_2, y_1) is given by

$$b = |y_2 - y_1|.$$

From the Pythagorean theorem, we obtain the following.

$$c^2 = a^2 + b^2$$

$$c^2 = (x_2 - x_1)^2 + (y_2 - y_1)^2$$

$$\text{For all real numbers } a, |a|^2 = a^2.$$

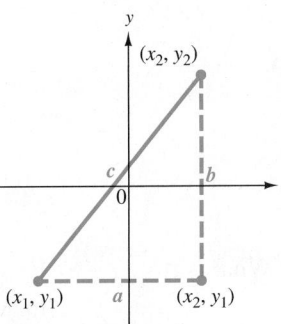

Figure 8

Choosing the principal square root gives the **distance formula.** In this formula, we use d (to denote distance) rather than c.

> **Distance Formula**
>
> The distance d between the points (x_1, y_1) and (x_2, y_2) is
>
> $$d = \sqrt{(x_2 - x_1)^2 + (y_2 - y_1)^2}.$$

EXAMPLE 9   **Using the Distance Formula**

Find the distance between the points $(-3, 5)$ and $(6, 4)$.

Designating the points as (x_1, y_1) and (x_2, y_2) is arbitrary. We choose $(x_1, y_1) = (-3, 5)$ and $(x_2, y_2) = (6, 4)$.

$$d = \sqrt{(x_2 - x_1)^2 + (y_2 - y_1)^2} \qquad \text{Distance formula}$$

$$d = \sqrt{[6 - (-3)]^2 + (4 - 5)^2} \qquad x_2 = 6, y_2 = 4, x_1 = -3, y_1 = 5$$

$$d = \sqrt{9^2 + (-1)^2} \qquad \boxed{\text{Substitute carefully.}}$$

$$d = \sqrt{82} \qquad \text{Leave in radical form.}$$

Work Problem ⑨ at the Side. ▶

⑨ Find the distance between each pair of points.

(a) $(2, -1)$ and $(5, 3)$

(b) $(-3, 2)$ and $(0, -4)$

Answers

9. (a) 5   (b) $3\sqrt{5}$

9.3 Exercises

CONCEPT CHECK *Choose the correct response.*

1. Which is the greatest perfect square factor of 128?

A. 12 **B.** 16 **C.** 32 **D.** 64

2. Which is the greatest perfect cube factor of $81a^7$?

A. $8a^3$ **B.** $27a^3$ **C.** $81a^6$ **D.** $27a^6$

3. Which radical can be simplified?

A. $\sqrt{21}$ **B.** $\sqrt{48}$ **C.** $\sqrt[3]{12}$ **D.** $\sqrt[4]{10}$

4. Which radical *cannot* be simplified?

A. $\sqrt[3]{30}$ **B.** $\sqrt[3]{27a^2b}$ **C.** $\sqrt{\dfrac{25}{81}}$ **D.** $\sqrt[4]{32}$

5. Which one of the following is *not* equal to $\sqrt{\frac{1}{2}}$? (Do not use calculator approximations.)

A. $\sqrt{0.5}$ **B.** $\sqrt{\dfrac{2}{4}}$ **C.** $\sqrt{\dfrac{3}{6}}$ **D.** $\dfrac{\sqrt{4}}{\sqrt{16}}$

6. Which one of the following is *not* equal to $\sqrt[3]{\frac{2}{5}}$? (Do not use calculator approximations.)

A. $\sqrt[3]{\dfrac{6}{15}}$ **B.** $\dfrac{\sqrt[3]{50}}{5}$ **C.** $\dfrac{\sqrt[3]{10}}{\sqrt[3]{25}}$ **D.** $\dfrac{\sqrt[3]{10}}{5}$

7. Explain why $\sqrt[3]{x} \cdot \sqrt[3]{x}$ is not equal to x. What is it equal to?

8. Explain why $\sqrt[4]{x} \cdot \sqrt[4]{x}$ is not equal to x, but *is* equal to $\sqrt{x}$, for $x \geq 0$.

Multiply using the product rule. Assume all variables represent positive real numbers. **See Examples 1 and 2.**

9. $\sqrt{5} \cdot \sqrt{6}$

10. $\sqrt{10} \cdot \sqrt{3}$

11. $\sqrt{14} \cdot \sqrt{x}$

12. $\sqrt{23} \cdot \sqrt{t}$

13. $\sqrt{14} \cdot \sqrt{3pqr}$

14. $\sqrt{7} \cdot \sqrt{5xt}$

15. $\sqrt[3]{7x} \cdot \sqrt[3]{2y}$

16. $\sqrt[3]{9x} \cdot \sqrt[3]{4y}$

17. $\sqrt[4]{11} \cdot \sqrt[4]{3}$

18. $\sqrt[4]{6} \cdot \sqrt[4]{9}$

19. $\sqrt[4]{2x} \cdot \sqrt[4]{3y^2}$

20. $\sqrt[4]{3y^2} \cdot \sqrt[4]{6yz}$

21. $\sqrt[3]{7} \cdot \sqrt[4]{3}$ **22.** $\sqrt[5]{8} \cdot \sqrt[6]{12}$ **23.** $\sqrt{12} \cdot \sqrt[3]{3}$ **24.** $\sqrt[4]{5} \cdot \sqrt[5]{4}$

Simplify. Assume that all variables represent positive real numbers. **See Example 3.**

25. $\sqrt{\dfrac{64}{121}}$ **26.** $\sqrt{\dfrac{16}{49}}$ **27.** $\sqrt{\dfrac{3}{25}}$ **28.** $\sqrt{\dfrac{13}{49}}$

29. $\sqrt{\dfrac{x}{25}}$ **30.** $\sqrt{\dfrac{k}{100}}$ **31.** $\sqrt{\dfrac{p^6}{81}}$ **32.** $\sqrt{\dfrac{w^{10}}{36}}$

33. $\sqrt[3]{-\dfrac{27}{64}}$ **34.** $\sqrt[3]{-\dfrac{216}{125}}$ **35.** $\sqrt[3]{\dfrac{r^2}{8}}$ **36.** $\sqrt[3]{\dfrac{t}{125}}$

37. $-\sqrt[4]{\dfrac{81}{x^4}}$ **38.** $-\sqrt[4]{\dfrac{625}{y^4}}$ **39.** $\sqrt[5]{\dfrac{1}{x^{15}}}$ **40.** $\sqrt[5]{\dfrac{32}{y^{20}}}$

Express each radical in simplified form. **See Example 4.**

41. $\sqrt{12}$ **42.** $\sqrt{18}$ **43.** $\sqrt{288}$ **44.** $\sqrt{72}$

45. $-\sqrt{32}$ **46.** $-\sqrt{48}$ **47.** $-\sqrt{28}$ **48.** $-\sqrt{24}$

49. $\sqrt{30}$ **50.** $\sqrt{46}$ **51.** $\sqrt[3]{128}$ **52.** $\sqrt[3]{24}$

53. $\sqrt[3]{-16}$

54. $\sqrt[3]{-250}$

55. $\sqrt[3]{40}$

56. $\sqrt[3]{375}$

57. $-\sqrt[4]{512}$

58. $-\sqrt[4]{1250}$

59. $\sqrt[5]{64}$

60. $\sqrt[5]{128}$

61. CONCEPT CHECK A student claimed that $\sqrt[3]{14}$ is not in simplified form, since $14 = 8 + 6$, and 8 is a perfect cube. Was his reasoning correct? Why or why not?

62. CONCEPT CHECK Why is $\sqrt[3]{k^4}$ not a simplified radical?

Express each radical in simplified form. Assume that all variables represent positive real numbers. **See Example 5.**

63. $\sqrt{72k^2}$

64. $\sqrt{18m^2}$

65. $\sqrt{144x^3y^9}$

66. $\sqrt{169s^5t^{10}}$

67. $\sqrt{121x^6}$

68. $\sqrt{256z^{12}}$

69. $-\sqrt[3]{27t^{12}}$

70. $-\sqrt[3]{64y^{18}}$

71. $-\sqrt{100m^8z^4}$

72. $-\sqrt{25t^6s^{20}}$

73. $-\sqrt[3]{-125a^6b^9c^{12}}$

74. $-\sqrt[3]{-216x^6y^{15}z^3}$

75. $\sqrt[4]{\dfrac{1}{16}r^8t^{20}}$

76. $\sqrt[4]{\dfrac{81}{256}t^{12}u^8}$

77. $\sqrt{50x^3}$

78. $\sqrt{300z^3}$

79. $-\sqrt{500r^{11}}$

80. $-\sqrt{200p^{13}}$

81. $\sqrt{13x^7y^8}$

82. $\sqrt{23k^9p^{14}}$

83. $\sqrt[3]{8z^6w^9}$ **84.** $\sqrt[3]{64a^{15}b^{12}}$ **85.** $\sqrt[3]{-16z^5t^7}$ **86.** $\sqrt[3]{-81m^4n^{10}}$

87. $\sqrt[4]{81x^{12}y^{16}}$ **88.** $\sqrt[4]{81t^8u^{28}}$ **89.** $-\sqrt[4]{162r^{15}s^{10}}$ **90.** $-\sqrt[4]{32k^5m^{10}}$

91. $\sqrt{\dfrac{y^{11}}{36}}$ **92.** $\sqrt{\dfrac{v^{13}}{49}}$ **93.** $\sqrt[3]{\dfrac{x^{16}}{27}}$ **94.** $\sqrt[3]{\dfrac{y^{17}}{125}}$

Simplify. Assume that $x \geq 0$. ***See Example 6.***

95. $\sqrt[4]{48^2}$ **96.** $\sqrt[4]{50^2}$ **97.** $\sqrt[4]{25}$ **98.** $\sqrt[6]{8}$

99. $\sqrt[10]{x^{25}}$ **100.** $\sqrt[12]{x^{44}}$ **101.** $\sqrt[10]{x^{16}}$ **102.** $\sqrt[12]{x^{38}}$

Simplify by first writing the radicals as radicals with the same index. Then multiply.
Assume that $x \geq 0$. ***See Example 7.***

103. $\sqrt[3]{4} \cdot \sqrt{3}$ **104.** $\sqrt[3]{5} \cdot \sqrt{6}$ **105.** $\sqrt[4]{3} \cdot \sqrt[3]{4}$

106. $\sqrt[3]{2} \cdot \sqrt[5]{3}$ **107.** $\sqrt{x} \cdot \sqrt[3]{x}$ **108.** $\sqrt[3]{x} \cdot \sqrt[4]{x}$

Find the unknown length in each right triangle. Simplify the answer if possible.
See Example 8.

109.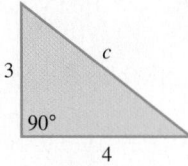

110.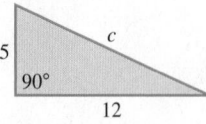

111.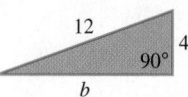

112.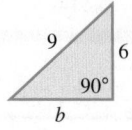

113.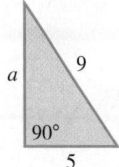

114.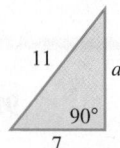

Find the distance between each pair of points. **See Example 9.**

115. $(6, 13)$ and $(1, 1)$

116. $(8, 13)$ and $(2, 5)$

117. $(-6, 5)$ and $(3, -4)$

118. $(-1, 5)$ and $(-7, 7)$

119. $(-8, 2)$ and $(-4, 1)$

120. $(-1, 2)$ and $(5, 3)$

121. $(4.7, 2.3)$ and $(1.7, -1.7)$

122. $(-2.9, 18.2)$ and $(2.1, 6.2)$

123. $\left(\sqrt{2}, \sqrt{6}\right)$ and $\left(-2\sqrt{2}, 4\sqrt{6}\right)$

124. $\left(\sqrt{7}, 9\sqrt{3}\right)$ and $\left(-\sqrt{7}, 4\sqrt{3}\right)$

125. $(x + y, y)$ and $(x - y, x)$

126. $(c, c - d)$ and $(d, c + d)$

▦ *Solve each problem.*

127. A Panasonic Smart Viera E50 LCD HDTV has a rectangular screen with a 36.5-in. width. Its height is 20.8 in. What is the length of the diagonal of the screen to the nearest tenth of an inch? (*Source:* Actual measurements of the author's television.)

36.5 in.

20.8 in.

128. The length of the diagonal of a box is given by

$$D = \sqrt{L^2 + W^2 + H^2},$$

where *L*, *W*, and *H* are the length, width, and height of the box. Find the length of the diagonal, *D*, of a box that is 4 ft long, 3 ft high, and 2 ft wide. Give the exact value, and then round to the nearest tenth of a foot.

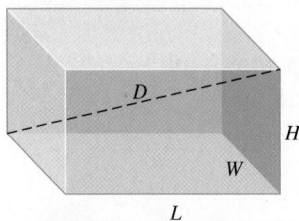

D

H

W

L

129. A formula from electronics dealing with impedance of parallel resonant circuits is

$$I = \frac{E}{\sqrt{R^2 + \omega^2 L^2}},$$

where the variables are in appropriate units. Find *I* if $E = 282$, $R = 100$, $L = 264$, and $\omega = 120\pi$. Give the answer to the nearest thousandth.

130. In the study of sound, one version of the law of tensions is

$$f_1 = f_2 \sqrt{\frac{F_1}{F_2}}.$$

Find f_1 to the nearest unit if $F_1 = 300$, $F_2 = 60$, and $f_2 = 260$.

The following letter appeared in the column "Ask Tom Why," written by Tom Skilling of the Chicago Tribune.

Dear Tom,
    I cannot remember the formula to calculate the distance to the horizon. I have a stunning view from my 14th floor condo, 150 feet above the ground. How far can I see?

Ted Fleischaker; Indianapolis, Ind.

Skilling's answer was as follows.

    To find the distance to the horizon in miles, take the square root of the height of your view in feet and multiply that result by 1.224. Your answer will be the number of miles to the horizon. (*Source: Chicago Tribune.*)

Use this information in Exercises 131 and 132.

131. (a) Write a formula for calculating the distance *d* to the horizon in terms of the height *h* of the view.

(b) Assuming Ted's eyes are 6 ft above the ground, the total height from the ground is $150 + 6 = 156$ ft. To the nearest tenth of a mile, how far can he see to the horizon?

132. Ted's neighbor Sheri lives on a floor that is 100 ft above the ground. Assuming that her eyes are 5 ft above the ground, to the nearest tenth of a mile, how far can she see to the horizon?

9.4 Adding and Subtracting Radical Expressions

OBJECTIVE

1 Simplify radical expressions involving addition and subtraction.

1 Add or subtract to simplify each radical expression.

(a) $3\sqrt{5} + 7\sqrt{5}$

(b) $2\sqrt{11} - \sqrt{11} + 3\sqrt{44}$

(c) $5\sqrt{12y} + 6\sqrt{75y}$, $y \geq 0$

(d) $3\sqrt{8} - 6\sqrt{50} + 2\sqrt{200}$

(e) $9\sqrt{5} - 4\sqrt{10}$

OBJECTIVE **1** **Simplify radical expressions involving addition and subtraction.** We do so by using the distributive property.

$$4\sqrt{2} + 3\sqrt{2}$$
$$= (4 + 3)\sqrt{2}$$
$$= 7\sqrt{2}$$

This is similar to simplifying $4x + 3x$ as $7x$.

$$2\sqrt{3} - 5\sqrt{3}$$
$$= (2 - 5)\sqrt{3}$$
$$= -3\sqrt{3}$$

This is similar to simplifying $2x - 5x$ as $-3x$.

CAUTION

Only radical expressions with the same index and the same radicand may be combined.

EXAMPLE 1 Adding and Subtracting Radicals

Add or subtract to simplify each radical expression.

(a) $3\sqrt{24} + \sqrt{54}$ ◁ Simplify each individual radical.

$$= 3\sqrt{4} \cdot \sqrt{6} + \sqrt{9} \cdot \sqrt{6} \quad \text{Product rule}$$
$$= 3 \cdot 2\sqrt{6} + 3\sqrt{6} \quad \text{Find the square roots.}$$
$$= 6\sqrt{6} + 3\sqrt{6} \quad \text{Multiply.}$$
$$= (6 + 3)\sqrt{6} \quad \text{Distributive property}$$
$$= 9\sqrt{6} \quad \text{Add.}$$

(b) $2\sqrt{20x} - \sqrt{45x}$, $x \geq 0$

$$= 2\sqrt{4} \cdot \sqrt{5x} - \sqrt{9} \cdot \sqrt{5x} \quad \text{Product rule}$$
$$= 2 \cdot 2\sqrt{5x} - 3\sqrt{5x} \quad \text{Find the square roots.}$$
$$= 4\sqrt{5x} - 3\sqrt{5x} \quad \text{Multiply.}$$
$$= \sqrt{5x} \quad \begin{array}{c}(4-3)\sqrt{5x} \\ = 1\sqrt{5x}, \text{ or } \sqrt{5x}\end{array} \quad \text{Combine like terms.}$$

(c) $2\sqrt{3} - 4\sqrt{5}$ The radicands differ and are already simplified, so this expression cannot be simplified further.

◁ Work Problem **1** at the Side.

CAUTION

In general, the root of a sum does not equal the sum of the roots. For example,

$$\sqrt{9 + 16} \neq \sqrt{9} + \sqrt{16},$$

since $\sqrt{9 + 16} = \sqrt{25} = 5$, but $\sqrt{9} + \sqrt{16} = 3 + 4 = 7$.

Answers

1. **(a)** $10\sqrt{5}$ **(b)** $7\sqrt{11}$
(c) $40\sqrt{3y}$ **(d)** $-4\sqrt{2}$
(e) cannot be simplified further

EXAMPLE 2   Adding and Subtracting Radicals

Simplify. Assume that all variables represent positive real numbers.

(a) $2\sqrt[3]{16} - 5\sqrt[3]{54}$ *(Remember to write the index with each radical.)*

$$= 2\sqrt[3]{8 \cdot 2} - 5\sqrt[3]{27 \cdot 2} \qquad \text{Factor.}$$

$$= 2\sqrt[3]{8} \cdot \sqrt[3]{2} - 5\sqrt[3]{27} \cdot \sqrt[3]{2} \qquad \text{Product rule}$$

$$= 2 \cdot 2 \cdot \sqrt[3]{2} - 5 \cdot 3 \cdot \sqrt[3]{2} \qquad \text{Find the cube roots.}$$

$$= 4\sqrt[3]{2} - 15\sqrt[3]{2} \qquad \text{Multiply.}$$

$$= (4 - 15)\sqrt[3]{2} \qquad \text{Distributive property}$$

$$= -11\sqrt[3]{2} \qquad \text{Subtract.}$$

(b) $2\sqrt[3]{x^2y} + \sqrt[3]{8x^5y^4}$

$$= 2\sqrt[3]{x^2y} + \sqrt[3]{8x^3y^3} \cdot \sqrt[3]{x^2y} \qquad \text{Factor; product rule}$$

$$= 2\sqrt[3]{x^2y} + 2xy\sqrt[3]{x^2y} \qquad \text{Find the cube root.}$$

(This result cannot be simplified further.)
$$= (2 + 2xy)\sqrt[3]{x^2y} \qquad \text{Distributive property}$$

················· **Work Problem 2 at the Side.** ▶

EXAMPLE 3   Adding and Subtracting Radicals with Fractions

Simplify. Assume that all variables represent positive real numbers.

(a) $2\sqrt{\dfrac{75}{16}} + 4\dfrac{\sqrt{8}}{\sqrt{32}}$

$$= 2\frac{\sqrt{25 \cdot 3}}{\sqrt{16}} + 4\frac{\sqrt{4 \cdot 2}}{\sqrt{16 \cdot 2}} \qquad \text{Quotient rule; factor.}$$

$$= 2\left(\frac{5\sqrt{3}}{4}\right) + 4\left(\frac{2\sqrt{2}}{4\sqrt{2}}\right) \qquad \text{Product rule; take square roots.}$$

$$= \frac{5\sqrt{3}}{2} + 2 \qquad \text{Multiply; } \frac{\sqrt{2}}{\sqrt{2}} = 1$$

$$= \frac{5\sqrt{3} + 4}{2} \qquad 2 = \frac{4}{2}; \frac{a}{c} + \frac{b}{c} = \frac{a+b}{c}$$

(b) $10\sqrt[3]{\dfrac{5}{x^6}} - 3\sqrt[3]{\dfrac{4}{x^9}}$

$$= 10\frac{\sqrt[3]{5}}{\sqrt[3]{x^6}} - 3\frac{\sqrt[3]{4}}{\sqrt[3]{x^9}} \qquad \text{Quotient rule}$$

$$= \frac{10\sqrt[3]{5}}{x^2} - \frac{3\sqrt[3]{4}}{x^3} \qquad \text{Simplify denominators.}$$

$$= \frac{10\sqrt[3]{5} \cdot x}{x^2 \cdot x} - \frac{3\sqrt[3]{4}}{x^3} \qquad \text{Write with a common denominator.}$$

$$= \frac{10x\sqrt[3]{5} - 3\sqrt[3]{4}}{x^3} \qquad \frac{a}{c} - \frac{b}{c} = \frac{a-b}{c}$$

················· **Work Problem 3 at the Side.** ▶

---

2 Simplify. Assume that all variables represent positive real numbers.

(a) $7\sqrt[3]{81} + 3\sqrt[3]{24}$

(b) $-2\sqrt[4]{32} - 7\sqrt[4]{162}$

(c) $\sqrt[3]{p^4q^7} - \sqrt[3]{64pq}$

3 Simplify. Assume that all variables represent positive real numbers.

(a) $2\sqrt{\dfrac{8}{9}} - 2\dfrac{\sqrt{27}}{\sqrt{108}}$

(b) $\sqrt{\dfrac{80}{y^4}} + \sqrt{\dfrac{81}{y^{10}}}$

Answers

2. (a) $27\sqrt[3]{3}$ **(b)** $-25\sqrt[4]{2}$
   (c) $(pq^2 - 4)\sqrt[3]{pq}$

3. (a) $\dfrac{4\sqrt{2} - 3}{3}$ **(b)** $\dfrac{4y^3\sqrt{5} + 9}{y^5}$

9.4 Exercises

 Download the MyDashBoard App

MyMathLab®

CONCEPT CHECK *Choose the correct response.*

1. Which sum can be simplified without first simplifying the individual radical expressions?

   A. $\sqrt{50} + \sqrt{32}$ **B.** $3\sqrt{6} + 9\sqrt{6}$

   C. $\sqrt[3]{32} - \sqrt[3]{108}$ **D.** $\sqrt[5]{6} - \sqrt[5]{192}$

2. Which difference can be simplified without first simplifying the individual radical expressions?

   A. $\sqrt{81} - \sqrt{18}$ **B.** $\sqrt[3]{8} - \sqrt[3]{16}$

   C. $4\sqrt[3]{7} - 9\sqrt[3]{7}$ **D.** $\sqrt{75} - \sqrt{12}$

3. **CONCEPT CHECK** Even though the indexes of the terms are not equal, the sum

$$\sqrt{64} + \sqrt[3]{125} + \sqrt[4]{16}$$

   can be simplified easily. What is this sum? Why can these terms be easily combined?

4. **CONCEPT CHECK** On an algebra quiz, Erin gave the difference $28 - 4\sqrt{2}$ as $24\sqrt{2}$. Her teacher did not give her any credit for this answer. **What Went Wrong?**

Simplify. Assume that all variables represent positive real numbers.
See Examples 1 and 2.

5. $\sqrt{36} - \sqrt{100}$

6. $\sqrt{25} - \sqrt{81}$

7. $-2\sqrt{48} + 3\sqrt{75}$

8. $4\sqrt{32} - 2\sqrt{8}$

9. $\sqrt[3]{16} + 4\sqrt[3]{54}$

10. $3\sqrt[3]{24} - 2\sqrt[3]{192}$

11. $\sqrt[4]{32} + 3\sqrt[4]{2}$

12. $\sqrt[4]{405} - 2\sqrt[4]{5}$

13. $6\sqrt{18} - \sqrt{32} + 2\sqrt{50}$

14. $5\sqrt{8} + 3\sqrt{72} - 3\sqrt{50}$

15. $5\sqrt{6} + 2\sqrt{10}$

16. $3\sqrt{11} - 5\sqrt{13}$

17. $2\sqrt{5} + 3\sqrt{20} + 4\sqrt{45}$

18. $5\sqrt{54} - 2\sqrt{24} - 2\sqrt{96}$

19. $8\sqrt{2x} - \sqrt{8x} + \sqrt{72x}$

20. $4\sqrt{18k} - \sqrt{72k} + \sqrt{50k}$

21. $3\sqrt{72m^2} - 5\sqrt{32m^2}$

22. $9\sqrt{27p^2} - 14\sqrt{108p^2}$

23. $-\sqrt[3]{54} + 2\sqrt[3]{16}$

24. $15\sqrt[3]{81} - 4\sqrt[3]{24}$

25. $2\sqrt[3]{27x} - 2\sqrt[3]{8x}$

26. $6\sqrt[3]{128m} + 3\sqrt[3]{16m}$

27. $\sqrt[3]{x^2y} - \sqrt[3]{8x^2y}$

28. $3\sqrt[3]{x^2y^2} - 2\sqrt[3]{64x^2y^2}$

29. $3x\sqrt[3]{xy^2} - 2\sqrt[3]{8x^4y^2}$

30. $6q^2\sqrt[3]{5q} - 2q\sqrt[3]{40q^4}$

31. $5\sqrt[4]{32} + 3\sqrt[4]{162}$

32. $2\sqrt[4]{512} + 4\sqrt[4]{32}$

33. $3\sqrt[4]{x^5y} - 2x\sqrt[4]{xy}$

34. $2\sqrt[4]{m^9p^6} - 3m^2p\sqrt[4]{mp^2}$

35. $2\sqrt[4]{32a^3} + 5\sqrt[4]{2a^3}$

36. $-\sqrt[4]{16r} + 5\sqrt[4]{r}$

Simplify. Assume that all variables represent positive real numbers. ***See Example 3.***

37. $\dfrac{2\sqrt{5}}{3} + \dfrac{\sqrt{5}}{6}$

38. $\dfrac{4\sqrt{3}}{3} + \dfrac{2\sqrt{3}}{9}$

39. $\sqrt{\dfrac{8}{9}} + \sqrt{\dfrac{18}{36}}$

40. $\sqrt{\dfrac{12}{16}} + \sqrt{\dfrac{48}{64}}$

41. $\dfrac{\sqrt{32}}{3} + \dfrac{2\sqrt{2}}{3} - \dfrac{\sqrt{2}}{\sqrt{9}}$

42. $\dfrac{\sqrt{27}}{2} - \dfrac{3\sqrt{3}}{2} + \dfrac{\sqrt{3}}{\sqrt{4}}$

43. $3\sqrt{\dfrac{50}{9}} + 8\dfrac{\sqrt{2}}{\sqrt{8}}$

44. $5\sqrt{\dfrac{288}{25}} + 21\dfrac{\sqrt{2}}{\sqrt{18}}$

45. $3\sqrt{\dfrac{50}{49}} - \dfrac{\sqrt{27}}{\sqrt{12}}$

46. $9\sqrt{\dfrac{48}{25}} - 2\dfrac{\sqrt{2}}{\sqrt{98}}$

47. $\sqrt{\dfrac{25}{x^8}} - \sqrt{\dfrac{9}{x^6}}$

48. $\sqrt{\dfrac{100}{y^4}} + \sqrt{\dfrac{81}{y^{10}}}$

49. $3\sqrt[3]{\dfrac{m^5}{27}} - 2m\sqrt[3]{\dfrac{m^2}{64}}$

50. $2a\sqrt[4]{\dfrac{a}{16}} - 5a\sqrt[4]{\dfrac{a}{81}}$

51. $3\sqrt[3]{\dfrac{2}{x^6}} - 4\sqrt[3]{\dfrac{5}{x^9}}$

52. $-4\sqrt[3]{\dfrac{4}{t^9}} + 3\sqrt[3]{\dfrac{9}{t^{12}}}$

Solve each problem. Give answers as simplified radical expressions.

53. Find the perimeter of the triangle.

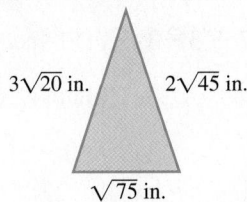

$3\sqrt{20}$ in. $2\sqrt{45}$ in.

$\sqrt{75}$ in.

54. Find the perimeter of the rectangle.

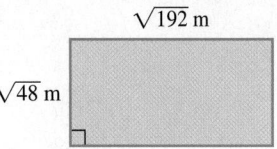

$\sqrt{192}$ m

$\sqrt{48}$ m

55. What is the perimeter of the rectangular screen?

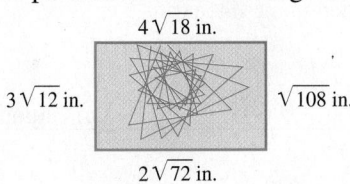

$4\sqrt{18}$ in.

$3\sqrt{12}$ in. $\sqrt{108}$ in.

$2\sqrt{72}$ in.

56. Find the area of the trapezoid.

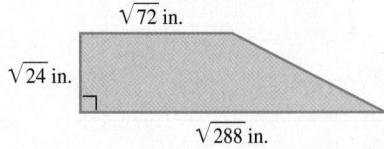

$\sqrt{72}$ in.

$\sqrt{24}$ in.

$\sqrt{288}$ in.

9.5 Multiplying and Dividing Radical Expressions

OBJECTIVES

1. Multiply radical expressions.
2. Rationalize denominators with one radical term.
3. Rationalize denominators with binomials involving radicals.
4. Write radical quotients in lowest terms.

OBJECTIVE **1** **Multiply radical expressions.** We multiply binomial expressions involving radicals by using the FOIL method from **Section 5.3.** Recall that the acronym **FOIL** refers to multiplying the **F**irst terms, **O**uter terms, **I**nner terms, and **L**ast terms of the binomials.

EXAMPLE 1 Multiplying Binomials Involving Radical Expressions

Multiply by using the FOIL method.

(a) $\left(\sqrt{5} + 3\right)\left(\sqrt{6} + 1\right)$

$$
\begin{array}{cccc}
\text{First} & \text{Outer} & \text{Inner} & \text{Last}
\end{array}
$$
$$= \sqrt{5} \cdot \sqrt{6} + \sqrt{5} \cdot 1 + 3 \cdot \sqrt{6} + 3 \cdot 1 \quad \text{FOIL method}$$
$$= \sqrt{30} + \sqrt{5} + 3\sqrt{6} + 3 \quad \text{This result cannot be simplified further.}$$

(b) $\left(7 - \sqrt{3}\right)\left(\sqrt{5} + \sqrt{2}\right)$

$$
\begin{array}{cccc}
\text{F} & \text{O} & \text{I} & \text{L}
\end{array}
$$
$$= 7\sqrt{5} + 7\sqrt{2} - \sqrt{3} \cdot \sqrt{5} - \sqrt{3} \cdot \sqrt{2}$$
$$= 7\sqrt{5} + 7\sqrt{2} - \sqrt{15} - \sqrt{6} \quad \text{Product rule}$$

(c) $\left(\sqrt{10} + \sqrt{3}\right)\left(\sqrt{10} - \sqrt{3}\right)$

$$= \sqrt{10} \cdot \sqrt{10} - \sqrt{10} \cdot \sqrt{3} + \sqrt{3} \cdot \sqrt{10} - \sqrt{3} \cdot \sqrt{3} \quad \text{FOIL method}$$
$$= 10 - 3 \quad \text{Product rule}$$
$$= 7 \quad \text{Subtract.}$$

The product $\left(\sqrt{10} + \sqrt{3}\right)\left(\sqrt{10} - \sqrt{3}\right) = \left(\sqrt{10}\right)^2 - \left(\sqrt{3}\right)^2$ is the difference of squares. (See **Section 5.4.**)

$$(x + y)(x - y) = x^2 - y^2 \quad \text{Here, } x = \sqrt{10} \text{ and } y = \sqrt{3}.$$

(d) $\left(\sqrt{7} - 3\right)^2$

$$= \left(\sqrt{7} - 3\right)\left(\sqrt{7} - 3\right) \quad a^2 = a \cdot a$$
$$= \sqrt{7} \cdot \sqrt{7} - 3\sqrt{7} - 3\sqrt{7} + 3 \cdot 3 \quad \text{FOIL method}$$
$$= 7 - 6\sqrt{7} + 9 \quad \text{Multiply. Combine like terms.}$$
$$= 16 - 6\sqrt{7} \quad \text{Add.}$$

Be careful. These terms cannot be combined.

(e) $\left(5 - \sqrt[3]{3}\right)\left(5 + \sqrt[3]{3}\right)$

$$= 5 \cdot 5 + 5\sqrt[3]{3} - 5\sqrt[3]{3} - \sqrt[3]{3} \cdot \sqrt[3]{3}$$
$$= 25 - \sqrt[3]{3^2} \quad \text{Multiply. Combine like terms.}$$
$$= 25 - \sqrt[3]{9} \quad \text{Apply the exponent.}$$

Remember to write the index 3 in *each* radical.

Continued on Next Page

(f) $\left(\sqrt{k} + \sqrt{y}\right)\left(\sqrt{k} - \sqrt{y}\right)$

$= \left(\sqrt{k}\right)^2 - \left(\sqrt{y}\right)^2$ Difference of squares

$= k - y, \quad k \geq 0 \text{ and } y \geq 0$

Note

In **Example 1(d),** we could have used the formula for the square of a binomial to obtain the same result.

$\left(\sqrt{7} - 3\right)^2$

$= \left(\sqrt{7}\right)^2 - 2\left(\sqrt{7}\right)(3) + 3^2 \quad (x-y)^2 = x^2 - 2xy + y^2$

$= 7 - 6\sqrt{7} + 9$ Apply the exponents. Multiply.

$= 16 - 6\sqrt{7}$ Add.

Work Problem ❶ at the Side. ▶

OBJECTIVE ❷ Rationalize denominators with one radical term. As defined earlier, a simplified radical expression will have no radical in the denominator. The origin of this agreement no doubt occurred before the days of high-speed calculation, when computation was a tedious process performed by hand.

For example, consider the radical expression $\frac{1}{\sqrt{2}}$. To find a decimal approximation by hand, it would be necessary to divide 1 by a decimal approximation for $\sqrt{2}$, such as 1.414. It would be much easier if the divisor were a whole number. This can be accomplished by multiplying $\frac{1}{\sqrt{2}}$ by 1 in the form $\frac{\sqrt{2}}{\sqrt{2}}$. *Multiplying by 1 in any form does not change the value of the original expression.*

$$\frac{1}{\sqrt{2}} \cdot \frac{\sqrt{2}}{\sqrt{2}} = \frac{\sqrt{2}}{2} \quad \text{Multiply by 1; } \frac{\sqrt{2}}{\sqrt{2}} = 1$$

Now the computation for the approximation would require dividing 1.414 by 2 to obtain 0.707, a much easier task.

With current technology, either form of this fraction can be approximated with the same number of keystrokes. See **Figure 9**, which shows how a calculator gives the same approximation for both forms of the expression.

```
1/√(2)
         .7071067812
√(2)/2
         .7071067812
```

Figure 9

Rationalizing a Denominator

A common way of "standardizing" the form of a radical expression is to have the denominator contain no radicals. The process of removing radicals from a denominator so that the denominator contains only rational numbers is called **rationalizing the denominator.** This is done by multiplying by a form of 1.

❶ Multiply by using the FOIL method.

(a) $\left(2 + \sqrt{3}\right)\left(1 + \sqrt{5}\right)$

$= 2 \cdot \underline{\quad} + 2 \cdot \underline{\quad}$

$+ \sqrt{3} \cdot \underline{\quad} + \sqrt{3} \cdot \underline{\quad}$

$= \underline{\hspace{3cm}}$

(b) $\left(4 + \sqrt{3}\right)\left(4 - \sqrt{3}\right)$

(c) $\left(\sqrt{13} - 2\right)^2$

(d) $\left(4 + \sqrt[3]{7}\right)\left(4 - \sqrt[3]{7}\right)$

(e) $\left(\sqrt{p} + \sqrt{s}\right)\left(\sqrt{p} - \sqrt{s}\right),$ $p \geq 0 \text{ and } s \geq 0$

Answers

1. **(a)** 1; $\sqrt{5}$; 1; $\sqrt{5}$; $2 + 2\sqrt{5} + \sqrt{3} + \sqrt{15}$
   (b) 13 **(c)** $17 - 4\sqrt{13}$
   (d) $16 - \sqrt[3]{49}$ **(e)** $p - s$

❷ Rationalize each denominator.

(a) $\dfrac{8}{\sqrt{3}}$

(b) $\dfrac{5\sqrt{6}}{\sqrt{5}}$

(c) $\dfrac{3}{\sqrt{48}}$

(d) $\dfrac{-16}{\sqrt{32}}$

❸ Simplify each radical. Assume that all variables represent positive real numbers.

(a) $\sqrt{\dfrac{8}{45}}$

(b) $\sqrt{\dfrac{72}{y}}$

(c) $\sqrt{\dfrac{200k^6}{y^7}}$

Answers

2. (a) $\dfrac{8\sqrt{3}}{3}$ (b) $\sqrt{30}$ (c) $\dfrac{\sqrt{3}}{4}$

   (d) $-2\sqrt{2}$

3. (a) $\dfrac{2\sqrt{10}}{15}$ (b) $\dfrac{6\sqrt{2y}}{y}$ (c) $\dfrac{10k^3\sqrt{2y}}{y^4}$

EXAMPLE 2 **Rationalizing Denominators with Square Roots**

Rationalize each denominator.

(a) $\dfrac{3}{\sqrt{7}}$

Multiply by $\dfrac{\sqrt{7}}{\sqrt{7}}$. This is an application of the identity property for multiplication.

$$\dfrac{3}{\sqrt{7}} = \dfrac{3 \cdot \sqrt{7}}{\sqrt{7} \cdot \sqrt{7}} = \dfrac{3\sqrt{7}}{7}$$

In the denominator, $\sqrt{7} \cdot \sqrt{7} = \sqrt{7 \cdot 7} = \sqrt{49} = 7$. The final denominator is now a rational number.

(b) $\dfrac{5\sqrt{2}}{\sqrt{5}} = \dfrac{5\sqrt{2} \cdot \sqrt{5}}{\sqrt{5} \cdot \sqrt{5}} = \dfrac{5\sqrt{10}}{5} = \sqrt{10}$

(c) $\dfrac{-6}{\sqrt{12}}$

Less work is involved if the radical in the denominator is simplified first.

$$\dfrac{-6}{\sqrt{12}} = \dfrac{-6}{\sqrt{4 \cdot 3}} = \dfrac{-6}{2\sqrt{3}} = \dfrac{-3}{\sqrt{3}}$$

Now rationalize the denominator.

$$\dfrac{-3}{\sqrt{3}} = \dfrac{-3 \cdot \sqrt{3}}{\sqrt{3} \cdot \sqrt{3}} = \dfrac{-3\sqrt{3}}{3} = -\sqrt{3}$$

◀ **Work Problem ❷ at the Side.**

EXAMPLE 3 **Rationalizing Denominators in Roots of Fractions**

Simplify each radical. In part (b), $p > 0$.

(a) $-\sqrt{\dfrac{18}{125}}$

$= -\dfrac{\sqrt{18}}{\sqrt{125}}$ Quotient rule

$= -\dfrac{\sqrt{9 \cdot 2}}{\sqrt{25 \cdot 5}}$ Factor.

$= -\dfrac{3\sqrt{2}}{5\sqrt{5}}$ Product rule

$= -\dfrac{3\sqrt{2} \cdot \sqrt{5}}{5\sqrt{5} \cdot \sqrt{5}}$ Multiply by $\dfrac{\sqrt{5}}{\sqrt{5}}$.

$= -\dfrac{3\sqrt{10}}{5 \cdot 5}$ Product rule

$= -\dfrac{3\sqrt{10}}{25}$ Multiply.

(b) $\sqrt{\dfrac{50m^4}{p^5}}$

$= \dfrac{\sqrt{50m^4}}{\sqrt{p^5}}$ Quotient rule

$= \dfrac{\sqrt{25m^4 \cdot 2}}{\sqrt{p^4 \cdot p}}$ Factor.

$= \dfrac{5m^2\sqrt{2}}{p^2\sqrt{p}}$ Product rule

$= \dfrac{5m^2\sqrt{2} \cdot \sqrt{p}}{p^2\sqrt{p} \cdot \sqrt{p}}$ Multiply by $\dfrac{\sqrt{p}}{\sqrt{p}}$.

$= \dfrac{5m^2\sqrt{2p}}{p^2 \cdot p}$ Product rule

$= \dfrac{5m^2\sqrt{2p}}{p^3}$ Multiply.

◀ **Work Problem ❸ at the Side.**

| EXAMPLE 4 | **Rationalizing Denominators with Higher Roots** |

Simplify.

(a) $\sqrt[3]{\dfrac{27}{16}}$

Use the quotient rule, and simplify the numerator and denominator.

$$\sqrt[3]{\frac{27}{16}} = \frac{\sqrt[3]{27}}{\sqrt[3]{16}} = \frac{3}{\sqrt[3]{8} \cdot \sqrt[3]{2}} = \frac{3}{2\sqrt[3]{2}}$$

Since $2 \cdot 4 = 8$, a perfect cube, multiply the numerator and denominator by $\sqrt[3]{4}$.

$$\frac{3}{2\sqrt[3]{2}} \qquad\qquad \sqrt[3]{\frac{27}{16}} = \frac{3}{2\sqrt[3]{2}}$$

$$= \frac{3 \cdot \sqrt[3]{4}}{2\sqrt[3]{2} \cdot \sqrt[3]{4}} \qquad \text{Multiply by } \sqrt[3]{4} \text{ in numerator and denominator.}$$
$$\text{This will give } \sqrt[3]{8} = 2 \text{ in the denominator.}$$

$$= \frac{3\sqrt[3]{4}}{2\sqrt[3]{8}} \qquad \text{Multiply.}$$

$$= \frac{3\sqrt[3]{4}}{2 \cdot 2} \qquad \sqrt[3]{8} = 2$$

$$= \frac{3\sqrt[3]{4}}{4} \qquad \text{Multiply.}$$

(b) $\sqrt[4]{\dfrac{5x}{z}}$

$$= \frac{\sqrt[4]{5x}}{\sqrt[4]{z}} \qquad\qquad \text{Quotient rule}$$

$$= \frac{\sqrt[4]{5x}}{\sqrt[4]{z}} \cdot \frac{\sqrt[4]{z^3}}{\sqrt[4]{z^3}} \qquad \text{Multiply by 1.}$$

> $\sqrt[4]{z} \cdot \sqrt[4]{z^3}$ will give $\sqrt[4]{z^4}$.

$$= \frac{\sqrt[4]{5xz^3}}{\sqrt[4]{z^4}} \qquad \text{Product rule}$$

$$= \frac{\sqrt[4]{5xz^3}}{z}, \quad x \geq 0, z > 0$$

CAUTION

In **Example 4(a)**, a typical error is to multiply the numerator and denominator by $\sqrt[3]{2}$, forgetting that $\sqrt[3]{2} \cdot \sqrt[3]{2} = \sqrt[3]{2^2}$ which does **not** equal 2. We need **three** factors of 2 to obtain 2^3 under the radical.

$$\sqrt[3]{2} \cdot \sqrt[3]{2} \cdot \sqrt[3]{2} = \sqrt[3]{2^3}, \quad \text{which does equal 2.}$$

Work Problem 4 at the Side. ▶

---

④ Simplify.

(a) $\sqrt[3]{\dfrac{15}{32}}$

(b) $\sqrt[3]{\dfrac{m^{12}}{n}}, \quad n \neq 0$

(c) $\sqrt[4]{\dfrac{6y}{w^2}}, \quad y \geq 0, w \neq 0$

Answers

4. (a) $\dfrac{\sqrt[3]{30}}{4}$ **(b)** $\dfrac{m^4\sqrt[3]{n^2}}{n}$ **(c)** $\dfrac{\sqrt[4]{6yw^2}}{w}$

OBJECTIVE ▶ **3** **Rationalize denominators with binomials involving radicals.** Recall the special product

$$(x + y)(x - y) = x^2 - y^2.$$

To rationalize a denominator that contains a binomial expression (one that contains exactly two terms) involving radicals, such as

$$\frac{3}{1 + \sqrt{2}},$$

we must use *conjugates*. The conjugate of $1 + \sqrt{2}$ is $1 - \sqrt{2}$. In general, $x + y$ and $x - y$ are **conjugates.**

Rationalizing a Binomial Denominator

Whenever a radical expression has a sum or difference with square root radicals in the denominator, rationalize the denominator by multiplying both the numerator and denominator by the conjugate of the denominator.

EXAMPLE 5 **Rationalizing Binomial Denominators**

Rationalize each denominator.

(a) $\dfrac{3}{1 + \sqrt{2}}$

> Again, we are multiplying by a form of 1.

$$= \frac{3\left(1 - \sqrt{2}\right)}{\left(1 + \sqrt{2}\right)\left(1 - \sqrt{2}\right)}$$

Multiply the numerator and denominator by $1 - \sqrt{2}$, the conjugate of the denominator.

$$= \frac{3\left(1 - \sqrt{2}\right)}{-1}$$

> The denominator is now a rational number.

$$\left(1 + \sqrt{2}\right)\left(1 - \sqrt{2}\right)$$
$$= 1^2 - \left(\sqrt{2}\right)^2$$
$$= 1 - 2, \text{ or } -1$$

$$= \frac{3}{-1}\left(1 - \sqrt{2}\right)$$

$$\frac{a \cdot b}{c} = \frac{a}{c} \cdot b$$

$$= -3\left(1 - \sqrt{2}\right), \quad \text{or} \quad -3 + 3\sqrt{2} \qquad \text{Distributive property}$$

(b) $\dfrac{5}{4 - \sqrt{3}}$

$$= \frac{5\left(4 + \sqrt{3}\right)}{\left(4 - \sqrt{3}\right)\left(4 + \sqrt{3}\right)}$$

Multiply the numerator and denominator by $4 + \sqrt{3}$.

$$= \frac{5\left(4 + \sqrt{3}\right)}{16 - 3}$$

Multiply in the denominator.

$$= \frac{5\left(4 + \sqrt{3}\right)}{13}$$

Subtract in the denominator.

We leave the numerator in factored form. This makes it easier to determine whether the expression is written in lowest terms.

· **Continued on Next Page**

(c) $\dfrac{\sqrt{2} - \sqrt{3}}{\sqrt{5} + \sqrt{3}}$

$= \dfrac{(\sqrt{2} - \sqrt{3})(\sqrt{5} - \sqrt{3})}{(\sqrt{5} + \sqrt{3})(\sqrt{5} - \sqrt{3})}$ Multiply the numerator and denominator by $\sqrt{5} - \sqrt{3}$.

$= \dfrac{\sqrt{10} - \sqrt{6} - \sqrt{15} + 3}{5 - 3}$ Multiply.

$= \dfrac{\sqrt{10} - \sqrt{6} - \sqrt{15} + 3}{2}$ Subtract in the denominator.

(d) $\dfrac{3}{\sqrt{5m} - \sqrt{p}}, \quad 5m \neq p, m > 0, p > 0$

$= \dfrac{3(\sqrt{5m} + \sqrt{p})}{(\sqrt{5m} - \sqrt{p})(\sqrt{5m} + \sqrt{p})}$ Multiply the numerator and denominator by $\sqrt{5m} + \sqrt{p}$.

$= \dfrac{3(\sqrt{5m} + \sqrt{p})}{5m - p}$ Multiply in the denominator.

················· **Work Problem ❺ at the Side.** ▶

OBJECTIVE ▶ ❹ **Write radical quotients in lowest terms.**

EXAMPLE 6   **Writing Radical Quotients in Lowest Terms**

Write each quotient in lowest terms.

(a) $\dfrac{6 + 2\sqrt{5}}{4}$

$= \dfrac{2(3 + \sqrt{5})}{2 \cdot 2}$ *This is a key step.*   Factor the numerator and denominator.

$= \dfrac{3 + \sqrt{5}}{2}$ Divide out the common factor.

Here is an alternative method for writing this expression in lowest terms.

$$\frac{6 + 2\sqrt{5}}{4} = \frac{6}{4} + \frac{2\sqrt{5}}{4} = \frac{3}{2} + \frac{\sqrt{5}}{2}, \quad \text{or} \quad \frac{3 + \sqrt{5}}{2}$$

(b) $\dfrac{5y - \sqrt{8y^2}}{6y}, \quad y > 0$

$= \dfrac{5y - 2y\sqrt{2}}{6y}$ $\sqrt{8y^2} = \sqrt{4y^2 \cdot 2} = 2y\sqrt{2}$

$= \dfrac{y(5 - 2\sqrt{2})}{6y}$ Factor the numerator.

$= \dfrac{5 - 2\sqrt{2}}{6}$ Divide out the common factor.

················· **Work Problem ❻ at the Side.** ▶

---

❺ Rationalize each denominator.

(a) $\dfrac{-4}{\sqrt{5} + 2}$

(b) $\dfrac{15}{\sqrt{7} + \sqrt{2}}$

(c) $\dfrac{\sqrt{3} + \sqrt{5}}{\sqrt{2} - \sqrt{7}}$

(d) $\dfrac{2}{\sqrt{k} + \sqrt{z}},$

$k \neq z, k > 0, z > 0$

❻ Write each quotient in lowest terms.

(a) $\dfrac{24 - 36\sqrt{7}}{16}$

(b) $\dfrac{2x + \sqrt{32x^2}}{6x}, \quad x > 0$

Answers

5. (a) $-4(\sqrt{5} - 2)$ **(b)** $3(\sqrt{7} - \sqrt{2})$

    (c) $\dfrac{-(\sqrt{6} + \sqrt{21} + \sqrt{10} + \sqrt{35})}{5}$

    (d) $\dfrac{2(\sqrt{k} - \sqrt{z})}{k - z}$

6. (a) $\dfrac{6 - 9\sqrt{7}}{4}$ **(b)** $\dfrac{1 + 2\sqrt{2}}{3}$

9.5 Exercises

MyMathLab®

CONCEPT CHECK *Match each part of a rule for a special product in Column I with the part it equals in Column II.*

I	**II**
1. $(x + \sqrt{y})(x - \sqrt{y})$	**A.** $x - y$
2. $(\sqrt{x} + y)(\sqrt{x} - y)$	**B.** $x + 2y\sqrt{x} + y^2$
3. $(\sqrt{x} + \sqrt{y})(\sqrt{x} - \sqrt{y})$	**C.** $x - y^2$
4. $(\sqrt{x} + \sqrt{y})^2$	**D.** $x - 2\sqrt{xy} + y$
5. $(\sqrt{x} - \sqrt{y})^2$	**E.** $x^2 - y$
6. $(\sqrt{x} + y)^2$	**F.** $x + 2\sqrt{xy} + y$

Multiply, and then simplify each product. Assume that all variables represent positive real numbers. **See Example 1.**

7. $\sqrt{3}\left(\sqrt{12} - 4\right)$

8. $\sqrt{5}\left(\sqrt{125} - 6\right)$

9. $\sqrt{2}\left(\sqrt{18} - \sqrt{3}\right)$

10. $\sqrt{5}\left(\sqrt{15} + \sqrt{5}\right)$

11. $\left(\sqrt{6} + 2\right)\left(\sqrt{6} - 2\right)$

12. $\left(\sqrt{7} + 8\right)\left(\sqrt{7} - 8\right)$

13. $\left(\sqrt{12} - \sqrt{3}\right)\left(\sqrt{12} + \sqrt{3}\right)$

14. $\left(\sqrt{18} + \sqrt{8}\right)\left(\sqrt{18} - \sqrt{8}\right)$

15. $\left(\sqrt{3} + 2\right)\left(\sqrt{6} - 5\right)$

16. $\left(\sqrt{7} + 1\right)\left(\sqrt{2} - 4\right)$

17. $\left(\sqrt{3x} + 2\right)\left(\sqrt{3x} - 2\right)$

18. $\left(\sqrt{6y} - 4\right)\left(\sqrt{6y} + 4\right)$

19. $\left(2\sqrt{x} + \sqrt{y}\right)\left(2\sqrt{x} - \sqrt{y}\right)$

20. $\left(\sqrt{p} + 5\sqrt{s}\right)\left(\sqrt{p} - 5\sqrt{s}\right)$

21. $\left(4\sqrt{x} + 3\right)^2$

22. $\left(5\sqrt{p} - 6\right)^2$

23. $\left(9 - \sqrt[3]{2}\right)\left(9 + \sqrt[3]{2}\right)$

24. $\left(7 + \sqrt[3]{6}\right)\left(7 - \sqrt[3]{6}\right)$

25. CONCEPT CHECK The correct answer to **Exercise 7** is $6 - 4\sqrt{3}$. Why is it not equal to $2\sqrt{3}$?

26. CONCEPT CHECK When we rationalize the denominator in the radical expression $\frac{1}{\sqrt{2}}$, we multiply both the numerator and denominator by $\sqrt{2}$. What property of real numbers covered in **Section 1.7** justifies this procedure?

Rationalize the denominator in each expression. Assume that all variables represent positive real numbers. ***See Example 2.***

27. $\dfrac{7}{\sqrt{7}}$

28. $\dfrac{11}{\sqrt{11}}$

29. $\dfrac{15}{\sqrt{3}}$

30. $\dfrac{12}{\sqrt{6}}$

31. $\dfrac{\sqrt{3}}{\sqrt{2}}$

32. $\dfrac{\sqrt{7}}{\sqrt{6}}$

33. $\dfrac{9\sqrt{3}}{\sqrt{5}}$

34. $\dfrac{3\sqrt{2}}{\sqrt{11}}$

35. $\dfrac{-6}{\sqrt{18}}$

36. $\dfrac{-5}{\sqrt{24}}$

37. $\dfrac{-8\sqrt{3}}{\sqrt{k}}$

38. $\dfrac{-4\sqrt{13}}{\sqrt{m}}$

39. $\dfrac{6\sqrt{3y}}{\sqrt{y^3}}$

40. $\dfrac{-8\sqrt{5y}}{\sqrt{y^5}}$

Simplify. Assume that all variables represent positive real numbers. ***See Examples 3 and 4.***

41. $\sqrt{\dfrac{7}{2}}$

42. $\sqrt{\dfrac{10}{3}}$

43. $-\sqrt{\dfrac{7}{50}}$

44. $-\sqrt{\dfrac{13}{75}}$

45. $\sqrt{\dfrac{24}{x}}$

46. $\sqrt{\dfrac{52}{y}}$

47. $-\sqrt{\dfrac{98r^3}{s}}$

48. $-\sqrt{\dfrac{150m^5}{n}}$

49. $\sqrt{\dfrac{288x^7}{y^9}}$

50. $\sqrt{\dfrac{242t^9}{u^{11}}}$

51. $\sqrt[3]{\dfrac{2}{3}}$

52. $\sqrt[3]{\dfrac{4}{5}}$

53. $\sqrt[3]{\dfrac{4}{9}}$

54. $\sqrt[3]{\dfrac{5}{16}}$

55. $-\sqrt[3]{\dfrac{2p}{r^2}}$

56. $-\sqrt[3]{\dfrac{6x}{y^2}}$

57. $\sqrt[4]{\dfrac{16}{x}}$

58. $\sqrt[4]{\dfrac{81}{y}}$

59. $\sqrt[4]{\dfrac{2y}{z}}$

60. $\sqrt[4]{\dfrac{7t}{s^2}}$

Rationalize the denominator in each expression. Assume that all variables represent positive real numbers and that no denominators are 0. **See Example 5.**

61. $\dfrac{2}{4 + \sqrt{3}}$

62. $\dfrac{6}{5 + \sqrt{2}}$

63. $\dfrac{6}{\sqrt{5} + \sqrt{3}}$

64. $\dfrac{12}{\sqrt{6} + \sqrt{3}}$

65. $\dfrac{-4}{\sqrt{3} - \sqrt{7}}$

66. $\dfrac{-3}{\sqrt{2} + \sqrt{5}}$

67. $\dfrac{1 - \sqrt{2}}{\sqrt{7} + \sqrt{6}}$

68. $\dfrac{-1 - \sqrt{3}}{\sqrt{6} + \sqrt{5}}$

69. $\dfrac{\sqrt{2} - \sqrt{3}}{\sqrt{6} - \sqrt{5}}$

70. $\dfrac{\sqrt{5} + \sqrt{6}}{\sqrt{3} - \sqrt{2}}$

71. $\dfrac{4}{\sqrt{x} - 2\sqrt{y}}$

72. $\dfrac{5}{3\sqrt{r} + \sqrt{s}}$

73. $\dfrac{\sqrt{x} - \sqrt{y}}{\sqrt{2x} + \sqrt{3y}}$

74. $\dfrac{\sqrt{a} + \sqrt{b}}{\sqrt{5a} - \sqrt{2b}}$

Write each quotient in lowest terms. Assume that all variables represent positive real numbers. **See Example 6.**

75. $\dfrac{25 + 10\sqrt{6}}{20}$

76. $\dfrac{12 - 6\sqrt{2}}{24}$

77. $\dfrac{16 + 4\sqrt{8}}{12}$

78. $\dfrac{12 + 9\sqrt{72}}{18}$

79. $\dfrac{6x + \sqrt{24x^3}}{3x}$

80. $\dfrac{11y + \sqrt{242y^5}}{22y}$

Summary Exercises *Performing Operations with Radicals and Rational Exponents*

Recall that a simplified radical satisfies the following conditions.

Conditions for a Simplified Radical

1. The radicand has no factor raised to a power greater than or equal to the index.

2. The radicand has no fractions.

3. No denominator contains a radical.

4. Exponents in the radicand and the index of the radical have greatest common factor 1.

CONCEPT CHECK   *Give the reason why each radical is not simplified.*

1. $\sqrt{\dfrac{2}{5}}$

2. $\sqrt[15]{x^5}$

3. $\dfrac{5}{\sqrt[3]{10}}$

4. $\sqrt[3]{x^5 y^6}$

Perform all indicated operations, and express each answer in simplest form with positive exponents. Assume that all variables represent positive real numbers.

5. $6\sqrt{10} - 12\sqrt{10}$

6. $\sqrt{7}\left(\sqrt{7} - \sqrt{2}\right)$

7. $\left(1 - \sqrt{3}\right)\left(2 + \sqrt{6}\right)$

8. $\sqrt{50} - \sqrt{98} + \sqrt{72}$

9. $\left(3\sqrt{5} + 2\sqrt{7}\right)^2$

10. $\dfrac{-3}{\sqrt{6}}$

11. $\dfrac{8}{\sqrt{7} + \sqrt{5}}$

12. $\sqrt[3]{16x^2} - \sqrt[3]{54x^2} + \sqrt[3]{128x^2}$

13. $\dfrac{1 - \sqrt{2}}{1 + \sqrt{2}}$

14. $\left(1 - \sqrt[3]{3}\right)\left(1 + \sqrt[3]{3} + \sqrt[3]{9}\right)$

15. $\left(\sqrt{5} + 7\right)\left(\sqrt{5} - 7\right)$

16. $\dfrac{1}{\sqrt{x} - \sqrt{5}}, \quad x \neq 5$

17. $\sqrt[3]{8a^3 b^5 c^9}$

18. $\dfrac{15}{\sqrt[3]{9}}$

19. $\dfrac{3}{\sqrt{5} + 2}$

20. $\sqrt{\dfrac{3}{5x}}$

21. $\dfrac{16\sqrt{3}}{5\sqrt{12}}$

22. $\dfrac{2\sqrt{25}}{8\sqrt{50}}$

23. $\dfrac{-10}{\sqrt[3]{10}}$

24. $\dfrac{\sqrt{6}+\sqrt{5}}{\sqrt{6}-\sqrt{5}}$

25. $\sqrt{12x}-\sqrt{75x}$

26. $\left(5-3\sqrt{3}\right)^2$

27. $\left(\sqrt{74}-\sqrt{73}\right)\left(\sqrt{74}+\sqrt{73}\right)$

28. $\sqrt[3]{\dfrac{13}{81}}$

29. $-t^2\sqrt[4]{t}+3\sqrt[4]{t^9}-t\sqrt[4]{t^5}$

30. $\dfrac{\sqrt{3}+\sqrt{7}}{\sqrt{6}-\sqrt{5}}$

31. $\dfrac{6}{\sqrt[4]{3}}$

32. $\dfrac{1}{1-\sqrt[3]{3}}$

33. $\sqrt[3]{\dfrac{x^2y}{x^{-3}y^4}}$

34. $\sqrt{12}-\sqrt{108}-\sqrt[3]{27}$

35. $\dfrac{x^{-2/3}y^{4/5}}{x^{-5/3}y^{-2/5}}$

36. $\left(\dfrac{x^{3/4}y^{2/3}}{x^{1/3}y^{5/8}}\right)^{24}$

37. $\left(125x^3\right)^{-2/3}$

38. $\left(3x^{-2/3}y^{1/2}\right)\left(-2x^{5/8}y^{-1/3}\right)$

39. $\dfrac{4^{1/2}+3^{1/2}}{4^{1/2}-3^{1/2}}$

40. $\left(\sqrt{6}-\sqrt{5}\right)^2\left(\sqrt{6}+\sqrt{5}\right)^2$

9.6 Solving Equations with Radicals

An equation that includes one or more radical expressions with a variable is a **radical equation.**

$$\sqrt{x-4}=8, \quad \sqrt{5x+12}=3\sqrt{2x-1}, \quad \sqrt[3]{6+x}=27$$

Radical equations

OBJECTIVES

1. Solve radical equations using the power rule.

2. Solve radical equations that require additional steps.

3. Solve radical equations with indexes greater than 2.

4. Use the power rule to solve a formula for a specified variable.

OBJECTIVE ▶ 1 Solve radical equations using the power rule. The equation $x = 1$ has only one solution. Its solution set is $\{1\}$. If we square both sides of this equation, we get $x^2 = 1$. This new equation has *two* solutions: -1 and 1. Notice that the solution of the original equation is also a solution of the squared equation. However, the squared equation has another solution, -1, that is *not* a solution of the original equation.

When solving equations with radicals, we use this idea of raising both sides to a power. This is an application of the **power rule.**

> **Power Rule for Solving an Equation with Radicals**
>
> If both sides of an equation are raised to the same power, all solutions of the original equation are also solutions of the new equation.

The power rule does not say that all solutions of the new equation are solutions of the original equation. They may or may not be. Solutions that do not satisfy the original equation are **extraneous solutions.** They must be rejected.

> **CAUTION**
>
> When the power rule is used to solve an equation, *every solution of the new equation* must *be checked in the original equation.*

1 Solve each equation.

(a) $\sqrt{r} = 3$

(b) $\sqrt{5x + 1} = 4$

EXAMPLE 1   Using the Power Rule

Solve $\sqrt{3x + 4} = 8$.

$$\left(\sqrt{3x+4}\right)^2 = 8^2 \qquad \text{Use the power rule and square each side.}$$

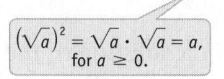

$$3x + 4 = 64 \qquad \text{Apply the exponents.}$$

$\left(\sqrt{a}\right)^2 = \sqrt{a} \cdot \sqrt{a} = a,$ for $a \geq 0$.

$$3x = 60 \qquad \text{Subtract 4.}$$

$$x = 20 \qquad \text{Divide by 3.}$$

To check, substitute the proposed solution in the *original* equation.

CHECK
$$\sqrt{3x + 4} = 8 \qquad \text{Original equation}$$

$$\sqrt{3 \cdot 20 + 4} \overset{?}{=} 8 \qquad \text{Let } x = 20.$$

$$\sqrt{64} \overset{?}{=} 8 \qquad \text{Evaluate the radicand.}$$

$$8 = 8 \ \checkmark \ \text{True}$$

Since 20 satisfies the *original* equation, the solution set is $\{20\}$.

Work Problem 1 at the Side. ▶

❷ Solve each equation.

Gs **(a)** $\sqrt{5x + 3} + 2 = 0$

Step 1 To isolate the radical on the left side, subtract _____ from each side.

$$\sqrt{} = \underline{}$$

Step 2 To apply the power rule, square each side.

$$\left(\sqrt{}\right)^2 = \left(\underline{}\right)^2$$

Step 3 Solve the equation from Step 2.

$$5x + 3 = \underline{}$$

$$5x = \underline{}$$

$$x = \underline{}$$

Step 4 Check the proposed solution in the original equation.

The result is a (*true / false*) statement, so the solution set is _____.

(b) $\sqrt{x - 9} - 3 = 0$

(c) $\sqrt{3x + 4} + 5 = 0$

The method used in the solution of the equation in **Example 1** can be generalized.

Solving an Equation with Radicals

Step 1 **Isolate the radical.** Make sure that one radical term is alone on one side of the equation.

Step 2 **Apply the power rule.** Raise each side of the equation to a power that is the same as the index of the radical.

Step 3 **Solve** the resulting equation. If it still contains a radical, repeat Steps 1 and 2.

Step 4 **Check** all proposed solutions in the *original* equation.

CAUTION

Remember to check (Step 4) or you may get an incorrect solution set.

EXAMPLE 2 Using the Power Rule

Solve $\sqrt{5x - 1} + 3 = 0$.

Step 1	$\sqrt{5x - 1} = -3$	To isolate the radical on one side, subtract 3 from each side.
Step 2	$\left(\sqrt{5x - 1}\right)^2 = (-3)^2$	Square each side.
Step 3	$5x - 1 = 9$	Apply the exponents.
	$5x = 10$	Add 1.
	$x = 2$	Divide by 5.

Step 4 **CHECK**

$\sqrt{5x - 1} + 3 = 0$	Original equation
$\sqrt{5 \cdot 2 - 1} + 3 \overset{?}{=} 0$	Let $x = 2$.
$\sqrt{9} + 3 \overset{?}{=} 0$	Evaluate the radicand.
$3 + 3 \overset{?}{=} 0$	Take the square root.
$6 = 0$	False

This false result shows that the *proposed solution* 2 is *not* a solution of the original equation. It is extraneous. The solution set is $\emptyset$.

Note

We could have determined after Step 1 that the equation in **Example 2** has no solution because the expression on the left cannot equal a negative number.

◀ **Work Problem** ❷ at the Side.

OBJECTIVE 2 Solve radical equations that require additional steps. Recall the following rule from **Section 5.4.**

$$(x + y)^2 = x^2 + 2xy + y^2$$

EXAMPLE 3 Using the Power Rule (Squaring a Binomial)

Solve $\sqrt{4 - x} = x + 2$.

Step 1 The radical is isolated on the left side of the equation.

Step 2 Square each side. On the right, $(x + 2)^2 = x^2 + 2(x)(2) + 2^2$.

$$\left(\sqrt{4 - x}\right)^2 = (x + 2)^2$$

$$4 - x = x^2 + 4x + 4$$

Remember the middle term.

Twice the product of 2 and x

Step 3 The new equation is quadratic, so write it in standard form.

$0 = x^2 + 5x$ Subtract 4. Add x.

$0 = x(x + 5)$ Factor.

$x = 0$ or $x + 5 = 0$ Zero-factor property

$x = -5$ Solve.

Set *each* factor equal to 0.

Step 4 Check each proposed solution in the original equation.

CHECK

$\sqrt{4 - x} = x + 2$	$\sqrt{4 - x} = x + 2$
$\sqrt{4 - 0} \stackrel{?}{=} 0 + 2$ Let $x = 0$.	$\sqrt{4 - (-5)} \stackrel{?}{=} -5 + 2$ Let $x = -5$.
$\sqrt{4} \stackrel{?}{=} 2$	$\sqrt{9} \stackrel{?}{=} -3$
$2 = 2$ ✓ True	$3 = -3$ False

The solution set is $\{0\}$. The other proposed solution, -5, is extraneous.

Work Problem 3 at the Side. ▶

EXAMPLE 4 Using the Power Rule (Squaring a Binomial)

Solve $\sqrt{x^2 - 4x + 9} = x - 1$.

Squaring gives $(x - 1)^2 = x^2 - 2(x)(1) + 1^2$ on the right.

$$\left(\sqrt{x^2 - 4x + 9}\right)^2 = (x - 1)^2$$

$$x^2 - 4x + 9 = x^2 - 2x + 1$$

Remember the middle term.

Twice the product of x and -1

$-2x = -8$ Subtract x^2 and 9. Add $2x$.

$x = 4$ Divide by -2.

CHECK $\sqrt{x^2 - 4x + 9} = x - 1$ Original equation

$\sqrt{4^2 - 4 \cdot 4 + 9} \stackrel{?}{=} 4 - 1$ Let $x = 4$.

$3 = 3$ ✓ True

The solution set is $\{4\}$.

Work Problem 4 at the Side. ▶

3 Solve.

GS **(a)** $\sqrt{3x - 5} = x - 1$

Step 1 The radical is isolated on the left side of the equation.

Step 2 To apply the power rule, square each side.

$\left(\sqrt{______}\right)^2 = (______)^2$

$3x - 5 = ______$

Step 3 The new equation is quadratic, so write it in standard form.

$0 = ______$

Factor on the right and use the zero-factor property to solve the quadratic equation.

$0 = (x - 2)(______)$

$x - 2 = 0$ or $______ = 0$

$x = 2$ or $x = ____$

Step 4 Check each proposed solution in the original equation.

The solution set is ____.

(b) $x + 1 = \sqrt{-2x - 2}$

4 Solve.

$$\sqrt{4x^2 + 2x - 3} = 2x + 7$$

Answers

3. **(a)** $3x - 5$; $x - 1$; $x^2 - 2x + 1$; $x^2 - 5x + 6$; $x - 3$; $x - 3$; 3; $\{2, 3\}$
(b) $\{-1\}$

4. $\{-2\}$

⑤ Solve each equation.

(a) $\sqrt{2x + 3} + \sqrt{x + 1} = 1$

(b) $\sqrt{3x + 1} - \sqrt{x + 4} = 1$

EXAMPLE 5 **Using the Power Rule (Squaring Twice)**

Solve $\sqrt{5x + 6} + \sqrt{3x + 4} = 2$.

Isolate one radical on one side of the equation by subtracting $\sqrt{3x + 4}$ from each side.

$$\sqrt{5x + 6} = 2 - \sqrt{3x + 4} \qquad \text{Subtract } \sqrt{3x + 4}.$$

$$\left(\sqrt{5x + 6}\right)^2 = \left(2 - \sqrt{3x + 4}\right)^2 \qquad \text{Square each side.}$$

$$5x + 6 = 4 - 4\sqrt{3x + 4} + (3x + 4) \qquad \text{Be careful here.}$$

Remember the middle term. — Twice the product of 2 and $-\sqrt{3x + 4}$

This equation still contains a radical. Isolate this radical term on the right.

$$5x + 6 = 8 + 3x - 4\sqrt{3x + 4} \qquad \text{Combine like terms.}$$

$$2x - 2 = -4\sqrt{3x + 4} \qquad \text{Subtract 8 and } 3x.$$

Divide *each* term by 2. $\quad x - 1 = -2\sqrt{3x + 4} \qquad$ Divide by 2 to make the numbers smaller.

$$(x - 1)^2 = \left(-2\sqrt{3x + 4}\right)^2 \qquad \text{Square each side again.}$$

$$x^2 - 2x + 1 = (-2)^2\left(\sqrt{3x + 4}\right)^2 \qquad \text{On the right, } (ab)^2 = a^2b^2.$$

$$x^2 - 2x + 1 = 4(3x + 4) \qquad \text{Apply the exponents.}$$

$$x^2 - 2x + 1 = 12x + 16 \qquad \text{Distributive property}$$

$$x^2 - 14x - 15 = 0 \qquad \text{Standard form}$$

$$(x + 1)(x - 15) = 0 \qquad \text{Factor.}$$

$$x + 1 = 0 \quad \text{or} \quad x - 15 = 0 \qquad \text{Zero-factor property}$$

$$x = -1 \quad \text{or} \quad x = 15 \qquad \text{Solve each equation.}$$

CHECK $\qquad \sqrt{5x + 6} + \sqrt{3x + 4} = 2 \qquad$ Original equation

$$\sqrt{5(-1) + 6} + \sqrt{3(-1) + 4} \overset{?}{=} 2 \qquad \text{Let } x = -1.$$

$$\sqrt{1} + \sqrt{1} \overset{?}{=} 2 \qquad \text{Evaluate the radicands.}$$

$$1 + 1 \overset{?}{=} 2 \qquad \text{Take square roots.}$$

$$2 = 2 \checkmark \qquad \text{True}$$

$$\sqrt{5x + 6} + \sqrt{3x + 4} = 2 \qquad \text{Original equation}$$

$$\sqrt{5(15) + 6} + \sqrt{3(15) + 4} \overset{?}{=} 2 \qquad \text{Let } x = 15.$$

$$\sqrt{81} + \sqrt{49} \overset{?}{=} 2 \qquad \text{Evaluate the radicands.}$$

$$9 + 7 \overset{?}{=} 2 \qquad \text{Take square roots.}$$

$$16 = 2 \qquad \text{False}$$

The proposed solution -1 is valid, but 15 is extraneous and must be rejected. Thus, the solution set is $\{-1\}$.

◀ **Work Problem ⑤ at the Side.**

Answers

5. (a) $\{-1\}$ (b) $\{5\}$

OBJECTIVE ▶ ③ **Solve radical equations with indexes greater than 2.** The power rule also applies to powers greater than 2.

> **EXAMPLE 6** **Using the Power Rule for a Power Greater than 2**
>
> Solve $\sqrt[3]{x + 5} = \sqrt[3]{2x - 6}$.
>
> $$\left(\sqrt[3]{x + 5}\right)^3 = \left(\sqrt[3]{2x - 6}\right)^3 \qquad \text{Cube each side.}$$
>
> $$x + 5 = 2x - 6 \qquad \left(\sqrt[3]{a}\right)^3 = a$$
>
> $$11 = x \qquad \text{Subtract } x. \text{ Add } 6.$$
>
> **CHECK** $\qquad \sqrt[3]{x + 5} = \sqrt[3]{2x - 6} \qquad$ Original equation
>
> $$\sqrt[3]{11 + 5} \stackrel{?}{=} \sqrt[3]{2 \cdot 11 - 6} \qquad \text{Let } x = 11.$$
>
> $$\sqrt[3]{16} = \sqrt[3]{16} \; \checkmark \qquad \text{True}$$
>
> The solution set is $\{11\}$.

························ **Work Problem ⑥ at the Side.** ▶

OBJECTIVE ▶ ④ **Use the power rule to solve a formula for a specified variable.**

> **EXAMPLE 7** **Solving a Formula from Electronics for a Variable**
>
> An important property of a radio-frequency transmission line is its **characteristic impedance,** represented by Z and measured in ohms. If L and C are the inductance and capacitance, respectively, per unit of length of the line, then these quantities are related by the formula $Z = \sqrt{\frac{L}{C}}$. Solve this formula for C.
>
> $$Z = \sqrt{\frac{L}{C}} \qquad \text{Our goal is to isolate } C \text{ on one side of the equality symbol.}$$
>
> $$Z^2 = \left(\sqrt{\frac{L}{C}}\right)^2 \qquad \text{Square each side.}$$
>
> $$Z^2 = \frac{L}{C} \qquad \left(\sqrt{a}\right)^2 = a$$
>
> $$CZ^2 = L \qquad \text{Multiply by } C.$$
>
> $$C = \frac{L}{Z^2} \qquad \text{Divide by } Z^2.$$

························ **Work Problem ⑦ at the Side** ▶

⑥ Solve each equation.

(a) $\sqrt[3]{2x + 7} = \sqrt[3]{3x - 2}$

(b) $\sqrt[4]{2x + 5} + 1 = 0$

⑦ Solve the formula for R.

$$Z = \sqrt{\frac{R}{T}}$$

Answers

6. (a) $\{9\}$ (b) $\emptyset$

7. $R = TZ^2$

9.6 Exercises

MyMathLab®

CONCEPT CHECK *Check each equation to see if the given value for x is a solution.*

1. $\sqrt{3x + 18} = x$

   (a) 6   **(b)** −3

2. $\sqrt{3x - 3} = x - 1$

   (a) 1   **(b)** 4

3. $\sqrt{x + 2} = \sqrt{9x - 2} - 2\sqrt{x - 1}$

   (a) 2   **(b)** 7

4. $\sqrt{8x - 3} = 2x$

   (a) $\dfrac{3}{2}$ **(b)** $\dfrac{1}{2}$

5. CONCEPT CHECK Is 9 a solution of the following equation?

$$\sqrt{x} = -3$$

If not, can there be a solution of this equation?

6. CONCEPT CHECK Before even attempting to solve

$$\sqrt{3x + 18} = x,$$

how can we be sure that the equation cannot have a negative solution?

*Solve each equation. **See Examples 1–4.***

7. $\sqrt{x - 2} = 3$

8. $\sqrt{x + 1} = 7$

9. $\sqrt{6x - 1} = 1$

10. $\sqrt{7x - 3} = 5$

11. $\sqrt{4x + 3} + 1 = 0$

12. $\sqrt{5x - 3} + 2 = 0$

13. $\sqrt{3k + 1} - 4 = 0$

14. $\sqrt{5z + 1} - 11 = 0$

15. $4 - \sqrt{x - 2} = 0$

16. $9 - \sqrt{4k + 1} = 0$

17. $\sqrt{9a - 4} = \sqrt{8a + 1}$

18. $\sqrt{4p - 2} = \sqrt{3p + 5}$

19. $2\sqrt{x} = \sqrt{3x + 4}$

20. $2\sqrt{m} = \sqrt{5m - 16}$

21. $3\sqrt{z - 1} = 2\sqrt{2z + 2}$

22. $5\sqrt{4x + 1} = 3\sqrt{10x + 25}$

23. $k = \sqrt{k^2 + 4k - 20}$

24. $p = \sqrt{p^2 - 3p + 18}$

25. $x = \sqrt{x^2 + 3x + 9}$

26. $z = \sqrt{z^2 - 4z - 8}$

27. $\sqrt{9 - x} = x + 3$

28. $\sqrt{5 - x} = x + 1$

29. $\sqrt{k^2 + 2k + 9} = k + 3$

30. $\sqrt{x^2 - 3x + 3} = x - 1$

31. $\sqrt{r^2 + 9r + 3} = -r$

32. $\sqrt{p^2 - 15p + 15} = p - 5$

33. $\sqrt{z^2 + 12z - 4} + 4 - z = 0$

34. $\sqrt{m^2 + 3m + 12} - m - 2 = 0$

35. CONCEPT CHECK A student wrote the following as his first step in solving $\sqrt{3x + 4} = 8 - x$.

$$3x + 4 = 64 + x^2$$

What Went Wrong? Solve the given equation correctly.

36. CONCEPT CHECK A student wrote the following as her first step in solving $\sqrt{5x + 6} = \sqrt{x + 3} + 3$.

$$5x + 6 = x + 3 + 9$$

What Went Wrong? Solve the given equation correctly.

*Solve each equation. **See Examples 5 and 6.***

37. $\sqrt[3]{2x + 5} = \sqrt[3]{6x + 1}$

38. $\sqrt[3]{p - 1} = 2$

39. $\sqrt[3]{a^2 + 5a + 1} = \sqrt[3]{a^2 + 4a}$

40. $\sqrt[3]{r^2 + 2r + 8} = \sqrt[3]{r^2}$

41. $\sqrt[3]{2m - 1} = \sqrt[3]{m + 13}$

42. $\sqrt[3]{2k - 11} - \sqrt[3]{5k + 1} = 0$

43. $\sqrt[4]{a + 8} = \sqrt[4]{2a}$

44. $\sqrt[4]{z + 11} = \sqrt[4]{2z + 6}$

45. $\sqrt[3]{x - 8} + 2 = 0$

46. $\sqrt[3]{r + 1} + 1 = 0$

47. $\sqrt[4]{2k - 5} + 4 = 0$

48. $\sqrt[4]{8z - 3} + 2 = 0$

49. $\sqrt{k + 2} - \sqrt{k - 3} = 1$

50. $\sqrt{r + 6} - \sqrt{r - 2} = 2$

51. $\sqrt{2r + 11} - \sqrt{5r + 1} = -1$

52. $\sqrt{3x - 2} - \sqrt{x + 3} = 1$

53. $\sqrt{3p + 4} - \sqrt{2p - 4} = 2$

54. $\sqrt{4x + 5} - \sqrt{2x + 2} = 1$

55. $\sqrt{3 - 3p} - 3 = \sqrt{3p + 2}$

56. $\sqrt{4x + 7} - 4 = \sqrt{4x - 1}$

57. $\sqrt{2\sqrt{x + 11}} = \sqrt{4x + 2}$

58. $\sqrt{1 + \sqrt{24 - 10x}} = \sqrt{3x + 5}$

For each equation, rewrite the expressions with rational exponents as radical expressions, and then solve using the procedures explained in this section.

59. $(2x - 9)^{1/2} = 2 + (x - 8)^{1/2}$

60. $(3w + 7)^{1/2} = 1 + (w + 2)^{1/2}$

61. $(2w - 1)^{2/3} - w^{1/3} = 0$

62. $(x^2 - 2x)^{1/3} - x^{1/3} = 0$

*Solve each formula for the indicated variable. **See Example 7.** (Source: Cooke, Nelson M., and Joseph B. Orleans, Mathematics Essential to Electricity and Radio, McGraw-Hill.)*

63. $Z = \sqrt{\dfrac{L}{C}}$ for L

64. $r = \sqrt{\dfrac{A}{\pi}}$ for A

65. $V = \sqrt{\dfrac{2K}{m}}$ for K

66. $V = \sqrt{\dfrac{2K}{m}}$ for m

67. $r = \sqrt{\dfrac{Mm}{F}}$ for M

68. $r = \sqrt{\dfrac{Mm}{F}}$ for F

The following formula is used to find the rotational rate N of a space station.

$$N = \frac{1}{2\pi}\sqrt{\frac{a}{r}}$$

Here, a is the acceleration and r represents the radius of the space station in meters. To find the value of r that will make N simulate the effect of gravity on Earth, the equation must be solved for r, using the required value of N. (Source: Kastner, Bernice, Space Mathematics, NASA.)

69. Solve the equation for the indicated variable.

    (a) for r

    (b) for a

70. If $a = 9.8$ m per sec^2, find the value of r (to the nearest tenth) using each value of N.

    (a) $N = 0.063$ rotation per sec

    (b) $N = 0.04$ rotation per sec

9.7 Complex Numbers

OBJECTIVES

1. Simplify numbers of the form $\sqrt{-b}$, where $b > 0$.

2. Recognize subsets of the complex numbers.

3. Add and subtract complex numbers.

4. Multiply complex numbers.

5. Divide complex numbers.

6. Simplify powers of i.

Recall that the set of real numbers includes many other number sets (the rational numbers, integers, and natural numbers, for example). In this section, we introduce a new set of numbers that includes the set of real numbers, as well as numbers that are even roots of negative numbers, like $\sqrt{-2}$.

OBJECTIVE ▶ **1** Simplify numbers of the form $\sqrt{-b}$, where $b > 0$. The equation $x^2 + 1 = 0$ has no real number solution since any solution must be a number whose square is -1. In the set of real numbers, all squares are nonnegative numbers because the product of two positive numbers or two negative numbers is positive and $0^2 = 0$. To provide a solution for the equation $x^2 + 1 = 0$, we introduce a new number i.

Imaginary Unit i

The **imaginary unit i** is defined as

$$i = \sqrt{-1}, \quad \text{and thus} \quad i^2 = -1.$$

In words, i is the principal square root of -1.

This definition of i makes it possible to define any square root of a negative number as follows.

Meaning of $\sqrt{-b}$

For any positive number b, $\qquad \sqrt{-b} = i\sqrt{b}$.

---

1 Write each number as a product of a real number and i.

(a) $\sqrt{-16}$

(b) $-\sqrt{-81}$

(c) $\sqrt{-7}$

(d) $\sqrt{-32}$

EXAMPLE 1 Simplifying Square Roots of Negative Numbers

Write each number as a product of a real number and i.

(a) $\sqrt{-100} = i\sqrt{100} = 10i$

(b) $-\sqrt{-36} = -i\sqrt{36} = -6i$

(c) $\sqrt{-2} = i\sqrt{2}$

(d) $\sqrt{-8} = i\sqrt{8} = i\sqrt{4 \cdot 2} = 2i\sqrt{2}$

◀ Work Problem **1** at the Side.

CAUTION

It is easy to mistake $\sqrt{2}i$ for $\sqrt{2i}$, with the i under the radical. For this reason, we usually write $\sqrt{2}i$ as $i\sqrt{2}$, as in the definition of $\sqrt{-b}$.

When finding a product such as $\sqrt{-4} \cdot \sqrt{-9}$, we cannot use the product rule for radicals because it applies only to nonnegative radicands.

For this reason, we change $\sqrt{-b}$ to the form $i\sqrt{b}$ before performing any multiplications or divisions.

Answers

1. (a) $4i$ **(b)** $-9i$ **(c)** $i\sqrt{7}$ **(d)** $4i\sqrt{2}$

EXAMPLE 2   **Multiplying Square Roots of Negative Numbers**

Multiply.

(a)
$$\sqrt{-4} \cdot \sqrt{-9}$$

> First write all square roots in terms of i.

$= i\sqrt{4} \cdot i\sqrt{9}$ $\sqrt{-b} = i\sqrt{b}$

$= i \cdot 2 \cdot i \cdot 3$ Take square roots.

$= 6i^2$ Multiply.

$= 6(-1)$ Substitute -1 for i^2.

$= -6$ Multiply.

(b)
$$\sqrt{-3} \cdot \sqrt{-7}$$

> First write all square roots in terms of i.

$= i\sqrt{3} \cdot i\sqrt{7}$ $\sqrt{-b} = i\sqrt{b}$

$= i^2\sqrt{3 \cdot 7}$ Product rule

$= (-1)\sqrt{21}$ Substitute -1 for i^2.

$= -\sqrt{21}$ $(-1)a = -a$

(c) $\sqrt{-2} \cdot \sqrt{-8}$

$= i\sqrt{2} \cdot i\sqrt{8}$

$= i^2\sqrt{2 \cdot 8}$

$= (-1)\sqrt{16}$

$= (-1)4$,   or -4

(d) $\sqrt{-5} \cdot \sqrt{6}$

$= i\sqrt{5} \cdot \sqrt{6}$

$= i\sqrt{30}$

······· **Work Problem 2 at the Side.** ▶

CAUTION

Use the definition of $\sqrt{-b}$ **before** the product rule for radicals.

$$\sqrt{-4} \cdot \sqrt{-9} = i\sqrt{4} \cdot i\sqrt{9} = -6,$$ Correct **(Example 2(a))**

but $\sqrt{-4(-9)} = \sqrt{36} = 6.$ Incorrect

Thus, $\sqrt{-4} \cdot \sqrt{-9} \neq \sqrt{-4(-9)}$.

EXAMPLE 3   **Dividing Square Roots of Negative Numbers**

Divide.

(a) $\dfrac{\sqrt{-75}}{\sqrt{-3}}$

> First write all square roots in terms of i.

$= \dfrac{i\sqrt{75}}{i\sqrt{3}}$

$= \sqrt{\dfrac{75}{3}}$ $\frac{i}{i} = 1$; Quotient rule

$= \sqrt{25}$ Divide.

$= 5$

(b) $\dfrac{\sqrt{-32}}{\sqrt{8}}$

$= \dfrac{i\sqrt{32}}{\sqrt{8}}$ $\sqrt{-32} = i\sqrt{32}$

$= i\sqrt{\dfrac{32}{8}}$ Quotient rule

$= i\sqrt{4}$ Divide.

$= 2i$

······· **Work Problem 3 at the Side.** ▶

2 Multiply.

(a) $\sqrt{-7} \cdot \sqrt{-5}$

(b) $\sqrt{-5} \cdot \sqrt{-10}$

(c) $\sqrt{-15} \cdot \sqrt{2}$

3 Divide.

(a) $\dfrac{\sqrt{-32}}{\sqrt{-2}}$

(b) $\dfrac{\sqrt{-27}}{\sqrt{-3}}$

(c) $\dfrac{\sqrt{-40}}{\sqrt{10}}$

Answers

2. (a) $-\sqrt{35}$ **(b)** $-5\sqrt{2}$ **(c)** $i\sqrt{30}$

3. (a) 4 **(b)** 3 **(c)** $2i$

4 Add.

(a) $(4 + 6i) + (-3 + 5i)$

(b) $(-1 + 8i) + (9 - 3i)$

OBJECTIVE **2** **Recognize subsets of the complex numbers.** A new set of numbers, the *complex numbers*, is defined as follows.

Complex Number

If a and b are real numbers, then any number of the form $a + bi$ is a **complex number.** In the complex number $a + bi$, the number a is the **real part** and b is the **imaginary part.***

For a complex number $a + bi$, if $b = 0$, then $a + bi = a$, which is a real number. ***Thus, the set of real numbers is a subset of the set of complex numbers.*** If $a = 0$ and $b \neq 0$, the complex number is said to be a **pure imaginary number.** For example, $3i$ is a pure imaginary number. A number such as $7 + 2i$ is a **nonreal complex number.**

A complex number written in the form $a + bi$ is in **standard form.** In this section, most answers will be given in standard form, but if a or b is 0, we consider answers such as a or bi to be in standard form.

The relationships among the sets of numbers are shown in **Figure 10.**

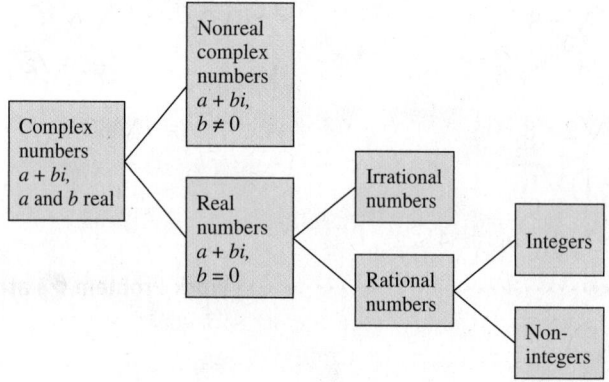

Figure 10

OBJECTIVE **3** **Add and subtract complex numbers.** The commutative, associative, and distributive properties for real numbers are also valid for complex numbers. ***Thus, to add complex numbers, we add their real parts and add their imaginary parts.***

EXAMPLE 4 **Adding Complex Numbers**

Add.

(a) $(2 + 3i) + (6 + 4i)$

$\quad = (2 + 6) + (3 + 4)i$ Commutative, associative, and distributive properties

$\quad = 8 + 7i$ Add real parts. Add imaginary parts.

(b) $5 + (9 - 3i)$

$\quad = (5 + 9) - 3i$ Associative property

$\quad = 14 - 3i$ Add real parts.

◀ **Work Problem** **4** at the Side.

Answers

4. (a) $1 + 11i$ **(b)** $8 + 5i$

*Some texts define bi as the imaginary part of the complex number $a + bi$.

To subtract complex numbers, we subtract their real parts and subtract their imaginary parts.

EXAMPLE 5 | **Subtracting Complex Numbers**

Subtract.

(a) $(6 + 5i) - (3 + 2i)$

$\quad = (6 - 3) + (5 - 2)i \qquad$ Properties of real numbers

$\quad = 3 + 3i \qquad\qquad$ Subtract real parts. Subtract imaginary parts.

(b) $(7 - 3i) - (8 - 6i)$　　　　　**(c)** $(-9 + 4i) - (-9 + 8i)$

$\quad = (7 - 8) + [-3 - (-6)]i \qquad\qquad = (-9 + 9) + (4 - 8)i$

$\quad = -1 + 3i \qquad\qquad\qquad\qquad\qquad = 0 - 4i$

$\qquad\qquad\qquad\qquad\qquad\qquad\qquad\qquad = -4i$

·················· **Work Problem ⑤ at the Side.** ▶

OBJECTIVE ▶ ④ Multiply complex numbers. We multiply complex numbers in the same way we multiply polynomials.

EXAMPLE 6 | **Multiplying Complex Numbers**

Multiply.

(a) $4i(2 + 3i)$

$\quad = 4i(2) + 4i(3i) \qquad$ Distributive property

$\quad = 8i + 12i^2 \qquad\qquad$ Multiply.

$\quad = 8i + 12(-1) \qquad$ Substitute -1 for i^2.

$\quad = -12 + 8i \qquad\qquad$ Standard form

(b) $(3 + 5i)(4 - 2i)$

$\quad = \underbrace{3(4)}_{\text{First}} + \underbrace{3(-2i)}_{\text{Outer}} + \underbrace{5i(4)}_{\text{Inner}} + \underbrace{5i(-2i)}_{\text{Last}} \qquad$ FOIL method

$\quad = 12 - 6i + 20i - 10i^2 \qquad$ Multiply.

$\quad = 12 + 14i - 10(-1) \qquad$ Combine imaginary terms. Substitute -1 for i^2.

$\quad = 12 + 14i + 10 \qquad\qquad$ Multiply.

$\quad = 22 + 14i \qquad\qquad\qquad$ Combine real terms.

(c) $(2 + 3i)(1 - 5i)$

$\quad = 2(1) + 2(-5i) + 3i(1) + 3i(-5i) \qquad$ FOIL method

$\quad = 2 - 10i + 3i - 15i^2 \qquad$ Multiply.

$\quad = 2 - 7i - 15(-1) \qquad$ *Use parentheses around -1 to avoid errors.* $\qquad i^2 = -1$

$\quad = 2 - 7i + 15 \qquad\qquad$ Multiply.

$\quad = 17 - 7i \qquad\qquad\qquad$ Combine real terms.

·················· **Work Problem ⑥ at the Side.** ▶

---

⑤ Subtract.

(a) $(7 + 3i) - (4 + 2i)$

(b) $(-6 - i) - (-5 - 4i)$

(c) $8 - (3 - 2i)$

⑥ Multiply.

(a) $6i(4 + 3i)$

(b) $(6 - 4i)(2 + 4i)$

(c) $(3 - 2i)(3 + 2i)$

Answers

5. (a) $3 + i$　**(b)** $-1 + 3i$　**(c)** $5 + 2i$

6. (a) $-18 + 24i$　**(b)** $28 + 16i$　**(c)** 13

7 Find each quotient.

(a) $\dfrac{2+i}{3-i}$

Multiply the numerator and denominator by the conjugate of the denominator.

$$= \frac{(2+i)(\underline{\hspace{1cm}})}{(3-i)(\underline{\hspace{1cm}})}$$

$$= \frac{6+2i+3i+\underline{\hspace{0.3cm}}^2}{\underline{\hspace{0.5cm}}^2-\underline{\hspace{0.5cm}}^2}$$

$$= \frac{6+5i-1}{9-(\underline{\hspace{0.5cm}})}$$

$$= \frac{5+5i}{10}$$

$$= \frac{\underline{\hspace{0.5cm}}(1+i)}{\underline{\hspace{0.5cm}} \cdot 2}$$

$$= \frac{1+i}{2}$$

$$= \underline{\hspace{0.5cm}} + \underline{\hspace{0.5cm}}$$

(b) $\dfrac{8-4i}{1-i}$

(c) $\dfrac{5}{3-2i}$

(d) $\dfrac{5-i}{i}$

Answers

7. **(a)** $3+i$; $3+i$; i; 3; i; -1; 5; 5; $\dfrac{1}{2}$; $\dfrac{1}{2}i$

(b) $6+2i$ **(c)** $\dfrac{15}{13}+\dfrac{10}{13}i$ **(d)** $-1-5i$

The two complex numbers $a + bi$ and $a - bi$ are **complex conjugates,** or simply *conjugates*, of each other. ***The product of a complex number and its conjugate is always a real number,*** as shown here.

$$(a + bi)(a - bi)$$

$$= a^2 - abi + abi - b^2i^2 \quad \text{FOIL method}$$

$$= a^2 - b^2(-1) \quad \text{Combine like terms; } i^2 = -1$$

$$= a^2 + b^2 \quad \boxed{\text{The product eliminates } i.}$$

For example, $(3 + 7i)(3 - 7i) = 3^2 + 7^2 = 9 + 49 = 58.$

OBJECTIVE **5** **Divide complex numbers.** The quotient of two complex numbers should be a complex number. To write the quotient as a complex number, we need to eliminate i in the denominator. We use conjugates and a process similar to that for rationalizing a denominator.

EXAMPLE 7 **Dividing Complex Numbers**

Find each quotient.

(a) $\dfrac{8+9i}{5+2i}$ $\qquad \boxed{\frac{5-2i}{5-2i}=1}$

$$= \frac{(8+9i)(5-2i)}{(5+2i)(5-2i)} \quad \begin{array}{l}\text{Multiply numerator and}\\\text{denominator by } 5-2i, \text{ the}\\\text{conjugate of the denominator.}\end{array}$$

$$= \frac{40-16i+45i-18i^2}{5^2+2^2} \quad \begin{array}{l}\text{In the denominator,}\\(a+bi)(a-bi)=a^2+b^2.\end{array}$$

$$= \frac{40+29i-18(-1)}{25+4} \quad \begin{array}{l}\text{In the numerator, combine}\\\text{imaginary terms; } i^2 = -1\end{array}$$

$$= \frac{58+29i}{29} \quad \text{Multiply. Combine real terms.}$$

$$= \frac{29(2+i)}{29} \quad \text{Factor the numerator.}$$

$$= 2 + i \quad \text{Lowest terms}$$

$\boxed{\begin{array}{l}\text{Factor first. Then}\\\text{divide out the}\\\text{common factor.}\end{array}}$

(b) $\dfrac{1+i}{i}$

$$= \frac{(1+i)(-i)}{i(-i)} \quad \begin{array}{l}\text{Multiply numerator and denominator by } -i,\\\text{the conjugate of } i.\end{array}$$

$$= \frac{-i-i^2}{-i^2} \quad \begin{array}{l}\text{Use the distributive property in the numerator.}\\\text{Multiply in the denominator.}\end{array}$$

$$= \frac{-i-(-1)}{-(-1)} \quad \text{Substitute } -1 \text{ for } i^2.$$

$$= \frac{-i+1}{1} \quad \boxed{\begin{array}{l}\text{Use parentheses}\\\text{to avoid errors.}\end{array}}$$

$$= 1 - i \quad \frac{a}{1} = a$$

◀ **Work Problem** **7** at the Side.

Calculator Tip

In **Examples 4–7,** we showed how complex numbers can be added, subtracted, multiplied, and divided algebraically. Many current models of graphing calculators can perform these operations. **Figure 11** shows how the computations in parts of **Examples 4–7** are displayed on a TI-83/84 Plus calculator. Be sure to use parentheses as shown.

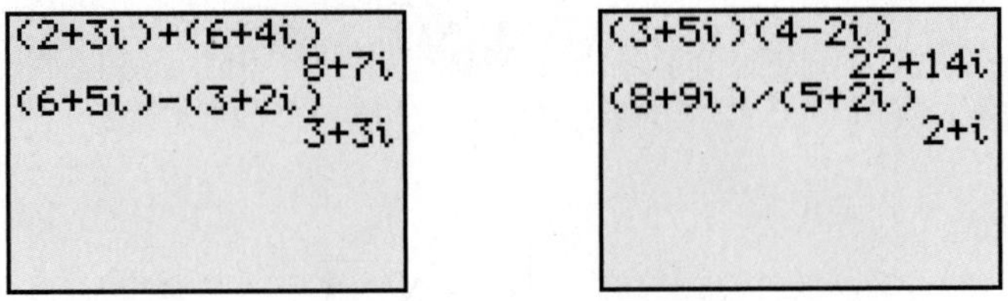

```
(2+3i)+(6+4i)
           8+7i
(6+5i)-(3+2i)
           3+3i
```

```
(3+5i)(4-2i)
        22+14i
(8+9i)/(5+2i)
           2+i
```

Figure 11

OBJECTIVE 6 Simplify powers of i. Because i^2 is defined to be -1, we can find greater powers of i as shown in the following examples.

$$i^3 = i \cdot i^2 = i(-1) = -i \qquad i^6 = i^2 \cdot i^4 = (-1) \cdot 1 = -1$$

$$i^4 = i^2 \cdot i^2 = (-1)(-1) = 1 \qquad i^7 = i^3 \cdot i^4 = (-i) \cdot 1 = -i$$

$$i^5 = i \cdot i^4 = i \cdot 1 = i \qquad i^8 = i^4 \cdot i^4 = 1 \cdot 1 = 1$$

Notice that the powers of i rotate through the four numbers i, -1, $-i$, and 1. Greater powers of i can be simplified by using the fact that $i^4 = 1$.

EXAMPLE 8   Simplifying Powers of i

Find each power of i.

(a) $i^{12} = (i^4)^3 = 1^3 = 1$

(b) $i^{39} = i^{36} \cdot i^3 = (i^4)^9 \cdot i^3 = 1^9 \cdot (-i) = -i$

(c) $i^{-2} = \dfrac{1}{i^2} = \dfrac{1}{-1} = -1$

(d) $i^{-1} = \dfrac{1}{i} = \dfrac{1(-i)}{i(-i)} = \dfrac{-i}{-i^2} = \dfrac{-i}{-(-1)} = \dfrac{-i}{1} = -i$

························· **Work Problem 8 at the Side.** ▶

---

8 Find each power of i.

(a) i^{21}

$$= i^{\underline{\quad}} \cdot i$$

$$= (i^4)^{\underline{\quad}} \cdot i$$

$$= \underline{\quad} \cdot i$$

$$= \underline{\quad}$$

(b) i^{36}

(c) i^{50}

(d) i^{-9}

Answers

8. **(a)** 20; 5; 1; i **(b)** 1 **(c)** -1 **(d)** $-i$

9.7 Exercises

FOR EXTRA HELP

 Download the MyDashBoard App

 MyMathLab®

CONCEPT CHECK *List all of the following sets to which each number belongs. A number may belong to more than one set.*

real numbers   pure imaginary numbers   nonreal complex numbers   complex numbers

1. $3 + 5i$

2. $-7i$

3. $\sqrt{2}$

4. $\dfrac{13}{3}$

5. $\sqrt{-49}$

6. $-\sqrt{-8}$

CONCEPT CHECK *Decide whether each expression is equal to 1, −1, i, or −i.*

7. $\sqrt{-1}$

8. $-i^2$

9. $\dfrac{1}{i}$

10. $(-i)^2$

Write each number as a product of a real number and i. Simplify all radical expressions. **See Example 1.**

11. $\sqrt{-169}$

12. $\sqrt{-225}$

13. $-\sqrt{-144}$

14. $-\sqrt{-196}$

15. $\sqrt{-5}$

16. $\sqrt{-21}$

17. $\sqrt{-48}$

18. $\sqrt{-96}$

Multiply or divide as indicated. **See Examples 2 and 3.**

19. $\sqrt{-15} \cdot \sqrt{-15}$

20. $\sqrt{-19} \cdot \sqrt{-19}$

21. $\sqrt{-3} \cdot \sqrt{-19}$

22. $\sqrt{-7} \cdot \sqrt{-15}$

23. $\sqrt{-4} \cdot \sqrt{-25}$

24. $\sqrt{-9} \cdot \sqrt{-81}$

25. $\sqrt{-3} \cdot \sqrt{11}$

26. $\sqrt{-5} \cdot \sqrt{13}$

27. $\dfrac{\sqrt{-300}}{\sqrt{-100}}$

28. $\dfrac{\sqrt{-40}}{\sqrt{-10}}$

29. $\dfrac{\sqrt{-75}}{\sqrt{3}}$

30. $\dfrac{\sqrt{-160}}{\sqrt{10}}$

Add or subtract as indicated. Write your answers in standard form. **See Examples 4 and 5.**

31. $(3 + 2i) + (-4 + 5i)$

32. $(7 + 15i) + (-11 + 14i)$

33. $(5 - i) + (-5 + i)$

34. $(-2 + 6i) + (2 - 6i)$

35. $(4 + i) - (-3 - 2i)$

36. $(9 + i) - (3 + 2i)$

37. $(-3 - 4i) - (-1 - 4i)$

38. $(-2 - 3i) - (-5 - 3i)$

39. $(-4 + 11i) + (-2 - 4i) + (7 + 6i)$

40. $(-1 + i) + (2 + 5i) + (3 + 2i)$

41. $\left[(7 + 3i) - (4 - 2i) \right] + (3 + i)$

42. $\left[(7 + 2i) + (-4 - i) \right] - (2 + 5i)$

CONCEPT CHECK *Fill in the blank with the correct response.*

43. Because $(4 + 2i) - (3 + i) = 1 + i$, using the definition of subtraction we can check this to find that

$$(1 + i) + (3 + i) = \underline{\hspace{1cm}}.$$

44. Because $\frac{-5}{2 - i} = -2 - i$, using the definition of division we can check this to find that

$$(-2 - i)(2 - i) = \underline{\hspace{1cm}}.$$

Multiply. ***See Example 6.***

45. $(3i)(27i)$

46. $(5i)(125i)$

47. $(-8i)(-2i)$

48. $(-32i)(-2i)$

49. $5i(-6 + 2i)$

50. $3i(4 + 9i)$

51. $(4 + 3i)(1 - 2i)$

52. $(7 - 2i)(3 + i)$

53. $(4 + 5i)^2$

54. $(3 + 2i)^2$

55. $(12 + 3i)(12 - 3i)$

56. $(6 + 7i)(6 - 7i)$

CONCEPT CHECK *Answer each of the following.*

57. Let a and b represent real numbers.

   (a) What is the conjugate of $a + bi$?

   (b) If we multiply $a + bi$ by its conjugate, we get $\underline{\hspace{1cm}} + \underline{\hspace{1cm}}$, which is always a real number.

58. By what complex number should we multiply the numerator and denominator of $\frac{2 + i\sqrt{2}}{2 - i\sqrt{2}}$ to write the quotient in standard form?

   A. $\sqrt{2}$ **B.** $i\sqrt{2}$

   C. $2 + i\sqrt{2}$ **D.** $2 - i\sqrt{2}$

*Write each quotient in the form a + bi. **See Example 7.***

59. $\dfrac{2}{1 - i}$

60. $\dfrac{29}{5 + 2i}$

61. $\dfrac{-7 + 4i}{3 + 2i}$

62. $\dfrac{-38 - 8i}{7 + 3i}$

63. $\dfrac{8i}{2 + 2i}$

64. $\dfrac{-8i}{1 + i}$

65. $\dfrac{2 - 3i}{2 + 3i}$

66. $\dfrac{-1 + 5i}{3 + 2i}$

*Find each power of i. **See Example 8.***

67. i^{18}

68. i^{26}

69. i^{89}

70. i^{45}

71. i^{96}

72. i^{48}

73. i^{-5}

74. i^{-17}

75. i^{-20}

76. i^{-27}

77. A student simplified i^{-18} as follows:

$$i^{-18} = i^{-18} \cdot i^{20} = i^{-18+20} = i^2 = -1.$$

Explain the mathematical justification for this correct work.

78. Explain why

$$(46 + 25i)(3 - 6i) \quad \text{and} \quad (46 + 25i)(3 - 6i)i^{12}$$

must be equal. (Do not actually perform the computation.)

Ohm's law *for the current I in a circuit with voltage E, resistance R, capacitance reactance X_c, and inductive reactance X_L is*

$$I = \frac{E}{R + (X_L - X_c)i}.$$

Use this law to work Exercises 79 and 80.

79. Find I if $E = 2 + 3i$, $R = 5$, $X_L = 4$, and $X_c = 3$.

80. Find E if $I = 1 - i$, $R = 2$, $X_L = 3$, and $X_c = 1$.

81. Show that $1 + 5i$ is a solution of

$$x^2 - 2x + 26 = 0.$$

82. Show that $3 + 2i$ is a solution of

$$x^2 - 6x + 13 = 0.$$

Chapter 9 *Summary*

Key Terms

9.1

square root A number a is a square root of b if $a^2 = b$.

principal square root The positive square root of a number is its principal square root.

Radical symbol ↘ Index ↙
$\sqrt[n]{a}$ ← Radicand
Radical

radicand The number or expression inside a radical symbol is the radicand.

radical A radical symbol with a radicand is a radical.

radical expression An algebraic expression containing a radical is a radical expression.

perfect square A number with a rational square root is a perfect square.

irrational number A real number that is not rational is an irrational number.

cube root A number a is a cube root of b if $a^3 = b$.

index (order) In a radical of the form $\sqrt[n]{a}$, the number n is the index, or order.

9.5

rationalizing the denominator The process of removing radicals from the denominator so that the denominator contains only rational quantities is called rationalizing the denominator.

conjugate The conjugate of $a + b$ is $a - b$.

9.6

radical equation A radical equation is an equation that includes one or more radical expressions with variables.

extraneous solution (of a radical equation) An extraneous solution of a radical equation is a solution found after applying the power rule that is not a solution of the original equation.

9.7

complex number A complex number is a number that can be written in the form $a + bi$, where a and b are real numbers.

real part The real part of $a + bi$ is a.

imaginary part The imaginary part of $a + bi$ is b.

pure imaginary number A complex number $a + bi$ with $a = 0$ and $b \neq 0$ is a pure imaginary number.

nonreal complex number A complex number $a + bi$ with $b \neq 0$ is a nonreal complex number.

standard form (of a complex number) A complex number is in standard form if it is written in the form $a + bi$.

complex conjugate The complex conjugate of $a + bi$ is $a - bi$.

New Symbols

$\sqrt{}$	radical symbol
$\sqrt[n]{a}$	radical; principal nth root of a
$\pm$	"positive or negative," or "plus or minus"
$\approx$	is approximately equal to
$a^{1/n}$	a to the power $\dfrac{1}{n}$
$a^{m/n}$	a to the power $\dfrac{m}{n}$
i	imaginary unit

Test Your Word Power

See how well you have learned the vocabulary in this chapter.

1 A **radicand** is
A. the index of a radical
B. the number or expression under the radical symbol
C. the positive root of a number
D. the radical symbol.

2 A **hypotenuse** is
A. either of the two shorter sides of a triangle
B. the shortest side of a triangle
C. the side opposite the right angle in a right triangle
D. the longest side in any triangle.

3 **Rationalizing the denominator** is the process of
A. eliminating fractions from a radical expression
B. changing the denominator of a fraction from a radical expression to a rational number
C. clearing a radical expression of radicals
D. multiplying radical expressions.

4 An **extraneous solution** is a solution
A. that does not satisfy the original equation

B. that makes an equation true
C. that makes an expression equal 0
D. that checks in the original equation.

5 A **complex number** is
A. a real number that includes a complex fraction
B. a zero multiple of i
C. a number of the form $a + bi$, where a and b are real numbers
D. the square root of -1.

Answers To Test Your Word Power

1. B; *Example:* In $\sqrt{3xy}$, $3xy$ is the radicand.

2. C; *Example:* In a right triangle where the sides measure 9, 12, and 15 units, the hypotenuse is the side with measure 15 units.

3. B; *Example:* To rationalize the denominator of $\dfrac{5}{\sqrt{3}+1}$, multiply both the numerator and denominator by $\sqrt{3}-1$ to obtain $\dfrac{5(\sqrt{3}-1)}{2}$.

4. A; *Example:* The proposed solution 2 is extraneous in $\sqrt{5x-1}+3=0$, as it leads to $6=0$, a false statement.

5. C; *Examples:* -5 (or $-5+0i$), $7i$ (or $0+7i$), and $\sqrt{2}-4i$.

Quick Review

Concepts	Examples

9.1 Radical Expressions and Graphs

$\sqrt[n]{a} = b$ means $b^n = a$.

$\sqrt[n]{a}$ is the principal nth root of a.

$\sqrt[n]{a^n} = |a|$ if n is even. $\sqrt[n]{a^n} = a$ if n is odd.

The two square roots of 64 are $\sqrt{64} = 8$, the principal square root, and $-\sqrt{64} = -8$.

$$\sqrt[4]{(-2)^4} = |-2| = 2 \qquad \sqrt[3]{-27} = -3$$

Functions Defined by Radical Expressions

The square root function $f(x) = \sqrt{x}$ and the cube root function $f(x) = \sqrt[3]{x}$ are two important functions defined by radical expressions.

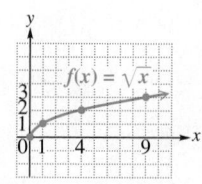

Square root function

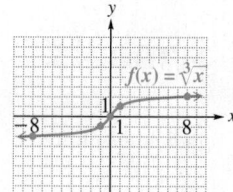

Cube root function

9.2 Rational Exponents

$a^{1/n} = \sqrt[n]{a}$ whenever $\sqrt[n]{a}$ exists.

If m and n are positive integers with m/n in lowest terms, then $a^{m/n} = (a^{1/n})^m$, provided that $a^{1/n}$ is a real number.

All of the usual definitions and rules for exponents are valid for rational exponents.

$$81^{1/2} = \sqrt{81} = 9 \qquad -64^{1/3} = -\sqrt[3]{64} = -4$$

$$8^{5/3} = (8^{1/3})^5 = 2^5 = 32 \qquad (y^{2/5})^{10} = y^4$$

$$5^{-1/2} \cdot 5^{1/4} = 5^{-1/2+1/4} = 5^{-1/4} = \frac{1}{5^{1/4}}$$

$$\frac{x^{-1/3}}{x^{-1/2}} = x^{-1/3-(-1/2)} = x^{-1/3+1/2} = x^{1/6}, \quad x > 0$$

Concepts	Examples

9.3 Simplifying Radical Expressions

Product and Quotient Rules for Radicals

If $\sqrt[n]{a}$ and $\sqrt[n]{b}$ are real numbers and n is a natural number, then

$$\sqrt[n]{a} \cdot \sqrt[n]{b} = \sqrt[n]{ab} \quad \text{and} \quad \sqrt[n]{\frac{a}{b}} = \frac{\sqrt[n]{a}}{\sqrt[n]{b}}, \quad b \neq 0.$$

Conditions for a Simplified Radical

1. The radicand has no factor raised to a power greater than or equal to the index.

2. The radicand has no fractions.

3. No denominator contains a radical.

4. Exponents in the radicand and the index of the radical have greatest common factor 1.

Pythagorean Theorem

If a and b are the lengths of the shorter sides of a right triangle and c is the length of the longest side, then

$$a^2 + b^2 = c^2.$$

The two shorter sides are the legs of the triangle, and the longest side is the hypotenuse. The hypotenuse is opposite the right angle.

Distance Formula

The distance d between the points (x_1, y_1) and (x_2, y_2) is

$$d = \sqrt{(x_2 - x_1)^2 + (y_2 - y_1)^2}.$$

$$\sqrt{3} \cdot \sqrt{7} = \sqrt{21} \qquad \sqrt[5]{x^3 y} \cdot \sqrt[5]{xy^2} = \sqrt[5]{x^4 y^3}$$

$$\frac{\sqrt{x^5}}{\sqrt{x^4}} = \sqrt{\frac{x^5}{x^4}} = \sqrt{x}, \quad x > 0$$

$$\sqrt{18} = \sqrt{9 \cdot 2} = 3\sqrt{2}$$

$$\sqrt[3]{54x^5 y^3} = \sqrt[3]{27x^3 y^3 \cdot 2x^2} = 3xy\sqrt[3]{2x^2}$$

$$\sqrt{\frac{7}{4}} = \frac{\sqrt{7}}{\sqrt{4}} = \frac{\sqrt{7}}{2}$$

$$\sqrt[9]{x^3} = x^{3/9} = x^{1/3}, \quad \text{or} \quad \sqrt[3]{x}$$

Find b for the triangle in the figure.

$$10^2 + b^2 = \left(2\sqrt{61}\right)^2$$
$$b^2 = 4(61) - 100$$
$$b^2 = 144$$
$$b = 12$$

Find the distance between $(3, -2)$ and $(-1, 1)$.

$$\sqrt{(-1 - 3)^2 + [1 - (-2)]^2}$$
$$= \sqrt{(-4)^2 + 3^2}$$
$$= \sqrt{16 + 9}$$
$$= \sqrt{25}, \quad \text{or} \quad 5$$

9.4 Adding and Subtracting Radical Expressions

Only radical expressions with the same index and the same radicand may be combined.

$$3\sqrt{17} + 2\sqrt{17} - 8\sqrt{17}$$
$$= (3 + 2 - 8)\sqrt{17}$$
$$= -3\sqrt{17}$$

$$\left.\begin{array}{l} \sqrt{15} + \sqrt{30} \\ \sqrt{3} + \sqrt[3]{9} \end{array}\right\} \text{ Cannot be combined}$$

9.5 Multiplying and Dividing Radical Expressions

Multiply binomial radical expressions by using the FOIL method. Special products from **Section 5.4** may apply.

$$\left(\sqrt{2} + \sqrt{7}\right)\left(\sqrt{3} - \sqrt{6}\right)$$
$$= \sqrt{6} - 2\sqrt{3} + \sqrt{21} - \sqrt{42} \qquad \sqrt{12} = 2\sqrt{3}$$

$$\left(\sqrt{5} - \sqrt{10}\right)\left(\sqrt{5} + \sqrt{10}\right) \qquad \left(\sqrt{3} - \sqrt{2}\right)^2$$
$$= 5 - 10 \qquad\qquad\qquad = 3 - 2\sqrt{3} \cdot \sqrt{2} + 2$$
$$= -5 \qquad\qquad\qquad\qquad = 5 - 2\sqrt{6}$$

(continued)

Concepts	Examples

9.5 **Multiplying and Dividing Radical Expressions** (*continued*)

$$\frac{\sqrt{7}}{\sqrt{5}} = \frac{\sqrt{7} \cdot \sqrt{5}}{\sqrt{5} \cdot \sqrt{5}} = \frac{\sqrt{35}}{5}$$

Rationalizing a Denominator

Rationalize the denominator by multiplying both the numerator and denominator by the same expression, one that will yield a rational number in the final denominator.

$$\frac{4}{\sqrt{5} - \sqrt{2}} = \frac{4(\sqrt{5} + \sqrt{2})}{(\sqrt{5} - \sqrt{2})(\sqrt{5} + \sqrt{2})}$$

$$= \frac{4(\sqrt{5} + \sqrt{2})}{5 - 2} = \frac{4(\sqrt{5} + \sqrt{2})}{3}$$

9.6 **Solving Equations with Radicals**

Solve $\sqrt{2x + 3} - x = 0$.

Solving an Equation with Radicals

Step 1 Isolate one radical on one side of the equation.

$$\sqrt{2x + 3} = x \qquad \text{Add } x.$$

Step 2 Raise each side of the equation to a power that is the same as the index of the radical.

$$(\sqrt{2x + 3})^2 = x^2 \qquad \text{Square each side.}$$
$$2x + 3 = x^2 \qquad \text{Apply the exponent.}$$

Step 3 Solve the resulting equation. If it still contains a radical, repeat Steps 1 and 2.

$$x^2 - 2x - 3 = 0 \qquad \text{Standard form}$$
$$(x + 1)(x - 3) = 0 \qquad \text{Factor.}$$

Step 4 Check all proposed solutions in the *original* equation.

$$x + 1 = 0 \quad \text{or} \quad x - 3 = 0 \qquad \text{Zero-factor property}$$

Proposed solutions that do not check are extraneous. They are not part of the solution set.

$$x = -1 \quad \text{or} \qquad x = 3 \qquad \text{Solve each equation.}$$

A check shows that 3 is a solution, but -1 is extraneous (as it leads to $2 = 0$, a false statement). The solution set is $\{3\}$.

9.7 **Complex Numbers**

$$i = \sqrt{-1}, \quad \text{and thus} \quad i^2 = -1.$$

For any positive number b, $\sqrt{-b} = i\sqrt{b}$.

To multiply radicals with negative radicands, first change each factor to the form $i\sqrt{b}$, and then multiply. The same procedure applies to quotients.

$$\sqrt{-25} = i\sqrt{25} = 5i$$

$$\sqrt{-3} \cdot \sqrt{-27}$$
$$= i\sqrt{3} \cdot i\sqrt{27} \qquad \sqrt{-b} = i\sqrt{b}$$
$$= i^2\sqrt{81}$$
$$= -1 \cdot 9, \quad \text{or} \quad -9 \qquad i^2 = -1$$

$$\frac{\sqrt{-18}}{\sqrt{-2}} = \frac{i\sqrt{18}}{i\sqrt{2}} = \sqrt{\frac{18}{2}} = \sqrt{9} = 3$$

Adding and Subtracting Complex Numbers

Add (or subtract) the real parts and add (or subtract) the imaginary parts.

$$(5 + 3i) + (8 - 7i) \qquad \qquad (5 + 3i) - (8 - 7i)$$
$$= 13 - 4i \qquad \qquad \qquad = -3 + 10i$$

Multiplying Complex Numbers

Multiply complex numbers by using the FOIL method.

$$(2 + i)(5 - 3i)$$
$$= 10 - 6i + 5i - 3i^2 \qquad \text{FOIL method}$$
$$= 10 - i - 3(-1) \qquad i^2 = -1$$
$$= 10 - i + 3 \qquad \text{Multiply.}$$
$$= 13 - i \qquad \text{Combine real terms.}$$

Dividing Complex Numbers

Divide complex numbers by multiplying the numerator and the denominator by the conjugate of the denominator.

$$\frac{2}{3 + i} = \frac{2(3 - i)}{(3 + i)(3 - i)} = \frac{2(3 - i)}{9 - i^2}$$

$$= \frac{2(3 - i)}{10} = \frac{2(3 - i)}{2 \cdot 5} = \frac{3 - i}{5} = \frac{3}{5} - \frac{1}{5}i$$

Chapter 9 *Review Exercises*

9.1 *Find each real number root. Use a calculator as necessary.*

1. $\sqrt{1764}$

2. $-\sqrt{289}$

3. $-\sqrt{-841}$

4. $\sqrt[3]{216}$

5. $\sqrt[5]{-32}$

6. $\sqrt{x^2}$

7. $\sqrt[3]{x^3}$

8. $\sqrt[4]{x^{20}}$

Graph each function. Give the domain and the range.

9. $f(x) = \sqrt{x-1}$

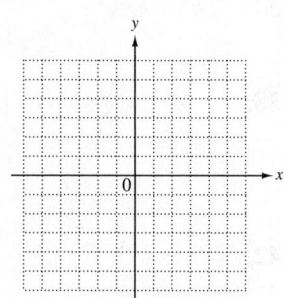

10. $f(x) = \sqrt[3]{x} - 2$

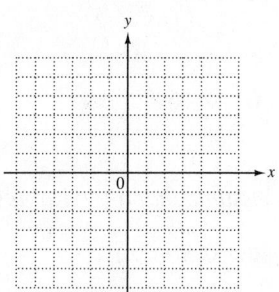

CONCEPT CHECK *Answer each question.*

11. Under what conditions is $\sqrt[n]{a}$ not a real number?

12. If a is negative and n is even, what can be said about $a^{1/n}$?

📷 *Use a calculator to find a decimal approximation for each radical. Round to the nearest thousandth.*

13. $\sqrt{40}$

14. $\sqrt{77}$

15. $\sqrt{310}$

📷 *Solve each problem.*

16. Use the formula for the time for one complete swing of a pendulum

$$t = 2\pi\sqrt{\frac{L}{g}}$$

to find the time to the nearest tenth of a second of a complete swing if the pendulum is 3 ft long and g is 32 ft per sec^2.

17. Use Heron's formula

$$A = \sqrt{s(s-a)(s-b)(s-c)},$$

where $s = \frac{1}{2}(a+b+c)$, to find the area of a triangle with sides of lengths 11, 13, and 20 in.

9.2 *Evaluate each exponential.*

18. $49^{1/2}$

19. $-8^{1/3}$

20. $(-16)^{1/4}$

21. Explain the relationship between the expressions $a^{m/n}$ and $\sqrt[n]{a^m}$.

Simplify each expression. Assume that all variables represent positive real numbers.

22. $16^{5/4}$

23. $-8^{2/3}$

24. $-\left(\dfrac{36}{25}\right)^{3/2}$

25. $\left(-\dfrac{1}{8}\right)^{-5/3}$

26. $\left(\dfrac{81}{10,000}\right)^{-3/4}$

27. $7^{1/3} \cdot 7^{5/3}$

28. $\dfrac{96^{2/3}}{96^{-1/3}}$

29. $\dfrac{k^{2/3}k^{-1/2}k^{3/4}}{2\left(k^2\right)^{-1/4}}$

30. Write $2^{4/5}$ as a radical.

Simplify each expression. Write answers in radical form. Assume that all variables represent positive real numbers.

31. $\sqrt{3^{18}}$

32. $\sqrt{7^9}$

33. $\sqrt[3]{m^5} \cdot \sqrt[3]{m^8}$

34. $\sqrt[4]{k^2} \cdot \sqrt[4]{k^7}$

35. $\sqrt[3]{\sqrt{m}}$

36. $\sqrt[4]{16y^5}$

37. $\sqrt[5]{y} \cdot \sqrt[3]{y}$

38. $\dfrac{\sqrt[3]{y^2}}{\sqrt[4]{y}}$

9.3 *Simplify each expression. Assume that all variables represent positive real numbers.*

39. $\sqrt{6} \cdot \sqrt{11}$

40. $\sqrt{5} \cdot \sqrt{r}$

41. $\sqrt[3]{6} \cdot \sqrt[3]{5}$

42. $\sqrt[4]{7} \cdot \sqrt[4]{3}$

43. $\sqrt{20}$

44. $-\sqrt{125}$

45. $\sqrt[3]{-108x^4y}$

46. $\sqrt[3]{64p^4q^6}$

47. $\sqrt{\dfrac{49}{81}}$

48. $\sqrt{\dfrac{y^3}{144}}$

49. $\sqrt[3]{\dfrac{m^{15}}{27}}$

50. $\sqrt[3]{\dfrac{r^2}{8}}$

51. $\dfrac{\sqrt[3]{2^4}}{\sqrt[4]{32}}$

52. $\dfrac{\sqrt{x}}{\sqrt[5]{x}}$

53. $\sqrt[4]{2} \cdot \sqrt{10}$

54. $\sqrt{5} \cdot \sqrt[3]{3}$

Find the distance between each pair of points.

55. $(2, 7)$ and $(-1, -4)$

56. $(-3, -5)$ and $(4, -3)$

9.4 *Perform the indicated operations. Assume that all variables represent positive real numbers.*

57. $2\sqrt{8} - 3\sqrt{50}$

58. $8\sqrt{80} - 3\sqrt{45}$

59. $-\sqrt{27y} + 2\sqrt{75y}$

60. $2\sqrt{54m^3} + 5\sqrt{96m^3}$

61. $3\sqrt[3]{54} + 5\sqrt[3]{16}$

62. $-6\sqrt[4]{32} + \sqrt[4]{512}$

9.5 *Multiply, and then simplify the products.*

63. $(\sqrt{3} + 1)(\sqrt{3} - 2)$

64. $(\sqrt{7} + \sqrt{5})(\sqrt{7} - \sqrt{5})$

65. $(3\sqrt{2} + 1)(2\sqrt{2} - 3)$

66. $(\sqrt{11} + 3\sqrt{5})(\sqrt{11} + 5\sqrt{5})$

67. $(\sqrt{13} - \sqrt{2})^2$

68. $(\sqrt{5} - \sqrt{7})^2$

Rationalize each denominator. Assume that all variables represent positive real numbers.

69. $\dfrac{-6\sqrt{3}}{\sqrt{2}}$

70. $\dfrac{3\sqrt{7p}}{\sqrt{y}}$

71. $-\sqrt[3]{\dfrac{9}{25}}$

72. $\sqrt[3]{\dfrac{108m^3}{n^5}}$

73. $\dfrac{1}{\sqrt{2} + \sqrt{7}}$

74. $\dfrac{-5}{\sqrt{6} - \sqrt{3}}$

9.6 *Solve each equation.*

75. $\sqrt{8x + 9} = 5$

76. $\sqrt{2z - 3} - 3 = 0$

77. $\sqrt{3m + 1} = -1$

78. $\sqrt{7z + 1} = z + 1$

79. $3\sqrt{m} = \sqrt{10m - 9}$

80. $\sqrt{p^2 + 3p + 7} = p + 2$

81. $\sqrt{x + 2} - \sqrt{x - 3} = 1$

82. $\sqrt[3]{5m - 1} = \sqrt[3]{3m - 2}$

83. $\sqrt[4]{x + 6} = \sqrt[4]{2x}$

9.7 *Write as a product of a real number and i.*

84. $\sqrt{-25}$

85. $\sqrt{-200}$

86. $\sqrt{-160}$

Perform the indicated operations. Write answers in standard form.

87. $(-2 + 5i) + (-8 - 7i)$

88. $(5 + 4i) - (-9 - 3i)$

89. $\sqrt{-5} \cdot \sqrt{-7}$

90. $\sqrt{-25} \cdot \sqrt{-81}$

91. $\dfrac{\sqrt{-72}}{\sqrt{-8}}$

92. $(2 + 3i)(1 - i)$

93. $(6 - 2i)^2$

94. $\dfrac{3 - i}{2 + i}$

95. $\dfrac{5 + 14i}{2 + 3i}$

Simplify each power of i.

96. i^{11}

97. i^{52}

98. i^{-13}

Mixed Review Exercises

Simplify. Assume that all variables represent positive real numbers.

99. $-\sqrt{169a^2b^4}$

100. $1000^{-2/3}$

101. $\dfrac{y^{-1/3} \cdot y^{5/6}}{y}$

102. $\dfrac{z^{-1/4}x^{1/2}}{z^{1/2}x^{-1/4}}$

103. $\sqrt[4]{k^{24}}$

104. $\sqrt[3]{54z^9t^8}$

105. $5i(3 - 7i)$

106. $\sqrt[3]{2} \cdot \sqrt[4]{5}$

107. $\left(7\sqrt{a} - 5\right)^2$

108. $-5\sqrt{18} + 12\sqrt{72}$

109. $8\sqrt[3]{x^3y^2} - 2x\sqrt[3]{y^2}$

110. $\left(\sqrt{5} - \sqrt{3}\right)\left(\sqrt{7} + \sqrt{3}\right)$

111. $\dfrac{-1}{\sqrt{12}}$

112. $\sqrt[3]{\dfrac{12}{25}}$

113. $\dfrac{2\sqrt{z}}{\sqrt{z} - 2}$

114. $\sqrt{-49}$

115. $(4 - 9i) + (-1 + 2i)$

116. $\dfrac{\sqrt{50}}{\sqrt{-2}}$

Solve each equation.

117. $\sqrt{x + 4} = x - 2$

118. $\sqrt{6 + 2x} - 1 = \sqrt{7 - 2x}$

Solve each problem.

119. Carpenters stabilize wall frames with a diagonal brace, as shown in the figure. The length of the brace is given by $L = \sqrt{H^2 + W^2}$.

   (a) Solve this formula for H.

   (b) If the bottom of the brace is attached 9 ft from the corner and the brace is 12 ft long, how far up the corner post should it be nailed? Give the answer to the nearest tenth of a foot.

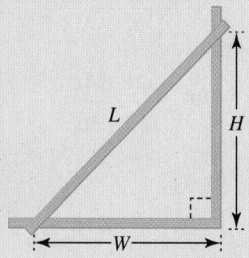

120. Find the perimeter of a triangular electronic highway road sign having the dimensions shown in the figure.

All

Traffic

$\sqrt{108}$ ft   Must Exit $2\sqrt{27}$ ft

Iowa

Highway 64

$\sqrt{50}$ ft

Chapter 9 **Test**  CHAPTER
Test Prep
VIDEO

The Chapter Test Prep Videos with test solutions are available
on DVD, in MyMathLab, and on You Tube — search
"LialCombinedAlg" and click on "Channels."

Find each root. Use a calculator as necessary.

1. $-\sqrt{841}$

2. $\sqrt[3]{-512}$

3. $125^{1/3}$

4. For $\sqrt{146.25}$, which choice gives the best estimate?

   A. 10   **B.** 11   **C.** 12   **D.** 13

Use a calculator to approximate each root to the nearest thousandth.

5. $\sqrt{478}$

6. $\sqrt[3]{-832}$

7. Graph the function $f(x) = \sqrt{x + 6}$, and give the domain
and the range.

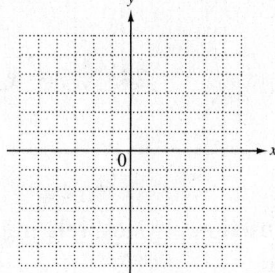

Simplify each expression. Assume that all variables represent positive real numbers.

8. $(-64)^{-4/3}$

9. $\dfrac{3^{2/5} x^{-1/4} y^{2/5}}{3^{-8/5} x^{7/4} y^{1/10}}$

10. $\sqrt{54x^5 y^6}$

11. $\sqrt[4]{32a^7 b^{13}}$

12. $\sqrt{2} \cdot \sqrt[3]{5}$
   (Express as a radical.)

13. $2\sqrt{300} + 5\sqrt{48}$

14. $3\sqrt{20} - 5\sqrt{80} + 4\sqrt{500}$

15. $\left(7\sqrt{5} + 4\right)\left(2\sqrt{5} - 1\right)$

16. $\left(\sqrt{3} - 2\sqrt{5}\right)^2$

17. $\dfrac{-5}{\sqrt{40}}$

18. $\dfrac{2}{\sqrt[3]{5}}$

19. $\dfrac{-4}{\sqrt{7} + \sqrt{5}}$

20. Write $\dfrac{6 + \sqrt{24}}{2}$ in lowest terms.

21. Find the distance between the points $(-3, 8)$ and $(2, 7)$.

22. Use the Pythagorean theorem to find the exact length of side b in the figure.

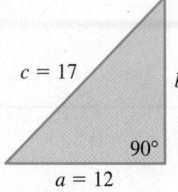

$c = 17$ b $90°$ $a = 12$

Solve each equation.

23. $\sqrt[3]{5x} = \sqrt[3]{2x - 3}$

24. $\sqrt{7 - x} + 5 = x$

25. $\sqrt{x + 4} - \sqrt{1 - x} = -1$

26. The following formula is used in physics, relating the velocity V of sound to the temperature T.

$$V = \dfrac{V_0}{\sqrt{1 - kT}}$$

(a) Find an approximation of V to the nearest tenth if $V_0 = 50$, $k = 0.01$, and $T = 30$. Use a calculator.

(b) Solve the formula for T.

Perform the indicated operations. Express answers in standard form.

27. $(-2 + 5i) - (3 + 6i) - 7i$

28. $(-4 + 2i)(3 - i)$

29. $\dfrac{7 + i}{1 - i}$

30. Simplify i^{35}.

Chapters R–9 *Cumulative Review Exercises*

Solve each equation or inequality.

1. $7 - (4 + 3t) + 2t = -6(t - 2) - 5$

2. $\frac{1}{3}x + \frac{1}{4}(x + 8) = x + 7$

3. $|6x - 9| = |-4x + 2|$

4. $-5 - 3(x - 2) < 11 - 2(x + 2)$

5. $1 + 4x > 5$ and $-2x > -6$

6. $-2 < 1 - 3x < 7$

7. Write the standard form of the equation of the line through the points $(-4, 6)$ and $(7, -6)$.

8. Choose the correct response: The lines with equations $2x + 3y = 8$ and $6y = 4x + 16$ are

    A. parallel   **B.** perpendicular   **C.** neither.

9. Consider the graph of $f(x) = -3x + 6$. Give the intercepts.

10. What is the slope of the line described in **Exercise 9?**

11. Graph the inequality $-2x + y < -6$.

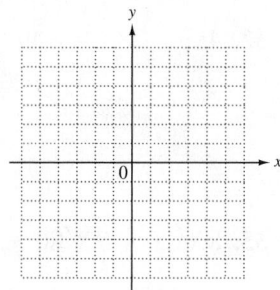

12. Find the measures of the marked angles.

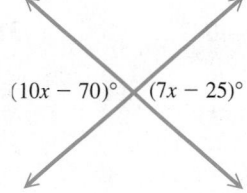

$(10x - 70)°$ $(7x - 25)°$

Solve each system.

13. $3x - y = 23$
    $2x + 3y = 8$

14. $5x + 2y = 7$
    $10x + 4y = 12$

15. $2x + y - z = 5$
    $6x + 3y - 3z = 15$
    $4x + 2y - 2z = 10$

16. In 2012, if we had sent five 2-oz letters and three 3-oz letters by first-class mail, it would have cost $5.80. Sending three 2-oz letters and five 3-oz letters would have cost $6.20. What was the 2012 postage rate for one 2-oz letter and for one 3-oz letter? (*Source:* U.S. Postal Service.)

Perform the indicated operations.

17. $(3k^3 - 5k^2 + 8k - 2) - (4k^3 + 11k + 7)$
    $+ (2k^2 - 5k)$

18. $(8x - 7)(x + 3)$

19. $\dfrac{8z^3 - 16z^2 + 24z}{8z^2}$

20. $\dfrac{6y^4 - 3y^3 + 5y^2 + 6y - 9}{2y + 1}$

Factor each polynomial completely.

21. $2p^2 - 5pq + 3q^2$

22. $18k^4 + 9k^2 - 20$

23. $x^3 + 512$

Perform each operation and express answers in lowest terms.

24. $\dfrac{y^2 + y - 12}{y^3 + 9y^2 + 20y} \div \dfrac{y^2 - 9}{y^3 + 3y^2}$

25. $\dfrac{1}{x + y} + \dfrac{3}{x - y}$

26. $\dfrac{x^2 - 12x + 36}{x^2 + 2x - 8} \cdot \dfrac{x^2 + 4x}{x^2 - 6x}$

Simplify each complex fraction.

27. $\dfrac{\dfrac{-6}{x - 2}}{\dfrac{8}{3x - 6}}$

28. $\dfrac{\dfrac{1}{a} - \dfrac{1}{b}}{\dfrac{a}{b} - \dfrac{b}{a}}$

29. $\dfrac{\dfrac{5r^2 s^3}{9}}{\dfrac{10r^4 s^5}{27}}$

Solve.

30. $2x^2 + 11x + 15 = 0$

31. $5t(t - 1) = 2(1 - t)$

32. $4x^2 - 28x = -49$

Simplify.

33. $27^{-5/3}$

34. $\dfrac{x^{-2/3}}{x^{-3/4}}, \quad x \neq 0$

35. $8\sqrt{20} + 3\sqrt{80} - 2\sqrt{500}$

36. $\dfrac{-9}{\sqrt{80}}$

37. $\dfrac{4}{\sqrt{6} - \sqrt{5}}$

38. $\dfrac{12}{\sqrt[3]{2}}$

39. Find the distance between the points $(-4, 4)$ and $(-2, 9)$.

40. Solve $\sqrt{8x - 4} - \sqrt{7x + 2} = 0$.

Solve each problem.

41. The current of a river runs at 3 mph. Brent's boat can travel 36 mi downstream in the same time that it takes to travel 24 mi upstream. Find the rate of the boat in still water.

42. How many liters of pure alcohol must be mixed with 40 L of 18% alcohol to obtain a 22% alcohol solution?

43. A jar containing only dimes and quarters has 29 coins with a face value of $4.70. How many of each denomination are there?

44. Brenda rides her bike 4 mph faster than her husband, Chuck. If Brenda can ride 48 mi in the same time that Chuck can ride 24 mi, what are their rates?

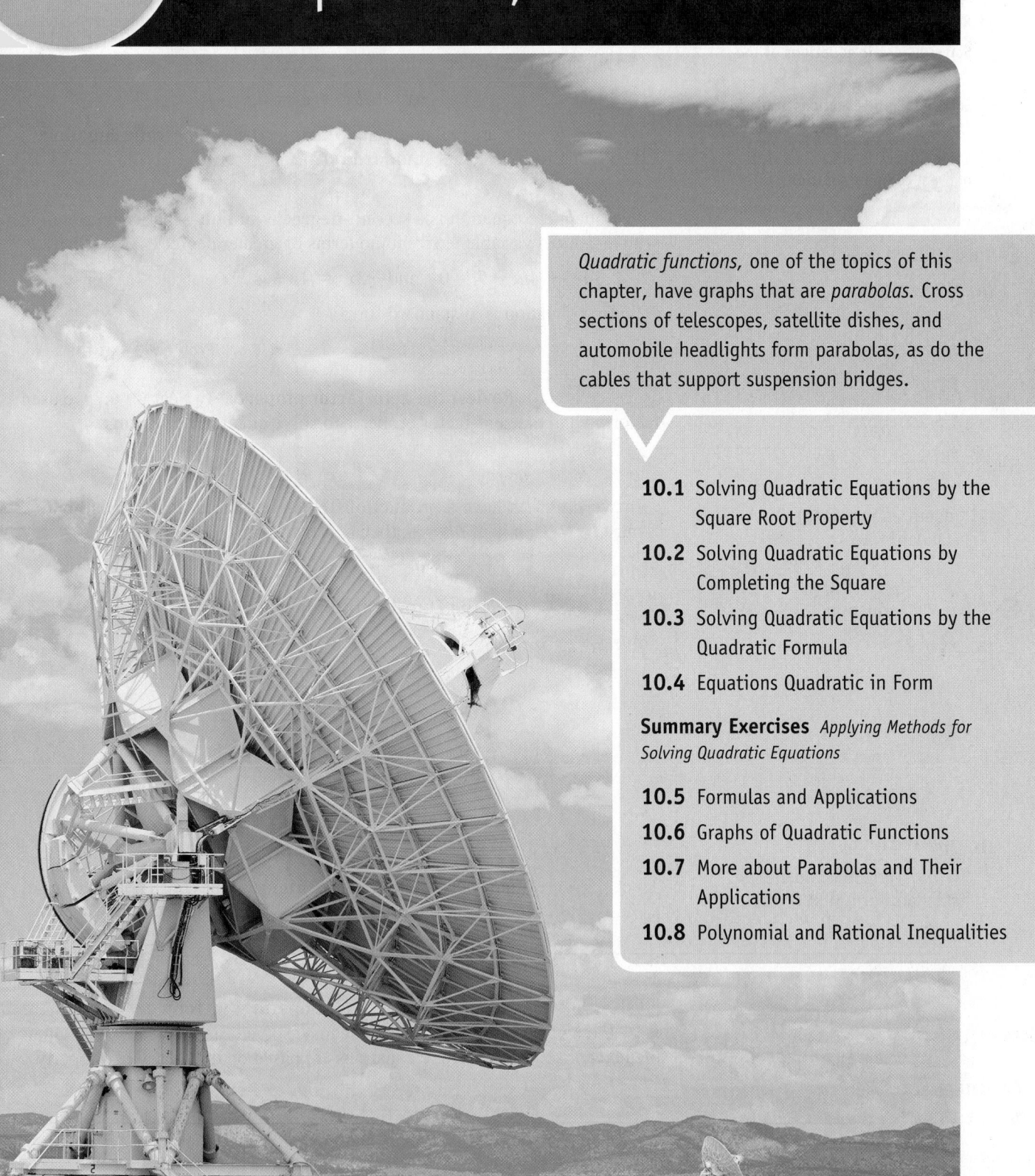

10

Quadratic Equations, Inequalities, and Functions

Quadratic functions, one of the topics of this chapter, have graphs that are *parabolas*. Cross sections of telescopes, satellite dishes, and automobile headlights form parabolas, as do the cables that support suspension bridges.

10.1 Solving Quadratic Equations by the Square Root Property

10.2 Solving Quadratic Equations by Completing the Square

10.3 Solving Quadratic Equations by the Quadratic Formula

10.4 Equations Quadratic in Form

Summary Exercises *Applying Methods for Solving Quadratic Equations*

10.5 Formulas and Applications

10.6 Graphs of Quadratic Functions

10.7 More about Parabolas and Their Applications

10.8 Polynomial and Rational Inequalities

10.1 Solving Quadratic Equations by the Square Root Property

OBJECTIVES

1 Review the zero-factor property.

2 Solve quadratic equations of the form $x^2 = k$, where $k > 0$.

3 Solve quadratic equations of the form $(ax + b)^2 = k$, where $k > 0$.

4 Solve quadratic equations with nonreal complex solutions.

1 Answer each question.

(a) Which of the following are quadratic equations?

   A. $x + 2y = 0$

   B. $x^2 - 8x + 16 = 0$

   C. $2t^2 - 5t = 3$

   D. $x^3 + x^2 + 4 = 0$

(b) Which quadratic equation identified in part (a) is in standard form?

2 Solve each equation.

(a) $x^2 + 3x + 2 = 0$

(b) $3m^2 = 3 - 8m$

   (*Hint:* Write the equation in standard form first.)

Answers

1. (a) B, C (b) B

2. (a) $\{-2, -1\}$ (b) $\left\{-3, \frac{1}{3}\right\}$

Recall from **Section 6.7** that a *quadratic equation* is defined as follows.

Quadratic Equation

An equation that can be written in the form

$$ax^2 + bx + c = 0,$$

where a, b, and c are real numbers, with $a \neq 0$, is a **quadratic equation.** The given form is called **standard form.**

A quadratic equation is a **second-degree equation**—that is, an equation with a squared variable term and no terms of greater degree.

$$4m^2 + 4m - 5 = 0 \quad \text{and} \quad 3x^2 = 4x - 8 \quad \text{Quadratic equations}$$

(The first equation is in standard form.)

◀ **Work Problem** 1 **at the Side.**

OBJECTIVE 1 **Review the zero-factor property.** In **Section 6.7** we used factoring and the zero-factor property to solve quadratic equations.

Zero-Factor Property

If two numbers have a product of 0, then at least one of the numbers must be 0. That is, if $ab = 0$, then $a = 0$ or $b = 0$.

EXAMPLE 1   **Using the Zero-Factor Property**

Solve $3x^2 - 5x - 28 = 0$.

$$3x^2 - 5x - 28 = 0 \quad \boxed{\text{The equation must be in standard form.}}$$

$$(3x + 7)(x - 4) = 0 \quad \text{Factor.}$$

$$3x + 7 = 0 \quad \text{or} \quad x - 4 = 0 \quad \text{Zero-factor property}$$

$$3x = -7 \quad \text{or} \quad x = 4 \quad \text{Solve each equation.}$$

$$x = -\frac{7}{3}$$

To check, substitute each solution in the original equation. The solution set is $\left\{-\frac{7}{3}, 4\right\}$.

◀ **Work Problem** 2 **at the Side.**

OBJECTIVE 2 **Solve quadratic equations of the form $x^2 = k$, where $k > 0$.** Not every quadratic equation can be solved easily by factoring. Other methods of solving quadratic equations are based on the following property.

Square Root Property

If x and k are complex numbers and $x^2 = k$, then

$$x = \sqrt{k} \quad \text{or} \quad x = -\sqrt{k}.$$

These steps justify the square root property.

$$x^2 = k$$

$$x^2 - k = 0 \qquad \text{Subtract } k.$$

$$\left(x - \sqrt{k}\right)\left(x + \sqrt{k}\right) = 0 \qquad \text{Factor.}$$

$$x - \sqrt{k} = 0 \quad \text{or} \quad x + \sqrt{k} = 0 \qquad \text{Zero-factor property}$$

$$\boldsymbol{x = \sqrt{k}} \quad \text{or} \quad \boldsymbol{x = -\sqrt{k}} \qquad \text{Solve each equation.}$$

Thus, the solutions of the equation $x^2 = k$ are $\sqrt{k}$ and $-\sqrt{k}$.

CAUTION

If $k > 0$, then using the square root property always produces *two* square roots, one positive and one negative.

EXAMPLE 2 **Using the Square Root Property**

Solve each equation.

(a) $x^2 = 16$

By the square root property, if $x^2 = 16$, then

$$x = \sqrt{16} = 4 \quad \text{or} \quad x = -\sqrt{16} = -4.$$

Check each solution by substituting it for x in the original equation. The solution set is $\{4, -4\}$, or $\{\pm 4\}$.

> This notation indicates two solutions, one positive and one negative.

(b) $r^2 = 5$

By the square root property, if $r^2 = 5$, then

$$r = \sqrt{5} \quad \text{or} \quad r = -\sqrt{5}.$$

> Don't forget the negative solution.

The solution set is $\{\sqrt{5}, -\sqrt{5}\}$, which may be written $\{\pm\sqrt{5}\}$.

(c)

$$m^2 - 20 = 0$$

$$m^2 = 20 \qquad \text{Add 20.}$$

$$m = \sqrt{20} \quad \text{or} \quad m = -\sqrt{20} \qquad \text{Square root property}$$

$$m = 2\sqrt{5} \quad \text{or} \quad m = -2\sqrt{5} \qquad \sqrt{20} = \sqrt{4} \cdot \sqrt{5} = 2\sqrt{5}$$

CHECK

$$m^2 - 20 = 0 \qquad \text{Original equation}$$

$$\left(2\sqrt{5}\right)^2 - 20 \stackrel{?}{=} 0 \qquad \text{Let } m = 2\sqrt{5}.$$

$$20 - 20 \stackrel{?}{=} 0 \qquad \left(2\sqrt{5}\right)^2 = 2^2 \cdot \left(\sqrt{5}\right)^2 = 4 \cdot 5 = 20$$

$$0 = 0 \checkmark \quad \text{True}$$

The check of $-2\sqrt{5}$ is similar. The solution set is $\{2\sqrt{5}, -2\sqrt{5}\}$.

········· **Continued on Next Page**

③ Solve each equation.

(a) $m^2 = 64$

(b) $p^2 = 7$

(c) $x^2 - 27 = 0$

(d) $3x^2 - 54 = 0$

(e) $2x^2 + 7 = 55$

④ Solve the problem.

An expert marksman can hold a silver dollar at forehead level, drop it, draw his gun, and shoot the coin as it passes waist level. If the coin falls about 4 ft, use the formula in **Example 3** to find the time that elapses between the dropping of the coin and the shot.

(d)
$$4x^2 - 48 = 0$$
$$4x^2 = 48 \qquad \text{Add 48.}$$
$$x^2 = 12 \qquad \text{Divide by 4.}$$
$$x = \sqrt{12} \quad \text{or} \quad x = -\sqrt{12} \qquad \text{Square root property}$$
$$x = 2\sqrt{3} \quad \text{or} \quad x = -2\sqrt{3} \qquad \sqrt{12} = \sqrt{4} \cdot \sqrt{3} = 2\sqrt{3}$$

The solutions are $2\sqrt{3}$ and $-2\sqrt{3}$. Check each in the original equation.

CHECK $\qquad 4x^2 - 48 = 0 \qquad$ Original equation

$$4\left(2\sqrt{3}\right)^2 - 48 \overset{?}{=} 0 \quad \text{Let } x =$$
$$4(12) - 48 \overset{?}{=} 0 \quad 2\sqrt{3}.$$

$\boxed{(2\sqrt{3})^2 = 2^2 \cdot (\sqrt{3})^2}$

$$48 - 48 \overset{?}{=} 0$$
$$0 = 0 \checkmark \text{ True}$$

$$4\left(-2\sqrt{3}\right)^2 - 48 \overset{?}{=} 0 \quad \text{Let } x =$$
$$4(12) - 48 \overset{?}{=} 0 \quad -2\sqrt{3}.$$
$$48 - 48 \overset{?}{=} 0$$
$$0 = 0 \checkmark \text{ True}$$

The solution set is $\left\{2\sqrt{3}, -2\sqrt{3}\right\}$.

(e)
$$5m^2 - 32 = 8$$
$$5m^2 = 40 \qquad \text{Add 32.}$$
$$m^2 = 8 \qquad \text{Divide by 5.}$$

$\boxed{\text{Don't stop here. Simplify the radicals.}}$ $\quad m = \sqrt{8} \quad \text{or} \quad m = -\sqrt{8} \qquad$ Square root property

$$m = 2\sqrt{2} \quad \text{or} \quad m = -2\sqrt{2} \qquad \sqrt{8} = \sqrt{4} \cdot \sqrt{2} = 2\sqrt{2}$$

The solution set is $\left\{2\sqrt{2}, -2\sqrt{2}\right\}$.

◀ **Work Problem ③ at the Side.**

EXAMPLE 3 **Using the Square Root Property in an Application**

Galileo Galilei developed a formula for freely falling objects,

$$d = 16t^2,$$

where d is the distance in feet that an object falls (disregarding air resistance) in t seconds, regardless of weight. The Leaning Tower of Pisa is about 180 ft tall. Use the formula to determine how long it would take an object dropped from the top of the tower to fall to the ground. (*Source:* www.brittanica.com)

$$d = 16t^2 \qquad \text{Galileo's formula}$$
$$180 = 16t^2 \qquad \text{Let } d = 180.$$
$$11.25 = t^2 \qquad \text{Divide by 16.}$$
$$t = \sqrt{11.25} \quad \text{or} \quad t = -\sqrt{11.25} \qquad \text{Square root property}$$

Time cannot be negative, so we discard $t = -\sqrt{11.25}$. Using a calculator, $\sqrt{11.25} \approx 3.4$ so $t \approx 3.4$. The object would fall to the ground in about 3.4 sec.

◀ **Work Problem ④ at the Side.**

Answers

3. (a) $\{8, -8\}$ **(b)** $\left\{\sqrt{7}, -\sqrt{7}\right\}$

(c) $\left\{3\sqrt{3}, -3\sqrt{3}\right\}$

(d) $\left\{3\sqrt{2}, -3\sqrt{2}\right\}$

(e) $\left\{2\sqrt{6}, -2\sqrt{6}\right\}$

4. 0.5 sec

OBJECTIVE ▶ 3 **Solve quadratic equations of the form $(ax + b)^2 = k$, where $k > 0$.** We can extend the square root property to solve equations in which the base is a binomial.

EXAMPLE 4 Extending the Square Root Property

Solve each equation.

(a) $(x - 5)^2 = 36$

$$x^2 \quad = \quad k$$

$$(x - 5)^2 = 36 \quad \begin{array}{l}\text{Substitute } (x - 5)^2 \text{ for } x^2 \text{ and } 36 \text{ for}\\ k \text{ in the square root property.}\end{array}$$

$x - 5 = \sqrt{36}$ or $x - 5 = -\sqrt{36}$ Square root property

$x - 5 = 6$ or $x - 5 = -6$ Take square roots.

$x = 11$ or $x = -1$ Add 5.

CHECK $(x - 5)^2 = 36$ Original equation

$(11 - 5)^2 \stackrel{?}{=} 36$ Let $x = 11$. | $(-1 - 5)^2 \stackrel{?}{=} 36$ Let $x = -1$.

$6^2 \stackrel{?}{=} 36$ | $(-6)^2 \stackrel{?}{=} 36$

$36 = 36$ ✓ True | $36 = 36$ ✓ True

Both solutions satisfy the original equation. The solution set is $\{-1, 11\}$.

(b) $(x + 1)^2 = 6$

$x + 1 = \sqrt{6}$ or $x + 1 = -\sqrt{6}$ Square root property

$x = -1 + \sqrt{6}$ or $x = -1 - \sqrt{6}$ Add -1.

CHECK $\left(-1 + \sqrt{6} + 1\right)^2 = \left(\sqrt{6}\right)^2 = 6$ ✓ Let $x = -1 + \sqrt{6}$.

$\left(-1 - \sqrt{6} + 1\right)^2 = \left(-\sqrt{6}\right)^2 = 6$ ✓ Let $x = -1 - \sqrt{6}$.

The solution set is $\left\{-1 + \sqrt{6}, -1 - \sqrt{6}\right\}$, or $\left\{-1 \pm \sqrt{6}\right\}$.

··· **Work Problem ❺ at the Side.** ▶

EXAMPLE 5 Extending the Square Root Property

Solve $(2x - 3)^2 = 18$.

$$(2x - 3)^2 = 18$$

$2x - 3 = \sqrt{18}$ or $2x - 3 = -\sqrt{18}$ Square root property

$2x = 3 + \sqrt{18}$ or $2x = 3 - \sqrt{18}$ Add 3.

$x = \dfrac{3 + \sqrt{18}}{2}$ or $x = \dfrac{3 - \sqrt{18}}{2}$ Divide by 2.

$x = \dfrac{3 + 3\sqrt{2}}{2}$ or $x = \dfrac{3 - 3\sqrt{2}}{2}$ $\sqrt{18} = \sqrt{9} \cdot \sqrt{2} = 3\sqrt{2}$

··· **Continued on Next Page**

---

❺ Solve each equation.

(a) $(x - 3)^2 = 16$

(b) $(x + 3)^2 = 25$

GS **(c)** $(x - 4)^2 = 3$

$x - 4 =$ _____ or $x - 4 =$ _____

$x =$ _____ or $x =$ _____

The solution set is _____ .

Answers

5. **(a)** $\{-1, 7\}$ **(b)** $\{-8, 2\}$
  (c) $\sqrt{3}; -\sqrt{3}; 4 + \sqrt{3}; 4 - \sqrt{3};$
  $\left\{4 + \sqrt{3}, 4 - \sqrt{3}\right\}$

6 Solve each equation.

(a) $(3k + 1)^2 = 2$

(b) $(2r + 3)^2 = 8$

We show the check for the first solution. The check for the second solution is similar.

CHECK $\qquad\qquad (2x - 3)^2 = 18 \qquad$ Original equation

$$\left[2\left(\frac{3 + 3\sqrt{2}}{2}\right) - 3\right]^2 \overset{?}{=} 18 \qquad \text{Let } x = \frac{3 + 3\sqrt{2}}{2}.$$

$$\left(3 + 3\sqrt{2} - 3\right)^2 \overset{?}{=} 18 \qquad \text{Multiply.}$$

$$\left(3\sqrt{2}\right)^2 \overset{?}{=} 18 \qquad \text{Simplify.}$$

$$18 = 18 \;\checkmark \quad \text{True}$$

> The $\pm$ symbol denotes two solutions.

The solution set is $\left\{\frac{3 + 3\sqrt{2}}{2}, \frac{3 - 3\sqrt{2}}{2}\right\}$, abbreviated $\left\{\frac{3 \pm \sqrt{2}}{2}\right\}$.

◀ **Work Problem 6 at the Side.**

OBJECTIVE **4** **Solve quadratic equations with nonreal complex solutions.** If $k < 0$ in the equation $x^2 = k$, then there will be two nonreal complex solutions.

EXAMPLE 6 **Solving for Nonreal Complex Solutions**

Solve each equation.

(a) $\qquad\qquad\qquad x^2 = -15$

$$x = \sqrt{-15} \quad \text{or} \quad x = -\sqrt{-15} \qquad \text{Square root property}$$

$$x = i\sqrt{15} \quad \text{or} \quad x = -i\sqrt{15} \qquad \sqrt{-a} = i\sqrt{a}$$

The solution set is $\left\{i\sqrt{15}, -i\sqrt{15}\right\}$.

(b) $\qquad\qquad (t + 2)^2 = -16$

$$t + 2 = \sqrt{-16} \quad \text{or} \quad t + 2 = -\sqrt{-16} \qquad \text{Square root property}$$

$$t + 2 = 4i \quad\quad \text{or} \quad t + 2 = -4i \qquad\quad \sqrt{-16} = 4i$$

$$t = -2 + 4i \quad \text{or} \qquad t = -2 - 4i \qquad \text{Add } -2.$$

The solution set is $\{-2 + 4i, -2 - 4i\}$.

◀ **Work Problem 7 at the Side.**

7 Solve each equation.

(a) $x^2 = -17$

(b) $(k + 5)^2 = -100$

Answers

6. (a) $\left\{\dfrac{-1 + \sqrt{2}}{3}, \dfrac{-1 - \sqrt{2}}{3}\right\}$

   (b) $\left\{\dfrac{-3 + 2\sqrt{2}}{2}, \dfrac{-3 - 2\sqrt{2}}{2}\right\}$

7. (a) $\left\{i\sqrt{17}, -i\sqrt{17}\right\}$

   (b) $\{-5 + 10i, -5 - 10i\}$

10.1 Exercises

  MyMathLab®

CONCEPT CHECK *Fill in each blank with the correct response.*

1. An equation in the form $ax^2 + bx + c = 0$, where a, b, and c are real numbers with $a \neq 0$, is a(n) _____ equation, also called a(n) _____ -degree equation. The greatest degree of the variable is _____ .

2. The equation $2x^2 - 7x + 4 = 0$ is a(n) _____ equation that is written in _____ form.

3. CONCEPT CHECK A student incorrectly solved $x^2 - x - 2 = 5$ as follows. **What Went Wrong?**

$$x^2 - x - 2 = 5$$
$$(x - 2)(x + 1) = 5 \quad \text{Factor.}$$
$$x - 2 = 5 \quad \text{or} \quad x + 1 = 5 \quad \text{Zero-factor property}$$
$$x = 7 \quad \text{or} \quad x = 4 \quad \text{Solve each equation.}$$

4. CONCEPT CHECK A student was asked to solve the quadratic equation

$$x^2 = 16$$

and did not get full credit for the solution set $\{4\}$. **What Went Wrong?**

Use the zero-factor property to solve each equation. (Hint: In Exercises 9 and 10, write the equation in standard form first.) **See Example 1.**

5. $x^2 + 3x + 2 = 0$

6. $x^2 + 8x + 15 = 0$

7. $3x^2 + 8x - 3 = 0$

8. $2x^2 + x - 6 = 0$

9. $2x^2 = 9x - 4$

10. $5x^2 = 11x - 2$

Solve each equation. **See Example 2.**

11. $x^2 = 81$

12. $z^2 = 225$

13. $t^2 = 17$

14. $k^2 = 19$

15. $m^2 = 32$

16. $x^2 = 54$

17. $r^2 - 3 = 0$

18. $x^2 - 13 = 0$

19. $t^2 - 48 = 0$

20. $p^2 - 50 = 0$

21. $3n^2 - 72 = 0$

22. $5z^2 - 200 = 0$

23. $2x^2 + 7 = 61$

24. $3x^2 - 8 = 64$

Solve each equation. **See Examples 4 and 5.**

25. $(x + 2)^2 = 25$

26. $(t + 8)^2 = 9$

27. $(x - 6)^2 = 49$

28. $(x - 4)^2 = 64$

29. $(x - 4)^2 = 3$ **30.** $(x + 3)^2 = 11$ **31.** $(t + 5)^2 = 48$

32. $(m - 6)^2 = 27$ **33.** $(3x + 2)^2 = 49$ **34.** $(5x + 3)^2 = 36$

35. $(3k - 1)^2 = 7$ **36.** $(2x - 5)^2 = 10$ **37.** $(4p + 1)^2 = 24$

38. $(5k - 2)^2 = 12$ **39.** $(3x + 1)^2 = 18$ **40.** $(5x + 6)^2 = 75$

Solve Exercises 41 and 42 using Galileo's formula, $d = 16t^2$. Round answers to the nearest tenth. **See Example 3.**

41. The sculpture of American presidents at Mount Rushmore National Memorial is 500 ft above the valley floor. How long would it take a rock dropped from the top of the sculpture to fall to the ground? (*Source:* www.travelsd.com)

42. The Gateway Arch in St. Louis, Missouri, is 630 ft tall. How long would it take an object dropped from the top of the arch to fall to the ground? (*Source:* www.gatewayarch.com)

Find all nonreal complex solutions of each equation. **See Example 6.**

43. $x^2 = -100$ **44.** $x^2 = -64$ **45.** $x^2 = -12$ **46.** $x^2 = -18$

47. $(r - 5)^2 = -3$ **48.** $(t + 6)^2 = -5$ **49.** $(6k - 1)^2 = -8$ **50.** $(4m - 7)^2 = -27$

10.2 Solving Quadratic Equations by Completing the Square

OBJECTIVE ▶ **1** **Solve quadratic equations by completing the square when the coefficient of the second-degree term is 1.** The methods we have studied so far are not enough to solve the equation

$$x^2 + 8x + 10 = 0.$$

If we could write the equation in the form $(x + 4)^2$ equals a constant, we could solve it with the square root property discussed in **Section 10.1**. To do that, we need to have a perfect square trinomial on one side of the equation.

Recall from **Section 6.5** that the perfect square trinomial

$$x^2 + 8x + 16 \quad \text{can be factored as} \quad (x + 4)^2.$$

In the trinomial, the coefficient of x (the first-degree term) is 8 and the constant term is 16. If we take half of 8 and square it, we get the constant term, 16.

$$\overset{\text{Coefficient of } x}{\underset{\downarrow}{}} \qquad \overset{\text{Constant}}{\underset{\downarrow}{}}$$
$$\left[\frac{1}{2}(8)\right]^2 = 4^2 = 16$$

Similarly, in $\quad x^2 + 12x + 36, \quad \left[\frac{1}{2}(12)\right]^2 = 6^2 = 36,$

and in $\quad m^2 - 6m + 9, \quad \left[\frac{1}{2}(-6)\right]^2 = (-3)^2 = 9.$

This relationship is true in general and is the idea behind rewriting a quadratic equation so that the square root property can be applied.

Work Problem ❶ at the Side. ▶

EXAMPLE 1   Rewriting a Quadratic Equation to Use the Square Root Property

Solve $x^2 + 8x + 10 = 0$.

This quadratic equation cannot be solved easily by factoring, and it is not in the correct form to solve using the square root property. To solve it by completing the square, we need a perfect square trinomial on the left side of the equation. To get this form, we first subtract 10 from each side.

$$x^2 + 8x + 10 = 0 \qquad \text{Original equation}$$
Only terms with variables remain on the left side. $\longrightarrow$ $x^2 + 8x = -10 \qquad \text{Subtract 10.}$

We must add a constant to get a perfect square trinomial on the left.

$$\underbrace{x^2 + 8x + \underline{\;?\;}}_{\substack{\text{Needs to be a perfect} \\ \text{square trinomial}}}$$

Take half the coefficient of the first-degree term, $8x$, and square the result.

$$\left[\frac{1}{2}(8)\right]^2 = 4^2 = 16 \leftarrow \text{Desired constant}$$

⋯⋯⋯⋯⋯⋯⋯⋯⋯⋯ Continued on Next Page

❶ Find the constant that must be added to make each expression a perfect square trinomial. Then factor the trinomial.

GS **(a)** $x^2 + 4x$

The coefficient of the first-degree term is ____.

$$\left[\frac{1}{2}(4)\right]^2 = \underline{\quad}^2 = \underline{\quad}$$

The perfect square trinomial would be

$$x^2 + 4x + \underline{\quad}.$$

It factors as _____.

(b) $t^2 - 2t + \underline{\quad}$

It factors as _____.

(c) $m^2 + 5m + \underline{\quad}$

It factors as _____.

(d) $x^2 - \frac{2}{3}x + \underline{\quad}$

It factors as _____.

Answers

1. **(a)** 4; 2; 4; 4; $(x + 2)^2$ **(b)** 1; $(t - 1)^2$
   (c) $\frac{25}{4}$; $\left(m + \frac{5}{2}\right)^2$ **(d)** $\frac{1}{9}$; $\left(x - \frac{1}{3}\right)^2$

2 Solve each equation.

(GS) (a) $n^2 + 6n + 4 = 0$

Isolate the terms with variables on the left side.

$$n^2 + 6n = \underline{\quad}$$

Take half the coefficient of the first-degree term and square the result.

$$\left[\frac{1}{2}(\underline{\quad})\right]^2 = \underline{\quad}^2 = \underline{\quad}$$

Add this result to each side of the equation, and complete the solution.

$$n^2 + 6n + 9 = -4 + \underline{\quad}$$

$$(\underline{\quad\quad})^2 = 5$$

$$n + 3 = \sqrt{5} \qquad \text{or} \quad n + 3 = \underline{\quad}$$

$$n = -3 + \sqrt{5} \text{ or } \qquad n = \underline{\quad}$$

The solution set is

$$\{\underline{\quad\quad}, \underline{\quad\quad}\}.$$

(b) $x^2 + 2x - 10 = 0$

3 Solve each equation by completing the square.

(a) $x^2 + 4x = 1$

(b) $z^2 + 6z - 3 = 0$

Answers

2. **(a)** $-4; 6; 3; 9; 9; n + 3; -\sqrt{5}; -3 - \sqrt{5};$
   $-3 + \sqrt{5}; -3 - \sqrt{5}$
   (b) $\{-1 + \sqrt{11}, -1 - \sqrt{11}\}$
3. **(a)** $\{-2 + \sqrt{5}, -2 - \sqrt{5}\}$
   (b) $\{-3 + 2\sqrt{3}, -3 - 2\sqrt{3}\}$

We add this constant, 16, to *each* side of the equation $x^2 + 8x = -10$.

$$x^2 + 8x + 16 = -10 + 16 \qquad \text{Add 16 to each side.}$$

> This is a key step.

$$(x + 4)^2 = 6 \qquad \begin{array}{l}\text{Factor on the left.}\\\text{Add on the right.}\end{array}$$

$$x + 4 = \sqrt{6} \qquad \text{or} \quad x + 4 = -\sqrt{6} \qquad \begin{array}{l}\text{Square root}\\\text{property}\end{array}$$

$$x = -4 + \sqrt{6} \quad \text{or} \qquad x = -4 - \sqrt{6} \quad \text{Add } -4.$$

CHECK $\qquad\qquad x^2 + 8x + 10 = 0 \qquad$ Original equation

$$(-4 + \sqrt{6})^2 + 8(-4 + \sqrt{6}) + 10 \overset{?}{=} 0 \qquad \text{Let } x = -4 + \sqrt{6}.$$

$$16 - 8\sqrt{6} + 6 - 32 + 8\sqrt{6} + 10 \overset{?}{=} 0 \qquad \text{Multiply.}$$

> Remember the middle term when squaring $-4 + \sqrt{6}$.

$$0 = 0 \checkmark \quad \text{True}$$

The check of the other solution is similar. The solution set is

$$\{-4 + \sqrt{6}, -4 - \sqrt{6}\}.$$

◀ **Work Problem 2** at the Side.

The process of changing the form of the equation in **Example 1** from

$$x^2 + 8x + 10 = 0 \quad \text{to} \quad (x + 4)^2 = 6$$

is called **completing the square**. Completing the square changes only the form of the equation. To see this, multiply out the left side of $(x + 4)^2 = 6$. Then write the equation in standard form to get $x^2 + 8x + 10 = 0$.

EXAMPLE 2	**Solving a Quadratic Equation by Completing the Square $(a = 1)$**

Solve $x^2 - 6x = 12$.

To complete the square on $x^2 - 6x$, take half the coefficient of x and square it.

$$\frac{1}{2}(-6) = -3 \quad \text{and} \quad (-3)^2 = 9$$

Coefficient of x

Add the result, **9**, to each side of the equation.

$$x^2 - 6x = 12 \qquad\qquad \text{Given equation}$$

$$x^2 - 6x + 9 = 12 + 9 \qquad \text{Add 9.}$$

$$(x - 3)^2 = 21 \qquad\qquad \text{Factor. Add.}$$

$$x - 3 = \sqrt{21} \qquad \text{or} \quad x - 3 = -\sqrt{21} \qquad \text{Square root property}$$

$$x = 3 + \sqrt{21} \quad \text{or} \qquad x = 3 - \sqrt{21} \qquad \text{Add 3.}$$

A check indicates that the solution set is $\{3 + \sqrt{21}, 3 - \sqrt{21}\}$.

◀ **Work Problem 3** at the Side.

Completing the Square to Solve $ax^2 + bx + c = 0$ (Where $a \neq 0$)

Step 1 **Be sure the second-degree (squared variable) term has coefficient 1.** If the coefficient of the second-degree term is 1, proceed to Step 2. If the coefficient of the second-degree term is not 1 but some other nonzero number a, then divide each side of the equation by a.

Step 2 **Write the equation in correct form** so that terms with variables are on one side of the equality symbol and the constant is on the other side.

Step 3 **Square half the coefficient of the first-degree term.**

Step 4 **Add the square to each side.**

Step 5 **Factor the perfect square trinomial.** One side should now be a perfect square trinomial. Factor it as the square of a binomial. Simplify the other side.

Step 6 **Solve the equation.** Apply the square root property to complete the solution.

EXAMPLE 3 **Solving a Quadratic Equation by Completing the Square ($a = 1$)**

Solve $k^2 + 5k - 1 = 0$.

Since the coefficient of the second-degree term is 1, begin with Step 2.

Step 2 $\qquad\qquad k^2 + 5k = 1$ Add 1 to each side.

Step 3 Take half the coefficient of the first-degree term and square the result.

$$\left[\frac{1}{2}(5)\right]^2 = \left(\frac{5}{2}\right)^2 = \frac{25}{4}$$

Step 4 $\qquad k^2 + 5k + \dfrac{25}{4} = 1 + \dfrac{25}{4}$ ◁ Add the square to each side of the equation.

Step 5 $\qquad\qquad \left(k + \dfrac{5}{2}\right)^2 = \dfrac{29}{4}$ Factor on the left. Add on the right.

Step 6 $\quad k + \dfrac{5}{2} = \sqrt{\dfrac{29}{4}}$ or $k + \dfrac{5}{2} = -\sqrt{\dfrac{29}{4}}$ Square root property

$\qquad k + \dfrac{5}{2} = \dfrac{\sqrt{29}}{2}$ or $k + \dfrac{5}{2} = -\dfrac{\sqrt{29}}{2}$ $\sqrt{\frac{a}{b}} = \frac{\sqrt{a}}{\sqrt{b}}$

$\qquad k = -\dfrac{5}{2} + \dfrac{\sqrt{29}}{2}$ or $k = -\dfrac{5}{2} - \dfrac{\sqrt{29}}{2}$ Add $-\frac{5}{2}$.

$\qquad k = \dfrac{-5 + \sqrt{29}}{2}$ or $k = \dfrac{-5 - \sqrt{29}}{2}$ $\frac{a}{c} \pm \frac{b}{c} = \frac{a \pm b}{c}$

Check that the solution set is $\left\{\dfrac{-5 + \sqrt{29}}{2}, \dfrac{-5 - \sqrt{29}}{2}\right\}$.

▸ **Work Problem ④ at the Side.** ▶

---

④ Solve each equation.

(a) $x^2 + x - 3 = 0$

(b) $r^2 + 3r - 1 = 0$

Answers

4. **(a)** $\left\{\dfrac{-1 + \sqrt{13}}{2}, \dfrac{-1 - \sqrt{13}}{2}\right\}$

    (b) $\left\{\dfrac{-3 + \sqrt{13}}{2}, \dfrac{-3 - \sqrt{13}}{2}\right\}$

5 Solve each equation by completing the square.

GS **(a)** $3x^2 + 5x - 2 = 0$

Step 1
Divide each side by ____.

_____

Step 2
Add ____ to each side.

_____

Steps 3 and 4
Take half the coefficient of x and square it. Add ____ to each side to complete the square.

$$x^2 + \frac{5}{3}x + \underline{} = \frac{2}{3} + \frac{25}{36}$$

Step 5
Factor and add.

$$(\underline{})^2 = \underline{}$$

Continue with Step 6 to complete the solution.

(b) $9x^2 + 18x = -5$

(c) $4t^2 - 24t + 11 = 0$

OBJECTIVE **2** **Solve quadratic equations by completing the square when the coefficient of the second-degree term is not 1.** If a quadratic equation has the form

$$ax^2 + bx + c = 0, \quad \text{where } a \neq 1,$$

then to obtain 1 as the coefficient of x^2, we first divide each side by a.

EXAMPLE 4 Solving a Quadratic Equation by Completing the Square ($a \neq 1$)

Solve $4x^2 + 16x = 9$.

Step 1 *Before completing the square, the coefficient of x^2 must be 1,* not 4. We get 1 as the coefficient of x^2 here by dividing each side by 4.

The coefficient of x^2 must be 1. $\longrightarrow$ $x^2 + 4x = \dfrac{9}{4}$ Divide by 4.

Step 2 The equation is already in the correct form, with the variable terms on one side and the constant on the other.

Step 3 Complete the square. Take half the coefficient of x and square it.

$$\frac{1}{2}(4) = 2 \quad \text{and} \quad 2^2 = 4$$

Step 4 $x^2 + 4x + 4 = \dfrac{9}{4} + 4$ Add 4 to each side.

Step 5 $(x + 2)^2 = \dfrac{25}{4}$ Factor; $\frac{9}{4} + 4 = \frac{9}{4} + \frac{16}{4} = \frac{25}{4}$.

Step 6 Solve the equation by using the square root property.

$$x + 2 = \sqrt{\frac{25}{4}} \quad \text{or} \quad x + 2 = -\sqrt{\frac{25}{4}} \qquad \text{Square root property}$$

$$x + 2 = \frac{5}{2} \quad \text{or} \quad x + 2 = -\frac{5}{2} \qquad \text{Take square roots.}$$

$$x = \frac{1}{2} \quad \text{or} \quad x = -\frac{9}{2} \qquad \text{Subtract } 2 = \frac{4}{2}.$$

CHECK Substitute each solution in the original equation.

$$4x^2 + 16x = 9 \qquad\qquad\qquad 4x^2 + 16x = 9$$

$$4\left(\frac{1}{2}\right)^2 + 16\left(\frac{1}{2}\right) \overset{?}{=} 9 \quad \text{Let } x = \tfrac{1}{2}. \qquad 4\left(-\frac{9}{2}\right)^2 + 16\left(-\frac{9}{2}\right) \overset{?}{=} 9$$

$$\text{Let } x = -\tfrac{9}{2}.$$

$$4\left(\frac{1}{4}\right) + 8 \overset{?}{=} 9 \qquad\qquad\qquad 4\left(\frac{81}{4}\right) - 72 \overset{?}{=} 9$$

$$1 + 8 \overset{?}{=} 9 \qquad\qquad\qquad\qquad 81 - 72 \overset{?}{=} 9$$

$$9 = 9 \; \checkmark \; \text{True} \qquad\qquad\qquad 9 = 9 \; \checkmark \; \text{True}$$

The two solutions $\frac{1}{2}$ and $-\frac{9}{2}$ check, so the solution set is $\left\{-\frac{9}{2}, \frac{1}{2}\right\}$.

◀ **Work Problem** **5** at the Side.

Answers

5. (a) 3; $x^2 + \frac{5}{3}x - \frac{2}{3} = 0$; $\frac{2}{3}$; $x^2 + \frac{5}{3}x = \frac{2}{3}$;

$\frac{25}{36}$; $\frac{25}{36}$; $x + \frac{5}{6}$; $\frac{49}{36}$; $\left\{-2, \frac{1}{3}\right\}$

(b) $\left\{-\frac{1}{3}, -\frac{5}{3}\right\}$ **(c)** $\left\{\frac{11}{2}, \frac{1}{2}\right\}$

EXAMPLE 5 Solving a Quadratic Equation by Completing the Square ($a \neq 1$)

Solve $2x^2 - 4x - 5 = 0$.

$$x^2 - 2x - \frac{5}{2} = 0 \qquad \text{Step 1}$$

$$x^2 - 2x = \frac{5}{2} \qquad \text{Step 2}$$

$$\left[\frac{1}{2}(-2)\right]^2 = (-1)^2 = 1 \qquad \text{Step 3}$$

$$x^2 - 2x + 1 = \frac{5}{2} + 1 \qquad \text{Step 4}$$

$$(x-1)^2 = \frac{7}{2} \qquad \text{Step 5}$$

$$x - 1 = \sqrt{\frac{7}{2}} \qquad \text{or} \quad x - 1 = -\sqrt{\frac{7}{2}} \qquad \text{Step 6}$$

$$x = 1 + \sqrt{\frac{7}{2}} \qquad \text{or} \qquad x = 1 - \sqrt{\frac{7}{2}} \qquad \text{Add 1.}$$

$$x = 1 + \frac{\sqrt{14}}{2} \qquad \text{or} \qquad x = 1 - \frac{\sqrt{14}}{2}$$

$$\sqrt{\frac{7}{2}} = \frac{\sqrt{7}}{\sqrt{2}} = \frac{\sqrt{7}}{\sqrt{2}} \cdot \frac{\sqrt{2}}{\sqrt{2}} = \frac{\sqrt{14}}{2}$$

Add the two terms in each solution as follows.

$$1 + \frac{\sqrt{14}}{2} = \frac{2}{2} + \frac{\sqrt{14}}{2} = \frac{2 + \sqrt{14}}{2}$$

$$1 - \frac{\sqrt{14}}{2} = \frac{2}{2} - \frac{\sqrt{14}}{2} = \frac{2 - \sqrt{14}}{2}$$

$$1 = \frac{2}{2}$$

Check that the solution set is $\left\{\frac{2 + \sqrt{14}}{2}, \frac{2 - \sqrt{14}}{2}\right\}$.

▸ **Work Problem ❻ at the Side.** ▶

EXAMPLE 6 Solving for Nonreal Complex Solutions

Solve each equation.

(a) $x^2 + 8x + 21 = 0$

$$x^2 + 8x = -21 \qquad \text{Subtract 21.}$$

$$x^2 + 8x + 16 = -21 + 16 \qquad \left[\frac{1}{2}(8)\right]^2 = 4^2 = 16; \text{ Add 16 to each side.}$$

$$(x + 4)^2 = -5 \qquad \text{Factor on the left. Add on the right.}$$

Because the constant term on the right side of the equation is negative, this equation will have nonreal complex solutions.

$$x + 4 = \sqrt{-5} \qquad \text{or} \quad x + 4 = -\sqrt{-5} \qquad \text{Square root property}$$

$$x + 4 = i\sqrt{5} \qquad \text{or} \quad x + 4 = -i\sqrt{5} \qquad \sqrt{-b} = i\sqrt{b}$$

$$x = -4 + i\sqrt{5} \qquad \text{or} \qquad x = -4 - i\sqrt{5} \qquad \text{Add } -4.$$

The solution set is $\left\{-4 + i\sqrt{5}, -4 - i\sqrt{5}\right\}$.

Continued on Next Page

---

❻ Solve each equation.

(a) $2r^2 - 4r + 1 = 0$

(b) $3z^2 - 6z - 2 = 0$

(c) $8x^2 - 4x - 2 = 0$

Answers

6. (a) $\left\{\frac{2 + \sqrt{2}}{2}, \frac{2 - \sqrt{2}}{2}\right\}$

(b) $\left\{\frac{3 + \sqrt{15}}{3}, \frac{3 - \sqrt{15}}{3}\right\}$

(c) $\left\{\frac{1 + \sqrt{5}}{4}, \frac{1 - \sqrt{5}}{4}\right\}$

7 Solve each equation.

(a) $x^2 + 2x + 7 = 0$

(b) $5t^2 - 15t + 12 = 0$

8 Solve each equation.

(a) $r(r - 3) = -1$

(b) $(x + 2)(x + 1) = 5$

(b) $4p^2 + 8p + 5 = 0$

> The coefficient of the second-degree term must be 1.

$p^2 + 2p + \dfrac{5}{4} = 0$ Divide by 4.

$p^2 + 2p = -\dfrac{5}{4}$ Subtract $\frac{5}{4}$.

$p^2 + 2p + 1 = -\dfrac{5}{4} + 1$ $\left[\frac{1}{2}(2)\right]^2 = 1^2 = 1;$
Add 1.

> There will be nonreal complex solutions because of $-\frac{1}{4}$.

$(p + 1)^2 = -\dfrac{1}{4}$ Factor on the left.
Add on the right.

$p + 1 = \sqrt{-\dfrac{1}{4}}$ or $p + 1 = -\sqrt{-\dfrac{1}{4}}$ Square root property

$p + 1 = \dfrac{1}{2}i$ or $p + 1 = -\dfrac{1}{2}i$ $\sqrt{-\frac{1}{4}} = \frac{1}{2}i$

$p = -1 + \dfrac{1}{2}i$ or $p = -1 - \dfrac{1}{2}i$ Add -1.

The solution set is $\left\{ -1 + \frac{1}{2}i, -1 - \frac{1}{2}i \right\}$.

◀ **Work Problem 7 at the Side.**

OBJECTIVE ▶ 3 Simplify the terms of an equation before solving.

EXAMPLE 7   Simplifying before Completing the Square

Solve $(x + 3)(x - 1) = 2$.

$(x + 3)(x - 1) = 2$ Given equation

$x^2 + 2x - 3 = 2$ Multiply using the FOIL method.

$x^2 + 2x = 5$ Add 3.

$x^2 + 2x + 1 = 5 + 1$ Add $\left[\frac{1}{2}(2)\right]^2 = 1^2 = 1$.

$(x + 1)^2 = 6$ Factor on the left. Add on the right.

$x + 1 = \sqrt{6}$ or $x + 1 = -\sqrt{6}$ Square root property

$x = -1 + \sqrt{6}$ or $x = -1 - \sqrt{6}$ Add -1.

The solution set is $\left\{ -1 + \sqrt{6}, -1 - \sqrt{6} \right\}$.

◀ **Work Problem 8 at the Side.**

Note

We will use completing the square in **Section 10.7** when we graph quadratic equations and in **Section 12.2** when we work with circles.

Answers

7. (a) $\left\{ -1 + i\sqrt{6}, -1 - i\sqrt{6} \right\}$

(b) $\left\{ \dfrac{3}{2} + \dfrac{\sqrt{15}}{10}i, \dfrac{3}{2} - \dfrac{\sqrt{15}}{10}i \right\}$

8. (a) $\left\{ \dfrac{3 + \sqrt{5}}{2}, \dfrac{3 - \sqrt{5}}{2} \right\}$

(b) $\left\{ \dfrac{-3 + \sqrt{21}}{2}, \dfrac{-3 - \sqrt{21}}{2} \right\}$

10.2 Exercises

 MyMathLab®

1. CONCEPT CHECK Which one of the two equations

$$(2x + 1)^2 = 5 \quad \text{and} \quad x^2 + 4x = 12$$

is more suitable for solving by the square root property? By completing the square?

2. CONCEPT CHECK What would be the first step in solving the equation

$$2x^2 + 8x = 9$$

by completing the square?

CONCEPT CHECK *Find the constant that must be added to make each expression a perfect square trinomial. Then factor the trinomial.*

3. $x^2 + 6x +$ ____

It factors as _____.

4. $x^2 + 14x +$ ____

It factors as _____.

5. $p^2 - 12p +$ ____

It factors as _____.

6. $x^2 - 20x +$ ____

It factors as _____.

7. $q^2 + 9q +$ ____

It factors as _____.

8. $t^2 + 3t +$ ____

It factors as _____.

Solve each equation by completing the square. See Examples 1–3.

9. $x^2 + 5x + 6 = 0$

10. $x^2 + 6x + 5 = 0$

11. $x^2 - 4x = -3$

12. $x^2 - 2x = 8$

13. $x^2 + 2x - 5 = 0$

14. $x^2 + 4x + 1 = 0$

15. $x^2 + 10x + 18 = 0$

16. $x^2 + 8x + 11 = 0$

17. $x^2 - 8x = -4$

18. $m^2 - 4m = 14$

19. $x^2 + x - 1 = 0$

20. $x^2 + x - 3 = 0$

21. $r^2 - 3r = 2$

22. $x^2 - 3x = 6$

Solve each equation by completing the square. See Examples 4, 5, and 7.

23. $4x^2 + 4x - 3 = 0$

24. $9x^2 + 3x - 2 = 0$

25. $3x^2 - 6x = 4$

26. $2x^2 - 6x = 3$

27. $2k^2 + 5k - 2 = 0$

28. $3r^2 + 2r - 2 = 0$

29. $3k^2 + 7k = 4$

30. $2k^2 + 5k = 1$

31. $(r - 3)(r - 5) = 2$

32. $(k - 1)(k - 7) = 1$

33. $-x^2 + 2x = -5$

34. $-r^2 + 3r = -2$

Find all nonreal complex solutions. **See Example 6.**

35. $m^2 + 4m + 13 = 0$

36. $t^2 + 6t + 10 = 0$

37. $3r^2 + 4r + 4 = 0$

38. $4x^2 + 5x + 5 = 0$

39. $-m^2 - 6m - 12 = 0$

40. $-k^2 - 5k - 10 = 0$

Relating Concepts (Exercises 41–46)   For Individual or Group Work

The Greeks had a method of completing the square geometrically in which they literally changed a figure into a square. For example, to complete the square for $x^2 + 6x$, we begin with a square of side x, as in the figure on the left. We add three rectangles of width 1 to the right side and the bottom to get a region with area $x^2 + 6x$. To fill in the corner (complete the square), we must add 9 1-by-1 squares, as shown on the right.

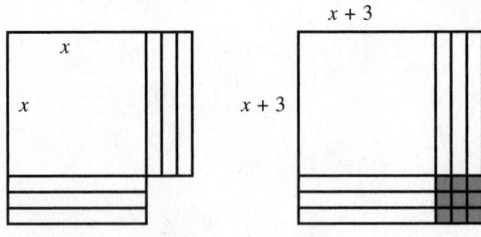

Work Exercises 41–46 in order.

41. What is the area of the original square?

42. What is the area of each strip?

43. What is the total area of the six strips?

44. What is the area of each small square in the corner of the second figure?

45. What is the total area of the small squares?

46. What is the area of the new "complete" square?

10.3 Solving Quadratic Equations by the Quadratic Formula

In this section, we complete the square to solve the general quadratic equation

$$ax^2 + bx + c = 0,$$

where a, b, and c are complex numbers and $a \neq 0$. The solution gives a formula for finding the solution of *any* specific quadratic equation.

OBJECTIVE ▶ **1** **Derive the quadratic formula.** We follow the steps given in **Section 10.2** to solve $ax^2 + bx + c = 0$ by completing the square (assuming $a > 0$).

OBJECTIVES

1 Derive the quadratic formula.

2 Solve quadratic equations by using the quadratic formula.

3 Use the discriminant to determine the number and type of solutions.

$$ax^2 + bx + c = 0 \qquad \text{General quadratic equation}$$

$$x^2 + \frac{b}{a}x + \frac{c}{a} = 0 \qquad \text{Divide by } a. \text{ (Step 1)}$$

$$x^2 + \frac{b}{a}x = -\frac{c}{a} \qquad \text{Subtract } \tfrac{c}{a}. \text{ (Step 2)}$$

$$\left[\frac{1}{2}\left(\frac{b}{a}\right)\right]^2 = \left(\frac{b}{2a}\right)^2 = \frac{b^2}{4a^2} \qquad \text{(Step 3)}$$

$$x^2 + \frac{b}{a}x + \frac{b^2}{4a^2} = -\frac{c}{a} + \frac{b^2}{4a^2} \qquad \text{Add } \tfrac{b^2}{4a^2} \text{ to each side. (Step 4)}$$

$$\left(x + \frac{b}{2a}\right)^2 = \frac{b^2}{4a^2} + \frac{-c}{a} \qquad \begin{array}{l}\text{Write the left side as a perfect square.}\\ \text{Rearrange the right side. (Step 5)}\end{array}$$

$$\left(x + \frac{b}{2a}\right)^2 = \frac{b^2}{4a^2} + \frac{-4ac}{4a^2} \qquad \text{Write with a common denominator.}$$

$$\left(x + \frac{b}{2a}\right)^2 = \frac{b^2 - 4ac}{4a^2} \qquad \text{Add fractions.}$$

$$x + \frac{b}{2a} = \sqrt{\frac{b^2 - 4ac}{4a^2}} \quad \text{or} \quad x + \frac{b}{2a} = -\sqrt{\frac{b^2 - 4ac}{4a^2}} \qquad \begin{array}{l}\text{Square root}\\ \text{property (Step 6)}\end{array}$$

We can simplify

$$\sqrt{\frac{b^2 - 4ac}{4a^2}} \quad \text{as} \quad \frac{\sqrt{b^2 - 4ac}}{\sqrt{4a^2}}, \quad \text{or} \quad \frac{\sqrt{b^2 - 4ac}}{2a}.$$

The right side of each equation can be expressed as follows.

$$x + \frac{b}{2a} = \frac{\sqrt{b^2 - 4ac}}{2a} \qquad \text{or} \qquad x + \frac{b}{2a} = -\frac{\sqrt{b^2 - 4ac}}{2a}$$

$$x = \frac{-b}{2a} + \frac{\sqrt{b^2 - 4ac}}{2a} \qquad \text{or} \qquad x = \frac{-b}{2a} - \frac{\sqrt{b^2 - 4ac}}{2a}$$

$$x = \frac{-b + \sqrt{b^2 - 4ac}}{2a} \qquad \text{or} \qquad x = \frac{-b - \sqrt{b^2 - 4ac}}{2a}$$

If $a < 0$, the same two solutions are obtained. The result is the **quadratic formula,** which is abbreviated as shown on the next page.

1 Identify the values of a, b, and c. (*Hint:* If necessary, first write the equation in standard form with 0 on the right side.) *Do not actually solve.*

(a) $-3x^2 + 9x - 4 = 0$

(b) $3x^2 = 6x + 2$

Quadratic Formula

The solutions of $ax^2 + bx + c = 0$ (where $a \neq 0$) are given by

$$x = \frac{-b \pm \sqrt{b^2 - 4ac}}{2a}.$$

CAUTION

In the quadratic formula, $x = \dfrac{-b \pm \sqrt{b^2 - 4ac}}{2a}$, *the square root is added to or subtracted from the value of* $-b$ *before dividing by 2a.*

OBJECTIVE ▶ **2** **Solve quadratic equations by using the quadratic formula.**
To use the quadratic formula, first write the equation in standard form

$$ax^2 + bx + c = 0.$$

Then identify the values of a, b, and c and substitute them into the quadratic formula.

◀ **Work Problem** ❶ **at the Side.**

EXAMPLE 1   **Using the Quadratic Formula (Rational Solutions)**

Solve $6x^2 - 5x - 4 = 0$.

This equation is in standard form, so we identify the values of a, b, and c. Here a, the coefficient of the second-degree term, is 6, and b, the coefficient of the first-degree term, is -5. The constant c is -4.

Substitute these values into the quadratic formula.

$$x = \frac{-b \pm \sqrt{b^2 - 4ac}}{2a} \qquad \text{Quadratic formula}$$

2 Solve $4x^2 - 11x - 3 = 0$ by using the quadratic formula.

$$x = \frac{-(-5) \pm \sqrt{(-5)^2 - 4(6)(-4)}}{2(6)} \qquad a = 6, b = -5, c = -4$$

> Use parentheses and substitute carefully to avoid errors.

$$x = \frac{5 \pm \sqrt{25 + 96}}{12}$$

$$x = \frac{5 \pm \sqrt{121}}{12} \qquad \text{Add under the radical.}$$

$$x = \frac{5 \pm 11}{12} \qquad \text{Take the square root.}$$

There are two solutions, one from the $+$ sign and one from the $-$ sign.

$$x = \frac{5 + 11}{12} = \frac{16}{12} = \frac{4}{3} \quad \text{or} \quad x = \frac{5 - 11}{12} = \frac{-6}{12} = -\frac{1}{2}$$

Check each solution in the original equation. The solution set is $\left\{-\frac{1}{2}, \frac{4}{3}\right\}$.

◀ **Work Problem** ❷ **at the Side.**

Answers

1. (a) $-3; 9; -4$ **(b)** $3; -6; -2$

2. $\left\{-\frac{1}{4}, 3\right\}$

Note

We could have used factoring to solve the equation in **Example 1.**

$$6x^2 - 5x - 4 = 0$$

$$(3x - 4)(2x + 1) = 0 \qquad \text{Factor.}$$

$$3x - 4 = 0 \quad \text{or} \quad 2x + 1 = 0 \qquad \text{Zero-factor property}$$

$$3x = 4 \quad \text{or} \qquad 2x = -1 \qquad \text{Solve each equation.}$$

$$x = \frac{4}{3} \quad \text{or} \qquad x = -\frac{1}{2} \qquad \text{Same solutions as in \textbf{Example 1}}$$

When solving quadratic equations, it is a good idea to try factoring first. If the equation cannot be factored or if factoring is difficult, then use the quadratic formula.

EXAMPLE 2 Using the Quadratic Formula (Irrational Solutions)

Solve $4x^2 = 8x - 1$.

Write the equation in standard form as $4x^2 - 8x + 1 = 0$. *This is a key step.*

$$x = \frac{-b \pm \sqrt{b^2 - 4ac}}{2a} \qquad \text{Quadratic formula}$$

$$x = \frac{-(-8) \pm \sqrt{(-8)^2 - 4(4)(1)}}{2(4)} \qquad a = 4, b = -8, c = 1$$

$$x = \frac{8 \pm \sqrt{64 - 16}}{8} \qquad \begin{array}{l}\text{Simplify in the numerator}\\\text{and denominator.}\end{array}$$

$$x = \frac{8 \pm \sqrt{48}}{8} \qquad \text{Subtract under the radical.}$$

$$x = \frac{8 \pm 4\sqrt{3}}{8} \qquad \sqrt{48} = \sqrt{16} \cdot \sqrt{3} = 4\sqrt{3}$$

$$x = \frac{4(2 \pm \sqrt{3})}{4(2)} \qquad \text{Factor.}$$
Factor first. Then divide out the common factor.

$$x = \frac{2 \pm \sqrt{3}}{2} \qquad \text{Write in lowest terms.}$$

The solution set is $\left\{ \frac{2 + \sqrt{3}}{2}, \frac{2 - \sqrt{3}}{2} \right\}$.

CAUTION

1. *Every quadratic equation must be written in standard form $ax^2 + bx + c = 0$ before we begin to solve it,* whether we use factoring or the quadratic formula.
2. *When writing solutions in lowest terms, factor first. Then divide out the common factor.* See the last two steps in **Example 2.**

Work Problem ❸ at the Side. ▶

---

❸ Solve each equation.

GS **(a)** $6x^2 + 4x - 1 = 0$

Here, $a = \underline{\quad}$, $b = 4$, and $c = \underline{\quad}$.

$$x = \frac{-b \pm \sqrt{b^2 - 4ac}}{2a}$$

$$x = \frac{-4 \pm \sqrt{\underline{\quad}^2 - 4(6)(\underline{\quad})}}{2(6)}$$

$$x = \frac{-4 \pm \sqrt{16 + \underline{\quad}}}{12}$$

$$x = \frac{-4 \pm \sqrt{40}}{12}$$

$$x = \frac{-4 \pm \underline{\quad}\sqrt{10}}{12}$$

$$x = \frac{\underline{\quad}(-2 \pm \sqrt{10})}{2(6)}$$

$$x = \underline{\quad}$$

The solution set is $\underline{\quad}$.

(b) $2x^2 + 19 = 14x$

Answers

3. **(a)** $6; -1; 4; -1; 24; 2; 2; \dfrac{-2 \pm \sqrt{10}}{6};$

$$\left\{ \frac{-2 + \sqrt{10}}{6}, \frac{-2 - \sqrt{10}}{6} \right\}$$

(b) $\left\{ \dfrac{7 + \sqrt{11}}{2}, \dfrac{7 - \sqrt{11}}{2} \right\}$

4 Solve each equation.

(a) $x^2 + x + 1 = 0$

(b) $(x + 2)(x - 6) = -17$

EXAMPLE 3 **Using the Quadratic Formula (Nonreal Complex Solutions)**

Solve $(9x + 3)(x - 1) = -8$.

This is a quadratic equation—when the first terms $9x$ and x are multiplied, we get a second-degree term, $9x^2$. We must write the equation in standard form.

$$(9x + 3)(x - 1) = -8$$

$$9x^2 - 6x - 3 = -8 \qquad \text{Multiply using the FOIL method.}$$

$$\text{Standard form} \rightarrow 9x^2 - 6x + 5 = 0 \qquad \text{Add 8.}$$

From the standard form of the equation, we identify $a = 9$, $b = -6$, and $c = 5$.

$$x = \frac{-b \pm \sqrt{b^2 - 4ac}}{2a} \qquad \text{Quadratic formula}$$

$$x = \frac{-(-6) \pm \sqrt{(-6)^2 - 4(9)(5)}}{2(9)} \qquad \text{Substitute.}$$

$$x = \frac{6 \pm \sqrt{-144}}{18} \qquad \text{Simplify.}$$

$$x = \frac{6 \pm 12i}{18} \qquad \sqrt{-144} = 12i$$

$$x = \frac{6(1 \pm 2i)}{6(3)} \qquad \text{Factor.}$$

$$x = \frac{1 \pm 2i}{3} \qquad \text{Write in lowest terms.}$$

$$x = \frac{1}{3} \pm \frac{2}{3}i \qquad \text{Standard form } a + bi \text{ for a complex number}$$

The solution set is $\left\{ \frac{1}{3} + \frac{2}{3}i, \frac{1}{3} - \frac{2}{3}i \right\}$.

◀ **Work Problem 4 at the Side.**

OBJECTIVE **3** **Use the discriminant to determine the number and type of solutions.** The solutions of the quadratic equation $ax^2 + bx + c = 0$ are given by

$$x = \frac{-b \pm \sqrt{b^2 - 4ac}}{2a}. \qquad \leftarrow \text{Discriminant}$$

If a, b, and c are integers, the type of solutions of a quadratic equation—that is, rational, irrational, or nonreal complex—is determined by the expression under the radical symbol,

$$b^2 - 4ac.$$

This expression is called the *discriminant* (because it distinguishes among the three types of solutions). By calculating the discriminant before solving a quadratic equation, we can predict the number and type of solutions of the equation.

Answers

4. (a) $\left\{ -\frac{1}{2} + \frac{\sqrt{3}}{2}i, -\frac{1}{2} - \frac{\sqrt{3}}{2}i \right\}$

(b) $\{2 + i, 2 - i\}$

Discriminant

The **discriminant** of $ax^2 + bx + c = 0$ is $b^2 - 4ac$. If a, b, and c are integers, then the number and type of solutions are determined as follows.

Discriminant	Number and Type of Solutions
Positive, and the square of an integer	Two rational solutions
Positive, but not the square of an integer	Two irrational solutions
Zero	One distinct rational solution
Negative	Two nonreal complex solutions

We can also calculate the discriminant to help decide how to solve a quadratic equation.

If the discriminant is a perfect square (including 0), then the equation can be solved by factoring. Otherwise, the quadratic formula should be used.

EXAMPLE 4   Using the Discriminant

Find the discriminant. Use it to predict the number and type of solutions for each equation. Then tell whether the equation can be solved by factoring or whether the quadratic formula should be used.

(a) $6x^2 - x - 15 = 0$

First identify the values of a, b, and c. Because $-x = -1x$, the value of b is -1. We find the discriminant by evaluating $b^2 - 4ac$.

$$b^2 - 4ac$$

> Use parentheses and substitute carefully.

$$= (-1)^2 - 4\,(6)\,(-15) \qquad a = 6,\, b = -1,\, c = -15$$

$$= 1 + 360 \qquad \text{Apply the exponent. Multiply.}$$

$$= 361, \quad \text{or} \quad 19^2, \quad \text{which is a perfect square.}$$

Since a, b, and c are integers and the discriminant 361 is a perfect square, there will be two rational solutions. The equation can be solved by factoring.

(b) $3x^2 - 4x = 5$

First write the equation in standard form.

$$3x^2 - 4x - 5 = 0 \qquad \text{Subtract 5.}$$

Now find the discriminant.

$$b^2 - 4ac \qquad \text{Discriminant}$$

$$= (-4)^2 - 4\,(3)\,(-5) \qquad a = 3,\, b = -4,\, c = -5$$

$$= 16 + 60 \qquad \text{Apply the exponent. Multiply.}$$

$$= 76 \qquad \text{Add.}$$

Because 76 is positive but *not* the square of an integer and a, b, and c are integers, the equation will have two irrational solutions and is best solved using the quadratic formula.

Continued on Next Page

5 Find the discriminant. Use it to predict the number and type of solutions for each equation. Then tell whether the equation can be solved by factoring or whether the quadratic formula should be used.

(a) $2x^2 + 3x = 4$

(b) $2x^2 + 3x + 4 = 0$

(c) $x^2 + 20x + 100 = 0$

(d) $15x^2 + 11x = 14$

(c) $4x^2 + x + 1 = 0$

> $x = 1x$, so $b = 1$.

$$b^2 - 4ac \qquad \text{Discriminant}$$
$$= 1^2 - 4\,(4)\,(1) \quad a = 4, b = 1, c = 1$$
$$= 1 - 16 \qquad \text{Apply the exponent. Multiply.}$$
$$= -15 \qquad \text{Subtract.}$$

Because the discriminant is negative and a, b, and c are integers, this quadratic equation will have two nonreal complex solutions. The quadratic formula should be used to solve it.

(d) $4x^2 + 9 = 12x$

First write the equation in standard form.

$$4x^2 - 12x + 9 = 0 \quad \text{Subtract } 12x.$$

Now find the discriminant.

$$b^2 - 4ac \qquad \text{Discriminant}$$
$$= (-12)^2 - 4\,(4)\,(9) \quad a = 4, b = -12, c = 9$$
$$= 144 - 144 \qquad \text{Apply the exponent. Multiply.}$$
$$= 0 \qquad \text{Subtract.}$$

Because the discriminant is 0, this quadratic equation will have one distinct rational solution. The equation can be solved by factoring.

◀ **Work Problem 5** at the Side.

Note

In **Section 10.7** we extend our use of the discriminant to determine the number of x-intercepts of the graph of a quadratic function.

Answers

5. (a) 41; two irrational solutions; quadratic formula
(b) -23; two nonreal complex solutions; quadratic formula
(c) 0; one distinct rational solution; factoring
(d) 961; two rational solutions; factoring

10.3 Exercises

CONCEPT CHECK *Answer each question in Exercises 1–4.*

1. The documentation for an early version of Microsoft *Word* for Windows used the following for the quadratic formula. Is this correct? If not, correct it.

$$x = -b \pm \frac{\sqrt{b^2 - 4ac}}{2a}$$ Is this correct?

2. One patron wrote the quadratic formula, as shown here, on a wall at the Cadillac Bar in Houston, Texas. Is this correct? If not, correct it.

$$x = \frac{-b\sqrt{b^2 - 4ac}}{2a}$$ Is this correct?

3. A student solved $5x^2 - 5x + 1 = 0$ incorrectly as follows. **What Went Wrong?**

$$x = \frac{5 \pm \sqrt{25 - 4(5)(1)}}{2(5)}$$ $a = 5, b = -5, c = 1$

$$x = \frac{5 \pm \sqrt{5}}{10}$$ Simplify.

$$x = \frac{1}{2} \pm \sqrt{5}$$ Write in lowest terms.

4. A student claimed that the equation

$$2x^2 - 5 = 0$$

cannot be solved using the quadratic formula because there is no first-degree x-term. Was the student correct? If not, give the values of a, b, and c, and solve the equation.

Use the quadratic formula to solve each equation. (All solutions for these equations are real numbers.) **See Examples 1 and 2.**

5. $x^2 - 8x + 15 = 0$

6. $x^2 + 3x - 28 = 0$

7. $2x^2 + 4x + 1 = 0$

8. $2x^2 + 3x - 1 = 0$

9. $2x^2 - 2x = 1$

10. $9x^2 + 6x = 1$

11. $x^2 + 18 = 10x$

12. $x^2 - 4 = 2x$

13. $4k^2 + 4k - 1 = 0$

14. $4r^2 - 4r - 19 = 0$

15. $2 - 2x = 3x^2$

16. $26r - 2 = 3r^2$

17. $\dfrac{x^2}{4} - \dfrac{x}{2} = 1$

(*Hint:* First clear the fractions.)

18. $p^2 + \dfrac{p}{3} = \dfrac{1}{6}$

(*Hint:* First clear the fractions.)

19. $-2t(t+2) = -3$

20. $-3x(x+2) = -4$

21. $(r-3)(r+5) = 2$

22. $(k+1)(k-7) = 1$

*Use the quadratic formula to solve each equation. (All solutions for these equations are nonreal complex numbers.) **See Example 3.***

23. $x^2 - 3x + 17 = 0$

24. $x^2 - 5x + 20 = 0$

25. $r^2 - 6r + 14 = 0$

26. $t^2 + 4t + 11 = 0$

27. $4x^2 - 4x = -7$

28. $9x^2 - 6x = -7$

29. $x(3x+4) = -2$

30. $p(2p+3) = -2$

31. $(2x-1)(8x-4) = -1$

32. $(x-1)(9x-3) = -2$

Find the discriminant. Use it to determine whether the solutions for each equation are

A. *two rational numbers,*

C. *two irrational numbers,*

B. *one distinct rational number,*

D. *two nonreal complex numbers.*

*Then tell whether the equation can be solved by factoring or whether the quadratic formula should be used. Do not actually solve. **See Example 4.***

33. $25x^2 + 70x + 49 = 0$

34. $4k^2 - 28k + 49 = 0$

35. $x^2 + 4x + 2 = 0$

36. $9x^2 - 12x - 1 = 0$

37. $3x^2 = 5x + 2$

38. $4x^2 = 4x + 3$

39. $3m^2 - 10m + 15 = 0$

40. $18x^2 + 60x + 82 = 0$

*Based on your answers in **Exercises 33–36**, solve the equation given in each exercise.*

41. Exercise 33

42. Exercise 34

43. Exercise 35

44. Exercise 36

10.4 Equations Quadratic in Form

OBJECTIVES

1. Solve an equation with fractions by writing it in quadratic form.

2. Use quadratic equations to solve applied problems.

3. Solve an equation with radicals by writing it in quadratic form.

4. Solve an equation that is quadratic in form by substitution.

OBJECTIVE ▸ 1 Solve an equation with fractions by writing it in quadratic form. A variety of nonquadratic equations can be written in the form of a quadratic equation and solved by using the methods of this chapter.

EXAMPLE 1 Solving an Equation with Fractions That Leads to a Quadratic Equation

Solve $\dfrac{1}{x} + \dfrac{1}{x-1} = \dfrac{7}{12}$.

Clear fractions by multiplying each side by the least common denominator, $12x(x-1)$. (The domain must be restricted to $x \neq 0$, $x \neq 1$.)

$$12x(x-1)\left(\frac{1}{x} + \frac{1}{x+1}\right) = 12x(x-1)\left(\frac{7}{12}\right) \quad \text{Multiply by the LCD.}$$

$$12x(x-1)\frac{1}{x} + 12x(x-1)\frac{1}{x-1} = 12x(x-1)\left(\frac{7}{12}\right) \quad \text{Distributive property}$$

$$12(x-1) + 12x = 7x(x-1) \quad \text{Multiply.}$$

$$12x - 12 + 12x = 7x^2 - 7x \quad \text{Distributive property}$$

$$24x - 12 = 7x^2 - 7x \quad \text{Combine like terms.}$$

$$7x^2 - 31x + 12 = 0 \quad \text{Standard form}$$

$$(7x - 3)(x - 4) = 0 \quad \text{Factor.}$$

$$7x - 3 = 0 \quad \text{or} \quad x - 4 = 0 \quad \text{Zero-factor property}$$

$$x = \frac{3}{7} \quad \text{or} \quad x = 4 \quad \text{Solve each equation.}$$

Check these solutions in the original equation. The solution set is $\left\{\frac{3}{7}, 4\right\}$.

··· **Work Problem ❶ at the Side. ▸**

OBJECTIVE ▸ 2 Use quadratic equations to solve applied problems. Some distance-rate-time (motion) problems lead to quadratic equations. We use the six-step problem-solving method from **Section 2.4.**

EXAMPLE 2 Solving a Motion Problem

A riverboat for tourists averages 12 mph in still water. It takes the boat 1 hr, 4 min to travel 6 mi upstream and return. Find the rate of the current.

Step 1 **Read** the problem carefully.

Step 2 **Assign a variable.**

Let x = the rate of the current.

The current slows down the boat when it travels upstream, so the rate of the boat traveling upstream is its rate in still water *less* the rate of the current, or $(12 - x)$ mph. See **Figure 1** on the next page.

··· **Continued on Next Page**

❶ Solve each equation. Check your solutions.

(a) $\dfrac{5}{m} + \dfrac{12}{m^2} = 2$

(b) $\dfrac{2}{x} + \dfrac{1}{x-2} = \dfrac{5}{3}$

(c) $\dfrac{4}{m-1} + 9 = -\dfrac{7}{m}$

Answers

1. (a) $\left\{-\dfrac{3}{2}, 4\right\}$ (b) $\left\{\dfrac{4}{5}, 3\right\}$

(c) $\left\{\dfrac{7}{9}, -1\right\}$

2 Solve each problem.

GS **(a)** In 4 hr, Kerrie can travel 15 mi upriver and come back. The rate of the current is 5 mph. Find the rate of her boat in still water.

Let $x =$ _____.

The rate traveling upriver (*against* the current) is ____ mph.

The rate traveling back downriver (*with* the current) is ____ mph.

Complete the table.

	d	r	t
Up	___	___	___
Down	___	___	___

Write an equation, and complete the solution.

(b) In $1\frac{3}{4}$ hr, Ken rows his boat 5 mi upriver and comes back. The rate of the current is 3 mph. How fast does Ken row?

Answers

2. **(a)** the rate of her boat in still water;
$x - 5$; $x + 5$;
row 1 of table: 15; $x - 5$; $\dfrac{15}{x - 5}$;
row 2 of table: 15; $x + 5$; $\dfrac{15}{x + 5}$;
$\dfrac{15}{x - 5} + \dfrac{15}{x + 5} = 4$; 10 mph

(b) 7 mph

Riverboat traveling *upstream*—the current slows it down.

Figure 1

Similarly, the current speeds up the boat as it travels downstream, so its rate downstream is $(12 + x)$ mph. Thus,

$$12 - x = \text{the rate upstream in miles per hour,}$$

and $12 + x =$ the rate downstream in miles per hour.

	d	r	t
Upstream	6	$12 - x$	$\frac{6}{12 - x}$
Downstream	6	$12 + x$	$\frac{6}{12 + x}$

Complete a table. Use the distance formula, $d = rt$, solved for time t, $t = \frac{d}{r}$, to write expressions for t.

Step 3   **Write an equation.** The total time is 1 hr, 4 min. We write it as a fraction.

$$1 + \frac{4}{60} = 1 + \frac{1}{15} = \frac{16}{15} \text{ hr}   \text{Total time}$$

Time upstream	+	time downstream	=	total time.
$\downarrow$		$\downarrow$		$\downarrow$
$\dfrac{6}{12 - x}$	$+$	$\dfrac{6}{12 + x}$	$=$	$\dfrac{16}{15}$

Step 4   **Solve** the equation. The LCD is $15(12 - x)(12 + x)$.

$$15(12 - x)(12 + x)\left(\frac{6}{12 - x} + \frac{6}{12 + x}\right)$$

$$= 15(12 - x)(12 + x)\left(\frac{16}{15}\right)$$

Multiply by the LCD.

$$15(12 + x)6 + 15(12 - x)6 = (12 - x)(12 + x)16$$

Distributive property;
Multiply.

$$90(12 + x) + 90(12 - x) = 16(144 - x^2)$$ Multiply.

$$1080 + 90x + 1080 - 90x = 2304 - 16x^2$$ Distributive property

$$2160 = 2304 - 16x^2$$ Combine like terms.

$$16x^2 = 144$$ Add $16x^2$. Subtract 2160.

$$x^2 = 9$$ Divide by 16.

$$x = 3 \quad \text{or} \quad x = -3$$ Square root property

Step 5   **State the answer.** The rate of the current cannot be -3, so the answer is 3 mph.

Step 6   **Check** that this value satisfies the original problem.

◀ **Work Problem** **2** at the Side.

> **CAUTION**
> As shown in **Example 2,** when a quadratic equation is used to solve an applied problem, sometimes only *one* answer satisfies the application. *Always check each answer in the words of the original problem.*

Recall from **Section 7.5** that a person's work rate is $\frac{1}{t}$ part of the job per hour, where t is the time in hours required to complete the job. Thus, the part of the job the person will do in x hours is $\frac{1}{t}x$.

EXAMPLE 3 **Solving a Work Problem**

It takes two carpet layers 4 hr to carpet a room. If each worked alone, one of them could do the job in 1 hr less time than the other. How long would it take each carpet layer to complete the job alone?

Step 1 **Read** the problem again. There will be two answers.

Step 2 **Assign a variable.**

Let $x =$ the number of hours for the slower carpet layer to complete the job.

Then $x - 1 =$ the number of hours for the faster carpet layer to complete the job.

The slower worker's rate is thus $\frac{1}{x}$, and the faster worker's rate is $\frac{1}{x - 1}$. Together they can do the job in 4 hr. Complete a table.

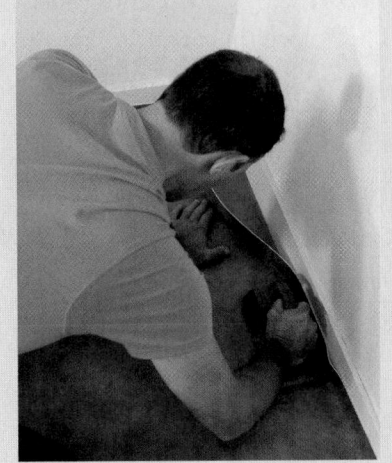

	Rate	Time Working Together	Fractional Part of the Job Done
Slower Worker	$\frac{1}{x}$	4	$\frac{1}{x}(4)$
Faster Worker	$\frac{1}{x - 1}$	4	$\frac{1}{x - 1}(4)$

Sum is 1 whole job.

Step 3 **Write an equation.** The sum of the fractional parts done by the workers should equal 1 (the whole job).

Part done by slower worker	+	part done by faster worker	=	1 whole job.
↓		↓		↓
$\frac{4}{x}$	+	$\frac{4}{x - 1}$	=	1

Step 4 **Solve** the equation from Step 3.

$$x(x - 1)\left(\frac{4}{x} + \frac{4}{x - 1}\right) = x(x - 1)(1)$$ Multiply by $x(x - 1)$, the LCD.

$$4(x - 1) + 4x = x(x - 1)$$ Distributive property

$$4x - 4 + 4x = x^2 - x$$ Distributive property

$$x^2 - 9x + 4 = 0$$ Standard form

This equation cannot be solved by factoring, so we use the quadratic formula.

Continued on Next Page

3 Solve each problem. Round answers to the nearest tenth.

(a) Carlos can complete a certain lab test in 2 hr less time than Jaime can. If they can finish the job together in 2 hr, how long would it take each of them working alone?

Let x = Jaime's time alone (in hours).

Then ____ = Carlos' time alone (in hours).

Complete the table.

	Rate	Time Working Together	Fractional Part of the Job Done
Carlos	____	____	____
Jaime	____	____	____

Write an equation, and complete the solution.

(b) Two chefs are preparing a banquet. One chef could prepare the banquet in 2 hr less time than the other. Together, they complete the job in 5 hr. How long would it take the faster chef working alone?

Answers

3. **(a)** $x - 2$;

  row 1 of table: $\dfrac{1}{x-2}$; 2; $\dfrac{2}{x-2}$;

  row 2 of table: $\dfrac{1}{x}$; 2; $\dfrac{2}{x}$;

  $\dfrac{2}{x-2} + \dfrac{2}{x} = 1$;

  Jaime: 5.2 hr; Carlos: 3.2 hr

  (b) 9.1 hr

$$x = \frac{-b \pm \sqrt{b^2 - 4ac}}{2a} \quad \text{Quadratic formula}$$

$$x = \frac{-(-9) \pm \sqrt{(-9)^2 - 4(1)(4)}}{2(1)} \quad \begin{array}{l}\text{In } x^2 - 9x + 4 = 0, a = 1, \\ b = -9, \text{ and } c = 4.\end{array}$$

$$x = \frac{9 \pm \sqrt{65}}{2} \quad \text{Simplify.}$$

$$x = \frac{9 + \sqrt{65}}{2} \approx 8.5 \quad \text{or} \quad x = \frac{9 - \sqrt{65}}{2} \approx 0.5 \quad \text{Use a calculator.}$$

Step 5 **State the answer.** Only the solution 8.5 makes sense in the original problem, because if $x = 0.5$, then

$$x - 1 = 0.5 - 1 = -0.5, \quad \text{Time cannot be negative.}$$

which cannot represent the time for the faster worker. The slower worker could do the job in about 8.5 hr and the faster in about

$$8.5 - 1 = 7.5 \text{ hr.}$$

Step 6 **Check** that these results satisfy the original problem.

◀ **Work Problem 3 at the Side.**

OBJECTIVE ▶ **3** **Solve an equation with radicals by writing it in quadratic form.**

EXAMPLE 4 Solving Radical Equations That Lead to Quadratic Equations

Solve each equation.

(a) $k = \sqrt{6k - 8}$

This equation is not quadratic. However, squaring each side of the equation gives a quadratic equation that can be solved by factoring.

$$\begin{array}{ll} k^2 = \left(\sqrt{6k - 8}\right)^2 & \text{Square each side.} \\ k^2 = 6k - 8 & \left(\sqrt{a}\right)^2 = a \\ k^2 - 6k + 8 = 0 & \text{Standard form} \\ (k - 4)(k - 2) = 0 & \text{Factor.} \\ k - 4 = 0 \quad \text{or} \quad k - 2 = 0 & \text{Zero-factor property} \\ k = 4 \quad \text{or} \quad k = 2 & \text{Proposed solutions} \end{array}$$

Recall that squaring each side of a radical equation can introduce extraneous solutions that do not satisfy the original equation. *All proposed solutions must be checked in the original (not the squared) equation.*

CHECK

$k = \sqrt{6k - 8}$	$k = \sqrt{6k - 8}$
$4 \overset{?}{=} \sqrt{6(4) - 8}$ Let $k = 4$.	$2 \overset{?}{=} \sqrt{6(2) - 8}$ Let $k = 2$.
$4 \overset{?}{=} \sqrt{16}$	$2 \overset{?}{=} \sqrt{4}$
$4 = 4$ ✓ True	$2 = 2$ ✓ True

Both solutions check, so the solution set is $\{2, 4\}$.

· · · · · · · · · · · · **Continued on Next Page**

(b)

$$x + \sqrt{x} = 6$$

$$\sqrt{x} = 6 - x \qquad \text{Isolate the radical on one side.}$$

$$(\sqrt{x})^2 = (6 - x)^2 \qquad \text{Square each side.}$$

$$x = 36 - 12x + x^2 \qquad (a - b)^2 = a^2 - 2ab + b^2$$

$$x^2 - 13x + 36 = 0 \qquad \text{Subtract } x. \text{ Interchange sides to write in standard form.}$$

$$(x - 4)(x - 9) = 0 \qquad \text{Factor.}$$

$$x - 4 = 0 \quad \text{or} \quad x - 9 = 0 \qquad \text{Zero-factor property}$$

$$x = 4 \quad \text{or} \quad x = 9 \qquad \text{Proposed solutions}$$

Check each proposed solution in the *original* equation.

CHECK

$x + \sqrt{x} = 6$		$x + \sqrt{x} = 6$	
$4 + \sqrt{4} \stackrel{?}{=} 6$	Let $x = 4$.	$9 + \sqrt{9} \stackrel{?}{=} 6$	Let $x = 9$.
$6 = 6$ ✓	True	$12 = 6$	False

Only the solution 4 checks, so the solution set is $\{4\}$.

·················· **Work Problem** ➍ **at the Side.** ▶

OBJECTIVE ➍ **Solve an equation that is quadratic in form by substitution.** A nonquadratic equation that can be written in the form

$$au^2 + bu + c = 0,$$

for $a \neq 0$ and an algebraic expression u, is **quadratic in form.**

Many equations that are quadratic in form can be solved more easily by defining and substituting a "temporary" variable u for an expression involving the variable in the original equation.

EXAMPLE 5 **Defining Substitution Variables**

Define a variable u, and write each equation in the form $au^2 + bu + c = 0$.

(a) $x^4 - 13x^2 + 36 = 0$

Look at the two terms involving the variable x, ignoring their coefficients. Try to find one variable expression that is the square of the other. Because $x^4 = (x^2)^2$, we can define $u = x^2$, and rewrite the original equation as a quadratic equation in u.

$$u^2 - 13u + 36 = 0 \qquad \text{Here, } u = x^2.$$

(b) $2(4m - 3)^2 + 7(4m - 3) + 5 = 0$

Because this equation involves both $(4m - 3)^2$ and $(4m - 3)$, we choose $u = 4m - 3$. Substituting u for $4m - 3$ gives a quadratic equation in u.

$$2u^2 + 7u + 5 = 0 \qquad \text{Here, } u = 4m - 3.$$

(c) $2x^{2/3} - 11x^{1/3} + 12 = 0$

We apply a power rule for exponents, $(a^m)^n = a^{mn}$ (**Section 5.2**). Because $(x^{1/3})^2 = x^{2/3}$, we define $u = x^{1/3}$. With this substitution, the original equation can be written as follows.

$$2u^2 - 11u + 12 = 0 \qquad \text{Here, } u = x^{1/3}.$$

·················· **Work Problem** ➎ **at the Side.** ▶

➍ Solve each equation. Check your solutions.

(a) $x = \sqrt{7x - 10}$

(b) $2x = \sqrt{x + 1}$

➎ Define a variable u, and write each equation in quadratic form $au^2 + bu + c = 0$.

(a) $2x^4 + 5x^2 - 12 = 0$

(b) $2(x + 5)^2 - 7(x + 5) + 6 = 0$

(c) $x^{4/3} - 8x^{2/3} + 16 = 0$

Answers

4. (a) $\{2, 5\}$ **(b)** $\{1\}$
5. (a) $u = x^2;\ 2u^2 + 5u - 12 = 0$
   (b) $u = x + 5;\ 2u^2 - 7u + 6 = 0$
   (c) $u = x^{2/3};\ u^2 - 8u + 16 = 0$

EXAMPLE 6	Solving Equations That Are Quadratic in Form

Solve each equation.

(a) $x^4 - 13x^2 + 36 = 0$

From **Example 5(a),** we write this equation in quadratic form by substituting u for x^2.

$$x^4 - 13x^2 + 36 = 0 \quad \text{◁ Think of this as a "disguised" quadratic equation.}$$

$$(x^2)^2 - 13x^2 + 36 = 0 \qquad x^4 = (x^2)^2$$

$$u^2 - 13u + 36 = 0 \qquad \text{Let } u = x^2.$$

$$(u - 4)(u - 9) = 0 \qquad \text{Factor.}$$

$$u - 4 = 0 \quad \text{or} \quad u - 9 = 0 \qquad \text{Zero-factor property}$$

$$\text{Don't stop here.} \quad ▷ \quad u = 4 \quad \text{or} \quad u = 9 \qquad \text{Solve each equation.}$$

$$x^2 = 4 \quad \text{or} \quad x^2 = 9 \qquad \text{Substitute } x^2 \text{ for } u.$$

$$x = \pm 2 \quad \text{or} \quad x = \pm 3 \qquad \text{Square root property}$$

The equation $x^4 - 13x^2 + 36 = 0$, a fourth-degree equation, has four solutions.* The solution set is

$$\{-3, -2, 2, 3\}.$$

Check each solution by substituting it for x in the original equation.

(b) $\qquad\qquad\qquad 4x^4 + 1 = 5x^2$

$$4x^4 - 5x^2 + 1 = 0 \qquad \text{Subtract } 5x^2.$$

$$4(x^2)^2 - 5x^2 + 1 = 0 \qquad x^4 = (x^2)^2$$

$$4u^2 - 5u + 1 = 0 \qquad \text{Let } u = x^2.$$

$$(4u - 1)(u - 1) = 0 \qquad \text{Factor.}$$

$$4u - 1 = 0 \quad \text{or} \quad u - 1 = 0 \qquad \text{Zero-factor property}$$

$$u = \frac{1}{4} \quad \text{or} \quad u = 1 \qquad \text{Solve each equation.}$$

$$\text{This is a key step.} \quad ▷ \quad x^2 = \frac{1}{4} \quad \text{or} \quad x^2 = 1 \qquad \text{Substitute } x^2 \text{ for } u.$$

$$x = \pm\frac{1}{2} \quad \text{or} \quad x = \pm 1 \qquad \text{Square root property}$$

Check that the solution set is $\left\{-1, -\frac{1}{2}, \frac{1}{2}, 1\right\}$.

(c) $\qquad\qquad\qquad x^4 = 6x^2 - 3$

$$x^4 - 6x^2 + 3 = 0 \qquad \text{Subtract } 6x^2. \text{ Add 3.}$$

$$(x^2)^2 - 6x^2 + 3 = 0 \qquad x^4 = (x^2)^2$$

$$u^2 - 6u + 3 = 0 \qquad \text{Let } u = x^2.$$

This equation cannot be solved by factoring, so we use the quadratic formula.

········ **Continued on Next Page**

---

*In general, an equation in which an nth-degree polynomial equals 0 has n solutions, although they may not all be distinct—that is, some may be repeated.

$$u = \frac{-(-6) \pm \sqrt{(-6)^2 - 4(1)(3)}}{2(1)}$$

In $u^2 - 6u + 3 = 0$, $a = 1$, $b = -6$, and $c = 3$.

$$u = \frac{6 \pm \sqrt{24}}{2}$$

Simplify.

$$u = \frac{6 \pm 2\sqrt{6}}{2}$$

$\sqrt{24} = \sqrt{4} \cdot \sqrt{6}$
$= 2\sqrt{6}$

$$u = \frac{2(3 \pm \sqrt{6})}{2}$$

Factor.

$$u = 3 \pm \sqrt{6}$$

Write in lowest terms.

| Find both square roots in each case. |

$x^2 = 3 + \sqrt{6}$ or $x^2 = 3 - \sqrt{6}$

Substitute x^2 for u.

$x = \pm\sqrt{3 + \sqrt{6}}$ or $x = \pm\sqrt{3 - \sqrt{6}}$

Square root property

The solution set contains four numbers, written as follows.

$$\left\{ \sqrt{3 + \sqrt{6}}, -\sqrt{3 + \sqrt{6}}, \sqrt{3 - \sqrt{6}}, -\sqrt{3 - \sqrt{6}} \right\}$$

Note

Some students prefer to solve equations like those in **Examples 6(a)** **and (b)** by factoring directly.

$$x^4 - 13x^2 + 36 = 0 \quad \text{**Example 6(a)** equation}$$
$$(x^2 - 9)(x^2 - 4) = 0 \quad \text{Factor.}$$
$$(x + 3)(x - 3)(x + 2)(x - 2) = 0 \quad \text{Factor again.}$$

Using the zero-factor property gives the same solutions that we obtained in **Example 6(a)**. Equations that cannot be solved by factoring (as in **Example 6(c)**) must be solved by substitution and the quadratic formula.

Work Problem 6 at the Side. ▶

The method used in **Example 6** can be generalized.

Solving an Equation That Is Quadratic in Form by Substitution

Step 1 **Define a temporary variable u,** based on the relationship between the variable expressions in the given equation. Substitute u in the original equation and rewrite the equation in the form $au^2 + bu + c = 0$.

Step 2 **Solve the quadratic equation obtained in Step 1** by factoring or the quadratic formula.

Step 3 **Replace u with the expression it defined in Step 1.**

Step 4 **Solve the resulting equations for the original variable.**

Step 5 **Check** all solutions by substituting them in the original equation.

---

6 Solve each equation. Check your solutions.

(a) $m^4 - 10m^2 + 9 = 0$

$(\underline{\quad})^2 - 10m^2 + 9 = 0$

Let $u = m^2$.

$\underline{\quad}^2 - 10\underline{\quad} + 9 = 0$

$(u - 9)(\underline{\quad\quad}) = 0$

$u - 9 = 0$ or $\underline{\quad\quad} = 0$

$u = 9$ or $u = \underline{\quad}$

Substitute m^2 for u.

$\underline{\quad} = 9$ or $m^2 = 1$

$m = \underline{\quad}$ or $m = \pm 1$

The solution set is $\underline{\quad\quad}$.

(b) $9k^4 - 37k^2 + 4 = 0$

(c) $x^4 - 4x^2 = -2$

Answers

6. **(a)** m^2; u; u; $u - 1$; $u - 1$; 1; m^2; ± 3;
$\{-3, -1, 1, 3\}$

(b) $\left\{ -2, -\dfrac{1}{3}, \dfrac{1}{3}, 2 \right\}$

(c) $\{\sqrt{2 + \sqrt{2}}, -\sqrt{2 + \sqrt{2}}, \sqrt{2 - \sqrt{2}},$
$-\sqrt{2 - \sqrt{2}}\}$

7 Solve each equation. Check your solutions.

(a) $5(r+3)^2 + 9(r+3) = 2$

(b) $4m^{2/3} = 3m^{1/3} + 1$

EXAMPLE 7 **Solving Equations That Are Quadratic in Form**

Solve each equation.

(a) $2(4m-3)^2 + 7(4m-3) + 5 = 0$

Step 1 Because of the repeated quantity $4m - 3$, substitute u for $4m - 3$ as in **Example 5(b).**

$$2(4m-3)^2 + 7(4m-3) + 5 = 0$$
$$2u^2 + 7u + 5 = 0 \qquad \text{Let } u = 4m - 3.$$

Step 2 $\qquad\qquad (2u+5)(u+1) = 0 \qquad$ Factor.

$$2u + 5 = 0 \qquad \text{or} \qquad u + 1 = 0 \qquad \text{Zero-factor property}$$

> Don't stop here.

$$u = -\frac{5}{2} \qquad \text{or} \qquad u = -1 \qquad \text{Solve for } u.$$

Step 3 $\quad 4m - 3 = -\dfrac{5}{2} \quad \text{or} \quad 4m - 3 = -1 \qquad$ Substitute $4m - 3$ for u.

Step 4 $\qquad 4m = \dfrac{1}{2} \qquad \text{or} \qquad 4m = 2 \qquad$ Add 3.

$$m = \frac{1}{8} \qquad \text{or} \qquad m = \frac{1}{2} \qquad \text{Divide by 4.}$$

Step 5 Check that the solution set of the original equation is $\left\{ \frac{1}{8}, \frac{1}{2} \right\}$.

(b) $2x^{2/3} - 11x^{1/3} + 12 = 0$

Step 1 From **Example 5(c),** $x^{2/3} = (x^{1/3})^2$, so substitute u for $x^{1/3}$.

$$2(x^{1/3})^2 - 11x^{1/3} + 12 = 0$$
$$2u^2 - 11u + 12 = 0 \qquad \text{Let } u = x^{1/3}.$$

Step 2 $\qquad\qquad (2u-3)(u-4) = 0 \qquad$ Factor.

$$2u - 3 = 0 \qquad \text{or} \quad u - 4 = 0 \qquad \text{Zero-factor property}$$

$$u = \frac{3}{2} \qquad \text{or} \qquad u = 4 \qquad \text{Solve for } u.$$

Step 3 $\qquad x^{1/3} = \dfrac{3}{2} \qquad \text{or} \qquad x^{1/3} = 4 \qquad$ Substitute $x^{1/3}$ for u.

Step 4 $\quad (x^{1/3})^3 = \left(\dfrac{3}{2}\right)^3 \text{ or } (x^{1/3})^3 = 4^3 \qquad$ Cube each side.

$$x = \frac{27}{8} \qquad \text{or} \qquad x = 64 \qquad \text{Apply the exponents.}$$

Step 5 Because the original equation involves variables with rational exponents, check that neither of these solutions is extraneous. The solution set is $\left\{ \frac{27}{8}, 64 \right\}$.

◀ **Work Problem** **7** **at the Side.**

CAUTION

A common error when solving problems like those in **Examples 6 and 7** is to stop too soon. *Once we have solved for u, we must remember to substitute and solve for the values of the original variable.*

Answers

7. **(a)** $\left\{ -5, -\dfrac{14}{5} \right\}$ **(b)** $\left\{ -\dfrac{1}{64}, 1 \right\}$

10.4 Exercises

CONCEPT CHECK *Based on the discussion and examples of this section, give the first step to solve each equation. Do not actually solve.*

1. $\dfrac{14}{x} = x - 5$

2. $\sqrt{1 + x} + x = 5$

3. $(r^2 + r)^2 - 8(r^2 + r) + 12 = 0$

4. $3t = \sqrt{16 - 10t}$

5. CONCEPT CHECK Study this incorrect "solution." **What Went Wrong?**

$$x = \sqrt{3x + 4}$$
$$x^2 = 3x + 4 \qquad \text{Square each side.}$$
$$x^2 - 3x - 4 = 0$$
$$(x - 4)(x + 1) = 0$$
$$x - 4 = 0 \quad \text{or} \quad x + 1 = 0$$
$$x = 4 \quad \text{or} \qquad x = -1$$

Solution set: $\{4, -1\}$

6. CONCEPT CHECK Study this incorrect "solution." **What Went Wrong?**

$$2(m - 1)^2 - 3(m - 1) + 1 = 0$$
$$2u^2 - 3u + 1 = 0 \qquad \text{Let } u = m - 1.$$
$$(2u - 1)(u - 1) = 0$$
$$2u - 1 = 0 \quad \text{or} \quad u - 1 = 0$$
$$u = \frac{1}{2} \quad \text{or} \qquad u = 1$$

Solution set: $\left\{\frac{1}{2}, 1\right\}$

Solve each equation. Check your solutions. **See Example 1.**

7. $1 - \dfrac{3}{x} - \dfrac{28}{x^2} = 0$

8. $4 - \dfrac{7}{r} - \dfrac{2}{r^2} = 0$

9. $3 - \dfrac{1}{t} = \dfrac{2}{t^2}$

10. $1 + \dfrac{2}{k} = \dfrac{3}{k^2}$

11. $\dfrac{1}{x} + \dfrac{2}{x + 2} = \dfrac{17}{35}$

12. $\dfrac{2}{m} + \dfrac{3}{m + 9} = \dfrac{11}{4}$

13. $\dfrac{2}{x + 1} + \dfrac{3}{x + 2} = \dfrac{7}{2}$

14. $\dfrac{4}{3 - p} + \dfrac{2}{5 - p} = \dfrac{26}{15}$

15. $\dfrac{3}{2x} - \dfrac{1}{2(x + 2)} = 1$

16. $\dfrac{4}{3x} - \dfrac{1}{2(x + 1)} = 1$

17. $\dfrac{6}{p} = 2 + \dfrac{p}{p + 1}$

18. $\dfrac{k}{2 - k} + \dfrac{2}{k} = 5$

CONCEPT CHECK *Answer each question.*

19. A boat travels 20 mph in still water, and the rate of the current is t mph.

(a) What is the rate of the boat when it travels upstream?

(b) What is the rate of the boat when it travels downstream?

20. It takes m hours to grade a set of papers.

(a) What is the grader's rate (in job per hour)?

(b) How much of the job will the grader do in 2 hr?

Complete any tables. Then solve each problem. **See Examples 2 and 3.**

21. On a windy day Yoshiaki found that he could travel 16 mi downstream and then 4 mi back upstream at top speed in a total of 48 min. What was the top speed of Yoshiaki's boat if the rate of the current was 15 mph? (Let x represent the rate of the boat in still water.)

	d	r	t
Upstream	4	$x - 15$	___
Downstream	16	___	___

22. Lekesha flew her plane for 6 hr at a constant rate. She traveled 810 mi with the wind, then turned around and traveled 720 mi against the wind. The wind speed was a constant 15 mph. Find the rate of the plane. (Let x represent the rate of the plane in still air.)

	d	r	t
With Wind	810	___	___
Against Wind	720	___	___

23. In Canada, Medicine Hat and Cranbrook are 300 km apart. Harry rides his Honda 20 km per hr faster than Sally rides her Yamaha. Find Harry's average rate if he travels from Cranbrook to Medicine Hat in $1\frac{1}{4}$ hr less time than Sally. (*Source: State Farm Road Atlas.*)

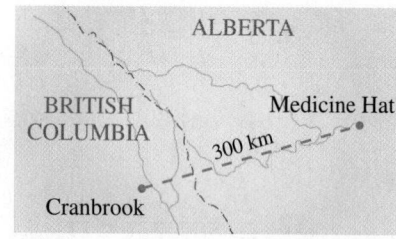

24. In California, the distance from Jackson to Lodi is about 40 mi, as is the distance from Lodi to Manteca. Rico drove from Jackson to Lodi, stopped in Lodi for a root beer, and then drove on to Manteca at 10 mph faster. Driving time for the entire trip was 88 min. Find his rate from Jackson to Lodi. (*Source: State Farm Road Atlas.*)

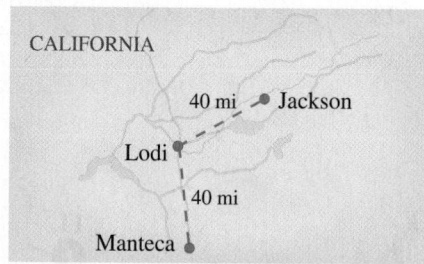

25. Working together, two people can cut a large lawn in 2 hr. One person can do the job alone in 1 hr less time than the other. How long (to the nearest tenth) would it take the faster person to do the job? (Let x represent the time of the faster person.)

	Rate	Time Working Together	Fractional Part of the Job Done
Faster Worker	___	2	___
Slower Worker	___	2	___

26. Working together, two people can clean an office building in 5 hr. One person takes 2 hr longer than the other person to clean the building alone. How long (to the nearest tenth) would it take the slower worker to clean the building alone? (Let x represent the time of the slower worker.)

	Rate	Time Working Together	Fractional Part of the Job Done
Faster Worker	___	___	___
Slower Worker	___	___	___

27. Rusty and Nancy Brauner are planting flats of spring flowers. Working alone, Rusty would take 2 hr longer than Nancy to plant the flowers. Working together, they do the job in 12 hr. How long (to the nearest tenth) would it have taken each person working alone?

28. Joel Spring can work through a stack of invoices in 1 hr less time than Noel White can. Working together they take $1\frac{1}{2}$ hr. How long (to the nearest tenth) would it take each person working alone?

29. A washing machine can be filled in 6 min if both the hot and cold water taps are fully opened. Filling the washer with hot water alone takes 9 min longer than filling it with cold water alone. How long does it take to fill the washer with cold water?

30. Two pipes together can fill a large tank in 2 hr. One of the pipes, used alone, takes 3 hr longer than the other to fill the tank. How long would each pipe take to fill the tank alone?

Solve each equation. Check your solutions. ***See Example 4.***

31. $z = \sqrt{5z - 4}$

32. $x = \sqrt{9x - 14}$

33. $2x = \sqrt{11x + 3}$

34. $4x = \sqrt{6x + 1}$

35. $3x = \sqrt{16 - 10x}$

36. $4t = \sqrt{8t + 3}$

37. $p - 2\sqrt{p} = 8$

38. $k + \sqrt{k} = 12$

39. $m = \sqrt{\dfrac{6 - 13m}{5}}$

40. $r = \sqrt{\dfrac{20 - 19r}{6}}$

41. $-x = \sqrt{\dfrac{8 - 2x}{3}}$

42. $-x = \sqrt{\dfrac{3x + 7}{4}}$

Solve each equation. Check your solutions. ***See Examples 5–7.***

43. $x^4 - 29x^2 + 100 = 0$

44. $x^4 - 37x^2 + 36 = 0$

45. $t^4 - 18t^2 + 81 = 0$

46. $x^4 - 8x^2 + 16 = 0$

47. $4k^4 - 13k^2 + 9 = 0$

48. $9x^4 - 25x^2 + 16 = 0$

49. $x^4 + 48 = 16x^2$

50. $z^4 = 17z^2 - 72$

51. $2x^4 - 9x^2 = -2$

52. $8x^4 + 1 = 11x^2$

53. $(x + 3)^2 + 5(x + 3) + 6 = 0$

54. $(k - 4)^2 + (k - 4) - 20 = 0$

55. $(t + 5)^2 + 6 = 7(t + 5)$

56. $3(m + 4)^2 - 8 = 2(m + 4)$

57. $2 + \dfrac{5}{3k - 1} = \dfrac{-2}{(3k - 1)^2}$

58. $3 - \dfrac{7}{2p + 2} = \dfrac{6}{(2p + 2)^2}$

59. $x^{2/3} + x^{1/3} - 2 = 0$

60. $x^{2/3} - 2x^{1/3} - 3 = 0$

61. $r^{2/3} + r^{1/3} - 12 = 0$

62. $3x^{2/3} - x^{1/3} - 24 = 0$

63. $4x^{4/3} - 13x^{2/3} + 9 = 0$

64. $9t^{4/3} - 25t^{2/3} + 16 = 0$

Relating Concepts (Exercises 65–70) For Individual or Group Work

Consider the following equation, which contains variable expressions in the denominators. **Work Exercises 65–70 in order.**

$$\dfrac{x^2}{(x - 3)^2} + \dfrac{3x}{x - 3} - 4 = 0$$

65. Why must 3 be excluded from the domain of this equation?

66. Multiply each side of the equation by the LCD, $(x - 3)^2$, and solve. There is only one solution—what is it?

67. Write the equation in a different manner so that it is quadratic in form using the expression $\frac{x}{x - 3}$.

68. Explain why the expression $\frac{x}{x - 3}$ cannot equal 1.

69. Solve the equation from **Exercise 67** by making the substitution

$$u = \dfrac{x}{x - 3}.$$

Two values should be obtained for u. Why is one of them impossible for this equation?

70. Solve the equation

$$x^2(x - 3)^{-2} + 3x(x - 3)^{-1} - 4 = 0$$

by letting $u = (x - 3)^{-1}$. Two values should be obtained for u. Why is one of them impossible for this equation?

Summary Exercises *Applying Methods for Solving Quadratic Equations*

We have introduced four methods for solving quadratic equations written in standard form $ax^2 + bx + c = 0$.

METHODS FOR SOLVING QUADRATIC EQUATIONS

Method	Advantages	Disadvantages
Factoring	This is usually the fastest method.	Not all polynomials are factorable. Some factorable polynomials are difficult to factor.
Square root property	This is the simplest method for solving equations of the form $(ax + b)^2 = c$.	Few equations are given in this form.
Completing the square	This method can always be used.	It requires more steps than other methods.
Quadratic formula	This method can always be used.	It may be more difficult than factoring. Sign errors may occur when evaluating $\sqrt{b^2 - 4ac}$.

CONCEPT CHECK   *Decide whether* factoring, the square root property, *or the quadratic formula is most appropriate for solving each quadratic equation. Do not actually solve.*

1. $(2x + 3)^2 = 4$

2. $4x^2 - 3x = 1$

3. $z^2 + 5z - 8 = 0$

4. $2k^2 + 3k = 1$

5. $3m^2 = 2 - 5m$

6. $p^2 = 5$

Solve each quadratic equation by the method of your choice. Check your solutions.

7. $p^2 = 47$

8. $6x^2 - x - 15 = 0$

9. $n^2 + 8n + 6 = 0$

10. $(x - 4)^2 = 49$

11. $\dfrac{9}{m} + \dfrac{5}{m^2} = 2$

12. $3m^2 = 3 - 8m$

13. $3x^2 - 9x + 4 = 0$

*14. $x^2 = -12$

15. $x\sqrt{2} = \sqrt{5x - 2}$

16. $12x^4 - 11x^2 + 2 = 0$

17. $(2k + 5)^2 = 12$

18. $\dfrac{2}{x} + \dfrac{1}{x - 2} - \dfrac{5}{3} = 0$

19. $t^4 + 14 = 9t^2$

20. $2x^2 + 4x = 5$

*21. $z^2 + z + 2 = 0$

22. $x^4 - 8x^2 = -1$

23. $4t^2 - 12t + 9 = 0$

24. $x\sqrt{3} = \sqrt{2 - x}$

25. $r^2 - 72 = 0$

26. $-3x^2 + 4x = -4$

27. $x^2 - 5x - 36 = 0$

28. $w^2 = 169$

*29. $3p^2 = 6p - 4$

30. $z = \sqrt{\dfrac{5z + 3}{2}}$

31. $2(3k - 1)^2 + 5(3k - 1) = -2$

*32. $\dfrac{4}{r^2} + 3 = \dfrac{1}{r}$

33. $x - \sqrt{15 - 2x} = 0$

34. $3 = \dfrac{1}{t + 2} + \dfrac{2}{(t + 2)^2}$

*35. $4k^4 + 5k^2 + 1 = 0$

36. $(x + 1)^{2/3} - (x + 1)^{1/3} = 2$

---

*This exercise requires knowledge of complex numbers.

10.5 Formulas and Applications

OBJECTIVE ▸ **1** **Solve formulas for variables involving squares and square roots.**

EXAMPLE 1 Solving for Variables Involving Squares or Square Roots

Solve each formula for the given variable.

(a) $w = \dfrac{kFr}{v^2}$ for v

$$w = \frac{kFr}{v^2} \quad \boxed{\text{The goal is to isolate } v \text{ on one side.}}$$

$$v^2 w = kFr \qquad \text{Multiply by } v^2.$$

$$v^2 = \frac{kFr}{w} \qquad \text{Divide by } w.$$

$$\boxed{\text{Include both positive and negative roots.}} \quad v = \pm\sqrt{\frac{kFr}{w}} \qquad \text{Square root property}$$

$$v = \frac{\pm\sqrt{kFr}}{\sqrt{w}} \cdot \frac{\sqrt{w}}{\sqrt{w}} \qquad \text{Rationalize the denominator.}$$

$$v = \frac{\pm\sqrt{kFrw}}{w} \qquad \sqrt{a} \cdot \sqrt{b} = \sqrt{ab};\ \sqrt{a} \cdot \sqrt{a} = a$$

In applications, the negative root is often rejected. However in problems like this, we indicate both roots.

(b) $d = \sqrt{\dfrac{4A}{\pi}}$ for A

$$d = \sqrt{\frac{4A}{\pi}} \quad \boxed{\text{The goal is to isolate } A \text{ on one side.}}$$

$$d^2 = \frac{4A}{\pi} \qquad \text{Square each side.}$$

$$\pi d^2 = 4A \qquad \text{Multiply by } \pi.$$

$$\frac{\pi d^2}{4} = A, \quad \text{or} \quad A = \frac{\pi d^2}{4} \qquad \begin{array}{l}\text{Divide by 4.}\\ \text{Interchange sides.}\end{array}$$

▸ **Work Problem** **1** **at the Side.** ▸

EXAMPLE 2 Solving for a Variable That Appears in First- and Second-Degree Terms

Solve $s = 2t^2 + kt$ for t.

Since the equation has terms with t^2 and t, we write it in standard form $ax^2 + bx + c = 0$, with t as the variable instead of x.

$$s = 2t^2 + kt$$

$$0 = 2t^2 + kt - s \qquad \text{Subtract } s.$$

$$2t^2 + kt - s = 0 \qquad \text{Standard form}$$

▸ **Continued on Next Page**

OBJECTIVES

1 Solve formulas for variables involving squares and square roots.

2 Solve applied problems using the Pythagorean theorem.

3 Solve applied problems using area formulas.

4 Solve applied problems using quadratic functions as models.

1 Solve each formula for the given variable.

(a) $A = \pi r^2$ for r

GS (b) $s = 30\sqrt{\dfrac{a}{p}}$ for a

$$s = 30\sqrt{\frac{a}{p}}$$

$$\frac{s}{\rule{1cm}{0.4pt}} = \sqrt{\frac{a}{p}}$$

Square each side.

$$\frac{s^2}{\rule{1cm}{0.4pt}} = \frac{\rule{0.6cm}{0.4pt}}{\rule{0.6cm}{0.4pt}}$$

$$\rule{1cm}{0.4pt} = a$$

(c) $S = \sqrt{\dfrac{pq}{n}}$ for p

Answers

1. (a) $r = \dfrac{\pm\sqrt{A\pi}}{\pi}$

(b) 30; 900; $\dfrac{a}{p}$; $\dfrac{ps^2}{900}$

(c) $p = \dfrac{nS^2}{q}$

❷ Solve $2t^2 - 5t + k = 0$ for t.

Now solve $2t^2 + kt - s = 0$ for t using the quadratic formula.

$$t = \frac{-k \pm \sqrt{k^2 - 4(2)(-s)}}{2(2)} \qquad \text{Let } a = 2, b = k, \text{ and } c = -s.$$

$$t = \frac{-k \pm \sqrt{k^2 + 8s}}{4} \qquad \text{Simplify.}$$

The solutions are $\quad t = \dfrac{-k + \sqrt{k^2 + 8s}}{4} \quad$ and $\quad t = \dfrac{-k - \sqrt{k^2 + 8s}}{4}$.

◀ **Work Problem ❷** at the Side.

Hypotenuse
c
Leg a
90°
Leg b
$a^2 + b^2 = c^2$
Pythagorean Theorem
Figure 2

OBJECTIVE ▶ **2** **Solve applied problems using the Pythagorean theorem.**
The Pythagorean theorem, represented by the equation

$$a^2 + b^2 = c^2,$$

is illustrated in **Figure 2** and was introduced in **Section 9.3.** It is used to solve applications involving right triangles.

EXAMPLE 3 Using the Pythagorean Theorem

Two cars left an intersection at the same time, one heading due north, the other due west. Some time later, they were exactly 100 mi apart. The car headed north had gone 20 mi farther than the car headed west. How far had each car traveled?

❸ Solve the problem.
A 13-ft ladder is leaning against a house. The distance from the bottom of the ladder to the house is 7 ft less than the distance from the top of the ladder to the ground. How far is the bottom of the ladder from the house?

Step 1 **Read** the problem carefully.

Step 2 **Assign a variable.**

Let $\quad x$ = the distance traveled by the car headed west.

Then $x + 20$ = the distance traveled by the car headed north.

See **Figure 3.** The cars are 100 mi apart, so the hypotenuse of the right triangle equals 100.

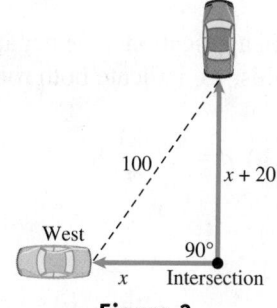

Figure 3

Step 3 **Write an equation.** Use the Pythagorean theorem.

$$a^2 + b^2 = c^2 \qquad \text{Pythagorean theorem}$$

$(x + y)^2 = x^2 + 2xy + y^2$ $\quad x^2 + (x + 20)^2 = 100^2 \qquad$ See **Figure 3.**

Step 4 **Solve.** $\quad x^2 + x^2 + 40x + 400 = 10{,}000 \qquad$ Square the binomial.

$$2x^2 + 40x - 9600 = 0 \qquad \text{Standard form}$$

$$x^2 + 20x - 4800 = 0 \qquad \text{Divide by 2.}$$

$$(x + 80)(x - 60) = 0 \qquad \text{Factor.}$$

$$x + 80 = 0 \qquad \text{or} \quad x - 60 = 0 \qquad \text{Zero-factor property}$$

$$x = -80 \quad \text{or} \qquad x = 60 \qquad \text{Solve for } x.$$

Step 5 **State the answer.** Distance cannot be negative, so discard the negative solution. The distances are 60 mi and $60 + 20 = 80$ mi.

Step 6 **Check.** Since $60^2 + 80^2 = 100^2$, the answers are correct.

◀ **Work Problem ❸** at the Side.

Answers

2. $t = \dfrac{5 + \sqrt{25 - 8k}}{4}, t = \dfrac{5 - \sqrt{25 - 8k}}{4}$

3. 5 ft

OBJECTIVE ▸ ③ **Solve applied problems using area formulas.**

EXAMPLE 4   Solving an Area Problem

A rectangular reflecting pool in a park is 20 ft wide and 30 ft long. The park gardener wants to plant a strip of grass of uniform width around the edge of the pool. She has enough seed to cover 336 ft². How wide will the strip be?

Step 1   **Read** the problem carefully.

Step 2   **Assign a variable.** The pool is shown in **Figure 4.**

> Let x = the unknown width of the grass strip.
>
> Then $20 + 2x$ = the width of the large rectangle (the width of the pool plus two grass strips),
>
> and $30 + 2x$ = the length of the large rectangle.

Step 3   **Write an equation.** Refer to **Figure 4.**

> $(30 + 2x)(20 + 2x)$ Area of large rectangle (length · width)
>
> $30 \cdot 20$,   or 600 Area of pool (in square feet)

The area of the large rectangle minus the area of the pool should equal 336 ft², the area of the grass strip.

Area of large   area of   area of
rectangle   −   pool   =   grass.
↓   ↓   ↓

$$(30 + 2x)(20 + 2x) - 600 = 336$$

Step 4   **Solve.** $600 + 100x + 4x^2 - 600 = 336$ Multiply.

$$4x^2 + 100x - 336 = 0 \quad \text{Standard form}$$

$$x^2 + 25x - 84 = 0 \quad \text{Divide by 4.}$$

$$(x + 28)(x - 3) = 0 \quad \text{Factor.}$$

$$x + 28 = 0 \quad \text{or} \quad x - 3 = 0 \quad \text{Zero-factor property}$$

$$x = -28 \quad \text{or} \quad x = 3 \quad \text{Solve each equation.}$$

Step 5   **State the answer.** The width cannot be −28 ft, so the grass strip will be 3 ft wide.

Step 6   **Check.** If $x = 3$, we can find the area of the large rectangle (which includes the grass strip).

$$(30 + 2 \cdot 3)(20 + 2 \cdot 3) = 36 \cdot 26 = 936 \text{ ft}^2 \quad \text{Area of pool and strip}$$

The area of the pool is $30 \cdot 20 = 600$ ft². So, the area of the grass strip is

$$936 - 600 = 336 \text{ ft}^2, \quad \text{as required.}$$

The answer is correct.

· Work Problem ❹ at the Side. ▶

OBJECTIVE ▸ ④ **Solve applied problems using quadratic functions as models.** Some applied problems can be modeled by *quadratic functions,* which for real numbers a, b, and c can be written in the form

$$f(x) = ax^2 + bx + c \quad \text{(where } a \neq 0\text{).}$$

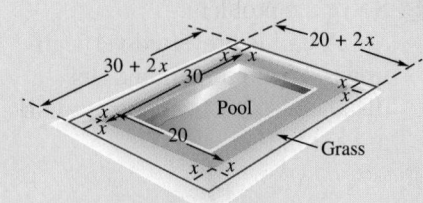

Figure 4

❹ Solve each problem.

  (a) Suppose the pool in **Example 4** is 20 ft by 40 ft and there is enough seed to cover 700 ft². How wide should the grass strip be?

  (b) A football practice field is 30 yd wide and 40 yd long. A strip of grass sod of uniform width is to be placed around the perimeter of the practice field. There is enough money budgeted for 296 sq yd of sod. How wide will the strip be?

Answers
4. (a) 5 ft   **(b)** 2 yd

5 Solve the problem.

A ball is projected vertically upward from the ground. Its distance in feet from the ground at t seconds is

$$s(t) = -16t^2 + 64t.$$

At what times will the ball be 32 ft from the ground? Use a calculator and round answers to the nearest tenth. (*Hint:* There are two answers.)

EXAMPLE 5 **Solving an Applied Problem Using a Quadratic Function**

If an object is projected upward from the top of a 144-ft building at 112 ft per sec, its position (in feet above the ground) is given by

$$s(t) = -16t^2 + 112t + 144,$$

where t is time in seconds after it was projected. When does it hit the ground?

When the object hits the ground, its distance above the ground is 0. We must find the value of t that makes $s(t) = 0$.

$0 = -16t^2 + 112t + 144$	Let $s(t) = 0$.
$0 = t^2 - 7t - 9$	Divide by -16.
$t = \dfrac{-(-7) \pm \sqrt{(-7)^2 - 4(1)(-9)}}{2(1)}$	Substitute into the quadratic formula. Let $a = 1$, $b = -7$, and $c = -9$.
$t = \dfrac{7 \pm \sqrt{85}}{2} \approx \dfrac{7 \pm 9.2}{2}$	Use a calculator.

The solutions are $t \approx 8.1$ or $t \approx -1.1$. Since time cannot be negative, discard the negative solution. The object will hit the ground about 8.1 sec after it is projected.

◀ **Work Problem 5** at the Side.

---

6 Use a calculator to evaluate

$$\frac{-4.61 \pm \sqrt{4.61^2 - 4(-0.002)(-116.5)}}{2(-0.002)}$$

for both solutions. Round to the nearest whole number. Which solution is valid for this problem?

EXAMPLE 6 **Using a Quadratic Function to Model the CPI**

The Consumer Price Index (CPI) is used to measure trends in prices for a "basket" of goods purchased by typical American families. This index uses a base period 1982–1984, which means that the index number for that period is 100. The quadratic function

$$f(x) = -0.002x^2 + 4.61x + 83.5$$

approximates the CPI for the years 1980–2010, where x is the number of years that have elapsed since 1980. (*Source:* Bureau of Labor Statistics.)

(a) Use the model to approximate the CPI for 1995.

For 1995, $x = 1995 - 1980 = 15$, so find $f(15)$.

$f(x) = -0.002x^2 + 4.61x + 83.5$	Given model
$f(15) = -0.002(15)^2 + 4.61(15) + 83.5$	Let $x = 15$.
$f(15) \approx 152$	Nearest whole number

The CPI for 1995 was about 152.

(b) In what year did the CPI reach 200?

Find the value of x that makes $f(x) = 200$.

$f(x) = -0.002x^2 + 4.61x + 83.5$	Given model
$200 = -0.002x^2 + 4.61x + 83.5$	Let $f(x) = 200$.
$0 = -0.002x^2 + 4.61x - 116.5$	Standard form

Now use $a = -0.002$, $b = 4.61$, and $c = -116.5$ in the quadratic formula.

◀ **Work Problem 6** at the Side.

The first solution is $x \approx 26$. Rounding up to the next whole number, the CPI first reached 200 in $1980 + 26 = 2006$. (Reject the solution $x \approx 2279$, which gives an invalid year.)

Answers

5. at 0.6 sec and at 3.4 sec
6. 26; 2279; 26

10.5 Exercises

CONCEPT CHECK *Answer each question.*

1. What is the first step in solving a formula like $gw^2 = 2r$ for w?

2. What is the first step in solving a formula like $gw^2 = kw + 24$ for w?

In Exercises 3 and 4, solve for m in terms of the other variables (where $m > 0$).

3.

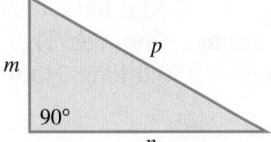

4.

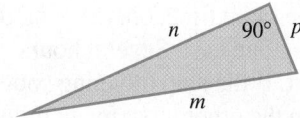

Solve each equation for the indicated variable. (Leave $\pm$ in the answers as needed.)
See Examples 1 and 2.

5. $d = kt^2$ for t

6. $S = 6e^2$ for e

7. $s = kwd^2$ for d

8. $S = \pi r^2 h$ for r

9. $I = \dfrac{ks}{d^2}$ for d

10. $R = \dfrac{k}{d^2}$ for d

11. $F = \dfrac{kA}{v^2}$ for v

12. $L = \dfrac{kd^4}{h^2}$ for h

13. $V = \pi r^2 h$ for r

14. $V = \dfrac{1}{3}\pi r^2 h$ for r

15. $At^2 + Bt = -C$ for t

16. $S = 2\pi rh + \pi r^2$ for r

17. $D = \sqrt{kh}$ for h

18. $F = \dfrac{k}{\sqrt{d}}$ for d

19. $p = \sqrt{\dfrac{k\ell}{g}}$ for ℓ

20. $p = \sqrt{\dfrac{k\ell}{g}}$ for g

Solve each problem. When appropriate, round answers to the nearest tenth.
See Example 3.

21. Find the lengths of the sides of the triangle.

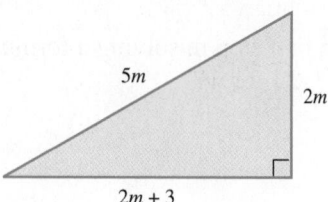

22. Find the lengths of the sides of the triangle.

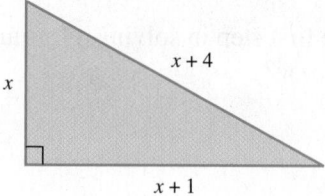

23. Two ships leave port at the same time, one heading due south and the other heading due east. Several hours later, they are 170 mi apart. If the ship traveling south traveled 70 mi farther than the other, how many miles did they each travel?

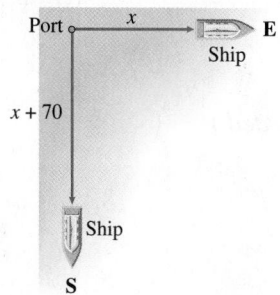

24. Faith Varnado is flying a kite that is 30 ft farther above her hand than its horizontal distance from her. The string from her hand to the kite is 150 ft long. How high is the kite?

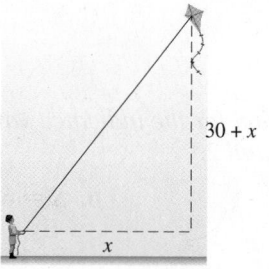

25. A game board is in the shape of a right triangle. The hypotenuse is 2 in. longer than the longer leg, and the longer leg is 1 in. less than twice as long as the shorter leg. How long is each side of the game board?

26. Manuel Bovi is planting a garden in the shape of a right triangle. The longer leg is 3 ft longer than the shorter leg. The hypotenuse is 3 ft longer than the longer leg. Find the lengths of the three sides of the garden.

Solve each problem. See Example 4.

27. A couple wants to buy a rug for a room that is 20 ft long and 15 ft wide. They want to leave an even strip of flooring uncovered around the edges of the room. How wide a strip will they have if they buy a rug with an area of 234 ft²?

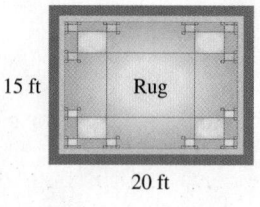

28. A club swimming pool is 30 ft wide and 40 ft long. The club members want an exposed aggregate border in a strip of uniform width around the pool. They have enough material for 296 ft². How wide can the strip be?

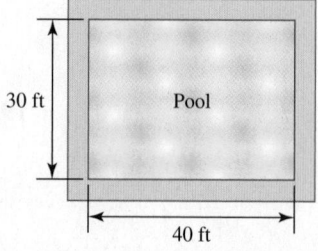

29. A rectangular piece of sheet metal has a length that is 4 in. less than twice the width. A square piece 2 in. on a side is cut from each corner. The sides are then turned up to form an uncovered box of volume 256 in.^{3}. Find the length and width of the original piece of metal.

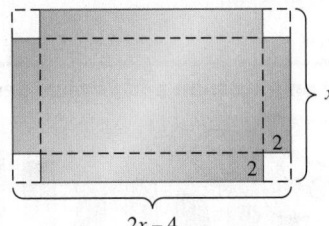

30. A rectangular piece of cardboard is 2 in. longer than it is wide. A square piece 3 in. on a side is cut from each corner. The sides are then turned up to form an uncovered box of volume 765 in.^{3}. Find the dimensions of the original piece of cardboard.

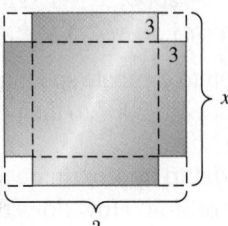

⊞ *Solve each problem. Round answers to the nearest tenth.* **See Example 5.**

31. A ball is projected upward from the ground. Its distance in feet from the ground in t seconds is given by

$$s(t) = -16t^2 + 128t.$$

At what times will the ball be 213 ft from the ground?

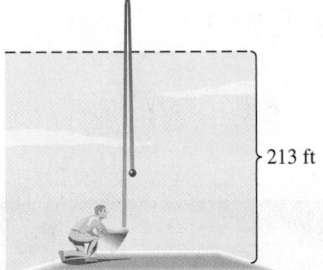

32. A toy rocket is launched from ground level. Its distance in feet from the ground in t seconds is given by

$$s(t) = -16t^2 + 208t.$$

At what times will the rocket be 550 ft from the ground?

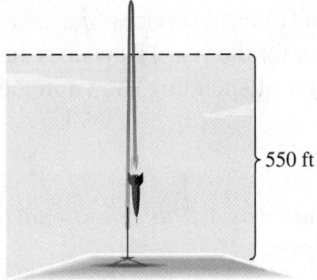

33. The following function gives the distance in feet a car going approximately 68 mph will skid in t seconds.

$$D(t) = 13t^2 - 100t$$

Find the time it would take for the car to skid 180 ft.

34. Refer to the function in **Exercise 33.** Find the time it would take for the car to skid 500 ft.

A ball is projected upward from ground level, and its distance in feet from the ground in t seconds is given by $s(t) = -16t^2 + 160t.$

35. After how many seconds does the ball reach a height of 400 ft? How would you describe in words its position at this height?

36. After how many seconds does it reach a height of 425 ft? How would you interpret the mathematical result here?

⊞ *Solve each problem using a quadratic equation.*

37. A certain bakery has found that the daily demand for blueberry muffins is $\frac{6000}{p}$, where p is the price of a muffin in cents. The daily supply is $3p - 410$. Find the price at which supply and demand are equal.

38. In one area the demand for Blu-ray discs is $\frac{1900}{P}$ per day, where P is the price in dollars per disc. The supply is $5P - 1$ per day. At what price, to the nearest cent, does supply equal demand?

Total spending (in billions of dollars) in the United States from all sources on physician and clinical services for the years 2000–2009 are shown in the bar graph and can be modeled by the quadratic function

$$f(x) = 0.4032x^2 + 27.69x + 289.8.$$

Here, $x = 0$ represents 2000, $x = 1$ represents 2001, and so on. Use the graph and the model to work Exercises 39–42. **See Example 6.**

39. (a) Use the graph to estimate spending on physician and clinical services in 2007 to the nearest $10 billion.

   (b) Use the model to approximate spending to the nearest $10 billion. How does this result compare to the estimate in part (a)?

40. Based on the model, in what year did spending on physician and clinical services first exceed $350 billion? (Round down for the year.) How does this result compare to the amount of spending shown in the graph?

41. Based on the model, in what year did spending on physician and clinical services first exceed $400 billion? (Round down for the year.) How does this result compare to the amount of spending shown in the graph?

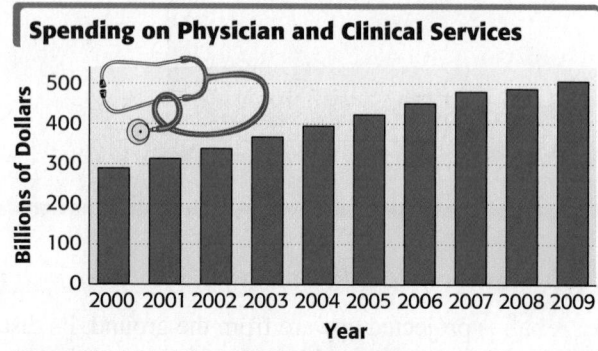

Spending on Physician and Clinical Services

Source: U.S. Centers for Medicare and Medicaid Services.

42. If these data were modeled by a *linear* function $f(x) = ax + b$, would the value of a be positive or negative? Explain.

William Froude was a 19th-century naval architect who used the expression

$$\frac{v^2}{g\ell}$$

in shipbuilding. This expression, known as the **Froude number,** *was also used by R. McNeill Alexander in his research on dinosaurs. (Source: "How Dinosaurs Ran," Scientific American.)*

In Exercises 43 and 44, find to the nearest tenth the value of v (in meters per second), given that $g = 9.8$ m per sec^2.

43. Rhinoceros: $\ell = 1.2$; Froude number $= 2.57$

44. Triceratops: $\ell = 2.8$; Froude number $= 0.16$

Recall from the **Section 7.5** *exercises that corresponding sides of similar triangles are proportional. Use this fact to find the lengths of the indicated sides of each pair of similar triangles. Check all possible solutions in both triangles. Sides of a triangle cannot be negative (and are not drawn to scale here).*

45. Side *AC*

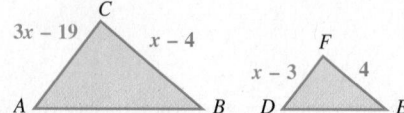

46. Side *RQ*

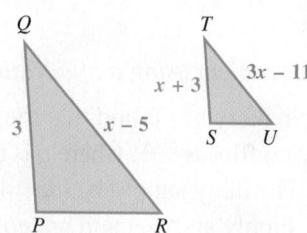

10.6 Graphs of Quadratic Functions

OBJECTIVE ▶ ① **Graph a quadratic function.** **Figure 5** gives a graph of the simplest *quadratic function* $y = x^2$.

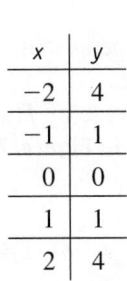

x	y
-2	4
-1	1
0	0
1	1
2	4

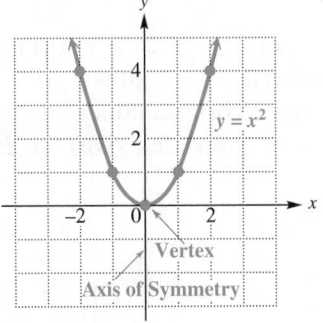

$y = x^2$

Vertex

Axis of Symmetry

Figure 5

This graph is a **parabola.** The point $(0, 0)$, the lowest point on the curve, is the **vertex** of this parabola. The vertical line through the vertex is the **axis of symmetry,** or simply the **axis,** of the parabola. Here, its equation is $x = 0$. A parabola is **symmetric about its axis**—that is, if the graph were folded along the axis, the two portions of the curve would coincide.

As **Figure 5** suggests, x can be any real number, so the domain of the function $y = x^2$ is $(-\infty, \infty)$. Since y is always nonnegative, the range is $[0, \infty)$.

In **Section 10.5,** we solved applications modeled by quadratic functions.

Quadratic Function

A function that can be written in the form

$$f(x) = ax^2 + bx + c$$

for real numbers a, b, and c, with $a \neq 0$, is a **quadratic function.**

The graph of any quadratic function is a parabola with a vertical axis.

Note

We use the variable y and function notation $f(x)$ interchangeably. Although the letter f is most often used to name quadratic functions, other letters can be used. We use the capital letter F to distinguish between different parabolas graphed on the same coordinate axes.

Parabolas, which are a type of *conic section* **(Chapter 12),** have a special reflecting property that makes them useful in the design of telescopes, radar equipment, solar furnaces, and automobile headlights. See **Figure 6.**

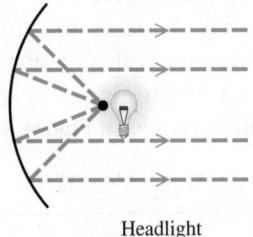

Headlight

Figure 6

① Graph each parabola. Give the vertex, axis of symmetry, domain, and range.

(a) $f(x) = x^2 + 3$

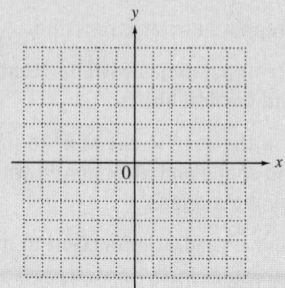

(b) $f(x) = x^2 - 1$

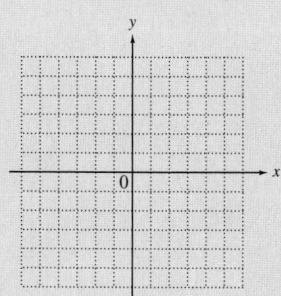

OBJECTIVE ► ② **Graph parabolas with horizontal and vertical shifts.**
Parabolas need not have their vertices at the origin, as does the graph of $f(x) = x^2$.

EXAMPLE 1 Graphing a Parabola (Vertical Shift)

Graph $F(x) = x^2 - 2$.

The graph of $F(x) = x^2 - 2$ has the same shape as that of $f(x) = x^2$, but is *shifted*, or *translated*, 2 units down, with vertex $(0, -2)$. Every function value is 2 less than the corresponding function value of $f(x) = x^2$. Plotting points on both sides of the vertex gives the graph in **Figure 7.**

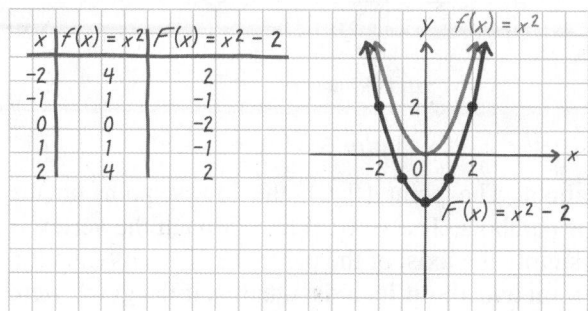

$F(x) = x^2 - 2$

Vertex: $(0, -2)$

Axis of symmetry: $x = 0$

Domain: $(-\infty, \infty)$

Range: $[-2, \infty)$

The graph of $f(x) = x^2$ is shown for comparison.

Figure 7

This parabola is symmetric about its axis $x = 0$, so the plotted points are "mirror images" of each other. Since x can be any real number, the domain is still $(-\infty, \infty)$. The value of y (or $F(x)$) is always greater than or equal to -2, so the range is $[-2, \infty)$.

Vertical Shift

The graph of $F(x) = x^2 + k$ is a parabola.

• The graph has the same shape as the graph of $f(x) = x^2$.

• The parabola is shifted k units up if $k > 0$, and $|k|$ units down if $k < 0$.

• The vertex of the parabola is $(0, k)$.

◄ **Work Problem ① at the Side.**

EXAMPLE 2 Graphing a Parabola (Horizontal Shift)

Graph $F(x) = (x - 2)^2$.

If $x = 2$, then $F(x) = 0$, which gives the vertex $(2, 0)$. The graph of $F(x) = (x - 2)^2$ has the same shape as that of $f(x) = x^2$, but is shifted 2 units to the right. We plot several points on one side of the vertex. Then we use symmetry about the axis $x = 2$ to find corresponding points on the other side of the vertex. The graph is shown in **Figure 8** on the next page.

Continued on Next Page

Answers

1. (a)

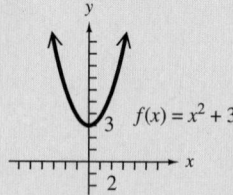

$f(x) = x^2 + 3$

vertex: $(0, 3)$; axis: $x = 0$;
domain: $(-\infty, \infty)$; range: $[3, \infty)$

(b)

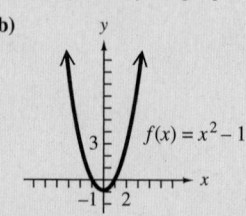

$f(x) = x^2 - 1$

vertex: $(0, -1)$; axis: $x = 0$;
domain: $(-\infty, \infty)$; range: $[-1, \infty)$

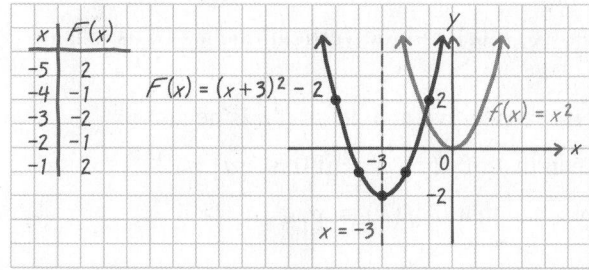

Figure 8

$F(x) = (x - 2)^2$

Vertex: $(2, 0)$

Axis of symmetry: $x = 2$

Domain: $(-\infty, \infty)$

Range: $[0, \infty)$

Horizontal Shift

The graph of $F(x) = (x - h)^2$ is a parabola.

- The graph has the same shape as the graph of $f(x) = x^2$.

- The parabola is shifted h units to the right if $h > 0$, and $|h|$ units to the left if $h < 0$.

- The vertex of the parabola is $(h, 0)$.

CAUTION

Errors frequently occur when horizontal shifts are involved. To determine the direction and magnitude of a horizontal shift, find the value that causes the expression $x - h$ to equal 0, as shown below.

$F(x) = (x - 5)^2$	$F(x) = (x + 5)^2$
Because +5 causes $x - 5$ to equal 0, the graph of $F(x)$ illustrates a shift of	Because −5 causes $x + 5$ to equal 0, the graph of $F(x)$ illustrates a shift of
5 units to the right.	**5 units to the left.**

Work Problem ❷ at the Side. ▶

EXAMPLE 3 Graphing a Parabola (Horizontal and Vertical Shifts)

Graph $F(x) = (x + 3)^2 - 2$.

This graph has the same shape as that of $f(x) = x^2$, but is shifted 3 units to the left (since $x + 3 = 0$ if $x = -3$) and 2 units down (because of the negative sign in -2). See **Figure 9**.

x	$F(x)$
−5	2
−4	−1
−3	−2
−2	−1
−1	2

Figure 9

$F(x) = (x + 3)^2 - 2$

Vertex: $(-3, -2)$

Axis of symmetry: $x = -3$

Domain: $(-\infty, \infty)$

Range: $[-2, \infty)$

---

❷ Graph each parabola. Give the vertex, axis of symmetry, domain, and range.

(a) $f(x) = (x - 3)^2$

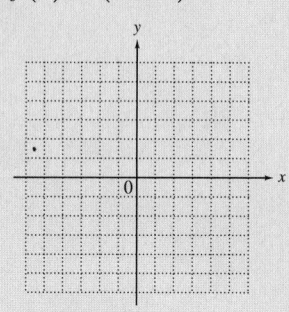

(b) $f(x) = (x + 2)^2$

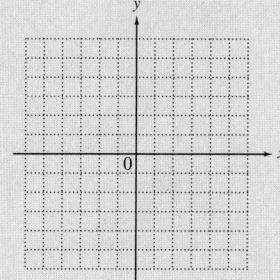

Answers

2. (a)

vertex: $(3, 0)$; axis: $x = 3$;
domain: $(-\infty, \infty)$; range: $[0, \infty)$

(b)

vertex: $(-2, 0)$; axis: $x = -2$;
domain: $(-\infty, \infty)$; range: $[0, \infty)$

❸ Graph each parabola. Give the vertex, axis of symmetry, domain, and range.

(a) $f(x) = (x + 2)^2 - 1$

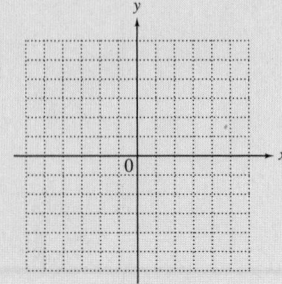

(b) $f(x) = (x - 2)^2 + 5$

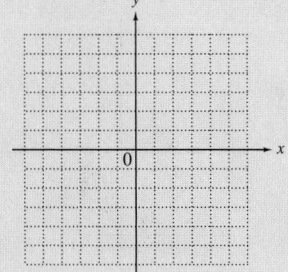

Answers

3. **(a)**

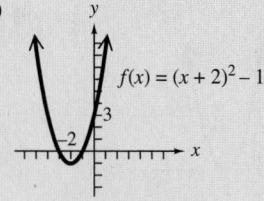

vertex: $(-2, -1)$; axis: $x = -2$;
domain: $(-\infty, \infty)$; range: $[-1, \infty)$

(b)

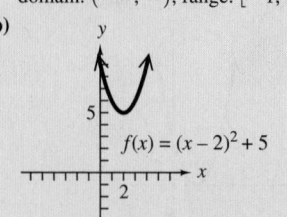

vertex: $(2, 5)$; axis: $x = 2$;
domain: $(-\infty, \infty)$; range: $[5, \infty)$

Vertex and Axis of a Parabola

The graph of $F(x) = (x - h)^2 + k$ is a parabola.

- The graph has the same shape as the graph of $f(x) = x^2$.
- The vertex of the parabola is (h, k).
- The axis of symmetry is the vertical line $x = h$.

◀ **Work Problem ❸ at the Side.**

OBJECTIVE ▶ ❸ **Use the coefficient of x^2 to predict the shape and direction in which a parabola opens.** Not all parabolas open up, and not all parabolas have the same shape as the graph of $f(x) = x^2$.

EXAMPLE 4 **Graphing a Parabola That Opens Down**

Graph $f(x) = -\frac{1}{2}x^2$.

This parabola is shown in **Figure 10.** The coefficient $-\frac{1}{2}$ affects the shape of the graph—the $\frac{1}{2}$ makes the parabola wider (since the values of $\frac{1}{2}x^2$ increase more slowly than those of x^2), and the negative sign makes the parabola open down. The graph is not shifted in any direction. Unlike the parabolas graphed in **Examples 1–3,** the vertex here has the *greatest* function value of any point on the graph.

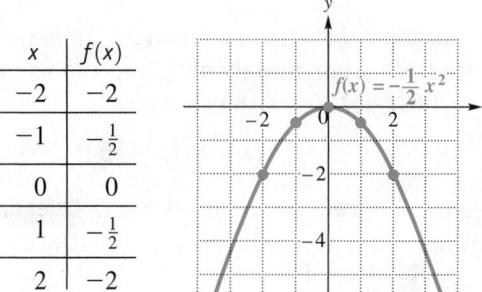

x	$f(x)$
-2	-2
-1	$-\frac{1}{2}$
0	0
1	$-\frac{1}{2}$
2	-2

$f(x) = -\frac{1}{2}x^2$
Vertex: $(0, 0)$
Axis of symmetry: $x = 0$
Domain: $(-\infty, \infty)$
Range: $(-\infty, 0]$

Figure 10

Some general characteristics of the graph of $F(x) = a(x - h)^2 + k$ are summarized as follows.

General Characteristics of $F(x) = a(x - h)^2 + k$

1. The graph of the quadratic function

$$F(x) = a(x - h)^2 + k \quad (\text{with } a \neq 0)$$

is a parabola with vertex (h, k) and the vertical line $x = h$ as axis of symmetry.

2. The graph opens up if a is positive and down if a is negative.

3. The graph is wider than that of $f(x) = x^2$ if $0 < |a| < 1$.

The graph is narrower than that of $f(x) = x^2$ if $|a| > 1$.

Work Problems ❹ and ❺ at the Side. ▶

EXAMPLE 5 Using the General Characteristics to Graph a Parabola

Graph $F(x) = -2(x + 3)^2 + 4$. Give the domain and the range.

   The parabola opens down (because $a < 0$) and is narrower than the graph of $f(x) = x^2$, since $|-2| = 2$ and $2 > 1$. This causes values of $F(x)$ to decrease more quickly than those of $f(x) = -x^2$. This parabola has vertex $(-3, 4)$ as shown in **Figure 11.** To complete the graph, we plotted the ordered pairs $(-4, 2)$ and, by symmetry, $(-2, 2)$. Symmetry can be used to find additional ordered pairs that satisfy the equation.

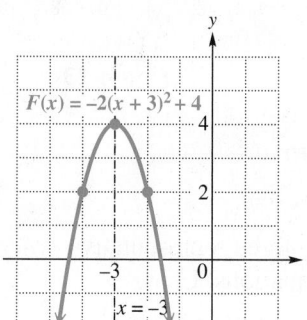

$F(x) = -2(x + 3)^2 + 4$
Vertex: $(-3, 4)$
Axis of symmetry: $x = -3$
Domain: $(-\infty, \infty)$
Range: $(-\infty, 4]$

Figure 11

········· Work Problem ❻ at the Side. ▶

OBJECTIVE ▶ ④ **Find a quadratic function to model data.**

EXAMPLE 6 Modeling the Number of Multiple Births

The number of higher-order multiple births (triplets or more) in the United States has declined in recent years, as shown by the data in the table. Here, x represents the number of years since 1996 and y represents the number of higher-order multiple births.

Year	x	y
1996	0	5939
2000	4	7325
2002	6	7401
2004	8	7275
2006	10	6540
2008	12	6268

Source: National Center for Health Statistics.

Find a quadratic function that models the data.

   A scatter diagram of the ordered pairs (x, y) is shown in **Figure 12** on the next page. The general shape suggested by the scatter diagram indicates that a parabola should approximate these points, as shown by the dashed curve in **Figure 13.** The equation for such a parabola would have a negative coefficient for x^2 since the graph opens down.

········· Continued on Next Page

---

❹ Decide whether each parabola opens up or down.

(a) $f(x) = -\dfrac{2}{3}x^2$

(b) $f(x) = \dfrac{3}{4}x^2 + 1$

(c) $f(x) = -2x^2 - 3$

(d) $f(x) = 3x^2 + 2$

❺ Decide whether each parabola in **Margin Problem 4** is wider or narrower than the graph of $f(x) = x^2$.

❻ Graph

$$f(x) = \dfrac{1}{2}(x - 2)^2 + 1.$$

Give the vertex, axis of symmetry, domain, and range.

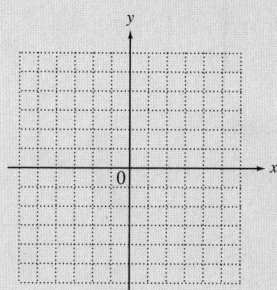

Answers

4. (a) down (b) up (c) down (d) up
5. (a) wider (b) wider (c) narrower
   (d) narrower
6.

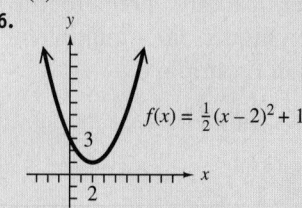

$f(x) = \frac{1}{2}(x - 2)^2 + 1$

vertex: $(2, 1)$; axis: $x = 2$;
domain: $(-\infty, \infty)$; range: $[1, \infty)$

7 Tell whether a linear or quadratic function would be a more appropriate model for each set of graphed data. If linear, tell whether the slope should be positive or negative. If quadratic, tell whether the coefficient a of x^2 should be positive or negative.

(a)   **AVERAGE DAILY E-MAIL VOLUME**

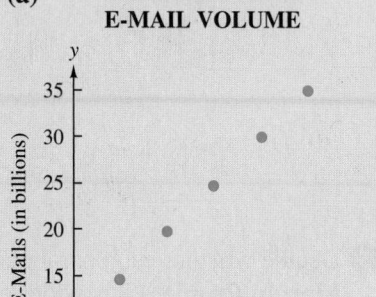

Source: General Accounting Office.

(b)   **MP3 PLAYER SALES IN U.S.**

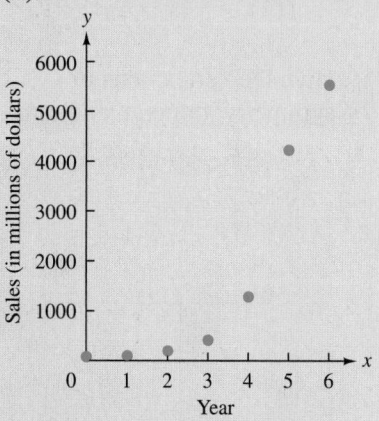

Source: Consumer Electronics Association.

8 Using the points $(4, 7325)$, $(8, 7275)$, and $(10, 6540)$, find another quadratic model for the data on higher-order multiple births in **Example 6.**

U.S. HIGHER-ORDER MULTIPLE BIRTHS

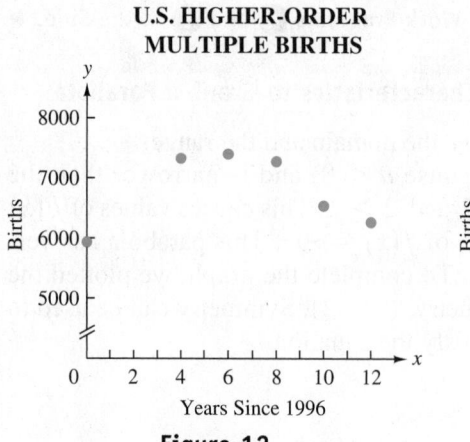

Figure 12

U.S. HIGHER-ORDER MULTIPLE BIRTHS

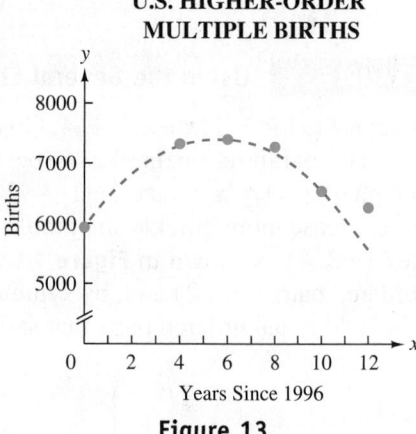

Figure 13

To find a quadratic function of the form

$$y = ax^2 + bx + c$$

that models, or *fits,* these data, we choose three representative ordered pairs and use them to write a system of three equations. Using

$$(0, 5939), \quad (6, 7401), \quad \text{and} \quad (12, 6268),$$

we substitute the x- and y-values from the ordered pairs into the quadratic form $ax^2 + bx + c = y$ to get three equations.

$$a(0)^2 + b(0) + c = \mathbf{5939} \quad \text{or} \quad c = 5939 \quad (1)$$
$$a(6)^2 + b(6) + c = \mathbf{7401} \quad \text{or} \quad 36a + 6b + c = 7401 \quad (2)$$
$$a(12)^2 + b(12) + c = \mathbf{6268} \quad \text{or} \quad 144a + 12b + c = 6268 \quad (3)$$

We can find the values of a, b, and c by solving this system of three equations in three variables using the methods of **Section 8.5.** From equation (1), $c = \mathbf{5939}$. Substitute 5939 for c in equations (2) and (3).

$$36a + 6b + \mathbf{5939} = 7401, \quad \text{or} \quad 36a + 6b = 1462 \quad (4)$$
$$144a + 12b + \mathbf{5939} = 6268, \quad \text{or} \quad 144a + 12b = 329 \quad (5)$$

We can eliminate b from this system of two equations in two variables by multiplying equation (4) by -2 and adding the result to equation (5).

$$72a = -2595$$
$$a = -36.04 \quad \text{Divide by 72. Use a calculator and round.}$$

We substitute -36.04 for a in equation (4) or (5) to find that $b = \mathbf{459.9}$. Using the values we have found for a, b, and c, our model is

$$y = -\mathbf{36.04}x^2 + \mathbf{459.9}x + \mathbf{5939}.$$

◀ **Work Problems 7** and **8** at the Side.

Note

If we had chosen three different ordered pairs of data in **Example 6,** a slightly different model would have resulted, as in **Margin Problem 8.**

🖩 **Calculator Tip**

The *quadratic regression* feature on a graphing calculator can be used to generate a quadratic model that fits given data. See your owner's manual.

Answers

7. (a) linear; positive   **(b)** quadratic; positive
8. $y = -59.17x^2 + 697.5x + 5482$

10.6 Exercises

FOR EXTRA HELP

 Download the MyDashBoard App

 MyMathLab®

CONCEPT CHECK *In Exercises 1 and 2, match each quadratic function in parts (a)–(d) with its graph from choices A–D.*

1. (a) $f(x) = (x + 2)^2 - 1$ **(b)** $f(x) = (x + 2)^2 + 1$ **(c)** $f(x) = (x - 2)^2 - 1$ **(d)** $f(x) = (x - 2)^2 + 1$

A.
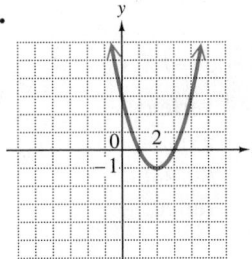

B.
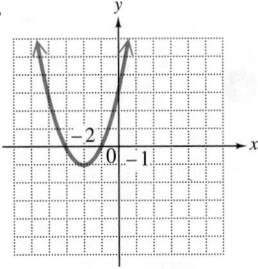

C.
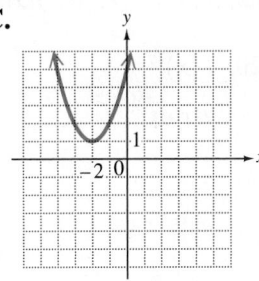

D.
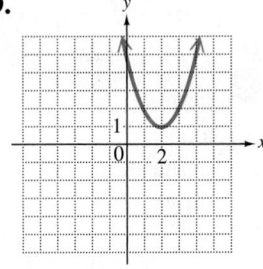

2. (a) $f(x) = -x^2 + 2$ **(b)** $f(x) = -x^2 - 2$ **(c)** $f(x) = -(x + 2)^2$ **(d)** $f(x) = -(x - 2)^2$

A.
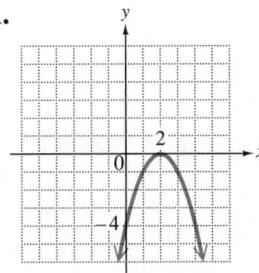

B.
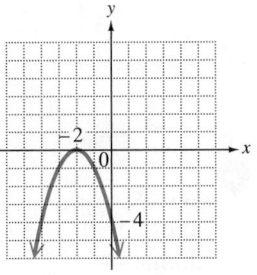

C.
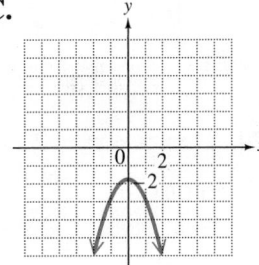

D.
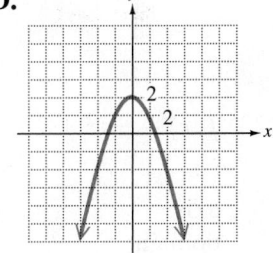

3. CONCEPT CHECK For $f(x) = a(x - h)^2 + k$, in what quadrant is the vertex if the values of h and k are as follows?

(a) $h > 0, k > 0$ **(b)** $h > 0, k < 0$

(c) $h < 0, k > 0$ **(d)** $h < 0, k < 0$

Consider the value of a.

(e) If $|a| > 1$, then the graph is (*narrower* / *wider*) than the graph of $f(x) = x^2$.

(f) If $0 < |a| < 1$, then the graph is (*narrower* / *wider*) than the graph of $f(x) = x^2$.

4. CONCEPT CHECK Match each quadratic function with the description of the parabola that is its graph.

(a) $f(x) = (x - 4)^2 - 2$ **A.** Vertex $(2, -4)$, opens down

(b) $f(x) = (x - 2)^2 - 4$ **B.** Vertex $(2, -4)$, opens up

(c) $f(x) = -(x - 4)^2 - 2$ **C.** Vertex $(4, -2)$, opens down

(d) $f(x) = -(x - 2)^2 - 4$ **D.** Vertex $(4, -2)$, opens up

*Identify the vertex of each parabola. **See Examples 1–4.***

5. $f(x) = -3x^2$ **6.** $f(x) = \frac{1}{2}x^2$ **7.** $f(x) = x^2 + 4$ **8.** $f(x) = x^2 - 4$

9. $f(x) = (x - 1)^2$ **10.** $f(x) = (x + 3)^2$ **11.** $f(x) = (x + 3)^2 - 4$

12. $f(x) = (x - 5)^2 - 8$ **13.** $f(x) = -(x - 5)^2 + 6$ **14.** $f(x) = -(x - 2)^2 + 1$

For each quadratic function, tell whether the graph opens up or down and whether the graph is wider, narrower, or the same shape as the graph of $f(x) = x^2$.
See Examples 4 and 5.

15. $f(x) = -\dfrac{2}{5}x^2$ **16.** $f(x) = -2x^2$ **17.** $f(x) = 3x^2 + 1$ **18.** $f(x) = \dfrac{2}{3}x^2 - 4$

*Graph each parabola. Plot at least two points in addition to the vertex. Give the vertex, axis of symmetry, domain, and range. **See Examples 1–5.***

19. $f(x) = -2x^2$
vertex:
axis:
domain:
range:

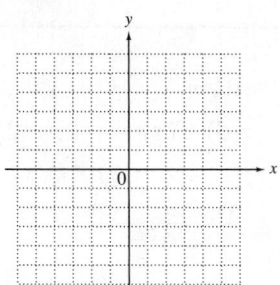

20. $f(x) = -\dfrac{1}{3}x^2$
vertex:
axis:
domain:
range:

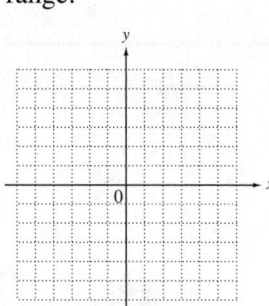

21. $f(x) = x^2 - 1$
vertex:
axis:
domain:
range:

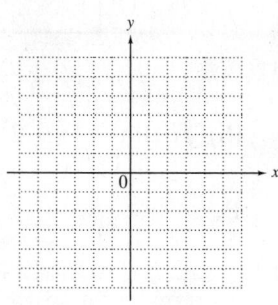

22. $f(x) = x^2 + 3$
vertex:
axis:
domain:
range:

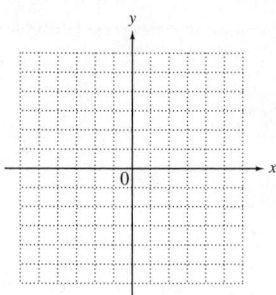

23. $f(x) = -x^2 + 2$
vertex:
axis:
domain:
range:

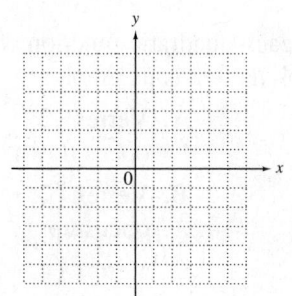

24. $f(x) = -x^2 - 2$
vertex:
axis:
domain:
range:

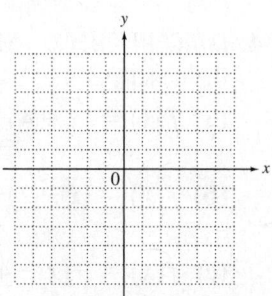

25. $f(x) = (x - 4)^2$
vertex:
axis:
domain:
range:

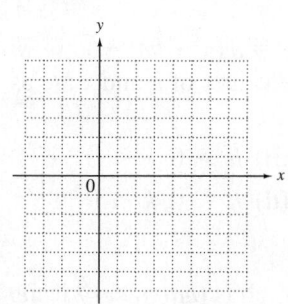

26. $f(x) = (x + 1)^2$
vertex:
axis:
domain:
range:

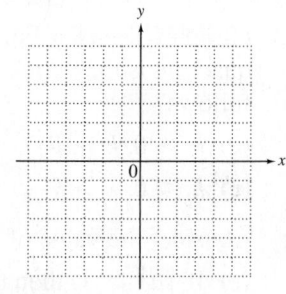

27. $f(x) = (x + 2)^2 - 1$
vertex:
axis:
domain:
range:

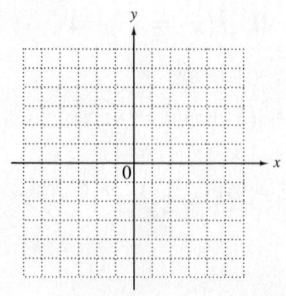

28. $f(x) = (x - 1)^2 + 2$
vertex:
axis:
domain:
range:

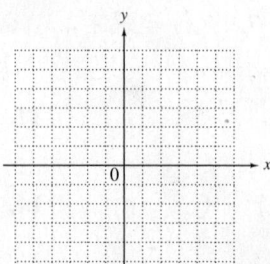

29. $f(x) = 2(x - 2)^2 - 3$
vertex:
axis:
domain:
range:

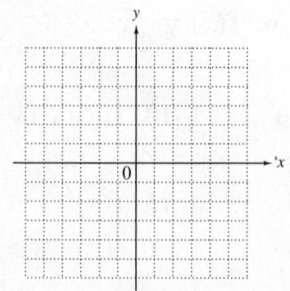

30. $f(x) = 3(x - 2)^2 + 1$
vertex:
axis:
domain:
range:

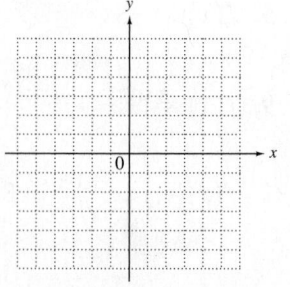

31. $f(x) = -2(x + 3)^2 + 4$ **32.** $f(x) = -2(x - 2)^2 - 3$ **33.** $f(x) = -\dfrac{2}{3}(x + 2)^2 + 1$ **34.** $f(x) = -\dfrac{1}{2}(x + 1)^2 + 2$

vertex: vertex: vertex: vertex:

axis: axis: axis: axis:

domain: domain: domain: domain:

range: range: range: range:

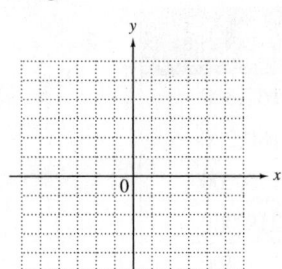 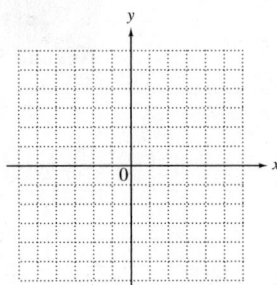 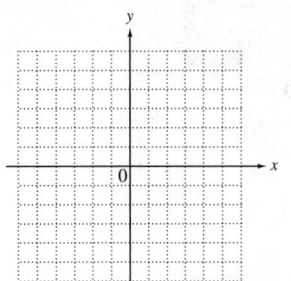 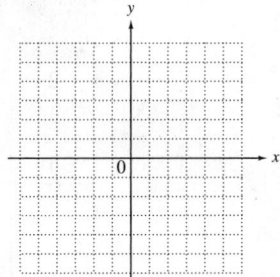

In Exercises 35–40, tell whether a linear *or* quadratic *function would be a more appropriate model for each set of graphed data. If linear, tell whether the slope should be* positive *or* negative. *If quadratic, tell whether the coefficient a of x^2 should be* positive *or* negative. **See Example 6.**

35.   PLASMA TV SALES IN U.S.   **36.**   AVERAGE DAILY VOLUME OF FIRST-CLASS MAIL   **37.**   SOCIAL SECURITY ASSETS*

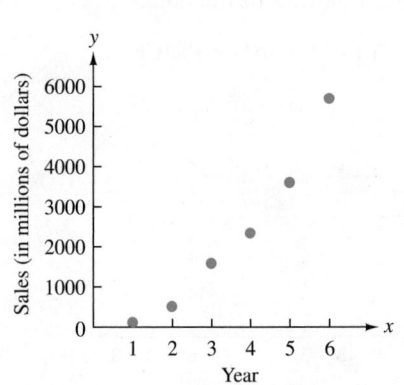 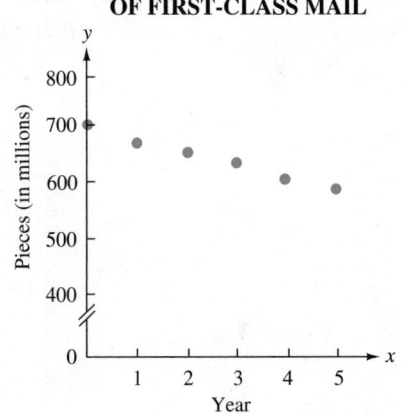 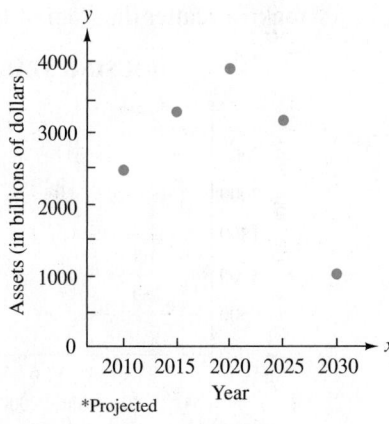

Source: Consumer Electronics Association.   ***Source:*** General Accounting Office.   ***Source:*** Social Security Administration.

38.   FOOD ASSISTANCE SPENDING IN IOWA   **39.**   AVERAGE MONTHLY BASIC CABLE RATE   **40.**   SMARTPHONE SALES IN U.S.

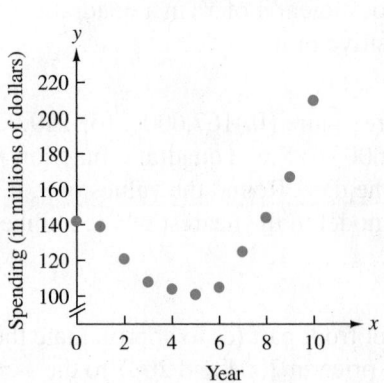 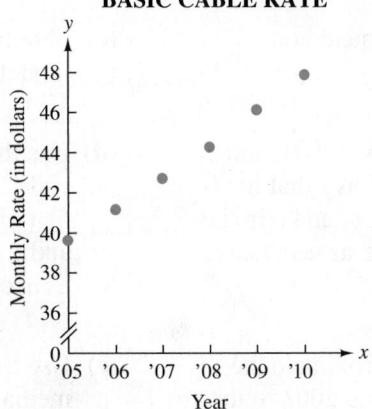 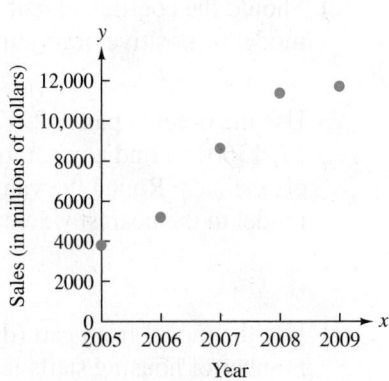

Source: Iowa Department of Human Services.   ***Source:*** SNL Kagan.   ***Source:*** Consumer Electronics Association.

Solve each problem. See Example 6.

41. The number (in thousands) of new, privately owned housing units started in the United States is shown in the table for the years 2002–2009. In the year column, 2 represents 2002, 3 represents 2003, and so on.

Year	Housing Starts (in thousands)
2	1700
3	1850
4	1960
5	2070
6	1800
7	1360
8	910
9	580

Source: U.S. Census Bureau.

(a) Use the ordered pairs (year, housing starts) to make a scatter diagram of the data.

HOUSING STARTS

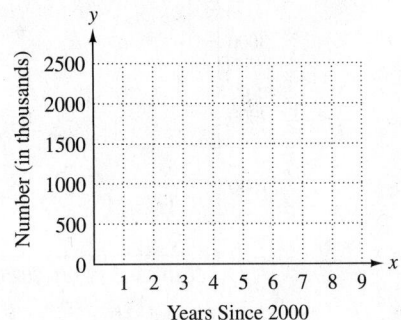

(b) Would a linear or quadratic function better model the data?

(c) Should the coefficient a of x^2 in a quadratic model be positive or negative?

(d) Use the ordered pairs (2, 1700), (4, 1960), and (7, 1360) to find a quadratic function f that models the data. Round the values of a, b, and c in the model to the nearest whole number, as necessary.

(e) Use the model from part (d) to approximate the number of housing starts in 2003 and 2008 to the nearest thousand. How well does the model approximate the actual data from the table?

42. Median sales prices (in dollars) for an existing single-family home in the United States over the period 2002–2009 are shown in the table. In the year column, 0 represents 2002, 1 represents 2003, and so on.

Year	Median Sales Price (in dollars)
0	167,600
1	180,200
2	195,200
3	219,000
4	221,900
5	217,900
6	196,600
7	172,100

Source: National Association of Realtors.

(a) Use the ordered pairs (year, median sales price) to make a scatter diagram of the data.

MEDIAN SALES PRICE

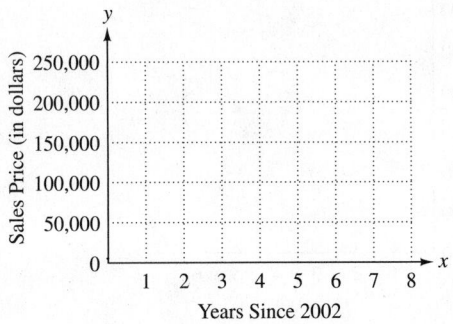

(b) Would a linear or quadratic function better model the data?

(c) Should the coefficient a of x^2 in a quadratic model be positive or negative?

(d) Use the ordered pairs (0, 167,600), (3, 219,000), and (6, 196,600) to find a quadratic function f that models the data. Round the values of a, b, and c in the model to the nearest whole number, as necessary.

(e) Use the model from part (d) to approximate the median sales price in 2004 and 2007 to the nearest hundred dollars. How well does the model approximate the actual data from the table?

10.7 More about Parabolas and Their Applications

OBJECTIVE ▶ 1 Find the vertex of a vertical parabola. When the equation of a parabola is given in the form $f(x) = ax^2 + bx + c$, there are two ways to locate the vertex.

1. Complete the square (**Examples 1 and 2**).

2. Use a formula derived by completing the square (**Example 3**).

EXAMPLE 1 Completing the Square to Find the Vertex ($a = 1$)

Find the vertex of the graph of $f(x) = x^2 - 4x + 5$.

To find the vertex, we need to write the expression $x^2 - 4x + 5$ in the form $(x - h)^2 + k$. We do this by completing the square on $x^2 - 4x$, as in **Section 10.2.** The process is slightly different here because we want to keep $f(x)$ alone on one side of the equation. Instead of adding the appropriate number to each side, we *add and subtract* it on the right.

$f(x) = x^2 - 4x + 5$

$f(x) = (x^2 - 4x\quad) + 5$ \qquad Group the variable terms.

This is equivalent to adding 0. → $\left[\frac{1}{2}(-4)\right]^2 = (-2)^2 = 4$ \quad Square half the coefficient of the first-degree term.

$f(x) = (x^2 - 4x + 4 - 4) + 5$ \quad Add and subtract 4.

$f(x) = (x^2 - 4x + 4) - 4 + 5$ \quad Bring -4 outside the parentheses.

$f(x) = (x - 2)^2 + 1$ \quad Factor. Combine like terms.

The vertex of this parabola is $(2, 1)$.

······· Work Problem **1** at the Side. ▶

1 Find the vertex of the graph of each quadratic function.

(a) $f(x) = x^2 - 6x + 7$

(b) $f(x) = x^2 + 4x - 9$

EXAMPLE 2 Completing the Square to Find the Vertex ($a \neq 1$)

Find the vertex of the graph of $f(x) = -3x^2 + 6x - 1$.

Because the x^2-term has a coefficient other than 1, we factor that coefficient out of the first two terms before completing the square.

$f(x) = -3x^2 + 6x - 1$

$f(x) = (-3x^2 + 6x) - 1$ \quad Group the variable terms.

$f(x) = -3(x^2 - 2x) - 1$ \quad Factor out -3.

$f(x) = -3(x^2 - 2x\quad) - 1$ \quad Prepare to complete the square.

→ $\left[\frac{1}{2}(-2)\right]^2 = (-1)^2 = 1$ \quad Square half the coefficient of the first-degree term.

$f(x) = -3(x^2 - 2x + 1 - 1) - 1$ \quad Add and subtract 1.

Now bring -1 outside the parentheses. Be sure to multiply it by -3.

$f(x) = -3(x^2 - 2x + 1) + (-3)(-1) - 1$ \quad Distributive property

$f(x) = -3(x^2 - 2x + 1) + 3 - 1$ \quad This is a key step.

$f(x) = -3(x - 1)^2 + 2$ \quad Factor. Combine like terms.

The vertex is $(1, 2)$.

······· Work Problem **2** at the Side. ▶

2 Find the vertex of the graph of each quadratic function.

(a) $f(x) = 2x^2 - 4x + 1$

(b) $f(x) = -\frac{1}{2}x^2 + 2x - 3$

Answers

1. (a) $(3, -2)$ (b) $(-2, -13)$
2. (a) $(1, -1)$ (b) $(2, -1)$

3 Use the vertex formula to find the vertex of the graph of each quadratic function.

(a) $f(x) = -2x^2 + 3x - 1$

We complete the square to derive a formula for the vertex of the graph of the quadratic function $f(x) = ax^2 + bx + c$ (where $a \neq 0$).

$$f(x) = ax^2 + bx + c \qquad \text{Standard form}$$

$$f(x) = (ax^2 + bx) + c \qquad \text{Group the terms with } x.$$

$$f(x) = a\left(x^2 + \frac{b}{a}x \quad\right) + c \qquad \text{Factor } a \text{ from the first two terms.}$$

$$\left[\frac{1}{2}\left(\frac{b}{a}\right)\right]^2 = \left(\frac{b}{2a}\right)^2 = \frac{b^2}{4a^2} \qquad \begin{array}{l}\text{Square half the} \\ \text{coefficient of the first-} \\ \text{degree term.}\end{array}$$

$$f(x) = a\left(x^2 + \frac{b}{a}x + \frac{b^2}{4a^2} - \frac{b^2}{4a^2}\right) + c \qquad \text{Add and subtract } \tfrac{b^2}{4a^2}.$$

$$f(x) = a\left(x^2 + \frac{b}{a}x + \frac{b^2}{4a^2}\right) + a\left(-\frac{b^2}{4a^2}\right) + c \qquad \text{Distributive property}$$

$$f(x) = a\left(x^2 + \frac{b}{a}x + \frac{b^2}{4a^2}\right) - \frac{b^2}{4a} + c \qquad -\tfrac{ab^2}{4a^2} = -\tfrac{b^2}{4a}$$

$$f(x) = a\left(x + \frac{b}{2a}\right)^2 + \frac{4ac - b^2}{4a} \qquad \begin{array}{l}\text{Factor. Rewrite terms with a} \\ \text{common denominator.}\end{array}$$

$$f(x) = a\underbrace{\left[x - \left(\frac{-b}{2a}\right)\right]}_{h}^2 + \underbrace{\frac{4ac - b^2}{4a}}_{k} \qquad \begin{array}{l} f(x) = a(x - h)^2 + k \\ \text{The vertex } (h, k) \text{ can be} \\ \text{expressed in terms of } a, b, \text{ and } c. \end{array}$$

(b) $f(x) = 4x^2 - x + 5$

The expression for k can be found by replacing x with $\frac{-b}{2a}$. Using function notation, if $y = f(x)$, then the y-value of the vertex is $f\left(\frac{-b}{2a}\right)$.

Vertex Formula

The graph of the quadratic function $f(x) = ax^2 + bx + c$ has vertex

$$\left(\frac{-b}{2a}, f\left(\frac{-b}{2a}\right)\right).$$

The axis of symmetry of the parabola is the line having equation

$$x = \frac{-b}{2a}.$$

EXAMPLE 3 **Using the Formula to Find the Vertex**

Use the vertex formula to find the vertex of the graph of $f(x) = x^2 - x - 6$.
The x-coordinate of the vertex of the parabola is given by $\frac{-b}{2a}$.

$$\frac{-b}{2a} = \frac{-(-1)}{2(1)} = \frac{1}{2} \quad \leftarrow \text{x-coordinate of vertex} \qquad a = 1, b = -1, \text{ and } c = -6$$

The y-coordinate is $f\left(\frac{-b}{2a}\right) = f\left(\frac{1}{2}\right)$.

$$f\left(\frac{1}{2}\right) = \left(\frac{1}{2}\right)^2 - \frac{1}{2} - 6 = \frac{1}{4} - \frac{1}{2} - 6 = -\frac{25}{4} \quad \leftarrow \text{y-coordinate of vertex}$$

The vertex is $\left(\frac{1}{2}, -\frac{25}{4}\right)$.

◄ **Work Problem 3** at the Side.

Answers

3. **(a)** $\left(\frac{3}{4}, \frac{1}{8}\right)$ **(b)** $\left(\frac{1}{8}, \frac{79}{16}\right)$

OBJECTIVE ➤ 2 **Graph a quadratic function.**

Graphing a Quadratic Function $f(x) = ax^2 + bx + c$
Step 1 **Determine whether the graph opens up or down.** If $a > 0$, then the parabola opens up. If $a < 0$, then it opens down.
Step 2 **Find the vertex.** Use either the vertex formula or completing the square.
Step 3 **Find any intercepts.** To find the x-intercepts (if any), solve $f(x) = 0$. To find the y-intercept, evaluate $f(0)$.
Step 4 **Complete the graph.** Plot the points found so far. Find and plot additional points as needed, using symmetry about the axis.

EXAMPLE 4   Graphing a Quadratic Function

Graph the quadratic function $f(x) = x^2 - x - 6$.

Step 1 From the equation, $a = 1$, so the graph of the function opens up.

Step 2 The vertex, $\left(\frac{1}{2}, -\frac{25}{4}\right)$, was found in **Example 3** by substituting the values $a = 1$, $b = -1$, and $c = -6$ in the vertex formula.

Step 3 Find any intercepts. Since the vertex, $\left(\frac{1}{2}, -\frac{25}{4}\right)$, is in quadrant IV and the graph opens up, there will be two x-intercepts. Let $f(x) = 0$ and solve.

$$f(x) = x^2 - x - 6$$
$$0 = x^2 - x - 6 \qquad \text{Let } f(x) = 0.$$
$$0 = (x - 3)(x + 2) \qquad \text{Factor.}$$
$$x - 3 = 0 \quad \text{or} \quad x + 2 = 0 \qquad \text{Zero-factor property}$$
$$x = 3 \quad \text{or} \qquad x = -2 \qquad \text{Solve each equation.}$$

The x-intercepts are $(3, 0)$ and $(-2, 0)$. Find the y-intercept by evaluating $f(0)$.

$$f(x) = x^2 - x - 6$$
$$f(0) = 0^2 - 0 - 6 \qquad \text{Let } x = 0.$$
$$f(0) = -6 \qquad \text{Apply the exponent. Subtract.}$$

The y-intercept is $(0, -6)$.

Step 4 Plot the points found so far and additional points as needed using symmetry about the axis, $x = \frac{1}{2}$. See **Figure 14.**

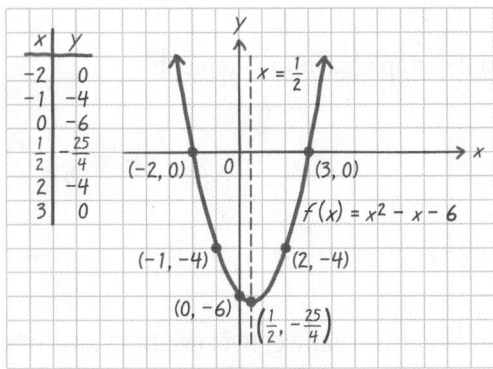

$f(x) = x^2 - x - 6$

Vertex: $\left(\frac{1}{2}, -\frac{25}{4}\right)$

Axis of symmetry: $x = \frac{1}{2}$

Domain: $(-\infty, \infty)$

Range: $\left[-\frac{25}{4}, \infty\right)$

Figure 14

⸽⸽ **Work Problem 4 at the Side.** ▶

---

4 Graph the quadratic function

$$f(x) = x^2 - 6x + 5.$$

Give the vertex, axis of symmetry, domain, and range.

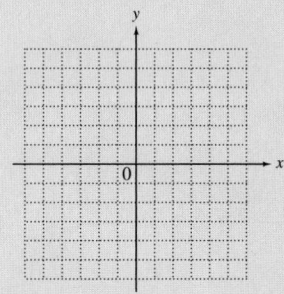

Answer

4.

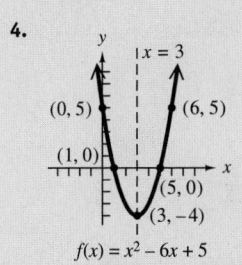

$f(x) = x^2 - 6x + 5$

vertex: $(3, -4)$; axis: $x = 3$; domain: $(-\infty, \infty)$; range: $[-4, \infty)$

5 Find the discriminant and use it to determine the number of x-intercepts of the graph of each quadratic function.

(a) $f(x) = 4x^2 - 20x + 25$

(b) $f(x) = 2x^2 + 3x + 5$

(c) $f(x) = -3x^2 - x + 2$

OBJECTIVE ▶ **3** **Use the discriminant to find the number of x-intercepts of a parabola with a vertical axis.** Recall from **Section 10.3** that

$$b^2 - 4ac \quad \text{Discriminant}$$

is the *discriminant* of the quadratic equation $ax^2 + bx + c = 0$ and that we can use it to determine the number of real solutions of a quadratic equation.

In a similar way, we can use the discriminant of a quadratic *function* to determine the number of x-intercepts of its graph. See **Figure 15**.

1. If the discriminant is positive, the parabola will have two x-intercepts.

2. If the discriminant is 0, there will be only one x-intercept, and it will be the vertex of the parabola.

3. If the discriminant is negative, the graph will have no x-intercepts.

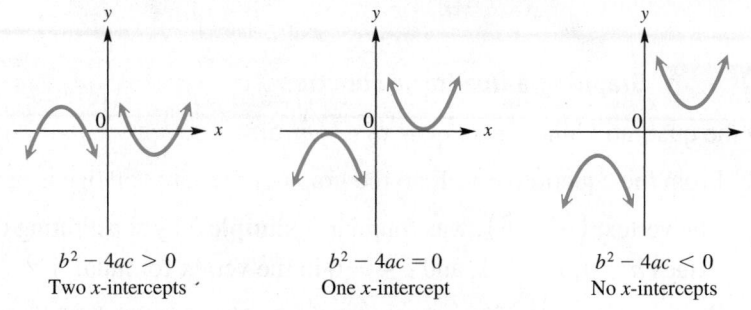

| $b^2 - 4ac > 0$ | $b^2 - 4ac = 0$ | $b^2 - 4ac < 0$ |
| Two x-intercepts | One x-intercept | No x-intercepts |

Figure 15

EXAMPLE 5 **Using the Discriminant to Determine the Number of x-Intercepts**

Find the discriminant and use it to determine the number of x-intercepts of the graph of each quadratic function.

(a) $f(x) = 2x^2 + 3x - 5$

$$b^2 - 4ac \qquad \text{Discriminant}$$
$$= 3^2 - 4(2)(-5) \qquad a = 2, \, b = 3, \, c = -5$$
$$= 9 - (-40) \qquad \text{Apply the exponent. Multiply.}$$
$$= 49 \qquad \text{Subtract.}$$

Because the discriminant is positive, the parabola has two x-intercepts.

(b) $f(x) = -3x^2 - 1$

$$b^2 - 4ac$$
$$= 0^2 - 4(-3)(-1) \qquad a = -3, \, b = 0, \, c = -1$$
$$= 0 - 12 \qquad \text{Apply the exponent. Multiply.}$$
$$= -12 \qquad \text{Subtract.}$$

The discriminant is negative, so the graph has no x-intercepts.

(c) $f(x) = 9x^2 + 6x + 1$

$$b^2 - 4ac$$
$$= 6^2 - 4(9)(1) \qquad a = 9, \, b = 6, \, c = 1$$
$$= 36 - 36 \qquad \text{Apply the exponent. Multiply.}$$
$$= 0 \qquad \text{Subtract.}$$

The parabola has only one x-intercept (its vertex) because the value of the discriminant is 0.

◀ **Work Problem** **5** at the Side.

OBJECTIVE ▶ ④ **Use quadratic functions to solve problems involving maximum or minimum value.** The vertex of the graph of a quadratic functions is either the highest or the lowest point on the parabola. It provides the following information.

1. The *y*-value of the vertex gives the maximum or minimum value of *y*.

2. The *x*-value tells where the maximum or minimum occurs.

Problem-Solving Hint

In many applied problems we must find the least or greatest value of some quantity. When we can express that quantity as a quadratic function, the value of *k* in the vertex (h, k) gives that optimum value.

EXAMPLE 6   **Finding Maximum Area**

A farmer has 120 ft of fencing to enclose a rectangular area next to a building. See **Figure 16.** Find the maximum area he can enclose and the dimensions of the field when the area is maximized.

Figure 16

Let *x* = the width of the rectangle.

$$x + x + \text{length} = 120 \qquad \text{Sum of the sides is 120 ft.}$$
$$2x + \text{length} = 120 \qquad \text{Combine like terms.}$$
$$\text{length} = \mathbf{120 - 2x} \qquad \text{Subtract } 2x.$$

The area $A(x)$ is given by the product of the length and width.

$$A(x) = (\mathbf{120 - 2x})x \qquad \text{Area = length · width}$$
$$A(x) = 120x - 2x^2 \qquad \text{Distributive property}$$

To determine the maximum area, use the vertex formula to find the vertex of the parabola given by $A(x) = 120x - 2x^2$. Write the equation in standard form.

$$A(x) = -2x^2 + 120x \qquad a = -2, \ b = 120, \ c = 0$$

Then

$$x = \frac{-b}{2a} = \frac{-120}{2(-2)} = \frac{-120}{-4} = \mathbf{30},$$

and

$$A(30) = -2(30)^2 + 120(30) = -2(900) + 3600 = \mathbf{1800}.$$

The graph is a parabola that opens down, and its vertex is $(\mathbf{30, 1800})$. Thus, the maximum area will be $\mathbf{1800}$ ft^2. This area will occur if *x*, the width of the rectangle, is $\mathbf{30}$ ft and the length is $120 - 2(30) = \mathbf{60}$ ft.

⋯⋯⋯⋯⋯⋯⋯⋯⋯⋯⋯⋯⋯ **Work Problem ⑥ at the Side.** ▶

⑥ Solve **Example 6** if the farmer has only 100 ft of fencing.

Answer

6. The field should be 25 ft by 50 ft with maximum area 1250 ft^2.

7 Solve the problem.

A toy rocket is launched from the ground so that its distance in feet above the ground after t seconds is

$$s(t) = -16t^2 + 208t.$$

Find the maximum height it reaches and the number of seconds it takes to reach that height.

CAUTION

Be careful when interpreting the meanings of the coordinates of the vertex. The first coordinate, x, gives the value for which the *function value*, y or $f(x)$, is a maximum or a minimum.

Read the problem carefully to determine whether to find the value of the independent variable, the function value, or both.

EXAMPLE 7 **Finding Maximum Height**

If air resistance is neglected, a projectile on Earth shot straight upward with an initial velocity of 40 m per sec will be at a height s in meters given by

$$s(t) = -4.9t^2 + 40t,$$

where t is the number of seconds elapsed after projection. After how many seconds will it reach its maximum height, and what is this maximum height?

For this function, $a = -4.9$, $b = 40$, and $c = 0$. Use the vertex formula.

$$t = \frac{-b}{2a} = \frac{-40}{2(-4.9)} \approx 4.1 \quad \text{Use a calculator.}$$

This indicates that the maximum height is attained at 4.1 sec. To find this maximum height, calculate $s(4.1)$.

$$s(t) = -4.9t^2 + 40t$$

$$s(4.1) = -4.9(4.1)^2 + 40(4.1) \quad \text{Let } t = 4.1.$$

$$s(4.1) \approx 81.6 \quad \text{Use a calculator.}$$

The projectile will attain a maximum height of approximately 81.6 m.

⋯⋯⋯⋯⋯⋯⋯⋯⋯⋯⋯⋯⋯⋯⋯⋯⋯⋯⋯⋯ ◀ **Work Problem** **7** **at the Side.**

OBJECTIVE ▶ **5** **Graph parabolas with horizontal axes.** If x and y are interchanged in the equation

$$y = ax^2 + bx + c,$$

the equation becomes

$$x = ay^2 + by + c.$$

Because of the interchange of the roles of x and y, these parabolas are horizontal (with horizontal lines as axes of symmetry).

General Characteristics of the Graph of a Horizontal Parabola

The graph of $x = ay^2 + by + c$ or $x = a(y - k)^2 + h$ is a parabola.

- The vertex of the parabola is (h, k).
- The axis of symmetry is the horizontal line $y = k$.
- The graph opens to the right if $a > 0$ and to the left if $a < 0$.

Answer

7. 676 ft; 6.5 sec

Graphing a Horizontal Parabola ($a = 1$)

Graph $x = (y - 2)^2 - 3$.

This graph has its vertex at $(-3, 2)$, since the roles of x and y are interchanged. It opens to the right (the positive x-direction) because $a = 1$ and $1 > 0$, and has the same shape as $y = x^2$ (but situated horizontally). Plotting a few additional points gives the graph shown in **Figure 17.**

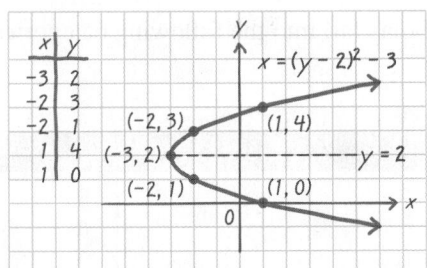

x	y
-3	2
-2	3
-2	1
1	4
1	0

$x = (y - 2)^2 - 3$

Vertex: $(-3, 2)$

Axis of symmetry: $y = 2$

Domain: $[-3, \infty)$

Range: $(-\infty, \infty)$

Figure 17

········· **Work Problem ❽ at the Side.** ▶

When a quadratic equation is given in the form $x = ay^2 + by + c$, we can complete the square on y to find the vertex.

Completing the Square to Graph a Horizontal Parabola ($a \neq 1$)

Graph $x = -2y^2 + 4y - 3$.

$x = -2y^2 + 4y - 3$

$x = (-2y^2 + 4y) - 3$ Group the variable terms.

$x = -2(y^2 - 2y \quad) - 3$ Factor out -2.

$x = -2(y^2 - 2y + 1 - 1) - 3$ Complete the square within the parentheses. Add and subtract 1.

$x = -2(y^2 - 2y + 1) + (-2)(-1) - 3$ Distributive property

> Be careful here.

$x = -2(y - 1)^2 - 1$ Factor. Simplify.

Because of the negative coefficient -2 in $x = -2(y - 1)^2 - 1$, the graph opens to the left (the negative x-direction). The graph is narrower than the graph of $y = x^2$ because $|-2| = 2$ and $2 > 1$. See **Figure 18.**

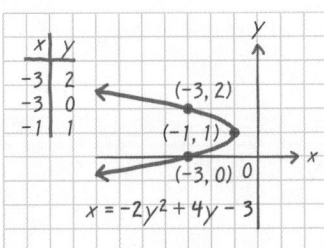

x	y
-3	2
-3	0
-1	1

$x = -2y^2 + 4y - 3$

Vertex: $(-1, 1)$

Axis of symmetry: $y = 1$

Domain: $(-\infty, -1]$

Range: $(-\infty, \infty)$

Figure 18

········· **Work Problem ❾ at the Side.** ▶

❽ Graph $x = (y + 1)^2 - 4$. Give the vertex, axis of symmetry, domain, and range.

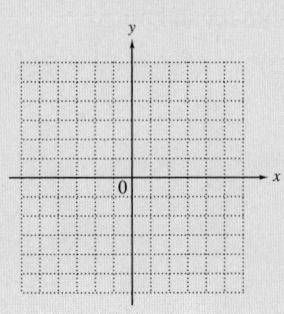

❾ Graph $x = -y^2 + 2y + 5$. Give the vertex, axis of symmetry, domain, and range.

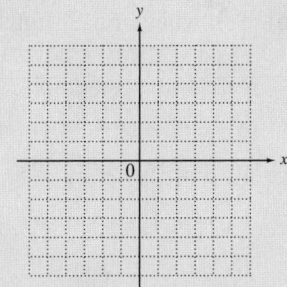

Answers

8.

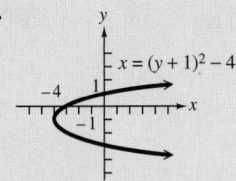

vertex: $(-4, -1)$; axis: $y = -1$;
domain: $[-4, \infty)$; range: $(-\infty, \infty)$

9.

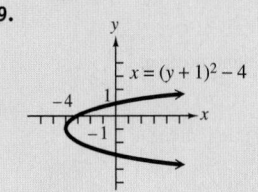

vertex: $(6, 1)$; axis: $y = 1$;
domain: $(-\infty, 6]$; range: $(-\infty, \infty)$

10 Find the vertex of each parabola. Tell whether the graph opens to the right or to the left. Give the domain and range.

(a) $x = 2y^2 - 6y + 5$

(b) $x = -3y^2 - 6y - 5$

11 Refer to the table on graphs of parabolas as needed.

(a) Tell whether each equation has a vertical or horizontal parabola as its graph.

A. $y = -x^2 + 20x + 80$

B. $x = 2y^2 + 6y + 5$

C. $x + 1 = (y + 2)^2$

D. $f(x) = (x - 4)^2$

(b) Which of the equations in part (a) represent functions?

> **CAUTION**
>
> *Quadratic equations solved for y (whose graphs are vertical parabolas) are examples of functions.* The horizontal parabolas given in **Examples 8 and 9** are *not* graphs of functions, because they do not satisfy the conditions of the vertical line test.

In summary, the graphs of parabolas fall into the following categories.

GRAPHS OF PARABOLAS

Equation	Graph
$y = ax^2 + bx + c$ or $y = a(x - h)^2 + k$	 These graphs represent functions.
$x = ay^2 + by + c$ or $x = a(y - k)^2 + h$	 These graphs are not graphs of functions.

◀ **Work Problems 10 and 11 at the Side.**

10.7 Exercises

 Download the MyDashBoard App ▶ MyMathLab®

CONCEPT CHECK *Answer each question.*

1. How can we determine just by looking at the equation of a parabola whether it has a vertical or a horizontal axis of symmetry?

2. Why can't the graph of a quadratic function be a parabola with a horizontal axis of symmetry?

Find the vertex of each parabola. **See Examples 1–3.**

3. $f(x) = x^2 + 8x + 10$

4. $f(x) = x^2 + 10x + 23$

5. $f(x) = -2x^2 + 4x - 5$

6. $f(x) = -3x^2 + 12x - 8$

7. $f(x) = x^2 + x - 7$

8. $f(x) = x^2 - x + 5$

Find the vertex of each parabola. For each equation, decide whether the graph opens up, down, to the left, or to the right, and whether it is wider, narrower, or the same shape as the graph of $y = x^2$. If it is a parabola with a vertical axis of symmetry, use the discriminant to determine the number of x-intercepts. **See Examples 1–3, 5, 8, and 9.**

9. $f(x) = 2x^2 + 4x + 5$

10. $f(x) = 3x^2 - 6x + 4$

11. $f(x) = -x^2 + 5x + 3$

12. $f(x) = -x^2 + 7x - 2$

13. $x = \frac{1}{3}y^2 + 6y + 24$

14. $x = \frac{1}{2}y^2 + 10y - 5$

Graph each parabola. Give the vertex, axis of symmetry, domain, and range.
See Examples 4, 8, and 9.

15. $f(x) = x^2 + 4x + 3$

vertex:
axis:
domain:
range:

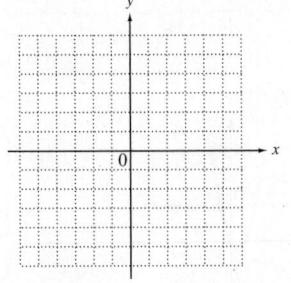

16. $f(x) = x^2 + 2x - 2$

vertex:
axis:
domain:
range:

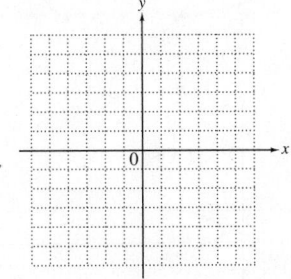

17. $f(x) = -2x^2 + 4x - 5$

vertex:
axis:
domain:
range:

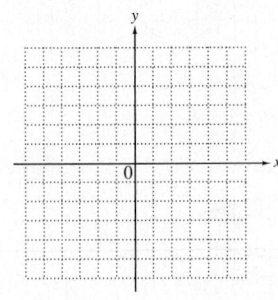

18. $f(x) = -3x^2 + 12x - 8$

vertex:
axis:
domain:
range:

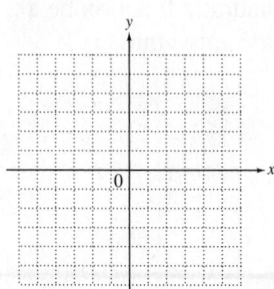

19. $x = (y + 2)^2 + 1$

vertex:
axis:
domain:
range:

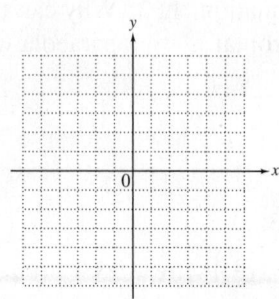

20. $x = (y + 3)^2 - 2$

vertex:
axis:
domain:
range:

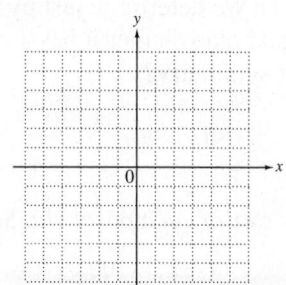

21. $x = -\dfrac{1}{5}y^2 + 2y - 4$

vertex:
axis:
domain:
range:

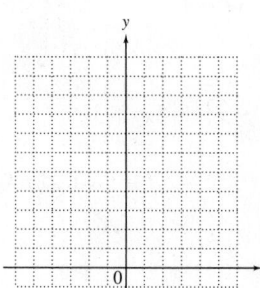

22. $x = -\dfrac{1}{2}y^2 - 4y - 6$

vertex:
axis:
domain:
range:

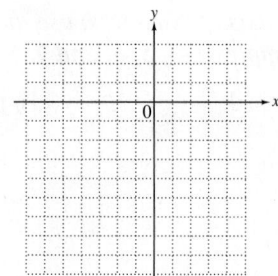

23. $x = 3y^2 + 12y + 5$

vertex:
axis:
domain:
range:

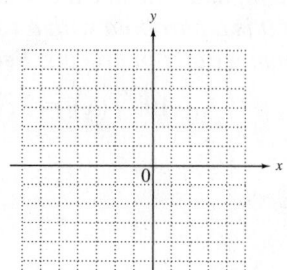

24. $x = 4y^2 + 16y + 11$

vertex:
axis:
domain:
range:

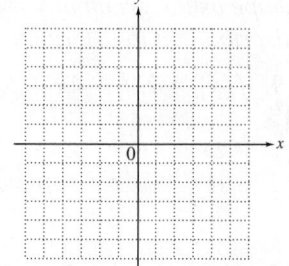

CONCEPT CHECK *Match each equation in Exercises 25–30 with its graph in choices A–F.*

25. $y = 2x^2 + 4x - 3$

26. $y = -x^2 + 3x + 5$

27. $y = -\dfrac{1}{2}x^2 - x + 1$

28. $x = y^2 + 6y + 3$

29. $x = -y^2 - 2y + 4$

30. $x = 3y^2 + 6y + 5$

A.

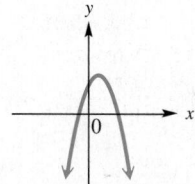

B.

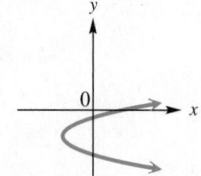

C.

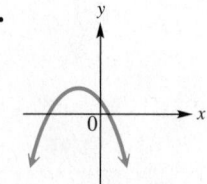

D.

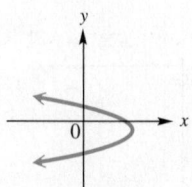

E.

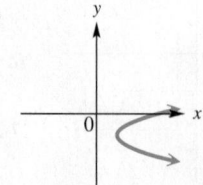

F.

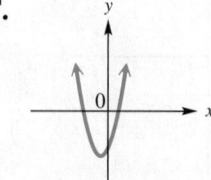

*Solve each problem. **See Examples 6 and 7.***

31. Find the pair of numbers whose sum is 60 and whose product is a maximum. (*Hint:* Let *x* and 60 − *x* represent the two numbers.)

32. Find the pair of numbers whose sum is 10 and whose product is a maximum.

33. Palo Alto College is planning to construct a rectangular parking lot on land bordered on one side by a highway. The plan is to use 640 ft of fencing to fence off the other three sides. What should the dimensions of the lot be if the enclosed area is to be a maximum? What is the maximum area?

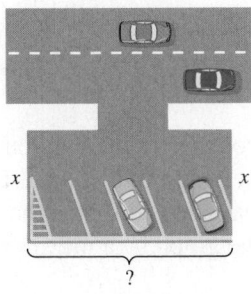

34. Bonnie Wolansky has 100 ft of fencing material to enclose a rectangular exercise run for her dog. One side of the run will border her house, so she will only need to fence three sides. What dimensions will give the enclosure the maximum area? What is the maximum area?

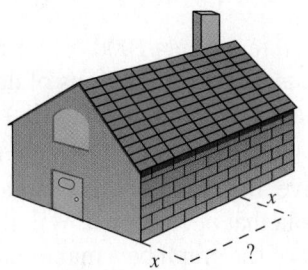

35. Klaus Loewy has a taco stand. He has found that his daily costs are approximated by

$$C(x) = x^2 - 40x + 610,$$

where $C(x)$ is the cost, in dollars, to sell *x* units of tacos. Find the number of units of tacos he should sell to minimize his costs. What is the minimum cost?

36. Mohammad Asghar has a frozen yogurt cart. His daily costs are approximated by

$$C(x) = x^2 - 70x + 1500,$$

where $C(x)$ is the cost, in dollars, to sell *x* units of frozen yogurt. Find the number of units of frozen yogurt he must sell to minimize his costs. What is the minimum cost?

37. A charter flight charges a fare of $200 per person, plus $4 per person for each unsold seat on the plane. If the plane holds 100 passengers and if *x* represents the number of unsold seats, find the following.

(a) A function $R(x)$ that describes the total revenue received for the flight (*Hint:* To find $R(x)$, multiply the number of people flying, 100 − *x*, by the price per ticket, 200 + 4*x*.)

(b) The number of unsold seats that will produce the maximum revenue

(c) The maximum revenue

38. A charter bus company charges a fare of $48 per person, plus $2 per person for each unsold seat on the bus. If the bus has 42 seats and if *x* represents the number of unsold seats, find the following.

(a) A function $R(x)$ that describes the total revenue from the trip (*Hint:* To find $R(x)$, multiply the total number riding, 42 − *x*, by the price per ticket, 48 + 2*x*.)

(b) The number of unsold seats that will produce the maximum revenue

(c) The maximum revenue

39. If an object on Earth is projected upward with an initial velocity of 32 ft per sec, then its height (in feet) after t seconds is given by

$$h(t) = -16t^2 + 32t.$$

Find the maximum height attained by the object and the number of seconds it takes to hit the ground.

40. A projectile on Earth is fired straight upward so that its distance (in feet) above the ground t seconds after firing is given by

$$s(t) = -16t^2 + 400t.$$

Find the maximum height it reaches and the number of seconds it takes to reach that height.

41. The total amount spent by Americans on clothing and footwear in the years 2000–2009 can be modeled by the quadratic function

$$f(x) = -4.979x^2 + 71.73x + 97.29,$$

where $x = 0$ represents 2000, $x = 1$ represents 2001, and so on, and $f(x)$ is in billions of dollars. (*Source:* U.S. Bureau of Economic Analysis.)

(a) Since the coefficient of x^2 in the model is negative, the graph of this quadratic function is a parabola that opens down. Will the y-value of the vertex of the graph be a maximum or a minimum?

(b) According to the model, in what year during this period was the amount spent on clothing and footwear a maximum? (Round down for the year.) Use the actual x-value of the vertex, to the nearest tenth, to find this amount. Round the answer to the nearest billion dollars.

42. The percent of the U.S. population that was foreign-born over the period 1930–2010 can be modeled by the quadratic function

$$f(x) = 0.0043x^2 - 0.3245x + 11.53,$$

where $x = 0$ represents 1930, $x = 10$ represents 1940, and so on. (*Source:* U.S. Census Bureau.)

(a) Since the coefficient of x^2 given in the model is positive, the graph of this quadratic function is a parabola that opens up. Will the y-value of the vertex of the graph be a maximum or a minimum?

(b) According to the model, in what year during this period was the percent of foreign-born population a minimum? (Round down for the year.) Use the actual x-value of the vertex, to the nearest tenth, to find this percent. Round the answer to the nearest tenth of a percent.

The graph shows how Social Security trust fund assets are expected to change, and suggests that a quadratic function would be a good fit to the data. The data are approximated by the function

$$f(x) = -20.57x^2 + 758.9x - 3140.$$

In the model, $x = 10$ represents 2010, $x = 15$ represents 2015, and so on, and $f(x)$ is in billions of dollars.

43. How could we have predicted that this quadratic model would have a negative coefficient for x^2, based only on the graph shown?

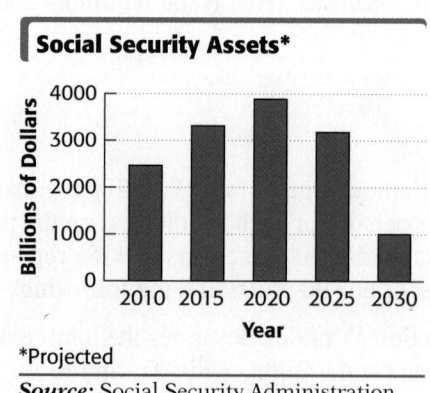

Social Security Assets*

*Projected

Source: Social Security Administration.

44. Algebraically determine the vertex of the graph, with coordinates to four significant digits.

45. Interpret the answer to **Exercise 44** as it applies to the application.

10.8 Polynomial and Rational Inequalities

Now we combine methods of solving linear inequalities and methods of solving quadratic equations to solve *quadratic inequalities*.

OBJECTIVES

1 Solve quadratic inequalities.

2 Solve polynomial inequalities of degree 3 or greater.

3 Solve rational inequalities.

> **Quadratic Inequality**
>
> A **quadratic inequality** can be written in the form
>
> $$ax^2 + bx + c < 0, \qquad ax^2 + bx + c > 0,$$
> $$ax^2 + bx + c \leq 0, \quad \text{or} \quad ax^2 + bx + c \geq 0,$$
>
> where a, b, and c are real numbers, with $a \neq 0$.

OBJECTIVE ▶ 1 **Solve quadratic inequalities.** One method for solving a quadratic inequality is by graphing the related quadratic function.

EXAMPLE 1 Solving Quadratic Inequalities by Graphing

Solve each inequality.

(a) $x^2 - x - 12 > 0$

We graph the related quadratic function $f(x) = x^2 - x - 12$. We are particularly interested in the x-intercepts, which are found as in **Section 10.7** by letting $f(x) = 0$ and solving the quadratic equation.

$$x^2 - x - 12 = 0$$
$$(x - 4)(x + 3) = 0 \qquad \text{Factor.}$$
$$x - 4 = 0 \quad \text{or} \quad x + 3 = 0 \qquad \text{Zero-factor property}$$
$$x = 4 \quad \text{or} \qquad x = -3 \leftarrow \text{The } x\text{-intercepts are } (4, 0) \text{ and } (-3, 0).$$

The graph, which opens up since the coefficient of x^2 is positive, is shown in **Figure 19(a).** Notice that x-values less than -3 or greater than 4 result in y-values *greater than* 0. Therefore, the solution set of $x^2 - x - 12 > 0$, written in interval notation, is

$$(-\infty, -3) \cup (4, \infty).$$

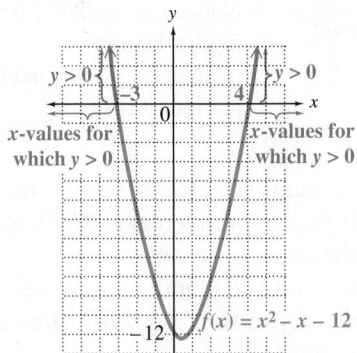

The graph is *above* the x-axis for
$(-\infty, -3) \cup (4, \infty)$.

(a)

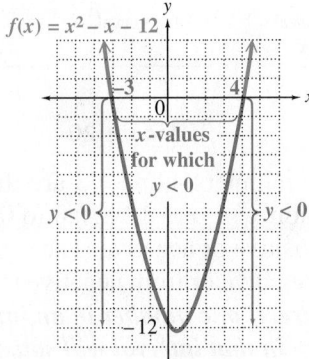

The graph is *below* the x-axis for
$(-3, 4)$.

(b)

Figure 19

················· **Continued on Next Page**

1 Use the graph to solve each quadratic inequality.

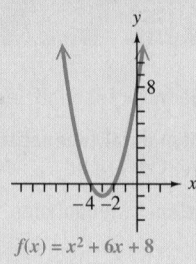

$f(x) = x^2 + 6x + 8$

(a) $x^2 + 6x + 8 > 0$

(b) $x^2 + 6x + 8 < 0$

2 Graph $f(x) = x^2 + 3x - 4$ and use the graph to solve each quadratic inequality.

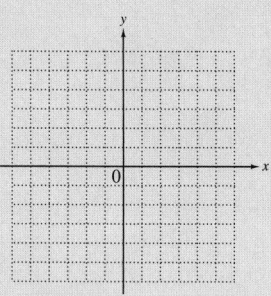

(a) $x^2 + 3x - 4 \geq 0$

(b) $x^2 + 3x - 4 \leq 0$

Answers

1. **(a)** $(-\infty, -4) \cup (-2, \infty)$ **(b)** $(-4, -2)$
2. **(a)** $(-\infty, -4] \cup [1, \infty)$ **(b)** $[-4, 1]$

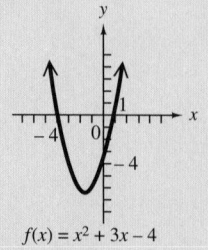

$f(x) = x^2 + 3x - 4$

(b) $x^2 - x - 12 < 0$

Here we want values of y that are *less than* 0. Referring to **Figure 19(b)** on the previous page, we notice from the graph that x-values between -3 and 4 result in y-values less than 0. Therefore, the solution set of the inequality $x^2 - x - 12 < 0$, written in interval notation, is $(-3, 4)$.

Note

If the inequalities in **Example 1** had used $\geq$ and $\leq$, the solution sets would have included the x-values of the intercepts, which make the quadratic expression equal to 0. They would have been written in interval notation as

$$(-\infty, -3] \cup [4, \infty) \quad \text{and} \quad [-3, 4].$$

Square brackets would indicate that the endpoints -3 and 4 are *included* in the solution sets.

◀ **Work Problems 1 and 2 at the Side.**

Another method for solving a quadratic inequality uses these basic ideas without actually graphing the related quadratic function.

EXAMPLE 2 **Solving a Quadratic Inequality Using Test Numbers**

Solve and graph the solution set of $x^2 - x - 12 > 0$.

First solve the quadratic equation $x^2 - x - 12 = 0$ by factoring, as in **Example 1(a)**.

$$x^2 - x - 12 = 0$$
$$(x - 4)(x + 3) = 0 \qquad \text{Factor.}$$
$$x - 4 = 0 \quad \text{or} \quad x + 3 = 0 \qquad \text{Zero-factor property}$$
$$x = 4 \quad \text{or} \qquad x = -3 \qquad \text{Solve each equation.}$$

The numbers 4 and -3 divide a number line into three intervals, as shown in **Figure 20**. *Be careful to put the lesser number on the left.*

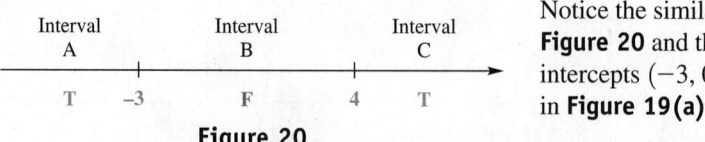

Figure 20

Notice the similarity between **Figure 20** and the x-axis with intercepts $(-3, 0)$ and $(4, 0)$ in **Figure 19(a)**.

The numbers 4 and -3 are the only numbers that make the quadratic expression $x^2 - x - 12$ equal to 0. All other numbers make the expression either positive or negative. The sign of the expression can change from positive to negative or from negative to positive only at a number that makes it 0. *Therefore, if one number in an interval satisfies the inequality, then all the numbers in that interval will satisfy the inequality.*

To see if the numbers in Interval A satisfy the inequality, choose any number from Interval A in **Figure 20** (that is, any number less than -3). Substitute this test number for x in the original inequality $x^2 - x - 12 > 0$.

······ **Continued on Next Page**

We choose -5 from Interval A, and substitute -5 for x.

$$x^2 - x - 12 > 0 \quad \text{Original inequality}$$

$$(-5)^2 - (-5) - 12 \overset{?}{>} 0 \quad \text{Let } x = -5.$$

> Use parentheses to avoid sign errors.

$$25 + 5 - 12 \overset{?}{>} 0 \quad \text{Simplify.}$$

$$18 > 0 \; \checkmark \; \text{True}$$

Because -5 satisfies the inequality, *all* numbers from Interval A are solutions.

Now try 0 from Interval B.

$$x^2 - x - 12 > 0 \quad \text{Original inequality}$$

$$0^2 - 0 - 12 \overset{?}{>} 0 \quad \text{Let } x = 0.$$

$$-12 > 0 \quad \text{False}$$

The numbers in Interval B are *not* solutions.

Work Problem ③ at the Side. ▶

In **Margin Problem 3,** the test number 5 satisfies the inequality, so *all* numbers in Interval C are also solutions.

Based on these results (shown by the colored letters in **Figure 20**), the solution set includes the numbers in Intervals A and C, as shown on the graph in **Figure 21.** The solution set is written in interval notation as

$$(-\infty, -3) \cup (4, \infty).$$

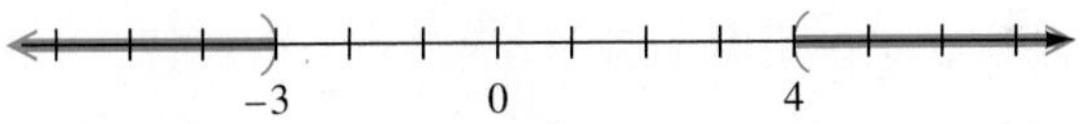

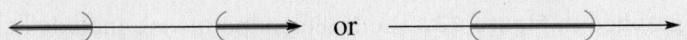

Figure 21

This agrees with the solution set we found in **Example 1(a).**

In summary, follow these steps to solve a quadratic inequality.

Solving a Quadratic Inequality

Step 1 **Write the inequality as an equation and solve it.**

Step 2 **Use the solutions from Step 1 to determine intervals.** Graph the numbers found in Step 1 on a number line. These numbers divide the number line into intervals.

Step 3 **Find the intervals that satisfy the inequality.** Substitute a test number from each interval into the original inequality to determine the intervals that satisfy the inequality. All numbers in those intervals are in the solution set. A graph of the solution set will usually look like one of these. (Square brackets might be used instead of parentheses.)

$\qquad$ or $\qquad$

Step 4 **Consider the endpoints separately.** The numbers from Step 1 are included in the solution set if the inequality is $\leq$ or $\geq$. They are not included if it is $<$ or $>$.

Work Problem ④ at the Side. ▶

③ Does the number 5 from Interval C satisfy $x^2 - x - 12 > 0$?

④ Solve each inequality, and graph the solution set.

(a) $x^2 + x - 6 > 0$

(b) $3m^2 - 13m - 10 \leq 0$

Answers

3. yes

4. (a) $(-\infty, -3) \cup (2, \infty)$

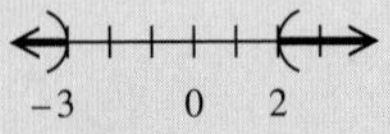

$-3 \quad 0 \quad 2$

(b) $\left[-\dfrac{2}{3}, 5\right]$

$-\dfrac{2}{3}$

$-1 \; 0 \qquad 3 \quad 5 \quad 7$

5 Solve each inequality.

(a) $(3x - 2)^2 > -2$

(b) $(3x - 2)^2 < -2$

6 Solve each inequality, and graph the solution set.

(a) $(x - 3)(x + 2)(x + 1) > 0$

⟶

(b) $(x - 5)(x + 1)(x - 3) \le 0$

⟶

Answers

5. (a) $(-\infty, \infty)$ **(b)** $\emptyset$
6. (a) $(-2, -1) \cup (3, \infty)$

-2 -1 0 1 2 3 4

(b) $(-\infty, -1] \cup [3, 5]$

-1 0 1 3 5

EXAMPLE 3	Solving Special Cases

Solve each inequality.

(a) $(2x - 3)^2 > -1$
  Because $(2x - 3)^2$ is never negative, it is *always* greater than -1. Thus, the solution set of $(2x - 3)^2 > -1$ is the set of all real numbers, $(-\infty, \infty)$.

(b) $(2x - 3)^2 < -1$
  Using similar reasoning as in part (a), there is no solution for this inequality. The solution set is $\emptyset$.

◀ **Work Problem 5** at the Side.

OBJECTIVE ▶ 2 Solve polynomial inequalities of degree 3 or greater. Higher-degree inequalities that have factorable polynomials are solved using a method similar to that of solving quadratic inequalities.

EXAMPLE 4	Solving a Third-Degree Polynomial Inequality

Solve and graph the solution set of $(x - 1)(x + 2)(x - 4) \le 0$.
  This is a *cubic* (third-degree) inequality rather than a quadratic inequality, but it can be solved using the preceding method by extending the zero-factor property to more than two factors (Step 1).

$$(x - 1)(x + 2)(x - 4) = 0 \qquad \text{Set the factored polynomial } \textit{equal} \text{ to } 0.$$

$$x - 1 = 0 \quad \text{or} \quad x + 2 = 0 \quad \text{or} \quad x - 4 = 0 \qquad \text{Zero-factor property}$$

$$x = 1 \quad \text{or} \quad x = -2 \quad \text{or} \quad x = 4 \qquad \text{Solve each equation.}$$

Locate the numbers -2, 1, and 4 on a number line, as in **Figure 22,** to determine the Intervals A, B, C, and D (Step 2).

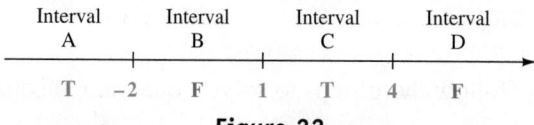

Figure 22

Substitute a test number from each interval in the *original* inequality to determine which intervals satisfy the inequality (Step 3).

Interval	Test Number	Test of Inequality	True or False?
A	-3	$-28 \le 0$	T
B	0	$8 \le 0$	F
C	2	$-8 \le 0$	T
D	5	$28 \le 0$	F

Use a table to organize this information. (Verify it.)

The numbers in Intervals A and C are in the solution set, which is written as

$$(-\infty, -2] \cup [1, 4],$$

and graphed in **Figure 23.** The three endpoints are included since the inequality symbol involves equality (Step 4).

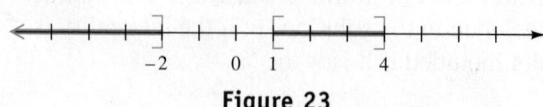

Figure 23

◀ **Work Problem 6** at the Side.

OBJECTIVE ▶ ③ **Solve rational inequalities.** **Rational inequalities** involve rational expressions and are solved similarly using the following steps.

Solving a Rational Inequality

Step 1 **Write the inequality** so that 0 is on one side and there is a single fraction on the other side.

Step 2 **Determine the numbers that make the numerator or denominator equal to 0.**

Step 3 **Divide a number line into intervals.** Use the numbers from Step 2.

Step 4 **Find the intervals that satisfy the inequality.** Test a number from each interval by substituting it into the *original* inequality.

Step 5 **Consider the endpoints separately.** Exclude any values that make the denominator 0.

EXAMPLE 5 **Solving a Rational Inequality**

Solve and graph the solution set of $\dfrac{-1}{x-3} > 1$.

Write the inequality so that 0 is on one side (Step 1).

$$\frac{-1}{x-3} - 1 > 0 \qquad \text{Subtract 1.}$$

$$\frac{-1}{x-3} - \frac{x-3}{x-3} > 0 \qquad \text{Use } x - 3 \text{ as the common denominator.}$$

> **Be careful with signs.**

$$\frac{-1-x+3}{x-3} > 0 \qquad \text{Write the left side as a single fraction.}$$

$$\frac{-x+2}{x-3} > 0 \qquad \text{Combine like terms in the numerator.}$$

The sign of $\frac{-x+2}{x-3}$ will change from positive to negative or negative to positive only at those numbers that make the numerator or denominator 0. The number 2 makes the numerator 0, and 3 makes the denominator 0 (Step 2). These two numbers, 2 and 3, divide a number line into three intervals. See **Figure 24** (Step 3).

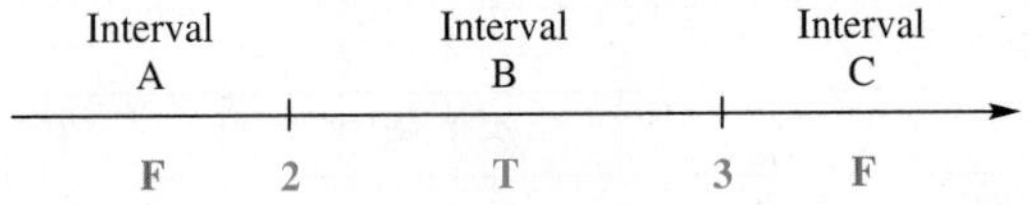

Figure 24

Testing a number from each interval in the *original* inequality, $\frac{-1}{x-3} > 1$, gives the results shown in the table (Step 4).

Interval	Test Number	Test of Inequality	True or False?
A	0	$\frac{1}{3} > 1$	F
B	2.5	$2 > 1$	T
C	4	$-1 > 1$	F

· **Continued on Next Page**

7 Solve each inequality, and graph the solution set.

(a) $\dfrac{2}{x-4} < 3$

(b) $\dfrac{5}{x+1} > 4$

8 Solve and graph the solution set of $\dfrac{x+2}{x-1} \le 5$.

Answers

7. (a) $(-\infty, 4) \cup \left(\dfrac{14}{3}, \infty\right)$

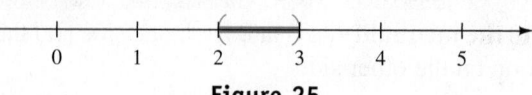

(b) $\left(-1, \dfrac{1}{4}\right)$

8. $(-\infty, 1) \cup \left[\dfrac{7}{4}, \infty\right)$

The solution set of $\dfrac{-1}{x-3} > 1$ is the interval $(2, 3)$. This interval does not include 3 since it would make the denominator of the original inequality 0. The number 2 is not included either, since the inequality symbol $>$ does not involve equality (Step 5). See **Figure 25**.

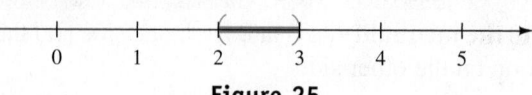

Figure 25

·· ◀ **Work Problem 7** at the Side.

> **CAUTION**
>
> *When solving a rational inequality, any number that makes the denominator 0 must be excluded from the solution set.*

EXAMPLE 6 **Solving a Rational Inequality**

Solve and graph the solution set of $\dfrac{x-2}{x+2} \le 2$.

Write the inequality so that 0 is on one side (Step 1).

$$\dfrac{x-2}{x+2} - 2 \le 0 \qquad \text{Subtract 2.}$$

$$\dfrac{x-2}{x+2} - \dfrac{2(x+2)}{x+2} \le 0 \qquad \text{Use } x+2 \text{ as the common denominator.}$$

$$\dfrac{x-2}{x+2} - \dfrac{2x+4}{x+2} \le 0 \qquad \text{Distributive property}$$

Be careful with signs. $\dfrac{x-2-2x-4}{x+2} \le 0 \qquad \text{Write as a single fraction.}$

$$\dfrac{-x-6}{x+2} \le 0 \qquad \text{Combine like terms in the numerator.}$$

The number -6 makes the numerator 0, and -2 makes the denominator 0 (Step 2). These two numbers determine three intervals (Step 3). Test a number from each interval in the *original* inequality $\frac{x-2}{x+2} \le 2$ (Step 4).

Interval	Test Number	Test of Inequality	True or False?
A	-8	$\frac{5}{3} \le 2$	T
B	-4	$3 \le 2$	F
C	0	$-1 \le 2$	T

The solution set is the interval

$$(-\infty, -6] \cup (-2, \infty).$$

The number -6 satisfies the original inequality, but -2 cannot be included as a solution since it makes the denominator 0 (Step 5). See **Figure 26**.

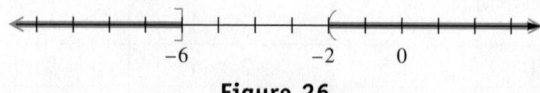

Figure 26

·· ◀ **Work Problem 8** at the Side.

10.8 Exercises

1. Explain how to determine whether to include or exclude endpoints when solving a quadratic or higher-degree inequality.

2. **CONCEPT CHECK** The solution set of the inequality $x^2 + x - 12 < 0$ is the interval $(-4, 3)$. Without actually performing any work, give the solution set of the inequality $x^2 + x - 12 \geq 0$.

*In Exercises 3–6, the graph of a quadratic function f is given. Use the graph to find the solution set of each equation or inequality. **See Example 1.***

3. (a) $x^2 - 4x + 3 = 0$
   (b) $x^2 - 4x + 3 > 0$
   (c) $x^2 - 4x + 3 < 0$

$f(x) = x^2 - 4x + 3$

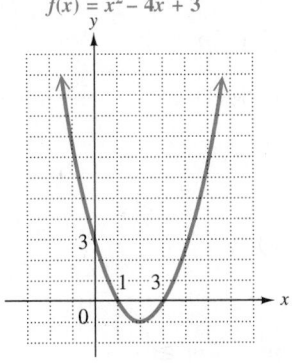

4. (a) $3x^2 + 10x - 8 = 0$
   (b) $3x^2 + 10x - 8 \geq 0$
   (c) $3x^2 + 10x - 8 < 0$

$f(x) = 3x^2 + 10x - 8$

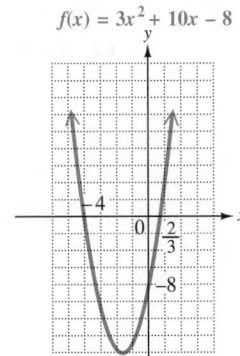

5. (a) $-2x^2 - x + 15 = 0$
   (b) $-2x^2 - x + 15 \geq 0$
   (c) $-2x^2 - x + 15 \leq 0$

$f(x) = -2x^2 - x + 15$

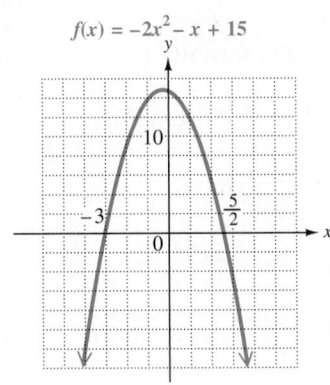

6. (a) $-x^2 + 3x + 10 = 0$
   (b) $-x^2 + 3x + 10 \geq 0$
   (c) $-x^2 + 3x + 10 \leq 0$

$f(x) = -x^2 + 3x + 10$

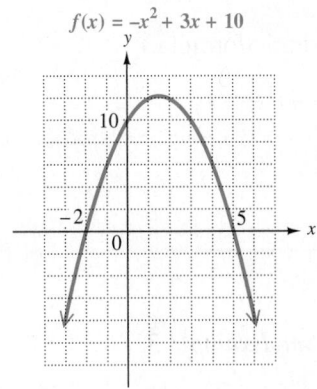

Solve each inequality, and graph the solution set. ***See Example 2.***

7. $(x + 1)(x - 5) > 0$

8. $(m + 6)(m - 2) > 0$

9. $(r + 4)(r - 6) < 0$

10. $(x + 4)(x - 8) < 0$

11. $x^2 - 4x + 3 \geq 0$

12. $m^2 - 3m - 10 \geq 0$

13. $10t^2 + 9t \geq 9$

14. $3r^2 + 10r \geq 8$

15. $9p^2 + 3p < 2$

16. $2x^2 + x < 15$

17. $6x^2 + x \geq 1$

18. $4m^2 + 7m \geq -3$

19. $x^2 - 6x + 6 \geq 0$
(*Hint:* Use the quadratic formula.)

20. $3k^2 - 6k + 2 \leq 0$
(*Hint:* Use the quadratic formula.)

Solve each inequality. ***See Example 3.***

21. $(4 - 3x)^2 \geq -2$

22. $(6p + 7)^2 \geq -1$

23. $(3x + 5)^2 \leq -4$

24. $(8t + 5)^2 \leq -5$

Solve each inequality, and graph the solution set. ***See Example 4.***

25. $(p-1)(p-2)(p-4) < 0$

⊳
┼─┼─┼─┼─┼─┼─┼─┼─┼─►

26. $(2r+1)(3r-2)(4r+7) < 0$

┼─┼─┼─┼─┼─┼─┼─┼─►

27. $(x-4)(2x+3)(3x-1) \geq 0$

┼─┼─┼─┼─┼─┼─┼─┼─┼─►

28. $(z+2)(4z-3)(2z+7) \geq 0$

┼─┼─┼─┼─┼─┼─┼─┼─►

Solve each inequality, and graph the solution set. ***See Examples 5 and 6.***

29. $\dfrac{x-1}{x-4} > 0$

┼─┼─┼─┼─┼─┼─┼─┼─►

30. $\dfrac{x+1}{x-5} > 0$

┼─┼─┼─┼─┼─┼─┼─┼─►

31. $\dfrac{2n+3}{n-5} \leq 0$

┼─┼─┼─┼─┼─┼─┼─┼─►

32. $\dfrac{3t+7}{t-3} \leq 0$

┼─┼─┼─┼─┼─┼─┼─┼─►

33. $\dfrac{8}{x-2} \geq 2$

⊳
┼─┼─┼─┼─┼─┼─┼─┼─►

34. $\dfrac{20}{x-1} \geq 1$

┼─┼─┼─┼─┼─┼─┼─┼─►

35. $\dfrac{3}{2t-1} < 2$

┼─┼─┼─┼─┼─┼─┼─┼─►

36. $\dfrac{6}{m-1} < 1$

┼─┼─┼─┼─┼─┼─┼─┼─►

37. $\dfrac{w}{w+2} \geq 2$

┼─┼─┼─┼─┼─┼─┼─┼─►

38. $\dfrac{m}{m+5} \geq 2$

┼─┼─┼─┼─┼─┼─┼─┼─►

39. $\dfrac{4k}{2k-1} < k$

┼─┼─┼─┼─┼─┼─┼─┼─►

40. $\dfrac{r}{r+2} < 2r$

┼─┼─┼─┼─┼─┼─┼─┼─►

41. $\dfrac{x-3}{x+2} \geq 2$

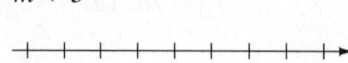

42. $\dfrac{m+4}{m+5} \geq 2$

43. $\dfrac{x-8}{x-4} \leq 3$

44. $\dfrac{2t-3}{t+1} \geq 4$

45. $\dfrac{2x-3}{x^2+1} \geq 0$

46. $\dfrac{9x-8}{4x^2+25} < 0$

Relating Concepts (Exercises 47–50) For Individual or Group Work

A toy rocket is projected vertically upward from the ground. Its distance s in feet above the ground after t seconds is given by the quadratic function

$$s(t) = -16t^2 + 256t.$$

Work Exercises 47–50 in order, *to see how quadratic equations and inequalities are related.*

47. At what times will the rocket be 624 ft above the ground? (*Hint:* Let $s(t) = 624$ and solve the quadratic *equation.*)

48. At what times will the rocket be more than 624 ft above the ground? (*Hint:* Let $s(t) > 624$ and solve the quadratic *inequality.*)

49. At what times will the rocket be at ground level? (*Hint:* Let $s(t) = 0$ and solve the quadratic *equation.*)

50. At what times will the rocket be less than 624 ft above the ground? (*Hint:* Let $s(t) < 624$, solve the quadratic *inequality,* and observe the solutions in **Exercises 48 and 49** to determine the least and greatest possible values of *t.*)

Chapter 10   *Summary*

Key Terms

10.1

quadratic equation A quadratic equation is an equation that can be written in the form $ax^2 + bx + c = 0$, where a, b, and c are real numbers, with $a \neq 0$. This form is called standard form.

second-degree equation An equation with a second-degree term and no terms of greater degree is a second-degree (or quadratic) equation.

10.3

quadratic formula The solutions of any quadratic equation $ax^2 + bx + c = 0$ (where $a \neq 0$) are given by the quadratic formula $x = \dfrac{-b \pm \sqrt{b^2 - 4ac}}{2a}$.

discriminant The discriminant is the expression $b^2 - 4ac$ under the radical in the quadratic formula.

10.4

quadratic in form A nonquadratic equation that can be written as a quadratic equation is quadratic in form.

10.6

quadratic function A function $f(x) = ax^2 + bx + c$, for real numbers a, b, and c, with $a \neq 0$, is a quadratic function.

parabola The graph of a quadratic function is a parabola.

vertex The point on a parabola that has the least y-value (if the parabola opens up) or the greatest y-value (if the parabola opens down) is the vertex of the parabola.

axis of symmetry The vertical (or horizontal) line through the vertex of a vertical (or horizontal) parabola is its axis of symmetry.

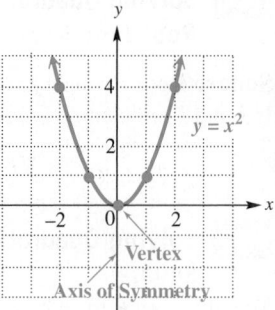

10.8

quadratic inequality A quadratic inequality is an inequality that can be written in the form $ax^2 + bx + c < 0$ or $ax^2 + bx + c > 0$ (or with $\leq$ or $\geq$), where a, b, and c are real numbers, with $a \neq 0$.

rational inequality An inequality that involves a rational expression is a rational inequality.

Test Your Word Power

See how well you have learned the vocabulary in this chapter.

1 The **quadratic formula** is
 A. a formula to find the number of solutions of a quadratic equation
 B. a formula to find the type of solutions of a quadratic equation
 C. the standard form of a quadratic equation
 D. a general formula for solving any quadratic equation.

2 A **quadratic function** is a function that can be written in the form
 A. $f(x) = mx + b$, for real numbers m and b
 B. $f(x) = \dfrac{P(x)}{Q(x)}$, where $Q(x) \neq 0$
 C. $f(x) = ax^2 + bx + c$, for real numbers a, b, and c $(a \neq 0)$
 D. $f(x) = \sqrt{x}$, for $x \geq 0$.

3 A **parabola** is the graph of
 A. any equation in two variables
 B. a linear equation
 C. an equation of degree 3
 D. a quadratic equation in two variables.

4 The **vertex** of a parabola is
 A. the point where the graph intersects the y-axis
 B. the point where the graph intersects the x-axis
 C. the lowest point on a parabola that opens up or the highest point on a parabola that opens down
 D. the origin.

5 The **axis of symmetry** of a parabola is
 A. either the x-axis or the y-axis
 B. the vertical line (of a vertical parabola) or the horizontal line (of a horizontal parabola) through the vertex
 C. the lowest or highest point on the graph of a parabola
 D. a line through the origin.

6 A parabola is **symmetric about its axis** since
 A. its graph is near the axis
 B. its graph is a mirror image on each side of the axis
 C. its graph looks different on each side of the axis
 D. its graph intersects the axis.

Answers To Test Your Word Power

1. D; *Example:* The solutions of $ax^2 + bx + c = 0$ (where $a \neq 0$) are given by

$$x = \frac{-b \pm \sqrt{b^2 - 4ac}}{2a}.$$

2. C; *Examples:* $f(x) = x^2 - 2$, $f(x) = (x + 4)^2 + 1$, $f(x) = x^2 - 4x + 5$

3. D; *Examples:* See the figures in the Quick Review for **Sections 10.6 and 10.7.**

4. C; *Example:* The graph of $y = (x + 3)^2$ has vertex $(-3, 0)$, which is the lowest point on the graph.

5. B; *Example:* The axis of symmetry of $y = (x + 3)^2$ is the vertical line $x = -3$.

6. B; *Example:* Since the graph of $y = (x + 3)^2$ is symmetric about its axis $x = -3$, the points $(-2, 1)$ and $(-4, 1)$ are on this graph.

Quick Review

Concepts	Examples

10.1 Solving Quadratic Equations by the Square Root Property

Square Root Property
If x and k are complex numbers and $x^2 = k$, then

$$x = \sqrt{k} \quad \text{or} \quad x = -\sqrt{k}.$$

Solve $(x - 1)^2 = 8$.

$$x - 1 = \sqrt{8} \qquad \text{or} \quad x - 1 = -\sqrt{8}$$
$$x = 1 + 2\sqrt{2} \quad \text{or} \qquad x = 1 - 2\sqrt{2}$$

Solution set: $\left\{ 1 + 2\sqrt{2}, 1 - 2\sqrt{2} \right\}$

10.2 Solving Quadratic Equations by Completing the Square

To solve $ax^2 + bx + c = 0$ (where $a \neq 0$) by completing the square, follow these steps.

Step 1 If $a \neq 1$, divide each side by a.

Step 2 Write the equation with the variable terms on one side and the constant on the other.

Step 3 Take half the coefficient of x and square it.

Step 4 Add the square to each side.

Step 5 Factor the perfect square trinomial, and write it as the square of a binomial. Simplify the other side.

Step 6 Use the square root property to complete the solution.

Solve $2x^2 - 4x - 18 = 0$.

$$x^2 - 2x - 9 = 0 \qquad \text{Divide by 2.}$$
$$x^2 - 2x = 9 \qquad \text{Add 9.}$$
$$\left[\tfrac{1}{2}(-2) \right]^2 = (-1)^2 = 1$$
$$x^2 - 2x + 1 = 9 + 1 \qquad \text{Add 1.}$$
$$(x - 1)^2 = 10 \qquad \text{Factor. Add.}$$
$$x - 1 = \sqrt{10} \qquad \text{or} \quad x - 1 = -\sqrt{10}$$
$$x = 1 + \sqrt{10} \quad \text{or} \qquad x = 1 - \sqrt{10}$$

Solution set: $\left\{ 1 + \sqrt{10}, 1 - \sqrt{10} \right\}$

10.3 Solving Quadratic Equations by the Quadratic Formula

Quadratic Formula
The solutions of $ax^2 + bx + c = 0$ (where $a \neq 0$) are given by

$$x = \frac{-b \pm \sqrt{b^2 - 4ac}}{2a}.$$

The Discriminant
If a, b, and c are integers, then the discriminant, $b^2 - 4ac$, of $ax^2 + bx + c = 0$ determines the number and type of solutions.

Discriminant	Number and Type of Solutions
Positive, the square of an integer	Two rational solutions
Positive, not the square of an integer	Two irrational solutions
Zero	One distinct rational solution
Negative	Two nonreal complex solutions

Solve $3x^2 + 5x + 2 = 0$.

$$x = \frac{-5 \pm \sqrt{5^2 - 4(3)(2)}}{2(3)} \qquad a = 3, b = 5, c = 2$$

$$x = \frac{-5 \pm 1}{6} \qquad \begin{array}{l}\text{Evaluate the radicand;}\\ \sqrt{1} = 1\end{array}$$

$$x = -\frac{2}{3} \quad \text{or} \quad x = -1$$

Solution set: $\left\{ -1, -\frac{2}{3} \right\}$

For $x^2 + 3x - 10 = 0$, the discriminant is

$$3^2 - 4(1)(-10)$$
$$= 49. \qquad \text{Two rational solutions}$$

For $x + 1 = 0$, the discriminant is

$$1^2 - 4(4)(1)$$
$$= -15. \qquad \text{Two nonreal complex solutions}$$

Concepts	Examples

10.4 Equations Quadratic in Form

A nonquadratic equation that can be written in the form

$$au^2 + bu + c = 0,$$

for $a \neq 0$ and an algebraic expression u, is quadratic in form. Substitute u for the expression, solve for u, and then solve for the variable in the expression.

Solve $3(x+5)^2 + 7(x+5) + 2 = 0$.

$$3u^2 + 7u + 2 = 0 \qquad \text{Let } u = x + 5.$$
$$(3u+1)(u+2) = 0 \qquad \text{Factor.}$$
$$u = -\frac{1}{3} \quad \text{or} \quad u = -2 \quad \text{Solve for } u.$$
$$x + 5 = -\frac{1}{3} \quad \text{or} \quad x + 5 = -2 \quad x + 5 = u$$
$$x = -\frac{16}{3} \quad \text{or} \quad x = -7 \quad \text{Subtract 5.}$$

Solution set: $\left\{-7, -\frac{16}{3}\right\}$

10.5 Formulas and Applications

To solve a formula for a second-degree variable, proceed as follows.

(a) If the variable appears only to the second power:
Isolate the second-degree variable on one side of the equation, and then use the square root property.

Solve $A = \dfrac{2mp}{r^2}$ for r.

$$r^2 A = 2mp \qquad \text{Multiply by } r^2.$$
$$r^2 = \frac{2mp}{A} \qquad \text{Divide by } A.$$
$$r = \pm\sqrt{\frac{2mp}{A}} \qquad \text{Square root property}$$
$$r = \frac{\pm\sqrt{2mpA}}{A} \qquad \text{Rationalize the denominator.}$$

(b) If the variable appears to the first and second powers:
Write the equation in standard form, and then use the quadratic formula.

Solve $m^2 + rm = t$ for m.
$$m^2 + rm - t = 0 \qquad \text{Standard form; } a = 1, b = r, c = -t$$
$$m = \frac{-r \pm \sqrt{r^2 - 4(1)(-t)}}{2(1)} = \frac{-r \pm \sqrt{r^2 + 4t}}{2}$$

10.6 Graphs of Quadratic Functions

1. The graph of the quadratic function
$$F(x) = a(x-h)^2 + k \quad (\text{where } a \neq 0)$$
is a parabola with vertex at (h, k) and the vertical line $x = h$ as axis of symmetry.
2. The graph opens up if a is positive and down if a is negative.
3. The graph is wider than the graph of $f(x) = x^2$ if $0 < |a| < 1$ and narrower if $|a| > 1$.

Graph $f(x) = -(x+3)^2 + 1$.

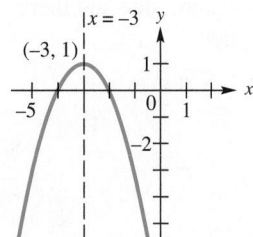

The graph opens down since $a < 0$.
Vertex: $(-3, 1)$
Axis of symmetry: $x = -3$
Domain: $(-\infty, \infty)$
Range: $(-\infty, 1]$

10.7 More about Parabolas and Their Applications

The vertex of the graph of $f(x) = ax^2 + bx + c$ (where $a \neq 0$) has coordinates as follows.

$$\left(\frac{-b}{2a}, f\left(\frac{-b}{2a}\right)\right)$$

Graphing a Quadratic Function

Step 1 Determine whether the graph opens up or down.

Step 2 Find the vertex.

Step 3 Find the x-intercepts (if any). Find the y-intercept.

Step 4 Find and plot additional points as needed.

Graph $f(x) = x^2 + 4x + 3$.

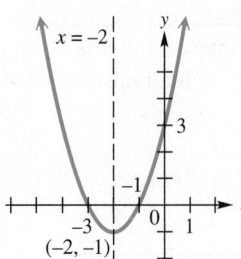

The graph opens up since $a > 0$.
Vertex: $(-2, -1)$
The solutions of $x^2 + 4x + 3 = 0$ are -1 and -3, so the x-intercepts are $(-1, 0)$ and $(-3, 0)$.
$f(0) = 3$, so the y-intercept is $(0, 3)$.
Axis of symmetry: $x = -2$
Domain: $(-\infty, \infty)$
Range: $[-1, \infty)$

(continued)

Concepts	Examples

10.7 **More about Parabolas and Their Applications (continued)**

Horizontal Parabolas

The graph of

$$x = ay^2 + by + c \quad \text{or} \quad x = a(y - k)^2 + h$$

is a horizontal parabola with vertex (h, k) and the horizontal line $y = k$ as axis of symmetry. The graph opens to the right if $a > 0$ and to the left if $a < 0$.

Horizontal parabolas do not represent functions.

Graph $x = 2y^2 + 6y + 5$.

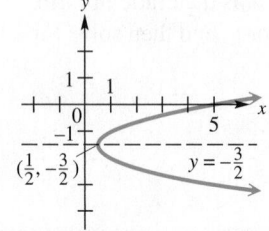

The graph opens to the right since $a > 0$.

Vertex: $\left(\frac{1}{2}, -\frac{3}{2}\right)$

Axis of symmetry: $y = -\frac{3}{2}$

Domain: $\left[\frac{1}{2}, \infty\right)$

Range: $(-\infty, \infty)$

10.8 **Polynomial and Rational Inequalities**

Solving a Quadratic (or Higher-Degree Polynomial) Inequality

Step 1 Write the inequality as an equation and solve.

Solve $2x^2 + 5x + 2 < 0$.

$$2x^2 + 5x + 2 = 0 \quad \text{Related equation}$$
$$(2x + 1)(x + 2) = 0 \quad \text{Factor.}$$
$$x = -\frac{1}{2} \quad \text{or} \quad x = -2$$

Step 2 Use the numbers found in Step 1 to divide a number line into intervals.

Intervals: $(-\infty, -2)$, $\left(-2, -\frac{1}{2}\right)$, $\left(-\frac{1}{2}, \infty\right)$

Step 3 Substitute a number from each interval into the original inequality to determine the intervals that belong in the solution set.

Step 4 Consider the endpoints separately.

Test values: $-3, -1, 0$

$x = -3$ from Interval A makes the original inequality false, $x = -1$ from Interval B makes it true, and from Interval C $x = 0$ makes it false. Choose the interval(s) which yield(s) a true statement. The solution set is the interval $\left(-2, -\frac{1}{2}\right)$.

Solving a Rational Inequality

Step 1 Write the inequality so that 0 is on one side and there is a single fraction on the other side.

Solve $\dfrac{x}{x + 2} \geq 4$.

$$\frac{x}{x + 2} - 4 \geq 0 \quad \text{Subtract 4.}$$
$$\frac{x}{x + 2} - \frac{4(x + 2)}{x + 2} \geq 0 \quad \text{Write with a common denominator.}$$
$$\frac{-3x - 8}{x + 2} \geq 0 \quad \text{Subtract fractions.}$$

Step 2 Determine the numbers that make the numerator or denominator 0.

$-\frac{8}{3}$ makes the numerator 0 and -2 makes the denominator 0.

Step 3 Use the numbers from Step 2 to divide a number line into intervals.

Step 4 Substitute a number from each interval into the original inequality to determine the intervals that belong in the solution set.

-4 from Interval A makes the original inequality false, $-\frac{7}{3}$ from Interval B makes it true, and 0 from Interval C makes it false.

Step 5 Consider the endpoints separately.

The solution set is the interval $\left[-\frac{8}{3}, -2\right)$. Note that $-\frac{8}{3}$ is included because the symbol $\geq$ involves equality, and -2 is excluded since it makes the denominator 0.

Chapter 10 Review Exercises

10.1–10.2 *Solve each equation by using the square root property or by completing the square.*

1. $t^2 = 121$

2. $p^2 = 3$

3. $(r - 3)^2 = 10$

4. $(2x + 5)^2 = 100$

*5. $(3k - 2)^2 = -25$

6. $x^2 + 4x = 15$

7. $2m^2 - 3m = -1$

8. $-x^2 + 5 = 2x$

9. $2x^2 - 3 = -8x$

10. The Singapore Flyer has a height of 165 m. Use the metric version of Galileo's formula, $d = 4.9t^2$ (where d is in meters and t is in seconds), to find how long it would take a wallet dropped from the top of the Singapore Flyer to reach the ground. Round your answer to the nearest tenth of a second. (*Source:* www.singaporeflyer.com)

10.3 *Solve each equation by using the quadratic formula.*

11. $2x^2 + x - 21 = 0$

12. $k^2 + 5k = 7$

13. $(t + 3)(t - 4) = -2$

*14. $2x^2 + 3x + 4 = 0$

*15. $3p^2 = 2(2p - 1)$

16. $m(2m - 7) = 3m^2 + 3$

Find the discriminant. Use it to predict whether the solutions for each equation are

A. *two rational numbers,*

B. *one distinct rational number,*

C. *two irrational numbers,*

D. *two nonreal complex numbers.*

Tell whether the equation can be solved by factoring or whether the quadratic formula should be used. Do not actually solve.

17. $x^2 + 5x + 2 = 0$

18. $4t^2 = 3 - 4t$

19. $9z^2 + 30z + 25 = 0$

20. $4x^2 = 6x - 8$

---

*This exercise requires knowledge of complex numbers.

10.4 *Solve each equation.*

21. $\dfrac{15}{x} = 2x - 1$

22. $\dfrac{1}{n} + \dfrac{2}{n+1} = 2$

23. $-2r = \sqrt{\dfrac{48 - 20r}{2}}$

24. $8(3x+5)^2 + 2(3x+5) - 1 = 0$

25. $2x^{2/3} - x^{1/3} - 28 = 0$

26. $p^4 - 5p^2 + 4 = 0$

Solve each problem. Round answers to the nearest tenth, as necessary.

27. Matthew Sudak drove 8 mi to pick up his cousin Jack, and then drove 11 mi to a mall at a rate 15 mph faster. If Matthew's total travel time was 24 min, what was his rate on the trip to pick up Jack?

28. An old machine processes a batch of checks in 1 hr more time than a new one. How long would it take the old machine to process a batch of checks that the two machines together process in 2 hr?

10.5 *Solve each formula for the indicated variable. (Leave $\pm$ in the answers as needed.)*

29. $k = \dfrac{rF}{wv^2}$ for v

30. $mt^2 = 3mt + 6$ for t

Solve each problem. Round answers to the nearest tenth, as necessary.

31. A large machine requires a part in the shape of a right triangle with a hypotenuse 9 ft less than twice the length of the longer leg. The shorter leg must be $\frac{3}{4}$ the length of the longer leg. Find the lengths of the three sides of the part.

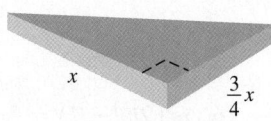

32. A square has an area of 256 cm². If the same amount is removed from one dimension and added to the other, the resulting rectangle has an area 16 cm² less. Find the dimensions of the rectangle.

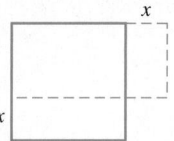

33. Nancy wants to buy a mat for a photograph that measures 14 in. by 20 in. She wants to have an even border around the picture when it is mounted on the mat. If the area of the mat she chooses is 352 in.², how wide will the border be?

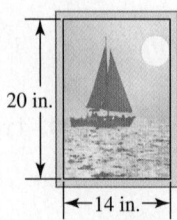

34. Lewis Tower in Philadelphia, Pennsylvania, is 400 ft high. Suppose that a ball is projected upward from the top of the Tower, and its position in feet above the ground is given by the quadratic function

$$f(t) = -16t^2 + 45t + 400,$$

where t is the number of seconds elapsed. How long will it take for the ball to reach a height of 200 ft above the ground? (*Source: World Almanac and Book of Facts.*)

10.6–10.7 *Identify the vertex of the graph of each parabola.*

35. $f(x) = -(x-1)^2$ **36.** $f(x) = (x-3)^2 + 7$ **37.** $y = -3x^2 + 4x - 2$ **38.** $x = (y-3)^2 - 4$

Graph each parabola. Give the vertex, axis of symmetry, domain, and range.

39. $y = 2(x-2)^2 - 3$ **40.** $f(x) = -2x^2 + 8x - 5$ **41.** $x = 2(y+3)^2 - 4$ **42.** $x = -\dfrac{1}{2}y^2 + 6y - 14$

vertex: vertex: vertex: vertex:

axis: axis: axis: axis:

domain: domain: domain: domain:

range: range: range: range:

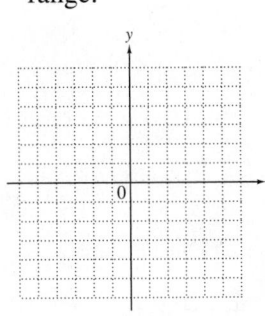 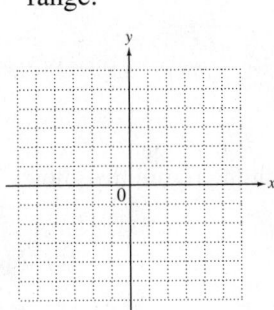 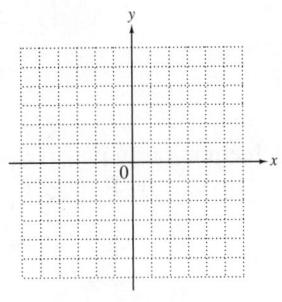 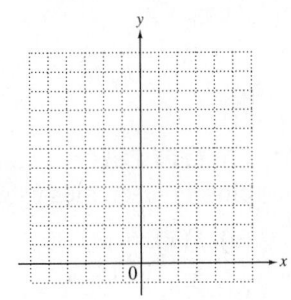

Solve each problem.

43. The height (in feet) of a projectile t seconds after being fired from ground level into the air is given by

$$f(t) = -16t^2 + 160t.$$

Find the number of seconds required for the projectile to reach maximum height. What is the maximum height?

44. Find the length and width of a rectangle having a perimeter of 200 m if the area is to be a maximum. What is the maximum area?

10.8 *Solve each inequality, and graph the solution set.*

45. $(x-4)(2x+3) > 0$ **46.** $x^2 + x \leq 12$ **47.** $(x+2)(x-3)(x+5) \leq 0$

48. $(4m+3)^2 \leq -4$ **49.** $\dfrac{6}{2z-1} < 2$ **50.** $\dfrac{3t+4}{t-2} \leq 1$

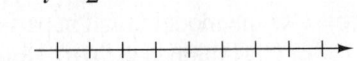

Mixed Review Exercises

Solve each equation or inequality.

51. $V = r^2 + R^2h$ for R
(Leave $\pm$ in the answer.)

*52. $3t^2 - 6t = -4$

53. $(x^2 - 2x)^2 = 11(x^2 - 2x) - 24$

54. $(r - 1)(2r + 3)(r + 6) < 0$

55. $(3k + 11)^2 = 7$

56. $S = \dfrac{Id^2}{k}$ for d
(Leave $\pm$ in the answer.)

57. $2x - \sqrt{x} = 6$

58. $6 + \dfrac{15}{s^2} = -\dfrac{19}{s}$

59. $\dfrac{-2}{x + 5} \le -5$

60. $(8x - 7)^2 \ge -1$

61. CONCEPT CHECK Match each equation in parts (a)–(f) with the figure that most closely resembles its graph in choices A–F.

(a) $g(x) = x^2 - 5$

(b) $h(x) = -x^2 + 4$

(c) $F(x) = (x - 1)^2$

(d) $G(x) = (x + 1)^2$

(e) $H(x) = (x - 1)^2 + 1$

(f) $K(x) = (x + 1)^2 + 1$

A.
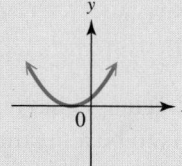

B.
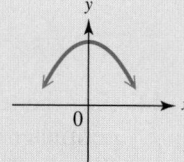

C.
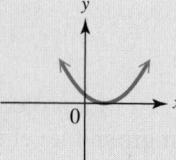

D.
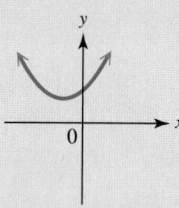

E.
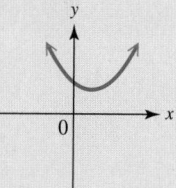

F.
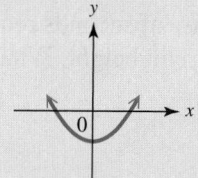

62. The numbers of music CDs (in millions) shipped by U.S. manufacturers for selected years are given in the table. Let $x = 0$ represent 2005, $x = 1$ represent 2006, and so on.

(a) Use the data for 2005, 2007, and 2009 in the quadratic form $ax^2 + bx + c = y$ to write a system of three equations.

(b) Solve the system in part (a) to get a quadratic function f that models the data.

(c) Use the model found in part (b) to approximate the number of CDs shipped in 2010. How does the answer compare to the actual data from the table?

CDs SHIPPED

Year	Number (in millions)
2005	705
2006	620
2007	511
2008	368
2009	293
2010	226

Source: Recording Industry Association of America.

*This exercise requires knowledge of complex numbers.

Chapter 10 **Test**  CHAPTER
Test Prep
VIDEO

The Chapter Test Prep Videos with test solutions are available
on DVD, in MyMathLab, and on YouTube—search
"LialCombinedAlg" and click on "Channels."

Solve by using either the square root property or by completing the square.

1. $t^2 = 54$

2. $(7x + 3)^2 = 25$

3. $x^2 + 2x = 1$

Solve by using the quadratic formula.

4. $2x^2 - 3x - 1 = 0$

***5.** $3t^2 - 4t = -5$

6. $3x = \sqrt{\dfrac{9x + 2}{2}}$

***7.** If k is a negative number, then which one of the following equations will
have two nonreal complex solutions?

A. $x^2 = 4k$ **B.** $x^2 = -4k$ **C.** $(x + 2)^2 = -k$ **D.** $x^2 + k = 0$

8. What is the discriminant for $2x^2 - 8x - 3 = 0$? How many and what type
of solutions does this equation have? (Do not actually solve.)

Solve by any method.

9. $3 - \dfrac{16}{x} - \dfrac{12}{x^2} = 0$

10. $4x^2 + 7x - 3 = 0$

11. $9x^4 + 4 = 37x^2$

12. $12 = (2n + 1)^2 + (2n + 1)$

13. $S = 4\pi r^2$ for r
(Leave $\pm$ in the answer.)

Solve each problem.

14. Andrew and Kent do desktop publishing. Kent can
prepare a certain prospectus 2 hr faster than Andrew.
If they work together, they can do the entire
prospectus in 5 hr. How long will it take each of them
working alone to prepare the prospectus? Round
answers to the nearest tenth of an hour.

15. Bryn Ruhberg paddled her canoe 10 mi upstream, and
then paddled back to her starting point. If the rate of
the current was 3 mph and the entire trip took $3\frac{1}{2}$ hr,
what was Bryn's rate?

16. Tyler McGinnis has a pool 24 ft long and 10 ft wide.
He wants to construct a concrete walk around the
pool. If he plans for the walk to be of uniform width
and cover 152 ft², what will the width of the walk be?

17. At a point 30 m from the base of a tower, the distance
to the top of the tower is 2 m more than twice the
height of the tower. Find the height of the tower.

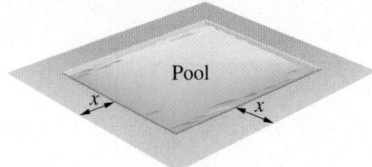

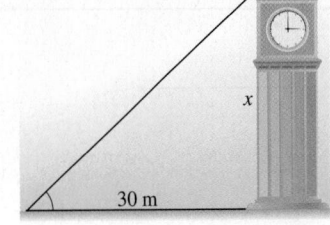

---

*This exercise requires knowledge of complex numbers.

18. Professor Bernstein has found that the number of students attending her intermediate algebra class is approximated by

$$S(x) = -x^2 + 20x + 80,$$

where x is the number of hours that the Campus Center is open daily.

(a) Find the number of hours that the center should be open so that the number of students attending class is a maximum.

(b) What is this maximum number of students?

19. The manager of Morgan's Department Store wants to construct a rectangular parking lot on land bordered on one side by a highway. The store has 280 ft of fencing that is to be used to fence off the other three sides. What should be the dimensions of the lot if the enclosed area is to be a maximum? What is the maximum area?

20. Which one of the following most closely resembles the graph of $f(x) = a(x - h)^2 + k$ if $a < 0, h > 0$, and $k < 0$?

A.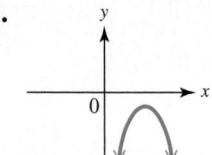

B.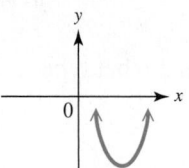

C.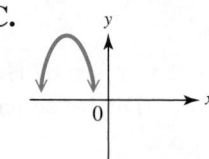

D.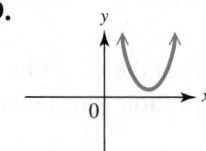

Graph each parabola. Give the vertex, axis of symmetry, domain, and range.

21. $f(x) = \dfrac{1}{2}x^2 - 2$

vertex:
axis:
domain:
range:

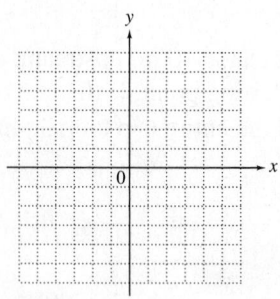

22. $f(x) = -x^2 + 4x - 1$

vertex:
axis:
domain:
range:

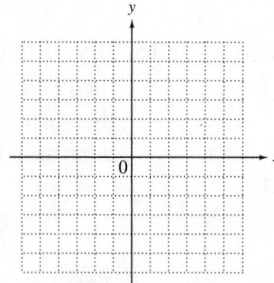

23. $x = 2y^2 + 8y + 3$

vertex:
axis:
domain:
range:

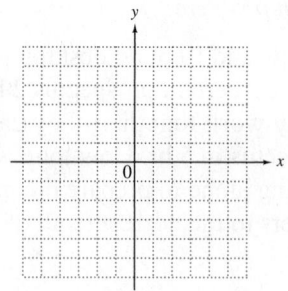

Solve. Graph each solution set.

24. $2x^2 + 7x > 15$

25. $\dfrac{5}{t - 4} \leq 1$

Chapters R–10 *Cumulative Review Exercises*

1. Let $S = \left\{-\frac{7}{3}, -2, -\sqrt{3}, 0, 0.7, \sqrt{12}, \sqrt{-8}, 7, \frac{32}{3}\right\}$. List the elements of S that are elements of each set.

(**a**) Integers (**b**) Rational numbers (**c**) Real numbers (**d**) Complex numbers

Solve each equation or inequality.

2. $-2x + 4 = 5(x - 4) + 17$

3. $-2x + 4 \leq -x + 3$

4. $|3x - 7| \leq 1$

5. $2x = \sqrt{\dfrac{5x + 2}{3}}$

6. $2x^2 - 4x - 3 = 0$

7. $z^2 - 2z = 15$

8. $\dfrac{3}{x - 3} - \dfrac{2}{x - 2} = \dfrac{3}{x^2 - 5x + 6}$

9. $p^4 - 10p^2 + 9 = 0$

10. Find the slope and y-intercept of the line with equation $2x - 4y = 7$.

11. Write the equation in standard form of the line through $(2, -1)$ and perpendicular to $-3x + y = 5$.

Graph each relation. Tell whether or not each is a function, and if it is, give its domain and range.

12. $4x - 5y = 15$

13. $4x - 5y < 15$

14. $y = -2(x - 1)^2 + 3$

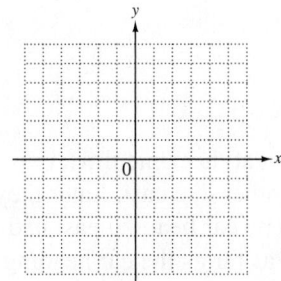

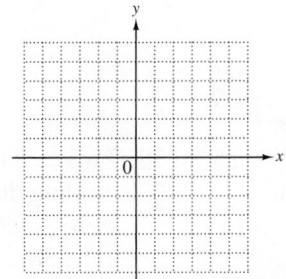

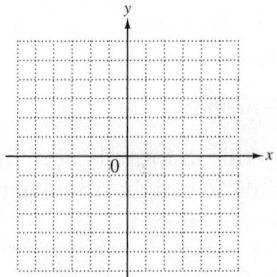

Solve each system of equations.

15. $2x - 4y = 10$
$9x + 3y = 3$

16. $x + y + 2z = 3$
$-x + y + z = -5$
$2x + 3y - z = -8$

Write with positive exponents only. Assume that variables represent positive real numbers.

17. $\left(\dfrac{x^{-3}y^2}{x^5y^{-2}}\right)^{-1}$

18. $\dfrac{(4x^{-2})^2(2y^3)}{8x^{-3}y^5}$

19. Multiply $(2t + 9)^2$.

20. Divide $4x^3 + 2x^2 - x + 26$ by $x + 2$.

Factor completely.

21. $16x - x^3$

22. $24m^2 + 2m - 15$

23. $9x^2 - 30xy + 25y^2$

Perform the indicated operations, and express answers in lowest terms. Assume that denominators represent nonzero real numbers.

24. $\dfrac{5t + 2}{-6} \div \dfrac{15t + 6}{5}$

25. $\dfrac{3}{2 - k} - \dfrac{5}{k} + \dfrac{6}{k^2 - 2k}$

26. $\dfrac{\dfrac{r}{s} - \dfrac{s}{r}}{\dfrac{r}{s} + 1}$

Simplify each radical expression.

27. $\sqrt[3]{\dfrac{27}{16}}$

28. $\dfrac{2}{\sqrt{7} - \sqrt{5}}$

Solve each problem.

29. Clark's rule, a formula used in reducing drug dosage according to weight from the recommended adult dosage to a child dosage, is

$$\dfrac{\text{weight of child in pounds}}{150} \times \text{adult dose} = \text{child's dose.}$$

Find a child's dosage if the child weighs 55 lb and the recommended adult dosage is 120 mg.

30. Two cars left an intersection at the same time, one heading due south and the other due east. Later they were exactly 95 mi apart. The car heading east had gone 38 mi less than twice as far as the car heading south. How far had each car traveled?

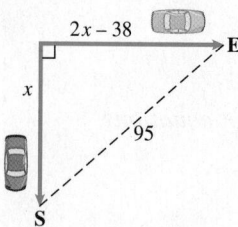

11 Inverse, Exponential, and Logarithmic Functions

The magnitudes of earthquakes, intensities of sounds, and population growth and decay are some examples of applications of *exponential* and *logarithmic functions*.

11.1 Operations on Functions and Composition

11.2 Inverse Functions

11.3 Exponential Functions

11.4 Logarithmic Functions

11.5 Properties of Logarithms

11.6 Common and Natural Logarithms

11.7 Exponential and Logarithmic Equations and Their Applications

Math in the Media *So, Did the Scarecrow Really Get a Brain?*

833

11.1 Operations on Functions and Composition

OBJECTIVES

1 Perform operations on functions.

2 Find the composition of functions.

OBJECTIVE ▶ 1 **Perform operations on functions.** The operations of addition, subtraction, multiplication, and division are also defined for functions. For example, the graph in **Figure 1** shows dollars (in billions) spent for general science and for space/other technologies in selected years.

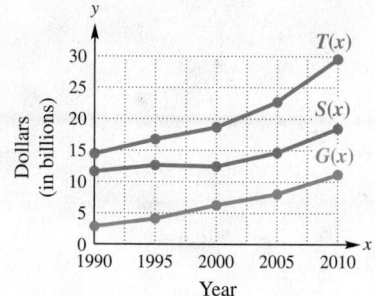

Science and Space Spending

Source: U.S. Office of Management and Budget.

Figure 1

$G(x)$ represents dollars spent for general science.

$S(x)$ represents dollars spent for space/other technologies.

$T(x)$ represents total expenditures for these two categories.

The total expenditures function can be found by *adding* the spending functions for the two individual categories.

$$T(x) = G(x) + S(x)$$

As another example, businesses use the equation "profit equals revenue minus cost," which can be written using function notation.

$$P(x) = R(x) - C(x) \qquad x \text{ is the number of items produced and sold.}$$

↑ ↑ ↑
Profit Revenue Cost
function function function

The profit function is found by *subtracting* the cost function from the revenue function.

We define the following **operations on functions.**

Operations on Functions

If $f(x)$ and $g(x)$ define functions, then

$$(f + g)(x) = f(x) + g(x), \qquad \text{Sum function}$$
$$(f - g)(x) = f(x) - g(x), \qquad \text{Difference function}$$
$$(fg)(x) = f(x) \cdot g(x), \qquad \text{Product function}$$

and
$$\left(\frac{f}{g}\right)(x) = \frac{f(x)}{g(x)}. \qquad \text{Quotient function}$$

In each case, the domain of the new function is the intersection of the domains of $f(x)$ and $g(x)$. Additionally, the domain of the quotient function must exclude any values of x for which $g(x) = 0$.

Section 11.1 Operations on Functions and Composition **835**

EXAMPLE 1 Adding and Subtracting Functions

Find each of the following for polynomial functions f and g as defined.

$$f(x) = x^2 - 3x + 7 \quad \text{and} \quad g(x) = -3x^2 - 7x + 7$$

(a) $(f + g)(x)$ — This notation does *not* indicate the distributive property.

$= f(x) + g(x)$ Use the definition.

$= (x^2 - 3x + 7) + (-3x^2 - 7x + 7)$ Substitute.

$= -2x^2 - 10x + 14$ Add the polynomials.

(b) $(f - g)(x)$

$= f(x) - g(x)$ Use the definition.

$= (x^2 - 3x + 7) - (-3x^2 - 7x + 7)$ Substitute.

$= (x^2 - 3x + 7) + (3x^2 + 7x - 7)$ Change to addition.

$= 4x^2 + 4x$ Add.

···· **Work Problem ❶ at the Side.** ▶

EXAMPLE 2 Adding and Subtracting Functions

Find each of the following for polynomial functions f and g as defined.

$$f(x) = 10x^2 - 2x \quad \text{and} \quad g(x) = 2x.$$

(a) $(f + g)(2)$

$(f + g)(2) = f(2) + g(2)$ Use the definition.

$\overbrace{f(x) = 10x^2 - 2x}\qquad \overbrace{g(x) = 2x}$

This is a key step. $= [10(2)^2 - 2(2)] \ + \ 2(2)$ Substitute.

$= [40 - 4] + 4$ Order of operations

$= 40$ Subtract, and then add.

Alternatively, we could first find $(f + g)(x)$.

$(f + g)(x) = f(x) + g(x)$ Use the definition.

$= (10x^2 - 2x) + 2x$ Substitute.

$= 10x^2$ Combine like terms.

$(f + g)(2) = 10(2)^2$ Substitute.

$= 40$ The result is the same.

(b) $(f - g)(x)$ and $(f - g)(1)$

$(f - g)(x) = f(x) - g(x)$ Use the definition.

$= (10x^2 - 2x) - 2x$ Substitute.

$= 10x^2 - 4x$ Combine like terms.

$(f - g)(1) = 10(1)^2 - 4(1)$ Substitute.

$= 6$ Perform the operations.

Confirm that $f(1) - g(1)$ gives the same result.

···· **Work Problem ❷ at the Side.** ▶

---

❶ For

$$f(x) = 3x^2 + 8x - 6$$

and $g(x) = -4x^2 + 4x - 8,$

find each of the following.

(GS) (a) $(f + g)(x)$

$= f(x) +$ ____

$= ($ _____ $) + ($ _____ $)$

$=$ _____

(b) $(f - g)(x)$

❷ For

$$f(x) = 18x^2 - 24x$$

and $g(x) = 3x,$

find each of the following.

(a) $(f + g)(x)$ and $(f + g)(-1)$

(b) $(f - g)(x)$ and $(f - g)(1)$

Answers

1. (a) $g(x)$; $3x^2 + 8x - 6$; $-4x^2 + 4x - 8$;
     $-x^2 + 12x - 14$
(b) $7x^2 + 4x + 2$
2. (a) $18x^2 - 21x$; 39
(b) $18x^2 - 27x$; -9

3 For
$$f(x) = 2x + 7$$
and $\quad g(x) = x^2 - 4,$
find $(fg)(x)$ and $(fg)(2)$.

EXAMPLE 3 Multiplying Functions

For $f(x) = 3x + 4$ and $g(x) = 2x^2 + x$, find $(fg)(x)$ and $(fg)(-1)$.

$$(fg)(x)$$
$$= f(x) \cdot g(x) \qquad \text{Use the definition.}$$
$$= (3x + 4)(2x^2 + x) \qquad \text{Substitute.}$$
$$= 6x^3 + 3x^2 + 8x^2 + 4x \qquad \text{FOIL method}$$
$$= 6x^3 + 11x^2 + 4x \qquad \text{Combine like terms.}$$

$$(fg)(-1)$$
$$= 6(-1)^3 + 11(-1)^2 + 4(-1) \qquad \text{Let } x = -1.$$
$$= -6 + 11 - 4 \qquad \begin{array}{l}\text{Apply the exponents.} \\ \text{Multiply.}\end{array}$$

> Be careful with signs.

$$= 1 \qquad \text{Add and subtract.}$$

(Another way to find $(fg)(-1)$ is to find $f(-1)$ and $g(-1)$ and then multiply the results. Verify this by showing that $f(-1) \cdot g(-1)$ equals 1. This follows from the definition.)

◀ **Work Problem 3 at the Side.**

CAUTION

Write the product $f(x) \cdot g(x)$ as $(fg)(x)$, **not** $f(g(x))$, which has a different mathematical meaning, as discussed later in this section.

---

4 For
$$f(x) = 2x^2 + 17x + 30$$
and $\quad g(x) = 2x + 5,$
find $\left(\frac{f}{g}\right)(x)$ and $\left(\frac{f}{g}\right)(-1)$.

EXAMPLE 4 Dividing Functions

For $f(x) = 2x^2 + x - 10$ and $g(x) = x - 2$, find $\left(\frac{f}{g}\right)(x)$ and $\left(\frac{f}{g}\right)(-3)$.

$$\left(\frac{f}{g}\right)(x)$$
$$= \frac{f(x)}{g(x)} \qquad \text{Use the definition.}$$
$$= \frac{2x^2 + x - 10}{x - 2} \qquad \text{Substitute.}$$
$$= \frac{(2x + 5)(x - 2)}{x - 2} \qquad \text{Factor.}$$

> $g(x) \neq 0$, so $x \neq 2$

$$= 2x + 5, \quad x \neq 2$$

$$\left(\frac{f}{g}\right)(-3)$$
$$= 2(-3) + 5 \qquad \text{Let } x = -3.$$
$$= -1 \qquad \text{Multiply and then add.}$$

Verify that the same value is found by evaluating $\frac{f(-3)}{g(-3)}$.

◀ **Work Problem 4 at the Side.**

Answers

3. $2x^3 + 7x^2 - 8x - 28; 0$

4. $x + 6, \quad x \neq -\dfrac{5}{2}; 5$

OBJECTIVE ▸ 2 Find the composition of functions. The diagram in **Figure 2** shows a function f that assigns, to each element x of set X, some element y of set Y. Suppose that a function g takes each element of set Y and assigns a value z of set Z. Then f and g together assign an element x in X to an element z in Z.

The result of this process is a new function h, which takes an element x in X and assigns it an element z in Z.

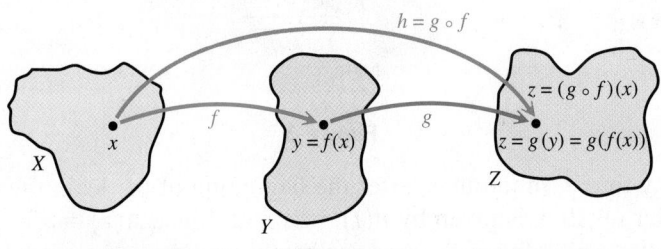

Figure 2

This function h is the *composition* of functions g and f, written $g \circ f$.

Composition of Functions

If f and g are functions, then the **composite function,** or **composition,** of g and f is

$$(g \circ f)(x) = g(f(x)).$$

The domain of $g \circ f$ is the set of all numbers x in the domain of f such that $f(x)$ is in the domain of g.

Read $g \circ f$ as "g of f."

As a real-life example of how composite functions occur, consider the following retail situation.

A $40 pair of blue jeans is on sale for 25% off. If we purchase the jeans before noon, the retailer offers an additional 10% off. What is the final sale price of the blue jeans?

We might be tempted to say that the jeans are 35% off and calculate

$$\$40 \, (0.35) = \$14,$$

giving a final sale price of $\Big\}$ Incorrect

$$\$40 - \$14 = \$26.$$

This is not correct. To find the final sale price, we must first find the price after taking 25% off, and then take an additional 10% off that price.

$\$40 \, (0.25) = \$10$, giving a sale price of $\$40 - \$10 = \$30$. — Take 25% off original price.

$\$30 \, (0.10) = \$3$, for a ***final sale price*** of $\$30 - \$3 = \$27$. — Take additional 10% off.

This is the idea behind composition of functions.

5 Let $f(x) = x - 4$ and $g(x) = x^2$.

GS **(a)** Find $(f \circ g)(3)$.

Apply the definition of composition of functions.

$(f \circ g)(3)$

$= f(\underline{}(\underline{}))$

Use the rule for $g(x)$.

$= f(\underline{}^2)$

Apply the exponent.

$= f(\underline{})$

Use the rule for $f(x)$.

$= \underline{} - 4$

$= \underline{}$

(b) Find $(g \circ f)(3)$.

As another example of composition, suppose an oil well off the Louisiana coast is leaking, with the leak spreading oil in a circular layer over the surface. See **Figure 3**.

$r(t)$

Figure 3

At any time t, in minutes, after the beginning of the leak, the radius of the circular oil slick is given by $r(t) = 5t$ feet. Since $A(r) = \pi r^2$ gives the area of a circle of radius r, the area can be expressed as a function of time by substituting $5t$ for r in $A(r) = \pi r^2$.

$$A(r) = \pi r^2$$

$$A(r(t)) = \pi (5t)^2$$

$$A(r(t)) = 25\pi t^2$$

The function $A(r(t))$ is a composite function of the functions A and r.

EXAMPLE 5 **Finding a Composite Function**

Let $f(x) = x^2$ and $g(x) = x + 3$. Find $(f \circ g)(4)$.

$(f \circ g)(4)$ Evaluate the "inside" function value first.

$= f(g(4))$ Definition

$= f(4 + 3)$ Use the rule for $g(x)$; $g(4) = 4 + 3$

$= f(7)$ Add.

Now evaluate the "outside" function.

$= 7^2$ Use the rule for $f(x)$; $f(7) = 7^2$

$= 49$ Square 7.

◀ **Work Problem** **5** at the Side.

If we interchange the order of the functions in **Example 5**, the composition of g and f is symbolized $g(f(x))$. To find $(g \circ f)(4)$, we let $x = 4$.

$(g \circ f)(4)$

$= g(f(4))$ Definition

$= g(4^2)$ Use the rule for $f(x)$; $f(4) = 4^2$

$= g(16)$ Square 4.

$= 16 + 3$ Use the rule for $g(x)$; $g(16) = 16 + 3$

$= 19$ Add.

Here we see that $(f \circ g)(4) \neq (g \circ f)(4)$ because $49 \neq 19$. In general,

$$(f \circ g)(x) \neq (g \circ f)(x).$$

EXAMPLE 6 **Finding Composite Functions**

Let $f(x) = 4x - 1$ and $g(x) = x^2 + 5$. Find each of the following.

(a) $\qquad\qquad (f \circ g)(2)$

$\qquad\qquad = f(g(2))$

$\qquad\qquad = f(2^2 + 5) \quad g(x) = x^2 + 5$

$\qquad\qquad = f(9) \qquad\quad$ Work inside the parentheses.

$\qquad\qquad = 4(9) - 1 \quad f(x) = 4x - 1$

$\qquad\qquad = 35 \qquad\quad$ Multiply. Subtract.

(b) $\quad (f \circ g)(x)$

$\qquad\quad = f(g(x)) \qquad\qquad$ Use $g(x)$ as the input for the function f.

$\qquad\quad = 4(g(x)) - 1 \qquad$ Use the rule for $f(x)$; $f(x) = 4x - 1$

$\qquad\quad = 4(x^2 + 5) - 1 \quad g(x) = x^2 + 5$

$\qquad\quad = 4x^2 + 20 - 1 \qquad$ Distributive property

$\qquad\quad = 4x^2 + 19 \qquad\quad$ Combine like terms.

(c) Find $(f \circ g)(2)$ again, this time using the rule obtained in part (b).

$\qquad\quad (f \circ g)(x) = 4x^2 + 19 \qquad$ From part (b)

$\qquad\quad (f \circ g)(2) = 4(2)^2 + 19 \qquad$ Let $x = 2$.

$\qquad\qquad\qquad\quad = 4(4) + 19 \qquad$ Square 2.

$\qquad\qquad\qquad\quad = 16 + 19 \qquad\quad$ Multiply.

Same result as in part (a) $\rightarrow\quad = 35 \qquad\qquad$ Add.

(d) $\quad (g \circ f)(x)$

$\qquad\quad = g(f(x)) \qquad\qquad$ Use $f(x)$ as the input for the function g.

$\qquad\quad = (f(x))^2 + 5 \qquad$ Use the rule for $g(x)$; $g(x) = x^2 + 5$

$\qquad\quad = (4x - 1)^2 + 5 \qquad f(x) = 4x - 1$

$\qquad\quad = 16x^2 - 8x + 1 + 5 \quad (x - y)^2 = x^2 - 2xy + y^2$

$\qquad\quad = 16x^2 - 8x + 6 \qquad$ Combine like terms.

Compare this result to that in part (b). Again,

$$(f \circ g)(x) \neq (g \circ f)(x).$$

$\cdots\cdots\cdots$ **Work Problem ❻ at the Side.** ▶

❻ Let $f(x) = 3x + 6$ and $g(x) = x^3$. Find each of the following.

(a) $(f \circ g)(2)$

(b) $(g \circ f)(2)$

(c) $(f \circ g)(x)$

(d) $(g \circ f)(x)$

Answers

6. **(a)** 30 **(b)** 1728 **(c)** $3x^3 + 6$
$\qquad$ **(d)** $(3x + 6)^3$

11.1 Exercises

MyMathLab®

CONCEPT CHECK *Find two polynomial functions defined by $f(x)$ and $g(x)$ such that each statement is true.*

1. $(f + g)(x) = 3x^3 - x + 3$

2. $(f - g)(x) = -x^2 + x - 5$

Let $f(x) = x^2 - 9$, $g(x) = 2x$, and $h(x) = x - 3$. Find each of the following.
See Examples 1–4.

3. $(f + g)(x)$

4. $(f - g)(x)$

5. $(f + g)(3)$

6. $(f - g)(-3)$

7. $(f - h)(x)$

8. $(f + h)(x)$

9. $(f - h)(-3)$

10. $(f + h)(-2)$

11. $(g - h)(-3)$

12. $(g + h)(1)$

13. $(g + h)\left(\dfrac{1}{4}\right)$

14. $(g + h)\left(\dfrac{1}{3}\right)$

15. $(fg)(x)$

16. $(fh)(x)$

17. $(fg)(2)$

18. $(fh)(1)$

19. $(fh)(-1)$

20. $(gh)(-3)$

21. $(fg)(-2)$

22. $(fh)(0)$

23. $(fg)\left(-\dfrac{1}{2}\right)$

24. $(fg)\left(-\dfrac{1}{3}\right)$

25. $\left(\dfrac{f}{g}\right)(x)$

26. $\left(\dfrac{f}{h}\right)(x)$

27. $\left(\dfrac{f}{g}\right)(2)$

28. $\left(\dfrac{f}{h}\right)(1)$

29. $\left(\dfrac{h}{g}\right)(x)$

30. $\left(\dfrac{g}{h}\right)(x)$

31. $\left(\dfrac{h}{g}\right)(3)$

32. $\left(\dfrac{f}{g}\right)(-1)$

Solve each problem. See Objective 1.

33. The cost in dollars to produce x t-shirts is $C(x) = 2.5x + 50$. The revenue in dollars from sales of x t-shirts is $R(x) = 10.99x$.

    (a) Write and simplify a function P that gives profit in terms of x.

    (b) Find the profit if 100 t-shirts are produced and sold.

34. The cost in dollars to produce x baseball caps is $C(x) = 4.3x + 75$. The revenue in dollars from sales of x caps is $R(x) = 25x$.

    (a) Write and simplify a function P that gives profit in terms of x.

    (b) Find the profit if 50 caps are produced and sold.

CONCEPT CHECK   *Let $f(x) = x^2$ and $g(x) = 2x - 1$. Match each expression in Column I with the description of how to evaluate it in Column II.*

I	II
35. $(f \circ g)(5)$	**A.** Square 5. Take the result and square it.
36. $(g \circ f)(5)$	**B.** Double 5 and subtract 1. Take the result and square it.
37. $(f \circ f)(5)$	**C.** Double 5 and subtract 1. Take the result, double it, and subtract 1.
38. $(g \circ g)(5)$	**D.** Square 5. Take the result, double it, and subtract 1.

Let $f(x) = x^2 + 4$, $g(x) = 2x + 3$, and $h(x) = x + 5$. Find each value or expression. ***See Examples 5 and 6.***

39. $(h \circ g)(4)$ ▶

40. $(f \circ g)(4)$

41. $(g \circ f)(6)$

42. $(h \circ f)(6)$

43. $(f \circ h)(-2)$

44. $(h \circ g)(-2)$

45. $(f \circ g)(x)$

46. $(g \circ h)(x)$ ▶

47. $(f \circ h)(x)$

48. $(g \circ f)(x)$

49. $(h \circ g)(x)$ ▶

50. $(h \circ f)(x)$

51. $(f \circ h)\left(\dfrac{1}{2}\right)$

52. $(h \circ f)\left(\dfrac{1}{2}\right)$

53. $(f \circ g)\left(-\dfrac{1}{2}\right)$

54. $(g \circ f)\left(-\dfrac{1}{2}\right)$

CONCEPT CHECK   *The tables give some selected ordered pairs for functions f and g.*

x	3	4	6
f(x)	1	3	9

x	2	7	1	9
g(x)	3	6	9	12

Tables like these can be used to evaluate composite functions. For example, to evaluate $(g \circ f)(6)$, use the first table to find $f(6) = 9$. Then use the second table to find

$$(g \circ f)(6) = g(f(6)) = g(9) = 12.$$

Find each of the following.

55. $(f \circ g)(2)$ ▶

56. $(f \circ g)(7)$

57. $(g \circ f)(3)$

58. $(g \circ f)(6)$

59. $(f \circ f)(4)$ ▶

60. $(g \circ g)(1)$

Solve each problem.

61. The function

$$f(x) = 12x$$

computes the number of inches in x feet, and the function

$$g(x) = 5280x$$

computes the number of feet in x miles. Find and simplify $(f \circ g)(x)$. What does it compute?

62. The function

$$f(x) = 60x$$

computes the number of minutes in x hours, and the function

$$g(x) = 24x$$

computes the number of hours in x days. Find and simplify $(f \circ g)(x)$. What does it compute?

63. The perimeter x of a square with sides of length s is given by the formula $x = 4s$.

(a) Solve for s in terms of x.

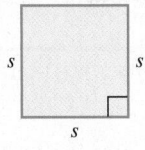

(b) If y represents the area of this square, write y as a function of the perimeter x.

(c) Use the composite function of part (b) to find the area of a square with perimeter 6.

64. The perimeter x of an equilateral triangle with sides of length s is given by the formula $x = 3s$.

(a) Solve for s in terms of x.

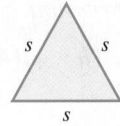

(b) The area y of an equilateral triangle with sides of length s is given by the formula $y = \frac{s^2\sqrt{3}}{4}$. Write y as a function of the perimeter x.

(c) Use the composite function of part (b) to find the area of an equilateral triangle with perimeter 12.

65. When a thermal inversion layer is over a city (as happens often in Los Angeles), pollutants cannot rise vertically but are trapped below the layer and must disperse horizontally.

　　Assume that a factory smokestack begins emitting a pollutant at 8 A.M. and that the pollutant disperses horizontally over a circular area. Suppose that t represents the time, in hours, since the factory began emitting pollutants ($t = 0$ represents 8 A.M), and assume that the radius of the circle of pollution is $r(t) = 2t$ miles. Let $A(r) = \pi r^2$ represent the area of a circle of radius r. Find and interpret $(A \circ r)(t)$.

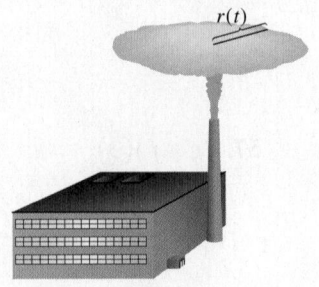

66. An oil well off the Gulf Coast is leaking, with the leak spreading oil over the surface as a circle. At any time t, in minutes, after the beginning of the leak, the radius of the circular oil slick on the surface is $r(t) = 4t$ feet. Let $A(r) = \pi r^2$ represent the area of a circle of radius r. Find and interpret $(A \circ r)(t)$.

11.2 Inverse Functions

In this chapter we will study two important types of functions, *exponential* and *logarithmic*. These functions are related: They are *inverses* of one another.

▦ **Calculator Tip**

A calculator with the following keys will be essential in this chapter.

 , or (LOG) , (eˣ) or (LN)

y^x , 10^x or LOG , e^x or LN

We will explain how these keys are used at appropriate places in the chapter.

> **OBJECTIVES**
>
> **1** Decide whether a function is one-to-one and, if it is, find its inverse.
>
> **2** Use the horizontal line test to determine whether a function is one-to-one.
>
> **3** Find the equation of the inverse of a function.
>
> **4** Graph f^{-1} from the graph of f.

OBJECTIVE ▶ **1** **Decide whether a function is one-to-one and, if it is, find its inverse.** Suppose we define the function

$$G = \{(-2, 2), (-1, 1), (0, 0), (1, 3), (2, 5)\}.$$

We can form another set of ordered pairs from G by interchanging the x- and y-values of each pair in G. We can call this set F, so

$$F = \{(2, -2), (1, -1), (0, 0), (3, 1), (5, 2)\}.$$

To show that these two sets are related, F is called the *inverse* of G. For a function f to have an inverse function, f must be *one-to-one*.

One-to-One Function

In a **one-to-one function,** each x-value corresponds to just one y-value, and each y-value corresponds to just one x-value.

The function shown in **Figure 4(a)** is not one-to-one because the y-value 7 corresponds to *two* x-values, 2 and 3. That is, the ordered pairs $(2, 7)$ and $(3, 7)$ both appear in the function. The function in **Figure 4(b)** is one-to-one.

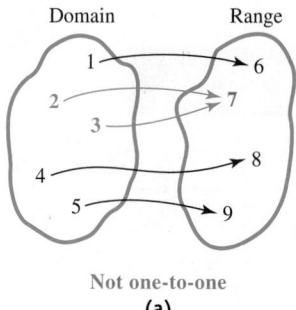

Not one-to-one	One-to-one
(a)	(b)

Figure 4

The *inverse* of any one-to-one function f is found by interchanging the components of the ordered pairs of f. The inverse of f is written f^{-1}. Read f^{-1} as **"the inverse of f"** or **"f-inverse."**

CAUTION

The symbol $f^{-1}(x)$ does not represent $\dfrac{1}{f(x)}$.

1 Find the inverse of each function that is one-to-one.

(a) $\{(1, 2), (2, 4), (3, 3), (4, 5)\}$

(b) $\{(0, 3), (-1, 2), (1, 3)\}$

(c) A Norwegian physiologist has developed a rule for predicting running times based on the time to run 5 km (5K). An example for one runner is shown here. (*Source:* Stephen Seiler, Agder College, Kristiansand, Norway.)

Distance	Time
1.5K	4:22
3K	9:18
5K	16:00
10K	33:40

Answers

1. **(a)** $\{(2, 1), (4, 2), (3, 3), (5, 4)\}$
   (b) not a one-to-one function
   (c)

Time	Distance
4:22	1.5K
9:18	3K
16:00	5K
33:40	10K

The definition of the inverse of a function follows.

Inverse of a Function

The **inverse** of a one-to-one function f, written f^{-1}, is the set of all ordered pairs of the form (y, x), where (x, y) belongs to f. *Since the inverse is formed by interchanging x and y, the domain of f becomes the range of f^{-1} and the range of f becomes the domain of f^{-1}.*

For inverses f and f^{-1}, it follows that for all x in their domains,

$$(f \circ f^{-1})(x) = x \quad \text{and} \quad (f^{-1} \circ f)(x) = x.$$

EXAMPLE 1 Finding the Inverses of One-to-One Functions

Find the inverse of each function that is one-to-one.

(a) $F = \{(-2, 1), (-1, 0), (0, 1), (1, 2), (2, 2)\}$

Each x-value in F corresponds to just one y-value. However, the y-value 2 corresponds to two x-values, 1 and 2. Also, the y-value 1 corresponds to both -2 and 0. Because some y-values correspond to more than one x-value, F is not one-to-one and does not have an inverse function.

(b) $G = \{(3, 1), (0, 2), (2, 3), (4, 0)\}$

Every x-value in G corresponds to only one y-value, and every y-value corresponds to only one x-value, so G is a one-to-one function. The inverse function is found by interchanging the x- and y-values in each ordered pair.

$$G^{-1} = \{(1, 3), (2, 0), (3, 2), (0, 4)\}$$

The domain and range of G become the range and domain, respectively, of G^{-1}.

(c) The U.S. Environmental Protection Agency sets air-quality standards for the United States. The table shows the number of days in which the air in the Los Angeles–Long Beach–Santa Ana metropolitan area failed to meet acceptable air-quality standards for the years 2005–2010.

Year	Number of Days Exceeding Standards
2005	29
2006	26
2007	18
2008	31
2009	18
2010	4

Source: U.S. Environmental Protection Agency.

Let f be the function defined in the table, with the years forming the domain and the number of days exceeding the air-quality standards forming the range. Then f is not one-to-one because in two different years (2007 and 2009), the number of days with unacceptable air quality was the same, 18.

◀ **Work Problem 1** at the Side.

OBJECTIVE ▶ **2** **Use the horizontal line test to determine whether a function is one-to-one.** By graphing a function and observing the graph, we can use the *horizontal line test* to tell whether the function is one-to-one.

Horizontal Line Test

A function is one-to-one if every horizontal line intersects the graph of the function at most once.

The horizontal line test follows from the definition of a one-to-one function. Any two points that lie on the same horizontal line have the same y-coordinate. No two ordered pairs that belong to a one-to-one function may have the same y-coordinate. Therefore, no horizontal line will intersect the graph of a one-to-one function more than once.

EXAMPLE 2   **Using the Horizontal Line Test**

Use the horizontal line test to determine whether each graph is the graph of a one-to-one function.

(a)

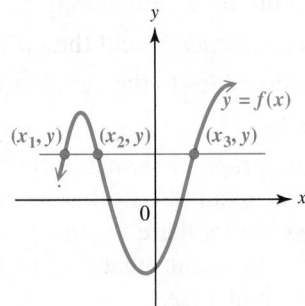

Figure 5(a)

(b)

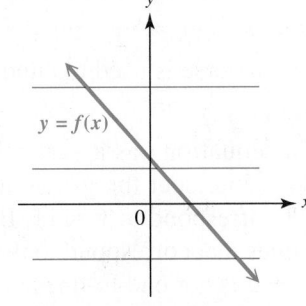

Figure 5(b)

Because the red horizontal line shown in **Figure 5(a)** intersects the graph in more than one point (actually three points), the function is not one-to-one.

Every horizontal line will intersect the graph in **Figure 5(b)** in exactly one point. This function is one-to-one.

⋯⋯⋯⋯⋯⋯⋯⋯⋯⋯⋯⋯⋯⋯⋯⋯⋯⋯ **Work Problem ❷ at the Side.** ▶

OBJECTIVE ▶ **3** **Find the equation of the inverse of a function.** The inverse of a one-to-one function is found by interchanging the x- and y-values of each of its ordered pairs. The equation of the inverse of a function defined by $y = f(x)$ is found in the same way.

Finding the Equation of the Inverse of $y = f(x)$

For a one-to-one function f defined by an equation $y = f(x)$, find the defining equation of the inverse as follows.

Step 1   Interchange x and y.

Step 2   Solve for y.

Step 3   Replace y with $f^{-1}(x)$.

---

❷ Use the horizontal line test to determine whether each graph is the graph of a one-to-one function.

(a)

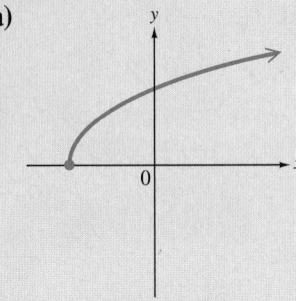

(b)

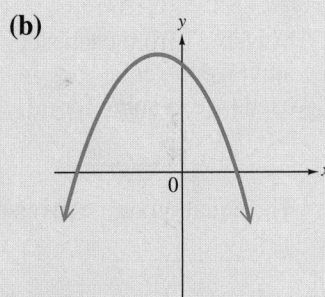

Answers

2. (a) one-to-one   **(b)** not one-to-one

❸ Decide whether each equation represents a one-to-one function. If so, find the equation for the inverse.

(a) $f(x) = 3x - 4$

GS **(b)** $f(x) = x^3 + 1$

This function (*is / is not*) one-to-one. As a result, it (*has / does not have*) an inverse.

Replace $f(x)$ with y.

$y = $ _____

Interchange x and y.

$x = $ _____

Subtract 1 from each side and take the cube root on each side to solve for y.

$y = $ _____

The equation of the inverse is _____.

(c) $f(x) = (x - 3)^2$

Answers

3. **(a)** one-to-one function;
$f^{-1}(x) = \dfrac{x + 4}{3}$, or $f^{-1}(x) = \dfrac{1}{3}x + \dfrac{4}{3}$

(b) is; has; $x^3 + 1$; $y^3 + 1$; $\sqrt[3]{x - 1}$;
$f^{-1}(x) = \sqrt[3]{x - 1}$

(c) not a one-to-one function

EXAMPLE 3 **Finding Equations of Inverses**

Decide whether each equation represents a one-to-one function. If so, find the equation for the inverse.

(a) $f(x) = 2x + 5$

The graph of $y = 2x + 5$ is a nonvertical line, so by the horizontal line test, f is a one-to-one function. Find the inverse as follows.

$$y = 2x + 5 \qquad \text{Let } y = f(x).$$
$$x = 2y + 5 \qquad \text{Interchange } x \text{ and } y. \text{ (Step 1)}$$
$$2y = x - 5 \qquad \text{Solve for } y. \text{ (Step 2)}$$
$$y = \frac{x - 5}{2}$$
$$f^{-1}(x) = \frac{x - 5}{2} \qquad \text{Replace } y \text{ with } f^{-1}(x). \text{ (Step 3)}$$
$$f^{-1}(x) = \frac{x}{2} - \frac{5}{2}, \quad \text{or} \quad f^{-1}(x) = \frac{1}{2}x - \frac{5}{2} \qquad \frac{a - b}{c} = \frac{a}{c} - \frac{b}{c}$$

Thus, f^{-1} is a linear function. In the function $y = 2x + 5$, the value of y is found by starting with a value of x, multiplying by 2, and adding 5. The equation $f^{-1}(x) = \frac{x - 5}{2}$ for the inverse has us *subtract* 5, and then *divide* by 2. An inverse is used to "undo" what a function does to the variable x.

(b) $y = x^2 + 2$

This equation has a vertical parabola as its graph, so some horizontal lines will intersect the graph at two points. For example, both $x = 3$ and $x = -3$ correspond to $y = 11$. Because of the x^2-term, there are many pairs of x-values that correspond to the same y-value. This means that the function $y = x^2 + 2$ is not one-to-one and does not have an inverse.

Alternatively, applying the steps for finding the equation of an inverse leads to the following.

$$y = x^2 + 2$$
$$x = y^2 + 2 \qquad \text{Interchange } x \text{ and } y.$$
$$y^2 = x - 2 \qquad \text{Solve for } y.$$
$$y = \pm\sqrt{x - 2} \qquad \text{Square root property}$$

The last step shows that there are two y-values for each choice of x in $(2, \infty)$, so the given function is not one-to-one. It does not have an inverse.

(c) $f(x) = (x - 2)^3$

A cubing function like this is one-to-one. (If we sketch its graph, we see that the conditions of the horizontal line test are satisfied.)

$$y = (x - 2)^3 \qquad \text{Replace } f(x) \text{ with } y.$$
$$x = (y - 2)^3 \qquad \text{Interchange } x \text{ and } y.$$
$$\sqrt[3]{x} = \sqrt[3]{(y - 2)^3} \qquad \text{Take the cube root on each side.}$$
$$\sqrt[3]{x} = y - 2 \qquad \sqrt[3]{a^3} = a$$
$$y = \sqrt[3]{x} + 2 \qquad \text{Solve for } y.$$
$$f^{-1}(x) = \sqrt[3]{x} + 2 \qquad \text{Replace } y \text{ with } f^{-1}(x).$$

◀ **Work Problem ❸ at the Side.**

OBJECTIVE ▶ ④ **Graph f^{-1} from the graph of f.** One way to graph the inverse of a function f whose equation is given is as follows.

1. Find several ordered pairs that belong to f.

2. Interchange x and y to obtain ordered pairs that belong to f^{-1}.

3. Plot those points, and sketch the graph of f^{-1} through them.

A simpler way is to select points on the graph of f and use symmetry to find corresponding points on the graph of f^{-1}.

For example, suppose the point (a, b) shown in **Figure 6** belongs to a one-to-one function f. Then the point (b, a) belongs to f^{-1}. The line segment connecting (a, b) and (b, a) is perpendicular to, and cut in half by, the line $y = x$. The points (a, b) and (b, a) are "mirror images" of each other with respect to $y = x$.

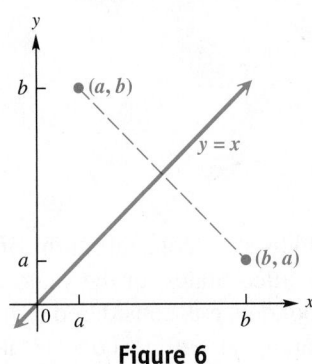

Figure 6

We can find the graph of f^{-1} from the graph of f by locating the mirror image of each point in f with respect to the line $y = x$.

EXAMPLE 4 | **Graphing the Inverse**

Graph the inverses of the functions labeled f shown in **Figure 7**.

In **Figure 7** the graphs of two functions labeled f are shown in blue. Their inverses are shown in red. In each case, the graph of f^{-1} is symmetric to the graph of f with respect to the line $y = x$.

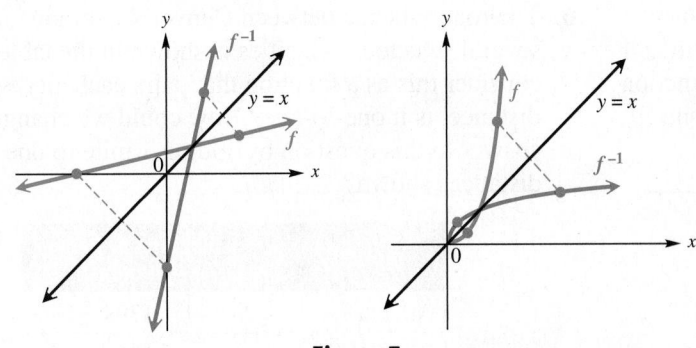

Figure 7

·················· **Work Problem ④ at the Side.** ▶

④ Use the given graphs to graph each inverse.

(a)

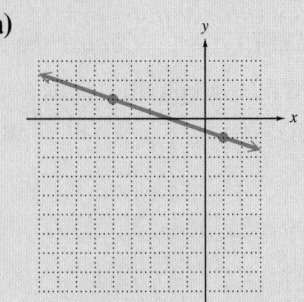

(b)

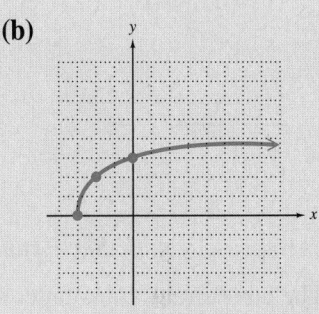

(c)

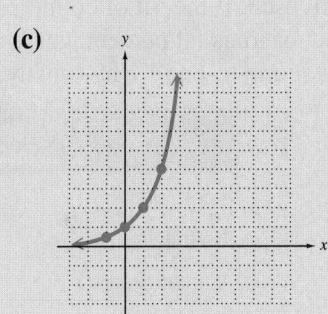

Answers

4. (a)   **(b)**

(c)

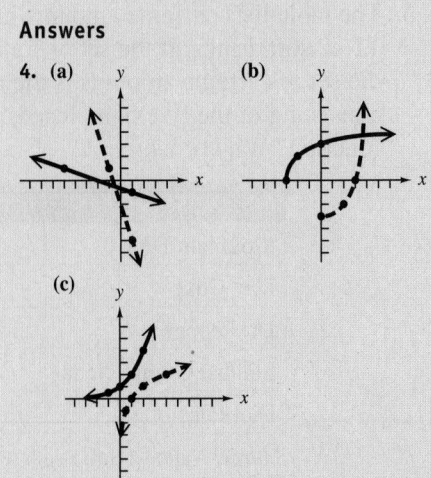

11.2 Exercises

CONCEPT CHECK *Choose the correct response.*

1. Which graph illustrates a one-to-one function?

A.

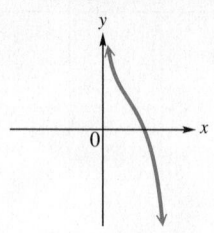

B.

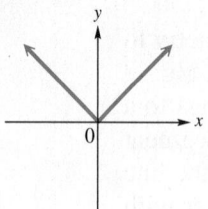

C.

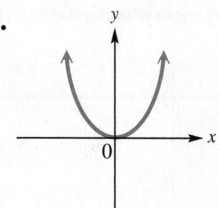

D.

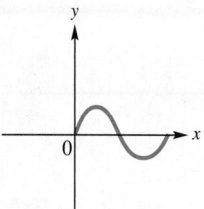

2. If a function is made up of ordered pairs in such a way that the same *y*-value appears in a correspondence with two different *x*-values, then which of the following applies?

 A. The function is one-to-one.

 B. The function is not one-to-one.

 C. Its graph does not pass the vertical line test.

 D. It has an inverse function associated with it.

Answer each question. ***See Example 1.***

3. The table shows trans fat content in a fast-food product in various countries, based on type of frying oil used. If the set of countries is the domain and the set of trans fat percentages is the range of a function, is it one-to-one? Why or why not?

Country	Percentage of Trans Fat in McDonald's Chicken
Scotland	14
France	11
United States	11
Peru	9
Russia	5
Denmark	1

Source: New England Journal of Medicine.

4. The table shows concentrations of carbon monoxide in the United States for the years 2004–2009. If this correspondence is considered to be a function that pairs each year with its concentration, is it one-to-one? If not, explain why.

Year	Concentration (in parts per million)
2004	2.5
2005	2.3
2006	2.2
2007	2.0
2008	1.9
2009	1.8

Source: E.P.A.

5. The table lists caffeine amounts in several popular 12-oz soft drinks. If the set of sodas is the domain and the set of caffeine amounts is the range of the function consisting of the five pairs listed, is it a one-to-one function? Why or why not?

Soda	Caffeine (in mg)
Mountain Dew	55
Diet Coke	45
Dr. Pepper	41
Sunkist Orange Soda	41
Diet Pepsi-Cola	36

Source: National Soft Drink Association.

6. The road mileage between Denver, Colorado, and several selected U.S. cities is shown in the table. If we consider this as a function that pairs each city with a distance, is it one-to-one? How could we change the answer to this question by adding 1 mile to one of the distances shown?

City	Distance to Denver (in miles)
Atlanta	1398
Indianapolis	1058
Kansas City, MO	600
Los Angeles	1059

CONCEPT CHECK   *Choose the correct response.*

7. Which function is one-to-one?

A. $f(x) = x$ **B.** $f(x) = x^2$

C. $f(x) = |x|$ **D.** $f(x) = -x^2 + 2x - 1$

8. If a function f is one-to-one and the point (p, q) lies on the graph of f, then which point *must* lie on the graph of f^{-1}?

A. $(-p, q)$ **B.** $(-q, -p)$

C. $(p, -q)$ **D.** (q, p)

If the function is one-to-one, find its inverse. ***See Examples 1 and 3.***

9. $\{(3, 6), (2, 10), (5, 12)\}$

10. $\left\{(-1, 3), (0, 5), \left(7, -\dfrac{1}{2}\right)\right\}$

11. $\{(-1, 3), (2, 7), (4, 3), (5, 8)\}$

12. $\{(-8, 6), (-4, 3), (0, 6)\}$

13. $\{(0, 4.5), (2, 8.6), (4, 12.7)\}$

14. $\{(1, 5.8), (2, 8.8), (3, 5.8)\}$

15. $f(x) = 2x + 4$

16. $f(x) = 3x + 1$

17. $g(x) = \sqrt{x - 3}, \quad x \geq 3$

18. $g(x) = \sqrt{x + 2}, \quad x \geq -2$

19. $f(x) = 3x^2 + 2$

20. $f(x) = -4x^2 - 1$

21. $f(x) = x^3 - 4$

22. $f(x) = x^3 - 3$

23. $f(x) = 5$

24. $f(x) = -7$

Let $f(x) = 2^x$. We will see in the next section that the function f is one-to-one. Find each value, always working part (a) before part (b).

25. **(a)** $f(3)$

 (b) $f^{-1}(8)$

26. **(a)** $f(4)$

 (b) $f^{-1}(16)$

27. **(a)** $f(0)$

 (b) $f^{-1}(1)$

28. **(a)** $f(-2)$

 (b) $f^{-1}\left(\dfrac{1}{4}\right)$

Graphs of selected functions are given in Exercises 29–34. **(a)** *Use the horizontal line test to determine whether each function is one-to-one.* **(b)** *If the function is one-to-one, graph its inverse with a dashed line (or curve) on the same set of axes.* (Remember: If f is one-to-one and $f(a) = b$, then $f^{-1}(b) = a$.) ***See Examples 3 and 4.***

29.

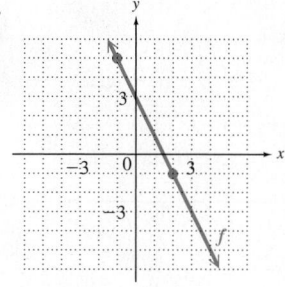

30.

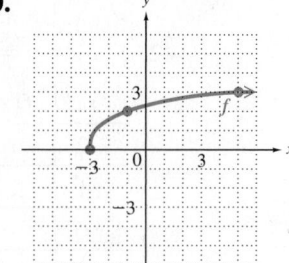

31.

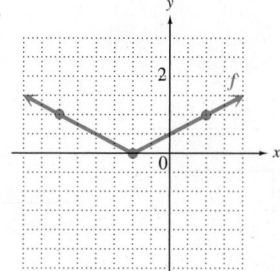

32.

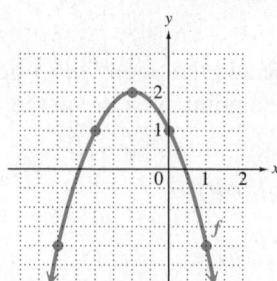

33.

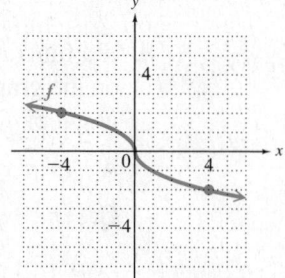

34.

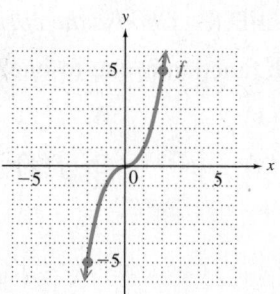

*Each function in Exercises 35–42 represents a one-to-one function. Graph the
function as a solid line (or curve), and then graph its inverse on the same set of axes
as a dashed line (or curve). In Exercises 39–42, complete the table so that graphing
the function will be easier. **See Example 4.***

35. $f(x) = 2x - 1$

36. $f(x) = 2x + 3$

37. $g(x) = -4x$

38. $g(x) = -2x$

39. $f(x) = y = \sqrt{x}, x \geq 0$

x	$f(x) = y$
0	
1	
4	

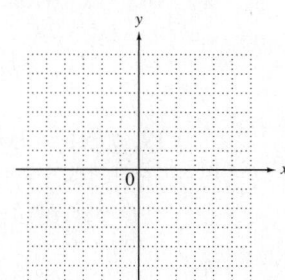

40. $f(x) = y = -\sqrt{x}, x \geq 0$

x	$f(x) = y$
0	
1	
4	

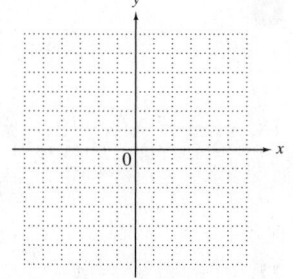

41. $f(x) = y = x^3 - 2$

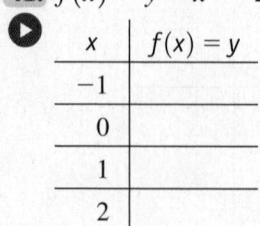

x	$f(x) = y$
−1	
0	
1	
2	

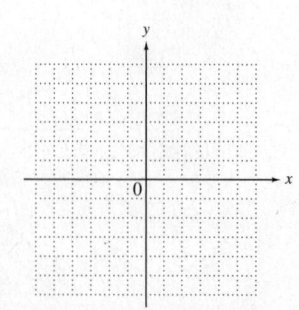

42. $f(x) = y = x^3 + 3$

x	$f(x) = y$
−2	
−1	
0	
1	

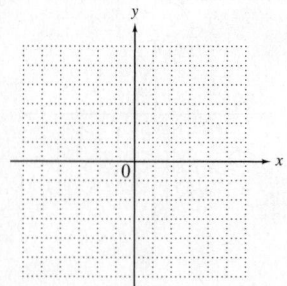

11.3 Exponential Functions

OBJECTIVES

1. Define exponential functions.
2. Graph exponential functions.
3. Solve exponential equations of the form $a^x = a^k$ for x.
4. Use exponential functions in applications involving growth or decay.

OBJECTIVE ▸ **1** **Define exponential functions.** In **Section 9.2**, we evaluated 2^x for rational values of x.

$$2^3 = 8, \quad 2^{-1} = \frac{1}{2}, \quad 2^{1/2} = \sqrt{2}, \quad 2^{3/4} = \sqrt[4]{2^3} = \sqrt[4]{8}$$ Examples of 2^x for rational x

In more advanced courses it is shown that 2^x exists for all real number values of x, both rational and irrational. The following definition of an exponential function assumes that a^x exists for all real numbers x.

Exponential Function

For $a > 0$, $a \neq 1$, and all real numbers x,

$$F(x) = a^x$$

is the **exponential function with base** a.

Note

The two restrictions on the value of a in the definition of an exponential function $F(x) = a^x$ are important.

1. The restriction $a > 0$ is necessary so that the function can be defined for all real numbers x. Letting a be negative ($a = -2$, for instance) and letting $x = \frac{1}{2}$ would give $(-2)^{1/2}$, which is not real.

2. The restriction $a \neq 1$ is necessary because 1 raised to any power is equal to 1. The function would then be the linear function $F(x) = 1$.

OBJECTIVE ▸ **2** **Graph exponential functions.** When graphing exponential functions of the form $F(x) = a^x$, pay particular attention to whether $a > 1$ or $0 < a < 1$.

EXAMPLE 1 **Graphing an Exponential Function ($a > 1$)**

Graph $f(x) = 2^x$. Then compare it to the graph of $F(x) = 5^x$.

Choose some values of x, and find the corresponding values of $f(x)$. Plotting these points and drawing a smooth curve through them gives the graph of $f(x) = 2^x$ shown in **Figure 8.**

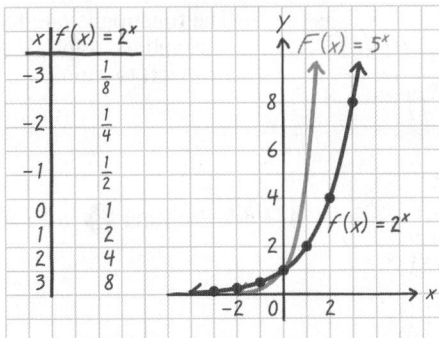

x	$f(x) = 2^x$
-3	$\frac{1}{8}$
-2	$\frac{1}{4}$
-1	$\frac{1}{2}$
0	1
1	2
2	4
3	8

Exponential function with base $a > 1$
Domain: $(-\infty, \infty)$
Range: $(0, \infty)$
The function is one-to-one, and its graph rises from left to right.

Figure 8

Continued on Next Page

1 Graph $f(x) = 10^x$.

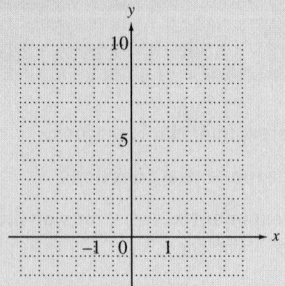

The graph of $f(x) = 2^x$ (repeated below) is typical of the graphs of exponential functions of the form $F(x) = a^x$, where $a > 1$.

> **The larger the value of a, the faster the graph rises.**

To see this, compare the graph of $F(x) = 5^x$ with the graph of $f(x) = 2^x$ in **Figure 8.** When graphing such functions, be sure to plot a sufficient number of points to see how rapidly the graph rises.

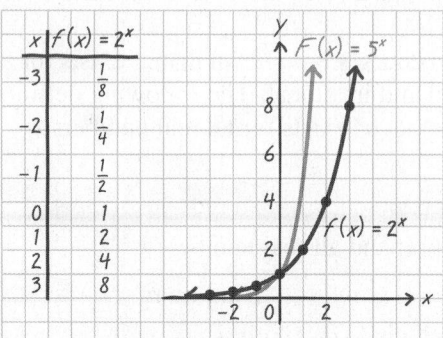

Figure 8 (repeated)

The vertical line test assures us that the graphs in **Figure 8** represent functions. **Figure 8** also shows an important characteristic of exponential functions where $a > 1$:

> **As x gets larger, y increases at a faster and faster rate.**

◀ **Work Problem 1** at the Side.

2 Graph $g(x) = \left(\dfrac{1}{10}\right)^x$.

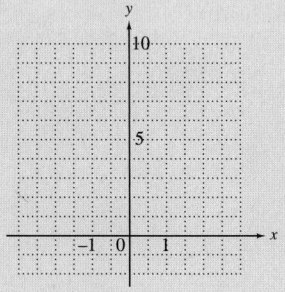

EXAMPLE 2   Graphing an Exponential Function ($0 < a < 1$)

Graph $g(x) = \left(\dfrac{1}{2}\right)^x$.

Again, find some points on the graph. The graph, shown in **Figure 9,** is very similar to that of $f(x) = 2^x$ (**Figure 8**) with the same domain and range, except that here **as x gets larger, y decreases.** This graph is typical of the graphs of exponential functions of the form $F(x) = a^x$, where $0 < a < 1$.

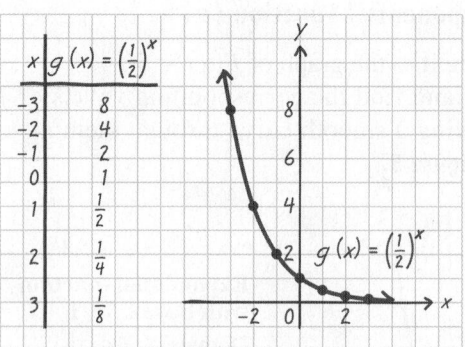

Exponential function with base $0 < a < 1$
Domain: $(-\infty, \infty)$
Range: $(0, \infty)$
The function is one-to-one, and its graph falls from left to right.

Figure 9

◀ **Work Problem 2** at the Side.

Answers

1.

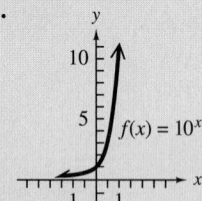

2.

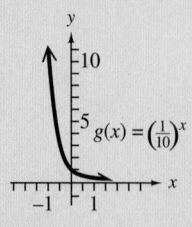

CAUTION

The graph of an exponential function of the form $f(x) = a^x$ approaches the x-axis, but does **not** touch it.

Characteristics of the Graph of $F(x) = a^x$

1. The graph contains the point $(0, 1)$.

2. The function is one-to-one.
   When $a > 1$, the graph *rises* from left to right. (See **Figure 8.**)
   When $0 < a < 1$, the graph *falls* from left to right. (See **Figure 9.**)
   In both cases, the graph goes from the second quadrant to the first.

3. The graph approaches the x-axis, but never touches it. (Recall from **Section 7.4** that such a line is an **asymptote.**)

4. The domain is $(-\infty, \infty)$, and the range is $(0, \infty)$.

❸ Graph $y = 2^{4x-3}$.

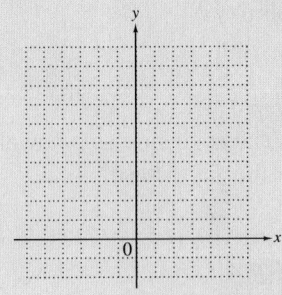

EXAMPLE 3 **Graphing a More Complicated Exponential Function**

Graph $f(x) = 3^{2x-4}$.

Find several ordered pairs. We let $x = 0$ and $x = 2$ and substitute to find values of $f(x)$, or y.

$y = 3^{2(0)-4}$ Let $x = 0$.	$y = 3^{2(2)-4}$ Let $x = 2$.
$y = 3^{-4}$, or $\dfrac{1}{81}$	$y = 3^0$, or 1

These ordered pairs, $\left(0, \frac{1}{81}\right)$ and $(2, 1)$, along with the other ordered pairs shown in the table, lead to the graph in **Figure 10.**

x	y
0	$\frac{1}{81}$
1	$\frac{1}{9}$
2	1
3	9

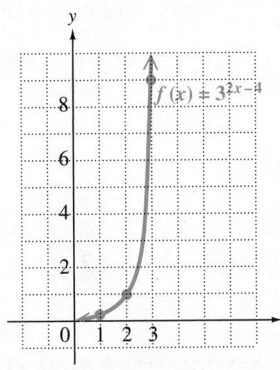

Figure 10

The graph of $f(x) = 3^{2x-4}$ is similar to the graph of $f(x) = 3^x$ except that it is shifted to the right and rises more rapidly.

································ Work Problem **❸** at the Side. ▶

OBJECTIVE ▶ ❸ Solve exponential equations of the form $a^x = a^k$ for x.
Until this chapter, we have solved only equations that had the variable as a base, like $x^2 = 8$. In these equations, all exponents have been constants. An **exponential equation** is an equation that has a variable in an exponent, such as

$$9^x = 27.$$

We can use the following property to solve many exponential equations.

Property for Solving an Exponential Equation

For $a > 0$ and $a \neq 1$, if $a^x = a^y$ then $x = y$.

This property would not necessarily be true if $a = 1$.

Answer

3.
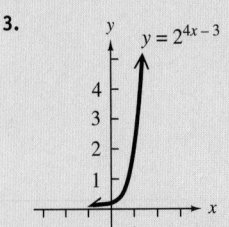

④ Solve each equation and check the solution.

(a) $25^x = 125$

Step 1
Write each side as a power of the same base.

$$(5\text{—})^x = 5\text{—}$$

Step 2
Use the power rule for exponents to simplify exponents on the left.

$$5\text{—} = 5^3$$

Step 3
Set the exponents equal.

$$\text{———} = \text{———}$$

Step 4
Solve for x.

$$x = \text{———}$$

Verify that the solution set is ———.

(b) $4^x = 32$

(c) $81^p = 27$

Solving an Exponential Equation

Step 1 **Each side must have the same base.** If the two sides of the equation do not have the same base, express each as a power of the same base.

Step 2 **Simplify exponents** if necessary, using the rules of exponents.

Step 3 **Set exponents equal** using the property given in this section.

Step 4 **Solve** the equation obtained in Step 3.

Note

These steps cannot be applied to an exponential equation like

$$3^x = 12$$

because Step 1 cannot easily be done. A method for solving such equations is given in **Section 11.7**.

EXAMPLE 4   **Solving an Exponential Equation**

Solve the equation $9^x = 27$.

$$9^x = 27$$
$$(3^2)^x = 3^3 \qquad \text{Write with the same base;}$$
$$\qquad\qquad\qquad 9 = 3^2 \text{ and } 27 = 3^3 \text{ (Step 1)}$$
$$3^{2x} = 3^3 \qquad \text{Power rule for exponents (Step 2)}$$
$$2x = 3 \qquad \text{If } a^x = a^y, \text{ then } x = y. \text{ (Step 3)}$$
$$x = \frac{3}{2} \qquad \text{Solve for } x. \text{ (Step 4)}$$

CHECK   Substitute $\frac{3}{2}$ for x.

$$9^x = 9^{3/2} = (9^{1/2})^3 = 3^3 = 27 \ \checkmark \quad \text{True}$$

The solution set is $\left\{\frac{3}{2}\right\}$.

◀ **Work Problem ④ at the Side.**

EXAMPLE 5   **Solving Exponential Equations**

Solve each equation.

(a) $4^{3x-1} = 16^{x+2}$ Be careful multiplying the exponents.

$$4^{3x-1} = (4^2)^{x+2} \qquad \text{Write with the same base; } 16 = 4^2$$
$$4^{3x-1} = 4^{2x+4} \qquad \text{Power rule for exponents}$$
$$3x - 1 = 2x + 4 \qquad \text{Set the exponents equal.}$$
$$x = 5 \qquad \text{Subtract } 2x. \text{ Add 1.}$$

Verify that the solution set is $\{5\}$.

············ **Continued on Next Page**

Answers

4. (a) $2; 3; 2x; 2x; 3; \frac{3}{2}; \left\{\frac{3}{2}\right\}$

(b) $\left\{\frac{5}{2}\right\}$ **(c)** $\left\{\frac{3}{4}\right\}$

(b) $6^x = \dfrac{1}{216}$

$6^x = \dfrac{1}{6^3}$ $216 = 6^3$

$6^x = 6^{-3}$ Write with the same base; $\frac{1}{6^3} = 6^{-3}$

$x = -3$ Set the exponents equal.

CHECK   Substitute -3 for x.

$$6^x = 6^{-3} = \frac{1}{6^3} = \frac{1}{216} \;\checkmark\; \text{True}$$

The solution set is $\{-3\}$.

(c) $\left(\dfrac{2}{3}\right)^x = \dfrac{9}{4}$

$\left(\dfrac{2}{3}\right)^x = \left(\dfrac{4}{9}\right)^{-1}$ $\frac{9}{4} = \left(\frac{4}{9}\right)^{-1}$

$\left(\dfrac{2}{3}\right)^x = \left[\left(\dfrac{2}{3}\right)^2\right]^{-1}$ Write with the same base.

$\left(\dfrac{2}{3}\right)^x = \left(\dfrac{2}{3}\right)^{-2}$ Power rule for exponents

$x = -2$ Set the exponents equal.

Check that the solution set is $\{-2\}$.

·· **Work Problem 5 at the Side.** ▶

OBJECTIVE 4 **Use exponential functions in applications involving growth or decay.**

EXAMPLE 6   **Applying an Exponential Growth Function**

The graph in **Figure 11** shows the concentration of carbon dioxide (in parts per million) in the air. This concentration is increasing exponentially.

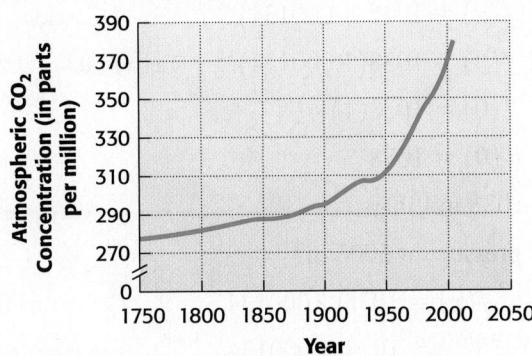

Carbon Dioxide in the Air

Source: Sacramento Bee; National Oceanic and Atmospheric Administration.

Figure 11

····················· **Continued on Next Page**

---

5 Solve each equation and check the solution.

GS **(a)** $25^{x-2} = 125^x$

   Step 1
   Write each side as a power of the same base.

$$(5\text{—})\text{—} = (5\text{—})^x$$

   Step 2
   Use the power rule for exponents to simplify exponents on each side.

$$5\text{—} = 5\text{—}$$

   Step 3
   Set the exponents equal.

$$\underline{\quad\quad} = \underline{\quad\quad}$$

   Step 4
   Solve for x.

$$x = \underline{\quad\quad}$$

   Verify that the solution set is $\underline{\quad}$.

(b) $3^{2x-1} = 27^{x+4}$

(c) $4^x = \dfrac{1}{32}$

(d) $\left(\dfrac{3}{4}\right)^x = \dfrac{16}{9}$

Answers

5. (a) $2; x - 2; 3; 2x - 4; 3x; 2x - 4; 3x;$
    $-4; \{-4\}$

   (b) $\{-13\}$ **(c)** $\left\{-\dfrac{5}{2}\right\}$ **(d)** $\{-2\}$

6 Use the exponential function in **Example 6** to approximate the carbon dioxide concentration in 1925.

The data graphed in **Figure 11** are approximated by the exponential function

$$f(x) = 266(1.001)^x,$$

where x is the number of years since 1750. Use this function and a calculator to approximate the concentration of carbon dioxide in parts per million, to the nearest unit, for each year.

(a) 1900

Because x represents the number of years since 1750,

$$x = 1900 - 1750 = 150.$$

$$f(x) = 266(1.001)^x \qquad \text{Given function}$$

$$f(150) = 266(1.001)^{150} \qquad \text{Let } x = 150.$$

$$f(150) \approx 309 \qquad \text{Use a calculator.}$$

The concentration in 1900 was 309 parts per million.

(b) 1950

$$f(x) = 266(1.001)^x \qquad \text{Given function}$$

$$f(200) = 266(1.001)^{200} \qquad x = 1950 - 1750 = 200$$

$$f(200) \approx 325 \qquad \text{Use a calculator.}$$

The concentration in 1950 was 325 parts per million.

◀ **Work Problem 6** at the Side.

7 Use the exponential function in **Example 7** to find the pressure at 8000 m.

EXAMPLE 7 **Applying an Exponential Decay Function**

The atmospheric pressure (in millibars) at a given altitude x, in meters, can be approximated by the exponential function

$$f(x) = 1038(1.000134)^{-x},$$

for values of x between 0 and 10,000. Because the base is greater than 1 and the coefficient of x in the exponent is negative, the function values decrease as x increases. This means that as the altitude increases, the atmospheric pressure decreases. (*Source:* Miller, A. and J. Thompson, *Elements of Meteorology,* Fourth Edition, Charles E. Merrill Publishing Company.)

(a) According to this function, what is the pressure at ground level?

$$f(x) = 1038(1.000134)^{-x} \qquad \text{Given function}$$

At ground level, $x = 0$.

$$f(0) = 1038(1.000134)^{-0} \qquad \text{Let } x = 0.$$

$$f(0) = 1038(1) \qquad a^0 = 1$$

$$f(0) = 1038$$

The pressure is 1038 millibars.

(b) What is the pressure at 5000 m?

$$f(x) = 1038(1.000134)^{-x} \qquad \text{Given function}$$

$$f(5000) = 1038(1.000134)^{-5000} \qquad \text{Let } x = 5000.$$

$$f(5000) \approx 531 \qquad \text{Use a calculator.}$$

The pressure is approximately 531 millibars.

◀ **Work Problem 7** at the Side.

Answers

6. 317 parts per million
7. approximately 355 millibars

11.3 Exercises

FOR EXTRA HELP

 MyMathLab®

Download the MyDashBoard App

CONCEPT CHECK *Choose the correct response.*

1. Which point lies on the graph of $f(x) = 2^x$?

   A. $(1, 0)$ **B.** $(2, 1)$

   C. $(0, 1)$ **D.** $\left(\sqrt{2}, \dfrac{1}{2}\right)$

2. The asymptote of the graph of $F(x) = a^x$

   A. is the x-axis.   **B.** is the y-axis.

   C. has equation $x = 1$.   **D.** has equation $y = 1$.

3. Which statement is true?

   A. The y-intercept of the graph of $f(x) = 10^x$ is $(0, 10)$.

   B. For any $a > 1$, the graph of $f(x) = a^x$ falls from left to right.

   C. The point $\left(\frac{1}{2}, \sqrt{5}\right)$ lies on the graph of $f(x) = 5^x$.

   D. The graph of $y = 4^x$ rises at a faster rate than the graph of $y = 10^x$.

4. Which statement is false?

   A. The domain of the function $f(x) = \left(\frac{1}{4}\right)^x$ is $(-\infty, \infty)$.

   B. The range of the function $f(x) = \left(\frac{1}{4}\right)^x$ is $(0, \infty)$.

   C. The function $f(x) = \left(\frac{1}{4}\right)^x$ is one-to-one.

   D. The graph of the function $f(x) = \left(\frac{1}{4}\right)^x$ has one x-intercept.

Graph each exponential function. ***See Examples 1–3.***

5. $f(x) = 3^x$

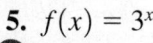

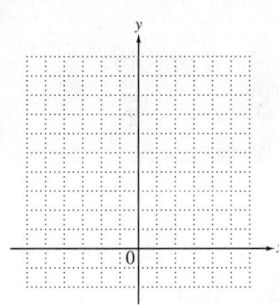

6. $f(x) = 5^x$

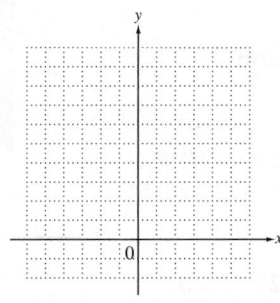

7. $g(x) = \left(\dfrac{1}{3}\right)^x$

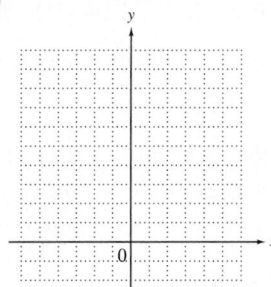

8. $g(x) = \left(\dfrac{1}{5}\right)^x$

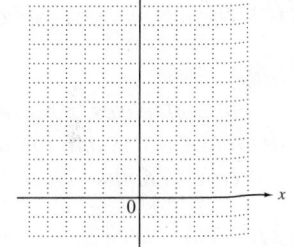

9. $f(x) = 4^{-x}$

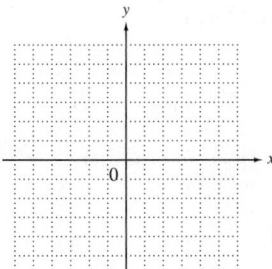

10. $f(x) = 6^{-x}$

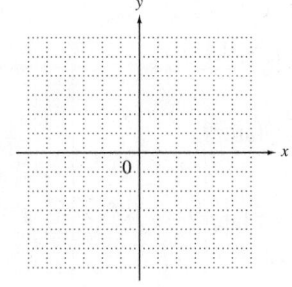

11. $y = 2^{2x-2}$

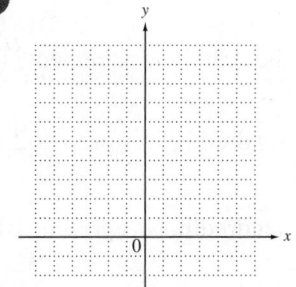

12. $y = 2^{2x+1}$

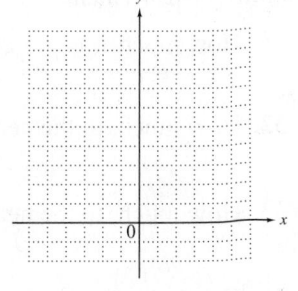

*Solve each equation. **See Examples 4 and 5.***

13. $6^t = 36$

14. $8^x = 64$

15. $100^x = 1000$
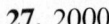

16. $8^x = 4$

17. $16^{2x+1} = 64^{x+3}$
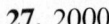

18. $9^{2x-8} = 27^{x-4}$

19. $5^x = \dfrac{1}{125}$

20. $3^x = \dfrac{1}{81}$

21. $5^x = 0.2$

22. $10^x = 0.1$

23. $\left(\dfrac{3}{2}\right)^x = \dfrac{8}{27}$

24. $\left(\dfrac{4}{3}\right)^x = \dfrac{27}{64}$

25. CONCEPT CHECK For an exponential function $f(x) = a^x$, if $a > 1$, then the graph (rises / falls) from left to right. If $0 < a < 1$, then the graph (rises / falls) from left to right.

26. CONCEPT CHECK Based on your answers in **Exercise 25**, make a conjecture (an educated guess) concerning whether an exponential function $f(x) = a^x$ is one-to-one. Then decide whether it has an inverse based on the concepts of **Section 11.2.**

*Solve each problem. **See Examples 6 and 7.***

A major scientific periodical published an article in 1990 dealing with the problem of global warming. The article was accompanied by a graph that illustrated two possible scenarios.

(a) The warming might be modeled by an exponential function of the form

$$y = (1.046 \times 10^{-38})(1.0444^x).$$

(b) The warming might be modeled by a linear function of the form

$$y = 0.009x - 17.67.$$

In both cases, x represents the year, and y represents the increase in degrees Celsius due to the warming. Use these functions to approximate the increase in temperature for each of the following years.

27. 2000

28. 2010

29. 2020

30. 2040

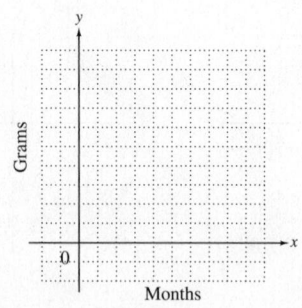

The amount of radioactive material in an ore sample is given by the function

$$A(t) = 100(3.2)^{-0.5t},$$

where $A(t)$ is the amount present, in grams, of the sample t months after the initial measurement.

31. How much was present at the initial measurement? (*Hint:* $t = 0$.)

32. How much, to the nearest hundredth, was present 2 months later?

33. How much, to the nearest hundredth, was present 10 months later?

34. Graph the function on the axes at the right.

11.4 Logarithmic Functions

The graph of $y = 2^x$ is the curve shown in blue in **Figure 12**. Because $y = 2^x$ defines a one-to-one function, it has an inverse function. Interchanging x and y gives

$$x = 2^y, \quad \text{the inverse of} \quad y = 2^x.$$

As we saw earlier in **Section 11.2**, the graph of the inverse is found by reflecting the graph of $y = 2^x$ about the line $y = x$. The graph of the inverse $x = 2^y$ is the curve shown in red in **Figure 12**.

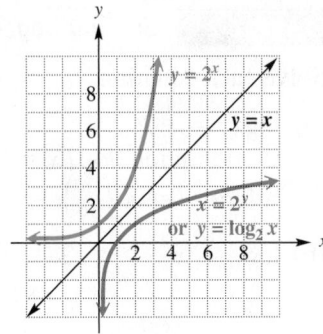

Figure 12

OBJECTIVES

1. Define a logarithm.
2. Convert between exponential and logarithmic forms.
3. Solve logarithmic equations of the form $\log_a b = k$ for a, b, or k.
4. Define and graph logarithmic functions.
5. Use logarithmic functions in applications involving growth or decay.

OBJECTIVE ▶ 1 Define a logarithm. We cannot solve the equation $x = 2^y$ for the dependent variable y with the methods presented up to now. We need the following definition to solve $x = 2^y$ for y.

Logarithm

For all positive numbers a, with $a \neq 1$, and all positive numbers x,

$$y = \log_a x \quad \text{means the same as} \quad x = a^y.$$

This key statement should be memorized. The abbreviation **log** is used for **logarithm**. Read $\log_a x$ as **"the logarithm of x with base a"** or **"the base a logarithm of x."** To remember the location of the base and the exponent in each form, refer to the following diagrams.

Logarithmic form: $\quad y = \log_a x \quad$ | $\quad$ Exponential form: $\quad x = a^y$

(Exponent / Base)

Meaning of $\log_a x$

A logarithm is an exponent. *The expression $\log_a x$ represents the exponent to which the base a must be raised to obtain x.*

OBJECTIVE ▶ 2 Convert between exponential and logarithmic forms. We use the definition of a logarithm to convert between these forms.

EXAMPLE 1   Converting between Exponential and Logarithmic Forms

The table shows several pairs of equivalent forms.

Exponential Form	Logarithmic Form
$3^2 = 9$	$\log_3 9 = 2$
$\left(\frac{1}{5}\right)^{-2} = 25$	$\log_{1/5} 25 = -2$
$10^5 = 100{,}000$	$\log_{10} 100{,}000 = 5$
$4^{-3} = \frac{1}{64}$	$\log_4 \frac{1}{64} = -3$

$y = \log_a x$
means
$x = a^y$.

···· **Work Problem ❶ at the Side.** ▶

---

❶ Complete the table.

Exponential Form	Logarithmic Form
$2^5 = 32$	
$100^{1/2} = 10$	
	$\log_8 4 = \frac{2}{3}$
	$\log_6 \frac{1}{1296} = -4$

Answers

1. $\log_2 32 = 5$; $\log_{100} 10 = \frac{1}{2}$;

$8^{2/3} = 4$; $6^{-4} = \frac{1}{1296}$

2 Solve each equation.

(a) $\log_3 27 = x$

(b) $\log_5 p = 2$

GS **(c)** $\log_2 (3x - 2) = 4$

Write the equation in exponential form.

$3x - 2 = \underline{\quad}^{\overline{\quad}}$

Apply the exponent.

$3x - 2 = \underline{\quad}$

Solve the equation for x.

$3x = \underline{\quad}$

$x = \underline{\quad}$

Check to confirm that the solution set is $\underline{\quad}$.

(d) $\log_m \dfrac{1}{16} = -4$

(e) $\log_x 12 = 3$

Answers

2. **(a)** $\{3\}$ **(b)** $\{25\}$ **(c)** 2; 4; 16; 18; 6; $\{6\}$
(d) $\{2\}$ **(e)** $\{\sqrt[3]{12}\}$

OBJECTIVE **3** **Solve logarithmic equations of the form $\log_a b = k$ for a, b, or k.** A **logarithmic equation** is an equation with a logarithm in at least one term.

EXAMPLE 2 **Solving Logarithmic Equations**

Solve each equation.

(a) $\log_4 x = -2$

By definition, $\log_4 x = -2$ is equivalent to $x = 4^{-2}$, and $4^{-2} = \frac{1}{16}$. The solution set is $\left\{ \frac{1}{16} \right\}$.

(b) $\log_{1/2} (3x + 1) = 2$

$$3x + 1 = \left(\frac{1}{2} \right)^2 \qquad \boxed{\text{This is a key step.}} \quad \text{Write in exponential form.}$$

$$3x + 1 = \frac{1}{4} \qquad \text{Apply the exponent.}$$

$$12x + 4 = 1 \qquad \text{Multiply each term by 4.}$$

$$x = -\frac{1}{4} \qquad \begin{array}{l}\text{Subtract 4, divide by 12, and}\\ \text{write in lowest terms.}\end{array}$$

CHECK $\log_{1/2} \left[3\left(-\dfrac{1}{4} \right) + 1 \right] \overset{?}{=} 2 \qquad$ Let $x = -\frac{1}{4}$.

$$\log_{1/2} \frac{1}{4} \overset{?}{=} 2 \qquad \text{Simplify within the parentheses.}$$

$$\left(\frac{1}{2} \right)^2 \overset{?}{=} \frac{1}{4} \qquad \text{Exponential form}$$

$$\frac{1}{4} = \frac{1}{4} \quad \checkmark \quad \text{True}$$

The solution set is $\left\{ -\frac{1}{4} \right\}$.

(c) $\log_x 3 = 2$

$$x^2 = 3 \qquad \text{Write in exponential form.}$$

$$x = \pm \sqrt{3} \qquad \text{Take square roots.}$$

Only the *principal* square root satisfies the equation since the base x must be a positive number. The solution set is $\left\{ \sqrt{3} \right\}$.

(d) $\log_{49} \sqrt[3]{7} = x$

$$49^x = \sqrt[3]{7} \qquad \text{Write in exponential form.}$$

$$(7^2)^x = 7^{1/3} \qquad \text{Write with the same base.}$$

$$7^{2x} = 7^{1/3} \qquad \text{Power rule for exponents}$$

$$2x = \frac{1}{3} \qquad \text{Set the exponents equal.}$$

$$x = \frac{1}{6} \qquad \text{Divide by 2 (or multiply by } \tfrac{1}{2}\text{).}$$

The solution set is $\left\{ \frac{1}{6} \right\}$.

◀ **Work Problem** **2** at the Side.

For any real positive number b, we know that $b^1 = b$ and $b^0 = 1$. Writing these two statements in logarithmic form gives the following properties of logarithms.

Properties of Logarithms

For any positive real number b, with $b \neq 1$, the following are true.

$$\log_b b = 1 \quad \text{and} \quad \log_b 1 = 0$$

EXAMPLE 3 **Using Properties of Logarithms**

Evaluate each logarithm.

(a) $\log_7 7 = 1$ $\quad \log_b b = 1$ $\qquad$ **(b)** $\log_{\sqrt{2}} \sqrt{2} = 1$

(c) $\log_9 1 = 0$ $\quad \log_b 1 = 0$ $\qquad$ **(d)** $\log_{0.2} 1 = 0$

·········· **Work Problem ③ at the Side.** ▶

OBJECTIVE ④ Define and graph logarithmic functions. Now we define the logarithmic function with base a.

Logarithmic Function

If a and x are positive numbers, with $a \neq 1$, then

$$G(x) = \log_a x$$

is the **logarithmic function with base** a.

EXAMPLE 4 **Graphing a Logarithmic Function ($a > 1$)**

Graph $f(x) = \log_2 x$.

By writing $y = f(x) = \log_2 x$ in exponential form as $x = 2^y$, we can identify ordered pairs that satisfy the equation. It is easier to choose values for y and find the corresponding values of x. Plotting the points in the table and connecting them with a smooth curve gives the graph in **Figure 13.** This graph is typical of logarithmic functions with base $a > 1$.

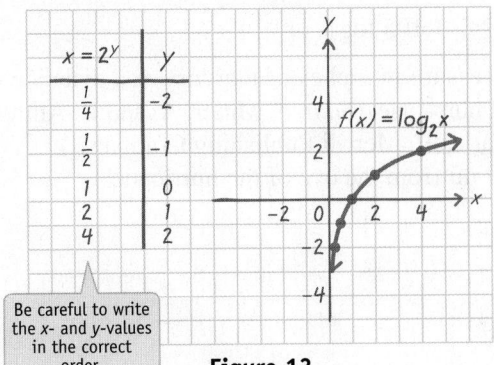

Be careful to write the x- and y-values in the correct order.

Figure 13

Logarithmic function with base $a > 1$

Domain: $(0, \infty)$

Range: $(-\infty, \infty)$

The function is one-to-one, and its graph rises from left to right.

·········· **Work Problem ④ at the Side.** ▶

---

③ Evaluate each logarithm.

(a) $\log_{2/5} \dfrac{2}{5}$

(b) $\log_\pi \pi$

(c) $\log_{0.4} 1$

(d) $\log_6 1$

④ Graph $y = \log_{10} x$.

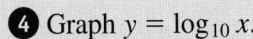

Answers

3. (a) 1 **(b)** 1 **(c)** 0 **(d)** 0

4.

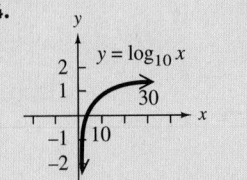

5 Graph $y = \log_{1/10} x$.

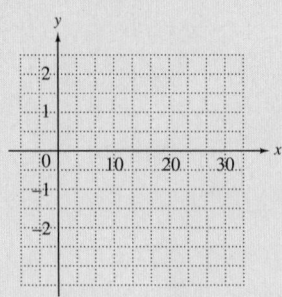

6 Solve the problem.

A population of mites in a laboratory is growing according to the logarithmic function

$$P(t) = 80 \log_{10}(t + 10),$$

where t is the number of days after a study is begun.

(a) Find the number of mites at the beginning of the study.

(b) Find the number present after 90 days.

Answers

5.

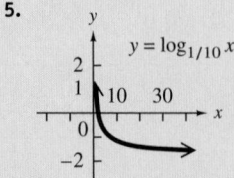

6. (a) 80 **(b)** 160

EXAMPLE 5 **Graphing a Logarithmic Function ($0 < a < 1$)**

Graph $g(x) = \log_{1/2} x$.

We write $y = g(x) = \log_{1/2} x$ in exponential form as $x = \left(\frac{1}{2}\right)^y$. Then we choose values for y and find the corresponding values of x. Plotting these points and connecting them with a smooth curve gives the graph in **Figure 14.** This graph is typical of logarithmic functions with $0 < a < 1$.

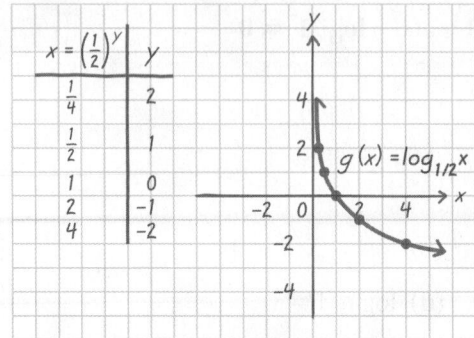

Logarithmic function with base $0 < a < 1$

Domain: $(0, \infty)$
Range: $(-\infty, \infty)$

The function is one-to-one, and its graph falls from left to right.

Figure 14

◀ **Work Problem 5** at the Side.

Characteristics of the Graph of $G(x) = \log_a x$

1. The graph contains the point $(1, 0)$.

2. The function is one-to-one.

   When $a > 1$, the graph *rises* from left to right, from the fourth quadrant to the first. (See **Figure 13.**)

   When $0 < a < 1$, the graph *falls* from left to right, from the first quadrant to the fourth. (See **Figure 14.**)

3. The graph approaches the y-axis, but never touches it. (The y-axis is an asymptote.)

4. The domain is $(0, \infty)$, and the range is $(-\infty, \infty)$.

OBJECTIVE ▶ **5** **Use logarithmic functions in applications involving growth or decay.**

EXAMPLE 6 **Solving an Application of a Logarithmic Function**

The logarithmic function

$$f(x) = 27 + 1.105 \log_{10}(x + 1)$$

approximates the barometric pressure in inches of mercury at a distance of x miles from the eye of a typical hurricane. (*Source:* Miller, A. and R. Anthes, *Meteorology,* Fifth Edition, Charles E. Merrill Publishing Company.)

Approximate the pressure 9 mi from the eye of the hurricane.

$$f(9) = 27 + 1.105 \log_{10}(9 + 1) \quad \text{Let } x = 9.$$
$$f(9) = 27 + 1.105 \log_{10} 10 \quad \text{Add inside the parentheses.}$$
$$f(9) = 27 + 1.105 (1) \quad \log_{10} 10 = 1$$
$$f(9) = 28.105 \quad \text{Add.}$$

The pressure 9 mi from the eye of the hurricane is 28.105 in.

◀ **Work Problem 6** at the Side.

11.4 Exercises

 MyMathLab®

1. CONCEPT CHECK Match the logarithm in Column I with its value in Column II. (*Example:* $\log_3 9$ is equal to 2 because 2 is the exponent to which 3 must be raised in order to obtain 9.)

I	**II**
(a) $\log_4 16$	**A.** -2
(b) $\log_3 81$	**B.** -1
(c) $\log_3\left(\dfrac{1}{3}\right)$	**C.** 2
(d) $\log_{10} 0.01$	**D.** 0
(e) $\log_5 \sqrt{5}$	**E.** $\dfrac{1}{2}$
(f) $\log_{13} 1$	**F.** 4

2. CONCEPT CHECK Match the logarithmic equation in Column I with the corresponding exponential equation from Column II.

I	**II**
(a) $\log_{1/3} 3 = -1$	**A.** $8^{1/3} = \sqrt[3]{8}$
(b) $\log_5 1 = 0$	**B.** $\left(\dfrac{1}{3}\right)^{-1} = 3$
(c) $\log_2 \sqrt{2} = \dfrac{1}{2}$	**C.** $4^1 = 4$
(d) $\log_{10} 1000 = 3$	**D.** $2^{1/2} = \sqrt{2}$
(e) $\log_8 \sqrt[3]{8} = \dfrac{1}{3}$	**E.** $5^0 = 1$
(f) $\log_4 4 = 1$	**F.** $10^3 = 1000$

Write in logarithmic form. ***See Example 1.***

3. $4^5 = 1024$

4. $3^6 = 729$

5. $\left(\dfrac{1}{2}\right)^{-3} = 8$

6. $\left(\dfrac{1}{6}\right)^{-3} = 216$

7. $10^{-3} = 0.001$

8. $10^{-1} = 0.1$

9. $\sqrt[4]{625} = 5$

10. $\sqrt[3]{343} = 7$

11. $8^{-2/3} = \dfrac{1}{4}$

12. $16^{-3/4} = \dfrac{1}{8}$

13. $5^0 = 1$

14. $7^0 = 1$

Write in exponential form. ***See Example 1.***

15. $\log_4 64 = 3$

16. $\log_2 512 = 9$

17. $\log_{12} 12 = 1$

18. $\log_{100} 100 = 1$

19. $\log_6 1 = 0$

20. $\log_\pi 1 = 0$

21. $\log_9 3 = \dfrac{1}{2}$

22. $\log_{64} 2 = \dfrac{1}{6}$

23. $\log_{1/4} \dfrac{1}{2} = \dfrac{1}{2}$

24. $\log_{1/8} \dfrac{1}{2} = \dfrac{1}{3}$

25. $\log_5 5^{-1} = -1$

26. $\log_{10} 10^{-2} = -2$

27. CONCEPT CHECK Match each logarithm in Column I with its value in Column II.

I	II
(a) $\log_8 8$	A. -1
(b) $\log_{16} 1$	B. 0
(c) $\log_{0.3} 1$	C. 1
(d) $\log_{\sqrt{7}} \sqrt{7}$	D. 0.1

28. When a student asked his teacher to explain how to evaluate

$$\log_9 3$$

without showing any work, his teacher told him, "Think radically." Explain what the teacher meant by this hint.

Solve each equation. See Examples 2 and 3.

29. $x = \log_{27} 3$

30. $x = \log_{125} 5$

31. $\log_x 9 = \dfrac{1}{2}$

32. $\log_x 5 = \dfrac{1}{2}$

33. $\log_x 125 = -3$

34. $\log_x 64 = -6$

35. $\log_{12} x = 0$

36. $\log_4 x = 0$

37. $\log_x x = 1$

38. $\log_x 1 = 0$

39. $\log_x \dfrac{1}{25} = -2$

40. $\log_x \dfrac{1}{10} = -1$

41. $\log_8 32 = x$

42. $\log_{81} 27 = x$

43. $\log_\pi \pi^4 = x$

44. $\log_{\sqrt{2}} \left(\sqrt{2} \right)^9 = x$

45. $\log_6 \sqrt{216} = x$

46. $\log_4 \sqrt{64} = x$

47. $\log_4 (2x + 4) = 3$

48. $\log_3 (2x + 7) = 4$

*If the point (p, q) is on the graph of $f(x) = a^x$ (for $a > 0$ and $a \neq 1$), then the point (q, p) is on the graph of $f^{-1}(x) = \log_a x$. Use this fact and refer to the graphs required in **Exercises 5–10** in **Section 11.3** to graph each logarithmic function. **See Examples 4 and 5.***

49. $y = \log_3 x$

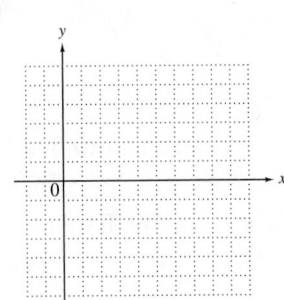

50. $y = \log_5 x$

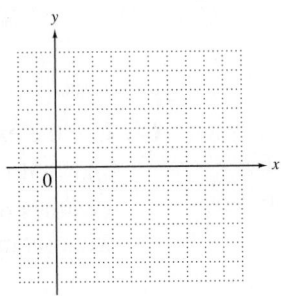

51. $y = \log_{1/3} x$

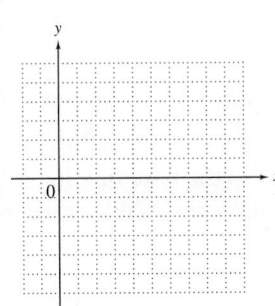

52. $y = \log_{1/5} x$

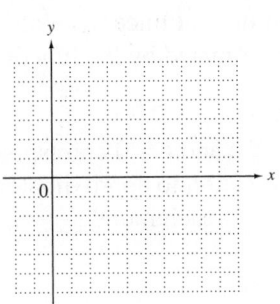

53. $y = \log_{1/4} x$

(*Hint:* $4^{-x} = \left(\frac{1}{4}\right)^x$.)

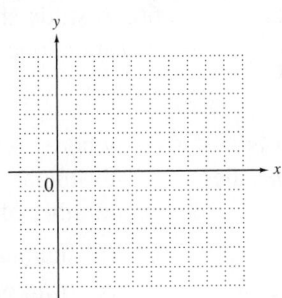

54. $y = \log_{1/6} x$

(*Hint:* $6^{-x} = \left(\frac{1}{6}\right)^x$.)

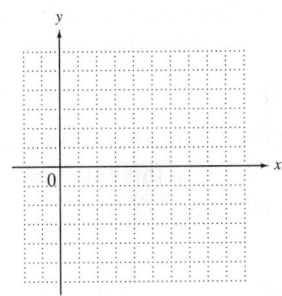

55. CONCEPT CHECK Compare the summary of characteristics of the graph of $F(x) = a^x$ in **Section 11.3** with the similar summary of characteristics of the graph of $G(x) = \log_a x$ in this section. Make a list of the characteristics that reinforce the concept that F and G are inverse functions.

56. CONCEPT CHECK The domain of $F(x) = a^x$ is $(-\infty, \infty)$, while the range is $(0, \infty)$. Therefore, since $G(x) = \log_a x$ defines the inverse of F, the domain of G is _____, while the range of G is _____.

CONCEPT CHECK *Use the graph to predict the value of $f(t)$ for each value of t.*

57. $t = 0$

58. $t = 10$

59. $t = 60$

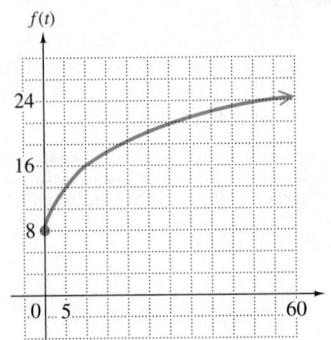

60. Show that the points determined in **Exercises 57–59** lie on the graph of

$$f(t) = 8 \log_5 (2t + 5).$$

CONCEPT CHECK *Answer each question.*

61. Why is 1 not allowed as a base for a logarithmic function?

62. The graphs of both $f(x) = 3^x$ and $g(x) = \log_3 x$ rise from left to right. Which one rises at a faster rate?

Write a short answer to each problem.

63. Explain why $\log_a 1$ is 0 for any value of a that is allowed as the base of a logarithm. Use a rule of exponents introduced earlier in your explanation.

64. Use the exponential key of your calculator to find approximations for the expression $\left(1 + \frac{1}{x}\right)^x$, using x-values of 1, 10, 100, 1000, and 10,000. Explain what seems to be happening as x gets larger and larger.

Solve each problem. See Example 6.

65. Sales (in thousands of units) of a new product are approximated by the function

$$S(t) = 100 + 30 \log_3 (2t + 1),$$

where t is the number of years after the product is introduced. Use this function to approximate the sales after each period of time.

(a) 1 yr

(b) 4 yr

(c) 13 yr

66. A study showed that the number of mice in an old abandoned house was approximated by the function

$$M(t) = 6 \log_4 (2t + 4),$$

where t is measured in months and $t = 0$ corresponds to January 2012. Use this function to approximate the number of mice in the house in each month.

(a) January 2012

(b) July 2012

(c) July 2014

*In the United States, the intensity of an earthquake is rated using the **Richter scale**. The Richter scale rating of an earthquake of intensity x is given by*

$$R = \log_{10} \frac{x}{x_0},$$

where x_0 is the intensity of an earthquake of a certain (small) size. The figure shows Richter scale ratings for major Southern California earthquakes since 1920. As the figure indicates, earthquakes "come in bunches" and the 1990s were an especially busy time.

67. The 1994 Northridge earthquake had a Richter scale rating of 6.7 and the 1992 Landers earthquake had a rating of 7.3. How much more powerful was the Landers earthquake than the Northridge earthquake?

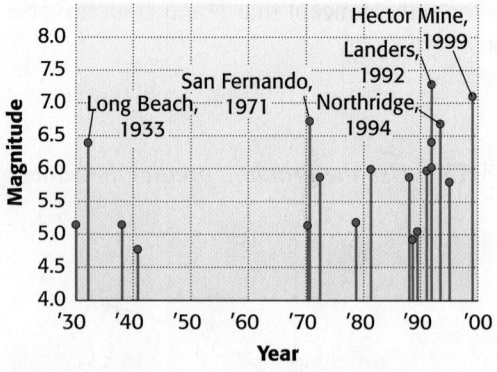

Major Southern California Earthquakes
(with magnitudes greater than 4.7)

Source: Caltech; U.S. Geological Survey.

68. Compare the smallest rated earthquake in the figure (at 4.8) with the Landers quake. How much more powerful was the Landers quake?

11.5 Properties of Logarithms

Logarithms have been used as an aid to numerical calculation for several hundred years. Today the widespread use of calculators has made the use of logarithms for calculation obsolete. However, logarithms are still very important in applications and in further work in mathematics.

OBJECTIVE ▶ 1 Use the product rule for logarithms. One way in which logarithms are used is to change a problem of multiplication into one of addition. For example, since we know that

$$\log_2 4 = 2, \quad \log_2 8 = 3, \quad \text{and} \quad \log_2 32 = 5,$$

we can make the following statements.

$$\log_2 32 = \log_2 4 + \log_2 8 \qquad 5 = 2 + 3$$
$$\log_2 (4 \cdot 8) = \log_2 4 + \log_2 8 \qquad 32 = 4 \cdot 8$$

This is an example of the product rule for logarithms.

> **Product Rule for Logarithms**
>
> If x, y, and b are positive real numbers, where $b \neq 1$, then the following is true.
>
> $$\log_b xy = \log_b x + \log_b y$$
>
> In words, the logarithm of a product is the sum of the logarithms of the factors.

To prove this rule, let $m = \log_b x$ and $n = \log_b y$, and recall that

$$\log_b x = m \quad \text{means} \quad b^m = x.$$
$$\log_b y = n \quad \text{means} \quad b^n = y.$$

Now consider the product xy.

$xy = b^m \cdot b^n$	Substitute.
$xy = b^{m+n}$	Product rule for exponents
$\log_b xy = m + n$	Convert to logarithmic form.
$\log_b xy = \log_b x + \log_b y$	Substitute for m and n.

The last statement is the result we wished to prove.

> **Note**
>
> The word statement of the product rule for logarithms above can be restated by replacing the word "logarithm" with the word "exponent." The rule then becomes the familiar rule for multiplying exponential expressions:
>
> The *exponent* of a product is equal to the sum of the *exponents* of the factors.

❶ Use the product rule to rewrite each expression.

(a) $\log_6 (5 \cdot 8)$

(b) $\log_4 3 + \log_4 7$

(c) $\log_8 8k, \quad k > 0$

(d) $\log_5 m^2, \quad m > 0$

❷ Use the quotient rule to rewrite each expression.

(a) $\log_7 \dfrac{9}{4}$

(b) $\log_3 p - \log_3 q,$
$p > 0, \quad q > 0$

(c) $\log_4 \dfrac{3}{16}$

(d) $\log_7 32 - \log_7 4$

EXAMPLE 1	Using the Product Rule

Use the product rule to rewrite each expression. Assume $x > 0$.

(a) $\log_5 (6 \cdot 9)$

$= \log_5 6 + \log_5 9$ Product rule

(b) $\log_7 8 + \log_7 12$

$= \log_7 (8 \cdot 12)$ Product rule

$= \log_7 96$ Multiply.

(c) $\log_3 (3x)$

$= \log_3 3 + \log_3 x$ Product rule

$= 1 + \log_3 x$ $\log_3 3 = 1$

(d) $\log_4 x^3$

$= \log_4 (x \cdot x \cdot x)$ $x^3 = x \cdot x \cdot x$

$= \log_4 x + \log_4 x + \log_4 x$ Product rule

$= 3 \log_4 x$ Combine like terms.

◀ **Work Problem ❶ at the Side.**

OBJECTIVE ▶ ❷ Use the quotient rule for logarithms.

Quotient Rule for Logarithms

If x, y, and b are positive real numbers, where $b \neq 1$, then the following is true.

$$\log_b \frac{x}{y} = \log_b x - \log_b y$$

In words, the logarithm of a quotient is the difference between the logarithm of the numerator and the logarithm of the denominator.

The proof of this rule is very similar to the proof of the product rule.

EXAMPLE 2	Using the Quotient Rule

Use the quotient rule to rewrite each expression. Assume $x > 0$.

(a) $\log_4 \dfrac{7}{9}$

$= \log_4 7 - \log_4 9$ Quotient rule

(b) $\log_5 6 - \log_5 x$

$= \log_5 \dfrac{6}{x}$ Quotient rule

(c) $\log_3 \dfrac{27}{5}$

$= \log_3 27 - \log_3 5$

$= 3 - \log_3 5$

(d) $\log_6 28 - \log_6 7$

$= \log_6 \dfrac{28}{7}$

$= \log_6 4$

◀ **Work Problem ❷ at the Side.**

CAUTION

There is no property of logarithms to rewrite the logarithm of a sum or difference. For example, we *cannot* write $\log_b (x + y)$ in terms of $\log_b x$ and $\log_b y$. Also,

$$\log_b \frac{x}{y} \neq \frac{\log_b x}{\log_b y}.$$

Answers

1. **(a)** $\log_6 5 + \log_6 8$ **(b)** $\log_4 21$
    (c) $1 + \log_8 k$ **(d)** $2 \log_5 m$
2. **(a)** $\log_7 9 - \log_7 4$ **(b)** $\log_3 \dfrac{p}{q}$
    (c) $\log_4 3 - 2$ **(d)** $\log_7 8$

OBJECTIVE ▶ 3 Use the power rule for logarithms. An exponential expression such as 2^3 means $2 \cdot 2 \cdot 2$. The base 2 is used as a factor 3 times. Similarly, the product rule can be extended to rewrite the logarithm of a power as the product of the exponent and the logarithm of the base.

$$\log_5 2^3$$
$$= \log_5 (2 \cdot 2 \cdot 2)$$
$$= \log_5 2 + \log_5 2 + \log_5 2$$
$$= 3 \log_5 2$$

$$\log_2 7^4$$
$$= \log_2 (7 \cdot 7 \cdot 7 \cdot 7)$$
$$= \log_2 7 + \log_2 7 + \log_2 7 + \log_2 7$$
$$= 4 \log_2 7$$

Furthermore, we saw in **Example 1(d)** that $\log_4 x^3 = 3 \log_4 x$. These examples suggest the power rule for logarithms.

Power Rule for Logarithms

If x and b are positive real numbers, where $b \neq 1$, and if r is any real number, then the following is true.

$$\log_b x^r = r \log_b x$$

In words, the logarithm of a number to a power equals the exponent times the logarithm of the number.

Here are some further examples of this rule.

$$\log_b m^5 = 5 \log_b m \quad \text{and} \quad \log_3 5^4 = 4 \log_3 5$$

To prove the power rule, let $\log_b x = m$.

$\log_b x = m$	
$b^m = x$	Convert to exponential form.
$(b^m)^r = x^r$	Raise to the power r.
$b^{mr} = x^r$	Power rule for exponents
$\log_b x^r = mr$	Convert to logarithmic form.
$\log_b x^r = rm$	Commutative property
$\log_b x^r = r \log_b x$	$m = \log_b x$

This is the statement to be proved.

As a special case of the power rule, let $r = \frac{1}{p}$, so

$$\log_b \sqrt[p]{x} = \log_b x^{1/p} = \frac{1}{p} \log_b x.$$

For example, using this result, with $x > 0$,

$$\log_b \sqrt[5]{x} = \log_b x^{1/5} = \frac{1}{5} \log_b x \quad \text{and} \quad \log_b \sqrt[3]{x^4} = \log_b x^{4/3} = \frac{4}{3} \log_b x.$$

Another special case is

$$\log_b \frac{1}{x} = \log_b x^{-1} = -\log_b x.$$

Note

For a review of rational exponents, refer to **Section 9.2.**

3 Use the power rule to rewrite each logarithm. Assume that $a > 0, b > 0, x > 0, a \neq 1$, and $b \neq 1$.

(a) $\log_3 5^2$

(b) $\log_a x^4$

(c) $\log_b \sqrt{8}$

(d) $\log_2 \sqrt[3]{2}$

(e) $\log_3 \dfrac{1}{x^5}$

4 Find the value of each logarithmic expression.

(a) $\log_{10} 10^3$

(b) $\log_2 8$

(c) $5^{\log_5 3}$

(d) $6^{\log_6 \sqrt{3}}$

EXAMPLE 3 **Using the Power Rule**

Use the power rule to rewrite each logarithm. Assume that $b > 0, x > 0$, and $b \neq 1$.

(a) $\log_5 4^2$

   $= 2 \log_5 4$ Power rule

(b) $\log_b x^5$

   $= 5 \log_b x$ Power rule

(c) $\log_b \sqrt{7}$

When using the power rule with logarithms of expressions involving radicals, begin by rewriting the radical expression with a rational exponent.

$$\log_b \sqrt{7}$$
$$= \log_b 7^{1/2} \quad \sqrt{x} = x^{1/2}$$
$$= \frac{1}{2} \log_b 7 \quad \text{Power rule}$$

(d) $\log_2 \sqrt[5]{x^2}$

   $= \log_2 x^{2/5}$ $\sqrt[5]{x^2} = x^{2/5}$

   $= \dfrac{2}{5} \log_2 x$ Power rule

(e) $\log_3 \dfrac{1}{x^4}$

   $= \log_3 x^{-4}$ Definition of negative exponent

   $= -4 \log_3 x$ Power rule

◀ **Work Problem 3** at the Side.

Two special properties involving both exponential and logarithmic expressions come directly from the fact that logarithmic and exponential functions are inverses of each other.

Special Properties

If $b > 0$ and $b \neq 1$, then the following are true.

$$b^{\log_b x} = x, \; x > 0 \quad \text{and} \quad \log_b b^x = x$$

To prove the first statement, let $y = \log_b x$.

$$y = \log_b x$$
$$b^y = x \qquad \text{Convert to exponential form.}$$
$$b^{\log_b x} = x \qquad \text{Replace } y \text{ with } \log_b x.$$

The proof of the second statement is similar.

EXAMPLE 4 **Using the Special Properties**

Find the value of each logarithmic expression.

(a) $\log_5 5^4 = 4$ $\log_b b^x = x$

(b) $\log_3 9$

   $= \log_3 3^2$ $9 = 3^2$

   $= 2$ $\log_b b^x = x$

(c) $4^{\log_4 10} = 10$ $b^{\log_b x} = x$

(d) $8^{\log_8 \sqrt{2}} = \sqrt{2}$ $b^{\log_b x} = x$

◀ **Work Problem 4** at the Side.

Answers

3. **(a)** $2 \log_3 5$ **(b)** $4 \log_a x$ **(c)** $\dfrac{1}{2} \log_b 8$

   (d) $\dfrac{1}{3}$ **(e)** $-5 \log_3 x$

4. **(a)** 3 **(b)** 3 **(c)** 3 **(d)** $\sqrt{3}$

We summarize the properties of logarithms.

Properties of Logarithms

If x, y, and b are positive real numbers, where $b \neq 1$, and r is any real number, then the following are true.

Product Rule $\log_b xy = \log_b x + \log_b y$

Quotient Rule $\log_b \dfrac{x}{y} = \log_b x - \log_b y$

Power Rule $\log_b x^r = r \log_b x$

Special Properties $b^{\log_b x} = x$ and $\log_b b^x = x$

OBJECTIVE ▶ ④ Use properties to write alternative forms of logarithmic expressions.

EXAMPLE 5   **Writing Logarithms in Alternative Forms**

Use the properties of logarithms to rewrite each expression if possible. Assume that all variables represent positive real numbers.

(a) $\log_4 4x^3$

$\quad = \log_4 4 + \log_4 x^3 \qquad$ Product rule

$\quad = 1 + 3 \log_4 x \qquad\quad \log_4 4 = 1;$ Power rule

(b) $\log_7 \sqrt{\dfrac{m}{n}}$

$\quad = \log_7 \left(\dfrac{m}{n}\right)^{1/2} \qquad$ Write the radical expression with a rational exponent.

$\quad = \dfrac{1}{2} \log_7 \dfrac{m}{n} \qquad\quad$ Power rule

$\quad = \dfrac{1}{2}\left(\log_7 m - \log_7 n\right) \qquad$ Quotient rule

(c) $\log_5 \dfrac{a^2}{bc}$

$\quad = \log_5 a^2 - \log_5 bc \qquad\qquad$ Quotient rule

$\quad = 2 \log_5 a - \log_5 bc \qquad\qquad$ Power rule

$\quad = 2 \log_5 a - \left(\log_5 b + \log_5 c\right) \qquad$ Product rule

$\quad = 2 \log_5 a - \log_5 b - \log_5 c \qquad$ Parentheses are necessary here.

(d) $4 \log_b m - \log_b n$

$\quad = \log_b m^4 - \log_b n \qquad$ Power rule

$\quad = \log_b \dfrac{m^4}{n} \qquad\qquad$ Quotient rule

···· **Continued on Next Page**

5 Use the properties of logarithms to rewrite each expression if possible. Assume that all variables represent positive real numbers.

(a) $\log_6 36m^5$

(b) $\log_2 \sqrt{9z}$

(c) $\log_q \dfrac{8r^2}{m-1}$, $m > 1, q \neq 1$

(d) $2 \log_a x + 3 \log_a y$, $a \neq 1$

(e) $\log_4 (3x + y)$

6 Decide whether each statement is *true* or *false*.

(a) $\log_6 36 - \log_6 6 = \log_6 30$

(b) $\log_4 (\log_2 16) = \dfrac{\log_6 6}{\log_6 36}$

(e) $\log_b (x + 1) + \log_b (2x + 1) - \dfrac{2}{3} \log_b x$

$= \log_b (x + 1) + \log_b (2x + 1) - \log_b x^{2/3}$ Power rule

$= \log_b \dfrac{(x + 1)(2x + 1)}{x^{2/3}}$ Product and quotient rules

$= \log_b \dfrac{2x^2 + 3x + 1}{x^{2/3}}$ Multiply in the numerator.

(f) $\log_8 (2p + 3r)$ cannot be rewritten using the properties of logarithms. ***There is no property of logarithms to rewrite the logarithm of a sum.***

◀ **Work Problem 5** at the Side.

EXAMPLE 6   **Deciding Whether Statements about Logarithms Are True**

Decide whether each statement is *true* or *false*.

(a) $\log_2 8 - \log_2 4 = \log_2 4$
Evaluate each side.

$\log_2 8 - \log_2 4$	Left side	$\log_2 4$	Right side
$= \log_2 2^3 - \log_2 2^2$	Write 8 and 4 as powers of 2.	$= \log_2 2^2$	Write 4 as a power of 2.
$= 3 - 2$	$\log_a a^x = x$	$= 2$	$\log_a a^x = x$
$= 1$	Subtract.		

The statement is false because $1 \neq 2$.

(b) $\log_3 (\log_2 8) = \dfrac{\log_7 49}{\log_8 64}$
Evaluate each side.

$\log_3 (\log_2 8)$	Left side	$\dfrac{\log_7 49}{\log_8 64}$	Right side
$= \log_3 (\log_2 2^3)$	Write 8 as a power of 2.	$= \dfrac{\log_7 7^2}{\log_8 8^2}$	Write 49 and 64 using exponents.
$= \log_3 3$	$\log_a a^x = x$	$= \dfrac{2}{2}$	$\log_a a^x = x$
$= 1$	$3 = 3^1$	$= 1$	Simplify.

The statement is true because $1 = 1$.

◀ **Work Problem 6** at the Side.

Answers

5. **(a)** $2 + 5 \log_6 m$ **(b)** $\log_2 3 + \dfrac{1}{2} \log_2 z$
  (c) $\log_q 8 + 2 \log_q r - \log_q (m - 1)$
  (d) $\log_a x^2 y^3$ **(e)** cannot be rewritten
6. **(a)** false   **(b)** false

11.5 Exercises

CONCEPT CHECK *Use the indicated rule of logarithms to complete each equation.*

1. $\log_{10}(7 \cdot 8) = $ _____ Product rule

2. $\log_{10}\dfrac{7}{8} = $ _____ Quotient rule

3. $3^{\log_3 4} = $ _____ Special property

4. $\log_{10} 3^6 = $ _____ Power rule

CONCEPT CHECK *Decide whether each statement of a logarithmic property is* true *or* false. *If it is* false, *correct it by changing the right side of the equation.*

5. $\log_b x + \log_b y = \log_b (x + y)$

6. $\log_b \dfrac{x}{y} = \log_b x - \log_b y$

7. $\log_b b^x = x$

8. $\log_b x^r = \log_b rx$

Use the properties of logarithms to express each logarithm as a sum or difference of logarithms, or as a single number if possible. Assume that all variables represent positive real numbers. **See Examples 1–5.**

9. $\log_7 \dfrac{4}{5}$

10. $\log_8 \dfrac{9}{11}$

11. $\log_2 8^{1/4}$

12. $\log_3 9^{3/4}$

13. $\log_4 \dfrac{3\sqrt{x}}{y}$

14. $\log_5 \dfrac{6\sqrt{z}}{w}$

15. $\log_3 \dfrac{\sqrt[3]{4}}{x^2 y}$

16. $\log_7 \dfrac{\sqrt[3]{13}}{pq^2}$

17. $\log_3 \sqrt{\dfrac{xy}{5}}$

18. $\log_6 \sqrt{\dfrac{pq}{7}}$

19. $\log_2 \dfrac{\sqrt[3]{x} \cdot \sqrt[5]{y}}{r^2}$

20. $\log_4 \dfrac{\sqrt[4]{z} \cdot \sqrt[5]{w}}{s^2}$

21. CONCEPT CHECK A student erroneously wrote

$$\log_a (x + y) = \log_a x + \log_a y.$$

When his teacher explained that this was wrong, the student claimed he had used the distributive property. **What Went Wrong?**

22. CONCEPT CHECK Why can't we determine a logarithm of 0? (*Hint:* Think of the definition of logarithm.)

Use the properties of logarithms to rewrite each expression as a single logarithm.
Assume that all variables are defined in such a way that the variable expressions
are positive, and bases are positive numbers not equal to 1. See Examples 1–5.

23. $\log_b x + \log_b y$

24. $\log_b 2 + \log_b z$

25. $3 \log_a m - \log_a n$

26. $5 \log_b x - \log_b y$

27. $(\log_a r - \log_a s) + 3 \log_a t$

28. $(\log_a p - \log_a q) + 2 \log_a r$

29. $3 \log_a 5 - 4 \log_a 3$

30. $3 \log_a 5 + \dfrac{1}{2} \log_a 9$

31. $\log_{10}(x + 3) + \log_{10}(x - 3)$

32. $\log_{10}(y + 4) + \log_{10}(y - 4)$

33. $3 \log_p x + \dfrac{1}{2} \log_p y - \dfrac{3}{2} \log_p z - 3 \log_p a$

34. $\dfrac{1}{3} \log_b x + \dfrac{2}{3} \log_b y - \dfrac{3}{4} \log_b s - \dfrac{2}{3} \log_b t$

Decide whether each statement is true *or* false. *See Example 6.*

35. $\log_2(8 + 32) = \log_2 8 + \log_2 32$

36. $\log_2(64 - 16) = \log_2 64 - \log_2 16$

37. $\log_3 7 + \log_3 7^{-1} = 0$

38. $\log_9 14 - \log_{14} 9 = 0$

39. $\log_6 60 - \log_6 10 = 1$

40. $\log_3 8 + \log_3 \dfrac{1}{8} = 0$

41. $\dfrac{\log_{10} 7}{\log_{10} 14} = \dfrac{1}{2}$

42. $\dfrac{\log_{10} 10}{\log_{10} 100} = \dfrac{1}{10}$

Relating Concepts (Exercises 43–48) For Individual or Group Work

Work Exercises 43–48 in order.

43. Evaluate $\log_3 81$.

44. Write the *meaning* of the expression $\log_3 81$.

45. Evaluate $3^{\log_3 81}$.

46. Write the *meaning* of the expression $\log_2 19$.

47. Evaluate $2^{\log_2 19}$.

48. Keeping in mind that a logarithm is an exponent, and using the results from **Exercises 43–47**, what is the simplest form of the expression $k^{\log_k m}$?

11.6 Common and Natural Logarithms

Logarithms are important in many applications in biology, engineering, economics, and social science. In this section we find numerical approximations for logarithms. Traditionally, base 10 logarithms were used most often because our number system is base 10. Logarithms with base 10 are called **common logarithms,** and

$$\log_{10} x \quad \text{is abbreviated as simply} \quad \log x,$$

where the base is understood to be 10.

OBJECTIVE **1** **Evaluate common logarithms using a calculator.**

> **📱 Calculator Tip**
>
> In **Example 1,** we give the results of evaluating some common logarithms using a calculator with a **LOG** key. (This may be a second function key on some calculators.) For simple scientific calculators, just enter the number, then press the **LOG** key. For graphing calculators, these steps are reversed.
>
> In this section, we give calculator approximations for logarithms to four decimal places.

EXAMPLE 1 **Evaluating Common Logarithms**

Evaluate each logarithm to four decimal places using a calculator.

(a) $\log 327.1 \approx 2.5147$ **(b)** $\log 437{,}000 \approx 5.6405$

(c) $\log 0.0615 \approx -1.2111$

In part (c), $\log 0.0615 \approx -1.2111$ is a negative number. *The common logarithm of a number between 0 and 1 is always negative* because the logarithm is the exponent on 10 that produces the number. In this case, we have

$$10^{-1.2111} \approx 0.0615.$$

If the exponent (the logarithm) were positive, the result would be greater than 1 because $10^0 = 1$. The graph in **Figure 15** illustrates these concepts.

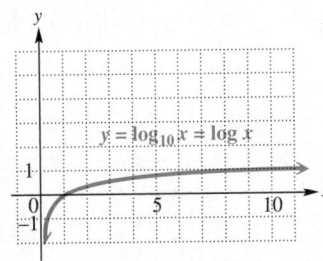

Figure 15

··········· **Work Problem ① at the Side.** ▶

OBJECTIVE **2** **Use common logarithms in applications.** In chemistry, pH is a measure of the acidity or alkalinity of a solution. Water, for example, has pH 7. In general, acids have pH numbers less than 7, and alkaline solutions have pH values greater than 7. The **pH** of a solution is defined as

$$\text{pH} = -\log [\text{H}_3\text{O}^+],$$

where $[\text{H}_3\text{O}^+]$ is the hydronium ion concentration in moles per liter. It is customary to round pH values to the nearest tenth.

OBJECTIVES

1 Evaluate common logarithms using a calculator.

2 Use common logarithms in applications.

3 Evaluate natural logarithms using a calculator.

4 Use natural logarithms in applications.

1 Evaluate each logarithm to four decimal places using a calculator.

(a) log 41,600

(b) log 43.5

(c) log 0.442

Answers

1. (a) 4.6191   **(b)** 1.6385   **(c)** −0.3546

② Solve the problem.

Find the pH of water with a hydronium ion concentration of 1.2×10^{-3}. If this water had been taken from a wetland, is the wetland a rich fen, a poor fen, or a bog?

③ Find the hydronium ion concentrations of solutions with the following pH values.

GS **(a)** 3.6

$$pH = -\log\left[H_3O^+\right]$$

Let pH = 3.6.

$$\underline{\quad\quad} = -\log\left[H_3O^+\right]$$

$$\log\left[H_3O^+\right] = \underline{\quad\quad}$$

Solve for $\left[H_3O^+\right]$ by writing the equation in exponential form.

$$\left[H_3O^+\right] = 10 - \underline{\quad}$$

$$\left[H_3O^+\right] \approx \underline{\quad\quad}$$

(b) 7.5

Figure 16 illustrates the pH scale.

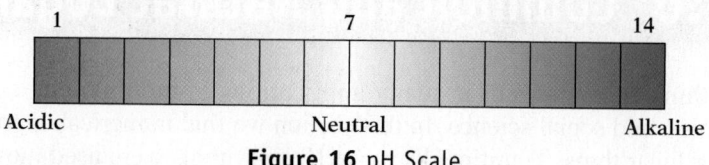

Acidic   Neutral   Alkaline

Figure 16 pH Scale

EXAMPLE 2   **Using pH in an Application**

Wetlands are classified as *bogs, fens, marshes,* and *swamps*. These classifications are based on pH values. A pH value between 6.0 and 7.5, such as that of Summerby Swamp in Michigan's Hiawatha National Forest, indicates that the wetland is a "rich fen." When the pH is between 3.0 and 6.0, the wetland is a "poor fen," and if the pH falls to 3.0 or less, it is a "bog." (*Source:* Mohlenbrock, R., "Summerby Swamp, Michigan," *Natural History.*)

Suppose that the hydronium ion concentration of a sample of water from a wetland is 6.3×10^{-3}. How would this wetland be classified?

Use the definition of pH.

$pH = -\log\left(6.3 \times 10^{-3}\right)$	Definition of pH
$pH = -\left(\log 6.3 + \log 10^{-3}\right)$	Product rule
$pH = -\left[0.7993 - 3(1)\right]$	Use a calculator to find log 6.3.
$pH = -0.7993 + 3$	Distributive property
$pH \approx 2.2$	Add.

Since the pH is less than 3.0, the wetland is a bog.

◀ Work Problem **②** at the Side.

EXAMPLE 3   **Finding Hydronium Ion Concentration**

Find the hydronium ion concentration of drinking water with pH 6.5.

$\mathbf{pH = -\log\left[H_3O^+\right]}$	
$\mathbf{6.5 = -\log\left[H_3O^+\right]}$	Let pH = 6.5.
$\log\left[H_3O^+\right] = -6.5$	Multiply by −1.

Solve for $\left[H_3O^+\right]$ by writing the equation in exponential form using base 10.

$\left[H_3O^+\right] = 10^{-6.5}$	Write in exponential form.
$\left[H_3O^+\right] \approx 3.2 \times 10^{-7}$	Use a calculator.

◀ Work Problem **③** at the Side.

The loudness of sound is measured in a unit called a **decibel**, abbreviated **dB.** To measure with this unit, we first assign an intensity of I_0 to a very faint sound, called the **threshold sound.** If a particular sound has intensity I, then the decibel level D of this louder sound is given by this formula.

$$D = 10 \log\left(\frac{I}{I_0}\right)$$

Any sound over 85 dB exceeds what hearing experts consider safe. Permanent hearing damage can be suffered at levels above 150 dB.

EXAMPLE 4   **Measuring the Loudness of Sound**

If music delivered through the earphones of an iPod has intensity I of $(3.162 \times 10^9)I_0$, find the decibel level.

$$D = 10 \log\left(\frac{I}{I_0}\right)$$

$$D = 10 \log\left(\frac{(3.162 \times 10^9)I_0}{I_0}\right) \qquad \text{Substitute the given value for } I.$$

$$D = 10 \log\left(3.162 \times 10^9\right)$$

$$D \approx 95 \qquad \begin{array}{l}\text{Use a calculator. Round to}\\ \text{the nearest unit.}\end{array}$$

·········· **Work Problem ④ at the Side.** ▶

OBJECTIVE ▸ **③ Evaluate natural logarithms using a calculator.** Logarithms used in applications are often **natural logarithms,** which have as base the number e. The letter e was chosen to honor the mathematician Leonhard Euler, who published extensive results on the number in 1748. Since it is an irrational number, its decimal expansion never terminates and never repeats.

Approximation for e

$$e \approx 2.718281828$$

🖩 **Calculator Tip**

A calculator key $\boxed{e^x}$ or the two keys $\boxed{\text{INV}}$ and $\boxed{\text{LN}}$ are used to approximate powers of e. For example, a calculator gives

$$e^2 \approx 7.389056099, \quad e^3 \approx 20.08553692, \quad \text{and} \quad e^{0.6} \approx 1.8221188.$$

Logarithms with base e are called natural logarithms because they occur in natural situations that involve growth or decay. The base e logarithm of x is written **ln x** (read "el en x"). A graph of $y = \ln x$, the equation that defines the natural logarithmic function, is given in **Figure 17.**

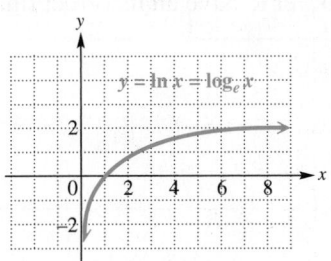

Figure 17

AVERAGE DECIBEL LEVELS FOR SOME COMMON SOUNDS

Decibel Level	Examples
60	Normal conversation
90	Rush hour traffic, lawn mower
100	Garbage truck, chain saw, pneumatic drill
120	Rock concert, thunderclap
140	Gunshot blast, jet engine
180	Rocket launching pad

Source: Deafness Research Foundation.

---

④ Find the decibel level to the nearest whole number of each of the following.

GS **(a)** A whisper with intensity I of $115 I_0$

$$D = 10 \log\left(\frac{I}{I_0}\right)$$

$$D = 10 \log\left(\frac{\rule{2em}{0.4pt}\,I_0}{I_0}\right)$$

$$D = 10 \log \rule{2em}{0.4pt}$$

$$D \approx \rule{2em}{0.4pt}$$

To the nearest whole number, the decibel level is ____ .

(b) A jet engine with intensity I of $(6.312 \times 10^{13})I_0$

Answers

4. (a) 115; 115; 21; 21 dB   (b) 138 dB

5 Find each logarithm to four decimal places.

(a) ln 0.01

(b) ln 27

(c) ln 529

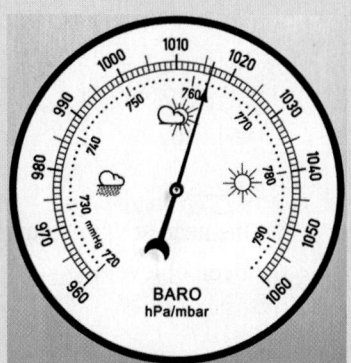

A barometer is an instrument that measures atmospheric pressure.

6 Use the natural logarithmic function in **Example 6** to approximate the altitude at 700 millibars of pressure. Round to the nearest hundred.

🖩 **Calculator Tip**

A calculator key labeled ⒧Ⓝ is used to evaluate natural logarithms. If a scientific calculator has an ⒠ˣ key, but not a key labeled ⒧Ⓝ, find a natural logarithm by entering the number, pressing the ⒤ⓃⓋ key, and then pressing the ⒠ˣ key. This works because $y = e^x$ defines the inverse function of $y = \ln x$ (or $y = \log_e x$).

EXAMPLE 5 **Finding Natural Logarithms**

Find each logarithm to four decimal places.

(a) $\ln 0.5841 \approx -0.5377$
   As with common logarithms, a number between 0 and 1 has a negative natural logarithm.

(b) $\ln 192.7 \approx 5.2611$ **(c)** $\ln 10.84 \approx 2.3832$

◄ **Work Problem** **5** at the Side.

OBJECTIVE **4** Use natural logarithms in applications.

EXAMPLE 6 **Applying Natural Logarithms**

The altitude in meters that corresponds to an atmospheric pressure of x millibars is given by the natural logarithmic function

$$f(x) = 51{,}600 - 7457 \ln x.$$

(*Source:* Miller, A. and J. Thompson, *Elements of Meteorology*, Fourth Edition, Charles E. Merrill Publishing Company.) Use this function to find the altitude when atmospheric pressure is 400 millibars. Round to the nearest hundred.

Let $x = 400$ and substitute for $f(x)$.

$f(x) = 51{,}600 - 7457 \ln x$ Given function

$f(\mathbf{400}) = 51{,}600 - 7457 \ln \mathbf{400}$ Let $x = 400$.

$f(400) \approx 6900$ Use a calculator.

Atmospheric pressure is 400 millibars at 6900 m.

◄ **Work Problem** **6** at the Side.

🖩 **Calculator Tip**

In Example 6, the final answer was obtained using a calculator *without* rounding the intermediate values. In general, it is best to wait until the final step to round the answer. Otherwise, a buildup of round-off error may cause the final answer to have an incorrect final decimal place digit.

Answers

5. (a) −4.6052 **(b)** 3.2958 **(c)** 6.2710
6. 2700 m

11.6 Exercises

FOR EXTRA HELP

Download the MyDashBoard App

MyMathLab®

CONCEPT CHECK *Choose the correct response.*

1. What is the base in the expression $\log x$?

   A. e **B.** 1   **C.** 10   **D.** x

2. What is the base in the expression $\ln x$?

   A. e **B.** 1   **C.** 10   **D.** x

3. Given $10^0 = 1$ and $10^1 = 10$, between what two consecutive integers is the value of $\log 5.6$?

   A. 5 and 6   **B.** 10 and 11

   C. 0 and 1   **D.** -1 and 0

4. Given $e^1 \approx 2.718$ and $e^2 \approx 7.389$, between what two consecutive integers is the value of $\ln 5.6$?

   A. 5 and 6   **B.** 2 and 3

   C. 1 and 2   **D.** 0 and 1

CONCEPT CHECK *Without using a calculator, give the value of each expression.*

5. $\log 10^{19.2}$

6. $\ln e^{\sqrt{2}}$

7. $10^{\log \sqrt{3}}$

8. $e^{\ln 75.2}$

▦ *Use a calculator for the remaining exercises in this set.*

Find each logarithm. Give approximations to four decimal places.
See Examples 1 and 5.

9. $\log 328.4$

10. $\log 457.2$

11. $\log 0.0326$

12. $\log 0.1741$

13. $\log (4.76 \times 10^9)$

14. $\log (2.13 \times 10^4)$

15. $\ln 7.84$

16. $\ln 8.32$

17. $\ln 0.0556$

18. $\ln 0.0217$

19. $\ln 10$

20. $\log e$

Suppose that water from a wetland is sampled and found to have the given hydronium ion concentration. Is the wetland a rich fen, *a* poor fen, *or a* bog? *See Example 2.*

21. 2.5×10^{-5}

22. 3.1×10^{-5}

23. 3.6×10^{-2}

24. 2.5×10^{-2}

25. 2.5×10^{-7}

26. 2.7×10^{-7}

Find the pH *of the substance with the given hydronium ion concentration.*
See Example 2.

27. Ammonia, 2.5×10^{-12}

28. Sodium bicarbonate, 4.0×10^{-9}

29. Tuna, 1.3×10^{-6}

30. Grapes, 5.0×10^{-5}

Use the formula for pH *to find the hydronium ion concentration of the substance with the given* pH. *See Example 3.*

31. Human gastric contents, 2.0

32. Human blood plasma, 7.4

33. Bananas, 4.6

34. Spinach, 5.4

Solve each problem. See Examples 4 and 6.

35. The time t in years for an amount increasing at a rate r (in decimal form) to double is given by

$$t = \frac{\ln 2}{\ln (1 + r)}.$$

This is called **doubling time.** Find the doubling time to the nearest tenth for an investment at each interest rate.

(a) 2% (or 0.02) **(b)** 5% (or 0.05)

36. The number of years, $N(r)$, since two independently evolving languages split off from a common ancestral language is approximated by

$$N(r) = -5000 \ln r,$$

where r is the percent of words (as a decimal) from the ancestral language common to both languages now. Find the number of years (to the nearest hundred) since the split for each percent of common words.

(a) 85% (or 0.85) **(b)** 35% (or 0.35)

37. Consumers can now enjoy movies at home in elaborate home-theater systems. Find the decibel level

$$D = 10 \log \left(\frac{I}{I_0} \right)$$

for each movie with the given intensity I.

(a) *Avatar;* $(5.012 \times 10^{10}) I_0$

(b) *Iron Man 2;* $10^{10} I_0$

(c) *Clash of the Titans;* $6{,}310{,}000{,}000 \, I_0$

38. Find the decibel level of each sound. (*Source:* The Canadian Society of Otolaryngology.)

(a) noisy restaurant: $I = 10^8 I_0$

(b) farm tractor: $I = (5.340 \times 10^9) I_0$

(c) snowmobile: $I = 31{,}622{,}776{,}600 \, I_0$

39. In the central Sierra Nevada of California, the percent of moisture p that falls as snow rather than rain is approximated reasonably well by

$$p(h) = 86.3 \ln h - 680,$$

where h is the altitude in feet.

(a) What percent of the moisture at 5000 ft falls as snow?

(b) What percent at 7500 ft falls as snow?

40. The **cost-benefit equation**

$$T = -0.642 - 189 \ln (1 - 0.01p)$$

describes the approximate tax T, in dollars per ton, that would result in a $p\%$ reduction in carbon dioxide emissions.

(a) What tax will reduce emissions 25%?

(b) Explain why the equation is not valid for $p = 0$.

11.7 Exponential and Logarithmic Equations and Their Applications

General methods for solving exponential and logarithmic equations depend on the properties that follow.

Properties for Solving Exponential and Logarithmic Equations

For all real numbers $b > 0$, $b \neq 1$, and any real numbers x and y, the following are true.

1. If $x = y$, then $b^x = b^y$.
2. If $b^x = b^y$, then $x = y$. (We used this property in **Section 11.3**.)
3. If $x = y$, and $x > 0$, $y > 0$, then $\log_b x = \log_b y$.
4. If $x > 0$, $y > 0$, and $\log_b x = \log_b y$, then $x = y$.

OBJECTIVES

1. Solve equations involving variables in the exponents.
2. Solve equations involving logarithms.
3. Solve applications involving compound interest.
4. Solve applications involving base e exponential growth and decay.
5. Use the change-of-base rule.

OBJECTIVE **1** Solve equations involving variables in the exponents.

EXAMPLE 1 Solving an Exponential Equation (Property 3)

Solve $3^x = 12$. Approximate the solution to three decimal places.

$$3^x = 12$$

$$\log 3^x = \log 12 \qquad \text{Property 3 (common logarithms)}$$

$$x \log 3 = \log 12 \qquad \text{Power rule}$$

$$\text{Exact solution} \longrightarrow x = \frac{\log 12}{\log 3} \qquad \text{Divide by log 3.}$$

$$\text{Decimal approximation} \longrightarrow x \approx 2.262 \qquad \text{Use a calculator.}$$

The solution set is $\{2.262\}$. Check with a calculator that $3^{2.262} \approx 12$.

·················· **Work Problem 1 at the Side.** ▶

CAUTION

Be careful: $\frac{\log 12}{\log 3}$ is **not** equal to log 4. Check to see that

$$\log 4 \approx 0.6021, \quad \text{but} \quad \frac{\log 12}{\log 3} \approx 2.262.$$

EXAMPLE 2 Solving an Exponential Equation (Base e)

Solve $e^{0.003x} = 40$. Approximate the solution to three decimal places.

$$\ln e^{0.003x} = \ln 40 \qquad \text{Property 3 (natural logarithms)}$$

$$0.003x \ln e = \ln 40 \qquad \text{Power rule}$$

$$0.003x = \ln 40 \qquad \ln e = \ln e^1 = 1$$

$$x = \frac{\ln 40}{0.003} \qquad \text{Divide by 0.003.}$$

$$x \approx 1229.626 \qquad \text{Use a calculator.}$$

The solution set is $\{1229.626\}$. Check that $e^{0.003(1229.626)} \approx 40$.

·················· **Work Problem 2 at the Side.** ▶

1 Solve each equation. Approximate the solutions to three decimal places.

(a) $2^x = 9$

(b) $10^x = 4$

2 Solve $e^{-0.01t} = 0.38$. Approximate the solution to three decimal places.

Answers

1. (a) $\{3.170\}$ (b) $\{0.602\}$
2. $\{96.758\}$

③ Solve $\log_3 (x + 1)^5 = 3$. Give the exact solution.

General Method for Solving an Exponential Equation

Take logarithms with the same base on both sides and then use the power rule of logarithms or the special property $\log_b b^x = x$. (See **Examples 1 and 2.**)

As a special case, if both sides can be written as exponentials with the same base, do so, and then set the exponents equal. (See **Section 11.3.**)

OBJECTIVE ▶ ② Solve equations involving logarithms. We use the definition of logarithm and the properties of logarithms to change equations to exponential form.

EXAMPLE 3 **Solving a Logarithmic Equation**

Solve $\log_2 (x + 5)^3 = 4$. Give the exact solution.

$$(x + 5)^3 = 2^4 \qquad \text{Convert to exponential form.}$$
$$(x + 5)^3 = 16 \qquad 2^4 = 16$$
$$x + 5 = \sqrt[3]{16} \qquad \text{Take the cube root on each side.}$$
$$x = -5 + \sqrt[3]{16} \qquad \text{Subtract 5.}$$
$$x = -5 + 2\sqrt[3]{2} \qquad \sqrt[3]{16} = \sqrt[3]{8 \cdot 2} = \sqrt[3]{8} \cdot \sqrt[3]{2} = 2\sqrt[3]{2}$$

CHECK $\qquad \log_2 (x + 5)^3 = 4 \qquad$ Original equation

$$\log_2 \left(-5 + 2\sqrt[3]{2} + 5\right)^3 \stackrel{?}{=} 4 \qquad \text{Let } x = -5 + 2\sqrt[3]{2}.$$
$$\log_2 \left(2\sqrt[3]{2}\right)^3 \stackrel{?}{=} 4 \qquad \text{Work inside the parentheses.}$$
$$\log_2 16 \stackrel{?}{=} 4 \qquad \left(2\sqrt[3]{2}\right)^3 = 2^3\left(\sqrt[3]{2}\right)^3 = 8 \cdot 2 = 16$$
$$2^4 \stackrel{?}{=} 16 \qquad \text{Write in exponential form.}$$
$$16 = 16 \checkmark \qquad \text{True}$$

A true statement results, so the solution set is $\left\{-5 + 2\sqrt[3]{2}\right\}$.

◀ **Work Problem ③ at the Side.**

EXAMPLE 4 **Solving a Logarithmic Equation (Property 4)**

Solve $\log_2 (x + 1) - \log_2 x = \log_2 7$.

$$\log_2 (x + 1) - \log_2 x = \log_2 7$$

Transform the left side to an expression with only *one* logarithm.

$$\log_2 \frac{x + 1}{x} = \log_2 7 \qquad \text{Quotient rule}$$
$$\frac{x + 1}{x} = 7 \qquad \text{Property 4}$$
$$x + 1 = 7x \qquad \text{Multiply by } x.$$
$$1 = 6x \qquad \text{Subtract } x.$$

This proposed solution must be checked.

$$\frac{1}{6} = x \qquad \text{Divide by 6.}$$

Answer

3. $\left\{-1 + \sqrt[5]{27}\right\}$

Continued on Next Page

CHECK

$$\log_2(x+1) - \log_2 x = \log_2 7 \qquad \text{Original equation}$$

$$\log_2\left(\frac{1}{6}+1\right) - \log_2\frac{1}{6} \overset{?}{=} \log_2 7 \qquad \text{Let } x = \frac{1}{6}.$$

$$\log_2\frac{\frac{7}{6}}{\frac{1}{6}} \overset{?}{=} \log_2 7 \qquad \text{Add within the parentheses;}$$
$$\text{Quotient rule}$$

$$\boxed{\frac{\frac{7}{6}}{\frac{1}{6}} = \frac{7}{6} \div \frac{1}{6} = \frac{7}{6} \cdot \frac{6}{1} = 7}$$

$$\log_2 7 = \log_2 7 \ \checkmark \qquad \text{True}$$

A true statement results, so the solution set is $\left\{\frac{1}{6}\right\}$.

·················· **Work Problem 4 at the Side.** ▶

CAUTION

The domain of $y = \log_b x$ is $(0, \infty)$. Keep the following in mind.

1. **It is always necessary to check that proposed solutions yield only logarithms of positive numbers in the original equation.**

2. **Do not reject a proposed solution just because it is nonpositive. Reject any value that leads to the logarithm of a nonpositive number.**

EXAMPLE 5 Solving a Logarithmic Equation

Solve $\log x + \log(x - 21) = 2$.

$$\log x + \log(x - 21) = 2$$

$$\log x(x - 21) = 2 \qquad \text{Product rule}$$

$$\boxed{\text{The base is 10.}} \quad x(x - 21) = 10^2 \qquad \text{Write in exponential form.}$$

$$x^2 - 21x = 100 \qquad \text{Distributive property; } 10^2 = 100$$

$$x^2 - 21x - 100 = 0 \qquad \text{Standard form}$$

$$(x - 25)(x + 4) = 0 \qquad \text{Factor.}$$

$$x - 25 = 0 \quad \text{or} \quad x + 4 = 0 \qquad \text{Zero-factor property}$$

$$x = 25 \quad \text{or} \qquad x = -4 \qquad \text{Solve each equation.}$$

The value -4 must be rejected as a solution since it leads to the logarithm of a negative number in the original equation.

$$\log(-4) + \log(-4 - 21) = 2 \qquad \text{The left side is undefined.}$$

Check that the only solution is 25, so the solution set is $\{25\}$.

·················· **Work Problem 5 at the Side.** ▶

Solving a Logarithmic Equation

Step 1 **Transform the equation so that a single logarithm appears on one side** using the product or quotient rule of logarithms.

Step 2 **(a) Use Property 4.**
If $\log_b x = \log_b y$, then $x = y$. (See **Example 4**.)

(b) Write the equation in exponential form.
If $\log_b x = k$, then $x = b^k$. (See **Examples 3 and 5**.)

---

4 Solve.

$$\log_8(2x + 5) + \log_8 3 = \log_8 33$$

---

5 Solve each equation.

(a) $\log_3 2x - \log_3(3x + 15) = -2$

Apply the quotient rule to get a single logarithm on the left.

$$\log_3 \frac{2x}{\underline{}} = -2$$

Write the equation in exponential form.

$$\frac{2x}{\underline{}} = 3^{\underline{}}$$

Find the value of the exponential expression.

$$\frac{2x}{3x + 15} = \frac{1}{\underline{}}$$

Solve the proportion.

$$x = \underline{}$$

Verify that this solution satisfies the equation, so the solution set is _____.

(b) $\log x + \log(x + 15) = 2$

Answers

4. $\{3\}$

5. **(a)** $3x + 15$; $3x + 15$; -2; 9; 1; $\{1\}$
(b) $\{5\}$

6 Find the value of $2000 deposited at 5% compounded annually for 10 yr.

OBJECTIVE **3** **Solve applications involving compound interest.** We have solved simple interest problems using the formula

$$I = prt. \quad \text{Simple interest formula}$$

In most cases, interest paid or charged is **compound interest** (interest paid on both principal and interest). The formula for compound interest is an application of exponential functions. *In this book, monetary amounts are given to the nearest cent.*

> **Compound Interest Formula (for a Finite Number of Periods)**
>
> If a principal of P dollars is deposited at an annual rate of interest r compounded (paid) n times per year, the account will contain
>
> $$A = P\left(1 + \frac{r}{n}\right)^{nt}$$
>
> dollars after t years. (In this formula, r is expressed as a decimal.)

7 Find the number of years, to the nearest hundredth, it will take for money deposited in an account paying 4% interest compounded semiannually to double.

EXAMPLE 6 Solving a Compound Interest Problem for A

How much money will there be in an account at the end of 5 yr if $1000 is deposited at 3% compounded quarterly? (Assume no withdrawals are made.)
   Because interest is compounded quarterly, $n = 4$.

$$A = P\left(1 + \frac{r}{n}\right)^{nt} \quad \text{Compound interest formula}$$

$$A = 1000\left(1 + \frac{0.03}{4}\right)^{4 \cdot 5} \quad \begin{array}{l}\text{Substitute } P = 1000, r = 0.03 \text{ (because}\\ 3\% = 0.03), n = 4, \text{ and } t = 5.\end{array}$$

$$A = 1000(1.0075)^{20} \quad \text{Simplify.}$$

$$A = 1161.18 \quad \text{Use a calculator.}$$

The account will contain $1161.18. (The actual amount of interest earned is $1161.18 − $1000 = $161.18. Why?)

◀ **Work Problem 6** at the Side.

EXAMPLE 7 Solving a Compound Interest Problem for t

Suppose inflation is averaging 3% per year. To the nearest hundredth, how many years will it take for prices to double? (This is called the **doubling time** of the money.)
   We want to find the number of years t for P dollars to grow to $2P$ dollars at a rate of 3% per year.

$$2P = P\left(1 + \frac{0.03}{1}\right)^{1t} \quad \begin{array}{l}\text{Substitute } A = 2P, r = 0.03, \text{ and } n = 1\\ \text{in the compound interest formula.}\end{array}$$

$$2 = (1.03)^t \quad \text{Divide by } P. \text{ Simplify.}$$

$$\log 2 = \log (1.03)^t \quad \text{Property 3}$$

$$\log 2 = t \log 1.03 \quad \text{Power rule}$$

$$t = \frac{\log 2}{\log 1.03} \quad \text{Divide by } \log 1.03. \text{ Interchange sides.}$$

$$t \approx 23.45 \quad \text{Use a calculator.}$$

Prices will double in 23.45 yr. To check, verify that $1.03^{23.45} \approx 2$.

◀ **Work Problem 7** at the Side.

Interest can be compounded annually, semiannually, quarterly, daily, and so on. If the number of compounding periods n is allowed to approach infinity, we have an example of **continuous compounding**.

Continuous Compound Interest Formula

If a principal of P dollars is deposited at an annual rate of interest r compounded continuously for t years, the final amount A on deposit is given by

$$A = Pe^{rt}.$$

EXAMPLE 8 Solving a Continuous Interest Problem

In **Example 6,** we found that $1000 invested for 5 yr at 3% interest compounded quarterly would grow to $1161.18.

(a) How much would this investment grow to if compounded continuously?

$A = Pe^{rt}$	Continuous compounding formula
$A = 1000e^{(0.03)5}$	Let $P = 1000$, $r = 0.03$, and $t = 5$.
$A = 1161.83$	Use a calculator.

The account will grow to $1161.83 (which is $0.65 more than the amount in **Example 6** when interest was compounded quarterly).

(b) How long would it take for the initial investment amount to double? Round to the nearest hundredth.
We must find the value of t that will cause A to be $2\,(\$1000) = \$2000$.

$A = Pe^{rt}$	Continuous compounding formula
$2000 = 1000e^{0.03t}$	Let $A = 2P = 2000$, $P = 1000$, and $r = 0.03$.
$2 = e^{0.03t}$	Divide by 1000.
$\ln 2 = 0.03t$	Take natural logarithms; $\ln e^k = k$.
$t = \dfrac{\ln 2}{0.03}$	Divide by 0.03. Interchange sides.
$t \approx 23.10$	Use a calculator.

It would take 23.10 yr for the original investment to double.

⋯⋯⋯⋯⋯⋯⋯⋯⋯⋯⋯⋯⋯⋯⋯⋯⋯⋯ Work Problem **8** at the Side. ▶

OBJECTIVE ▶ **4** **Solve applications involving base e exponential growth and decay.**

EXAMPLE 9 Solving an Exponential Decay Application

After a plant or animal dies, the amount of radioactive carbon-14 that is present disintegrates according to the natural logarithmic function

$$y = y_0 e^{-0.000121t},$$

where t is time in years, y is the amount of the sample at time t, and y_0 is the initial amount present at $t = 0$.

(a) If an initial sample contains $y_0 = 10$ g of carbon-14, how many grams, to the nearest hundredth, will be present after 3000 yr?

$$y = 10e^{-0.000121(3000)} \approx 6.96 \text{ g} \qquad \text{Let } y_0 = 10 \text{ and } t = 3000 \text{ in the formula.}$$

⋯⋯⋯⋯⋯⋯⋯⋯⋯⋯⋯⋯⋯⋯⋯⋯⋯⋯ **Continued on Next Page**

8 Solve each problem.

(a) How much will $2500 grow to at 4% interest compounded continuously for 3 yr?

(b) How long would it take for the initial investment in part (a) to double? Round to the nearest hundredth.

Answers

8. (a) $2818.74 **(b)** 17.33 yr

9 Radioactive strontium decays according to the natural logarithmic function

$$y = y_0 e^{-0.0239t},$$

where t is time in years.

(a) If an initial sample contains $y_0 = 12$ g of radioactive strontium, how many grams, to the nearest hundredth, will be present after 35 yr?

(b) What is the half-life of radioactive strontium to the nearest year?

10 Use the change-of-base rule to find each logarithm to four decimal places.

(a) $\log_3 17$, using common logarithms

(b) $\log_3 17$, using natural logarithms

(b) How long would it take to the nearest year for the initial sample to decay to half of its original amount? (This is called the **half-life.**)
Let $y = \frac{1}{2}(10) = 5$, and solve for t.

$$5 = 10e^{-0.000121t} \qquad \text{Substitute in } y = y_0 e^{kt}.$$

$$\frac{1}{2} = e^{-0.000121t} \qquad \text{Divide by 10.}$$

$$\ln \frac{1}{2} = -0.000121t \qquad \text{Take natural logarithms; } \ln e^k = k.$$

$$t = \frac{\ln \frac{1}{2}}{-0.000121} \qquad \text{Divide by } -0.000121. \text{ Interchange sides.}$$

$$t \approx 5728 \qquad \text{Use a calculator.}$$

The half-life is 5728 yr.

◀ Work Problem **9** at the Side.

OBJECTIVE **5** **Use the change-of-base rule.** In **Section 11.6,** we used a calculator to approximate the values of common logarithms (base 10) or natural logarithms (base e). The rule that follows is used to convert logarithms from one base to another.

Change-of-Base Rule

If $a > 0, a \neq 1, b > 0, b \neq 1,$ and $x > 0,$ then the following is true.

$$\log_a x = \frac{\log_b x}{\log_b a}$$

Any positive number other than 1 can be used for base b in the change-of-base rule, but usually the only practical bases are e and 10 because calculators give logarithms for these two bases.
To derive the change-of-base rule, let $\log_a x = m.$

$$\log_a x = m$$

$$a^m = x \qquad \text{Change to exponential form.}$$

$$\log_b(a^m) = \log_b x \qquad \text{Property 3}$$

$$m \log_b a = \log_b x \qquad \text{Power rule}$$

$$(\log_a x)(\log_b a) = \log_b x \qquad \text{Substitute for } m.$$

$$\log_a x = \frac{\log_b x}{\log_b a} \qquad \text{Divide by } \log_b a.$$

The last step gives the change-of-base rule.

EXAMPLE 10 **Using the Change-of-Base Rule**

Find $\log_5 12$ to four decimal places.
Use common logarithms and the change-of-base rule.

$$\log_5 12 = \frac{\log 12}{\log 5} \qquad \boxed{\text{Either common or natural logarithms can be used.}}$$

$$\log_5 12 \approx 1.5440 \qquad \text{Use a calculator.}$$

◀ Work Problem **10** at the Side.

Answers

9. (a) 5.20 g **(b)** 29 yr
10. (a) 2.5789 **(b)** 2.5789

11.7 Exercises

MyMathLab®

CONCEPT CHECK *Tell whether common logarithms or natural logarithms would be a better choice to use for solving each equation. Do not actually solve.*

1. $10^{0.0025x} = 75$

2. $10^{3x+1} = 13$

3. $e^{x-2} = 24$

4. $e^{-0.28x} = 30$

Many of the problems in the remaining exercises require a scientific calculator.

Solve each equation. Give solutions to three decimal places. **See Example 1.**

5. $7^x = 5$

6. $4^x = 3$

7. $9^{-x+2} = 13$

8. $6^{-x+1} = 22$

9. $3^{2x} = 14$

10. $5^{0.3x} = 11$

11. $2^{x+3} = 5^x$

12. $6^{x+3} = 4^x$

13. $2^{x+3} = 3^{x-4}$

14. $4^{x-2} = 5^{3x+2}$

15. $4^{2x+3} = 6^{x-1}$

16. $3^{2x+1} = 5^{x-1}$

Solve each equation. Use natural logarithms. Give solutions to three decimal places. **See Example 2.**

17. $e^{0.012x} = 23$

18. $e^{0.006x} = 30$

19. $e^{-0.205x} = 9$

20. $e^{-0.103x} = 7$

21. $\ln e^{3x} = 9$

22. $\ln e^{5x} = 20$

23. $\ln e^{0.45x} = \sqrt{7}$

24. $\ln e^{0.04x} = \sqrt{3}$

25. $\ln e^{2x} = \pi$

26. $\ln e^{-x} = \pi$

27. $e^{\ln 2x} = e^{\ln(x+1)}$

28. $e^{\ln(6-x)} = e^{\ln(4+2x)}$

Solve each equation. Give exact solutions. **See Example 3.**

29. $\log_3(6x + 5) = 2$

30. $\log_5(12x - 8) = 3$

31. $\log_2(2x - 1) = 5$

32. $\log_6(4x + 2) = 2$

33. $\log_7(x + 1)^3 = 2$

34. $\log_4(x - 3)^3 = 4$

Solve each equation. Give exact solutions. ***See Examples 4 and 5.***

35. $\log (6x + 1) = \log 3$

36. $\log (2x - 3) = \log 12$

37. $\log_5 (3x + 2) - \log_5 x = \log_5 4$

38. $\log_2 (x + 5) - \log_2 (x - 1) = \log_2 3$

39. $\log 4x - \log (x - 3) = \log 2$

40. $\log (-x) + \log 3 = \log (2x - 15)$

41. $\log_2 x + \log_2 (x - 7) = 3$

42. $\log_3 x + \log_3 (2x + 5) = 1$

43. $\log 5x - \log (2x - 1) = \log 4$

44. $\log (2x + 1) - \log 10x = \log 10$

45. $\log_2 x + \log_2 (x - 6) = 4$

46. $\log_2 x + \log_2 (x + 4) = 5$

Solve each problem. ***See Examples 6–8.***

47. Suppose that $2000 is deposited at 4% compounded quarterly.

    (a) How much money will there be in an account at the end of 6 yr? (Assume no withdrawals are made.)

    (b) To two decimal places, how long will it take for the account to grow to $3000?

48. Suppose that $3000 is deposited at 3.5% compounded quarterly.

    (a) How much money will there be in an account at the end of 7 yr? (Assume no withdrawals are made.)

    (b) To two decimal places, how long will it take for the account to grow to $5000?

49. What will be the amount A in an account with initial principal $4000 if interest is compounded continuously at an annual rate of 3.5% for 6 yr?

50. Refer to **Exercise 48.** Does the money grow to a larger value under those conditions, or when invested for 7 yr at 3% compounded continuously?

51. How long, to the nearest hundredth of a year, would it take an initial principal P to double if it is invested at 4.5% compounded continuously?

52. How long, to the nearest hundredth of a year, would it take $4000 to double at 3.25% compounded continuously?

Solve each problem. See Example 9.

53. A sample of 400 g of lead-210 decays to polonium-210 according to the function
$$A(t) = 400e^{-0.032t},$$
where t is time in years. How much lead, to the nearest hundredth of a gram, will be left in the sample after 25 yr?

54. How long, to the nearest hundredth of a year, will it take the initial sample of lead in **Exercise 53** to decay to half of its original amount?

Use the change-of-base rule (with either common or natural logarithms) to find each logarithm to four decimal places. See Example 10.

55. $\log_6 13$

56. $\log_7 19$

57. $\log_{\sqrt{2}} \pi$

58. $\log_\pi \sqrt{2}$

59. $\log_{21} 0.7496$

60. $\log_{19} 0.8325$

61. $\log_{1/2} 5$

62. $\log_{1/3} 7$

Relating Concepts (Exercises 63–66)   For Individual or Group Work

*In **Section 11.3**, we solved an equation such as $5^x = 125$ as follows.*

$5^x = 125$	Original equation
$5^x = 5^3$	$125 = 5^3$
$x = 3$	Set exponents equal.

Solution set: $\{3\}$

The method described in this section can also be used to solve this equation.
Work Exercises 63–66 in order, *to see how this is done.*

63. Take common logarithms on both sides, and write this equation.

64. Apply the power rule for logarithms on the left.

65. Write the equation so that x is alone on the left.

66. Use a calculator to find the decimal form of the solution. What is the solution set?

Chapter 11 *Summary*

Key Terms

11.1

composition (composite function) If f and g are functions, then the composition of g and f is defined by $(g \circ f)(x) = g(f(x))$ for all x in the domain of f such that $f(x)$ is in the domain of g.

11.2

one-to-one function A one-to-one function is a function in which each x-value corresponds to just one y-value and each y-value corresponds to just one x-value.

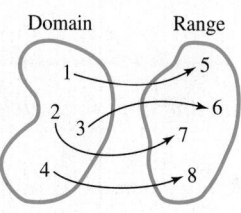

Domain   Range

inverse of a function f If f is a one-to-one function, the inverse of f is the set of all ordered pairs of the form (y, x), where (x, y) belongs to f.

11.3

exponential equation An equation involving an exponential, where the variable is in the exponent, is an exponential equation.

11.4

logarithm A logarithm is an exponent. The expression $\log_a x$ represents the exponent on the base a that gives the number x.

logarithmic equation A logarithmic equation is an equation with a logarithm in at least one term.

11.6

common logarithm A common logarithm is a logarithm with base 10.

natural logarithm A natural logarithm is a logarithm with base e.

New Symbols

$(f \circ g)(x) = f(g(x))$	composite function of f and g
f^{-1}	inverse of f
$\log_a x$	logarithm of x with base a

$\log x$	common (base 10) logarithm of x
$\ln x$	natural (base e) logarithm of x
e	a constant, approximately 2.718281828

Test Your Word Power

See how well you have learned the vocabulary in this chapter.

1 In a **one-to-one function**
- **A.** each x-value corresponds to only two y-values
- **B.** each x-value corresponds to one or more y-values
- **C.** each x-value is the same as each y-value
- **D.** each x-value corresponds to only one y-value and each y-value corresponds to only one x-value.

2 If f is a one-to-one function, then the **inverse** of f is
- **A.** the set of all solutions of f
- **B.** the set of all ordered pairs formed by interchanging the coordinates of the ordered pairs of f

- **C.** an equation involving an exponential expression
- **D.** the set of all ordered pairs that are the opposite (negative) of the coordinates of the ordered pairs of f.

3 An **exponential function** is a function defined by an expression of the form
- **A.** $f(x) = ax^2 + bx + c$, for real numbers a, b, c ($a \neq 0$)
- **B.** $f(x) = \log_a x$, for a and x positive numbers ($a \neq 1$)
- **C.** $f(x) = a^x$, for all real numbers x ($a > 0$, $a \neq 1$)
- **D.** $f(x) = \sqrt{x}$, for $x \geq 0$.

4 A **logarithm** is
- **A.** an exponent
- **B.** a base
- **C.** an equation
- **D.** a radical expression.

5 A **logarithmic function** is a function defined by an expression of the form
- **A.** $f(x) = ax^2 + bx + c$, for real numbers a, b, c ($a \neq 0$)
- **B.** $f(x) = \log_a x$, for a and x positive numbers ($a \neq 1$)
- **C.** $f(x) = a^x$, for all real numbers x ($a > 0$, $a \neq 1$)
- **D.** $f(x) = \sqrt{x}$, for $x \geq 0$.

Answers to Test Your Word Power

1. D; *Example:* The function $f = \{(0, 2), (1, -1), (3, 5), (-2, 3)\}$ is one-to-one.

2. B; *Example:* The inverse of the one-to-one function f defined in Answer 1 is $f^{-1} = \{(2, 0), (-1, 1), (5, 3), (3, -2)\}$.

3. C; *Examples:* $f(x) = 4^x$, $g(x) = \left(\frac{1}{2}\right)^x$

4. A; *Example:* $\log_a x$ is the exponent to which a must be raised to obtain x. For instance, $\log_3 9 = 2$ since $3^2 = 9$.

5. B; *Examples:* $y = \log_3 x$, $y = \log_{1/3} x$

Quick Review

Concepts	Examples

11.1 Operations on Functions and Composition

Operations on Functions

If $f(x)$ and $g(x)$ define functions, then

$$(f + g)(x) = f(x) + g(x),$$
$$(f - g)(x) = f(x) - g(x),$$
$$(fg)(x) = f(x) \cdot g(x),$$

and
$$\left(\frac{f}{g}\right)(x) = \frac{f(x)}{g(x)}, \quad g(x) \neq 0.$$

Composition of f and g

$$(f \circ g)(x) = f(g(x))$$

If $f(x) = x^2$ and $g(x) = 2x + 1$, then

$$(f + g)(x) = f(x) + g(x) = x^2 + 2x + 1,$$
$$(f - g)(x) = f(x) - g(x) = x^2 - 2x - 1,$$
$$(fg)(x) = f(x) \cdot g(x) = 2x^3 + x^2,$$

and
$$\left(\frac{f}{g}\right)(x) = \frac{f(x)}{g(x)} = \frac{x^2}{2x + 1}, \quad x \neq -\frac{1}{2}.$$

Again, let $f(x) = x^2$ and $g(x) = 2x + 1$.

$$
\begin{array}{l|l}
(f \circ g)(x) = f(g(x)) & (g \circ f)(x) = g(f(x)) \\
\quad = (2x + 1)^2 & \quad = g(x^2) \\
\quad = 4x^2 + 4x + 1 & \quad = 2x^2 + 1
\end{array}
$$

11.2 Inverse Functions

Horizontal Line Test

A function is one-to-one if every horizontal line intersects the graph of the function at most once.

Inverse Functions

For a one-to-one function $y = f(x)$, the equation of the inverse function f^{-1} is found by interchanging x and y, solving for y, and replacing y with $f^{-1}(x)$.

In general, the graph of f^{-1} is the mirror image of the graph of f with respect to the line $y = x$.

Find f^{-1} if $f(x) = 2x - 3$. The graph of f is a slanted straight line, and thus f is one-to-one by the horizontal line test.

Interchange x and y in the equation $y = 2x - 3$.

$$x = 2y - 3$$

Solve for y.
$$y = \frac{x + 3}{2}$$

Therefore, $\quad f^{-1}(x) = \frac{x + 3}{2}$, or $f^{-1}(x) = \frac{1}{2}x + \frac{3}{2}$.

11.3 Exponential Functions

For $a > 0$, $a \neq 1$, $F(x) = a^x$ is the exponential function with base a.

Graph of $F(x) = a^x$

1. The graph contains the point $(0, 1)$.

2. When $a > 1$, the graph rises from left to right.
   When $0 < a < 1$, the graph falls from left to right.

3. The x-axis is an asymptote.

4. The domain is $(-\infty, \infty)$, and the range is $(0, \infty)$.

$F(x) = 3^x$ is the exponential function with base 3.

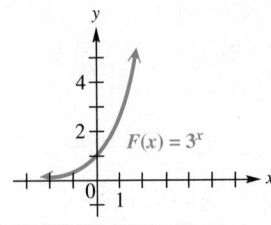

Concepts	Examples

11.4 Logarithmic Functions

$y = \log_a x$ means $x = a^y$.

For $b > 0, b \neq 1$, $\log_b b = 1$ and $\log_b 1 = 0$.

For $a > 0, a \neq 1, x > 0, G(x) = \log_a x$ is the logarithmic function with base a.

Graph of $G(x) = \log_a x$

1. The graph contains the point $(1, 0)$.
2. When $a > 1$, the graph rises from left to right.
   When $0 < a < 1$, the graph falls from left to right.
3. The y-axis is an asymptote.
4. The domain is $(0, \infty)$, and the range is $(-\infty, \infty)$.

$y = \log_2 x$ means $x = 2^y$.

$$\log_3 3 = 1 \qquad \log_5 1 = 0$$

$G(x) = \log_3 x$ is the logarithmic function with base 3.

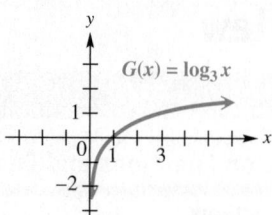

11.5 Properties of Logarithms

Product Rule
$$\log_b xy = \log_b x + \log_b y$$

Quotient Rule
$$\log_b \frac{x}{y} = \log_b x - \log_b y$$

Power Rule
$$\log_b x^r = r \log_b x$$

Special Properties
$$b^{\log_b x} = x \quad \text{and} \quad \log_b b^x = x$$

$\log_2 3m = \log_2 3 + \log_2 m$ Product rule

$\log_5 \dfrac{9}{4} = \log_5 9 - \log_5 4$ Quotient rule

$\log_{10} 2^3 = 3 \log_{10} 2$ Power rule

$6^{\log_6 10} = 10 \qquad \log_3 3^4 = 4$ Special properties

11.6 Common and Natural Logarithms

Common logarithms (base 10) are used in applications such as pH, sound level, and intensity of an earthquake. Use the $\boxed{\text{LOG}}$ key of a calculator to evaluate common logarithms.

Use the formula pH $= -\log [H_3O^+]$ to find the pH (to one decimal place) of grapes with hydronium ion concentration 5.0×10^{-5}.

$$\text{pH} = -\log(5.0 \times 10^{-5}) \qquad \text{Substitute.}$$
$$\text{pH} = -(\log 5.0 + \log 10^{-5}) \qquad \text{Property of logarithms}$$
$$\text{pH} \approx 4.3 \qquad \text{Evaluate.}$$

Natural logarithms (base e) are most often used in applications of growth and decay, such as time for money invested to double, decay of chemical compounds, and biological growth. Use the $\boxed{\text{LN}}$ key or both the $\boxed{\text{INV}}$ and $\boxed{e^x}$ keys to evaluate natural logarithms.

Use the formula for doubling time (in years)

$$t = \frac{\ln 2}{\ln(1 + r)}$$

to find the doubling time, to the nearest hundredth of a year, for an interest rate of 4% compounded annually.

$$t = \frac{\ln 2}{\ln(1 + 0.04)} \qquad \text{Substitute.}$$
$$t \approx 17.67 \qquad \text{Evaluate.}$$

The doubling time is 17.67 yr.

Concepts

Examples

11.7 Exponential and Logarithmic Equations and Their Applications

To solve exponential equations, use these properties (where $b > 0, b \neq 1$).

1. If $b^x = b^y$, then $x = y$.

Solve. $\qquad 2^{3x} = 2^5$

$\qquad\qquad 3x = 5 \qquad$ Set the exponents equal.

$\qquad\qquad x = \dfrac{5}{3} \qquad$ Divide by 3.

The solution set is $\left\{\frac{5}{3}\right\}$.

2. If $x = y \ (x > 0, y > 0)$, then $\log_b x = \log_b y$.

Solve. $\qquad 5^x = 8$

$\qquad\quad \log 5^x = \log 8 \qquad$ Take common logarithms.

$\qquad\quad x \log 5 = \log 8 \qquad$ Power rule

$\qquad\qquad x = \dfrac{\log 8}{\log 5} \qquad$ Divide by log 5.

$\qquad\qquad x \approx 1.2920 \qquad$ Use a calculator.

The solution set is $\{1.2920\}$.

To solve logarithmic equations, use these properties (where $b > 0, b \neq 1, x > 0, y > 0$). First use the properties of **Section 11.5,** if necessary, to write the equation in the proper form.

1. If $\log_b x = \log_b y$, then $x = y$.

Solve. $\qquad \log_3 2x = \log_3(x + 1)$

$\qquad\qquad 2x = x + 1 \qquad$ Property 1

$\qquad\qquad x = 1 \qquad$ Subtract x.

The solution set is $\{1\}$.

2. If $\log_b x = y$, then $b^y = x$.

Solve.

$\log x + \log(x + 15) = 2$

$\qquad \log x(x + 15) = 2 \qquad$ Product rule

$\qquad \log(x^2 + 15x) = 2 \qquad$ Distributive property

$\qquad\qquad x^2 + 15x = 10^2 \qquad$ Write in exponential form.

$\qquad\qquad x^2 + 15x = 100 \qquad 10^2 = 100$

$\qquad x^2 + 15x - 100 = 0 \qquad$ Standard form

$\qquad (x + 20)(x - 5) = 0 \qquad$ Factor.

$\qquad x + 20 = 0 \quad$ or $\quad x - 5 = 0 \qquad$ Zero-factor property

$\qquad\qquad x = -20 \quad$ or $\qquad x = 5 \qquad$ Solve each equation.

The value -20 must be rejected as a solution since it leads to the logarithm of at least one negative number in the original equation. Check that the only solution is 5, so the solution set is $\{5\}$.

Change-of-Base Rule

If $a > 0, a \neq 1, b > 0, b \neq 1, x > 0$, then the following is true.

$$\log_a x = \frac{\log_b x}{\log_b a}$$

Approximate $\log_3 37$ to four decimal places.

$$\log_3 37 = \frac{\ln 37}{\ln 3} = \frac{\log 37}{\log 3} \approx 3.2868$$

Chapter 11   Review Exercises

11.1

1. For $f(x) = 2x + 3$ and $g(x) = 5x^2 - 3x + 2$, find each of the following.

    (a) $(f + g)(x)$ **(b)** $(f - g)(x)$ **(c)** $(f + g)(-1)$ **(d)** $(f - g)(-1)$

2. For $f(x) = 12x^2 - 3x$ and $g(x) = 3x$, find each of the following.

    (a) $(fg)(x)$ **(b)** $\left(\dfrac{f}{g}\right)(x)$ **(c)** $(fg)(-1)$ **(d)** $\left(\dfrac{f}{g}\right)(2)$

Let $f(x) = 3x^2 + 2x - 1$ and $g(x) = 5x + 7$. Find each of the following.

3. (a) $(g \circ f)(3)$ **4. (a)** $(f \circ g)(-2)$ **5. (a)** $(f \circ g)(x)$

    (b) $(f \circ g)(3)$ **(b)** $(g \circ f)(-2)$ **(b)** $(g \circ f)(x)$

6. Based on your answers to **Exercises 3–5,** discuss whether composition of functions is a commutative operation.

11.2   *Determine whether each graph is the graph of a one-to-one function.*

7. 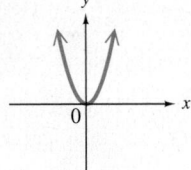   **8.** 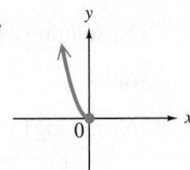   **9.**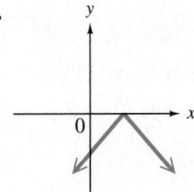

Determine whether each function is one-to-one. If it is, find its inverse.

10. $f(x) = -3x + 7$ **11.** $f(x) = \sqrt[3]{6x - 4}$ **12.** $\{(-1, 1), (0, 0), (1, 1), (2, 4)\}$

Each function graphed is one-to-one. Graph its inverse on the same set of axes as a dashed line or curve.

13. 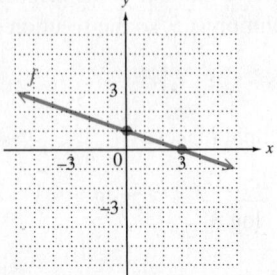   **14.** 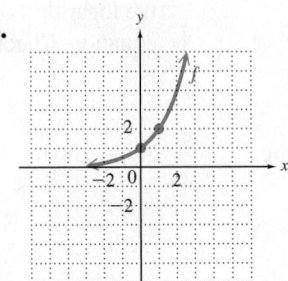   **15.**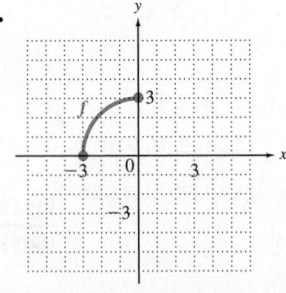

11.3 *Graph each function.*

16. $f(x) = 4^x$

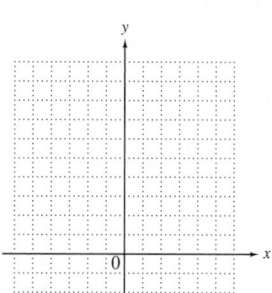

17. $f(x) = \left(\dfrac{1}{4}\right)^x$

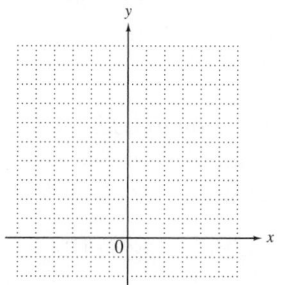

18. $f(x) = 4^{x+1}$

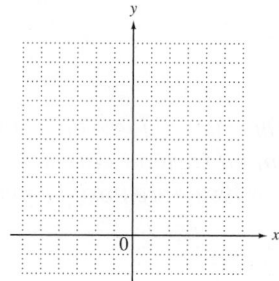

Solve each equation.

19. $4^{3x} = 8^{x+4}$

20. $\left(\dfrac{1}{27}\right)^{x-1} = 9^{2x}$

21. $5^x = 1$

22. $\left(\dfrac{2}{5}\right)^x = \dfrac{125}{8}$

📱 *In the remainder of the Chapter Review, many exercises will require a scientific calculator. We do not mark each such exercise.*

23. A 2008 report predicted that the U.S. Hispanic population will increase from 46.9 million in 2008 to 132.8 million in 2050. (*Source:* U.S. Census Bureau.) Assuming an exponential growth pattern, the population is approximated by

$$f(x) = 46.9e^{0.0247x},$$

where x represents the number of years since 2008. Use this function to approximate, to the nearest tenth, the Hispanic population in each year.

(a) 2015 **(b)** 2030

11.4 **CONCEPT CHECK** *Work each problem.*

24. Convert each equation to the indicated form.

(a) Write in exponential form: $\log_5 625 = 4$.

(b) Write in logarithmic form: $5^{-2} = 0.04$.

25. Fill in the blanks with the correct responses:

The value of $\log_2 32$ is _____. This means that if we raise _____ to the _____ power, the result is _____.

Graph each function.

26. $g(x) = \log_4 x$
(*Hint:* See **Exercise 16.**)

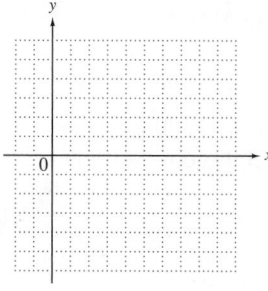

27. $g(x) = \log_{1/4} x$
(*Hint:* See **Exercise 17.**)

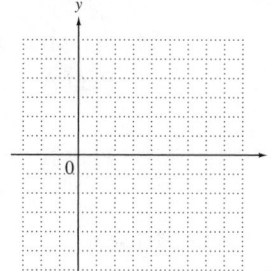

28. $g(x) = \ln x$

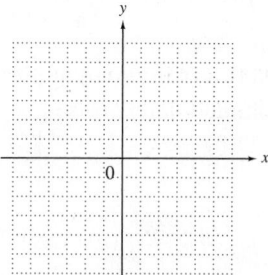

Solve each equation.

29. $\log_8 64 = x$

30. $\log_7 \dfrac{1}{49} = x$

31. $\log_4 x = \dfrac{3}{2}$

32. $\log_b b^2 = 2$

11.5 *Apply the properties of logarithms to express each logarithm as a sum or difference of logarithms. Assume that all variables represent positive real numbers.*

33. $\log_4 3x^2$

34. $\log_5 \dfrac{a^3 b^2}{c^4}$

35. $\log_4 \dfrac{\sqrt{x} \cdot w^2}{z}$

36. $\log_2 \dfrac{p^2 r}{\sqrt{z}}$

Use the properties of logarithms to rewrite each expression as a single logarithm. Assume that all variables are defined in such a way that the variable expressions are positive, and bases are positive numbers not equal to 1.

37. $2 \log_a 7 - 4 \log_a 2$

38. $3 \log_a 5 + \dfrac{1}{3} \log_a 8$

39. $\log_b 3 + \log_b x - 2 \log_b y$

40. $\log_3 (x + 7) - \log_3 (4x + 6)$

11.6 *Evaluate each logarithm. Give approximations to four decimal places.*

41. $\log 28.9$

42. $\log 0.257$

43. $\ln 28.9$

44. $\ln 0.257$

Find the pH of each substance with the given hydronium ion concentration.

45. Milk, 4.0×10^{-7}

46. Crackers, 3.8×10^{-9}

47. If vinegar has pH 2.2, what is its hydronium ion concentration?

11.7 *Solve each equation. Give solutions to three decimal places.*

48. $3^x = 9.42$

49. $2^{x-1} = 15$

50. $e^{0.06x} = 3$

Solve each equation. Give exact solutions.

51. $\log_3 (9x + 8) = 2$

52. $\log_5 (x + 6)^3 = 2$

53. $\log_3 (p + 2) - \log_3 p = \log_3 2$

54. $\log (2x + 3) - \log x = 1$

55. $\log_4 x + \log_4 (8 - x) = 2$

56. $\log_2 x + \log_2 (x + 15) = 4$

Solve each problem.

57. How much would be in an account after 3 yr if $6500.00 was invested at 3% annual interest, compounded daily? (Use $n = 365$.)

58. Which is a better plan?

Plan A: Invest $1000.00 at 4% compounded quarterly for 3 yr

Plan B: Invest $1000.00 at 3.9% compounded monthly for 3 yr

Use the change-of-base rule (with either common or natural logarithms) to find each logarithm. Give approximations to four decimal places.

59. $\log_{16} 13$

60. $\log_4 12$

61. $\log_{\sqrt{6}} \sqrt{13}$

62. $\log_{1/4} 17$

Mixed Review Exercises

Evaluate.

63. $\log_2 128$

64. $\log_{12} 1$

65. $\log_{2/3} \dfrac{27}{8}$

66. $5^{\log_5 36}$

67. $e^{\ln 4}$

68. $10^{\log e}$

69. $\log_3 3^{-5}$

70. $\ln e^{5.4}$

Find each logarithm. Give approximations to four decimal places.

71. $\log 385$

72. $\ln 0.68$

73. $\log_2 25$

74. $\log_{1/3} 14$

Solve.

75. $\log_3 (x + 9) = 4$

76. $\log_2 32 = x$

77. $\log_x \dfrac{1}{81} = 2$

78. $27^x = 81$

79. $2^{2x-3} = 8$

80. $\log_3 (x + 1) - \log_3 x = 2$

81. $\log (3x - 1) = \log 10$

82. $5^{x+2} = 25^{2x+1}$

83. $\log_4 (x + 2) - \log_4 x = 3$

84. $\ln (x^2 + 3x + 4) = \ln 2$

*A machine purchased for business use **depreciates,** or loses value, over a period of years. The value of the machine at the end of its useful life is its **scrap value.** By one method of depreciation the scrap value, S, is given by*

$$S = C(1 - r)^n,$$

where C is the original cost, n is the useful life in years, and r is the constant percent of depreciation.

85. Find the scrap value, to the nearest dollar, of a machine costing $30,000, having a useful life of 12 yr and a constant annual rate of depreciation of 15%.

86. A machine has a "half-life" of 6 yr. Find the constant annual rate of depreciation to the nearest unit of percent.

*One measure of the diversity of species in an ecological community is the **index of diversity,** given by the logarithmic expression*

$$-(p_1 \ln p_1 + p_2 \ln p_2 + \ldots + p_n \ln p_n),$$

where $p_1, p_2, \ldots, p_n$ are the proportions of a sample belonging to each of n species in the sample. (Source: Ludwig, John and James Reynolds, Statistical Ecology: A Primer on Methods and Computing, *New York, John Wiley and Sons.)*

   Approximate the index of diversity to the nearest thousandth if a sample of 100 from a community produces the following numbers.

87. 90 of one species, 10 of another

88. 60 of one species, 40 of another

Chapter 11 Test

The Chapter Test Prep Videos with test solutions are available on DVD, in MyMathLab, and on You**Tube**—search "LialCombinedAlg" and click on "Channels."

1. For $f(x) = -2x^2 + 5x - 6$ and $g(x) = 7x - 3$, find each of the following.

   (a) $(f + g)(x)$ (b) $(f - g)(x)$

   (c) $(f - g)(-2)$

2. For $f(x) = x^2 + 3x + 2$ and $g(x) = x + 1$, find each of the following.

   (a) $(fg)(-2)$ (b) $\left(\dfrac{f}{g}\right)(x)$

   (c) $\left(\dfrac{f}{g}\right)(-2)$

3. For $f(x) = 3x + 5$ and $g(x) = x^2 + 2$, find each of the following.

   (a) $(f \circ g)(-2)$ (b) $(f \circ g)(x)$

   (c) $(g \circ f)(x)$

4. Decide whether each function is one-to-one.

   (a) $f(x) = x^2 + 9$ (b)

   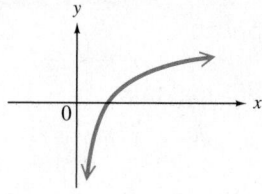

5. Find $f^{-1}(x)$ for the one-to-one function

   $$f(x) = \sqrt[3]{x + 7}.$$

6. The graph of a one-to-one function f is given. Graph f^{-1} with a dashed curve on the same set of axes.

   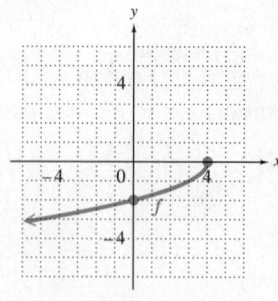

7. Graph $y = 6^x$.

   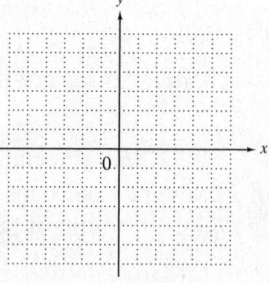

8. Graph $y = \log_6 x$.

   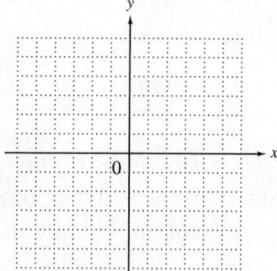

Solve each equation. Give exact solutions.

9. $5^x = \dfrac{1}{625}$

10. $2^{3x-7} = 8^{2x+2}$

11. The atmospheric pressure (in millibars) at a given altitude x (in meters) is approximated by

    $$f(x) = 1013e^{-0.0001341x}.$$

    Use this function to approximate the atmospheric pressure at each altitude.

    (a) 2000 m   (b) 10,000 m

12. Write in logarithmic form: $4^{-2} = 0.0625$.

13. Write in exponential form: $\log_7 49 = 2$.

Solve each equation.

14. $\log_{1/2} x = -5$

15. $x = \log_9 3$

16. $\log_x 16 = 4$

Use properties of logarithms to express each logarithm as a sum or difference of logarithms. Assume that variables represent positive real numbers.

17. $\log_3 x^2 y$

18. $\log_5 \left(\dfrac{\sqrt{x}}{yz} \right)$

Use properties of logarithms to rewrite each expression as a single logarithm. Assume that variables represent positive real numbers, and bases are positive numbers not equal to 1.

19. $3 \log_b s - \log_b t$

20. $\dfrac{1}{4} \log_b r + 2 \log_b s - \dfrac{2}{3} \log_b t$

21. Use a calculator to approximate each logarithm to four decimal places.

    (a) $\log 21.3$ **(b)** $\ln 0.43$ **(c)** $\log_6 45$

22. Solve $3^x = 78$, giving the solution to four decimal places.

23. Solve $\log_8 (x + 5) + \log_8 (x - 2) = \log_8 8$.

24. Suppose that $10,000 is invested at 4.5% annual interest, compounded quarterly.

    (a) How much will be in the account in 5 yr if no money is withdrawn?

    (b) How long, to the nearest tenth of a year, will it take for the initial principal to double?

25. Suppose that $15,000 is invested at 5% annual interest, compounded continuously.

    (a) How much will be in the account in 5 yr if no money is withdrawn?

    (b) How long, to the nearest tenth of a year, will it take for the initial principal to double?

26. Use the change-of-base rule to express $\log_3 19$ as described.

    (a) in terms of common logarithms

    (b) in terms of natural logarithms

    (c) approximated to four decimal places

Chapters R–11 *Cumulative Review Exercises*

Let $S = \left\{ -\frac{9}{4}, -2, -\sqrt{2}, 0, 0.6, \sqrt{11}, \sqrt{-8}, 6, \frac{30}{3} \right\}$. *List the elements of S that are elements of each set.*

1. Integers

2. Rational numbers

3. Irrational numbers

Solve each equation or inequality.

4. $7 - (3 + 4x) + 2x = -5(x - 1) - 3$

5. $2x + 2 \leq 5x - 1$

6. $|2x - 5| = 9$

7. $|4x + 2| > 10$

8. The graph indicates that the number of international travelers to the United States increased from 50,977 thousand in 2006 to 59,745 thousand in 2010.

   (a) Is this the graph of a function?

   (b) What is the slope of the line in the graph? Interpret the slope in the context of international travelers to the United States.

International Travelers to the U.S.

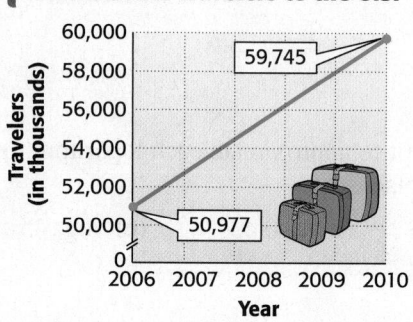

Source: U.S. Department of Commerce.

Solve each system of equations.

9. $5x - 3y = 14$
$2x + 5y = 18$

10. $x + 2y + 3z = 11$
$3x - y + z = 8$
$2x + 2y - 3z = -12$

Perform the indicated operations.

11. $(2p + 3)(3p - 1)$

12. $(4k - 3)^2$

13. $(3m^3 + 2m^2 - 5m) - (8m^3 + 2m - 4)$

14. Divide $6t^4 + 17t^3 - 4t^2 + 9t + 4$ by $3t + 1$.

Factor completely.

15. $5z^3 - 19z^2 - 4z$

16. $16a^2 - 25b^4$

17. $8c^3 + d^3$

Perform the indicated operations.

18. $\dfrac{(5p^3)^4(-3p^7)}{2p^2(4p^4)}$

19. $\dfrac{x^2 - 9}{x^2 + 7x + 12} \div \dfrac{x - 3}{x + 5}$

20. $\dfrac{2}{k + 3} - \dfrac{5}{k - 2}$

Simplify.

21. $\sqrt{288}$

22. $\dfrac{-8^{4/3}}{8^2}$

23. $2\sqrt{32} - 5\sqrt{98}$

24. Solve $\sqrt{2x + 1} - \sqrt{x} = 1$.

25. Multiply $(5 + 4i)(5 - 4i)$.

26. Simplify i^{-21}.

Solve each equation or inequality.

27. $3x^2 = x + 1$

28. $x^2 + 2x - 8 > 0$

29. $x^4 - 5x^2 + 4 = 0$

Solve.

30. $5^{x+3} = \left(\dfrac{1}{25}\right)^{3x+2}$

31. $\log_5 x + \log_5(x + 4) = 1$

32. Write $\log_5 125 = 3$ in exponential form.

33. Rewrite the following using the product, quotient, and power rules for logarithms.

$$\log \frac{x^3\sqrt{y}}{z}$$

Graph.

34. $y = \dfrac{1}{3}(x - 1)^2 + 2$

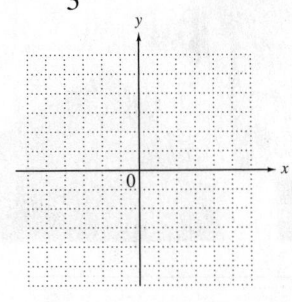

35. $f(x) = 2^x$

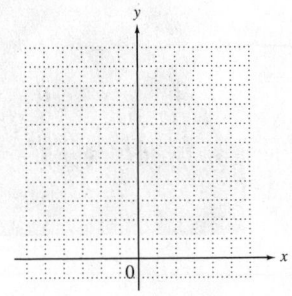

36. $f(x) = \log_3 x$

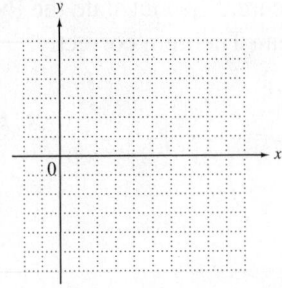

Math in the Media

SO, DID THE SCARECROW REALLY GET A BRAIN?

Probably the most famous mathematical statement in the history of motion pictures is heard in the 1939 classic *The Wizard of Oz*. Ray Bolger's character, the Scarecrow, wants a brain. When the Wizard grants him his "Th.D." (Doctor of Thinkology), the Scarecrow replies with a statement that has made mathematics teachers shudder for over 70 years.

Scarecrow: *The sum of the square roots of any two sides of an isosceles triangle is equal to the square root of the remaining side.*

His statement is quite impressive and sounds like the formula for the *Pythagorean Theorem* in **Section 9.3.** Let's see why it is incorrect.

1. To what kind of triangle does the Scarecrow refer in his statement? To what kind of triangle does the Pythagorean Theorem actually refer?

2. In the Scarecrow's statement, he refers to square roots. In applying the formula for the Pythagorean Theorem, do you find square roots of the sides? If not, what do you find?

3. An isosceles triangle has two sides of equal length. Draw an isosceles triangle with two sides of length 9 units and remaining side of length 4 units. Now show that this triangle does not satisfy the Scarecrow's statement.

   (This is called a *counterexample* and is sufficient to show that his statement is false in general.)

4. Use wording similar to that of the Scarecrow, but state the Pythagorean Theorem correctly.

12 Nonlinear Functions, Conic Sections, and Nonlinear Systems

An *ellipse*, one of a group of curves known as *conic sections*, has a special reflecting property responsible for "whispering galleries" like that in the Old House Chamber of the U.S. Capitol. We investigate ellipses in this chapter.

12.1 Additional Graphs of Functions

12.2 The Circle and the Ellipse

12.3 The Hyperbola and Other Functions Defined by Radicals

12.4 Nonlinear Systems of Equations

12.5 Second-Degree Inequalities and Systems of Inequalities

903

12.1 Additional Graphs of Functions

OBJECTIVES

1. Recognize graphs of the absolute value, reciprocal, and square root functions, and graph their translations.

2. Recognize and graph step functions.

OBJECTIVE 1 Recognize graphs of the absolute value, reciprocal, and square root functions, and graph their translations. The elementary function $f(x) = |x|$ is the **absolute value function**. This function pairs each real number with its absolute value. Its graph is shown in **Figure 1**.

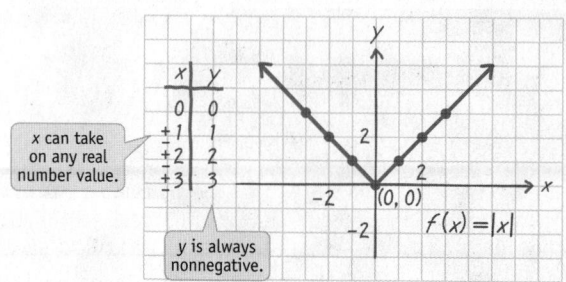

Absolute value function
$$f(x) = |x|$$
Domain: $(-\infty, \infty)$
Range: $[0, \infty)$

Figure 1

The **reciprocal function** $f(x) = \frac{1}{x}$, was introduced in **Section 7.4**. Its graph is shown in **Figure 2**. Since x can never equal 0, as x gets closer and closer to 0, $\frac{1}{x}$ approaches either ∞ or $-\infty$. Also, $\frac{1}{x}$ can never equal 0, and as x approaches ∞ or $-\infty$, $\frac{1}{x}$ approaches 0. The axes are **asymptotes** for the function.

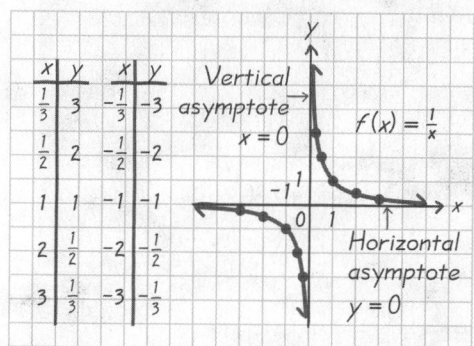

Reciprocal function
$$f(x) = \frac{1}{x}$$
Domain: $(-\infty, 0) \cup (0, \infty)$
Range: $(-\infty, 0) \cup (0, \infty)$

Figure 2

The **square root function** $f(x) = \sqrt{x}$, was introduced in **Section 9.1**. Its graph is shown in **Figure 3**.

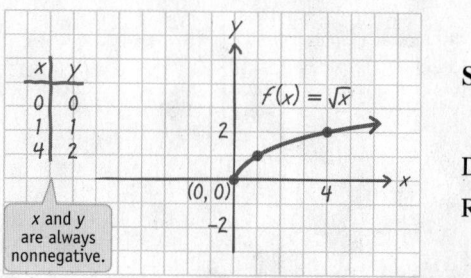

Square root function
$$f(x) = \sqrt{x}$$
Domain: $[0, \infty)$
Range: $[0, \infty)$

Figure 3

The graphs of these elementary functions can be shifted, or translated, just as we saw with the graph of $f(x) = x^2$ in **Section 10.6.**

EXAMPLE 1 **Applying a Horizontal Shift**

Graph $f(x) = |x - 2|$. Give the domain and range.

The graph of $y = (x - 2)^2$ is obtained by shifting the graph of $y = x^2$ two units to the right. In a similar manner, the graph of $f(x) = |x - 2|$ is found by shifting the graph of $y = |x|$ two units to the right, as shown in **Figure 4.**

x	y
0	2
1	1
2	0
3	1
4	2

Compare this table of values to that with **Figure 1.**

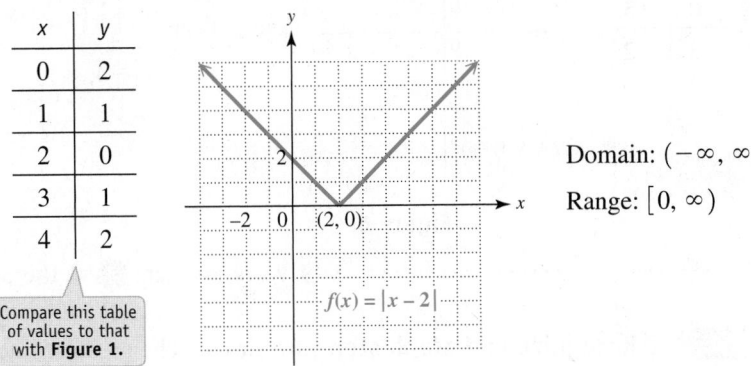

Domain: $(-\infty, \infty)$

Range: $[0, \infty)$

Figure 4

·····**Work Problem ❶ at the Side.** ▶

As seen in **Example 1,** the graph of

$$y = f(x + h)$$

is a *horizontal* translation of the graph of $y = f(x)$. In **Example 2,** we use the fact that the graph of

$$y = f(x) + k$$

is a *vertical* translation of the graph of $y = f(x)$.

EXAMPLE 2 **Applying a Vertical Shift**

Graph $f(x) = \frac{1}{x} + 3$. Give the domain and range.

The graph of this function is found by shifting the graph of $y = \frac{1}{x}$ three units up. See **Figure 5.**

x	y		x	y
$\frac{1}{3}$	6		$-\frac{1}{3}$	0
$\frac{1}{2}$	5		$-\frac{1}{2}$	1
1	4		-1	2
2	3.5		-2	2.5

Compare this table of values to that with **Figure 2.**

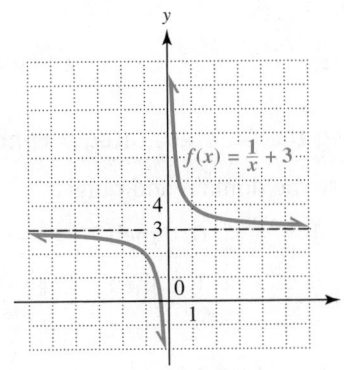

Domain:
$(-\infty, 0) \cup (0, \infty)$

Range:
$(-\infty, 3) \cup (3, \infty)$

Vertical
asymptote: $x = 0$

Horizontal
asymptote: $y = 3$

Figure 5

·····**Work Problem ❷ at the Side.** ▶

---

❶ Graph $f(x) = \sqrt{x + 4}$. Give the domain and range.

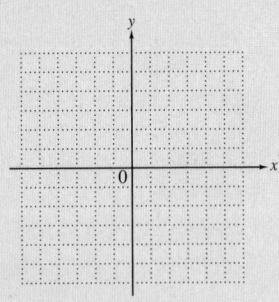

❷ Graph $f(x) = \frac{1}{x} - 2$. Give the domain and range.

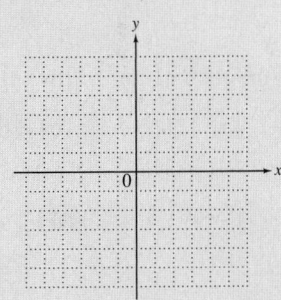

Answers

1.

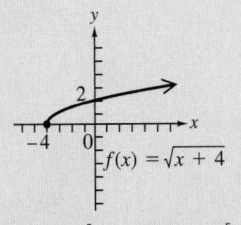

domain: $[-4, \infty)$; range: $[0, \infty)$

2.

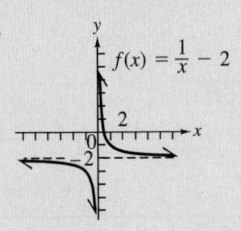

domain: $(-\infty, 0) \cup (0, \infty)$;
range: $(-\infty, -2) \cup (-2, \infty)$

❸ Graph $f(x) = |x + 2| + 1$.
Give the domain and range.

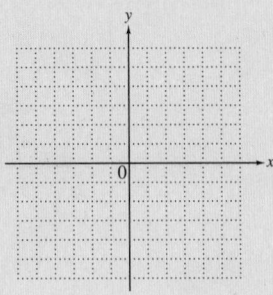

❹ Find each of the following.

(a) $[\![18]\!]$ **(b)** $[\![8.7]\!]$

(c) $[\![-5]\!]$ **(d)** $[\![-6.9]\!]$

(e) $\left[\!\!\left[1\dfrac{1}{2} \right]\!\!\right]$ **(f)** $[\![\pi]\!]$

Answers

3.

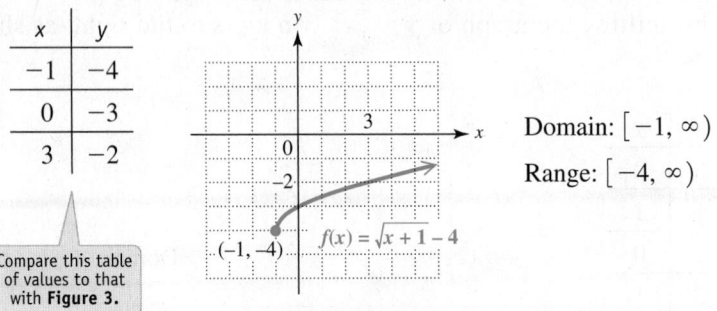

$f(x) = |x + 2| + 1$

domain: $(-\infty, \infty)$; range: $[1, \infty)$

4. (a) 18   **(b)** 8   **(c)** -5 **(d)** -7
   (e) 1   **(f)** 3

EXAMPLE 3   **Applying Both Horizontal and Vertical Shifts**

Graph $f(x) = \sqrt{x + 1} - 4$. Give the domain and range.

    The graph of $y = (x + 1)^2 - 4$ is obtained by shifting the graph of $y = x^2$ one unit to the left and four units down. Following this pattern, we shift the graph of $y = \sqrt{x}$ one unit to the left and four units down to get the graph of $f(x) = \sqrt{x + 1} - 4$. See **Figure 6.**

x	y
-1	-4
0	-3
3	-2

Compare this table of values to that with **Figure 3.**

$f(x) = \sqrt{x + 1} - 4$

$(-1, -4)$

Domain: $[-1, \infty)$

Range: $[-4, \infty)$

Figure 6

◀ **Work Problem ❸ at the Side.**

OBJECTIVE ❷ Recognize and graph step functions. The greatest integer function is defined as follows.

$f(x) = [\![x]\!]$

The **greatest integer function,** written $f(x) = [\![x]\!]$, pairs every real number x with the greatest integer less than or equal to x.

EXAMPLE 4   **Finding the Greatest Integer**

Evaluate each expression.

(a) $[\![8]\!] = 8$ **(b)** $[\![-1]\!] = -1$ **(c)** $[\![0]\!] = 0$ If x is an integer, then $[\![x]\!] = x$.

(d) $[\![7.45]\!] = 7$ The greatest integer *less than or equal to* 7.45 is 7.

(e) $[\![-2.6]\!] = -3$

-2.6

$-3\ -2\ -1\ \ 0$

    Think of a number line with -2.6 graphed on it. Since -3 is to the *left of* (and is, therefore, *less than*) -2.6, the greatest integer less than or equal to -2.6 is -3, **not** -2.

◀ **Work Problem ❹ at the Side.**

EXAMPLE 5   **Graphing the Greatest Integer Function**

Graph $f(x) = [\![x]\!]$. Give the domain and range.

For $[\![x]\!]$,   if $-1 \le x < 0$,   then $[\![x]\!] = -1$;

          if $\ \ 0 \le x < 1$,   then $[\![x]\!] = 0$;

          if $\ \ 1 \le x < 2$,   then $[\![x]\!] = 1$;

          if $\ \ 2 \le x < 3$,   then $[\![x]\!] = 2$;

          if $\ \ 3 \le x < 4$,   then $[\![x]\!] = 3$,   and so on.

····Continued on Next Page

Thus, the graph, as shown in **Figure 7,** consists of a series of horizontal line segments. In each one, the left endpoint is included and the right endpoint is excluded. These segments continue indefinitely following this pattern to the left and right. The appearance of the graph is the reason that this function is called a **step function.**

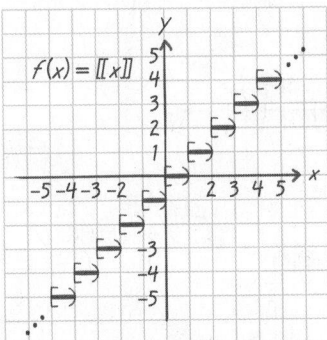

Figure 7

Greatest integer function

$$f(x) = [\![x]\!]$$

Domain: $(-\infty, \infty)$

Range: $\{\ldots, -3, -2, -1, 0, 1, 2, 3, \ldots\}$
(the set of integers)

The dots indicate that the graph continues indefinitely in the same pattern.

The graph of a step function also may be shifted. For example, the graph of $h(x) = [\![x - 2]\!]$ is the same as the graph of $f(x) = [\![x]\!]$ shifted two units to the right. Similarly, the graph of $g(x) = [\![x]\!] + 2$ is the graph of $f(x)$ shifted two units up.

················· **Work Problem 5 at the Side.** ▶

EXAMPLE 6 **Applying a Greatest Integer Function**

An overnight delivery service charges $25 for a package weighing up to 2 lb. For each additional pound or fraction of a pound there is an additional charge of $3. Let $D(x)$, or y, represent the cost to send a package weighing x pounds. Graph $D(x)$ for x in the interval $(0, 6]$.

For x in the interval $(0, 2]$, $y = 25$.

For x in the interval $(2, 3]$, $y = 25 + 3 = 28$.

For x in the interval $(3, 4]$, $y = 28 + 3 = 31$.

For x in the interval $(4, 5]$, $y = 31 + 3 = 34$.

For x in the interval $(5, 6]$, $y = 34 + 3 = 37$.

The graph, which is that of a step function, is shown in **Figure 8.**

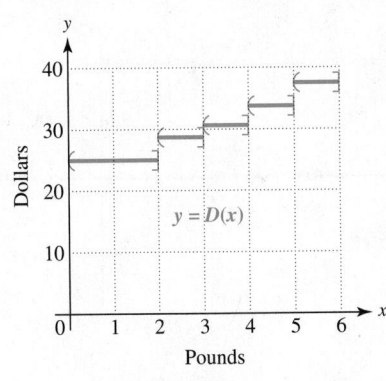

Figure 8

················· **Work Problem 6 at the Side.** ▶

5 Graph $f(x) = [\![x + 1]\!]$. Give the domain and range.

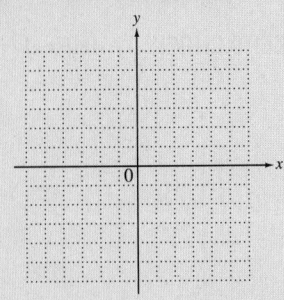

6 Assume that the post office charges $0.80 per oz (or fraction of an ounce) to mail a letter to Europe. Graph the ordered pairs (ounces, cost) for x in the interval $(0, 4]$.

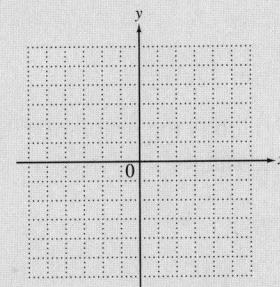

Answers

5.

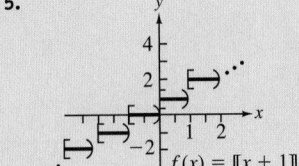

domain: $(-\infty, \infty)$;
range: $\{\ldots, -2, -1, 0, 1, 2, \ldots\}$

6.

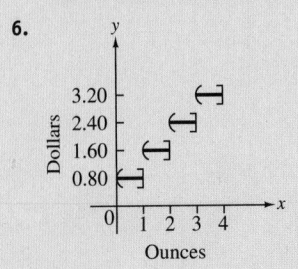

12.1 Exercises

 FOR EXTRA HELP   Download the MyDashBoard App   MyMathLab®

CONCEPT CHECK *For Exercises 1–6, refer to the basic graphs in A–F.*

A.
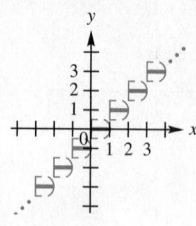

B.
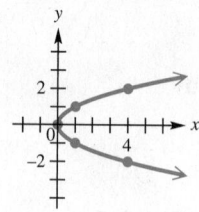

C.
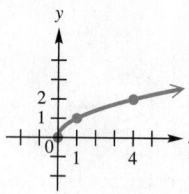

D.
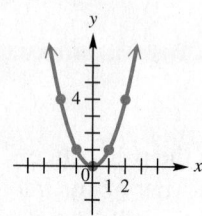

E.
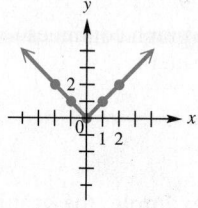

F.
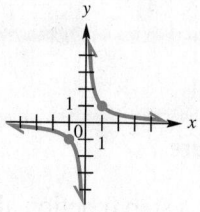

1. Which is the graph of $f(x) = |x|$? The lowest point on its graph has coordinates (____ , ____).

2. Which is the graph of $f(x) = x^2$? Give the domain and range.

3. Which is the graph of $f(x) = [\![x]\!]$? Give the domain and range.

4. Which is the graph of $f(x) = \sqrt{x}$? Give the domain and range.

5. Which is not the graph of a function? Why?

6. Which is the graph of $f(x) = \frac{1}{x}$? The lines with equations $x = 0$ and $y = 0$ are its _____.

Graph each function. Give the domain and range. ***See Examples 1–3.***

7. $f(x) = |x + 1|$
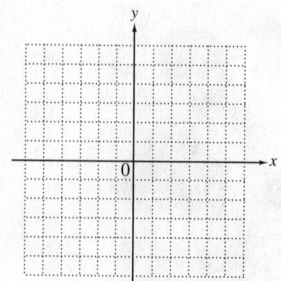

8. $f(x) = |x - 1|$

9. $f(x) = \frac{1}{x} + 1$
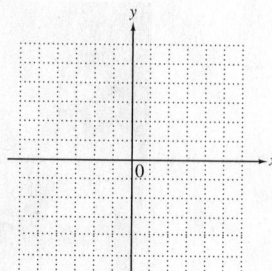

10. $f(x) = \frac{1}{x} - 1$
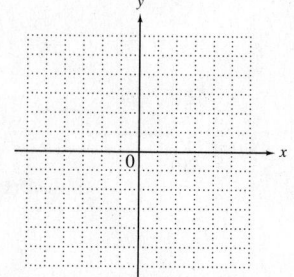

11. $f(x) = \sqrt{x-2}$

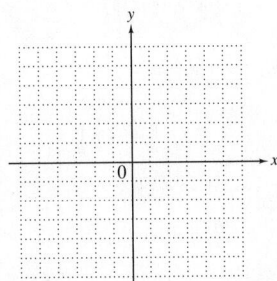

12. $f(x) = \sqrt{x+5}$

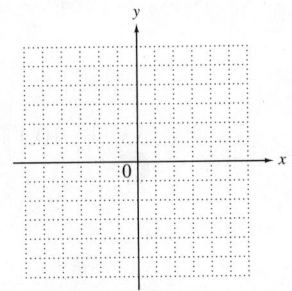

13. $f(x) = \dfrac{1}{x-2}$

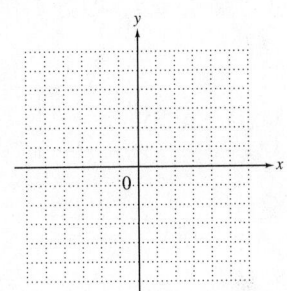

14. $f(x) = \dfrac{1}{x+2}$

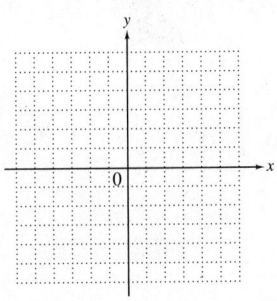

15. $f(x) = \sqrt{x+3} - 3$

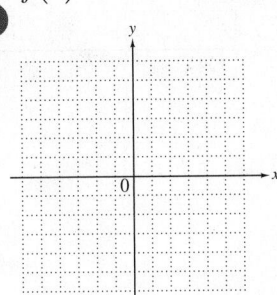

16. $f(x) = \sqrt{x-2} + 2$

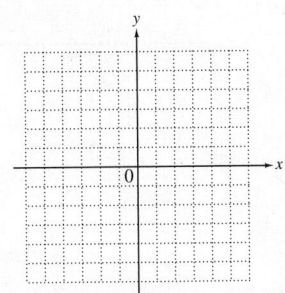

17. $f(x) = |x-3| + 1$

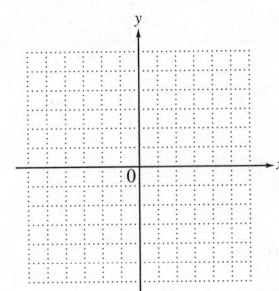

18. $f(x) = |x+1| - 4$

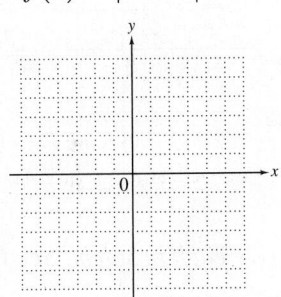

CONCEPT CHECK *Without actually plotting points, match each function defined by the absolute value expression with its graph.*

19. $f(x) = |x-2| + 2$

A.

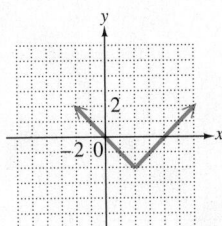

20. $f(x) = |x+2| + 2$

B.

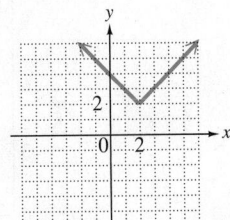

21. $f(x) = |x-2| - 2$

C.

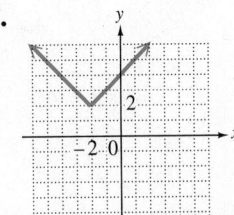

22. $f(x) = |x+2| - 2$

D.

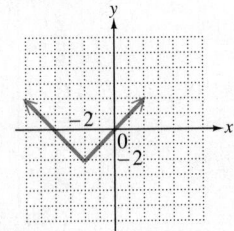

23. CONCEPT CHECK How is the graph of
$$f(x) = \frac{1}{x-3} + 2$$
obtained from the graph of $g(x) = \frac{1}{x}$?

24. CONCEPT CHECK How is the graph of
$$f(x) = \frac{1}{x+5} - 3$$
obtained from the graph of $g(x) = \frac{1}{x}$?

Evaulate each expression. **See Example 4.**

25. $[\![3]\!]$ **26.** $[\![28]\!]$ **27.** $[\![4.5]\!]$ **28.** $[\![7.6]\!]$ **29.** $\left[\!\!\left[\dfrac{1}{2}\right]\!\!\right]$

30. $\left[\!\!\left[\dfrac{3}{4}\right]\!\!\right]$ **31.** $[\![-14]\!]$ **32.** $[\![-10]\!]$ **33.** $[\![-10.1]\!]$ **34.** $[\![-6.5]\!]$

Graph each step function. **See Example 5.**

35. $f(x) = [\![x - 3]\!]$ **36.** $g(x) = [\![x + 2]\!]$ **37.** $f(x) = [\![x]\!] - 1$ **38.** $f(x) = [\![x]\!] + 1$

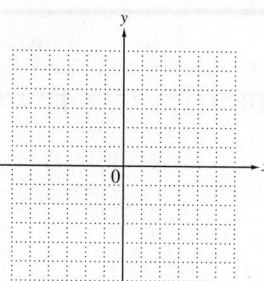   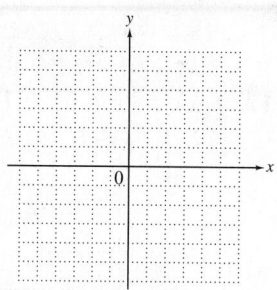   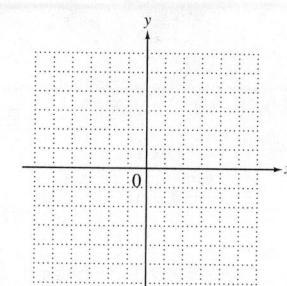   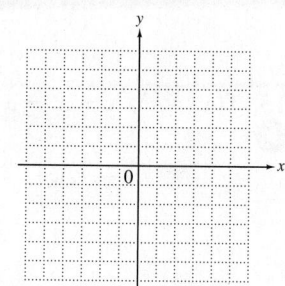

Solve each problem. **See Example 6.**

39. In 2012, postage rates were $0.45 for the first ounce, plus $0.20 for each additional ounce. Assume that each letter carried one $0.45 stamp and as many $0.20 stamps as necessary. Graph the function $y = p(x) =$ the number of stamps on a letter weighing x ounces. Use the interval $(0, 5]$. (*Source:* www.usps.com)

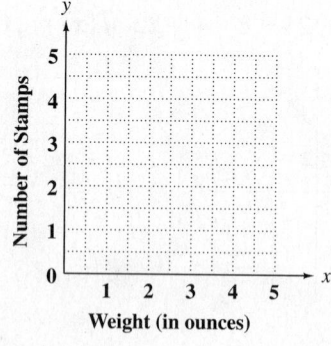

40. The cost of parking a car at an airport hourly parking lot is $3 for the first half-hour and $2 for each additional half-hour or fraction thereof. Graph the function $y = f(x) =$ the cost of parking a car for x hours. Use the interval $(0, 2]$.

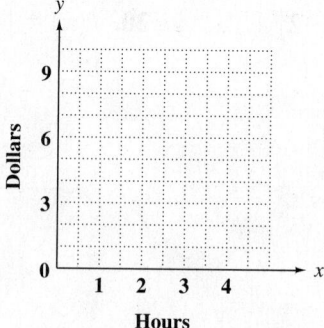

41. A certain long-distance carrier provides service between Podunk and Nowhereville. If x represents the number of minutes for the call, where $x > 0$, then the function

$$f(x) = 0.40 [\![x]\!] + 0.75$$

gives the total cost of the call in dollars. Find the cost of a 5.5-minute call.

42. Total rental cost in dollars for a power washer, where x represents the number of hours with $x > 0$, can be represented by the function

$$f(x) = 12 [\![x]\!] + 25.$$

Find the cost of a $7\frac{1}{2}$ hr rental.

12.2 The Circle and the Ellipse

When an infinite cone is intersected by a plane, the resulting figure is a **conic section.** A parabola is one example of a conic section. Circles, ellipses, and hyperbolas may also result. See **Figure 9.**

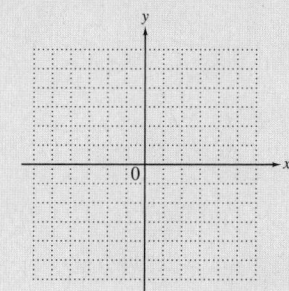

OBJECTIVES

1. Write an equation of a circle given the center and radius.
2. Determine the center and radius of a circle given its equation.
3. Recognize the equation of an ellipse.
4. Graph ellipses.

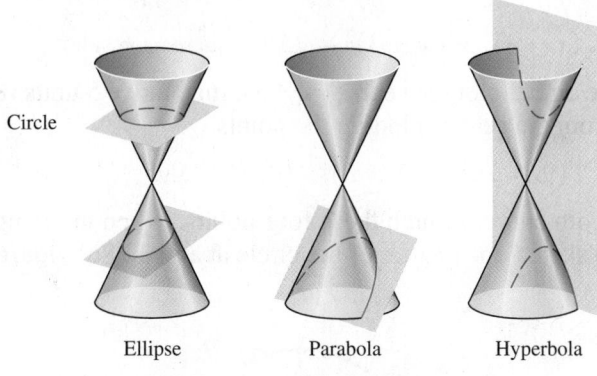

Figure 9

OBJECTIVE 1 **Write an equation of a circle given the center and radius.** A **circle** is the set of all points in a plane that lie a fixed distance from a fixed point. The fixed point is the **center,** and the fixed distance is the **radius.** We use the distance formula from **Section 9.3** to find an equation of a circle.

EXAMPLE 1   Writing an Equation of a Circle and Graphing It

Write an equation of the circle with radius 3 and center at $(0, 0)$, and graph it.

If the point (x, y) is on the circle, then the distance from (x, y) to the center $(0, 0)$ is 3.

$$\sqrt{(x_2 - x_1)^2 + (y_2 - y_1)^2} = d \qquad \text{Distance formula}$$

$$\sqrt{(x - 0)^2 + (y - 0)^2} = 3 \qquad \text{Let } x_1 = 0,\, y_1 = 0,\, \text{and } d = 3.$$

$$\left(\sqrt{x^2 + y^2}\right)^2 = 3^2 \qquad \text{Square each side.}$$

$$x^2 + y^2 = 9 \qquad \left(\sqrt{a}\right)^2 = a$$

An equation of this circle is $x^2 + y^2 = 9$. The graph is shown in **Figure 10.**

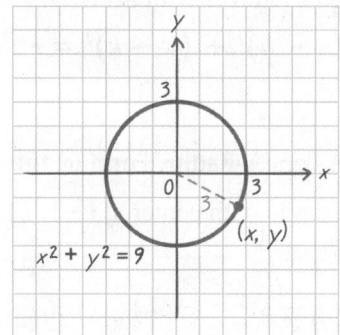

Figure 10

1 Write an equation of the circle with radius 4 and center $(0, 0)$. Sketch its graph.

Answer

1. $x^2 + y^2 = 16$

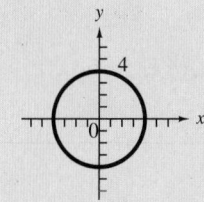

···················· **Work Problem** 1 **at the Side.** ▶

② Write an equation of the circle with center at $(3, -2)$ and radius 3, and graph it.

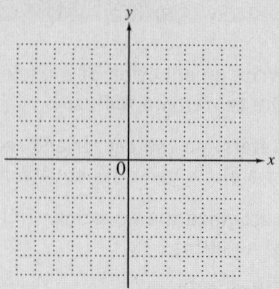

A circle may not be centered at the origin, as seen in the next example.

EXAMPLE 2 **Writing an Equation of a Circle and Graphing It**

Write an equation of the circle with center at $(4, -3)$ and radius 5, and graph it.

$$\sqrt{(x_2 - x_1)^2 + (y_2 - y_1)^2} = d \quad \text{Distance formula}$$

$$\sqrt{(x - 4)^2 + [y - (-3)]^2} = 5 \quad \text{Let } x_1 = 4, y_1 = -3, \text{ and } d = 5.$$

$$(x - 4)^2 + (y + 3)^2 = 25 \quad \text{Square each side.}$$

To graph the circle, plot the center $(4, -3)$, then move 5 units right, left, up, and down from the center, plotting the points

$$(9, -3), \quad (-1, -3), \quad (4, 2), \quad \text{and} \quad (4, -8).$$

Draw a smooth curve through these four points. When graphing by hand, it is helpful to sketch one quarter of the circle at a time. See **Figure 11.**

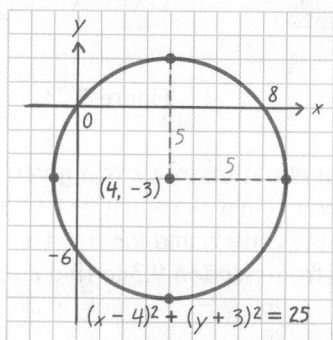

Figure 11

◀ Work Problem **②** at the Side.

③ Write an equation of the circle with center at $(-5, 4)$ and radius $\sqrt{6}$.

Examples 1 and 2 suggest the form of an equation of a circle with radius r and center at (h, k). If (x, y) is a point on the circle, then the distance from the center (h, k) to the point (x, y) is r. By the distance formula,

$$\sqrt{(x - h)^2 + (y - k)^2} = r.$$

Squaring each side gives the **center-radius form** of the equation of a circle.

Equation of a Circle (Center-Radius Form)

An equation of a circle of radius r with center at (h, k) is

$$(x - h)^2 + (y - k)^2 = r^2.$$

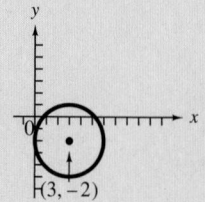
EXAMPLE 3 **Using the Center-Radius Form of the Equation of a Circle**

Write an equation of the circle with center at $(-1, 2)$ and radius $\sqrt{7}$.

$$(x - h)^2 + (y - k)^2 = r^2 \quad \text{Center-radius form}$$

$$[x - (-1)]^2 + (y - 2)^2 = \left(\sqrt{7}\right)^2 \quad \text{Let } h = -1, k = 2, \text{ and } r = \sqrt{7}.$$

Pay attention to signs here. $\quad (x + 1)^2 + (y - 2)^2 = 7 \quad \text{Simplify; } \left(\sqrt{a}\right)^2 = a$

◀ Work Problem **③** at the Side.

OBJECTIVE **2** **Determine the center and radius of a circle given its equation.** In the equation found in **Example 2,** multiplying out $(x - 4)^2$ and $(y + 3)^2$ and then combining like terms gives the following.

$$(x - 4)^2 + (y + 3)^2 = 25$$
$$x^2 - 8x + 16 + y^2 + 6y + 9 = 25$$
$$x^2 + y^2 - 8x + 6y = 0$$

This general form suggests that an equation with both x^2- and y^2-terms that have equal coefficients may represent a circle.

EXAMPLE 4   **Completing the Square to Find the Center and Radius**

Find the center and radius of the circle $x^2 + y^2 + 2x + 6y - 15 = 0$, and graph it.

Since the equation has an x^2-term and a y^2-term with equal coefficients, its graph might be that of a circle. To find the center and radius, complete the squares on x and y.

$$x^2 + y^2 + 2x + 6y = 15$$
Transform so that the constant is on the right.

$$(x^2 + 2x \quad) + (y^2 + 6y \quad) = 15$$
Rewrite in anticipation of completing the square.

$$\left[\frac{1}{2}(2)\right]^2 = 1 \qquad \left[\frac{1}{2}(6)\right]^2 = 9$$
Square half the coefficient of each middle term.

$$(x^2 + 2x + 1) + (y^2 + 6y + 9) = 15 + 1 + 9$$
Complete the squares on both x and y.

$$(x + 1)^2 + (y + 3)^2 = 25$$
Factor on the left. Add on the right.

$$[x - (-1)]^2 + [y - (-3)]^2 = 5^2$$
Center-radius form

The graph is a circle with center at $(-1, -3)$ and radius 5. See **Figure 12.**

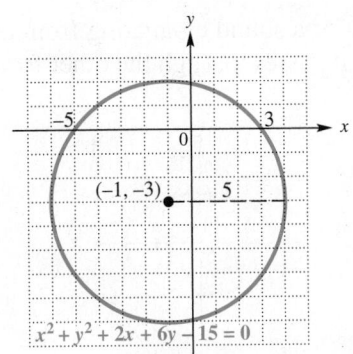

Figure 12

··········· **Work Problem** **4** **at the Side.** ▶

Note

Consider the following.

1. If the procedure of **Example 4** leads to an equation of the form
$$(x - h)^2 + (y - k)^2 = 0,$$
then the graph is the single point (h, k).

2. If the constant on the right side of the equation is *negative,* then the equation has *no graph.*

---

4 Find the center and radius of the circle with equation
$$x^2 + y^2 - 10x + 4y + 20 = 0,$$
and graph it.

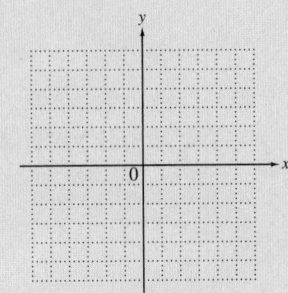

Answer

4. center: $(5, -2)$; radius: 3

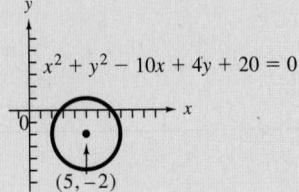

OBJECTIVE **3** **Recognize the equation of an ellipse.** An **ellipse** is the set of all points in a plane the *sum* of whose distances from two fixed points is constant. These fixed points are the **foci** (singular: *focus*). **Figure 13** shows an ellipse whose foci are $(c, 0)$ and $(-c, 0)$, with x-intercepts $(a, 0)$ and $(-a, 0)$ and y-intercepts $(0, b)$ and $(0, -b)$. It can be shown in more advanced courses that $c^2 = a^2 - b^2$ for an ellipse of this type. The origin is the **center** of the ellipse.

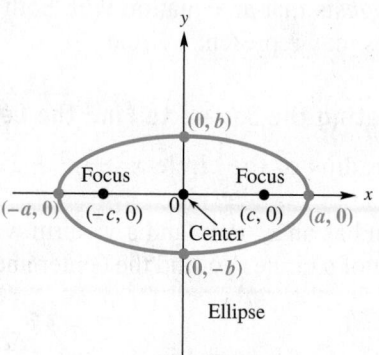

Figure 13

An ellipse centered at the origin has the following equation.

Equation of an Ellipse

The ellipse whose x-intercepts are $(a, 0)$ and $(-a, 0)$ and whose y-intercepts are $(0, b)$ and $(0, -b)$ has an equation of the form

$$\frac{x^2}{a^2} + \frac{y^2}{b^2} = 1.$$

Note that a circle is a special case of an ellipse, where $a^2 = b^2$.

When a ray of light or a sound emanating from one focus of an ellipse bounces off the ellipse, it passes through the other focus. See **Figure 14.**

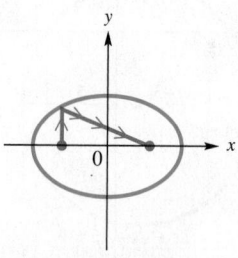

Reflecting property
of an ellipse

Figure 14

As mentioned in the chapter introduction, this reflecting property is responsible for whispering galleries. In the Old House Chamber of the U.S. Capitol, John Quincy Adams was able to listen in on his opponents' conversations—his desk was positioned at one of the foci beneath the ellipsoidal ceiling and his opponents were located across the room at the other focus.

The paths of Earth and other planets around the sun are approximately ellipses. The sun is at one focus and a point in space is at the other. Orbits of communication satellites and other space vehicles are also elliptical.

OBJECTIVE ▶ ④ **Graph ellipses.** To graph an ellipse centered at the origin, we plot the four intercepts and then sketch the ellipse through those points.

EXAMPLE 5 Graphing Ellipses

Graph each ellipse.

(a) $\dfrac{x^2}{49} + \dfrac{y^2}{36} = 1$

Here, $a^2 = 49$, so $a = 7$, and the x-intercepts for this ellipse are $(7, 0)$ and $(-7, 0)$. Similarly, $b^2 = 36$, so $b = 6$, and the y-intercepts for this ellipse are $(0, 6)$ and $(0, -6)$. Plotting the intercepts and sketching the ellipse through them gives the graph in **Figure 15.**

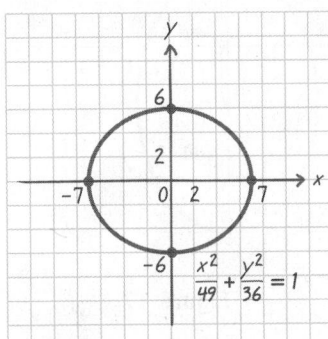

Figure 15

(b) $\dfrac{x^2}{36} + \dfrac{y^2}{121} = 1$

The x-intercepts for this ellipse are $(6, 0)$ and $(-6, 0)$, and the y-intercepts are $(0, 11)$ and $(0, -11)$. Join these intercepts with the smooth curve of an ellipse. See **Figure 16.**

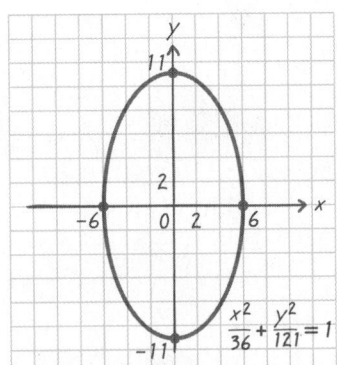

Figure 16

·············· Work Problem ❺ at the Side. ▶

> **CAUTION**
>
> Hand-drawn graphs of ellipses are smooth curves and show symmetry with respect to the center.

❺ Graph each ellipse.

(a) $\dfrac{x^2}{4} + \dfrac{y^2}{25} = 1$

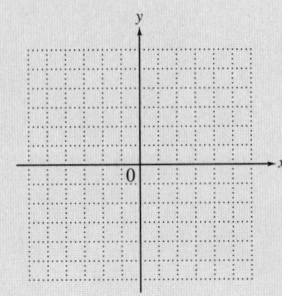

(b) $\dfrac{x^2}{64} + \dfrac{y^2}{49} = 1$

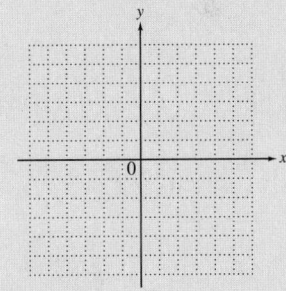

Answers

5. (a)

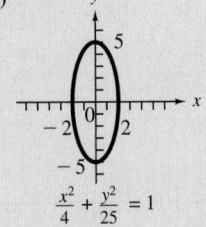

$\dfrac{x^2}{4} + \dfrac{y^2}{25} = 1$

(b)

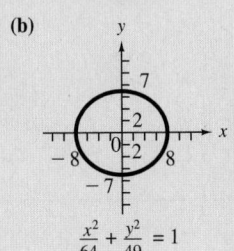

$\dfrac{x^2}{64} + \dfrac{y^2}{49} = 1$

6 Graph

$$\frac{(x+4)^2}{16} + \frac{(y-1)^2}{36} = 1.$$

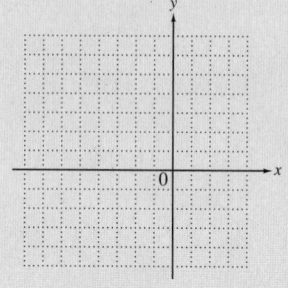

EXAMPLE 6 **Graphing an Ellipse Shifted Horizontally and Vertically**

Graph $\dfrac{(x-2)^2}{25} + \dfrac{(y+3)^2}{49} = 1.$

Just as $(x-2)^2$ and $(y+3)^2$ would indicate that the center of a circle would be $(2, -3)$, so it is with this ellipse. **Figure 17** shows that the graph goes through the four points

$$(2, 4), \quad (7, -3), \quad (2, -10), \quad \text{and} \quad (-3, -3).$$

The x-values of these points are found by adding $\pm a = \pm 5$ to 2, and the y-values come from adding $\pm b = \pm 7$ to -3.

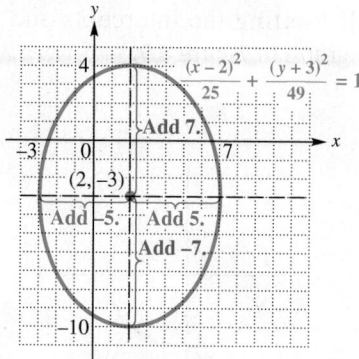

Figure 17

◀ **Work Problem** **6** **at the Side.**

Note

Graphs of circles and ellipses are not graphs of functions. Of the conic sections studied up to this point, only the vertical parabola

$$f(x) = ax^2 + bx + c$$

is the graph of a function.

Answer

6.

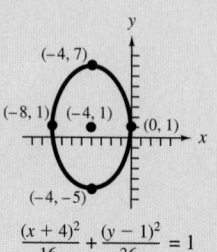

$$\frac{(x+4)^2}{16} + \frac{(y-1)^2}{36} = 1$$

12.2 Exercises

 MyMathLab®

Download the MyDashBoard App

CONCEPT CHECK *Match each equation with the correct graph.*

1. $(x - 3)^2 + (y - 2)^2 = 25$

A.

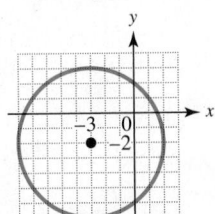

B.

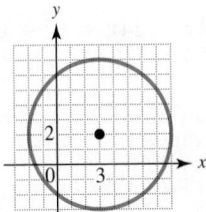

2. $(x - 3)^2 + (y + 2)^2 = 25$

C.

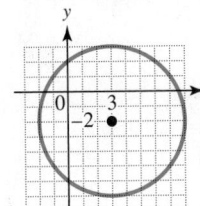

D.

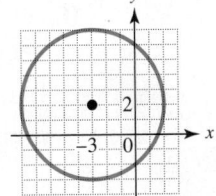

3. $(x + 3)^2 + (y - 2)^2 = 25$

4. $(x + 3)^2 + (y + 2)^2 = 25$

5. See Example 1. Consider the circle whose equation is

$$x^2 + y^2 = 25.$$

(a) What are the coordinates of its center?

(b) What is its radius?

(c) Sketch its graph.

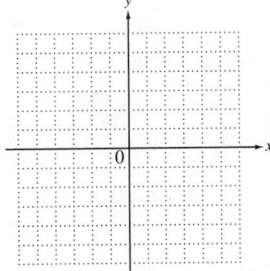

6. Explain why a set of points defined by a circle does not satisfy the definition of a function.

Write the equation of a circle satisfying the given conditions. **See Examples 2 and 3.**

7. Center: $(-4, 3)$; radius: 2

8. Center: $(5, -2)$; radius: 4

9. Center: $(-8, -5)$; radius: $\sqrt{5}$

10. Center: $(-12, 13)$; radius: $\sqrt{7}$

Find the center and radius of each circle. (Hint: In Exercises 15 and 16, divide each side by a common factor.) **See Example 4.**

11. $x^2 + y^2 + 4x + 6y + 9 = 0$

12. $x^2 + y^2 - 8x - 12y + 3 = 0$

13. $x^2 + y^2 + 10x - 14y - 7 = 0$

14. $x^2 + y^2 - 2x + 4y - 4 = 0$

15. $3x^2 + 3y^2 - 12x - 24y + 12 = 0$

16. $2x^2 + 2y^2 + 20x + 16y + 10 = 0$

Graph each circle. Identify the center and the radius. ***See Examples 1, 2, and 4.***

17. $x^2 + y^2 = 4$ **18.** $x^2 + y^2 = 9$ **19.** $3x^2 = 48 - 3y^2$ **20.** $2y^2 = 10 - 2x^2$

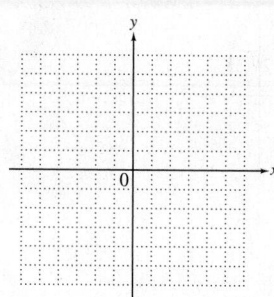 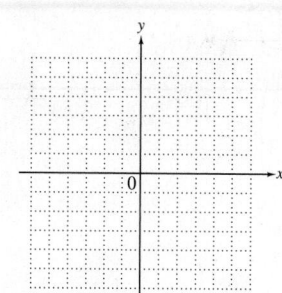 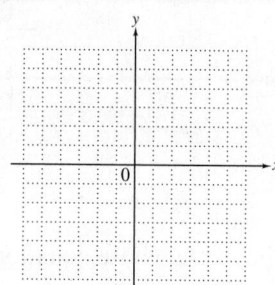 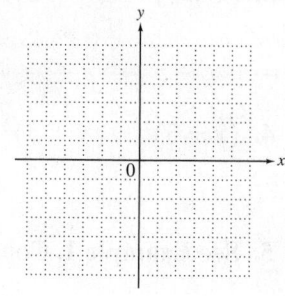

21. $(x - 1)^2 + (y + 2)^2 = 16$ **22.** $(x + 3)^2 + (y - 2)^2 = 9$ **23.** $x^2 + y^2 + 2x + 2y - 23 = 0$

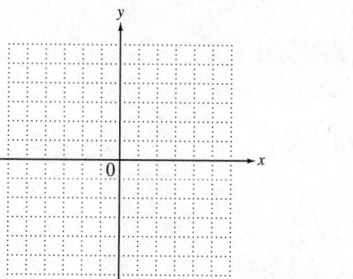 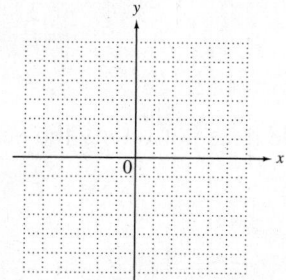 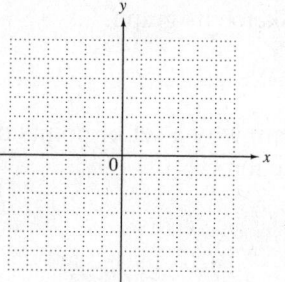

24. $x^2 + y^2 - 4x - 6y + 9 = 0$ **25.** $x^2 + y^2 + 6x - 6y + 9 = 0$ **26.** $x^2 + y^2 - 4x + 6y + 4 = 0$

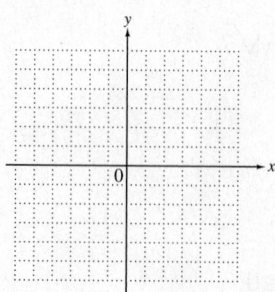 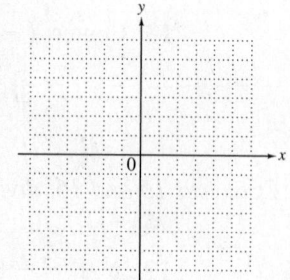 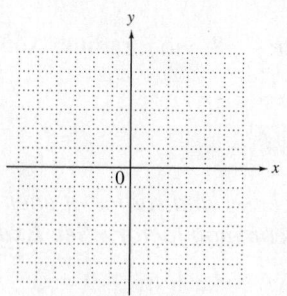

Write a short answer to each problem.

27. A circle can be drawn on a piece of posterboard by fastening one end of a length of string with a thumbtack, pulling the string taut with a pencil, and tracing a curve, as shown in the figure. Explain why this method works.

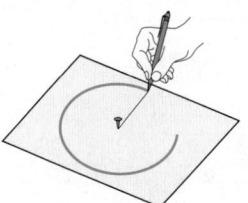

28. An ellipse can be drawn on a piece of posterboard by fastening two ends of a length of string with thumbtacks, pulling the string taut with a pencil, and tracing a curve, as shown in the figure. Explain why this method works.

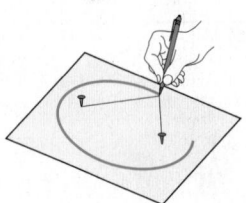

Graph each ellipse. See Examples 5 and 6.

29. $\dfrac{x^2}{9} + \dfrac{y^2}{25} = 1$

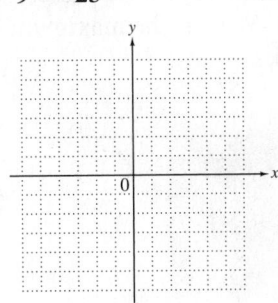

30. $\dfrac{x^2}{9} + \dfrac{y^2}{16} = 1$

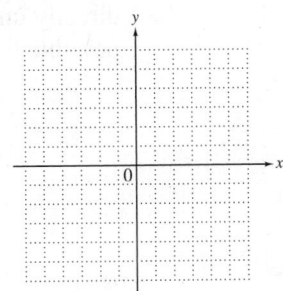

31. $\dfrac{x^2}{36} + \dfrac{y^2}{16} = 1$

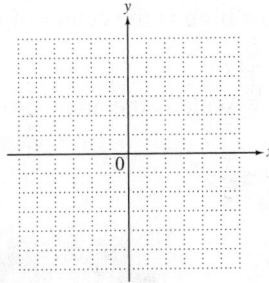

32. $\dfrac{x^2}{9} + \dfrac{y^2}{4} = 1$

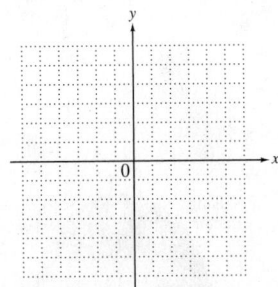

33. $\dfrac{x^2}{25} + \dfrac{y^2}{4} = 1$

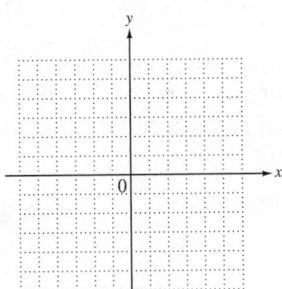

34. $\dfrac{x^2}{16} + \dfrac{y^2}{9} = 1$

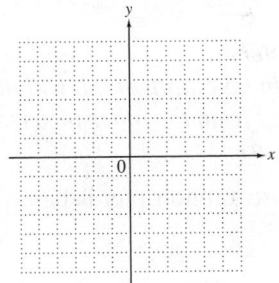

35. $\dfrac{(x-2)^2}{16} + \dfrac{(y-1)^2}{9} = 1$

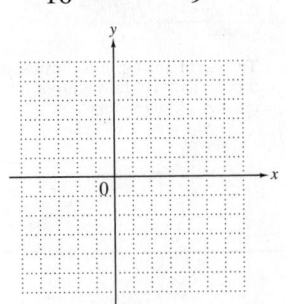

36. $\dfrac{(x-3)^2}{9} + \dfrac{(y+2)^2}{4} = 1$

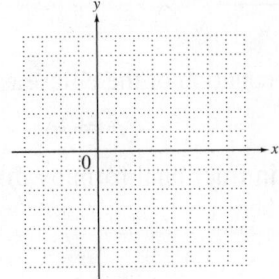

37. $\dfrac{(x+3)^2}{4} + \dfrac{(y-1)^2}{25} = 1$

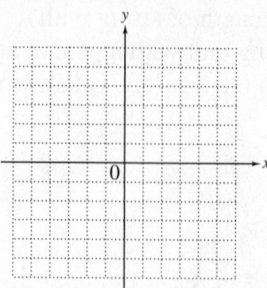

38. $\dfrac{(x+2)^2}{16} + \dfrac{(y+1)^2}{25} = 1$

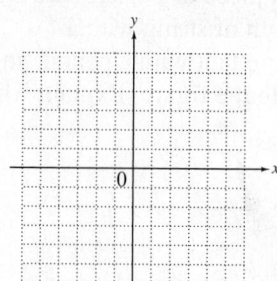

Solve each problem.

39. An arch has the shape of half an ellipse. The equation of the complete ellipse, where x and y are in meters, is $100x^2 + 324y^2 = 32{,}400$.

 (a) How high is the center of the arch?

 (b) How wide is the arch across the bottom?

NOT TO SCALE

40. A one-way street passes under an overpass, which is in the form of the top half of an ellipse, as shown in the figure. Suppose that a truck 12 ft wide passes directly under the overpass. What is the maximum possible height of this truck?

15 ft

20 ft

NOT TO SCALE

*A **lithotripter** is a machine used to crush kidney stones using shock waves. The patient is placed in an elliptical tub with the kidney stone at one focus of the ellipse. A beam is projected from the other focus to the tub, so that it reflects to hit the kidney stone. See the figure.*

41. Suppose a lithotripter is based on the ellipse with equation

$$\frac{x^2}{36} + \frac{y^2}{9} = 1.$$

How far from the center of the ellipse must the kidney stone and the source of the beam be placed? (*Hint:* Use the fact that $c^2 = a^2 - b^2$, since $a > b$ here.)

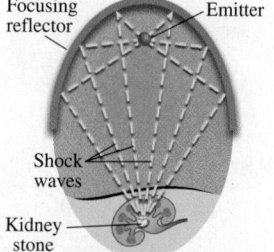

Focusing reflector

Emitter

Shock waves

Kidney stone

The top of an ellipse is illustrated in this depiction of how a lithotripter crushes a kidney stone.

42. Rework **Exercise 41** if the equation of the ellipse is

$$9x^2 + 4y^2 = 36.$$

(*Hint:* Write the equation in fractional form by dividing each term by 36, and use $c^2 = b^2 - a^2$, since $b > a$ here.)

12.3 The Hyperbola and Other Functions Defined by Radicals

OBJECTIVES

1. Recognize the equation of a hyperbola.
2. Graph hyperbolas by using asymptotes.
3. Identify conic sections by their equations.
4. Graph generalized square root functions.

OBJECTIVE ▶ 1 Recognize the equation of a hyperbola. A **hyperbola** is the set of all points in a plane such that the absolute value of the *difference* of the distances from two fixed points (the *foci*) is constant. The graph of a hyperbola has two parts, or *branches,* and two intercepts (or *vertices*) that lie on its axis, called the **transverse axis.**

The hyperbola in **Figure 18** has a horizontal transverse axis, with foci $(c, 0)$ and $(-c, 0)$ and x-intercepts $(a, 0)$ and $(-a, 0)$. (A hyperbola with vertical transverse axis would have its intercepts on the y-axis.)

A hyperbola centered at the origin has one of the following equations. (It is shown in more advanced courses that for a hyperbola, $c^2 = a^2 + b^2$.)

Equations of Hyperbolas

A hyperbola with x-intercepts $(a, 0)$ and $(-a, 0)$ has equation

$$\frac{x^2}{a^2} - \frac{y^2}{b^2} = 1.$$

A hyperbola with y-intercepts $(0, b)$ and $(0, -b)$ has equation

$$\frac{y^2}{b^2} - \frac{x^2}{a^2} = 1.$$

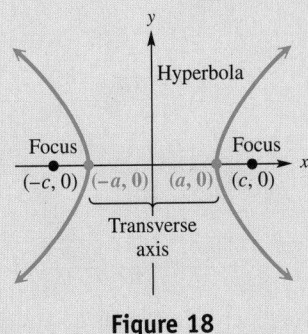

Figure 18

OBJECTIVE ▶ 2 Graph hyperbolas by using asymptotes. The two branches of the graph of a hyperbola approach a pair of intersecting straight lines, which are its *asymptotes.* (See **Figures 19 and 20** on the next page.)

Asymptotes of Hyperbolas

The extended diagonals of the rectangle, called the **fundamental rectangle,** with vertices (corners) at the points (a, b), $(-a, b)$, $(-a, -b)$, and $(a, -b)$ are the **asymptotes** of the hyperbolas

$$\frac{x^2}{a^2} - \frac{y^2}{b^2} = 1 \quad \text{and} \quad \frac{y^2}{b^2} - \frac{x^2}{a^2} = 1.$$

To graph a hyperbola, follow these steps.

Graphing a Hyperbola

Step 1 **Find the intercepts.** Locate the intercepts at $(a, 0)$ and $(-a, 0)$ if the x^2-term has a positive coefficient, or at $(0, b)$ and $(0, -b)$ if the y^2-term has a positive coefficient.

Step 2 **Find the fundamental rectangle.** Its vertices will be at the points (a, b), $(-a, b)$, $(-a, -b)$, and $(a, -b)$.

Step 3 **Sketch the asymptotes.** The extended diagonals of the fundamental rectangle are the asymptotes of the hyperbola. They have equations $y = \pm\frac{b}{a}x$.

Step 4 **Draw the graph.** Sketch each branch of the hyperbola through an intercept, approaching (but not touching) the asymptotes.

1 Graph $\dfrac{x^2}{4} - \dfrac{y^2}{25} = 1$.

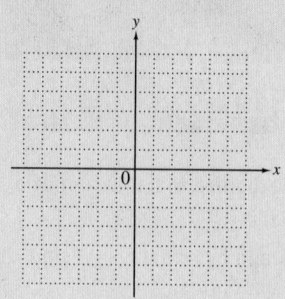

2 Graph $\dfrac{y^2}{81} - \dfrac{x^2}{64} = 1$.

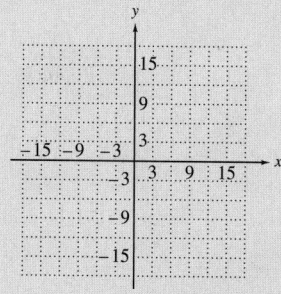

Answers

1.

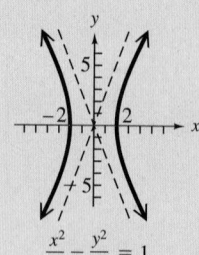

$$\frac{x^2}{4} - \frac{y^2}{25} = 1$$

2.

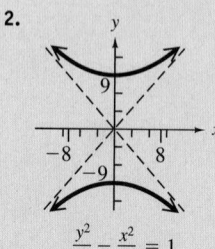

$$\frac{y^2}{81} - \frac{x^2}{64} = 1$$

EXAMPLE 1   Graphing a Horizontal Hyperbola

Graph $\dfrac{x^2}{16} - \dfrac{y^2}{25} = 1$.

Step 1   Here $a = 4$ and $b = 5$. The x-intercepts are $(4, 0)$ and $(-4, 0)$.

Step 2   As shown in **Figure 19**, the vertices of the fundamental rectangle are the four points

$$(a, b) \quad (-a, b) \quad (-a, -b) \qquad (a, -b)$$
$$\downarrow\downarrow \qquad \downarrow\downarrow \qquad \downarrow\downarrow \qquad\qquad \downarrow\downarrow$$
$$(4, 5), \quad (-4, 5), \quad (-4, -5), \quad \text{and} \quad (4, -5).$$

Steps 3 and 4   The equations of the asymptotes are $y = \pm\dfrac{b}{a}x$, or $y = \pm\dfrac{5}{4}x$. The hyperbola approaches these lines as x and y get larger and larger in absolute value.

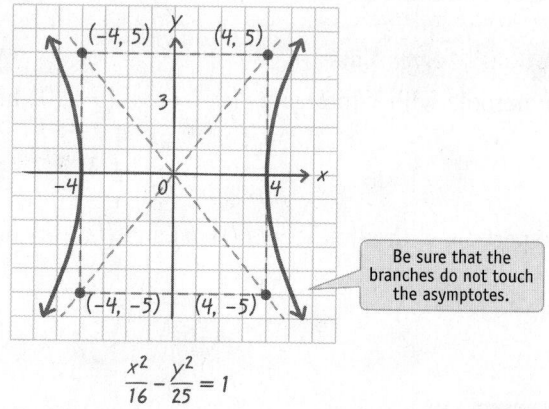

Be sure that the branches do not touch the asymptotes.

$$\frac{x^2}{16} - \frac{y^2}{25} = 1$$

Figure 19

◀ **Work Problem 1 at the Side.**

EXAMPLE 2   Graphing a Vertical Hyperbola

Graph $\dfrac{y^2}{49} - \dfrac{x^2}{16} = 1$.

This hyperbola has y-intercepts $(0, 7)$ and $(0, -7)$. The asymptotes are the extended diagonals of the fundamental rectangle with vertices at

$$(4, 7), \quad (-4, 7), \quad (-4, -7), \quad \text{and} \quad (4, -7).$$

Their equations are $y = \pm\dfrac{7}{4}x$. See **Figure 20.**

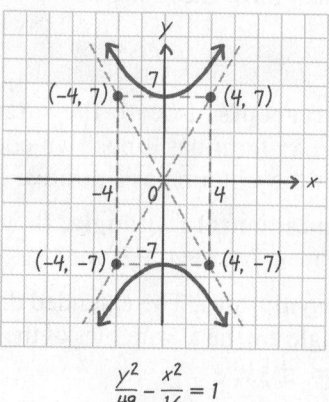

$$\frac{y^2}{49} - \frac{x^2}{16} = 1$$

Figure 20

◀ **Work Problem 2 at the Side.**

OBJECTIVE ▶ **3** **Identify conic sections by their equations.** Rewriting a second-degree equation in one of the forms given for ellipses, hyperbolas, circles, or parabolas makes it possible to identify the graph of the equation.

SUMMARY OF CONIC SECTIONS

Equation	Graph	Description	Identification
$y = ax^2 + bx + c$ or $y = a(x - h)^2 + k$	 Parabola	It opens up if $a > 0$, down if $a < 0$. The vertex is (h, k).	It has an x^2-term. y is not squared.
$x = ay^2 + by + c$ or $x = a(y - k)^2 + h$	 Parabola	It opens to the right if $a > 0$, to the left if $a < 0$. The vertex is (h, k).	It has a y^2-term. x is not squared.
$(x - h)^2 + (y - k)^2 = r^2$	 Circle	The center is (h, k), and the radius is r.	x^2- and y^2-terms have the same positive coefficient.
$\dfrac{x^2}{a^2} + \dfrac{y^2}{b^2} = 1$	 Ellipse	The x-intercepts are $(a, 0)$ and $(-a, 0)$. The y-intercepts are $(0, b)$ and $(0, -b)$.	x^2- and y^2-terms have different positive coefficients.
$\dfrac{x^2}{a^2} - \dfrac{y^2}{b^2} = 1$	 Hyperbola	The x-intercepts are $(a, 0)$ and $(-a, 0)$. The asymptotes are found from (a, b), $(a, -b)$, $(-a, -b)$, and $(-a, b)$.	x^2 has a positive coefficient. y^2 has a negative coefficient.
$\dfrac{y^2}{b^2} - \dfrac{x^2}{a^2} = 1$	 Hyperbola	The y-intercepts are $(0, b)$ and $(0, -b)$. The asymptotes are found from (a, b), $(a, -b)$, $(-a, -b)$, and $(-a, b)$.	y^2 has a positive coefficient. x^2 has a negative coefficient.

❸ Identify the graph of each equation.

(a) $3x^2 = 27 - 4y^2$

(b) $6x^2 = 100 + 2y^2$

(c) $3x^2 = 27 - 4y$

(d) $3x^2 = 27 - 3y^2$

EXAMPLE 3 **Identifying the Graphs of Equations**

Identify the graph of each equation.

(a) $9x^2 = 108 + 12y^2$

Both variables are squared, so the graph is either an ellipse or a hyperbola. (This situation also occurs for a circle, which is a special case of the ellipse.) To see which one it is, rewrite the equation so that both the x^2-term and y^2-term are on one side of the equation and 1 is on the other.

$$9x^2 - 12y^2 = 108 \quad \text{Subtract } 12y^2.$$

$$\frac{x^2}{12} - \frac{y^2}{9} = 1 \quad \text{Divide by 108.}$$

Because of the subtraction symbol, the graph of this equation is a hyperbola.

(b) $x^2 = y - 3$

Only one of the two variables, x, is squared, so this is the vertical parabola with equation $y = x^2 + 3$.

(c) $x^2 = 9 - y^2$

Write the variable terms on the same side of the equation.

$$x^2 + y^2 = 9 \quad \text{Add } y^2.$$

The graph of this equation is a circle with center at the origin and radius 3.

◀ **Work Problem ❸ at the Side.**

OBJECTIVE ▶ ④ Graph generalized square root functions. Recall from **Section 3.6** that no vertical line will intersect the graph of a function in more than one point. Thus, horizontal parabolas and all circles, ellipses, and hyperbolas with horizontal or vertical axes are examples of graphs that do not satisfy the conditions of a function. However, by considering only a part of the graph of each of these, we have the graph of a function, as seen in **Figure 21.**

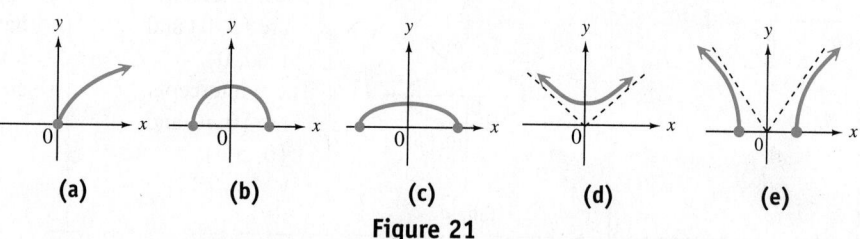

| (a) | (b) | (c) | (d) | (e) |

Figure 21

In parts (a)–(d) of **Figure 21,** the top portion of a conic section is shown (parabola, circle, ellipse, and hyperbola, respectively). In part (e), the top two portions of a hyperbola are shown. In each case, the graph is that of a function since the graph satisfies the conditions of the vertical line test.

In **Sections 9.1** and **12.1** we worked with the square root function $f(x) = \sqrt{x}$. To find equations for the types of graphs shown in **Figure 21,** we extend its definition.

Generalized Square Root Function

For an algebraic expression in x defined by u, with $u \geq 0$, a function of the form

$$f(x) = \sqrt{u}$$

is a **generalized square root function.**

Answers

3. **(a)** ellipse **(b)** hyperbola **(c)** parabola
  (d) circle

EXAMPLE 4	Graphing a Semicircle

Graph $f(x) = \sqrt{25 - x^2}$. Give the domain and range.

$$f(x) = \sqrt{25 - x^2} \qquad \text{Given function}$$

$$y = \sqrt{25 - x^2} \qquad \text{Replace } f(x) \text{ with } y.$$

$$y^2 = \left(\sqrt{25 - x^2}\right)^2 \qquad \text{Square each side.}$$

$$y^2 = 25 - x^2 \qquad \left(\sqrt{a}\right)^2 = a$$

$$x^2 + y^2 = 25 \qquad \text{Add } x^2.$$

This is the graph of a circle with center at $(0, 0)$ and radius 5. *Since $f(x)$, or y, represents a principal square root in the original equation, $f(x)$ must be nonnegative. This restricts the graph to the upper half of the circle.* See **Figure 22.** Use the graph and the vertical line test to verify that it is indeed a function. The domain is $[-5, 5]$, and the range is $[0, 5]$.

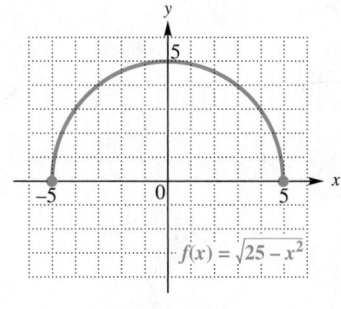

Figure 22

····· Work Problem ❹ at the Side. ▶

EXAMPLE 5	Graphing a Portion of an Ellipse

Graph $\dfrac{y}{6} = -\sqrt{1 - \dfrac{x^2}{16}}$. Give the domain and range.

$$\frac{y}{6} = -\sqrt{1 - \frac{x^2}{16}} \qquad \text{Given equation}$$

$$\left(\frac{y}{6}\right)^2 = \left(-\sqrt{1 - \frac{x^2}{16}}\right)^2 \qquad \text{Square each side.}$$

$$\frac{y^2}{36} = 1 - \frac{x^2}{16} \qquad \text{Apply the exponents.}$$

$$\frac{x^2}{16} + \frac{y^2}{36} = 1 \qquad \text{Add } \tfrac{x^2}{16}.$$

This is the equation of an ellipse. The x-intercepts are $(4, 0)$ and $(-4, 0)$, and the y-intercepts are $(0, 6)$ and $(0, -6)$. *Since $\frac{y}{6}$ equals a negative square root in the original equation, y must be nonpositive, restricting the graph to the lower half of the ellipse.* See **Figure 23.** The domain is $[-4, 4]$, and the range is $[-6, 0]$.

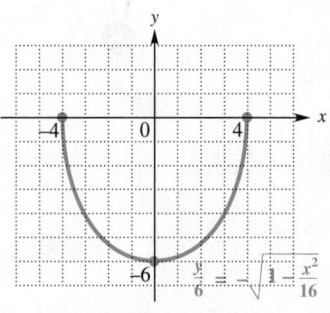

Figure 23

····· Work Problem ❺ at the Side. ▶

❹ Graph $f(x) = \sqrt{36 - x^2}$. Give the domain and range.

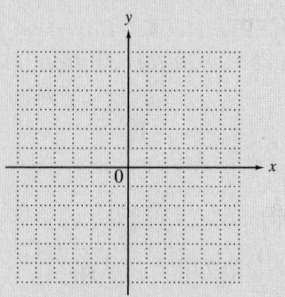

❺ Graph

$$\frac{y}{3} = -\sqrt{1 - \frac{x^2}{4}}.$$

Give the domain and range.

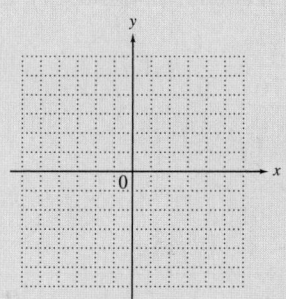

Answers

4.

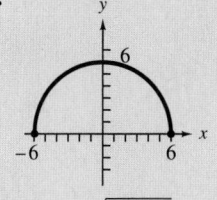

$f(x) = \sqrt{36 - x^2}$

domain: $[-6, 6]$; range: $[0, 6]$

5.

$\dfrac{y}{3} = -\sqrt{1 - \frac{x^2}{4}}$

domain: $[-2, 2]$; range: $[-3, 0]$

12.3 Exercises

FOR EXTRA HELP  Download the MyDashBoard App  MyMathLab®

CONCEPT CHECK *Based on the discussions of ellipses in the previous section and of hyperbolas in this section, match each equation with its graph.*

1. $\dfrac{x^2}{25} + \dfrac{y^2}{9} = 1$

2. $\dfrac{x^2}{9} + \dfrac{y^2}{25} = 1$

3. $\dfrac{x^2}{9} - \dfrac{y^2}{25} = 1$

4. $\dfrac{x^2}{25} - \dfrac{y^2}{9} = 1$

A.
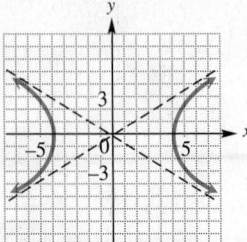

B.
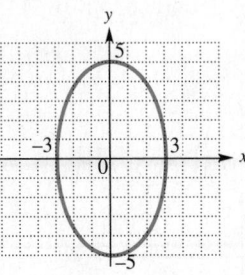

C.
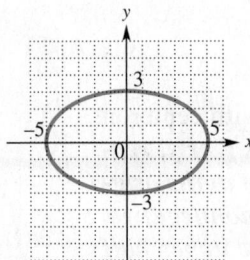

D.
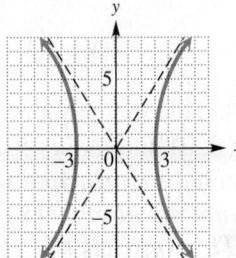

Graph each hyperbola. ***See Examples 1 and 2.***

5. $\dfrac{x^2}{16} - \dfrac{y^2}{9} = 1$

6. $\dfrac{x^2}{25} - \dfrac{y^2}{4} = 1$

7. $\dfrac{y^2}{4} - \dfrac{x^2}{25} = 1$

8. $\dfrac{y^2}{9} - \dfrac{x^2}{4} = 1$

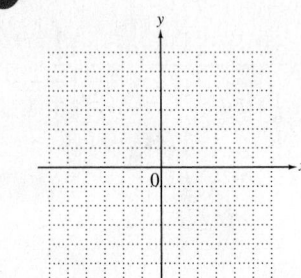

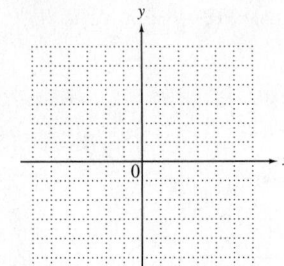

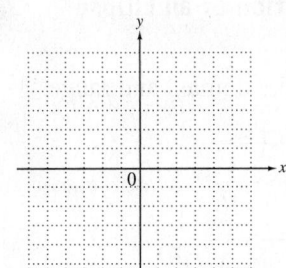

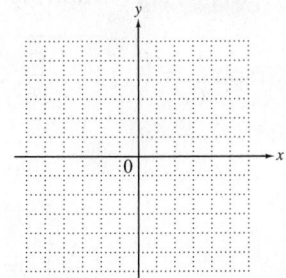

9. $\dfrac{x^2}{25} - \dfrac{y^2}{36} = 1$

10. $\dfrac{x^2}{49} - \dfrac{y^2}{16} = 1$

11. $\dfrac{y^2}{9} - \dfrac{x^2}{9} = 1$

12. $\dfrac{y^2}{16} - \dfrac{x^2}{16} = 1$

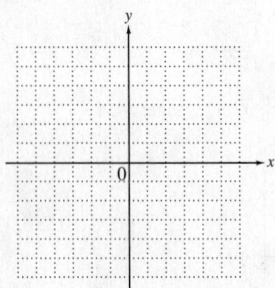

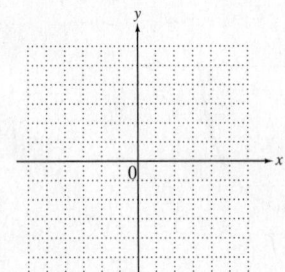

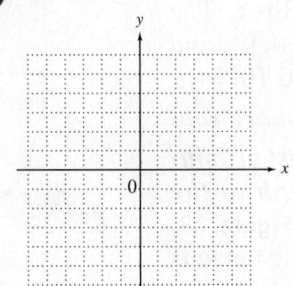

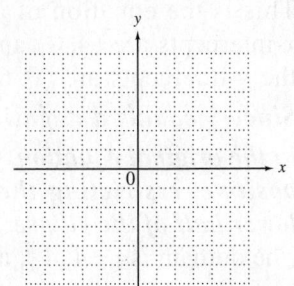

Identify the graph of each equation as a parabola, circle, ellipse, *or* hyperbola, *and sketch it.* **See Example 3.**

13. $x^2 - y^2 = 16$

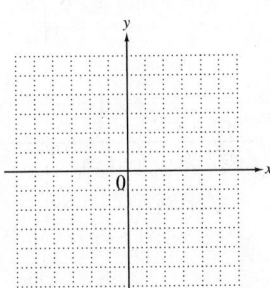

14. $x^2 + y^2 = 16$

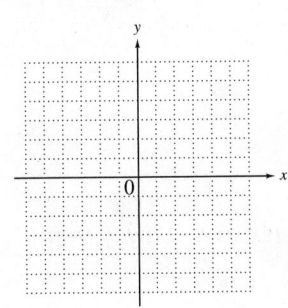

15. $4x^2 + y^2 = 16$

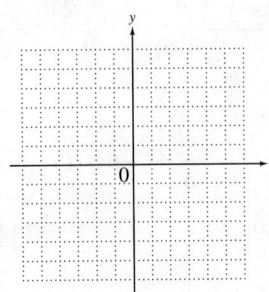

16. $x^2 - 2y = 0$

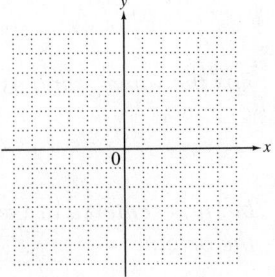

17. $y^2 = 36 - x^2$

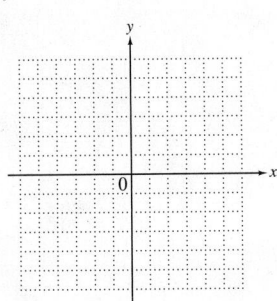

18. $9x^2 + 25y^2 = 225$

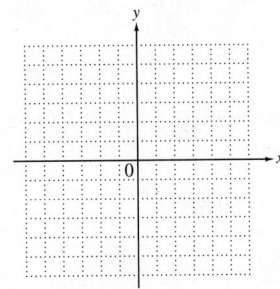

19. $x^2 + 9y^2 = 9$

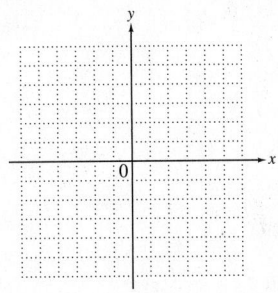

20. $y^2 = 4 + x^2$

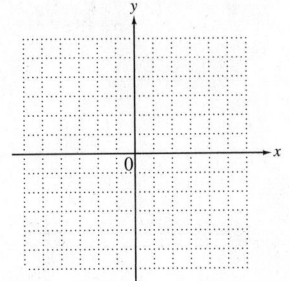

Graph each function defined by a radical expression. Give the domain and range. **See Examples 4 and 5.**

21. $f(x) = \sqrt{9 - x^2}$

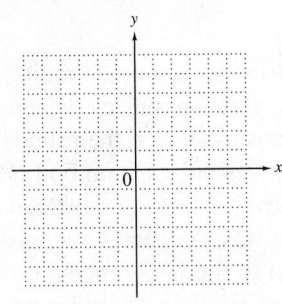

22. $f(x) = \sqrt{16 - x^2}$

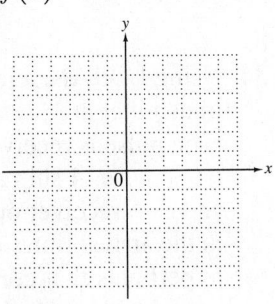

23. $f(x) = -\sqrt{25 - x^2}$

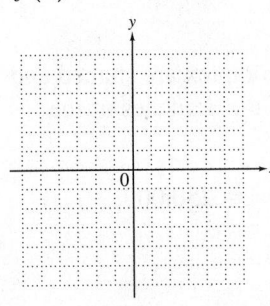

24. $f(x) = -\sqrt{36 - x^2}$

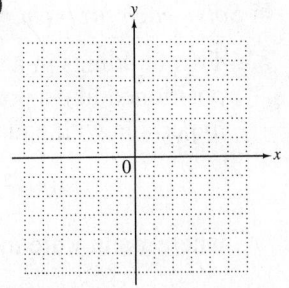

25. $y = \sqrt{\dfrac{x + 4}{2}}$

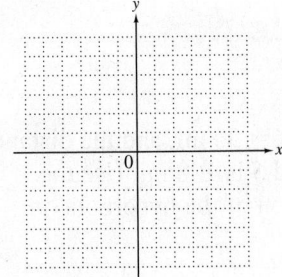

26. $\dfrac{y}{3} = \sqrt{1 + \dfrac{x^2}{9}}$

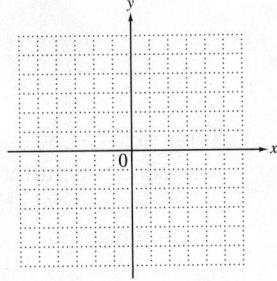

27. $y = -2\sqrt{\dfrac{9 - x^2}{9}}$

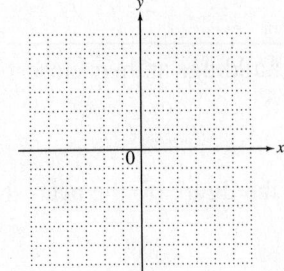

28. $y = -3\sqrt{1 - \dfrac{x^2}{25}}$

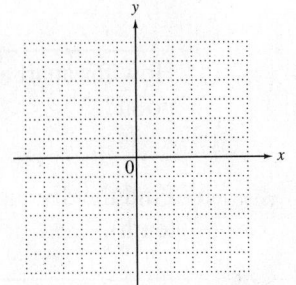

In Section 12.2, Example 6, we saw that the center of an ellipse may be shifted away from the origin. The same process can be applied to hyperbolas. For example, the hyperbola shown at the right,

$$\frac{(x+5)^2}{4} - \frac{(y-2)^2}{9} = 1,$$

has the same graph as

$$\frac{x^2}{4} - \frac{y^2}{9} = 1,$$

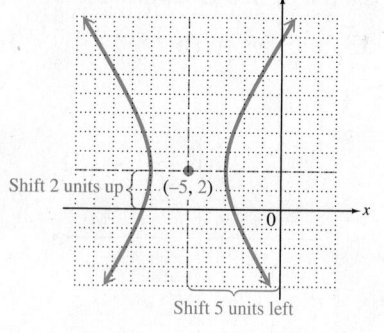

but it is centered at $(-5, 2)$. Graph each hyperbola with center shifted away from the origin.

29. $\dfrac{(x-2)^2}{4} - \dfrac{(y+1)^2}{9} = 1$ **30.** $\dfrac{(x+3)^2}{16} - \dfrac{(y-2)^2}{4} = 1$ **31.** $\dfrac{y^2}{36} - \dfrac{(x-2)^2}{49} = 1$ **32.** $\dfrac{(y-5)^2}{9} - \dfrac{x^2}{25} = 1$

  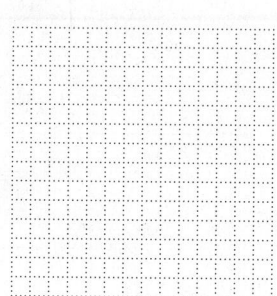 

Solve each problem.

33. Two buildings in a sports complex are shaped and positioned like a portion of the branches of the hyperbola with equation

$$400x^2 - 625y^2 = 250{,}000,$$

where x and y are in meters.

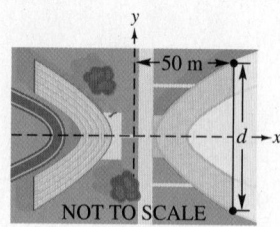

NOT TO SCALE

(a) How far apart are the buildings at their closest point?

(b) Find the distance d in the figure (to the nearest tenth).

34. Using LORAN, a location-finding system, a radio transmitter at M sends out a series of pulses. When each pulse is received at transmitter S, it then sends out a pulse. A ship at P receives pulses from both M and S. A receiver on the ship measures the difference in the arrival times of the pulses. A special map gives hyperbolas that correspond to the differences in arrival times (which give the distances d_1 and d_2 in the figure). The ship can then be located as lying on a branch of a particular hyperbola.

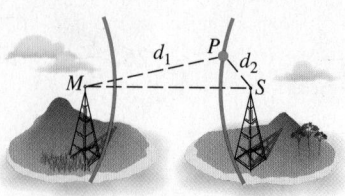

Suppose $d_1 = 80$ mi and $d_2 = 30$ mi, and the distance between transmitters M and S is 100 mi. Use the definition to find an equation of the hyperbola on which the ship is located.

12.4 Nonlinear Systems of Equations

An equation in which some terms have more than one variable or a variable of degree 2 or greater is a **nonlinear equation.** A **nonlinear system of equations** includes at least one nonlinear equation.

When solving a nonlinear system, it helps to visualize the types of graphs of the equations of the system to determine the possible number of points of intersection. For example, if a system includes two equations where the graph of one is a circle and the graph of the other is a line, then there may be zero, one, or two points of intersection, as illustrated in **Figure 24.**

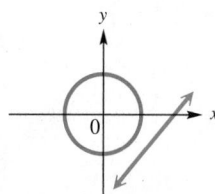
No points of intersection

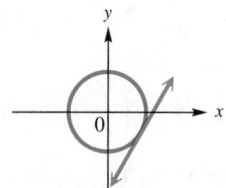
One point of intersection

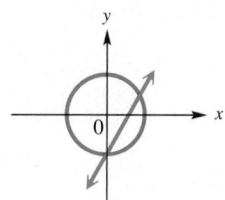
Two points of intersection

Figure 24

> **OBJECTIVES**
>
> **1** Solve a nonlinear system by substitution.
>
> **2** Use the elimination method to solve a nonlinear system with two second-degree equations.
>
> **3** Solve a nonlinear system that requires a combination of methods.

OBJECTIVE ▶ 1 Solve a nonlinear system by substitution. We can usually solve a nonlinear system by the substitution method **(Sections 4.2 and 8.4)** if one of the equations is linear.

EXAMPLE 1 Solving a Nonlinear System by Substitution

Solve the system.
$$x^2 + y^2 = 9 \quad (1)$$
$$2x - y = 3 \quad (2)$$

The graph of (1) is a circle and the graph of (2) is a line. The graphs could intersect in zero, one, or two points, as shown in **Figure 24.** We begin by solving the linear equation (2) for one of its two variables.

$$2x - y = 3 \qquad (2)$$
$$y = 2x - 3 \quad \text{Solve for } y. \quad (3)$$

Then we substitute $2x - 3$ for y in the nonlinear equation (1).

$$x^2 + y^2 = 9 \qquad (1)$$
$$x^2 + (2x - 3)^2 = 9 \qquad \text{Let } y = 2x - 3.$$
$$x^2 + 4x^2 - 12x + 9 = 9 \qquad \text{Square } 2x - 3.$$
$$5x^2 - 12x = 0 \qquad \text{Combine like terms. Subtract 9.}$$
$$x(5x - 12) = 0 \qquad \text{Factor. The GCF is } x.$$

> Set *both* factors equal to 0.

$$x = 0 \quad \text{or} \quad 5x - 12 = 0 \qquad \text{Zero-factor property}$$
$$x = \frac{12}{5} \qquad \text{Solve the equation.}$$

Let $x = 0$ in equation (3) to get $y = -3$. If $x = \frac{12}{5}$ in equation (3), then $y = \frac{9}{5}$. The solution set of the system is $\left\{(0, -3), \left(\frac{12}{5}, \frac{9}{5}\right)\right\}$. The graph in **Figure 25** confirms the two points of intersection.

·········· **Work Problem ❶ at the Side.** ▶

❶ Solve each system.

(a) $x^2 + y^2 = 10$
$$x = y + 2$$

(b) $x^2 - 2y^2 = 8$
$$y + x = 6$$

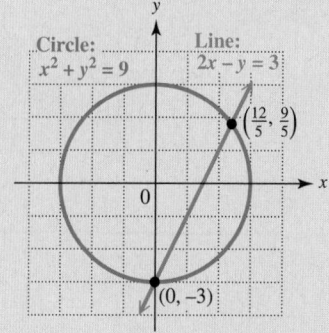

Figure 25

Answers

1. (a) $\{(3, 1), (-1, -3)\}$
  (b) $\{(4, 2), (20, -14)\}$

2 Solve each system.

(a) $xy = 8$

$x + y = 6$

(b) $xy + 10 = 0$

$4x + 9y = -2$

EXAMPLE 2 Solving a Nonlinear System by Substitution

Solve the system.

$$6x - y = 5 \quad (1)$$

$$xy = 4 \quad (2)$$

The graph of (1) is a line. It can be shown by plotting points that the graph of (2) is a hyperbola. Visualizing a line and a hyperbola indicates that there may be zero, one, or two points of intersection.

Solving $xy = 4$ for x gives $x = \frac{4}{y}$. We substitute $\frac{4}{y}$ for x in equation (1).

$$6x - y = 5 \quad (1)$$

$$6\left(\frac{4}{y}\right) - y = 5 \qquad \text{Let } x = \frac{4}{y}.$$

$$\frac{24}{y} - y = 5 \qquad \text{Multiply.}$$

$$24 - y^2 = 5y \qquad \text{Multiply by } y, \ y \neq 0.$$

$$y^2 + 5y - 24 = 0 \qquad \text{Standard form}$$

$$(y - 3)(y + 8) = 0 \qquad \text{Factor.}$$

$$y - 3 = 0 \quad \text{or} \quad y + 8 = 0 \qquad \text{Zero-factor property.}$$

$$y = 3 \quad \text{or} \qquad y = -8 \qquad \text{Solve each equation.}$$

We substitute these results into $x = \frac{4}{y}$ to obtain the corresponding values of x.

If $y = 3$, then $x = \dfrac{4}{3}$.

If $y = -8$, then $x = -\dfrac{1}{2}$.

The solution set of the system is

$$\left\{ \left(\frac{4}{3}, 3\right), \left(-\frac{1}{2}, -8\right) \right\}.$$

See **Figure 26**.

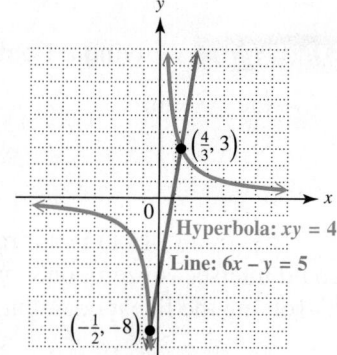

Figure 26

· ◀ **Work Problem 2** at the Side.

CAUTION

Be sure to write the *x*-coordinates first in the ordered-pair solutions of a nonlinear system.

OBJECTIVE ▶ 2 **Use the elimination method to solve a nonlinear system with two second-degree equations.** If a system consists of two second-degree equations, then there may be zero, one, two, three, or four solutions. **Figure 27** shows a case where a system consisting of a circle and a parabola has four solutions, all made up of ordered pairs of real numbers.

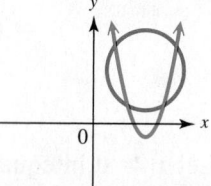

This system has four solutions, since there are four points of intersection.

Figure 27

Answers

2. (a) $\{(4, 2), (2, 4)\}$

(b) $\left\{ (-5, 2), \left(\dfrac{9}{2}, -\dfrac{20}{9}\right) \right\}$

The elimination method (**Sections 4.3 and 8.4**) is often used when both equations are second degree.

EXAMPLE 3   Solving a Nonlinear System by Elimination

Solve the system.

$$x^2 + y^2 = 9 \qquad (1)$$
$$2x^2 - y^2 = -6 \qquad (2)$$

The graph of (1) is a circle, while the graph of (2) is a hyperbola. By analyzing the possibilities, we conclude that there may be zero, one, two, three, or four points of intersection. Adding the two equations will eliminate y.

$$
\begin{array}{llll}
x^2 + y^2 = & 9 & (1) & \\
2x^2 - y^2 = & -6 & (2) & \\
\hline
3x^2 \quad = & 3 & & \text{Add.} \\
x^2 = 1 & & & \text{Divide by 3.} \\
x = 1 \quad \text{or} \quad x = -1 & & & \text{Square root property}
\end{array}
$$

Each value of x gives corresponding values for y when substituted into one of the original equations. Using equation (1) is easier since the coefficients of the x^2- and y^2-terms are 1.

$x^2 + y^2 = 9$ (1)	$x^2 + y^2 = 9$ (1)
$1^2 + y^2 = 9$ Let $x = 1$.	$(-1)^2 + y^2 = 9$ Let $x = -1$.
$y^2 = 8$	$y^2 = 8$
$y = \sqrt{8}$ or $y = -\sqrt{8}$	$y = \sqrt{8}$ or $y = -\sqrt{8}$
$y = 2\sqrt{2}$ or $y = -2\sqrt{2}$	$y = 2\sqrt{2}$ or $y = -2\sqrt{2}$

The solution set is

$$\left\{ \left(1, 2\sqrt{2}\right), \left(1, -2\sqrt{2}\right), \left(-1, 2\sqrt{2}\right), \left(-1, -2\sqrt{2}\right) \right\}.$$

Figure 28 shows the four points of intersection.

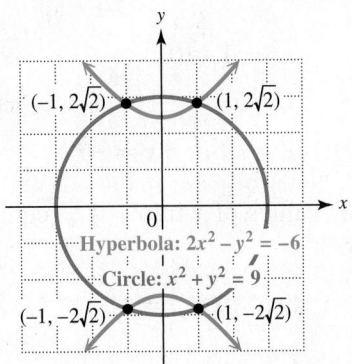

Figure 28

· · · · · · · · · · · · · · · **Work Problem ❸ at the Side.** ▶

---

❸ Solve each system.

(a) $x^2 + y^2 = 41$

$\quad x^2 - y^2 = 9$

(b) $\quad x^2 + 3y^2 = 40$

$\quad 4x^2 - \quad y^2 = 4$

Answers

3. (a) $\{(5, 4), (5, -4), (-5, 4), (-5, -4)\}$

(b) $\{ (2, 2\sqrt{3}), (2, -2\sqrt{3}),$
$(-2, 2\sqrt{3}), (-2, -2\sqrt{3}) \}$

OBJECTIVE **3** **Solve a nonlinear system that requires a combination of methods.**

EXAMPLE 4 **Solving a Nonlinear System by a Combination of Methods**

Solve the system.

$$x^2 + 2xy - y^2 = 7 \quad (1)$$
$$x^2 - y^2 = 3 \quad (2)$$

While we have not graphed equations like (1), its graph is a hyperbola. The graph of (2) is also a hyperbola. Two hyperbolas may have zero, one, two, three, or four points of intersection. We use the elimination method here in combination with the substitution method.

$$x^2 + 2xy - y^2 = \ 7 \quad (1)$$
$$-x^2 \qquad + y^2 = -3 \quad \text{Multiply (2) by } -1.$$

The x^2- and y^2-terms were eliminated.

$$2xy \qquad = \ 4 \quad \text{Add.}$$

Next, we solve $2xy = 4$ for one of the variables. We choose y.

$$2xy = 4$$
$$y = \frac{2}{x} \quad (3)$$

Now, we substitute $y = \frac{2}{x}$ into one of the original equations.

$$x^2 - y^2 = 3 \qquad\qquad \text{The substitution is easier in (2).}$$

$$x^2 - \left(\frac{2}{x}\right)^2 = 3 \qquad\qquad \text{Let } y = \frac{2}{x}.$$

$$x^2 - \frac{4}{x^2} = 3 \qquad\qquad \text{Square } \frac{2}{x}.$$

$$x^4 - 4 = 3x^2 \qquad\qquad \text{Multiply by } x^2, x \neq 0.$$

$$x^4 - 3x^2 - 4 = 0 \qquad\qquad \text{Subtract } 3x^2.$$

$$(x^2 - 4)(x^2 + 1) = 0 \qquad\qquad \text{Factor.}$$

$$x^2 - 4 = 0 \quad \text{or} \quad x^2 + 1 = 0 \qquad \text{Zero-factor property}$$

$$x^2 = 4 \quad \text{or} \qquad x^2 = -1 \quad \text{Solve each equation.}$$

$$x = 2 \ \text{ or } \ x = -2 \quad x = i \ \text{ or } \ x = -i$$

Substituting these four values of x into $y = \frac{2}{x}$ (equation (3)) gives the corresponding values for y.

If $x = 2$, then $y = \frac{2}{2} = 1$.

If $x = -2$, then $y = \frac{2}{-2} = -1$.

Multiply by the complex conjugate of the denominator. $i(-i) = 1$

If $x = i$, then $y = \frac{2}{i} = \frac{2}{i} \cdot \frac{-i}{-i} = -2i$.

If $x = -i$, then $y = \frac{2}{-i} = \frac{2}{-i} \cdot \frac{i}{i} = 2i$.

Continued on Next Page

If we substitute the x-values we found into equation (1) or (2) instead of into equation (3), we get extraneous solutions. ***It is always wise to check all solutions in both of the given equations.*** There are four ordered pairs in the solution set, two with real values and two with nonreal complex values. The solution set is

$$\{(2, 1), (-2, -1), (i, -2i), (-i, 2i)\}.$$

The graph of the system, shown in **Figure 29,** shows only the two real intersection points because the graph is in the real number plane. In general, if solutions contain nonreal complex numbers as components, they do not appear on the graph.

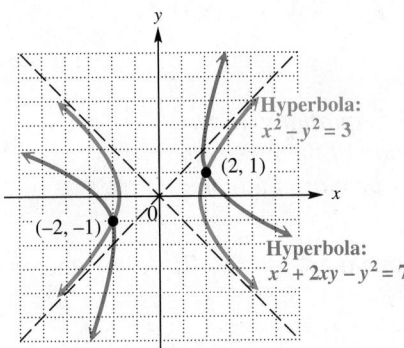

Figure 29

···················· Work Problem ➍ at the Side. ▶

Note

It is not essential to visualize the number of points of intersection of the graphs in order to solve a nonlinear system. Sometimes we are unfamiliar with the graphs or, as in **Example 4,** there are nonreal complex solutions that do not appear as points of intersection in the real plane. Visualizing the geometry of the graphs is only an aid to solving these systems.

➍ Solve each system.

(a) $x^2 + xy + y^2 = 3$

$\qquad x^2 + y^2 = 5$

(b) $x^2 + 7xy - 2y^2 = -8$

$\qquad -2x^2 + 4y^2 = 16$

12.4 Exercises

CONCEPT CHECK *Each sketch represents the graphs of a pair of equations in a system. How many ordered pairs of real numbers are in each solution set?*

1.
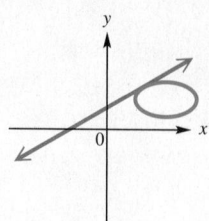

2.
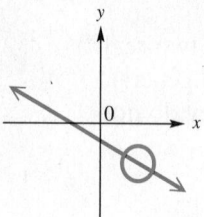

3.
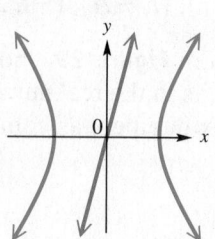

4.
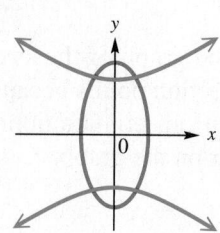

CONCEPT CHECK *Suppose that a nonlinear system is composed of equations whose graphs are those described, and the number of points of intersection of the two graphs is as given. Make a sketch satisfying these conditions. (There may be more than one way to do this.)*

5. A line and a circle; no points

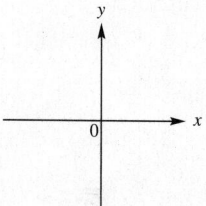

6. A line and a circle; one point

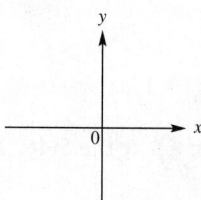

7. A line and an ellipse; two points

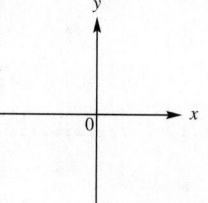

8. A line and a hyperbola; no points

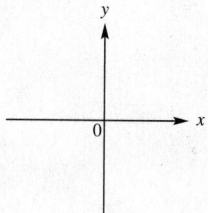

9. A circle and an ellipse; four points

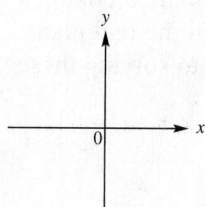

10. A parabola and an ellipse; one point

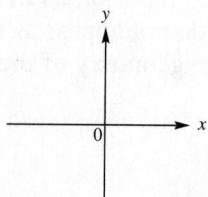

Solve each system by the substitution method. **See Examples 1 and 2.**

11. $y = 4x^2 - x$
$y = x$

12. $y = x^2 + 6x$
$3y = 12x$

13. $y = x^2 + 6x + 9$
$x + y = 3$

14. $y = x^2 + 8x + 16$
$x - y = -4$

15. $x^2 + y^2 = 2$
$2x + y = 1$

16. $2x^2 + 4y^2 = 4$
$x = 4y$

17. $xy = 4$
$3x + 2y = -10$

18. $xy = -5$
$2x + y = 3$

19. $xy = -3$
$x + y = -2$

20. $xy = 12$
$x + y = 8$

21. $y = 3x^2 + 6x$
$y = x^2 - x - 6$

22. $y = 2x^2 + 1$
$y = 5x^2 + 2x - 7$

23. $2x^2 - y^2 = 6$
$y = x^2 - 3$

24. $x^2 + y^2 = 4$
$y = x^2 - 2$

Solve each system using the elimination method or a combination of the elimination and substitution methods. **See Examples 3 and 4.**

25. $3x^2 + 2y^2 = 12$
$x^2 + 2y^2 = 4$

26. $6x^2 + y^2 = 9$
$3x^2 + 4y^2 = 36$

27. $5x^2 - 2y^2 = -13$
$3x^2 + 4y^2 = 39$

28. $2x^2 + y^2 = 28$
$4x^2 - 5y^2 = 28$

29. $xy = 6$
$3x^2 - y^2 = 12$

30. $xy = 5$
$2y^2 - x^2 = 5$

31. $2x^2 + 2y^2 = 8$
$3x^2 + 4y^2 = 24$

32. $5x^2 + 5y^2 = 20$
$x^2 + 2y^2 = 2$

33. $x^2 + xy + y^2 = 15$
$x^2 + y^2 = 10$

34. $2x^2 + 3xy + 2y^2 = 21$
$x^2 + y^2 = 6$

Solve each problem by using a nonlinear system.

35. The area of a rectangular rug is 84 ft^2 and its perimeter is 38 ft. Find the length and width of the rug.

36. Find the length and width of a rectangular room whose perimeter is 50 m and whose area is 100 m^2.

12.5 Second-Degree Inequalities and Systems of Inequalities

OBJECTIVES

1. Graph second-degree inequalities.
2. Graph the solution set of a system of inequalities.

1 Graph $x^2 + y^2 \geq 9$.

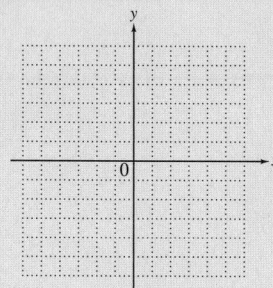

2 Graph $y \geq (x + 1)^2 - 5$.

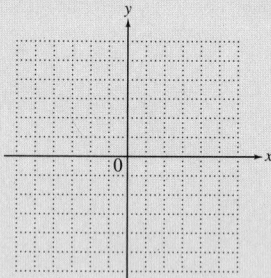

Answers

1.   2.

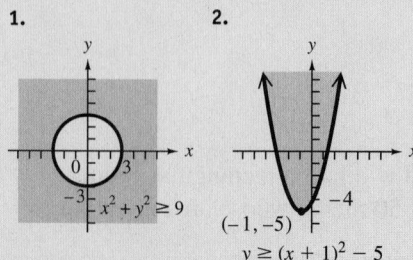

OBJECTIVE 1 Graph second-degree inequalities. A **second-degree inequality** is an inequality with at least one variable of degree 2 and no variable with degree greater than 2.

EXAMPLE 1 Graphing a Second-Degree Inequality

Graph $x^2 + y^2 \leq 36$.

The boundary of the inequality $x^2 + y^2 \leq 36$ is the graph of the equation $x^2 + y^2 = 36$, a circle with radius 6 and center at the origin, as shown in **Figure 30.**

The inequality $x^2 + y^2 \leq 36$ will include either the points outside the circle or the points inside the circle, as well as the boundary. To decide which region to shade, we substitute any test point not on the circle.

$$x^2 + y^2 < 36$$

$$0^2 + 0^2 \overset{?}{<} 36 \qquad \text{Use } (0, 0) \text{ as a test point.}$$

$$0 < 36 \checkmark \quad \text{True}$$

Since a true statement results, the original inequality includes the points *inside* the circle, the shaded region in **Figure 30,** and the boundary.

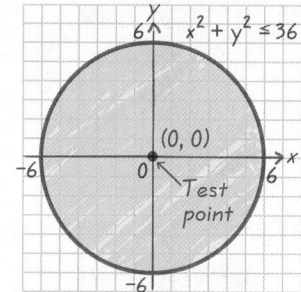

Figure 30

◀ Work Problem **1** at the Side.

EXAMPLE 2 Graphing a Second-Degree Inequality

Graph $y < -2(x - 4)^2 - 3$.

The boundary, $y = -2(x - 4)^2 - 3$, is a parabola that opens down with vertex at $(4, -3)$.

$$y < -2(x - 4)^2 - 3 \qquad \text{Original inequality}$$

$$0 \overset{?}{<} -2(0 - 4)^2 - 3 \qquad \text{Use } (0, 0) \text{ as a test point.}$$

$$0 \overset{?}{<} -32 - 3 \qquad \text{Simplify.}$$

$$0 < -35 \qquad \text{False}$$

Because the final inequality is a false statement, the points in the region containing $(0, 0)$ do not satisfy the inequality. In **Figure 31,** the parabola is drawn as a dashed curve since the points of the parabola itself do not satisfy the inequality, and the region inside (or below) the parabola is shaded.

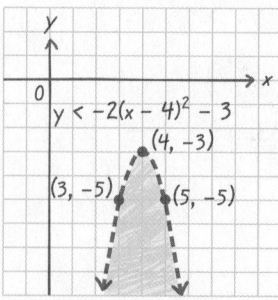

Figure 31

◀ Work Problem **2** at the Side.

> **Note**
>
> Since the substitution is easy, the origin is the test point of choice unless the graph actually passes through $(0, 0)$.

EXAMPLE 3 Graphing a Second-Degree Inequality

Graph $16y^2 \le 144 + 9x^2$.

Rewrite the inequality as follows.

$$16y^2 - 9x^2 \le 144 \qquad \text{Subtract } 9x^2.$$

$$\frac{y^2}{9} - \frac{x^2}{16} \le 1 \qquad \text{Divide by 144.}$$

This form shows that the boundary is the following hyperbola.

$$\frac{y^2}{9} - \frac{x^2}{16} = 1$$

Since the graph is a vertical hyperbola, the desired region will be either the region between the branches or the regions above the top branch and below the bottom branch. Choose $(0, 0)$ as a test point. Substituting into the original inequality leads to $0 \le 144$, a true statement, so the region between the branches containing $(0, 0)$ is shaded, as shown in **Figure 32**.

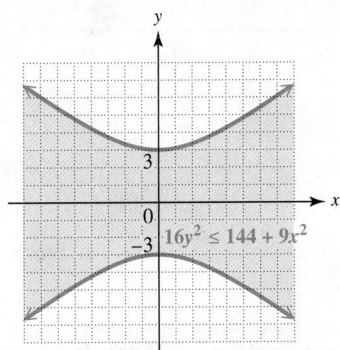

Figure 32

········· Work Problem ❸ at the Side. ▶

OBJECTIVE ❷ **Graph the solution set of a system of inequalities.** If two or more inequalities are considered at the same time, we have a **system of inequalities**. To find the solution set of the system, we find the intersection of the graphs (solution sets) of the inequalities in the system.

EXAMPLE 4 Graphing a System of Two Inequalities

Graph the solution set of the system.

$$2x + 3y > 6$$
$$x^2 + y^2 < 16$$

Begin by graphing the solution set of $2x + 3y > 6$. The boundary line is the graph of $2x + 3y = 6$ and is a dashed line because the symbol $>$ does not include equality. The test point $(0, 0)$ leads to a false statement in $2x + 3y > 6$, so shade the region above the line, as shown in **Figure 33** on the next page.

········· Continued on Next Page

---

❸ Graph $x^2 + 4y^2 > 36$.

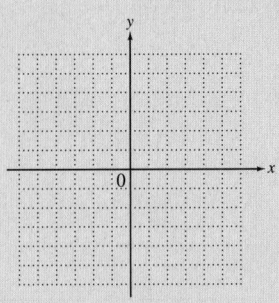

Answer

3.

$x^2 + 4y^2 > 36$

4 Graph the solution set of the system.

$$x^2 + y^2 \leq 25$$

$$x + y \leq 3$$

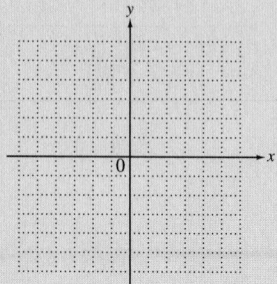

5 Graph the solution set of the system.

$$3x - 4y \geq 12$$

$$x + 3y > 6$$

$$y \leq 2$$

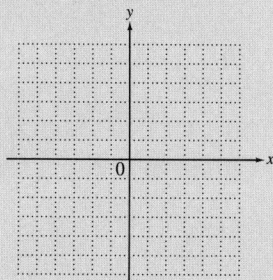

The graph of $x^2 + y^2 < 16$ is the interior of a dashed circle centered at the origin with radius 4. This is shown in **Figure 34.**

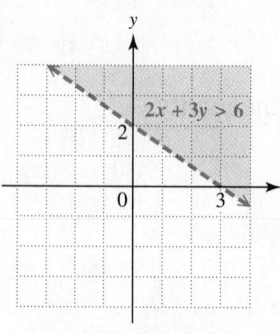

Figure 33

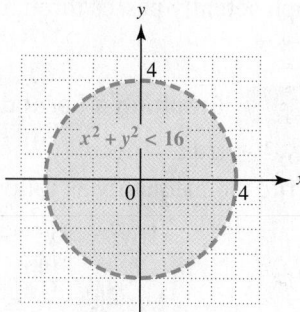

Figure 34

The graph of the solution set of the system is the intersection of the graphs of the two inequalities. The overlapping region in **Figure 35** is the solution set.

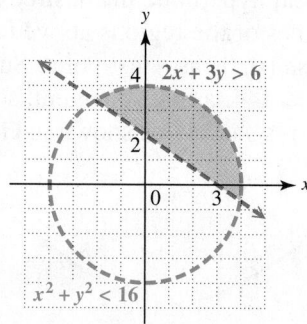

Figure 35

◀ **Work Problem ④ at the Side.**

EXAMPLE 5 **Graphing a System of Three Inequalities**

Graph the solution set of the system.

$$x + y < 1$$

$$y \leq 2x + 3$$

$$y \geq -2$$

Graph each inequality separately, on the same axes. The graph of $x + y < 1$ consists of all points below the dashed line $x + y = 1$. The graph of $y \leq 2x + 3$ is the region that lies below or along the solid line $y = 2x + 3$. Finally, the graph of $y \geq -2$ is the region above or along the solid horizontal line $y = -2$.

The graph of the system, the intersection of these three graphs, is the triangular region enclosed by the three boundary lines in **Figure 36,** including two of its boundaries.

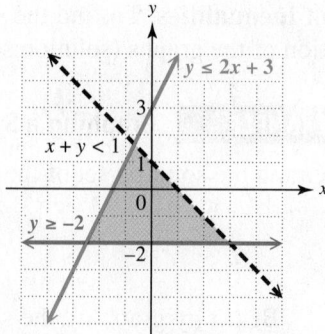

Figure 36

◀ **Work Problem ⑤ at the Side.**

Answers

4.

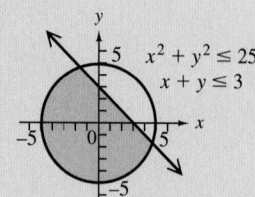

5.

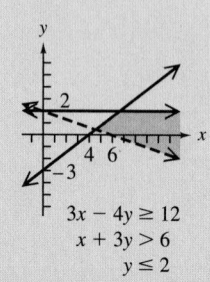

EXAMPLE 6	Graphing a System of Three Inequalities

Graph the solution set of the system.

$$y \geq x^2 - 2x + 1$$
$$2x^2 + y^2 > 4$$
$$y < 4$$

The graph of $y = x^2 - 2x + 1$ is a parabola with vertex at $(1, 0)$. Those points above (or in the interior of) the parabola satisfy the condition $y > x^2 - 2x + 1$. Thus, points on the parabola or in the interior are in the solution set of $y \geq x^2 - 2x + 1$.

The graph of the equation $2x^2 + y^2 = 4$ is an ellipse. We draw it as a dashed curve. To satisfy the inequality $2x^2 + y^2 > 4$, a point must lie outside the ellipse.

The graph of $y < 4$ includes all points below the dashed line $y = 4$. Finally, the graph of the system is the shaded region in **Figure 37,** which lies outside the ellipse, inside or on the boundary of the parabola, and below the line $y = 4$.

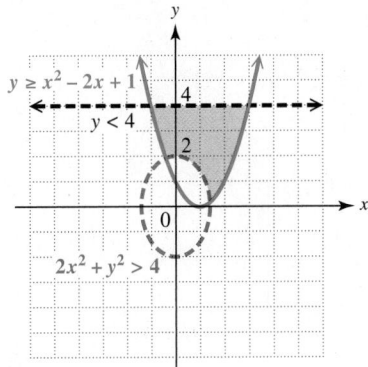

Figure 37

.... Work Problem **6** at the Side. ▶

6 Graph the solution set of the system.

$$y > x^2 + 1$$
$$\frac{x^2}{9} + \frac{y^2}{4} \geq 1$$
$$y \leq 5$$

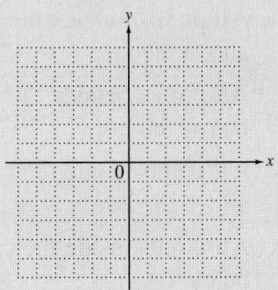

Answer

6.

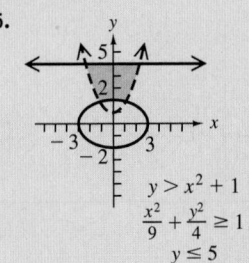

$$y > x^2 + 1$$
$$\frac{x^2}{9} + \frac{y^2}{4} \geq 1$$
$$y \leq 5$$

12.5 Exercises

1. CONCEPT CHECK Which is a description of the graph of the solution set of the following system?

$$x^2 + y^2 < 25$$
$$y > -2$$

A. All points outside the circle $x^2 + y^2 = 25$ and above the line $y = -2$

B. All points outside the circle $x^2 + y^2 = 25$ and below the line $y = -2$

C. All points inside the circle $x^2 + y^2 = 25$ and above the line $y = -2$

D. All points inside the circle $x^2 + y^2 = 25$ and below the line $y = -2$

2. CONCEPT CHECK The graph of the system

$$y > x^2 + 1$$
$$\frac{x^2}{9} + \frac{y^2}{4} > 1$$
$$y < 5$$

consists of all points (*above/below*) the parabola $y = x^2 + 1$, (*inside/outside*) the ellipse $\frac{x^2}{9} + \frac{y^2}{4} = 1$, and (*above/below*) the line $y = 5$.

CONCEPT CHECK *Match each nonlinear inequality with its graph.*

3. $y \geq x^2 + 4$

4. $y \leq x^2 + 4$

5. $y < x^2 + 4$

6. $y > x^2 + 4$

A.

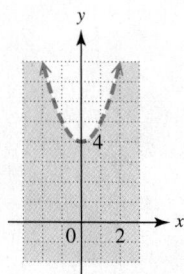

B.

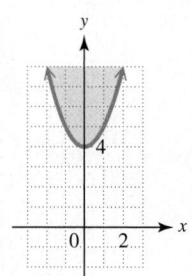

C.

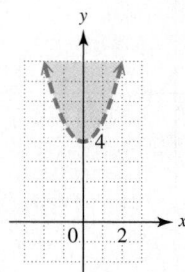

D.

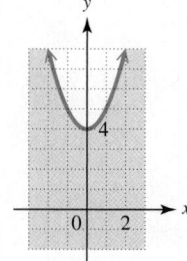

Graph each inequality. **See Examples 1–3.**

7. $y > x^2 - 1$

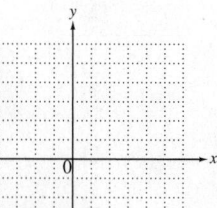

8. $y \geq x^2 - 2$

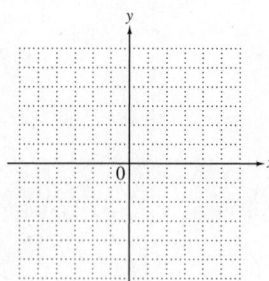

9. $y^2 \leq 4 - 2x^2$

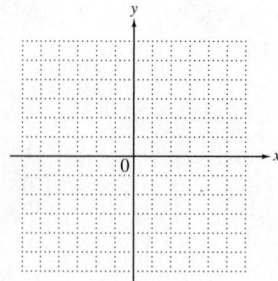

10. $2y^2 \geq 8 - x^2$

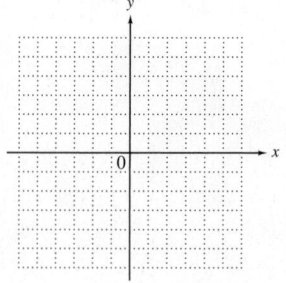

11. $x^2 \leq 16 - y^2$

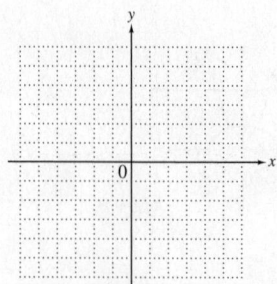

12. $x^2 > 4 - y^2$

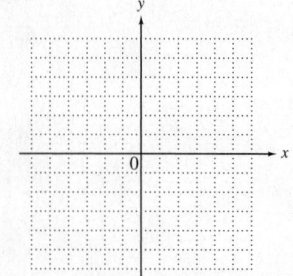

13. $x^2 \leq 16 + 4y^2$

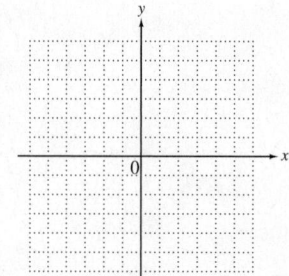

14. $y^2 > 4 + x^2$

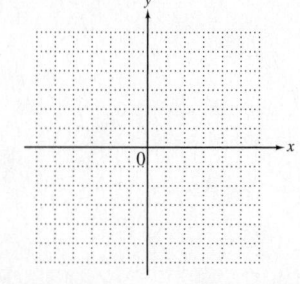

15. $9x^2 < 16y^2 - 144$

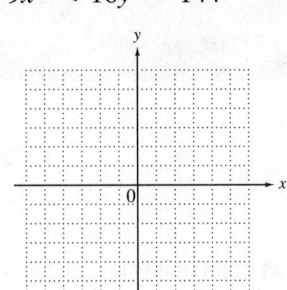

16. $9x^2 > 16y^2 + 144$

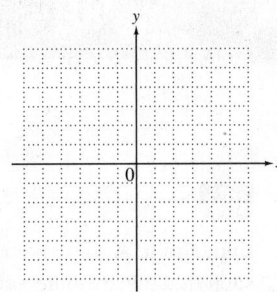

17. $4y^2 \leq 36 - 9x^2$

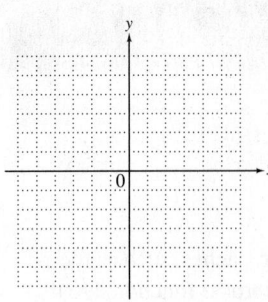

18. $x^2 - 4 \geq -4y^2$

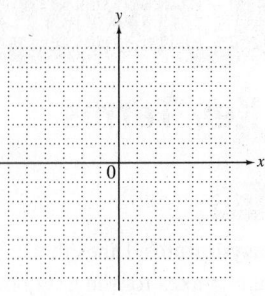

19. $x \geq y^2 - 8y + 14$

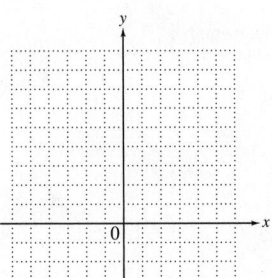

20. $x \leq -y^2 + 6y - 7$

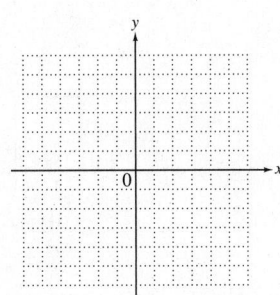

21. $25x^2 \leq 9y^2 + 225$

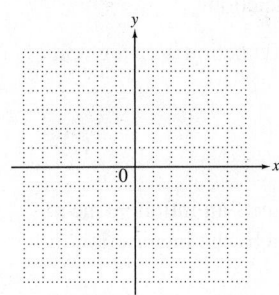

22. $y^2 - 16x^2 \leq 16$

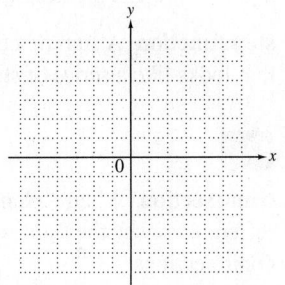

Graph each system of inequalities. ***See Examples 4–6.***

23. $3x - y > -6$
$\quad 4x + 3y > 12$

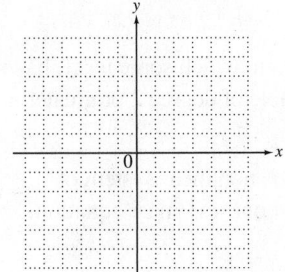

24. $2x + 5y < 10$
$\quad x - 2y < 4$

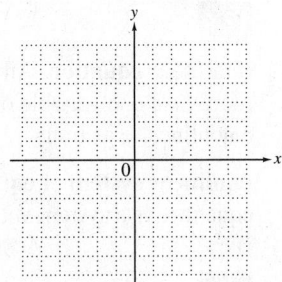

25. $4x - 3y \leq 0$
$\quad x + y \leq 5$

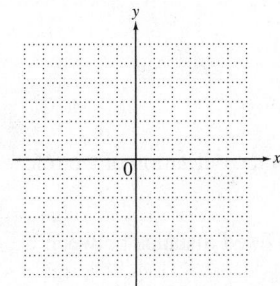

26. $5x - 3y \leq 15$
$\quad 4x + y \geq 4$

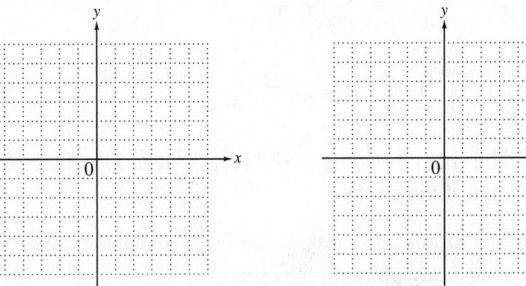

27. $x^2 - y^2 \geq 9$
$\quad \dfrac{x^2}{16} + \dfrac{y^2}{9} \leq 1$

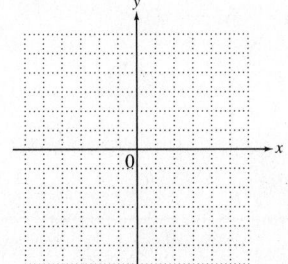

28. $y^2 - x^2 \geq 4$
$\quad -5 \leq y \leq 5$

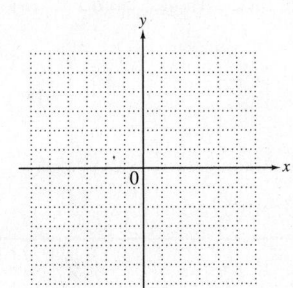

29. $y < x^2$
$\quad y > -2$
$\quad x + y < 3$
$\quad 3x - 2y > -6$

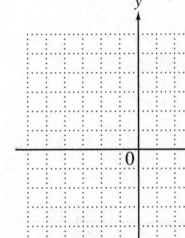

30. $y \leq -x^2$
$\quad y \geq x - 3$
$\quad y \leq -1$
$\quad x < 1$

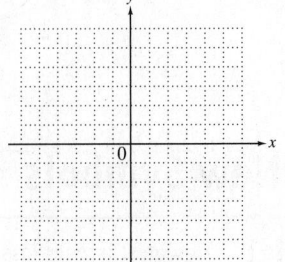

Chapter 12 *Summary*

Key Terms

12.1

asymptotes Lines that a graph approaches, such as the x- and y-axes for the graph of the reciprocal function, are the asymptotes of the graph.

greatest integer function The function $f(x) = [\![x]\!]$, where the symbol $[\![x]\!]$ represents the greatest integer less than or equal to x, is the greatest integer function.

step function A step function is a function with a graph that looks like a series of steps.

12.2

conic section When a plane intersects an infinite cone at different angles, the figures formed by the intersections are conic sections.

circle A circle is the set of all points in a plane that lie a fixed distance from a fixed point.

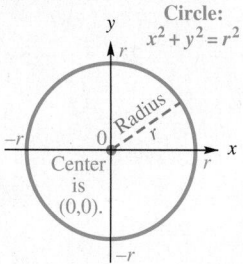

center The fixed point discussed in the definition of a circle is the center of the circle.

radius The radius of a circle is the fixed distance between the center and any point on the circle.

ellipse An ellipse is the set of all points in a plane the sum of whose distances from two fixed points (**foci**) is constant.

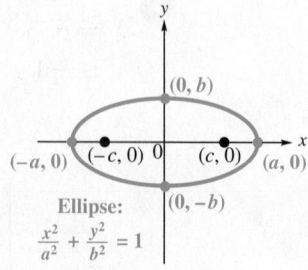

12.3

hyperbola A hyperbola is the set of all points in a plane such that the absolute value of the difference of the distances from two fixed points (foci) is constant.

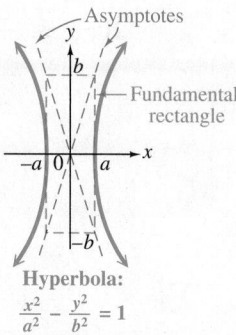

asymptotes of a hyperbola The two intersecting lines that the branches of a hyperbola approach are the asymptotes of the hyperbola.

fundamental rectangle The asymptotes of a hyperbola are the extended diagonals of its fundamental rectangle.

12.4

nonlinear equation An equation in which some terms have more than one variable or a variable of degree 2 or greater is a nonlinear equation.

nonlinear system of equations A nonlinear system of equations is a system with at least one nonlinear equation.

12.5

second-degree inequality A second-degree inequality is an inequality with at least one variable of degree 2 and no variable with degree greater than 2.

system of inequalities A system of inequalities consists of two or more inequalities to be solved at the same time.

New Symbols

$[\![x]\!]$ greatest integer less than or equal to x

Test Your Word Power

See how well you have learned the vocabulary in this chapter.

① **Conic sections** are
  A. graphs of first-degree equations
  B. the result of two or more intersecting planes
  C. graphs of first-degree inequalities
  D. figures that result from the intersection of an infinite cone with a plane.

② A **circle** is the set of all points in a plane
  A. such that the absolute value of the difference of the distances from two fixed points is constant
  B. that lie a fixed distance from a fixed point
  C. the sum of whose distances from two fixed points is constant
  D. that make up the graph of any second-degree equation.

③ An **ellipse** is the set of all points in a plane
  A. such that the absolute value of the difference of the distances from two fixed points is constant
  B. that lie a fixed distance from a fixed point
  C. the sum of whose distances from two fixed points is constant
  D. that make up the graph of any second-degree equation.

④ A **hyperbola** is the set of all points in a plane
  A. such that the absolute value of the difference of the distances from two fixed points is constant
  B. that lie a fixed distance from a fixed point
  C. the sum of whose distances from two fixed points is constant
  D. that make up the graph of any second-degree equation.

⑤ A **nonlinear equation** is an equation
  A. in which some terms have more than one variable or a variable of degree 2 or greater
  B. in which the terms have only one variable
  C. of degree 1
  D. of a linear function.

⑥ A **nonlinear system of equations** is a system
  A. with at least one linear equation
  B. with two or more inequalities
  C. with at least one nonlinear equation
  D. with at least two linear equations.

Answers to Test Your Word Power

1. D; *Example:* Parabolas, circles, ellipses, and hyperbolas are conic sections.

2. B; *Example:* See the graph of $x^2 + y^2 = 9$ in **Figure 10** of **Section 12.2.**

3. C; *Example:* See the graph of $\frac{x^2}{49} + \frac{y^2}{36} = 1$ in **Figure 15** of **Section 12.2.**

4. A; *Example:* See the graph of $\frac{x^2}{16} - \frac{y^2}{25} = 1$ in **Figure 19** of **Section 12.3.**

5. A; *Examples:* $y = x^2 + 8x + 16$, $xy = 5$, $2x^2 - y^2 = 6$

6. C; *Example:* $x^2 + y^2 = 2$
   $2x + y = 1$

Quick Review

Concepts	Examples

12.1 Additional Graphs of Functions

In addition to the squaring function, some other elementary functions include the following.

• Absolute value function $f(x) = |x|$

• Reciprocal function $f(x) = \frac{1}{x}$

• Square root function $f(x) = \sqrt{x}$

• Greatest integer function $f(x) = [\![x]\!]$, which is a step function

Their graphs can be translated, as shown in the first three examples at the right.

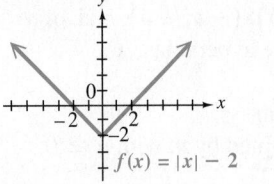
$f(x) = |x| - 2$

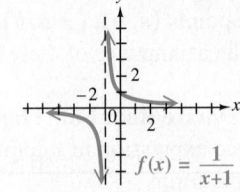
$f(x) = \frac{1}{x+1}$

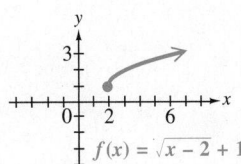
$f(x) = \sqrt{x-2} + 1$

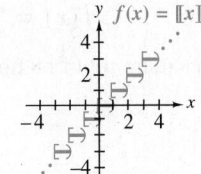
$f(x) = [\![x]\!]$

Concepts	Examples

12.2 The Circle and the Ellipse

Circle

The circle with radius r and center at (h, k) has an equation of the form

$$(x - h)^2 + (y - k)^2 = r^2.$$

The circle with equation $(x + 2)^2 + (y - 3)^2 = 25$, which can be written $\left[x - (-2)^2\right] + (y - 3)^2 = 5^2$, has center $(-2, 3)$ and radius 5.

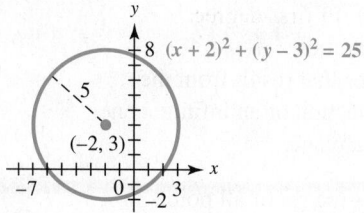

Ellipse

The ellipse whose x-intercepts are $(a, 0)$ and $(-a, 0)$ and whose y-intercepts are $(0, b)$ and $(0, -b)$ has an equation of the form

$$\frac{x^2}{a^2} + \frac{y^2}{b^2} = 1.$$

Graph $\dfrac{x^2}{9} + \dfrac{y^2}{4} = 1$.

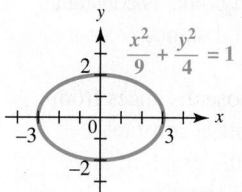

12.3 The Hyperbola and Other Functions Defined by Radicals

Hyperbola

A hyperbola with x-intercepts $(a, 0)$ and $(-a, 0)$ has an equation of the form

$$\frac{x^2}{a^2} - \frac{y^2}{b^2} = 1.$$

A hyperbola with y-intercepts $(0, b)$ and $(0, -b)$ has an equation of the form

$$\frac{y^2}{b^2} - \frac{x^2}{a^2} = 1.$$

The extended diagonals of the fundamental rectangle with vertices at the points $(a, b), (-a, b), (-a, -b),$ and $(a, -b)$ are the asymptotes of these hyperbolas.

Graph $\dfrac{x^2}{4} - \dfrac{y^2}{4} = 1$.

The graph has x-intercepts $(2, 0)$ and $(-2, 0)$.

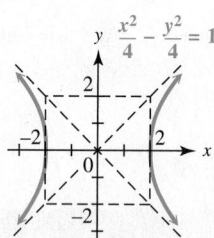

The fundamental rectangle has vertices at $(2, 2), (-2, 2),$ $(-2, -2),$ and $(2, -2)$.

Generalized Square Root Function

For an algebraic expression in x defined by u, with $u \geq 0$, a function of the form

$$f(x) = \sqrt{u}$$

is a generalized square root function.

Graph $y = -\sqrt{4 - x^2}$.

Square each side and rearrange terms.

$$x^2 + y^2 = 4$$

This equation has a circle as its graph. However, graph only the lower half of the circle, since the original equation indicates that y cannot be positive.

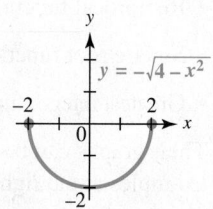

Concepts	Examples

12.4 Nonlinear Systems of Equations

Solving a Nonlinear System
A nonlinear system can be solved by the substitution method, the elimination method, or a combination of the two.

Solve the system.

$$x^2 + 2xy - y^2 = 14 \quad (1)$$
$$x^2 - y^2 = -16 \quad (2)$$

Multiply equation (2) by -1 and use elimination.

$$
\begin{aligned}
x^2 + 2xy - y^2 &= 14 \\
-x^2 \qquad\quad + y^2 &= 16 \\
\hline
2xy \qquad\quad &= 30 \\
xy &= 15
\end{aligned}
$$

Solve for y to obtain $y = \frac{15}{x}$, and substitute into equation (2).

$$x^2 - y^2 = -16 \quad (2)$$

$$x^2 - \left(\frac{15}{x}\right)^2 = -16 \quad \text{Let } y = \frac{15}{x}.$$

$$x^2 - \frac{225}{x^2} = -16 \quad \text{Apply the exponent.}$$

$$x^4 + 16x^2 - 225 = 0 \quad \text{Multiply by } x^2. \text{ Add } 16x^2.$$

$$(x^2 - 9)(x^2 + 25) = 0 \quad \text{Factor.}$$

$$x^2 - 9 = 0 \quad \text{or} \quad x^2 + 25 = 0 \quad \text{Zero-factor property}$$

$$x = \pm 3 \quad \text{or} \quad x = \pm 5i \quad \text{Solve each equation.}$$

Find corresponding y-values to obtain the solution set.

$$\{(3, 5), (-3, -5), (5i, -3i), (-5i, 3i)\}$$

12.5 Second-Degree Inequalities and Systems of Inequalities

Graphing a Second-Degree Inequality
To graph a second-degree inequality, graph the corresponding equation as a boundary and use test points to determine which region(s) form the solution set. Shade the appropriate region(s).

Graphing a System of Inequalities
The solution set of a system of inequalities is the intersection of the solution sets of the individual inequalities.

Graph $y \geq x^2 - 2x + 3$.

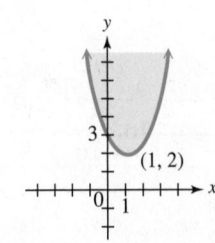

(1, 2)

Graph the solution set of the system.

$$3x - 5y > -15$$
$$x^2 + y^2 \leq 25$$

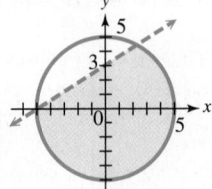

Chapter 12 *Review Exercises*

12.1 *Graph each function. Give the domain and range.*

1. $f(x) = |x + 4|$

2. $f(x) = \dfrac{1}{x - 4}$

3. $f(x) = \sqrt{x} + 3$

4. $f(x) = [\![-x]\!]$

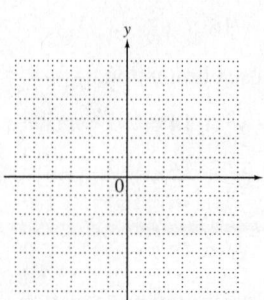
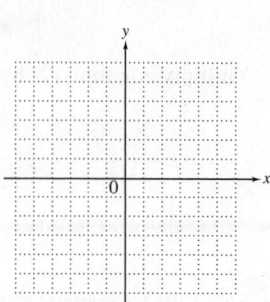
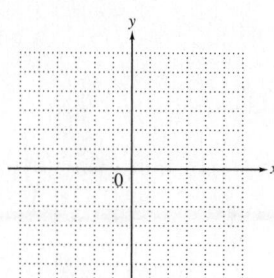
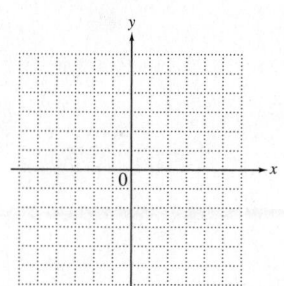

Find each of the following.

5. $[\![12]\!]$

6. $\left[\!\left[2\dfrac{1}{4} \right]\!\right]$

7. $[\![-4.75]\!]$

12.2 *Write an equation for each circle.*

8. Center $(-2, 4)$, $r = 3$

9. Center $(-1, -3)$, $r = 5$

10. Center $(4, 2)$, $r = 6$

Find the center and radius of each circle.

11. $x^2 + y^2 + 6x - 4y - 3 = 0$

12. $x^2 + y^2 - 8x - 2y + 13 = 0$

13. $2x^2 + 2y^2 + 4x + 20y = -34$

14. $4x^2 + 4y^2 - 24x + 16y = 48$

Graph each equation.

15. $x^2 + y^2 = 16$

16. $\dfrac{x^2}{16} + \dfrac{y^2}{9} = 1$

17. $\dfrac{x^2}{36} + \dfrac{y^2}{25} = 1$

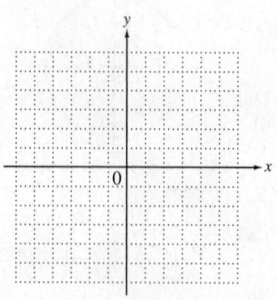
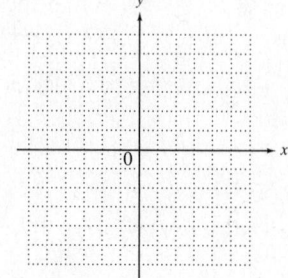
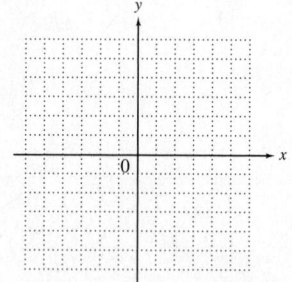

18. A satellite is in an elliptical orbit around Earth with perigee altitude of 160 km and apogee altitude of 16,000 km. See the figure. (*Source*: Kastner, Bernice, *Space Mathematics*, NASA.) Find the equation of the ellipse.

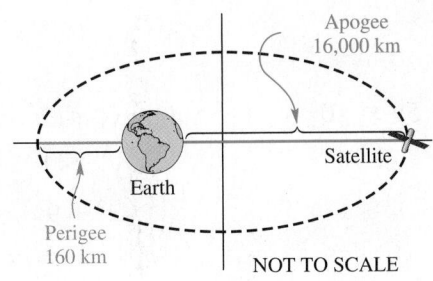

19. The Roman Colosseum is an ellipse with $a = 310$ ft and $b = 256.5$ ft.

(a) Find the distance between the foci of this ellipse to the nearest tenth of a foot.

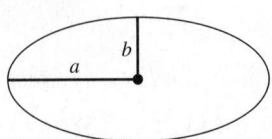

(b) A formula for the approximate perimeter of an ellipse is

$$C \approx 2\pi\sqrt{\frac{a^2 + b^2}{2}},$$

where a and b are the lengths given above. Use this formula to find the approximate perimeter of the Roman Colosseum.

12.3 *Graph each equation.*

20. $\dfrac{x^2}{16} - \dfrac{y^2}{25} = 1$

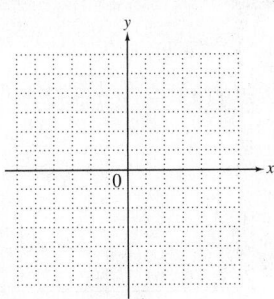

21. $\dfrac{y^2}{25} - \dfrac{x^2}{4} = 1$

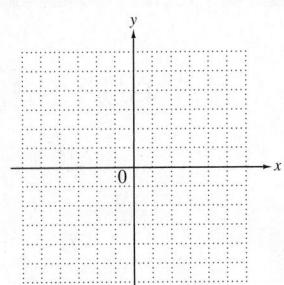

22. $f(x) = -\sqrt{16 - x^2}$

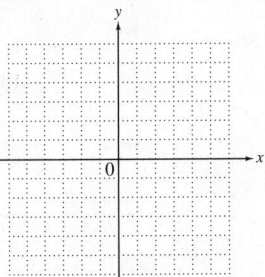

Identify the graph of each equation as a parabola, circle, ellipse, *or* hyperbola.

23. $x^2 + y^2 = 64$

24. $y = 2x^2 - 3$

25. $y^2 = 2x^2 - 8$

26. $y^2 = 8 - 2x^2$

27. $x = y^2 + 4$

28. $x^2 - y^2 = 64$

12.4 *Solve each system.*

29. $2y = 3x - x^2$
$\quad x + 2y = -12$

30. $y + 1 = x^2 + 2x$
$\quad y + 2x = 4$

31. $x^2 + 3y^2 = 28$
$\quad y - x = -2$

32. $xy = 8$
$\quad x - 2y = 6$

33. $x^2 + y^2 = 6$
$\quad x^2 - 2y^2 = -6$

34. $3x^2 - 2y^2 = 12$
$\quad x^2 + 4y^2 = 18$

CONCEPT CHECK *Answer each question.*

35. How many solutions are possible for a system of two equations whose graphs are a circle and a line?

36. How many solutions are possible for a system of two equations whose graphs are a parabola and a hyperbola?

12.5 *Graph each inequality or system of inequalities.*

37. $9x^2 \geq 16y^2 + 144$

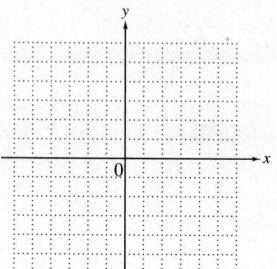

38. $4x^2 + y^2 \geq 16$

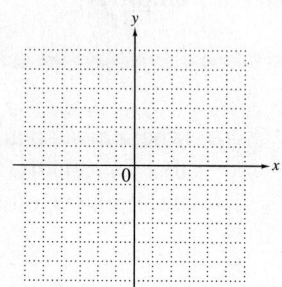

39. $2x + 5y \leq 10$
$3x - y \leq 6$

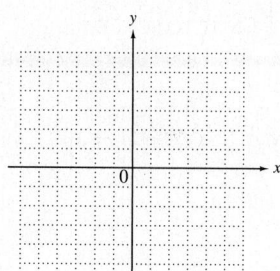

40. $|x| \leq 2$
$|y| > 1$
$4x^2 + 9y^2 \leq 36$

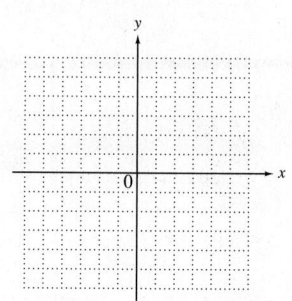

Mixed Review Exercises

Graph.

41. $x^2 + y^2 = 25$

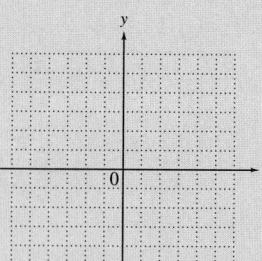

42. $x^2 + 9y^2 = 9$

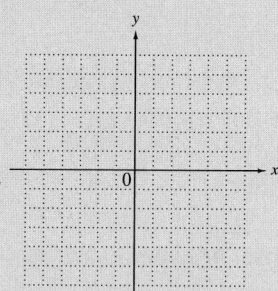

43. $x^2 - 9y^2 = 9$

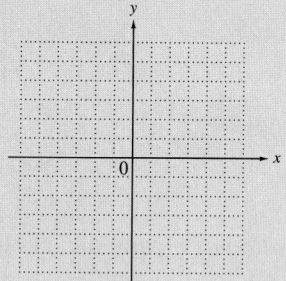

44. $f(x) = \sqrt{x + 2}$

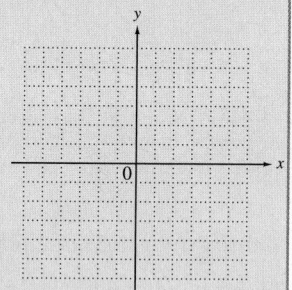

45. $f(x) = [\![x]\!] - 1$

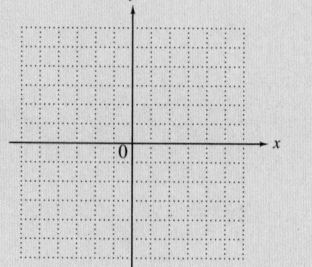

46. $y < -(x + 2)^2 + 1$

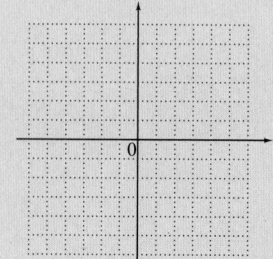

47. $4y > 3x - 12$
$x^2 < 16 - y^2$

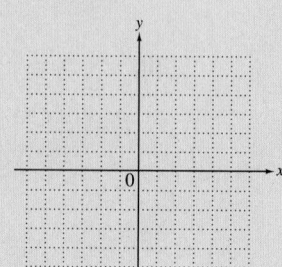

48. $9x^2 \leq 4y^2 + 36$
$x^2 + y^2 \leq 16$

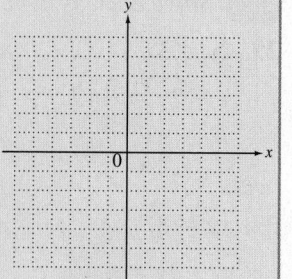

 CHAPTER
Test Prep
VIDEO

The Chapter Test Prep Videos with test solutions are available on DVD, in MyMathLab, and on You Tube —*search "LialCombinedAlg" and click on "Channels."*

Fill in each blank with the correct response.

1. For the reciprocal function $f(x) = \frac{1}{x}$, _____ is the only real number not in the domain.

2. The range of the square root function $f(x) = \sqrt{x}$ is _____.

3. The range of $f(x) = [\![x]\!]$, the greatest integer function, is _____.

4. Match each function in parts (a)–(d) with its graph from choices A–D.

   (a) $f(x) = \sqrt{x} - 2$ **(b)** $f(x) = \sqrt{x} + 2$ **(c)** $f(x) = \sqrt{x+2}$ **(d)** $f(x) = \sqrt{x-2}$

   A. 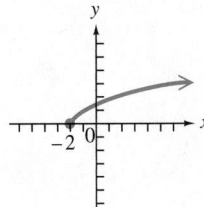 **B.** 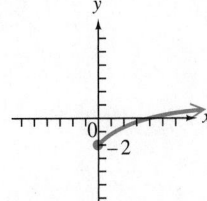 **C.** 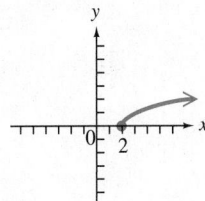 **D.**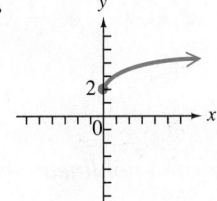

5. Sketch the graph of $f(x) = |x - 3| + 4$. Give the domain and range.

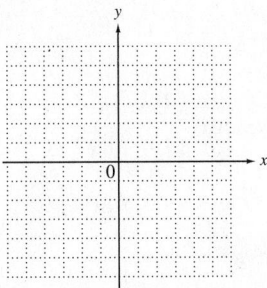

6. Find the center and radius of the circle whose equation is $(x - 2)^2 + (y + 3)^2 = 16$. Sketch the graph.

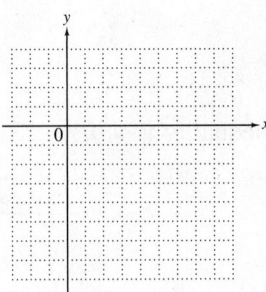

7. Find the center and radius of the circle whose equation is $x^2 + y^2 + 8x - 2y = 8$.

Graph.

8. $f(x) = \sqrt{9 - x^2}$

9. $4x^2 + 9y^2 = 36$

10. $16y^2 - 4x^2 = 64$

11. $\dfrac{y}{2} = -\sqrt{1 - \dfrac{x^2}{9}}$

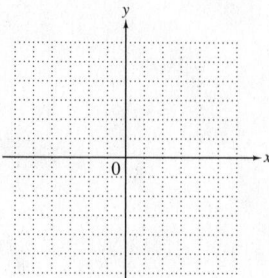 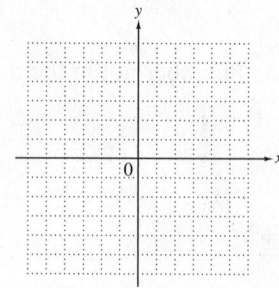 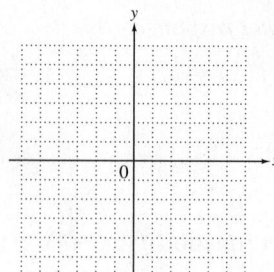 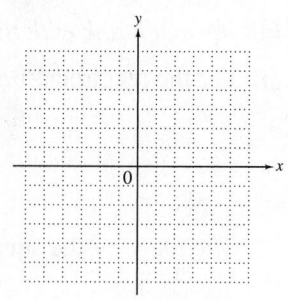

Identify the graph of each equation as a parabola, hyperbola, ellipse, *or* circle.

12. $6x^2 + 4y^2 = 12$

13. $16x^2 = 144 + 9y^2$

14. $4y^2 + 4x = 9$

15. $y^2 = 20 - x^2$

Solve each nonlinear system.

16. $2x - y = 9$
$xy = 5$

17. $x - 4 = 3y$
$x^2 + y^2 = 8$

18. $x^2 + y^2 = 25$
$x^2 - 2y^2 = 16$

19. Graph the inequality $y < x^2 - 2$.

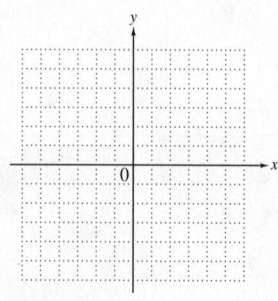

20. Graph the system $\begin{array}{l} x^2 + 25y^2 \leq 25 \\ x^2 + y^2 \leq 9 \end{array}$.

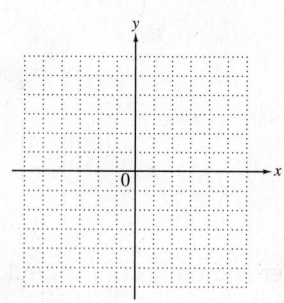

Chapters R–12 *Cumulative Review Exercises*

Solve.

1. $4 - (2x + 3) + x = 5x - 3$

2. $-4k + 7 \geq 6k + 1$

3. $|5m| - 6 = 14$

4. Find the slope of the line through $(2, 5)$ and $(-4, 1)$.

5. Find an equation in standard form of the line passing through the point $(-3, -2)$ and perpendicular to the graph of $2x - 3y = 7$.

Solve each system.

6. $\begin{aligned} 3x - \ y &= 12 \\ 2x + 3y &= -3 \end{aligned}$

7. $\begin{aligned} x + y - 2z &= 9 \\ 2x + y + \ z &= 7 \\ 3x - y - \ z &= 13 \end{aligned}$

8. $\begin{aligned} xy &= -5 \\ 2x + y &= 3 \end{aligned}$

Perform the indicated operations.

9. $(5y - 3)^2$

10. $(2r + 7)(6r - 1)$

11. $\dfrac{8x^4 - 4x^3 + 2x^2 + 13x + 8}{2x + 1}$

Factor.

12. $12x^2 - 7x - 10$

13. $z^4 - 1$

14. $a^3 - 27b^3$

Perform each operation.

15. $\dfrac{y^2 - 4}{y^2 - y - 6} \div \dfrac{y^2 - 2y}{y - 1}$

16. $\dfrac{5}{c + 5} - \dfrac{2}{c + 3}$

17. $\dfrac{p}{p^2 + p} + \dfrac{1}{p^2 + p}$

Solve.

18. Kareem and Jamal want to clean their office. Kareem can do the job alone in 3 hr, while Jamal can do it alone in 2 hr. How long will it take them if they work together?

Simplify. Assume that all variables represent positive real numbers.

19. $\dfrac{(2a)^{-2}a^4}{a^{-3}}$

20. $4\sqrt[3]{16} - 2\sqrt[3]{54}$

21. $\dfrac{3\sqrt{5x}}{\sqrt{2x}}$

22. $\dfrac{5 + 3i}{2 - i}$

Solve.

23. $2\sqrt{k} = \sqrt{5k + 3}$

24. $10q^2 + 13q = 3$

25. $3k^2 - 3k - 2 = 0$

26. $2(x^2 - 3)^2 - 5(x^2 - 3) = 12$

27. $\log(x + 2) + \log(x - 1) = 1$

28. $F = \dfrac{kwv^2}{r}$ for v
(Leave $\pm$ in the answer.)

29. If $f(x) = x^2 + 2x - 4$ and $g(x) = 3x + 2$,
find the following.

    (a) $(g \circ f)(1)$ **(b)** $(f \circ g)(x)$

30. If $f(x) = x^3 + 4$, find $f^{-1}(x)$.

31. Evaluate.

    (a) $3^{\log_3 4}$ **(b)** $e^{\ln 7}$

32. Use properties of logarithms to write the
following as a single logarithm.

$$2 \log(3x + 7) - \log 4$$

Graph.

33. $f(x) = -3x + 5$

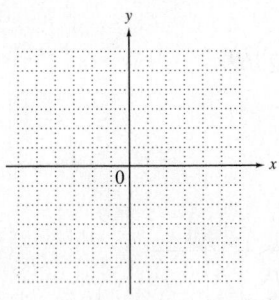

34. $f(x) = -2(x - 1)^2 + 3$

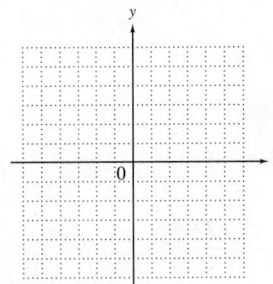

35. $\dfrac{x^2}{25} + \dfrac{y^2}{16} \leq 1$

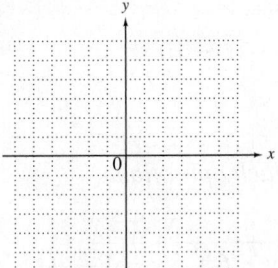

36. $f(x) = \sqrt{x - 2}$

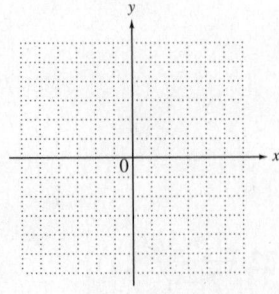

37. $\dfrac{x^2}{4} - \dfrac{y^2}{16} = 1$

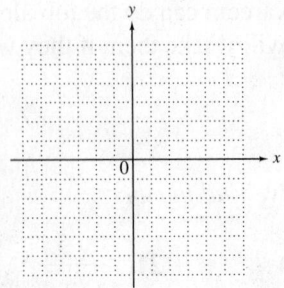

38. $f(x) = 3^x$

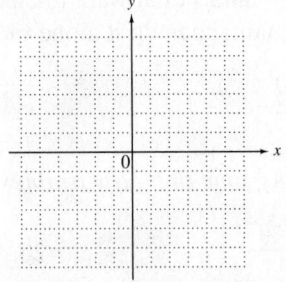

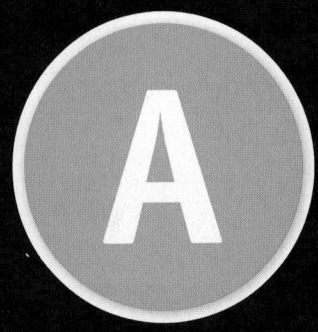

Appendix: Review of Exponents, Polynomials, and Factoring

Transition from Introductory to Intermediate Algebra

OBJECTIVE ① Review the basic rules for exponents.

OBJECTIVES

① Review the basic rules for exponents.

② Review addition, subtraction, and multiplication of polynomials.

③ Review factoring techniques.

Definitions and Rules for Exponents (Sections 5.2 and 5.5)

If no denominators are 0, for any integers m and n, the following hold.

			Examples
Product rule		$a^m \cdot a^n = a^{m+n}$	$7^4 \cdot 7^3 = 7^7$
Zero exponent		$a^0 = 1 \quad (a \neq 0)$	$(-3)^0 = 1$
Negative exponent		$a^{-n} = \dfrac{1}{a^n}$	$5^{-3} = \dfrac{1}{5^3}$
Quotient rule		$\dfrac{a^m}{a^n} = a^{m-n}$	$\dfrac{2^2}{2^5} = 2^{-3} = \dfrac{1}{2^3}$
Power rules	(a)	$(a^m)^n = a^{mn}$	$(4^2)^3 = 4^6$
	(b)	$(ab)^m = a^m b^n$	$(3k)^4 = 3^4 k^4$
	(c)	$\left(\dfrac{a}{b}\right)^m = \dfrac{a^m}{b^m}$	$\left(\dfrac{2}{3}\right)^{10} = \dfrac{2^{10}}{3^{10}}$
Negative-to-positive rules		$\dfrac{a^{-m}}{b^{-n}} = \dfrac{b^n}{a^m}$	$\dfrac{5^{-3}}{3^{-5}} = \dfrac{3^5}{5^3}$
		$\left(\dfrac{a}{b}\right)^{-m} = \left(\dfrac{b}{a}\right)^m$	$\left(\dfrac{4}{7}\right)^{-2} = \left(\dfrac{7}{4}\right)^2$

EXAMPLE 1   Applying Definitions and Rules for Exponents

Simplify. Assume that all variables represent nonzero real numbers.

(a) $(x^2 y^{-3})(x^{-5} y^7)$

$= (x^{2+(-5)})(y^{-3+7})$ Product rule

$= x^{-3} y^4$

$= \dfrac{1}{x^3} y^4, \quad \text{or} \quad \dfrac{y^4}{x^3}$ Definition of negative exponent

(b) $(-5)^0 + (-5^0)$ $-5^0 = -1 \cdot 5^0$
$= -1 \cdot 1$
$= -1$

$= 1 + (-1)$

$= 0$ Add.

Continued on Next Page

953

1 Simplify. Assume that all variables represent nonzero real numbers.

(a) $(a^{-4}bc^2)(a^2b^{-2}c^4)$

(b) $4^0 + (-4)^0$

(c) $\dfrac{(x^3y^{-2})^3}{(x^4y^{-3})^2}$

(d) $\left(\dfrac{2x^2y^{-2}}{x^{-4}y}\right)^{-3}$

(e) $(3ab^4c^2)^3(b^2c)^2$

2 Add or subtract as indicated.

(a) $(5x^3 - 3x^2 + x + 4)$
   $+ (2x^3 - x^2 - 3x - 1)$

(b) $(7y^2 - 11y + 8)$
   $- (-3y^2 + 4y + 6)$

(c) $\dfrac{(t^5s^{-4})^2}{(t^{-3}s^5)^3}$

$= \dfrac{t^{10}s^{-8}}{t^{-9}s^{15}}$ Power rules (a) and (b)

$= \dfrac{t^{10}t^9}{s^{15}s^8}$ Definition of negative exponent

$= \dfrac{t^{19}}{s^{23}}$ Product rule

(d) $\left(\dfrac{-3x^{-4}y}{x^5y^{-4}}\right)^{-2}$

$= \left(\dfrac{x^5y^{-4}}{-3x^{-4}y}\right)^2$ Negative-to-positive rule

$= \dfrac{x^{10}y^{-8}}{9x^{-8}y^2}$ Power rules (a), (b), and (c)

$= \dfrac{x^{18}}{9y^{10}}$ Quotient rule

(e) $(2x^2y^3z)^2(x^4y^2)^3$

$= (4x^4y^6z^2)(x^{12}y^6)$ Power rules (a) and (b)

$= 4x^{16}y^{12}z^2$ Product rule

◄ **Work Problem 1 at the Side.**

OBJECTIVE 2 Review addition, subtraction, and multiplication of polynomials.

Adding and Subtracting Polynomials (Section 5.1)

To add polynomials, add like terms.

To subtract polynomials, change all signs on the subtrahend (second polynomial) and add the result to the minuend (first polynomial).

EXAMPLE 2   Adding and Subtracting Polynomials

Add or subtract as indicated.

(a) $(-4x^3 + 3x^2 - 8x + 2) + (5x^3 - 8x^2 + 12x - 3)$

$= (-4 + 5)x^3 + (3 - 8)x^2 + (-8 + 12)x + (2 - 3)$

$= x^3 - 5x^2 + 4x - 1$

(b) $-4(x^2 + 3x - 6) - (2x^2 - 3x + 7)$

$= -4x^2 - 12x + 24 - 2x^2 + 3x - 7$

$= -6x^2 - 9x + 17$

(c) Subtract.

$\begin{array}{l} 2t^2 - 3t - 4 \\ -8t^2 + 4t - 1 \end{array} \longrightarrow \begin{array}{l} 2t^2 - 3t - 4 \\ \underline{8t^2 - 4t + 1} \\ 10t^2 - 7t - 3 \end{array}$ Change signs. Add.

◄ **Work Problem 2 at the Side.**

Multiplying Polynomials (Sections 5.3 and 5.4)

To multiply two polynomials, multiply each term of the second polynomial by each term of the first polynomial and add the products. In particular, when multiplying two binomials, use the FOIL method.

Also recall the following special product rules.

$$(a + b)^2 = a^2 + 2ab + b^2$$
$$(a - b)^2 = a^2 - 2ab + b^2$$
$$(a + b)(a - b) = a^2 - b^2$$

Answers

1. (a) $\dfrac{c^6}{a^2b}$ (b) 2 (c) x (d) $\dfrac{y^9}{8x^{18}}$

   (e) $27a^3b^{16}c^8$

2. (a) $7x^3 - 4x^2 - 2x + 3$

   (b) $10y^2 - 15y + 2$

EXAMPLE 3	**Multiplying Polynomials**

Find each product.

(a) $(4y - 1)(3y + 2)$

$\qquad = 4y(3y) + 4y(2) - 1(3y) - 1(2)$ $\qquad$ FOIL method

$\qquad = 12y^2 + 8y - 3y - 2$ $\qquad$ Multiply.

$\qquad = 12y^2 + 5y - 2$ $\qquad$ Combine like terms.

(b) $(3x + 5y)(3x - 5y)$

$\qquad = (3x)^2 - (5y)^2$ $\qquad$ $(a + b)(a - b) = a^2 - b^2$

$\qquad = 9x^2 - 25y^2$ $\qquad$ $(ab)^2 = a^2b^2$

(c) $(2t + 3)^2$

$\qquad = (2t)^2 + 2(2t)(3) + 3^2$
$\qquad \quad (a + b)^2 = a^2 + 2ab + b^2$

$\qquad = 4t^2 + 12t + 9$

(d) $(5x - 1)^2$

$\qquad = (5x)^2 - 2(5x)(1) + 1^2$
$\qquad \quad (a - b)^2 = a^2 - 2ab + b^2$

$\qquad = 25x^2 - 10x + 1$

(e) $(3x + 2)(9x^2 - 6x + 4)$

$$\begin{array}{r} 9x^2 - 6x + 4 \\ 3x + 2 \\ \hline 18x^2 - 12x + 8 \\ 27x^3 - 18x^2 + 12x \qquad\qquad\quad \\ \hline 27x^3 \qquad\qquad\qquad\quad + 8 \end{array}$$

Multiply vertically.

$\leftarrow 2(9x^2 - 6x + 4)$

$\leftarrow 3x(9x^2 - 6x + 4)$

Add like terms.

The product is a sum of cubes, $27x^3 + 8$.

· **Work-Problem ❸ at the Side.** ▶

OBJECTIVE ▶ ❸ Review factoring techniques. Factoring, which involves writing a polynomial as a product, was covered in **Chapter 6.**

Guidelines for Factoring a Polynomial (Chapter 6)

Question 1   **Is there a common factor other than 1?** If so, factor it out.

Question 2   **How many terms are in the polynomial?**

Two terms: Check to see whether it is a difference of squares or a sum or difference of cubes. If so, factor as in **Section 6.5.**

Three terms: Is it a perfect square trinomial? In this case, factor as in **Section 6.5.**

  If the trinomial is not a perfect square, check to see whether the coefficient of the second-degree term is 1. If so, use the method of **Section 6.2.**

  If the coefficient of the second-degree term is not 1, use the general factoring methods of **Sections 6.3 and 6.4.**

Four terms: Try to factor the polynomial by grouping, using the methods of **Sections 6.1 and 6.3.**

Question 3   **Can any factors be factored further?** If so, factor them.

---

❸ Find each product.

(a) $(2x + 5)(3x - 2)$

(b) $(2t + 7y)(2t - 7y)$

(c) $(6r - 5)^2$

(d) $(3x + 7)^2$

(e) $(2x - 3)(4x^2 + 6x + 9)$

Answers

3. (a) $6x^2 + 11x - 10$ **(b)** $4t^2 - 49y^2$
   (c) $36r^2 - 60r + 25$ **(d)** $9x^2 + 42x + 49$
   (e) $8x^3 - 27$

4 Factor each polynomial completely.

(a) $10s^3t^6 + 5s^9t^2$

(b) $12x^2 - 5x - 2$

(c) $49z^2 - 36$

(d) $4x^2 - 28x + 49$

(e) $8p^3 + 125$

(f) $xy + 6y + xz + 6z$

EXAMPLE 4 **Factoring Polynomials**

Factor each polynomial completely.

(a) $6x^2y^3 - 12x^3y^2$

$$= 6x^2y^2(y) - 6x^2y^2(2x) \quad \text{$6x^2y^2$ is the greatest common factor.}$$

$$= 6x^2y^2(y - 2x) \quad \text{Distributive property}$$

(b) $3x^2 - x - 2$

To find the factors, find two terms that multiply to give $3x^2$ ($3x$ and x) and two terms that multiply to give -2 ($+2$ and -1). Make sure that the sum of the outer and inner products in the factored form is $-x$.

$$3x^2 - x - 2 \quad \text{factors as} \quad (3x + 2)(x - 1).$$

CHECK Multiply $(3x + 2)(x - 1)$ to obtain $3x^2 - x - 2$. ✓

(c) $100t^2 - 81$

$$= (10t)^2 - 9^2 \quad \text{Difference of squares}$$

$$= (10t + 9)(10t - 9) \quad a^2 - b^2 = (a + b)(a - b)$$

(d) $4x^2 + 20xy + 25y^2$

The first and last terms are both perfect squares.

$$4x^2 = (2x)^2 \quad \text{and} \quad 25y^2 = (5y)^2$$

To factor as a perfect square trinomial, take twice the product of the two terms in the binomial $2x + 5y$.

$$2(2x)(5y) = 20xy$$

Twice ⟶ First term ⟶ Last term

Since $20xy$ is the middle term of the trinomial, the trinomial is a perfect square.

$$4x^2 + 20xy + 25y^2 \quad \text{factors as} \quad (2x + 5y)^2.$$

(e) $1000x^3 - 27$

$$= (10x)^3 - 3^3 \quad \text{Difference of cubes}$$

$$= (10x - 3)[(10x)^2 + 10x(3) + 3^2]$$

$$\qquad a^3 - b^3 = (a - b)(a^2 + ab + b^2)$$

$$= (10x - 3)(100x^2 + 30x + 9) \quad (10x)^2 = 10^2x^2 = 100x^2$$

(f) $6xy - 3x + 4y - 2$

Since there are four terms, try factoring by grouping.

$$6xy - 3x + 4y - 2$$

$$= (6xy - 3x) + (4y - 2) \quad \text{Group the terms.}$$

$$= 3x(2y - 1) + 2(2y - 1) \quad \text{Factor each group.}$$

$$= (2y - 1)(3x + 2) \quad \text{Factor out $2y - 1$.}$$

◀ **Work Problem** **4** at the Side.

Answers

4. (a) $5s^3t^2(2t^4 + s^6)$ **(b)** $(4x + 1)(3x - 2)$
(c) $(7z + 6)(7z - 6)$ **(d)** $(2x - 7)^2$
(e) $(2p + 5)(4p^2 - 10p + 25)$
(f) $(x + 6)(y + z)$

Appendix A Exercises

 MyMathLab®

*Apply the definitions and rules for exponents to simplify each expression. Write the final answers using only positive exponents. Assume that all variables represent positive real numbers. **See Example 1.***

1. $(a^4b^{-3})(a^{-6}b^2)$

2. $(t^{-3}s^{-5})(t^8s^{-2})$

3. $(5x^{-2}y)^2(2xy^4)^2$

4. $(7x^{-3}y^4)^3(2x^{-1}y^{-4})^2$

5. $-6^0 + (-6)^0$

6. $(-12)^0 - 12^0$

7. $\dfrac{(2w^{-1}x^2y^{-1})^3}{(4w^5x^{-2}y)^2}$

8. $\dfrac{(5p^{-3}q^2r^{-4})^2}{(10p^4q^{-1}r^5)^{-1}}$

9. $\left(\dfrac{-4a^{-2}b^4}{a^3b^{-1}}\right)^{-3}$

10. $\left(\dfrac{r^{-3}s^{-8}}{-6r^2s^{-4}}\right)^{-2}$

11. $(7x^{-4}y^2z^{-2})^{-2}(7x^4y^{-1}z^3)^2$

12. $(3m^{-5}n^2p^{-4})^3(3m^4n^{-3}p^5)^{-2}$

*Add or subtract as indicated. **See Example 2.***

13. $(2a^4 + 3a^3 - 6a^2 + 5a - 12)$
    $+ (-8a^4 + 8a^3 - 14a^2 + 21a - 3)$

14. $(-6r^4 - 3r^3 + 12r^2 - 9r + 9)$
    $+ (8r^4 - 13r^3 - 14r^2 - 10r - 3)$

15. $(6x^3 - 12x^2 + 3x - 4) - (-2x^3 + 6x^2 - 3x + 12)$

16. $(10y^3 - 4y^2 + 8y + 7) - (7y^3 + 5y^2 - 2y - 13)$

17. Add.
    $5x^2y + 2xy^2 + y^3$
    $\underline{-4x^2y - 3xy^2 + 5y^3}$

18. Add.
    $6ab^3 - 2a^2b^2 + 3b^5$
    $\underline{8ab^3 + 12a^2b^2 - 8b^5}$

19. $3\left(5x^2 - 12x + 4\right) - 2\left(9x^2 + 13x - 10\right)$

20. $-4\left(2t^3 - 3t^2 + 4t - 1\right)$
$\qquad - 3\left(-8t^3 + 3t^2 - 2t + 9\right)$

21. Subtract.

$\quad 6x^3 - 2x^2 + 3x - 1$
$\underline{-4x^3 + 2x^2 - 6x + 3}$

22. Subtract.

$\quad -9y^3 - 2y^2 + 3y - 8$
$\underline{-8y^3 + 4y^2 + 3y + 1}$

Find each product. ***See Example 3.***

23. $(3x + 1)(2x - 7)$

24. $(5z + 3)(2z - 3)$

25. $(4x - 1)(x - 2)$

26. $(7t - 3)(t - 4)$

27. $(4t + 3)(4t - 3)$

28. $(6x + 1)(6x - 1)$

29. $\left(2y^2 + 4\right)\left(2y^2 - 4\right)$

30. $\left(3b^3 + 2t\right)\left(3b^3 - 2t\right)$

31. $(4x - 3)^2$

32. $(9t + 2)^2$

33. $(6r + 5y)^2$

34. $(8m - 3n)^2$

35. $(c + 2d)\left(c^2 - 2cd + 4d^2\right)$

36. $(f + 3g)\left(f^2 - 3fg + 9g^2\right)$

37. $(4x - 1)\left(16x^2 + 4x + 1\right)$

38. $(5r - 2)\left(25r^2 + 10r + 4\right)$

39. $(7t + 5s)\left(2t^2 + 5st - s^2\right)$

40. $(8p + 3q)\left(2p^2 - 4pq + q^2\right)$

Factor each polynomial completely. See Example 4.

41. $8x^3y^4 + 12x^2y^3 + 36xy^4$

42. $10m^5n + 4m^2n^3 + 18m^3n^2$

43. $x^2 - 2x - 15$

44. $x^2 + x - 12$

45. $2x^2 - 9x - 18$

46. $3x^2 + 2x - 8$

47. $36t^2 - 25$

48. $49r^2 - 9$

49. $16t^2 + 24t + 9$

50. $25t^2 + 90t + 81$

51. $4m^2p - 12mnp + 9n^2p$

52. $16p^2r - 40pqr + 25q^2r$

53. $x^3 + 1$

54. $x^3 + 27$

55. $8t^3 + 125$

56. $27s^3 + 64$

57. $t^6 - 125$

58. $w^6 - 27$

59. $5xt + 15xr + 2yt + 6yr$

60. $3am + 18mb + 2an + 12nb$

61. $6ar + 12br - 5as - 10bs$

62. $7mt + 35ms - 2nt - 10ns$

63. $t^4 - 1$

64. $r^4 - 81$

Appendix: Solving Systems of Linear Equations by Matrix Methods

OBJECTIVES

1. Define a matrix.
2. Write the augmented matrix of a system.
3. Use row operations to solve a system with two equations.
4. Use row operations to solve a system with three equations.
5. Use row operations to solve special systems.

OBJECTIVE 1 Define a matrix. An ordered array of numbers such as

$$\text{Rows} \begin{bmatrix} 2 & 3 & 5 \\ 7 & 1 & 2 \end{bmatrix} \overset{\text{Columns}}{} \text{Matrix}$$

is a **matrix.** The numbers are the **elements** of the matrix. *Matrices* (the plural of *matrix*) are named according to the number of **rows** and **columns** they contain. The rows are read horizontally, and the columns are read vertically. This matrix is a 2×3 (read "two by three") matrix because it has 2 rows and 3 columns. The number of rows followed by the number of columns gives the **dimensions** of the matrix.

$$\begin{bmatrix} -1 & 0 \\ 1 & -2 \end{bmatrix} \quad \begin{matrix} 2 \times 2 \\ \text{matrix} \end{matrix} \qquad \begin{bmatrix} 8 & -1 & -3 \\ 2 & 1 & 6 \\ 0 & 5 & -3 \\ 5 & 9 & 7 \end{bmatrix} \quad \begin{matrix} 4 \times 3 \\ \text{matrix} \end{matrix}$$

A **square matrix** has the same number of rows as columns. The 2×2 matrix above is a square matrix.

We now discuss a matrix method of solving linear systems that is a structured way of using the elimination method from **Chapters 4 and 8.** The advantage of this new method is that it can be done by a graphing calculator or a computer.

OBJECTIVE 2 Write the augmented matrix of a system. To solve a linear system using matrices, we begin by writing an *augmented matrix* for the system. An **augmented matrix** has a vertical bar that separates the columns of the matrix into two groups. For example, to solve the system

$$x - 3y = 1$$
$$2x + y = -5,$$

we start by writing the augmented matrix

$$\begin{bmatrix} 1 & -3 & | & 1 \\ 2 & 1 & | & -5 \end{bmatrix}. \qquad \text{Augmented matrix}$$

Place the coefficients of the variables to the left of the bar, and the constants to the right.

A matrix is just a shorthand way of writing a system of equations, so the rows of an augmented matrix can be treated the same as the equations of a system of equations.

Exchanging the position of two equations in a system does not change the system. Also, multiplying any equation in a system by a nonzero number does not change the system. Comparable changes to the augmented matrix of a system of equations produce new matrices that correspond to systems with the same solutions as the original system.

The following **row operations** produce new matrices that lead to systems having the same solutions as the original system.

System of equations:

$$x - 3y = 1$$
$$2x + y = -5$$

Augmented matrix:

$$\begin{bmatrix} 1 & -3 & | & 1 \\ 2 & 1 & | & -5 \end{bmatrix}$$

Coefficients of the variables ⟶   The bar separates the coefficients from the constants. ⟶   Constants

---

Matrix Row Operations

1. Any two rows of the matrix may be interchanged.

2. The elements of any row may be multiplied by any nonzero real number.

3. Any row may be transformed by adding to the elements of the row the product of a real number and the corresponding elements of another row.

Example of row operation 1

$$\begin{bmatrix} 2 & 3 & 9 \\ 4 & 8 & -3 \\ 1 & 0 & 7 \end{bmatrix} \quad \text{becomes} \quad \begin{bmatrix} 1 & 0 & 7 \\ 4 & 8 & -3 \\ 2 & 3 & 9 \end{bmatrix}$$

Interchange row 1 and row 3.

Example of row operation 2

$$\begin{bmatrix} 2 & 3 & 9 \\ 4 & 8 & -3 \\ 1 & 0 & 7 \end{bmatrix} \quad \text{becomes} \quad \begin{bmatrix} 6 & 9 & 27 \\ 4 & 8 & -3 \\ 1 & 0 & 7 \end{bmatrix}$$

Multiply the numbers in row 1 by 3.

Example of row operation 3

$$\begin{bmatrix} 2 & 3 & 9 \\ 4 & 8 & -3 \\ 1 & 0 & 7 \end{bmatrix} \quad \text{becomes} \quad \begin{bmatrix} 0 & 3 & -5 \\ 4 & 8 & -3 \\ 1 & 0 & 7 \end{bmatrix}$$

Multiply the numbers in row 3 by -2. Add them to the corresponding numbers in row 1.

The third row operation corresponds to the way we eliminated a variable from a pair of equations to solve a system by the elimination method in **Sections 4.3 and 8.4.**

OBJECTIVE ▶ 3 Use row operations to solve a system with two equations.
Row operations can be used to rewrite a matrix until it is the matrix of a system whose solution is easy to find. The goal is a matrix in the form

$$\begin{bmatrix} 1 & a & | & b \\ 0 & 1 & | & c \end{bmatrix} \quad \text{or} \quad \begin{bmatrix} 1 & a & b & | & c \\ 0 & 1 & d & | & e \\ 0 & 0 & 1 & | & f \end{bmatrix}$$

for systems with two or three equations, respectively. Notice that there are 1s down the diagonal from upper left to lower right and 0s below the 1s. A matrix written this way is said to be in **row echelon form.**

① Use row operations to solve the system.

$$x - 2y = 9$$
$$3x + y = 13$$

EXAMPLE 1 Using Row Operations to Solve a System with Two Variables

Use row operations to solve the system.

$$x - 3y = 1$$
$$2x + y = -5$$

We start by writing the augmented matrix of the system.

$$\begin{bmatrix} 1 & -3 & | & 1 \\ 2 & 1 & | & -5 \end{bmatrix}$$ Write the augmented matrix.

Our goal is to use the various row operations to change this matrix into one that leads to a system that is easier to solve. It is best to work by columns.

We start with the first column and make sure that there is a 1 in the first row, first column position. There is already a 1 in this position.

Next, we introduce 0 in every position below the first. To get a 0 in row two, column one, we add to the numbers in row two the result of multiplying each number in row one by -2. (We abbreviate this as $-2R_1 + R_2$.) Row one remains unchanged.

$$\begin{bmatrix} 1 & -3 & | & 1 \\ 2 + 1(-2) & 1 + -3(-2) & | & -5 + 1(-2) \end{bmatrix}$$

↑ Original number from row two   ↑ -2 times number from row one

1 in the first position of column one → $\begin{bmatrix} 1 & -3 & | & 1 \\ 0 & 7 & | & -7 \end{bmatrix}$ $-2R_1 + R_2$
0 in every position below the first →

Now we go to column two. The number 1 is needed in row two, column two. We use the second row operation, multiplying each number of row two by $\frac{1}{7}$.

Stop here—this matrix is in row echelon form. → $\begin{bmatrix} 1 & -3 & | & 1 \\ 0 & 1 & | & -1 \end{bmatrix}$ $\frac{1}{7}R_2$

This augmented matrix leads to the system of equations

$$1x - 3y = 1 \qquad\qquad x - 3y = 1$$
$$0x + 1y = -1, \quad \text{or} \quad y = -1.$$

From the second equation, $y = -1$, we substitute -1 for y in the first equation to find x.

$$x - 3y = 1$$
$$x - 3(-1) = 1 \qquad \text{Let } y = -1.$$
$$x + 3 = 1 \qquad \text{Multiply.}$$
$$x = -2 \qquad \text{Subtract 3.}$$

The solution set of the system is $\{(-2, -1)\}$. Check this solution by substitution in both equations of the system.

Write the values of x and y in the correct order.

◀ **Work Problem ①** at the Side.

Answer

1. $\{(5, -2)\}$

OBJECTIVE ▶ **4** **Use row operations to solve a system with three equations.**

EXAMPLE 2 Using Row Operations to Solve a System with Three Variables

Use row operations to solve the system.

$$x - y + 5z = -6$$
$$3x + 3y - z = 10$$
$$x + 3y + 2z = 5$$

Start by writing the augmented matrix of the system.

$$\begin{bmatrix} 1 & -1 & 5 & | & -6 \\ 3 & 3 & -1 & | & 10 \\ 1 & 3 & 2 & | & 5 \end{bmatrix} \quad \text{Write the augmented matrix.}$$

This matrix already has 1 in row one, column one. Next get 0s in the rest of column one. First, add to row two the results of multiplying each number of row one by -3.

$$\begin{bmatrix} 1 & -1 & 5 & | & -6 \\ 0 & 6 & -16 & | & 28 \\ 1 & 3 & 2 & | & 5 \end{bmatrix} \quad -3R_1 + R_2$$

Now add to the numbers in row three the results of multiplying each number of row one by -1.

$$\begin{bmatrix} 1 & -1 & 5 & | & -6 \\ 0 & 6 & -16 & | & 28 \\ 0 & 4 & -3 & | & 11 \end{bmatrix} \quad -1R_1 + R_3$$

Obtain 1 in row two, column two by multiplying each number in row two by $\frac{1}{6}$.

$$\begin{bmatrix} 1 & -1 & 5 & | & -6 \\ 0 & 1 & -\frac{8}{3} & | & \frac{14}{3} \\ 0 & 4 & -3 & | & 11 \end{bmatrix} \quad \frac{1}{6}R_2$$

To obtain 0 in row three, column two, add to row three the results of multiplying each number in row two by -4.

$$\begin{bmatrix} 1 & -1 & 5 & | & -6 \\ 0 & 1 & -\frac{8}{3} & | & \frac{14}{3} \\ 0 & 0 & \frac{23}{3} & | & -\frac{23}{3} \end{bmatrix} \quad -4R_2 + R_3$$

Obtain 1 in row three, column three by multiplying each number in row three by $\frac{3}{23}$.

$$\begin{bmatrix} 1 & -1 & 5 & | & -6 \\ 0 & 1 & -\frac{8}{3} & | & \frac{14}{3} \\ 0 & 0 & 1 & | & -1 \end{bmatrix} \quad \frac{3}{23}R_3$$

This final matrix leads to the system of equations given at the top of the next page.

·········· **Continued on Next Page**

2 Use row operations to solve the system.

$$2x - y + z = 7$$
$$x - 3y - z = 7$$
$$-x + y - 5z = -9$$

3 Use row operations to solve each system.

(a) $x - y = 2$
$-2x + 2y = 2$

(b) $x - y = 2$
$-2x + 2y = -4$

$$x - y + 5z = -6$$
$$y - \frac{8}{3}z = \frac{14}{3}$$
$$z = -1$$

Substitute -1 for z in the second equation, $y - \frac{8}{3}z = \frac{14}{3}$, to find that $y = 2$. Finally, substitute 2 for y and -1 for z in the first equation,

$$x - y + 5z = -6,$$

to determine that $x = 1$. The solution set of the original system is $\{(1, 2, -1)\}$. Check by substitution.

◀ **Work Problem 2** at the Side.

OBJECTIVE **5** **Use row operations to solve special systems.**

EXAMPLE 3 **Recognizing Inconsistent Systems or Dependent Equations**

Use row operations to solve each system.

(a) $\begin{array}{r} 2x - 3y = 8 \\ -6x + 9y = 4 \end{array}$ $\longrightarrow$ $\left[\begin{array}{cc|c} 2 & -3 & 8 \\ -6 & 9 & 4 \end{array}\right]$ Write the augmented matrix.

$\left[\begin{array}{cc|c} 1 & -\frac{3}{2} & 4 \\ -6 & 9 & 4 \end{array}\right]$ $\frac{1}{2}R_1$

$\left[\begin{array}{cc|c} 1 & -\frac{3}{2} & 4 \\ 0 & 0 & 28 \end{array}\right]$ $6R_1 + R_2$

The corresponding system of equations is

$$x - \frac{3}{2}y = 4$$

$$0 = 28, \quad \text{False}$$

which has no solution and is inconsistent. The solution set is $\emptyset$.

(b) $\begin{array}{r} -10x + 12y = 30 \\ 5x - 6y = -15 \end{array}$ $\longrightarrow$ $\left[\begin{array}{cc|c} -10 & 12 & 30 \\ 5 & -6 & -15 \end{array}\right]$ Write the augmented matrix.

$\left[\begin{array}{cc|c} 1 & -\frac{6}{5} & -3 \\ 5 & -6 & -15 \end{array}\right]$ $-\frac{1}{10}R_1$

$\left[\begin{array}{cc|c} 1 & -\frac{6}{5} & -3 \\ 0 & 0 & 0 \end{array}\right]$ $-5R_1 + R_2$

The corresponding system of equations is

$$x - \frac{6}{5}y = -3$$

$$0 = 0, \quad \text{True}$$

which has dependent equations. We use the second equation of the original system, which is in standard form, to express the solution set.

$$\{(x, y) \mid 5x - 6y = -15\}$$

◀ **Work Problem 3** at the Side.

Answers

2. $\{(2, -2, 1)\}$
3. **(a)** $\emptyset$ **(b)** $\{(x, y) \mid x - y = 2\}$

Appendix B Exercises

MyMathLab®

Download the MyDashBoard App

1. CONCEPT CHECK Consider the matrix $\begin{bmatrix} -2 & 3 & 1 \\ 0 & 5 & -3 \\ 1 & 4 & 8 \end{bmatrix}$, and answer the following.

(a) What are the elements of the second row?

(b) What are the elements of the third column?

(c) Is this a square matrix? Why?

(d) Give the matrix obtained by interchanging the first and third rows.

(e) Give the matrix obtained by multiplying the first row by $-\frac{1}{2}$.

(f) Give the matrix obtained by multiplying the third row by 3 and adding it to the first row.

2. CONCEPT CHECK Give the dimensions of each matrix.

(a) $\begin{bmatrix} 3 & -7 \\ 4 & 5 \\ -1 & 0 \end{bmatrix}$

(b) $\begin{bmatrix} 4 & 9 & 0 \\ -1 & 2 & -4 \end{bmatrix}$

(c) $\begin{bmatrix} 6 & 3 \\ -2 & 5 \\ 4 & 10 \\ 1 & -11 \end{bmatrix}$

GS *Complete the steps in the matrix solution of each system by filling in the blanks. Give the final system and the solution set.* **See Example 1.**

3. $4x + 8y = 44$
$2x - y = -3$

$\begin{bmatrix} 4 & 8 & | & 44 \\ 2 & -1 & | & -3 \end{bmatrix}$

$\begin{bmatrix} 1 & \underline{} & | & \underline{} \\ 2 & -1 & | & -3 \end{bmatrix}$ $\frac{1}{4}R_1$

$\begin{bmatrix} 1 & 2 & | & 11 \\ 0 & \underline{} & | & \underline{} \end{bmatrix}$ $-2R_1 + R_2$

$\begin{bmatrix} 1 & 2 & | & 11 \\ 0 & 1 & | & \underline{} \end{bmatrix}$ $-\frac{1}{5}R_2$

4. $2x - 5y = -1$
$3x + y = 7$

$\begin{bmatrix} 2 & -5 & | & -1 \\ 3 & 1 & | & 7 \end{bmatrix}$

$\begin{bmatrix} 1 & -\frac{5}{2} & | & \underline{} \\ 3 & 1 & | & 7 \end{bmatrix}$ $\frac{1}{2}R_1$

$\begin{bmatrix} 1 & -\frac{5}{2} & | & -\frac{1}{2} \\ 0 & \underline{} & | & \underline{} \end{bmatrix}$ $-3R_1 + R_2$

$\begin{bmatrix} 1 & -\frac{5}{2} & | & -\frac{1}{2} \\ 0 & 1 & | & \underline{} \end{bmatrix}$ $\frac{2}{17}R_2$

Use row operations to solve each system. **See Examples 1 and 3.**

5. $x + y = 5$
$x - y = 3$

6. $x + 2y = 7$
$x - y = -2$

7. $2x + 4y = 6$
$3x - y = 2$

8. $4x + 5y = -7$
$x - y = 5$

9. $3x + 4y = 13$
$2x - 3y = -14$

10. $5x + 2y = 8$
$3x - y = 7$

11. $-4x + 12y = 36$
$x - 3y = 9$

12. $2x - 4y = 8$
$-3x + 6y = 5$

13. $2x + y = 4$
$4x + 2y = 8$

14. $3x + 4y = -1$
$6x + 8y = -2$

15. $\dfrac{1}{2}x + \dfrac{1}{3}y = 0$
$\dfrac{2}{3}x + \dfrac{3}{4}y = 0$

16. $1.2x + 0.3y = 0$
$2.9x - 0.6y = 0$

GS *Complete the steps in the matrix solution of each system by filling in the blanks. Give the final system and the solution set. **See Example 2.***

17. $x + y - z = -3$
$2x + y + z = 4$
$5x - y + 2z = 23$

$$\begin{bmatrix} 1 & 1 & -1 & | & -3 \\ 2 & 1 & 1 & | & 4 \\ 5 & -1 & 2 & | & 23 \end{bmatrix}$$

$$\begin{bmatrix} 1 & 1 & -1 & | & -3 \\ 0 & \underline{\quad} & \underline{\quad} & | & \underline{\quad} \\ 0 & \underline{\quad} & \underline{\quad} & | & \underline{\quad} \end{bmatrix} \begin{matrix} \\ -2R_1 + R_2 \\ -5R_1 + R_3 \end{matrix}$$

$$\begin{bmatrix} 1 & 1 & -1 & | & -3 \\ 0 & 1 & \underline{\quad} & | & \underline{\quad} \\ 0 & -6 & 7 & | & 38 \end{bmatrix} \begin{matrix} \\ -1R_2 \\ \\ \end{matrix}$$

$$\begin{bmatrix} 1 & 1 & -1 & | & -3 \\ 0 & 1 & -3 & | & -10 \\ 0 & 0 & \underline{\quad} & | & \underline{\quad} \end{bmatrix} \begin{matrix} \\ \\ 6R_2 + R_3 \end{matrix}$$

$$\begin{bmatrix} 1 & 1 & -1 & | & -3 \\ 0 & 1 & -3 & | & -10 \\ 0 & 0 & 1 & | & \underline{\quad} \end{bmatrix} \begin{matrix} \\ \\ -\frac{1}{11}R_3 \end{matrix}$$

18. $2x + y + 2z = 11$
$2x - y - z = -3$
$3x + 2y + z = 9$

$$\begin{bmatrix} 2 & 1 & 2 & | & 11 \\ 2 & -1 & -1 & | & -3 \\ 3 & 2 & 1 & | & 9 \end{bmatrix}$$

$$\begin{bmatrix} 1 & \underline{\quad} & \underline{\quad} & | & \underline{\quad} \\ 2 & -1 & -1 & | & -3 \\ 3 & 2 & 1 & | & 9 \end{bmatrix} \begin{matrix} \frac{1}{2}R_1 \\ \\ \end{matrix}$$

$$\begin{bmatrix} 1 & \frac{1}{2} & 1 & | & \frac{11}{2} \\ 0 & \underline{\quad} & \underline{\quad} & | & \underline{\quad} \\ 0 & \underline{\quad} & \underline{\quad} & | & \underline{\quad} \end{bmatrix} \begin{matrix} \\ -2R_1 + R_2 \\ -3R_1 + R_3 \end{matrix}$$

$$\begin{bmatrix} 1 & \frac{1}{2} & 1 & | & \frac{11}{2} \\ 0 & 1 & \underline{\quad} & | & \underline{\quad} \\ 0 & \frac{1}{2} & -2 & | & -\frac{15}{2} \end{bmatrix} \begin{matrix} \\ -\frac{1}{2}R_2 \\ \\ \end{matrix}$$

$$\begin{bmatrix} 1 & \frac{1}{2} & 1 & | & \frac{11}{2} \\ 0 & 1 & \frac{3}{2} & | & 7 \\ 0 & 0 & \underline{\quad} & | & \underline{\quad} \end{bmatrix} \begin{matrix} \\ \\ -\frac{1}{2}R_2 + R_3 \end{matrix}$$

$$\begin{bmatrix} 1 & \frac{1}{2} & 1 & | & \frac{11}{2} \\ 0 & 1 & \frac{3}{2} & | & 7 \\ 0 & 0 & 1 & | & \underline{\quad} \end{bmatrix} \begin{matrix} \\ \\ -\frac{4}{11}R_3 \end{matrix}$$

*Use row operations to solve each system. **See Examples 2 and 3.***

19. $x + y - 3z = 1$
▶ $2x - y + z = 9$
$3x + y - 4z = 8$

20. $2x + 4y - 3z = -18$
$3x + y - z = -5$
$x - 2y + 4z = 14$

21. $x + y - z = 6$
$2x - y + z = -9$
$x - 2y + 3z = 1$

22. $x + 3y - 6z = 7$
$2x - y + 2z = 0$
$x + y + 2z = -1$

23. $x - y = 1$
$y - z = 6$
$x + z = -1$

24. $x + y = 1$
$2x - z = 0$
$y + 2z = -2$

25. $4x + 8y + 4z = 9$
$x + 3y + 4z = 10$
$5x + 10y + 5z = 12$

26. $x + 2y + 3z = -2$
$2x + 4y + 6z = -5$
$x - y + 2z = 6$

27. $x - 2y + z = 4$
$3x - 6y + 3z = 12$
$-2x + 4y - 2z = -8$

28. $x + 3y + z = 1$
$2x + 6y + 2z = 2$
$3x + 9y + 3z = 3$

29. $5x + 3y - z = 0$
$2x - 3y + z = 0$
$x + 4y - 2z = 0$

30. $4x + 5y - z = 0$
$7x - 5y + z = 0$
$x + 3y - 2z = 0$

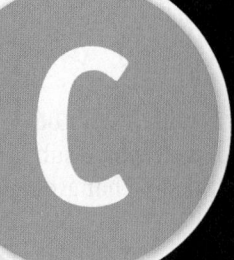

Appendix: Synthetic Division

We begin by reviewing the terminology for the parts of a division problem. The *divisor* is the quantity we are dividing by, the *dividend* is the quantity we are dividing into, and the *quotient* is the result of the division.

$$\underset{\text{Divisor}\longrightarrow}{247)\overline{\underset{\longleftarrow\text{ Dividend}}{385{,}814}}}\overset{\overset{\displaystyle 1\ 562}{\longleftarrow\text{ Quotient}}}{}$$

OBJECTIVES

1 Use synthetic division to divide by a polynomial of the form $x - k$.

2 Use the remainder theorem to evaluate a polynomial.

3 Decide whether a given number is a solution of an equation.

OBJECTIVE ▸ 1 Use synthetic division to divide by a polynomial of the form $x - k$. If a polynomial in x is divided by a binomial of the form $x - k$, a shortcut method can be used. For an illustration, look at the division on the left below.

$$
\begin{array}{r}
3x^2 + 9x + 25 \\
x - 3)\overline{3x^3 + 0x^2 - 2x + 5} \\
\underline{3x^3 - 9x^2} \\
9x^2 - 2x \\
\underline{9x^2 - 27x} \\
25x + 5 \\
\underline{25x - 75} \\
80
\end{array}
\qquad
\begin{array}{r}
3 \quad\ 9 \quad 25 \\
1 - 3)\overline{3 \quad\ 0 \quad -2 \quad\ 5} \\
\underline{3 \ -9} \\
9 \ -2 \\
\underline{9 \ -27} \\
25 \quad\ 5 \\
\underline{25 \ -75} \\
80
\end{array}
$$

On the right above, the same division is shown written without the variables. This is why it is *essential* to use 0 as a placeholder in synthetic division. All the numbers in color on the right are repetitions of the numbers directly above them, so they are omitted to condense the work, as shown on the left below.

$$
\begin{array}{r}
3 \quad\ 9 \quad 25 \\
1 - 3)\overline{3 \quad 0 \ -2 \quad\ 5} \\
\underline{-9} \\
9 \ -2 \\
\underline{-27} \\
25 \quad\ 5 \\
\underline{-75} \\
80
\end{array}
\qquad
\begin{array}{r}
3 \quad\ 9 \quad 25 \\
1 - 3)\overline{3 \quad 0 \ -2 \quad\ 5} \\
\underline{-9} \\
9 \\
\underline{-27} \\
25 \\
\underline{-75} \\
80
\end{array}
$$

The numbers in color on the left are again repetitions of the numbers directly above them. They too are omitted, as shown on the right above. If we bring the 3 in the dividend down to the beginning of the bottom row, the top row can be omitted since it duplicates the bottom row.

1 Divide, using synthetic division.

(a) $\dfrac{3x^2 + 10x - 8}{x + 4}$

(b) $(2x^2 + 3x - 5) \div (x + 1)$

$$
\begin{array}{r}
1 - 3\overline{)3 \quad\; 0 \quad -2 \quad\;\; 5} \\
-9 \;\; -27 \;\; -75 \\
\hline
3 \quad\;\; 9 \quad\;\; 25 \quad\;\; 80
\end{array}
$$

We omit the 1 at the upper left, since it represents $1x$, which will *always* be the first term in the divisor. Also, to simplify the arithmetic, we replace subtraction in the second row by addition. We compensate for this by changing the -3 at the upper left to its additive inverse, 3.

Additive inverse of -3 ⟶
$$
\begin{array}{r}
3\overline{)3 \quad\;\; 0 \quad\; -2 \quad\quad 5} \\
9 \quad\;\; 27 \quad\; 75 \quad \longleftarrow \text{Change signs.} \\
\hline
3 \quad\;\; 9 \quad\;\; 25 \quad\;\; 80 \quad \longleftarrow \text{Remainder} \\
\downarrow \quad\;\; \downarrow \quad\;\; \downarrow \quad\;\; \downarrow
\end{array}
$$

The quotient is read from the bottom row. $3x^2 + 9x + 25 + \dfrac{80}{x - 3}$

The first three numbers in the bottom row are the coefficients of the quotient polynomial with degree 1 less than the degree of the dividend. The last number gives the remainder.

Synthetic Division

This shortcut method is called **synthetic division**. *It is used only when dividing a polynomial $P(x)$ by a binomial of the form $x - k$.*

EXAMPLE 1   Using Synthetic Division

Use synthetic division to divide $5x^2 + 16x + 15$ by $x + 2$.
   We change $x + 2$ into the form $x - k$ by writing it as

$$x + 2 = x - (-2), \quad \text{where } k = -2.$$

Now write the coefficients of $5x^2 + 16x + 15$, placing -2 to the left.

$x + 2$ leads to -2. ⟶ $-2\overline{)5 \quad 16 \quad 15}$ ← Coefficients

$$
\begin{array}{r}
-2\overline{)5 \quad 16 \quad 15} \\
\downarrow \; -10 \\
5
\end{array}
$$
Bring down the 5, and multiply: $-2 \cdot 5 = -10$.

$$
\begin{array}{r}
-2\overline{)5 \quad 16 \quad\;\; 15} \\
-10 \; -12 \\
\hline
5 \quad\; 6
\end{array}
$$
Add 16 and -10, getting 6, and multiply -2 and 6 to get -12.

$$
\begin{array}{r}
-2\overline{)5 \quad 16 \quad\;\; 15} \\
-10 \; -12 \\
\hline
5 \quad\; 6 \quad\;\; 3
\end{array}
$$
← Remainder

Add 15 and -12, getting 3.

The result is read from the bottom row.

$$\frac{5x^2 + 16x + 15}{x + 2} = 5x + 6 + \frac{3}{x + 2}$$

◄ **Work Problem 1** at the Side.

Answers

1. (a) $3x - 2$ (b) $2x + 1 + \dfrac{-6}{x + 1}$

EXAMPLE 2 Using Synthetic Division with a Missing Term

Use synthetic division to find $(-4x^5 + x^4 + 6x^3 + 2x^2 + 50) \div (x - 2)$.

$$
\begin{array}{r|rrrrrr}
2) & -4 & 1 & 6 & 2 & 0 & 50 \\
  &   & -8 & -14 & -16 & -28 & -56 \\
\hline
  & -4 & -7 & -8 & -14 & -28 & -6
\end{array}
$$

Use the steps given earlier, first inserting a 0 for the missing x-term.

Read the result from the bottom row.

$$\frac{-4x^5 + x^4 + 6x^3 + 2x^2 + 50}{x - 2}$$

$$= -4x^4 - 7x^3 - 8x^2 - 14x - 28 + \frac{-6}{x - 2}$$

·········· **Work Problem 2 at the Side.** ▶

OBJECTIVE 2 Use the remainder theorem to evaluate a polynomial. We can use synthetic division to evaluate polynomials. For example, in the synthetic division of **Example 2,** where the polynomial was divided by $x - 2$, the remainder was -6.

Replacing x in the polynomial with 2 gives the following.

$$-4x^5 + x^4 + 6x^3 + 2x^2 + 50 \qquad \text{From Example 2}$$
$$= -4 \cdot 2^5 + 2^4 + 6 \cdot 2^3 + 2 \cdot 2^2 + 50 \qquad \text{Replace } x \text{ with 2.}$$
$$= -4 \cdot 32 + 16 + 6 \cdot 8 + 2 \cdot 4 + 50 \qquad \text{Evaluate the powers.}$$
$$= -128 + 16 + 48 + 8 + 50 \qquad \text{Multiply.}$$
$$= -6 \qquad \text{Add.}$$

This is the same number as the remainder. Thus, dividing by $x - 2$ produced a remainder equal to the result when x is replaced with 2. This always happens, as the following **remainder theorem** states. This result is proved in more advanced courses.

Remainder Theorem

If the polynomial $P(x)$ is divided by $x - k$, then the remainder is equal to $P(k)$.

EXAMPLE 3 Using the Remainder Theorem

Let $P(x) = 2x^3 - 5x^2 - 3x + 11$. Find $P(-2)$.

Use the remainder theorem, and divide $P(x)$ by $x - (-2)$.

$$
\begin{array}{r|rrrr}
\text{Value of } k \rightarrow -2) & 2 & -5 & -3 & 11 \\
 & & -4 & 18 & -30 \\
\hline
 & 2 & -9 & 15 & -19 \leftarrow \text{Remainder}
\end{array}
$$

Thus, $P(-2) = -19$.

·········· **Work Problem 3 at the Side.** ▶

2 Divide, using synthetic division.

(a) $\dfrac{3x^3 - 2x + 21}{x + 2}$

(b) $(-4x^4 + 3x^3 + 18x + 2) \div (x - 2)$

3 Let $P(x) = x^3 - 5x^2 + 7x - 3$. Use synthetic division to find each value.

(a) $P(1)$ (Divide by $x - 1$.)

(b) $P(-2)$

Answers

2. (a) $3x^2 - 6x + 10 + \dfrac{1}{x+2}$

   (b) $-4x^3 - 5x^2 - 10x - 2 + \dfrac{-2}{x-2}$

3. (a) 0 (b) -45

4 Use synthetic division to decide whether 2 is a solution of each equation.

(a) $3x^3 - 11x^2 + 17x - 14$
$= 0$

(b) $4x^5 - 7x^4 - 11x^2 + 2x + 6$
$= 0$

OBJECTIVE **3** **Decide whether a given number is a solution of an equation.** We can use the remainder theorem to do this.

EXAMPLE 4 **Using the Remainder Theorem**

Use synthetic division to decide whether -5 is a solution of the equation.

$$2x^4 + 12x^3 + 6x^2 - 5x + 75 = 0$$

If synthetic division gives a remainder of 0, then -5 is a solution. Otherwise, it is not.

$$
\begin{array}{r}
\text{Proposed solution} \rightarrow -5)\overline{2 \quad\quad 12 \quad\quad 6 \quad -5 \quad\quad 75} \\
\underline{-10 \quad -10 \quad 20 \quad -75} \\
2 \quad\quad 2 \quad -4 \quad 15 \quad\quad 0 \leftarrow \text{Remainder}
\end{array}
$$

Since the remainder is 0, the polynomial has a value of 0 when $x = -5$. Therefore, -5 is a solution of the given equation.

◀ **Work Problem** **4** **at the Side.**

The synthetic division in **Example 4** shows that $x - (-5)$ divides the polynomial with 0 remainder. Thus

$$x - (-5) = x + 5$$

is a *factor* of the polynomial and

$$2x^4 + 12x^3 + 6x^2 - 5x + 75$$

factors as

$$(x + 5)(2x^3 + 2x^2 - 4x + 15).$$

The second factor is the quotient polynomial found in the last row of the synthetic division.

Answers

4. (a) yes (b) no

Appendix C Exercises

Download the MyDashBoard App

MyMathLab®

CONCEPT CHECK *Choose the letter of the correct setup to perform synthetic division on the indicated quotient.*

1. $\dfrac{x^2 + 3x - 6}{x - 2}$

   A. $-2\overline{)1\quad 3\quad -6}$ **B.** $-2\overline{)-1\quad -3\quad 6}$

   C. $2\overline{)1\quad 3\quad -6}$ **D.** $2\overline{)-1\quad -3\quad 6}$

2. $\dfrac{x^3 - 3x^2 + 2}{x - 1}$

   A. $1\overline{)1\quad -3\quad 2}$ **B.** $-1\overline{)1\quad -3\quad 2}$

   C. $1\overline{)1\quad -3\quad 0\quad 2}$ **D.** $1\overline{)-1\quad 3\quad 0\quad -2}$

Use synthetic division to find each quotient. **See Examples 1 and 2.**

3. $\dfrac{x^2 - 6x + 5}{x - 1}$

4. $\dfrac{x^2 - 4x - 21}{x + 3}$

5. $\dfrac{4m^2 + 19m - 5}{m + 5}$

6. $\dfrac{3k^2 - 5k - 12}{k - 3}$

7. $\dfrac{2a^2 + 8a + 13}{a + 2}$

8. $\dfrac{4y^2 - 5y - 20}{y - 4}$

9. $(p^2 - 3p + 5) \div (p + 1)$

10. $(z^2 + 4z - 6) \div (z - 5)$

11. $\dfrac{4a^3 - 3a^2 + 2a - 3}{a - 1}$

12. $\dfrac{5p^3 - 6p^2 + 3p + 14}{p + 1}$

13. $(x^5 - 2x^3 + 3x^2 - 4x - 2) \div (x - 2)$

14. $(2y^5 - 5y^4 - 3y^2 - 6y - 23) \div (y - 3)$

15. $(-4r^6 - 3r^5 - 3r^4 + 5r^3 - 6r^2 + 3r) \div (r - 1)$

16. $(-3t^5 + 2t^4 - 5t^3 + 6t^2 - 3t - 2) \div (t - 2)$

17. $(-3y^5 + 2y^4 - 5y^3 - 6y^2 - 1) \div (y + 2)$

18. $(m^6 + 2m^4 - 5m + 11) \div (m - 2)$

19. $\dfrac{y^3 + 1}{y - 1}$

20. $\dfrac{z^4 + 81}{z - 3}$

Use the remainder theorem to find $P(k)$. ***See Example 3.***

21. $P(x) = 2x^3 - 4x^2 + 5x - 3; \quad k = 2$

22. $P(x) = x^3 + 3x^2 - x + 5; \quad k = -1$

23. $P(r) = -r^3 - 5r^2 - 4r - 2; \quad k = -4$

24. $P(z) = -z^3 + 5z^2 - 3z + 4; \quad k = 3$

25. $P(x) = 2x^3 - 4x^2 + 5x - 33; \quad k = 3$

26. $P(x) = x^3 - 3x^2 + 4x - 4; \quad k = 2$

Use synthetic division to decide whether the given number is a solution of each equation.
See Example 4.

27. $x^3 - 2x^2 - 3x + 10 = 0; \quad x = -2$

28. $x^3 - 3x^2 - x + 10 = 0; \quad x = -2$

29. $m^4 + 2m^3 - 3m^2 + 8m - 8 = 0; \quad m = -2$

30. $r^4 - r^3 - 6r^2 + 5r + 10 = 0; \quad r = -2$

31. $3x^3 + 2x^2 - 2x + 11 = 0; \quad x = -2$

32. $3z^3 + 10z^2 + 3z - 9 = 0; \quad z = -2$

33. Explain why it is important to insert 0s as placeholders for missing terms before performing synthetic division.

34. Explain why a 0 remainder in synthetic division of $P(x)$ by k indicates that k is a solution of the equation $P(x) = 0$.

Answers to Selected Exercises

In this section we provide the answers that we think most students will obtain when they work the exercises using the methods explained in the text. If your answer does not look exactly like the one given here, it is not necessarily wrong. In many cases there are equivalent forms of the answer that are correct. For example, if the answer section shows $\frac{3}{4}$ and your answer is 0.75, you have obtained the correct answer but written it in a different (yet equivalent) form. Unless the directions specify otherwise, 0.75 is just as valid an answer as $\frac{3}{4}$.

In general, if your answer does not agree with the one given in the text, see whether it can be transformed into the other form. If it can, then it is the correct answer. If you still have doubts, talk with your instructor.

CHAPTER R Prealgebra Review

SECTION R.1 (pages 11–14)

1. true **2.** true **3.** false; This is an improper fraction. Its value is 1.
4. false; The number 1 is neither prime nor composite. **5.** false; The fraction $\frac{17}{51}$ can be simplified to $\frac{1}{3}$. **6.** false; The reciprocal of $\frac{8}{2} = 4$ is $\frac{2}{8} = \frac{1}{4}$.
7. false; *Product* indicates multiplication, so the product of 8 and 2 is 16.
8. false; *Difference* indicates subtraction, so the difference between 12 and 2 is 10. **9.** prime **11.** composite **13.** composite **15.** prime
17. $2 \cdot 3 \cdot 5$ **19.** $2 \cdot 2 \cdot 3 \cdot 3 \cdot 7$ **21.** $2 \cdot 2 \cdot 31$ **23.** 29 **25.** $\frac{1}{2}$
27. $\frac{5}{6}$ **29.** $\frac{1}{5}$ **31.** $\frac{6}{5}$ **33.** $1\frac{5}{7}$ **35.** $6\frac{5}{12}$ **37.** $7\frac{6}{11}$ **39.** $\frac{13}{5}$ **41.** $\frac{83}{8}$
43. $\frac{51}{5}$ **45.** A **46.** C **47.** $\frac{24}{35}$ **49.** $\frac{6}{25}$ **51.** $\frac{6}{5}$, or $1\frac{1}{5}$ **53.** $\frac{232}{15}$, or
$15\frac{7}{15}$ **55.** $\frac{10}{3}$, or $3\frac{1}{3}$ **57.** 12 **59.** $\frac{1}{16}$ **61.** $\frac{84}{47}$, or $1\frac{37}{47}$ **63.** $\frac{2}{3}$
65. $\frac{8}{9}$ **67.** $\frac{27}{8}$, or $3\frac{3}{8}$ **69.** $\frac{17}{36}$ **71.** $\frac{11}{12}$ **73.** $\frac{4}{3}$, or $1\frac{1}{3}$ **75.** 6 cups
77. $618\frac{3}{4}$ ft **79.** $\frac{9}{16}$ in. **81.** $\frac{7}{16}$ in. **83.** $16\frac{5}{8}$ yd **85.** 8 cakes (There
will be some sugar left over.) **87.** $\frac{1}{15}$ **89.** about $4\frac{1}{2}$ million, or 4,500,000

SECTION R.2 (pages 23–24)

1. (a) 6 (b) 9 (c) 1 (d) 7 (e) 4 **2.** Answers will vary. One
example is 5243.0164. **3.** (a) 46.25 (b) 46.2 (c) 46 (d) 50
4. (a) 0.889 (b) 0.556 (c) 0.976 (d) 0.864 **5.** C **6.** D
7. B **8.** C **9.** $\frac{4}{10}$ **11.** $\frac{64}{100}$ **13.** $\frac{138}{1000}$ **15.** $\frac{3805}{1000}$ **17.** 139;
143.094 **19.** 27; 25.61 **21.** 10; 15.33 **23.** 82; 81.716 **25.** 17;
15.211 **27.** 0.006; 0.006 **29.** 90; 116.48 **31.** 8; 7.15 **33.** 2; 2.05
35. 6000; 5711.6 **37.** 0.2; 0.162 **39.** 0.125 **41.** 1.25 **43.** $0.\overline{5}$;
0.556 **45.** $0.1\overline{6}$; 0.167 **47.** To convert a decimal to a percent, move the
decimal point two places to the right and attach a percent symbol (%).
48. To convert a percent to a decimal, move the decimal point two places to the
left and drop the percent symbol (%). **49.** 0.54 **51.** 1.17 **53.** 0.024
55. 0.0625 **57.** 0.008 **59.** 73% **61.** 0.4% **63.** 128% **65.** 30%
67. 600% **69.** 80% **71.** $18.\overline{18}$% **73.** 175% **75.** $216.\overline{6}$%
77. $15 **79.** 70%

CHAPTER 1 The Real Number System

SECTION 1.1 (pages 33–35)

1. true **2.** false; $+$, $-$, $\cdot$, and $\div$ are operation symbols.
3. false; Using the rules for order of operations gives
$4 + 3(8 - 2) = 4 + 3(6) = 4 + 18 = 22$. **4.** false; $3^3 = 3 \cdot 3 \cdot 3 = 27$
5. false; The correct translation is $4 = 16 - 12$. **6.** false; The correct
translation is $6 = 10 - 4$. **7.** The 4 would be applied last because we
work first inside the parentheses. **8.** exponents **9.** 49 **11.** 144 **13.** 64
15. 1000 **17.** 81 **19.** 1024 **21.** $\frac{16}{81}$ **23.** 0.000064
25. ②①; 45; 58 **27.** ②①③; 12; 8; 13 **29.** ①②; 32
31. ①②; 19 **33.** $\frac{49}{30}$ **35.** 12 **37.** 36.14 **39.** 26 **41.** 4
43. 95 **45.** 12 **47.** 14 **49.** $\frac{19}{2}$ **51.** $3 \cdot (6 + 4) \cdot 2 = 60$
52. $2 \cdot (8 - 1) \cdot 3 = 42$ **53.** $10 - (7 - 3) = 6$
54. $15 - (10 - 2) = 7$ **55.** $(8 + 2)^2 = 100$ **56.** $(4 + 2)^2 = 36$
57. false **59.** true **61.** true **63.** false **65.** false **67.** true
69. $15 = 5 + 10$ **71.** $9 > 5 - 4$ **73.** $16 \neq 19$ **75.** $2 \leq 3$
77. Seven is less than nineteen; true **79.** One-third is not equal to
three-tenths; true **81.** Eight is greater than or equal to eleven; false
83. $30 > 5$ **85.** $3 \leq 12$ **87.** $1.3 \leq 2.5$ **89.** $\frac{3}{4} < \frac{4}{5}$
91. (a) $14.7 - 40 \cdot 0.13$ (b) 9.5 (c) 8.075; walking (5 mph)
(d) $14.7 - 55 \cdot 0.11$; 8.65; 7.3525, swimming **93.** Alaska, Texas,
California, Idaho **95.** Alaska, Texas, California, Idaho, Missouri

SECTION 1.2 (pages 41–44)

1. B **2.** C **3.** A **4.** B, C **5.** 11 **6.** 10 **7.** $13 + x$; 16
8. expression; equation **9.** The equation would be $5x - 9 = 49$.
10. $2x^3 = 2 \cdot x \cdot x \cdot x$, while $2x \cdot 2x \cdot 2x = (2x)^3$. **11.** Answers will
vary. Two such pairs are $x = 0$, $y = 6$ and $x = 1$, $y = 4$. For each pair, the
value of $2x + y$ is 6. **12.** The exponent 2 applies only to the base x. The
exponential must be evaluated before the product. **13.** (a) 64 (b) 144
15. (a) $\frac{7}{8}$ (b) $\frac{13}{12}$ **17.** (a) 9.569 (b) 14.353 **19.** (a) 52 (b) 114
21. (a) 12 (b) 33 **23.** (a) 6 (b) $\frac{9}{5}$ **25.** (a) $\frac{4}{3}$ (b) $\frac{13}{6}$
27. (a) $\frac{2}{7}$ (b) $\frac{16}{27}$ **29.** (a) 12 (b) 55 **31.** (a) 1 (b) $\frac{28}{17}$
33. (a) 3.684 (b) 8.841 **35.** $12x$ **37.** $x - 2$ **39.** $7 - \frac{1}{3}x$
41. $2x - 6$ **43.** $\frac{12}{x + 3}$ **45.** $6(x - 4)$ **47.** An equation cannot be *evaluated*. An equation that contains a variable can be *solved* to find the value
or values of the variable that make it a true statement. **48.** An expression
cannot be *solved*. It indicates a series of operations to be performed. An
equation can be solved. **49.** no **51.** yes **53.** yes **55.** no **57.** yes
59. yes **61.** $x + 8 = 18$ **63.** $2x + 5 = 5$ **65.** $16 - \frac{3}{4}x = 13$
67. $3x = 2x + 8$ **69.** expression **71.** equation **73.** equation
75. expression **77.** 69 yr **78.** 74 yr **79.** 77 yr **80.** 79 yr

SECTION 1.3 (pages 52–54)

1. 0 **2.** integers **3.** positive **4.** right **5.** quotient; denominator
6. irrational **7.** 2,632,000 **9.** −7067

11.
$-6\ -5\qquad 0\qquad 3$

13.
$-6\ -4\ -2\ \ 0\qquad 3\ 4$

$-3\tfrac{4}{5}\quad -\tfrac{13}{8}\quad \tfrac{1}{4}\quad 2\tfrac{1}{2}$

15.
$-4\qquad 0\qquad 3$

17. (a) 3, 7 (b) 0, 3, 7

(c) $-9, 0, 3, 7$ (d) $-9, -1\tfrac{1}{4}, -\tfrac{3}{5}, 0, 3, 5.9, 7$ (e) $-\sqrt{7}, \sqrt{5}$

(f) All are real numbers. **19.** (a) 11 (b) 0, 11 (c) 0, 11, −6

(d) $\tfrac{7}{9}, -2.\overline{3}, 0, -8\tfrac{3}{4}, 11, -6$ (e) $\sqrt{3}, \pi$ (f) All are real numbers.

21. 4 **22.** One example is 2.85. There are others. **23.** 0
24. One example is 4. There are others. **25.** One example is $\sqrt{13}$. There are others. **26.** 0 **27.** true **28.** false **29.** true **30.** true **31.** false
32. true *In Exercises 33–38, answers will vary.* **33.** $\tfrac{1}{2}, \tfrac{5}{8}, 1\tfrac{3}{4}$

34. $-1, -\tfrac{3}{4}, -5$ **35.** $-3\tfrac{1}{2}, -\tfrac{2}{3}, \tfrac{3}{7}$ **36.** $\tfrac{1}{2}, -\tfrac{2}{3}, \tfrac{2}{7}$

37. $\sqrt{5}, \pi, -\sqrt{3}$ **38.** $\tfrac{2}{3}, \tfrac{5}{6}, \tfrac{5}{2}$ **39.** −11 **41.** −21 **43.** −100

45. $-\tfrac{2}{3}$ **47.** false **49.** true **51.** (a) 2 (b) 2 **53.** (a) −6 (b) 6

55. (a) $\tfrac{3}{4}$ (b) $\tfrac{3}{4}$ **57.** (a) −4.95 (b) 4.95 **59.** (a) A (b) A

(c) B (d) B **60.** 5; 5; 5; −5 **61.** 7 **63.** −12 **65.** $-\tfrac{2}{3}$ **67.** 9

69. false **71.** true **73.** Energy, 2008 to 2009 **75.** 2009 to 2010

SECTION 1.4 (pages 60–63)

1. negative; −5 **2.** zero (0) **3.** negative; −2

4. −3; 5 **5.** Add −2 and 5. **6.** Add −6 and 2. **7.** Add −1 and −3.
8. Add −8 and 4. **9.** 2 **11.** −3 **13.** −10 **15.** −13 **17.** −15.9
19. −5; 5 **21.** 13 **23.** 0 **25.** −8 **27.** $\tfrac{3}{10}$ **29.** $\tfrac{1}{2}$ **31.** $-\tfrac{3}{4}$

33. −1.6 **35.** −8.7 **37.** −11; −14; −25 **39.** $-\tfrac{1}{4}$, or −0.25
41. true **43.** false **45.** true **47.** false **49.** true **51.** false
53. $-5 + 12 + 6$; 13 **55.** $[-19 + (-4)] + 14$; −9

57. $[-4 + (-10)] + 12$; −2 **59.** $\left[\tfrac{5}{7} + \left(-\tfrac{9}{7}\right)\right] + \tfrac{2}{7}$; $-\tfrac{2}{7}$ **61.** −$62

63. −184 m **65.** 17 **67.** 37 yd **69.** 120°F **71.** −$107 **73.** −6

SECTION 1.5 (pages 68–72)

1. −8; −6; 2 **2.** 5; −4 **3.** $7 - 12$; $12 - 7$ **4.** additive; inverse;
opposite **5.** −4 **6.** −22 **7.** −3; −10 **9.** −16 **11.** 4; 11

13. 19 **15.** −4 **17.** 5 **19.** 0 **21.** $\tfrac{3}{4}$ **23.** $-\tfrac{11}{8}$ **25.** $\tfrac{15}{8}$ **27.** 13.6

29. −11.9 **31.** 8 **33.** −2.8 **35.** −6.3 **37.** −28 **39.** $\tfrac{37}{12}$

41. −42.04 **43.** $4 - (-8)$; 12 **45.** $-2 - 8$; −10

47. $[9 + (-4)] - 7$; −2 **49.** $[8 - (-5)] - 12$; 1 **51.** −69°F

53. 14,776 ft **55.** −$80 **57.** −176.9°F **59.** $1045.55

61. 469 B.C. **63.** $323.83 **65.** 14 ft **67.** 40,776 ft **69.** 4.4%
70. Americans spent more money than they earned, which means
that they had to dip into savings or borrow money. **71.** $1530 billion
73. $2900 **75.** −$20,900 **77.** 136 ft **79.** negative **80.** positive
81. positive **82.** negative **83.** positive **84.** negative

SECTION 1.6 (pages 81–84)

1. greater than 0 **2.** less than 0 **3.** less than 0 **4.** less than 0
5. greater than 0 **6.** less than 0 **7.** −28 **9.** 30 **11.** 0 **13.** $\tfrac{5}{6}$

15. −2.38 **17.** $\tfrac{3}{2}$ **19.** −3 **21.** −2 **23.** 16 **25.** 0 **27.** undefined

29. $\tfrac{3}{2}$ **31.** C **32.** A **33.** 30; 10; 3 **35.** 7 **37.** 4 **39.** −3 **41.** −1

43. $\tfrac{7}{4}$ **45.** 68 **47.** −228 **49.** 1 **51.** 0 **53.** −6 **55.** undefined

57. $-12 + 4(-7)$; −40 **59.** $-1 - 2(-8)(2)$; 31

61. $-3[3 - (-7)]$; −30 **63.** $\tfrac{3}{10}[-2 + (-28)]$; −9 **65.** $\dfrac{-20}{-8 + (-2)}$; 2

67. $\dfrac{-18 + (-6)}{2(-4)}$; 3 **69.** $\dfrac{-\tfrac{2}{3}\left(-\tfrac{1}{5}\right)}{\tfrac{1}{7}}$; $\tfrac{14}{15}$ **71.** $9x = -36$ **73.** $\tfrac{x}{4} = -1$

75. $x - \tfrac{9}{11} = 5$ **77.** $\tfrac{6}{x} = -3$ **79.** 29 **80.** The incorrect answer, 92,
was obtained by performing all of the operations in order from left to right
rather than following the rules for order of operations. The multiplications
and divisions need to be done in order, before the additions and
subtractions. **81.** 42 **82.** 5 **83.** $8\tfrac{2}{5}$ **84.** $8\tfrac{2}{5}$ **85.** 2 **86.** $-12\tfrac{1}{2}$

SUMMARY EXERCISES Performing Operations with Real Numbers (pages 85–86)

1. −16 **2.** 4 **3.** 0 **4.** −24 **5.** −17 **6.** 76 **7.** −18 **8.** 90
9. 38 **10.** 4 **11.** −5 **12.** 5 **13.** $-\tfrac{7}{2}$, or $-3\tfrac{1}{2}$ **14.** 4 **15.** 13

16. $\tfrac{5}{4}$, or $1\tfrac{1}{4}$ **17.** 9 **18.** $\tfrac{37}{10}$, or $3\tfrac{7}{10}$ **19.** 0 **20.** 25 **21.** 14

22. 0 **23.** −4 **24.** $\tfrac{6}{5}$, or $1\tfrac{1}{5}$ **25.** −1 **26.** $\tfrac{52}{37}$, or $1\tfrac{15}{37}$

27. $\tfrac{17}{16}$, or $1\tfrac{1}{16}$ **28.** $-\tfrac{2}{3}$ **29.** 3.33 **30.** 1.02 **31.** −13 **32.** 0

33. 24 **34.** −7 **35.** 37 **36.** −3 **37.** −1 **38.** $\tfrac{1}{2}$ **39.** $-\tfrac{5}{13}$ **40.** 5

41. undefined **42.** 0

SECTION 1.7 (pages 93–96)

1. (a) B (b) F (c) C (d) I (e) B (f) D, F (g) B (h) A
(i) G (j) H **2.** order; grouping **3.** yes **4.** yes **5.** no **6.** no
7. no **8.** no **9.** (foreign sales) clerk; foreign (sales clerk)
10. (defective merchandise) counter; defective (merchandise counter)
11. −15; commutative property **13.** 3; commutative property
15. 6; associative property **17.** 7; associative property
19. Subtraction is not associative. **20.** Division is not associative.
21. row 1: $-5, \tfrac{1}{5}$; row 2: $10, -\tfrac{1}{10}$; row 3: $\tfrac{1}{2}, -2$; row 4: $-\tfrac{3}{8}, \tfrac{8}{3}$;

row 5: $-x, \tfrac{1}{x}$ ($x \neq 0$); row 6: $y, -\tfrac{1}{y}$ ($y \neq 0$); opposite; the same

22. identity property **23.** commutative property
25. associative property **27.** inverse property

29. inverse property   **31.** identity property

33. commutative property   **35.** distributive property

37. identity property   **39.** distributive property

41. The expression following the first equality symbol should be $-3(4) - 3(-6)$. This simplifies to $-12 + 18$, which equals 6.

42. We must multiply $\frac{3}{4}$ by 1 in the form of a fraction, $\frac{3}{3}$: $\frac{3}{4} \cdot \frac{3}{3} = \frac{9}{12}$.

43. $7 + r$ **45.** s **47.** $-6x + (-6)7$; $-6x - 42$

49. $w + [5 + (-3)]$; $w + 2$ **51.** 6700   **53.** 2   **55.** 0.77   **57.** $4t + 12$

59. $-8r - 24$ **61.** y; -4; $-5y + 20$ **63.** $-16y - 20z$ **65.** $8(z + w)$

67. $5(3 + 17)$; 100   **69.** $7(2v + 5r)$ **71.** $24r + 32s - 40y$

73. $-24x - 9y - 12z$ **75.** $-4t - 5m$ **77.** $5c + 4d$ **79.** $3q - 5r + 8s$

SECTION 1.8 (pages 101–103)

1. B   **2.** C   **3.** A   **4.** B   **5.** $15x$ **7.** $13b$ **9.** $4r + 11$

11. $5 + 2x - 6y$ **13.** $-7 + 3p$ **15.** $-\frac{4}{3}y - 10$ **17.** -12 **19.** 5

21. 1   **23.** -1 **25.** 74   **27.** $-\frac{3}{8}$ **29.** $\frac{1}{2}$ **31.** $\frac{2}{5}$ **33.** -1.28

35. like   **37.** unlike   **39.** like   **41.** unlike   **43.** x; -3; $2x$; 6; $1 - 2x$

45. $-\frac{1}{3}t - \frac{28}{3}$ **47.** $-4.1r + 4.2$ **49.** $-2y^2 + 3y^3$ **51.** $-19p + 16$

53. $-2x + 4$ **55.** $-\frac{14}{3}x - \frac{22}{3}$ **57.** $-\frac{3}{2}y + 16$ **59.** $-16y + 63$

61. $4r + 15$ **63.** $12k - 5$ **65.** $-2k - 3$ **67.** $4x - 7$

69. $-23.7y - 12.6$ **71.** $(x + 3) + 5x$; $6x + 3$

73. $(13 + 6x) - (-7x)$; $13 + 13x$ **75.** $2(3x + 4) - (-4 + 6x)$; 12

77. Wording may vary. One example is "the difference between 9 times a number and the sum of the number and 2."   **79.** The student made a sign error when applying the distributive property. $7x - 2(3 - 2x)$ means $7x - 2(3) - 2(-2x)$, which simplifies to $7x - 6 + 4x$, or $11x - 6$.   **80.** The student incorrectly started by adding $3 + 2$. As the first step, 2 must be multiplied by $4x - 5$. Thus, $3 + 2(4x - 5)$ equals $3 + 8x - 10$, which simplifies to $8x - 7$.   **81.** $1000 + 5x$ (dollars)

82. $750 + 3y$ (dollars)   **83.** $1000 + 5x + 750 + 3y$ (dollars)

84. $1750 + 5x + 3y$ (dollars)

Chapter 1 REVIEW EXERCISES (pages 109–113)

1. 625   **2.** 0.00000081   **3.** 0.009261   **4.** $\frac{125}{8}$ **5.** 27   **6.** 200   **7.** 7

8. 4   **9.** $13 < 17$ **10.** $5 + 2 \neq 10$ **11.** Six is less than fifteen.

12. Answers will vary. One example is $2 + 5 \geq \frac{16}{2}$.   **13.** 30   **14.** 60

15. 14   **16.** 13   **17.** $x + 6$ **18.** $8 - x$ **19.** $6x - 9$ **20.** $12 + \frac{3}{5}x$

21. yes   **22.** no   **23.** $2x - 6 = 10$ **24.** $4x = 8$ **25.** equation

26. expression

27.
$$\begin{array}{c} \overset{-\frac{1}{2}}{} \quad \overset{2.5}{} \\ \xleftarrow{\;\;} +\!+\!\bullet\!+\!\bullet\!+\!\bullet\!+\!\bullet\!+\!\bullet\!+ \xrightarrow{\;\;} \\ -6 \;\; -4 \;\; -2 \;\;\; 0 \;\;\; 2 \;\;\;\; 4 \;\; 5 \;\; 6 \end{array}$$

28.
$$\begin{array}{c} \overset{-3\frac{1}{4}}{}\, \overset{-1\frac{1}{8}}{}\;\; \overset{\frac{5}{6}}{}\;\; \overset{\frac{14}{5}}{} \\ \xleftarrow{\;} +\!+\!+\!+\!\bullet\!+\!\bullet\!+\!+\!\bullet\!+\!\bullet\!+\!+ \xrightarrow{\;} \\ -6 \;\; -4 \;\; -2 \;\;\; 0 \;\;\; 2 \;\;\; 4 \;\;\; 6 \end{array}$$

29. -10 **30.** -9 **31.** $-\frac{3}{4}$ **32.** $-|23|$ **33.** true   **34.** true   **35.** true

36. false   **37.** -3 **38.** -19 **39.** -7 **40.** 9   **41.** -6 **42.** -4

43. -17 **44.** $-\frac{29}{36}$ **45.** -10 **46.** -19 **47.** $(-31 + 12) + 19$; 0

48. $[-4 + (-8)] + 13$; 1   **49.** $26.25   **50.** $-10°F$ **51.** -11

52. -1 **53.** 7   **54.** $-\frac{43}{35}$ **55.** 10.31   **56.** -12 **57.** 2   **58.** 1

59. $-4 - (-6)$; 2   **60.** $[4 + (-8)] - 5$; -9

61. $[18 - (-23)] - 15$; 26   **62.** $19 - (-7 - 12)$; 38   **63.** 38

64. 12,392.69   **65.** -126 thousand   **66.** 267 thousand

67. 681 thousand   **68.** -857 thousand   **69.** 36   **70.** -105

71. $\frac{1}{2}$ **72.** 10.08   **73.** -20 **74.** -10 **75.** -24

76. -35 **77.** 4   **78.** -20 **79.** $-\frac{3}{4}$ **80.** 11.3   **81.** -1

82. undefined   **83.** 1   **84.** 0   **85.** -18 **86.** -18 **87.** 125

88. -423 **89.** $-4(5) - 9$; -29 **90.** $\frac{5}{6}[12 + (-6)]$; 5

91. $\frac{12}{8 + (-4)}$; 3   **92.** $\frac{-20(12)}{15 - (-15)}$; -8 **93.** $\frac{x}{x + 5} = -2$

94. $8x - 3 = -7$ **95.** identity property   **96.** identity property

97. inverse property   **98.** inverse property   **99.** distributive property

100. associative property   **101.** associative property   **102.** commutative property   **103.** $(7 + 1)y$; $8y$ **104.** $-12 \cdot 4 - 12(-t)$; $-48 + 12t$

105. $3(2s + 4y)$; $6s + 12y$ **106.** $-1(-4r) + (-1)(5s)$; $4r - 5s$

107. $17p^2$ **108.** $16r^2 + 7r$ **109.** $-19k + 54$ **110.** $5s - 6$

111. $-45t - 23$ **112.** $-45t^2 - 23.4t$ **113.** -6 **114.** $\frac{25}{36}$ **115.** -26

116. $\frac{8}{3}$ **117.** $-\frac{1}{24}$ **118.** $\frac{7}{2}$ **119.** 2   **120.** 77.6   **121.** $-1\frac{1}{2}$

122. 11   **123.** $-\frac{28}{15}$ **124.** 24   **125.** -11 **126.** 16   **127.** $-47°F$

128. 27 ft

Chapter 1 TEST (pages 114–115)

1. true   **2.** false   **3.**
$$\xleftarrow{\;} +\!+\!\bullet\!+\!\bullet\!+\!\bullet\!+\!+\!+\!\bullet\!+ \xrightarrow{\;}$$
$$-3 \;\; -1 \; 0 \; 1 \qquad 4$$
4. $-|-8|$ (or -8)

5. -1.277 **6.** $\frac{-6}{2 + (-8)}$; 1   **7.** negative   **8.** 4   **9.** $-2\frac{5}{6}$ **10.** 6

11. 2   **12.** 108   **13.** 11   **14.** $\frac{30}{7}$ **15.** -70 **16.** 3   **17.** $178°F$

18. 15   **19.** $-$1.42 trillion   **20.** D   **21.** A   **22.** E   **23.** B   **24.** C

25. $21x$ **26.** $-9x^2 - 6x - 8$ **27.** identity and distributive properties

28. **(a)** -18 **(b)** -18 **(c)** The distributive property tells us that the two methods produce equal results.

CHAPTER 2 Equations, Inequalities, and Applications

SECTION 2.1 (pages 123–124)

1. equality; expression   **2.** linear; $=$ **3.** equivalent   **4.** added to; subtracted from   **5.** C   **6.** A, B   **7.** **(a)** expression; $x + 15$

(b) expression; $m + 7$ **(c)** equation; $\{-1\}$ **(d)** equation; $\{-17\}$

8. Replace the variable(s) in the original equation with the proposed solution. A true statement will result if the proposed solution is correct.

9. $\{12\}$ **11.** $\{-3\}$ **13.** $\{4\}$ **15.** $\{-9\}$ **17.** $\left\{-\frac{3}{4}\right\}$ **19.** $\{6.3\}$

21. $\{-16.9\}$ **23.** $\{-10\}$ **25.** $\{-13\}$ **27.** $\left\{\frac{4}{15}\right\}$ **29.** $\{7\}$

31. $\{-3\}$ **33.** $\{-4\}$ **35.** $\{3\}$ **37.** $\{-2\}$ **39.** $\{4\}$ **41.** $\{-16\}$
43. $\{2\}$ **45.** $\{2\}$ **47.** $\{-4\}$ **49.** $\{4\}$ **51.** $\{0\}$ **53.** $\left\{\dfrac{7}{15}\right\}$
55. $\{7\}$ **57.** $\{-4\}$ **59.** $\{13\}$ **61.** $\{29\}$ **63.** $\{18\}$ **65.** Answers will vary. One example is $x - 6 = -8$. **66.** Answers will vary. One example is $x + \dfrac{1}{2} = 1$.

SECTION 2.2 (pages 130–131)

1. (a) and **(c)**: multiplication property of equality; **(b)** and **(d)**: addition property of equality **2.** C **3.** $\dfrac{3}{2}$ **4.** $\dfrac{5}{4}$ **5.** 10 **6.** 100 **7.** $-\dfrac{2}{9}$
8. $-\dfrac{3}{8}$ **9.** -1 **10.** -1 **11.** 6 **12.** 7 **13.** -4 **14.** -13
15. 0.12 **16.** 0.21 **17.** -1 **18.** -1 **19.** B **20.** A **21.** $\{6\}$
23. $\left\{\dfrac{15}{2}\right\}$ **25.** $\{-5\}$ **27.** $\left\{-\dfrac{18}{5}\right\}$ **29.** $\{12\}$ **31.** 2; 2; 0; $\{0\}$
33. $\{-12\}$ **35.** $\{40\}$ **37.** $\{-12.2\}$ **39.** $\{-48\}$ **41.** $\{72\}$
43. $\{-35\}$ **45.** $\{14\}$ **47.** $\left\{-\dfrac{27}{35}\right\}$ **49.** $\{3\}$ **51.** $\{-5\}$ **53.** $\{20\}$
55. $\{7\}$ **57.** 1; $\{-12\}$ **59.** $\{0\}$ **61.** $\{-6\}$ **63.** Answers will vary. One example is $\dfrac{3}{2}x = -6$. **64.** Answers will vary. One example is $100x = 17$. **65.** $-4x = 10$; $-\dfrac{5}{2}$ **67.** $\dfrac{x}{-5} = 2$; -10

SECTION 2.3 (pages 141–144)

1. addition; subtract **2.** left; like **3.** distributive; parentheses
4. multiplication; $\dfrac{4}{3}$ **5.** fractions; 6 **6.** decimals; 10 **7. (a)** identity; B
(b) conditional; A **(c)** contradiction; C **8.** D **9.** A **10.** D **11.** $\{4\}$
13. $\{-5\}$ **15.** $\left\{\dfrac{5}{2}\right\}$ **17.** $\left\{-\dfrac{1}{2}\right\}$ **19.** $\{5\}$ **21.** $\{1\}$ **23.** $\left\{-\dfrac{5}{3}\right\}$
25. $\{2\}$ **27.** $\left\{-\dfrac{5}{3}\right\}$ **29.** $\varnothing$ **31.** $\{0\}$ **33.** $\{-1\}$ **35.** $\varnothing$
37. {all real numbers} **39.** $\{0\}$ **41.** 14; $\{7\}$ **43.** $\{12\}$ **45.** $\{11\}$
47. $\{0\}$ **49.** $\left\{\dfrac{3}{25}\right\}$ **51.** 100; $\{60\}$ **53.** $\{4\}$ **55.** $\{5000\}$
57. $\left\{-\dfrac{72}{11}\right\}$ **59.** $\{0\}$ **61.** $\varnothing$ **63.** {all real numbers} **65.** $\{-6\}$
67. $\{15\}$ **69.** $12 - q$ **71.** $\dfrac{9}{z}$ **73.** $x + 29$ **75.** $a + 12$; $a - 2$
77. $25r$ **79.** $\dfrac{t}{5}$ **81.** $3x + 2y$

SUMMARY EXERCISES Applying Methods for Solving Linear Equations (pages 146–147)

1. $\{-5\}$ **2.** $\{4\}$ **3.** $\{-5.1\}$ **4.** $\{25\}$ **5.** $\{-25\}$ **6.** $\{-6\}$
7. $\{-3\}$ **8.** $\{-16\}$ **9.** $\{0\}$ **10.** $\left\{-\dfrac{96}{5}\right\}$ **11.** {all real numbers}
12. $\{23.7\}$ **13.** $\{7\}$ **14.** $\{0\}$ **15.** $\{5\}$ **16.** $\{1\}$ **17.** $\{-16\}$
18. $\varnothing$ **19.** $\{6\}$ **20.** $\{3\}$ **21.** $\varnothing$ **22.** $\left\{\dfrac{7}{3}\right\}$ **23.** $\{25\}$
24. $\{-10.8\}$ **25.** $\{3\}$ **26.** $\{7\}$ **27.** $\{2\}$ **28.** {all real numbers}
29. $\left\{-\dfrac{2}{7}\right\}$ **30.** $\{10\}$ **31.** $\left\{\dfrac{14}{17}\right\}$ **32.** $\left\{-\dfrac{5}{2}\right\}$
33. {all real numbers} **34.** $\{64\}$

SECTION 2.4 (pages 156–161)

1. *Step 1:* Read the problem carefully; *Step 2:* Assign a variable to represent the unknown; *Step 3:* Write an equation; *Step 4:* Solve the equation; *Step 5:* State the answer; *Step 6:* Check the answer.
2. Some examples are *is, are, was,* and *were*. **3.** D; There cannot be a fractional number of cars. **4.** D; A day cannot have more than 24 hr.
5. A; Distance cannot be negative. **6.** C; Time cannot be negative.
7. 1; 16 (or 14); -7 (or -9) **8.** odd; 2; 11 (or 15); even; 2; 10 (or 14)
9. complementary; supplementary **10.** $90°$; $180°$ **11.** yes, $90°$; yes, $45°$
12. $x - 1$; $x - 2$ **13.** $8 \cdot (x + 6) = 104$; 7 **15.** $5x + 2 = 4x + 5$; 3
17. $3x - 2 = 5x + 14$; -8 **19.** $3(x - 2) = x + 6$; 6
21. $3x + (x + 7) = -11 - 2x$; -3 **23.** *Step 1:* We are asked to find the number of drive-in movie screens in the two states; *Step 2:* the number of screens in Pennsylvania; *Step 3:* x; $x + 3$; *Step 4:* 28; *Step 5:* 28; 28; 31; *Step 6:* 3; screens in New York; 31; 59 **25.** Democrats: 193; Republicans: 242
27. Bon Jovi: \$108.2 million; Roger Waters: \$89.5 million **29.** wins: 57; losses: 25 **31.** orange: 97 mg; pineapple: 25 mg **33.** 420 lb
35. active: 225 mg; inert: 25 mg **37.** 1950 Denver nickel: \$16.00; 1944 Philadelphia nickel: \$12.00 **39.** onions: 81.3 kg; grilled steak: 536.3 kg
41. American: 18; United: 11; Southwest: 26 **43.** $x + 5$; $x + 9$; shortest piece: 15 in.; middle piece: 20 in.; longest piece: 24 in. **45.** gold: 46; silver: 29; bronze: 29 **47.** 36 million mi **49.** A and B: $40°$; C: $100°$
51. 68, 69 **53.** 146, 147 **55.** 10, 12 **57.** 17, 19 **59.** 10, 11 **61.** 18
63. $18°$ **65.** $20°$ **67.** $39°$ **69.** $50°$

SECTION 2.5 (pages 168–173)

1. The perimeter of a plane geometric figure is the distance around the figure.
2. The area of a plane geometric figure is the measure of the surface covered or enclosed by the figure. **3.** area **4.** area **5.** perimeter
6. perimeter **7.** area **8.** area **9.** area **10.** area **11.** $P = 26$
13. $A = 64$ **15.** $b = 4$ **17.** $t = 5.6$ **19.** $h = 7$ **21.** $r = 2.6$
23. $A = 50.24$ **25.** $V = 150$ **27.** $V = 52$ **29.** $V = 7234.56$
31. $I = \$600$ **33.** $p = \$550$ **35.** 0.025; $t = 1.5$ yr **37.** length: 18 in.; width: 9 in. **39.** length: 14 m; width: 4 m **41.** shortest: 5 in.; medium: 7 in.; longest: 8 in. **43.** two equal sides: 7 m; third side: 10 m
45. perimeter: 5.4 m; area: 1.8 m^2 **47.** 10 ft **49.** about 154,000 ft^2
51. 194.48 ft^2; 49.42 ft **53.** 23,800.10 ft^2 **55.** length: 36 in.; maximum volume: 11,664 in.^{3} **57.** $48°$, $132°$ **59.** $55°$, $35°$ **61.** $51°$, $51°$
63. $105°$, $105°$ **65.** $t = \dfrac{d}{r}$ **67.** $H = \dfrac{V}{LW}$ **69.** $b = P - a - c$
71. $r = \dfrac{C}{2\pi}$ **73.** $r = \dfrac{I}{pt}$ **75.** $h = \dfrac{2A}{b}$ **77.** $h = \dfrac{3V}{\pi r^2}$ **79.** $W = \dfrac{P - 2L}{2}$
81. $m = \dfrac{y - b}{x}$ **83.** $y = \dfrac{C - Ax}{B}$ **85.** $r = \dfrac{M - C}{C}$ **87.** $a = \dfrac{P - 2b}{2}$
We give one possible answer for Exercises 89–95. There are other correct forms.
89. $y = -6x + 4$ **91.** $y = 5x - 2$ **93.** $y = \dfrac{3}{5}x - 3$ **95.** $y = \dfrac{1}{3}x - 4$

SECTION 2.6 (pages 180–185)

1. compare; A, D **2.** ratios; proportion; cross products **3. (a)** C
(b) D **(c)** B **(d)** A **4.** C, E **5.** $\dfrac{6}{7}$ **7.** $\dfrac{18}{55}$ **9.** $\dfrac{5}{16}$ **11.** $\dfrac{4}{15}$ **13.** $\dfrac{6}{5}$
15. 10 lb; \$0.749 **17.** 64 oz; \$0.047 **19.** 32 oz; \$0.531 **21.** 32 oz; \$0.056
23. $\{35\}$ **25.** $\{7\}$ **27.** $\left\{\dfrac{8}{5}\right\}$ **29.** $\{2\}$ **31.** $\{-1\}$ **33.** $\{5\}$

35. $\left\{-\dfrac{31}{5}\right\}$ **37.** \$30.00   **39.** \$56.85   **41.** 50,000 fish   **43.** 4 ft

45. $\dfrac{87}{100}$; 209.67 million drivers   **47.** $2\dfrac{5}{8}$ cups   **49.** \$391.53   **51.** 9; $x = 4$

53. $x = 8$ **55.** $x = 3$; $y = 5.5$ **57.** side of triangle labeled Chair:

18 ft; side of triangle labeled Pole: 12 ft; One proportion is $\dfrac{x}{12} = \dfrac{18}{4}$; 54 ft

59. (a) 2625 mg   **(b)** $\dfrac{125\,\text{mg}}{5\,\text{mL}} = \dfrac{2625\,\text{mg}}{x\,\text{mL}}$ **(c)** 105 mL   **61.** \$238

63. \$278   **65.** C   **66.** A   **67.** 109.2   **69.** 700   **71.** 425   **73.** 8%

75. 120%   **77.** \$119.25; \$675.75   **79.** 80%   **81.** \$3000   **83.** \$304

85. 14,825,000; · ; 153,889,000; 9.6%   **87.** 79.3%   **89.** 892%   **91.** 30

92. (a) $5x = 12$ **(b)** $\left\{\dfrac{12}{5}\right\}$ **93.** $\left\{\dfrac{12}{5}\right\}$ **94.** Both methods give the

same solution set.

SUMMARY EXERCISES   Applying Problem-Solving Techniques (pages 186–187)

1. 48   **2.** 80°   **3.** 4   **4.** 104°, 104°   **5.** 3   **6.** 18, 20   **7.** 140°, 40°

8. 36 quart cartons   **9.** 24.34 in.; 727.28 in.²   **10.** 6.2%

11. Barcelona: \$7.9 million; Real Madrid: \$7.4 million; New York

Yankees: \$6.8 million   **12.** 12.42 cm   **13.** $16\dfrac{2}{3}\%$ **14.** 510 calories

15. 4000 calories   **16.** Charl Schwartzel: 274; Adam Scott: 276

17. 24 oz; \$0.074   **18.** 14 gold medals   **19.** \$49.21   **20.** 71.2%

SECTION 2.7 (pages 199–202)

1. < (or >); > (or <); ≤ (or ≥); ≥ (or ≤)   **2.** false   **3.** $(0, \infty)$

4. $(-\infty, \infty)$ **5.** $x > -4$ **6.** $x \geq -4$ **7.** $x \leq 4$ **8.** $x < 4$

9. $-1 < x \leq 2$ **10.** $-1 \leq x < 2$

11. $(-\infty, 4]$

13. $(-3, \infty)$

15. $[8, 10]$

17. $(0, 10]$

19. $[1, \infty)$

21. $[5, \infty)$

23. $(-\infty, -6)$

25. It must be reversed when multiplying or dividing by a negative number.

26. Divide by -5 and reverse the direction of the inequality symbol to get $x < -4$.

27. $(-\infty, 6)$

29. $[-10, \infty)$

31. $(-\infty, -3)$

33. $(-\infty, 0]$

35. $(20, \infty)$

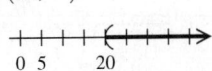

37. $[-3, \infty)$

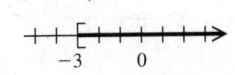

39. $(-\infty, -3]$

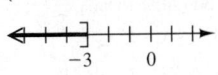

41. $(-1, \infty)$

43. $[-5, \infty)$

45. $(-\infty, 1)$

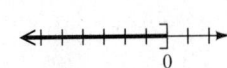 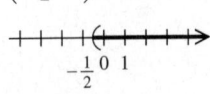

47. $(-\infty, 0]$

49. $\left(-\dfrac{1}{2}, \infty\right)$

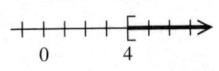 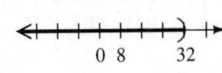

51. $[4, \infty)$

53. $(-\infty, 32)$

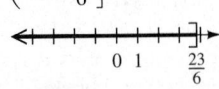 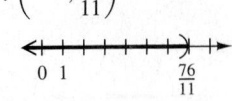

55. $\left(-\infty, \dfrac{23}{6}\right]$

57. $\left(-\infty, \dfrac{76}{11}\right)$

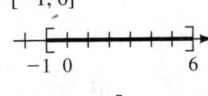 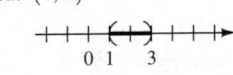

59. $[-1, 6]$

61. $(1, 3)$

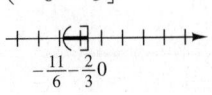 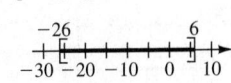

63. $\left(-\dfrac{11}{6}, -\dfrac{2}{3}\right]$

65. $[-26, 6]$

67. $[-3, 6]$

69. $x \geq 16$ **70.** $x < 1$

71. $x > 8$ **72.** $x \geq 12$

73. $x \leq 20$ **74.** $x > 40$

75. 88 or more   **77.** 80 or more   **79.** all numbers greater than 16

81. It has never exceeded 40°C.   **83.** 32 or greater   **85.** 12 min

87. $\{4\}$

88. (a) $(-\infty, 4)$

(b) $(4, \infty)$

89. It is the set of all real numbers.

90. The graph would be the set of all real numbers; $(-\infty, \infty)$

Chapter 2 REVIEW EXERCISES (pages 207–210)

1. $\{9\}$ **2.** $\{4\}$ **3.** $\{-6\}$ **4.** $\left\{\dfrac{3}{2}\right\}$ **5.** $\{20\}$ **6.** $\left\{-\dfrac{61}{2}\right\}$ **7.** $\{15\}$

8. $\{0\}$ **9.** $\emptyset$ **10.** $\{$all real numbers$\}$ **11.** $-\dfrac{7}{2}$ **12.** 20

13. Hawaii: 6425 mi²; Rhode Island: 1212 mi²   **14.** Seven Falls: 300 ft;

Twin Falls: 120 ft   **15.** 80°   **16.** 11, 13   **17.** $h = 11$ **18.** $A = 28$

19. $r = 4.75$ **20.** $V = 3052.08$ **21.** $h = \dfrac{A}{b}$ **22.** $h = \dfrac{2A}{b + B}$

23. (Other correct forms of the answers are possible.)   **(a)** $y = -x + 11$

(b) $y = \dfrac{3}{2}x - 6$ **24.** 135°, 45°   **25.** 100°, 100°   **26.** perimeter: 326.5 ft;

area: 6538.875 ft²   **27.** diameter: 46.78 ft; radius: 23.39 ft   **28.** $\dfrac{3}{2}$ **29.** $\dfrac{5}{14}$

30. $\frac{3}{4}$ **31.** $\left\{\frac{7}{2}\right\}$ **32.** $\left\{-\frac{8}{3}\right\}$ **33.** $\left\{\frac{25}{19}\right\}$ **34.** 40% means $\frac{40}{100}$, or $\frac{2}{5}$.

It is the same as the ratio 2 to 5. **35.** $6\frac{2}{3}$ lb **36.** 36 oz **37.** 375 km

38. 18 oz; $0.249 **39.** 6 **40.** 175% **41.** $33\frac{1}{3}$% **42.** 2500

43. $27,160 **44.** $350.46

45. $[-4, \infty)$

46. $(-\infty, 7)$

47. $[-5, 6)$

48. $\left[\frac{1}{2}, \infty\right)$

49. $[-3, \infty)$

50. $(-\infty, 2)$

51. $[3, \infty)$

52. $[46, \infty)$

53. $(-\infty, -5)$

54. $(-\infty, -37)$

55. $\left[-2, \frac{3}{2}\right)$

56. $(1, 5]$

57. 88 or more **58.** all numbers less than or equal to $-\frac{1}{3}$ **59.** {7}

60. $r = \frac{d}{2}$ **61.** $(-\infty, 2)$ **62.** {−9} **63.** {70} **64.** $\left\{\frac{13}{4}\right\}$ **65.** ∅

66. {all real numbers} **67.** DiGiorno: $668.7 million; Red Baron: $268.8 million **68.** 2.0 in. **69.** gold: 14; silver: 7; bronze: 5 **70.** *D*: 22°; *E*: 44°; *F*: 114° **71.** 44 m **72.** 70 ft **73.** 160 oz; $0.062 **74.** 24°, 66° **75.** 92 or more **76.** $197.50

Chapter 2 TEST (pages 211–212)

1. {6} **2.** {−6} **3.** $\left\{\frac{13}{4}\right\}$ **4.** {−10.8} **5.** {21}

6. {all real numbers} **7.** {30} **8.** ∅ **9.** wins: 62; losses: 20
10. Hawaii: 4021 mi²; Maui: 728 mi²; Kauai: 551 mi² **11.** 24, 26 **12.** 50°
13. (a) $W = \frac{P - 2L}{2}$ (b) 18 **14.** $y = \frac{5}{4}x - 2$ (Other correct forms of the answer are possible.) **15.** 100°, 80° **16.** 75°, 75° **17.** {6} **18.** {−29}
19. 16 oz; $0.249 **20.** 2300 mi **21.** $26.82; $122.18 **22.** 40%
23. (a) $x < 0$ (b) $-2 < x \le 3$

24. $(-\infty, 11)$

25. $[-3, \infty)$

26. $(-\infty, 4]$

27. $(-2, 6]$

28. 83 or more

---

1. $\frac{3}{8}$ **2.** $\frac{3}{4}$ **3.** $\frac{31}{20}$ **4.** $\frac{551}{40}$, or $13\frac{31}{40}$ **5.** 6 **6.** $\frac{6}{5}$ **7.** 34.03

8. 27.31 **9.** 30.51 **10.** 56.3 **11.** 35 yd **12.** $7\frac{1}{2}$ cups **13.** $3\frac{3}{8}$ in.

14. $2599.94 **15.** true **16.** true **17.** 7 **18.** 1 **19.** 13 **20.** −40

21. −12 **22.** undefined **23.** −6 **24.** 28 **25.** 1 **26.** 0 **27.** $\frac{73}{18}$

28. −64 **29.** −134 **30.** $-\frac{29}{6}$ **31.** distributive property

32. commutative property **33.** inverse property **34.** identity property

35. $7p - 14$ **36.** $2k - 11$ **37.** {7} **38.** {−4} **39.** {−1}

40. $\left\{-\frac{3}{5}\right\}$ **41.** {2} **42.** {−13} **43.** {26} **44.** {−12}

45. $c = P - a - b - B$ **46.** $s = \frac{P}{4}$

47. $(-\infty, 2]$

48. $(-\infty, 1)$

49. $230.50 **50.** $7125 **51.** 30 cm **52.** 16 in.

CHAPTER 3 Graphs of Linear Equations and Inequalities in Two Variables; Functions

SECTION 3.1 (pages 224–229)

1. does; do not **2.** II **3.** *y* **4.** 3 **5.** 6 **6.** −4 **7.** No. For two ordered pairs to be equal, their *x*-values must be equal and their *y*-values must be equal. Here we have $4 \ne -1$ and $-1 \ne 4$. **8.** Substituting $\frac{1}{3}$ for x in $y = 6x + 2$ gives $y = 6\left(\frac{1}{3}\right) + 2$. Here $6\left(\frac{1}{3}\right) = 2$, so $y = 2 + 2$, or 4. Because $6\left(\frac{1}{7}\right) = \frac{6}{7}$, calculating y for $x = \frac{1}{7}$ requires working with fractions.
9. negative; negative **10.** negative; positive **11.** positive; negative
12. positive; positive **13.** 2007, 2008, 2009, 2010 **15.** 2004: about 171 billion lb; 2010: about 193 billion lb **16.** U.S. milk production increased about 22 billion lb during these years, from about 171 billion lb to about 193 billion lb. **17.** 2009 to 2010; about 1.5 million
19. The number of cars imported each year was decreasing. **21.** yes
23. yes **25.** no **27.** yes **29.** no **31.** no **33.** 11 **35.** $-\frac{7}{2}$ **37.** −4
39. −5 **41.** 4, 6, −6; (0, 4), (6, 0), (−6, 8) **43.** 3, −5, −15; (0, 3), (−5, 0), (−15, −6) **45.** −9, −9, −9; (−9, 6), (−9, 2), (−9, −3)
47. −6, −6, −6; (8, −6), (4, −6), (−2, −6) **49.** 8, 8, 8; (8, 8), (8, 3), (8, 0) **51.** −2, −2, −2; (9, −2), (2, −2), (0, −2) **53.** (2, 4); I
55. (−5, 4); II **57.** (3, 0); no quadrant **59.** If $xy < 0$, then either $x < 0$ and $y > 0$ or $x > 0$ and $y < 0$. If $x < 0$ and $y > 0$, then the point lies in quadrant II. If $x > 0$ and $y < 0$, then the point lies in quadrant IV.
60. If $xy > 0$, then either $x > 0$ and $y > 0$ or $x < 0$ and $y < 0$. If $x > 0$ and $y > 0$, then the point lies in quadrant I. If $x < 0$ and $y < 0$, then the point lies in quadrant III.
61–72.

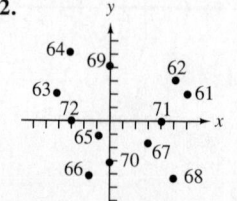

73. −3, 6, −2, 4

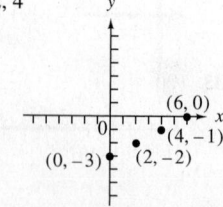

75. $-3, 4, -6, -\dfrac{4}{3}$

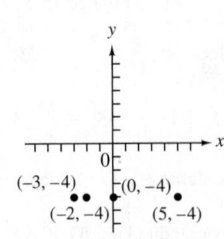

77. $-4, -4, -4, -4$

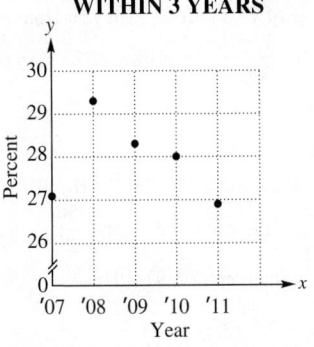

79. The points in each graph appear to lie on a straight line.

80. (a) They are the same (all -4).   **(b)** They are the same (all 5).

81. (a) $(5, 45)$ **(b)** $(6, 50)$ **83. (a)** $(2007, 27.1)$, $(2008, 29.3)$, $(2009, 28.3)$, $(2010, 28.0)$, $(2011, 26.9)$ **(b)** $(2000, 32.4)$ means that 32.4 percent of 2-year college students in 2000 received a degree within 3 years.

(c) 2-YEAR COLLEGE STUDENTS COMPLETING A DEGREE WITHIN 3 YEARS

(d) With the exception of the point for 2007, the points lie approximately on a straight line. Rates at which 2-year college students complete a degree within 3 years were generally decreasing.

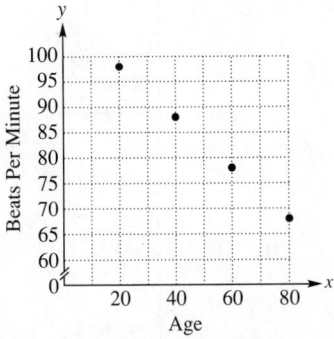

85. (a) 98, 88, 78, 68 **(b)** $(20, 98)$, $(40, 88)$, $(60, 78)$, $(80, 68)$

(c) TARGET HEART RATE ZONE (Lower Limit)

87. between 98 and 157 beats per minute; between 88 and 141 beats per minute

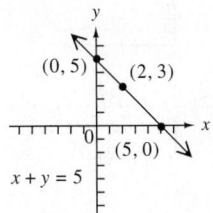

Yes the points lie in a linear pattern.

SECTION 3.2 (pages 238–243)

1. By; C; 0   **2.** line; solution   **3. (a)** A **(b)** C **(c)** D **(d)** B

4. A, C, D   **5.** x-intercept: $(4, 0)$; y-intercept: $(0, -4)$

6. x-intercept: $(-5, 0)$; y-intercept: $(0, 5)$ **7.** x-intercept: $(-2, 0)$; y-intercept: $(0, -3)$ **8.** x-intercept: $(4, 0)$; y-intercept: $(0, 3)$

9. 5, 5, 3

11. 1, 3, -1

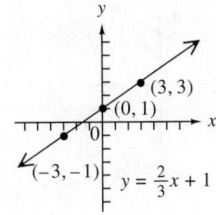

13. $-6, -2, -5$

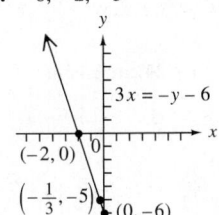

15. 0; $(0, -8)$; 0; $(8, 0)$

17. $(0, -8)$; $(12, 0)$

19. $(0, 0)$; $(0, 0)$

21.

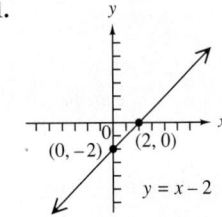

23.

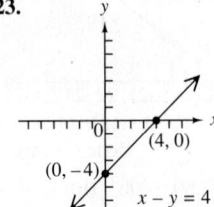

25.

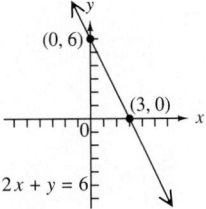

27.

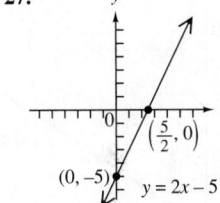

29.

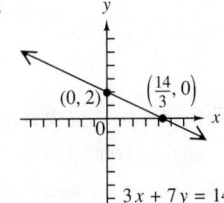

31.

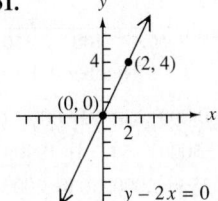

33.

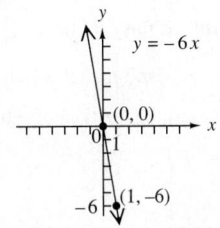

35.

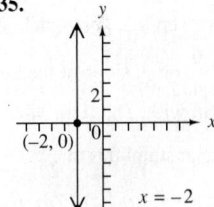

37.

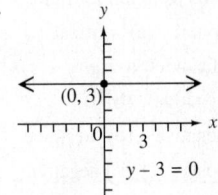

39.

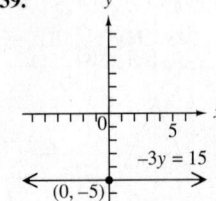

41. (a) D **(b)** C **(c)** B **(d)** A   **42.** $y = 0$; $x = 0$

In Exercises 43–46, descriptions may vary.

43. The graph is a line with x-intercept $(-3, 0)$ and y-intercept $(0, 9)$.

44. The graph is a vertical line with x-intercept $(11, 0)$.   **45.** The graph is a horizontal line with y-intercept $(0, -2)$.   **46.** The graph passes through the origin $(0, 0)$ and the points $(2, 1)$ and $(4, 2)$.

47. (a) 151.5 cm, 159.3 cm, 174.9 cm
(b) (20, 151.5), (22, 159.3), (26, 174.9)

(c)

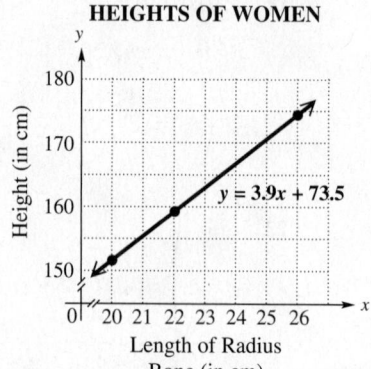

HEIGHTS OF WOMEN

(d) 24 cm; 24 cm

49. (a) \$62.50; \$100 **(b)** 200 **(c)** (50, 62.50), (100, 100), (200, 175)

(d)

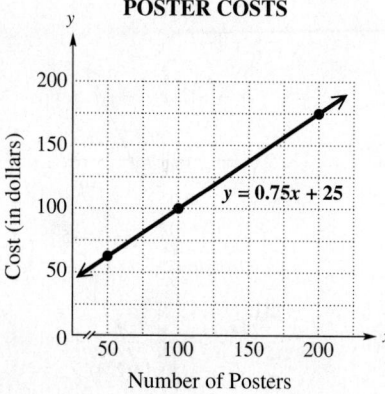

POSTER COSTS

51. (a) \$30,000 **(b)** \$15,000 **(c)** \$5000 **(d)** After 5 yr, the SUV has a value of \$5000. **53. (a)** 1990: 23.8 lb; 2000: 29.0 lb; 2009: 33.8 lb
(b) 1990: 25 lb; 2000: 30 lb; 2009: 33 lb **(c)** The values are quite close.

SECTION 3.3 (pages 254–259)

1. steepness; vertical; horizontal **2.** ratio; y; rise; x; run **3. (a)** 6
(b) 4 **(c)** $\frac{6}{4}$, or $\frac{3}{2}$; slope of the line **(d)** Yes, it doesn't matter which point we start with. The slope would be expressed as the quotient of -6 and -4, which simplifies to $\frac{3}{2}$.

4. (a) C **(b)** A **(c)** D **(d)** B **5.** 4 **7.** $-\frac{1}{2}$

9. Answers will vary.

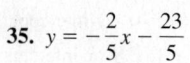

10. (a) falls from left to right
(b) horizontal **(c)** vertical
(d) rises from left to right
11. (a) negative **(b)** 0
12. (a) negative **(b)** negative
13. (a) positive **(b)** negative
14. (a) positive **(b)** 0 **15. (a)** 0
(b) negative **16. (a)** 0
(b) positive **17.** $\frac{8}{27}$ **18.** $\frac{3}{10}$

19. $-\frac{2}{3}$ **20.** $-\frac{1}{4}$ **21.** Because the student found the difference $3 - 5 = -2$ in the numerator, he or she should have subtracted in the same order in the denominator to get $-1 - 2 = -3$. The correct slope is $\frac{-2}{-3} = \frac{2}{3}$.

22. Slope is defined as $\frac{\text{change in } y}{\text{change in } x}$, but the student found $\frac{\text{change in } x}{\text{change in } y}$. The correct slope is $\frac{-5}{8} = -\frac{5}{8}$. **23.** $\frac{5}{4}$ **25.** $\frac{3}{2}$ **27.** -3 **29.** 0

31. undefined **33.** $\frac{1}{4}$ **35.** $-\frac{1}{2}$ **37.** 5 **39.** $\frac{1}{4}$ **41.** $\frac{3}{2}$ **43.** $-\frac{3}{2}$

45. 0 **47.** undefined **49.** 1 **51.** $\frac{4}{3}$; $\frac{4}{3}$; parallel **53.** $\frac{5}{3}$; $\frac{3}{5}$; neither

55. $\frac{3}{5}$; $-\frac{5}{3}$; perpendicular **57.** 0.42 **59.** 0.42% **61.** negative; decreased

63. $\frac{7}{10}$ **65.** $-\$4000$ per yr; The value of the machine is decreasing \$4000 each year during these years. **66.** \$50 per month; The amount saved is increasing \$50 each month during these months. **67.** 0% per yr (or no change); The percent of pay raise is not changing—it is 3% each year during these years. **68.** positive; negative **69. (a)** -6 theaters per yr **(b)** The negative slope means that the number of drive-in theaters decreased by an average of 6 each year from 2005 to 2011. **71. (a)** 17.5 **(b)** The number of subscribers increased by an average of 17.5 million each year from 2006 to 2010. **73.** $\frac{1}{3}$ **74.** $\frac{1}{3}$ **75.** $\frac{1}{3}$ **76.** $\frac{1}{3} = \frac{1}{3} = \frac{1}{3}$ is true. **77.** They are collinear. **78.** They are not collinear.

SECTION 3.4 (pages 270–275)

1. (a) D **(b)** C **(c)** B **(d)** A **2.** A, B, D **3. (a)** C **(b)** B
(c) A **(d)** D **4. (a)** the y-axis **(b)** the x-axis **5.** slope: $\frac{5}{2}$;
y-intercept: $(0, -4)$ **7.** slope: -1; y-intercept: $(0, 9)$ **9.** slope: $\frac{1}{5}$;
y-intercept: $\left(0, -\frac{3}{10}\right)$ **11.** $y = 4x - 3$ **13.** $y = -x - 7$ **15.** $y = 3$

17. $x = 0$ **19.** $y = 3x - 3$ **20.** $y = 2x - 4$ **21.** $y = -x + 3$

22. $y = -x - 2$ **23.** $y = -\frac{1}{2}x + 2$ **24.** $y = \frac{3}{2}x - 3$

25.

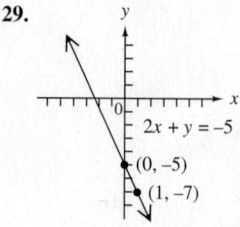

27.

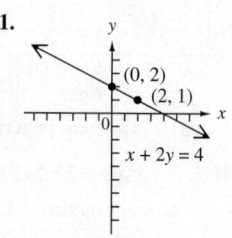

29.

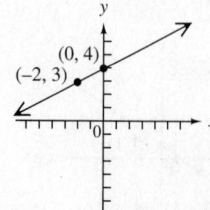

31.

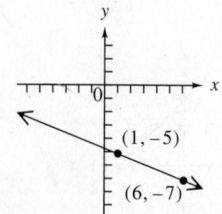

33. $y = \frac{1}{2}x + 4$

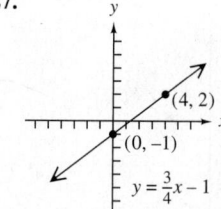

35. $y = -\frac{2}{5}x - \frac{23}{5}$

37. $y = 2$

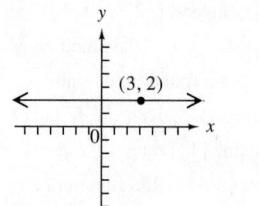

39. $x = 3$ (no slope-intercept form)

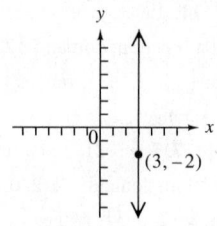

41. $y = \dfrac{2}{3}x$

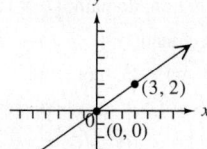

43. $y = 2x - 7$ **45.** $y = -2x - 4$

47. $y = \dfrac{2}{3}x + \dfrac{19}{3}$ **49. (a)** $y = x - 3$

(b) $x - y = 3$ **51. (a)** $y = -\dfrac{5}{7}x - \dfrac{54}{7}$

(b) $5x + 7y = -54$ **53. (a)** $y = -\dfrac{2}{3}x - 2$

(b) $2x + 3y = -6$ **55. (a)** $y = \dfrac{1}{3}x + \dfrac{4}{3}$

(b) $x - 3y = -4$ **57.** $y = 5$ **59.** $x = 9$

61. $y = -\dfrac{3}{2}$ **63.** $y = 8$ **65.** $x = 0.5$ **67.** $y = -2x - 3$

69. $y = 4x - 5$ **71.** $y = \dfrac{3}{4}x - \dfrac{9}{2}$ **73.** $y = 45x$; $(0, 0), (5, 225), (10, 450)$

75. $y = 5.00x$; $(0, 0), (5, 25.00), (10, 50.00)$ **77. (a)** $y = 41x + 99$

(b) $(5, 304)$; The cost of a 5-month membership is \$304.   **(c)** \$591

79. (a) $y = 60x + 36$ **(b)** $(5, 336)$; The cost of the plan for 5 months

is \$336.   **(c)** \$1476   **81. (a)** \$400   **(b)** \$0.25   **(c)** $y = 0.25x + 400$

(d) \$425   **(e)** 1500   **83. (a)** $y = -729x + 13{,}855$; The number of U.S.

travelers to Canada decreased by 729 thousand per yr from 2006 to 2009.

(b) 10,939 thousand

85. (a) $(1, 2294), (2, 2372), (3, 2558), (4, 2727), (5, 2963)$

(b) yes

**AVERAGE ANNUAL COSTS AT
2-YEAR COLLEGES**

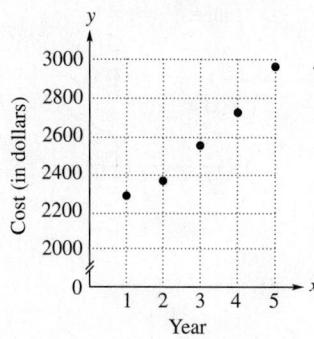

(c) $y = 197x + 1978$ **(d)** \$3160

**SUMMARY EXERCISES   Applying Graphing and
Equation-Writing Techniques for Lines (pages 276–277)**

1. (a) B   **(b)** D   **(c)** A   **(d)** C   **(e)** F   **(f)** E   **2.** A, B

3.

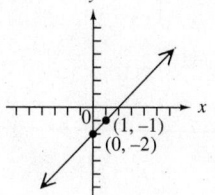

4.

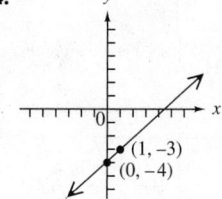

5.

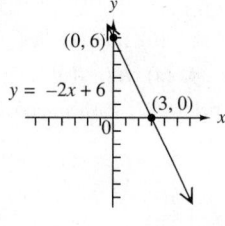

6.

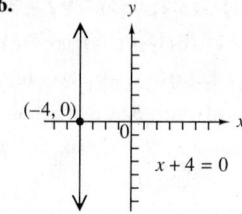

7.

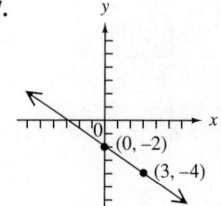

8.

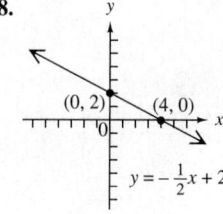

9.

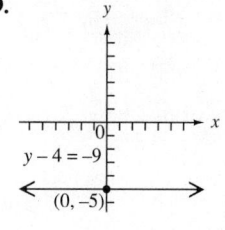

10.

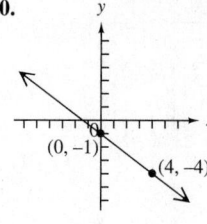

11.

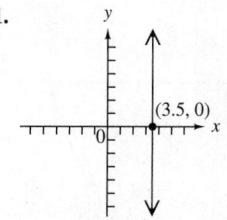

12.

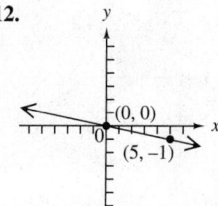

13.

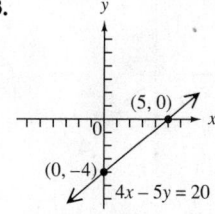

14.

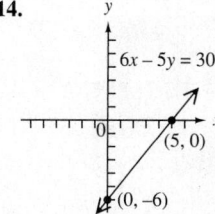

15.

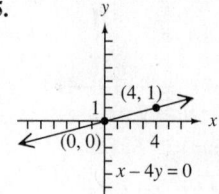

16.

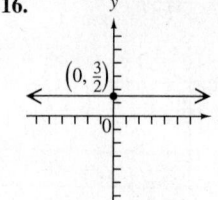

17.

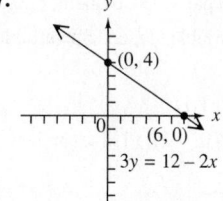

18.

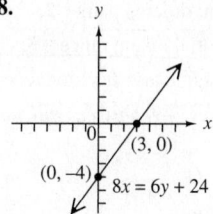

19. $y = -3x - 6$ **20.** $y = -4x - 3$ **21.** $y = \dfrac{3}{5}x$ **22.** $x = 0$

23. $y = 0$ **24.** $y = -2x - 4$ **25.** $y = \dfrac{5}{3}x + 5$ **26.** $y = 3x + 11$

SECTION 3.5 (pages 282–283)

1. (a) yes (b) yes (c) no (d) yes **2.** (a) no (b) no (c) yes
(d) yes **3.** (a) no (b) no (c) no (d) yes **4.** (a) yes (b) yes
(c) yes (d) no **5.** solid; below **6.** dashed; below **7.** dashed; above
8. solid; above **9.** $\leq$ **10.** $\geq$ **11.** $>$ **12.** $<$

13.

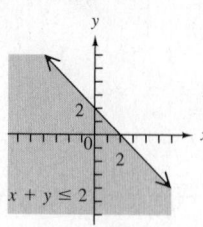

$x + y \leq 2$

15.

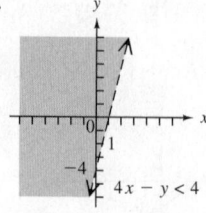

$4x - y < 4$

17.

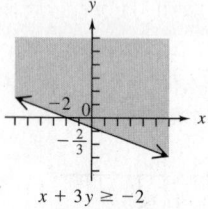

$x + 3y \geq -2$

19.

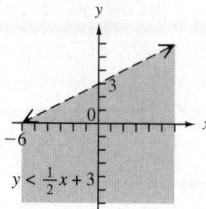

$y < \frac{1}{2}x + 3$

21.

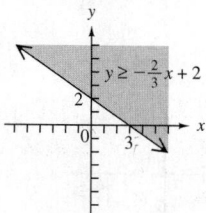

$y \geq -\frac{2}{3}x + 2$

23.

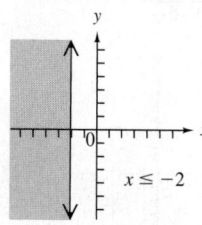

$x \leq -2$

25.

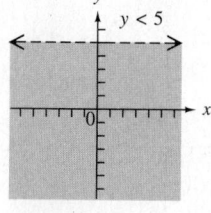

$y < 5$

27.

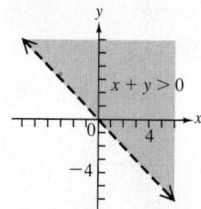

$x + y > 0$

29.

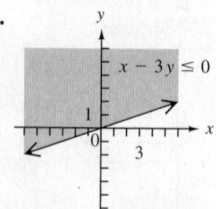

$x - 3y \leq 0$

31.

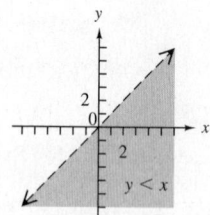

$y < x$

SECTION 3.6 (pages 290–292)

1. relation; ordered pairs **2.** function; ordered pairs **3.** domain; range
4. does not; domain; range **5.** independent variable; dependent variable
6. vertical line test; function; one

In Exercises 7–9, answers will vary.

7.

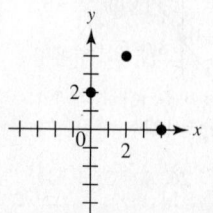

8. $\{(-1, -3), (0, -1), (1, 1), (3, 3)\}$

9.

x	y
-3	-4
-3	1
2	0

10. No, the same x-value, -3, is paired with two different y-values, -4
and 1. **11.** function; domain: $\{5, 3, 4, 7\}$; range: $\{1, 2, 9, 3\}$
13. not a function; domain: $\{2, 0\}$; range: $\{4, 2, 6\}$ **15.** function;
domain: $\{-3, 4, -2\}$; range: $\{1, 7\}$ **17.** not a function; domain:
$\{1, 0, 2\}$; range: $\{1, -1, 0, 4, -4\}$ **19.** function; domain: $\{2, 5, 11, 17, 3\}$;
range: $\{1, 7, 20\}$ **21.** not a function; domain: $\{1\}$; range: $\{5, 2, -1, -4\}$
23. function; domain: $\{4, 2, 0, -2\}$; range: $\{-3\}$ **25.** function;
domain: $\{-2, 0, 3\}$; range: $\{2, 3\}$ **27.** function; domain: $(-\infty, \infty)$;
range: $(-\infty, \infty)$ **29.** not a function; domain: $(-\infty, 0]$; range: $(-\infty, \infty)$
31. function; domain: $(-\infty, \infty)$; range: $(-\infty, 4]$ **33.** not a function;
domain: $[-4, 4]$; range: $[-3, 3]$ **35.** not a function; domain: $(-\infty, 3]$;
range: $(-\infty, \infty)$ **37.** function; $(-\infty, \infty)$ **39.** function; $(-\infty, \infty)$
41. function; $(-\infty, \infty)$ **43.** not a function; $[0, \infty)$ **45.** not a function;
$(-\infty, \infty)$ **47.** function; $(-\infty, \infty)$ **49.** function; $(-\infty, 0) \cup (0, \infty)$
51. function; $(-\infty, 4) \cup (4, \infty)$ **53.** function; $(-\infty, 0) \cup (0, \infty)$
55. (a) yes (b) domain: $\{2006, 2007, 2008, 2009, 2010\}$;
range: $\{39.6, 40.5, 40.3, 43.0\}$ (c) Answers will vary. Two possible
answers are $(2006, 39.6)$ and $(2010, 39.6)$.

SECTION 3.7 (pages 297–300)

1. $f(x)$; function; domain; x; f of x (or "f at x") **2.** B **3.** 4 **5.** -11
7. -59 **9.** 3 **11.** 2.75 **13.** $-3p + 4$ **15.** $3x + 4$ **17.** $-3x - 2$
19. $-6t + 1$ **21.** $-\dfrac{p^2}{9} + \dfrac{4p}{3} + 1$ **23.** (a) -1 (b) -1 **25.** (a) 2
(b) 3 **27.** (a) 15 (b) 10 **29.** (a) 4 (b) 1 **31.** (a) 3
(b) -3 **33.** (a) -3 (b) 2 **35.** (a) 2 (b) 0 (c) -1
37. (a) $f(x) = -\dfrac{1}{3}x + 4$ (b) 3 **39.** (a) $f(x) = 3 - 2x^2$ (b) -15
41. (a) $f(x) = \dfrac{4}{3}x - \dfrac{8}{3}$ (b) $\dfrac{4}{3}$ **43.** line; -2; linear; $-2x + 4$; -2; 3; -2
44. $ax + b$; line; slope; y-intercept; $(-\infty, \infty)$
45. domain: $(-\infty, \infty)$;   **47.** domain: $(-\infty, \infty)$;
range: $(-\infty, \infty)$ range: $(-\infty, \infty)$

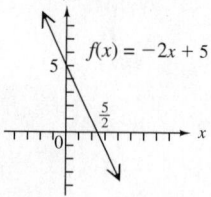

$f(x) = -2x + 5$

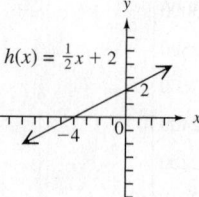

$h(x) = \frac{1}{2}x + 2$

49. domain: $(-\infty, \infty)$;   **51.** domain: $(-\infty, \infty)$;
range: $\{-4\}$ range: $\{0\}$

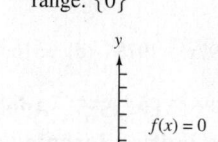

$g(x) = -4$

$f(x) = 0$

53. **(a)** \$0; \$2.50; \$5.00; \$7.50 **(b)** $2.50x$ **(c)**

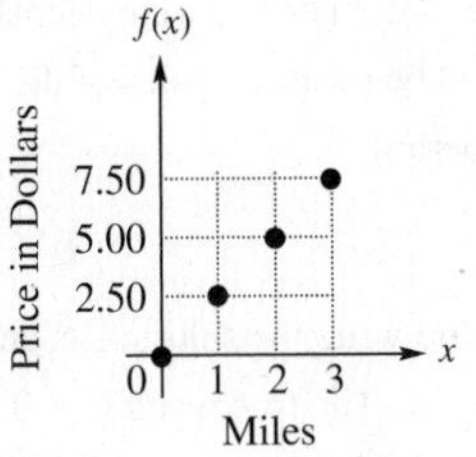

55. **(a)** $f(x) = 12x + 100$ **(b)** 1600; The cost to print 125 t-shirts is \$1600. **(c)** 75; $f(75) = 1000$; The cost to print 75 t-shirts is \$1000. **57.** **(a)** 1.1 **(b)** 4 **(c)** -1.2; $(0, 3.5)$ **(d)** $f(x) = -1.2x + 35$ **59.** **(a)** $[0, 100]$; $[0, 3000]$ **(b)** 25 hr; 25 hr **(c)** 2000 gal **(d)** $g(0) = 0$; The pool is empty at time 0.

Chapter 3 REVIEW EXERCISES (pages 305–308)

1. (2006, 52.5), (2011, 55.4) **2.** In the year 2000, 52.0% of first-year college students at two-year public institutions returned for a second year. **3.** 2008 to 2009; about 53.7% **4.** 2007 to 2008; 2009 to 2010; about 2% increase each time **5.** -1; 2; 1 **6.** 2; $\frac{3}{2}$; $\frac{14}{3}$ **7.** no **8.** yes

9.

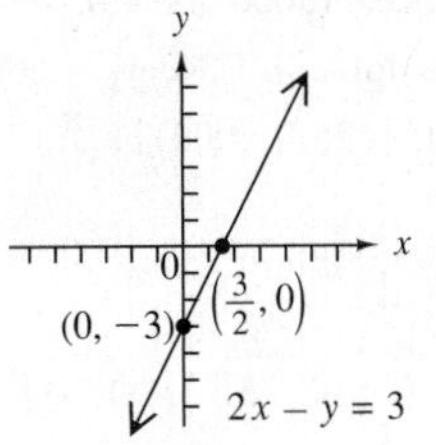

10.

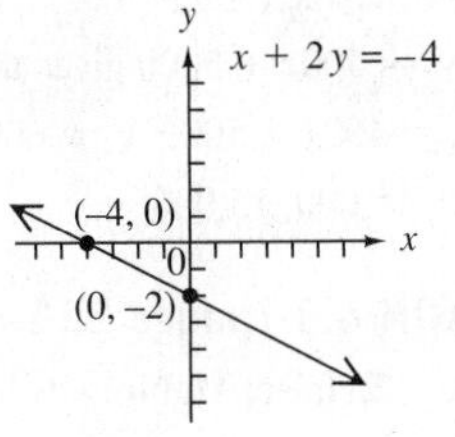

11.

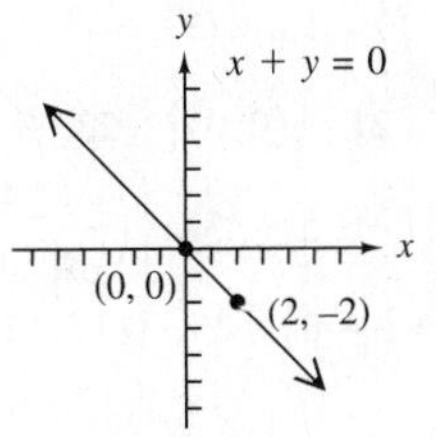

12.

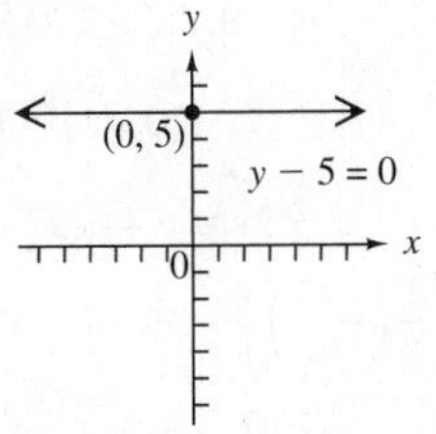

13. $-\frac{1}{2}$ **14.** undefined **15.** 0 **16.** 3 **17.** $-\frac{1}{3}$ **18.** **(a)** 2 **(b)** $\frac{1}{3}$

19. parallel **20.** perpendicular **21.** 12 ft **22.** \$1446 per yr

23. $y = -x + \frac{2}{3}$ **24.** $y = -\frac{1}{3}x + 1$ **25.** $y = x - 7$

26. $y = -\frac{1}{4}x + \frac{3}{2}$; $x + 4y = 6$ **27.** $y = 4x - 26$; $4x - y = 26$

28. $y = -\frac{5}{2}x + 1$; $5x + 2y = 2$ **29.** $y = 12$ **30.** $x = 2$ **31.** $x = 0.3$

32. $y = 4$ **33.** $y = 87.95x + 20$; \$1339.25 **34.** $y = 47x + 159$; \$723

35.

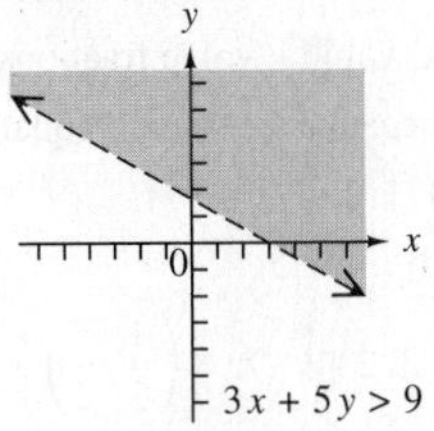

36.

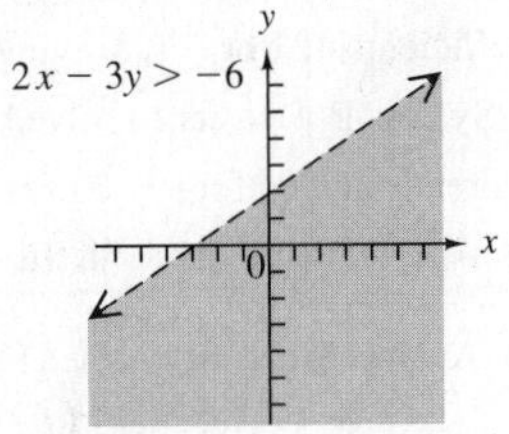

37.

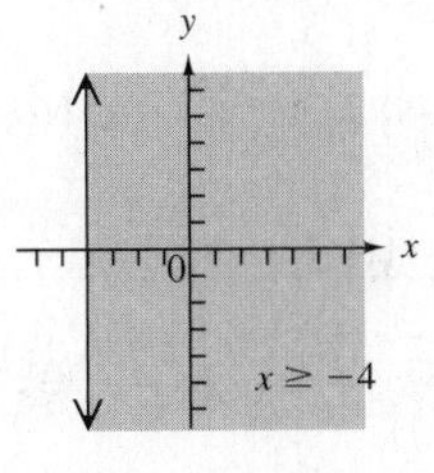

38.

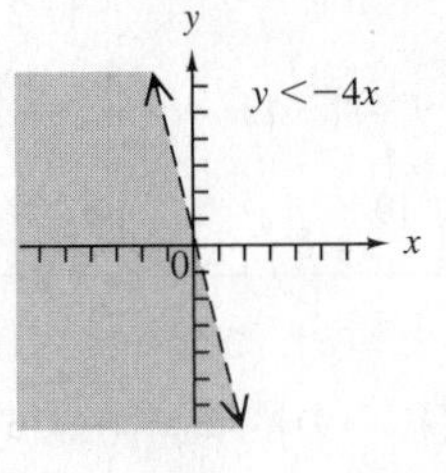

39. domain: $\{-4, 1\}$; range: $\{2, -2, 5, -5\}$; not a function **40.** domain: $\{9, 11, 4, 17, 25\}$; range: $\{32, 47, 69, 14\}$; function **41.** domain: $[-4, 4]$; range: $[0, 2]$; function **42.** domain; $(-\infty, 0]$; range: $(-\infty, \infty)$; not a function **43.** function; linear function; domain: $(-\infty, \infty)$ **44.** not a function; domain: $(-\infty, \infty)$ **45.** not a function; domain: $[0, \infty)$ **46.** function; domain: $(-\infty, 36) \cup (36, \infty)$ **47.** -6 **48.** -15 **49.** $-2p^2 + 3p - 6$ **50.** $-2k^2 - 3k - 6$ **51.** C **52.** A **53.** $f(x) = 2x^2$; 18 **54.** It is a horizontal line. **55.** A **56.** C, D **57.** A, B, D **58.** D **59.** C **60.** B **61.** **(a)** $y = -\frac{1}{4}x - \frac{5}{4}$ **(b)** $x + 4y = -5$ **62.** **(a)** $y = -3x + 30$ **(b)** $3x + y = 30$ **63.** **(a)** $y = -\frac{4}{7}x - \frac{23}{7}$ **(b)** $4x + 7y = -23$ **64.** A, B, D **65.** D **66.** **(a)** yes **(b)** domain: $\{1950, 1960, 1970, 1980, 1990, 2000, 2010\}$; range: $\{68.2, 69.7, 70.8, 73.7, 75.4, 77.0, 78.7\}$ **(c)** Answers will vary. Two possible ordered pairs are (1960, 69.7) and (2010, 78.7). **(d)** 78.7; In 2010, life expectancy at birth was 78.7 yr. **(e)** 1990

Chapter 3 TEST (pages 309–310)

1. (0, 6); (2, 0)

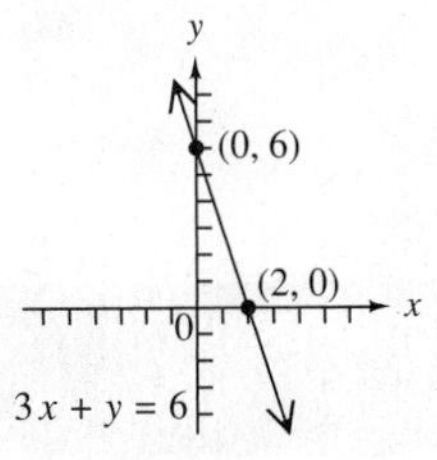

2. (0, 0); (0, 0)

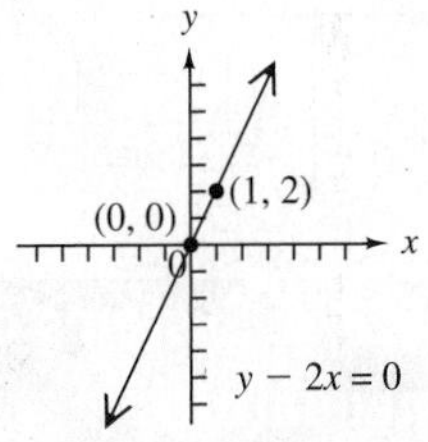

3. none; $(-3, 0)$

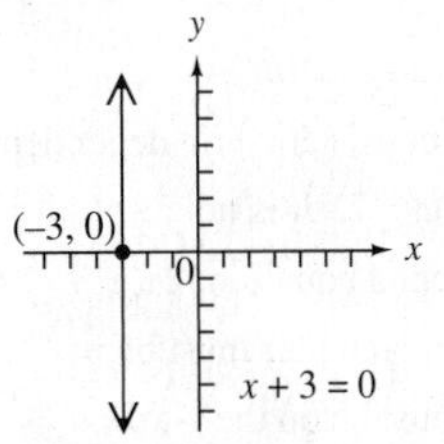

4. (0, 1); none

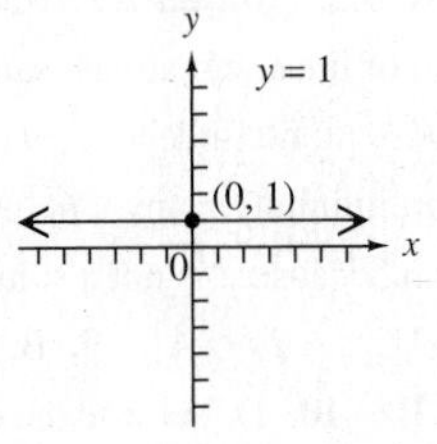

5. $-\frac{8}{3}$ **6.** -2 **7.** undefined **8.** $\frac{5}{2}$ **9.** -900 farms per yr; The number of farms decreased, on the average, by about 900 each year from 1980 to 2010. **10.** **(a)** $y = -\frac{1}{2}x + 2$ **(b)** $x + 2y = 4$ **11.** **(a)** $y = -5x + 19$ **(b)** $5x + y = 19$ **12.** $y = 14$ **13.** $x = 5$ **14.** **(a)** $y = -\frac{3}{5}x - \frac{11}{5}$ **(b)** $y = -\frac{1}{2}x - \frac{3}{2}$

15.

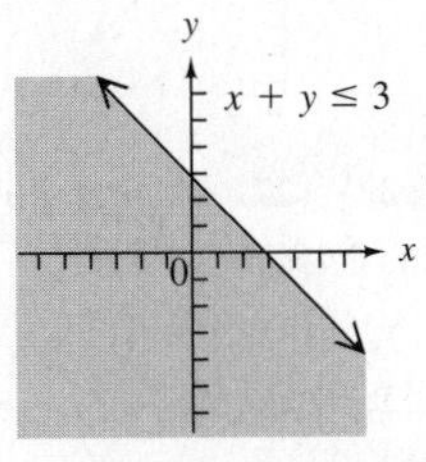

16.

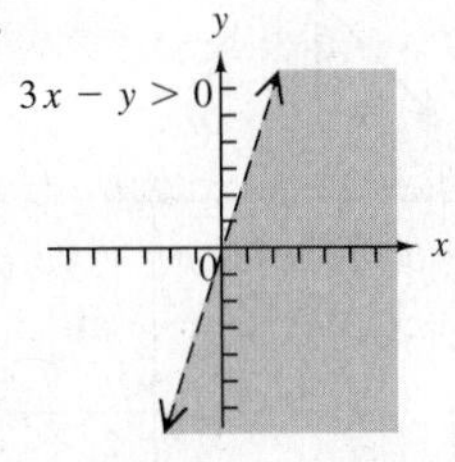

17. C; domain: $(-\infty, \infty)$; range: $[0, \infty)$ **18.** C; domain: $\{0, 3, 6\}$; range: $\{1, 2, 3\}$ **19.** 0; $-a^2 + 2a - 1$ **20.** The slope is negative since sales are decreasing. **21.** (0, 209), (11, 123); -7.8 **22.** $y = -7.8x + 209$ **23.** 154.4 thousand; The equation gives a number that is lower than the actual sales. **24.** In 2011, worldwide snowmobile sales were 123 thousand.

Chapters R–3 CUMULATIVE REVIEW EXERCISES (pages 311–312)

1. $\dfrac{301}{40}$, or $7\dfrac{21}{40}$ **2.** 6 **3.** 7 **4.** $\dfrac{73}{18}$, or $4\dfrac{1}{18}$ **5.** true **6.** -43

7. distributive property **8.** $-p + 2$ **9.** $h = \dfrac{3V}{\pi r^2}$ **10.** $\{-1\}$ **11.** $\{2\}$

12. $\{-13\}$ **13.** $(-2.6, \infty)$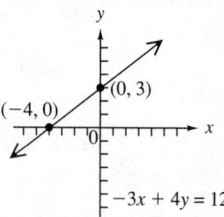

14. $(0, \infty)$ **15.** $(-\infty, -4]$

16. high school diploma: $32,552; bachelor's degree: $53,976 **17.** 13 mi
18. **(a)** 85.63, 77.82, 76.09 **(b)** $(20, 85.63), (38, 77.82), (42, 76.09)$
(c) In 1992, the winning time was approximately 80.42 sec.
19. **(a)** 19.8 million **(b)** 13.2 million **(c)** 7.8 million
20. $(-4, 0); (0, 3)$ **21.** $\dfrac{3}{4}$

22.

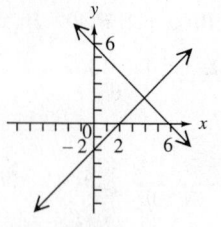

23. 6 **24.** perpendicular
25. $y = 3x - 11$ **26.** $y = 4$

CHAPTER 4 Systems of Linear Equations and Inequalities

SECTION 4.1 (pages 319–322)

1. system of linear equations; same **2.** ordered pair; true
3. inconsistent; no; independent **4.** solution; consistent **5.** dependent;
consistent; infinitely many **6.** parallel; solution **7.** It is not a solution of
the system because it is not a solution of the second equation, $2x + y = 4$.
8. $\{(x, y) \mid 3x - 2y = 4\}$ **9.** B, because the ordered pair must be in
quadrant II. **10.** D, because the ordered pair must be on the y-axis,
with $y < 0$. **11.** no **13.** yes **15.** yes **17.** no
We show the graphs here only for Exercises 19–23.
19. $\{(4, 2)\}$ **21.** $\{(0, 4)\}$

23. $\{(4, -1)\}$ **25.** $\{(1, 3)\}$ **27.** $\{(0, 2)\}$
29. Ø (inconsistent system)
31. $\{(x, y) \mid 5x - 3y = 2\}$
(dependent equations)
33. $\{(4, -3)\}$
35. $\{(x, y) \mid 2x - y = 4\}$
(dependent equations)
37. Ø (inconsistent system) **39.** **(a)** neither **(b)** intersecting lines
(c) one solution **41.** **(a)** dependent **(b)** one line **(c)** infinite number
of solutions **43.** **(a)** inconsistent **(b)** parallel lines **(c)** no solution

45. **(a)** neither **(b)** intersecting lines **(c)** one solution
47. **(a)** 40 **(b)** 30 **49.** 1980–2000 **51.** 2006; about 600 (million)
units **53.** The slope would be negative. Sales of CDs were decreasing
during this period. **54.** The slope would be positive. Sales of digital
downloads were increasing during this period.

SECTION 4.2 (pages 329–330)

1. The student must find the value of y and write the solution as an ordered pair. The solution set is $\{(3, 0)\}$. **2.** The true result $0 = 0$ means
that the system has an infinite number of solutions. The solution set is
$\{(x, y) \mid x + y = 4\}$. **3.** A false statement, such as $0 = 3$, occurs.
4. A true statement, such as $0 = 0$, occurs. **5.** $\{(3, 9)\}$ **7.** $\{(7, 3)\}$
9. $\{(-2, 4)\}$ **11.** $\{(-4, 8)\}$ **13.** $\{(3, -2)\}$ **15.** $\{(x, y) \mid 3x - y = 5\}$
17. Ø **19.** $\{(x, y) \mid 2x - y = -12\}$ **21.** $\left\{\left(\dfrac{1}{3}, -\dfrac{1}{2}\right)\right\}$ **23.** $\{(2, -3)\}$
25. $\{(2, -4)\}$ **27.** $\{(-4, 2)\}$ **29.** $\{(7, -3)\}$ **31.** $\{(2, 3)\}$
33. To find the total cost, multiply the number of bicycles (x) by the
cost per bicycle (400 dollars) and add the fixed cost (5000 dollars).
Thus $y_1 = 400x + 5000$ gives this total cost (in dollars). **34.** $y_2 = 600x$
35. $y_1 = 400x + 5000$, $y_2 = 600x$; solution set: $\{(25, 15,000)\}$
36. 25; 15,000; 15,000

SECTION 4.3 (pages 335–337)

1. true **2.** false; Multiply by -3. **3.** true **4.** true **5.** $\{(-1, 3)\}$
7. $\{(-1, -3)\}$ **9.** $\{(-2, 3)\}$ **11.** $\left\{\left(\dfrac{1}{2}, 4\right)\right\}$ **13.** $\{(3, -6)\}$
15. $\{(7, 4)\}$ **17.** $\{(0, 4)\}$ **19.** $\{(-4, 0)\}$ **21.** $\{(0, 0)\}$ **23.** Ø
25. $\{(x, y) \mid x - 3y = -4\}$ **27.** $\{(0, 7)\}$ **29.** $\{(-6, 5)\}$ **31.** $\left\{\left(-\dfrac{6}{5}, \dfrac{4}{5}\right)\right\}$
33. $\left\{\left(\dfrac{1}{8}, -\dfrac{5}{6}\right)\right\}$ **35.** Ø **37.** $\{(x, y) \mid 2x + y = 0\}$ **39.** $\{(11, 15)\}$
41. $\left\{\left(13, -\dfrac{7}{5}\right)\right\}$ **43.** $\{(6, -4)\}$ **45.** $5.66 = 2001a + b$
46. $7.89 = 2010a + b$ **47.** $2001a + b = 5.66, 2010a + b = 7.89$;
solution set: $\{(0.248, -490.590)\}$ **48.** **(a)** $y = 0.248x - 490.590$
(b) $7.39; This is a bit more than the actual figure.

SUMMARY EXERCISES Applying Techniques for Solving Systems of Linear Equations (pages 338–339)

1. **(a)** Use substitution since the second equation is solved for y.
(b) Use elimination since the coefficients of the y-terms are opposites.
(c) Use elimination since the equations are in standard form with no
coefficients of 1 or -1. Solving by substitution would involve fractions.
2. System B is easier to solve by substitution because the second equation
is already solved for y. **3.** **(a)** $\{(1, 4)\}$ **(b)** $\{(1, 4)\}$
(c) Answers will vary. **4.** **(a)** $\{(-5, 2)\}$ **(b)** $\{(-5, 2)\}$
(c) Answers will vary. **5.** $\{(2, 6)\}$ **6.** $\{(-3, 2)\}$ **7.** $\left\{\left(\dfrac{1}{3}, \dfrac{1}{2}\right)\right\}$
8. Ø **9.** $\{(3, 0)\}$ **10.** $\left\{\left(\dfrac{3}{2}, -\dfrac{3}{2}\right)\right\}$ **11.** $\{(x, y) \mid 3x + y = 7\}$
12. $\{(9, 4)\}$ **13.** $\left\{\left(\dfrac{45}{31}, \dfrac{4}{31}\right)\right\}$ **14.** $\{(4, -5)\}$ **15.** Ø **16.** $\{(0, 0)\}$
17. $\left\{\left(\dfrac{19}{3}, -5\right)\right\}$ **18.** $\left\{\left(\dfrac{22}{13}, -\dfrac{23}{13}\right)\right\}$ **19.** $\{(-12, -60)\}$
20. $\{(2, -3)\}$ **21.** $\{(24, -12)\}$ **22.** $\{(-2, 1)\}$ **23.** $\{(-35, 13)\}$
24. $\{(10, -9)\}$ **25.** $\{(-4, 6)\}$ **26.** $\{(5, 3)\}$

SECTION 4.4 (pages 346–351)

1. D **2.** A **3.** B **4.** B **5.** D **6.** D **7.** C **8.** C

9. the second number; $x - y = 48$; The two numbers are 73 and 25.

11. *The Phantom of the Opera:* 9803; *Cats:* 7485 **13.** *Harry Potter and the Deathly Hallows Part 2:* $381 million; *Transformers: Dark of the Moon:* $352 million **15.** Terminal Tower: 708 ft; Key Tower: 947 ft

17. variables; width; Equation (1): $x = 38 + y$; length; twice; Equation (2): $2x + 2y = 188$; length: 66 yd; width: 28 yd **19.** **(a)** 45 units **(b)** Do not produce—the product will lead to a loss. **21.** table entries: (third column) $1x$ (or x), $10y$; 46 ones; 28 tens **23.** 5 DVDs of *Moneyball;* 2 Blu-ray discs of *The Concert for George* **25.** table entries: (second column) 4%, or 0.04; (third column) $0.04y$; Equation (1): $x = 2y$; Equation (2): $0.05x + 0.04y = 350$; $2500 at 4%; $5000 at 5%

27. U2: $92; Taylor Swift: $72 **29.** table entries: (third column) $0.40x$, $0.70y$, $0.50(120)$ (or 60); 80 L of 40% solution; 40 L of 70% solution

31. table entries: (second column) 3, 4; (third column) $6x$, $3y$, $4(90)$ (or 360); 30 lb at $6 per lb; 60 lb at $3 per lb **33.** nuts: 40 lb; raisins: 20 lb **35.** table entries: (third column) 4.5, 4.5; (fourth column) $4.5x$, $4.5y$; Equation (1): $4.5x + 4.5y = 495$; Equation (2): $x = 10 + y$; 60 mph; 50 mph **37.** bicycle: 13.5 mph; car: 49.3 mph **39.** car leaving Cincinnati: 55 mph; car leaving Toledo: 70 mph **41.** Roberto: 17.5 mph; Juana: 12.5 mph **43.** table entries: (third column) 3, 3; (fourth column) 24; boat: 10 mph; current: 2 mph **45.** plane: 470 mph; wind: 30 mph

SECTION 4.5 (pages 355–356)

1. C **2.** A **3.** B **4.** D

5.
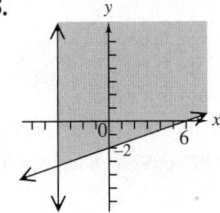

6.
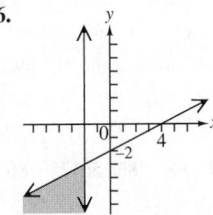

7.
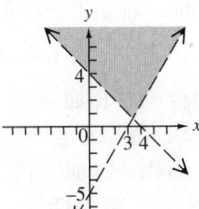

8.
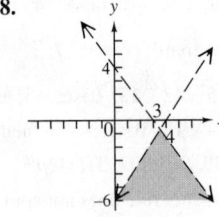

9.
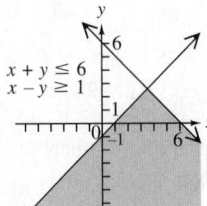

11.
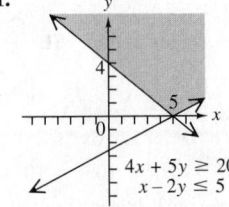

13.
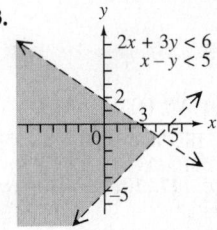

15.
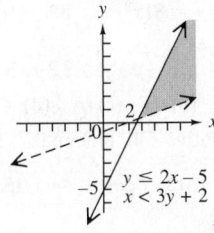

17.
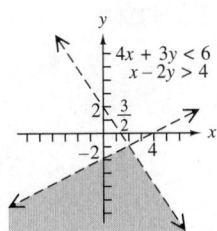

19.
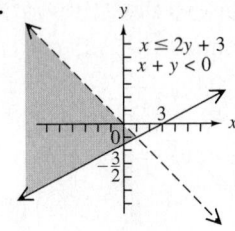

21.
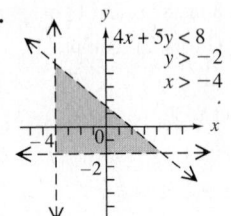

23.
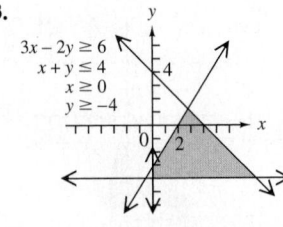

25. There are infinitely many points in any of the shaded regions. There are, however, points that are not in the shaded regions, and thus there are ordered pairs that are not solutions.

Chapter 4 REVIEW EXERCISES (pages 360–362)

1. yes **2.** no

We do not show the graphs for Exercises 3–6.

3. $\{(3, 1)\}$ **4.** $\{(0, -2)\}$ **5.** $\{(x, y) \mid x - 2y = 2\}$ **6.** $\emptyset$

7. $\{(2, 1)\}$ **8.** $\{(3, 5)\}$ **9.** $\{(6, 4)\}$ **10.** $\emptyset$ **11.** $\{(7, 1)\}$

12. $\{(-5, -2)\}$ **13.** $\{(-4, 3)\}$ **14.** $\{(x, y) \mid 3x - 4y = 9\}$

15. $\{(9, 2)\}$ **16.** $\left\{\left(\dfrac{10}{7}, -\dfrac{9}{7}\right)\right\}$ **17.** $\{(8, 9)\}$ **18.** $\{(2, 1)\}$

19. $\{(7, -2)\}$ **20.** $\{(-4, 2)\}$ **21.** Pizza Hut: 7542 locations; Domino's: 4929 locations **22.** *Reader's Digest:* 5.7 million; *People:* 3.6 million **23.** table entries: (first column) x; (second column) 0.90; (third column) $0.90y$, 100 (1) (or 100); 25 lb of $1.30 candy; 75 lb of $0.90 candy **24.** table entries: (first column) y, 20; (second column) 20; (third column) $10x$; 13 twenties; 7 tens **25.** length: 27 m; width: 18 m **26.** plane: 250 mph; wind: 20 mph **27.** table entries: (second column) 0.04; (third column) $0.03x$, $0.04y$, 650; $7000 at 3%; $11,000 at 4% **28.** table entries: (second column) 0.70; (third column) $0.40x$, $0.70y$, $0.50(90)$ (or 45); 60 L of 40% solution; 30 L of 70% solution **29.** B **30.** B

31.
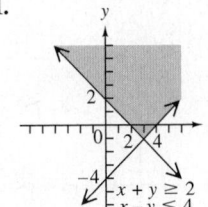

32.
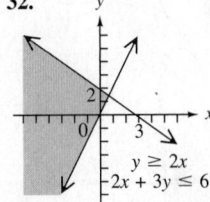

33.
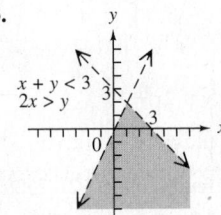

34. $\{(2, 0)\}$ **35.** $\{(-4, 15)\}$
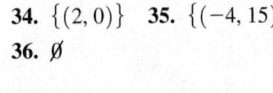
36. $\emptyset$

37.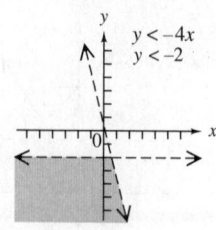

$x + y < 5$
$x - y \geq 2$

38.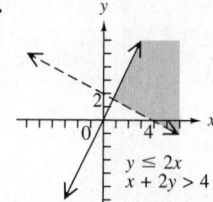

$y \leq 2x$
$x + 2y > 4$

25.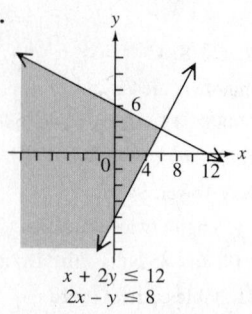

$x + 2y \leq 12$
$2x - y \leq 8$

39.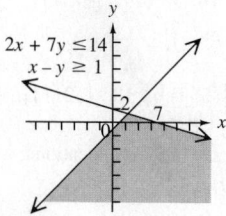

$y < -4x$
$y < -2$

40. 8 in., 8 in., and 13 in.
41. Giants: 21; Patriots: 17
42. (a) years 0–6 **(b)** year 6; about $650

CHAPTER 5 Exponents and Polynomials

SECTION 5.1 (pages 373–376)

1. 7; 5 **2.** two **3.** 8 **4.** is not **5.** 26 **6.** $5x^9$ **7.** 1; 6 **9.** 1; 1

11. 1; $\dfrac{1}{5}$ **13.** 2; -19, -1 **15.** 3; 1, -8, $\dfrac{2}{3}$ **17.** $2m^5$ **19.** $-r^5$

21. $\dfrac{2}{3}x^4$ **23.** cannot be simplified; $0.2m^5 - 0.5m^2$ **25.** $-5x^5$

27. $5p^9 + 4p^7$ **29.** $-2y^2$ **31.** already simplified; 4; binomial

33. already simplified; $6m^5 + 5m^4 - 7m^3 - 3m^2$; 5; none of these

35. $x^4 + \dfrac{1}{3}x^2 - 4$; 4; trinomial **37.** 7; 0; monomial

39. $1.5x^2 - 0.5x$; 2; binomial **41. (a)** -1 **(b)** 5 **43. (a)** 19 **(b)** -2

45. (a) 36 **(b)** -12 **47. (a)** -124 **(b)** 5 **49.** $5m^2 + 3m$

51. $4x^4 - 4x^2$ **53.** $\dfrac{7}{6}x^2 - \dfrac{2}{15}x + \dfrac{5}{6}$ **55.** $12m^3 - 13m^2 + 6m + 11$

57. $2.9x^3 - 3.5x^2 - 1.5x - 9$ **59.** $8r^2 + 5r - 12$ **61.** $5m^2 - 14m + 6$

63. $4x^3 + 2x^2 + 5x$ **65.** $-18y^5 + 7y^4 + 5y^3 + 3y^2 + y$

67. $-2m^3 + 7m^2 + 8m - 9$ **69.** $-11x^2 - 3x - 3$ **71.** $2x^2 + 8x$

73. $8x^2 + 8x + 6$ **75.** $8t^2 + 8t + 13$ **77.** $13a^2b - 7a^2 - b$

79. $c^4d - 5c^2d^2 + d^2$ **81.** $12m^3n - 11m^2n^2 - 4mn^2$ **83. (a)** $23y + 5t$
(b) $25°, 67°, 88°$ **85.** 5; 175 **86.** 87 ft; $(1, 87)$ **87.** 6; $27 **88.** 2.5; 130

SECTION 5.2 (pages 383–384)

1. false **2.** true **3.** false **4.** true **5.** 1 **6.** three squared; five cubed; seven to the fourth power **7.** t^7 **9.** $\left(\dfrac{1}{2}\right)^5$ **11.** $(-8p)^2$ **13.** base: 3; exponent: 5; 243 **15.** base: -3; exponent: 5; -243 **17.** base: $-6x$; exponent: 4 **19.** base: x; exponent: 4 **21.** 5^8 **23.** 4^{12} **25.** $(-7)^9$

27. t^{24} **29.** $-56r^7$ **31.** $42p^{10}$ **33.** The product rule does not apply.

35. The product rule does not apply. **37.** 4^6 **39.** t^{20} **41.** $343r^3$

43. 5^{12} **45.** -8^{15} **47.** $5^5x^5y^5$ **49.** $8q^3r^3$ **51.** $\dfrac{1}{8}$ **53.** $\dfrac{a^3}{b^3}$ **55.** $\dfrac{9^8}{5^8}$

57. $-8x^6y^3$ **59.** $9a^6b^4$ **61.** $\dfrac{5^5}{2^5}$ **63.** $\dfrac{9^5}{8^3}$ **65.** $2^{12}x^{12}$ **67.** -6^5p^5

69. $6^5x^{10}y^{15}$ **71.** x^{21} **73.** $4w^4x^{26}y^7$ **75.** $-r^{18}s^{17}$ **77.** $\dfrac{125a^6b^{15}}{c^{18}}$

79. $25m^6p^{14}q^5$ **81.** $16x^{10}y^{16}z^{10}$ **83.** Using power rule (a), simplify as follows: $(10^2)^3 = 10^{2 \cdot 3} = 10^6 = 1{,}000{,}000$. **84.** The 4 is used as an *exponent* on 3. It is *not* multiplied by 3. The correct simplification is $3^4 \cdot x^8 \cdot y^{12} = 81x^8y^{12}$. **85.** $30x^7$ **87.** $6p^7$

SECTION 5.3 (pages 389–391)

1. (a) B **(b)** D **(c)** A **(d)** C **2. (a)** C **(b)** A **(c)** B **(d)** D

3. distributive **4.** binomials **5.** $15pq^2$ **7.** $-18m^3n^2$ **9.** $9y^{10}$

11. $-8x^{10}$ **13.** $-6m^2 - 4m$ **15.** $6p - \dfrac{9}{2}p^2 + 9p^4$ **17.** $10y^9 + 4y^6 + 6y^5$

Chapter 4 TEST (pages 363–364)

1. (a) no **(b)** no **(c)** yes **2.** $\{(2, -3)\}$ **3.** It has no solution.
4. $\{(1, -6)\}$ **5.** $\{(-35, 35)\}$ **6.** $\{(5, 6)\}$ **7.** $\{(-1, 3)\}$
8. $\{(-1, 3)\}$ **9.** $\{(0, 0)\}$ **10.** Ø **11.** $\{(x, y) \mid 3x - y = 6\}$
12. $\{(-15, 6)\}$ **13.** Memphis and Atlanta: 394 mi; Minneapolis and Houston: 1176 mi **14.** Statue of Liberty: 3.8 million; National World War II Memorial: 4.0 million **15.** 20 L of 15% solution; 30 L of 40% solution **16.** slower car: 45 mph; faster car: 60 mph

17.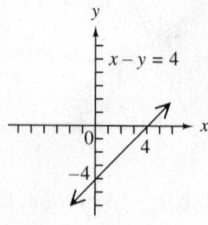

$2x + 7y \leq 14$
$x - y \geq 1$

18.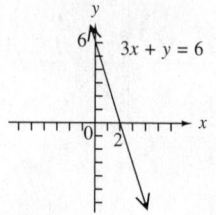

$2x - y > 6$
$4y + 12 \geq -3x$

Chapters R–4 CUMULATIVE REVIEW EXERCISES (pages 365–366)

1. $-1, 1, -2, 2, -4, 4, -5, 5, -8, 8, -10, 10, -20, 20, -40, 40$ **2.** 1
3. commutative property **4.** distributive property **5.** inverse property
6. 46 **7.** $T = \dfrac{PV}{k}$ **8.** $\left\{-\dfrac{13}{11}\right\}$ **9.** $\left\{\dfrac{9}{11}\right\}$ **10.** $(-18, \infty)$

11. $\left(-\dfrac{11}{2}, \infty\right)$ **12.** length: 9.50 in.; width: 7.31 in.

13.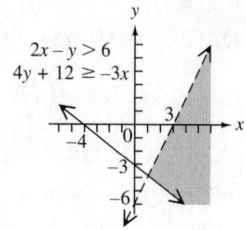

$x - y = 4$

14.

$3x + y = 6$

15. $-\dfrac{4}{3}$ **16.** $-\dfrac{1}{4}$ **17.** $y = \dfrac{1}{2}x + 3$ **18.** $y = 2x + 1$

19. (a) $x = 9$ **(b)** $y = -1$ **20.** $\{(-1, 6)\}$ **21.** $\{(3, -4)\}$
22. Ø **23.** table entries: (third column) 2; (fourth column) 2y, 2528; 405 adults, 49 children **24.** 19 in., 19 in., 15 in.

19. $6y^6 + 4y^4 + 2y^3$ **21.** $28r^5 - 32r^4 + 36r^3$ **23.** $6a^4 - 12a^3b + 15a^2b^2$

25. $3m^2$; $2mn$; n^3; $21m^5n^2 + 14m^4n^3 - 7m^3n^5$ **27.** $12x^3 + 26x^2 + 10x + 1$

29. $6r^3 + 5r^2 - 12r + 4$ **31.** $20m^4 - m^3 - 8m^2 - 17m - 15$

33. $5x^4 - 13x^3 + 20x^2 + 7x + 5$ **35.** $3x^5 + 18x^4 - 2x^3 - 8x^2 + 24x$

37. first row: x^2; $4x$; second row: $3x$; 12; Product: $x^2 + 7x + 12$

39. first row: $2x^3$; $6x^2$; $4x$; second row: x^2; $3x$; 2; Product: $2x^3 + 7x^2 + 7x + 2$

41. $m^2 + 12m + 35$ **43.** $n^2 + n - 6$ **45.** $8r^2 - 10r - 3$ **47.** $9x^2 - 4$

49. $9q^2 + 6q + 1$ **51.** $15xy - 40x + 21y - 56$ **53.** $6t^2 + 23st + 20s^2$

55. $-0.3t^2 + 0.22t + 0.24$ **57.** $x^2 - \dfrac{5}{12}x - \dfrac{1}{6}$ **59.** $\dfrac{15}{16} - \dfrac{1}{4}r - 2r^2$

61. $2x^3 + x^2 - 15x$ **63.** $6y^5 - 21y^4 - 45y^3$ **65.** $-200r^7 + 32r^3$

67. (a) $3y^2 + 10y + 7$ (b) $8y + 16$ **69.** $30x + 60$

70. $30x + 60 = 600$; $\{18\}$ **71.** 10 yd by 60 yd **72.** 140 yd

73. \$2100 **74.** \$1260

SECTION 5.4 (pages 396–398)

1. square; twice; product; square **2.** positive; negative **3.** difference;

squares **4.** conjugates **5.** (a) $4x^2$ (b) $12x$ (c) 9 (d) $4x^2 + 12x + 9$

7. $p^2 + 4p + 4$ **9.** $z^2 - 10z + 25$ **11.** $x^2 - \dfrac{3}{2}x + \dfrac{9}{16}$

13. $v^2 + 0.8v + 0.16$ **15.** $16x^2 - 24x + 9$ **17.** $100z^2 + 120z + 36$

19. $4p^2 + 20pq + 25q^2$ **21.** $0.64t^2 + 1.12ts + 0.49s^2$

23. $25x^2 + 4xy + \dfrac{4}{25}y^2$ **25.** $9t^3 - 6t^2 + t$ **27.** $48t^3 + 24t^2 + 3t$

29. $-16r^2 + 16r - 4$ **31.** (a) $49x^2$ (b) 0 (c) $-9y^2$ (d) $49x^2 - 9y^2$;

Because 0 is the identity element for addition, it is not necessary to write

"$+ 0$." **33.** $q^2 - 4$ **35.** $r^2 - \dfrac{9}{16}$ **37.** $s^2 - 6.25$ **39.** $4w^2 - 25$

41. $100x^2 - 9y^2$ **43.** $4x^4 - 25$ **45.** $49x^2 - \dfrac{9}{49}$ **47.** $9p^3 - 49p$

49. $m^3 - 15m^2 + 75m - 125$ **51.** $y^3 + 6y^2 + 12y + 8$

53. $8a^3 + 12a^2 + 6a + 1$ **55.** $81r^4 - 216r^3t + 216r^2t^2 - 96rt^3 + 16t^4$

57. $3x^5 - 27x^4 + 81x^3 - 81x^2$

59. $-8x^6y - 32x^5y^2 - 48x^4y^3 - 32x^3y^4 - 8x^2y^5$ **61.** $\dfrac{1}{2}m^2 - 2n^2$

63. $9a^2 - 4$ **65.** $\pi x^2 + 4\pi x + 4\pi$ **67.** $x^3 + 6x^2 + 12x + 8$

69. $(a + b)^2$ **70.** a^2 **71.** $2ab$ **72.** b^2 **73.** $a^2 + 2ab + b^2$

74. They both represent the area of the entire large square. **75.** 1225

76. $30^2 + 2(30)(5) + 5^2$ **77.** 1225 **78.** They are equal.

SECTION 5.5 (pages 406–407)

1. negative **2.** positive **3.** negative **4.** negative **5.** positive

6. positive **7.** 0 **8.** 0 **9.** (a) B (b) C (c) D (d) B (e) E (f) B

10. (a) C (b) F (c) F (d) B (e) E (f) B **11.** 1 **13.** 1 **15.** -1

17. 0 **19.** 0 **21.** 2 **23.** $\dfrac{1}{64}$ **25.** 16 **27.** $\dfrac{49}{36}$ **29.** $\dfrac{1}{81}$ **31.** $\dfrac{8}{15}$

33. $-\dfrac{7}{18}$ **35.** $\dfrac{1}{9}$ **37.** $\dfrac{1}{6^5}$, or $\dfrac{1}{7776}$ **39.** 216 **41.** $2r^4$ **43.** $\dfrac{25}{64}$ **45.** $\dfrac{p^5}{q^8}$

47. r^9 **49.** $\dfrac{x^5}{6}$ **51.** $3y^2$ **53.** x^3 **55.** $\dfrac{yz^2}{4x^3}$ **57.** $a + b$ **59.** 343

61. $\dfrac{1}{x^2}$ **63.** $\dfrac{64x}{9}$ **65.** $\dfrac{x^2z^4}{y^2}$ **67.** $6x$ **69.** $\dfrac{1}{m^{10}n^5}$ **71.** $\dfrac{5}{16x^5}$ **73.** $\dfrac{36q^2}{m^4p^2}$

75. The student attempted to use the quotient rule with unequal bases.

The correct way to simplify is $\dfrac{16^3}{2^2} = \dfrac{(2^4)^3}{2^2} = \dfrac{2^{12}}{2^2} = 2^{12-2} = 2^{10} = 1024$.

76. The student incorrectly assumed that the negative exponent indicated

a negative number. The correct way to simplify is $5^{-4} = \dfrac{1}{5^4} = \dfrac{1}{625}$.

SUMMARY EXERCISES Applying the Rules for Exponents (pages 408–409)

1. positive **2.** negative **3.** negative **4.** negative **5.** positive

6. negative **7.** positive **8.** negative **9.** $\dfrac{6^{12}x^{24}}{5^{12}}$ **10.** $\dfrac{r^6s^{12}}{729t^6}$

11. $100{,}000x^7y^{14}$ **12.** $-128a^{10}b^{15}c^4$ **13.** $\dfrac{729w^3x^9}{y^{12}}$ **14.** $\dfrac{x^4y^6}{16}$

15. c^{22} **16.** $\dfrac{1}{k^4t^{12}}$ **17.** $\dfrac{11}{30}$ **18.** $y^{12}z^3$ **19.** $\dfrac{x^6}{y^5}$ **20.** 0 **21.** $\dfrac{1}{z^2}$

22. $\dfrac{9}{r^2s^2t^{10}}$ **23.** $\dfrac{300x^3}{y^3}$ **24.** $\dfrac{3}{5x^6}$ **25.** x^8 **26.** $\dfrac{y^{11}}{x^{11}}$ **27.** $\dfrac{a^6}{b^4}$ **28.** $6ab$

29. $\dfrac{61}{900}$ **30.** 1 **31.** $\dfrac{343a^6b^9}{8}$ **32.** 1 **33.** -1 **34.** 0 **35.** $\dfrac{27y^{18}}{4x^8}$

36. $\dfrac{1}{a^8b^{12}c^{16}}$ **37.** $\dfrac{x^{15}}{216z^9}$ **38.** $\dfrac{q}{8p^6r^3}$ **39.** x^6y^6 **40.** 0 **41.** $\dfrac{343}{x^{15}}$

42. $\dfrac{9}{x^6}$ **43.** $5p^{10}q^9$ **44.** $\dfrac{7}{24}$ **45.** $\dfrac{r^{14}t}{2s^2}$ **46.** 1 **47.** $8p^{10}q$ **48.** $\dfrac{1}{mn^3p^3}$

49. -1 **50.** $\dfrac{3}{40}$

SECTION 5.6 (pages 412–413)

1. $6x^2 + 8$; 2; $3x^2 + 4$ **2.** 0 **3.** $3x^2 + 4$; 2 (These may be reversed.);

$6x^2 + 8$ **4.** is not **5.** To use the method of this section, the divisor must

be a monomial. This is true of the first problem, but not the second.

6. 8; 13; no **7.** 2 **8.** -30 **9.** $2m^2 - m$ **11.** $30x^3 - 10x + 5$

13. $-4m^3 + 2m^2 - 1$ **15.** $4t^4 - 2t^2 + 2t$ **17.** $a^4 - a + \dfrac{2}{a}$

19. $-2x^3 + \dfrac{2x^2}{3} - x$ **21.** $-9x^2 + 5x + 1$ **23.** $\dfrac{4x^2}{3} + x - \dfrac{2}{3x}$

25. $27r^4$; $36r^3$; $6r^2$; $3r$; 2; $9r^3 - 12r^2 - 2r + 1 - \dfrac{2}{3r}$ **27.** $-m^2 + 3m - \dfrac{4}{m}$

29. $\dfrac{12}{x} - \dfrac{6}{x^2} + \dfrac{14}{x^3} - \dfrac{10}{x^4}$ **31.** $-4b^2 + 3ab - \dfrac{5}{a}$ **33.** $6x - 2 + \dfrac{1}{x}$

35. $15x^5 - 35x^4 + 35x^3$ **36.** $-72y^6 + 60y^5 - 24y^4 + 36y^3 - 84y^2$

SECTION 5.7 (pages 418–419)

1. The divisor is $2x + 5$. The quotient is $2x^3 - 4x^2 + 3x + 2$. **2.** Stop

when the degree of the remainder is less than the degree of the divisor, or

when the remainder is 0. **3.** Divide $12m^2$ by $2m$ to get $6m$. **4.** Multiply

$6m$ by $2m - 3$ to get $12m^2 - 18m$. **5.** $x + 2$ **7.** $2y - 5$

9. $p - 4 + \dfrac{44}{p + 6}$ **11.** $r - 5$ **13.** $2a - 14 + \dfrac{74}{2a + 3}$ **15.** $4x^2 - 7x + 3$

17. $3y^2 - 2y + 2$ **19.** $2x^2 - 2x + 3 + \dfrac{-1}{x + 1}$ **21.** $3k - 4 + \dfrac{2}{k^2 - 2}$

23. $x^2 + 1$ **25.** $x^2 + 1$ **27.** $2p^2 - 5p + 4 + \dfrac{6}{3p^2 + 1}$ **29.** $x^3 + 6x - 7$

31. $2x^2 + \dfrac{3}{5}x + \dfrac{1}{5}$ **33.** $(x^2 + x - 3)$ units **35.** 33 **36.** 33 **37.** They

are the same. **38.** The answers should agree.

SECTION 5.8 (pages 424–426)

1. (a) C (b) A (c) B (d) D **2.** (a) A (b) C (c) B (d) D

3. in scientific notation **4.** in scientific notation **5.** not in scientific

notation; 5.6×10^6 **6.** not in scientific notation; 3.4×10^4 **7.** not in

scientific notation; 4×10^{-3} **8.** not in scientific notation; 7×10^{-4}

9. not in scientific notation; 8×10^1 **10.** not in scientific notation; 9×10^2

11. A number is written in scientific notation if it is the product of a number

whose absolute value is between 1 and 10 (inclusive of 1) and a power of 10.

12. To multiply by a positive power of 10, move the decimal point to the right as many places as the exponent on 10. With a negative power of 10, move the decimal point to the left as many places as the absolute value of the exponent on 10. **13.** 5.876×10^9 **15.** 8.235×10^4 **17.** 7×10^{-6} **19.** -2.03×10^{-3} **21.** 750,000 **23.** 5,677,000,000,000 **25.** 1,000,000,000,000 **27.** -6.21 **29.** 0.00078 **31.** 0.000000005134 **33.** 6×10^{11}; 600,000,000,000 **35.** 1.5×10^7; 15,000,000 **37.** 8×10^{-3}; 0.008 **39.** 2.4×10^2; 240 **41.** 6.3×10^{-2}; 0.063 **43.** 6.426×10^4; 64,260 **45.** 3×10^{-4}; 0.0003 **47.** 4×10^1; 40 **49.** 1.3×10^{-5}; 0.000013 **51.** 5×10^2; 500 **53.** 2.6×10^{-3}; 0.0026 **55.** 7.205×10^{-6}; 0.000007205 **57.** 0.000002 **59.** 4.2×10^{42} **61.** 1.5×10^{17} mi **63.** $3209 **65.** $52,733 **67.** 3.59×10^2 sec, or 359 sec **69.** $86.40

Chapter 5 REVIEW EXERCISES (pages 430–433)

1. $22m^2$; degree 2; monomial **2.** $p^3 - p^2 + 4p + 2$; degree 3; none of these **3.** already in descending powers; degree 5; none of these **4.** $-8y^5 - 7y^4 + 9y$; degree 5; trinomial **5.** $-5a^3 + 4a^2$ **6.** $2r^3 - 3r^2 + 9r$ **7.** $11y^2 - 10y + 9$ **8.** $-13k^4 - 15k^2 - 4k - 6$ **9.** $10m^3 - 6m^2 - 3$ **10.** $-y^2 - 4y + 26$ **11.** $10p^2 - 3p - 11$ **12.** $7r^4 - 4r^3 - 1$ **13.** 4^{11} **14.** -5^{11} **15.** $-72x^7$ **16.** $10x^{14}$ **17.** 19^5x^5 **18.** -4^7y^7 **19.** $5p^4t^4$ **20.** $\dfrac{7^6}{5^6}$ **21.** $27x^6y^9$ **22.** t^{42} **23.** $36x^{16}y^4z^{16}$ **24.** $\dfrac{8m^9n^3}{p^6}$ **25.** $125x^6$ **26.** The product rule for exponents does not apply here because we want the sum of 7^2 and 7^4, not their product. **27.** $10x^2 + 70x$ **28.** $-6p^5 + 15p^4$ **29.** $6r^3 + 8r^2 - 17r + 6$ **30.** $8y^3 + 27$ **31.** $5p^5 - 2p^4 - 3p^3 + 25p^2 + 15p$ **32.** $x^2 + 3x - 18$ **33.** $6k^2 - 9k - 6$ **34.** $12p^2 - 48pq + 21q^2$ **35.** $2m^4 + 5m^3 - 16m^2 - 28m + 9$ **36.** $a^2 + 8a + 16$ **37.** $9p^2 - 12p + 4$ **38.** $4r^2 + 20rs + 25s^2$ **39.** $r^3 + 6r^2 + 12r + 8$ **40.** $8x^3 - 12x^2 + 6x - 1$ **41.** $4z^2 - 49$ **42.** $36m^2 - 25$ **43.** $25a^2 - 36b^2$ **44.** $4x^4 - 25$ **45.** three; two **46.** $(a+b)^2 = (a+b)(a+b) = a^2 + 2ab + b^2$. The term $2ab$ is not in $a^2 + b^2$. **47.** 2 **48.** $\dfrac{1}{32}$ **49.** $\dfrac{25}{36}$ **50.** $-\dfrac{3}{16}$ **51.** 36 **52.** x^2 **53.** $\dfrac{1}{p^{12}}$ **54.** r^4 **55.** 2^8 **56.** $\dfrac{1}{9^6}$ **57.** 5^8 **58.** $\dfrac{1}{8^{12}}$ **59.** $\dfrac{1}{m^2}$ **60.** y^7 **61.** r^{13} **62.** $25m^6$ **63.** $\dfrac{y^{12}}{8}$ **64.** $\dfrac{1}{a^3b^5}$ **65.** $72r^5$ **66.** $\dfrac{8n^{10}}{3m^{13}}$ **67.** $\dfrac{5y^2}{3}$ **68.** $-2x^2y$ **69.** $-y^3 + 2y - 3$ **70.** $p - 3 + \dfrac{5}{2p}$ **71.** $-x^9 + 2x^8 - 4x^3 + 7x$ **72.** $-2m^2n + mn^2 + \dfrac{6n^3}{5}$ **73.** $2r + 7$ **74.** $4m + 3 + \dfrac{5}{3m-5}$ **75.** $2a + 1 + \dfrac{-8a+12}{5a^2-3}$ **76.** $k^2 + 2k + 4 + \dfrac{-2k-12}{2k^2+1}$ **77.** 4.8×10^7 **78.** 2.8988×10^{10} **79.** 6.5×10^{-5} **80.** 8.24×10^{-8} **81.** 24,000 **82.** 78,300,000 **83.** 0.000000897 **84.** 0.00000000000995 **85.** 8×10^2; 800 **86.** 4×10^6; 4,000,000 **87.** 2.5×10^{-2}; 0.025 **88.** 1×10^{-2}; 0.01 **89.** 2.796×10^{10} calculations; 1.6776×10^{12} calculations **90.** about 3.3 **91.** 0 **92.** $\dfrac{243}{p^3}$ **93.** $\dfrac{1}{49}$ **94.** $49 - 28k + 4k^2$ **95.** $y^2 + 5y + 1$ **96.** $\dfrac{1296r^8s^4}{625}$ **97.** $-8m^7 - 10m^6 - 6m^5$ **98.** 32 **99.** $5xy^3 - \dfrac{8y^2}{5} + 3x^2y$ **100.** $\dfrac{r^2}{6}$ **101.** $8x^3 + 12x^2y + 6xy^2 + y^3$ **102.** $\dfrac{3}{4}$ **103.** $a^3 - 2a^2 - 7a + 2$ **104.** $8y^3 - 9y^2 + 5$ **105.** $10r^2 + 21r - 10$ **106.** $144a^2 - 1$ **107.** $2x^2 + x - 6$; $6x - 2$ **108.** $20x^4 + 8x^2$; $25x^8 + 20x^6 + 4x^4$ **109.** The friend wrote the second term of the quotient as $-12x$ rather than $-2x$. Here is the correct method: $\dfrac{6x^2 - 12x}{6} = \dfrac{6x^2}{6} - \dfrac{12x}{6} = x^2 - 2x.$ **110.** $2mn + 3m^4n^2 - 4n$

Chapter 5 TEST (pages 434–435)

1. $-7x^2 + 8x$; 2; binomial **2.** $4n^4 + 13n^3 - 10n^2$; 4; trinomial **3.** $4t^4 + t^3 - 6t^2 - t$ **4.** $-2y^2 - 9y + 17$ **5.** $-12t^2 + 5t + 8$ **6.** -32 **7.** $\dfrac{216}{m^6}$ **8.** $-27x^5 + 18x^4 - 6x^3 + 3x^2$ **9.** $2r^3 + r^2 - 16r + 15$ **10.** $t^2 - 5t - 24$ **11.** $8x^2 + 2xy - 3y^2$ **12.** $25x^2 - 20xy + 4y^2$ **13.** $100v^2 - 9w^2$ **14.** $x^3 + 3x^2 + 3x + 1$ **15.** $12x + 36$; $9x^2 + 54x + 81$ **16. (a)** positive **(b)** positive **(c)** negative **(d)** positive **(e)** zero **(f)** negative **17.** $\dfrac{1}{625}$ **18.** 2 **19.** $\dfrac{7}{12}$ **20.** 8^5 **21.** x^2y^6 **22.** $4y^2 - 3y + 2 + \dfrac{5}{y}$ **23.** $-3xy^2 + 2x^3y^2 + 4y^2$ **24.** $2x + 9$ **25.** $3x^2 + 6x + 11 + \dfrac{26}{x-2}$ **26. (a)** 3.44×10^{11} **(b)** 5.57×10^{-6} **27. (a)** 29,600,000 **(b)** 0.0000000607 **28. (a)** 1×10^3; 5.89×10^{12} **(b)** 5.89×10^{15} mi

Chapters R–5 CUMULATIVE REVIEW EXERCISES (pages 436–437)

1. $\dfrac{19}{24}$ **2.** $-\dfrac{1}{20}$ **3.** 3.72 **4.** 62.006 **5.** $1836 **6.** -8 **7.** 24 **8.** $\dfrac{1}{2}$ **9.** -4 **10.** associative property **11.** inverse property **12.** distributive property **13.** $\{10\}$ **14.** $\left\{\dfrac{13}{4}\right\}$ **15.** $\varnothing$ **16.** $r = \dfrac{d}{t}$ **17.** $\{-5\}$ **18.** $\{-12\}$ **19.** $\{20\}$ **20.** {all real numbers} **21.** mouse: 160; elephant: 10 **22.** 4 **23.** $[10, \infty)$ **24.** $\left(-\infty, -\dfrac{14}{5}\right)$ **25.** $[-4, 2)$ **26.** $(0, 2)$ and $(-3, 0)$

27.

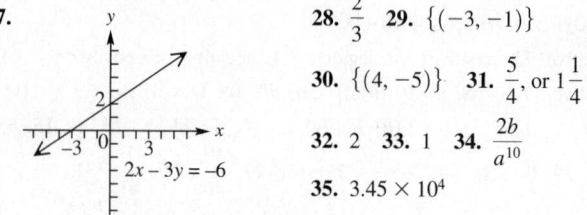

28. $\dfrac{2}{3}$ **29.** $\{(-3, -1)\}$ **30.** $\{(4, -5)\}$ **31.** $\dfrac{5}{4}$, or $1\dfrac{1}{4}$ **32.** 2 **33.** 1 **34.** $\dfrac{2b}{a^{10}}$ **35.** 3.45×10^4 **36.** $11x^3 - 14x^2 - x + 14$ **37.** $18x^7 - 54x^6 + 60x^5$ **38.** $63x^2 + 57x + 12$ **39.** $25x^2 + 80x + 64$ **40.** $y^2 - 2y + 6$

CHAPTER 6 Factoring and Applications

SECTION 6.1 (pages 447–449)

1. product; multiplying **2.** common factor; is; divides **3.** 4 **5.** 4 **7.** 6 **9.** 1 **11.** 8 **13.** $10x^3$ **15.** xy^2 **17.** 6 **19.** $6m^3n^2$ **21.** factored **22.** factored **23.** not factored **24.** not factored

25. The correct factored form is $18x^3y^2 + 9xy = 9xy(2x^2y + 1)$. If a polynomial has two terms, the product of the factors must have two terms. $9xy(2x^2y) = 18x^3y^2$ is just one term.   **26.** First, we verify that we have factored completely. Then we multiply the factors. The product should be the original polynomial.   **27.** $3m^2$ **29.** $2z^4$ **31.** $2mn^4$ **33.** $y + 2$
35. $a - 2$ **37.** $2 + 3xy$ **39.** $x(x - 4)$ **41.** $3t(2t + 5)$ **43.** $m^2(m - 1)$
45. $-6x^2(2x + 1)$ **47.** $5y^6(13y^4 + 7)$ **49.** no common factor (except 1)
51. $8mn^3(1 + 3m)$ **53.** $-2x(2x^2 - 5x + 3)$ **55.** $13y^2(y^6 + 2y^2 - 3)$
57. $9qp^3(5q^3p^2 + 4p^3 + 9q)$ **59.** $(x + 2)(c + d)$ **61.** $(2a + b)(a^2 - b)$
63. $(p + 4)(q - 1)$ **65.** not in factored form; $(7t + 4)(8 + x)$
66. not in factored form; $(5x - 1)(3r + 7)$ **67.** in factored form
68. in factored form   **69.** not in factored form   **70.** not in factored form
71. $(5 + n)(m + 4)$ **73.** $(2y - 7)(3x + 4)$ **75.** $(y + 3)(3x + 1)$
77. $(z + 2)(7z - a)$ **79.** $(3r + 2y)(6r - x)$ **81.** $(w + 1)(w^2 + 9)$
83. $(a + 2)(3a^2 - 2)$ **85.** $(4m - p^2)(4m^2 - p)$ **87.** $(y + 3)(y + x)$
89. $(z - 2)(2z - 3w)$ **91.** $(5 - 2p)(m + 3)$ **93.** $(3r + 2y)(6r - t)$
95. commutative property   **96.** $2x(y - 4) - 3(y - 4)$ **97.** No, because it is not a product. It is the difference between $2x(y - 4)$ and $3(y - 4)$.
98. $(y - 4)(2x - 3)$, or $(2x - 3)(y - 4)$; yes

SECTION 6.2 (pages 455–456)

1. a and b must have different signs.   **2.** a and b must have the same sign.
3. C   **4.** Factor out the greatest common factor, $2x$.   **5.** $a^2 + 13a + 36$
6. $y^2 - 4y - 21$ **7.** A prime polynomial is one that cannot be factored using only integers in the factors.   **8.** To check a factored form, we multiply the factors. We should get the trinomial we started with. If we forget to factor out any common factors, the check will *not* indicate it.   **9.** 1 and 12, -1 and -12, 2 and 6, -2 and -6, 3 and 4, -3 and -4; The pair with a sum of 7 is 3 and 4.   **11.** 1 and -24, -1 and 24, 2 and -12, -2 and 12, 3 and -8, -3 and 8, 4 and -6, -4 and 6; The pair with a sum of -5 is 3 and -8.   **13.** $p + 6$ **15.** $x + 11$ **17.** $x - 8$ **19.** $y - 5$ **21.** $x + 11$
23. $y - 9$ **25.** $(y + 8)(y + 1)$ **27.** $(b + 3)(b + 5)$
29. $(m + 5)(m - 4)$ **31.** $(x + 8)(x - 5)$ **33.** prime
35. $(y - 5)(y - 3)$ **37.** $(z - 8)(z - 7)$ **39.** $(r - 6)(r + 5)$
41. $(a - 12)(a + 4)$ **43.** $(r + 2a)(r + a)$ **45.** $(x + y)(x + 3y)$
47. $(t + 2z)(t - 3z)$ **49.** $(v - 5w)(v - 6w)$ **51.** $(a + 5b)(a - 3b)$
53. $4(x + 5)(x - 2)$ **55.** $2t(t + 1)(t + 3)$ **57.** $-2x^4(x - 3)(x + 7)$
59. $a^3(a + 4b)(a - b)$ **61.** $5m^2(m^3 + 5m^2 - 8)$
63. $mn(m - 6n)(m - 4n)$

SECTION 6.3 (pages 459–460)

1. B   **2.** D   **3.** $(m + 6)(m + 2)$ **5.** $(a + 5)(a - 2)$
7. $(2t + 1)(5t + 2)$ **9.** $(3z - 2)(5z - 3)$ **11.** $(2s - t)(4s + 3t)$
13. $(3a + 2b)(5a + 4b)$ **15.** **(a)** 2; 12; 24; 11   **(b)** 3; 8 (Order is irrelevant.)   **(c)** $3m$; $8m$ **(d)** $2m^2 + 3m + 8m + 12$
(e) $(2m + 3)(m + 4)$ **(f)** $(2m + 3)(m + 4) = 2m^2 + 11m + 12$
17. $(2x + 1)(x + 3)$ **19.** $(4r - 3)(r + 1)$ **21.** $(4m + 1)(2m - 3)$
23. $(3m + 1)(7m + 2)$ **25.** $(2b + 1)(3b + 2)$ **27.** $(4y - 3)(3y - 1)$
29. $(4 + x)(4 + 3x)$, or $(x + 4)(3x + 4)$ **31.** $3(4x - 1)(2x - 3)$
33. $2m(m - 4)(m + 5)$ **35.** $-4z^3(z - 1)(8z + 3)$.
37. $(3p + 4q)(4p - 3q)$ **39.** $(3a - 5b)(2a + b)$ **41.** The student stopped too soon. He needs to factor out the common factor $4x - 1$ to get $(4x - 1)(4x - 5)$ as the correct answer.   **42.** The student forgot to include the common factor $3k$ in her answer. The correct answer is $3k(k - 5)(k + 1)$.

SECTION 6.4 (pages 465–466)

1. B   **2.** A   **3.** A   **4.** B   **5.** A   **6.** A   **7.** $2a + 5b$
9. $x^2 + 3x - 4$; $x + 4$, $x - 1$ **11.** $2z^2 - 5z - 3$; $2z + 1$, $z - 3$
13. $(4x + 4)$ cannot be a factor because its terms have a common factor of 4, but those of the polynomial do not. The correct factored form is $(4x - 3)(3x + 4)$.   **14.** The student forgot to factor out the common factor 2 from the terms of the trinomial. The completely factored form is $2(2x - 1)(x + 3)$.   **15.** $(3a + 7)(a + 1)$ **17.** $(2y + 3)(y + 2)$
19. $(3m - 1)(5m + 2)$ **21.** $(3s - 1)(4s + 5)$ **23.** $(5m - 4)(2m - 3)$
25. $(4w - 1)(2w - 3)$ **27.** $(4y + 1)(5y - 11)$ **29.** prime
31. $2(5x + 3)(2x + 1)$ **33.** $-q(5m + 2)(8m - 3)$
35. $3n^2(5n - 3)(n - 2)$ **37.** $-y^2(5x - 4)(3x + 1)$
39. $(5a + 3b)(a - 2b)$ **41.** $(4s + 5t)(3s - t)$
43. $m^4n(3m + 2n)(2m + n)$ **45.** $-1(x + 7)(x - 3)$
47. $-1(3x + 4)(x - 1)$ **49.** $-1(a + 2b)(2a + b)$ **51.** $5 \cdot 7$
52. $(-5)(-7)$ **53.** The product of $3x - 4$ and $2x - 1$ is $6x^2 - 11x + 4$.
54. The product of $4 - 3x$ and $1 - 2x$ is $6x^2 - 11x + 4$.   **55.** The factors in **Exercise 53** are the opposites of the factors in **Exercise 54**.
56. $(3 - 7t)(5 - 2t)$

SECTION 6.5 (pages 473–476)

1. 1; 4; 9; 16; 25; 36; 49; 64; 81; 100; 121; 144; 169; 196; 225; 256; 289; 324; 361; 400   **2.** 1; 16; 81; 256; 625   **3.** A, D   **4.** The binomial $4x^2 + 16$ can be factored as $4(x^2 + 4)$. *After* any common factor is removed, a sum of squares (like $x^2 + 4$ here) *cannot* be factored.
5. $(y + 5)(y - 5)$ **7.** $(x + 12)(x - 12)$ **9.** prime   **11.** prime
13. $(3r + 2)(3r - 2)$ **15.** $4(3x + 2)(3x - 2)$
17. $(14p + 15)(14p - 15)$ **19.** $(4r + 5a)(4r - 5a)$ **21.** prime
23. $(p^2 + 7)(p^2 - 7)$ **25.** $(x^2 + 1)(x + 1)(x - 1)$
27. $(p^2 + 16)(p + 4)(p - 4)$ **29.** B, C   **30.** No, it is not a perfect square since the middle term would have to be 30y.   **31.** $(w + 1)^2$
33. $(x - 4)^2$ **35.** prime   **37.** $2(x + 6)^2$ **39.** $(2x + 3)^2$
41. $(4x - 5)^2$ **43.** $(7x - 2y)^2$ **45.** $(8x + 3y)^2$ **47.** $-2h(5h - 2y)^2$
49. $\left(p + \dfrac{1}{3}\right)\left(p - \dfrac{1}{3}\right)$ **51.** $\left(2m + \dfrac{3}{5}\right)\left(2m - \dfrac{3}{5}\right)$
53. $(x + 0.8)(x - 0.8)$ **55.** $\left(t + \dfrac{1}{2}\right)^2$ **57.** $(x - 0.5)^2$
59. 1; 8; 27; 64; 125; 216; 343; 512; 729; 1000   **60.** 3   **61.** C, D
62. A, D   **63.** $(x + 1)(x^2 - x + 1)$ **65.** $(x - 1)(x^2 + x + 1)$
67. $(p + q)(p^2 - pq + q^2)$ **69.** $(y - 6)(y^2 + 6y + 36)$
71. $(k + 10)(k^2 - 10k + 100)$ **73.** $(3x - 1)(9x^2 + 3x + 1)$
75. $(5x + 2)(25x^2 - 10x + 4)$ **77.** $(y - 2x)(y^2 + 2xy + 4x^2)$
79. $(3x - 4y)(9x^2 + 12xy + 16y^2)$
81. $(2p + 9q)(4p^2 - 18pq + 81q^2)$ **83.** $2(2t - 1)(4t^2 + 2t + 1)$
85. $5(2w + 3)(4w^2 - 6w + 9)$ **87.** $(x + y^2)(x^2 - xy^2 + y^4)$
89. $(5k - 2m^3)(25k^2 + 10km^3 + 4m^6)$ **91.** $(x^3 - 1)(x^3 + 1)$
92. $(x - 1)(x^2 + x + 1)(x + 1)(x^2 - x + 1)$
93. $(x^2 - 1)(x^4 + x^2 + 1)$ **94.** $(x - 1)(x + 1)(x^4 + x^2 + 1)$
95. The result in **Exercise 92** is completely factored.   **96.** Show that $x^4 + x^2 + 1 = (x^2 + x + 1)(x^2 - x + 1)$.   **97.** difference of squares
98. $(x - 3)(x^2 + 3x + 9)(x + 3)(x^2 - 3x + 9)$

SECTION 6.6 (pages 479–480)

1. (a) B **(b)** D **(c)** A **(d)** C **(e)** A, B **2. (a)** C **(b)** E **(c)** C
(d) B **(e)** A, B **3.** $8m^3(4m^6 + 2m^2 + 3)$ **5.** $7k(2k + 5)(k - 2)$
7. $(m + n)(m - 4n)$ **9.** $10nr(10nr + 3r^2 - 5n)$ **11.** $(4 + m)(5 + 3n)$
13. $(y^2 + 9)(y + 3)(y - 3)$ **15.** $(2p - 5)(4p^2 + 10p + 25)$
17. $(p - 12)^2$ **19.** $(4z - 1)^2$ **21.** prime **23.** $(p + 4)(p^2 - 4p + 16)$
25. $(2a + 1)(a^2 - 7)$ **27.** $(4r + 3m)^2$ **29.** prime **31.** $4(2k - 3)^2$
33. $(4k - 3h)(2k + h)$ **35.** $(5z - 6)(2z + 1)$ **37.** $(3y - 1)(3y + 5)$
39. prime **41.** $(2 - q)(2 - 3p)$ **43.** $8(5z + 4)(25z^2 - 20z + 16)$
45. $(10a + 9y)(10a - 9y)$ **47.** $(a + 4)^2$ **49.** prime **51.** $(5a - 7b)^2$
53. $(m + 3)(2m - 5n)$ **55.** $2(3m - 10)(9m^2 + 30m + 100)$

SECTION 6.7 (pages 487–489)

1. $ax^2 + bx + c$ **2.** standard **3.** factor **4.** 0; zero; factor
5. (a) linear **(b)** quadratic **(c)** quadratic **(d)** linear **6.** Because
$(x - 9)^2 = (x - 9)(x - 9)$, applying the zero-factor property leads to two
solutions of 9. Thus, 9 is a double solution. **7.** Set each *variable* factor
equal to 0, to get $2x = 0$ or $3x - 4 = 0$. The solution set is $\left\{0, \dfrac{4}{3}\right\}$.
8. The variable x is another factor to set equal to 0, so the solution set is
$\left\{0, \dfrac{1}{7}\right\}$. **9.** $\{-5, 2\}$ **11.** $\left\{3, \dfrac{7}{2}\right\}$ **13.** $\left\{-\dfrac{1}{2}, \dfrac{1}{6}\right\}$ **15.** $\left\{-\dfrac{5}{6}, 0\right\}$
17. $\left\{0, \dfrac{4}{3}\right\}$ **19.** $\{9\}$ **21.** $\{-2, -1\}$ **23.** $\{1, 2\}$ **25.** $\{-8, 3\}$
27. $\{-1, 3\}$ **29.** $\{-2, -1\}$ **31.** $\{-4\}$ **33.** $\left\{-2, \dfrac{1}{3}\right\}$
35. $\left\{-\dfrac{4}{3}, \dfrac{1}{2}\right\}$ **37.** $\left\{-\dfrac{2}{3}\right\}$ **39.** $\{-3, 3\}$ **41.** $\left\{-\dfrac{7}{4}, \dfrac{7}{4}\right\}$
43. $\{-11, 11\}$ **45.** $\{0, 7\}$ **47.** $\left\{0, \dfrac{1}{2}\right\}$ **49.** $\{2, 5\}$ **51.** $\left\{-4, \dfrac{1}{2}\right\}$
53. $\left\{-12, \dfrac{11}{2}\right\}$ **55.** $\{-2, 0, 2\}$ **57.** $\left\{-\dfrac{7}{3}, 0, \dfrac{7}{3}\right\}$ **59.** $\left\{-\dfrac{5}{2}, \dfrac{1}{3}, 5\right\}$
61. $\left\{-\dfrac{7}{2}, -3, 1\right\}$ **63.** $\{-5, 0, 4\}$ **65.** $\{-3, 0, 5\}$ **67.** $\{-1, 3\}$
69. (a) 64; 144; 4; 6 **(b)** No time has elapsed, so the object hasn't fallen
(been released) yet. **70.** Time cannot be negative.

SECTION 6.8 (pages 496–500)

1. Read; variable; equation; Solve; answer; Check; original **2.** Only 6 is
reasonable since a square cannot have a side of negative length.
3. *Step 3:* $(2x + 1)(x + 1)$; *Step 4:* 4; $-\dfrac{11}{2}$; *Step 5:* 9; 5; *Step 6:* $9 \cdot 5$
5. *Step 3:* 192; 4x; *Step 4:* 6; -8; *Step 5:* 8; 6; *Step 6:* $8 \cdot 6$; 192
7. length: 14 cm; width: 12 cm **9.** base: 12 in.; height: 5 in.
11. length: 15 in.; width: 12 in. **13.** height: 13 in.; width: 10 in.
15. mirror: 7 ft; painting: 9 ft **17.** 20, 21 **19.** $-3, -2$ or 4, 5
21. $-3, -1$ or 7, 9 **23.** $-2, 0, 2$ or 6, 8, 10 **25.** 7, 9, 11 **27.** 12 cm
29. 12 mi **31.** 8 ft **33. (a)** 1 sec **(b)** $\dfrac{1}{2}$ sec and $1\dfrac{1}{2}$ sec **(c)** 3 sec
(d) The negative solution, -1, does not make sense since t represents time,
which cannot be negative. **35.** 112 ft **37.** 256 ft **39. (a)** 109 million;
The result using the model is the same as the actual number from the table
for 2000. **(b)** 14 **(c)** 270 million; The result is a little less than
286 million, the actual number for 2008. **(d)** 393 million

Chapter 6 REVIEW EXERCISES (pages 504–508)

1. $15(t + 3)$ **2.** $30z(2z^2 + 1)$ **3.** $11x^2(4x + 5)$
4. $50m^2n^2(2n - mn^2 + 3)$ **5.** $(x - 4)(2y + 3)$ **6.** $(2y + 3)(3y + 2x)$
7. $(x + 3)(x + 7)$ **8.** $(y - 5)(y - 8)$ **9.** $(q + 9)(q - 3)$
10. $(r - 8)(r + 7)$ **11.** prime **12.** $3(x^2 + 2x + 2)$
13. $(r + 8s)(r - 12s)$ **14.** $(p + 12q)(p - 10q)$
15. $-8p(p + 2)(p - 5)$ **16.** $3x^2(x + 2)(x + 8)$
17. $(m + 3n)(m - 6n)$ **18.** $(y - 3z)(y - 5z)$
19. $p^5(p - 2q)(p + q)$ **20.** $-3r^3(r + 3s)(r - 5s)$
21. r and $6r$, $2r$ and $3r$ **22.** Factor out z. **23.** $(2k - 1)(k - 2)$
24. $(3r - 1)(r + 4)$ **25.** $(3r + 2)(2r - 3)$ **26.** $(5z + 1)(2z - 1)$
27. prime **28.** $4x^3(3x - 1)(2x - 1)$ **29.** $-3(x + 2)(2x - 5)$
30. $rs(5r + 6s)(2r + s)$ **31.** $-5y(3y + 2)(2y - 1)$ **32.** prime
33. $-mn(3m + 5)(m - 8)$ **34.** $(2a - 5b)(7a + 4b)$ **35.** B
36. D **37.** $(n + 8)(n - 8)$ **38.** $(5b + 11)(5b - 11)$
39. $(7y + 5w)(7y - 5w)$ **40.** $36(2p + q)(2p - q)$ **41.** prime
42. $(z + 5)^2$ **43.** $(3t - 7)^2$ **44.** $(4m + 5n)^2$
45. $(5x - 1)(25x^2 + 5x + 1)$ **46.** $(10p + 3)(100p^2 - 30p + 9)$
47. $\left\{-\dfrac{3}{4}, 1\right\}$ **48.** $\{-7, -3, 4\}$ **49.** $\left\{0, \dfrac{5}{2}\right\}$ **50.** $\{-3, -1\}$
51. $\{1, 4\}$ **52.** $\{3, 5\}$ **53.** $\left\{-\dfrac{4}{3}, 5\right\}$ **54.** $\left\{-\dfrac{8}{9}, \dfrac{8}{9}\right\}$ **55.** $\{0, 8\}$
56. $\{-1, 6\}$ **57.** $\{7\}$ **58.** $\{6\}$ **59.** $\left\{-\dfrac{2}{5}, -2, -1\right\}$ **60.** $\{-3, 3\}$
61. $\left\{-\dfrac{3}{8}, 0, \dfrac{3}{8}\right\}$ **62.** $\left\{-\dfrac{3}{4}, -\dfrac{1}{2}, \dfrac{1}{3}\right\}$ **63.** $\left\{\dfrac{9}{5}\right\}$ **64.** $\{-3, 10\}$
65. length: 10 ft; width: 4 ft **66.** 5 ft **67.** 6, 7 or $-5, -4$
68. $-5, -4, -3$ or 5, 6, 7 **69.** 26 mi **70.** 6 m **71.** width: 10 m;
length: 17 m **72. (a)** 256 ft **(b)** 1024 ft **73.** 602 thousand;
The result is a little higher than the 592 thousand given in the table.
74. 882 thousand; The estimate may be unreliable because the
conditions that prevailed in the years 2001–2009 may have changed,
causing either a greater increase or a greater decrease predicted by the
model for the number of such vehicles. **75.** D **76.** The terms of
$2x + 8$ have a common factor of 2. The completely factored form is
$2(x + 4)(3x - 4)$. **77.** $(3m + 4p)(5m - 4)$
78. $8abc(3b^2c - 7ac^2 + 9ab)$ **79.** $(z - x)(z - 10x)$
80. $(3k + 5)(k + 2)$ **81.** $(y^2 + 25)(y + 5)(y - 5)$
82. $3m(2m + 3)(m - 5)$ **83.** prime **84.** $8(z + 2y)(z^2 - 2zy + 4y^2)$
85. $-1(2r + 3q)(6r - 5q)$ **86.** $(10a + 3)(10a - 3)$ **87.** $(7t + 4)^2$
88. $\{0, 7\}$ **89.** $\{-5, 2\}$ **90.** $\left\{-\dfrac{2}{5}\right\}$ **91.** 15 m, 36 m, 39 m
92. length: 6 m; width: 4 m

Chapter 6 TEST (pages 509–510)

1. D **2.** $6x(2x - 5)$ **3.** $m^2n(2mn + 3m - 5n)$ **4.** $(2x + y)(a - b)$
5. $(x - 7)(x - 2)$ **6.** $(2x + 3)(x - 1)$ **7.** $(3x + 1)(2x - 7)$
8. $3(x + 1)(x - 5)$ **9.** $(5z - 1)(2z - 3)$ **10.** prime **11.** prime
12. $(y + 7)(y - 7)$ **13.** $(9a + 11b)(9a - 11b)$ **14.** $(x + 8)^2$
15. $(2x - 7y)^2$ **16.** $-2(x + 1)^2$ **17.** $4t(t + 4)^2$
18. $(x^2 + 9)(x + 3)(x - 3)$ **19.** $(x - 8)(x^2 + 8x + 64)$
20. $8(k + 2)(k^2 - 2k + 4)$ **21.** $\{-3, 9\}$ **22.** $\left\{\dfrac{1}{2}, 6\right\}$
23. $\left\{-\dfrac{2}{5}, \dfrac{2}{5}\right\}$ **24.** $\{10\}$ **25.** $\{0, 3\}$ **26.** $\left\{-8, -\dfrac{5}{2}, \dfrac{1}{3}\right\}$
27. 6 ft by 9 ft **28.** $-2, -1$ **29.** 17 ft **30.** 85 billion pieces

Chapters R–6 CUMULATIVE REVIEW EXERCISES (pages 511–512)

1. $\{0\}$ **2.** $\{0.05\}$ **3.** $\{6\}$ **4.** $t = \dfrac{A - P}{Pr}$ **5.** second column: 38%, 12%; third column: 230, 205 **6.** gold: 9; silver: 15; bronze: 13 **7.** $669 **8.** 110° and 70° **9. (a)** negative; positive **(b)** negative; negative **10.** $\left(-\dfrac{3}{4}, 0\right)$, $(0, 3)$ **11.** 4

12.

(0, 3) $y = 4x + 3$ $\left(-\dfrac{3}{4}, 0\right)$

13. (a) 8; A slope of (approximately) 8 means that the retail sales of prescription drugs increased by about $8 billion per year. **(b)** (2008, 250)
14. $\{(-1, 2)\}$ **15.** $\varnothing$ **16.** 4
17. $\dfrac{16}{9}$ **18.** 1 **19.** 256 **20.** $\dfrac{1}{p^2}$
21. $\dfrac{1}{m^6}$ **22.** $-4k^2 - 4k + 8$ **23.** $45x^2 + 3x - 18$ **24.** $9p^2 + 12p + 4$
25. $4x^3 + 6x^2 - 3x + 10$ **26.** $(2a - 1)(a + 4)$
27. $(2m + 3)(5m + 2)$ **28.** $(4t + 3v)(2t + v)$ **29.** $(2p - 3)^2$
30. $(5r + 9t)(5r - 9t)$ **31.** $2pq(3p + 1)(p + 1)$ **32.** $\left\{-\dfrac{2}{3}, \dfrac{1}{2}\right\}$
33. $\{0, 8\}$ **34.** $\left\{\dfrac{4}{7}\right\}$ **35.** 5 m, 12 m, 13 m

Chapter 7 Rational Expressions and Functions

SECTION 7.1 (pages 521–524)

1. $\dfrac{2}{15}$ **2.** $\dfrac{4}{15}$ **3.** $\dfrac{9}{10}$ **4.** $\dfrac{25}{28}$ **5.** $\dfrac{3}{4}$ **6.** $\dfrac{7}{12}$ **7. (a)** C **(b)** A **(c)** D
(d) B **(e)** E **(f)** F **8.** B, E, F **9.** 7; $\{x \mid x \neq 7\}$
11. $-\dfrac{1}{7}$; $\left\{x \mid x \neq -\dfrac{1}{7}\right\}$ **13.** 0; $\{x \mid x \neq 0\}$ **15.** $-2, \dfrac{3}{2}$; $\left\{x \mid x \neq -2, \dfrac{3}{2}\right\}$
17. none; $\{x \mid x \text{ is a real number}\}$ **19.** none; $\{x \mid x \text{ is a real number}\}$
21. numerator: x^2, $4x$; denominator: x, 4; It simplifies to x. **22.** D **23.** B
24. B, D **25.** x **27.** $\dfrac{x - 3}{x + 5}$ **29.** $\dfrac{x + 3}{2x(x - 3)}$ **31.** already in lowest terms
33. $\dfrac{6}{7}$ **35.** $\dfrac{z}{6}$ **37.** $\dfrac{2}{t - 3}$ **39.** $\dfrac{x - 3}{x + 1}$ **41.** $\dfrac{4x + 1}{4x + 3}$ **43.** $a^2 - ab + b^2$
45. $\dfrac{c + 6d}{c - d}$ **47.** $\dfrac{a + b}{a - b}$ **49.** -1 *In Exercises 51–55, there are other acceptable ways to express each answer.* **51.** $-x - 2$ **53.** $-x - y$
55. $-\dfrac{x + y}{x - y}$ **57.** $-\dfrac{1}{2}$ **59.** already in lowest terms **61.** $\dfrac{x + 4}{x - 2}$
63. $\dfrac{2x + 3}{x + 2}$ **65.** $-\dfrac{35}{8}$ **67.** $\dfrac{7x}{6}$ **69.** $-\dfrac{p + 5}{2p}$ (There are other ways.)
71. $\dfrac{-m(m + 7)}{m + 1}$ (There are other ways.) **73.** -2 **75.** $\dfrac{x + 4}{x - 4}$
77. $\dfrac{2x - 3}{2(x - 3)}$ **79.** $\dfrac{2x + 3y}{2x - 3y}$ **81.** $\dfrac{k + 5p}{2k + 5p}$ **83.** $(k - 1)(k - 2)$

SECTION 7.2 (pages 531–533)

1. $\dfrac{4}{5}$ **2.** $\dfrac{7}{8}$ **3.** $-\dfrac{1}{18}$ **4.** $-\dfrac{1}{12}$ **5.** $\dfrac{31}{36}$ **6.** $\dfrac{23}{30}$ **7.** $\dfrac{9}{t}$ **9.** $\dfrac{2}{x}$ **11.** 1
13. $x - 5$ **15.** $\dfrac{1}{p + 3}$ **17.** $a - b$ **19.** $72x^4y^5$ **21.** $z(z - 2)$

23. $2(y + 4)$ **25.** $(x + 9)^2(x - 9)$ **27.** $(m + n)(m - n)$
29. $x(x - 4)(x + 1)$ **31.** $(t + 5)(t - 2)(2t - 3)$ **33.** $2y(y + 3)(y - 3)$
35. The expression $\dfrac{x - 4x - 1}{x + 2}$ is incorrect. The third term in the numerator should be $+ 1$, since the $-$ sign should be distributed over *both* $4x$ and -1. The answer should be $\dfrac{-3x + 1}{x + 2}$. **36.** The expressions are equivalent. Multiplying $\dfrac{3}{5 - y}$ by 1 in the form $\dfrac{-1}{-1}$ gives $\dfrac{-3}{y - 5}$. **37.** $\dfrac{31}{3t}$ **39.** $\dfrac{5 - 22x}{12x^2y}$
41. $\dfrac{1}{x(x - 1)}$ **43.** $\dfrac{5a^2 - 7a}{(a + 1)(a - 3)}$ **45.** 3 **47.** $\dfrac{3}{x - 4}$, or $\dfrac{-3}{4 - x}$
49. $\dfrac{w + z}{w - z}$, or $\dfrac{-w - z}{z - w}$ **51.** $\dfrac{-13}{12(3 + x)}$ **53.** $\dfrac{2(2x - 1)}{x - 1}$ **55.** $\dfrac{7}{y}$
57. $\dfrac{6}{x - 2}$ **59.** $\dfrac{3x - 2}{x - 1}$ **61.** $\dfrac{4x - 7}{x^2 - x + 1}$ **63.** $\dfrac{2x + 1}{x}$
65. $\dfrac{x}{(x - 2)^2(x - 3)}$ **67.** $\dfrac{10x + 23}{(x + 2)^2(x + 3)}$ **69.** $\dfrac{2x(x + 12y)}{(x + 2y)(x - y)(x + 6y)}$
71. $c(x) = \dfrac{10x}{49(101 - x)}$

SECTION 7.3 (pages 538–539)

1. complex; numerator; both **2.** reciprocal; denominator
3. LCD; identity **4. (a)** $\dfrac{1}{20}$ **(b)** -2 **5.** $\dfrac{1}{6}$ **6.** $\dfrac{9}{5}$ **7.** $\dfrac{4}{15}$ **8.** $-\dfrac{16}{15}$
9. $\dfrac{2x}{x - 1}$ **11.** $\dfrac{2(k + 1)}{3k - 1}$ **13.** $\dfrac{5x^2}{9z^3}$ **15.** $\dfrac{1 + x}{-1 + x}$ **17.** $\dfrac{6x + 1}{7x - 3}$
19. $\dfrac{y + x}{y - x}$ **21.** $4x$ **23.** $\dfrac{y + 4}{2}$ **25.** $x + 4y$ **27.** $\dfrac{a + b}{ab}$ **29.** xy
31. $\dfrac{3y}{2}$ **33.** $\dfrac{x^2 + 5x + 4}{x^2 + 5x + 10}$ **35.** $\dfrac{x^2y^2}{y^2 + x^2}$ **37.** $\dfrac{y^2 + x^2}{xy^2 + x^2y}$, or $\dfrac{y^2 + x^2}{xy(y + x)}$
39. $\dfrac{2xy - 3x}{x + 3y^2}$ **41.** $\dfrac{1}{2xy}$

SECTION 7.4 (pages 545–547)

1. (a) equation **(b)** expression **(c)** expression **(d)** equation
2. "Solve" refers to finding the solution set of an equation. This is an expression. "Solve" should be replaced by "Simplify" or "Add."
3. (a) 0 **(b)** $\{x \mid x \neq 0\}$ **5. (a)** $-1, 2$ **(b)** $\{x \mid x \neq -1, 2\}$
7. (a) $-\dfrac{5}{3}, 0, -\dfrac{3}{2}$ **(b)** $\left\{x \mid x \neq -\dfrac{5}{3}, 0, -\dfrac{3}{2}\right\}$ **9. (a)** $4, \dfrac{7}{2}$
(b) $\left\{x \mid x \neq 4, \dfrac{7}{2}\right\}$ **11. (a)** $0, 1, -3, 2$ **(b)** $\{x \mid x \neq 0, 1, -3, 2\}$
13. $\{1\}$ **15.** $\{-6, 4\}$ **17.** $\{-7, 3\}$ **19.** $\left\{-\dfrac{7}{12}\right\}$ **21.** $\varnothing$ **23.** $\{-3\}$
25. $\{5\}$ **27.** $\{0\}$ **29.** $\{5\}$ **31.** $\left\{\dfrac{27}{56}\right\}$ **33.** $\{-10\}$ **35.** $\varnothing$ **37.** $\{0\}$
39. $\{-3, -1\}$ **41.** $\varnothing$ **43.** $\varnothing$ **45.** $\left\{x \mid x \neq -\dfrac{3}{2}, \dfrac{3}{2}\right\}$
47. $x = 0$; $y = 0$ **49.** $x = 0$; $y = 0$ **51.** $x = 2$; $y = 0$

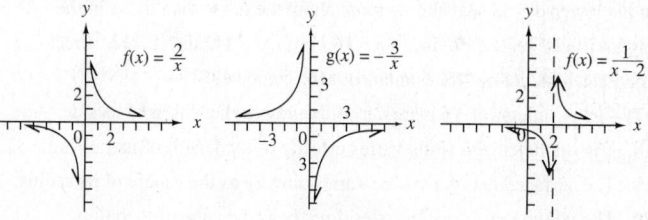

53. (a) 0 **(b)** 1.6 **(c)** 4.1 **(d)** The waiting time also increases.

SUMMARY EXERCISES Simplifying Rational Expressions vs. Solving Rational Equations (pages 548–549)

1. equation; $\{20\}$ **2.** expression; $\dfrac{2(x+5)}{5}$ **3.** expression; $-\dfrac{22}{7x}$

4. expression; $\dfrac{y+x}{y-x}$ **5.** equation; $\left\{\dfrac{1}{2}\right\}$ **6.** equation; $\{7\}$

7. expression; $\dfrac{43}{24x}$ **8.** equation; $\{1\}$ **9.** expression; $\dfrac{5x-1}{-2x+2}$, or

$\dfrac{5x-1}{-2(x-1)}$ **10.** expression; $\dfrac{25}{4(r+2)}$ **11.** expression; $\dfrac{x^2+xy+2y^2}{(x+y)(x-y)}$

12. expression; $\dfrac{24p}{p+2}$ **13.** expression; $-\dfrac{5}{36}$ **14.** equation; $\{0\}$

15. expression; $\dfrac{b+3}{3}$ **16.** expression; $\dfrac{5}{3z}$ **17.** expression;

$\dfrac{2x+10}{x(x-2)(x+2)}$ **18.** equation; $\left\{\dfrac{1}{7},2\right\}$ **19.** expression; $\dfrac{-1}{x-3}$, or $\dfrac{1}{3-x}$

20. expression; $\dfrac{t-2}{8}$ **21.** equation; $\varnothing$ **22.** expression; $\dfrac{13x+28}{2x(x+4)(x-4)}$

23. expression; $\dfrac{-x}{3x+5y}$ **24.** expression; $\dfrac{k(2k^2-2k+5)}{(k-1)(3k^2-2)}$

25. equation; $\{-10\}$ **26.** equation; $\{-13\}$ **27.** expression; $\dfrac{3y+2}{y+3}$

28. equation; $\left\{\dfrac{5}{4}\right\}$ **29.** equation; $\varnothing$ **30.** expression; $\dfrac{2z-3}{2z+3}$

SECTION 7.5 (pages 559–564)

1. A **2.** B **3.** D **4.** D **5.** 65.625 **7.** $\dfrac{25}{4}$ **9.** Multiply each side

by $a-b$. **10.** Factor out r on the left. **11.** $G=\dfrac{Fd^2}{Mm}$ **13.** $a=\dfrac{bc}{c+b}$

15. $v=\dfrac{PVt}{pT}$ **17.** $r=\dfrac{nE-IR}{In}$ **19.** $b=\dfrac{2A}{h}-B$, or $b=\dfrac{2A-Bh}{h}$

21. $r=\dfrac{eR}{E-e}$ **23.** $R=\dfrac{D}{1-DT}$ **25.** 21 girls, 7 boys **26.** \$0.72

27. 1.75 in. **29.** 5.4 in. **31.** 93 games **33.** 25,000 fish

35. 6.6 more gallons **37.** 2.4 mL **39.** $x=\dfrac{7}{2}$; $AC=8$; $DF=12$

41. 18.809 min **43.** 314.248 m per min **45.** table entries: (fourth column)

$\dfrac{10}{12+x},\dfrac{6}{12-x}$; 3 mph **47.** 10 mph **49.** 1020 mi **51.** 480 mi

53. 190 mi **55.** table entries: (fourth column) $\dfrac{1}{15}x,\dfrac{1}{12}x$; $6\dfrac{2}{3}$ min

57. table entries: (second column) $\dfrac{1}{20},\dfrac{1}{x}$; (fourth column) $\dfrac{1}{20}(12)$ or

$\dfrac{3}{5},\dfrac{1}{x}(12)$ or $\dfrac{12}{x}$; 30 hr **59.** 20 hr **61.** $2\dfrac{4}{5}$ hr

SECTION 7.6 (pages 571–574)

1. direct **2.** direct **3.** inverse **4.** inverse **5.** direct **6.** inverse
7. (a) increases; decreases (b) decreases; increases **8.** The customers
in the lower-priced seats know more about the game than those in the
higher priced seats. **9.** inverse **10.** inverse **11.** direct **12.** direct
13. joint **14.** joint **15.** combined **16.** combined
17. The perimeter of a square varies directly as the length of its side.
18. The diameter of a circle varies directly as the length of its radius.
19. The surface area of a sphere varies directly as the square of its radius.
20. The volume of a sphere varies directly as the cube of its radius.

21. The area of a triangle varies jointly as the length of its base and its height.
22. The volume of a cone varies jointly as the square of its radius and

its height. **23.** 36 **25.** $\dfrac{16}{9}$ **27.** 0.625 **29.** $\dfrac{16}{5}$ **31.** $222\dfrac{2}{9}$

33. $\$4.59\dfrac{9}{10}$ **35.** 8 lb **37.** about 450 cm³ **39.** 256 ft

41. $13\dfrac{1}{3}$ amperes **43.** $21\dfrac{1}{3}$ foot-candles **45.** \$420 **47.** 11.8 lb

49. 448.1 lb **51.** approximately 68,600 calls

Chapter 7 REVIEW EXERCISES (pages 579–582)

1. -6; $\{x\,|\,x\neq-6\}$ **2.** 2, 5; $\{x\,|\,x\neq2,5\}$ **3.** 9; $\{x\,|\,x\neq9\}$ **4.** $\dfrac{x}{2}$

5. $\dfrac{5m+n}{5m-n}$ **6.** $\dfrac{-1}{2+r}$ **7.** $\dfrac{3y^2(2y+3)}{2y-3}$ **8.** $\dfrac{-3(w+4)}{w}$ **9.** $\dfrac{z(z+2)}{z+5}$

10. 1 **11.** $96b^5$ **12.** $9r^2(3r+1)$ **13.** $(3x-1)(2x+5)(3x+4)$

14. $3(x-4)^2(x+2)$ **15.** $\dfrac{16z-3}{2z^2}$ **16.** 12 **17.** $\dfrac{71}{30(a+2)}$

18. $\dfrac{13r^2+5rs}{(5r+s)(2r-s)(r+s)}$ **19.** $\dfrac{3+2t}{4-7t}$ **20.** -2 **21.** $\dfrac{1}{3q+2p}$

22. $\dfrac{y+x}{xy}$ **23.** $\{-3\}$ **24.** $\{-2\}$ **25.** $\{0\}$ **26.** $\varnothing$ **27.** (a) equation;

$\{-24\}$ (b) expression; $\dfrac{24+x}{6x}$ **28.** Although her algebra was correct,
3 is not a solution because it is not in the domain of the variable in the
equation. Thus, $\varnothing$ is correct. **29.** C; $x=0$; $y=0$

30. $x=-1$; $y=0$ **31.** $\dfrac{15}{2}$ **32.** 2 **33.** $c=\dfrac{ab}{b-a}$, or $c=\dfrac{-ab}{a-b}$

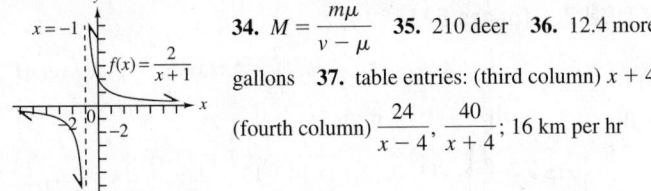

34. $M=\dfrac{m\mu}{v-\mu}$ **35.** 210 deer **36.** 12.4 more
gallons **37.** table entries: (third column) $x+4$;
(fourth column) $\dfrac{24}{x-4},\dfrac{40}{x+4}$; 16 km per hr

38. table entries: (second column) $\dfrac{1}{8},\dfrac{1}{12}$; (fourth column) $\dfrac{1}{8}x,\dfrac{1}{12}x$; $4\dfrac{4}{5}$ min

39. C **40.** 6 **41.** 430 mm **42.** 36 ft³ **43.** $\dfrac{1}{x-2y}$ **44.** $\dfrac{x+5}{x+2}$

45. $\dfrac{6m+5}{3m^2}$ **46.** $\dfrac{11}{3-x}$, or $\dfrac{-11}{x-3}$ **47.** $\dfrac{x^2-6}{2(2x+1)}$ **48.** $\dfrac{s^2+t^2}{st(s-t)}$

49. $\dfrac{3-5x}{6x+1}$ **50.** $\dfrac{acd+b^2d+bc^2}{bcd}$ **51.** $\dfrac{1}{3}$ **52.** $\dfrac{k-3}{36k^2+6k+1}$

53. $\dfrac{5a^2+4ab+12b^2}{(a+3b)(a-2b)(a+b)}$ **54.** $\dfrac{x(9x+1)}{3x+1}$ **55.** $\left\{\dfrac{1}{3}\right\}$

56. $r=\dfrac{AR}{R-A}$, or $r=\dfrac{-AR}{A-R}$ **57.** $\{1,4\}$ **58.** $\left\{-\dfrac{14}{3}\right\}$ **59.** $3\dfrac{3}{5}$ hr

60. 2.4 mi **61.** 5.59 vibrations per sec **62.** 12 ft²

Chapter 7 TEST (pages 583–584)

1. $-2,\dfrac{4}{3}$; $\left\{x\,\middle|\,x\neq-2,\dfrac{4}{3}\right\}$ **2.** $\dfrac{2x-5}{x(3x-1)}$ **3.** $\dfrac{3(x+3)}{4}$ **4.** $\dfrac{y+4}{y-5}$

5. -2 **6.** $\dfrac{x+5}{x}$ **7.** $t^2(t+3)(t-2)$ **8.** $\dfrac{7-2t}{6t^2}$ **9.** $\dfrac{13x+35}{(x-7)(x+7)}$

10. $\dfrac{4}{x+2}$ **11.** $\dfrac{11x+21}{(x-3)^2(x+3)}$ **12.** $\dfrac{72}{11}$ **13.** $-\dfrac{1}{a+b}$ **14.** $\dfrac{2y^2+x^2}{xy(y-x)}$

15. (a) expression; $\dfrac{11(x-6)}{12}$ (b) equation; $\{6\}$ **16.** $\left\{\dfrac{1}{2}\right\}$

17. $\{5\}$ **18.** $\ell = \dfrac{2S}{n} - a$, or $\ell = \dfrac{25 - na}{n}$

19. $x = -1; y = 0$ **20.** $3\dfrac{3}{14}$ hr **21.** 15 mph **22.** 48,000 fish

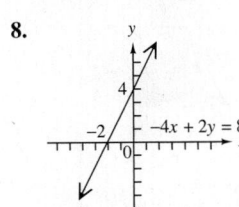

23. (a) 3 units **(b)** 0 **24.** 200 amps
$f(x) = \dfrac{-2}{x+1}$
25. 0.8 lb

Chapters R–7 CUMULATIVE REVIEW EXERCISES (pages 585–586)

1. $\left\{-\dfrac{15}{4}\right\}$ **2.** $\{11\}$ **3.** $\left(-\infty, \dfrac{240}{13}\right]$ **4.** \$4000 at 4%; \$8000 at 3%

5. 6 m **6. (a)** $-\dfrac{3}{2}$ **(b)** $\dfrac{3}{4}$ **7. (a)** $y = -\dfrac{3}{2}x + \dfrac{1}{2}$ **(b)** $y = \dfrac{3}{4}x - \dfrac{7}{4}$

8.

9.

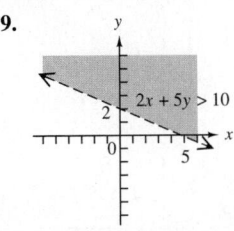

$-4x + 2y = 8$

$2x + 5y > 10$

10. (a) $f(x) = \dfrac{5}{3}x - \dfrac{8}{3}$ **(b)** -1 **11. (a)** yes **(b)** domain: $(-\infty, \infty)$;
range: $(-\infty, \infty)$ **12.** $\{(-1, 3)\}$ **13.** $\{(-1, 2)\}$

14. $\{(x, y) \mid 2x - y = 6\}$ **15.** $\dfrac{a^{10}}{b^{10}}$ **16.** $\dfrac{m}{n}$ **17.** $4y^2 - 7y - 6$

18. $x^2 + 4x - 7$ **19.** $12x^2 + 5x - 3$ **20.** $49t^6 - 64$

21. $16x^2 + 40x + 25$ **22.** $(2x + 5)(x - 9)$ **23.** $25(2t + 1)(2t - 1)$

24. $(2p + 5)(4p^2 - 10p + 25)$ **25.** $\dfrac{a(a - b)}{2(a + b)}$ **26.** 3 **27.** $\dfrac{2(x + 2)}{2x - 1}$

28. $\left\{-\dfrac{7}{3}, 1\right\}$ **29.** $\{-4\}$ **30.** $q = \dfrac{fp}{p - f}$, or $q = \dfrac{-fp}{f - p}$

CHAPTER 8 Equations, Inequalities, and Systems Revisited

SECTION 8.1 (pages 598–601)

1. algebraic expression; does; is not **2.** linear equation; =; first-degree
equation; one **3.** true; solution; solution set **4.** solution set; conditional
equation **5.** identity; all real numbers **6.** contradiction; empty set $\emptyset$
7. A, C **8.** B is nonlinear because the variable is squared. D is nonlinear
because there is a variable in the second denominator. **9.** equation
11. expression **13.** equation **15.** $\{-1\}$ **17.** $\{-4\}$ **19.** $\{-7\}$
21. $\{0\}$ **23.** $\left\{-\dfrac{7}{8}\right\}$ **25.** $\left\{-\dfrac{1}{2}\right\}$ **27.** $\{-2\}$ **29.** $\{-1\}$ **31.** $\{4\}$
33. $\{0\}$ **35.** $\{3\}$ **37.** $\{2000\}$ **39.** $\{25\}$ **41.** $\{3\}$ **43.** $\emptyset$;
contradiction **45.** $\{0\}$; conditional equation **47.** $\{$all real numbers$\}$;
identity **49.** D **50.** C **51.** B **52.** A **53.** F **54.** E
55. $(-\infty, 7]$ **57.** $[5, \infty)$

59. $(-5, \infty)$

61. $(-4, \infty)$

63. $(-\infty, -40]$

65. $(7, \infty)$

67. $\left(-\infty, -\dfrac{15}{2}\right)$

69. $(-\infty, -7)$

71. $\left[\dfrac{1}{2}, \infty\right)$

73. $(3, \infty)$

75. $(1, 11)$

77. $[-5, 6]$

79. $(-6, -4)$

81. $\left[-\dfrac{1}{3}, \dfrac{1}{9}\right)$

SECTION 8.2 (pages 607–610)

1. true **2.** false; The intersection is $\{5\}$. **3.** false; The union is
$(-\infty, 6) \cup (6, \infty)$. **4.** true **5.** $\{4\}$, or D **7.** $\emptyset$ **9.** $\{1, 2, 3, 4, 5, 6\}$,
or A **11.** $\{1, 3, 5, 6\}$

13.

14.

15.

16.

17.

18.

19. Answers will vary. One example is: The intersection of two streets is
the region common to *both* streets. **20.** If the word is *and*, use intersection. If the word is *or*, use union.

21. $(-3, 2)$

23. $(-\infty, 2]$

25. $\emptyset$

27. $[5, 9]$

29. $(-\infty, 4]$

31. $(-\infty, 8]$

33. $[-2, \infty)$

35. $(-\infty, \infty)$

37. $(-\infty, -5) \cup (5, \infty)$

39. $(-\infty, -1] \cup (2, \infty)$

41. $(-\infty, 2) \cup (2, \infty)$

43. $[-4, -1]$ **45.** $[-9, -6]$
47. $(-\infty, 3)$ **49.** $[3, 9)$

51. intersection; $(-5, -1)$

53. union; $(-\infty, 4)$

55. intersection; $[4, 12]$

57. union; $(-\infty, 0] \cup [2, \infty)$

59. {Tuition and fees} **61.** {Tuition and fees, Board rates, Dormitory charges} **63.** Mario, Joe **64.** none of them **65.** none of them **66.** Luigi, Than **67.** Mario, Joe **68.** Luigi, Mario, Than, Joe

SECTION 8.3 (pages 617–623)

1. E; C; D; B; A **2.** E; D; A; C; B **3.** (a) one (b) two (c) none
4. Use *or* for the = statement and the > statement. Use *and* for the < statement. **5.** $\{-12, 12\}$ **7.** $\{-5, 5\}$ **9.** $\{-6, 12\}$ **11.** $\{-5, 4\}$
13. $\left\{-3, \dfrac{11}{2}\right\}$ **15.** $\left\{-\dfrac{19}{2}, \dfrac{9}{2}\right\}$ **17.** $\{-10, -2\}$ **19.** $\left\{-8, \dfrac{32}{3}\right\}$
21. $(-\infty, -3) \cup (3, \infty)$

23. $(-\infty, -4] \cup [4, \infty)$

25. $(-\infty, -10) \cup (6, \infty)$

27. $\left(-\infty, -\dfrac{7}{3}\right] \cup [3, \infty)$

29. $(-\infty, -2) \cup (8, \infty)$

31. (a)

(b)

32. (a)

(b)

33. $[-3, 3]$

35. $(-4, 4)$

37. $[-10, 6]$

39. $\left(-\dfrac{7}{3}, 3\right)$

41. $[-2, 8]$

43. $(-\infty, -2) \cup (10, \infty)$

45. $\{-6, -1\}$

47. $\left[-\dfrac{10}{3}, 4\right]$

49. $\left[-4, -\dfrac{4}{3}\right]$

51. $\{-5, 5\}$ **53.** $\{-5, -3\}$ **55.** $(-\infty, -3) \cup (2, \infty)$ **57.** $[-10, 0]$

59. $\{-1, 3\}$ **61.** $\left\{-3, \dfrac{5}{3}\right\}$ **63.** $\left\{-\dfrac{1}{3}, -\dfrac{1}{15}\right\}$ **65.** $\left\{-\dfrac{5}{4}\right\}$
67. $(-\infty, \infty)$ **69.** $\emptyset$ **71.** $\left\{-\dfrac{1}{4}\right\}$ **73.** $\emptyset$ **75.** $(-\infty, \infty)$
77. $\left\{-\dfrac{3}{7}\right\}$ **79.** $(-\infty, \infty)$ **81.** $\left(-\infty, -\dfrac{7}{10}\right) \cup \left(-\dfrac{7}{10}, \infty\right)$
83. $(-\infty, \infty)$ **85.** $|x - 1000| \le 100$; $900 \le x \le 1100$ **87.** 810.5 ft
89. Williams Tower, Bank of America Center, Texaco Heritage Plaza, Enterprise Plaza, Centerpoint Energy Plaza, Continental Center I, Fulbright Tower

In Relating Concepts Exercises 91–104, we give answers only for the odd-numbered problems.

91.

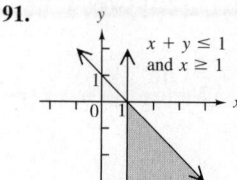

93.

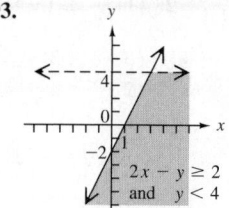

95.

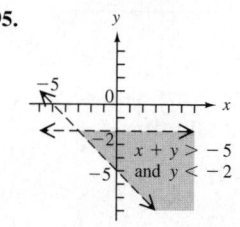

97.

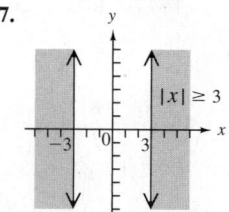

99.

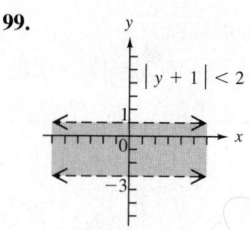

101.

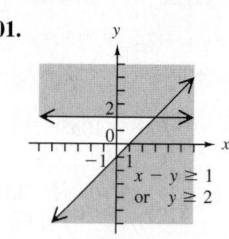

103.

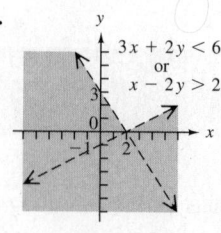

SUMMARY EXERCISES   Solving Linear and Absolute Value Equations and Inequalities (pages 624–625)

1. $\{12\}$ **2.** $\{-5, 7\}$ **3.** $\{7\}$ **4.** $\left\{-\dfrac{2}{5}\right\}$ **5.** $\emptyset$ **6.** $(-\infty, -1]$
7. $\left[-\dfrac{2}{3}, \infty\right)$ **8.** $\{-1\}$ **9.** $\{-3\}$ **10.** $\left\{1, \dfrac{11}{3}\right\}$ **11.** $(-\infty, 5]$
12. $(-\infty, \infty)$ **13.** $\{2\}$ **14.** $(-\infty, -8] \cup [8, \infty)$ **15.** $\emptyset$
16. $(-\infty, \infty)$ **17.** $(-5.5, 5.5)$ **18.** $\left\{\dfrac{13}{3}\right\}$ **19.** $\left\{-\dfrac{96}{5}\right\}$
20. $(-\infty, 32]$ **21.** $(-\infty, -24)$ **22.** $\left\{\dfrac{3}{8}\right\}$ **23.** $\left\{\dfrac{7}{2}\right\}$ **24.** $(-6, 8)$

25. {all real numbers}   26. $(-\infty, 5)$ 27. $(-\infty, -4) \cup (7, \infty)$

28. {24}   29. $\left\{-\dfrac{1}{5}\right\}$ 30. $\left(-\infty, -\dfrac{5}{2}\right]$ 31. $\left[-\dfrac{1}{3}, 3\right]$ 32. $[1, 7]$

33. $\left\{-\dfrac{1}{6}, 2\right\}$ 34. {−3}   35. $(-\infty, -1] \cup \left[\dfrac{5}{3}, \infty\right)$ 36. $\left[\dfrac{3}{4}, \dfrac{15}{8}\right]$

37. $\left\{-\dfrac{5}{2}\right\}$ 38. {60}   39. $\left[-\dfrac{9}{2}, \dfrac{15}{2}\right]$ 40. $(1, 9)$ 41. $(-\infty, \infty)$

42. $\left\{\dfrac{1}{3}, 9\right\}$ 43. {all real numbers}   44. $\left\{-\dfrac{10}{9}\right\}$ 45. {−2}

46. $\emptyset$ 47. $(-\infty, -1) \cup (2, \infty)$ 48. $[-3, -2]$

SECTION 8.4 (pages 633–634)

1. D; The ordered pair solution must be in quadrant IV, since that is where the graphs of the equations intersect.   2. B; The ordered pair solution must be on the *x*-axis, with $x < 0$, since that is where the graphs of the equations intersect.   3. **(a)** B **(b)** C **(c)** A **(d)** D   4. **(a)** Use substitution since the second equation is solved for *y*.   **(b)** Use elimination since the coefficients of the *y*-terms are opposites.   **(c)** Use elimination since the equations are in standard form with no coefficients of 1 or −1. Solving by substitution would involve fractions.

5. $\{(-2, -3)\}$ 7. $\{(0, 1)\}$

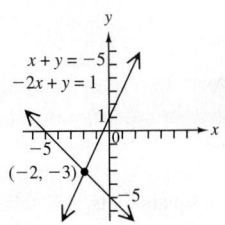

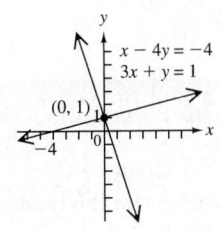

9. $\{(1, 2)\}$ 11. $\{(2, 3)\}$ 13. $\left\{\left(\dfrac{22}{9}, \dfrac{22}{3}\right)\right\}$ 15. $\{(5, 4)\}$

17. $\left\{\left(-5, -\dfrac{10}{3}\right)\right\}$ 19. $\{(2, 6)\}$ 21. $\{(x, y) \mid 2x - y = 0\}$; dependent equations   23. $\emptyset$; inconsistent system   25. $\{(2, -4)\}$ 27. $\{(3, -1)\}$

29. $\{(2, -3)\}$ 31. $\left\{\left(\dfrac{3}{2}, -\dfrac{3}{2}\right)\right\}$ 33. $\{(x, y) \mid 7x + 2y = 6\}$; dependent equations   35. $\{(2, -4)\}$ 37. $\emptyset$; inconsistent system

SECTION 8.5 (pages 643–646)

1. Answers will vary. Some possible answers are   **(a)** two perpendicular walls and the ceiling in a normal room,   **(b)** the floors of three different levels of an office building, and   **(c)** three consecutive pages of this book (since they intersect in the spine).   2. Answers will vary. Three possibilities are $\left(1, 1, -\dfrac{1}{2}\right), \left(0, 0, \dfrac{1}{2}\right)$, and $(2, 5, -3)$.   3. The statement means that when −1 is substituted for *x*, 2 is substituted for *y*, and 3 is substituted for *z* in the three equations, the resulting three statements are true.   4. B

5. $\{(3, 2, 1)\}$ 7. $\{(1, 4, -3)\}$ 9. $\{(1, 0, 3)\}$ 11. $\left\{\left(1, \dfrac{3}{10}, \dfrac{2}{5}\right)\right\}$

13. $\{(0, 2, -5)\}$ 15. $\left\{\left(\dfrac{20}{59}, -\dfrac{33}{59}, \dfrac{35}{59}\right)\right\}$ 17. $\{(4, 5, 3)\}$

19. $\{(2, 2, 2)\}$ 21. $\{(-1, 0, 0)\}$ 23. $\left\{\left(\dfrac{8}{3}, \dfrac{2}{3}, 3\right)\right\}$

25. $\emptyset$; inconsistent system   27. $\{(x, y, z) \mid x - y + 4z = 8\}$; dependent equations   29. $\{(x, y, z) \mid 2x + y - z = 6\}$; dependent equations   31. $\{(0, 0, 0)\}$ 33. $\emptyset$; inconsistent system   35. $\{(3, 0, 2)\}$ 37. $x + y + z = 180$; angle measures: 70°, 30°, 80°   39. first: 20°; second: 70°; third: 90°   41. shortest: 12 cm; middle: 25 cm; longest: 33 cm   43. Independent: 36%; Democrat: 34%; Republican: 28%   45. $16 tickets: 1170; $23 tickets: 985; $40 tickets: 130   47. bookstore A: 140; bookstore B: 280; bookstore C: 380   49. wins: 49; losses: 29; overtime losses: 4

Chapter 8 REVIEW EXERCISES (pages 652–655)

1. $\left\{-\dfrac{9}{5}\right\}$ 2. {10}   3. $\left\{\dfrac{1}{3}\right\}$ 4. {300}   5. {all real numbers}; identity   6. $\emptyset$; contradiction   7. {0}; conditional equation   8. {all real numbers}; identity

9. $(-9, \infty)$

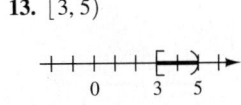

10. $(-\infty, -3]$

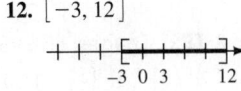

11. $[-3, \infty)$

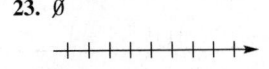

12. $[-3, 12]$

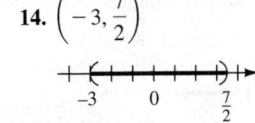

13. $[3, 5)$

14. $\left(-3, \dfrac{7}{2}\right)$

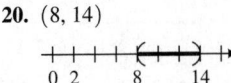

15. $\{a, c\}$ 16. $\{a\}$ 17. $\{a, c, e, f, g\}$ 18. $\{a, b, c, d, e, f, g\}$

19. $(4, 7)$

20. $(8, 14)$

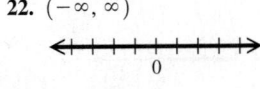

21. $(-\infty, -3] \cup (5, \infty)$

22. $(-\infty, \infty)$

23. $\emptyset$

24. $(-\infty, -2] \cup [7, \infty)$

25. $(-3, 4)$ 26. $(-\infty, 2)$ 27. $(4, \infty)$ 28. $(1, \infty)$ 29. $\{-7, 7\}$

30. $\{-11, 7\}$ 31. $\left\{-\dfrac{1}{3}, 5\right\}$ 32. $\emptyset$ 33. $\{0, 7\}$ 34. $\left\{-\dfrac{3}{2}, \dfrac{1}{2}\right\}$

35. $\left\{-\dfrac{3}{4}, \dfrac{1}{2}\right\}$ 36. $\left\{-\dfrac{8}{5}\right\}$

37. $(-12, 12)$

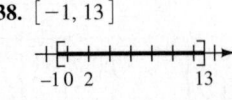

38. $[-1, 13]$

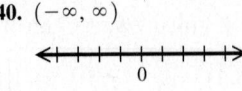

39. $[-3, -2]$

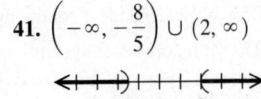

40. $(-\infty, \infty)$

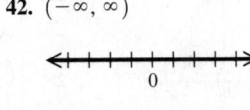

41. $\left(-\infty, -\dfrac{8}{5}\right) \cup (2, \infty)$

42. $(-\infty, \infty)$

43. $\{(2, 2)\}$ 44. C   45. $\left\{\left(-\dfrac{8}{9}, -\dfrac{4}{3}\right)\right\}$ 46. $\{(0, 4)\}$

47. $\{(2,4)\}$ **48.** $\{(-1,2)\}$ **49.** $\{(-6,3)\}$ **50.** $\left\{\left(\dfrac{68}{13}, -\dfrac{31}{13}\right)\right\}$

51. $\{(x,y)\,|\,3x - y = -6\}$; dependent equations **52.** $\varnothing$; inconsistent system **53.** $\{(0,0)\}$ **54.** $\{(1,-5,3)\}$ **55.** $\varnothing$; inconsistent system **56.** $\{(1,2,3)\}$ **57.** $85°, 35°, 60°$ **58.** Mantle: 54; Maris: 61; Berra: 22

59. $\left\{\dfrac{7}{6}\right\}$ **60.** $\varnothing$ **61.** $[-4,5)$ **62.** $(-\infty, 2]$

63. $(-\infty, -1) \cup \left(\dfrac{11}{7}, \infty\right)$ **64.** $\{-5, 15\}$ **65.** $[-16, 10]$

66. $(-\infty, \infty)$ **67.** $\left\{-4, -\dfrac{2}{3}\right\}$ **68.** $\left[-\dfrac{1}{3}, \dfrac{13}{3}\right]$

69. [number line] 0, 6, 8 **70.** [number line] -1, 0, 7

71. $\left\{\left(\dfrac{82}{23}, -\dfrac{4}{23}\right)\right\}$ **72.** $\{(3,-1)\}$ **73.** $\{(5,3)\}$ **74.** $\{(0,4)\}$

75. $\{(x,y,z)\,|\,3x - 4y + z = 8\}$; dependent equations **76.** $\{(0,0,0)\}$

77. U.S.A.: 37; Germany: 30; Canada: 26 **78. (a)** $\varnothing$ **(b)** $(-\infty, \infty)$ **(c)** $\varnothing$

Chapter 8 TEST (pages 656–657)

1. $\{-19\}$ **2.** $\{5\}$ **3.** $\{4\}$ **4.** $\varnothing$; contradiction **5.** $\{$all real numbers$\}$; identity **6.** $\{0\}$; conditional equation

7. $[2, \infty)$ [number line] 0, 2

8. $[1, \infty)$ [number line] 0, 1

9. $(-\infty, 28)$ [number line] 0, 7, 28

10. $[-3, 3]$ [number line] -3, 0, 3

11. $\{1, 5\}$ **12.** $\{1, 2, 5, 7, 9, 12\}$

13. $[2, 9)$ [number line] 0, 2, 9

14. $(-\infty, 3) \cup [6, \infty)$ [number line] 0, 3, 6

15. $\left[-\dfrac{5}{2}, 1\right]$ [number line] $-\dfrac{5}{2}$, 0, 1

16. $\left(-\infty, -\dfrac{7}{6}\right) \cup \left(\dfrac{17}{6}, \infty\right)$ [number line] $-\dfrac{7}{6}$, 0, 1, $\dfrac{17}{6}$

17. $\left(\dfrac{1}{3}, \dfrac{7}{3}\right)$ [number line] $0\frac{1}{3}$, $\dfrac{7}{3}$

18. $\varnothing$ [number line]

19. $\left\{-\dfrac{5}{3}, 3\right\}$ **20.** $\left\{-\dfrac{5}{7}, \dfrac{11}{3}\right\}$ **21.** $\varnothing$

22. (a) $\varnothing$ **(b)** $(-\infty, \infty)$ **(c)** $\varnothing$

23. $\{(6,1)\}$ [graph] $x + y = 7$, $x - y = 5$, (6, 1)

24. $\left\{\left(-\dfrac{9}{4}, \dfrac{5}{4}\right)\right\}$

25. $\{(x,y)\,|\,12x - 5y = 8\}$; dependent equations **26.** $\{(0, -2)\}$ **27.** $\varnothing$; inconsistent system

28. $\left\{\left(-\dfrac{2}{3}, \dfrac{4}{5}, 0\right)\right\}$ **29.** $\{(3, -2, 1)\}$

30. 60 oz of Orange Pekoe; 30 oz of Irish Breakfast; 10 oz of Earl Grey

Chapters R–8 CUMULATIVE REVIEW EXERCISES (page 658–659)

1. $-2m + 6$ **2.** $4m - 3$ **3.** $2x^2 + 5x + 4$ **4.** -24 **5.** 204

6. undefined **7.** 10 **8.** $\left\{\dfrac{7}{6}\right\}$ **9.** $\{-1\}$ **10.** $\left(-\infty, \dfrac{15}{4}\right]$

11. $\left(-\dfrac{1}{2}, \infty\right)$ **12.** $(2, 3)$ **13.** $(-\infty, 2) \cup (3, \infty)$ **14.** $\left\{-\dfrac{16}{5}, 2\right\}$

15. $(-11, 7)$ **16.** $(-\infty, -2] \cup [7, \infty)$ **17.** $h = \dfrac{V}{lw}$

18. table entries: (fourth column) $550x$, $500x$; 2 hr

19. [graph] $4x + 2y = -8$

20. -1 **21.** 0 **22.** -1 **23.** $\left(-\dfrac{7}{2}, 0\right)$

24. $(0, 7)$ **25.** $\{(1, 5)\}$

26. $\{(1, 1, 0)\}$ **27.** $\dfrac{y}{18x}$ **28.** $\dfrac{5my^4}{3}$

29. $x^3 + 12x^2 - 3x - 7$

30. $49x^2 + 42xy + 9y^2$

31. $10p^3 + 7p^2 - 28p - 24$ **32.** $(2w + 7z)(8w - 3z)$ **33.** $(2y - 9)^2$

34. $(2p + 3)(4p^2 - 6p + 9)$ **35.** $\left\{-4, -\dfrac{3}{2}, 1\right\}$ **36.** $\left\{\dfrac{1}{3}\right\}$

37. $\{-2, 1\}$ **38.** $\dfrac{4}{q}$ **39.** $\dfrac{3r + 28}{7r}$ **40.** $\dfrac{7}{15(q - 4)}$

41. $\dfrac{7(2z + 1)}{24}$ **42.** $\dfrac{195}{29}$

CHAPTER 9 Roots, Radicals, and Root Functions

SECTION 9.1 (pages 670–674)

1. true **2.** false; A negative number has no real square roots. **3.** false; Zero has only one square root. **4.** true **5.** true **6.** false; A positive number has just one real cube root. **7.** E **8.** F **9.** D **10.** B **11.** C **12.** E **13.** C **14.** B **15.** C **16.** D **17. (a)** not a real number **(b)** negative **(c)** 0 **18. (a)** a must be positive ($a > 0$). **(b)** a must be negative ($a < 0$). **(c)** a must be 0 ($a = 0$). **19.** $-3, 3$

21. $-8, 8$ **23.** $-13, 13$ **25.** $-\dfrac{5}{14}, \dfrac{5}{14}$ **27.** $-30, 30$ **29.** $64; 8$

31. 1 **33.** 7 **35.** -16 **37.** $-\dfrac{12}{11}$ **39.** 0.8 **41.** not a real number

43. not a real number **45.** 100 **47.** 19 **49.** $\dfrac{2}{3}$ **51.** $3x^2 + 4$

53. rational; 5 **55.** irrational; 5.385 **57.** rational; -8 **59.** irrational; -17.321 **61.** not a real number **63.** irrational; 34.641 **65.** 6 **67.** -4

69. -8 **71.** 6 **73.** -2 **75.** not a real number **77.** 3 **79.** not a real number **81.** $\dfrac{4}{3}$ **83.** $-\dfrac{1}{2}$ **85.** 0.1

87. domain: $[-3, \infty)$; range: $[0, \infty)$ **89.** domain: $[0, \infty)$; range: $[-2, \infty)$

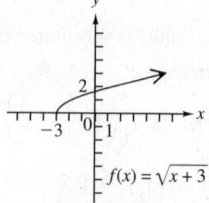

$f(x) = \sqrt{x + 3}$

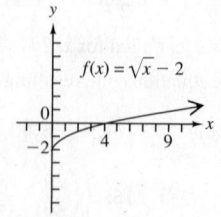

$f(x) = \sqrt{x} - 2$

91. domain: $(-\infty, \infty)$;
range: $(-\infty, \infty)$

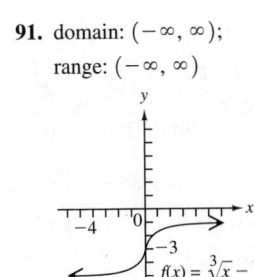

$f(x) = \sqrt[3]{x} - 3$

93. 12 **95.** 10 **97.** 2 **99.** -9
101. -5 **103.** $|x|$ **105.** $|z|$ **107.** x
109. x^5 **111.** $|x|^5$ (or $|x^5|$) **113.** 97.381
115. 16.863 **117.** -9.055 **119.** 7.507
121. 3.162 **123.** 1.885
125. (a) 1,183,000 cycles per sec
(b) 118,000 cycles per sec **127.** 10 mi
129. 392,000 mi^2

SECTION 9.2 (pages 681–684)

1. C **2.** E **3.** A **4.** J **5.** H **6.** G **7.** B **8.** F **9.** D **10.** I

11. 13 **13.** 9 **15.** 2 **17.** $\dfrac{8}{9}$ **19.** -3 **21.** not a real number

23. 1000 **25.** 27 **27.** -1024 **29.** 16 **31.** $\dfrac{1}{8}$ **33.** $\dfrac{1}{512}$ **35.** $\dfrac{9}{25}$

37. $\sqrt{12}$ **39.** $\left(\sqrt[4]{8}\right)^3$ **41.** $\left(\sqrt[8]{9q}\right)^5 - \left(\sqrt[3]{2x}\right)^2$ **43.** $\dfrac{1}{\left(\sqrt{2m}\right)^3}$

45. $\left(\sqrt[3]{2y + x}\right)^2$ **47.** $\dfrac{1}{\left(\sqrt[3]{3m^4 + 2k^2}\right)^2}$

49. $\sqrt{a^2 + b^2} = \sqrt{3^2 + 4^2} = 5;\ a + b = 3 + 4 = 7;\ 5 \neq 7$

50. The statement is true for this particular choice of values for a and b. However, it is not true *in general*. For example, let $a = 3$, $b = 4$, and $n = 2$.

51. 64 **53.** 64 **55.** x^{10} **57.** $\sqrt[6]{x^5}$ **59.** $\sqrt[15]{t^8}$ **61.** 9 **63.** 4 **65.** y

67. $x^{5/12}$ **69.** $k^{2/3}$ **71.** $x^3 y^8$ **73.** $\dfrac{1}{x^{10/3}}$ **75.** $\dfrac{1}{m^{1/4} n^{3/4}}$ **77.** p^2

79. $\dfrac{c^{11/3}}{b^{11/4}}$ **81.** $\dfrac{q^{5/3}}{9p^{7/2}}$ **83.** $p + 2p^2$ **85.** $k^{7/4} - k^{3/4}$ **87.** $6 + 18a$

89. $5 + \dfrac{5}{m^3}$ **91.** $y^{3/2}$ **93.** $\dfrac{1}{k^{2/3}}$ **95.** $x^{1/3} z^{5/6}$ **97.** $k^{1/6}$ **99.** $y^{1/30}$

101. $x^{5/27}$ **103.** 72 in.; 6.0 ft **105.** $-12.3°$; The table gives $-12°$.
107. $4.2°$; The table gives $4°$.

SECTION 9.3 (pages 692–697)

1. D **2.** D **3.** B **4.** A **5.** D **6.** D **7.** Because there are only two factors of $\sqrt[3]{x}$, $\sqrt[3]{x} \cdot \sqrt[3]{x} = \left(\sqrt[3]{x}\right)^2$, or $\sqrt[3]{x^2}$. **8.** Because there are only two factors, $\sqrt[4]{x} \cdot \sqrt[4]{x} = \left(\sqrt[4]{x}\right)^2 = x^{2/4} = x^{1/2} = \sqrt{x}$, for $x \geq 0$. **9.** $\sqrt{30}$
11. $\sqrt{14x}$ **13.** $\sqrt{42pqr}$ **15.** $\sqrt[3]{14xy}$ **17.** $\sqrt[4]{33}$ **19.** $\sqrt[4]{6xy^2}$

21. cannot be multiplied using the product rule directly **23.** cannot be multiplied using the product rule directly **25.** $\dfrac{8}{11}$ **27.** $\dfrac{\sqrt{3}}{5}$ **29.** $\dfrac{\sqrt{x}}{5}$

31. $\dfrac{p^3}{9}$ **33.** $-\dfrac{3}{4}$ **35.** $\dfrac{\sqrt[3]{r^2}}{2}$ **37.** $-\dfrac{3}{x}$ **39.** $\dfrac{1}{x^3}$ **41.** $2\sqrt{3}$ **43.** $12\sqrt{2}$

45. $-4\sqrt{2}$ **47.** $-2\sqrt{7}$ **49.** cannot be simplified further **51.** $4\sqrt[3]{2}$
53. $-2\sqrt[3]{2}$ **55.** $2\sqrt[3]{5}$ **57.** $-4\sqrt[4]{2}$ **59.** $2\sqrt[5]{2}$ **61.** His reasoning was incorrect. Here 8 is a term, not a factor. **62.** It is not simplified because the power of k is greater than the index of the radical. The simplified form is $k\sqrt[3]{k}$. **63.** $6k\sqrt{2}$ **65.** $12xy^4\sqrt{xy}$ **67.** $11x^3$ **69.** $-3t^4$

71. $-10m^4 z^2$ **73.** $5a^2 b^3 c^4$ **75.** $\dfrac{1}{2} r^2 t^5$ **77.** $5x\sqrt{2x}$ **79.** $-10r^5\sqrt{5r}$
81. $x^3 y^4\sqrt{13x}$ **83.** $2z^2 w^3$ **85.** $-2zt^2\sqrt[3]{2z^2 t}$ **87.** $3x^3 y^4$

89. $-3r^3 s^2\sqrt[4]{2r^3 s^2}$ **91.** $\dfrac{y^5\sqrt{y}}{6}$ **93.** $\dfrac{x^5\sqrt[3]{x}}{3}$ **95.** $4\sqrt{3}$ **97.** $\sqrt{5}$

99. $x^2\sqrt{x}$ **101.** $x\sqrt[6]{x^3}$ **103.** $\sqrt[6]{432}$ **105.** $\sqrt[12]{6912}$ **107.** $\sqrt[6]{x^5}$

109. 5 **111.** $8\sqrt{2}$ **113.** $2\sqrt{14}$ **115.** 13 **117.** $9\sqrt{2}$ **119.** $\sqrt{17}$

121. 5 **123.** $6\sqrt{2}$ **125.** $\sqrt{5y^2 - 2xy + x^2}$ **127.** 42.0 in.
129. 0.003 **131.** (a) $d = 1.224\sqrt{h}$ (b) 15.3 mi

SECTION 9.4 (pages 700–701)

1. B **2.** C **3.** 15; Each radical expression simplifies to a whole number.
4. We cannot group $28 - 4$ here. The terms 28 and $4\sqrt{2}$ are not like terms and cannot be combined. The difference cannot be combined.
5. -4 **7.** $7\sqrt{3}$ **9.** $14\sqrt[3]{2}$ **11.** $5\sqrt[4]{2}$ **13.** $24\sqrt{2}$ **15.** cannot be simplified further **17.** $20\sqrt{5}$ **19.** $12\sqrt{2x}$ **21.** $-2m\sqrt{2}$ **23.** $\sqrt[3]{2}$
25. $2\sqrt[3]{x}$ **27.** $-\sqrt[3]{x^2 y}$ **29.** $-x\sqrt[3]{xy^2}$ **31.** $19\sqrt[4]{2}$ **33.** $x\sqrt[4]{xy}$

35. $9\sqrt[4]{2a^3}$ **37.** $\dfrac{5\sqrt{5}}{6}$ **39.** $\dfrac{7\sqrt{2}}{6}$ **41.** $\dfrac{5\sqrt{2}}{3}$ **43.** $5\sqrt{2} + 4$

45. $\dfrac{30\sqrt{2} - 21}{14}$ **47.** $\dfrac{5 - 3x}{x^4}$ **49.** $\dfrac{m\sqrt[3]{m^2}}{2}$ **51.** $\dfrac{3x\sqrt[3]{2} - 4\sqrt[3]{5}}{x^3}$

53. $\left(12\sqrt{5} + 5\sqrt{3}\right)$ in. **55.** $\left(24\sqrt{2} + 12\sqrt{3}\right)$ in.

SECTION 9.5 (pages 708–710)

1. E **2.** C **3.** A **4.** F **5.** D **6.** B **7.** $6 - 4\sqrt{3}$ **9.** $6 - \sqrt{6}$
11. 2 **13.** 9 **15.** $3\sqrt{2} - 5\sqrt{3} + 2\sqrt{6} - 10$ **17.** $3x - 4$
19. $4x - y$ **21.** $16x + 24\sqrt{x} + 9$ **23.** $81 - \sqrt[3]{4}$
25. Because 6 and $4\sqrt{3}$ are not like terms, they cannot be combined.
26. identity property for multiplication **27.** $\sqrt{7}$ **29.** $5\sqrt{3}$ **31.** $\dfrac{\sqrt{6}}{2}$

33. $\dfrac{9\sqrt{15}}{5}$ **35.** $-\sqrt{2}$ **37.** $\dfrac{-8\sqrt{3k}}{k}$ **39.** $\dfrac{6\sqrt{3}}{y}$ **41.** $\dfrac{\sqrt{14}}{2}$

43. $-\dfrac{\sqrt{14}}{10}$ **45.** $\dfrac{2\sqrt{6x}}{x}$ **47.** $-\dfrac{7r\sqrt{2rs}}{s}$ **49.** $\dfrac{12x^3\sqrt{2xy}}{y^5}$ **51.** $\dfrac{\sqrt[3]{18}}{3}$

53. $\dfrac{\sqrt[3]{12}}{3}$ **55.** $-\dfrac{\sqrt[3]{2pr}}{r}$ **57.** $\dfrac{2\sqrt[4]{x^3}}{x}$ **59.** $\dfrac{\sqrt[4]{2yz^3}}{z}$ **61.** $\dfrac{2\left(4 - \sqrt{3}\right)}{13}$

63. $3\left(\sqrt{5} - \sqrt{3}\right)$ **65.** $\sqrt{3} + \sqrt{7}$ **67.** $\sqrt{7} - \sqrt{6} - \sqrt{14} + 2\sqrt{3}$

69. $2\sqrt{3} + \sqrt{10} - 3\sqrt{2} - \sqrt{15}$ **71.** $\dfrac{4\left(\sqrt{x} + 2\sqrt{y}\right)}{x - 4y}$

73. $\dfrac{x\sqrt{2} - \sqrt{3xy} - \sqrt{2xy} + y\sqrt{3}}{2x - 3y}$ **75.** $\dfrac{5 + 2\sqrt{6}}{4}$ **77.** $\dfrac{4 + 2\sqrt{2}}{3}$

79. $\dfrac{6 + 2\sqrt{6x}}{3}$

SUMMARY EXERCISES   Performing Operations with Radicals and Rational Exponents (pages 711–712)

1. The radicand is a fraction, $\dfrac{2}{5}$. **2.** The exponent in the radicand and the index of the radical have greatest common factor 5. **3.** The denominator contains a radical, $\sqrt[3]{10}$. **4.** The radicand has two factors, x and y, that are raised to powers greater than the index, 3. **5.** $-6\sqrt{10}$ **6.** $7 - \sqrt{14}$
7. $2 + \sqrt{6} - 2\sqrt{3} - 3\sqrt{2}$ **8.** $4\sqrt{2}$ **9.** $73 + 12\sqrt{35}$ **10.** $\dfrac{-\sqrt{6}}{2}$
11. $4\left(\sqrt{7} - \sqrt{5}\right)$ **12.** $3\sqrt[3]{2x^2}$ **13.** $-3 + 2\sqrt{2}$ **14.** -2 **15.** -44
16. $\dfrac{\sqrt{x} + \sqrt{5}}{x - 5}$ **17.** $2abc^3\sqrt[3]{b^2}$ **18.** $5\sqrt[3]{3}$ **19.** $3\left(\sqrt{5} - 2\right)$
20. $\dfrac{\sqrt{15x}}{5x}$ **21.** $\dfrac{8}{5}$ **22.** $\dfrac{\sqrt{2}}{8}$ **23.** $-\sqrt[3]{100}$ **24.** $11 + 2\sqrt{30}$
25. $-3\sqrt{3x}$ **26.** $52 - 30\sqrt{3}$ **27.** 1 **28.** $\dfrac{\sqrt[3]{117}}{9}$ **29.** $t^2\sqrt[4]{t}$
30. $3\sqrt{2} + \sqrt{15} + \sqrt{42} + \sqrt{35}$ **31.** $2\sqrt[4]{27}$ **32.** $\dfrac{1 + \sqrt[3]{3} + \sqrt[3]{9}}{-2}$
33. $\dfrac{x\sqrt[3]{x^2}}{y}$ **34.** $-4\sqrt{3} - 3$ **35.** $xy^{6/5}$ **36.** $x^{10} y$ **37.** $\dfrac{1}{25x^2}$
38. $\dfrac{-6y^{1/6}}{x^{1/24}}$ **39.** $7 + 4 \cdot 3^{1/2}$, or $7 + 4\sqrt{3}$ **40.** 1

SECTION 9.6 (pages 718–721)

1. (a) yes (b) no **2.** (a) yes (b) yes **3.** (a) yes (b) no
4. (a) yes (b) yes **5.** no; There is no solution. The radical expression, which is nonnegative, cannot equal a negative number. **6.** Since the radical on the left side cannot be negative, and it must equal x, x cannot be negative.
7. $\{11\}$ **9.** $\left\{\dfrac{1}{3}\right\}$ **11.** $\varnothing$ **13.** $\{5\}$ **15.** $\{18\}$ **17.** $\{5\}$ **19.** $\{4\}$
21. $\{17\}$ **23.** $\{5\}$ **25.** $\varnothing$ **27.** $\{0\}$ **29.** $\{0\}$ **31.** $\left\{-\dfrac{1}{3}\right\}$ **33.** $\varnothing$
35. We cannot just square each term. The right side should be $(8-x)^2 = 64 - 16x + x^2$. The correct first step is $3x + 4 = 64 - 16x + x^2$, and the solution set is $\{4\}$. **36.** We cannot just square each term. The right side should be $x + 3 + 2\sqrt{x+3} \cdot 3 + 9$. The correct first step is $5x + 6 = x + 3 + 2\sqrt{x+3} \cdot 3 + 9$, and the solution set is $\{6\}$.
37. $\{1\}$ **39.** $\{-1\}$ **41.** $\{14\}$ **43.** $\{8\}$ **45.** $\{0\}$ **47.** $\varnothing$
49. $\{7\}$ **51.** $\{7\}$ **53.** $\{4, 20\}$ **55.** $\varnothing$ **57.** $\left\{\dfrac{5}{4}\right\}$ **59.** $\{9, 17\}$
61. $\left\{\dfrac{1}{4}, 1\right\}$ **63.** $L = CZ^2$ **65.** $K = \dfrac{V^2 m}{2}$ **67.** $M = \dfrac{r^2 F}{m}$
69. (a) $r = \dfrac{a}{4\pi^2 N^2}$ (b) $a = 4\pi^2 N^2 r$

SECTION 9.7 (pages 728–730)

1. nonreal complex, complex **2.** pure imaginary, nonreal complex, complex **3.** real, complex **4.** real, complex **5.** pure imaginary, nonreal complex, complex **6.** pure imaginary, nonreal complex, complex
7. i **8.** 1 **9.** $-i$ **10.** -1 **11.** $13i$ **13.** $-12i$ **15.** $i\sqrt{5}$
17. $4i\sqrt{3}$ **19.** -15 **21.** $-\sqrt{57}$ **23.** -10 **25.** $i\sqrt{33}$ **27.** $\sqrt{3}$
29. $5i$ **31.** $-1 + 7i$ **33.** 0 **35.** $7 + 3i$ **37.** -2 **39.** $1 + 13i$
41. $6 + 6i$ **43.** $4 + 2i$ **44.** -5 **45.** -81 **47.** -16 **49.** $-10 - 30i$
51. $10 - 5i$ **53.** $-9 + 40i$ **55.** 153 **57.** (a) $a - bi$ (b) $a^2; b^2$
58. C **59.** $1 + i$ **61.** $-1 + 2i$ **63.** $2 + 2i$ **65.** $-\dfrac{5}{13} - \dfrac{12}{13}i$
67. -1 **69.** i **71.** 1 **73.** $-i$ **75.** 1 **77.** Since $i^{20} = (i^4)^5 = 1^5 = 1$, the student multiplied by 1, which is justified by the identity property for multiplication. **78.** $i^{12} = (i^4)^3 = 1^3 = 1$, so by the identity property for multiplication, the two products must be equal. **79.** $\dfrac{1}{2} + \dfrac{1}{2}i$
81. $(1 + 5i)^2 - 2(1 + 5i) + 26$ will simplify to 0 when the operations are applied.

Chapter 9 REVIEW EXERCISES (pages 735–738)

1. 42 **2.** -17 **3.** not a real number **4.** 6 **5.** -2 **6.** $|x|$ **7.** x
8. $|x|^5$ (or $|x^5|$)

9. domain: $[1, \infty)$; range: $[0, \infty)$

10. domain: $(-\infty, \infty)$; range: $(-\infty, \infty)$

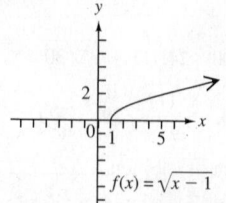

$f(x) = \sqrt{x - 1}$

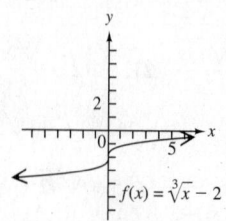

$f(x) = \sqrt[3]{x} - 2$

11. n must be even, and a must be negative. **12.** It is not a real number.
13. 6.325 **14.** 8.775 **15.** 17.607 **16.** 1.9 sec **17.** 66 in.^{2} **18.** 7

19. -2 **20.** not a real number **21.** By a power rule for exponents and the definition of $x^{1/n}$, $a^{m/n} = (a^m)^{1/n} = \sqrt[n]{a^m}$. **22.** 32 **23.** -4 **24.** $-\dfrac{216}{125}$
25. -32 **26.** $\dfrac{1000}{27}$ **27.** 49 **28.** 96 **29.** $\dfrac{k^{17/12}}{2}$ **30.** $\sqrt[5]{2^4}$, or $\sqrt[5]{16}$
31. 3^9 **32.** $7^4\sqrt{7}$ **33.** $m^4\sqrt[3]{m}$ **34.** $k^2\sqrt[4]{k}$ **35.** $\sqrt[6]{m}$ **36.** $2y\sqrt[4]{y}$
37. $\sqrt[15]{y^8}$ **38.** $\sqrt[12]{y^5}$ **39.** $\sqrt{66}$ **40.** $\sqrt{5r}$ **41.** $\sqrt[3]{30}$ **42.** $\sqrt[4]{21}$
43. $2\sqrt{5}$ **44.** $-5\sqrt{5}$ **45.** $-3x\sqrt[3]{4xy}$ **46.** $4pq^2\sqrt[3]{p}$ **47.** $\dfrac{7}{9}$
48. $\dfrac{y\sqrt{y}}{12}$ **49.** $\dfrac{m^5}{3}$ **50.** $\dfrac{\sqrt[3]{r^2}}{2}$ **51.** $\sqrt[12]{2}$ **52.** $\sqrt[10]{x^3}$ **53.** $\sqrt[4]{200}$
54. $\sqrt[6]{1125}$ **55.** $\sqrt{130}$ **56.** $\sqrt{53}$ **57.** $-11\sqrt{2}$ **58.** $23\sqrt{5}$
59. $7\sqrt{3y}$ **60.** $26m\sqrt{6m}$ **61.** $19\sqrt[3]{2}$ **62.** $-8\sqrt[4]{2}$ **63.** $1 - \sqrt{3}$
64. 2 **65.** $9 - 7\sqrt{2}$ **66.** $86 + 8\sqrt{55}$ **67.** $15 - 2\sqrt{26}$
68. $12 - 2\sqrt{35}$ **69.** $-3\sqrt{6}$ **70.** $\dfrac{3\sqrt{7py}}{y}$ **71.** $-\dfrac{\sqrt[3]{45}}{5}$ **72.** $\dfrac{3m\sqrt[3]{4n}}{n^2}$
73. $\dfrac{\sqrt{2} - \sqrt{7}}{-5}$ **74.** $\dfrac{-5(\sqrt{6} + \sqrt{3})}{3}$ **75.** $\{2\}$ **76.** $\{6\}$ **77.** $\varnothing$
78. $\{0, 5\}$ **79.** $\{9\}$ **80.** $\{3\}$ **81.** $\{7\}$ **82.** $\left\{-\dfrac{1}{2}\right\}$ **83.** $\{6\}$
84. $5i$ **85.** $10i\sqrt{2}$ **86.** $4i\sqrt{10}$ **87.** $-10 - 2i$ **88.** $14 + 7i$
89. $-\sqrt{35}$ **90.** -45 **91.** 3 **92.** $5 + i$ **93.** $32 - 24i$ **94.** $1 - i$
95. $4 + i$ **96.** $-i$ **97.** 1 **98.** $-i$ **99.** $-13ab^2$ **100.** $\dfrac{1}{100}$
101. $\dfrac{1}{y^{1/2}}$ **102.** $\dfrac{x^{3/4}}{z^{3/4}}$ **103.** k^6 **104.** $3z^3t^2\sqrt[4]{2t^2}$ **105.** $35 + 15i$
106. $\sqrt[12]{2000}$ **107.** $49a - 70\sqrt{a} + 25$ **108.** $57\sqrt{2}$
109. $6x\sqrt[3]{y^2}$ **110.** $\sqrt{35} + \sqrt{15} - \sqrt{21} - 3$ **111.** $-\dfrac{\sqrt{3}}{6}$
112. $\dfrac{\sqrt[3]{60}}{5}$ **113.** $\dfrac{2\sqrt{z}(\sqrt{z} + 2)}{z - 4}$ **114.** $7i$ **115.** $3 - 7i$ **116.** $-5i$
117. $\{5\}$ **118.** $\left\{\dfrac{3}{2}\right\}$ **119.** (a) $H = \sqrt{L^2 - W^2}$ (b) 7.9 ft
120. $\left(12\sqrt{3} + 5\sqrt{2}\right)$ ft

Chapter 9 TEST (pages 739–740)

1. -29 **2.** -8 **3.** 5 **4.** C **5.** 21.863 **6.** -9.405
7. domain: $[-6, \infty)$; range: $[0, \infty)$

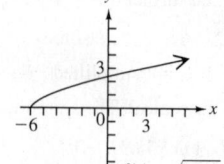

$f(x) = \sqrt{x + 6}$

8. $\dfrac{1}{256}$ **9.** $\dfrac{9y^{3/10}}{x^2}$ **10.** $3x^2 y^3\sqrt{6x}$
11. $2ab^3\sqrt[4]{2a^3 b}$ **12.** $\sqrt[6]{200}$ **13.** $40\sqrt{3}$
14. $26\sqrt{5}$ **15.** $66 + \sqrt{5}$
16. $23 - 4\sqrt{15}$ **17.** $\dfrac{-\sqrt{10}}{4}$ **18.** $\dfrac{2\sqrt[3]{25}}{5}$
19. $-2\left(\sqrt{7} - \sqrt{5}\right)$ **20.** $3 + \sqrt{6}$
21. $\sqrt{26}$ **22.** $\sqrt{145}$ **23.** $\{-1\}$

24. $\{6\}$ **25.** $\{-3\}$ **26.** (a) 59.8 (b) $T = \dfrac{V_0^2 - V^2}{-V^2 k}$, or $T = \dfrac{V^2 - V_0^2}{V^2 k}$ **27.** $-5 - 8i$ **28.** $-10 + 10i$ **29.** $3 + 4i$
30. $-i$

Chapters R–9 CUMULATIVE REVIEW EXERCISES (pages 741–742)

1. $\left\{\dfrac{4}{5}\right\}$ **2.** $\{-12\}$ **3.** $\left\{\dfrac{11}{10}, \dfrac{7}{2}\right\}$ **4.** $(-6, \infty)$ **5.** $(1, 3)$ **6.** $(-2, 1)$

7. $12x + 11y = 18$ **8.** C **9.** x-intercept: $(2, 0)$; y-intercept: $(0, 6)$

10. -3

11.

12. Both angles measure $80°$.

13. $\{(7, -2)\}$ **14.** $\varnothing$

15. $\{(x, y, z) \mid 2x + y - z = 5\}$

16. 2-oz letter: $0.65; 3-oz letter: $0.85

17. $-k^3 - 3k^2 - 8k - 9$

18. $8x^2 + 17x - 21$ **19.** $z - 2 + \dfrac{3}{z}$

20. $3y^3 - 3y^2 + 4y + 1 + \dfrac{-10}{2y + 1}$ **21.** $(2p - 3q)(p - q)$

22. $(3k^2 + 4)(6k^2 - 5)$ **23.** $(x + 8)(x^2 - 8x + 64)$ **24.** $\dfrac{y}{y + 5}$

25. $\dfrac{4x + 2y}{(x + y)(x - y)}$ **26.** $\dfrac{x - 6}{x - 2}$ **27.** $-\dfrac{9}{4}$ **28.** $-\dfrac{1}{a + b}$ **29.** $\dfrac{3}{2r^2 s^2}$

30. $\left\{-3, -\dfrac{5}{2}\right\}$ **31.** $\left\{-\dfrac{2}{5}, 1\right\}$ **32.** $\left\{\dfrac{7}{2}\right\}$ **33.** $\dfrac{1}{243}$ **34.** $x^{1/12}$

35. $8\sqrt{5}$ **36.** $\dfrac{-9\sqrt{5}}{20}$ **37.** $4(\sqrt{6} + \sqrt{5})$ **38.** $6\sqrt[3]{4}$ **39.** $\sqrt{29}$

40. $\{6\}$ **41.** 15 mph **42.** $\dfrac{80}{39}$ L, or $2\dfrac{2}{39}$ L **43.** dimes: 17; quarters: 12

44. Brenda: 8 mph; Chuck: 4 mph

CHAPTER 10 Quadratic Equations, Inequalities, and Functions

SECTION 10.1 (pages 749–750)

1. quadratic; second; two **2.** quadratic; standard **3.** The zero-factor property requires a product equal to 0. The first step should have been to write the equation with 0 on one side. **4.** The equation is also true for -4. The solution set is $\{-4, 4\}$. **5.** $\{-2, -1\}$ **7.** $\left\{-3, \dfrac{1}{3}\right\}$

9. $\left\{\dfrac{1}{2}, 4\right\}$ **11.** $\{9, -9\}$ **13.** $\{\sqrt{17}, -\sqrt{17}\}$ **15.** $\{4\sqrt{2}, -4\sqrt{2}\}$

17. $\{\sqrt{3}, -\sqrt{3}\}$ **19.** $\{4\sqrt{3}, -4\sqrt{3}\}$ **21.** $\{2\sqrt{6}, -2\sqrt{6}\}$

23. $\{3\sqrt{3}, -3\sqrt{3}\}$ **25.** $\{-7, 3\}$ **27.** $\{-1, 13\}$

29. $\{4 + \sqrt{3}, 4 - \sqrt{3}\}$ **31.** $\{-5 + 4\sqrt{3}, -5 - 4\sqrt{3}\}$ **33.** $\left\{-3, \dfrac{5}{3}\right\}$

35. $\left\{\dfrac{1 + \sqrt{7}}{3}, \dfrac{1 - \sqrt{7}}{3}\right\}$ **37.** $\left\{\dfrac{-1 + 2\sqrt{6}}{4}, \dfrac{-1 - 2\sqrt{6}}{4}\right\}$

39. $\left\{\dfrac{-1 + 3\sqrt{2}}{3}, \dfrac{-1 - 3\sqrt{2}}{3}\right\}$ **41.** 5.6 sec **43.** $\{10i, -10i\}$

45. $\{2i\sqrt{3}, -2i\sqrt{3}\}$ **47.** $\{5 + i\sqrt{3}, 5 - i\sqrt{3}\}$

49. $\left\{\dfrac{1}{6} + \dfrac{\sqrt{2}}{3}i, \dfrac{1}{6} - \dfrac{\sqrt{2}}{3}i\right\}$

SECTION 10.2 (pages 757–758)

1. square root property for $(2x + 1)^2 = 5$; completing the square for $x^2 + 4x = 12$ **2.** Divide each side by 2. **3.** 9; $(x + 3)^2$

4. 49; $(x + 7)^2$ **5.** 36; $(p - 6)^2$ **6.** 100; $(x - 10)^2$

7. $\dfrac{81}{4}$; $\left(q + \dfrac{9}{2}\right)^2$ **8.** $\dfrac{9}{4}$; $\left(t + \dfrac{3}{2}\right)^2$ **9.** $\{-3, -2\}$ **11.** $\{1, 3\}$

13. $\{-1 + \sqrt{6}, -1 - \sqrt{6}\}$ **15.** $\{-5 + \sqrt{7}, -5 - \sqrt{7}\}$

17. $\{4 + 2\sqrt{3}, 4 - 2\sqrt{3}\}$ **19.** $\left\{\dfrac{-1 + \sqrt{5}}{2}, \dfrac{-1 - \sqrt{5}}{2}\right\}$

21. $\left\{\dfrac{3 + \sqrt{17}}{2}, \dfrac{3 - \sqrt{17}}{2}\right\}$ **23.** $\left\{-\dfrac{3}{2}, \dfrac{1}{2}\right\}$

25. $\left\{\dfrac{3 + \sqrt{21}}{3}, \dfrac{3 - \sqrt{21}}{3}\right\}$ **27.** $\left\{\dfrac{-5 + \sqrt{41}}{4}, \dfrac{-5 - \sqrt{41}}{4}\right\}$

29. $\left\{\dfrac{-7 + \sqrt{97}}{6}, \dfrac{-7 - \sqrt{97}}{6}\right\}$ **31.** $\{4 + \sqrt{3}, 4 - \sqrt{3}\}$

33. $\{1 + \sqrt{6}, 1 - \sqrt{6}\}$ **35.** $\{-2 + 3i, -2 - 3i\}$

37. $\left\{-\dfrac{2}{3} + \dfrac{2\sqrt{2}}{3}i, -\dfrac{2}{3} - \dfrac{2\sqrt{2}}{3}i\right\}$ **39.** $\{-3 + i\sqrt{3}, -3 - i\sqrt{3}\}$

41. x^2 **42.** x **43.** $6x$ **44.** 1 **45.** 9 **46.** $(x + 3)^2$, or $x^2 + 6x + 9$

SECTION 10.3 (pages 765–766)

1. No. The fraction bar should extend under the term $-b$. The correct formula is $x = \dfrac{-b \pm \sqrt{b^2 - 4ac}}{2a}$. **2.** No. The patron forgot the $\pm$ symbol in the numerator. (See the correct formula in the **Exercise 1** answer.) **3.** The last step is wrong. Because 5 is not a common factor of the terms in the numerator, the fraction cannot be simplified further. The solutions are $\dfrac{5 \pm \sqrt{5}}{10}$.

4. The quadratic formula can be used to solve *any* quadratic equation. Since the equation can be written as $2x^2 + 0x - 5 = 0$, it follows that $a = 2, b = 0$, and $c = -5$. The solution set is $\left\{\dfrac{\sqrt{10}}{2}, -\dfrac{\sqrt{10}}{2}\right\}$.

5. $\{3, 5\}$ **7.** $\left\{\dfrac{-2 + \sqrt{2}}{2}, \dfrac{-2 - \sqrt{2}}{2}\right\}$ **9.** $\left\{\dfrac{1 + \sqrt{3}}{2}, \dfrac{1 - \sqrt{3}}{2}\right\}$

11. $\{5 + \sqrt{7}, 5 - \sqrt{7}\}$ **13.** $\left\{\dfrac{-1 + \sqrt{2}}{2}, \dfrac{-1 - \sqrt{2}}{2}\right\}$

15. $\left\{\dfrac{-1 + \sqrt{7}}{3}, \dfrac{-1 - \sqrt{7}}{3}\right\}$ **17.** $\{1 + \sqrt{5}, 1 - \sqrt{5}\}$

19. $\left\{\dfrac{-2 + \sqrt{10}}{2}, \dfrac{-2 - \sqrt{10}}{2}\right\}$ **21.** $\{-1 + 3\sqrt{2}, -1 - 3\sqrt{2}\}$

23. $\left\{\dfrac{3}{2} + \dfrac{\sqrt{59}}{2}i, \dfrac{3}{2} - \dfrac{\sqrt{59}}{2}i\right\}$ **25.** $\{3 + i\sqrt{5}, 3 - i\sqrt{5}\}$

27. $\left\{\dfrac{1}{2} + \dfrac{\sqrt{6}}{2}i, \dfrac{1}{2} - \dfrac{\sqrt{6}}{2}i\right\}$ **29.** $\left\{-\dfrac{2}{3} + \dfrac{\sqrt{2}}{3}i, -\dfrac{2}{3} - \dfrac{\sqrt{2}}{3}i\right\}$

31. $\left\{\dfrac{1}{2} + \dfrac{1}{4}i, \dfrac{1}{2} - \dfrac{1}{4}i\right\}$ **33.** 0; B; factoring **35.** 8; C; quadratic formula **37.** 49; A; factoring **39.** -80; D; quadratic formula

41. $\left\{-\dfrac{7}{5}\right\}$ **43.** $\{-2 + \sqrt{2}, -2 - \sqrt{2}\}$

SECTION 10.4 (pages 775–778)

1. Multiply by the LCD, x. **2.** Isolate the radical term on one side.

3. Substitute a variable for $r^2 + r$. **4.** Square each side.

5. The proposed solution -1 does not check. The solution set is $\{4\}$.

6. The solutions given are for u. Each must be set equal to $m - 1$ and solved for m. The correct solution set is $\left\{\dfrac{3}{2}, 2\right\}$. **7.** $\{-4, 7\}$

9. $\left\{-\dfrac{2}{3}, 1\right\}$ **11.** $\left\{-\dfrac{14}{17}, 5\right\}$ **13.** $\left\{-\dfrac{11}{7}, 0\right\}$

15. $\left\{\dfrac{-1 + \sqrt{13}}{2}, \dfrac{-1 - \sqrt{13}}{2}\right\}$ **17.** $\left\{\dfrac{2 + \sqrt{22}}{3}, \dfrac{2 - \sqrt{22}}{3}\right\}$

19. (a) $(20 - t)$ mph **(b)** $(20 + t)$ mph **20. (a)** $\dfrac{1}{m}$ job per hr

(b) $\dfrac{2}{m}$ job **21.** table entries: (third column) $x + 15$; (fourth column)

$\dfrac{4}{x - 15}, \dfrac{16}{x + 15}$; 25 mph **23.** 80 km per hr **25.** table entries: (second

column) $\dfrac{1}{x}, \dfrac{1}{x + 1}$; (fourth column) $\dfrac{2}{x}, \dfrac{2}{x + 1}$; 3.6 hr **27.** Rusty: 25.0 hr;

Nancy: 23.0 hr **29.** 9 min **31.** $\{1, 4\}$ **33.** $\{3\}$ **35.** $\left\{\dfrac{8}{9}\right\}$

37. $\{16\}$ **39.** $\left\{\dfrac{2}{5}\right\}$ **41.** $\{-2\}$ **43.** $\{-5, -2, 2, 5\}$ **45.** $\{-3, 3\}$

47. $\left\{-\dfrac{3}{2}, -1, 1, \dfrac{3}{2}\right\}$ **49.** $\{-2\sqrt{3}, -2, 2, 2\sqrt{3}\}$

51. $\left\{\dfrac{\sqrt{9 + \sqrt{65}}}{2}, -\dfrac{\sqrt{9 + \sqrt{65}}}{2}, \dfrac{\sqrt{9 - \sqrt{65}}}{2}, -\dfrac{\sqrt{9 - \sqrt{65}}}{2}\right\}$

53. $\{-6, -5\}$ **55.** $\{-4, 1\}$ **57.** $\left\{-\dfrac{1}{3}, \dfrac{1}{6}\right\}$ **59.** $\{-8, 1\}$

61. $\{-64, 27\}$ **63.** $\left\{-\dfrac{27}{8}, -1, 1, \dfrac{27}{8}\right\}$ **65.** It would cause both

denominators to equal 0, and division by 0 is undefined. **66.** $\dfrac{12}{5}$

67. $\left(\dfrac{x}{x - 3}\right)^2 + 3\left(\dfrac{x}{x - 3}\right) - 4 = 0$ **68.** The numerator can never equal

the denominator, since the denominator is 3 less than the numerator.

69. $\left\{\dfrac{12}{5}\right\}$; The values for u are -4 and 1. The value 1 is impossible

because it leads to a contradiction $\left(\text{since } \dfrac{x}{x - 3} \text{ is never equal to } 1\right)$.

70. $\left\{\dfrac{12}{5}\right\}$; The values for u are $\dfrac{1}{x}$ and $\dfrac{-4}{x}$. The value $\dfrac{1}{x}$ is impossible,

since $\dfrac{1}{x} \neq \dfrac{1}{x - 3}$ for all x.

SUMMARY EXERCISES Applying Methods for Solving Quadratic Equations (pages 779–780)

1. square root property **2.** factoring **3.** quadratic formula
4. quadratic formula **5.** factoring **6.** square root property

7. $\{\sqrt{47}, -\sqrt{47}\}$ **8.** $\left\{-\dfrac{3}{2}, \dfrac{5}{3}\right\}$ **9.** $\{-4 + \sqrt{10}, -4 - \sqrt{10}\}$

10. $\{-3, 11\}$ **11.** $\left\{-\dfrac{1}{2}, 5\right\}$ **12.** $\left\{-3, \dfrac{1}{3}\right\}$

13. $\left\{\dfrac{9 + \sqrt{33}}{6}, \dfrac{9 - \sqrt{33}}{6}\right\}$ **14.** $\{2i\sqrt{3}, -2i\sqrt{3}\}$ **15.** $\left\{\dfrac{1}{2}, 2\right\}$

16. $\left\{-\dfrac{\sqrt{6}}{3}, -\dfrac{1}{2}, \dfrac{1}{2}, \dfrac{\sqrt{6}}{3}\right\}$ **17.** $\left\{\dfrac{-5 + 2\sqrt{3}}{2}, \dfrac{-5 - 2\sqrt{3}}{2}\right\}$

18. $\left\{\dfrac{4}{5}, 3\right\}$ **19.** $\{-\sqrt{7}, -\sqrt{2}, \sqrt{2}, \sqrt{7}\}$

20. $\left\{\dfrac{-2 + \sqrt{14}}{2}, \dfrac{-2 - \sqrt{14}}{2}\right\}$ **21.** $\left\{-\dfrac{1}{2} + \dfrac{\sqrt{7}}{2}i, -\dfrac{1}{2} - \dfrac{\sqrt{7}}{2}i\right\}$

22. $\left\{\sqrt{4 + \sqrt{15}}, -\sqrt{4 + \sqrt{15}}, \sqrt{4 - \sqrt{15}}, -\sqrt{4 - \sqrt{15}}\right\}$

23. $\left\{\dfrac{3}{2}\right\}$ **24.** $\left\{\dfrac{2}{3}\right\}$ **25.** $\{6\sqrt{2}, -6\sqrt{2}\}$ **26.** $\left\{-\dfrac{2}{3}, 2\right\}$

27. $\{-4, 9\}$ **28.** $\{13, -13\}$ **29.** $\left\{1 + \dfrac{\sqrt{3}}{3}i, 1 - \dfrac{\sqrt{3}}{3}i\right\}$

30. $\{3\}$ **31.** $\left\{-\dfrac{1}{3}, \dfrac{1}{6}\right\}$ **32.** $\left\{\dfrac{1}{6} + \dfrac{\sqrt{47}}{6}i, \dfrac{1}{6} - \dfrac{\sqrt{47}}{6}i\right\}$

33. $\{3\}$ **34.** $\left\{-\dfrac{8}{3}, -1\right\}$ **35.** $\left\{-i, i, -\dfrac{1}{2}i, \dfrac{1}{2}i\right\}$ **36.** $\{-2, 7\}$

SECTION 10.5 (pages 785–788)

1. Solve for w^2 by dividing each side by g. **2.** Write it in standard form
(with 0 on one side, in descending powers of w). **3.** $m = \sqrt{p^2 - n^2}$

5. $t = \dfrac{\pm \sqrt{dk}}{k}$ **7.** $d = \dfrac{\pm \sqrt{skw}}{kw}$ **9.** $d = \dfrac{\pm \sqrt{skl}}{l}$ **11.** $v = \dfrac{\pm \sqrt{kAF}}{F}$

13. $r = \dfrac{\pm \sqrt{V\pi h}}{\pi h}$ **15.** $t = \dfrac{-B \pm \sqrt{B^2 - 4AC}}{2A}$ **17.** $h = \dfrac{D^2}{k}$

19. $\ell = \dfrac{p^2 g}{k}$ **21.** 2.3, 5.3, 5.8 **23.** eastbound ship: 80 mi; southbound
ship: 150 mi **25.** 8 in., 15 in., 17 in. **27.** 1 ft **29.** 20 in. by 12 in.
31. 2.4 sec and 5.6 sec **33.** 9.2 sec **35.** It reaches its *maximum* height
at 5 sec because this is the only time it reaches 400 ft. **36.** Because the
discriminant is negative, the ball never reaches a height of 425 ft.
37. \$1.50 **39. (a)** \$490 billion **(b)** \$500 billion; The model gives a
slightly greater value than the graph. **41.** 2003; The graph indicates that
spending first exceeded \$400 billion in 2005. **42.** The value of a would
be positive, because it would be the slope of a line that rises from left to
right. **43.** 5.5 m per sec **45.** 5 or 14

SECTION 10.6 (pages 795–798)

1. (a) B **(b)** C **(c)** A **(d)** D **2. (a)** D **(b)** C **(c)** B **(d)** A
3. (a) I **(b)** IV **(c)** II **(d)** III **(e)** narrower **(f)** wider
4. (a) D **(b)** B **(c)** C **(d)** A **5.** $(0, 0)$ **7.** $(0, 4)$ **9.** $(1, 0)$
11. $(-3, -4)$ **13.** $(5, 6)$ **15.** down; wider **17.** up; narrower
19. vertex: $(0, 0)$; axis: $x = 0$; **21.** vertex: $(0, -1)$; axis: $x = 0$;
domain: $(-\infty, \infty)$; range: $(-\infty, 0]$ domain: $(-\infty, \infty)$; range: $[-1, \infty)$

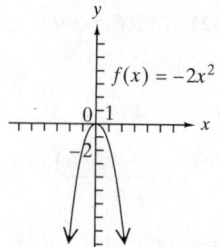

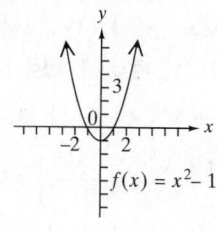

23. vertex: $(0, 2)$; axis: $x = 0$; **25.** vertex: $(4, 0)$; axis: $x = 4$;
domain: $(-\infty, \infty)$; range: $(-\infty, 2]$ domain: $(-\infty, \infty)$; range: $[0, \infty)$

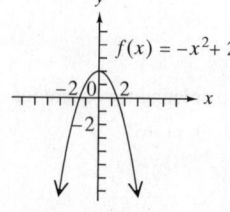

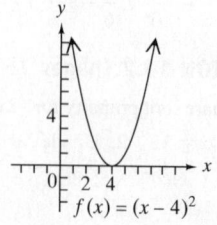

27. vertex: $(-2, -1)$; axis: $x = -2$; **29.** vertex: $(2, -3)$; axis: $x = 2$;
domain: $(-\infty, \infty)$; range: $[-1, \infty)$ domain: $(-\infty, \infty)$; range: $[-3, \infty)$

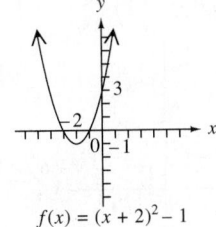

$f(x) = (x + 2)^2 - 1$

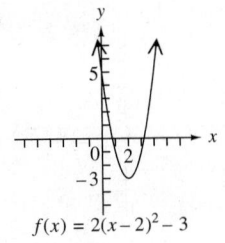

$f(x) = 2(x - 2)^2 - 3$

31. vertex: $(-3, 4)$; axis: $x = -3$; **33.** vertex: $(-2, 1)$; axis: $x = -2$;
domain: $(-\infty, \infty)$; range: $(-\infty, 4]$ domain: $(-\infty, \infty)$; range: $(-\infty, 1]$

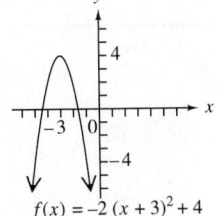

$f(x) = -2(x + 3)^2 + 4$

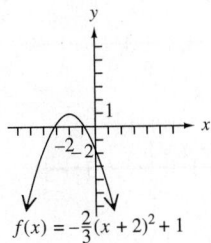

$f(x) = -\frac{2}{3}(x + 2)^2 + 1$

35. quadratic; positive **37.** quadratic; negative **39.** linear; positive

41. (a)

HOUSING STARTS

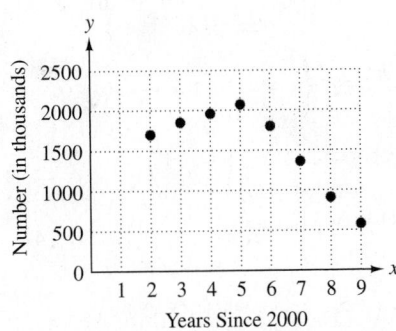

Years Since 2000

(b) quadratic **(c)** negative **(d)** $f(x) = -66x^2 + 526x + 912$
(e) 2003: 1896 thousand; 2008: 896 thousand; The model approximates the data fairly well.

SECTION 10.7 (pages 807–810)

1. If x is squared, it has a vertical axis. If y is squared, it has a horizontal axis. **2.** A parabola with a horizontal axis of symmetry fails the conditions of the vertical line test. **3.** $(-4, -6)$ **5.** $(1, -3)$

7. $\left(-\frac{1}{2}, -\frac{29}{4}\right)$ **9.** $(-1, 3)$; up; narrower; no x-intercepts

11. $\left(\frac{5}{2}, \frac{37}{4}\right)$; down; same; two x-intercepts

13. $(-3, -9)$; to the right; wider

15. vertex: $(-2, -1)$; axis: $x = -2$; **17.** vertex: $(1, -3)$; axis: $x = 1$;
domain: $(-\infty, \infty)$; range: $[-1, \infty)$ domain: $(-\infty, \infty)$; range: $(-\infty, -3]$

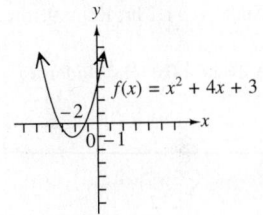

$f(x) = x^2 + 4x + 3$

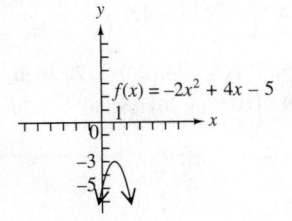

$f(x) = -2x^2 + 4x - 5$

19. vertex: $(1, -2)$; axis: $y = -2$; **21.** vertex: $(1, 5)$; axis: $y = 5$;
domain: $[1, \infty)$; range: $(-\infty, \infty)$ domain: $(-\infty, 1]$; range: $(-\infty, \infty)$

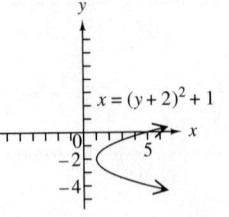

$x = (y + 2)^2 + 1$

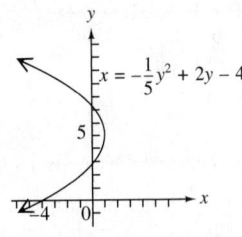

$x = -\frac{1}{5}y^2 + 2y - 4$

23. vertex: $(-7, -2)$; axis: $y = -2$; **25.** F **26.** A **27.** C **28.** B
domain: $[-7, \infty)$; range: $(-\infty, \infty)$ **29.** D **30.** E **31.** 30 and 30
33. 160 ft by 320 ft; 51,200 ft^2
35. 20 units; $210
37. (a) $R(x) = 20,000 + 200x - 4x^2$
(b) 25 **(c)** $22,500
39. 16 ft; 2 sec **41. (a)** maximum
(b) 2007; $356 billion

$x = 3y^2 + 12y + 5$

43. The coefficient of x^2 is negative because a parabola that models the data must open down. **45.** In 2018 Social Security assets will reach their maximum value of $3860 billion.

SECTION 10.8 (pages 817–820)

1. Include the endpoints if the symbol is $\geq$ or $\leq$. Exclude the endpoints if the symbol is $>$ or $<$. **2.** $(-\infty, -4] \cup [3, \infty)$ **3. (a)** $\{1, 3\}$

(b) $(-\infty, 1) \cup (3, \infty)$ **(c)** $(1, 3)$ **5. (a)** $\left\{-3, \frac{5}{2}\right\}$ **(b)** $\left[-3, \frac{5}{2}\right]$

(c) $\left(-\infty, -3\right] \cup \left[\frac{5}{2}, \infty\right)$

7. $(-\infty, -1) \cup (5, \infty)$ **9.** $(-4, 6)$

11. $(-\infty, 1] \cup [3, \infty)$ **13.** $\left(-\infty, -\frac{3}{2}\right] \cup \left[\frac{3}{5}, \infty\right)$

15. $\left(-\frac{2}{3}, \frac{1}{3}\right)$ **17.** $\left(-\infty, -\frac{1}{2}\right] \cup \left[\frac{1}{3}, \infty\right)$

19. $\left(-\infty, 3 - \sqrt{3}\right] \cup \left[3 + \sqrt{3}, \infty\right)$ **21.** $(-\infty, \infty)$ **23.** $\emptyset$

25. $(-\infty, 1) \cup (2, 4)$ **27.** $\left[-\frac{3}{2}, \frac{1}{3}\right] \cup [4, \infty)$

29. $(-\infty, 1) \cup (4, \infty)$ **31.** $\left[-\frac{3}{2}, 5\right)$

33. $(2, 6]$

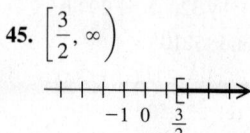

35. $\left(-\infty, \dfrac{1}{2}\right) \cup \left(\dfrac{5}{4}, \infty\right)$

37. $[-4, -2)$

39. $\left(0, \dfrac{1}{2}\right) \cup \left(\dfrac{5}{2}, \infty\right)$

41. $[-7, -2)$

43. $(-\infty, 2] \cup (4, \infty)$

45. $\left[\dfrac{3}{2}, \infty\right)$

47. 3 sec and 13 sec **48.** between 3 sec and 13 sec **49.** at 0 sec (the time when it is initially projected) and at 16 sec (the time when it hits the ground) **50.** between 0 and 3 sec and between 13 and 16 sec

Chapter 10 REVIEW EXERCISES (pages 825–828)

1. $\{11, -11\}$ **2.** $\left\{\sqrt{3}, -\sqrt{3}\right\}$ **3.** $\left\{3 + \sqrt{10}, 3 - \sqrt{10}\right\}$

4. $\left\{-\dfrac{15}{2}, \dfrac{5}{2}\right\}$ **5.** $\left\{\dfrac{2}{3} + \dfrac{5}{3}i, \dfrac{2}{3} - \dfrac{5}{3}i\right\}$ **6.** $\left\{-2 + \sqrt{19}, -2 - \sqrt{19}\right\}$

7. $\left\{\dfrac{1}{2}, 1\right\}$ **8.** $\left\{-1 + \sqrt{6}, -1 - \sqrt{6}\right\}$

9. $\left\{\dfrac{-4 + \sqrt{22}}{2}, \dfrac{-4 - \sqrt{22}}{2}\right\}$ **10.** 5.8 sec **11.** $\left\{-\dfrac{7}{2}, 3\right\}$

12. $\left\{\dfrac{-5 + \sqrt{53}}{2}, \dfrac{-5 - \sqrt{53}}{2}\right\}$ **13.** $\left\{\dfrac{1 + \sqrt{41}}{2}, \dfrac{1 - \sqrt{41}}{2}\right\}$

14. $\left\{-\dfrac{3}{4} + \dfrac{\sqrt{23}}{4}i, -\dfrac{3}{4} - \dfrac{\sqrt{23}}{4}i\right\}$ **15.** $\left\{\dfrac{2}{3} + \dfrac{\sqrt{2}}{3}i, \dfrac{2}{3} - \dfrac{\sqrt{2}}{3}i\right\}$

16. $\left\{\dfrac{-7 + \sqrt{37}}{2}, \dfrac{-7 - \sqrt{37}}{2}\right\}$ **17.** 17; C; quadratic formula

18. 64; A; factoring **19.** 0; B; factoring **20.** −92; D; quadratic formula

21. $\left\{-\dfrac{5}{2}, 3\right\}$ **22.** $\left\{-\dfrac{1}{2}, 1\right\}$ **23.** $\{-4\}$ **24.** $\left\{-\dfrac{11}{6}, -\dfrac{19}{12}\right\}$

25. $\left\{-\dfrac{343}{8}, 64\right\}$ **26.** $\{-2, -1, 1, 2\}$ **27.** 40 mph **28.** 4.6 hr

29. $v = \dfrac{\pm\sqrt{rFkw}}{kw}$ **30.** $t = \dfrac{3m \pm \sqrt{9m^2 + 24m}}{2m}$ **31.** 9 ft, 12 ft, 15 ft

32. 12 cm by 20 cm **33.** 1 in. **34.** 5.2 sec **35.** $(1, 0)$ **36.** $(3, 7)$

37. $\left(\dfrac{2}{3}, -\dfrac{2}{3}\right)$ **38.** $(-4, 3)$

39. vertex: $(2, -3)$; axis: $x = 2$; domain: $(-\infty, \infty)$; range: $[-3, \infty)$ **40.** vertex: $(2, 3)$; axis: $x = 2$; domain: $(-\infty, \infty)$; range: $(-\infty, 3]$

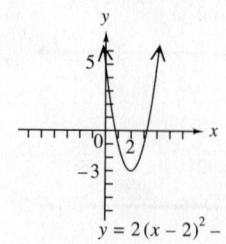

$y = 2(x - 2)^2 - 3$

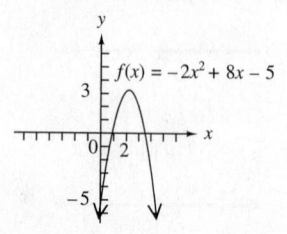

$f(x) = -2x^2 + 8x - 5$

41. vertex: $(-4, -3)$; axis: $y = -3$; domain: $[-4, \infty)$; range: $(-\infty, \infty)$ **42.** vertex: $(4, 6)$; axis: $y = 6$; domain: $(-\infty, 4]$; range: $(-\infty, \infty)$

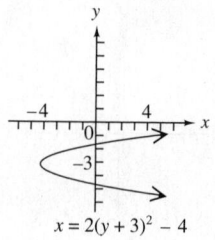

$x = 2(y + 3)^2 - 4$

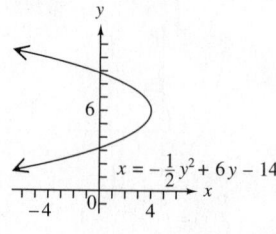

$x = -\dfrac{1}{2}y^2 + 6y - 14$

43. 5 sec; 400 ft **44.** length: 50 m; width: 50 m; maximum area: 2500 m²

45. $\left(-\infty, -\dfrac{3}{2}\right) \cup (4, \infty)$ **46.** $[-4, 3]$

47. $(-\infty, -5] \cup [-2, 3]$ **48.** Ø

49. $\left(-\infty, \dfrac{1}{2}\right) \cup (2, \infty)$ **50.** $[-3, 2)$

51. $R = \dfrac{\pm\sqrt{Vh - r^2h}}{h}$ **52.** $\left\{1 + \dfrac{\sqrt{3}}{3}i, 1 - \dfrac{\sqrt{3}}{3}i\right\}$

53. $\{-2, -1, 3, 4\}$ **54.** $(-\infty, -6) \cup \left(-\dfrac{3}{2}, 1\right)$

55. $\left\{\dfrac{-11 + \sqrt{7}}{3}, \dfrac{-11 - \sqrt{7}}{3}\right\}$ **56.** $d = \dfrac{\pm\sqrt{SkI}}{I}$ **57.** $\{4\}$

58. $\left\{-\dfrac{5}{3}, -\dfrac{3}{2}\right\}$ **59.** $\left(-5, -\dfrac{23}{5}\right]$ **60.** $(-\infty, \infty)$

61. (a) F (b) B (c) C (d) A (e) E (f) D
62. (a) $c = 705$; $4a + 2b + c = 511$; $16a + 4b + c = 293$
(b) $f(x) = -3x^2 - 91x + 705$ (c) 175 million; The result using the model is lower than the actual data.

Chapter 10 TEST (pages 829–830)

1. $\left\{3\sqrt{6}, -3\sqrt{6}\right\}$ **2.** $\left\{-\dfrac{8}{7}, \dfrac{2}{7}\right\}$ **3.** $\left\{-1 + \sqrt{2}, -1 - \sqrt{2}\right\}$

4. $\left\{\dfrac{3 + \sqrt{17}}{4}, \dfrac{3 - \sqrt{17}}{4}\right\}$ **5.** $\left\{\dfrac{2}{3} + \dfrac{\sqrt{11}}{3}i, \dfrac{2}{3} - \dfrac{\sqrt{11}}{3}i\right\}$

6. $\left\{\dfrac{2}{3}\right\}$ **7.** A **8.** 88; two irrational solutions **9.** $\left\{-\dfrac{2}{3}, 6\right\}$

10. $\left\{\dfrac{-7 + \sqrt{97}}{8}, \dfrac{-7 - \sqrt{97}}{8}\right\}$ **11.** $\left\{-2, -\dfrac{1}{3}, \dfrac{1}{3}, 2\right\}$

12. $\left\{-\dfrac{5}{2}, 1\right\}$ **13.** $r = \dfrac{\pm\sqrt{\pi S}}{2\pi}$ **14.** Andrew: 11.1 hr; Kent: 9.1 hr

15. 7 mph **16.** 2 ft **17.** 16 m **18.** (a) 10 hr (b) 180 students
19. 140 ft by 70 ft; 9800 ft² **20.** A

21. vertex: $(0, -2)$; axis: $x = 0$; domain: $(-\infty, \infty)$; range: $[-2, \infty)$

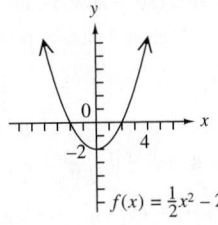

$f(x) = \frac{1}{2}x^2 - 2$

22. vertex: $(2, 3)$; axis: $x = 2$; domain: $(-\infty, \infty)$; range: $(-\infty, 3]$

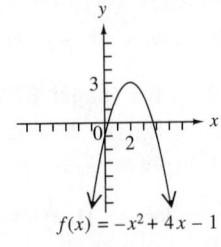

$f(x) = -x^2 + 4x - 1$

23. vertex: $(-5, -2)$; axis: $y = -2$; domain: $[-5, \infty)$; range: $(-\infty, \infty)$

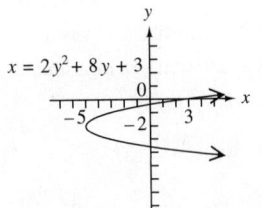

$x = 2y^2 + 8y + 3$

24. $(-\infty, -5) \cup \left(\dfrac{3}{2}, \infty\right)$

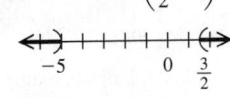

25. $(-\infty, 4) \cup [9, \infty)$

Chapters R–10 CUMULATIVE REVIEW EXERCISES (pages 831–832)

1. (a) $-2, 0, 7$ **(b)** $-\dfrac{7}{3}, -2, 0, 0.7, 7, \dfrac{32}{3}$ **(c)** All are real except $\sqrt{-8}$. **(d)** All are complex numbers. **2.** $\{1\}$ **3.** $[1, \infty)$

4. $\left[2, \dfrac{8}{3}\right]$ **5.** $\left\{\dfrac{2}{3}\right\}$ **6.** $\left\{\dfrac{2 + \sqrt{10}}{2}, \dfrac{2 - \sqrt{10}}{2}\right\}$ **7.** $\{-3, 5\}$

8. $\varnothing$ **9.** $\{-3, -1, 1, 3\}$ **10.** slope: $\dfrac{1}{2}$; y-intercept: $\left(0, -\dfrac{7}{4}\right)$

11. $x + 3y = -1$

12. function; domain: $(-\infty, \infty)$; range: $(-\infty, \infty)$

13. not a function

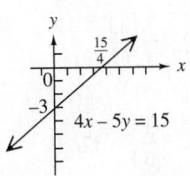

$4x - 5y = 15$

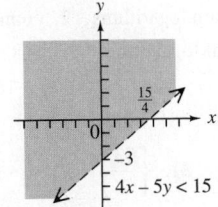

$4x - 5y < 15$

14. function; domain: $(-\infty, \infty)$; range: $(-\infty, 3]$

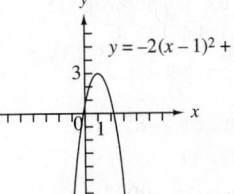

$y = -2(x-1)^2 + 3$

15. $\{(1, -2)\}$ **16.** $\{(3, -4, 2)\}$

17. $\dfrac{x^8}{y^4}$ **18.** $\dfrac{4}{xy^2}$

19. $4t^2 + 36t + 81$

20. $4x^2 - 6x + 11 + \dfrac{4}{x+2}$

21. $x(4 + x)(4 - x)$

22. $(4m - 3)(6m + 5)$

23. $(3x - 5y)^2$ **24.** $-\dfrac{5}{18}$

25. $-\dfrac{8}{k}$ **26.** $\dfrac{r - s}{r}$ **27.** $\dfrac{3\sqrt[3]{4}}{4}$ **28.** $\sqrt{7} + \sqrt{5}$ **29.** 44 mg

30. southbound car: 57 mi; eastbound car: 76 mi

CHAPTER 11 Inverse, Exponential, and Logarithmic Functions

SECTION 11.1 (pages 840–842)

1. $f(x)$ and $g(x)$ can be any two polynomials that have a sum of $3x^3 - x + 3$, such as $f(x) = 3x^3 + 1$ and $g(x) = -x + 2$.

2. $f(x)$ and $g(x)$ can be any two polynomials whose difference is $-x^2 + x - 5$, such as $f(x) = 2x^2 + 3x - 2$ and $g(x) = 3x^2 + 2x + 3$.

3. $x^2 + 2x - 9$ **5.** 6 **7.** $x^2 - x - 6$ **9.** 6 **11.** 0 **13.** $-\dfrac{9}{4}$

15. $2x^3 - 18x$ **17.** -20 **19.** 32 **21.** 20 **23.** $\dfrac{35}{4}$ **25.** $\dfrac{x^2 - 9}{2x}$, $x \neq 0$

27. $-\dfrac{5}{4}$ **29.** $\dfrac{x - 3}{2x}$, $x \neq 0$ **31.** 0 **33. (a)** $P(x) = 8.49x - 50$

(b) \$799 **35.** B **36.** D **37.** A **38.** C **39.** 16 **41.** 83 **43.** 13

45. $4x^2 + 12x + 13$ **47.** $x^2 + 10x + 29$ **49.** $2x + 8$ **51.** $\dfrac{137}{4}$ **53.** 8

55. 1 **56.** 9 **57.** 9 **58.** 12 **59.** 1 **60.** 12 **61.** $(f \circ g)(x) = 63{,}360x$;

It computes the number of inches in x miles. **63. (a)** $s = \dfrac{x}{4}$ **(b)** $y = \dfrac{x^2}{16}$

(c) 2.25 **65.** $(A \circ r)(t) = 4\pi t^2$; This is the area of the circular layer as a function of time.

SECTION 11.2 (pages 848–850)

1. A **2.** B **3.** It is not one-to-one. France and the United States are paired with the same trans fat percentage, 11. **4.** It is one-to-one.

5. This function is not one-to-one because two sodas in the list have 41 mg of caffeine. **6.** Yes. By adding 1 to 1058, two distances would be the same. The function would not be one-to-one. **7.** A **8.** D

9. $\{(6, 3), (10, 2), (12, 5)\}$ **11.** not one-to-one **13.** $\{(4.5, 0), (8.6, 2), (12.7, 4)\}$ **15.** $f^{-1}(x) = \dfrac{x - 4}{2}$, or $f^{-1}(x) = \dfrac{1}{2}x - 2$

17. $g^{-1}(x) = x^2 + 3, x \geq 0$ **19.** not one-to-one **21.** $f^{-1}(x) = \sqrt[3]{x + 4}$

23. not one-to-one **25. (a)** 8 **(b)** 3 **27. (a)** 1 **(b)** 0

29. (a) one-to-one

(b)

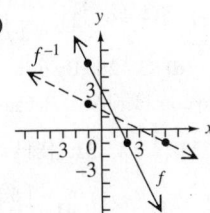

31. (a) not one-to-one

33. (a) one-to-one

(b)

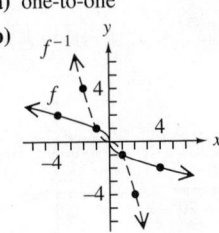

35.

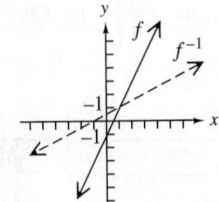

37.

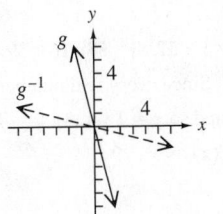

39. 0, 1, 2

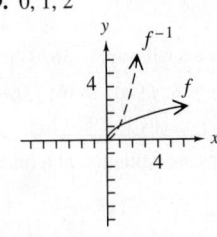

41. $-3, -2, -1, 6$

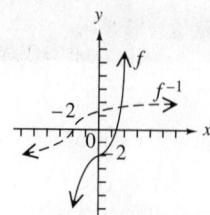

SECTION 11.3 (pages 857–858)

1. C **2.** A **3.** C **4.** D

5.

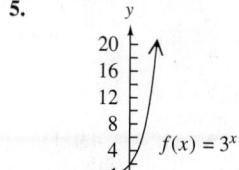

$f(x) = 3^x$

7.

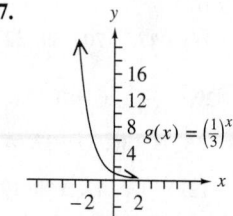

$g(x) = \left(\frac{1}{3}\right)^x$

9.

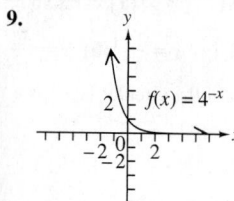

$f(x) = 4^{-x}$

11.

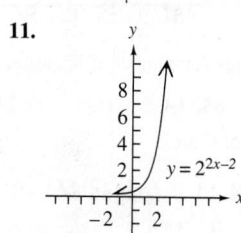

$y = 2^{2x-2}$

13. $\{2\}$ **15.** $\left\{\frac{3}{2}\right\}$ **17.** $\{7\}$ **19.** $\{-3\}$ **21.** $\{-1\}$ **23.** $\{-3\}$

25. rises; falls **26.** It is one-to-one and thus has an inverse. **27.** (a) 0.6°C
(b) 0.3°C **29.** (a) 1.4°C **(b)** 0.5°C **31.** 100 g **33.** 0.30 g

SECTION 11.4 (pages 863–866)

1. (a) C **(b)** F **(c)** B **(d)** A **(e)** E **(f)** D **2.** (a) B **(b)** E **(c)** D
(d) F **(e)** A **(f)** C **3.** $\log_4 1024 = 5$ **5.** $\log_{1/2} 8 = -3$

7. $\log_{10} 0.001 = -3$ **9.** $\log_{625} 5 = \frac{1}{4}$ **11.** $\log_8 \frac{1}{4} = -\frac{2}{3}$ **13.** $\log_5 1 = 0$

15. $4^3 = 64$ **17.** $12^1 = 12$ **19.** $6^0 = 1$ **21.** $9^{1/2} = 3$ **23.** $\left(\frac{1}{4}\right)^{1/2} = \frac{1}{2}$

25. $5^{-1} = 5^{-1}$ **27.** (a) C **(b)** B **(c)** B **(d)** C **28.** By using the
word "radically," the teacher meant for him to consider roots. Because
3 is the square (2nd) root of 9, $\log_9 3 = \frac{1}{2}$. **29.** $\left\{\frac{1}{3}\right\}$ **31.** $\{81\}$

33. $\left\{\frac{1}{5}\right\}$ **35.** $\{1\}$ **37.** $\{x \mid x > 0, x \neq 1\}$ **39.** $\{5\}$ **41.** $\left\{\frac{5}{3}\right\}$

43. $\{4\}$ **45.** $\left\{\frac{3}{2}\right\}$ **47.** $\{30\}$

49.

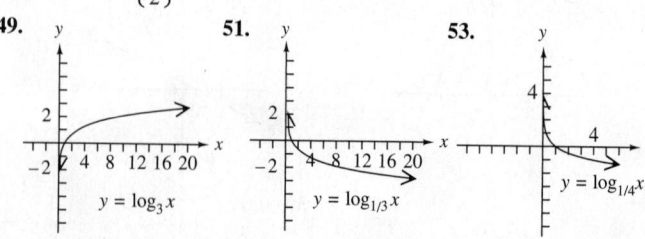

$y = \log_3 x$

51.

$y = \log_{1/3} x$

53.

$y = \log_{1/4} x$

55. Answers will vary. **56.** $(0, \infty)$; $(-\infty, \infty)$ **57.** 8 **58.** 16 **59.** 24
60. $f(0) = 8$; $f(10) = 16$; $f(60) = 24$ **61.** Since every real number
power of 1 equals 1, if $y = \log_1 x$, then $x = 1^y$ and so $x = 1$ for every y. This
contradicts the definition of a function. **62.** $f(x) = 3^x$

63. $x = \log_a 1$ is equivalent to $a^x = 1$. The only value of x that makes
$a^x = 1$ is 0. (Recall that $a \neq 1$.) **64.** The expression gets closer and closer
to a number that is approximately 2.718. **65.** (a) 130 thousand units
(b) 160 thousand units **(c)** 190 thousand units **67.** about 4 times as powerful

SECTION 11.5 (pages 873–874)

1. $\log_{10} 7 + \log_{10} 8$ **2.** $\log_{10} 7 - \log_{10} 8$ **3.** 4 **4.** $6 \log_{10} 3$ **5.** false;
$\log_b x + \log_b y = \log_b xy$ **6.** true **7.** true **8.** false; $\log_b x^r = r \log_b x$

9. $\log_7 4 - \log_7 5$ **11.** $\frac{1}{4} \log_2 8$, or $\frac{3}{4}$ **13.** $\log_4 3 + \frac{1}{2} \log_4 x - \log_4 y$

15. $\frac{1}{3} \log_3 4 - 2 \log_3 x - \log_3 y$ **17.** $\frac{1}{2} \log_3 x + \frac{1}{2} \log_3 y - \frac{1}{2} \log_3 5$

19. $\frac{1}{3} \log_2 x + \frac{1}{5} \log_2 y - 2 \log_2 r$ **21.** In the notation $\log_a (x + y)$, the
parentheses do not indicate multiplication. They indicate that $x + y$ is the
result of raising a to some power. **22.** No number allowed as a logarithmic
base can be raised to a power with a result of 0. **23.** $\log_b xy$

25. $\log_a \frac{m^3}{n}$ **27.** $\log_a \frac{rt^3}{s}$ **29.** $\log_a \frac{125}{81}$ **31.** $\log_{10} (x^2 - 9)$

33. $\log_p \frac{x^3 y^{1/2}}{z^{3/2} a^3}$ **35.** false **37.** true **39.** true **41.** false **43.** 4

44. It is the exponent to which 3 must be raised in order to obtain 81.
45. 81 **46.** It is the exponent to which 2 must be raised in order to
obtain 19. **47.** 19 **48.** m

SECTION 11.6 (pages 879–880)

1. C **2.** A **3.** C **4.** C **5.** 19.2 **6.** $\sqrt{2}$ **7.** $\sqrt{3}$ **8.** 75.2
9. 2.5164 **11.** -1.4868 **13.** 9.6776 **15.** 2.0592 **17.** -2.8896
19. 2.3026 **21.** poor fen **23.** bog **25.** rich fen **27.** 11.6 **29.** 5.9
31. 1.0×10^{-2} **33.** 2.5×10^{-5} **35.** (a) 35.0 yr **(b)** 14.2 yr
37. (a) 107 dB **(b)** 100 dB **(c)** 98 dB **39.** (a) 55% **(b)** 90%

SECTION 11.7 (pages 887–889)

1. common logarithms **2.** common logarithms **3.** natural logarithms
4. natural logarithms **5.** $\{0.827\}$ **7.** $\{0.833\}$ **9.** $\{1.201\}$
11. $\{2.269\}$ **13.** $\{15.967\}$ **15.** $\{-6.067\}$ **17.** $\{261.291\}$
19. $\{-10.718\}$ **21.** $\{3\}$ **23.** $\{5.879\}$ **25.** $\{1.571\}$ **27.** $\{1\}$

29. $\left\{\frac{2}{3}\right\}$ **31.** $\left\{\frac{33}{2}\right\}$ **33.** $\{-1 + \sqrt[3]{49}\}$ **35.** $\left\{\frac{1}{3}\right\}$ **37.** $\{2\}$

39. $\varnothing$ **41.** $\{8\}$ **43.** $\left\{\frac{4}{3}\right\}$ **45.** $\{8\}$ **47.** (a) $2539.47

(b) 10.19 yr **49.** $4934.71 **51.** 15.40 yr **53.** 179.73 g **55.** 1.4315

57. 3.3030 **59.** -0.0947 **61.** -2.3219 **63.** $\log 5^x = \log 125$

64. $x \log 5 = \log 125$ **65.** $x = \frac{\log 125}{\log 5}$ **66.** $\frac{\log 125}{\log 5} = 3; \{3\}$

Chapter 11 REVIEW EXERCISES (pages 894–897)

1. (a) $5x^2 - x + 5$ **(b)** $-5x^2 + 5x + 1$ **(c)** 11 **(d)** -9
2. (a) $36x^3 - 9x^2$ **(b)** $4x - 1, x \neq 0$ **(c)** -45 **(d)** 7
3. (a) 167 **(b)** 1495 **4.** (a) 20 **(b)** 42 **5.** (a) $75x^2 + 220x + 160$
(b) $15x^2 + 10x + 2$ **6.** No, composition of functions is not commutative.
For example, the results of **Exercise 5** show that $(f \circ g)(x) \neq (g \circ f)(x)$
in this case. **7.** not one-to-one **8.** one-to-one **9.** not one-to-one

10. $f^{-1}(x) = \frac{x - 7}{-3}$, or $f^{-1}(x) = -\frac{1}{3}x + \frac{7}{3}$ **11.** $f^{-1}(x) = \frac{x^3 + 4}{6}$

12. not one-to-one

13.

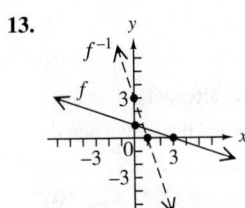

14.

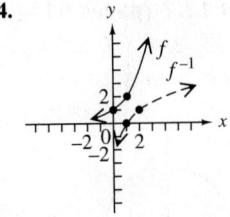

15.

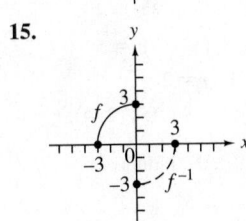

16.

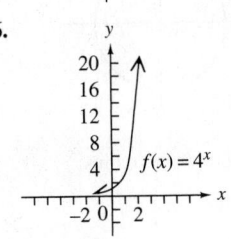

17.

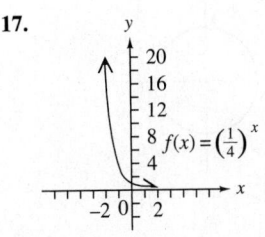

18.

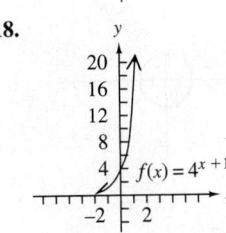

19. $\{4\}$ **20.** $\left\{\dfrac{3}{7}\right\}$ **21.** $\{0\}$ **22.** $\{-3\}$ **23. (a)** 55.8 million

(b) 80.8 million   **24. (a)** $5^4 = 625$ **(b)** $\log_5 0.04 = -2$ **25.** 5; 2; fifth; 32

26.

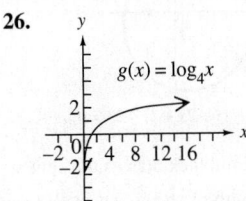

27.

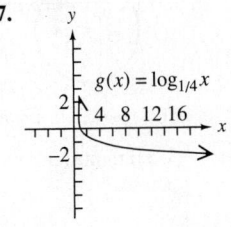

28.

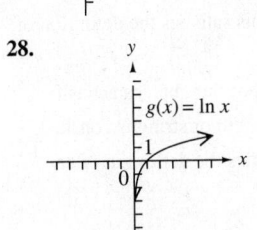

29. $\{2\}$ **30.** $\{-2\}$ **31.** $\{8\}$

32. $\{b \mid b > 0, b \neq 1\}$

33. $\log_4 3 + 2\log_4 x$

34. $3\log_5 a + 2\log_5 b - 4\log_5 c$

35. $\dfrac{1}{2}\log_4 x + 2\log_4 w - \log_4 z$

36. $2\log_2 p + \log_2 r - \dfrac{1}{2}\log_2 z$ **37.** $\log_a \dfrac{49}{16}$ **38.** $\log_a 250$ **39.** $\log_b \dfrac{3x}{y^2}$

40. $\log_3 \dfrac{x+7}{4x+6}$ **41.** 1.4609   **42.** -0.5901 **43.** 3.3638   **44.** -1.3587

45. 6.4   **46.** 8.4   **47.** 6.3×10^{-3} **48.** $\{2.042\}$ **49.** $\{4.907\}$

50. $\{18.310\}$ **51.** $\left\{\dfrac{1}{9}\right\}$ **52.** $\{-6 + \sqrt[3]{25}\}$ **53.** $\{2\}$ **54.** $\left\{\dfrac{3}{8}\right\}$

55. $\{4\}$ **56.** $\{1\}$ **57.** \$7112.11   **58.** Plan A would pay \$2.92 more.

59. 0.9251   **60.** 1.7925   **61.** 1.4315   **62.** -2.0437 **63.** 7   **64.** 0

65. -3 **66.** 36   **67.** 4   **68.** e **69.** -5 **70.** 5.4   **71.** 2.5855

72. -0.3857 **73.** 4.6439   **74.** -2.4022 **75.** $\{72\}$ **76.** $\{5\}$

77. $\left\{\dfrac{1}{9}\right\}$ **78.** $\left\{\dfrac{4}{3}\right\}$ **79.** $\{3\}$ **80.** $\left\{\dfrac{1}{8}\right\}$ **81.** $\left\{\dfrac{11}{3}\right\}$ **82.** $\{0\}$

83. $\left\{\dfrac{2}{63}\right\}$ **84.** $\{-2, -1\}$ **85.** \$4267   **86.** 11%   **87.** 0.325

88. 0.673

Chapter 11 TEST (pages 898–899)

1. (a) $-2x^2 + 12x - 9$ **(b)** $-2x^2 - 2x - 3$ **(c)** -7 **2. (a)** 0

(b) $x + 2,\ x \neq -1$ **(c)** 0   **3. (a)** 23   **(b)** $3x^2 + 11$ **(c)** $9x^2 + 30x + 27$

4. (a) not one-to-one   **(b)** one-to-one   **5.** $f^{-1}(x) = x^3 - 7$

6.

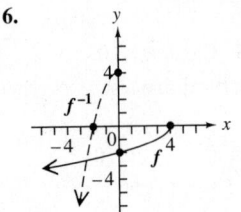

7.

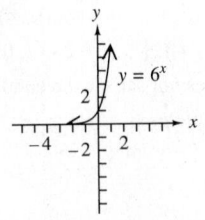

8.

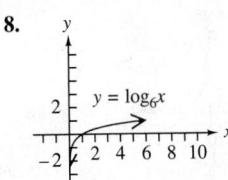

9. $\{-4\}$ **10.** $\left\{-\dfrac{13}{3}\right\}$

11. (a) 775 millibars

(b) 265 millibars

12. $\log_4 0.0625 = -2$

13. $7^2 = 49$ **14.** $\{32\}$

15. $\left\{\dfrac{1}{2}\right\}$ **16.** $\{2\}$ **17.** $2\log_3 x + \log_3 y$

18. $\dfrac{1}{2}\log_5 x - \log_5 y - \log_5 z$ **19.** $\log_b \dfrac{s^3}{t}$ **20.** $\log_b \dfrac{r^{1/4}\,s^2}{t^{2/3}}$

21. (a) 1.3284   **(b)** -0.8440 **(c)** 2.1245   **22.** $\{3.9656\}$ **23.** $\{3\}$

24. (a) \$12,507.51   **(b)** 15.5 yr   **25. (a)** \$19,260.38   **(b)** 13.9 yr

26. (a) $\dfrac{\log 19}{\log 3}$ **(b)** $\dfrac{\ln 19}{\ln 3}$ **(c)** 2.6801

Chapters R–11 CUMULATIVE REVIEW EXERCISES (pages 900–901)

1. $-2, 0, 6, \dfrac{30}{3}$ (or 10)   **2.** $-\dfrac{9}{4}, -2, 0, 0.6, 6, \dfrac{30}{3}$ (or 10)

3. $-\sqrt{2}, \sqrt{11}$ **4.** $\left\{-\dfrac{2}{3}\right\}$ **5.** $[1, \infty)$ **6.** $\{-2, 7\}$

7. $(-\infty, -3) \cup (2, \infty)$ **8. (a)** yes   **(b)** 2192; The number of travelers increased by an average of 2192 thousand per year during the period 2006–2010.   **9.** $\{(4, 2)\}$ **10.** $\{(1, -1, 4)\}$ **11.** $6p^2 + 7p - 3$

12. $16k^2 - 24k + 9$ **13.** $-5m^3 + 2m^2 - 7m + 4$ **14.** $2t^3 + 5t^2 - 3t + 4$

15. $z(5z + 1)(z - 4)$ **16.** $(4a + 5b^2)(4a - 5b^2)$

17. $(2c + d)(4c^2 - 2cd + d^2)$ **18.** $-\dfrac{1875p^{13}}{8}$ **19.** $\dfrac{x+5}{x+4}$

20. $\dfrac{-3k - 19}{(k+3)(k-2)}$ **21.** $12\sqrt{2}$ **22.** $-\dfrac{1}{4}$ **23.** $-27\sqrt{2}$ **24.** $\{0, 4\}$

25. 41   **26.** $-i$ **27.** $\left\{\dfrac{1 + \sqrt{13}}{6}, \dfrac{1 - \sqrt{13}}{6}\right\}$ **28.** $(-\infty, -4) \cup (2, \infty)$

29. $\{-2, -1, 1, 2\}$ **30.** $\{-1\}$ **31.** $\{1\}$ **32.** $5^3 = 125$

33. $3\log x + \dfrac{1}{2}\log y - \log z$

34.

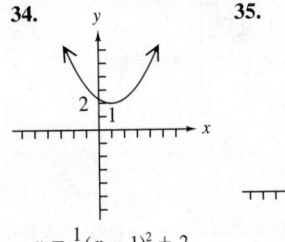

35.

36.

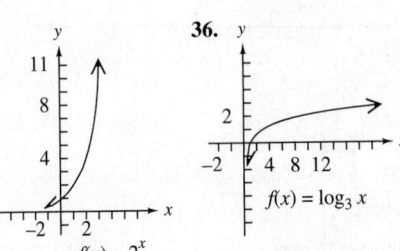

CHAPTER 12 Nonlinear Functions, Conic Sections, and Nonlinear Systems

SECTION 12.1 (pages 908–910)

1. E; 0; 0 **2.** D; $(-\infty, \infty)$; $[0, \infty)$
3. A; $(-\infty, \infty)$; $\{\ldots, -2, -1, 0, 1, 2, \ldots\}$ **4.** C; $[0, \infty)$; $[0, \infty)$
5. B; It does not satisfy the conditions of the vertical line test.
6. F; asymptotes
7. domain: $(-\infty, \infty)$; range: $[0, \infty)$

9. domain: $(-\infty, 0) \cup (0, \infty)$; range: $(-\infty, 1) \cup (1, \infty)$

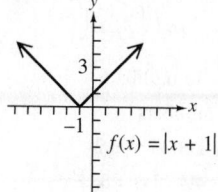
$f(x) = |x + 1|$

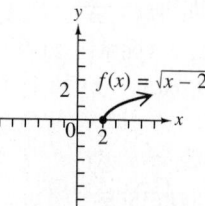
$f(x) = \frac{1}{x} + 1$

11. domain: $[2, \infty)$; range: $[0, \infty)$

13. domain: $(-\infty, 2) \cup (2, \infty)$; range: $(-\infty, 0) \cup (0, \infty)$

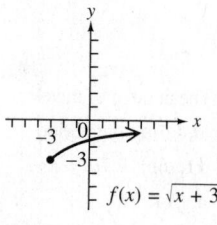
$f(x) = \sqrt{x - 2}$

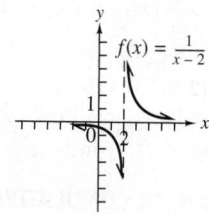
$f(x) = \frac{1}{x - 2}$

15. domain: $[-3, \infty)$; range: $[-3, \infty)$

17. domain: $(-\infty, \infty)$; range: $[1, \infty)$

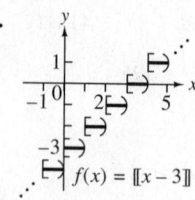
$f(x) = \sqrt{x + 3} - 3$

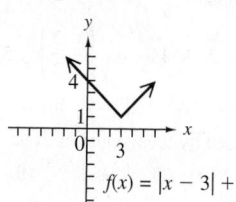
$f(x) = |x - 3| + 1$

19. B **20.** C **21.** A **22.** D **23.** Shift the graph of $g(x) = \frac{1}{x}$ three units to the right and two units up. **24.** Shift the graph of $g(x) = \frac{1}{x}$ five units to the left and three units down. **25.** 3 **27.** 4 **29.** 0
31. -14 **33.** -11

35.

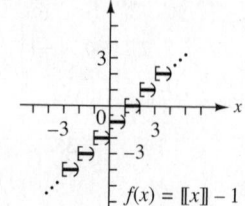
$f(x) = [\![x - 3]\!]$

37.

$f(x) = [\![x]\!] - 1$

39.

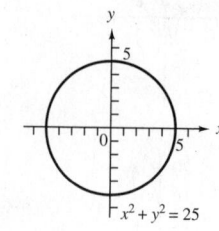

41. $2.75

SECTION 12.2 (pages 917–920)

1. B **2.** C **3.** D **4.** A
5. (a) $(0, 0)$ (b) 5
(c)

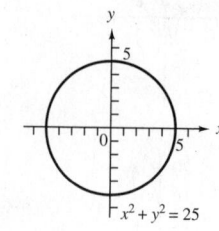
$x^2 + y^2 = 25$

6. There will always be domain values that yield more than one range value. A circle fails the conditions of the vertical line test.
7. $(x + 4)^2 + (y - 3)^2 = 4$
9. $(x + 8)^2 + (y + 5)^2 = 5$
11. $(-2, -3)$; $r = 2$
13. $(-5, 7)$; $r = 9$ **15.** $(2, 4)$; $r = 4$
19. center: $(0, 0)$; radius: 4

17. center: $(0, 0)$; radius: 2

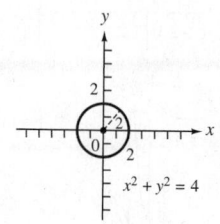
$x^2 + y^2 = 4$

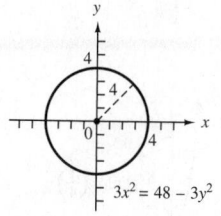
$3x^2 = 48 - 3y^2$

21. center: $(1, -2)$; radius: 4

23. center: $(-1, -1)$; radius: 5

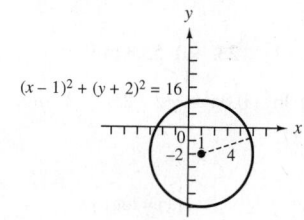
$(x - 1)^2 + (y + 2)^2 = 16$

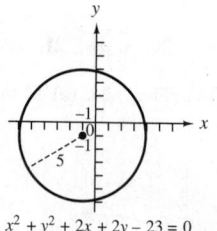
$x^2 + y^2 + 2x + 2y - 23 = 0$

25. center: $(-3, 3)$; radius: 3

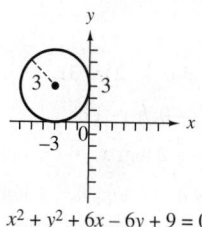
$x^2 + y^2 + 6x - 6y + 9 = 0$

27. The thumbtack acts as the center and the length of string acts as the radius. This satisfies the definition of a circle.
28. The two thumbtacks act as foci and the length of string is constant. This satisfies the definition of an ellipse.

29.

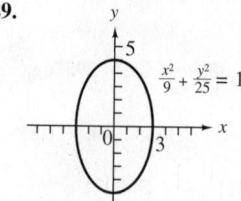
$\frac{x^2}{9} + \frac{y^2}{25} = 1$

31.

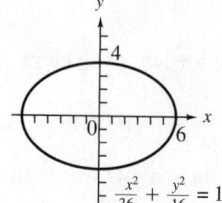
$\frac{x^2}{36} + \frac{y^2}{16} = 1$

33.

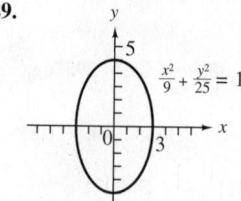
$\frac{x^2}{25} + \frac{y^2}{4} = 1$

35.

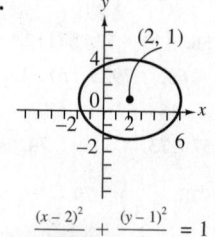
$\frac{(x - 2)^2}{16} + \frac{(y - 1)^2}{9} = 1$

37.

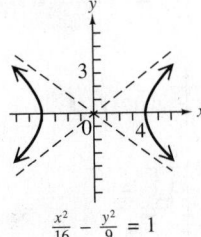

$$\frac{(x+3)^2}{4} + \frac{(y-1)^2}{25} = 1$$

39. (a) 10 m **(b)** 36 m

41. $3\sqrt{3}$ units

SECTION 12.3 (pages 926–928)

1. C **2.** B **3.** D **4.** A

5.

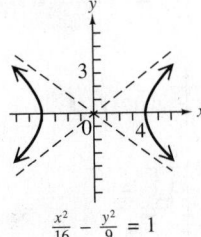

$$\frac{x^2}{16} - \frac{y^2}{9} = 1$$

7.

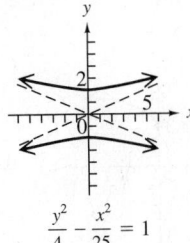

$$\frac{y^2}{4} - \frac{x^2}{25} = 1$$

9.

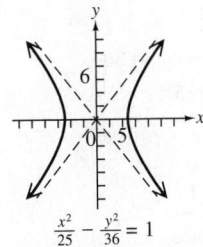

$$\frac{x^2}{25} - \frac{y^2}{36} = 1$$

11.

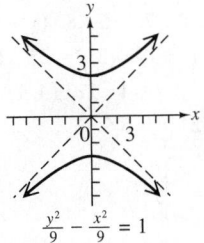

$$\frac{y^2}{9} - \frac{x^2}{9} = 1$$

13. hyperbola

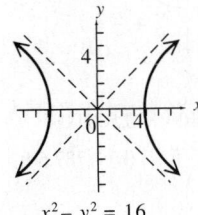

$$x^2 - y^2 = 16$$

15. ellipse

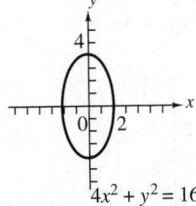

$$4x^2 + y^2 = 16$$

17. circle

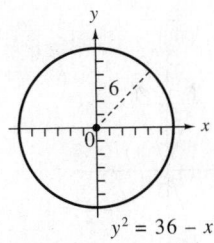

$$y^2 = 36 - x^2$$

19. ellipse

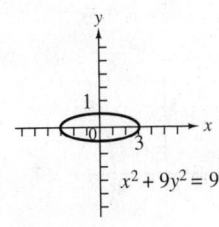

$$x^2 + 9y^2 = 9$$

21. domain: $[-3, 3]$; range: $[0, 3]$

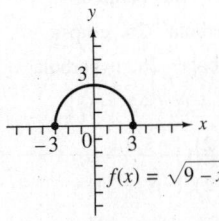

$$f(x) = \sqrt{9 - x^2}$$

23. domain: $[-5, 5]$; range: $[-5, 0]$

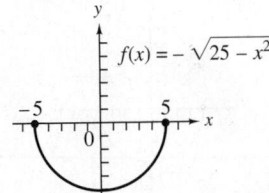

$$f(x) = -\sqrt{25 - x^2}$$

25. domain: $[-4, \infty)$; range: $[0, \infty)$

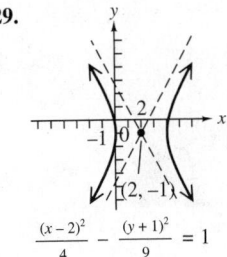

$$y = \sqrt{\frac{x+4}{2}}$$

27. domain: $[-3, 3]$; range: $[-2, 0]$

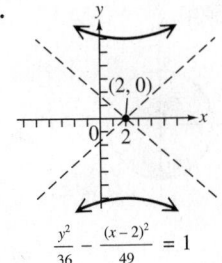

$$y = -2\sqrt{\frac{9 - x^2}{9}}$$

29.

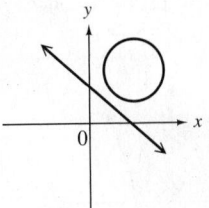

$$\frac{(x-2)^2}{4} - \frac{(y+1)^2}{9} = 1$$

31.

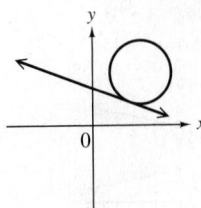

$$\frac{y^2}{36} - \frac{(x-2)^2}{49} = 1$$

33. (a) 50 m **(b)** 69.3 m

SECTION 12.4 (pages 934–935)

1. one **2.** two **3.** none **4.** four

In Exercises 5–10, answers may vary.

5.

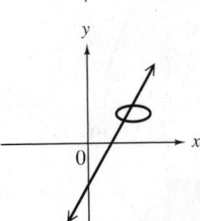

6.

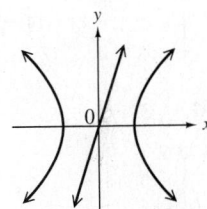

7.

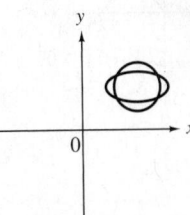

8.

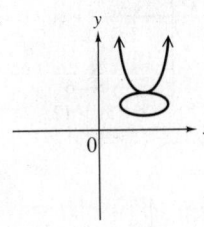

9.

10.

11. $\left\{(0, 0), \left(\frac{1}{2}, \frac{1}{2}\right)\right\}$ **13.** $\{(-6, 9), (-1, 4)\}$

15. $\left\{\left(-\frac{1}{5}, \frac{7}{5}\right), (1, -1)\right\}$ **17.** $\left\{(-2, -2), \left(-\frac{4}{3}, -3\right)\right\}$

19. $\{(-3, 1), (1, -3)\}$ **21.** $\left\{\left(-\frac{3}{2}, -\frac{9}{4}\right), (-2, 0)\right\}$

23. $\left\{\left(-\sqrt{3}, 0\right), \left(\sqrt{3}, 0\right), \left(-\sqrt{5}, 2\right), \left(\sqrt{5}, 2\right)\right\}$

25. $\{(-2, 0), (2, 0)\}$ **27.** $\{(1, 3), (1, -3), (-1, 3), (-1, -3)\}$

29. $\left\{\left(i\sqrt{2}, -3i\sqrt{2}\right), \left(-i\sqrt{2}, 3i\sqrt{2}\right), \left(-\sqrt{6}, -\sqrt{6}\right), \left(\sqrt{6}, \sqrt{6}\right)\right\}$

31. $\left\{\left(-2i\sqrt{2}, -2\sqrt{3}\right), \left(-2i\sqrt{2}, 2\sqrt{3}\right), \left(2i\sqrt{2}, -2\sqrt{3}\right), \left(2i\sqrt{2}, 2\sqrt{3}\right)\right\}$ **33.** $\left\{\left(-\sqrt{5}, -\sqrt{5}\right), \left(\sqrt{5}, \sqrt{5}\right)\right\}$

35. length: 12 ft; width: 7 ft

SECTION 12.5 (pages 940–941)

1. C   2. above; outside; below   3. B   4. D   5. A   6. C

7.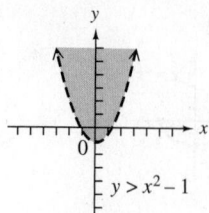
$y > x^2 - 1$

9.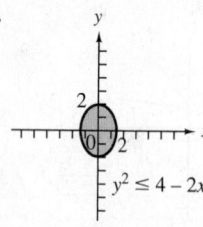
$y^2 \le 4 - 2x^2$

11.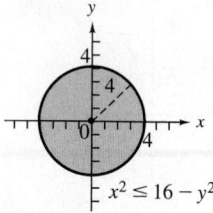
$x^2 \le 16 - y^2$

13.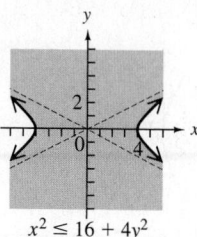
$x^2 \le 16 + 4y^2$

15.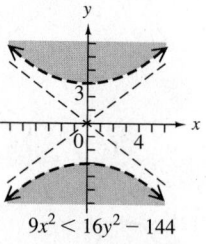
$9x^2 < 16y^2 - 144$

17.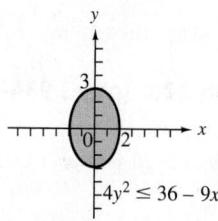
$4y^2 \le 36 - 9x^2$

19.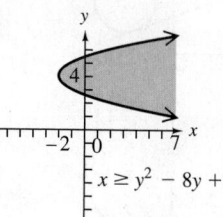
$x \ge y^2 - 8y + 14$

21.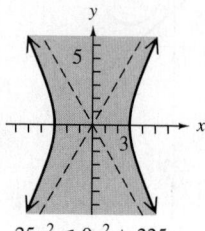
$25x^2 \le 9y^2 + 225$

23.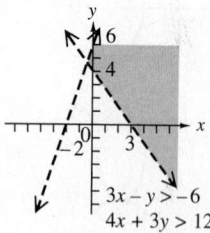
$3x - y > -6$
$4x + 3y > 12$

25.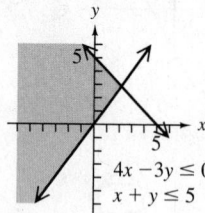
$4x - 3y \le 0$
$x + y \le 5$

27.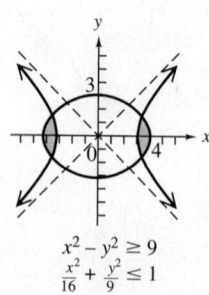
$x^2 - y^2 \ge 9$
$\frac{x^2}{16} + \frac{y^2}{9} \le 1$

29.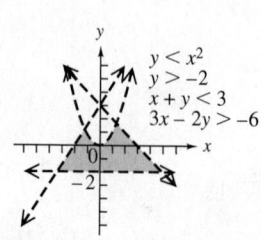
$y < x^2$
$y > -2$
$x + y < 3$
$3x - 2y > -6$

Chapter 12 REVIEW EXERCISES (pages 946–948)

1. domain: $(-\infty, \infty)$;
range: $[0, \infty)$

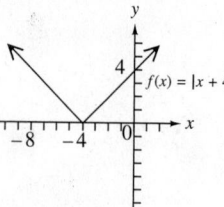
$f(x) = |x + 4|$

2. domain: $(-\infty, 4) \cup (4, \infty)$;
range: $(-\infty, 0) \cup (0, \infty)$

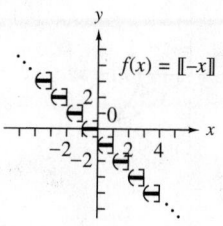
$f(x) = \frac{1}{x - 4}$

3. domain: $[0, \infty)$;
range: $[3, \infty)$

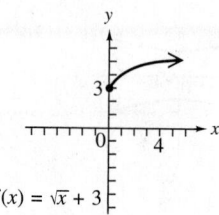
$f(x) = \sqrt{x} + 3$

4. domain: $(-\infty, \infty)$;
range: $\{\dots, -2, -1, 0, 1, 2, \dots\}$

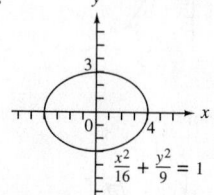
$f(x) = [\![-x]\!]$

5. 12   6. 2   7. -5 8. $(x + 2)^2 + (y - 4)^2 = 9$
9. $(x + 1)^2 + (y + 3)^2 = 25$ 10. $(x - 4)^2 + (y - 2)^2 = 36$
11. $(-3, 2); r = 4$ 12. $(4, 1); r = 2$ 13. $(-1, -5); r = 3$
14. $(3, -2); r = 5$

15.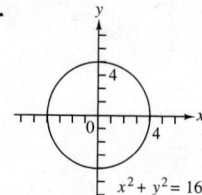
$x^2 + y^2 = 16$

16.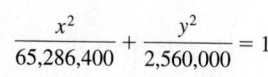
$\frac{x^2}{16} + \frac{y^2}{9} = 1$

17.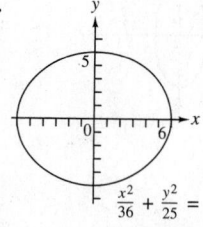
$\frac{x^2}{36} + \frac{y^2}{25} = 1$

18. $\dfrac{x^2}{65,286,400} + \dfrac{y^2}{2,560,000} = 1$

19. (a) 348.2 ft   (b) 1787.6 ft

20.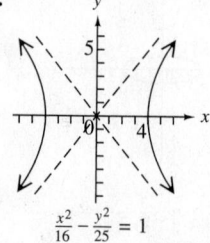
$\frac{x^2}{16} - \frac{y^2}{25} = 1$

21.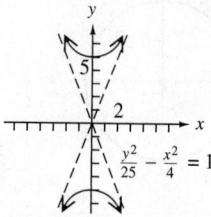
$\frac{y^2}{25} - \frac{x^2}{4} = 1$

22.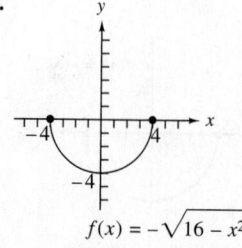
$f(x) = -\sqrt{16 - x^2}$

23. circle   24. parabola
25. hyperbola   26. ellipse
27. parabola   28. hyperbola
29. $\{(6, -9), (-2, -5)\}$
30. $\{(1, 2), (-5, 14)\}$
31. $\{(4, 2), (-1, -3)\}$
32. $\{(-2, -4), (8, 1)\}$

33. $\left\{ \left(-\sqrt{2}, 2 \right), \left(-\sqrt{2}, -2 \right), \left(\sqrt{2}, -2 \right), \left(\sqrt{2}, 2 \right) \right\}$

34. $\left\{ \left(-\sqrt{6}, -\sqrt{3} \right), \left(-\sqrt{6}, \sqrt{3} \right), \left(\sqrt{6}, -\sqrt{3} \right), \left(\sqrt{6}, \sqrt{3} \right) \right\}$

35. 0, 1, or 2   **36.** 0, 1, 2, 3, or 4

37.

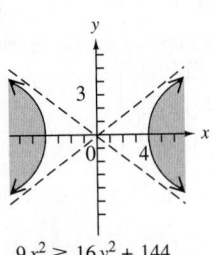

$9x^2 \geq 16y^2 + 144$

38.

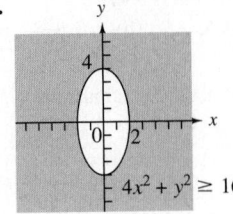

$4x^2 + y^2 \geq 16$

39.

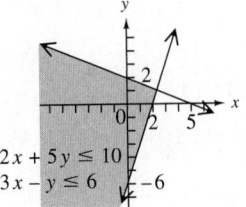

$2x + 5y \leq 10$
$3x - y \leq 6$

40.

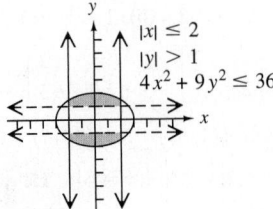

$|x| \leq 2$
$|y| > 1$
$4x^2 + 9y^2 \leq 36$

41.

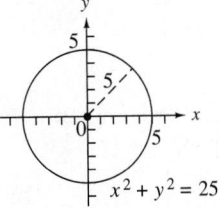

$x^2 + y^2 = 25$

42.

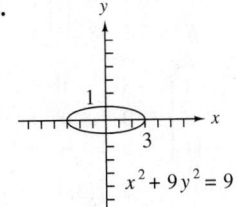

$x^2 + 9y^2 = 9$

43.

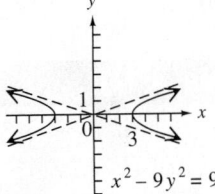

$x^2 - 9y^2 = 9$

44.

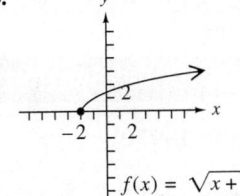

$f(x) = \sqrt{x + 2}$

45.

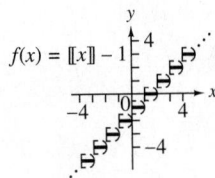

$f(x) = [\![x]\!] - 1$

46.

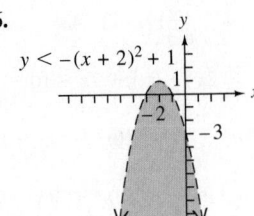

$y < -(x + 2)^2 + 1$

47.

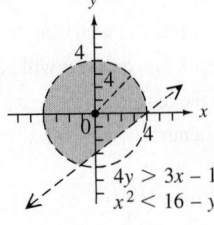

$4y > 3x - 12$
$x^2 < 16 - y^2$

48.

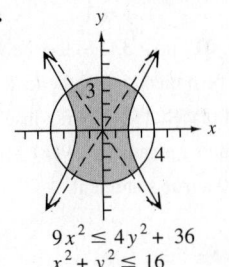

$9x^2 \leq 4y^2 + 36$
$x^2 + y^2 \leq 16$

Chapter 12 TEST (pages 949–950)

1. 0   **2.** $[0, \infty)$ **3.** $\{\ldots, -2, -1, 0, 1, 2, \ldots\}$

4. (a) C   **(b)** A   **(c)** D   **(d)** B

5. domain: $(-\infty, \infty)$; range: $[4, \infty)$ **6.** center: $(2, -3)$; radius: 4

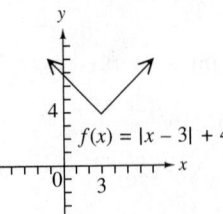

$f(x) = |x - 3| + 4$

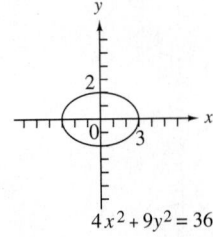

$(x - 2)^2 + (y + 3)^2 = 16$

7. center: $(-4, 1)$; radius: 5

8.

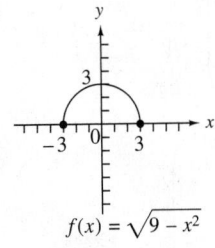

$f(x) = \sqrt{9 - x^2}$

9.

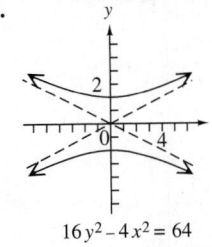

$4x^2 + 9y^2 = 36$

10.

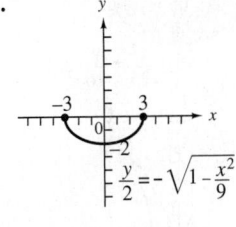

$16y^2 - 4x^2 = 64$

11.

$\dfrac{y}{2} = -\sqrt{1 - \dfrac{x^2}{9}}$

12. ellipse   **13.** hyperbola   **14.** parabola   **15.** circle

16. $\left\{ \left(-\dfrac{1}{2}, -10 \right), (5, 1) \right\}$ **17.** $\left\{ (-2, -2), \left(\dfrac{14}{5}, -\dfrac{2}{5} \right) \right\}$

18. $\left\{ \left(-\sqrt{22}, -\sqrt{3} \right), \left(-\sqrt{22}, \sqrt{3} \right), \left(\sqrt{22}, -\sqrt{3} \right), \left(\sqrt{22}, \sqrt{3} \right) \right\}$

19.

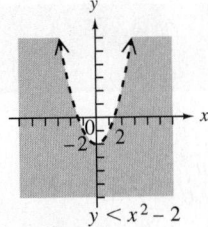

$y < x^2 - 2$

20.

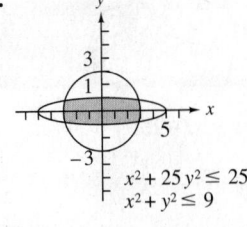

$x^2 + 25y^2 \leq 25$
$x^2 + y^2 \leq 9$

Chapters R–12 CUMULATIVE REVIEW EXERCISES
(pages 951–952)

1. $\left\{ \dfrac{2}{3} \right\}$ **2.** $\left(-\infty, \dfrac{3}{5} \right]$ **3.** $\{-4, 4\}$ **4.** $\dfrac{2}{3}$ **5.** $3x + 2y = -13$

6. $\{(3, -3)\}$ **7.** $\{(4, 1, -2)\}$ **8.** $\left\{ (-1, 5), \left(\dfrac{5}{2}, -2 \right) \right\}$

9. $25y^2 - 30y + 9$ **10.** $12r^2 + 40r - 7$

11. $4x^3 - 4x^2 + 3x + 5 + \dfrac{3}{2x + 1}$ **12.** $(3x + 2)(4x - 5)$

13. $(z^2 + 1)(z + 1)(z - 1)$ **14.** $(a - 3b)(a^2 + 3ab + 9b^2)$

15. $\dfrac{y - 1}{y(y - 3)}$ **16.** $\dfrac{3c + 5}{(c + 5)(c + 3)}$ **17.** $\dfrac{1}{p}$ **18.** $1\dfrac{1}{5}$ hr   **19.** $\dfrac{a^5}{4}$

20. $2\sqrt[3]{2}$ **21.** $\dfrac{3\sqrt{10}}{2}$ **22.** $\dfrac{7}{5} + \dfrac{11}{5}i$ **23.** $\varnothing$ **24.** $\left\{ \dfrac{1}{5}, -\dfrac{3}{2} \right\}$

25. $\left\{\dfrac{3 + \sqrt{33}}{6}, \dfrac{3 - \sqrt{33}}{6}\right\}$ **26.** $\left\{-\dfrac{\sqrt{6}}{2}, \dfrac{\sqrt{6}}{2}, -\sqrt{7}, \sqrt{7}\right\}$

27. $\{3\}$ **28.** $v = \dfrac{\pm \sqrt{rFkw}}{kw}$ **29. (a)** -1 **(b)** $9x^2 + 18x + 4$

30. $f^{-1}(x) = \sqrt[3]{x - 4}$ **31. (a)** 4   **(b)** 7   **32.** $\log \dfrac{(3x + 7)^2}{4}$

33.

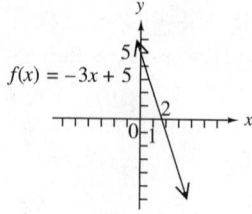

34.

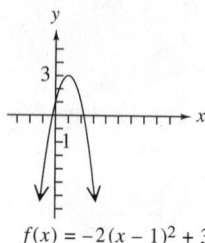

$f(x) = -2(x - 1)^2 + 3$

35.

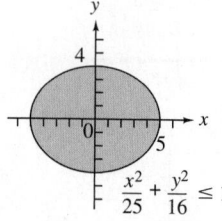

$\dfrac{x^2}{25} + \dfrac{y^2}{16} \le 1$

36.

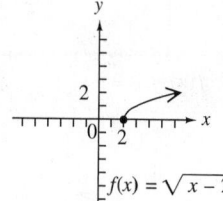

$f(x) = \sqrt{x - 2}$

37.

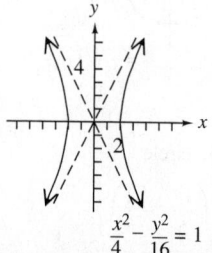

$\dfrac{x^2}{4} - \dfrac{y^2}{16} = 1$

38.

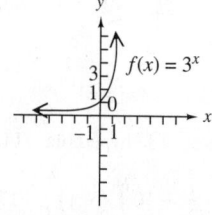

$f(x) = 3^x$

APPENDIX A Review of Exponents, Polynomials, and Factoring

(pages 957–959)

1. $\dfrac{1}{a^2 b}$ **3.** $\dfrac{100y^{10}}{x^2}$ **5.** 0   **7.** $\dfrac{x^{10}}{2w^{13}y^5}$ **9.** $\dfrac{a^{15}}{-64b^{15}}$ **11.** $\dfrac{x^{16}z^{10}}{y^6}$

13. $-6a^4 + 11a^3 - 20a^2 + 26a - 15$ **15.** $8x^3 - 18x^2 + 6x - 16$

17. $x^2 y - xy^2 + 6y^3$ **19.** $-3x^2 - 62x + 32$ **21.** $10x^3 - 4x^2 + 9x - 4$

23. $6x^2 - 19x - 7$ **25.** $4x^2 - 9x + 2$ **27.** $16t^2 - 9$ **29.** $4y^4 - 16$

31. $16x^2 - 24x + 9$ **33.** $36r^2 + 60ry + 25y^2$ **35.** $c^3 + 8d^3$

37. $64x^3 - 1$ **39.** $14t^3 + 45st^2 + 18s^2t - 5s^3$

41. $4xy^3(2x^2 y + 3x + 9y)$ **43.** $(x + 3)(x - 5)$ **45.** $(2x + 3)(x - 6)$

47. $(6t + 5)(6t - 5)$ **49.** $(4t + 3)^2$ **51.** $p(2m - 3n)^2$

53. $(x + 1)(x^2 - x + 1)$ **55.** $(2t + 5)(4t^2 - 10t + 25)$

57. $(t^2 - 5)(t^4 + 5t^2 + 25)$ **59.** $(5x + 2y)(t + 3r)$

61. $(6r - 5s)(a + 2b)$ **63.** $(t^2 + 1)(t + 1)(t - 1)$

APPENDIX B Solving Systems of Linear Equations by Matrix Methods

(pages 965–966)

1. (a) $0, 5, -3$ **(b)** $1, -3, 8$ **(c)** yes; The number of rows is the same as the number of columns (three).   **(d)** $\begin{bmatrix} 1 & 4 & 8 \\ 0 & 5 & -3 \\ -2 & 3 & 1 \end{bmatrix}$

(e) $\begin{bmatrix} 1 & -\frac{3}{2} & -\frac{1}{2} \\ 0 & 5 & -3 \\ 1 & 4 & 8 \end{bmatrix}$ **(f)** $\begin{bmatrix} 1 & 15 & 25 \\ 0 & 5 & -3 \\ 1 & 4 & 8 \end{bmatrix}$

2. (a) 3×2 **(b)** 2×3 **(c)** 4×2

3. $\begin{bmatrix} 1 & 2 & | & 11 \\ 2 & -1 & | & -3 \end{bmatrix}$; $\begin{bmatrix} 1 & 2 & | & 11 \\ 0 & -5 & | & -25 \end{bmatrix}$; $\begin{bmatrix} 1 & 2 & | & 11 \\ 0 & 1 & | & 5 \end{bmatrix}$; $x + 2y = 11$;

$y = 5$; $\{(1, 5)\}$ **5.** $\{(4, 1)\}$ **7.** $\{(1, 1)\}$ **9.** $\{(-1, 4)\}$ **11.** $\varnothing$

13. $\{(x, y) \mid 2x + y = 4\}$ **15.** $\{(0, 0)\}$

17. $\begin{bmatrix} 1 & 1 & -1 & | & -3 \\ 0 & -1 & 3 & | & 10 \\ 0 & -6 & 7 & | & 38 \end{bmatrix}$; $\begin{bmatrix} 1 & 1 & -1 & | & -3 \\ 0 & 1 & -3 & | & -10 \\ 0 & -6 & 7 & | & 38 \end{bmatrix}$;

$\begin{bmatrix} 1 & 1 & -1 & | & -3 \\ 0 & 1 & -3 & | & -10 \\ 0 & 0 & -11 & | & -22 \end{bmatrix}$; $\begin{bmatrix} 1 & 1 & -1 & | & -3 \\ 0 & 1 & -3 & | & -10 \\ 0 & 0 & 1 & | & 2 \end{bmatrix}$; $x + y - z = -3$;

$y - 3z = -10$; $z = 2$; $\{(3, -4, 2)\}$ **19.** $\{(4, 0, 1)\}$

21. $\{(-1, 23, 16)\}$ **23.** $\{(3, 2, -4)\}$ **25.** $\varnothing$

27. $\{(x, y, z) \mid x - 2y + z = 4\}$ **29.** $\{(0, 0, 0)\}$

APPENDIX C Synthetic Division

(pages 971–972)

1. C   **2.** C   **3.** $x - 5$ **5.** $4m - 1$ **7.** $2a + 4 + \dfrac{5}{a + 2}$

9. $p - 4 + \dfrac{9}{p + 1}$ **11.** $4a^2 + a + 3$

13. $x^4 + 2x^3 + 2x^2 + 7x + 10 + \dfrac{18}{x - 2}$

15. $-4r^5 - 7r^4 - 10r^3 - 5r^2 - 11r - 8 + \dfrac{-8}{r - 1}$

17. $-3y^4 + 8y^3 - 21y^2 + 36y - 72 + \dfrac{143}{y + 2}$

19. $y^2 + y + 1 + \dfrac{2}{y - 1}$ **21.** 7   **23.** -2 **25.** 0   **27.** yes

29. no   **31.** no   **33.** Since the variables are not present, a missing term will not be noticed in synthetic division. As a result, the quotient will be wrong if placeholders are not inserted.   **34.** By the remainder theorem, a 0 remainder means that $P(k) = 0$. That is, k is a number that makes $P(x) = 0$ a true statement.

Solutions to Selected Exercises

CHAPTER 1 The Real Number System

SECTION 1.1 (pages 33–35)

33. $\dfrac{1}{4} \cdot \dfrac{2}{3} + \dfrac{2}{5} \cdot \dfrac{11}{3}$

$= \dfrac{1}{6} + \dfrac{22}{15}$ Multiply; $\dfrac{2}{12} = \dfrac{1}{6}$

$= \dfrac{5}{30} + \dfrac{44}{30}$ LCD = 30

$= \dfrac{49}{30}$ Add.

45. $\left(\dfrac{3}{2}\right)^2\left[\left(11 + \dfrac{1}{3}\right) - 6\right]$

$= \dfrac{9}{4}\left[\left(11 + \dfrac{1}{3}\right) - 6\right]$ Apply the exponent.

$= \dfrac{9}{4}\left[\left(\dfrac{33}{3} + \dfrac{1}{3}\right) - 6\right]$ LCD = 3

$= \dfrac{9}{4}\left(\dfrac{34}{3} - 6\right)$ Add inside parentheses.

$= \dfrac{9}{4}\left(\dfrac{34}{3} - \dfrac{18}{3}\right)$ LCD = 3

$= \dfrac{9}{4}\left(\dfrac{16}{3}\right)$ Subtract inside parentheses.

$= \dfrac{144}{12}$,   or   12   Multiply. Divide.

65. $\dfrac{9(7-1) - 8 \cdot 2}{4(6-1)} > 3$

$\dfrac{9(6) - 8 \cdot 2}{4(5)} > 3$ Work inside parentheses.

$\dfrac{54 - 16}{20} > 3$ Multiply.

$\dfrac{38}{20} > 3$ Subtract.

$1\dfrac{9}{10} > 3$ Write as a mixed number.

This statement is *false*.

SECTION 1.2 (pages 41–44)

31. (a) $\dfrac{3x + y^2}{2x + 3y}$

$= \dfrac{3(2) + 1^2}{2(2) + 3(1)}$ Let $x = 2$ and $y = 1$.

$= \dfrac{3(2) + 1}{4 + 3}$

$= \dfrac{6 + 1}{7}$

$= \dfrac{7}{7}$,   or   1

(b) $\dfrac{3x + y^2}{2x + 3y}$

$= \dfrac{3(1) + 5^2}{2(1) + 3(5)}$ Let $x = 1$ and $y = 5$.

$= \dfrac{3 + 25}{2 + 15}$ Multiply.

$= \dfrac{28}{17}$ Add.

33. (a) $0.841x^2 + 0.32y^2$

$= 0.841 \cdot 2^2 + 0.32 \cdot 1^2$
   Let $x = 2$, $y = 1$.

$= 0.841 \cdot 4 + 0.32 \cdot 1$
   Apply the exponents.

$= 3.364 + 0.32$ Multiply.

$= 3.684$ Add.

(b) $0.841x^2 + 0.32y^2$

$= 0.841 \cdot 1^2 + 0.32 \cdot 5^2$
   Let $x = 1$, $y = 5$.

$= 0.841 \cdot 1 + 0.32 \cdot 25$

$= 0.841 + 8$

$= 8.841$

57. $\dfrac{z + 4}{2 - z} = \dfrac{13}{5}$

$\dfrac{\dfrac{1}{3} + 4}{2 - \dfrac{1}{3}} \overset{?}{=} \dfrac{13}{5}$ Let $z = \dfrac{1}{3}$.

$\dfrac{\dfrac{1}{3} + \dfrac{12}{3}}{\dfrac{6}{3} - \dfrac{1}{3}} \overset{?}{=} \dfrac{13}{5}$ Write with LCD = 3.

$\dfrac{\dfrac{13}{3}}{\dfrac{5}{3}} \overset{?}{=} \dfrac{13}{5}$ Add and subtract.

$\dfrac{13}{3} \cdot \dfrac{3}{5} \overset{?}{=} \dfrac{13}{5}$ Definition of division

$\dfrac{13}{5} = \dfrac{13}{5}$ ✓   True

The true result shows that $\dfrac{1}{3}$ is a solution.

SECTION 1.3 (pages 52–54)

45. In order to compare these two numbers, write them with a common denominator.

$-\dfrac{2}{3} = -\dfrac{8}{12}$ and $-\dfrac{1}{4} = -\dfrac{3}{12}$

Since $\dfrac{8}{12} > \dfrac{3}{12}$,

$-\dfrac{2}{3}$ is farther to the left of 0 on a number line than $-\dfrac{1}{4}$, so $-\dfrac{2}{3}$ is the lesser number.

73. A decrease is represented by a negative number, and the amount of decrease is represented by its absolute value. There are three negative numbers in the table: -43.6, -16.2, and -0.6. Of these, -43.6 has the greatest absolute value, so Energy, 2008 to 2009 represents the greatest decrease.

SECTION 1.4 (pages 60–63)

39. $\left(-\dfrac{1}{2} + 0.25\right) + \left(-\dfrac{3}{4} + 0.75\right)$

$= \left(-\dfrac{1}{2} + \dfrac{1}{4}\right) + \left(-\dfrac{3}{4} + \dfrac{3}{4}\right)$

$= \left(-\dfrac{2}{4} + \dfrac{1}{4}\right) + 0$

$= -\dfrac{1}{4}$,   or -0.25

59. "$\dfrac{2}{7}$ more than the sum of $\dfrac{5}{7}$ and $-\dfrac{9}{7}$" is interpreted and evaluated as follows.

$\left[\dfrac{5}{7} + \left(-\dfrac{9}{7}\right)\right] + \dfrac{2}{7}$

$= -\dfrac{4}{7} + \dfrac{2}{7}$ Work inside brackets.

$= -\dfrac{2}{7}$ Add.

SECTION 1.5 (pages 68–72)

39. $\left(-\dfrac{3}{8} - \dfrac{2}{3}\right) - \left(-\dfrac{9}{8} - 3\right)$

$= \left[-\dfrac{3}{8} + \left(-\dfrac{2}{3}\right)\right] - \left[-\dfrac{9}{8} + (-3)\right]$

$= \left[-\dfrac{9}{24} + \left(-\dfrac{16}{24}\right)\right]$ Get a common denominator for each pair of fractions.

$\quad - \left[-\dfrac{9}{8} + \left(-\dfrac{24}{8}\right)\right]$

$= -\dfrac{25}{24} - \left(-\dfrac{33}{8}\right)$ Add inside brackets.

$= -\dfrac{25}{24} - \left(-\dfrac{99}{24}\right)$ Get a common denominator.

$= -\dfrac{25}{24} + \dfrac{99}{24}$ Definition of subtraction

$= \dfrac{74}{24}$ Add.

$= \dfrac{37}{12}$ Lowest terms

41. $[-12.25 - (8.34 + 3.57)] - 17.88$

$= [-12.25 - 11.91] - 17.88$

$= [-12.25 + (-11.91)] - 17.88$

$= -24.16 - 17.88$

$= -24.16 + (-17.88)$

$= -42.04$

59. Sum of checks:

$\$35.84 + \$26.14 + \$3.12$

$= \$61.98 + \$3.12$

$= \$65.10$

Sum of deposits:

$\$85.00 + \$120.76 = \$205.76$

Final balance:

$=$ Beginning balance

$-$ checks $+$ deposits

$= \$904.89 - \$65.10 + \$205.76$

$= \$839.79 + \$205.76$

$= \$1045.55$

Her account balance at the end of August was $\$1045.55$.

SECTION 1.6 (pages 81–84)

43. $\dfrac{4(2^3 - 5) - 5(-3^3 + 21)}{3[6 - (-2)]}$

$= \dfrac{4(8 - 5) - 5(-27 + 21)}{3[6 - (-2)]}$

$= \dfrac{4(3) - 5(-6)}{3[8]}$

$= \dfrac{12 + 30}{24}$

$= \dfrac{42}{24}, \text{ or } \dfrac{7}{4}$

49. $\left(\dfrac{5}{6}x + \dfrac{3}{2}y\right)\left(-\dfrac{1}{3}a\right)$

$= \left[\dfrac{5}{6}(6) + \dfrac{3}{2}(-4)\right]\left[-\dfrac{1}{3}(3)\right]$ $\begin{array}{l}\text{Let } x = 6, \\ y = -4, \\ \text{and } a = 3.\end{array}$

$= [5 + (-6)](-1)$

$= (-1)(-1)$

$= 1$

69. "The product of $-\dfrac{2}{3}$ and $-\dfrac{1}{5}$, divided by $\dfrac{1}{7}$,"

is interpreted and evaluated as follows.

$\dfrac{-\dfrac{2}{3}\left(-\dfrac{1}{5}\right)}{\dfrac{1}{7}}$

$= \dfrac{\dfrac{2}{15}}{\dfrac{1}{7}}$ Multiply in the numerator.

$= \dfrac{2}{15} \cdot \dfrac{7}{1}$ Definition of division

$= \dfrac{14}{15}$ Multiply.

SECTION 1.7 (pages 93–96)

73. $-3(8x + 3y + 4z)$

$= -3(8x) + (-3)(3y) + (-3)(4z)$
    Distributive property

$= -24x - 9y - 12z$ Multiply.

79. $-(-3q + 5r - 8s)$

$= -1(-3q + 5r - 8s)$

$= 3q - 5r + 8s$

SECTION 1.8 (pages 101–103)

45. $-\dfrac{4}{3} + 2t + \dfrac{1}{3}t - 8 - \dfrac{8}{3}t$

$= \left(2t + \dfrac{1}{3}t - \dfrac{8}{3}t\right) + \left(-\dfrac{4}{3} - 8\right)$

$= \left(2 + \dfrac{1}{3} - \dfrac{8}{3}\right)t + \left(-\dfrac{4}{3} - 8\right)$

$= \left(\dfrac{6}{3} + \dfrac{1}{3} - \dfrac{8}{3}\right)t + \left(-\dfrac{4}{3} - \dfrac{24}{3}\right)$

$= -\dfrac{1}{3}t - \dfrac{28}{3}$

47. $-5.3r + 4.9 - (2r + 0.7) + 3.2r$

$= -5.3r + 4.9 - 2r - 0.7 + 3.2r$

$= (-5.3r - 2r + 3.2r) + (4.9 - 0.7)$

$= (-5.3 - 2 + 3.2)r + (4.9 - 0.7)$

$= -4.1r + 4.2$

CHAPTER 2 Equations, Inequalities, and Applications

SECTION 2.1 (pages 123–124)

33. $10x + 4 = 9x$

$10x + 4 - 9x = 9x - 9x$ Subtract $9x$.

$1x + 4 = 0$

$x + 4 - 4 = 0 - 4$ Subtract 4.

$x = -4$

CHECK $10x + 4 = 9x$

$10(-4) + 4 \overset{?}{=} 9(-4)$ Let $x = -4$.

$-36 = -36$ ✓   True

Solution set: $\{-4\}$

53. $\dfrac{5}{7}x + \dfrac{1}{3} = \dfrac{2}{5} - \dfrac{2}{7}x + \dfrac{2}{5}$

$\dfrac{5}{7}x + \dfrac{1}{3} = \dfrac{4}{5} - \dfrac{2}{7}x$

$\dfrac{5}{7}x + \dfrac{1}{3} + \dfrac{2}{7}x = \dfrac{4}{5} - \dfrac{2}{7}x + \dfrac{2}{7}x$ Add $\dfrac{2}{7}x$.

$\dfrac{7}{7}x + \dfrac{1}{3} = \dfrac{4}{5}$ Combine like terms.

$1x + \dfrac{1}{3} - \dfrac{1}{3} = \dfrac{4}{5} - \dfrac{1}{3}$ Subtract $\dfrac{1}{3}$.

$x = \dfrac{12}{15} - \dfrac{5}{15}$ LCD = 15

$x = \dfrac{7}{15}$ Subtract.

CHECK $\dfrac{5}{7}x + \dfrac{1}{3} = \dfrac{2}{5} - \dfrac{2}{7}x + \dfrac{2}{5}$

$\dfrac{5}{7}\left(\dfrac{7}{15}\right) + \dfrac{1}{3} \overset{?}{=} \dfrac{2}{5} - \dfrac{2}{7}\left(\dfrac{7}{15}\right) + \dfrac{2}{5}$

   Let $x = \dfrac{7}{15}$.

$\dfrac{1}{3} + \dfrac{1}{3} \overset{?}{=} \dfrac{4}{5} - \dfrac{2}{15}$ Multiply.

$\dfrac{2}{3} \overset{?}{=} \dfrac{12}{15} - \dfrac{2}{15}$ LCD = 15

$\dfrac{2}{3} = \dfrac{10}{15}, \text{ or } \dfrac{2}{3}$ ✓ True

Solution set: $\left\{\dfrac{7}{15}\right\}$

61. $10(-2x + 1) = -19(x + 1)$

$-20x + 10 = -19x - 19$
    Distributive property

$-20x + 10 + 19x = -19x - 19 + 19x$
    Add $19x$.

$-x + 10 = -19$
    Combine like terms.

$-x + 10 - 10 = -19 - 10$
    Subtract 10.

$-x = -29$

$x = 29$

CHECK $10(-2x + 1) = -19(x + 1)$

$10[-2(29) + 1] \overset{?}{=} -19(29 + 1)$
    Let $x = 29$.

$10[-57] \overset{?}{=} -19(30)$

$-570 = -570$ ✓ True

Solution set: $\{29\}$

SECTION 2.2 (pages 130–131)

53. $\dfrac{2}{5}x - \dfrac{3}{10}x = 2$

$\dfrac{4}{10}x - \dfrac{3}{10}x = 2$ LCD = 10

$\dfrac{1}{10}x = 2$ Combine like terms.

$10 \cdot \dfrac{1}{10}x = 10 \cdot 2$ Multiply by 10.

$x = 20$

CHECK $\dfrac{2}{5}x - \dfrac{3}{10}x = 2$

$\dfrac{2}{5}(20) - \dfrac{3}{10}(20) \overset{?}{=} 2$ Let $x = 20$.

$8 - 6 \overset{?}{=} 2$

$2 = 2$ ✓   True

Solution set: $\{20\}$

61. $0.9w - 0.5w + 0.1w = -3$

$$0.5w = -3 \qquad \text{Combine like terms.}$$

$$\frac{0.5w}{0.5} = \frac{-3}{0.5} \qquad \text{Divide by 0.5.}$$

$$w = -6$$

A check confirms this solution.

Solution set: $\{-6\}$

SECTION 2.3 (pages 141–144)

55. $0.02(5000) + 0.03x = 0.025(5000 + x)$

Multiply both sides by 1000, *not* 100.

$$1000\big[0.02(5000) + 0.03x\big]$$
$$= 1000\big[0.025(5000 + x)\big]$$

$$20(5000) + 30x = 25(5000 + x)$$

$$100{,}000 + 30x = 125{,}000 + 25x$$
$$\text{Distributive property}$$

$$5x + 100{,}000 = 125{,}000$$
$$\text{Subtract } 25x.$$

$$5x = 25{,}000$$
$$\text{Subtract } 100{,}000.$$

$$x = 5000 \qquad \text{Divide by 5.}$$

CHECK Substituting 5000 for x in the original equation results in a true statement,

$$250 = 250. \checkmark$$

Solution set: $\{5000\}$

59. $-(6k - 5) - (-5k + 8) = -3$

$$-1(6k - 5) - 1(-5k + 8) = -3$$

$$-6k + 5 + 5k - 8 = -3$$

$$-k - 3 = -3$$

$$-k = 0$$

$$k = 0$$

CHECK Substituting 0 for k in the original equation results in a true statement,

$$-3 = -3. \checkmark$$

Solution set: $\{0\}$

SECTION 2.4 (pages 156–161)

21. *Step 1*

Read the problem again.

Step 2

Let $x =$ the unknown number. Then $3x$ is three times the number, $x + 7$ is 7 more than the number, and the sum is

$$3x + (x + 7).$$

$2x$ is twice the number, and

$$-11 - 2x$$

is the difference between -11 and twice the number.

Step 3

$$3x + (x + 7) = -11 - 2x$$

Step 4

$$4x + 7 = -11 - 2x$$

$$6x + 7 = -11 \qquad \text{Add } 2x.$$

$$6x = -18 \qquad \text{Subtract 7.}$$

$$x = -3 \qquad \text{Divide by 6.}$$

Step 5

The number is -3.

Step 6

Check that -3 is the correct answer by substituting this result into the words of the original problem. The sum of three times this number and 7 more than the number is

$$3(-3) + (-3 + 7) = -5.$$

The difference between -11 and twice this number is

$$-11 - 2(-3) = -5.$$

The values are equal, so the number -3 is the correct answer.

49. *Step 1*

Read the problem again.

Step 2

Let $x =$ the measures of angles A and B.

$x + 60 =$ the measure of angle C.

Step 3

The sum of the measures of the angles of any triangle is $180°$, so

$$x + x + (x + 60) = 180.$$

Step 4

$$3x + 60 = 180$$

$$3x = 120 \qquad \text{Subtract 60.}$$

$$x = 40 \qquad \text{Divide by 3.}$$

Step 5

Angles A and B have measures of $40°$, and angle C has a measure of $40 + 60 = 100°$.

Step 6

The answer checks since

$$40 + 40 + 100 = 180.$$

SECTION 2.5 (pages 168–173)

55. Let L represent the length of the box. The girth is $4 \cdot 18 = 72$ in. Since the length plus the girth is 108, we have

$$L + 72 = 108$$

$$L = 36 \text{ in.} \qquad \text{Subtract 72.}$$

The maximum volume of the box is

$$V = LWH$$

$$V = 36(18)(18) \qquad \text{Substitute.}$$

$$V = 11{,}664 \text{ in.}^3.$$

59. The two angles are complementary, so the sum of their measures is $90°$.

$$(8x - 1) + 5x = 90$$

$$13x - 1 = 90 \qquad \text{Combine like terms.}$$

$$13x = 91 \qquad \text{Add 1.}$$

$$x = 7 \qquad \text{Divide by 13.}$$

Since $x = 7$, we have

$$8x - 1 = 8(7) - 1 = 56 - 1 = 55$$

and $\quad 5x = 5(7) = 35.$

The angles measure $55°$ and $35°$.

85. $M = C(1 + r)$ for r

$$M = C + Cr \qquad \text{Distributive property}$$

$$M - C = Cr \qquad \text{Subtract } C.$$

$$\frac{M - C}{C} = \frac{Cr}{C} \qquad \text{Divide by } C.$$

$$\frac{M - C}{C} = r, \quad \text{or} \quad r = \frac{M - C}{C}$$

SECTION 2.6 (pages 180–185)

47. Let $x =$ the number of cups of cleaner. Set up a proportion with one ratio involving the number of cups of cleaner and the other involving the number of gallons of water.

$$\frac{x \text{ cups}}{\frac{1}{4}\text{ cup}} = \frac{10\frac{1}{2}\text{ gallons}}{1 \text{ gallon}}$$

$$x \cdot 1 = \frac{1}{4}\left(10\frac{1}{2}\right) \qquad \text{Cross products}$$

$$x = \frac{1}{4}\left(\frac{21}{2}\right) \qquad \begin{array}{l}\text{Write an}\\ \text{improper fraction.}\end{array}$$

$$x = \frac{21}{8}, \quad \text{or} \quad 2\frac{5}{8}$$

The amount of cleaner needed is $2\frac{5}{8}$ cups.

89. The increase in value was

$$\$6200 - \$625 = \$5575.$$

We can state the problem as follows:

"What percent of \$625 is \$5575?"

This translates into a percent equation, where p is the percent as a decimal.

$$p \cdot 625 = 5575$$

$$p = \frac{5575}{625} \qquad \text{Divide by 625.}$$

$$p = 8.92, \quad \text{or} \quad 892\%$$

The percent increase in the value of the coin was 892%.

SECTION 2.7 (pages 199–202)

53. $\dfrac{2}{3}(p + 3) > \dfrac{5}{6}(p - 4)$

$$6\left(\frac{2}{3}\right)(p + 3) > 6\left(\frac{5}{6}\right)(p - 4)$$
$$\text{Multiply by 6, the LCD.}$$

$$4(p + 3) > 5(p - 4)$$

(continued)

$$4p + 12 > 5p - 20$$

Distributive property

$$-p + 12 > -20$$

Subtract $5p$.

$$-p > -32$$

Subtract 12.

$$\frac{-p}{-1} < \frac{-32}{-1}$$

Divide by -1. Reverse the direction of the symbol.

$$p < 32$$

Solution set: $(-\infty, 32)$

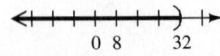

81. The Celsius temperature C must give a Fahrenheit temperature F that has *never exceeded* 104°, which translates as *is less than or equal to* 104°.

$$F \le 104$$

$$\frac{9}{5}C + 32 \le 104$$

$$\frac{9}{5}C \le 72 \qquad \text{Subtract 32.}$$

$$\frac{5}{9}\left(\frac{9}{5}C\right) \le \frac{5}{9}(72) \qquad \text{Multiply by } \frac{5}{9}.$$

$$C \le 40$$

The temperature of Providence, Rhode Island, has never exceeded 40°C.

CHAPTER 3 Graphs of Linear Equations and Inequalities in Two Variables; Functions

SECTION 3.1 (pages 224–229)

29. Is $(5, -6)$ a solution of the equation $x = -6$?

Since y does not appear in the equation, we just substitute 5 for x.

$$x = -6$$

$$5 \stackrel{?}{=} -6 \qquad \text{Let } x = 5.$$

The result is false, so $(5, -6)$ is not a solution of the equation $x = -6$.

49. The given equation $x - 8 = 0$ may be written as $x = 8$. For any value of y, the value of x will always be 8. The completed table of values follows.

x	y	Ordered pairs
8	8	$\rightarrow (8, 8)$
8	3	$\rightarrow (8, 3)$
8	0	$\rightarrow (8, 0)$

SECTION 3.2 (pages 238–243)

39. $-3y = 15$

$$y = \frac{15}{-3}, \text{ or } -5 \qquad \text{Divide by } -3.$$

For any value of x, the value of y is -5. Three ordered pairs are $(-2, -5)$, $(0, -5)$, and $(1, -5)$. Plot these points and draw a line through them. The graph is a horizontal line.

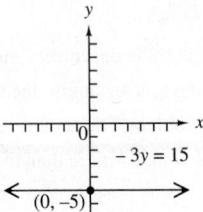

SECTION 3.3 (pages 254–259)

17. The slope (or grade) of the hill is the ratio of the rise to the run, or the ratio of the vertical change to the horizontal change. Since the rise is 32 and the run is 108, the slope is

$$\frac{32}{108} = \frac{8 \cdot 4}{27 \cdot 4} = \frac{8}{27}. \qquad \text{Lowest terms}$$

35. First label the given points.

$$\underbrace{(x_1, y_1)}_{} \qquad \underbrace{(x_2, y_2)}_{}$$

$$\left(-\frac{7}{5}, \frac{3}{10}\right) \quad \text{and} \quad \left(\frac{1}{5}, -\frac{1}{2}\right)$$

Then apply the slope formula.

$$\text{slope } m = \frac{\text{change in } y}{\text{change in } x}$$

$$= \frac{y_2 - y_1}{x_2 - x_1}$$

$$= \frac{-\frac{1}{2} - \frac{3}{10}}{\frac{1}{5} - \left(-\frac{7}{5}\right)} \qquad \text{Substitute.}$$

$$= \frac{-\frac{5}{10} - \frac{3}{10}}{\frac{1}{5} + \frac{7}{5}} \qquad \begin{array}{l}\text{Get a common} \\ \text{denominator in} \\ \text{the numerator.} \\ \text{Simplify the} \\ \text{denominator.}\end{array}$$

$$= \frac{-\frac{8}{10}}{\frac{8}{5}} \qquad \begin{array}{l}\text{Subtract in} \\ \text{the numerator.} \\ \text{Add in the} \\ \text{denominator.}\end{array}$$

$$= -\frac{8}{10} \div \frac{8}{5} \qquad \frac{a}{b} = a \div b$$

$$= -\frac{8}{10} \cdot \frac{5}{8} \qquad \begin{array}{l}\text{Multiply by the} \\ \text{reciprocal.}\end{array}$$

$$= -\frac{1}{2} \qquad \begin{array}{l}\text{Multiply. Write in} \\ \text{lowest terms.}\end{array}$$

65. Use the points $(0, 20)$ and $(4, 4)$ from the graph to find the average rate of change.

$$\frac{\text{change in } y}{\text{change in } x} = \frac{4 - 20}{4 - 0} = \frac{-16}{4} = -4$$

The average rate of change is $-\$4000$ per year. The value of the machine is decreasing $\$4000$ each year during these years.

SECTION 3.4 (pages 270–275)

19. The rise is 3 and the run is 1, so the slope m is given by

$$m = \frac{\text{rise}}{\text{run}} = \frac{3}{1} = 3.$$

The y-intercept is $(0, -3)$, so $b = -3$. The equation of the line, written in slope-intercept form $y = mx + b$, is

$$y = 3x - 3. \qquad m = 3, b = -3$$

55. (a) First, find the slope of the line.

$$\underbrace{(x_1, y_1)}_{} \qquad\qquad \underbrace{(x_2, y_2)}_{}$$

$$\left(\frac{1}{2}, \frac{3}{2}\right) \quad \text{and} \quad \left(-\frac{1}{4}, \frac{5}{4}\right)$$

$$m = \frac{\frac{5}{4} - \frac{3}{2}}{-\frac{1}{4} - \frac{1}{2}} = \frac{\frac{5}{4} - \frac{6}{4}}{-\frac{1}{4} - \frac{2}{4}} = \frac{-\frac{1}{4}}{-\frac{3}{4}}$$

$$= -\frac{1}{4} \div \left(-\frac{3}{4}\right) = -\frac{1}{4}\left(-\frac{4}{3}\right) = \frac{1}{3}$$

Now use the point $\left(\frac{1}{2}, \frac{3}{2}\right)$ for (x_1, y_1) and $m = \frac{1}{3}$ in the point-slope form.

$$y - y_1 = m(x - x_1)$$

$$y - \frac{3}{2} = \frac{1}{3}\left(x - \frac{1}{2}\right) \qquad \text{Substitute.}$$

$$y - \frac{3}{2} = \frac{1}{3}x - \frac{1}{6} \qquad \begin{array}{l}\text{Distributive} \\ \text{property}\end{array}$$

$$y = \frac{1}{3}x - \frac{1}{6} + \frac{9}{6} \qquad \text{Add } \frac{3}{2} = \frac{9}{6}.$$

$$y = \frac{1}{3}x + \frac{4}{3} \qquad \begin{array}{l}\text{Combine like} \\ \text{terms; } \frac{8}{6} = \frac{4}{3}\end{array}$$

This last equation is written in slope-intercept form $y = mx + b$.

(b) Clear the fractions in the final equation in part (a) by multiplying each term by 3.

$$y = \frac{1}{3}x + \frac{4}{3} \qquad \text{From part (a)}$$

$$3y = x + 4 \qquad \text{Multiply by 3.}$$

$$-x + 3y = 4 \qquad \text{Subtract } x.$$

$$x - 3y = -4 \qquad \text{Multiply by } -1.$$

This last equation is written in standard form $Ax + By = C$.

65. Through $(0.5, 0.2)$; vertical

A vertical line through the point (a, b) has equation $x = a$. Since the x-value in $(0.5, 0.2)$ is 0.5, the equation of this line is $x = 0.5$.

71. Solve the given equation for y.

$$3x = 4y + 5$$

$$3x - 4y = 5 \qquad \text{Subtract } 4y.$$

$$-4y = -3x + 5 \qquad \text{Subtract } 3x.$$

$$y = \frac{3}{4}x - \frac{5}{4} \qquad \text{Divide by } -4.$$

The slope is $\frac{3}{4}$. A line parallel to this line has the same slope. Use the point-slope form with $m = \frac{3}{4}$ and $(x_1, y_1) = (2, -3)$.

$$y - y_1 = m(x - x_1)$$

$$y - (-3) = \frac{3}{4}(x - 2) \qquad \text{Substitute.}$$

$$y + 3 = \frac{3}{4}x - \frac{3}{2}$$

$$y = \frac{3}{4}x - \frac{3}{2} - \frac{6}{2} \qquad \begin{array}{l}\text{Subtract} \\ 3 = \frac{6}{2}.\end{array}$$

$$y = \frac{3}{4}x - \frac{9}{2} \qquad \begin{array}{l}\text{Combine} \\ \text{like terms.}\end{array}$$

SECTION 3.6 (pages 290–292)

35. Any vertical line drawn through the shaded region would intersect many points of the graph, so this relation is not the graph of a function. The graph extends indefinitely to the left from 3, so the domain is $(-\infty, 3]$. The graph extends indefinitely upward and downward, so the range is $(-\infty, \infty)$.

53. $xy = 1$

Divide both sides of the equation by x to rewrite $xy = 1$ as $y = \frac{1}{x}$. Note that x can never equal 0, or the denominator would equal 0. Thus, the domain is

$$(-\infty, 0) \cup (0, \infty).$$

Each nonzero x-value gives exactly one y-value. Therefore, $xy = 1$ defines y as a function of x.

SECTION 3.7 (pages 297–300)

21. $g(x) = -x^2 + 4x + 1$

$$g\left(\frac{p}{3}\right) = -\left(\frac{p}{3}\right)^2 + 4\left(\frac{p}{3}\right) + 1 \quad \text{Let } x = \frac{p}{3}.$$

$$g\left(\frac{p}{3}\right) = -\frac{p^2}{9} + \frac{4p}{3} + 1$$

41. (a) Solve the given equation for y.

$$4x - 3y = 8$$

$$-3y = -4x + 8 \qquad \text{Subtract } 4x.$$

$$y = \frac{4}{3}x - \frac{8}{3} \qquad \text{Divide by } -3.$$

Since $y = f(x)$,

$$f(x) = \frac{4}{3}x - \frac{8}{3}.$$

(b) $f(3) = \frac{4}{3}(3) - \frac{8}{3} \qquad \text{Let } x = 3.$

$$f(3) = \frac{12}{3} - \frac{8}{3} \qquad \text{Multiply.}$$

$$f(3) = \frac{4}{3} \qquad \text{Subtract.}$$

59. Refer to the graph given with **Exercise 59.**

(a) The independent variable is t, the number of hours, and the possible values are in the set $[0, 100]$. The dependent variable is g, the number of gallons, and the possible values are in the set $[0, 3000]$.

(b) The graph rises for the first 25 hr, so the water level increases for 25 hr. The graph falls for $t = 50$ to $t = 75$, so the water level decreases for $75 - 50 = 25$ hr.

(c) There are 2000 gal in the pool when $t = 90$.

(d) $g(0)$ is the number of gallons in the pool at time $t = 0$. Here, $g(0) = 0$, which means the pool is empty at time 0.

CHAPTER 4 Systems of Linear Equations and Inequalities

SECTION 4.1 (pages 319–322)

27. $2x - 3y = -6$

$$y = -3x + 2$$

To graph $2x - 3y = -6$, find the intercepts.

$$2x - 3(0) = -6 \qquad \text{Let } y = 0.$$

$$2x = -6$$

$$x = -3$$

The x-intercept is $(-3, 0)$.

$$2(0) - 3y = -6 \qquad \text{Let } x = 0.$$

$$-3y = -6$$

$$y = 2$$

The y-intercept is $(0, 2)$.

Plot the intercepts, $(-3, 0)$ and $(0, 2)$, and draw the line through them.

To graph the second line, which is in slope-intercept form, start by plotting the y-intercept, $(0, 2)$. From this point, move 3 units down and 1 unit to the right (because the slope is -3) to reach the point $(1, -1)$.

Draw the line through $(0, 2)$ and $(1, -1)$.

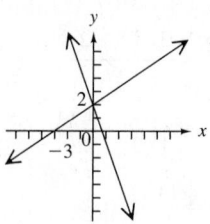

The lines intersect at their common y-intercept, $(0, 2)$.

Solution set: $\{(0, 2)\}$

35. $4x - 2y = 8 \qquad (1)$

$$2x = y + 4 \qquad (2)$$

Graph the line $4x - 2y = 8$ using its intercepts, $(2, 0)$ and $(0, -4)$.

Graph $2x = y + 4$ using its intercepts, $(2, 0)$ and $(0, -4)$. Since both equations have the same intercepts, they are equations of the same line.

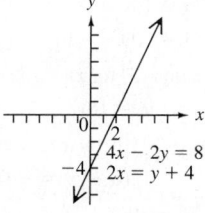

There are an infinite number of solutions, so we use set-builder notation to write the solution set. Rewrite equation (2) in standard form.

$$2x - y = 4$$

The solution set is

$$\{(x, y) \mid 2x - y = 4\}.$$

The given system consists of dependent equations.

SECTION 4.2 (pages 329–330)

21. $6x - 8y = 6 \qquad (1)$

$$2y = -2 + 3x \qquad (2)$$

Solve equation (2) for y.

$$2y = -2 + 3x$$

$$y = \frac{3x - 2}{2} \qquad (3)$$

Substitute $\frac{3x - 2}{2}$ for y in equation (1).

$$6x - 8y = 6 \qquad (1)$$

$$6x - 8\left(\frac{3x - 2}{2}\right) = 6$$

$$6x - 4(3x - 2) = 6$$

$$6x - 12x + 8 = 6$$

$$-6x + 8 = 6$$

$$-6x = -2$$

$$x = \frac{-2}{-6}, \quad \text{or} \quad \frac{1}{3}$$

(continued)

To find y, let $x = \frac{1}{3}$ in equation (3).

$$y = \frac{3x - 2}{2} = \frac{3\left(\frac{1}{3}\right) - 2}{2} = \frac{1 - 2}{2} = -\frac{1}{2}$$

Solution set: $\left\{\left(\frac{1}{3}, -\frac{1}{2}\right)\right\}$

27. $\frac{x}{5} + 2y = \frac{16}{5}$ (1)

$\frac{3x}{5} + \frac{y}{2} = -\frac{7}{5}$ (2)

Multiply each side of equation (1) by 5.

$$5\left(\frac{x}{5} + 2y\right) = 5\left(\frac{16}{5}\right)$$

$$x + 10y = 16 \quad (3)$$

Multiply each side of equation (2) by 10.

$$10\left(\frac{3x}{5} + \frac{y}{2}\right) = 10\left(-\frac{7}{5}\right)$$

$$6x + 5y = -14 \quad (4)$$

We now have an equivalent system that does not include fractions.

$$x + 10y = 16 \quad (3)$$
$$6x + 5y = -14 \quad (4)$$

Solve equation (3) for x.

$$x = 16 - 10y \quad (5)$$

Substitute $16 - 10y$ for x in equation (4).

$$6x + 5y = -14 \quad (4)$$
$$6(16 - 10y) + 5y = -14$$
$$96 - 60y + 5y = -14$$
$$-55y = -110$$
$$y = 2$$

To find x, let $y = 2$ in equation (5).

$$x = 16 - 10y \quad (5)$$
$$x = 16 - 10(2)$$
$$x = -4$$

Solution set: $\{(-4, 2)\}$

SECTION 4.3 (pages 335–337)

25. $-x + 3y = 4$ (1)

$-2x + 6y = 8$ (2)

Multiply equation (1) by -2 and add the result to equation (2).

$$2x - 6y = -8 \quad (3)$$
$$\underline{-2x + 6y = 8} \quad (2)$$
$$0 = 0 \quad \text{True}$$

The true result, $0 = 0$, indicates that the equations of the original system are dependent. To obtain an equation in standard form $Ax + By = C$, with $A > 0$, multiply equation (1) by -1.

$$x - 3y = -4$$

Solution set: $\{(x, y) \mid x - 3y = -4\}$

37. $6x + 3y = 0$ (1)

$-18x - 9y = 0$ (2)

Multiply equation (1) by 3 and add the result to equation (2).

$$18x + 9y = 0 \quad (3)$$
$$\underline{-18x - 9y = 0} \quad (2)$$
$$0 = 0 \quad \text{True}$$

The true result, $0 = 0$, means that the system has an infinite number of solutions. To obtain an equation for the solution set in which the coefficients have greatest common factor 1, divide both sides of equation (1) by 3 to obtain the equivalent equation

$$2x + y = 0.$$

Solution set: $\{(x, y) \mid 2x + y = 0\}$

SECTION 4.4 (pages 346–351)

19. $C = 85x + 900$ (1)

$R = 105x$ (2)

No more than 38 units can be sold.

(a) Since $C = R$, substitute $105x$ for C and solve for x.

$$105x = 85x + 900$$
$$20x = 900 \quad \text{Subtract } 85x.$$
$$x = 45 \quad \text{Divide by 20.}$$

The break-even quantity is 45 units.

(b) Since no more than 38 units can be sold and the break-even quantity is 45, do not produce the product ($45 > 38$). Doing so will lead to a loss.

37. *Step 1*

Read the problem again.

Step 2

Assign variables.

Let $x = $ the average rate of the bicycle;

$y = $ the average rate of the car.

	r	t	d
Bicycle	x	7.5	$7.5x$
Car	y	7.5	$7.5y$

↑
Use the formula $d = rt$.

Step 3

The total distance traveled by the bicycle and the car is 471 mi.

$$7.5x + 7.5y = 471 \quad (1)$$

The car traveled 35.8 mph faster than the bicycle.

$$y = x + 35.8 \quad (2)$$

These equations form the following system.

$$7.5x + 7.5y = 471 \quad (1)$$
$$y = x + 35.8 \quad (2)$$

Step 4

Because equation (2) is already solved for y, the substitution method is a good choice. Substitute $x + 35.8$ for y in equation (1).

$$7.5x + 7.5y = 471 \quad (1)$$
$$7.5x + 7.5(x + 35.8) = 471$$
$$\text{Let } y = x + 35.8.$$
$$7.5x + 7.5x + 268.5 = 471$$
$$\text{Distributive property}$$
$$15x + 268.5 = 471$$
$$\text{Combine like terms.}$$
$$15x = 202.5$$
$$\text{Subtract 268.5.}$$
$$x = 13.5$$
$$\text{Divide by 15.}$$

To find the value of y substitute 13.5 for x in equation (2).

$$y = x + 35.8 \quad (2)$$
$$y = 13.5 + 35.8 \quad \text{Let } x = 13.5.$$
$$y = 49.3$$

Step 5

The average rate of the bicycle is 13.5 mph, and the average rate of the car is 49.3 mph.

Step 6

In 7.5 hr, the total distance traveled by the bicycle and the car is

$$7.5(13.5) + 7.5(49.3) = 471 \text{ mi.}$$

Since $49.3 - 13.5 = 35.8$, the car traveled 35.8 mph faster than the bicycle, as stated.

SECTION 4.5 (pages 355–356)

19. $x \leq 2y + 3$

$x + y < 0$

Graph $x = 2y + 3$ as a solid line through $(3, 0)$ and $(5, 1)$. Using $(0, 0)$ as a test point results in the true statement $0 \leq 3$, so shade the region containing the origin. Graph $x + y = 0$ as a dashed line through $(0, 0)$ and $(1, -1)$. We cannot use $(0, 0)$ as a test point because it is on the boundary line. Using $(1, 0)$ results in the false statement $1 < 0$, so shade the region *not* containing $(1, 0)$.

The solution set of the system is the intersection of the two shaded regions. It includes the portion of the line $x = 2y + 3$ that bounds the region but not the portion of the line $x + y = 0$.

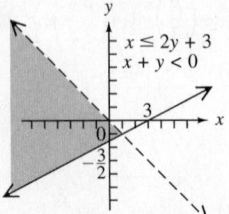

23. $3x - 2y \geq 6$
$\quad\ x + \ y \leq 4$
$\quad\quad\ x \geq 0$
$\quad\quad\ y \geq -4$

Graph $3x - 2y = 6$, $x + y = 4$, $x = 0$,
and $y = -4$ as solid lines, as shown in
the graph at the top of the next column.
All four inequalities are true for $(2, -2)$.
Shade the region bounded by the four lines,
which contains the test point $(2, -2)$. The
solution set is the shaded region.

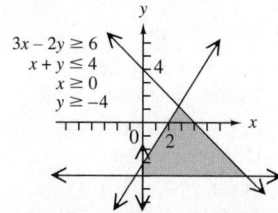

CHAPTER 5 Exponents and Polynomials

SECTION 5.1 (pages 373–376)

21. $\dfrac{1}{2}x^4 + \dfrac{1}{6}x^4$

$\quad = \left(\dfrac{1}{2} + \dfrac{1}{6}\right)x^4$ Distributive property

$\quad = \left(\dfrac{3}{6} + \dfrac{1}{6}\right)x^4$ Write fractions with a common denominator.

$\quad = \dfrac{4}{6}x^4$ Add fractions.

$\quad = \dfrac{2}{3}x^4$ Lowest terms

37. $0.8x^4 - 0.3x^4 - 0.5x^4 + 7$
$\quad = (0.8 - 0.3 - 0.5)x^4 + 7$
$\quad = 0x^4 + 7$
$\quad = 7$

Since 7 can be written as $7x^0$, the degree of
the simplified polynomial is 0. It has one
term, so it is a monomial.

53. Add.

$\quad \dfrac{2}{3}x^2 + \dfrac{1}{5}x + \dfrac{1}{6}$

$\quad \dfrac{1}{2}x^2 - \dfrac{1}{3}x + \dfrac{2}{3}$

Rewrite so that the fractions in each column
have a common denominator.

$\quad \dfrac{4}{6}x^2 + \dfrac{3}{15}x + \dfrac{1}{6}$

$\quad \dfrac{3}{6}x^2 - \dfrac{5}{15}x + \dfrac{4}{6}$

$\quad \dfrac{7}{6}x^2 - \dfrac{2}{15}x + \dfrac{5}{6}$ Add column by column.

71. Use the formula for the perimeter of a
square, $P = 4s$, with $s = \dfrac{1}{2}x^2 + 2x$.

$\quad P = 4s$

$\quad P = 4\left(\dfrac{1}{2}x^2 + 2x\right)$ Substitute.

$\quad P = 4\left(\dfrac{1}{2}x^2\right) + 4(2x)$ Distributive property

$\quad P = 2x^2 + 8x$ Multiply.

SECTION 5.2 (pages 383–384)

43. $(-5^2)^6$

$\quad = (-1 \cdot 5^2)^6$ Write the factor -1.

$\quad = (-1)^6 \cdot (5^2)^6$ Power rule (b)

$\quad = 1 \cdot 5^{2 \cdot 6}$ Power rule (a)

$\quad = 5^{12}$ Multiply.

77. $\left(\dfrac{5a^2b^5}{c^6}\right)^3 \quad (c \neq 0)$

$\quad = \dfrac{(5a^2b^5)^3}{(c^6)^3}$ Power rule (c)

$\quad = \dfrac{5^3(a^2)^3(b^5)^3}{(c^6)^3}$ Power rule (b)

$\quad = \dfrac{125a^6b^{15}}{c^{18}}$ Power rule (a)

79. $(-5m^3p^4q)^2(p^2q)^3$

$\quad = (-1 \cdot 5m^3p^4q)^2(p^2q)^3$ Write the factor -1.

$\quad = (-1)^2 \cdot 5^2 \cdot (m^3)^2 \cdot$
$\quad\quad\quad (p^4)^2 \cdot q^2 \cdot (p^2)^3 \cdot q^3$ Power rule (b)

$\quad = 1 \cdot 25 \cdot m^6p^8q^2p^6q^3$ Power rule (a)

$\quad = 25m^6p^{8+6}q^{2+3}$ Product rule

$\quad = 25m^6p^{14}q^5$

SECTION 5.3 (pages 389–391)

57. $\left(x - \dfrac{2}{3}\right)\left(x + \dfrac{1}{4}\right)$

$\quad\quad\ \ \mathbf{F} \quad\quad \mathbf{O} \quad\quad \mathbf{I} \quad\quad \mathbf{L}$

$\quad = x(x) + x\left(\dfrac{1}{4}\right) + \left(-\dfrac{2}{3}\right)x + \left(-\dfrac{2}{3}\right)\dfrac{1}{4}$

$\quad = x^2 + \dfrac{1}{4}x - \dfrac{2}{3}x - \dfrac{1}{6}$

$\quad = x^2 + \left(\dfrac{3}{12}x - \dfrac{8}{12}x\right) - \dfrac{1}{6}$

$\quad = x^2 - \dfrac{5}{12}x - \dfrac{1}{6}$

SECTION 5.4 (pages 396–398)

11. $\left(x - \dfrac{3}{4}\right)^2$

$\quad = x^2 - 2(x)\left(\dfrac{3}{4}\right) + \left(\dfrac{3}{4}\right)^2$
$\quad\quad\quad\quad\quad\quad (a - b)^2 = a^2 - 2ab + b^2$

$\quad = x^2 - \dfrac{3}{2}x + \dfrac{9}{16}$

21. $(0.8t + 0.7s)^2$

$\quad = (0.8t)^2 + 2(0.8t)(0.7s) + (0.7s)^2$
$\quad\quad\quad\quad\quad (a + b)^2 = a^2 + 2ab + b^2$

$\quad = 0.64t^2 + 1.12ts + 0.49s^2$

29. $-(4r - 2)^2$

First square the binomial.

$\quad (4r - 2)^2$

$\quad = (4r)^2 - 2(4r)(2) + 2^2$

$\quad = 16r^2 - 16r + 4$

Now multiply by -1.

$\quad -1(16r^2 - 16r + 4)$

$\quad = -16r^2 + 16r - 4$

43. $(2x^2 - 5)(2x^2 + 5)$

$\quad = (2x^2)^2 - 5^2$ $\quad \begin{array}{l}(a - b)(a + b) \\ = a^2 - b^2\end{array}$

$\quad = 4x^4 - 25$ $\quad (2x^2)^2 = 2^2x^4 = 4x^4$

SECTION 5.5 (pages 406–407)

17. $(-2)^0 - 2^0$

$\quad = 1 - 1$ $\quad a^0 = 1$

$\quad = 0$ Subtract.

33. $-2^{-1} + 3^{-2}$

$\quad = -(2^{-1}) + 3^{-2}$ In -2^{-1}, the base is 2.

$\quad = -\dfrac{1}{2^1} + \dfrac{1}{3^2}$ $\quad a^{-n} = \dfrac{1}{a^n}$

$\quad = -\dfrac{1}{2} + \dfrac{1}{9}$ Apply the exponents.

$\quad = -\dfrac{9}{18} + \dfrac{2}{18}$ Write fractions with a common denominator.

$\quad = -\dfrac{7}{18}$ Add.

69. $\dfrac{(m^7n)^{-2}}{m^{-4}n^3}$

$\quad = \dfrac{(m^7)^{-2}n^{-2}}{m^{-4}n^3}$ Power rule (b)

$\quad = \dfrac{m^{7(-2)}n^{-2}}{m^{-4}n^3}$ Power rule (a)

$\quad = \dfrac{m^{-14}n^{-2}}{m^{-4}n^3}$

$\quad = m^{-14-(-4)}n^{-2-3}$ Quotient rule

$\quad = m^{-10}n^{-5}$

$\quad = \dfrac{1}{m^{10}n^5}$ $\quad a^{-n} = \dfrac{1}{a^n}$

SECTION 5.6 (pages 412–413)

23. $\dfrac{-3x^3 - 4x^4 + 2x}{-3x^2}$

$= \dfrac{-4x^4 - 3x^3 + 2x}{-3x^2}$ Write in descending powers.

$= \dfrac{-4x^4}{-3x^2} - \dfrac{3x^3}{-3x^2} + \dfrac{2x}{-3x^2}$ Divide each term by $-3x^2$.

$= \dfrac{4x^2}{3} + x - \dfrac{2}{3x}$ Quotient rule; be careful with signs.

Notice how the third term $-\dfrac{2}{3x}$ is written with x in the denominator, which is **not** the same as $-\dfrac{2}{3}x$ or $-\dfrac{2x}{3}$. In $-\dfrac{2}{3x}$, we are *dividing* by x. In $-\dfrac{2}{3}x$ and $-\dfrac{2x}{3}$, we are *multiplying* by x. Applying the quotient rule to the term $\dfrac{2x}{-3x^2}$ gives the following.

$\dfrac{2x}{-3x^2} = \dfrac{2x^1}{-3x^2} = -\dfrac{2}{3}x^{1-2} = -\dfrac{2}{3}x^{-1}$

$= -\dfrac{2}{3}\left(\dfrac{1}{x}\right) = -\dfrac{2}{3x}$

$\left(\dfrac{4}{3}x^2 \text{ is an acceptable form for the first term,}\right.$

$\left.\dfrac{4x^2}{3}. \text{ Why?}\right)$

33. Use the formula for the area of a rectangle, $A = LW$, with $A = 12x^2 - 4x + 2$ and $W = 2x$.

$A = LW$

$12x^2 - 4x + 2 = L(2x)$ Substitute for A and W.

$\dfrac{12x^2 - 4x + 2}{2x} = L$ Divide each side by $2x$.

$\dfrac{12x^2}{2x} - \dfrac{4x}{2x} + \dfrac{2}{2x} = L$ Divide each term by $2x$.

$6x - 2 + \dfrac{1}{x} = L$ $\dfrac{2}{2x} = \dfrac{2}{2} \cdot \dfrac{1}{x}$

$= 1 \cdot \dfrac{1}{x} = \dfrac{1}{x}$

$6x - 2 + \dfrac{1}{x}$ represents the length of the rectangle.

SECTION 5.7 (pages 418–419)

23. $(x^4 - x^2 - 2) \div (x^2 - 2)$

Use 0 as the coefficient of the missing x^3- and x-terms in the dividend and the missing x-term in the divisor.

$$
\begin{array}{r}
x^2 + 1 \\
x^2 + 0x - 2 \overline{) x^4 + 0x^3 - x^2 + 0x - 2} \\
\underline{x^4 + 0x^3 - 2x^2} \\
x^2 + 0x - 2 \\
\underline{x^2 + 0x - 2} \\
0
\end{array}
$$

The remainder is 0. The answer is the quotient, $x^2 + 1$.

29. $\dfrac{2x^5 + x^4 + 11x^3 - 8x^2 - 13x + 7}{2x^2 + x - 1}$

$$
\begin{array}{r}
x^3 + 6x - 7 \\
2x^2 + x - 1 \overline{) 2x^5 + x^4 + 11x^3 - 8x^2 - 13x + 7} \\
\underline{2x^5 + x^4 - x^3} \\
12x^3 - 8x^2 - 13x \\
\underline{12x^3 + 6x^2 - 6x} \\
-14x^2 - 7x + 7 \\
\underline{-14x^2 - 7x + 7} \\
0
\end{array}
$$

The remainder is 0. The answer is the quotient, $x^3 + 6x - 7$.

SECTION 5.8 (pages 424–426)

51. $\dfrac{4 \times 10^5}{8 \times 10^2}$

$= \dfrac{4}{8} \times \dfrac{10^5}{10^2}$

$= 0.5 \times 10^3$ Divide; quotient rule

$= (5 \times 10^{-1}) \times 10^3$ Write 0.5 in scientific notation.

$= 5 \times (10^{-1} \times 10^3)$ Associative property

$= 5 \times 10^2$ Product rule

$= 500$ Write without exponents.

55. $\dfrac{(1.65 \times 10^8)(5.24 \times 10^{-2})}{(6 \times 10^4)(2 \times 10^7)}$

$= \dfrac{1.65 \times 5.24}{6 \times 2} \times \dfrac{10^8 \times 10^{-2}}{10^4 \times 10^7}$

Associative and commutative properties

$= 0.7205 \times \dfrac{10^6}{10^{11}}$ Product rule

$= 0.7205 \times 10^{-5}$ Quotient rule

$= (7.205 \times 10^{-1}) \times 10^{-5}$

Write 0.7205 in scientific notation.

$= 7.205 \times (10^{-1} \times 10^{-5})$

Associative property

$= 7.205 \times 10^{-6}$ Product rule

$= 0.000007205$

Write without exponents.

CHAPTER 6 Factoring and Applications

SECTION 6.1 (pages 447–449)

79. $18r^2 + 12ry - 3xr - 2xy$

$= (18r^2 + 12ry) + (-3xr - 2xy)$

Group the terms.

$= 6r(3r + 2y) - x(3r + 2y)$

Factor each group.

$= (3r + 2y)(6r - x)$

Factor out the common factor, $3r + 2y$.

87. $y^2 + 3x + 3y + xy$

$= y^2 + 3y + xy + 3x$

Rearrange terms.

$= (y^2 + 3y) + (xy + 3x)$

Group the terms.

$= y(y + 3) + x(y + 3)$

Factor each group.

$= (y + 3)(y + x)$

Factor out the common factor, $y + 3$.

SECTION 6.2 (pages 455–456)

41. $a^2 - 8a - 48$

Find two integers whose product is -48 and whose sum is -8. Since c is negative, one integer must be positive and one must be negative.

Factors of -48	Sums of Factors
$-1, 48$	47
$1, -48$	-47
$-2, 24$	22
$2, -24$	-22
$-3, 16$	13
$3, -16$	-13
$-4, 12$	8
$4, -12$	$-8 \leftarrow$
$-6, 8$	2
$6, -8$	-2

Thus,

$a^2 - 8a - 48$ factors as $(a + 4)(a - 12)$.

57. $-2x^6 - 8x^5 + 42x^4$

First, factor out the negative common factor, $-2x^4$.

$-2x^6 - 8x^5 + 42x^4$

$= -2x^4(x^2 + 4x - 21)$

Now factor $x^2 + 4x - 21$.

Factors of -21	Sums of Factors
$1, -21$	-20
$-1, 21$	20
$3, -7$	-4
$-3, 7$	$4 \leftarrow$

Thus,

$x^2 + 4x - 21$ factors as $(x - 3)(x + 7)$.

The completely factored form is
$$-2x^6 - 8x^5 + 42x^4$$
$$= -2x^4(x - 3)(x + 7).$$

61. $5m^5 + 25m^4 - 40m^2$

First, factor out the GCF, $5m^2$.
$$5m^5 + 25m^4 - 40m^2$$
$$= 5m^2(m^3 + 5m^2 - 8)$$

The trinomial $m^3 + 5m^2 - 8$ cannot be factored further, so the polynomial is in completely factored form.

63. $m^3n - 10m^2n^2 + 24mn^3$

First, factor out the GCF, mn.
$$m^3n - 10m^2n^2 + 24mn^3$$
$$= mn(m^2 - 10mn + 24n^2)$$

The expressions $-6n$ and $-4n$ have a product of $24n^2$ and a sum of $-10n$. The completely factored form is
$$m^3n - 10m^2n^2 + 24mn^3$$
$$= mn(m - 6n)(m - 4n).$$

SECTION 6.3 (pages 459–460)

29. $16 + 16x + 3x^2$

Write the trinomial in descending powers.
$$3x^2 + 16x + 16$$

To factor, find two integers whose product is $3(16) = 48$ and whose sum is 16. The integers are 12 and 4.
$$3x^2 + 16x + 16$$
$$= 3x^2 + 12x + 4x + 16$$
$$= (3x^2 + 12x) + (4x + 16)$$
Group the terms.
$$= 3x(x + 4) + 4(x + 4)$$
Factor each group.
$$= (x + 4)(3x + 4)$$
Factor out the common factor, $x + 4$.

35. $-32z^5 + 20z^4 + 12z^3$

First factor out the negative common factor, $-4z^3$.
$$-32z^5 + 20z^4 + 12z^3$$
$$= -4z^3(8z^2 - 5z - 3)$$

To factor $8z^2 - 5z - 3$, find two integers whose product is $8(-3) = -24$ and whose sum is -5. These integers are -8 and 3. Now rewrite the given trinomial and factor it.
$$-32z^5 + 20z^4 + 12z^3$$
$$= -4z^3(8z^2 - 5z - 3)$$
$$= -4z^3(8z^2 - 8z + 3z - 3)$$
$$\quad -5z = -8z + 3z$$
$$= -4z^3[(8z^2 - 8z) + (3z - 3)]$$
Group the terms.
$$= -4z^3[8z(z - 1) + 3(z - 1)]$$
Factor each group.
$$= -4z^3[(z - 1)(8z + 3)]$$
Factor out the common factor, $z - 1$.
$$= -4z^3(z - 1)(8z + 3)$$

SECTION 6.4 (pages 465–466)

43. $6m^6n + 7m^5n^2 + 2m^4n^3$

Factor out the GCF, m^4n.
$$6m^6n + 7m^5n^2 + 2m^4n^3$$
$$= m^4n(6m^2 + 7mn + 2n^2)$$

Now factor $6m^2 + 7mn + 2n^2$ by trial and error. Possible factors of $6m^2$ are $6m$ and m or $3m$ and $2m$. Possible factors of $2n^2$ are $2n$ and n.
$$(3m + 2n)(2m + n)$$
$$= 6m^2 + 7mn + 2n^2 \quad \text{Correct}$$

The completely factored form is
$$6m^6n + 7m^5n^2 + 2m^4n^3$$
$$= m^4n(3m + 2n)(2m + n).$$

SECTION 6.5 (pages 473–476)

51. $4m^2 - \dfrac{9}{25}$

Because $4m^2 = (2m)^2$ and $\dfrac{9}{25} = \left(\dfrac{3}{5}\right)^2$,

$4m^2 - \dfrac{9}{25}$ is a difference of squares.

$$4m^2 - \frac{9}{25}$$
$$= (2m)^2 - \left(\frac{3}{5}\right)^2$$
$$= \left(2m + \frac{3}{5}\right)\left(2m - \frac{3}{5}\right)$$

57. $x^2 - 1.0x + 0.25$

The first and last terms are perfect squares, x^2 and $(-0.5)^2$. The trinomial is a perfect square, since the middle term is
$$2 \cdot x \cdot (-0.5) = -1.0x.$$
$$x^2 - 1.0x + 0.25$$
$$= (x)^2 - 2(x)(0.5) + (0.5)^2$$
$$= (x - 0.5)^2$$

SECTION 6.7 (pages 487–489)

57. $\qquad 9y^3 - 49y = 0$

To factor the polynomial, begin by factoring out the greatest common factor.
$$y(9y^2 - 49) = 0$$
Now factor $9y^2 - 49$ as the difference of two squares.
$$y(3y + 7)(3y - 7) = 0$$
Set each of the three factors equal to 0 and solve.

$y = 0$ or $3y + 7 = 0$ or $3y - 7 = 0$
$$\qquad\qquad 3y = -7 \quad \text{or} \quad 3y = 7$$
$$\qquad\qquad y = -\frac{7}{3} \quad \text{or} \quad y = \frac{7}{3}$$

Solution set: $\left\{-\dfrac{7}{3}, 0, \dfrac{7}{3}\right\}$

59. $(2r + 5)(3r^2 - 16r + 5) = 0$
$$(2r + 5)(3r - 1)(r - 5) = 0$$
Factor $3r^2 - 16r + 5$.

Set each factor equal to 0 and solve.

$2r + 5 = 0$ or $3r - 1 = 0$ or $r - 5 = 0$
$2r = -5$ or $\qquad 3r = 1$ or $\qquad r = 5$
$r = -\dfrac{5}{2}$ or $\qquad r = \dfrac{1}{3}$

Solution set: $\left\{-\dfrac{5}{2}, \dfrac{1}{3}, 5\right\}$

65. $\qquad\qquad r^4 = 2r^3 + 15r^2$

Write with 0 on one side of the equation.
$$r^4 - 2r^3 - 15r^2 = 0$$
Subtract $2r^3$ and $15r^2$.
$$r^2(r^2 - 2r - 15) = 0 \quad \text{GCF} = r^2$$
$$r \cdot r \cdot (r + 3)(r - 5) = 0 \quad \text{Factor.}$$

Set each of the factors equal to 0, and solve the resulting equations. (Since r is a factor twice, it gives a double solution.)

$r = 0$ or $r + 3 = 0$ or $r - 5 = 0$
$$\qquad\qquad r = -3 \quad \text{or} \quad r = 5$$

Solution set: $\{-3, 0, 5\}$

SECTION 6.8 (pages 496–500)

15. Let $\quad x =$ the length of a side of the square painting.

Then $x - 2 =$ the length of a side of the square mirror.

Since the formula for the area of a square is $A = s^2$, the area of the painting is x^2, and the area of the mirror is $(x - 2)^2$. The difference between their areas is 32.

$$x^2 - (x - 2)^2 = 32$$
$$x^2 - (x^2 - 4x + 4) = 32$$
$$x^2 - x^2 + 4x - 4 = 32$$
$$4x - 4 = 32$$
$$4x = 36$$
$$x = 9$$

The length of a side of the painting is 9 ft. The length of a side of the mirror is
$$9 - 2 = 7 \text{ ft.}$$

As a check,
$$9^2 - 7^2 = 81 - 49 = 32, \quad \text{as required.}$$

SOLUTIONS

23. Let $\quad x$ = the least even integer.

Then $x + 2$ = the next even integer and

$\quad x + 4$ = the third even integer.

$x^2 + (x + 2)^2 = (x + 4)^2$

$x^2 + x^2 + 4x + 4 = x^2 + 8x + 16$

$\qquad$ Square the binomials.

$2x^2 + 4x + 4 = x^2 + 8x + 16$

$\qquad$ Combine like terms.

$x^2 - 4x - 12 = 0$

$\qquad$ Standard form

$(x + 2)(x - 6) = 0$

$\qquad$ Factor.

$x + 2 = 0 \quad$ or $\quad x - 6 = 0$

$\qquad$ Zero-factor property

$x = -2 \quad$ or $\quad x = 6$

If $x = -2$, then $x + 2 = 0$ and $x + 4 = 2$.

If $x = 6$, then $x + 2 = 8$ and $x + 4 = 10$.

The integers are $-2, 0$, and 2 or $6, 8$, and 10.

CHECK

$(-2)^2 + 0^2 \overset{?}{=} 2^2 \qquad\qquad 6^2 + 8^2 \overset{?}{=} 10^2$

$4 + 0 \overset{?}{=} 4 \qquad\qquad 36 + 64 \overset{?}{=} 100$

$4 = 4 \checkmark \qquad\qquad 100 = 100 \checkmark$

$\quad$ True $\qquad\qquad\qquad\quad$ True

CHAPTER 7 Rational Expressions and Functions

SECTION 7.1 (pages 521–524)

15. $f(x) = \dfrac{3x + 1}{2x^2 + x - 6}$

Set the denominator equal to zero and solve.

$2x^2 + x - 6 = 0$

$(x + 2)(2x - 3) = 0 \quad$ Factor.

$x + 2 = 0 \quad$ or $\quad 2x - 3 = 0$

$\qquad$ Zero-factor property

$x = -2 \quad$ or $\qquad 2x = 3$

$\qquad$ Solve each equation.

$x = \dfrac{3}{2}$

The numbers -2 and $\dfrac{3}{2}$ are not in the domain

of the function. In set-builder notation, the

domain is

$\left\{ x \,\middle|\, x \neq -2, \dfrac{3}{2} \right\}.$

45. $\dfrac{2c^2 + 2cd - 60d^2}{2c^2 - 12cd + 10d^2}$

$= \dfrac{2(c^2 + cd - 30d^2)}{2(c^2 - 6cd + 5d^2)}$

$\qquad$ Factor out the GCF in the numerator and denominator.

$= \dfrac{2(c + 6d)(c - 5d)}{2(c - d)(c - 5d)}$

$\qquad$ Factor trinomials in the numerator and denominator.

$= \dfrac{c + 6d}{c - d}$

$\qquad$ Lowest terms

83. $\left(\dfrac{6k^2 - 13k - 5}{k^2 + 7k} \div \dfrac{2k - 5}{k^3 + 6k^2 - 7k} \right)$

$\quad \cdot \dfrac{k^2 - 5k + 6}{3k^2 - 8k - 3}$

Factor k from the denominator of the divisor.

Multiply by the reciprocal.

$= \left[\dfrac{6k^2 - 13k - 5}{k^2 + 7k} \cdot \dfrac{k(k^2 + 6k - 7)}{2k - 5} \right]$

$\quad \cdot \dfrac{k^2 - 5k + 6}{3k^2 - 8k - 3}$

$= \dfrac{(3k + 1)(2k - 5)}{k(k + 7)} \cdot \dfrac{k(k + 7)(k - 1)}{2k - 5}$

$\quad \cdot \dfrac{(k - 2)(k - 3)}{(3k + 1)(k - 3)}$

$\qquad$ Factor numerators and denominators.

$= (k - 1)(k - 2) \qquad$ Lowest terms

SECTION 7.2 (pages 531–533)

17. $\dfrac{a^3}{a^2 + ab + b^2} - \dfrac{b^3}{a^2 + ab + b^2}$

$= \dfrac{a^3 - b^3}{a^2 + ab + b^2}$

$\qquad$ Subtract the numerators. Keep the common denominator.

$= \dfrac{(a - b)(a^2 + ab + b^2)}{a^2 + ab + b^2}$

$\qquad$ Factor the difference of cubes in the numerator.

$= a - b \qquad$ Lowest terms

53. $\dfrac{4x}{x - 1} - \dfrac{2}{x + 1} - \dfrac{4}{x^2 - 1}$

$x^2 - 1 = (x + 1)(x - 1)$, the LCD.

$\dfrac{4x}{x - 1} - \dfrac{2}{x + 1} - \dfrac{4}{x^2 - 1}$

$= \dfrac{4x(x + 1)}{(x - 1)(x + 1)} - \dfrac{2(x - 1)}{(x + 1)(x - 1)}$

$\quad - \dfrac{4}{(x + 1)(x - 1)}$

$\qquad$ Fundamental property

$= \dfrac{4x(x + 1) - 2(x - 1) - 4}{(x + 1)(x - 1)}$

$\qquad$ Subtract numerators.

$= \dfrac{4x^2 + 4x - 2x + 2 - 4}{(x - 1)(x + 1)}$

$\qquad$ Distributive property

$= \dfrac{4x^2 + 2x - 2}{(x - 1)(x + 1)}$

$\qquad$ Combine like terms in the numerator.

$= \dfrac{2(2x^2 + x - 1)}{(x - 1)(x + 1)}$

$\qquad$ Factor out the GCF in the numerator.

$= \dfrac{2(2x - 1)(x + 1)}{(x - 1)(x + 1)} \qquad$ Factor.

$= \dfrac{2(2x - 1)}{x - 1} \qquad$ Lowest terms

61. $\dfrac{4}{x + 1} + \dfrac{1}{x^2 - x + 1} - \dfrac{12}{x^3 + 1}$

$x^3 + 1 = (x + 1)(x^2 - x + 1)$, the LCD.

$\dfrac{4}{x + 1} + \dfrac{1}{x^2 - x + 1}$

$- \dfrac{12}{(x + 1)(x^2 - x + 1)}$

$= \dfrac{4(x^2 - x + 1)}{(x + 1)(x^2 - x + 1)}$

$+ \dfrac{1 \cdot (x + 1)}{(x^2 - x + 1)(x + 1)}$

$- \dfrac{12}{(x + 1)(x^2 - x + 1)}$

$\qquad$ Fundamental property

$= \dfrac{4(x^2 - x + 1) + (x + 1) - 12}{(x + 1)(x^2 - x + 1)}$

$\qquad$ Add and subtract numerators.

$= \dfrac{4x^2 - 4x + 4 + x + 1 - 12}{(x + 1)(x^2 - x + 1)}$

$\qquad$ Distributive property

$= \dfrac{4x^2 - 3x - 7}{(x + 1)(x^2 - x + 1)}$

$\qquad$ Combine like terms.

$= \dfrac{(4x - 7)(x + 1)}{(x + 1)(x^2 - x + 1)} \qquad$ Factor.

$= \dfrac{4x - 7}{x^2 - x + 1} \qquad$ Lowest terms

69. $\dfrac{5x}{x^2 + xy - 2y^2} - \dfrac{3x}{x^2 + 5xy - 6y^2}$

$\qquad$ Factor each denominator.

$x^2 + xy - 2y^2 = (x + 2y)(x - y)$

$x^2 + 5xy - 6y^2 = (x + 6y)(x - y)$

The LCD is $(x + 2y)(x - y)(x + 6y)$.

$\dfrac{5x}{(x + 2y)(x - y)} - \dfrac{3x}{(x + 6y)(x - y)}$

$= \dfrac{5x(x + 6y)}{(x + 2y)(x - y)(x + 6y)}$

$- \dfrac{3x(x + 2y)}{(x + 6y)(x - y)(x + 2y)}$

$\qquad$ Fundamental property

$= \dfrac{5x(x + 6y) - 3x(x + 2y)}{(x + 6y)(x - y)(x + 2y)}$

$\qquad$ Subtract numerators.

$= \dfrac{5x^2 + 30xy - 3x^2 - 6xy}{(x + 2y)(x - y)(x + 6y)}$

$\qquad$ Distributive property

$= \dfrac{2x^2 + 24xy}{(x + 2y)(x - y)(x + 6y)}$

$\qquad$ Combine like terms.

$= \dfrac{2x(x + 12y)}{(x + 2y)(x - y)(x + 6y)}$

$\qquad$ Factor out the GCF.

SECTION 7.3 (pages 538–539)

13. $\dfrac{\dfrac{4z^2x^4}{9}}{\dfrac{12x^2z^5}{15}}$ ⎦ These are single fractions. We use Method 1.

$= \dfrac{4z^2x^4}{9} \div \dfrac{12x^2z^5}{15}$ Write as a division problem.

$= \dfrac{4z^2x^4}{9} \cdot \dfrac{15}{12x^2z^5}$ Multiply by the reciprocal of the divisor.

$= \dfrac{60z^2x^4}{108x^2z^5}$ Multiply.

$= \dfrac{5 \cdot 12 \cdot z^2 \cdot x^2 \cdot x^2}{9 \cdot 12 \cdot x^2 \cdot z^2 \cdot z^3}$ Factor.

$= \dfrac{5x^2}{9z^3}$ Lowest terms

25. $\dfrac{\dfrac{x^2 - 16y^2}{xy}}{\dfrac{1}{y} - \dfrac{4}{x}}$ We use Method 2.

Multiply the numerator and denominator by xy, the LCD of all the fractions.

$= \dfrac{\left(\dfrac{x^2 - 16y^2}{xy}\right)xy}{\left(\dfrac{1}{y} - \dfrac{4}{x}\right)xy}$

$= \dfrac{x^2 - 16y^2}{x - 4y}$ Distributive property

$= \dfrac{(x + 4y)(x - 4y)}{x - 4y}$ Factor the difference of squares in the numerator.

$= x + 4y$ Lowest terms

33. $\dfrac{\dfrac{x + 2}{x} + \dfrac{1}{x + 2}}{\dfrac{5}{x} + \dfrac{x}{x + 2}}$ We use Method 2.

Multiply the numerator and denominator by $x(x + 2)$, the LCD of all the fractions.

$= \dfrac{x(x + 2)\left(\dfrac{x + 2}{x} + \dfrac{1}{x + 2}\right)}{x(x + 2)\left(\dfrac{5}{x} + \dfrac{x}{x + 2}\right)}$

$= \dfrac{x(x + 2)\left(\dfrac{x + 2}{x}\right) + x(x + 2)\left(\dfrac{1}{x + 2}\right)}{x(x + 2)\left(\dfrac{5}{x}\right) + x(x + 2)\left(\dfrac{x}{x + 2}\right)}$ Distributive property

$= \dfrac{(x + 2)(x + 2) + x}{5(x + 2) + x^2}$ Multiply.

$= \dfrac{x^2 + 4x + 4 + x}{5x + 10 + x^2}$ Multiply.

$= \dfrac{x^2 + 5x + 4}{x^2 + 5x + 10}$ Combine like terms.

41. $\dfrac{x^{-1} + 2y^{-1}}{2y + 4x}$

$= \dfrac{\dfrac{1}{x} + \dfrac{2}{y}}{2y + 4x}$ Write with positive exponents.

Multiply the numerator and denominator by xy, the LCD of all the fractions.

$= \dfrac{xy\left(\dfrac{1}{x} + \dfrac{2}{y}\right)}{xy(2y + 4x)}$

$= \dfrac{y + 2x}{2xy(y + 2x)}$ Distributive property; factor $2y + 4x$ as $2(y + 2x)$.

$= \dfrac{1}{2xy}$ Lowest terms

SECTION 7.4 (pages 545–547)

21. $\dfrac{3x + 1}{x - 4} = \dfrac{6x + 5}{2x - 7}$

The domain excludes 4 and $\dfrac{7}{2}$.

Multiply each side by the LCD, $(x - 4)(2x - 7)$.

$(x - 4)(2x - 7)\left(\dfrac{3x + 1}{x - 4}\right)$

$= (x - 4)(2x - 7)\left(\dfrac{6x + 5}{2x - 7}\right)$

$(2x - 7)(3x + 1) = (x - 4)(6x + 5)$

$6x^2 - 19x - 7 = 6x^2 - 19x - 20$

$-7 = -20$ False

The false statement indicates that the original equation has no solution.

Solution set: $\varnothing$

31. $\dfrac{9}{x} + \dfrac{4}{6x - 3} = \dfrac{2}{6x - 3}$

The domain excludes 0 and $\dfrac{1}{2}$.

Multiply by the LCD, $x(6x - 3)$.

$x(6x - 3)\left(\dfrac{9}{x} + \dfrac{4}{6x - 3}\right)$

$= x(6x - 3)\left(\dfrac{2}{6x - 3}\right)$

$9(6x - 3) + 4x = 2x$

$54x - 27 + 4x = 2x$

$56x = 27$

$x = \dfrac{27}{56}$

Note that $\dfrac{27}{56}$ is in the domain. Substitute $\dfrac{27}{56}$ for x in the original equation to check the solution.

Solution set: $\left\{\dfrac{27}{56}\right\}$

45. $\dfrac{4x - 7}{4x^2 - 9} = \dfrac{-2x^2 + 5x - 4}{4x^2 - 9} + \dfrac{x + 1}{2x + 3}$

$\dfrac{4x - 7}{(2x + 3)(2x - 3)}$

$= \dfrac{-2x^2 + 5x - 4}{(2x + 3)(2x - 3)} + \dfrac{x + 1}{2x + 3}$ Factor.

The domain excludes $-\dfrac{3}{2}$ and $\dfrac{3}{2}$.

Multiply by the LCD, $(2x + 3)(2x - 3)$.

$4x - 7 = -2x^2 + 5x - 4$
         $+ (2x - 3)(x + 1)$

$4x - 7 = -2x^2 + 5x - 4 + 2x^2 - x - 3$

$4x - 7 = 4x - 7$ True

This equation is true for every real number value of x, but we have already determined that $-\dfrac{3}{2}$ and $\dfrac{3}{2}$ are excluded from the domain. Thus, every real number except $-\dfrac{3}{2}$ and $\dfrac{3}{2}$ is a solution.

Solution set: $\left\{x \mid x \neq -\dfrac{3}{2}, \dfrac{3}{2}\right\}$

SECTION 7.5 (pages 559–564)

33. Let $x =$ the number of fish in the lake. Write and solve a proportion.

$\dfrac{\text{total in lake}}{\text{tagged in lake}} = \dfrac{\text{total in sample}}{\text{tagged in sample}}$

$\dfrac{x}{500} = \dfrac{400}{8}$

$\dfrac{x}{500} = 50$

$x = 500(50)$

$x = 25{,}000$

There are 25,000 fish in the lake.

51. *Step 1*

Read the problem again.

Step 2

Let $x =$ the distance in miles from San Francisco to the secret rendezvous. Make a table.

	d	r	t
First Trip	x	200	$\dfrac{x}{200}$
Return Trip	x	300	$\dfrac{x}{300}$

Step 3

Time there	plus	time back	equals	4 hr.
$\dfrac{x}{200}$	$+$	$\dfrac{x}{300}$	$=$	4

(continued)

Step 4

Multiply by the LCD, 600.

$$600\left(\frac{x}{200}+\frac{x}{300}\right)=600\,(4)$$

$$3x+2x=2400$$

$$5x=2400$$

$$x=480$$

Step 5

The distance is 480 mi.

Step 6

Check.

480 mi at 200 mph takes $\dfrac{480}{200}$, or 2.4 hr.

480 mi at 300 mph takes $\dfrac{480}{300}$, or 1.6 hr.

The total time is 4 hr, as required.

61. Let $x=$ the time from Mimi's arrival home to the time the place is a shambles.

	Rate	Time to Mess up House	Fractional Part of the Job Done
Hortense and Mort	$-\dfrac{1}{7}$	x	$-\dfrac{1}{7}x$
Mimi	$\dfrac{1}{2}$	x	$\dfrac{1}{2}x$

Notice that Hortense and Mort's rate is negative since they are "undoing" the messing up by cleaning the house.

$$\begin{array}{ccccc}\text{Part done}\\\text{by Hortense}\\\text{and Mort}\end{array}+\begin{array}{c}\text{part done}\\\text{by Mimi}\end{array}=\begin{array}{c}\text{1 whole job}\\\text{of messing}\\\text{up.}\end{array}$$

$$-\frac{1}{7}x\quad+\quad\frac{1}{2}x\quad=\quad1$$

Multiply by the LCD, 14.

$$14\left(-\frac{1}{7}x\right)+14\left(\frac{1}{2}x\right)=14\,(1)$$

$$-2x+7x=14$$

$$5x=14$$

$$x=\frac{14}{5},\quad\text{or}\quad2\frac{4}{5}$$

It would take $\dfrac{14}{5}$ hr, or $2\dfrac{4}{5}$ hr after Mimi got home for the house to be a shambles.

SECTION 7.6 (pages 571–574)

43. Let $I=$ the illumination produced by a light source and $d=$ the distance from the source. I varies inversely as d^2, so

$$I=\frac{k}{d^2}$$

for some constant k. Since $I=768$ when $d=1$, substitute these values in the equation and solve for k.

$$I=\frac{k}{d^2}$$

$$768=\frac{k}{1^2}$$

$$768=k$$

So $I=\dfrac{768}{d^2}$. Now let $d=6$.

$$I=\frac{768}{d^2}=\frac{768}{36}=\frac{64}{3},\text{ or }21\frac{1}{3}$$

The illumination produced by the light source is $21\dfrac{1}{3}$ foot-candles.

49. Let $F=$ the force, $w=$ the weight of the car, $s=$ the speed, and $r=$ the radius. The force varies inversely as the radius and jointly as the weight and the square of the speed, so for some constant k,

$$F=\frac{kws^2}{r}.$$

Let $F=242$, $w=2000$, $r=500$, and $s=30$.

$$242=\frac{k\,(2000)\,(30)^2}{500}$$

$$k=\frac{242\,(500)}{2000\,(900)}$$

$$k=\frac{121}{1800}$$

So $F=\dfrac{121ws^2}{1800r}$.

Let $r=750$, $s=50$, and $w=2000$.

$$F=\frac{121\,(2000)\,(50)^2}{1800\,(750)}\approx448.1$$

Approximately 448.1 lb of force would be needed.

CHAPTER 8 Equations, Inequalities, and Systems Revisited

SECTION 8.1 (pages 598–601)

29. $2\,[w-(2w+4)+3]=2\,(w+1)$

$\quad 2\,[w-2w-4+3]=2\,(w+1)$

$\qquad\qquad$ Distributive property

$\quad\quad 2\,[-w-1]=2\,(w+1)$

$\qquad\qquad$ Combine like terms.

$\quad\quad -2w-2=2w+2$

$\qquad\qquad$ Distributive property

$\quad\quad\quad -2=4w+2$

$\qquad\qquad$ Add 2w.

$\quad\quad\quad -4=4w$

$\qquad\qquad$ Subtract 2.

$\quad\quad\quad -1=w$

$\qquad\qquad$ Divide by 4.

CHECK Substitute -1 for w in the original equation.

Solution set: $\{-1\}$

67. $\dfrac{2k-5}{-4}>5$

Multiply each side by -4 and reverse the direction of the inequality symbol.

$$-4\left(\frac{2k-5}{-4}\right)<-4\,(5)$$

$$2k-5<-20\qquad\text{Multiply.}$$

$$2k<-15\qquad\text{Add 5.}$$

$$k<-\frac{15}{2}\qquad\text{Divide by 2.}$$

Check that the solution set is the interval $\left(-\infty,-\dfrac{15}{2}\right)$.

77. $-6\le2\,(z+2)\le16$

$\quad -6\le\ \ 2z+4\ \ \le16$

$\qquad\qquad$ Distributive property

$\quad -10\le\quad 2z\quad\le12$

$\qquad\qquad$ Subtract 4 from each part.

$\quad\ -5\le\quad z\quad\le6$

$\qquad\qquad$ Divide each part by 2.

Check that the solution set is the interval $[-5,6]$.

SECTION 8.2 (pages 607–610)

41. $4x - 8 > 0 \quad$ or $\quad 4x - 1 < 7$

$\qquad 4x > 8 \quad$ or $\qquad 4x < 8$

$\qquad x > 2 \quad$ or $\qquad x < 2$

The graph of the solution set contains all numbers either greater than 2 or less than 2. This is all real numbers except 2. The solution set is $(-\infty, 2) \cup (2, \infty)$.

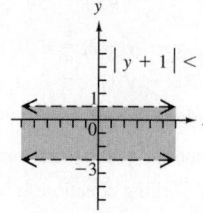

43. $(-\infty, -1] \cap [-4, \infty)$

The intersection is the set of numbers less than or equal to -1 and greater than or equal to -4. The numbers common to *both* original sets are between, and including, -4 and -1. The interval form is $[-4, -1]$.

SECTION 8.3 (pages 617–623)

19. $\left| 1 - \dfrac{3}{4}k \right| = 7$

$1 - \dfrac{3}{4}k = 7 \quad$ or $\quad 1 - \dfrac{3}{4}k = -7$

$-\dfrac{3}{4}k = 6 \quad$ or $\qquad -\dfrac{3}{4}k = -8$

$\qquad\qquad\qquad\qquad\qquad$ Subtract 1.

$k = -8 \quad$ or $\qquad k = \dfrac{32}{3}$

$\qquad\qquad\qquad$ Multiply by $-\dfrac{4}{3}$.

Solution set: $\left\{ -8, \dfrac{32}{3} \right\}$

49. $\qquad |-3x - 8| \le 4$

$\qquad -4 \le -3x - 8 \le 4$

$\qquad\; 4 \le \quad -3x \quad \le 12 \qquad$ Add 8.

Divide each part by -3 and reverse the direction of the inequality symbols.

$$\dfrac{4}{-3} \ge x \ge \dfrac{12}{-3}$$

$$-\dfrac{4}{3} \ge x \ge -4$$

Rewrite in order based on a number line.

$$-4 \le x \le -\dfrac{4}{3}$$

Solution set: $\left[-4, -\dfrac{4}{3} \right]$

65. $|2p - 6| = |2p + 11|$

$\quad 2p - 6 = 2p + 11$

$\qquad\; -6 = 11 \qquad$ False

$\quad$ *No solution* ($\varnothing$)

or

$\quad 2p - 6 = -(2p + 11)$

$\quad 2p - 6 = -2p - 11 \qquad$ Distributive property

$\qquad\qquad\qquad\qquad\qquad$ Add $2p$ and add 6.

$\qquad 4p = -5$

$\qquad\quad p = -\dfrac{5}{4} \qquad$ Divide by 4.

Solution set: $\varnothing \cup \left\{ -\dfrac{5}{4} \right\} = \left\{ -\dfrac{5}{4} \right\}$

81. $|10z + 7| > 0$

Since the absolute value of an expression is always nonnegative, there is only one possible value of z that makes this statement false. Solving the equation $10z + 7 = 0$ will give that value of z.

$$10z + 7 = 0$$

$$10z = -7$$

$$z = -\dfrac{7}{10}$$

The solution set of the inequality includes *all values other than* $-\dfrac{7}{10}$, which makes the absolute value expression equal 0.

Solution set: $\left(-\infty, -\dfrac{7}{10} \right) \cup \left(-\dfrac{7}{10}, \infty \right)$

99. $|y + 1| < 2$ can be rewritten as follows.

$$-2 < y + 1 < 2$$

$$-3 < \quad y \quad < 1$$

The boundaries are the dashed horizontal lines $y = -3$ and $y = 1$. Since y is between -3 and 1, the graph includes all points between the lines.

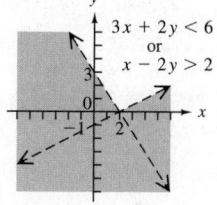

103. $3x + 2y < 6 \quad$ or $\quad x - 2y > 2$

Graph $3x + 2y = 6$, which has intercepts $(2, 0)$ and $(0, 3)$, as a dashed line. Test $(0, 0)$, which yields $0 < 6$, a true statement. Shade the region that includes $(0, 0)$. Graph $x - 2y = 2$, which has intercepts $(2, 0)$ and $(0, -1)$, as a dashed line. Test $(0, 0)$, which yields $0 > 2$, a false statement. Shade the region that does not include $(0, 0)$.

The required graph of the union includes all the shaded regions, that is, all the points that satisfy either inequality.

SECTION 8.4 (pages 633–634)

21. $\qquad\quad y = 2x \qquad (1)$

$\qquad 4x - 2y = 0 \qquad (2)$

From equation (1), substitute $2x$ for y in equation (2).

$\qquad 4x - 2y = 0 \qquad (2)$

$\quad 4x - 2(2x) = 0 \qquad$ Let $y = 2x$.

$\qquad 4x - 4x = 0 \qquad$ Multiply.

$\qquad\qquad\quad 0 = 0 \qquad$ True

The equations are dependent, and the solution set is the set of all points on the line. We use one of the equations of the system to write the solution set in set-builder notation. We give the equation in standard form with coefficients that are integers having greatest common factor 1 and positive coefficient of x. Thus, we use equation (2) and divide each term by the common factor 2 to get the solution set.

$$\{ (x, y) \mid 2x - y = 0 \}$$

35.
$$\frac{x}{2} + \frac{y}{3} = -\frac{1}{3} \quad (1)$$

$$\frac{x}{2} + 2y = -7 \quad (2)$$

Clear the fractions by multiplying equation (1) by -6 and equation (2) by 6. (We multiply equation (2) by 6 instead of 2 so that when the fractions are cleared, the x-terms are opposites.) Then add the results to eliminate x.

$$
\begin{array}{lll}
-3x - 2y = 2 & \text{Multiply (1) by } -6. \ (3) \\
\underline{3x + 12y = -42} & \text{Multiply (2) by 6.} \\
10y = -40 & \text{Add.} \\
y = -4 & \text{Divide by 10.}
\end{array}
$$

To find x, substitute -4 for y in equation (3).

$$
\begin{array}{lll}
-3x - 2y = 2 & (3) \\
-3x - 2(-4) = 2 & \text{Let } y = -4. \\
-3x + 8 = 2 & \text{Multiply.} \\
-3x = -6 & \text{Subtract 8.} \\
x = 2 & \text{Divide by } -3.
\end{array}
$$

The solution $(2, -4)$ checks in both equations (1) and (2).
Solution set: $\{(2, -4)\}$

SECTION 8.5 (pages 643–646)

15.
$$
\begin{array}{ll}
x + 2y + 3z = 1 & (1) \\
-x - y + 3z = 2 & (2) \\
-6x + y + z = -2 & (3)
\end{array}
$$

Step 1
Since x in equation (1) has coefficient 1, we choose it as the focus variable and (1) as the working equation.

Step 2
Add equations (1) and (2) to eliminate x.

$$
\begin{array}{ll}
x + 2y + 3z = 1 & (1) \\
\underline{-x - y + 3z = 2} & (2) \\
y + 6z = 3 & \text{Add.} \quad (4)
\end{array}
$$

Step 3
Multiply working equation (1) by 6 and add to equation (3) to eliminate x again.

$$
\begin{array}{ll}
6x + 12y + 18z = 6 & \text{Multiply (1) by 6.} \\
\underline{-6x + y + z = -2} & (3) \\
13y + 19z = 4 & \text{Add.} \quad (5)
\end{array}
$$

Step 4
The equations that resulted in Steps 2 and 3 form a system in y and z.

$$
\begin{array}{ll}
y + 6z = 3 & (4) \\
13y + 19z = 4 & (5)
\end{array}
$$

Solve this system of equations (4) and (5).

$$
\begin{array}{ll}
-13y - 78z = -39 & \text{Multiply (4) by } -13. \\
\underline{13y + 19z = 4} & (5) \\
-59z = -35 & \text{Add.} \\
z = \frac{35}{59} & \text{Divide by } -59.
\end{array}
$$

Substitute $\frac{35}{59}$ for z in equation (4) to find y.
Be careful—the arithmetic gets messy.

$$
\begin{array}{ll}
y + 6z = 3 & (4) \\
y + 6\left(\frac{35}{59}\right) = 3 & \text{Let } z = \frac{35}{59}. \\
y + \frac{210}{59} = 3 & \text{Multiply.} \\
y = \frac{177}{59} - \frac{210}{59} & 3 = \frac{177}{59} \\
y = -\frac{33}{59} & \text{Subtract.}
\end{array}
$$

Step 5
Substitute $-\frac{33}{59}$ for y and $\frac{35}{59}$ for z in working equation (1) to find focus variable x.

$$x + 2y + 3z = 1 \quad (1)$$

$$x + 2\left(-\frac{33}{59}\right) + 3\left(\frac{35}{59}\right) = 1$$

$$x - \frac{66}{59} + \frac{105}{59} = 1$$

$$x + \frac{39}{59} = 1$$

$$x = \frac{59}{59} - \frac{39}{59}$$

$$x = \frac{20}{59}$$

Step 6
The solution $\left(\frac{20}{59}, -\frac{33}{59}, \frac{35}{59}\right)$ checks when substituted in equations (1), (2), and (3).
Solution set: $\left\{\left(\frac{20}{59}, -\frac{33}{59}, \frac{35}{59}\right)\right\}$

31.
$$
\begin{array}{ll}
x + y - 2z = 0 & (1) \\
3x - y + z = 0 & (2) \\
4x + 2y - z = 0 & (3)
\end{array}
$$

We choose z as the focus variable and (2) as the working equation. Eliminate z by adding equations (2) and (3).

$$
\begin{array}{ll}
3x - y + z = 0 & (2) \\
\underline{4x + 2y - z = 0} & (3) \\
7x + y = 0 & (4)
\end{array}
$$

To get another equation without z, multiply working equation (2) by 2 and add the result to equation (1).

$$
\begin{array}{ll}
6x - 2y + 2z = 0 & \text{Multiply (2) by 2.} \\
\underline{x + y - 2z = 0} & (1) \\
7x - y = 0 & (5)
\end{array}
$$

Add equations (4) and (5) to find x.

$$
\begin{array}{ll}
7x + y = 0 & (4) \\
\underline{7x - y = 0} & (5) \\
14x = 0 & \text{Add.} \\
x = 0 & \text{Divide by 14.}
\end{array}
$$

Substitute 0 for x in equation (4) to find y.

$$
\begin{array}{ll}
7x + y = 0 & (4) \\
7(0) + y = 0 & \text{Let } x = 0. \\
0 + y = 0 & \text{Multiply.} \\
y = 0 & \text{Add.}
\end{array}
$$

Substitute 0 for x and 0 for y in working equation (2) to find focus variable z.

$$
\begin{array}{ll}
3x - y + z = 0 & (2) \\
3(0) - 0 + z = 0 & \text{Substitute.} \\
z = 0
\end{array}
$$

The solution $(0, 0, 0)$ checks when substituted in equations (1), (2), and (3).
Solution set: $\{(0, 0, 0)\}$

CHAPTER 9 Roots, Radicals, and Root Functions

SECTION 9.1 (pages 670–674)

15. The length $\sqrt{98}$ is closer to $\sqrt{100} = 10$ than to $\sqrt{81} = 9$. The width $\sqrt{26}$ is closer to $\sqrt{25} = 5$ than to $\sqrt{36} = 6$. Use the estimates $L = 10$ and $W = 5$ in the area formula $A = LW$ to find an estimate of the area.

$$A \approx 10 \cdot 5 = 50$$

Choice C is the best estimate.

111. $\sqrt[6]{x^{30}}$

$$= \sqrt[6]{(x^5)^6}$$

$$= |x^5|, \quad \text{or} \quad |x|^5 \quad \text{(6 is even.)}$$

129. Let $a = 850$, $b = 925$, and $c = 1300$. First find the semiperimeter s.

$$s = \frac{1}{2}(a + b + c)$$

$$s = \frac{1}{2}(850 + 925 + 1300)$$

$$s = \frac{3075}{2}$$

$$s = 1537.5$$

Now find the area A using Heron's formula and a calculator.

$$A = \sqrt{s(s-a)(s-b)(s-c)}$$
$$A = \sqrt{1537.5\,(687.5)\,(612.5)\,(237.5)}$$
$$A \approx 392{,}128.8$$

The area of the Bermuda Triangle is about $392{,}000$ mi^2.

SECTION 9.2 (pages 681–684)

47. $(3m^4 + 2k^2)^{-2/3}$

$$= \frac{1}{(3m^4 + 2k^2)^{2/3}}$$
$$= \frac{1}{\left[(3m^4 + 2k^2)^{1/3}\right]^2}$$
$$= \frac{1}{\left(\sqrt[3]{3m^4 + 2k^2}\right)^2}$$

57. $\sqrt[3]{x} \cdot \sqrt{x}$

$\quad = x^{1/3} \cdot x^{1/2}$ Convert to rational exponents.

$\quad = x^{1/3+1/2}$ Product rule

$\quad = x^{2/6+3/6}$ Find a common denominator.

$\quad = x^{5/6}$ Add exponents.

$\quad = \sqrt[6]{x^5}$ Write as a radical.

79. $\left(\dfrac{b^{-3/2}}{c^{-5/3}}\right)^2 \left(b^{-1/4}c^{-1/3}\right)^{-1}$

$\quad = \left(\dfrac{c^{5/3}}{b^{3/2}}\right)^2 \left(b^{1/4}c^{1/3}\right)$ Definition of negative exponent; power rule

$\quad = \dfrac{c^{10/3}}{b^3}\left(b^{1/4}c^{1/3}\right)$ Power rule

$\quad = \dfrac{c^{10/3}b^{1/4}c^{1/3}}{b^3}$ Multiply.

$\quad = c^{10/3+1/3}b^{1/4-3}$ Product and quotient rules

$\quad = c^{11/3}b^{-11/4}$ $\dfrac{1}{4} - 3 = \dfrac{1}{4} - \dfrac{12}{4} = -\dfrac{11}{4}$

$\quad = \dfrac{c^{11/3}}{b^{11/4}}$

81. $\left(\dfrac{p^{-1/4}q^{-3/2}}{3^{-1}p^{-2}q^{-2/3}}\right)^{-2}$

$\quad = \dfrac{p^{1/2}q^3}{3^2 p^4 q^{4/3}}$ Power rule

$\quad = \dfrac{p^{1/2-4}q^{3-4/3}}{9}$ Quotient rule

$\quad = \dfrac{p^{1/2-8/2}q^{9/3-4/3}}{9}$ Write exponents with a common denominator.

$\quad = \dfrac{p^{-7/2}q^{5/3}}{9}$ Subtract exponents.

$\quad = \dfrac{q^{5/3}}{9p^{7/2}}$

99. $\sqrt[3]{\sqrt[5]{\sqrt{y}}}$

$\quad = \sqrt[3]{\sqrt[5]{y^{1/2}}}$
$\quad = \sqrt[3]{(y^{1/2})^{1/5}}$
$\quad = (y^{1/10})^{1/3}$
$\quad = y^{1/30}$

SECTION 9.3 (pages 692–697)

73. $-\sqrt[3]{-125a^6b^9c^{12}}$

$\quad = -\sqrt[3]{(-5a^2b^3c^4)^3}$
$\quad = -(-5a^2b^3c^4)$
$\quad = 5a^2b^3c^4$

89. $-\sqrt[4]{162r^{15}s^{10}}$

$\quad = -\sqrt[4]{81r^{12}s^8(2r^3s^2)}$
$\quad = -\sqrt[4]{81r^{12}s^8} \cdot \sqrt[4]{2r^3s^2}$
$\quad = -3r^3s^2 \sqrt[4]{2r^3s^2}$

93. $\sqrt[3]{\dfrac{x^{16}}{27}}$

$\quad = \dfrac{\sqrt[3]{x^{15} \cdot x^1}}{\sqrt[3]{27}}$
$\quad = \dfrac{\sqrt[3]{x^{15}} \cdot \sqrt[3]{x}}{\sqrt[3]{27}}$
$\quad = \dfrac{x^5\sqrt[3]{x}}{3}$

123. Let $(x_1, y_1) = \left(\sqrt{2}, \sqrt{6}\right)$ and $(x_2, y_2) = \left(-2\sqrt{2}, 4\sqrt{6}\right)$.

$d = \sqrt{(x_2 - x_1)^2 + (y_2 - y_1)^2}$
$\quad = \sqrt{\left(-2\sqrt{2} - \sqrt{2}\right)^2 + \left(4\sqrt{6} - \sqrt{6}\right)^2}$
$\quad = \sqrt{\left(-3\sqrt{2}\right)^2 + \left(3\sqrt{6}\right)^2}$
$\quad = \sqrt{9 \cdot 2 + 9 \cdot 6}$
$\quad = \sqrt{18 + 54}$
$\quad = \sqrt{72}$
$\quad = \sqrt{36} \cdot \sqrt{2}$
$d = 6\sqrt{2}$

SECTION 9.4 (pages 700–701)

13. $6\sqrt{18} - \sqrt{32} + 2\sqrt{50}$

$\quad = 6\sqrt{9 \cdot 2} - \sqrt{16 \cdot 2} + 2\sqrt{25 \cdot 2}$
$\quad = 6\sqrt{9} \cdot \sqrt{2} - \sqrt{16} \cdot \sqrt{2}$
$\qquad + 2\sqrt{25} \cdot \sqrt{2}$
$\quad = 6 \cdot 3\sqrt{2} - 4\sqrt{2} + 2 \cdot 5\sqrt{2}$
$\quad = 18\sqrt{2} - 4\sqrt{2} + 10\sqrt{2}$
$\quad = 24\sqrt{2}$

29. $3x\sqrt[3]{xy^2} - 2\sqrt[3]{8x^4y^2}$

$\quad = 3x\sqrt[3]{xy^2} - 2\sqrt[3]{8x^3} \cdot \sqrt[3]{xy^2}$
$\quad = 3x\sqrt[3]{xy^2} - 2 \cdot 2x \cdot \sqrt[3]{xy^2}$
$\quad = 3x\sqrt[3]{xy^2} - 4x\sqrt[3]{xy^2}$
$\quad = (3x - 4x)\sqrt[3]{xy^2}$
$\quad = -x\sqrt[3]{xy^2}$

49. $3\sqrt[3]{\dfrac{m^5}{27}} - 2m\sqrt[3]{\dfrac{m^2}{64}}$

$\quad = \dfrac{3\sqrt[3]{m^5}}{\sqrt[3]{27}} - \dfrac{2m\sqrt[3]{m^2}}{\sqrt[3]{64}}$
$\quad = \dfrac{3\sqrt[3]{m^3} \cdot \sqrt[3]{m^2}}{3} - \dfrac{2m\sqrt[3]{m^2}}{4}$
$\quad = \dfrac{m\sqrt[3]{m^2}}{1} - \dfrac{m\sqrt[3]{m^2}}{2}$
$\quad = \dfrac{2m\sqrt[3]{m^2} - m\sqrt[3]{m^2}}{2}$
$\quad = \dfrac{m\sqrt[3]{m^2}}{2}$

55. $4\sqrt{18} + \sqrt{108} + 2\sqrt{72} + 3\sqrt{12}$

$\quad = 4\sqrt{9} \cdot \sqrt{2} + \sqrt{36} \cdot \sqrt{3}$
$\qquad + 2\sqrt{36} \cdot \sqrt{2} + 3\sqrt{4} \cdot \sqrt{3}$
$\quad = 4 \cdot 3\sqrt{2} + 6\sqrt{3} + 2 \cdot 6\sqrt{2}$
$\qquad + 3 \cdot 2\sqrt{3}$
$\quad = 12\sqrt{2} + 6\sqrt{3} + 12\sqrt{2} + 6\sqrt{3}$
$\quad = 24\sqrt{2} + 12\sqrt{3}$

The perimeter is $\left(24\sqrt{2} + 12\sqrt{3}\right)$ in.

SECTION 9.5 (pages 708–710)

9. $\sqrt{2}\left(\sqrt{18} - \sqrt{3}\right)$

$\quad = \sqrt{2} \cdot \sqrt{18} - \sqrt{2} \cdot \sqrt{3}$
$\quad = \sqrt{36} - \sqrt{6}$
$\quad = 6 - \sqrt{6}$

21. $\left(4\sqrt{x} + 3\right)^2$

$\quad = \left(4\sqrt{x}\right)^2 + 2\left(4\sqrt{x}\right)(3) + 3^2$
$\quad = 16x + 24\sqrt{x} + 9$

39. $\dfrac{6\sqrt{3y}}{\sqrt{y^3}}$

$\quad = \dfrac{6\sqrt{3y} \cdot \sqrt{y}}{\sqrt{y^3} \cdot \sqrt{y}}$
$\quad = \dfrac{6\sqrt{3y^2}}{\sqrt{y^4}}$
$\quad = \dfrac{6y\sqrt{3}}{y^2}$
$\quad = \dfrac{6\sqrt{3}}{y}$

73. $\dfrac{\sqrt{x} - \sqrt{y}}{\sqrt{2x} + \sqrt{3y}}$

Multiply both the numerator and denominator by the conjugate of the denominator, $\sqrt{2x} - \sqrt{3y}$.

$$= \dfrac{\left(\sqrt{x} - \sqrt{y}\right)\left(\sqrt{2x} - \sqrt{3y}\right)}{\left(\sqrt{2x} + \sqrt{3y}\right)\left(\sqrt{2x} - \sqrt{3y}\right)}$$

$$= \dfrac{\sqrt{2x^2} - \sqrt{3xy} - \sqrt{2xy} + \sqrt{3y^2}}{\left(\sqrt{2x}\right)^2 - \left(\sqrt{3y}\right)^2}$$

$$= \dfrac{x\sqrt{2} - \sqrt{3xy} - \sqrt{2xy} + y\sqrt{3}}{2x - 3y}$$

SECTION 9.6 (pages 718–721)

21. $3\sqrt{z-1} = 2\sqrt{2z+2}$

$\left(3\sqrt{z-1}\right)^2 = \left(2\sqrt{2z+2}\right)^2$
    Square each side.

$9(z - 1) = 4(2z + 2)$

$9z - 9 = 8z + 8$
        Distributive property

$z = 17$ Subtract $8z$. Add 9.

A check confirms that 17 is a solution of the original equation.

Solution set: $\{17\}$

33. $\sqrt{z^2 + 12z - 4} + 4 - z = 0$

$\sqrt{z^2 + 12z - 4} = z - 4$
        Isolate the radical.

$\left(\sqrt{z^2 + 12z - 4}\right)^2 = (z - 4)^2$
        Square each side.

$z^2 + 12z - 4 = z^2 - 8z + 16$
        Simplify.

$20z = 20$

$z = 1$

Substituting 1 for z makes the left side of the original equation positive, but the right side is zero, so 1 is not a solution. It is extraneous.

Solution set: $\varnothing$

43. $\sqrt[4]{a + 8} = \sqrt[4]{2a}$

Raise each side to the fourth power.

$\left(\sqrt[4]{a + 8}\right)^4 = \left(\sqrt[4]{2a}\right)^4$

$a + 8 = 2a$

$8 = a$

A check confirms that 8 is a solution of the original equation.

Solution set: $\{8\}$

57. $\sqrt{2\sqrt{x} + 11} = \sqrt{4x + 2}$

$\left(\sqrt{2\sqrt{x} + 11}\right)^2 = \left(\sqrt{4x + 2}\right)^2$
        Square each side.

$2\sqrt{x} + 11 = 4x + 2$

$\left(2\sqrt{x} + 11\right)^2 = (4x + 2)^2$
        Square again.

$4(x + 11) = 16x^2 + 16x + 4$

$4x + 44 = 16x^2 + 16x + 4$

$16x^2 + 12x - 40 = 0$ Standard form

$4x^2 + 3x - 10 = 0$ Divide by 4.

$(x + 2)(4x - 5) = 0$ Factor.

$x + 2 = 0$ or $4x - 5 = 0$
        Zero-factor property

$x = -2$ or $x = \dfrac{5}{4}$

CHECK

Let $x = -2$ in the original equation.

$\sqrt{2\sqrt{-2} + 11} \overset{?}{=} \sqrt{4(-2) + 2}$

$\sqrt{2\sqrt{9}} \overset{?}{=} \sqrt{-8 + 2}$

$\sqrt{6} = \sqrt{-6}$ False

Let $x = \dfrac{5}{4}$ in the original equation.

$\sqrt{2\sqrt{\dfrac{5}{4}} + 11} \overset{?}{=} \sqrt{4\left(\dfrac{5}{4}\right) + 2}$

$\sqrt{2\sqrt{\dfrac{49}{4}}} \overset{?}{=} \sqrt{5 + 2}$

$\sqrt{2\left(\dfrac{7}{2}\right)} = \sqrt{7}$

$\sqrt{7} = \sqrt{7}$ ✓   True

Solution set: $\left\{\dfrac{5}{4}\right\}$

61. $(2w - 1)^{2/3} - w^{1/3} = 0$

$\sqrt[3]{(2w - 1)^2} - \sqrt[3]{w} = 0$
        Write with radicals.

$\sqrt[3]{(2w - 1)^2} = \sqrt[3]{w}$
        Add $\sqrt[3]{w}$.

$\left(\sqrt[3]{(2w - 1)^2}\right)^3 = \left(\sqrt[3]{w}\right)^3$
        Cube each side.

$(2w - 1)^2 = w$

$4w^2 - 4w + 1 = w$
        Square on the left.

$4w^2 - 5w + 1 = 0$
        Standard form

$(4w - 1)(w - 1) = 0$

$4w - 1 = 0$ or $w - 1 = 0$

$w = \dfrac{1}{4}$ or $w = 1$

A check confirms that $\dfrac{1}{4}$ and 1 are both solutions of the original equation.

Solution set: $\left\{\dfrac{1}{4}, 1\right\}$

65. Solve $V = \sqrt{\dfrac{2K}{m}}$ for K.

$V^2 = \left(\sqrt{\dfrac{2K}{m}}\right)^2$ Square each side.

$V^2 = \dfrac{2K}{m}$

$\dfrac{V^2 m}{2} = K$,   or $K = \dfrac{V^2 m}{2}$

        Multiply by $\dfrac{m}{2}$.

SECTION 9.7 (pages 728–730)

41. $\left[(7 + 3i) - (4 - 2i)\right] + (3 + i)$
        Work inside the brackets first.

$= \left[(7 - 4) + (3 + 2)i\right] + (3 + i)$

$= (3 + 5i) + (3 + i)$

$= (3 + 3) + (5 + 1)i$

$= 6 + 6i$

53. $(4 + 5i)^2$

$= 4^2 + 2(4)(5i) + (5i)^2$

$= 16 + 40i + 25i^2$

$= 16 + 40i + 25(-1)$ $i^2 = -1$

$= 16 + 40i - 25$

$= -9 + 40i$

79. $I = \dfrac{E}{R + (X_L - X_c)i}$

Substitute $2 + 3i$ for E, 5 for R, 4 for X_L, and 3 for X_c.

$I = \dfrac{2 + 3i}{5 + (4 - 3)i}$

$I = \dfrac{2 + 3i}{5 + i}$

$I = \dfrac{(2 + 3i)(5 - i)}{(5 + i)(5 - i)}$

$I = \dfrac{10 - 2i + 15i - 3i^2}{5^2 - i^2}$

$I = \dfrac{10 + 13i + 3}{25 + 1}$

$I = \dfrac{13 + 13i}{26}$

$I = \dfrac{13(1 + i)}{13 \cdot 2}$

$I = \dfrac{1 + i}{2}$

$I = \dfrac{1}{2} + \dfrac{1}{2}i$

CHAPTER 10 Quadratic Equations, Inequalities, and Functions

SECTION 10.1 (pages 749–750)

49.
$$(6k - 1)^2 = -8$$

$6k - 1 = \sqrt{-8}$ or $6k - 1 = -\sqrt{-8}$

Square root property

$6k - 1 = \sqrt{-4} \cdot \sqrt{2}$

or $6k - 1 = -\sqrt{-4} \cdot \sqrt{2}$

Product rule

$6k - 1 = 2i\sqrt{2}$ or $6k - 1 = -2i\sqrt{2}$

$\sqrt{-4} = 2i$

$6k = 1 + 2i\sqrt{2}$ or $6k = 1 - 2i\sqrt{2}$

Add 1.

$$k = \frac{1 + 2i\sqrt{2}}{6} \quad \text{or} \quad k = \frac{1 - 2i\sqrt{2}}{6}$$

Divide by 6.

$$k = \frac{1}{6} + \frac{2\sqrt{2}}{6}i \quad \text{or} \quad k = \frac{1}{6} - \frac{2\sqrt{2}}{6}i$$

Standard form

$$k = \frac{1}{6} + \frac{\sqrt{2}}{3}i \quad \text{or} \quad k = \frac{1}{6} - \frac{\sqrt{2}}{3}i$$

Lowest terms

Solution set: $\left\{\dfrac{1}{6} + \dfrac{\sqrt{2}}{3}i, \dfrac{1}{6} - \dfrac{\sqrt{2}}{3}i\right\}$

SECTION 10.2 (pages 757–758)

19.
$$x^2 + x - 1 = 0$$
$$x^2 + x = 1 \quad \text{Add 1.}$$

Take half of 1, the coefficient of x, square it, and add the result to each side.

$$x^2 + x + \frac{1}{4} = 1 + \frac{1}{4}$$

Add $\left[\frac{1}{2}(1)\right]^2 = \frac{1}{4}$.

$$\left(x + \frac{1}{2}\right)^2 = \frac{5}{4} \quad \text{Factor and add.}$$

$x + \dfrac{1}{2} = \sqrt{\dfrac{5}{4}}$ or $x + \dfrac{1}{2} = -\sqrt{\dfrac{5}{4}}$

$x + \dfrac{1}{2} = \dfrac{\sqrt{5}}{2}$ or $x + \dfrac{1}{2} = -\dfrac{\sqrt{5}}{2}$

$x = -\dfrac{1}{2} + \dfrac{\sqrt{5}}{2}$ or $x = -\dfrac{1}{2} - \dfrac{\sqrt{5}}{2}$

$x = \dfrac{-1 + \sqrt{5}}{2}$ or $x = \dfrac{-1 - \sqrt{5}}{2}$

Solution set: $\left\{\dfrac{-1 + \sqrt{5}}{2}, \dfrac{-1 - \sqrt{5}}{2}\right\}$

25.
$$3x^2 - 6x = 4$$
$$x^2 - 2x = \frac{4}{3} \quad \text{Divide by 3.}$$
$$x^2 - 2x + 1 = \frac{4}{3} + 1$$

Add $\left[\frac{1}{2}(-2)\right]^2 = 1$.

$$(x - 1)^2 = \frac{7}{3} \quad \text{Factor and add.}$$

$x - 1 = \sqrt{\dfrac{7}{3}}$ or $x - 1 = -\sqrt{\dfrac{7}{3}}$

$x - 1 = \dfrac{\sqrt{7}}{\sqrt{3}}$ or $x - 1 = -\dfrac{\sqrt{7}}{\sqrt{3}}$

$x - 1 = \dfrac{\sqrt{21}}{3}$ or $x - 1 = -\dfrac{\sqrt{21}}{3}$

Rationalize denominators by multiplying by $\dfrac{\sqrt{3}}{\sqrt{3}}$.

$x = 1 + \dfrac{\sqrt{21}}{3}$ or $x = 1 - \dfrac{\sqrt{21}}{3}$

$x = \dfrac{3}{3} + \dfrac{\sqrt{21}}{3}$ or $x = \dfrac{3}{3} - \dfrac{\sqrt{21}}{3}$

$x = \dfrac{3 + \sqrt{21}}{3}$ or $x = \dfrac{3 - \sqrt{21}}{3}$

Solution set: $\left\{\dfrac{3 + \sqrt{21}}{3}, \dfrac{3 - \sqrt{21}}{3}\right\}$

39. $-m^2 - 6m - 12 = 0$

Multiply each side by -1.

$m^2 + 6m + 12 = 0$

$m^2 + 6m = -12 \quad \text{Subtract 12.}$

Complete the square.

$$\left[\frac{1}{2}(6)\right]^2 = 3^2 = 9$$

$m^2 + 6m + 9 = -12 + 9$

$(m + 3)^2 = -3$

$m + 3 = \sqrt{-3}$ or $m + 3 = -\sqrt{-3}$

$m + 3 = i\sqrt{3}$ or $m + 3 = -i\sqrt{3}$

$m = -3 + i\sqrt{3}$ or $m = -3 - i\sqrt{3}$

Solution set: $\left\{-3 + i\sqrt{3}, -3 - i\sqrt{3}\right\}$

SECTION 10.3 (pages 765–766)

17.
$$\frac{x^2}{4} - \frac{x}{2} = 1$$

First clear fractions by multiplying each side by the LCD, 4.

$$4\left(\frac{x^2}{4} - \frac{x}{2}\right) = 4(1)$$
$$x^2 - 2x = 4$$
$$x^2 - 2x - 4 = 0$$

Here $a = 1$, $b = -2$, and $c = -4$.

Substitute in the quadratic formula.

$$x = \frac{-b \pm \sqrt{b^2 - 4ac}}{2a}$$

$$x = \frac{-(-2) \pm \sqrt{(-2)^2 - 4(1)(-4)}}{2(1)}$$

$$x = \frac{2 \pm \sqrt{4 + 16}}{2}$$

$$x = \frac{2 \pm \sqrt{20}}{2}$$

$$x = \frac{2 \pm 2\sqrt{5}}{2}$$

$$x = \frac{2(1 \pm \sqrt{5})}{2}$$

$$x = 1 \pm \sqrt{5}$$

Solution set: $\left\{1 + \sqrt{5}, 1 - \sqrt{5}\right\}$

19. $-2t(t + 2) = -3$

$-2t^2 - 4t = -3$

$-2t^2 - 4t + 3 = 0$

Here $a = -2$, $b = -4$, and $c = 3$.

$$t = \frac{-b \pm \sqrt{b^2 - 4ac}}{2a}$$

$$t = \frac{-(-4) \pm \sqrt{(-4)^2 - 4(-2)(3)}}{2(-2)}$$

$$t = \frac{4 \pm \sqrt{16 + 24}}{-4}$$

$$t = \frac{4 \pm \sqrt{40}}{-4}$$

$$t = \frac{4 \pm 2\sqrt{10}}{-4}$$

$$t = \frac{2(2 \pm \sqrt{10})}{-2 \cdot 2}$$

$$t = \frac{2 \pm \sqrt{10}}{-2} \cdot \frac{-1}{-1}$$

$$t = \frac{-2 \mp \sqrt{10}}{2}$$

$$t = \frac{-2 \pm \sqrt{10}}{2}$$

Solution set: $\left\{\dfrac{-2 + \sqrt{10}}{2}, \dfrac{-2 - \sqrt{10}}{2}\right\}$

SOLUTIONS

29. $x(3x + 4) = -2$

$3x^2 + 4x = -2$

$3x^2 + 4x + 2 = 0$

Here $a = 3$, $b = 4$, and $c = 2$.

$x = \dfrac{-b \pm \sqrt{b^2 - 4ac}}{2a}$

$x = \dfrac{-4 \pm \sqrt{4^2 - 4(3)(2)}}{2(3)}$

$x = \dfrac{-4 \pm \sqrt{16 - 24}}{6}$

$x = \dfrac{-4 \pm \sqrt{-8}}{6}$

$x = \dfrac{-4 \pm 2i\sqrt{2}}{6}$

$x = \dfrac{2(-2 \pm i\sqrt{2})}{2 \cdot 3}$

$x = \dfrac{-2 \pm i\sqrt{2}}{3}$

Solution set: $\left\{ -\dfrac{2}{3} + \dfrac{\sqrt{2}}{3}i, \ -\dfrac{2}{3} - \dfrac{\sqrt{2}}{3}i \right\}$

SECTION 10.4 (pages 775–778)

15. $\dfrac{3}{2x} - \dfrac{1}{2(x+2)} = 1$

Multiply by the LCD, $2x(x + 2)$.

$2x(x+2)\left(\dfrac{3}{2x} - \dfrac{1}{2(x+2)} \right)$

$= 2x(x+2) \cdot 1$

$3(x + 2) - x(1) = 2x(x + 2)$

$3x + 6 - x = 2x^2 + 4x$

$0 = 2x^2 + 2x - 6$

$0 = x^2 + x - 3$

Use $a = 1$, $b = 1$, and $c = -3$ in the quadratic formula.

$x = \dfrac{-b \pm \sqrt{b^2 - 4ac}}{2a}$

$x = \dfrac{-1 \pm \sqrt{1^2 - 4(1)(-3)}}{2(1)}$

$x = \dfrac{-1 \pm \sqrt{1 + 12}}{2}$

$x = \dfrac{-1 \pm \sqrt{13}}{2}$

Use a calculator to check both proposed solutions. Both solutions check.

Solution set: $\left\{ \dfrac{-1 + \sqrt{13}}{2}, \dfrac{-1 - \sqrt{13}}{2} \right\}$

23. Let x = Harry's average rate.

Then $x - 20$ = Sally's average rate.

	d	r	t
Harry	300	x	$\frac{300}{x}$
Sally	300	$x - 20$	$\frac{300}{x - 20}$

It takes Harry $1\dfrac{1}{4}$ hr, or $\dfrac{5}{4}$ hr, less time than Sally.

$\dfrac{300}{x} = \dfrac{300}{x - 20} - \dfrac{5}{4}$

Multiply by the LCD, $4x(x - 20)$.

$4x(x - 20)\left(\dfrac{300}{x} \right) = 4x(x - 20)\left(\dfrac{300}{x - 20} - \dfrac{5}{4} \right)$

$1200(x - 20) = 4x(300) - x(x - 20) \cdot 5$

$1200x - 24{,}000 = 1200x - 5x^2 + 100x$

$5x^2 - 100x - 24{,}000 = 0$

$x^2 - 20x - 4800 = 0$ Divide by 5.

$(x - 80)(x + 60) = 0$

$x - 80 = 0$ or $x + 60 = 0$

$x = 80$ or $x = -60$

Reject $x = -60$. Harry's average rate is 80 km per hr.

39. $m = \sqrt{\dfrac{6 - 13m}{5}}$

$m^2 = \dfrac{6 - 13m}{5}$

     Square each side.

$5m^2 = 6 - 13m$

     Multiply by 5.

$5m^2 + 13m - 6 = 0$ Standard form

$(5m - 2)(m + 3) = 0$ Factor.

$5m - 2 = 0$ or $m + 3 = 0$

     Zero-factor property

$m = \dfrac{2}{5}$ or $m = -3$

CHECK   If $m = \dfrac{2}{5}$, then $\dfrac{2}{5} = \sqrt{\dfrac{4}{25}}$. ✓

               True

If $m = -3$, then $-3 = \sqrt{9}$.

                False

Solution set: $\left\{ \dfrac{2}{5} \right\}$

57. $2 + \dfrac{5}{3k - 1} = \dfrac{-2}{(3k - 1)^2}$

Let $u = 3k - 1$, so $u^2 = (3k - 1)^2$.

$2 + \dfrac{5}{u} = -\dfrac{2}{u^2}$

Multiply by the LCD, u^2.

$u^2\left(2 + \dfrac{5}{u} \right) = u^2\left(-\dfrac{2}{u^2} \right)$

$2u^2 + 5u = -2$

$2u^2 + 5u + 2 = 0$

$(2u + 1)(u + 2) = 0$

$2u + 1 = 0$ or $u + 2 = 0$

$u = -\dfrac{1}{2}$ or $u = -2$

To find k, substitute $3k - 1$ for u.

$3k - 1 = -\dfrac{1}{2}$ or $3k - 1 = -2$

$3k = \dfrac{1}{2}$ or $3k = -1$

$k = \dfrac{1}{6}$ or $k = -\dfrac{1}{3}$

CHECK   If $k = \dfrac{1}{6}$, then $2 - 10 = -8$. ✓

                 True

If $k = -\dfrac{1}{3}$, then $2 - \dfrac{5}{2} = -\dfrac{1}{2}$. ✓

                 True

Solution set: $\left\{ -\dfrac{1}{3}, \dfrac{1}{6} \right\}$

SECTION 10.5 (pages 785–788)

29. Let x be the width of the sheet metal. Then the length is $2x - 4$.

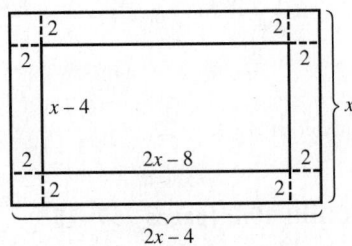

By cutting out 2-in. squares from each corner, we get a rectangle with width $(x - 4)$ in. and length

$(2x - 4) - 4 = (2x - 8)$ in.

The uncovered box then has height 2 in., length $(2x - 8)$ in., and width $(x - 4)$ in. Use the formula $V = LWH$ or $V = HLW$.

$256 = 2(2x - 8)(x - 4)$

$256 = 4(x - 4)(x - 4)$ Factor out 2.

$64 = (x - 4)^2$ Divide by 4.

$(x - 4)^2 = 64$

Use the square root property.

$x - 4 = 8$ or $x - 4 = -8$

$x = 12$ or $x = -4$

Since x represents width, discard the negative solution. The width is 12 in., and the length is

$2(12) - 4 = 20$ in.

37. Supply and demand are equal when

$3p - 410 = \dfrac{6000}{p}$.

To solve for p, multiply both sides by p.

$3p^2 - 410p = 6000$

$3p^2 - 410p - 6000 = 0$

Use the quadratic formula with $a = 3$, $b = -410$, and $c = -6000$.

$$p = \frac{-(-410) \pm \sqrt{(-410)^2 - 4(3)(-6000)}}{2(3)}$$

$$p = \frac{410 \pm \sqrt{168,100 + 72,000}}{6}$$

$$p = \frac{410 \pm \sqrt{240,100}}{6}$$

$$p = \frac{410 \pm 490}{6}$$

$$p = \frac{900}{6} = 150 \quad \text{or} \quad p = \frac{-80}{6} = -\frac{40}{3}$$

Discard the negative solution. The supply and demand are equal when the price is 150 cents, or \$1.50.

45. Write a proportion.

$$\frac{4}{x-4} = \frac{x-3}{3x-19}$$

Multiply by the LCD, $(x-4)(3x-19)$.

$$(x-4)(3x-19)\left(\frac{4}{x-4}\right)$$
$$= (x-4)(3x-19)\left(\frac{x-3}{3x-19}\right)$$

$$4(3x-19) = (x-4)(x-3)$$
$$12x - 76 = x^2 - 7x + 12$$
$$x^2 - 19x + 88 = 0$$
$$(x-8)(x-11) = 0$$
$$x - 8 = 0 \quad \text{or} \quad x - 11 = 0$$
$$x = 8 \quad \text{or} \qquad x = 11$$

If $x = 8$, then

$$3x - 19 = 3(8) - 19 = 5.$$

If $x = 11$, then

$$3x - 19 = 3(11) - 19 = 14.$$

Thus, $AC = 5$ or $AC = 14$.

SECTION 10.6 (pages 795–798)

33. $f(x) = -\frac{2}{3}(x+2)^2 + 1$

Because $a = -\frac{2}{3}$, the graph opens down and is wider than the graph of $y = x^2$. Because $h = -2$ and $k = 1$, the graph is shifted 2 units to the left and 1 unit up. The vertex is at $(-2, 1)$ and the axis of symmetry is $x = -2$. Two other points on the graph are $\left(-4, -\frac{5}{3}\right)$ and $\left(0, -\frac{5}{3}\right)$. We can substitute any value for x, so the domain is $(-\infty, \infty)$. The value of y is always less than or equal to 1, so the range is $(-\infty, 1]$.

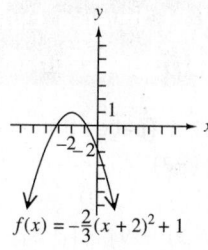

$$f(x) = -\frac{2}{3}(x+2)^2 + 1$$

SECTION 10.7 (pages 807–810)

21. $x = -\frac{1}{5}y^2 + 2y - 4$

The roles of x and y are reversed, so this is a parabola with a horizontal axis.

Step 1

Since $a = -\frac{1}{5}$ and $-\frac{1}{5} < 0$, the graph opens to the left. It is wider than the graph of $y = x^2$ because $\left|-\frac{1}{5}\right| = \frac{1}{5}$ and $\frac{1}{5} < 1$.

Step 2

The y-coordinate of the vertex is

$$\frac{-b}{2a} = \frac{-2}{2\left(-\frac{1}{5}\right)} = \frac{-2}{-\frac{2}{5}} = -2 \cdot \left(-\frac{5}{2}\right) = 5.$$

The x-coordinate of the vertex is

$$-\frac{1}{5}(5)^2 + 2(5) - 4 = -5 + 10 - 4 = 1.$$

Thus, the vertex is $(1, 5)$. Since the graph opens left, the axis of symmetry goes through the y-coordinate of the vertex—its equation is $y = 5$.

Step 3

To find the x-intercept, let $y = 0$. If $y = 0$, then $x = -4$, so the x-intercept is $(-4, 0)$. To find the y-intercepts, let $x = 0$.

$$0 = -\frac{1}{5}y^2 + 2y - 4$$

$$0 = y^2 - 10y + 20 \quad \text{Multiply by } -5.$$

Since $y^2 - 10y + 20$ does not factor, let $a = 1$, $b = -10$, and $c = 20$ in the quadratic formula.

$$y = \frac{-(-10) \pm \sqrt{(-10)^2 - 4(1)(20)}}{2(1)}$$

$$y = \frac{10 \pm \sqrt{20}}{2}$$

$$y = \frac{10 \pm 2\sqrt{5}}{2}$$

$$y = \frac{2(5 \pm \sqrt{5})}{2}$$

$$y = 5 \pm \sqrt{5}$$

The y-intercepts are approximately $(0, 7.2)$ and $(0, 2.8)$.

Step 4

For an additional point on the graph, let $y = 7$ (two units above the axis) to get $x = \frac{1}{5}$. So the point $\left(\frac{1}{5}, 7\right)$ is on the graph. By symmetry, the point $\left(\frac{1}{5}, 3\right)$ (two units below the axis) is on the graph.

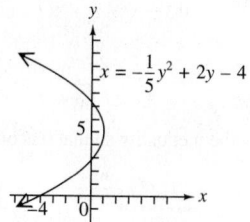

$$x = -\frac{1}{5}y^2 + 2y - 4$$

From the graph, we see that the domain is $(-\infty, 1]$ and the range is $(-\infty, \infty)$.

SECTION 10.8 (pages 817–820)

19. $x^2 - 6x + 6 \geq 0$

Solve the equation

$$x^2 - 6x + 6 = 0.$$

Since $x^2 - 6x + 6$ does not factor, let $a = 1$, $b = -6$, and $c = 6$ in the quadratic formula.

$$x = \frac{-(-6) \pm \sqrt{(-6)^2 - 4(1)(6)}}{2(1)}$$

$$x = \frac{6 \pm \sqrt{12}}{2}$$

$$x = \frac{6 \pm 2\sqrt{3}}{2}$$

$$x = \frac{2(3 \pm \sqrt{3})}{2}$$

$$x = 3 \pm \sqrt{3}$$

$$x = 3 + \sqrt{3} \approx 4.7 \quad \text{or}$$

$$x = 3 - \sqrt{3} \approx 1.3$$

$$\begin{array}{ccc} A & B & C \\ \hline & 3 - \sqrt{3} & \quad 3 + \sqrt{3} \end{array}$$

Test a number from each interval in the inequality

$$x^2 - 6x + 6 \geq 0.$$

Interval A: Let $x = 0$.

$$0^2 - 6(0) + 6 \overset{?}{\geq} 0$$

$$6 \geq 0 \quad \text{True}$$

Interval B: Let $x = 3$.

$$3^2 - 6(3) + 6 \overset{?}{\geq} 0$$

$$-3 \geq 0 \quad \text{False}$$

Interval C: Let $x = 5$.

$$5^2 - 6(5) + 6 \overset{?}{\geq} 0$$

$$1 \geq 0 \quad \text{True}$$

(continued)

The solution set includes the numbers in Intervals A and C, including $3 - \sqrt{3}$ and $3 + \sqrt{3}$ because equality is included in the symbol $\geq$.

Solution set:

$$\left(-\infty, 3 - \sqrt{3}\,\right] \cup \left[3 + \sqrt{3}, \infty\right)$$

37.

$$\frac{w}{w + 2} \geq 2$$

Write the inequality so that 0 is on one side.

$$\frac{w}{w + 2} - 2 \geq 0$$

$$\frac{w}{w + 2} - \frac{2(w + 2)}{w + 2} \geq 0$$

$$\frac{w - 2w - 4}{w + 2} \geq 0$$

$$\frac{-w - 4}{w + 2} \geq 0$$

The number -4 makes the numerator 0, and -2 makes the denominator 0. These two numbers determine three intervals.

Test a number from each interval in the inequality

$$\frac{w}{w + 2} \geq 2.$$

Interval A: Let $w = -5$.

$$\frac{-5}{-3} \overset{?}{\geq} 2$$

$$\frac{5}{3} \geq 2 \quad \text{False}$$

Interval B: Let $w = -3$.

$$\frac{-3}{-1} \overset{?}{\geq} 2$$

$$3 \geq 2 \quad \text{True}$$

Interval C: Let $w = 0$.

$$\frac{0}{2} \overset{?}{\geq} 2$$

$$0 \geq 2 \quad \text{False}$$

The solution set includes numbers in Interval B, including -4, but excluding -2 which makes the fraction undefined.

Solution set: $[-4, -2)$

CHAPTER 11 Inverse, Exponential, and Logarithmic Functions

SECTION 11.1 (pages 840–842)

9. $(f - h)(-3)$

$$= f(-3) - h(-3)$$

$$= \left[(-3)^2 - 9\right] - \left[(-3) - 3\right]$$

$$= (9 - 9) - (-6)$$

$$= 0 + 6$$

$$= 6$$

Alternatively, we could evaluate the polynomial in **Exercise 7**, $x^2 - x - 6$, using $x = -3$.

33. Since "profit equals revenue minus cost,"

$$P(x) = R(x) - C(x).$$

(a) Substitute $10.99x$ for $R(x)$ and $2.5x + 50$ for $C(x)$.

$$P(x) = 10.99x - (2.5x + 50)$$
$$\quad \text{Substitute for } R(x) \text{ and } C(x).$$

$$P(x) = 10.99x - 2.5x - 50$$
$$\quad \text{Distributive property}$$

$$P(x) = 8.49x - 50$$
$$\quad \text{Combine like terms.}$$

(b) If 100 t-shirts are produced and sold, let $x = 100$ in the function from part (a).

$$P(x) = 8.49x - 50 \quad \text{See part (a).}$$
$$P(100) = 8.49(100) - 50$$
$$\quad \text{Let } x = 100.$$
$$P(100) = 849 - 50 \quad \text{Multiply.}$$
$$P(100) = 799 \quad \text{Subtract.}$$

If 100 t-shirts are produced and sold, profit is $799.

51. $(f \circ h)\left(\dfrac{1}{2}\right) = f\left(h\left(\dfrac{1}{2}\right)\right) \quad \text{Definition}$

$$= f\left(\frac{1}{2} + 5\right) \quad h(x) = x + 5$$

$$= f\left(\frac{11}{2}\right) \quad 5 = \frac{10}{2}$$

$$= \left(\frac{11}{2}\right)^2 + 4 \quad f(x) = x^2 + 4$$

$$= \frac{121}{4} + \frac{16}{4} \quad \text{LCD} = 4$$

$$= \frac{137}{4} \quad \text{Add.}$$

61. $(f \circ g)(x) = f(g(x))$

$$= f(5280x) \quad g(x) = 5280x$$

$$= 12(5280x) \quad f(x) = 12x$$

$$= 63{,}360x \quad \text{Multiply.}$$

$(f \circ g)(x)$ computes the number of inches in x miles.

SECTION 11.2 (pages 848–850)

17. Write $g(x) = \sqrt{x - 3}$ as $y = \sqrt{x - 3}$.

Since $x \geq 3$, $y \geq 0$. The graph of g is half of a horizontal parabola that opens to the right. The graph passes the horizontal line test, so g is one-to-one. To find the inverse, interchange x and y to get

$$x = \sqrt{y - 3}.$$

Note that now $y \geq 3$, so $x \geq 0$. Solve for y.

$$x^2 = y - 3 \quad \text{Square each side.}$$

$$x^2 + 3 = y \quad \text{Add 3.}$$

Replace y with $g^{-1}(x)$.

$$g^{-1}(x) = x^2 + 3, \quad x \geq 0$$

27. (a) To find $f(0)$, substitute 0 for x.

$$f(x) = 2^x, \text{ so } f(0) = 2^0 = 1.$$

(b) Since f is one-to-one and $f(0) = 1$, it follows that $f^{-1}(1) = 0$.

41. $f(x) = y = x^3 - 2$

Complete the table of values.

x	y
-1	-3
0	-2
1	-1
2	6

Plot these points, and connect them with a solid smooth curve.

Interchange the values of x and y to make a table of values for f^{-1}.

x	y
-3	-1
-2	0
-1	1
6	2

Plot these points, and connect them with a dashed smooth curve. Use the fact that the graph of f^{-1} is symmetric to the graph of f with respect to the line $y = x$.

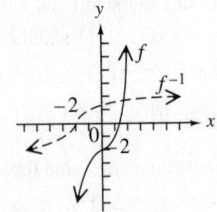

SECTION 11.3 (pages 857–858)

17. $16^{2x+1} = 64^{x+3}$

Write each side as a power of 4.

$$(4^2)^{2x+1} = (4^3)^{x+3}$$
$$4^{4x+2} = 4^{3x+9}$$

Set the exponents equal.

$$4x + 2 = 3x + 9$$
$$x = 7$$

CHECK $16^{2x+1} = 64^{x+3}$

$$16^{2(7)+1} \overset{?}{=} 64^{7+3} \qquad \text{Let } x = 7.$$
$$16^{15} \overset{?}{=} 64^{10}$$
$$(4^2)^{15} \overset{?}{=} (4^3)^{10}$$
$$4^{30} = 4^{30} \checkmark \qquad \text{True}$$

Solution set: $\{7\}$

21. $5^x = 0.2$

$$5^x = \frac{2}{10} \qquad \text{Write the decimal as a fraction.}$$

$$5^x = \frac{1}{5} \qquad \text{Write the fraction in lowest terms.}$$

$$5^x = 5^{-1} \qquad \text{Write with the same base.}$$

$$x = -1 \qquad \text{Set the exponents equal.}$$

Check by substituting -1 for x in the original equation.

Solution set: $\{-1\}$

SECTION 11.4 (pages 863–866)

45. $\log_6 \sqrt{216} = x$

$$6^x = \sqrt{216} \qquad \text{Write in exponential form.}$$

$$6^x = 216^{1/2}$$

$$6^x = (6^3)^{1/2} \qquad \text{Write with the same base.}$$

$$6^x = 6^{3/2} \qquad (a^m)^n = a^{mn}$$

$$x = \frac{3}{2} \qquad \text{Set the exponents equal.}$$

Solution set: $\left\{\dfrac{3}{2}\right\}$

67. $R = \log_{10} \dfrac{x}{x_0}$

Change to exponential form.

$$10^R = \frac{x}{x_0}, \quad \text{so} \quad x = x_0 \, 10^R.$$

Let $R = 6.7$ for the Northridge earthquake, x_1.

$$x_1 = x_0 10^{6.7}$$

Let $R = 7.3$ for the Landers earthquake, x_2.

$$x_2 = x_0 10^{7.3}$$

The ratio of x_2 to x_1 is

$$\frac{x_2}{x_1} = \frac{x_0 10^{7.3}}{x_0 10^{6.7}} = 10^{0.6} \approx 3.98.$$

The Landers earthquake was about 4 times as powerful as the Northridge earthquake.

SECTION 11.5 (pages 873–874)

15. $\log_3 \dfrac{\sqrt[3]{4}}{x^2 y}$

$$= \log_3 \frac{4^{1/3}}{x^2 y} \qquad \text{Write the radical expression with a rational exponent.}$$

$$= \log_3 4^{1/3} - \log_3 (x^2 y) \qquad \text{Quotient rule}$$

$$= \log_3 4^{1/3} - (\log_3 x^2 + \log_3 y) \qquad \text{Product rule}$$

$$= \log_3 4^{1/3} - \log_3 x^2 - \log_3 y$$

$$= \frac{1}{3}\log_3 4 - 2\log_3 x - \log_3 y \qquad \text{Power rule}$$

33. $3\log_p x + \dfrac{1}{2}\log_p y - \dfrac{3}{2}\log_p z - 3\log_p a$

$$= \log_p x^3 + \log_p y^{1/2} - \log_p z^{3/2} - \log_p a^3 \qquad \text{Power rule}$$

$$= \left(\log_p x^3 + \log_p y^{1/2}\right) - \left(\log_p z^{3/2} + \log_p a^3\right) \qquad \text{Group the terms into sums.}$$

$$= \log_p x^3 y^{1/2} - \log_p z^{3/2} a^3 \qquad \text{Product rule}$$

$$= \log_p \frac{x^3 y^{1/2}}{z^{3/2} a^3} \qquad \text{Quotient rule}$$

SECTION 11.6 (pages 879–880)

39. $p(h) = 86.3 \ln h - 680$

(a) $p(5000) = 86.3 \ln 5000 - 680$

$$p(5000) \approx 55 \qquad \text{Use a calculator.}$$

The percent of moisture at 5000 ft that falls as snow rather than rain is 55%.

(b) $p(7500) = 86.3 \ln 7500 - 680$

$$p(7500) \approx 90 \qquad \text{Use a calculator.}$$

The percent of moisture at 7500 ft that falls as snow rather than rain is 90%.

SECTION 11.7 (pages 887–889)

13.

$$2^{x+3} = 3^{x-4}$$

$$\log 2^{x+3} = \log 3^{x-4} \qquad \text{Property 3 (common logs)}$$

$$(x+3)\log 2 = (x-4)\log 3 \qquad \text{Power rule}$$

$$x\log 2 + 3\log 2 = x\log 3 - 4\log 3 \qquad \text{Distributive property}$$

$$x\log 2 - x\log 3 = -3\log 2 - 4\log 3 \qquad \text{Write } x\text{-terms on one side.}$$

$$x(\log 2 - \log 3) = -3\log 2 - 4\log 3 \qquad \text{Factor out } x.$$

$$x = \frac{-3\log 2 - 4\log 3}{\log 2 - \log 3} \qquad \text{Divide by } \log 2 - \log 3.$$

$$x \approx 15.967 \qquad \text{Use a calculator.}$$

Check that $2^{15.967+3} \approx 3^{15.967-4}$.

Solution set: $\{15.967\}$

23. $\ln e^{0.45x} = \sqrt{7}$

$$0.45x = \sqrt{7} \qquad \ln e^k = k$$

$$x = \frac{\sqrt{7}}{0.45} \qquad \text{Divide by 0.45.}$$

$$x \approx 5.879 \qquad \text{Use a calculator.}$$

Solution set: $\{5.879\}$

39. $\log 4x - \log(x-3) = \log 2$

$$\log \frac{4x}{x-3} = \log 2 \qquad \text{Quotient rule}$$

$$\frac{4x}{x-3} = 2 \qquad \text{Property 4}$$

$$4x = 2(x-3) \qquad \text{Multiply by } x-3.$$

$$4x = 2x - 6 \qquad \text{Distributive property}$$

$$2x = -6 \qquad \text{Subtract } 2x.$$

$$x = -3 \qquad \text{Divide by 2.}$$

Reject $x = -3$, which yields an equation in which the logarithms of negative numbers appear.

Solution set: $\emptyset$

CHAPTER 12 Nonlinear Functions, Conic Sections, and Nonlinear Systems

SECTION 12.1 (pages 908–910)

39. For any portion of the first ounce, the cost will be one \$0.45 stamp. If the weight exceeds one ounce (up to two ounces), an additional \$0.20 stamp is required. The following table summarizes the weight of a letter, x, and the number of stamps required, $p(x)$, on the interval $(0, 5]$.

x	$(0, 1]$	$(1, 2]$	$(2, 3]$	$(3, 4]$	$(4, 5]$
$p(x)$	1	2	3	4	5

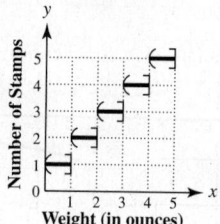

SECTION 12.2 (pages 917–920)

19.
$$3x^2 = 48 - 3y^2$$
$$3x^2 + 3y^2 = 48 \quad \text{Add } 3y^2.$$
$$x^2 + y^2 = 16 \quad \text{Divide by 3.}$$
$$x^2 + y^2 = 4^2$$

This is an equation of a circle with center $(0, 0)$ and radius 4. See the graph.

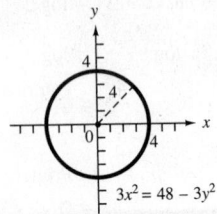

39. (a) $100x^2 + 324y^2 = 32,400$
$$\frac{x^2}{324} + \frac{y^2}{100} = 1 \quad \text{Divide by 32,400.}$$
$$\frac{x^2}{18^2} + \frac{y^2}{10^2} = 1$$

The height in the center is the y-coordinate of the upper y-intercept. The height is 10 m.

(b) The width of the ellipse is the distance between x-intercepts, $(-18, 0)$ and $(18, 0)$. The width across the bottom of the arch is
$$18 + 18 = 36 \text{ m}.$$

SECTION 12.3 (pages 926–928)

25.
$$y = \sqrt{\frac{x + 4}{2}}$$
$$y^2 = \frac{x + 4}{2} \quad \text{Square each side.}$$
$$2y^2 = x + 4 \quad \text{Multiply by 2.}$$
$$2y^2 - 4 = x \quad \text{Subtract 4.}$$
$$2(y - 0)^2 - 4 = x$$

This is a parabola that opens to the right with vertex $(-4, 0)$. However, y is non-negative in the original equation, so only the top half of the parabola is included in the graph.

x	y
-4	0
-2	1
0	$\sqrt{2}$
4	2

The domain is $[-4, \infty)$, and the range is $[0, \infty)$.

33. (a) $400x^2 - 625y^2 = 250,000$
$$\frac{x^2}{625} - \frac{y^2}{400} = 1 \quad \text{Divide by 250,000.}$$
$$\frac{x^2}{25^2} - \frac{y^2}{20^2} = 1$$

The x-intercepts are $(25, 0)$ and $(-25, 0)$. The distance between the buildings is the distance between the x-intercepts. The buildings are
$$25 + 25 = 50 \text{ m}$$
apart at their closest point.

(b) At $x = 50$, $y = \dfrac{d}{2}$, so $d = 2y$.
$$400(50)^2 - 625y^2 = 250,000$$
$$1,000,000 - 625y^2 = 250,000$$
$$-625y^2 = -750,000$$
$$y^2 = 1200$$
$$y = \sqrt{1200}$$

The distance d is
$$2\sqrt{1200} \approx 69.3 \text{ m}.$$

SECTION 12.4 (pages 934–935)

23.
$$2x^2 - y^2 = 6 \quad (1)$$
$$y = x^2 - 3 \quad (2)$$
Solve equation (2) for x^2.
$$x^2 = y + 3 \quad (3)$$
Substitute $y + 3$ for x^2 in equation (1).
$$2x^2 - y^2 = 6 \quad (1)$$
$$2(y + 3) - y^2 = 6$$
$$2y + 6 - y^2 = 6$$
$$0 = y^2 - 2y$$
$$0 = y(y - 2)$$
$$y = 0 \quad \text{or} \quad y = 2$$
Using equation (3) with $y = 0$,
$$x^2 = 3, \quad \text{so} \quad x = \pm\sqrt{3}.$$
Using equation (3) with $y = 2$,
$$x^2 = 5, \quad \text{so} \quad x = \pm\sqrt{5}.$$
All four ordered pairs check in *both* equations.
Solution set:
$$\{(-\sqrt{3}, 0), (\sqrt{3}, 0), (-\sqrt{5}, 2), (\sqrt{5}, 2)\}$$

35. Let $W =$ the width, and $L =$ the length.
The formula for the area of a rectangle is
$A = LW$, or $LW = A$.
$$LW = 84 \quad (1)$$
The perimeter of a rectangle is given by
$P = 2L + 2W$, or $2L + 2W = P$.
$$2L + 2W = 38 \quad (2)$$
Solve equation (2) for L to get
$$L = 19 - W. \quad (3)$$

Substitute $19 - W$ for L in equation (1).
$$LW = 84 \quad (1)$$
$$(19 - W)W = 84$$
$$19W - W^2 = 84$$
$$-W^2 + 19W - 84 = 0$$
$$W^2 - 19W + 84 = 0 \quad \text{Multiply by } -1.$$
$$(W - 7)(W - 12) = 0$$
$$W - 7 = 0 \quad \text{or} \quad W - 12 = 0$$
$$W = 7 \quad \text{or} \qquad W = 12$$

Using equation (3), with $W = 7$,
$$L = 19 - 7 = 12.$$

If $W = 12$, then $L = 7$, which are the same two numbers. Length must be greater than width, so the length is 12 ft and the width is 7 ft.

Index

A

Absolute value
  adding numbers with different signs, 57–58
  evaluating, 51
  explanation of, 51, 105
  simplifying roots using, 668
Absolute value equations
  explanation of, 611, 647
  solving, 612–616, 649–650
Absolute value functions, 904, 943
Absolute value inequalities
  explanation of, 611, 647
  solving, 613–616, 649–650
Addition
  associative property of, 87–88, 92, 98, 108
  with brackets, 58
  commutative property of, 87, 88, 92, 98, 108
  of complex numbers, 724–725, 734
  of decimals, 16
  of exponential expressions, 379
  of fractions, 6–7
  identity element for, 89, 106
  identity property of, 88–89, 92, 108
  inverse property of, 89–90, 92, 108
  of like terms, 368
  with number line, 56–57
  phrases that indicate, 59
  of polynomial functions, 835
  of polynomials, 370–371, 428, 954
  of radical expressions, 698–699, 733
  of rational expressions, 525–530, 577
  of real numbers, 56–59, 107
  of signed numbers, 57–58
  word phrases for, 59
Addition property
  of equality, 118–122, 204, 589
  of inequality, 189–190, 594
Additive identity element, 89, 106
Additive inverse, 50, 89–90
Agreement on domain, 288
Algebraic expressions
  distinguishing between equations and, 40, 80
  evaluating, 37–38
  examples of, 588
  explanation of, 37, 105
  simplifying, 98–100, 108
  symbols for inequality, 594
  translating word phrases into, 38–39, 59, 66–67, 79–80, 100, 140
Algebraic fractions. See Rational expressions
Alternative forms of logarithms, 871–872
Ampere, 573
Angles
  complementary, 153–156, 203
  measure of, 154–155, 165
  right, 153, 203
  straight, 153, 165, 203
  supplementary, 153–156, 203
  vertical, 165, 203
Apogee, 569
Applications
  of exponents, 420–423, 429
  of fractions, 9
  from geometry, 162–167
  of inequalities, 193–198
  of linear equations, 148–155, 205, 237, 268–289
  of linear systems, 340–345, 359
  with mixed numbers, 9
  of negative numbers, 47
  of percents, 178–179, 342–343
  of proportions, 177
  of Pythagorean theorem, 493–494
  of quadratic equations, 490–495, 503
  for scientific notation, 423
  steps to solve, 148–155, 490
Applied problems. See also Problem-solving strategies
  with distance, rate, and time, 553–556
  methods to solve, 803–804

  modeled by quadratic functions, 783–784
  Pythagorean theorem to solve, 782
  with work rate, 556–558
"Approximately equal to" symbol, 731
Approximations, for roots, 669
Area
  of circles, 169
  explanation of, 162, 203
  of rectangles, 162, 417
  of trapezoids, 162, 168
  of triangles, 163–164, 168, 674
Area formula
  explanation of, 162–164, 382
  use of, 162, 164, 417, 490–491
Associative property
  of addition, 87–88, 92, 98, 108
  distinguishing between commutative property and, 87–88
  of multiplication, 87–88, 92, 108
  use of, 87–88, 98
Asymptotes
  explanation of, 514, 942
  horizontal, 544, 575
  of hyperbola, 921–922, 942
  of reciprocal function, 904
  vertical, 544, 575
Augmented matrix, 960–961
Average rate of change, 252–253
Axes
  of a coordinate system, 221, 301
  of parabola, 789, 792, 821
  x-, 221, 301
  y-, 221, 301

B

Bar graphs
  examples of, 216
  explanation of, 216, 301
  interpretation of, 216
Base
  explanation of, 28, 105, 177, 377
  of exponents, 377
  of percentage, 177
Binomials
  explanation of, 369, 427
  factoring, 477–478
  factoring out as the greatest common factor, 444
  finding greater powers of, 395
  multiplication of, 387–388, 702–703
  square of, 392–393, 428, 715
Body mass index (BMI), 570
Boundary lines
  explanation of, 278, 301
  graphs of linear inequalities with vertical, 281
Brackets, 30, 58, 65–66
Breaching, 113
Break-even quantity, 348

C

Calculator tips
  for complex numbers, 727
  for logarithms, 875, 877–878
  for matrices, 960
  for negative numbers, 58
  for order of operations, 31
  for percents, 177
  for quadratic regression, 794
  for roots, 669
  for scientific notation, 422
  for square roots, 663, 665
  for systems of equations, 627
Cartesian coordinate system, 221, 301, 303
Celsius-Fahrenheit conversions, 67, 167, 202
Center
  of circle, 911–913, 942
  of ellipse, 914
Center-radius form of circle, 912

Change-of-base rule, 886, 893
Characteristic impedance, 717
Circle(s)
  center of, 911–913, 942
  center-radius form of, 912
  circumference of, 169, 565
  equation of, 911–912, 944
  explanation of, 942
  formulas for, 169
  graphs of, 911–913, 923
  radius of, 565, 911–913, 942
Circle graphs, 10
Circumference of a circle, 169, 565
Classifying polynomials, 369
Coefficients
  numerical, 98–99, 106, 368
  of polynomials, 368
Collinear points, 259
Columns, of matrix, 960
Combined variation, 570, 575
Combining like terms, 99–100, 122
Common denominators, 6. See also Least common denominators (LCDs)
Common factors
  explanation of, 2–3, 440
  factoring out binomial, 444
  factoring out negative, 443
  factoring trinomials with, 457, 464
  greatest. See Greatest common factor (GCF)
  looking for, 454
  quadratic equations with, 483
Common logarithms
  applications of, 875–877
  evaluating, 875
  explanation of, 875, 890, 892
Commutative property
  of addition, 87–88, 92, 98, 108
  distinguishing between associative property and, 88
  of multiplication, 87, 88, 92, 108
  use of, 87–88, 98
Complementary angles
  explanation of, 203
  solving problems involving, 153–155
Completing the square
  explanation of, 752, 822
  to find vertex, 799–800
  to graph horizontal parabolas, 805
  simplifying before, 756
  solving quadratic equations by, 751–756, 779
Complex conjugates, 726, 731
Complex fractions
  explanation of, 534, 575
  simplifying, 534–537, 577
Complex numbers
  addition of, 724–725, 734
  on calculators, 727
  conjugates of, 726
  division of, 726–727, 734
  explanation of, 724, 731, 734
  imaginary part of, 724
  multiplication of, 725–726, 734
  nonreal, 724, 731, 748, 755–756
  powers of i and, 727
  real part of, 724, 731
  simplifying, 722–723
  standard form of, 724, 731
  subsets of, 724
  subtraction of, 724–725, 734
Composite functions, 837–839, 890, 891
Composite numbers, 1
Composition. See Composite functions
Compound inequalities
  explanation of, 602, 647
  solving with and, 602–604, 649
  solving with or, 605–606, 649
Compound interest formulas, 884–885
Compound interest problems, 884–885
Conditional equations, 136, 203, 592–593, 647

Conic sections
    examples of, 789, 911
    explanation of, 911, 942
    summary of, 923
Conjugates, 393, 427, 706, 726, 731
Consecutive integers
    even, 152, 203, 491
    explanation of, 203
    odd, 152–153, 203, 491
    solving problems involving, 152–153, 491–493
Consistent systems, 317, 357, 627, 647
Constant, explanation of, 37, 105
Constant functions, 296, 302
Constant of variation, 565, 575
Consumer Price Index (CPI), 183, 784
Continuously compounded interest, 885
Contradictions, 137, 140, 203, 592–593, 647
Coordinate system, rectangular, 221, 301, 303
Coordinates
    explanation of, 47, 105, 221, 301
    of vertex, 789, 804
Cost-benefit equation, 880
Cost per unit, 174–175
Costs
    fixed, 274
    solving problems about, 341–342
    variable, 274
Counterexample, 902
Cross products
    explanation of, 175, 203
    solving equations using, 176, 552
Cube root function, 667, 732
Cube roots
    explanation of, 731
    finding, 665–666
Cubes
    difference of, 471–472, 477, 503
    factoring sum of, 472
    of numbers, 28
    perfect, 665
    sum of, 472, 477

D

Data, equations of lines that describe, 268
dB (decibel), 877
Decimal places, 17
Decimals
    addition of, 16
    converting percents and, 20–21
    division of, 17–19
    equivalents, 22
    explanation of, 15
    linear equations with, 128
    multiplication of, 17, 19
    place value in, 15
    repeating, 20
    rounding of, 19, 20
    solving equations with, 128, 139, 592
    solving linear systems with, 327–328
    subtraction of, 16
    writing fractions as, 19–20
    written as fractions, 15–16
Degree
    of polynomials, 369, 427
    symbol for, 153
    of a term, 369, 427
Denominators
    adding fractions with different, 7
    adding fractions with same, 6
    common, 6
    of complex fractions, 534–537
    explanation of, 1, 2–3
    least common. See Least common denominators (LCDs)
    of rational expressions, 525, 526–530
    rationalizing, 703–707, 731
Dependent equations
    explanation of, 317, 357, 627, 647
    recognizing, 964
    solving a linear system with, 326
    solving a system of, 631
    in three variables, 639
Dependent variable, 284, 301
Depreciation, 242, 897
Descartes, René, 221
Descending powers, 369, 427

Difference
    of cubes, 471–472, 477, 503
    of squares, 467–468, 477, 503
    in subtraction, 8, 39, 64, 105
Direct variation, 565–567, 575, 578
Discriminant
    explanation of, 802, 821, 822
    use of, 762–764, 802
Distance, rate, and time problems, 168, 343–345, 553–556
Distance formula, 690–691, 733
Distributive property
    explanation of, 90–92, 108
    to simplify terms in equations, 122
    to solve linear equations, 590–591
    to solve linear inequalities, 596
    to subtract rational expressions, 528–530
    use of, 99, 368
Diversity, index of, 897
Dividend, 17, 967
Division
    of complex numbers, 726–727, 734
    of decimals, 17–19
    definition of, 75–77
    of fractions, 5, 9
    of polynomial functions, 836
    of polynomials, 410–411, 414–417, 429, 967–969
    by powers of ten, 19
    of radical expressions, 703–707, 733–734
    of rational expressions, 520, 576
    of real numbers, 75–76, 108
    with scientific notation, 422–423
    of signed numbers, 76–77
    synthetic, 967–970
    word phrases for, 79–80
    by zero, 76
Divisors, 17, 967
Domain
    agreement on, 288
    explanation of, 286–287, 302, 304
    of functions, 286–287, 302, 304
    of rational equations, 540, 575
    of rational expressions, 514–515, 540, 575
    of rational functions, 514–515
    of relations, 286–287, 302, 304
Double negative rule, 50
Doubling time of money, 880, 884

E

e, 877
Earthquake intensities, 438
Elements
    of a matrix, 960
    of a set, 46
Elimination method
    explanation of, 331
    solving linear systems with, 331–334, 359, 629–631, 636–638, 650–651
    solving nonlinear systems with, 930–931
Elimination number, 116
Ellipse(s)
    center of, 914
    equation of, 914, 944
    explanation of, 911, 942
    foci of, 914
    graphs of, 914–916, 923, 925
    intercepts of, 914
Empty set, 137
Equality
    addition property of, 118–122, 204, 589
    multiplication property of, 126–129, 204, 589
    symbols for, 32, 39–40
Equation(s)
    absolute value, 611–616, 647, 649–650
    of circle, 911–912, 944
    conditional, 136, 203, 592–593, 647
    contradiction, 137, 140, 592–593, 647
    cost-benefit, 880
    dependent. See Dependent equations
    of ellipse, 914, 944
    equivalent, 118, 126, 203, 588, 647
    explanation of, 39, 105
    exponential. See Exponential equations
    expressions vs., 40, 80, 548, 588
    factoring to solve, 779

first-degree, 647
graphs of, 924
of horizontal lines, 265–266
of hyperbola, 921, 944
identifying functions from, 289
identity, 137, 140, 592–593, 647
independent, 317, 357, 627
with infinitely many solutions, 136–137
of inverse function, 845–846
linear in one variable. See Linear equations in one variable
linear in two variables. See Linear equations in two variables
linear systems of, 626, 635, 647
of lines, 248–249, 260–269
logarithmic. See Logarithmic equations
mathematical models as, 44, 222–223, 237, 268–269, 494–495
with no solutions, 136–137
nonlinear, 588, 929, 942
nonlinear systems of, 929–933, 942, 945
power rules for, 713–717, 731
quadratic. See Quadratic equation(s)
radical. See Radical equations
rational, 540–543, 577–578
second-degree, 744, 821, 923, 930–931
simplifying terms in, 122, 129
solution of, 39, 105, 118
solving literal, 166–167
solving using cross products, 176
solving with decimals, 139
solving with fractions, 138–139
square root property of, 744–748
systems of. See Systems of equations
translating word sentences to, 40, 80
variation, 565–570
of vertical lines, 265–266
working, 636
Equilibrium demand, 322
Equilibrium supply, 322
Equivalent equations, 118, 126, 203, 588, 647
Equivalent forms for positive numbers, 77
Equivalent inequalities, 594, 647
Estimation, of decimals, 19
Euler, Leonhard, 877
Even consecutive integers, 152–153, 203, 491–493
Even index, 665
Exponent(s)
    applications of, 420–423, 429
    base of, 377
    explanation of, 28, 105, 377
    integer, 399–405, 429
    negative, 400–402, 429, 537
    negative-to-positive rules for, 953
    positive, 402
    power rules for, 379–382, 428, 953
    product rule for, 377–379, 381–382, 428, 953
    quotient rule for, 402–404, 429, 953
    rational, 675–680, 732
    rules for, 404, 953
    scientific notation and, 420–423
    zero, 399, 429, 953
Exponential equations
    explanation of, 853, 890
    growth and decay problems, 885–886
    properties of, 881
    solving, 853–855, 881–882, 893
Exponential expressions
    addition vs. multiplication of, 379
    explanation of, 28, 105, 377, 427
Exponential form to logarithmic form, 859
Exponential functions
    applications of, 855–856
    explanation of, 851, 891
    graphs of, 851–853, 891
Exponential growth and decay problems, 855–856, 885–886
Expressions
    algebraic. See Algebraic expressions
    equations vs., 40, 80, 548, 588
    evaluating, 78, 107
    exponential. See Exponential expressions
    radical. See Radical expressions
    rational. See Rational expressions
    with related unknown quantities, 140
    simplifying, 98–100, 108

Extraneous solutions, 713, 731
Extremes of a proportion, 175, 203

F

f (x) notation. *See* Function notation
Factor tree, 2
Factored form
  explanation of, 440, 501
  prime, 2, 440
Factoring
  binomials, 477–478
  difference of cubes, 471–472, 477
  difference of squares, 467–468, 477, 503
  explanation of, 440, 501
  by grouping, 444–446, 457–458, 502
  perfect square trinomials, 469–470, 478, 503
  polynomials, 442–446, 450–454, 457–458, 461–464, 467–472, 477–478, 955–956
  to solve equations, 779
  to solve quadratic equations, 481–486, 503
  sum of cubes, 472, 477
  trinomials, 450–454, 457–458, 461–464, 478, 501–503
Factors
  common. *See* Common factors
  distinguishing between terms and, 98
  explanation of, 1, 501
  greatest common. *See* Greatest common factor (GCF)
  of numbers, 1, 440
Fahrenheit-Celsius conversions, 67, 167, 202
First-degree equations, 647. *See also* Linear Equations; Linear equations in one variable; Linear equations in two variables
Fixed cost, 274
Foci (*sing: Focus*)
  of ellipse, 914
  of hyperbola, 921
Focus variable, 636
FOIL method
  explanation of, 387, 427, 428, 702
  factoring trinomials using, 450, 461–464, 503
  inner product of, 387, 427
  multiplying binomials by, 387–388, 428
  multiplying complex numbers by, 725
  outer product of, 387, 427
  squaring by, 392
Formula(s)
  for body mass index, 570
  for circumference of circle, 169, 565
  compound interest, 884–885
  distance, 690–691, 733
  to evaluate variables, 162
  explanation of, 203
  finding values of unknown variables in, 550
  from geometry, 162–167
  involving squares and square roots, 781–782
  for Pythagorean theorem, 493–494, 503, 689–690, 733, 782, 902
  quadratic, 759–764, 821, 822
  rotational rate N of space station, 721
  slope, 244–247
  solving for specified variables in, 164, 166–169, 205, 550–551
  vertex, 800
Fourth root, 665–666
Fraction(s)
  addition of, 6–7, 9
  applications of, 9–10
  complex, 534–537, 575, 577
  division of, 5, 9
  equivalents, 22
  explanation of, 1
  improper, 1, 3–4
  linear equations with, 127, 591
  lowest terms of, 2–3, 89
  multiplication of, 4
  proper, 1
  reciprocals of, 5
  simplifying, 2–3
  solving equations with, 138–139
  solving linear systems with, 327
  subtraction of, 6–8
  writing decimals as, 15–16
  written as decimals, 19–20

Fraction bar, 1, 29–30
Froude, William, 788
Froude number, 788
Function(s)
  absolute value, 904, 943
  composite, 837–839, 890, 891
  concept of, 376
  constant, 296, 302
  cube root, 667, 732
  defined by radical expressions, 666–667, 732
  domain of, 286–287, 302, 304
  equation of the inverse, 845–846
  evaluating, 293–294
  explanation of, 284–286, 301, 304, 489
  exponential, 851–856, 891
  exponential decay, 855–856
  exponential growth, 855–856
  greatest integer, 906–907, 942
  identifying from equations, 289
  identifying from graphs, 287–288
  inverse. *See* Inverse functions
  linear, 296, 302, 304
  logarithmic, 861–862, 892
  one-to-one, 843–848, 890, 891
  operations on, 834–836, 891
  polynomial. *See* Polynomial functions
  quadratic. *See* Quadratic functions
  range of, 286–287, 302, 304
  rational. *See* Rational functions
  reciprocal, 543, 904, 943
  relations as, 285–286
  square root, 666, 732, 904–905, 924–925, 944
  step, 907, 942
  translation of, 904–906
  variations in definition of, 289
  vertical line test for, 287–288
Function notation
  explanation of, 304
  use of, 293–295
Fundamental property of rational numbers, 515–518
Fundamental rectangle of hyperbola, 921–922, 942

G

Gains and losses, 59
Galilei, Galileo, 481, 489, 746
GCF. *See* Greatest common factor (GCF)
Geometry
  formulas and applications from, 162–166
  of systems of linear equations, 635–636
Girth, 574
Googol, 425
Graph(s)
  of absolute value function, 904
  bar, 216, 301
  circle, 10
  of circles, 911–913, 923
  of cube root functions, 666–667
  of elementary functions, 904–906
  of ellipses, 914–916, 923, 925
  of equations, 924
  explanation of, 231, 301
  of exponential functions, 851–853, 891
  of greatest integer functions, 907, 943
  of hyperbolas, 921–922
  identifying functions from, 287–288
  of inequalities, 188–189
  interpretation of, 216–217
  of intervals on number lines, 188–189
  of inverse functions, 847
  line, 217–218, 235–236, 261–262, 301
  of linear equations, 231–237, 303, 626–627
  of linear inequalities, 188–189, 278–281, 304, 352–354
  of linear systems, 314–318, 626–627, 635–636
  of logarithmic functions, 861–862
  of numbers, 47–48
  of ordered pairs, 222
  of parabolas, 790–792, 804–806, 823–824, 923
  pie, 10
  of quadratic functions, 789–794, 801, 823
  of radical functions, 666–667
  of rational functions, 543–544, 575, 578
  of rational numbers, 48
  of reciprocal functions, 904
  of relations, 287–288
  of second-degree inequalities, 936–939

of semicircles, 925
  to solve quadratic inequalities, 811–812
  of square root functions, 666, 904–905, 924–925, 943
  of systems of inequalities, 937–939, 945
Greater than, 31–32, 188
Greater than or equal to, 31–32, 188
Greatest common factor (GCF)
  explanation of, 440, 501
  factoring out, 442–444, 454
  for numbers, 440–441
  steps to find, 442, 502
  of variable terms, 441–442
Greatest integer functions
  explanation of, 942
  graphs of, 907, 943
  method for applying, 906–907
Grouping
  factoring by, 444–446, 502
  factoring trinomials by, 457–458
  with negative signs, 446
Grouping symbols
  explanation of, 29–30
  subtraction with, 65–66
Growth and decay applications, 855–856, 885–886

H

Half-life, 886
Heron's formula, 674
Horizontal asymptotes, 544, 575
Horizontal line test, 845, 891
Horizontal lines
  equations of, 265–267, 304
  graphs of, 236
  slope of, 247–248
Horizontal parabolas, 804–806, 824
Horizontal shift
  of ellipse, 916
  explanation of, 791
  method for applying, 905–906
  of parabola, 790–792
Hyperbola(s)
  asymptotes of, 921–922, 942
  equations of, 921, 944
  explanation of, 911, 942
  foci of, 921
  fundamental rectangle of, 942
  graphs of, 921–922
  intercepts of, 921
Hypotenuse
  explanation of, 689
  of a right triangle, 493–494, 501

I

i, 722, 727, 731
Identities, 137, 140, 203, 592–593, 647
Identity element
  for addition, 89, 106
  for multiplication, 89, 106
Identity equations, 137, 140
Identity properties
  of addition, 88–89, 92, 108
  of multiplication, 88–89, 92, 108
Imaginary numbers, 724, 731
Imaginary part, 724, 731
Imaginary unit, 722, 727, 731
Impedance, characteristic, 717
Improper fractions, 1, 3–4
Inconsistent systems
  explanation of, 317, 357, 627, 647
  solving, 631, 632
  substitution method to solve, 325–326
Independent equations, 317, 357, 627
Independent variable, 284, 301
Index
  of diversity, 897
  of a radical, 665, 689, 698, 717, 731
Inequality(ies)
  absolute value, 611–616, 647, 649–650
  addition property of, 189–190, 594
  compound, 602–606, 647, 649
  equivalent, 594, 647
  explanation of, 203, 594, 647
  graphs of, 188–189
  linear in one variable, 188, 203, 206, 594–596
  linear in two variables, 304, 352–354

Inequality(ies) (continued)
    multiplication property of, 191–192, 595
    polynomial, 814
    quadratic, 811–814, 821, 824
    rational, 815–816, 821, 824
    second-degree, 936–939, 942, 945
    solutions to, 192–198
    symbols of, 31, 32, 188, 278
    systems of, 937–939, 945
    three-part, 194–196, 203, 597
Infinity symbol, 188
Inner products, 387, 427
Integers
    consecutive, 152–153, 203, 491–493
    explanation of, 46, 105
    as exponents, 399–405, 429
Intensities, 438
Intercepts
    of ellipse, 914
    of hyperbola, 921
    of a line, 232–234
    method for finding, 232
Interest
    compound, 884–885
    formula for, 169
Intersection of sets, 602, 606, 647
Interval notation, 188, 203
Intervals, 188–189, 203
Inverse functions
    equations of, 845–846
    explanation of, 843–844, 890, 891
    graphs of, 847
    horizontal line test and, 845
    notation for, 890
Inverse properties, 89–90, 92, 108
Inverse variation, 567–569, 575, 578
Irrational numbers, 48, 105, 664, 731
Irrational square roots, 664–665

J

Joint variation, 569–570, 575, 578

L

Least common denominators (LCDs)
    in equations, 137–139
    explanation of, 6, 526, 575
    method to find, 526
Legs of a right triangle, 493–494, 501, 689
Leonardo da Pisa, 662
Less than
    interpretation of, 66–67, 279
    symbol for, 31–32, 188
Less than or equal to, 31–32, 188
Light-year, 435
Like terms
    addition of, 368
    combining, 99–100, 122
    explanation of, 99, 106, 427
Line graphs, 217–218, 235–236, 261–262, 301
Linear equations in one variable
    applications of, 148–155, 205
    with decimals, 128, 592
    explanation of, 118, 203, 588, 647
    with fractions, 127, 591
    simplifying terms in, 129
    solution set of, 588
    solutions to, 133–140, 205
    solving, 588–593, 648
    types of, 592–593
Linear equations in three variables, 635–642
Linear equations in two variables
    applications of, 237, 268
    explanation of, 218, 301
    graphs of, 231–237, 303
    intercepts of, 233–234
    slope of a line, 244–253, 303
    standard form of, 264–265, 267
    systems of, 626–632
Linear functions, 296, 302, 304
Linear inequalities in one variable
    addition property of, 594
    explanation of, 188, 203, 594, 647
    graphs of, 188–189
    multiplication property of, 595

solutions to, 206
solving, 649
steps to solve, 596
with three parts, 597
Linear inequalities in two variables
    boundary line of, 278, 281
    explanation of, 278, 301
    graphs of, 278–281, 304
    system of, 352–354
Linear systems. See Systems of linear equations
Linear systems of equations, 626–627, 635, 639, 647.
    See also Systems of linear equations; Systems of
    linear equations in three variables; Systems of
    linear equations in two variables
Lines. See also Number lines
    boundary, 278, 301
    equations of, 249, 260–269
    graphs of, 217–218, 235–236, 261–262, 301
    horizontal, 236, 247–248, 265–266, 304
    parallel, 250–251, 266–267, 301
    perpendicular, 250–251, 266–267, 301
    slopes of, 244–253, 303
    vertical, 236, 248, 265–266, 304
Literal equations, 166–167
Lithotripter, 920
Logarithmic equations
    explanation of, 859, 890
    properties of, 881
    solving, 860–861, 882–883, 893
Logarithmic form to exponential form, 859
Logarithmic functions
    applications of, 862
    with base a, 861–862
    explanation of, 861, 892
    graphs of, 861–862, 892
Logarithms
    alternative forms of, 871–872
    on calculators, 875, 877–878
    change-of-base rule for, 886, 893
    common, 875–877, 890, 892
    explanation of, 859, 890
    natural, 877–878, 890, 892
    power rules for, 869–871, 892
    product rule for, 867–868, 871, 892
    properties of, 861, 867–872, 892
    quotient rule for, 868, 871, 892
Loudness of sound, 877
Lowest terms
    of fractions, 2–3, 89
    of rational expressions, 515–518, 576, 707

M

Magic number, 116
Math in the Media
    magic number in sports, 116
    Pythagorean theorem, 902
    reforming the grade system, 660
    Richter scale, 438
Mathematical expressions. See Algebraic expressions
Mathematical models
    explanation of, 44
    with linear equations, 222–223, 237, 268–269
    quadratic, 494–495
Matrices
    augmented, 960–961
    on calculators, 960
    columns of, 960
    elements of, 960
    explanation of, 960
    row echelon form of, 961
    row operations on, 961–964
    rows of, 960
    square, 960
Matrix method for solving systems, 960–964
Maximum value problems, 803–804
Means of a proportion, 175, 203
Minimum value problems, 803–804
Minuend, 64, 105
Mixed numbers
    applications with, 8–10
    converting between improper fractions and, 3–4
    explanation of, 3
Mixture problems, 342–343
Models. See Mathematical models
Money, doubling time of, 880, 884

Monomials
    dividing polynomials by, 410–411, 429
    explanation of, 369, 427
Motion problems, 343–345, 767–768
Multiplication
    associative property of, 87, 88, 92, 108
    of binomials, 387–388, 702–703
    commutative property of, 87, 88, 92, 108
    of complex numbers, 725–726, 734
    of decimals, 17, 19
    of exponential expressions, 379
    FOIL method of, 387–388, 428
    of fractions, 4–5
    identity element for, 89, 106
    identity property of, 88–89, 92, 108
    inverse property of, 89–90, 92, 108
    of polynomial functions, 836
    of polynomials, 385–388, 428, 954–955
    by powers of ten, 19
    properties of, 73–74
    of radical expressions, 702–707, 733–734
    of rational expressions, 518–519, 576
    of real numbers, 73–74, 108
    with scientific notation, 422–423
    of signed numbers, 73–74
    word phrases for, 79
Multiplication property
    of equality, 126–129, 204, 589
    of inequality, 191–192, 595
    of zero, 73
Multiplicative identity element, 89, 106
Multiplicative inverses, 75, 89–90. See also Reciprocals
Multivariable polynomials, 372

N

Natural logarithms
    applications of, 878
    evaluating, 877–878
    explanation of, 877, 890, 892
Natural numbers, 46, 105
Negative exponents
    changing to positive exponents, 402
    explanation of, 400–401, 429, 953
    simplifying rational expressions with, 537
Negative infinity symbol, 188
Negative numbers
    addition with, 56–57
    applications of, 47
    calculator tip for, 58
    explanation of, 46, 105
    multiplication with, 73–74
    square roots of, 722–723
Negative slope, 247
Negative square roots, 662, 664
Negative-to-positive rules, 429
Nonlinear equations, 588, 929, 942
Nonlinear systems of equations
    explanation of, 929, 942
    solving, 929–933, 945
Nonreal complex numbers, 724, 731, 748, 755–756
Not equal, 31
Notation
    function, 293–295, 304
    interval, 188, 203
    scientific. See Scientific notation
    set-builder, 46, 105
    subscript, 245–246, 301
nth roots, 665–666, 668, 675
Null set, 137
Number(s)
    absolute value of, 51, 106
    complex, 722–727, 731, 734
    composite, 1
    factored form of, 440
    factors of, 1
    fraction. See Fractions
    Froude, 788
    graphs of, 47–48
    greatest common factor of, 440–442
    imaginary, 731
    integer, 46, 105
    irrational. See Irrational numbers
    mixed, 1, 3–4, 8–9
    natural, 46, 105
    negative. See Negative numbers

nonreal complex, 724, 731, 748, 755–756
opposite of, 46, 50
ordering of, 49, 107
positive. *See* Positive numbers
prime, 1–2
prime factored form of, 2
rational, 47, 105
real. *See* Real numbers
signed. *See* Signed numbers
whole, 1, 46, 105, 414–415
Number lines
addition with, 50, 56–57
explanation of, 46, 105
graphing intervals on, 188–189
irrational numbers on, 48
linear inequalities on, 188–189
ordering real numbers on, 49, 107
subtraction with, 64
Numerators, 1, 2–3
Numerical coefficients, 98–99, 106, 368

O

Odd consecutive integers, 153, 203, 491
Ohm's law, 730
One-to-one functions
explanation of, 843–844, 890, 891
horizontal line test for, 845
inverse of, 843–848
Operations
on functions, 834–836, 891
order of, 29–31, 107
set, 602, 604, 606, 649
with signed numbers, 56–58, 64–66, 73–77
Opposite
explanation of, 46, 89, 105
of a number, 46, 50
of real numbers, 50, 89
Opposites, quotients of, 518
Order
of operations, 29–31, 77, 107
of radicals, 665
of real numbers, 49, 107
Ordered pairs
completing, 219
explanation of, 218, 301, 303
graphs of, 222–223
plotting, 221–223, 231–232, 301
solutions as, 218–219, 314–315
as solutions of linear systems, 626
Ordered triple, 635
Origin
explanation of, 221, 301
graphing inequalities with boundary lines through, 281
graphing line through, 235
Outer products, 387, 427

P

Pairs, ordered. *See* Ordered pairs
Parabola(s)
applications of, 803–804
axis of, 789, 792, 821
explanation of, 789, 821
graphs of, 790–792, 804–806, 824, 923
horizontal, 804–806, 824
horizontal shift of, 790–792
symmetry of, 789
vertex of, 789, 792, 799–800, 802, 821
vertical, 799–800, 802, 806
vertical shift of, 790–792
x-intercepts, number of, 802
Parallel lines
explanation of, 301
slopes of, 250–251
writing equations of, 266–267
Parentheses, 92
Percents and percentages
applications of, 178–179, 342–343
base of, 177
calculator tip for, 177
converting decimals and, 20–21
equivalents, 22
explanation of, 177
solutions to, 206
Perfect cubes, 665

Perfect fourth powers, 665
Perfect square trinomials, 469–470, 478, 501, 503
Perfect squares, 469–470, 664, 731
Perigee, 569
Perimeter
explanation of, 13, 163, 203
formulas for, 162, 168
Perpendicular lines
explanation of, 301
slopes of, 250–251
writing equations of, 266–267
pH, 875–876
Pi (π), 48
Pie chart, 10
Place value in decimals, 15
Plane, 221, 301
Plotting ordered pairs, 221–223, 231–232, 301
Point-slope form, 263–264, 267, 304
Polynomial(s)
addition of, 370–371, 428, 954
binomial, 369
classifying, 369
coefficients of, 368
degree of, 369, 427
descending powers of, 369, 427
divided by monomials, 410–411, 429
divided by polynomials, 414–417, 429
division of, 410–411, 414–417, 429, 967–969
evaluating, 370
explanation of, 427
factoring, 442–446, 450–454, 457–458, 461–464, 467–472, 477–478, 955–956
monomial, 369
multiplication of, 385–388, 428, 954–955
multivariable, 372
prime, 453, 501
subtraction of, 371–372, 428, 954
synthetic division of, 967–969
terms of, 416–417
trinomial, 369
in *x*, 369
Polynomial functions
addition of, 835
division of, 836
evaluating, 376
multiplication of, 836
subtraction of, 835
Polynomial inequalities, 814
Positive exponents, 402
Positive numbers
adding with negative numbers, 58
equivalent forms for, 77
explanation of, 46, 105
subtracting, 64
writing in scientific notation, 421
Positive or negative (plus or minus) symbol, 731
Positive slope, 247
Positive square root, 662
Power(s). *See also* Exponents
descending, 369, 427
explanation of, 28, 105, 377
of *i*, 727
Power rules
for exponents, 379–382, 428
for logarithms, 869–871, 892
for radical equations, 713–717, 731
Powers of ten
division by, 19
explanation of, 15
multiplication by, 19
Prime factored form, 2, 440–441
Prime numbers, 1–2
Prime polynomials, 453, 501
Principal square root, 662, 731
Problem-solving strategies, 148–155. *See also*
Applied problems
Product
explanation of, 1, 73, 105
factored form as, 448
of sum and difference of two terms, 393–395
Product rule
for exponents, 377–379, 381–382, 428, 953
for logarithms, 867–868, 871, 892
for radicals, 685, 722, 733
Proper fractions, 1
Properties of one, 2, 3

Proportions
applications of, 177
cross products of, 175–176, 203
explanation of, 175–176, 203, 551, 575
extremes of, 175, 203
finding unknown in, 176
means of, 175, 203
solutions of, 176, 206
solving applications using, 551–552
terms of, 175
Pure imaginary numbers, 724, 731
Pyramids, 169
Pythagoras, 493
Pythagorean theorem
explanation of, 493, 503
use of, 493–494, 689–690, 733, 782, 902

Q

Quadrants, 221, 301
Quadratic equation(s)
applications of, 490–495, 503, 767–770
with common factor, 483
completing the square to solve, 751–756, 752–756, 779
discriminant of, 762–764, 802, 821, 822
explanation of, 481, 501, 744, 821
factoring to solve, 481–486, 503
methods for solving, 482–486, 779
quadratic formula to solve, 760–762
solving equation that leads to, 767
square root property to solve, 744–748
standard form of, 481, 501
Quadratic formula
derivation of, 759–760
explanation of, 821, 822
solving quadratic equations using, 760–762, 779
use of discriminant and, 762–764
Quadratic functions
applied problems modeled by, 783–784
discriminant of, 802
explanation of, 789, 821
graphs of, 789–794, 801, 823
Quadratic inequalities
explanation of, 811, 821
solving by graphing, 811–812
special cases of, 814
steps to solve, 813
using test numbers to solve, 812–813
Quadratic regression, 794
Quotient
explanation of, 5, 75, 105, 967
of opposites, 518
radical, 707
translating words and phrases, 79
Quotient rule
for exponents, 402–403, 429, 953
for logarithms, 868, 871, 892
for radicals, 686, 733

R

Radical equations
explanation of, 731
power rules for, 713–717, 731
Radical expressions
addition of, 698–699, 733
division of, 702–707, 733–734
explanation of, 662, 666, 731, 732
functions defined by, 666–667, 732
multiplication of, 702–707, 733–734
simplifying, 685–691, 698–699, 733
squaring, 663
subtraction of, 698–699, 733
Radical symbol, 731
Radicals
converting between rational exponents and, 678
explanation of, 662, 731
index of, 665, 689, 698, 717, 731
order of, 665
product rule for, 685, 722, 733
quotient rule for, 686, 733
simplifying, 686–689, 733
solving equations with, 713–717, 734, 770–771
Radicands, 662, 665, 731
Radius, 565, 911–913, 942

Range
  explanation of, 286–287, 302, 304
  of functions, 286–287, 302, 304
  of relations, 286–287, 302, 304
Rates of change, 252–253
Rational equations, 540–543, 577–578
Rational exponents
  converting between radicals and, 678
  evaluating, 675–677
  explanation of, 732
  negative, 676
  rules for, 679–680
Rational expressions
  addition of, 525–530, 577
  applications of, 550–558, 578
  division of, 520, 576
  domain of, 514–515, 540, 575
  equations with, 540–543, 577–578
  explanation of, 514, 575
  lowest terms of, 515–518, 576, 707
  multiplication of, 518–519, 576
  reciprocals of, 519–520
  simplifying with negative exponents, 537
  solving equations with, 540–543
  subtraction of, 525–530, 577
Rational functions
  discontinuous graphs of, 543, 575
  domain of, 514–515, 575
  explanation of, 514–515, 567, 575
  graphs of, 543–544, 578
Rational inequalities
  explanation of, 815, 821
  solving, 815–816, 824
Rational numbers, 47, 105, 515–518
Rationalizing the denominator
  with binomials involving radicals, 706–707
  explanation of, 731
  with one radical term, 703–705
Ratios
  explanation of, 174–175, 203, 551, 575
  writing, 174, 206
Real numbers. *See also* Numbers
  addition of, 56–59, 107
  division of, 75–76, 108
  explanation of, 48, 105
  multiplication of, 73–75, 108
  opposite of, 50, 89
  order of, 49, 107
  properties of, 87–92, 108
  reciprocals of, 75, 89
  subtraction of, 64–67, 108
Real part, 724, 731
Reciprocal function, 543, 904, 943
Reciprocals
  to apply definition of division, 75–77
  explanation of, 5, 75, 105
  of fractions, 5
  negative exponents and, 400–402
  of rational expressions, 519–520
  of real numbers, 75, 89
Rectangles
  area of, 162, 417
  length of, 163
  perimeter of, 164, 168
  width of, 163
Rectangular coordinate system, 221, 301, 303
Relation(s)
  domain of, 286–287, 302, 304
  explanation of, 284–286, 301, 304
  as functions, 285–286
  graphs of, 287–288
  range of, 286–287, 302, 304
Remainder theorem, 969
Repeating decimals, 20
Resonant frequency, 669
Richter scale, 438, 866
Right angles, 153, 203
Right triangle
  hypotenuse of, 493–494, 501
  legs of, 493–494, 501
Rise
  comparing run to, 244–246
  explanation of, 244–245, 301
Roots
  approximations for, 669
  on calculators, 669

cube, 665, 667, 731
  evaluating, 662–669
  fourth, 665–666
  *n*th, 668
  simplifying, 687
  square. *See* Square roots
Rotational rate N of space station formula, 721
Rounding, of decimals, 19, 20
Row echelon form of matrix, 961
Row operations on matrix, 961–964
Rows, of matrix, 960
Rule for rounding, 19
Run
  comparing rise to, 244–246
  explanation of, 244–245, 301

S

Scatter diagrams, 223, 269, 301
Scientific notation
  applications for, 423
  calculations using, 422–423
  calculator tip for, 422
  explanation of, 420, 427
  use of, 420–423, 429
  writing positive numbers in, 421
Scrap value, 897
Second-degree equations. *See* Quadratic equations
Second-degree inequalities
  explanation of, 936, 942
  graphs of, 936–939, 945
Semicircles, 925
Semiperimeter, 674
Set(s)
  elements of, 46
  empty, 137
  intersection of, 602, 647–648
  null, 137
  of numbers, 724
  solution. *See* Solution set(s)
  union of, 604, 647–468
Set-builder notation, 46, 105
Set operations, 602, 604, 606, 649
Signed numbers
  addition of, 56–58
  division of, 76–77
  explanation of, 46, 105
  multiplication of, 73–74
  subtraction of, 64–65
  summary of operations with, 58, 64, 74, 76
Similar triangles, 182, 562, 788
Simple interest formula, 169
Slope
  from an equation, 249
  as average rate of change, 252–253
  explanation of, 244–246, 301
  formula, 244–247
  of horizontal lines, 247–248
  of a line, 244–253, 303
  negative, 247
  of parallel lines, 250–251
  of perpendicular lines, 250–251
  positive, 247
  of vertical lines, 248
Slope-intercept form
  explanation of, 260–263, 267, 304
  use of, 316
Solution(s)
  of equations, 39, 105, 118, 588, 647, 970
  explanation of, 39, 105, 118
  extraneous, 713, 731
Solution set(s)
  explanation of, 118, 203, 314, 357, 588
  of system of equations, 626, 647
  of system of inequalities, 937–939
Sound, loudness of, 877
Specified variables, solving for, 164, 166–169, 205, 550–551
Spheres, volume of, 169
Square brackets, 30
Square matrix, explanation of, 960
Square root functions
  explanation of, 666, 732, 924, 944
  graphs of, 666, 904–905, 924–925, 943
Square root property
  explanation of, 822

extending, 747–748
  rewriting quadratic equations to use, 751–752
  solving quadratic equations with, 744–748, 779
Square roots
  calculator tip for, 663, 665
  explanation of, 662, 731
  finding, 662
  irrational, 664–665
  negative, 662
  of negative numbers, 722–723
  positive, 662
  principal, 662, 731
  simplifying, 668, 722
  solving for variables involving, 781–782
Squares
  of binomials, 392–393, 428, 715
  completing, 751–756
  difference of, 467–468, 477, 503
  of numbers, 28
  perfect, 469–470, 664, 731
  solving for variables involving, 781–782
Standard form
  of complex numbers, 724, 731
  of a linear equation, 218, 264–265, 267
  of a quadratic equation, 481–483, 501
Step functions, 907, 942
Straight angles, 153, 165, 203
Study Skills
  analyzing test results, 132
  managing time, 125
  preparing for math final exam, 230
  reading math textbooks, 25–26
  reviewing a chapter, 97
  tackling homework, 45
  taking lecture notes, 36
  taking math tests, 104
  using study cards, 55, 145
Subscript notation, 245–246
Substitution method
  explanation of, 323
  to solve equation quadratic in form, 771–774
  solving linear systems with, 323–328, 627–629, 650–651
  solving nonlinear systems with, 929–930
Subtraction
  application of, 67
  of complex numbers, 724–725, 734
  of decimals, 16
  definition of, 64–65
  of fractions, 6–8
  with grouping symbols, 65–66
  interpreting expressions involving, 66–67
  with number line, 64
  phrases that indicate, 66–67
  of polynomial functions, 835
  of polynomials, 371–372, 428, 954
  of radical expressions, 698–699, 733
  of rational expressions, 525–530, 577
  of real numbers, 64–67, 108
  of signed numbers, 64–65
  of variable terms, 120–121
  word phrases for, 66–67
Subtrahend, 64, 105
Sum
  in addition, 6, 38, 105
  of cubes, 472, 477
  and difference of two terms, 393–394, 429
Supplementary angles
  explanation of, 153, 203
  solving problems involving, 153–155
Supply and demand, 322
Symbol(s)
  for approximately equal to, 731
  for degree, 153
  for equality and inequality, 31–32, 39, 188, 278, 594
  for grouping, 29, 30, 65–66
  for infinity, 188
  for negative infinity, 188
  for positive or negative, 745
  for radicals, 662
  for subtraction, 65–66
  word phrases converted to, 32
Symmetry, of parabola, 789
Synthetic division
  to determine solutions of equations, 970
  to divide by binomial, 967–969

to evaluate polynomials, 969
explanation of, 968
Systems of equations. *See also* Systems of linear equations
on calculators, 627
consistent, 627, 647
explanation of, 626, 647
inconsistent, 627, 647
nonlinear, 942
solving problems with three variables, 641–642
special cases of, 631–632
Systems of inequalities
explanation of, 937
graphs of, 937–939, 945
solution set of, 937–939
Systems of linear equations
applications of, 340–345, 359
consistent, 317, 357
with decimals, 328
elimination method to solve, 331–334, 359
explanation of, 314, 357, 626
with fractions, 327
graphing method to solve, 314–318
inconsistent, 317, 325–326, 357
with infinitely many solutions, 317
matrix method to solve, 960–964
methods to solve, 314–318, 323–328, 331–334, 338–339, 358–359
with no solution, 317
solution set of, 314
substitution method to solve, 323–328, 627–629
Systems of linear equations in three variables
elimination method to solve, 636–638, 651
explanation of, 635
geometry of, 635–636
graphs of, 635–636
inconsistent, 640
matrix method to solve, 960–964
with missing terms, 638–639
special cases of, 639–640
Systems of linear equations in two variables
consistent, 627
elimination method to solve, 629–631, 650
graphs of, 626–627
inconsistent, 627
matrix method to solve, 960–964
ordered pair as solution of, 626
special cases of, 631–632
steps to solve, 626–632, 630, 650
substitution method to solve, 627–629, 650
Systems of linear inequalities
explanation of, 357
graphing to solve, 352–354
solutions of, 352–354, 357, 359
Systems of nonlinear equations, 929–933, 942

T

Table of values, 220–221, 301
Temperature conversion, 67, 167, 202
Terms
combining, 99–100
degree of, 369, 427

distinguishing between factors and, 98
explanation of, 98, 106, 203, 427
like, 99–100, 106, 368, 427
numerical coefficient of, 98–99, 106, 368
of polynomials, 416–417
product of sum and difference of, 393–395, 429
of proportions, 175
unlike, 99
Three-dimensional objects, 169
Three-part inequalities, 194–196, 203, 597
Threshold sound, 877
Traffic intensity, 547
Translating sentences into equations, 40, 80
Translating word phrases into expressions, 38–39, 59, 66–67, 79–80, 100, 140
Translations
of functions, 904–906
of parabola, 789
Trapezoids, 168
Triangles
area formula for, 164
area of, 674
perimeter of, 163–164, 168
right, 493–494, 501
similar, 182, 562, 788
Trillion, 426
Trinomials
explanation of, 369, 427
factoring, 450–454, 457–458, 461–464, 469–470, 478, 501–503
perfect square, 469–470, 478, 501
Triple, ordered, 635

U

Union of sets, 604, 606, 647
Unit pricing, 174–175
Unlike terms, 99

V

Values
absolute, 51, 57, 105
table of, 220–221, 301
Variable(s)
dependent, 284, 301
evaluating expressions with, 37–38, 107
explanation of, 37, 105
focus, 636
formulas to evaluate, 162
greatest common factor for, 441–444
independent, 284, 301
simplifying radicals involving, 688
solving application problems, 641–642
solving for specified, 166–167, 205, 550–551
solving for squared, 781–782
Variable cost, 274
Variation
combined, 570, 575
constant of, 565, 575
direct, 565–567, 575, 578
inverse, 567–569, 575, 578
joint, 569–570, 578
Variation equations, 565–570

Vertex
coordinates of, 789, 804
explanation of, 789, 792, 821
formula, 800
of parabolas, 789, 792, 799–800, 802, 821
Vertical angles, 165, 203
Vertical asymptotes, 544, 575
Vertical line test, 287–288
Vertical lines
equations of, 265–267, 304
graphs of, 235–236
slope of, 248
Vertical parabolas
explanation of, 806
vertex of, 799–800, 802
x-intercepts of, 802
Vertical shift
of ellipse, 916
explanation of, 790
method for applying, 905–906
of parabola, 790–792
Volume
explanation of, 169
formulas for, 169

W

Whole numbers
division of, 414–415
explanation of, 1, 46, 105
Windchill factor, 684
Word phrases
for addition, 59
converted to symbols, 32
for division, 79–80
for multiplication, 79
as ratios, 174
for subtraction, 66–67
translating to algebraic expressions, 38–39, 59, 66–67, 79–80, 100, 140
Work rates, 556–558, 769–770
Working equation, 636

X

x, polynomial in, 368
x-axis, 221, 301
x-intercepts
explanation of, 232, 301
of parabola, number of, 802

Y

y-axis, 221, 301
y-intercepts, 232, 260–262, 301

Z

Zero
division by, 76
multiplication property of, 73
Zero exponent, 399, 429, 953
Zero-factor property
explanation of, 481, 503
use of, 481–486, 744

Credits

Geometry Formulas

Square

Perimeter: $P = 4s$

Area: $A = s^2$

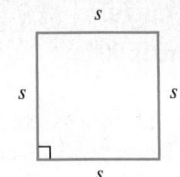

Rectangular Solid

Volume: $V = LWH$

Surface area: $A = 2HW + 2LW + 2LH$

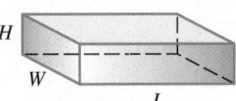

Rectangle

Perimeter: $P = 2L + 2W$

Area: $A = LW$

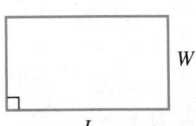

Cube

Volume: $V = e^3$

Surface area: $S = 6e^2$

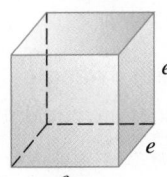

Triangle

Perimeter: $P = a + b + c$

Area: $A = \dfrac{1}{2}bh$

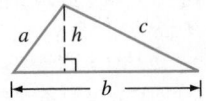

Right Circular Cylinder

Volume: $V = \pi r^2 h$

Surface area: $S = 2\pi rh + 2\pi r^2$

(Includes both circular bases)

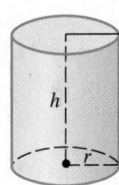

Parallelogram

Perimeter: $P = 2a + 2b$

Area: $A = bh$

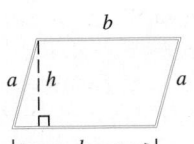

Cone

Volume: $V = \dfrac{1}{3}\pi r^2 h$

Surface area: $S = \pi r \sqrt{r^2 + h^2} + \pi r^2$

(Includes circular base)

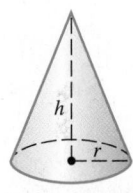

Trapezoid

Perimeter: $P = a + b + c + B$

Area: $A = \dfrac{1}{2}h(b + B)$

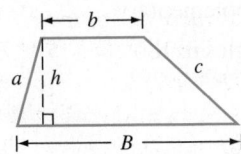

Right Pyramid

Volume: $V = \dfrac{1}{3}Bh$

B = area of the base

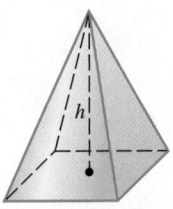

Circle

Diameter: $d = 2r$

Circumference: $C = 2\pi r$

$C = \pi d$

Area: $A = \pi r^2$

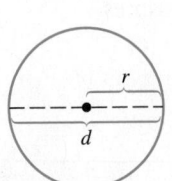

Sphere

Volume: $V = \dfrac{4}{3}\pi r^3$

Surface area: $S = 4\pi r^2$

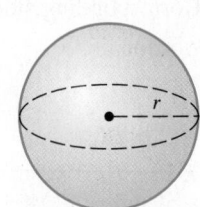

Triangles and Angles

Right Triangle
Triangle has one 90° (right) angle.

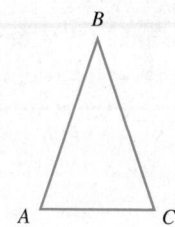

Pythagorean Theorem (for right triangles)
$a^2 + b^2 = c^2$

Right Angle
Measure is 90°.

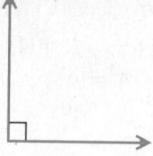

Isosceles Triangle
Two sides are equal.

$AB = BC$

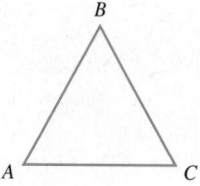

Straight Angle
Measure is 180°.

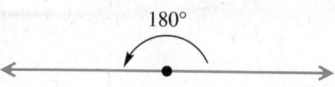

Equilateral Triangle
All sides are equal.

$AB = BC = CA$

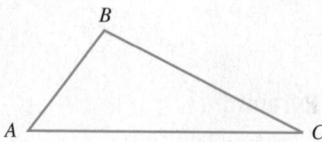

Complementary Angles
The sum of the measures of two complementary angles is 90°.

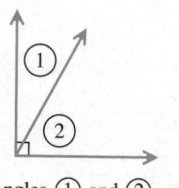

Angles ① and ② are complementary.

Sum of the Angles of Any Triangle
$A + B + C = 180°$

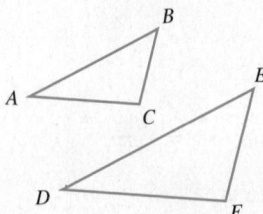

Supplementary Angles
The sum of the measures of two supplementary angles is 180°.

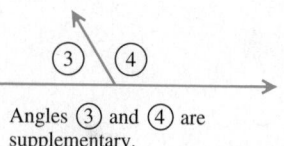

Angles ③ and ④ are supplementary.

Similar Triangles
Corresponding angles are equal. Corresponding sides are proportional.

$A = D, B = E, C = F$

$$\frac{AB}{DE} = \frac{AC}{DF} = \frac{BC}{EF}$$

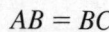

Vertical Angles
Vertical angles have equal measures.

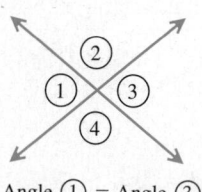

Angle ① = Angle ③
Angle ② = Angle ④